HANDBOOK OF RESEARCH ON HERBAL LIVER PROTECTION

Hepatoprotective Plants

HANDBOOK OF RESEARCH ON HERBAL LIVER PROTECTION

Hepatoprotective Plants

T. Pullaiah
Maddi Ramaiah

First edition published 2021

Apple Academic Press Inc.
1265 Goldenrod Circle, NE,
Palm Bay, FL 32905 USA
4164 Lakeshore Road, Burlington,
ON, L7L 1A4 Canada

CRC Press
6000 Broken Sound Parkway NW,
Suite 300, Boca Raton, FL 33487-2742 USA
2 Park Square, Milton Park,
Abingdon, Oxon, OX14 4RN UK

First issued in paperback 2021

Library and Archives Canada Cataloguing in Publication

CIP data on file with Canada Library and Archives

Library of Congress Cataloging-in-Publication Data

Names: Pullaiah, T., author. | Ramaiah, Maddi, 1985- author.

Title: Handbook of research on herbal liver protection : hepatoprotective plants / T. Pullaiah, Maddi Ramaiah.

Description: First edition. | Palm Bay, FL : Apple Academic Press, 2021. | Includes bibliographical references and index. | Summary: "This important volume provides a comprehensive overview of hepatotoxicity and medicinal plants used for protecting the liver and curing liver toxicity and liver diseases. To date, there has been no extensive resource on the plants that are used in this capacity, both in traditional medicine and in modern medicine. This book, Handbook of Research on Herbal Liver Protection: Hepatoprotective Plants, fills that gap. It presents information on the medicinal plants used in traditional medicine (both codified and noncodified) and in ethnomedicine, including the plant parts used and methods of use and dosages. The phytochemicals extracted from medicinal plants, screened and used in modern medicine for liver protection and curing liver problems, are given in detail. The volume discusses the medicinal plants screened for hepatoprotection, and the methods of screening are given as well. Methods of assay for screening the medicinal plants are also presented. Key features: Provides complete information on plants that show hepatoprotective properties Lists and discusses the phytochemicals useful for liver protection and cures Considers traditional uses and ethnomedicinal plants for liver protection Details the plant parts and the extracts that livers protection properties and active principles showing hepatoprotection. This volume will be a single point of reference for researchers who require complete information on hepatoprotective plants. It is also useful for information on liver toxicity/diseases prevention and cures as well as on plants and phytochecmicals used for liver protection. It will be useful to researchers working on liver disorders and diseases, physicians treating the patients with liver problems, for hospitals, pharmacy institutions, pharmacy students, pharmacy researchers, and ethnobotanists and others working in traditional medicine"-- Provided by publisher.

Identifiers: LCCN 2020040906 (print) | LCCN 2020040907 (ebook) | ISBN 9781771889186 (hardcover) | ISBN 9781003043171 (ebook)

Subjects: MESH: Liver Diseases--prevention & control | Phytochemicals--analysis | Chemical and Drug Induced Liver Injury--prevention & control | Plant Extracts--analysis | Plants, Medicinal--toxicity | Medicine, Traditional

Classification: LCC RC846 (print) | LCC RC846 (ebook) | NLM WI 710 | DDC 616.3/620610724--dc23

LC record available at https://lccn.loc.gov/2020040906

LC ebook record available at https://lccn.loc.gov/2020040907

ISBN: 978-1-77188-918-6 (hbk)
ISBN: 978-1-77463-780-7 (pbk)
ISBN: 978-1-00304-317-1 (ebk)

About the Authors

T. Pullaiah, PhD, is a former Professor at the Department of Botany at Sri Krishnadevaraya University in Andhra Pradesh, India, where he has taught for more than 35 years. He has held several positions at the university, including Dean, Faculty of Biosciences, Head of the Department of Botany, Head of the Department of Biotechnology, and Member of Academic Senate. He was President of the Indian Botanical Society (2014), President of the Indian Association for Angiosperm Taxonomy (2013), and Fellow of Andhra Pradesh Akademi of Sciences. He was awarded the Panchanan Maheshwari Gold Medal, the Dr. G. Panigrahi Memorial Lecture Award of the Indian Botanical Society and the Prof. Y.D. Tyagi Gold Medal of the Indian Association for Angiosperm Taxonomy, and a best teacher award from the Government of Andhra Pradesh. Under his guidance, 54 students obtained their doctoral degrees. He has authored over 50 books, edited 20 books, and published over 330 research papers, including reviews and book chapters. His books include *Advances in Cell and Molecular Diagnostics* (Elsevier), *Ethnobotany of India* (5 volumes, Apple Academic Press), *Red Sanders: Silviculture and Conservation* (Springer), *Monograph on Brachystelma and Ceropegia in India* (CRC Press), *Flora of Andhra Pradesh* (5 volumes), *Flora of Eastern Ghats* (7 volumes), *Flora of Telangana* (3 volumes), *Encyclopaedia of World Medicinal Plants* (7 volumes, 2nd edition), and *Encyclopaedia of Herbal Antioxidants* (3 volumes). He was also a member of the Species Survival Commission of the International Union for Conservation of Nature (IUCN). Professor Pullaiah received his PhD from Andhra University, India, attended Moscow State University, Russia, and worked as postdoctoral fellow during 1976–1978.

Maddi Ramaiah, PhD, is Professor and Head, Department of Pharmacognosy, Hindu College of Pharmacy, Amaravathi Road, Guntur, Andhra Pradesh, India. He has 11 years of teaching and research experience and has published more than 55 research and review papers in various reputed indexed national and international journals. He has also presented more than 15 research and review posters at national and international conferences. He is a textbook and book chapter author. He attended many conferences, seminars, symposia, workshops, and QIP programs. He has been an active advisor member, editorial board member, and scientific reviewer for more than 50 reputed national and international journals, including Elsevier publications. He is a professional member of the Association of Pharmaceutical Teachers of India, Association of Pharmacy Professionals, Society for Ethnopharmacology, and the Society of Pharmacognosy. He received his Doctor of Philosophy degree in Pharmaceutical Sciences from Andhra University, Visakhapatnam, Andhra Pradesh, India.

Contents

Abbreviations

AA	acrylamide
AAE	ascorbic acid equivalent
ABEO	*A. biebersteinii* essential oil
AC	ammonium chloride
ACBE	*Acacia confusa* bark extract
ACE	alcoholic extract
AcHz	acetylhydrazine
ACP	acid phosphatase
ADR	anticancer drug
ADS	antioxidant defense systems
AEAR	aqqueous extract of *Asparagus racemosus* root
AEF	aqueous extract of formula
AEJP	aqueous extract of *J. phoenicea* berries
AEMP	aqueous extract of *Melothria perpusilla*
AESG	aqueous extract of leaves of *S. grandiflora*
AF	acetone fraction
AFP	alfa fetoprotein
AG	Amalkadi Ghrita
AHH	aryl hydrocarbon hydroxylase
AI	*A. indica*
ALDH	aldehyde dehydrogenase
ALI	acute liver injury
ALP	alkaline phosphatase
ALT	alanine transaminase
AM	*Acacia mellifera*
AMPK	AMP-activated protein kinase
ANIT	alpha-naphthylisothiocyanate
AP	*Andrographis paniculata*
APAP	acetaminophen
APIG	apigenin-7-glucoside (APIG)
APME	*Artemisia pallens* methanol extract
APS	*Astragalus* polysaccharides
AQ	aromatic water
AQTB	aqueous extract of *T. bellirica*

ASE	*Amaranthus spinosus* extract
ASE	*Angelica sinensis* extract
AST	aspartate aminotranferase
ATE	*Amaranthus tricolor* extract
ATEs	*Acer tegmentosum* extracts
ATP	Adenosine triphosphate
ATR	atractyloside
ATT	antitubercular treatment
BAEE	*Borago officinalis* aerial ethanolic extract
BCA	Biochanin A
BCG	Bacille Calmette–Guerin
BDL	bile duct ligation
BELE	*Berchemia lineata* ethanol extract
BG	*Ballota glandulosissima*
BHEE	biherbal ethanolic extract
Bil	bilirubin
BPA	bisphenol A
BR	*Boesenbergia rotunda*
BRE	*Boschniakia rossica* extract
BRN	bilirubin
BSA	bovine serum albumin
BUN	blood urea nitrogen
BVLE	*B. variegata* leaves extract
BW	body weight
CA	caffeic acid
CA	*Centella asiatica*
CaR	*C. arisanense* roots
CAT	catalyse
CATR	carboxyatractyloside
CBIL	conjugated bilirubin
CBL	*Croton bonplandianus*
CCl_4	carbon tetrachloride
CCME	*Caesalpinia crista* methanolic extract
CCW	*Crossostephium chinensis* water extract
CD	conjugated dienes
CE	chloroform extract
CGA	chlorogenic acid
CGRE	*Cichorium glandulosum* root extract
CGX	Chunggan extract
ChEE	*Citrus hystrix* peel ethanolic extract

CHOL	cholesterol
CHY	Chyavanaprash
CISE	chestnut (*Castanea crenata*) inner shell extract
CMC	carboxy methyl cellulose
CPA	cyclophosphamide
CPCESA	Committee for the Purpose of Control and Supervision of Animals
CPE	*C. papaya* fruit extract
CPL	*C. papaya* leaf
CPP	*Cissus pteroclada*
CPRO	cold-pressed *R. officinalis* oil
CR	Coptidis rhizome
CRAE	Coptidis rhizome aqueous extract
CS	*Cuscuta chinensis*
CSE	*Carum copticum* seeds
CSE	cassia seed ethanol extract
CSP	*Crocus sativus* petals
CX	*Curcuma xanthorrhiza*
CYP2E1	cytochrome P450 2E1
DADS	diallyl disulfide
DBL	direct bilirubin
DE	dry extract
DEN	diethylnitrosamine
DHQ	dihydroquercetin
DL	dried latex
DLAE	*Dicranopteris linearis* aqueous leaf extract
DLLE	*Diospyros lotus* leaf extract
DME	*Digera muricata* methanol extract
DMN	dimethylnitrosamine
DOP	*Dendrobium officinale*
DOX	doxorubicin
EAF	Ethyl acetate fraction
EAP	ethyl acetate partition
EATE	ethyl acetate fraction of *Tagetes erecta*
EEAA	ethanolic extract of *Aquilaria agallocha*
EECM	ethanol extract of *Citrus macroptera*
EEOS	ethanolic extract of *Oxalis stricta*
EEPT	ethanolic extract of aerial parts of *Piper trioicum*
EETP	ethanol extract of *Tephrosia purpurea*
EETS	ethanolic extract of flowers of *Tecoma stans*

ELAP	ethanolic leaf extract *Andrographis paniculata*
EP	*Eclipta prostrata*
ERZO	ethanolic extract of rhizomes of *Zingiber officinale*
ESEt	ethanolic seeds extract
FBG	fasting blood glucose
FI	*Flagellaria indica*
FIE	*F. indica* extract
FRAP	ferric reducing antioxidant power
G-6-Pase	glucose-6-phosphatase
GA	gallic acid
GAGL	gluonic acid gamma-lactone
GAK	ground apricot kernel
GalN	galactosamine
GbE	*G. biloba* extract
GBP	*G. biloba* phytosomes
GCA	*Gentiana cruciata* aerial parts
GCL	glutamate-cysteine ligase
GGT	gamma glutamyl transferase
GGTP	gamma-glutamate transpeptidase
GPT	glutamic-pyruvate transaminase
GPx	glutathione peroxidase
GRD	glutathione reductase
GSH	glutathione level
GSO	grape seed oil
GST	glutathione *S*-transferase
H_2O_2	hydrogen peroxide
HA	hyaluronic acid
HC	*Heinsia crinita*
HDL	high-density lipoprotein
HO	heme oxygenase
HO-1	heme oxygenase-1
HP	hydroperoxides
HP	hydroxy proline
HPLC	high-performance liquid chromatography
HS	*Hygrophila schulli*
HSC	hepatic stellate cells
HT	*Helianthus tuberosus*
HTG	hepatic triglycerides
HV	Himoliv
i.p.	intraperitoneal

IBD	infectious bursal disease
IBU	Ibuprofen
IL-6	interleukin-6
ILI	immunological liver injury
INH	induced hepatotoxicity
INH	isoniazid
iNOS	inducible nitric oxide synthase
IT	*Indigofera tinctoria*
JNK	Jun N-terminal kinase
LB	*Lagenaria breviflora*
LC	*Lysimachia clethroides*
LD	lindane
LDH	lactate dehydrogenase
LDL	low-density lipoprotein
LDL-C	low-dentistry lipoprotein-cholesterol
LGO	lemongrass essential oil
LN	*Lippia nodiflora*
LOX	lipooxygenase
LP	lipid peroxidation
LPO	lipid peroxide
LPS	lipopolysaccharide
LSEE	*L. sativum* ethanolic extract
LTH	liver tissue homogenate
LXRα	liver X receptor α
MAF	methanol active fraction
MAPK	mitogen-activated protein kinase
MC	*Micromeria croatica*
MCHL	*M. cecropioides* hydromethanolic leaf
MD	Majoon-e-Dabeed-ul-ward
MDA	malanodialdehyde
MDME	microsomal drug metabolizing enzymes
ME	methanolic extract
MEBT	methanolic extract of *Berberis tinctoria*
MECA	methanolic extract of *Cyperus articulatus*
MECB	methanolic extract of *Caesalpinia bonduc*
MECI	methanolic extract of *Clerodendrum infortunatum*
MECM	methanolic extract of *Cucurbita maxima*
MECS	methanolic extract of *C. spinosa* leaves
MECT	methanolic extract of *Cyperus tegetum* rhizome
MEDP	methanolic extract of *Dipteracanthus patulus*

MEGG	methanolic extract of *Gardenia gummifera* root
MELC	methanolic extract of *Luffa cylindrica* leaves
MEME	methanolic extract of *Merremia emarginata*
MEMM	methanolic extract of *M. malabathricum* leaves
MEZJ	methanolic extract of *Ziziphus jujuba* fruits
MMP	mitochondrial membrane potential
MO	*Moringa oleifera*
MPO	myeloperoxidase
MTX	methotrexate
MUSE	*M. uniflorum* seed extract
NAC	*N*-acetylcysteine
NAFLD	nonalcoholic fatty liver disease
NAPQI	*N*-acetyl-*p*-benzoquinone imine
NB	nitrobenzene
ND	Newcastle disease
NDEA	*N*-nitrosodiethylamine
NDMA	*N*-nitrosodimethylamine
NF	nuclear factor
NILE	*Nerium oleander* leaf extract
NL	*Nauclea latifolia*
NL	*Newbouldia laevis*
NL	*Nymphaea lotus*
NO	nitric oxide
O_2^-	superoxide anions
OATP	organic anion-transporting polypeptide
OCME	*Oxalis corniculata* methanol extract
ODFR	oxygen-derived free radicals
OECD/OCDE	Organisation for Economic Co-operation and Development
ORAC	oxygen radical absorbance capacity
OV	*Origanum vulgare*
PA	*Phyllanthus amarus*
PC	protein carbonyl
PCC	protein carbonyl content
PCEE	*Piper cubeba* fruits ethanol extract
PCM	paracetamol
PCO	protein carbonyl
PDM	petroleum ether, diethyl ether, and methanol
PE	*P. emblica*
PEE	petroleum ether extract
PFDB	polyphenolic fraction of *Desmostachya bipinnata* root

PG	*Punica granatum*
PG	Panchagavya Ghrita
PHF	polyherbal formulation
PN	*Phyllanthus niruri*
POD	peroxidase
POE	*Polygonum orientale*
PPARγ	peroxisome proliferator activated receptor gamma
PPE	purple potato extract
PX	Poloxamer
PZA	pyrazinamide
RBCs	red blood cells
RGDS	red grape dried seeds
RIF	rifampicin
ROE	*Rosmarinus officinalis* leaves extract
ROS	reactive oxidative stress (ROS)
RSE	*Rhodiola sachalinensis*
SAFE	*Sonneratia apetala* fruit extract
SBE	stem bark extract
SBLN	serum bilirubin
SCPE	*S. chinensis* pollen extract
SD	Sprague Dawley
SMA	smooth muscle actin
SML	*S. mombin* leaf
SMS	*S. mombin* stem
SOD	superoxide dismutase
SP	Seabuckthorn polysaccharide
SPP	seed powder
SRE	*S. rhombifolia*
Ss-Bb	soyasaponin Bb
STE	*Solanum trilobatum* extract
STZ	streptozotocin
TAA	thioacetamide
TAC	total antioxidant capacity
TAG	total triglycerol
TAM	Tamoxifen citrate
TAP	turmeric antioxidant protein
TBA	thiobarbituric acid
t-BHP	*tert*-butyl hydroperoxide
TBL	total bilirubin
TC	total cholesterol

TCA	tricarboxylic acid
TCW	tender coconut water
TFG	*Trigonella foenum-graecum*
TFs	total flavonoids
TGF	transforming growth factor
TGF-β	transforming growth factor-beta
TGs	triglycerides
TL	total lipids
TLR	toll-like receptor
TMCA	trimethylcolchicinic acid
TNF	tumor necrosis factor
TNF-α	tumor necrosis factor-α
TOE	*Taraxacum officinale* leaf extract
TP	total lipoprotein
TPC	total phenolic content
TPE	*Tanacetum parthenium* extract
TPs	taxus polyprenols
TPTN	total protein
TREE	turnip root ethanolic extract
TrxR	thioredoxin reductase
TSB	*T. undulata* stem bark
TSTA	total saponins of *T. affinis*
UA	ursolic acid
UGT	UDP-glucuronosyl transferase
UPME	*Uraria picta* methanolic extract
VEGF	vascular endothelial growth factor
VLDL	very low density lipoprotein
VLDL-C	very low-dentistry lipoprotein-cholesterol
VN	*Vitex negundo*
WBCs	white blood cells
WEPT	water extract of pu-erh tea
XO	xanthine oxidase
ZSC	*Ziziphus spina-christi*
ZZDHD	Zhi-Zi-Da-Huang decoction

Preface

Liver is one of the most vital organs present in humans and vertebrate animals. Various toxic chemicals that are being used in food and water are causing toxicity to this important organ. There is a necessity to know about the methods of protecting this vital organ and the medicinal plants that are being used to protect it.

This book, *Handbook of Research on Herbal Liver Protection,* gives an overview of the hepatotoxicity and medicinal plants used for protecting the liver and for curing liver toxicity and diseases. There is no comprehensive source of information on the plants that are used, both in traditional medicine and in modern medicine. This book is an answer for this. It gives information on medicinal plants used in traditional medicine (both codified and noncodified) and ethnomedicine. Plant parts used, method of use and dosage, and references are given. Phytochemicals extracted from medicinal plants, screened and used in modern medicine for liver protection, and curing liver problems are given in detail. Details of medicinal plants screened for hepatoprotection and methods of screening are given along with the reference. Methods of assay for screening the medicinal plants are also given.

This book is a single point of reference for researchers who require complete information on hepatoprotective plants. The book provides complete details on screening methods and assays of hepatoprotectants. It is also useful to get information on liver toxicity/diseases prevention and cures. Simply, it will be a valuable desk reference on herbal liver protection.

The book will be useful for researchers working on liver disorders and diseases, physicians treating the patients with liver problems, hospitals, pharmacy institutions, pharmacy students, pharmacy researchers, and ethnobotanists or people working on traditional medicine.

Since it is a voluminous subject, we might have missed some references. Readers are requested to bring such omissions to our notice so that the same can be included in future editions.

I wish to express our appreciation and help rendered by Sandra Sickels, Rakesh Kumar, and staff of Apple Academic Press. Their patience and perseverance have made this book a reality and is greatly appreciated.

—T. Pullaiah, PhD
Maddi Ramaiah, PhD

CHAPTER 1

Liver: Activities, Functions, Problems, and Diseases

ABSTRACT

The liver is a vital organ present in vertebrates and some other animals and it is the largest organ in the body and constitutes about 2% of the body weight. Liver is involved in the carrying of major physiological processes such as maintenance of growth, immunity, nutrition, energy metabolism, reproduction, synthesis and secretion of bile, albumin, prothrombin, and many more. It is a body detoxification center. Consumption of toxic chemicals, alcohol, infections, etc., may affect the liver functions and their physiological process. Most of the hepatotoxic chemicals damage liver cells mainly by inducing lipid peroxidation and other oxidative damages in the liver. Liver fibrosis and liver cirrhosis are chronic degenerative liver disorders. Examples of liver diseases include hepatitis, alcoholic liver disease, fatty liver disease, liver cancer, etc. This chapter gives a detailed note on various activities, functions, problems, and diseases of the liver.

LIVER ACTIVITIES, FUNCTIONS, PROBLEMS AND DISEASES

The liver is a vital organ present in vertebrates and some other animals, and it is typically the largest visceral organ. Liver is the largest organ in the body, contributing about 2% of the body weight in the average human. It is connected with most of the physiological processes that include growth, immunity, nutrition, energy metabolism, reproduction, synthesis and secretion of bile, albumin, prothrombin, and the reduction of the compliments, which are the major effectors of the hormonal branch of the immune system (Dey and Saha, 2013). It has a great capacity to detoxicate toxic substances and synthesize useful principles (Shanani, 1999; Subramoniam and Pushpangadan, 1999). It helps in the maintenance, performance, and regulation of homeostasis of the body. It is involved with almost all the biochemical pathways to growth, fight against disease, nutrient supply, energy provision, and reproduction.

In addition, it aids the metabolism of carbohydrates, proteins, and fats; detoxification; secretion of bile; and storage of vitamins (Ahsan et al., 2009). The role played by this organ in the removal of substances from the portal circulation makes it susceptible to first and persistent attack by offending foreign compounds, culminating in liver dysfunction (Bodakhe and Ram, 2007).

Furthermore, detoxification of a variety of drugs and xenobiotics occurs in the liver. The bile secreted by the liver has, among other things, an important role in digestion. Liver diseases are among the most serious ailments. They may be classified as acute or chronic hepatitis (noninflammatory diseases) and cirrhosis (degenerative disorder resulting in fibrosis of the liver). Liver diseases are mainly caused by toxic chemicals (certain antibiotics, chemotherapeutics, peroxidized oil, aflatoxin, carbon tetrachloride, chlorinated hydrocarbons, etc.), excess consumption of alcohol, infections, and autoimmune disorder.

Liver diseases remain one of the major threats to public health and are a worldwide problem (Asha and Pushpangadan, 1998). They are caused by chemicals like acetaminophen (in large doses), excess consumption of alcohol, infections, and autoimmune disorders. Most of the hepatotoxic chemicals damage liver cells mainly by inducing lipid peroxidation and other oxidative damages (Recknagel, 1983; Wendel et al., 1987; Dianzani et al., 1991). Acetaminophen, a mild analgesic and antipyretic drug, developed in the last century, causes serious liver necrosis in humans and in experimental animals if taken in large doses (Lin et al., 1995; Mitchell et al., 1973; Hinson 1980; Mitchell, 1988). While alcohol is one of the main causes of end-stage liver disease worldwide, alcoholic liver disease is the second most common reason for liver transplantation in the United States (Mandayam et al., 2004). Due to increased frequency of drinking and change of diet construction, such as the increase of fat content, the incidence of liver diseases has increased in China, becoming another important risk factor for morbidity and mortality in addition to viral hepatitis (Zhuang and Zhang, 2003). The spectrum of alcoholic liver disease ranges from fatty liver to alcoholic hepatitis and ultimately fibrosis and cirrhosis (Tuma and Sorrell, 2004). Liver cirrhosis is the ninth leading cause of death in the United States (Kim et al., 2002). Chronic liver diseases are the fifth most frequent cause of death in the European Union, as they entail multiple risks, such as portal hypertension, ascites, spontaneous bacterial peritonitis, hepatorenal and hepatopulmonary syndromes, hepatic encephalopathy, and, of course, hepatocellular carcinoma (Bosetti et al., 2007, 2008).

The liver plays a major role in transforming and clearing chemicals and is susceptible to the toxicity of these agents. Some medicinal agents, when taken in overdoses and sometimes even when introduced within therapeutic ranges, may injure the organ. Chemicals that cause liver injury are called hepatotoxins (Maity and Ahmad, 2012). Chemicals often cause subclinical injury to the liver which manifests only as abnormal liver enzyme tests. Drug-induced hepatotoxicity represents a major clinical problem accounting for 50% of all cases of acute liver failure. Although the majority of cases of acute liver failure are due to intentional or unintentional misuse, 16% are idiosyncratic (Bell and Cahalsani, 2009). Some of the inorganic compounds producing hepatotoxicity are arsenic, phosphorus, copper, and iron. The organic agents include certain naturally occurring plant toxins such as pyrrolizidine alkaloids, mycotoxins, and bacterial toxins. In addition, exposure to hepatotoxic compounds may be occupational, environmental, or domestic that could be accidental, homicidal, or suicidal ingestion (Sumanth, 2007).

Most of the hepatotoxic chemicals damage liver cells mainly by inducing lipid peroxidation and other oxidative damages in the liver. Enhanced lipid peroxidation produced during the liver microsomal metabolism of ethanol may result in hepatitis and cirrhosis (Smuckler, 1975). It has been estimated that about 90% of acute hepatitis is due to viruses. The major viral agents involved are hepatitis B, A, C, D (delta agents), E, and G. Of these, Hepatitis B infection often results in chronic liver diseases and cirrhosis of the liver. Primary liver cancer has also shown to be produced by these viruses. It has been estimated that approximately 14–16 million people are infected with this virus in Southeast Asia region and about 6% of the total population in the region are carriers of this virus. A vaccine has become available for immunization against Hepatitis B virus. Hepatitis C and E infections are also common in countries of the Southeast Asia region (WHO, 2013).

The liver is known as the chemical factory of the human body which regulates the metabolism of food, xenobiotics, and drugs. Healthy habits commonly protect and regenerate hepatic cells if any damage or injury to the liver occurs accidentally, by environmental factors or even by the consumption of the drugs. But chronic alcoholism causes severe damage to the liver cells which may develop hepatitis, alcoholic steatonecrosis because of excessive production of cytokines—tumor necrosis factor α (TNF-α), interleukins (IL) (IL6 and IL8) (Carmo et al., 2017). During metabolism alcohol is converted to acetaldehyde, a highly toxic substance, and also it liberates acetate anions. If the acetate anions react with hydrogen atoms of the body, they generate highly reactive free radicals causing reactive oxidative stress (ROS). Acute free radical accumulation depletes the equilibrium between ROS production and ROS removal (Kumar, 2012). The enzymes cytochrome P450 2E1 (CYP2E1) and catalase become active and play an important role in the conversion of alcohol to acetaldehyde. Small amount of alcohol is removed by interacting with fatty acids to form reactive compounds called fatty acid ethyl esters. However, the chronic accumulation of these reactive species certainly contributes to hepatic tissue damage and hepatitis (Guo and Ren, 2010).

Liver fibrosis, and ultimately liver cirrhosis, is the common end-stage of all chronic liver diseases. At the beginning of fibrogenesis stands a chronic inflammatory condition. But it is not the virus- or toxin-induced hepatocellular damage that primarily causes tissue destruction and the formation of granulation tissue, but the activation of immunocompetent cells (e.g., Kupffer-cells) and the release of proinflammatory cytokines, such as TNF-α, IL-6, and IL-12. These mediators and the accumulation of potentially toxic free fatty acids generate highly ROS, which exposes the hepatocyte to an oxidative stress, which, primarily via peroxidation of membrane lipids and DNA damage, leads to hepatocellular injury. In the meantime, it comes to an activation of mesenchymal cells, resulting in an increased synthesis and interstitial deposition of extracellular matrix components (Gressner et al., 2008). These mesenchymal cells, hepatic stellate cells (HSC), also known as Ito cells, are pericytes found in the perisinusoidal space of the liver also known as the space of Disse. Following liver injury, HSC undergo "activation" that connotes a transition from quiescent vitamin A-rich cells into proliferative, fibrogenic, and contractile myofibroblasts. This pathway has long been, and probably still is, considered as the "canonical" pathway in the pathogenic understanding of liver fibrogenesis. The major phenotypic changes after activation include proliferation, contractility, fibrogenesis, matrix degradation, chemotaxis, retinoid loss, and white blood cell chemoattraction (Friedman, 2008).

The liver, because of its strategic anatomical location and its large metabolic conversions, is exposed to many kinds of xenobiotics and therapeutic agents. Moreover, the rapidly growing morbidity and mortality from liver diseases are largely attributable to the increasing number of chemical compounds and environmental pollution. Unfortunately, so far, in the modern era of medicine, there is no specific treatment to counter the menacing impact of these dreaded diseases (Chatterjee, 2000). The therapeutic regimen followed in all these cases up to the present moment is by and large symptomatic and at best palliative, but it still confronts the practitioner with a formidable task. Due to this fact, efforts to find suitable palliative and/or curative agents for the treatment of liver diseases in natural products of plants and mineral origin are being made (Bhandarkar and Khan, 2004). Liver damage can also be caused by drugs, particularly paracetamol and drugs used to treat cancer. Liver injury induced by paracetamol is the best characterized system of xenobiotics-induced hepatotoxicity in human beings (Lee et al., 2004). Large doses of paracetamol, a widely used analgesic-antipyretic drug, is known to cause hepatotoxicity in man and laboratory animals. The modern medicines have little to offer for alleviation of hepatic ailments, whereas most important representatives are of phytoconstituents (Handa and Kapoor, 1989). With the increased use of chemicals and alcohol besides the growing incidence of viruses and autoimmune diseases, the incidence of liver injury is growing for which conventional drugs used for treatment are often inadequate.

Hepatoxicity is the inflammation of the liver, in general associated with various drugs used in the modern medicine, different chemicals, toxins, and viruses (Ravikumar et al., 2005; Stierum et al., 2005).

Liver diseases which are still a global health problem may be classified as acute or chronic hepatitis (inflammatory liver diseases), hepatosis (noninflammatory diseases), and cirrhosis (degenerative disorder resulting in liver fibrosis). Accumulation of toxins in the bloodstream results in hepatic encephalopathy, and these toxins are normally removed by the liver. Unfortunately, treatments of choice for liver diseases are controversial because conventional or synthetic drugs for the treatment of these diseases are insufficient and sometimes cause serious side effects (Kumar et al., 2011).

Hepatic problems along with heart problems are the major causes of death across the world. Roger and Pamplona (2001) and Ellahi et al. (2014) reported that two million people die annually from liver-related disorders with 60,000 occurring from hepatitis B alone. The WHO fact sheets (2005) also reveals that more than 170 million people have long-term liver infections with hepatitis C virus (Bartholomew et al., 2014).

Cirrhosis was the 12th leading cause of death in the United States in 2007 and represented a large economic burden with the national cost of treatment ranging from $14 million and above, in addition to the $2 billion as an indirect cost due to loss of work productivity and reduction in the health-related quality of life. Depending on the disease etiology, this has been on the progressive rise and is expected to significantly increase to $2.5 million and $10.6 million, respectively, by the year 2028 (Teschke et al., 2012; Christopher and Taosheng, 2017). Unwanted side effects experienced with most of the orthodox drugs are another major reason for a search of an alternative source of drugs against liver diseases.

HEPATOTOXICITY

Drugs continue to be pulled from the market with disturbing regularity because of late discovery of hepatotoxicity (Tripathi et al., 2010). The mechanism of hepatic injury has been proposed to involve two pathways-direct hepatotoxicity and adverse immune reactions. In most instances, hepatic injury is initiated by the bioactivation of drugs to chemically reactive metabolites, which have the ability to interact with cellular macromolecules such as proteins, lipids, and nucleic acids, leading to protein dysfunction, lipid peroxidation, DNA damage, and oxidative stress. Additionally, these reactive metabolites may induce disruption of ionic gradients and intracellular calcium stores, resulting in mitochondrial dysfunction and loss of energy production. Its dysfunction releases an excessive amount of oxidants which in turn damages hepatic cells (Tarantino et al., 2009). Hepatic cellular dysfunction and death also have the ability to initiate immunological reactions, including both innate and adaptive immune responses. Stress and damage to hepatocytes result in the release of signals that stimulate the activation of other cells, particularly those of the innate immune system, including Kupffer cells, natural killer cells, and NKT cells. These cells contribute to the progression of liver injury by producing proinflammatory mediators and secreting chemokines to further recruit inflammatory cells to the liver. It has been demonstrated that various inflammatory cytokines, such as TNF-α, interferon-γ, and IL-1β, produced during hepatic injury are involved in promoting tissue damage (Gao et al., 2009). However, innate immune cells are also the main source of IL-10, IL-6, and certain prostaglandins, all of which have been shown to play a hepatoprotective role (Wang et al., 2004). It is, therefore the delicate balance of inflammatory and hepatoprotective mediators produced after activation of the innate immune system that determines an individual's susceptibility and adaptation to hepatic injury.

In India, pulmonary tuberculosis is one of the major causes for adult deaths. Isoniazid and Rifampicin (RIF), the first line drugs used for tuberculosis chemotherapy, are associated with hepatotoxicity. The rate of hepatotoxicity has been reported to be much higher in developing countries like India (8%–30%) compared to that in advanced countries (2%–3%) with a similar dose schedule. It has been established that oxidative stress is one of the mechanisms for RIF-induced hepatic injury.

In spite of the tremendous advances in modern medicine, there is no effective drug available that stimulates liver function, offers protection to the liver from damage, or helps in the regeneration of hepatic cells (Chattopadhyay, 2003). Despite the tremendous scientific advancement in the field of gastroenterology over the recent years, there is not even a single effective allopathic medication available for the treatment of liver disorders. It is therefore necessary to search for alternative drugs for the treatment of liver diseases to replace currently used drugs of doubtful efficacy and safety.

KEYWORDS

- **liver**
- **liver functions**
- **liver health**
- **hepatotoxicity**
- **liver diseases**

REFERENCES

Ahsan, M.R.; Islam, K.M.; Bulbul, I.J. Hepatoprotective activity of methanol extract of some medicinal plants against carbon tetrachloride-induced hepatotoxicity in rats. *Eur. J. Sci. Res.* **2009**, *37*(2), 302–310.

Asha, V.V.; Pushpangadan, P. Preliminary evaluation of the anti-hepatotoxic activity of *Phyllanthus kozhikodianus*, *Phyllanthus maderspatensis* and *Solanum indicum*. *Fitoterapia* **1998**, *59*, 255–259.

Bartholomew, I.C.; Brai, R.A.; Adisa, A.; Odetola, A. Hepatoprotective properties of aqueous leaf extract of *Persea americana* Mill. (Lauraceae) “avocado” against CCl4-induced damage in rats. *Afr. J. Trad. Compl. Alternat. Med.* **2014**, *11*(2), 237–244.

Bell, L.N.; Chalasani, N. Epidemiology of idiosyncratic drug-induced liver injury. *Semin. Liver Dis.* **2009**, *29*, 337–347.

Bhandarkar, M.; Khan, A. Antihepatotoxic effect of *Nymphaea stellata* Willd against Carbon tetrachloride hepatic damage in albino rats. *J. Ethnopharmacol.* **2004**, *91*, 61–64.

Bodakhe, S.H.; Ram, A. Hepatoprotective properties of *Bauhinia variegata* bark extract. *Yakugaku Zasshi.* **2007**, *127*, 1503–1507.

Bosetti, C.; Levi, F.; Boffetta, P.; Lucchini, F.; Negri, E.; La Vecchia, C. Trends in mortality from hepatocellular carcinoma in Europe, 1980–2004. *Hepatology* **2008**, *48*, 137–145.

Bosetti, C.; Levi, F.; Lucchini, F.; Zatonski, W.A.; Negri, E.; La Vecchia, C. Worldwide mortality from cirrhosis: An update to 2002. *J. Hepatol.* **2007**, *46*, 827–839.

Carmo, R.F.; Cavalcanti, M.S.M.; Moura, P. Role of Interleukin-22 in chronic liver injury. *Cytokine* **2017**, *98*, 107–114.

Chatterjee, T.K. *Medicinal plants with hepatoprotective properties in herbal opinions*, Vol. III. Calcutta: Books & Allied, **2000**. p. 155.

Chattopadhyay, R.R. Possible mechanism of hepatoprotective activity of *Azadirachta indica* leaf extract: Part II. *J. Ethnopharmacol.* **2003**, *89*, 217–219.

Christopher, T.B.; Taosheng, C. Hepatotoxicity of herbal supplements mediated by modulation of cytochrome P450. *Int. J. Mol. Sci.* **2017**, *18*(11), 2353.

Dey, P.; Saha, M.R. Heptotoxicity and the present herbal hepatoprotective scenario. *Int. J. Pharm.* **2013**, *7*, 265–273.

Dianzani, M.U.; Muzia, G.; Biocca, M.E.; Canuto, R.A. Lipid peroxidation in fatty liver induced by caffeine in rats. *Int. J. Tissue React.* **1991**, *13*, 79–85.

Ellahi, B.I.; Salman, M.I.; Sheikh, S.A.; Summra, E. Hepatoprotective and hepatocurative properties of alcoholic extract of *Carthamus oxyacantha* seeds. *Afr. J. Plant Sci.* **2014**, *8*(1), 34–41.

Friedman, S.L. Hepatic stellate cells: Protean, multifunctional, and enigmatic cells of the liver. *Physiol. Rev.* **2008**, *88*, 125–172.

Gao, B.; Radaeva, S.; Park, O. Liver natural killer and natural killer T cells: Immunobiology and emerging roles in liver diseases. *J. Leukoc. Biol.* **2009**, *86*, 513–528.

Gressner, O.A.; Rizk, M.S.; Kovalenko, E.; Weiskirchen, R.; Gressner, A.M. Changing the pathogenetic roadmap of liver fibrosis? Where did it start; Where will it go? *J. Gastroenterol. Hepatol.* **2008**, *23*, 1024–1035.

Guo, R.; Ren, J. Alcohol and acetaldehyde in public health: From marvel to menace. *Int. J. Environ. Res. Public Health* **2010**, *7*(4),1285–1301.

Handa, S.S.; Kapoor, V.K. *International book of pharmacognosy.* New Delhi: Vallabh Prakashan, **1989**. p. 125.

Hinson, J.A. Biochemical toxicology of acetaminophen. *Rev. Biochem. Toxicol.* **1980**, *2*, 103–129.

Kim, W. R.; Brown, R. S.; Terrault, N. A.; El-Serag, H. Burden of liver disease in the United States: Summary of workshop. *Hepatology* **2002**, *36*(1), 227–242.

Kumar, A. A review on hepatoprotective herbal drugs. *Int. J. Res. Pharm. Chem.* **2012**, *2*(1), 92–102.

Kumar, C.H.; Ramesh, A.; Kumar, J.N.S.; Ishaq, B.M. A review on hepatoprotective activity of medicinal plants. *Int. J. Pharm. Sci. Res.* **2011**, *2*, 501–515.

Lee, K.J.; Woo, E.R.; Choi, C.Y. Protective effect of acteoside on carbon tetrachloride induced hepatotoxicity. *Life Sci.* **2004**, *74*, 1051–1064.

Lin, C.C.; Tsai, C.C.; Yen, M.H. The evaluation of hepatoprotective effects of Taiwan folk medicine "Teng–Khia–U." *J. Ethnopharmacol.* **1995**, *45*, 113–123.

Maity, T.; Ahmad, A. Protective effect of *Mikania scandens* (L.) Willd. against isoniazid induced hepatotoxicity in rats. *Int. J. Pharm. Pharm. Sci.* **2012**, *4*, 466–469.

Mandayam, S.; Jamal, M.M.; Morgan, T.R. Epidemiology of alcoholic liver disease. *Semin. Liver Dis.* **2004**, *24*, 217–232.

Mitchell, J.R. Acetaminophen toxicity. *N. Engl. J. Med.* **1988**, *319*, 1601–1602.

Mitchell, L.D.; Jollow, D.J.; Potter, W.Z.; Davis, D.C.; Gillette, J.R.; Brodie, B.B. Acetaminophen-induced hepatic necrosis. I. Role of drugs metabolism. *J. Pharmacol. Exp. Ther.* **1973**, *187*, 185–194.

Ravikumar, V.; Shivashangari, K.S.; Devaki, T. Hepatoprotective activity of *Tridax procumbens* against D-galatosamine/lipopolysaccharide-induced hepatitis in rats. *J. Ethnopharmacol.* **2005**, *101*, 55–60.

Recknagel, R.O. A new direction in the study of carbon tetrachloride hepatotoxicity. *Life Sci.* **1983**, *33*, 401–408.

Roger, D.; Pamplona, R. Liver toxicity. *Encyclopedia of medicinal plants*. London: Editorial Safeliz, **2001**. pp. 392–395.

Shanani, S. Evaluation of hepatoprotective efficacy of APCL-A polyherbal formulation *in vivo* in rats. *Indian Drugs* **1999**, *36*, 628–631.

Smuckler, E.A. Alcoholic drink: Its production and effects. *Fed. Proc.* **1975**, *34*, 2038–2044.

Stierum, R.; Heijne, W.; Keinhus, A.; vanOmen, B.; Groten, J. Toxicogenomics concept and applications to study hepatic effects of food additives and chemicals. *Toxicol. Appl. Pharmacol.* **2005**, *207*(2), 179–188.

Subramoniam, A.; Pushpangadan, P. Development of phytomedicine for liver diseases. *Indian J. Pharmacol.* **1999**, *31*, 166–175.

Sumanth M. Screening models for hepatoprotective agents. *Pharm. Rev.* **2007**, *5*, 2.

Tarantino, G.; Di Minno, M.N.; Capone, D. Drug-induced liver injury: Is it somehow foreseeable? *World J. Gastroenterol.* **2009**, *15*, 2817–2833.

Teschke, R.; Wolff, A.; Frenzel, C.; Schulze, J.; Eickhoff, A. Herbal hepatotoxicity: A tabular compilation of reported cases. *Liver Int.* **2012**, *32*(10), 1543–1556.

Tripathi, M.; Singh, B.K.; Mishra, C.; Raisuddin, S.; Kakkar, P. Involvement of mitochondria mediated pathways in hepatoprotection conferred by *Fumaria parviflora* Lam. extract against nimesulide induced apoptosis in vitro. *Toxicol. In Vitro* **2010**, *24*, 495–508.

Tuma, D.J.; Sorrell, M. Alcohol and alcoholic liver disease. *Semin. Liver Dis.* **2004**, *24*, 215

Wang, H.; Lafdil, F.; Kong, X.; Gao, B. Signal transducer and activator of transcription 3 in liver diseases: A novel therapeutic target. *Int. J. Biol. Sci.* **2004**, *7*, 536–550.

Wendcl, A.; Feurensteins, S.; Konz, K.H. Acute paracetamol intoxication of starved mice leads to lipid peroxidation in vivo. *Biochem. Pharmacol.* **1987**, *28*, 2051–2053.

WHO. *Fact sheets.* World Health Organisation. **2005**.

WHO. *Regional strategy for the prevention and control of viral hepatitis.* World Health Organisation. Regional Office, New Delhi. **2013**.

Zhuang, H.; Zhang, J.H. Epidemiology of alcoholic liver disease. *Chin. J. Gastroenterol.* **2003**, *8*, 294–297.

CHAPTER 2

Hepatotoxicity and Hepatoprotective Assay

ABSTRACT

The liver is the largest and vital organ and performs various physiological and metabolic functions in the body. Hepatotoxicity is generally applicable to liver damage caused by chemicals. Drug-induced liver injury is one of the causes of acute and chronic liver disease. This chapter gives in-depth information on hepatotoxic agents and their evaluation methods. Also it explained the evaluation of hepatoprotective activity and biochemical parameters estimated.

INTRODUCTION

The therapeutic value, efficacy, and toxicity of drugs may be evaluated in animals (rats, mice, etc.) experimentally made sick, followed by clinical trials. Detailed biochemical and other in vitro assays are obligatory to establish the mechanism of action. Both in vivo and in vitro test systems are employed to assess antihepatotoxic and hepatoprotective activity. These systems measure the ability of the test drug to prevent or cure liver toxicity (induced by various hepatotoxins) in experimental animals (Surendran et al., 2011). Several authors like Delgado-Montemayor et al. (2015) have given the models of hepatoprotective activity assessment.

EVALUATION OF HEPATOTOXIC AGENTS

IN VITRO MODELS

Fresh hepatocyte preparations and primary cultured hepatocytes are cultured to study the antihepatotoxic activity of drugs. Hepatocytes are treated with hepatotoxin and the effect of the test drug on the same is evaluated. The activities of the transaminases released into the medium are determined. An augmented activity of marker transaminases in the medium indicates liver damage. Parameters such as hepatocytes multiplication, morphology, macromolecular synthesis, and oxygen consumption are determined (Pradeep et al., 2009).

IN VIVO MODELS

A toxic dose or repeated doses of a known hepatotoxin is administered to induce liver damage in experimental animals. The test substance is administered along with, prior to and/or after the toxin treatment. Liver damage and recovery from damage are assessed by quantifying serum marker enzymes, bilirubin (Bil), bile flow, histopathological changes, and biochemical changes in the liver. An augmented level of liver marker enzymes such as glutamate pyruvate transaminase, glutamate oxaloacetate transaminase, and alkaline phosphatase (ALP) in the serum indicates liver damage (Amat et al., 2010). The therapeutic efficacy of a drug against diverse hepatotoxins differs especially when their mechanism of action varies. Consequently, the efficacy of each drug has to be tested against hepatotoxins which act by varied methods.

HEPATOTOXIC AGENTS

More than 600 chemicals can cause damage in the liver, one of which is carbon tetrachloride (CCl_4). Ahmad and Tabassum (2012) discussed various experimental models used for the study of antihepatotoxic agents. Several chemicals have been known to induce hepatotoxicity. CCl_4, galactosamine (GalN), D-GalN/lipopolysaccharide, thioacetamide, antitubercular drugs, paracetamol, arsenic, etc., are used to induce experimental hepatotoxicity in laboratory animals.

CCL_4

The progress in industrialization has blessed mankind with a technologically superior lifestyle but poor management of industrial waste has in turn poisoned nature. One such chemical is CCl_4, which is a potent environmental toxin emitted from chemical industries and its presence in the atmosphere is increasing at an alarming rate. The presence of CCl_4 in the human body is reported to cause liver damage through free-radical-mediated inflammatory processes. Kupffer cells present in the liver are potentially more sensitive to oxidative stress than hepatocytes. Kuffer cells produced tumor necrosis factor-α in response to reactive oxygen species (ROS) that might further cause inflammation or apoptosis. Liver injury due to CCl_4 in rats was first reported in 1965 and has been widely and successfully used by many investigators. CCl_4 is metabolized by cytochrome P-450 in endoplasmic reticulum and mitochondria with the formation of CCl_3O^-, a reactive oxidative free radical, which initiates lipid peroxidation (LPO). Administration of a single dose of CCl_4 to a rat produces, within 24 h, a centrilobular necrosis and fatty changes. The poison reaches its maximum concentration in the liver within 3 h of administration. Thereafter, the level falls and by 24 h there is no CCl_4 left in the liver. The development of necrosis is associated with leakage of hepatic enzymes into the serum. A dose of CCl_4 that induces hepatotoxicity ranges from 0.1 to 3 mL/kg administered intraperitoneally.

CCl_4 is one of the most commonly used hepatotoxins in the experimental study of liver diseases (Johnston and Kroening, 1998). The hepatotoxic effects of CCl_4 are largely due to its active metabolite, trichloromethyl radical (Johnston and Kroening, 1998; Srivastava et al., 1990). These activated radicals bind covalently to the macromolecules and induce peroxidative degradation of membrane lipids of endoplasmic reticulum rich in polyunsaturated fatty acids. This leads to the formation of lipid peroxides, which in turn give products like malondialdehyde (MDA) that cause damage to the membrane. This lipid peroxidative degradation of biomembranes is one of the principal causes of hepatotoxicity of CCl_4 (Cotran et al., 1994; Kaplowitz et al., 1986). This is evidenced by an elevation in the serum marker enzymes namely aspartate aminotransferase (AST), alanine aminotransferase (ALT), and ALP.

The effect of CCl_4 administration on liver mitochondrial function was studied in rats by Padma and Setty (1999). The following changes were observed in mitochondria due to the administration of CCl_4. A decrease in the rate of respiration, respiratory control ratio, and phosphate/oxygen ratio was observed using glutamate and malate or succinate as substrates. A decrease in the activities of NADH dehydrogenase (35%), succinate dehydrogenase (76%), and cytochrome *c* oxidase (51%) was found. The rate of electron transfer through site I, site II, and site III was studied independently and found to be significantly decreased. A decrease in the content of cytochrome aa3 (34%) was seen. A significant decrease in the levels of phospholipids particularly cardiolipin and a significant increase in the lipid peroxide level was observed. The CCl_4 induced toxicity may be partly due to the LPO and partly due to the effect on protein synthesis.

THIOACETAMIDE

Thioacetamide interferes with the movement of RNA from the nucleus to cytoplasm which may cause membrane injury. A metabolite of thioacetamide (perhaps S-oxide) is responsible for hepatic injury. Thioacetamide reduces the number of viable hepatocytes as well as the rate of oxygen consumption. It also decreases the volume of bile and its content, that is, bile salts, cholic acid, and deoxycholic acid. The dose of thioacetamide is 100 mg/kg, subcutaneously (Amin et al., 2012).

ACETAMINOPHEN (PARACETAMOL)

Nonsteroidal anti-inflammatory drugs are widely used for the alleviation of pain, fever, and inflammation. They are the most widely prescribed medications in the world and are used by millions of patients on a daily basis. However, excessive consumption of nonsteroidal anti-inflammatory drugs has been related to severe side effects caused by oxidative stress, resulting in considerable morbidity and mortality (Vonkeman and van de Laar, 2010). Acetaminophen (also known as paracetamol), a nonprescription drug, is a safe and effective analgesic and antipyretic drug when used at therapeutic doses. However, an acute or cumulative overdose can cause severe liver injury that may progress to acute liver failure. In fact, acetaminophen is the most common cause of acute liver failure in developed countries (Jaeschke and Bajt, 2005). The liver is the main organ involved in the metabolism of acetaminophen. At therapeutic doses, acetaminophen is eliminated via glucuronidation and sulfation reactions. However, at high doses, the conjugation pathways are saturated, and part of the drug is converted by cytochrome P450 2E1 (CYP2E1) to the highly reactive metabolite *N*-acetyl-*p*-benzoquinone imine (NAPQI) that reacts with sulfhydryl groups. Reduced glutathione (GSH) initially traps NAPQI, and the GSH adduct is excreted. However, when GSH is depleted, NAPQI reacts with cellular proteins, including a number of mitochondrial proteins, to form NAPQI adducts. Consequences of this process are the inhibition of mitochondrial respiration and ATP depletion, as well as mitochondrial oxidative stress (Jaeschke et al., 2003, 2011; McGill and Jaeschke, 2013). This results in increased susceptibility to liver injury by ROS, including hydrogen peroxide (H_2O_2), superoxide anions (O_2^-), and hydroxyl radicals (·OH). In addition to reducing the GSH level, the acetaminophen overdose also reduces the antioxidant enzyme activities, increases LPO, and causes hepatic DNA fragmentation, which ultimately leads to cellular necrosis (Arai et al., 2014; Das et al., 2010; González-Ponce et al., 2016). Interestingly, acetaminophen-induced liver injury is one of the most widely used models to evaluate the hepatoprotective potential of natural products (Jaeschke et al., 2013).

Acetaminophen (paracetamol) administration causes necrosis of the centrilobular hepatocytes characterized by nuclear pyknosis and eosinophilic cytoplasm followed by the large excessive hepatic lesion. The covalent binding of NAPQI, an oxidative product of paracetamol to sulfhydryl groups of protein, results in lipid peroxidative degradation of GSH level and thereby, produces cell necrosis in the liver. The dose of paracetamol is 1 g/kg post oral.

ACRYLAMIDE (AA)

AA is a water-soluble vinyl monomer used in the production and synthesis of polyacrylamides. Monomeric AA has been shown to cause diverse toxic effects in experimental animals. AA is carcinogenic to laboratory rodents and is described by the International Agency for Research of Cancer as a probable carcinogen to humans. In the human body, AA is oxidized to the epoxide glycidamide (2,3-epoxypropionamide) via an enzymatic reaction involving CYP2E1. AA undergoes biotransformation by conjugation with GSH and is probably being the major route of detoxification. A daily dose of 6 mg/kg, intraperitoneal (i.p.) for 15 days produces hepatotoxicity in female Sprague–Dawley rats (Khan et al., 2011).

ADRIAMYCIN

Adriamycin (doxorubicin) is an antibiotic isolated from *Streptomyces peucetius* subspecies *cesius*. Adriamycin is considered to be one of the most compelling drugs against a wide range of tumors. However, its clinical potential is contraindicated due to severe cytotoxic side effects. Based on the in vitro model of toxicity using isolated hepatocytes and liver microsomes, adriamycin has been shown to undergo redox cycling between semiquinone and quinone radicals during its oxidative metabolism. A single dose of 10 mg/kg body weight of adriamycin is given to rats to induce hepatotoxicity (Sakr and Abo-El-Yazid, 2011).

ALCOHOL

Liver is among the organs most susceptible to the toxic effects of ethanol. Alcohol consumption is known to cause fatty infiltration, hepatitis, and cirrhosis. Fat infiltration is a reversible phenomenon that occurs when alcohol replaces fatty acids in the mitochondria. Hepatitis and cirrhosis may occur because of enhanced lipid peroxidative reaction during the microsomal metabolism of ethanol. The mechanisms responsible for the effects of alcohol, an increase in hepatic LPO leads to alteration in membrane phospholipid composition. The effects of ethanol have been suggested to be a result of the enhanced generation of oxyfree radicals during its oxidation in the liver. The peroxidation of membrane lipids results in loss of membrane structure and integrity. These results in elevated levels of glutamyl transpeptidase; a membrane-bound enzyme in serum. Ethanol inhibits glutathione peroxidase, decrease the activity of catalase, superoxide dismutase, along with increase of antioxidant enzymes superoxide dismutase, glutathione peroxidase are speculated to be due to the damaging effects of free radicals produced following ethanol exposure or alternatively could be due to a direct effect of acetaldehyde, formed by oxidation of ethanol. Continuous administration of ethanol (7.9 g/kg body weight/day) for a period of 6 weeks induces liver damage in rats (Ilaiyaraja and Khanum, 2011).

ALPHA-NAPHTHYLISOTHIOCYANATE (ANIT)

ANIT injures bile duct epithelium and hepatic parenchymal cells in rats. It is commonly believed that ANIT undergoes bioactivation by hepatic, cytochrome P450-dependent mixed-function oxidases. Rats intoxicated once with ANIT (75 mg/kg, i.p.) show liver cell damage and biliary cell damage with cholestasis at 24 h, but not 12 h, after intoxication (Ohta et al., 2007).

ANTITUBERCULAR DRUGS

Drug-induced hepatotoxicity is a potentially serious adverse effect of the currently used antitubercular therapeutic regimens containing isoniazid (INH), rifampicin, and pyrazinamide. Adverse effects of antitubercular therapy are sometimes potentiated by multiple drug regimens. Thus, though INH, rifampicin, and pyrazinamide each in itself are potentially hepatotoxic, when given in combination, their toxic effect is enhanced. INH is metabolized to monoacetyl hydrazine, which is further metabolized to a toxic product by cytochrome P450 leading to hepatotoxicity. Patients on concurrent rifampicin therapy have an increased incidence of hepatitis (Grant and Rokey, 2012). This has been postulated due to rifampicin-induced cytochrome P450 enzyme-induction, causing an increased production of the toxic metabolites from acetylhydrazine (AcHz). Rifampicin also increases the metabolism of INH to isonicotinic acid and hydrazine, both of which are hepatotoxic. The plasma half-life of AcHz (metabolite of INH) is shortened by rifampicin and AcHz is quickly converted to its active metabolites by increasing the oxidative elimination rate of AcHz, which is related to the higher incidence of liver necrosis caused by INH and rifampicin in combination. Rifampicin induces the hydrolysis pathway of INH metabolism into the hepatotoxic metabolite hydrazine. Pharmacokinetic interactions exist between rifampicin and pyrazinamide in tuberculosis patients when these drugs are administered concomitantly. Pyrazinamide decreases the blood level of rifampicin by decreasing its bioavailability and increasing its clearance. Pyrazinamide, in combination with INH and rifampicin, appears to be associated with an increased incidence of hepatotoxicity (Rao et al., 2012). INH and rifampicin (50 mg/kg, i.p.) induce hepatotoxicity in rats.

INH is one of the most important antituberculosis drugs and it undergoes hydrolysis in the liver via an enzymatic reaction with CYP2E1, resulting in the formation of hepatotoxic compounds.

CADMIUM

Cadmium, a heavy metal well known to be highly toxic to both humans and animals is distributed widely. The toxic effects of cadmium exposure are testicular atrophy, renal dysfunction, hepatic damage, hypertension, central nervous system injury, and anemia. Cadmium may induce oxidative damage in different tissues by enhancing the peroxidation of membrane lipids in tissues and altering the antioxidant systems of the cells. The peroxidative damage to the cell membrane may cause injury to cellular components due to the interaction of metal ions with the cell organelles (Renugadevi and Prabhu, 2010). Cadmium depletes GSH and protein-bound sulfhydryl groups resulting in enhanced production of ROS such as O_2^-, ·OH, and H_2O_2. These ROS result in increased lipid. Cadmium is given orally (3 mg/kg body weight/day) as cadmium chloride ($CdCl_2$) for 3 weeks to induce hepatotoxicity in rats (Murugavel and Pari, 2007).

ERYTHROMYCIN

Erythromycin estolate is a potent macrolide antibiotic, generates free radicals, and has been reported to induce liver toxicity. Erythromycin when given as erythromycin stearate (100 mg/kg body weight for 14 days) (Sambo et al., 2009) or erythromycin estolate (800 mg/kg/day for 15 days) to albino rats produces hepatotoxicity in them (Pari and Uma 2003).

GalN

D-GalN is a well-established hepatotoxicant that induces a diffuse type of liver injury closely resembling human viral hepatitis. D-GalN by its property of generating free radicals causes severe damage to the membrane and affects almost all organs of the human body. It presumably disrupts the synthesis of essential uridylate nucleotides resulting in organelle injury and ultimately cell death. The depletion of those nucleotides would impede the normal synthesis of RNA and consequently would produce a decline in protein synthesis. This mechanism of toxicity brings about an increase in cell membrane permeability leading to enzyme leakage and eventually cell death. The cholestasis caused by GalN may be from its damaging effects on bile ducts or ductules or canalicular membrane of hepatocytes. GalN decreases the bile flow and its content, that is, bile salts, cholic acid, and deoxycholic acid. GalN reduces the number of viable hepatocytes as well as the rate of oxygen consumption. Hepatic injury is induced by i.p. single-dose injection of D-GalN (800 mg/kg) (Kiso et al., 1983; Taye et al., 2009).

AZATHIOPRINE

As an antimetabolite, Azathioprine inhibits the de novo and salvage pathways of purine synthesis. Intraperitoneal injection of this drug results in not only lymphocyte suppression but also toxicity to bone marrow, gastrointestinal tract, and liver. This Azathioprine-induced hepatotoxicity was found to be associated with oxidative damage. Plants with antioxidative properties have been traditionally used to prevent diseases associated with free radicals. Typically, the administration of Azathioprine induces oxidative stress through depleting the activities of antioxidants and elevating the level of MDA in the liver. This escalates levels of ALT and AST in serum.

IBUPROFEN (IBU)

Baghdadi et al. (2016) have given the uses of IBU in human health and its effect on the liver. IBU is used mostly for inflammatory diseases such as rheumatoid arthritis, menstrual cramps, toothache, and dysmenorrhea (Chen et al., 2005; Mahalakshmi et al., 2010). IBU has the same action as that of aspirin and shares the same side effects to a lesser extent (Thangapandian, 2012). IBU has been used through different traditions and considered to have cooling, stimulant, carminative, digestive properties, used to treat cystitis (Aboelsoud, 2010), headaches, measles, rectal prolapse, prevent cancer (Li and Ben, 1999), relief of anxiety and insomnia, and may have potential sedative, hypotensive, and muscle relaxant effects (Emamghoreishi et al., 2005), diuretic, antipyretic, stomachic, aphrodisiac, laxative, and anthelmintic properties (Deepa and Anuradha, 2011). On the other hand, IBU was commonly known to produce hepatotoxicity, inducing cholestatic hepatitis (Manov et al., 2006). Besides, IBU was also found to be most active in impairing gluconeogenesis from lactate, and in impairing albumin synthesis in vitro (Castell et al., 1988). A lower dose of IBU also induces more histopathological lesions in rats' liver than chlorpromazine, paracetamol, and others (Schoonen et al., 2007). Various theories have been proposed for the mechanism by which IBU damages the liver (Teoh and Farrell, 2003). It was reported to cause changes in the architecture of the hepatic cell, cell permeability, and to create ionic imbalance resulting in increased intracellular calcium concentration. Consequently, mitochondrial activity was inhibited, leading to the death of hepatic cells (Saraf et al., 1994).

LEAD

Lead is known to disrupt the biological systems by altering the molecular interactions, cell signaling, and cellular function. Exposure to even low levels of lead may have potentially hazardous effects on brain, liver, kidneys, and testes. Autopsy studies of lead-exposed humans indicate that among soft tissue, liver is the largest repository (33%) of lead, followed by the kidney. Lead-induced hepatic damage is mostly rooted in LPO and disturbance of the pro-oxidant–antioxidant balance by the generation of ROS. Hepatotoxicity can be induced by using lead acetate (550 ppm for 21 days in drinking water) (Haleagrahara et al., 2010) or lead nitrate (5 mg/kg body weight daily for 30 days) (Sharma and Pandey, 2010).

METHOTREXATE (MTX)

MTX is a folic acid competitor, used for chemotherapy in the treatment of diverse malignancies (acute lymphoblastic leukemia) as well as diverse inflammatory diseases (Coleshowers et al., 2010). MTX has also been used successfully in psoriasis since 1951, and over the past 10 years in psoriatic and rheumatoid arthritis. It is on the World Health Organization's list of essential medicines. The efficacy of MTX is restricted by its toxicity and severe side effects including hepatitis, liver cirrhosis and fibrosis, hepatocytes hypertrophy, hepatocellular necrosis, and death (Coleshowers et al., 2010). MTX toxicity has serious drawbacks on the hematopoietic system and liver enzymes (Coleshowers et al., 2010). It is toxic to the liver, kidneys, respiratory, and reproductive systems for long-term therapy even at very low doses. MTX causes LPO indicated by high levels of thiobarbituric acid (TBA) reactive substances. It also lowered the levels of antioxidant enzymes like superoxide dismutase, catalase, and glutathione reductase, indicating oxidative stress and distressing the antioxidant enzyme defense system (Coleshowers et al., 2010; Montessar et al., 2017).

MICROCYSTIN

Microcystin-LR, a cyclic heptapeptide synthesized by the blue-green algae, *Microcystis aeruginosa*, is a potent hepatotoxin. Pathological examination of livers from mice and rats that received microcystin-LR revealed severe, peracute, diffuse, centrilobular hepatocellular necrosis, and hemorrhage. Mice receiving sublethal doses of microcystin (20 μg/kg) for 28 weeks developed neoplastic liver nodules (Clark et al., 2007).

PHALLOIDIN

Phalloidin, is one of the main toxins of *Amanita phalloides*. It induces hepatotoxicity in rats at an intravenous dose of 50 μg/100 g body weight. Phalloidin also induces a cytolytic lesion. Phalloidin causes severe liver damage characterized by marked cholestasis, which is due in part to irreversible polymerization of actin filaments. Liver uptake of this toxin through the transporter OATP1B1 is inhibited by the bile acid derivative BALU-1, which does not inhibit the sodium-dependent bile acid transporter sodium taurocholate co-transporting polypeptide (NTCP) (Herraez et al., 2009).

TAMOXIFEN (TAM)

TAM citrate is a nonsteroidal antiestrogen drug used in the treatment and prevention of hormone-dependent breast cancer. In high dose, it is a known liver carcinogen in rats, due to oxygen radical overproduction and LPO via formation of lipid peroxy radicals. An i.p. dose of 45 mg/kg/day of TAM citrate in 0.1 mL dimethyl sulfoxide and normal saline for 6 days induce hepatotoxicity in rats (El-Beshbishy et al., 2010).

Tert-BUTYL HYDROPEROXIDE (t-BHP)

Hepatotoxicity and oxidative stress is induced in male rats at various times (0–24 h) after *t*-BHP (0, 0.2, 0.5, 1 or 3 mmol/kg, i.p.) treatment. Serum hepatotoxicity parameters have been reported to increase from 2 h following 1 mmol/kg *t*-BHP and maximum values at 8 h. The elevation of hepatotoxic parameters and plasma MDA has been observed from 0.5 to 1 mmol/kg *t*-BHP, respectively, in a dose-dependent manner. Being a short-chain analog of lipid peroxide, *t*-BHP is metabolized into free-radical intermediates by cytochrome P450 in hepatocytes, which initiate LPO, GSH depletion, and cell damage (Oh et al., 2012).

BIOCHEMICAL PARAMETERS

The hepatoprotective activity can be assessed using various biochemical parameters like ALT, AST, ALP, serum Bil, total protein, and serum antioxidant enzymes along with histopathological studies of liver tissue which are given below.

EXPERIMENTAL ANIMALS

Male or female Wistar albino rats, weighing between 150 and 200 g are to be used for the hepatoprotective activity. The animals are housed in polypropylene cages and maintained at 24 ± 2 °C under 12 h light/dark cycle are fed *ad libitum* with standard pellet diet and should have free access to water. They are initially acclimatized for the study protocol and study protocol is to be approved by the Institutional Ethical Committee as per the requirements of the Committee for the Purpose of Control and Supervision of Experiments on Animals (CPCESA), New Delhi or as per the Organisation for Economic Co-operation and Development (OECD/OCDE) Guidelines or NIH Guidelines (1985). Before conducting the experiment, ethical clearance has to be obtained from the Institutional Animal Ethical Committee of the Institution where the work is being carried out.

ACUTE ORAL TOXICITY STUDIES

Wistar albino rats 150–200 g (male), are maintained under standard husbandry conditions, and are used for all sets of experiments. The acute oral toxicity study is to be carried out as per the guidelines set by the OECD, received draft guidelines 423, received from CPCSEA, Ministry of Social Justice and Empowerment, Government of India. Animals are allowed to take standard laboratory feed and tap water.

EVALUATION OF HEPATOPROTECTIVE ACTIVITY

In the paracetamol-induced liver injury model, paracetamol (2 g/kg) suspension prepared using 0.1% Tween 80, is administered to all animals except the animal of the normal control group. Silymarin (100 mg/kg, p.o.) is used as a standard. The animals were segregated into six groups of six each. Group 1, which served as normal control receiving 1.5% Tween 80. Group 2 received paracetamol (2 g/kg, p.o.) single dose on 6th day. Group 3 received paracetamol (2 g/kg, p.o.) single dose and Silymarin (100 mg/kg, p.o.) simultaneously for 7 days. Group 4 received paracetamol (2 g/kg, p.o.) single dose and alcoholic extract (200 mg/kg, p.o.) simultaneously for 7 days. Group 5 received paracetamol (2 g/kg, p.o) single dose and chloroform extract (200 mg/kg, p.o.) simultaneously for 7 days. Group 6 received paracetamol (2 g/kg, p.o.) single dose and aqueous extract (200 mg/kg, p.o.) simultaneously for 7 days. On the seventh day of the start of respective treatment, the rats are anesthetized by light ether anesthesia and the blood is withdrawn from retro orbital plexus. It is allowed to coagulate for 30 min and serum is separated by centrifugation at 2500 rpm. The serum is used to estimate serum glutamate pyruvate transaminase, serum glutamate oxaloacetate transaminase, and ALP.

COLLECTION OF BLOOD SAMPLES

Prior to the termination of the studies, rat feed is withdrawn in order to fast the rats and their beddings changed. However, drinking water is provided *ad libitum*. On the termination day, rats are anesthetized using inhalational diethyl ether after which blood samples for hepatic function tests are obtained directly from the heart chamber using a 21G needle mounted on a 5 mL syringe plunger and collected into plain sample bottle.

COLLECTION OF LIVERS FOR HEPATIC TISSUE OXIDATIVE STRESS MARKERS

After blood collection through cardiac puncture, a deep longitudinal incision is made into the ventral surface of the rat abdomen. The livers are identified and carefully dissected out *en bloc* from each rat. The right lobe of the liver is rinsed in ice cold 1.15% KCl solution in order to preserve the oxidative enzyme activities of the liver before being stored in a clean sample bottle which itself is in an ice-pack filled cooler. This is to prevent the breakdown of the hepatic antioxidant biomarkers.

DETERMINATION OF LIVER AND RENAL TISSUE SUPEROXIDE DISMUTASE ACTIVITY

Superoxide dismutase activity is determined by its ability to inhibit the auto-oxidation of epinephrine by the increase in absorbance at 480 nm as described by Sun and Zigman (1978). The reaction mixture (3 mL) contains 2.95 mL 0.05 M sodium carbonate buffer (pH 10.2), 0.02 mL of liver homogenates, and 0.03 mL of epinephrine in 0.005 N HCl is used to initiate the reaction. The reference curette contained 2.95 mL buffer, 0.03 mL of substrate (epinephrine), and 0.02 mL of water. Enzyme activity is calculated by measuring the change in absorbance at 480 nm for 5 min.

DETERMINATION OF LIVER TISSUE CATALASE ACTIVITY

Hepatic tissue catalase activity is determined according to Kakkar et al. (1984) by measuring the decrease in absorbance at 240 nm due to the decomposition of H_2O_2 in a UV recording spectrophotometer. The reaction mixture (3 mL) contained 0.1 mL of serum in phosphate buffer (50 mM, pH 7.0) and 2.9 mL of 30 mM of H_2O_2 in the phosphate buffer (pH 7.0). An extinction coefficient for H_2O_2 at 240 nm of 40.0 M^{-1} cm^{-1} according to Aebi (1984) is used for calculation. The specific activity of catalase is expressed as moles of H_2O_2 reduced per min per mg protein.

DETERMINATION OF LIVER TISSUE REDUCED GSH ACTIVITY

The reduced GSH content in the liver tissue is estimated according to the method described by Sedlak and Lindsay (1968). To the homogenate 10% tricarboxylicacid (TCA) is added and centrifuged. 1 mL of the supernatant is treated with 0.5 mL of Ellman's reagent (19.8 mg of 5,5-dithiobisnitro benzoic acid in100 mL of 0.1% sodium nitrate) and 3.0 mL of phosphate buffer (0.2 M, pH 8.0). The absorbance is read at 412 nm.

DETERMINATION OF LIVER TISSUE MDA ACTIVITY

MDA an index of LPO is determined using the method of Buege and Aust (1978). 1 mL of supernatant is added to 2 mL of (1:1:1 ratio) TCA–TBA–HCl reagent (TBA 0.37%, 0.24 N HCl, and 15% TCA). TCA–TBA reagent boiled at 100 °C for 15 min, and allowed to cool. Flocculent material is removed by centrifuging at 3000 rpm for 10 min. The supernatant is removed and the absorbance is read at 532 nm against a blank. MDA is calculated using the molar extinction for MDA–TBA complex of $1.56 \times 10^5 m^{-1} cm^{-1}$.

DETERMINATION OF SERUM HEPATIC FUNCTION PARAMETERS

SERUM ASPARTATE TRANSAMINASE AND ALANINE TRANSAMINASE DETERMINATION

The method of Reitman and Frankel (1957) is used. Into a test tube, 0.1 mL substrate (D,L-aspartate, 0.2 mol/L and α-ketoglutaric acid, 1.8 mmol/L in phosphate buffer, pH 7.5) solution is pipetted and placed in a 37 °C water bath to warm 0.2 mL plasma is added and shaken gently to mix. Exactly one hour after adding plasma, 1.0 mL color reagent (2,4-dinitrophenylhydrazine approximately 20 mg/100 mL, in 10% HCl solution) is added and mixed gently and left at room temperature (18–26 °C). 20 min after adding color reagent, 10 mL 0.40 N sodium hydroxide solution is added and mixed by inversion. After 5 min, absorbance is read at 340 nm using water as a reference. AST activity in Sigma-Frankel units/mL is determined from the calibration curve. The same procedure is carried out for ALT except that procedures are started 30 min after starting AST. Substrate for ALT is L-alanine (0.2 mol/L) and α-ketoglutaric acid

(1.8 mmol/L) in phosphate buffer, pH 7.5 (Reitman and Frankel,1957).

SERUM ALKALINE PHOSPHATASE DETERMINATION

Alkaline phosphatase is determined using the colorimetric endpoint method of Tietz et al. (1983) and adapted by Teco Diagnostics Kits. The principle is based on the fact that alkaline phosphatase acts upon the AMP-buffered sodium thymolphthalein monophosphate. The addition of an alkaline reagent stops enzyme activity and simultaneously develops a blue chorogem, which is measured photometrically. For each sample, 0.5 mL of alkaline phosphatase substrate is dispensed into labeled test tubes and equilibrated to 37 °C for 3 min. At timed intervals 0.5 mL of standard, control, and sample were added to their respective test tube and mixed gently, deionized water is used as blank. The samples are incubated for exactly 10 min at 37 °C following the same sequences, 2.5 mL of alkaline phosphatase color developer at timed intervals are added. The wavelength of the spectrometer is set at 590 nm.

BILIRUBIN

Bilirubin (Bil) is the breakdown product of normal heme—a part of hemoglobin in red blood cells—catabolism of aged erythrocytes. Bilirubin, loosely bound to albumin in plasma to form a soluble species taken up from the Disse spaces of liver sinusoids into hepatocytes, where it is esterified at its propionyl sites with glucuronic acid under the catalytic activity of uridinediphosphoglucuronate 1A1 transferase enzymes. Esterified Bil is excreted into bile as water-soluble Bil diglucuronide. Serum concentration of Bil is a marker of the liver's ability to take up Bil from the plasma into the hepatocyte, conjugate it with glucuronic acid, and excrete Bil glucuronides into bile. An elevated level of serum conjugated Bil implies regurgitation of Bil glucuronides from hepatocytes back into plasma, usually because of intrahepatic or extrahepatic obstruction to bile outflow and cholestasis. The liver has substantial reserve capacity, and normal serum Bil levels can be maintained until there is enough injury to reduce the liver's capacity to clear Bil from plasma. Serum concentration of Bil is very specific for potentially serious liver damage and is an important indicator of the loss of liver function (Feild et al., 2008).

Reduction in the level of serum Bil is a strong indication of restoring normal liver function (Alqasoumi and Abdel-Kader, 2012).

GAMMA GLUTAMYL TRANSPEPTIDASE (GGT)

Serum GGT (also gamma-glutamyl-transferase) is specific to liver injury and a more sensitive marker for cholestatic damage than ALP. GGT may be elevated with even minor, subclinical levels of liver dysfunction. GGT is raised in alcohol toxicity following several days of moderate ingestion. Rifampin, phenytoin, or barbiturates, all resulted in the elevation of GGT level. An isolated GGT elevation in these situations does not indicate hepatocellular injury. The GGT level will return to normal after discontinuation of the offending agent. Hepatic dysfunction should be considered if the GGT elevation is associated with other abnormalities in liver biochemistry. Hepatoprotective agents will reduce the elevated level of GGT (Alqasoumi and Abdel-Kader, 2012).

DETERMINATION OF SERUM LIPIDS, ALBUMIN, PROTEINS, AND FASTING BLOOD GLUCOSE

Serum total protein is estimated by Biuret method (Treitz, 1970) while that of albumin is determined by bromocresol green (Waterborg, 2002). The total Bil and the conjugated Bil are determined by Jendrassike–Grof method (Zelenka et al., 2008). The serum total cholesterol and its fractions are determined by the Lierbermane–Burchard quantitative test (Cox and García-Palmieri, 1990). Fasting blood glucose is determined by the glucose oxidase method of Trinder (1969), using One Touch Basic Blood Glucose Monitoring System.

HISTOPATHOLOGICAL STUDIES

One animal from the treated groups showing maximal activity as indicated by improved biochemical parameters from each test, positive control, hepatotoxin, and control groups are utilized for this purpose. The animals are sacrificed and the abdomen is cut open to remove the liver. Then 5 mm thick piece of the liver is fixed in Bouin's solution (mixture of 75 mL of saturated picric acid, 25 mL of 40% formaldehyde, and 5 mL of glacial acetic acid) for 12 h and then embedded in paraffin, using conventional methods and cut into 5 μm thick sections and stained, using hematoxylin–eosin dye, and finally observed under microscope for histopathological changes in liver architecture, and their photomicrographs were taken.

KEYWORDS

- **hepatotoxicity**
- **hepatotoxic agents**
- **hepatoprotective assay**
- **hepatoprotective activity**

REFERENCES

Aebi, H. Catalase *in vitro*. *Methods Enzymol.* **1984**, *105*, 121–126.

Ahmad, F.; Tabassum, N. Experimental models used for the study of antihepatotoxic agents. *J. Acute Dis.* **2012**, *1*(2), 85–89.

Alqasoumi, S.I.; Abdel-Kader, M.S. Screening of some traditionally used plants for their hepatoprotective effect. In: Phytochemicals as Nutraceuticals—Global Approaches to their Role in Nutrition and Health. Rao, V. (ed.), InTech., Rijeka, Croatia. http://www.intechopen.com/books **2012**.

Amat, N.; Upur, H.; Blazeković, B. *In vivo* hepatoprotective activity of the aqueous extract of *Artemisia absinthium* L. against chemically and immunologically induced liver injuries in mice. *J. Ethnopharmacol.* **2010**, *131*, 478–484.

Amin, Z.A.; Bilgen, M.; Alshawsh, M.A.; Ali, H.M.; Hadi, A.H.; Abdulla, M.A. Protective role of *Phyllanthus niruri* extract against thioacetamide-induced liver cirrhosis in rat model. *Evid. Based Complement. Altern. Med.* **2012**, *2012*, 241583.

Arai, T.; Koyama, M.; Kitamura, D.; Mizuta, R. Acrolein, a highly toxic aldehyde generated under oxidative stress in vivo, aggravates the mouse liver damage after acetaminophen overdose. *Biomed. Res.* **2014**, *35*, 389–395.

Awang, D. Milk Thistle. *Can. Pharm. J.* **1993**, *23*, 749–754.

Baghdadi, H.H.; El-Demerdash, F.M.; Radwan, E.H.; Hussein, S. The protective effect of *Coriandrum sativum* L. oil against liver toxicity induced by ibuprofen in rats. *J. Biosci. Appl. Res.* **2016**, *2*, 197–202.

Buege, J.A.; Aust, S.D. Microsomal lipid peroxidation. *Meth. Enzymol.* **1978**, *52*, 302–310.

Coleshowers, C.L.; Oguntibeju, O.O.; Ukpong, M.; Truter, E.J. Effects of methotrexate on antioxidant enzyme status in a rodent model. *Med. Tech. South Afr.*, **2010,** *24* (1), 5–9.

Clark,S.P.; Davis, M.A.; Ryan, T.P.; Searfoss, G.H.; Hooser, S.B. Hepatic gene expression changes in mice associated with prolonged sublethal microcystin exposure. *Toxicol. Pathol.* **2007**, *35*, 594– 605.

Cox, R.A.; García-Palmieri, M.R. Cholesterol, triglycerides, and associated lipoproteins. In: Walker, H.K.; Hall, W.D.; Hurst, J.W., (eds.) Clinical Methods: The History, Physical, and Laboratory Examinations. 3rd ed. Massachusetts, Boston: Butterworths; Chapter 31. Available from: http://www.ncbi.nlm.nih.gov/books/NBK351/ **1990**.

Cotran, R.S.; Kumar, V.; Robbins, S.L. Cell injury and cellular death. In, Robbin's *Pathologic Basis of Disease,* 5th ed., Prism Book Pvt. Ltd., **1994**.pp. 379–430.

Das, J.; Ghosh, J.; Manna, P.; Sil, P.C. Acetaminophen induced acute liver failure via oxidative stress and JNK activation: protective role of taurine by the suppression of cytochrome P450 2E1. *Free Radic. Res.* **2010**, *44*, 340–355.

Delgado-Montemayor, C.; Cordero-Perez, P.; Salazar-Aranda, R.; Waksman-Minsky, N. Models of hepaoprotective activity assessment. *Med. Univ.* **2015**, *17*(69), 222–228.

El-Beshbishy, H.A.; Mohamadin, A.M.; Nagy, A.A.; Abdel-Naim, A.B. Amelioration of tamoxifen-induced liver injury in rats by grape seed extract, black seed extract and curcumin. *Indian J. Exp. Biol.* **2010**, *48*, 280–288.

González-Ponce, H.A.; Martínez-Saldaña, M.C.; Rincón-Sánchez, A.R.; Sumaya-Martínez, M.T.; Buist-Homan, M.; Faber, K.N.; Moshage, H.; Jaramillo-Juárez, F. Hepatoprotective effect of *Opuntia robusta* and *Opuntia streptacantha* fruits against acetaminophen-induced acute liver damage. *Nutrients* **2016,** *8*, 607.

Grant, L.M.; Rockey, D.C. Drug-induced liver injury. *Curr. Opin. Gastroenterol.* **2012,** *28*, 198–202.

Haleagrahara, N.; Jackie, T.; Chakravarthi, S.; Rao, M.; Kulur, A. Protective effect of *Etlingera elatior* (torch ginger) extract on lead acetate-induced hepatotoxicity in rats. *J. Toxicol. Sci.* **2010**, *35*, 663–671.

Herraez, E.; Macias, R.I.; Vazquez-Tato, J.; Hierro, C.; Monte, M.J.; Marin, J.J. Protective effect of bile acid derivatives in phalloidin induced rat liver toxicity. *Toxicol. Appl. Pharmacol.* **2009**, *239*, 212–218.

Ilaiyaraja, N.; Khanum, F. Amelioration of alcohol-induced hepatotoxicity and oxidative stress in rats by *Acorus calamus*. *J. Diet Suppl.* **2011,** *8*, 331–345.

Jaeschke, H.; Bajt, M.L. Intracellular signaling mechanisms of acetaminophen-induced liver cell death. *Toxicol. Sci.* **2005**, *89*, 31–41.

Jaeschke, H.; Knight, T.R.; Bajt, M.L. The role of oxidant stress and reactive nitrogen species in acetaminophen hepatotoxicity. *Toxicol. Lett.* **2003**, *144*, 279–288.

Jaeschke, H.; McGill, M.R.; Williams, C.D.; Ramachandran, A. Current issues with acetaminophen hepatotoxicity—A clinically relevant model to test the efficacy of natural products. *Life Sci.* **2011**, *88*, 737–745.

Jaeschke, H.; Williams, C.D.; McGill, M.R.; Xie, Y.; Ramachandran, A. Models of drug-induced liver injury for evaluation of phytotherapeutics and other natural products. *Food Chem. Toxicol.* **2013**, *55*, 279–289.

Johnston, D.E.; Kroening, C. Mechanism of early carbon tetrachloride toxicity in cultured rat hepatocytes. *Pharmacol. Toxicol.* **1998**, *83*, 231–239.

Kakkar, P.; Das, B.; Viswanathan, P.N. A modified spectrophotometric assay of superoxide dismutase. *Indian J. Biochem. Biophys.* **1984**, *21*, 130–132.

Kaplowitz, N.; Aw, T.Y.; Simon, F.R.; Stolz, A. Drug-induced hepatotoxicity. *Ann. Int. Med.* **1986**, *104*, 826–839.

Khan, M.R.; Afzaal, M.; Saeed, N.; Shabbir, M. Protective potential of methanol extract of *Digera muricata* on acrylamide induced hepatotoxicity in rats. *Afr. J. Biotechnol.* **2011**, *10*, 8456–8464.

Kiso, Y.; Tohkin, M.; Hikino, H. Assay method for antihepatotoxic activity using galactosamine-induced cytotoxicity in primary cultured hepatocytes. *J. Nat. Prod.* **1983**, *46*, 841–847.

McGill, M.R.; Jaeschke, H. Metabolism and disposition of acetaminophen: Recent advances in relation to hepatotoxicity and diagnosis. *Pharm. Res.* **2013**, *30*, 2174–2187.

Montasser, A.O.S.; Saleh, H.; Ahmed-Farid, O.A.; Saad, A.; Marie, M.S. Protective effects of *Balanites aegyptiaca* extract, melatonin and ursodeoxycholic acid against hepatotoxicity induced by Methotrexate in male rats. *Asian Pac. J. Trop. Med.* **2017**, *10*(6), 557–565.

Murugavel, P.; Pari, L. Effects of diallyl tetrasulfide on cadmium-induced oxidative damage in the liver of rats. *Hum. Exp. Toxicol.* **2007**, *26*, 527–534.

NIH Guide for the Care and Use of Laboratory Animals. Revised NIH Publication Number. 85-23, U.S. Department of Health, Education and Welfare. Research Triangle Park, North Carolina. **1985**.

Oh, J.M.; Jung, Y.S.; Jeon, B.S.; Yoon, B.I.; Lee, K.S.; Kim, B.H.; et al. Evaluation of hepatotoxicity and oxidative stress in rats treated with *tert*-butyl hydroperoxide. *Food Chem. Toxicol.* **2012**, *50*, 1215–1221.

Ohta, Y.; Kongo-Nishimura, M.; Hayashi, T.; Kitagawa, A.; Matsura, T.; Yamada, K. Saikokeishito extract exerts a therapeutic effect on alpha-naphthylisothiocyanate-induced liver injury in rats through attenuation of enhanced neutrophil infiltration and oxidative stress in the liver tissue. *J. Clin. Biochem. Nutr.* **2007**, *40*, 31–41.

Padma, P.; Setty, O.H. Protective effect of *Phyllanthus fraternus* against carbon tetrachloride-induced mitochondrial dysfunction. *Life Sci.* **1999**, *64*, 2411–2417.

Pari, L.; Uma, A. Protective effect of *Sesbania grandiflora* against erythromycin estolate-induced hepatotoxicity. *Therapie* **2003**, *58*, 439–443.

Pradeep, H.A.; Khan, S.; Ravikumar, K.; Ahmed, M.F.; Rao, M.S.; Kiranmai, M.;

et al. Hepatoprotective evaluation of *Anogeissus latifolia: in vitro* and *in vivo* studies. *World J. Gastroenterol.* **2009**, *15*, 4816–4822.

Rao, C.V.; Rawat, A.K.; Singh, A.P.; Singh, A.; Verma, N. Hepatoprotective potential of ethanolic extract of *Ziziphus oenoplia* (L.) Mill roots against antitubercular drugs induced hepatotoxicity in experimental models. *Asian Pac. J. Trop. Med.* **2012**, *5*, 283–288.

Renugadevi, J.; Prabu, S.M. Cadmium-induced hepatotoxicity in rats and the protective effect of naringenin. *Exp. Toxicol. Pathol.* **2010**, *62*, 171–181.

Reitman, S.; Frankel, S. Determination of serum glutamate-oxaloacetic and glutamic pyruvic acid transaminase. *Am. J. Clin. Pathol.* **1957**, *28*, 56–66.

Sakr, S.A.; Abo-El-Yazid, S.M. Effect of fenugreek seed extract on adriamycin-induced hepatotoxicity and oxidative stress in albino rats. *Toxicol. Ind. Health* **2011**, *28*(10), 876–885.

Sambo, N.; Garba, S.H.; Timothy, H. Effect of the aqueous extract of *Psidium guajava* on erythromycin-induced liver damage in rats. *Niger. J. Physiol. Sci.* **2009**, *24*, 171–176.

Sedlak, J.; Lindsay, R.H. Estimation of total, protein-bound and nonprotein sulfhydryl-groups in tissue with Ellman's reagent. *Anal. Biochem.* **1968**, *25*, 1192–1205.

Sharma, V.; Pandey, D. Protective role of *Tinospora cordifolia* against lead-induced hepatotoxicity. *Toxicol. Int.* **2010**, *17*, 12–17.

Srivastava, S.P.; Chen, N.O.; Holtzman, J.L. The *in vitro* NADPH dependent inhibition by CCl_4 of the ATP-dependent calcium uptake of hepatic microsomes from male rats. Studies on the mechanism of inactivation of the hepatic microsomal calcium pump by the CCl_3 radical. *J. Biol. Chem.* **1990**, *265*, 8392–8399.

Sun, M.; Zigman, S. An improved spectrophotometric assay of superoxide dismutase based on epinephrine autoxidation. *Anal. Biochem.* **1978**, *90*, 81–89.

Surendran, S.; Eswaran, M.B.; Vijayakumar, M.; Rao, C.V. *In vitro* and *in vivo* hepatoprotective activity of *Cissampelos pareira* against carbon-tetrachloride induced hepatic damage. *Indian J. Exp. Biol.* **2011**, *49*, 939–945.

Taye, A.; El-Moselhy, M.A.; Hassan, M.K.; Ibrahim, H.M.; Mohammed, A.F. Hepatoprotective effect of pentoxifylline against D-galactosamine-induced hepatotoxicity in rats. *Ann. Hepatol.* **2009**, *8*(4), 364–337.

Treitz, N.W. *Fundamentals of Clinical Chemistry with Clinical Correlation.* Philadelphia: W.B. Sanders; **1970**. pp. 280–284.

Trietz, N.W.; Rinker, A.D.; Shaw, L.M. International Federation of Clinical Chemistry. IFCC methods for the measurement of catalytic concentration of enzymes. Part5. IFCC method for alkaline phosphatase (orthophosphoric-monoester phosphohydrolase, alkaline optimum, EC 3.1.3.1). IFCC document stage 2, draft 1, 1983-03 with a view to an IFCC recommendation. *Clin. Chim. Acta.* **1983**, *135*, 339–367.

Trinder, P. Determination of blood glucose using 4-aminophenzone as oxygenacceptor. *J. Clin. Path.* **1969**, *22*, 246–248.

Vonkeman, H.E.; van de Laar, M.A. Nonsteroidal anti-inflammatory drugs: adverse effects and their prevention. *Semin. Arthritis Rheum.* **2010**, *39*, 294–312.

Waterborg, H.H. *The Lowry method for protein quantitation. The Protein Protocols Handbook.* Berlin, Heidelberg: Springer; 2002, pp. 7–9.

Zelenka, J.; Lenícek, M.; Muchova, L.; Jirsa, M.; Kudla, M.; Balaz, P.; Zadinova, M.; Ostrow, J.D.; Wong, R.J.; Vítek, L. High sensitive method for quantitative determination of bilirubin in biological fluids and tissues. *J. Chromatogr. B. Analyt. Technol. Biomed. Life Sci.* **2008**, *867*, 37–42.

CHAPTER 3

Hepatoprotective Plants

ABSTRACT

Traditional medicine, which includes plants, has been increasingly used worldwide, especially in people with chronic conditions. The use of medicinal plants for the treatment of liver diseases has a long history, beginning with the Ayurvedic treatment, and spreading to the Chinese, European, as well as other systems of traditional medicines. There are more than 300 preparations in Indian system of medicine for the treatment of jaundice and chronic liver diseases. In this chapter, plant species screened for hepatoprotective activity is explained in a simple way and it may be useful to herbal practioners, research scientists, and students for screening of hepatoprotective drugs.

INTRODUCTION

Since ancient times, mankind has made use of plants in the treatment of various ailments because their toxicity factors appear to have lower side effects (Elberry et al., 2011). Many of the currently available drugs were derived either directly or indirectly from medicinal plants. Numerous medicinal plants and their formulations are being used for liver disorders in ethnomedical practices and in traditional system of medicine in different parts of the World. This situation arose from the fact that conventional drugs used in the treatment of liver diseases are often unavailable, inaccessible, and unaffordable particularly to the rural poor that suffer most, the burden of the disease (Jain et al., 2013). The inadequacy of herbs used in curing of liver diseases and other dysfunctions caused by allopathic drugs is enough reason to focus on systematic scientific research to evaluate some species of plants that are traditionally claimed to possess hepatoprotective activities.

Recent interest in natural therapies and alternative medicines has made researchers pay attention to traditional herbal medicine. In the past decade, attention has been centered on scientific evaluation of traditional drugs with plant origin for the treatment of various diseases. Due to their effectiveness, with presumably minimal side effects in terms of treatment as well as relatively low costs, herbal drugs are widely prescribed, even when their biologically active constituents are not fully identified (Levy et al., 2004).

The utility of natural therapies for liver diseases has a long history. Despite the fact that most recommendations are not based on documented evidence, some of these combinations do have active constituents with confirmed antioxidant, anti-inflammatory, anticarcinogenic, antifibrotic, or antiviral properties. Although a large number of these plants and formulations have been investigated, the studies were mostly unsatisfactory. For instance, the therapeutic values, in most of these studies, were assessed against a few chemicals-induced subclinical levels of liver damages in rodents. The reasons that make us arrive at such a conclusion are lack of standardization of the herbal drugs, the limited number of randomized placebo-controlled clinical trials, and paucity of traditional toxicologic evaluations (Thyagarajan et al., 2002).

MEDICINAL PLANTS SCREENED FOR HEPATOPROTECGIVE ACTIVITY

The use of natural remedies for the treatment of liver diseases has a long history, beginning with the ayurvedic treatment, and spreading to the Chinese, European, as well as other systems of traditional medicines. Pharmacological validation of each hepatoprotective plant should include efficacy evaluation against liver diseases induced by various agents (Ding et al., 2012). Hundreds of plants have been so far examined to be taken for a wide spectrum of liver diseases (Asadi-Samani et al., 2013, 2014). Natural products, including herbal extracts, could significantly contribute to recovery processes of the intoxicated liver. According to reliable scientific information obtained from the research on medicinal plants, plants such as *Silybum marianum*, *Glycyrrhiza glabra*, *Phyllanthus* species (*P. amarus, P. niruri, P. emblica*), and *Picrorhiza kurroa* have been widely and most of the time fruitfully applied for the treatment of liver disorders, exerting their effects via antioxidant-related properties (McBride et al., 2012; Tatiya et al., 2012; Shukla et al., 1991; Hu et al., 2008).

More than 87 medicinal plants have been used in different combinations in the preparation of 33 patented herbal formulations in India (Evans, 1996; Sharma et al., 1991). Herbal formulations (Liv. 52, Livergen, Livokin, Octogen, Stimuliv, and Tefroliv) have been found to produce marked beneficial effects in the studied pharmacological, biochemical, and histological parameters against acute liver toxicity in mice model induced by paracetamol (PCM) (Girish et al., 2009). The potency of any hepatoprotective agent is dependent on its ability to either reduce the harmful effects or maintain the normal hepatic physiological mechanism, which have been caused by a hepatotoxin.

Plant drugs are known to play a major role in the management of liver diseases. There are many plants and their extracts that have been shown to possess hepatoprotective activities. There are more than 300 preparations in Indian system of medicine for the treatment of jaundice and chronic liver diseases. About 600 commercial herbal formulations with claimed hepatoprotective activity are being sold all over the globe. The active phytochemical fraction that imparts hepatoprotective activity has been identified in many plants. These phytochemicals can be isolated and developed as single-ingredient drugs, with quality and standards of modern medicine. The major problem faced with herbal products is their standardization and their quality assurance. There can be batch-to-batch variations in their efficacies as a result of natural and genetic alterations, seasonal changes, differences in soil and climatic conditions, and nutritional status of the medicinal plant. Pharmacological validation of each hepatoprotective plant should include efficacy evaluation against liver diseases induced by various agents. The most effective drugs for each kind of liver disease have to be selected by separate efficacy evaluations. To treat liver disease of known, unknown, or multiple causes, a combination of different herbs with active fractions (or purified compounds) has to be developed. They may prove to be useful in the treatment of infective, toxic, and degenerative diseases of the liver (Girish and Pradhan, 2012).

Several reviews have been published on hepatoprotective plants and phytochemicals useful in liver protection. Adewusi and Afolayan (2010) in their review reported 107 plants and 58 compounds isolated from higher plants useful in the treatment of liver diseases. Dey et al. (2013) reviewed the hepatotoxicity and the present scenario. They gave a list of plants screened for hepatoprotective activity. Lawal et al. (2016) gave an excellent review on African natural products with potential antioxidants and hepatoprotectives properties. A total 1076 plants species representing 287 families and 132 isolated compounds were found. Alqasoumi and Abdel-Kader (2012) reviewed the publications hepatoprotective activity and listed plants screened for hepatoprotection. Al-Asmari et al. (2014) reviewed the hepatoprotective plants used in Saudi Traditional medicine. Gupta et al. (2015) reviewed the hepatoprotective properties of Triphala and its constituents. Gressner's (2012) review focuses on established components of Western food, such as Curry, Coffee, or Chocolate, in terms of their hepatoprotective effects and discusses the biochemical background of the epidemiological observations. Reviews on hepatoprotective plants include Khan et al. (2016), Al-Snafi et al. (2019), etc.

In the following sections, plant species screened for hepatoptorective activity is given. Plant species are arranged in alphabetical sequency.

Abelmoschus manihot (L.) Medik. FAMILY: MALVACEAE

The decoction of the flowers of *Abelmoschus manihot* is traditionally used for the treatment of jaundice and various types of chronic and acute hepatitis in Anhui and Jiangsu Provinces of China. Phytochemical studies have indicated that total flavonoids extracted from flowers of *A. manihot* (TFA) were the major constituents of the flowers. Ai et al. (2013) investigated the hepatoprotective effect of the plant extracts against CCl_4-induced hepatocyte damage *in vitro* and liver injury *in vivo*. A concentration-dependent increase in the percentage viability was observed when CCl_4-exposed hepatocytes were treated with different concentrations of total flavonoids. Levels of alanine transaminase (ALT), aspartate aminotranferase (AST), and alkaline phosphatase (ALP) in the medium were significantly decreased. In the animal studies, total flavonoids showed significant protection with the depletion of ALT, AST, ALP, and gamma-glutamyltransferase (γ-GT) in serum as was raised by the induction of CCl_4. Moreover, total flavonoids decreased the malanodialdehyde (MDA) level and elevated the content of GSH in the liver as compared to those in the CCl_4 group. Furthermore, activities of antioxidative enzymes, including SOD, GPx, Catalase (CAT), and GST, were enhanced dose dependently with total flavonoids. Meanwhile, the inflammatory mediators (e.g., TNF-α, IL-1β, and NO) were inhibited by total flavonoids treatment both at the serum and mRNA levels. Additionally, histological analyses also showed that total flavonoids reduced the extent of liver lesions induced by CCl_4. Yan et al. (2015) investigated the protective effects and mechanisms of TFA on α-naphthylisothiocyanate (ANIT)-induced cholestatic liver injury in rats. The oral administration of TFA to ANIT-treated rats could reduce the increases in serum levels of ALT, AST, LDH, ALP, gamma glutamyl transpeptidase (GGT), total bilirubin (TBIL), DBIL, and TBA. Decreased bile flow by ANIT was restored with TFA treatment. Concurrent administration of TFA reduced the severity of polymorphonuclear neutrophil infiltration and other histological damages, which were consistent with the serological tests. Hepatic MDA and GSH contents in liver tissue were reduced, while SOD and GST activities, which had been suppressed by ANIT, were elevated in the groups pretreated with TFA. With TFA intervention, levels of TNF-α and NO in liver were decreased. Additionally, TFA was found to increase the expression of liver BSEP, MRP2, and NTCP in both protein and mRNA levels in ANIT-induced liver injury with cholestasis.

Abelmoschus moschatus Medik. FAMILY: MALVACEAE

Singh et al. (2012) evaluated the hepatoprotective activity of *Abelmoschus moschatus* seed extract against PCM- and ethanol-induced hepatotoxicity. Paracetamol-induced hepatotoxicity resulted in an increase in serum AST, ALT, ALP activity and bilirubin level. Paracetamol hepatotoxicity was manifested by an increase in lipid peroxidation (LP), depletion of reduced GSH and CAT activity in liver tissue. Administration of ethanolic as well as aqueous plants extract (300 mg/kg BW [BW] of rat) protects the PCM-induced LPO, restored altered serum marker enzymes and antioxidant level toward normal.

Abrus precatorius L. FAMILY: FABACEAE

Wakawa and Franklyne (2015) evaluated the protective effects of aqueous leaf extract of *Abrus precatorius* against CCl_4-induced liver injury in rats. A significant decrease was observed in both the groups treated with 200 and 400 mg/kg BW of the leaf extract on the levels of the enzymes and nonenzyme markers of liver injury and LPO as well as relative organ weight, with no significant changes in the mean final BW of the treated groups. This result showed that the leaf extract of *A. precatorius* contains phytochemical(s) that is (are) protective to the liver against CCl_4-induced injury in rats.

Abutilon bidentatum Hochst. ex A. Rich. FAMILY: MALVACEAE

Yasmin et al. (2011) evaluated the hepatoprotective activity of aqueous methanolic extracts of aerial parts of *Abutilon bidentatum* on CCl_4 and PCM-induced liver damage in rabbits. Administration of CCl_4 to rabbits produced hepatotoxicity showed by significant increase in the serum levels of SGOT, SGPT, ALP, and direct bilirubin (DBL) in comparison to control group. The aqueous methanolic extract of *A. bidentatum* at different dosages, administered orally for three days, showed a significant decrease in the serum enzymes when compared to the CCl_4 compared control groups. The biochemical observations were supplemented with histopathological examination of rabbit liver sections.

***Abutilon indicum* (L.) Sweet FAMILY: MALVACEAE**

The aqueous extract of *Abutilon indicum* was tested by Porchezhian and Ansari (2005) for hepatoprotective activity against CCl_4- and PCM-induced hepatotoxicities in rats. *A. indicum* exhibited significant hepatoprotective activity by reducing CCl_4- and PCM-induced change in biochemical parameters that was evident by enzymatic examination. The plant extract may interfere with free-radical formation, which may conclude in hepatoprotective action. Acute toxicity studies revealed that the LD_{50} value is more than the dose of 4 g/kg BW.

***Acacia catechu* L.f. Family: Fabaceae**

Ray et al. (2006) evaluated the hepatoprotective effect of the ethyl acetate extract of *Acacia catechu* in experimental animal models. Highly significant hepatoprotective activity was observed when the extract of *A. catechu* (250 mg/kg) was administered prophylactically for seven days. Sheshidhar et al. (2013) evaluated the hepatoprotective effect of ethanolic extract of *A. catechu* in paracetamol (PCM)-induced liver damage in experimental rats. Pretreatment with *Acacia* extract significantly prevented the biochemical enzyme levels and histological changes induced by PCM in the liver. The effects of *A. catechu* were comparable to that of the standard drug Silymarin.

Lakshmi et al. (2018) investigated the hepatoprotective effects and the possible mechanism of *A. catechu* in APAP-induced hepatotoxicity using female Wistar rat model. The seed (400 mg/kg BW) and bark (400 mg/kg BW) extract's treated groups exhibited hepatoprotective effects and was compared with well-known clinical antidote *N*-acetylcysteine (NAC). When groups treated with APAP, significant increase of liver weight/BW ratio, liver function enzymes such as ALT, ALP, and AST and decrease of antioxidant enzymes such as GSH and SOD were observed. The histopathology of APAP treated groups also showed moderate degree of sinusoidal congestion, centrilobular necrosis with polymorph nuclear cells infiltration, marked vacuolations, and congestion. However, pretreatment with seed or bark extract groups decreased LPO accumulation, reduced the liver function enzymes, and increased antioxidant defense enzymes. Moreover, histopathology of seed extract treated groups showed normal architecture whereas bark extract treated groups exhibited mild degree of vacuolations in the hepatocytes with minimal sinusoidal congestion. Taken together, their study concluded that *A. catechu* seed extract to be a more promising agent for protecting liver from APAP-induced hepatotoxicity.

Acacia confusa Merr. FAMILY: FABACEAE

The hepatoprotective effects of *Acacia confusa* bark extract (ACBE) and its active constituent gallic acid (GA) were evaluated by Tung et al. (2009) against CCl_4-induced hepatotoxicity in rats. CCl_4-induced hepatic pathological damage and significantly increased the levels of AST, ALT, and MDA in plasma, and cytochrome P4502E1 (CYP2E1) protein expression in hepatic samples, and decreased the activities of SOD, glutathione peroxidase (GPx) and CAT in erythrocytes. Treatment with ACBE, GA, or Silymarin could decrease significantly the AST, ALT, and MDA levels in plasma, and CYP2E1 expression in liver tissues, and increase the activities of SOD and GPx in erythrocyte when compared with CCl_4-treated group. Liver histopathology also showed that ACBE, GA, or Silymarin could significantly reduce the incidence of liver lesions induced by CCl_4.

Acacia mellifera (Vahl) Benth. FAMILY: FABACEAE

Arbab et al. (2015) investigated the hepatoprotective and anti-HBV efficacy of *Acacia mellifera* (AM) leaves extracts. The crude ethanolic extract, including organic and AFs (AFs), was tested for cytotoxicity on HepG2 and HepG2.2.15 cells (IC_{50} = 684 μg/mL). Of these, the ethyl acetate and AFs showed the most promising, dose-dependent hepatoprotection in DCFH-toxicated cells at 48 h. In CCl_4-injured rats, oral administration of AM ethanol extract (250 and 500 mg/kg BW) for three weeks significantly normalized the sera aminotransferases, ALP, bilirubin, cholesterol, triglycerides (TGs), and lipoprotein levels and elevated tissue nonprotein sulfydryl and total protein (TP). The histopathology of dissected livers also revealed that AM cured the tissue lesions. The phytochemical screening of the fractions showed the presence of alkaloids, flavonoids, tannins, sterols, and saponins. Further, anti-HBV potential of the fractions was evaluated on HepG2.2.15 cells. Of these, the *n*-butanol and AFs exhibited the best inhibitory effects on HBsAg and HBeAg expressions in dose- and time-dependent manner. Taken together, while the ethyl acetate and AFs exhibited the most promising antioxidant/hepatoprotective and anti-HBV activity, respectively, the *n*-butanol partition showed both activities.

Acacia modesta Wall. FAMILY: FABACEAE

Rahaman and Chaudhry (2015) investigated the mechanism of protective effect of *Acacia modesta* bark against PCM-induced hepatotoxicity in mice. The reduction in level of ALT, AST, and ALP showed that *A. modesta* extract has hepatoprotective constituents. While serum albumin (ALB) and TP level, which were reduced on treatment with APAP, were normalized by *A. modesta* extract. It proved as a good antioxidant than ascorbic acid. Phytochemical analysis showed that antioxidant activity is due to tannins and saponins extract.

Acalypha indica L. FAMILY: EUPHORBIACEAE

Kumar et al. (2013) evaluated the hepatoprotective activity of methanol extract (ME) and methanolic fraction of methanol extract of *Acalypha indica* against thioacetamide-induced toxicity in albino rats. Administration of thioacetamide (100 mg/kg i.p.) induced a marked increase in the serum levels of SGOT, SGPT, ALP, TBL, and CHL; and decrease in the levels of TN and ALB, indicating parenchymal cell necrosis. ME at dose levels of 300 mg/kg and methanol fraction of methanol extract at dose levels of 250 mg/kg has restored the altered parameters significantly as observed in case of Silymarin treated group.

Dwijayanti et al. (2015) evaluated the hepatoprotective effects of the combination of *A. indica* (AI) and *Centella asiatica* (CA) against hypoxia. Seven groups of Spraque Dawley rats are used as follows: negative control without hypoxic condition, six other groups were put in hypoxic chamber and treated with water, AI, CA, two groups of AI–CA combination, and vitamin C, respectively. After 7 days of treatment, MDA levels were measured in plasma and liver. One of the AI-CA combination groups showed significant protective effect against hypoxia in liver but not in plasma.

Acalypha racemosa Wall. ex Baill. FAMILY: EUPHORBIACEAE

The effects of simultaneous treatment of CCl_4 (i.p.) with 60 mg/kg (p.o.) of aqueous extract of leaves of *Acalypha racemosa* on rat liver were evaluated by Iniaghe et al. (2008). Administration of CCl_4 alone to rats significantly increased TBL concentration and the activities of ALT and AST in the serum while it significantly reduced serum TP and ALB concentrations when compared with controls which received distilled water (p.o.). Also it significantly increased liver MDA content when compared with control. However, simultaneous treatment of CCl_4 with 60 mg/kg of the aqueous extract significantly reversed these changes. Results of MDA content of liver homogenates suggest that a probable mechanism of action of the extract is antioxidation. Histopathological studies were carried out on the liver to confirm the observed changes.

Acalypha wilkesiana Muell.-Arg.
FAMILY: EUPHORBIACEAE

The potential of aqueous extract of the leaves of *Acalypha wilkesiana*, to protect against CCl_4-induced liver damage was investigated in Wistar albino rats by Ikewuchi et al. (2011). On fractionation and gas chromatographic analysis of the crude aqueous extract, 39 known alkaloids were detected, consisting mainly of akuamidine (69.027%), voacangine (26.226%), echitamine (1.974%), echitamidine (0.599%), lupanine (0.521%), and augustamine (0.278%). Compared to test control, the treatment dose dependently produced significantly lower ALP, AST, and ALT activities. Histopathological studies on the liver sections showed that pretreatment with the extract protected against CCl_4-induced fatty degeneration of hepatocytes; thus, confirming the results of the biochemical studies. The above results imply that the treatment with the plant extract protects the liver against CCl_4-induced hepatotoxicity; therefore, justifying the use of *A. wilkesiana* in African traditional health care for the management of liver problems.

Acantholimon gilliati Turril
FAMILY: PLUMBAGINACEAE

Acantholimon gilliati is used in a variety of diseases including hepatic ailments in the west region of Iran. Lashgari et al. (2017) investigated the hepatoprotective effect of methanolic extract of *A. gilliati* on formaldehyde-induced liver injury in adult male mice. Formaldehyde-induced liver damage both in histology and function. The levels of ALT, AST, and ALP enzymes were significantly increased in formaldehyde treated group. Administration of methanolic extract in all experimental groups significantly reduced serum levels of ALT and ALP. However, AST was reduced significantly just in groups III and IV who were treated with doses of 5 and 10 mg of *A. gilliati.* Similarly ME in doses of 5 and 10 mg protected liver histology against formaldehyde. Results showed that the ME of the *A. gilliati* in the lower doses has a protective effect on both histology and function of liver.

Acathopanax senticosus (Rupr. et Maxim.) Harms.
FAMILY: ARALIACEAE

Acathopanax senticosus is a popular folk medicine used as a nutrient for hepatitis and cancer in Taiwan. The antioxidant activity of the crude extract and the hepatoprotective activities on CCl_4- or APAP-induced toxicity in the rat liver were evaluated by Lin and Huang (2000). *A. senticosus* exerts some antioxidant effects. On a CCl_4- or APAP-intoxicated model, the levels of

AST and ALT were increased by CCl_4 or APAP administration and reduced by treatment with the plant extract. Histological changes around the hepatic central vein were also recovered by treatments. However, treatments with larger doses of the crude extract of *A. senticosus* enhanced liver damage. This result suggests that even if *A. senticosus* had hepatoprotective activity in small doses, treatment with larger doses would possibly induce some cell toxicity.

Acanthospermum hispidum DC. FAMILY: ASTERACEAE

The EtOH extract from the aerial parts of *Acanthospermum hispidum* (400 mg/kg) was evaluated by Himaja (2015) for 28 days against liver injury induced by anti-TB drugs (RIF 40 mg/kg, INH 27 mg/kg, PZA 66 and EMB 53 mg/kg) on Wistar rats. This plant extract reduced the level of liver enzymes in serum and microscopic analysis revealed that liver tissue was regenerated (Himaja 2015).

Acanthus ilicifolius L. FAMILY: ACANTHACEAE

The alcoholic extract (ACE) of *Acanthus ilicifolius* leaves inhibited the formation of oxygen-derived free radicals (ODFR) *in vitro* with IC_{50} of 550, 2750, 670, and 600 μg/mL (Fe(2+)/ascorbate system), 980 μg/mL (Fe(3+)/ADP/ascorbate system) for superoxide radical production, hydroxyl radical generation, nitric oxide (NO) radical formation, and lipid peroxide (LPO) formation, respectively (Babu et al., 2001). The oral administration of the extract (250 and 500 mg/kg) significantly reduced CCl_4-induced hepatotoxicity in rats, as judged from the serum and tissue activity of marker enzymes SGOT, SGPT, and ALP. These results were comparable with those obtained with curcumin (100 mg/kg, p.o.) (Babu et al., 2001).

Acer tegmentosum Maxim. FAMILY: SAPINDACEAE

Cho et al. (2015) investigated the antidiabetic, alcohol metabolism, anti-inflammatory, and hepatoprotective effects of *Acer tegmentosum* extracts (ATEs). 2,2-Diphenyl-1-picrylhydrazyl (DPPH) radical scavenging and SOD activities of ATE were about 89% and 82.9% at 0.5 μg/mL, respectively. Alcohol dehydrogenase and acetaldehyde dehydrogenase activities were 118.0% and 177% at 2 mg/mL, respectively. α-Glucosidase inhibitory activity of ATE was 75% higher at 50 μg/mL and remarkably increased in a dose-dependent manner. Nitric oxide productions in macrophage RAW 264.7 cells stimulated by lipopolysaccharide (LPS) was reduced to 16.7% by addition of ATE at 1 mg/

mL. ATE showed significant protective effects against tacrine-induced cytotoxicity in Hep G2 cells at 100 μg/mL. Based on these results, authors concluded that ATE may be used as a major pharmacological agent and antidiabetic, antihepatitis, and anti-inflammatory remedy.

Achillea biebersteinii Afan. FAMILY: ASTERACEAE

Hepatoprotective effect of *Achillea biebersteinii* ethanol extract was evaluated by Hosbas et al. (2011) against CCl_4-induced subacute hepatotoxicity in rats. Increasing doses of ethanol extract reduced plasma ALP levels. On the contrary, all doses of the extract increased plasma AST and ALT levels. Ethanol extract did not show any significant hepatoprotective activity at 250 and 500 mg/kg doses, whereas a weak activity was observed at 750 mg/kg. However, at all the tested doses, ethanol extract caused improvement in antioxidant defense potential (liver GSH, CAT levels). Moreover, total phenolic content (TPC) of *A. biebersteinii* was found to be 6.21 ± 0.004 mg GA equivalent (GAE)/g extract.

Al-Said et al. (2016) evaluated the efficiency of *A. biebersteinii* essential oil (ABEO) (0.2 mL/kg) in the amelioration of CCl_4-induced hepatotoxicity in rodent model. Around 44 components (92.0%) of the total oil have been identified by GC-MS analysis where α-terpinene and -cymene were the most abundant. The high serum enzymatic (GOT, GPT, GGT, and ALP) and bilirubin concentrations as well as the level of MDA, NP-SH, and TP contents in liver tissues were significantly reinstated toward normalization by the ABEO. Histopathological study further confirmed these findings. In addition, ABEO showed mild antioxidant activity in DPPH radical scavenging and β-carotene-linoleic acid assays.

Achillea millefolium L. FAMILY: ASTERACEAE

The crude extract of *Achillea millefolium* (Am.Cr) was studied for its possible hepatoprotective effect against D-galactosamine (D-GalN) and LPS-induced hepatitis in mice and antispasmodic effect in isolated gut preparations to rationalize some of the folklore uses (Yaeesh et al., 2006). Co-administration of D-GalN (700 mg/kg) and LPS (25 μg/kg) produced 100% mortality in mice. Pretreatment of animals with Am.Cr (300 mg/kg) reduced the mortality to 40%. Co-administration of D-GalN (700 mg/kg) and LPS (1 μg/kg) significantly raised the plasma ALT and AST levels compared with values in the control group. Pretreatment of mice with Am.Cr (150–600 mg/kg) significantly prevented the toxins induced rise in plasma ALT and

AST. The hepatoprotective effect of Am.Cr was further verified by histopathology of the liver, which showed improved architecture, absence of parenchymal congestion, decreased cellular swelling and apoptotic cells, compared with the toxin group of animals.

Achyrocline satureioides (Lam.) DC. FAMILY: ASTERACEAE

The hepatoprotective activity was evaluated by Kadarian et al. (2002) in the bromobenzene-(BB-)-induced hepatotoxicity model in mice through the measurement of the serum levels of ALT and AST, TBARS, and GSH levels. The aqueous extract of the aerial parts of *A. satureioides* administered before BB, at the dose of 300 mg/kg, p.o., demonstrated significant inhibition in the BB increase of liver ALT and AST and in the BB-induced increase of liver TBARS content. Moreover, it was able to significantly increase the depleted levels of liver GSH.

Acrocarpus fraxinifolius Arn. FAMILY: FABACEAE

Abd El-Ghffar et al. (2017) investigated the components of the *n*-hexane extract of *Acrocarpus fraxinifolius* and its hepatoprotective activity against PCM-induced hepatotoxicity in rats. Chromatographic analysis revealed the presence of 36 components. Major compounds were α-tocopherol, labda-8 (20)-13-dien-15-oic acid, lupeol, phytol, and squalene. In the acute oral toxicity study, the mortality rates and behavioral signs of toxicity were zero in all groups (doses from 0 to 5 g/kg BW of *A. fraxinifolius*). LD_{50} was found to be greater than 5 g/kg of the extract. Only the high dose (500 mg/kg BW) of extract significantly alleviated the liver relative weight and biomarkers, as serum AST, ALT, ALP, lipid profiles, bilirubin profiles, and hepatic LPO, and increased BW, serum protein profile, and hepatic TAC in PCM-induced hepatotoxicity in rats.

Adansonia digitata L. FAMILY: MALVACEAE

The fruit pulp of *Adansonia digitata*, commonly known as baobab is an important human nutrition source in East, Central, and West Africa (Beckier, 1983; Szolnoki, 1985). The aqueous extract of *A. digitata* pulp was tested for hepatoprotective activity against liver injury by CCl_4 in rats. The aqueous extract exhibited significant hepatoprotective activity and consumption of the fruit may play an important part in human resistance to liver damage in areas where the plant is consumed (Didibe et al., 1996). The mechanism of liver protection may be due to the presence of triterpenoids, β-sitosterol, β-amyrin palmitate, and ursolic acid in the fruit pulp of *A. digitata* (Al-Qarawi et al., 2003).

Mohamed et al. (2015) investigated the hepatoprotective effect of *A. digitata* fruits pulp methanolic extract on CCl_4-induced hepatotoxicity in rats. The two doses of the plant extract (100 and 200 mg/kg) showed dose-dependent hepatoprotective effect on CCl_4-induced hepatotoxicity, as evident by the significant reduction in serum levels of AST, ALT, ALP, and bilirubin along with the improved histopathological liver sections compared to CCl_4-treated animals.

The ME of the fruit pulp of *A. digitata* was examined by Hanafy et al. (2016) for its hepatoprotective activity against liver damage induced by APAP in rats. Treatment of the rats with the ME of the fruit pulp of *A. digitata* prior to administration of APAP significantly reduced the disturbance in liver function. Moreover, histopathological evaluation was performed in order to assess liver case regarding inflammatory infiltration or necrosis. Animals were observed for any symptoms of toxicity after administration of extract of the fruit pulp of *A. digitata* to ensure safety of the fruit extract. Decreased levels of GSH, SOD, and CAT, observed in APAP-treated rats, are an indication of tissue damage produced by free radicals. The increase in the concentration of these antioxidant enzymes in liver tissues of Silymarin- or extract-treated animals indicates antioxidant effect of Silymarin and extract. The histopathological findings confirmed the biochemical results.

Adenanthera pavonina **L.**
FAMILY: FABACEAE

Mujahid et al. (2013) evaluated the hepatoprotective action of the leaves of *Adenanthera pavonina* against isoiazid (INH) and rifampicin (RIF)-induced liver damage in experimental animals. The methanolic extract of *A. pavonina* was safe up to a dose of 2000 mg/kg. The significantly elevated serum enzymatic activities of SGOT, SGPT, ALP, bilirubin, and LDH due to INH + RIF treatment were restored to near normal in a dose-dependent manner after the treatment with methanolic extract of leaves of *A. pavonina*. Also the increased level of TP and ALB toward normal by extract of *A. pavonina* leaves. In the antioxidant studies, a significant increase in the levels of GSH, CAT, and SOD was observed. In addition, methanolic extract also significantly prevented the elevation of hepatic MDA formation in the liver of INH + RIF intoxicated rats in a dose-dependent manner. The biochemical observations were supplemented with histopathological examination of rat liver section.

Aegiceras corniculatum (L.) Blanco FAMILY: PRIMULACEAE

Roome et al. (2008) evaluated the antioxidant, anti-inflammatory, and hepatoprotective potential of *Aegiceras corniculatum*. The *n*-hexane, ethyl acetate, and MEs, derived from *A. corniculatum* stems, scavenged superoxide anions ($O_2^{\cdot}$) and hydroxyl radicals (•OH) in nitro blue tetrazolium reduction and deoxyribose degradation assays, respectively. All the extracts inhibited the process of LPO at its initiation step. Additionally, in rat liver microsomes *n*-hexane and ethyl acetate extracts also caused termination of radical chain reaction supporting their scavenging action toward lipid peroxy radicals (LOO•). Moreover, increased production of $O_2^{\cdot}$ in human neutrophils, stimulated by phorbol-12-myristate-13-acetate and/or opsonized zymosan were also suppressed (IC50 approximately 3–20 μg/mL), thereby, revealing the ability of plant extracts to antagonize the oxidative stress via interference with nicotinamide adenine dinucleotide phosphate (NADPH) oxidase metabolic pathway. These *in vitro* results coincide with the reduction in the glucose oxidase-induced paw edema in mice in the presence of ethyl acetate and MEs (10, 50, and 100 mg/kg, i.p.). Plant extracts (250, 500, and 1000 mg/kg, p.o.) also significantly protected the CCl_4-induced oxidative tissue injury in rat liver. This was reflected by approximately 60% decline in the levels of serum aminotransferase enzymes.

Aegle marmelos Roxb. FAMILY: RUTACEAE

Singanan et al. (2007) evaluated the hepatoprotective effect Bael leaves (*Aegle marmelos*) in alcohol-induced liver injury in albino rat. The observed values of TBARS in healthy, alcohol intoxicated, and herbal drug treated animals were 123.35, 235.68, and 141.85 μg/g tissue, respectively. The results were compared with the standard herbal drug Silymarin (133.04 μg/g tissue). The experimental results indicate that the Bael leaves have excellent hepatoprotective effect. A similar experimental result was also observed in other biochemical parameters.

Singh and Rao (2008) investigated the hepatoprotective effect of pulp/seeds of fruits of *A. marmelos* against CCl_4-induced hepatotoxicity in rats. Treatment with aqueous extract of fruit pulp/seeds significantly reduced CCl_4-induced elevation in plasma enzyme and bilirubin concentration in rats. Parmar et al. (2009, 2010) evaluated the hepatoprotective activity of leaves of *A. marmelos* against CCl_4 and PCM-induced toxicity in male Wistar rats. Toxin-induced hepatotoxicity resulted in an increase in serum AST, ALT, ALP activity, and bilirubin level accompanied by significant decrease in ALB level. Co-administration of the aqueous extract protects the toxin-induced LPO, restored latered serum marker enzymes and antioxidant level toward near normal.

Sumitha and Thirunalasundari (2011) investigated the hepatoprotective effect of crude ethanolic extract of the leaves of *A. marmelos* (AMEE) in CCl_4-induced toxicity in mice. The ethanolic extract at a dose of 500 mg/kg BW when given orally exhibited a significant protective effect evidenced by lowering the levels of enzymes like SGPT, SGOT, ALP, bilirubin, total cholesterol (TC), TGs, low-density lipoprotein (LDL), and very LDL but there was an increase in the level of high-density lipoprotein (HDL) as compared to CCl_4-induced group. It was also proved by increased level of antioxidant enzymes in *A. marmelos* treated group. Lipid peroxidation was significantly lowered in CCl_4 treated group followed by plant extract treated group. These biochemical observations were supported by histopathological examination of liver sections.

Singh et al. (2014) evaluated the hepatoprotective activity of aqueous extract of *A. marmelos* (AEAM) against cyclophosphamide (CPA)-induced liver damage in mice. SGOT, SGPT, ALP, acid phosphatase, bilirubin, cholesterol levels, and LPO were significantly increased, accompanied by a significant decrease in the level of ALD in CPA induced hepatotoxic group of mice compared to the control. However, significant amelioration in these parameters was found in AEAM treated groups of mice. CPA treatment markedly decreased the level of SOD and CAT in the liver as well as white blood cells (WBCs) and red blood cells (RBCs) counts, which were significantly enhanced by AEAM treatment. Histopathological examinations have also confirmed the protective efficacy of AEAM. The phytochemical screening of the extract revealed the presence of alkaloids, saponins, tannins, flavonoids, and phenols, which may have hepatoprotective role.

Aerva javanica **(Burm. f.) Juss. FAMILY: AMARANTHACEAE**

Aqueous methanolic extract (70%) of *Aerva javanica* had been tested by Shauka et al. (2015) for its hepatoprotective activity on albino rats. It showed that *A. javanica* extract applied against PCM intoxication reduced the increased levels of liver marker enzymes. Phytochemical analysis of *A. javanica* extract confirmed the presence of glycosides, flavonoids, Saponins, Terpenes, and Tannins. In another study, Arbab et al. (2016) investigated the hepatoprotective efficacy of *A. javanica* against CCl_4-induced liver injury in rats. It was observed that hepatocyte recovery was 90.2% by treatment with *A. javanica* extract.

Aerva lanata (L.) Juss. FAMILY: AMARANTHACEAE

Manokaran et al. (2008) evaluated the hepatoprotective activity of hydroalcoholic extract of *Aerva lanata* against PCM-induced liver damage in rats. The plant extract was effective in protecting liver against the injury induced by PCM in rats. This was evident from significant reduction in serum enzymes ALT, AST, ALP, and bilirubin.

Aerva sanguinolenta (L.) Blume FAMILY: AMARANTHACEAE

Lalee et al. (2012) investigated the hepatoprotective effects of ethanolic extract of *Aerva sanguinolenta* by oral route to adult male Wistar albino rats. From the result, it was found out that the ethanolic extract of the plant has hepatoprotective activity and that is comparable to that of Silymarin. Here hepatoprotective activity of ethanolic extract of *A. sanguinolenta* leaves may be due to the presence of polyphenolic compounds. Besides *A. sanguinolenta* contains flavonoid and tannin which are also known as natural antioxidants due to their electron donating property which either scavenge the principal propagating radicals or halt the radical chain.

Aeschynomene aspera L. FAMILY: FABACEAE

The hepatoprotective activity of benzene and alcoholic extracts of root of *Aeschynomene aspera* was investigated in rats for CCl_4-induced hepatotoxicity by Kumaresan and Pandae (2011). The extracts did not produce any mortality even at 5000 mg/kg while LD 50 of benzene and alcoholic extracts was found to be 100 and 200 mg/kg, respectively. Hepatotoxicity was induced in rats by intraperitoneal injection of CCl_4 (1 mL/kg/day diluted with olive oil (1:1) for 3 days). In benzene and alcoholic extracts-treated animals, the toxicity effect of CCl_4 was controlled significantly by restoration of the levels of serum bilirubin (SB) and enzymes as compared to the normal and standard drug Silymarin-treated groups. Histology of liver sections of the animals treated with the extracts showed the presence of normal hepatic cords, absence of necrosis and fatty infiltration which further evidence the hepatoprotective activity.

Aeschynomene elaphroxylon (Guill. & Perr.) Taub. FAMILY: FABACEAE

Hashem et al. (2019) investigated metabolic profile of *Aeschynomene elaphroxylon* extracts of flowers, leaves, and bark adopting ultra-performance liquid chromatography (UPLC)-Orbitrap high-resolution mass spectrometry (HRMS) analysis to determine their bioactive metabolites, and it was designed to investigate the potential hepatoprotective activity of *A. elaphroxylon* flowers and bark extracts against CCl_4-induced hepatic fibrosis in rats. Forty-nine compounds of various classes were detected in the three extracts, with triterpenoid saponins as the major detected metabolite. Flowers and bark extracts presented similar chemical profile while leaves extract was quite different. The antioxidant activities of the flowers, leaves, and bark extracts were measured by *in vitro* assays as Fe^{+3} reducing antioxidant power and oxygen radical absorbance capacity. It revealed that flowers and bark extracts had relatively high antioxidant activity as compared to leaves extract. Based on the metabolic profile and *in vitro* antioxidant activity, flowers, and bark ethanolic extracts were chosen for alleviation of hepatotoxicity induced by CCl_4 in rats. Flowers and bark ethanolic extracts exerted a significant hepatoprotective effect through reduction in the activities of ALT, AST, and ALB, the tested extracts reduced oxidative stress by increasing GSH content and reducing the MDA level. Furthermore, the extracts decreased levels of pro-inflammatory TNF-α. Moreover, this study revealed the potentiality of *A. elaphroxylon* in ameliorating the CCl_4-induced hepatic fibrosis in rats.

Aframomum melegueta K. Schum. FAMILY: ZINGIBERACEAE

The protective effect of aqueous plant extract of *Aframomum melegueta* on ethanol-induced toxicity was investigated in male Wistar rats by Nwozo and Oyinloye (2011). The rats were treated with 45% ethanol (4.8 g/kg BW) for 16 days to induce alcoholic diseases in the liver. The activities of ALT, AST, and TG were monitored and the histological changes in liver examined in order to evaluate the protective effects of the plant extract. Hepatic MDA and reduced GSH, as well as SOD and GSH-*S*-transferase activities were determined for the antioxidant status. Chronic ethanol administration resulted in a significant elevation of serum ALT and TG levels, as well as a decrease in reduced GSH and SOD which was dramatically attenuated by the co-administration of the plant extract. Histological changes were related to these indices. Co-administration of the plant extract suppressed the elevation of LPO, restored the reduced glutathion, and enhanced the SOD activity. These results highlight the ability of *A. melegueta* to ameliorate oxidative damage in the liver and the observed effects are associated with its antioxidant activities.

The hepatotoxic effects of the seeds of *A. melegueta*, a spice, were studied by Nwaehujor et al. (2014) in Sprague Dawley (SD) rats. Serum levels of AST significantly increased progressively in extract-treated rats compared to the control from day 7 till the termination of the study (day 21). However, serum ALT, ALP, and TB levels of test rats were only significantly elevated relative to the normal on days 14 and 21 of the investigation. The serum ALB levels of extract-treated rats were however, comparable with that of the normal rats throughout the study period. Histopathology of the rat livers revealed mild focal necrosis of hepatocytes at day 7, moderate multifocal areas of hepatic necrosis at day 14 and severe, diffused necrosis of hepatocytes at day 21 of treatment with the extract. The results demonstrated that the methanol seed extract of *A. melegueta* was potent in inducing liver toxicity at the tested dose (300 mg/kg). Maximal caution should therefore be imbibed in prolonged excessive use of the plant seeds as spice in delicacies.

Agrimonia eupatoria L. FAMILY: ROSACEAE

In popular medicine, *A. euptoria* is employed for the treatment of several disorders, for example, inflammations. The aqueous extract of *A. eupatoria* is full of several phenolic compounds and its ethyl acetate fraction (EAF) has exhibited antioxidant activity and lower toxicity (Correia et al., 2007). *A. eupatoria* is rich in coumarins, flavonoids, tannins, terpenoids, and phenolic compounds including protocatechuic acid, coumaric acid, chlorogenic acid, quercitrin, and GA (Gião et al., 2009). *A. eupatoria* caused effects on the liver cells in a preliminary study. The hepatoprotective effects of *A. eupatoria* on hepatocarcinogenesis induced by diethylnitrosamine (DEN) and $CC1_4$ were studied in the *in vivo* models. There is evidence on the biologic actions of *A. eupatoria* and its benefits for liver tumor therapy. The hepatoprotective effects of *A. eupatoria* water extract against ethanol-induced liver injury have been already shown. Animals were treated orally with *A. eupatoria* extract at 10, 30, 100, and 300 mg/kg/day doses. After ethanol chronic consumption, serum aminotransferase activities and proinflammatory cytokines pronouncedly increased, although attenuated by *A. eupatoria* extract. The cytochrome P450 activity and LPO also increased after ethanol consumption while GSH concentration decreased. *A. eupatoria* extract ameliorated chronic ethanol-induced liver injury, and protection likely relates to the suppression of oxidative stress and toll-like receptor (TLR)-mediated inflammatory signaling (Yoon et al., 2012). Hepatoprotective effects of aqueous extract of *A. eupatoria* were investigated in experimental liver damaged models. Hepatoprotective effects of the plant were monitored by reducing serum AST and ALT levels (Kang et al., 2012).

Ajuga nipponensis **Makino**
FAMILY: LAMIACEAE

Hsieh et al. (2016) investigated the hepatoprotective activity of *Ajuga nipponensis*. Maximum yields of flavonoids (7.87 ± 0.10) mg/g and ecdysterones (0.73 ± 0.02) mg/g could be obtained when the extraction time was 50 min, the extraction temperature was 60 °C, and the ratio of sample to 70% (v/v) ethanol was 1:20 (w/w). The antioxidant property of *A. nipponensis* was correlated to the concentration of its extracts. At 5 mg/mL, *A. nipponensis* extract scavenged 84.8% of DPPH radical and had absorbance values of 2.43 ± 0.04 reducing power. Upon CCl_4-induced liver injury, glutamic oxaloacetic transaminase and glutamic pyruvic transaminase (GPT) decreased significantly after the mice were treated with *A. nipponensis*. Histological researches also explained that *A. nipponensis* reduced the extent of liver lesions induced by CCl_4.

Albizia harveyi **E. Fourn.**
FAMILY: FABACEAE

Profiling the polyphenols in the ME from the bark of *Albizia harveyi* was performed by Sobeh et al. (2017). The phytochemical analysis identified 39 compounds, the majority of them were flavon-3-ol derivatives and condensed tannins. Total phenolic content, determined by the Folin–Ciocalteu method amounted to 489 mg GAEs/g extract. The extract showed promising antioxidant activities with an EC_{50} of 3.6 µg/mL and 18.32 mM $FeSO_4$ equivalent/mg extract in radical scavenging assay and ferric reducing antioxidant power (FRAP) assays, respectively. The hepatoprotectivee potential of the extract in rats was determined *in vivo* in a D-GalN-induced liver toxicity model. A dose of 100 mg/kg (BW) of the bark extract reduced levels of AST, gamma-glutamyl transferase (GGT), and TB by 35%–7%, 65.3%, and 23.8%, respectively, whereas GSH was increased by 59.1%. These effects were similar to Silymarin which was used as positive control.

Albizia lebbeck (L.) Benth. FAMILY: FABACEAE

Patel et al. (2010) evaluated hepatoprotective properties of *Albizzia lebbeck* against CCl_4-induced hepatotoxicity in rats. The test plant has shown dose-dependent antioxidant activity in all the models of the study. Pretreatment with test extract (200 and 400 mg/kg) prevented the depletion of tissue GSH, LPO, and reduced the elevated levels of all the biochemical markers of hepatotoxicity, indicating that the test extract possess hepatoprotective property. The histopathological study exhibited near to normal liver architecture as compared to control. The result of this study suggests that 70% ethanolic extract of bark of *A. lebbeck* possesses antioxidant and hepatoprotective effects in rats.

Shirode et al. (2012) evaluated the hepatoprotective activity of 70% ethanolic extract of leaves of *A. lebbeck* (70% EELAL) in experimental liver damage induced by thioacetamide (100 mg/kg, s.c.) in albino rats. The extract at the dose of 100 and 200 mg/kg produced significant protective effect as indicated by decreased in the activity of serum enzymes, bilirubin, tissue LPO and physical parameters and increased levels of tissue GSH in a dose-dependent manner. The effects of extract were comparable to that of standard drug Silymarin. Histopathological observation also confirmed these findings.

Albizzia procera (Roxb.) Benth. FAMILY: FABACEAE

Sivakrishnan and Kottaimuthu (2014) evaluated the hepatoprotective effect of *Albizzia procera* against PCM-induced liver injury in rats. Paracetamol treatment lead to the elevated levels of liver marker enzymes and histological observations. However, treatment with *A. procera* significantly reversed the above changes compared to the control group as observed in the PCM-changed rats.

Alcea rosea L. [Syn.: *Althaea rosea* (L.) Cav.] FAMILY: MALVACEAE

Hussain et al. (2014) evaluated the hepatoprotective effects of *Alcea rosea* against APAP-induced hepatotoxicity in mice. Acetaminophen significantly increased serum levels of liver enzyme markers whereas the extract of *A. rosea* significantly reduced serum levels of elevated liver enzyme markers in dose-dependent manner compared to APAP-treated mice group. Histopathological examination of liver tissues also supported the protective effects of *A. rosea* on liver enzyme markers.

Alchornea cordifolia (Schumach. & Thonn) Müll.-Arg. FAMILY: EUPHORBIACEAE

The hepatoprotective activity of the leaf extract of *Alchornea cordifolia*, a Nigerian plant on APAP-induced toxicity *in vivo* has been reported (Olaleye et al., 2006). The antioxidative properties revealed TPC of 0.22 mg/mL and reducing power of 0.062 mg/mL as compared to vitamin E with a reducing power of 0.042 mg/mL. The authors concluded that the hepatoprotective activity of this plant on APAP-induced liver damage is connected to its antioxidative properties.

The effect of combined ethanolic leaf extract of *Alchornea cordifolia* and *Costus afer* orally administered at two doses (50 and 100 mg/kg) on some biochemical parameters of rats with PCM-induced hepatotoxicity was evaluated by Arhoghro et al. (2015). The results obtained showed that PCM treated group (group B) showed a significant reduction in weight compared to group A, C, and D. Analysis of hepatic enzymes also showed an increase in serum marker enzymes of hepatic damage AST, ALT, and ALP, after PCM administration in group B. The combined ethanolic leaf extracts of *Alchornea cordifolia* and *Costus afer* brought the serum levels of these enzymes in group C and D, respectively, close to the control. In conclusion, combined ethanolic leaf extract of *Alchornea cordifolia* and *Costus afer* has hepatoprotective effects on PCM-induced hepatoxicity in Wistar rats.

Olaleye and Rocha (2008) investigated the hepatoprotective activity of *Alchornea cordifolia* on APAP-induced liver damage in mice. Paracetamol caused liver damage as evident by statistically significant increased in plasma activities of AST and ALT. There were general statistically significant losses in the activities of SOD, GPx, CAT, and delta-ALA-D and an increase in TBARS in the liver of PCM-treated group compared with the control group. However, the extract of *A. cordifolia* was able to counteract these effects.

Osadebe et al. (2012) investigated the hepatoprotective and antioxidant activities of *A. cordifolia* leaf extract by CCl_4-induced liver damage in rats. The ethyl acetate and chloroform fractions, at a dose of 300 mg/kg, produced significant hepatoprotection by decreasing the activities of the serum enzymes and bilirubin while there were marked scavenging of the DPPH free radicals by the fractions. The effects were comparable to those of the standard drugs used for the respective experiments, Silymarin and ascorbic acid. Alkaloids, flavonoids, saponins, and tannins were detected in the phytochemical screening.

Alhagi maurorum Medik. FAMILY: FABACEAE

The hepatroprotective effect of *Alhagi maurorum* aerial parts ethanol extract was studied using Wistar albino rats. Liver injury induced in rats by CCl_4. The normal appearance of hepatocytes and correction of SGOT, SGPT, ALP, and TB, indicated a good protection of the extract from CCl_4 hepatotoxicity. The results were compared with Silymarin, the reference hepatoprotective drug (Alqasoumi et al., 2009). Administration of 660 mg/kg of the ethanolic *A. maurorum* extract to mice showed a significant decrease in the level of transaminases in animals treated with a combination of ethanolic *A. maurorum* extract plus CCl_4 or APAP as compared to animals receiving CCl_4 or APAP alone. Histopatological investigation also confirmed that *A. maurorum* extract protects liver against damage-induced either by CCl_4 or APAP (Abdellatif et al., 2014; Gargoum et al., 2013).

Aqueous methanolic extract (500 mg/kg) of *A. maurorum* was tested on rabbits intoxicated by administration of PCM (250 mg/kg). It was concluded that there was a significant reduction in AST, ALT, ALP, and SB after treatment with *A. maurorum* extract. Confirmation of hepatoprotective activity of *A. maurorum* was strengthened by histopathological studies. It was further confirmed that tannins, saponins, alkaloids, and flavonoids present in *A. maurorum* extract showed hepatoprotective activity against PCM-induced toxicity in rabbits (Rehman et al., 2015). In another study, hepatoprotective activity of *A. maurorum* was investigated against CCl_4 and APAP-induced toxicity in mice. A significant decrease in level of liver marker enzymes in the serum was found which confirmed the hepatoprotective potential of *A. maurorum* extract (Huda et al., 2013).

Allanblackia gabonensis (Pellegr.) Bamps. FAMILY: CLUSIACEAE

Allanblackia gabonensis is used in traditional medicine to treat some inflammatory diseases. The aqueous suspension of the stem bark of *A. gabonensis* showed significant hepatho-nephroprotective activity against APAP-induced liver and kidney disorders in rats. In this, study, the pretreatment with 100 and 200 mg/kg significantly reduced the serum level of MDA, increase in enzymatic antioxidant activities (SOD and CAT) and nonenzymatic antioxidant (GSH) levels (Vouffo et al., 2012). The stem bark of this plant has been known to elaborate the following compounds xanthones, benzophenone, flavonoide, and phytosterol (Azebaze et al., 2008). In addition, *A. gabonensis* possess significant analgesic and anti-inflammatory activities (Ymele et al., 2011) which may be contributing to its hepatoprotective activities.

***Allium sativum* L. FAMILY: ALLIACEAE**

A. sativum (garlic) is one of the world's most known medicines that have been used for flavoring and as a medical herb mainly due to its prophylactic and therapeutic capacities. Garlic has known nutritional properties, particularly for its bioactive components, and is used as antidiabetic, anti-inflammatory, antihypertension, antimicrobial, antiatherosclerotic, and hepatoprotective in different diet-oriented therapeutical regimes to heal various lifestyle-associated disorders (Amagase et al., 2001). Garlic and its supplements are taken in many cultures for their hypolipidemic, antiplatelet, and procirculatory effects. Sapogenins, saponins, sulphuric compounds, and flavonoids have been detected in different species of *Allium* genus (Kazemi et al., 2010). Additional biological effects attributed to garlic extract may be due to S-allylcysteine, S-allylmercaptocysteine, and N (alpha)-fructosyl arginine that are formed throughout the extraction (Amagase et al., 2001). Most garlic's beneficial effects are due to organosulfate molecule allicin (Touloupakis and Ghanotakis, 2010). The hepatoprotective effect of garlic extracts on Cd-induced oxidative damage in rats has been reported. *A. sativum* extract decreased hepatic activities of ALT, AST, and ALP and simultaneously increased the plasma activities of ALT and AST. Cd-induced oxidative damage in rat liver is predisposed to decreasing by moderate dose of *A. sativum* extracts probably through reduced LPO and improved antioxidant defense system that could not prevent and protect Cd-induced hepatotoxicity (Obioha et al., 2009). *A. sativum* chemical compounds have curative effects on iron liver excess (Ghorbel et al., 2011). In another study, the hepatoprotective effects by *A. sativum*, ginger (*Zingiber officinale*), and vitamin E against $CC1_4$-induced liver damage were examined in male Wistar albino rats. Serum ALT, AST, and ALP levels decreased significantly 24 h after $CC1_4$ administration in rats pretreated with garlic, ginger, vitamin E, and various mixtures of garlic and ginger compared to the rats treated with only $CC1_4$. LPO expressed by serum MDA was assayed to assess the severity of liver damage by $CC1_4$, including the extent of hepatoprotection by garlic, ginger, and vitamin E. MDA concentration was significantly decreased in rats pretreated with garlic, ginger, vitamin E, and various mixtures of garlic and ginger compared to the rats administered by $CC1_4$ alone. Histological examination of the liver was indicative of severe infiltration of inflammatory cells in rats treated with CCl_4 alone although the change in the normal architecture of the hepatic cells decreased considerably in pretreated rats (Patrick-Iwuanyanwu et al., 2007).

The hepatoprotective activity of *A. sativum* extract at a dose of 300 mg/kg BW, administered intraperitoneally for 14 days before the induction of D-GalN and LPS (D-GalN/LPS) was investigated against D-GalN/LPS-induced hepatitis in rats. The pretreatment with aqueous *A. sativum* extract helped the altered parameters (ALT, AST, ALP, LDH, gamma glutamyl transferase (GGT), bilirubin, LPO, tumor necrosis factor (TNF), and myeloperoxidase (MPO) activity level, TC, TGs, free fatty acids, and antioxidant enzyme activities) reach to nearly normal control values. Aqueous *A. sativum* extract could afford a significant protection in the DGalN/LPS-induced hepatic damage easing (El-Beshbishy, 2008). An investigation of chemopreventive effects of *A. sativum* extract and Silymarin on *N*-nitrosodiethylamine (NDEA) and CCl_4-induced hepatotoxicity in male albino rats indicated synergistic effect of Silymarin and *A. sativum*, and their hepatoprotective features against hepatotoxicity (Park et al., 2008).

Ajayi et al. (2009) studied the hepatoprotective and some hematological effects of *Allium sativum* (garlic) and vitamin C on experimental rats that were exposed to lead (Pb) for one week. Garlic and vitamin C produced significant reduction in the levels of ALT, ALP, and PCV while the level of AST increases significantly. The level of Hb increases significantly in rats treated with garlic and reduces significantly in rats treated with vitamin C.

The main part of therapeutic effect of *A. sativum* is attributed to antioxidant compounds, probably associated with its phenolic compounds and flavonoid substances. *A. sativum* (irrespective of the processes such as baking) has phenolic, flavonoid, and flavenol compounds as well as allicin (Taji et al., 2012). The effect of fresh *A. sativum* on inhibiting the oxidation was higher compared to three-month *A. sativum*. Phenolic compounds of the fresh *A. sativum* were higher than the three-month *A. sativum*. The amount of allicin was 15 and 8 μg/mL in fresh and three-month dated aqueous *A. sativum* extract, respectively.

The aqueous extract of *A. sativum* bulbs (fresh garlic homogenate, 0.25 g/kg/d) generates a hepatoprotective effect against the subacute liver damage induced by the mixture of INH/RIF (50 mg/d, each) administered by oral via, half an hour before anti-TB drugs over a period of 28 days in Wistar rats. The results showed that ALT, AST and TB levels were reduced in animals receiving the garlic extract and RIF/INH, with respect to the group where only RIF/INH was administered. The authors also observed an increase in the GSH level and a low level of LPO; the effect observed was attributed to the presence of thiosulfinates, steroids, terpenes, flavonoids, and other phenols present in garlic (Pal et al., 2006). *A. sativum* (250 mg/kg, oral via), administered by 28 days also protects from liver injury caused only with INH (50 mg/kg); the effect observed was similar to that of Silymarin (200 mg/kg) employed as a positive control (Nasim et al., 2011).

Comparison of hepatoprotective role of garlic and Silymarin in antituberculosis drug (isoniazid)-induced hepatotoxicity was studied by Ilyas et al. (2011). In isoniazid-treated animals there was abnormal rise in the levels of biochemical markers (ALT, AST, ALP, and total bilirubin [TBIL]). In animals treated with isoniazid + Silymarin, the biochemical markers were near normal levels and in isnoazid + garlic extract treated animals the levels of biochemical markers were within normal limits.

The antihepatic toxicity of garlic was investigated experimentally in rats, $CrCl_3$ alone increased serum levels of AST and ALT. However, garlic inhibited the hepatotoxicity of $CrCl_3$, and the concomitant use of garlic and $CrCl_3$ decreased the levels of AST and ALT when garlic is used in a dose of 60 and 120 mg/kg and $CrCl_3$ is 8 mg/kg (Al-Snafi, 2013).

***Allium stipitatum* Regel (Syn: *Allium hirtifolium* Boiss.) FAMILY: ALLIACEAE**

Allium stipitatum (Syn.: *A. hirtifolium*), commonly known as Persian shallot (*Moosir* in Persian) is endemic to Iran. Based on available pharmaceutical investigations, antioxidant and hepatoprotective effects of *A. stipitatum* have also been demonstrated. In addition, *A. hirtifolium* extracts had antioxidant properties comparable to or slightly higher than garlic extracts (Kazemi et al., 2010). The commonly known phytochemical compounds identified in *A. stipitatum* are saponins, sapogenins, sulfur-containing compounds (e.g., thiosulfinates), and flavonoids including shallomin, quercetin, and kaempferol (Kazemi et al., 2010). Alliin, alliinase, allicin, *S*-allyl-cysteine, diallyldisulfide, diallyltrisulfide, and methylallyltrisulfide are the most important biological secondary metabolites of *A. stipitatum* (Azadi et al., 2009). Disulfide and trisulfide compounds are among the most important compounds existing in *A. stipitatum* (Rose et al., 2005). Researches have shown that both the corn and the flower of shallot contain a high density of glycosidic flavonols. Linolenic, linoleic, palmitic, palmitoleic, stearic, and oleic acids have been identified in *A. hirtifolium* oil, as well (Fattorusso et al., 2002).

Treating rats with hydroalcoholic extract of *A. stipitatum* could protect liver cells against oxidant effects of alloxan, and consequently caused a significant reduction in serum concentration of ALP, ALT, and AST. Biochemical results have confirmed the usefulness of *A. stipitatum* extract in decreasing the destructive effects of alloxan on liver tissue, and consequently decreasing the enzymes' leakage into cytosol, which is possibly achieved by herbal antioxidant compounds including flavonoids (Kazemi et al., 2012). It was also reported that consumption of *A. stipitatum* caused a reduction in AST level compared to the group with a hypercholesterolemic diet (Asgari et al., 2012). A research on the effect of hydroalcoholic *A. stipitatum* extract on the level of liver enzymes in streptozotocin-induced diabetic rats indicated that hydroalcoholic extract of *A. stipitatum* could significantly decrease serum levels of liver enzymes (AST, ALT, ALP, and LDH) in a dose-dependent manner. Antioxidant micronutrients in the extract of *A. stipitatum* may also restore liver damages. Shallomin and other active constituents of *A. stipitatum* did not produce any adverse effect on the organs such as liver and kidney (Amin et al., 2012).

Alocasia indica Spach. FAMILY: ARACEAE

Oral administration of hydroalcoholic extract of *Alocacia indica* (250 and 500 mg/kg) effectively inhibited CCl_4 and PCM-induced changes in the serum marker enzymes, cholesterol, serum protein, and ALB in a dose-dependent manner as compared to the normal and the standard drug Silymarin-treated groups. Hepatic steatosis, fatty infiltration, hydropic degeneration, and necrosis observed in CCl_4 and PCM-treated groups were completely absent in histology of the liver sections of the animals treated with the extracts (Mulla et al., 2009).

Aloe barbadensis Mill. (Syn. *Aloe vera* (L.) Burm. f.) FAMILY: LILIACEAE

The protective effect of fresh *Aloe barbadensis* (Syn.: *Aloe vera*) leaves extract on lindane (LD)-induced hepatoxicity and genotoxicity was studied by Etim et al. (2006). Serum levels of hepatic enzyme markers: SGPT, SGOT, GGT, and ALP were determined after oral administration of *A. barbadensis* leaves extract and lindane. The level of polychromatic erythrocytes was also observed. The pretreatment with *A. barbadensis* leaves extract at concentration of 1.0 mL/kg BW significantly decreased the serum levels of GPT, GOT, GGT, and ALP induced by 100 mg/kg BW of lindane. The level of polychromatic erythrocytes observed was not statistically significant when compared to control.

Chandan et al. (2007) evaluated the hepatoprotective activity of *A. barbadensis*. The shade dried aerial parts of *A. barbadensis* were extracted with petroleum ether (AB-1), chloroform (AB-2), and methanol (AB-3). The plant marc was extracted with distilled water (AB-4). All the extracts were evaluated for hepatoprotective activity on limited test models as hexobarbitone sleep time, zoxazolamine paralysis time, and marker biochemical parameters. AB-1 and AB-2 were observed to be devoid of any hepatoprotective activity. Out of two active extracts (AB-3 and AB-4), the most active AB-4 was studied in detail. AB-4 showed significant hepatoprotective activity against CCl_4-induced hepatotoxicity as evident by restoration of serum transaminases, ALP, bilirubin, and TGs. Hepatoprotective potential was confirmed by the restoration of LPO, GSH, glucose-6-phosphatase, and microsomal aniline hydroxylase and amidopyrine *N*-demethylase toward near normal. Histopathology of the liver tissue further supports the biochemical findings confirming the hepatoprotective potential of AB-4. This study showed that the aqueous extract of *A. barbadensis* is significantly capable of restoring integrity of hepatocytes indicated by improvement in physiological parameters, excretory capacity (bromosulphalen [BSP] retention) of hepatocytes and also by stimulation of bile flow secretion. AB-4 did not show any sign of toxicity up to an oral dose of 2 g/kg in mice (Chandan et al., 2007).

Alqasoumi et al. (2008a) evaluated the hepatoprotective activity of the ethanolic extract of *Aloe barbadensis* (=*A. vera*) using CCl_4 in Wistar albino rats. Treatment with *A. vera* extract *failed to restore* the normal appearance of hepatocytes.

Parmar et al. (2009) evaluated the hepatoprotective activity of *A. barbadensis* in CCl_4-induced hepatotoxicity in male albino rats. Animals exposed to CCl_4 showed significant increase in AST, ALT activities; ALP activity and TB level accompanied by significant decrease in TC, ALB and protein levels. Hepatic TBARS level exhibited an increase while there was a significant depletion in CAT and reduced GSH levels following exposure to CCl_4. Co-administration of aqueous plant extracts of *A. barbadensis*, with CCl_4 provided significant protection to most of the above-mentioned biochemical variables.

Parmar et al. (2010) evaluated the hepatoprotective activity of leaves of *Aloe barbadensis* (Syn.: *A. vera*) against PCM-induced toxicity in male Wistar rats. Paracetamol-induced hepatotoxicity resulted in an increase in serum AST, ALT, ALP activity, and bilirubin level accompanied by significant decrease in albumn level. Co-administration of the aqueous extract protects the PM-induced lipid peroxidatiom, restored latered serum marker enzymes, and antioxidant level toward near normal.

Werawatganon et al. (2014) examined the antioxidative, anti-inflammatory effects of *A. barbadensis* (*A. vera*) in mice with APAP-induced hepatitis. In APAP group, ALT, hepatic MDA and the number of interleukin (IL)-12 and IL-18 positive stained cells were significantly increased when compared to control group, whereas hepatic GSH was significantly decreased when compared to control group. The mean level of ALT, hepatic MDA, the number of IL-12 and IL-18 positive stained cells, and hepatic GSH in *A. barbadensis*-treated group were improved as compared with APAP group. Moreover, in the APAP group, the liver showed extensive hemorrhagic hepatic necrosis at all zones while in *Aloe vera*-treated group, the liver architecture was improved histopathology.

In a study conducted to evaluate hepatoprotective activity of *A. barbadensis (A. vera)* gel against PCM-induced hepatotoxicity in albino rats, seven day treatment with *A. barbadensis* significantly reduced the levels of AST, ALT, and ALP significantly and restored the depleted liver thiol levels significantly (Nayak et al., 2011). In another study done for the evaluation of hepatotherapeutic effect of *A. barbadensis* in alcohol-induced liver damage in albino rats, the resultant hepatic dysfunction was abrogated by *A. barbadensis* extract. Histopathological examination revealed that *A. barbadensis* treatment maintained hepatic architecture similar to that seen in the control group. This study shows that aqueous extract of *A. barbadensis* is hepatotherapeutic and thus lends credence to the use of the plant in folklore medicine in the management of alcohol-induced hepatic dysfunction (Saka et al., 2011).

Bhatt et al. (2014, 2015) evaluated the hepatoprotective activity of *Aloe barbadensis* (Syn.: *A. vera*). A total of 110 male and female patients of age group between 15 and 65 years, diagnosed clinically and biochemically as the case of acute viral hepatitis and ready to give consent were recruited in the study from SVBP Hospital, LLRM Medical College, Meerut, UP, India. Subjects who fulfilled selection criteria were randomized into two groups. Fifty patients belonging to control group received the conventional treatment for acute viral hepatitis while 60 patients enrolled in treated group were given the conventional treatment for acute viral hepatitis supplemented with *A. barbadensis* juice (Patanjali Ayurved Ltd., Haridwar) in dose of 20 mL BD orally. Every patient was followed-up for 6 weeks. Serum bilirubin, ALT, AST, and ALP levels were measured initially and at the end of 2, 4, and 6 weeks. Intragroup comparison using repeated measure ANOVA demonstrated a statistically significant decrease in the above-mentioned parameters for both treated and control groups at all intervals of time. Intergroup comparison done by Student's-*t* test, revealed statistically significant difference in all the mentioned parameters between treated and control groups at all intervals of times. Evaluation of hepatoprotective activity of *A. barbadensis* and its comparison with Liv. 52 against hepatotoxicity induced by CCl_4 in albino rats, it was seen that the hepatoprotective activity of *Aloe vera* was comparable to Liv. 52 and the aqueous extract of *A. barbaadensis* exhibited dose-dependent hepato-protection, both biochemically and histologically.

Hena et al. (2016) evaluated the hepatoprotective activity of ethanol and aqueous extracts of *Aloe barbadensis* (Syn.: *Aloe vera*) against PCM-induced liver damage in albino Wistar rats. They reported that pretreatment of rats with ethanol and aqueous extracts prior to PCM administration caused a significant reduction in the values of SGOT, SGPT, ALP, and bilirubin almost comparable to the Silymarin. The hepatoprotective effect was confirmed by histopathological examination of the liver tissue of control and treated animal.

Alpinia galanga **(L.) Willd.** **FAMILY: ZINGIBERACEAE**

Hemabarathy et al. (2009) evaluated the hepatoprotective effect of the crude extract of *Alpinia galanga* against PCM-induced hepatotoxicity in rats. Supplementation with the extract of *A. galanga* maintained serum protein and liver SOD levels similar to that of the normal group. Significant decrease in liver MDA levels as compared with the group treated with PCM was observed in groups treated with the extract. Histological analysis showed significant reduction in the number of necrotic cells in both groups supplemented with the extract.

Alternanthera sessilis **(L.) DC.** **FAMILY: AMARANTHACEAE**

The hepatoprotective effects of *Alternanthera sessilis* were investigated by Lin et al. (1994) in three kinds of experimental animal model. Acute hepatitis was induced by various chemicals such as CCl_4 or APAP in mice and D(+)-galactosamine in rats. When treated with *A. sessilis* (300 mg/kg, p.o.) at 2, 6, and 10 h, a reduction in elevation of SGOT and SGPT levels could be observed at 24 h after administration of the three hepatotoxins. These serological observations were also confirmed by histopathological examinations including centrilobular necrosis, eosinophilic bodies, pyknotic nuclei, microvesicular degeneration of hepatocytes and others. The liver microscopic examination showed a noted improvement in groups receiving *A. sessilis.*

Alysicarpus vaginalis (L.) DC. FAMILY: FABACEAE

Rathi et al. (2015) evaluated the hepatoprotective activity of ethanol extract of *Alysicarpus vaginalis* aerial parts in nitrobenzene (NB)-induced hepatic injury in Wistar rats. Liver injury was induced in rats by single oral administration of NB (50 mg/kg BW). One day after NB induction, the rats were treated with ethanolic extract of *A. vaginalis* orally at the doses of 200 mg/kg BW daily for 30 days. After 30 days, serum biochemical parameters, antioxidant enzyme status and histopathological parameters were analyzed. The results indicated that the ethanol extract of *A. vaginalis* the doses of 200 mg/kg orally significantly and dose dependently reduced and normalized the serum marker enzymes, and increased the antioxidant enzyme status as compared to that of NB control group. Furthermore, it was confirmed by histopathological studies.

Amaranthus caudatus L. FAMILY: AMARANTHACEAE

Methanol extract of whole plant of *Amaranthus caudatus* (MEAC) was screened by Kumar et al. (2011) for hepatoprotective potency against PCM-induced liver damage in Wistar rats. MEAC at 200 and 400 mg/kg significantly normalized the PCM-induced biochemical changes compared with PCM-treated group; increased ALT, AST, TB, and DB levels and decreased serum ALB were significantly reversed by the MEAC treatment (200 and 400 mg/kg). Treatment with MEAC (200 and 400 mg/kg) significantly prevented the rise of MDA and TP levels, and prevented the reduction of GSH, CAT, and TT levels significantly compared with PCM-treated group. Histopathological examination of the liver sections also proved the hepatoprotective activity of MEAC.

Amaranthus spinosus L. FAMILY: AMARANTHACEAE

The hepatoprotective and antioxidant activity of 50% ethanolic extract of whole plant of *Amaranthus spinosus* (ASE) was evaluated by Zeashan et al. (2008) against CCl_4-induced hepatic damage in rats. The ASE at a dose of 100, 200, and 400 mg/kg were administered orally once daily for 14 days. The substantially elevated serum enzymatic levels of serum AST, ALT, ALP, and TB were restored toward normalization significantly by the ASE in a dose-dependent manner. Higher dose exhibited significant hepatoprotective activity against CCl_4-induced hepatotoxicity in rats. The biochemical observations were supplemented with histopathological examination of rat liver sections. Meanwhile, *in vivo* antioxidant activities as MDA, hydroperoxides (HP), reduced GSH, SOD, and CAT were also screened which were also found significantly positive in a dose-dependent manner. The results of this study strongly indicate that whole plants of *A. spinosus* have potent hepatoprotective activity against CCl_4-induced hepatic damage in experimental animals. This study suggests that possible mechanism of this activity may be due to the presence of flavonoids and phenolics in the ASE which may be responsible for hepatoprotective activity.

Around 50% ethanolic extract of *A. spinosus* (whole plant) was evaluated for *in vitro* antioxidant and hepatoprotective activity by Zeashan et al. (2009). Ethanolic extract (6, 7, 8, 9, and 10 μg/mL) was able to normalise the levels of biochemical parameters in isolated rat hepatocytes intoxicated with CCl_4. A dose dependent increase in percentage viability was observed in CCl_4 intoxicated HepG2 cells.

Amaranthus tricolor L. FAMILY: AMARANTHACEAE

The ethanolic extract of *Amaranthus tricolor* (ATE) leaves was tested for its efficacy against CCl_4-induced liver toxicity in rats by Al-Dosari (2010). Oral administration of ATE for 3 weeks significantly reduced the elevated levels of serum GOT, GPT, GGT, ALP, bilirubin, cholesterol, LDL, very low-density lipoprotein (VLDL), TG, and MDA induced by CCl_4. Moreover, ATE treatment was also found to significantly increase the activities of NP-SH and TP in liver tissue. These biochemical findings have been supported by the evaluation of the liver histopathology in rats. The prolongation of narcolepsy induced by pentobarbital was shortened significantly by the extract. The acute toxicity test showed that no morbidity or mortality was caused by the extract. The observed hepatoprotective effect appears to be due to the antioxidant properties of *A. tricolor*.

The pharmacological investigation of Aneja et al. (2013) focuses on evaluation of the efficacy of aqueous extract of roots of *Amaranthus tricolor* for their protection against PCM overdose-induced hepatotoxicity. The aqueous extract of roots of *A. tricolor* was prepared and phytochemical screening was done. The biochemical investigation, namely, SGOT, SGPT, ALP, and TBIL was done against PCM-induced hepatotoxicity in Wistar albino rats. The histopathological studies of liver were also done. The phytochemical screening of the aqueous extract showed the presence of alkaloids, carbohydrates, flavanoids, amino acids, proteins, fixed oil, saponins and tannins, and phenolic compounds. Pretreatment with the aqueous extract of root significantly prevented the physical, biochemical, histological, and functional changes induced by PCM in the liver. The extract showed significant hepatoprotective effects as evidenced by decreased serum enzyme activities like SGPT, SGOT, ALP, and TB, which was supported by histopathological studies of liver. The aqueous extract showed significant hepatoprotective activity comparable with standard drug Silymarin. From these results, it was concluded that the *A. tricolor* has potential effectiveness in treating liver damage in a dose-dependent manner.

Amaranthus viridis L. FAMILY: AMARANTHACEAE

Heaptoprotective activity of methanolic extract of whole plant of *Amaranthus viridis* (MeAv) in PCM-induced hepatotoxicity was evaluated in Wistar rats by Kumar et al. (2011). MeAv significantly decreased the elevated liver marker enzymes (SGPT, SGOT), bilirubin (TBIL and DBL) and restored ALB, TP levels. A histopathological study also showed liver protective property of MeAv. In animal model, the antioxidant studies, the MeAv has notably retained the MDA, reduced GSH, CAT, and total thiols.

Ambrosia maritima L. FAMILY: ASTERACEAE

The hepatoprotective activity of the aqueous-methanolic extract of *Ambrosia maritima* was investigated by Ahmed and Khater (2001) against APAP (PCM, 4-hydroxy acetanilide)-induced hepatic damage. Acetaminophen at the dose of 640 mg/kg produced liver damage in rats as manifested by the significant rise in serum levels of AST, ALT, and ALP to 1178.5 +118.05; 607.5 + 32.6 and 274.16 + 8.89 IU/L (*n* = 10), respectively, compared with respective control values of 97.83 + 3.23; 46.0 + 3.92 and 168.67 + 7.86 IU/L. Pretreatment of rats with the plant extract (100 and 200 mg/kg) lowered significantly the respective serum AST to 203.3 + 5.74 and 157.1 + 8.78 IU/L, ALT to 138.67 + 7.7 and 87.5 + 3.6 IU/L and ALP levels to 238.0 + 5.89 and 206.5 + 7.5 IU/L, respectively. Treatment of rats with APAP led to a marked increase in LPO as measured by MDA (42%). This was associated with a significant reduction of the hepatic antioxidant system, for example, reduced GSH (65%), glutathione reductase (GR) (35%), total glutathione peroxidase (GSH-Px) (32%), and GST (16%). These biochemical alterations resulting from APAP administration were inhibited by pretreatment with *A. maritima* extract.

Ammi majus L. FAMILY: APIACEAE

Ammi majus is a local medicinal plant with fruits that are contraindicated in nursing, pregnancy, tuberculosis (TB), liver and kidney diseases, human immunodeficiency virus, and other autoimmune diseases (Asadi-Samani et al., 2015). It is commonly used for skin disorders such as psoriasis and vitiligo. *A. majus* is contraindicated in diseases associated with photosensitivity, cataract, invasive squamous-cell cancer, known sensitivity to xanthotoxin, and in children under the age of 12 (Selim and Ouf, 2012). *A. majus* concomitantly accumulates various 7-*O*-prenylated umbelliferones as the predominant coumarines (Hübner et al., 2003). It is considered as a source of 6-hydroxy-7-methoxy coumarine, which is known as the major coumarin. *A. majus* with confirmed antioxidant effect could be used in diabetic nephropathy and myocardial injury, thanks to different active compounds such as quercetine, kaempferol, and marmesinin that inhibit cytochrome P450 such as xanthotoxin bergapten, imperatorin, and isoimpinellin. Treatment of rats with different doses of *A. majus* seeds' extract could cause hepatoprotective effects against CCl_4-induced liver damage, in a dose-dependent fashion (Mutlag et al., 2011).

Amorphophallus campanulatus Roxb. FAMILY: ARACEAE

The hepatoprotective activity of ethanolic and aqueous extracts of *Amorphophallus campanulatus* tubers were evaluated by Jain et al. (2009) against CCl_4-induced hepatic damage in rats. The extracts at a dose of 500 mg/kg were administered orally once daily. The substantially elevated serum enzymatic levels were significantly restored toward normalization by the extracts. Silymarin was used as a standard reference and exhibited significant hepatoprotective activity against CCl_4-induced haptotoxicity in rats. The biochemical observations were supplemented with histopathological examination of rat liver sections. The results of this study strongly indicate that *A. campanulatus* tubers have potent hepatoprotective action against CCl_4-induced hepatic damage in rats. The ethanolic extract was found hepatoprotective more potent than the aqueous extract. The antioxidant activity was also screened and found positive for both ethanolic and aqueous extracts. This study suggests that a possible mechanism of this activity may be due to free-radical scavenging potential caused by the presence of flavonoids in the extracts (Jain et al., 2009).

Singh et al. (2011) evaluated the hepatoprotective effect of the dried tuber of *A. campanulatus* (MAC) against APAP-induced hepatic injury in albino rats. Pretreatment with MAC reduced the biochemical markers of hepatic injury like SGPT, SGOT, ALP, bilirubin (BRN), and TP showed hepatoprotective activity. There is increase in the levels of SOD, CAT, and GPx shows that the plant may possess hepatoprotective and antioxidant property.

Bais and Mali (2013) investigated the protective effect of *A. campanulatus* tuber extracts against H_2O_2-induced oxidative damage in human erythrocytes and leucocytes. There was increase in the CAT, SOD, GPx, and reduction of GSH and LPO levels in H_2O_2 group compared with control group. The extracts of tuber of *A. campanulatus* treated groups showed effective reduction of CAT, SOD, GPx, and increased the GSH and LPO levels as compared with H_2O_2 group on human erythrocytes and leucocytes. The ME was found more effective than others.

Anacardium occidentale L. FAMILY: ANACARDIACEAE

Ikyembe et al. (2014) investigated the hepatoprotective activity of pretreatment with methanolic leaf extract of *Anacardium occidentale* against CCl4-induced hepatotoxicity in Wistar rats. Histological and biochemical examinations of the liver showed CCl_4-induced hepatotoxicity. Pretreatment with the extract, especially at dose 500 mg/kg, revealed hepatoprotective effect against chemically induced acute liver toxicity by preserving the histoarchitecture of the liver and significantly reducing of the serum marker enzymes.

Ananas comosus (L.) Merr. FAMILY: BROMELIACEAE

Mohamad et al. (2015) investigated the reversing effects of pineapple vinegar on PCM-induced liver damage in murine model. Pineapple juice was fermented via anaerobic and aerobic fermentation to produce pineapple vinegar. Pineapple vinegar (0.08 and 2 mL/kg BW) and synthetic vinegar were used to treat PCM-induced liver damage in mice. Pineapple vinegar contained 169.67±0.05 μg GAE/mL of TPC, with 862.61±4.38 μg/mL GA as the main component. Oral administration of pineapple vinegar at 2 mL/kg BW reduced serum enzyme biomarker levels, including

AST, ALT, ALP, and TG after 7 days of PCM treatment. Liver antioxidant levels such as hepatic GSH, SOD, LPO, and FRAP were restored after the treatment. Pineapple vinegar reduced the expressions of inducible enitric oxide synthase (iNOS) and NF-kB and the level of NO significantly. Pineapple vinegar also downregulated liver cytochrome P450 protein expression. Oral administration of pineapple vinegar at 0.08 and 2 mL/kg BW reduced serum enzyme biomarker levels, restored liver antioxidant levels, reduced inflammatory factor expressions, and downregulated liver cytochrome P450 protein expression in PCM-induced liver damage in mice.

Pineapple fruit (*Ananas comosus*) can inhibit the activity of cytochrome 2E1 (CYP2E1) (Yantih et al., 2017). Extracts of ethanol and water from pineapple fruit can decrease ALT and AST levels in rats, the increased ALT and AST levels are directly proportional to the damage of liver function. Yantih et al. (2017) evaluated the hepatoprotective activity of pineapple juice in INH-induced rats. Pineapple juice exhibited hepatoprotective activity, as it decreased the ALT and AST levels in the rats after 4 weeks of treatment. Pineapple juice, protected the rats' livers by inhibiting the central venous diameter widening, although the data analysis showed that the liver function in these rats was not as good as that in the positive controls.

Anchusa strigosa **Banks & Sol.**
FAMILY: BORAGINACEAE

The aqueous and ethanolic extracts of *Anchusa strigosa* were studied to inhibit aryl hydrocarbon hydroxylase (AHH) activity and 3H-benzo (a) pyrene (3H-BP) binding to rat liver microsomal protein. The aqueous extracts showed no inhibitory effect while the ethanolic extracts exhibited strong inhibitory effect on both AHH and 3H-BP binding to the microsomal protein (Alwan et al., 1989).

Andrographis lineata **Nees**
FAMILY: ACANTHACEAE

Sangameswaran et al. (2008) evaluated the hepatoprotective effect of *Andrographis lineata* extracts in CCl_4-induced liver injury in rats. Male Wistar rats with chronic liver damage, induced by subcutaneous injection of 50% v/v CCl_4 in liquid paraffin at a dose of 3 mL/kg on alternate days for a period of 4 weeks, were treated with methanol and aqueous extracts of *A. lineata* orally at a dose of 845 mg/kg/day. The biochemical parameters such as SGOT, SGPT, SB, and ALP were estimated to assess the liver function. Histopathological studies of the liver were also carried out to confirm the biochemical changes. Histopathological examinations of liver tissue corroborated well with the biochemical changes. The activities of extracts were comparable to a standard drug. Hepatic steatosis, hydropic degeneration, and necrosis were observed in the CCl_4-treated group, while these were completely absent in the standard and extract treated groups. *A. lineata* extracts exhibited hepatoprotective action against CCl_4-induced liver injury.

Andrographis paniculata (Burm. f.) Nees FAMILY: ACANTHACEAE

Alcoholic extract of the leaves of *Andrographis paniculata* (AAP) was obtained by cold maceration. A dose of 300 mg/kg (1/6 of LD50) of the extract was selected to study hepatoprotective action against CCl_4-induced liver damage. The extract was found to be effective in preventing liver damage which was evident by morphological, biochemical, and functional parameters (Rana and Avadhoot, 1991).

Andrographolide active constituent of *A. paniculata* antagonized the toxic effects of PCM on certain enzymes (SGOT, SGPT, and ALP) in serum as well as in isolated hepatic cells as tested by trypan blue exclusion and oxygen uptake tests, in a significant dose-dependent (0.75–12 mg/kg, p.o. × 7 days) manner. Neoandrographolide increase GSH, GSH 5-transferase, GPx, SOD, and LPO level (Dahanukar et al., 2000).

Administration of an alcohol extract of *A. paniculata* (25 mg/kg) and two of its constituent diterpenes, andrographolide and neoandrographolide (6 mg/kg/day for 2 weeks) showed significant antihepatotoxic action in *P. berghei* K173-induced hepatic damage in *M. natalensis* (Chander et al., 2008). The increased levels of serum lipoprotein-X, ALP, GOT, GPT, and bilirubin were markedly reduced by *A. paniculata* and its diterpenes. In the liver, these preparations decreased the levels of LPO products and facilitated the recovery of SOD and glycogen. The protective effects of andrographolide were comparable to those of neoandrographolide.

Nagalekshmi et al. (2011) investigated the ability of the extract of *A. paniculata* to offer protection against acute hepatotoxicity induced by PCM (150 mg/kg) in Swiss albino mice. Oral administration of *A. paniculata* extract (100–200 mg/kg) offered a significant dose-dependent protection against PCM-induced hepatotoxicity as assessed in terms of biochemical and histopathological parameters. The PCM-induced elevated levels of serum marker enzymes such as SGPT, SGOT, ALP, and bilirubin in peripheral blood serum and distorted hepatic tissue architecture along with increased levels of LPOs and the reduction of SOD, CAT, reduced GSH and GPx in liver tissue. Administration of the plant extract after PCM insult restored the levels of these parameters to control (untreated) levels.

The effect of *A. paniculata* extract was studied on CCl_4-induced hepatic damage in rats by Vetriselvan et al. (2011). The degree of protection was measured by physical and biochemical changes. Pretreatment with extract significantly prevented the physical and biochemical changes induced by CCl_4 in the liver. Aqueous extract of *A. paniculata* at the doses of 100, 200 mg/kg caused a significant inhibition in the levels of SGOT, SGPT, ALP, LDH, Tot. Bil, Dir.Bil. toward the respective normal range and this is indication of stabilization plasma membrane as well as repair of hepatic tissue damage caused by CCl_4.

Vetriselvan and Subasini (2012) also evaluated the hepatoprotective activity of aqueous extract of *A. paniculata* in ethanol-induced hepatotoxicity in albino Wistar rats. Results revealed that there is a tremendous elevation in the liver enzymes such as AST and ALT associated with ethanol administration. The aqueous extract of *A. paniculata* showed significant reduction in the liver tissue abnormalities and liver enzymes. In this present study, it has shown that *A. paniculata* demonstrate a strong hepatoprotective activity against ethanol-induced rats.

Trivedi and Rawal (2001) reported that *A. paniculata* treatment prevents BHC-induced increase in the activities of enzymes γ-gutamyl tranpeptidase, GSH-*S*-transferase and LPO. Administration of *A. paniculata* showed protective effects in the activity of SOD, CAT, GSH peroxidise, GR, as well as the level of GSH. The activity of lipid peroxidise was also decreased.

Sivaraj et al. (2011) investigated the protective effect of the leaf extract of *A. paniculata* (Ap) against ethanol-induced liver toxicity in male albino rats. The ethanol induced animals the liver marker enzymes like ALT, AST, ALP, and bilirubin were significantly elevated when compared to the normal animals. After administration of aqueous extract of Ap, the elevated levels of marker enzymes were significantly decreased. The antioxidant enzymes were decreased significantly in ethanol-induced animals after administration of plant extract the decreased levels were increased significantly. The aqueous leaf extract of *A. paniculata* could protect the liver against ethanol-induced liver toxicity by possibly reducing the rate of LPO and increasing the antioxidant defense mechanism in rats.

Hepatoprotective effect of the aqueous leaf extract of *A. paniculata* was investigated by Nasir et al. (2013) against CCl_4-induced hepatic injury in rats. Significant increase of serum levels of ALT, AST, ALP, TBL, DBL, total cholesterol (CHL),

TGs, LDL, VLDL, and MDA in CCl_4 intoxicated rats were restored to normal levels when treated with the extract and CCl_4. Significant decrease of serum levels of TP, ALB, HDL and reduced GSH in CCl_4 intoxicated rats were restored to normal levels when treated with the extract and CCl_4. The LD_{50} of the leaf extract was greater than 3000 mg/kg.

Bardi et al. (2014) investigated the hepatoprotective effects of ethanolic *A. paniculata* leaf extract (ELAP) on thioacetamide-induced hepatotoxicity in rats. An acute toxicity study proved that ELAP is not toxic in rats. To examine the effects of ELAP *in vivo*, male *SD* rats were given intraperitoneal injections of vehicle 10% Tween-20, 5 mL/kg (normal control) or 200 mg/kg TAA thioacetamide (to induce liver cirrhosis) three times per week. Rats treated with ELAP exhibited significantly lower liver/BW ratios and smoother, more normal liver surfaces compared with the cirrhosis group. Histopathology showed minimal disruption of hepatic cellular structure, minor fibrotic septa, a low degree of lymphocyte infiltration, and minimal collagen deposition after ELAP treatment. Immunohistochemistry indicated that ELAP-induced downregulation of proliferating cell nuclear antigen. Also, hepatic antioxidant enzymes and oxidative stress parameters in ELAP-treated rats were comparable to Silymarin-treated rats. ELAP administration reduced levels of altered serum liver biomarkers. ELAP fractions were noncytotoxic to WRL-68 cells, but possessed antiproliferative activity on HepG2 cells, which was confirmed by a significant elevation of LDH, reactive oxygen species (ROS), cell membrane permeability, cytochrome *c*, and caspase-8, -9, and, -3/7 activity in HepG2 cells. A reduction of mitochondrial membrane potential (MMP) was also detected in ELAP-treated HepG2 cells. The hepatoprotective effect of 500 mg/kg of ELAP is proposed to result from the reduction of thioacetamide-induced toxicity, normalizing ROS levels, inhibiting cellular proliferation, and inducing apoptosis in HepG2 cells.

Angelica sinensis (Oliv.) Diels FAMILY: APIACEAE

Cao et a. (2014) evaluated the hepatoprotective effects of *Angelica sinensis* extract (ASE) against CCl_4-induced hepatotoxicity in Jian carp (*Cyprinus carpio* var. *Jian*). Results showed that the increases of SGPT and SGOT induced by CCl_4 were significantly inhibited by pre-treating the fish with 0.1%, 0.5%, and 1.0% ASE in the diets. The elevation of LDH and the reductions of the TP and ALB in the serum induced by CCl_4 were also inhibited by pretreatments with 0.5% and 1.0% ASE. In the liver tissue, pretreatment with 1.0% ASE significantly inhibited the MDA formation and the reductions of the total antioxidant capacity (T-AOC), SOD, GSH, and CYP3A mRNA expression induced by CCl_4. Comet assay showed that tail moment, olive tail moment, tail length, and tail deoxyribonucleic acid (DNA%) were positively changed in fish pretreated with 0.5% and 1.0% ASE. CCl_4-induced histological changes were obviously reduced by 0.5% and 1.0% ASE. Overall results proved the hepatoprotective effect of ASE in a dose-dependent manner and support the use of ASE (1.0%) as a hepatoprotective and antioxidant agent in fish.

Anisochilus carnosus (L.) Wall. FAMILY: LAMIACEAE

Kumar et al. (2010) investigated the hepatoprotective activity of alcoholic and aqueous extract of leaves of *Anisochilus carnosus* against rifampicin-induced hepatotoxicity.

Preliminary phytochemical studies revealed the presence of carbohydrates, glycosides, phenolic compound, tannins, alkaloids, and flavonoids. The leaf extract of *A. carnosus* was found to be nontoxic. Treatment of rats with Rifampicin produce an increase in the weight and volume of rat liver. Aqueous and alcoholic extract showed significant decrease in liver weight and volume compared to control group. Rifampicin administration resulted increase of SGOT, SGPT, ALP, glutamate transpeptidase (GGTP), and bilirubin direct, while TP was found to decrease compare to normal control group. Pretreatment with Silymarin, aqueous and, alcoholic extract significantly prevented the biochemical changes induced by rifampicin. Aqueous extract treatment offered greater hepatoprotective effect than alcoholic extract. Hepatocytes of control group showed a normal histology of liver. Rifampicin-treated group showed severe and macro-vasicular fatty changes, severe inflammation, and fatty degeneration. Silymarin-treated group showed normal hepatocytes and no evidence of hepatic damage. Alcoholic extract treated group showed mild fatty degeneration, mild chronic inflammation, and also show mild focal rearrangements of cells. Aqueous extract treated group showed normal lobular architecture of the liver with no evidence of inflammation.

Reshi et al. (2010) evaluated the hepatoprotective activity of leaf and leaf callus extracts of *A. carnosus* against alcohol-induced toxicity using HepG2 cell line. Ethanolic leaf extract pretreated HepG2 cells showed 94% cell viability compared to the standard Silymarin pretreated

HepG2 cells which showed 81% cell viability. Leaf callus extracts also exhibited significant hepatoprotective activity where ethanolic callus extract pretreated HepG2 cells showed 86% viability after intoxication with alcohol. HepG2 cell viability percentage was dose dependent. Phytochemical studies revealed the presence of different secondary metabolites in leaf and leaf callus extracts. The bio-efficacy study confirms the presence of secondary metabolites of hepatoprotective nature in leaf and leaf callus of *A. carnosus*.

An ethanolic extract of stems of *A. carnosus* was studied by Venkatesh et al. (2011) for hepatoprotective activity against CCl_4-induced hepatotoxicity in Albino Wistar rats. Ethanolic extract of *A. carnosus* was administered to the experimental rats at two dose levels, 200 and 400 mg/kg BW. The hepatoprotective effect of the extract was evaluated by the assay of liver function biochemical parameters like SGPT, SGOT, ALP, TB, and TP. In ethanolic extract treated animals, the toxic effect of CCl_4 was controlled significantly as compared to the normal and the standard drug Silymarin-treated group.

Annona muricata L. FAMILY: ANNONACEAE

Adewole and Ojewole (2008) investigated the possible protective effects of *Annona muricata* leaf aqueous extract (AME) in rat experimental paradigms of diabetes mellituss. Treatment of Groups B and C rats with STZ (70 mg/kg i.p.) resulted in hyperglycemia, hypoinsulinemia, and increased TBARS, ROS, TC, TG, and LDL levels. STZ treatment also significantly decreased CAT, GSH, SOD, GSH-Px activities, and HDL levels. AME-treated Groups C and D rats showed significant decrease in elevated blood glucose, ROS, TBARS, TC, TG, and LDL. Furthermore, AME treatment significantly increased antioxidant enzymes' activities, as well as serum insulin levels. The findings of this laboratory animal study suggest that *A. muricata* extract has a protective, beneficial effect on hepatic tissues subjected to STZ-induced oxidative stress, possibly by decreasing LPO and indirectly enhancing production of insulin and endogenous antioxidants.

Arthur et al. (2012) evaluated the hepatoprotective activity and antijaundice property of aqueous extract of *A. muricata* against CCl_4 and APAP-induced hepatotoxicity in Sprage Dawley rat model. The activity of all the marker enzymes registered significant increases in CCl_4- and APAP-treated rats, decreases in cholesterol and TG concentration, and increase in total and indirect bilirubin, an indication of hepatic jaundice. *A. muricata* at all doses significantly restored liver function toward normal levels which compared well against Silymarin control. Histopathological analysis of liver sections confirmed biochemical investigations. The results indicate that leaves of *A. muricata* possess hepatoprotective activity and can treat hepatic jaundice.

Usunomena (2014) evaluated the protective role of ethanolic leaf extract of *A. muricata* on dimethylnitrosamine (DMN)-induced hepatotoxicity in rat model. Pretreatment with *A. muricata* ethanolic leaf extract produced a significant decrease in the ALT and TC level and an increase in TP, globulin, ALB, and hematological parameters compared to DMN alone group.

Okoye et al. (2016) explored the hepatocurative effects of aqueous stem bark extract of *A. muricata*. Significant increase in serum levels of ALP, AST, ALT, and LDH was observed in animals treated with APAP. Liver microscopy also showed evidence of cholestasis and necrosis. It showed minimal change in tissue architecture with concomitant significant decrease in serum levels of ALP, AST, ALT, and LDH when administered with stem bark extract. The biochemical analysis and liver microscopy of this study suggest that aqueous stem bark extract of *A. muricata* possess significant hepatoprotective, anticholestasis and antisinusoidal congestion properties.

The hepatocurative effect of aqueous leaf extract of *A. muricata* on PCM-induced hepatotoxicity in rats was studied by Omeodu et al. (2017). Significant increase of serum levels of ALT, AST, GGT, CHOL, and TGs in PCM poisoned rats, with significantly, decreased serum levels of TP and HDL were recorded. These findings confirmed induction of hepatotoxicity by the *A. muricata* leaf extract was found to significant reduce the serum levels of AST, ALT, CHOL, and TGs and significantly increased the serum concentration of HDL. These indicate the possible hepatocurative effects of aqueous leaf extract of *A. muricata* on PCM-induced liver toxicity in rats.

Annona reticulata L. FAMILY: ANNONACEAE

Bijesh and Pramod (2014) evaluated the hepatoprotective effect of ethanolic extract of *Annona reticulata* leaves using Chang liver cell line. Simvastatin was used to induce the hepatotoxicity in the Chang liver cell lines. The extract was evaluated for hepatoprotective effect in simvastatin-induced hepatotoxicity in Chang liver cells. The percentage viability of in the cell line was evaluated by the MTT assay [(3-(4,5-dimethylthiazole-2 yl)-2,5 diphenyl tetrazolium bromide). The ethanol extract treatment significantly improved viability of Chang liver cell line which was evident in morphology. The hepatoprotective activity was confirmed by estimating level of super oxide dismutase enzyme and LPO assay. The ethanol extract significantly increased super oxide dismutase enzyme level and decreased LPO.

Annona squamosa L. FAMILY: ANNONACEAE

Raj et al. (2009) evaluated the hepatoprotective effect of custard apple (*Annona squamosa*) in DEN-induced Swiss albino mice. The levels of SGOT, SGPT, ALP, total and direct bilirubin (both in serum and tissue), acid phosphatise (ACP), alfa fetoprotein (AFP) (only in serum) are increased in DEN-administered mice and decreased in DEN+*A. squamosa* extract groups. Total proteins decreased in DEN-administered mice and increased in DEN+ *A. squamosa* treated groups. Histopathology also confirmed the hepatoprotective protect.

The MeOH extract of *A. squamosa* leaves showed a hepatoprotective effect on RIF/INH-induced liver damage (100 mg/kg each) in Wistar rats; treatment was administered during 21 days orally (Thattakudian et al., 2011). The extract of *A. squamosa* at 250 and 500 mg/kg reduced ALT, AST, ALP, GGT, protein, and TB levels. The GSH levels increased in group treated with extract and Silymarin. This hepatoprotective effect showed that the extract was similar to that exhibited by Silymarin (Thattakudian et al., 2011).

Saleem et al. (2008) demonstrated the hepatoprotective effect of alcoholic and water extract of *A. squamosa* (custard apple) hepatotoxic animals with a view to explore its use for the treatment of hepatotoxicity in human. These extracts were used to study the hepatoprotective effect in isoniazid + rifampicin-induced hepatotoxic model. There was a significant decrease in TB accompanied by significant increase in the level of TP and also significant decrease in ALP, AST, ALT, and γ-GT in the treatment group as compared to the hepatotoxic group. In the histopathological study, the hepatotoxic group showed hepatocytic necrosis and inflammation in the centrilobular region with portal triaditis. The treatment group showed minimal inflammation with moderate portal triaditis and their lobular architecture was normal. It should be concluded that the extracts of *A. squamosa* were not able to revert completely hepatic injury induced by isoniazid.

Rajeshkumar et al. (2015) evaluated the hepatoprotective activity of *A. squamosa* leaves aqueous extract induced with PCM. The increasing of enzymes and bilirubin, TGs, and cholesterol and decreasing protein shows the liver damage. These levels are changed into normal range indicates efficiency of our plant drug. The histopathalogical results also confirmed the hepatoprotective capability.

Anoectochilus formosanus Hayata FAMILY: ORCHIDACEAE

Aqueous extracts of fresh whole plant of *Anoectochilus formosanus* at a dose of 130 mg/kg showed inhibition of chronic hepatitis (induced by CCl_4) in mice by reducing SGPT and hepatic hydroxyproline (HYP) level. It also diminished the hypoalbuminemia and splenomegaly. In an *in vitro* study, the LD_{50} values for H_2O_2-induced cytotoxicity in normal liver cells were significantly higher after kinsenoside (isolated from the aqueous extract) pretreatment at the dose 20–40 µg/mL (Wu et al., 2007).

Anogeissus latifolia (Roxb. ex DC.) Wall. FAMILY: COMBRETACEAE

Pradeep et al. (2009) evaluated the hepatoprotective activity of a hydroalcoholic extract of the bark of *Anogeissus latifolia*; *in vitro* in primary rat hepatocyte monolayer culture and *in vivo* in the liver of Wistar rats intoxicated by CCl_4. *In vitro*: primary hepatocyte monolayer cultures were treated with CCl_4 and extract of *A. latifolia.* A protective activity could be demonstrated in the CCl_4 damaged primary monolayer culture. *In vivo*: Hydroalcoholic extract of *A. latifolia* (300 mg/kg) was found to have protective activity in rats with CCl_4-induced liver damage as judged from serum marker enzyme activity.

Aphanamixis polystachya (Wall.) R. Parker FAMILY: MELIACEAE

Gole and Dasgupta (2002) evaluated the antihepatotoxic activity of *Aphanamixis polystachya* on CCl_4-induced liver injury in a rat model. The crude leaf extract significantly inhibits the enhanced AST, ALT, ALP, ACP, and LDH activities released from the CCl_4-intoxicated animals. It also ameliorated the depressed value of serum ALB and the enhanced value of TB in plasma caused by CCl_4 intoxication.

Apium graveolens L. FAMILY: APIACEAE

Apium graveolens, commonly known as celery, is an edible plant. Data obtained from literature reveal that *A. graveolens* has many pharmacological properties such as antifungal, antihypertensive, antihyperlipidemic, diuretic, and anticancer (Mansi et al., 2009; Asif et al., 2011; Nagella et al., 2012; Fazal and Singla, 2012). This plant has also been shown to have medicinal features including hyperlipidemic effects as well as antioxidative and hepatoprotective activities (Mansi et al., 2009).

The active constituents are isoimperatorin, isoquercitrin, linoleicacid, coumarins (seselin, osthenol, apigravin, and celerin), furanocoumarins (including bergapten), flavonoids (apigenin, apiin), phenolic compounds, choline, and unidentified alkaloids (Asif et al., 2011). *A. graveolens* is full of betacarotene, folic acid, vitamin C, sodium, magnesium, silica, potassium, chlorophyll, and fiber. The essential oil contains deltalimonene and various sesquiterpene (Asif et al., 2011; Nagella et al., 2012).

Seeds of *A. graveolens* are used in Indian systems of medicine for the treatment of liver ailments. The antihepatotoxic effect of methanolic extracts of the seeds was studied by Singh and Handa (1995) on rat liver damage induced by a single dose of PCM (3 g/kg, p.o.) or thioacetamide (100 mg/kg, s.c.) by monitoring several liver function tests, namely, serum transaminases (SGOT and SGPT), ALP, sorbitol dehydrogenase, glutamate dehydrogenase, and bilirubin in serum. Furthermore, hepatic tissues were processed for assay of TGs and histopathological alterations simultaneously. A significant hepatoprotective activity of the methanolic extract of the seeds was reported.

The different extracts of *A. graveolens* were tested by Ahmed et al. (2002) for their hepatoprotective activity against CCl_4-induced hepatotoxicity in albino rats. The degree of protection was measured by using biochemical parameters like serum transaminases (SGOT and SGPT), ALP, TP and ALB. The methanolic extracts showed the most significant hepatoprotective activity comparable with standard drug Silymarin. Other extracts namely petroleum ether and acetone also exhibited a potent activity.

Seeds of *A. graveolens* are used in Iranian medicine for liver ailments and disorders, have effects on liver, and exhibit hepatoprotective activities. Examining the antihepatotoxic effect of *A. graveolens* seeds' methanolic extracts on rats' liver showed a significant hepatoprotective activity (Asif et al., 2011). The roots open obstruction of the liver and spleen, and help in dropsy and jaundice treatment (Fazal and Singla, 2012). Due to apigenin-related anti-inflammatory and antioxidant properties, *A. graveolens* seeds could counteract the prooxidant effect of 2-acetylaminofluorine through scavenging superoxide radicals, consequently declining hepatic GST and decreasing release of γ-GGT in serum; as a result, *A. graveolens* could be assumed as a potent plant against experimentally induced hepatocarcinogenesis in rats (Sultana et al., 2005). In addition, different extracts of the plant were examined for their hepatoprotective activity against CCl_4-induced hepatotoxicity in albino rats. Wu and Chen (2008) reported that the extracts of *A. graveolens* root significantly decreased CCl_4-induced acute hepatic injury, increasing the activities of AST and ALT and preventing CCl_4-induced acute liver injury.

Apocynum venetum L. FAMILY: APOCYNACEAE

A water extract (500 mg/kg/day, one week administration) of the leaves of *Apocynum venetum* showed protective effects against CCl_4 (30 microliters/mouse) or D-galactosamine (D-GalN, 700 mg/kg)/LPS (LPS, 20 µg/kg)-induced liver injury in mice (Xiong et al., 2000). Tumor necrosis factor-alpha (TNF-alpha) secreted from LPS-stimulated macrophages is the most crucial mediator in the D-GalN/LPS-induced liver injury model. The extract had no significant inhibition on the increase of serum TNF-alpha, but exhibited a complete inhibition at the concentration of 100 µg/mL on TNF-alpha (100 ng/mL)-induced cell death in D-GalN (0.5 mM)-sensitized mouse hepatocytes. Further activity-guided fractionation resulted in the isolation of fifteen flavonoids, namely, (−)-epicatechin (1), (−)-epigallocatechin (2), isoquercetin (3), hyperin (4), (+)-catechin (5), (+)-gallocatechin (6), kaempferol-6′-*O*-acetate (7), isoquercetin-6′-*O*-acetate (8), catechin-[8,7-e]-4 alpha-(3,4-dihydroxpyhenyl)-dihydro-2(3H)-pyranone (9), apocynin B (10), apocynin A (11), cinchonain Ia (12), apocynin C (13), apocynin D (14) and quercetin (15). All the compounds showed inhibitory effects on TNF-alpha-induced cell death with different intensities. The flavonol glycosides 3, 4, 7, and 8 and the phenylpropanoid-substituted flavan-3-ols 11 and 12 showed potent inhibitory effects on TNF-alpha-induced cell death with IC50 values of 37.5, 14.5, 31.2, 55.1, 71.9, and 41.2 µM, respectively. In contrast, the clinically used 5 and its analogs 1, 2, and 6 showed apparent activity only at 80 µM. These flavonoids appeared to be the hepatoprotective principles of the leaves of *A. venetum*.

Zhang et al. (2018) investigated the protective effects of the total flavonoids prepared from the leaves of *A. venetum* on CCl_4-induced hepatotoxicity in a cultured HepG2 cell line and in mice. Cell exposed to 0.4% CCl_4 (v/v) for 6 h led to a significant decrease in cell viability, increased LDH leakage, and intracellular ROS. CCl_4 also induced cell marked apoptosis, which was accompanied by the loss of MMP. Pretreatment with total flavonoids at concentrations of 25, 50, and 100 µg/mL effectively relieved CCl_4-induced cellular damage in a dose-dependent manner. *In vivo*, total flavonoids (100, 200, and 400 mg/kg BW) were administered via gavage daily for 14 days before CCl_4 treatment. The high serum ALT and AST levels induced by CCl_4 were dose dependently suppressed by pretreatment of total flavonoids (200 and 400 mg/kg BW). Histological analysis also supported the results obtained from serum assays. Furthermore, flavonoids could prevent CCl_4-caused oxidative damage by decreasing the MDA formation and increasing antioxidant enzymes (CAT, SOD, and GSH-Px) activities in liver tissues (Zhang et al., 2018).

Aquilaria agallocha Roxb. FAMILY: THYMELAEACEAE

Rahman et al. (2013) evaluated hepatoprotective effect of ethanolic extract of *Aquilaria agallocha* (EEAA) leaves induced by CCl_4 hepatotoxicity in rat model by estimated serum hepatic enzyme levels and hisopathological study of liver tissues of rats. EEAA at dose 200 and 400 mg/kg BW administered per oral for 10 days in rats and compared with standard Silymarin at dose 100 mg/kg orally. The results showed significant decline in serum ALT, AST, and ALP levels treated groups which increased due to CCl_4-induced liver damage compared with standard drug. Histopathological study of liver tissues revealed hepatoprotective activity of EEAA.

Alam et al. (2017) evaluated the hepatoprotective potential of ethanolic extract of *A. agallocha* leaves against PCM-induced hepatotoxicity in SD rats. Hepatoprotective potential was assessed by various biochemical parameters such as ALT, AST, ALP, LDH, bilirubin, cholesterol, TP, and ALB. Group IV rats showed significant decrease in ALT, AST, ALP, LDH, cholesterol, bilirubin, liver wt., and relative liver wt. levels while significant increase in final BW, TP and ALB levels as compared to group II rats. Hepatoprotective potential of AAE 400 mg/kg/day was comparable to that of standard drug Silymarin 100 mg/kg/day. Results of the study were well supported by the histopathological observations. This study confirms that *A. agallocha* leaves possesses hepatoprotective potential comparable to that of standard drug Silymarin as it exhibited comparable protective potential against PCM-induced hepatotoxicity in SD rats.

Arachniodes exilis (Hance) Ching. FAMILY: DRYOPTERIDACEAE

Zhou et al. (2010) investigated the antioxidant and hepatoprotective activity of ethanol extract of *Arachniodes exilis*. Antioxidant activity was evaluated by different assays, including reducing power, LPO, DPPH, 2,2′-azinobis-3-ethylbenzothiazoline-6-sulfonic acid (ABTS), superoxide anion, hydroxyl radicals, and hydrogen peroxide. The hepatoprotective activity of ethanol extract was studied in mice liver damage induced by CCl_4 by monitoring biochemical parameters. The extract showed potent activities on reducing power, LPO, DPPH, ABTS, superoxide anion, hydroxyl radical, and hydrogen peroxide. Also oral administration of *A. exilis* at different doses resulted in significant improvement on the levels of SGOT, SGPT, MDA, and SOD. The results indicate that this plant possesses potential antioxidant and hepatoprotective properties and has therapeutic potential for the treatment of liver diseases.

Arctium lappa L. FAMILY: ASTERACEAE

Arctium lappa was shown to suppress the CCl_4 or APAP-intoxicated mice as well as the ethanol plus CCl_4-induced rat liver damage. The underlying hepatoprotective ability of *A. lappa* could be related to the decrease of oxidative stress on hepatocytes by increasing GSH, cytochrome P-450 content and NADPH-cytochrome C reductase activity and by decreasing MDA content; hence alleviating the severity of liver damage based on histopathological observations (Lin et al., 2000, 2002).

Ahangarpour et al. (2017) evaluated the antidiabetic and hypolipidemic properties of *A. lappa* root extract on nicotinamide-streptozotocin-induced type 2 diabetes in mice. Induction of diabetes decreased the level of insulin, leptin, and HDL and increased the level of other lipids, glucose, and hepatic enzymes significantly. Administration of both doses of the extract significantly decreased the level of TG, VLDL, glucose and ALP in diabetic mice. Insulin levels increased in animals treated with 200 mg/kg and HDL and leptin levels increased in animals treated with 300 mg/kg of the extract.

Ardisia solanacea Roxb. FAMILY: PRIMULACEAE

Samal (2013) investigated the hepatoprotective activity of alcoholic extract of *Ardisia solanacea* leaves against CCl_4-induced hepatotoxicity. Alcoholic extract showed the presence of phenolic compound and flavanoids. Alteration in the levels of biochemical markers of hepatic damage like SGOT, SGPT, ALP, bilirubin, and protein were tested in both CCl_4 treated and untreated groups. CCl_4 (1 mL) has enhanced the SGOT, SGPT, ALP, and TB where decrease in TP level in liver. Treatment of alcoholic extract of *A. solanacea* (200 mg/kg) has brought back the altered levels of biochemical markers to the near normal levels in the dose-dependent manner. Moreover, it prevented CCl_4-induced prolongation in pentobarbital sleeping time confirming hepatoprotectivity.

Areca catechu L. FAMILY: ARECACEAE

Aqueous extracts from seeds of *Areca catechu* were investigated by Pithayanukul et al. (2009) for their hepatoprotective potential by studying their antioxidant capacity using four different methods by determining their *in vitro* anti-inflammatory activity against 5-lipoxygenase, and by evaluating their hepatoprotective potential against liver injury induced by CCl_4 in rats. *A. catechu* extract exhibited potent antioxidant and anti-inflammatory activities. Treatment of rats with extract reversed oxidative damage in hepatic tissues induced by CCl_4.

Artemisia absinthium L. FAMILY: ASTERACEAE

The effect of aqueous-methanolic extract of *Artemisia absinthium* was investigated by Gilani and Janbaz (1995) against APAP- and CCl_4-induced hepatic damage. Acetaminophen produced 100% mortality at the dose of 1 g/kg in mice while pretreatment of animals with plant extract (500 mg/kg) reduced the death rate to 20%. Pretreatment of rats with plant extract (500 mg/kg, orally twice daily for two days) prevented the APAP (640 mg/kg) as well as CCl_4 (1.5 mL/kg) induced rise in serum transaminases (GOT and GPT). Post-treatment with three successive doses of extract (500 mg/kg, 6 h) restricted the hepatic damage induced by APAP but CCl_4-induced hepatotoxicity was not altered. Plant extract (500 mg/kg) caused significant prolongation in pentobarbital (75 mg/kg)-induced sleep as well as increased strychnine-induced lethality in mice suggestive of inhibitory effect on microsomal drug metabolizing enzymes (MDME).

Amat et al. (2010) evaluated *in vivo* hepatoprotective activity of the aqueous extract of *A. absinthium* L., which has been used for the treatment of liver disorders in Traditional Uighur Medicine. Liver injury was induced chemically, by a single CCl_4 administration (0.1% in olive oil, 10 mL/kg, i.v.), or immunologically, by injection of endotoxin (LPS, 10 μg, i.v.) in Bacille Calmette–Guerin (BCG)-primed mice. The levels of AST, ALT, TNF-alpha (TNF-α) and interleukin-1 (IL-1) in mouse sera, as well as SOD, GPx, and MDA in mouse liver tissues were measured. The biochemical observations were supplemented by histopathological examination. The obtained results demonstrated that the pretreatment with AEAA significantly and dose dependently prevented chemically or immunologically induced increase in serum levels of hepatic enzymes. Furthermore, aqueous extract significantly reduced the LPO in the liver tissue and restored activities of defense antioxidant enzymes SOD and GPx toward normal levels. In the BCG/LPS model, increase in the levels of important pro-inflammatory mediators TNF-alpha and IL-1 was significantly suppressed by aqueous extract pretreatment. Histopathology of the liver tissue showed that AEAA attenuated the hepatocellular necrosis and led to reduction of inflammatory cells infiltration. Phytochemical analyses revealed the presence of sesquiterpene lactones, flavonoids, phenolic acids, and tannins in the aqueous extract.

Mohammadian et al. (2016) evaluated the hepatoprotective effects of *A. absinthium* on some factors reflecting the development of oxidative toxic stress in plasma. Levels of ALT, AST, and TTG were decreased in the animals treated with 10 mg/kg/day of *A. absinthium* compared to the control. ALT and AST in 50 mg/kg group was observed compared with control group. Also, total thiol groups increased in *Artemisia* 50 mg/kg group compared to control group. Results suggest that alcoholic extract of *Artemisia* can ameliorate liver toxicity in rats through reducing the serum levels of ALT, AST, and oxidative damage.

Artemisia annua L. FAMILY: ASTERACEAE

The significant hepatoprotective activities of *Artemisia annua* extracts seemed to be strongly connected to their content of hydroxycinnamoyl quinic acids and flavonoids (El-Askary et al., 2019).

Artemisia aucheri Boiss. FAMILY: ASTERACEAE

Rezaei et al. (2013) investigated the effects of *Artemisia aucheri* extract on thioacetamide-induced hepatotoxicity in Wistar rats. Significant decreases in aminotransferase and ALP activities and significant increases in the concentration of ALB and TP in groups treated with the extract compared with thioacetamide-treated group were observed. The results indicate that protective effects of *Artemisia* extract against the thioacetamide-induced hepatotoxicity may be due to its ability to block the bioactivation of thioacetamide, primarily by inhibiting the activity of Cyp_{450} and free radicals. *Artemisia* possesses quercetin. Studies have demonstrated that quercetin inhibits LPO and as an antioxidant can inhibit LPO.

Artemisia campestris L. FAMILY: ASTERACEAE

Aniya et al. (2000) evaluated the antioxidant and hepatoprotective activity of *Artemisia campestris.* A water extract of *A. campestris* showed a strong scavenging action of DPPH, hydroxyl, and superoxide anion radicals. When the extract was given intraperitoneally to mice prior to CCl_4 treatment, CCl_4-induced liver toxicity, as seen by an elevation of serum AST and ALT activities, was significantly reduced. Depression of the elevation of serum enzyme levels after CCl_4 treatment was also observed by oral administration of the extract. In that case, CCl_4-derived LPO in the liver was decreased by the extract treatment. These results suggest that the extract of *A. campestris* scavenges radicals formed by CCl_4 treatment resulting in protection against CCl_4-induced liver toxicity.

Artemisia capillaris Thunb. FAMILY: ASTERACEAE

Artemisia capillaris, also called "InJin" in Korean, has been widely used to treat various hepatic disorders in traditional Oriental medicine. Choi et al. (2013) evaluated the hepatoprotective effect of *A. capillaris* (aqueous extract, WAC) on alcoholic liver injury. Alcohol+pyrazole (PRZ) treatment drastically increased the serum levels of AST, ALT, and MDA levels in serum and liver tissues while these changes were significantly ameliorated by WAC administration. The prominent microvesicular steatosis and mild necrosis in hepatic histopathology were induced by alcohol- PRZ treatment, but notably attenuated by WAC administration. Moreover, the alcohol-PRZ treatment-induced depletions of the antioxidant components including GSH content, total antioxidant capacity (TAC), activities of glutathione peroxidase (GSH-Px), reductase (GSH-Rd), CAT, and SOD were significantly ameliorated by WAC administration (except GSH-Rd). These results were in accordance with the modulation of NF-E2-related factor (Nrf2) and heme oxygenase-1 (HO-1) gene expression. Alcohol-PRZ treatment increased the levels of TNF-α and transforming growth factor-beta (TGF-β) in hepatic tissues. However they were significantly normalized by WAC administration. In addition, WAC administration significantly attenuated the alterations of aldehyde dehydrogenase (ALDH) level in serum and hepatic gene expressions of ALDH and alcohol dehydrogenase. These results support the relevance in clinical use of *A. capillaris* for alcohol-associated hepatic disorders. The underlying mechanisms may involve both enhancement of antioxidant activities and modulation of proinflammatory cytokines.

Artemisia dracunculus L. FAMILY: ASTERACEAE

Zarezade et al. (2018) investigated the antioxidant and hepatoprotective activity of the hydroalcoholic extract of aerial parts of *Artemisia dracunculus* (HAAD) against CCl_4-induced hepatotoxicity in rats. Total phenolic content was 197.22 ± 3.73 mg GAE/g HAAD dry weight. HAAD indicated powerful activity in FRAP, DPPH, and ABTS tests. Acute toxicity study showed that the extract had an LD_{50} of >5000 mg/kg. The oral treatment with HAAD exhibited a significant decrease in the levels of AST, ALT, ALP, and TB and an increase in the level of TP. The extract significantly diminished MDA levels. The activities of the antioxidant enzymes were significantly augmented in rats pretreated with HAAD 200 mg/kg. Histopathological examination demonstrated lower liver damage in HAAD-treated groups as compared to CCl_4 groups.

Artemisia maritima L. FAMILY: ASTERACEAE

The hepatoprotective activity of the aqueous-methanolic extract of *Artemisia maritima* was investigated by Janbaz and Gilani (1995) against APAP (PCM, 4-hydroxy acetanilide)- and CCl_4-induced hepatic damage. Acetaminophen produced 100% mortality at the dose of 1 g/kg in mice, while pretreatment of animals with the plant extract (500 mg/kg) reduced the death rate to 20%. Acetaminophen at the dose of 640 mg/kg produced liver damage in rats as manifested by the significant rise in serum levels of SGOT and SGPT compared to respective control values. Pretreatment of rats with the plant extract (500 mg/kg) lowered significantly the respective SGOT and SGPT levels. Similarly, a hepatotoxic dose of CCl_4 (1.5 mL/kg, orally) raised significantly the serum GOT and GPT levels compared to respective control values. The same dose of plant extract (500 mg/kg) was able to prevent significantly the CCl_4-induced rise in serum transaminases. Moreover, it prevented CCl_4-induced prolongation in pentobarbital sleeping time confirming hepatoprotectivity and validates the traditional use of this plant against liver damage (Janbaz and Gilani, 1995).

Artemisia pallens Walls ex DC. FAMILY: ASTERACEAE

Honmore et al. (2015) evaluated the protective effects of *Artemisia pallens* methanol extract (APME) in APAP-induced hepatic-toxicity. Pretreatment with APME (200 and 400 mg/kg, p.o.) significantly decreased AST, ALT, bilirubin, blood urea nitrogen (BUN), and serum creatinine as compared with APAP-treated rat. Decreased level of serum ALB, serum uric acid, and HDL were significantly restored by APME (200 and 400 mg/kg, p.o.) pretreatment. Administration of APME (200 and 400 mg/kg, p.o.) significantly reduced the elevated level of cholesterol, LDL, LDH, TG, and VLDL. It also significantly restored the altered level of hepatic and renal antioxidant enzymes SOD and GSH. The increased level of MDA and NO in hepatic as well as renal tissue was significantly decreased by APME (200 and 400 mg/kg, p.o.) administration. Histological alternation induced by APAP in liver and kidney was also reduced by the APME (200 and 400 mg/kg, p.o.) pretreatment.

Artemisia sacrorum Ledeb. FAMILY: ASTERACEAE

Artemisia sacrorum has long been used as one kind of oriental folk medicine to treat some liver diseases. Yuan et al. (2010) investigated the hepatoprotective effects of 50% ethanol eluate precipitation of *A. sacrorum* (EEP) on APAP-induced toxicity in mice. Pretreated with EEP prior to the administration of APAP significantly prevented the increases of AST, ALT, and TNF-alpha levels in sera, and suppressed the GSH depletion, MDA accumulation in liver tissues markedly. In addition, EEP prevented APAP-induced apoptosis and necrosis, as indicated by liver histopathological analysis, immunohistochemical analysis, and DNA laddering. Furthermore, according to the results from Western blot analysis, EEP decreased APAP-induced caspase-3 and caspase-8 protein expressions in mouse livers markedly.

Artemisia scoparia Waldst. et Kit. FAMILY: ASTERACEAE

The hepatoprotective activity of hydroalcoholic extract of aerial parts *Artemisia scoparia* was investigated against CCl_4 (Gilani and Janbaz, 1994; Noguchi et al., 1982) and PCM (Gilani and Janbaz, 1993)-induced liver damage. The extract dose dependently attenuated hepatotoxin-induced biochemical parameters (rise in serum AST and ALT) and prolongation of phenobarbital-induced sleeping time clearly indicating its hepatoprotective action. Hepatotoxins like CCl_4 and PCM significantly reduced the activity of drug metabolizing enzymes in liver, leading to the slowing of drug metabolism resulting in increased level of drugs such as barbiturates which results in prolongation of their pharmacological activity (sleeping time). Reversal of barbiturate-induced sleeping time suggests hepatoprotective effect of *A. scoparia* (Gilani and Janbaz, 1994). Recent pharmacological studies also showed anti-inflammatory (Habib and Waheed, 2013) and antioxidant (Singh et al., 2009) activities of *A. scoparia* which may contribute to its hepatoprotective activity. Although the plant is recognised as antihelmintic, its mammalian toxicity is negligible

(Negahban et al., 2004). Some cases of dermatitis and allergic reaction have been reported (Anonymous). Phytochemical studies on aerial part of *A. scoparia* showed the presence of hyperin, eupafolin, pedalitin, 5,7,2′,4′-tetrahydroxy-6,5′-dimethoxyflavone, camphor, 1,8-,beta-caryophyllene, cirsilineol, cirsimaritin, arcapillin, and cirsiliol (Al-Asmari et al., 2014).

Artemisia vulgaris L. FAMILY: ASTERACEAE

The effect of a crude extract of the aerial parts of *Artemisia vulgaris* was investigated by Gilani et al. (2005) against D-GalN and LPS-induced hepatitis in mice. Co-administration of D-GalN (700 mg[sol]kg) and LPS (1 μg[sol]kg) significantly raised the plasma levels of ALT and AST in mice in the toxin group compared with the values in the control group. Pretreatment of mice with different doses of Av.Cr (150–600 mg[sol] kg) significantly reduced the toxin-induced rise in plasma ALT and AST. The hepatoprotective effect was further verified by histopathology of the liver, which showed improved architecture, absence of parenchyma congestion, decreased cellular swelling and apoptotic cells, compared with the findings in the toxin group of animals.

Oral administration during one month of 1 mL/kg of the decoction extract from *A. vulgaris* (prepared with 1 g leaves/mL water) was given on male and female Wistar rats with hepatotoxicity damage induced by antitubercular drugs (RIF, 54 mg/kg/d; INH, 27 mg/kg, d; PZA, 135 mg/kg/d). *A. vulgaris* leaves were collected at different seasonal time. The result showed that *A. vulgaris* collected in the May–June period exerts a better hepatoprotective effect in this model (Mitra et al., 2016).

Aspalathus linearis (Burm. f.) R. Dahlgren FAMILY: FABACEAE

Hepatoprotective properties of rooibos tea (*Aspalathus linearis*) were investigated in a rat model of liver injury induced by CCl_4 by Ulicná et al. (2003). Rooibos tea, like *N*-acetyl-L-cysteine which was used for the comparison, showed histological regression of steatosis and cirrhosis in the liver tissue with a significant inhibition of the increase of liver tissue concentrations of MDA, triacylglycerols and cholesterol. Simultaneously, rooibos tea significantly suppressed mainly the increase in plasma activities of aminotransferases (ALT, AST), ALP and bilirubin concentrations, which are considered as markers of liver functional state. The antifibrotic effect in the experimental model of hepatic cirrhosis of rats suggests the use of rooibos tea as a plant hepatoprotector in the diet of patients with hepatopathies.

Asparagus albus L. FAMILY: LILIACEAE

The antioxidant and hepatoprotective activities of hot aqueous extract from *Asparagus albus* leaf against CCl_4-induced liver damage in rats were investigated by Serairi-Beji et al. (2017). The total phenolic, flavonoid, and condensed tannin contents of hot aqueous extract from *A. albus* leaf were determined. The antioxidant activity of hot aqueous extract from *A. albus* leaf was evaluated using the antioxidant capacity, DPPH free-radical-scavenging ability and reducing power assays. Different polyphenolic compounds, namely GA, vanillic acid, 3,4 dimethoxybenzoic acid, catechin, rutin, and quercetin were identified. Oral administration of hot aqueous extract from *A. albus* leaf to male Wistar, intoxicated with CCl_4, demonstrated a significant protective effect by lowering the levels of hepatic marker enzymes (AST and lactate transaminases) and by improving the histological architecture of the rat liver. The hot aqueous extract from *A. albus* leaf attenuated oxidative stress by restoring the activities of SOD, CAT, and GPx.

Asparagus racemosus Willd. FAMILY: LILIACEAE

Aqueous extract of *Asparagus racemosus* root (AEAR) was evaluated by Rahiman et al. (2011) for its hepatoprotective activities in rats. The plant extract (150 and 250 mg/kg, p.o.) showed a remarkable hepatoprotective and antioxidant activity against PCM-induced hepatotoxicity as judged from the wet liver weight, serum marker enzymes, antioxidant levels and histopathological studies on liver tissues. Paracetamol-induced a significant rise in wet liver weight, AST, ALT, ALP, TB with a reduction of SOD and CAT. Treatment of rats with different doses of plant extracts (150 and 250 mg/kg) significantly altered serum marker enzymes and antioxidant levels to near normal against PCM-treated rats. The activity of the extracts was comparable to the standard drug, Silymarin (100 mg/kg, p.o.). Kumar et al. (2011) reported hepatoprotective activity of ethanol extract of *A. racemosus* root against PCM-induced hepatic damage. It was observed that extracts of *A. racemosus* has reversal effects on the levels of the above-mentioned parameters in PCM hepatotoxicity.

Asphodeline lutea (L.) Rchb. FAMILY: ASPHODELACEAE

Lazarova et al. (2016) investigated the effect of *Asphodeline lutea* dry root extract (ALE) administered alone and against CCl_4-induced liver injury *in vitro/in vivo*. Hepatoprotective potential was investigated by *in vivo*/*in vitro* assays in Wistar rats as well as antioxidant properties. At concentrations ranging from 10 to 200 μg/mL of ALE significant cytotoxic effects on isolated hepatocytes were found. ALE showed some toxicity in Wistar rats discerned by increased ALT, ALP activities, and MDA quantity, decreased GSH (reduced GSH) levels without affecting the activity of the antioxidant enzymes GPx, GR, and GST (GSH-*S*-transferase activity). The antioxidant and hepatoprotective potential of ALE was also observed *in vitro*/*in vivo* against CCl_4-induced liver injury, where ALE normalizes all the examined parameters perturbated by CCl_4 administration. In addition, ALE preserved the decreased cytochrome P450 level and ethylmorphine-*N*-demethylase activity without affecting aniline 4-hydroxylase activity. ALE is rich in anthraquinones, naphthalenes, and caffeic acid. The pro-oxidant effects of ALE could be due to naphthalene and anthraquinone bioactivation pathways involving toxic metabolites.

Astragalus kahiricus DC. FAMILY: FABACEAE

Allam et al. (2013) evaluated the hepatoprotective activity of the ethanol extract of *Astragalus kahiricus* roots against ethanol-induced liver apoptosis and it showed very promising hepatoprotective actions through different mechanisms. The extract counteracted the ethanol-induced liver enzymes leakage and GSH depletion. In addition, it demonstrated antiapoptotic effects against caspase-3 activation and DNA fragmentation that were confirmed by liver histopathological examination. Moreover, the phytochemical study of this extract led to the isolation of four cycloartane-type triterpenes identified as astrasieversianin II (1), astramembrannin II, (2) astrasieversianin XIV, (3) and cycloastragenol (4). The structures of these isolates were established by high-resolution electrospray ionization mass spectrometry (HRESI)-MS and 1D and 2D NMR experiments. The antimicrobial, antimalarial, and cytotoxic activities of the isolates were further evaluated, but none of them showed any activity.

Astragalus monspessulanus L. FAMILY: FABACEAE

Simeonova et al. (2015) investigated the hepatoprotective potential of *n*-butanolic extract of *Astragalus monspessulanus* against *in-vitro*/*in-vivo* CCl_4-induced liver damage in rats. Silymarin was used as a positive control. The *in-vitro* experiments were carried out in primary isolated rat hepatocytes first incubated with CCl_4 (86 μmol/L). Cell preincubation with the extract (1 and 10 μg/mL) significantly ameliorated the CCl_4-induced liver damage. *In-vivo* rats were challenged orally with CCl_4 (10% solution in olive oil) alone and after 7 days pretreatment with the extract. Extract pretreatment normalized the activities of the antioxidant enzymes and the levels of GSH and MDA. These data are supported by the histopathological examination.

Astragalus spruneri **Boiss.** **FAMILY: FABACEAE**

Kondeva-Burdina et al. (2018) evaluated the effect of a defatted extract (EAS) and three flavonoids, isolated from *Astragalus spruneri* using *in vitro*/*in vivo* models of liver injury. The EAS was characterized by high-performance liquid chromatography (HPLC) and flavonoids (14 mg/g dw) and saponins (8 mg/g dw) were proved. The flavonoids were isolated from the same extract and partially identified by LC-MS. In *in vitro* models of nonenzyme induced (Fe^{2+}/AA) LPO in isolated liver microsomes and CCl_4-induced metabolic bioactivation and *t*-BuOOH-induced oxidative stress in isolated rat hepatocytes, both EAS and the flavonoids exerted similar to silybin (positive control) an antioxidant and cytoprotective activity, discerned by decreased MDA production in the microsomes and by preserved cell viability and GSH levels as well as by decreased LDH activity and MDA quantity in isolated rat hepatocytes. The antioxidant and hepatoprotective effect of EAS has been confirmed *in vivo* against CCl_4-induced liver injury in rats. EAS restored the GSH levels and the activity of the antioxidant enzymes CAT and SOD, affected by CCl_4 administration, as well as decreased the production of MDA. The effect of EAS was commensurable with those of Silymarin.

Atalantia ceylanica **(Arn.) Oliv.** **FAMILY: RUTACEAE**

Decoction prepared from leaves of *Atalantia ceylanica* is used in traditional medicine in Sri Lanka for the treatment of various liver ailments since ancient times. Lyophilized powder of the water extract of *A. ceylanica* leaves was investigated by Fernando and Soysa (2014) for its phytochemical constituents, antioxidant and hepatoprotective activity *in-vitro*. Hepatotoxicity was induced on porcine liver slices with ethanol to study hepatoprotective activity. Porcine liver slices were incubated at 37 °C with different concentrations of the water extract of *A. ceylanica* in the presence of ethanol for 2 h. The hepatoprotective effects were quantified by the leakage of ALT, AST, and LDH to the medium. TBARS assay was performed to examine the anti-LPO activity caused by the plant extract.

The mean ± SD (n =9) for the levels of total phenolics and flavonoids were 4.87 ± 0.89 w/w% of GAEs and 16.48 ± 0.63 w/w% of (−)-epigallocatechin gallate equivalents, respectively. The decoction demonstrated high antioxidant activity. The mean ± SD values of EC_{50} were 131.2 ± 36.1, 48.4 ± 12.1, 263.5 ± 28.3 and 87.70 ± 6.06 µg/mL for DPPH, hydroxyl radical, NO scavenging assays and ferric ion reducing power assay, respectively. A significant decrease was observed in ALT, AST, and LDH release from porcine liver slices treated *with A. ceylanica* extract at a concentration of 2 mg/mL in the presence of ethanol (5 M) compared to that of ethanol (5 M) treated slices. Furthermore, a reduction in LPO was also observed in liver slices treated with the leaf extract of *A. ceylanica* (2 mg/mL) compared to that of ethanol-induced liver toxicity.

Autranella congolensis (De Willd.) A.Chev. FAMILY: SAPOTACEAE

Njouendou et al. (2014) assessed the hehepatoprotective activity of the extract of *Autranella congolensis* at two different doses (100 and 200 mg/kg) on rat model of thioacetamide-induced hepatotoxicity by investigating biochemical and histopathological markers after 29 days of treatment. Elevated SGOT, SGPT, ALP, TB, and DBL observed in thioacetamide toxic groups were restored toward the normal values, when animals received extract treatment at 100 and 200 mg/kg. Administration of thioacetamide lowered significantly the level of liver antioxidants markers SOD, CAT, GSH, and increased the level of LPO, which were moderated in group of animals treated with both extracts at 100 and 200 mg/kg. The activity of liver microsome CYP2E1 was significantly high when animals received thioacetamide treatment; both extracts reduced the increase in enzyme activity. The histological and biochemical changes exhibited by *Autranella congolensis* extract on toxic animals were comparable to those obtained with standard Silymarin. The hepatoprotective effects of AC extract observed in this study showed that these two medicinal plants are potential sources of antihepatotoxic drugs.

Avicennia marina Forssk. FAMILY: ACANTHACEAE

Mayuresh and Sunita (2016) evaluated hepatoprotective activity of standardized ethyl acetate extract of the leaves of *Avicennia marina* against CCl_4-induced liver intoxication. The extract provided concentration-dependent percent protection which was evident by the changes in the levels of biochemical parameters (SGOT, SGPT, ALP, TPN, TG,

CHO, HDL, LDL, direct and TB, and liver glycogen content) in albino Wistar rats. The observations of biochemical parameters, antioxidant assays and histopathology, endorse an overall promising effect against liver disorders.

Azadirachta indica A. Juss. FAMILY: MELIACEAE

The effect of *Azadirachta indica* leaf extract on serum enzyme levels (SGOT, SGPT, acid phosphatase, and ALP) elevated by PCM in rats was studied with a view to observe any possible hepatoprotective effect of this plant. It is stipulated that the extract treated group was protected from hepatic cell damage caused by PCM induction. The findings were further confirmed by histopathological study of liver. The antihepatotoxic action of picroliv seems likely due to an alteration in the biotransformation of the toxic substances resulting in decreased formation of reactive metabolites (Chattopadhyay et al., 1992).

The shade dried leaves of *A. indica* were extracted successively by petroleum ether (60–80 °C) and ethanol. The ethanol extract after removal of solvent was studied for CCl_4-induced hepatic changes using 0.5 g and 1 g/kg oral dose in albino rats. The changes were assessed by serum enzyme profile that include SGOT, SGPT, ALP, bilirubin (B) and hepatic triglycerides (HTG) levels, histological changes in liver and pentobarbitone sleeping time as a functional parameter. There was significant reversal of biochemical, histological, and functional changes induced in rats by ethanol extract treatment (Mujumdar et al., 1998).

Chattopadhyay and Bandyopadhyay (2005) investigated the effects of *A. indica* leaf extract on antioxidant enzymes to elucidate the possible mechanism of its hepatoprotective activity. Administration of *A. indica* leaf extract significantly enhanced the hepatic level of GSH-dependent enzymes and SOD and CAT activity suggesting that the hepatoprotective effect of the extract on PCM-induced hepatoxicity may be due to its antioxidant activity. Chemical analysis of the *A. indica* leaf extract revealed that the extract contains following six compounds. (i) Quercetin-3-*O*-β-D-glucoside, (ii) Myricetin-3-*O*-rutinoside, (iii) Quercetin-3-*O*-rutinoside, (iv) Kaempferol-3-*O*-rutinoside, (v) Kaempferol-3-*O*-β-D-glucoside, and (vi) Quercetin-3-*O*-a L-rhamnoside. It is well documented that the compounds quercetin, rutin, vitamins C and E are strong antioxidants. It is presumed that the quercetin and rutin compounds of *A. indica* leaf extract may be responsible for its hepatoprotective activity.

The effects of *A. indica* leaf extract on blood and liver GSH, Na+K(+)-ATPase activity and thiobarbutiric acid reactive substances against PCM-induced hepatic damage in rats have been studied by Chattopadhyay (2003) with a view to elucidate possible mechanism behind its hepatoprotective action. It was interesting to observe that *A. indica* leaf extract has reversal effects on the levels of the above-mentioned parameters in PCM hepatotoxicity.

Kale et al. (2003) evaluated the effect of aqueous extract of *A. indica* leaves on hepatotoxicity induced by antitubercular drugs in rats. The aqueous extract was administered with anti-TB drugs (INH/RIF/PAZ 27/54/135 mg/kg/d) during 30 days on Wistar rats by oral via. The authors found that this extract prevented the biochemical changes in serum; the levels of bilirubin, protein, ALT, AST, and ALP were similar to those of the Silymarin control group (Kale et al., 2003).

Maruthappan and Sree (2009) evaluated the hepatoprotective activity of *A. indica* using a ethyl alcohol-induced liver injury in Wistar rats and probe into its mechanism of action. Levels of serum marker enzymes (AST, ALT, and ALP) were significantly increased in ethyl alcohol-treated rats. Simultaneously, *A. indica* leaf powder and Silymarin, standard drug significantly suppressed mainly the increase in plasma activities of AST, ALT, and ALP concentration, which are considered as markers of liver functional state. The significant decrease in the liver TP and ALB levels after ethyl alcohol treatment, which were reversed with *A. indica.* These effects were comparable to Silymarin. In order to probe the possible mechanism by which *A. indica* prevents hepatic damage caused by ethyl alcohol, investigation on levels of TBARS was found to be elevated and significant decrease in GR, SOD, and CAT content of liver after treatment with ethyl alcohol, which were significantly reversed by *A. indica*.

The antioxidant and hepatoprotective activity of fresh juice of young (tender) stem bark of *A. indica* were evaluated by Gomase et al. (2010) against CCl_4-induced hepatic damage in albino rats. Pharmacological assay showed that fresh juice of young stem bark extract of *A. indica* was good hepatotoxic agent at a dose level of 500 mg/kg. The plant extract has decreased the enzyme level of SGOT, SGPT, ALP, bilirubin by the dose of 500 mg/kg.

Johnson et al. (2015) investigated the hepatoprotective effect of ethanolic leaf extract of *A. indica* against APAP-induced hepatotoxicity in SD male albino rats. Animals treated with Silymarin, and *A. indica* extracts significantly have reduced WBC count compared to PCM control group. HGB, RBC, and HCT values in all the groups administered with Silymarin, vitamin C, and *A. indica* extracts were significantly increased when compared to the PCM-intoxicated animals without treatment. Treatment with Silymarin, vitamins C and *A. indica* extracts showed effective hepatoprotective effect as evidence in the decrease in the plasma levels of liver biomarker enzymes and reduction in oxidative stress parameters. Histopathological evaluation of the liver architecture also revealed that all the treated animals have reduced the incidence of PCM-induced liver lesions.

Nwobodo et al. (2016) investigated the effect of leaves aqueous extract of *A. indica* on plasma levels of vitamins C and E, liver enzymes ALT, AST, and ALP in Wistar rats with hepatotoxicity. There were significant differences between the initial and final mean weights of the animals. There was no change in ALP levels. Vitamins C and E in liver homogenate were decreased in group B treated with PCM and increased in group D treated with PCM and *A. indica* extract. Vitamins C and E were decreased in group B and increased in group D. The observed decrease in liver enzymes and increase in vitamins C and E in group D suggest that the extract enhances vitamins C and E levels, and may be hepatoprotective.

Baccharis dracunculifolia DC. FAMILY: ASTERACEAE

Rezende et al. (2014) investigated the *in vivo* protective effects of *Baccharis dracunculifolia* leaves extract (BdE) against CCl_4 - and APAP-induced hepatotoxicity. Pretreatment with BdE significantly reduced the damage caused by CCl_4 and APAP on the serum markers of hepatic injury, AST, ALT, and ALP. Results were confirmed by histopathological analysis. Phytochemical analysis, performed by HPLC, showed that BdE was rich in *N*-coumaric acid derivatives, caffeoylquinic acids and flavonoids. BdE also showed DPPH antioxidant activity (EC50 of 15.75 ± 0.43 µg/mL), and high total phenolic (142.90 ± 0.77 mg GAE/g) and flavonoid (51.47 ± 0.60 mg RE/g) contents.

Baccharoides anthelmintica (L.) Moench *(Centratherum anthelminticum* (L.) Gamble; *Vernonia anthelmintica* (L.) Willd.) FAMILY: ASTERACEAE

Qureshi et al. (2016) evaluated the hepatoprotective effect of ethanolic seeds extract (ESEt) of *Baccharoides anthelmintica* (Syn.: *Centratherum anthelminticum*) (black cumin) in CCl_4-induced liver injury. The test doses (600 and 800 mg/kg) of same extract were found effective in their respective test groups by improving the body and liver weights, serum alanine and AST, γ-glutamyltranspeptidase, ALP, TPs, ALB, TB, especially indirect bilirubin and uric acid levels as compared to CCl_4-induced hepatotoxic control group. In addition, decreased percent inhibitions of antioxidant parameters including CAT, SOD, and reduced GSH accompanied with increased percent inhibition of LPO observed in both test groups. Histopathological studies also proved the liver regenerating property of ESEt by showing decrease in fatty deposition, necrosis, and inflammation around the central vein of liver lobules.

Bacopa monnieri L. FAMILY: PLANTAGINACEAE

Hepatoprotective activity of ethanolic extract of whole plant of *Bacopa monnieri* has been studied against NB (Menon et al., 2010) and morphine (Sumathy et al., 2001)-induced liver toxicity. The extract significantly attenuated hepatotoxin-induced changes in biochemical parameters (sera AST, ALT, and APT) and histopathological changes in liver tissues. Ethanolic extract of *B. monnieri* also showed significant antioxidant (Sumathy et al., 2001) and anti-inflammatory (Channa et al., 2006) activities, which may contribute to its hepatoprotective activity. Acute toxicity studies showed no deterious effect in pharmacological doses. The single dose LD50 was found to be 2400 mg/kg BW in rats. In a chronic toxicity study in rats, *B. monnieri* was found to be well tolerated up to the dose of 500 mg/kg BW for 3 months (Allan et al., 2007). Phytochemical analysis on plant of *B. monnieri* showed the presence of alkaloid (brahmine), bacosides, nicotine, herpestine, D-mannitol, hersaponin, stigmosterol, beta-sitosterol, and bacosaponins (Sumathy et al., 2001; Weisner, n.d.). Bacoside, a major constituent of brahmi, has been shown to possess significant anticancer activity against liver tumors in rats (Sudharani et al., 2011).

Balanites roxburghii Planch. (Syn.: *B. aegyptiaca auct.)* FAMILY: BALANITACEAE

The effect of ethanolic extracts of bark of *Balanites roxburghii (B. aegyptiaca)* has been investigated against PCM (Ali et al., 2001) and CCl_4 (Jaiprakash et al., 2003)-induced hepatotoxicity in rats. The extract dose dependently attenuated the hepatotoxin-induced biochemical (serum AST, ALT, ALP, and bilirubin) and histopathological changes in liver which was comparable with Silymarin. The extract also reversed toxin-induced prolongation of pentobarbital sleeping time in rats. The purified fractions of *B. aegyptiaca* possess significant antioxidant (Suky et al., 2011) and anti-inflammatory (Gaur et al., 2008) activities which may contribute to its hepatoprotective action. Phytochemical studies on

B. aegyptiaca showed the presence of flavonoids, saponins, quercetin 3-glucoside, quercetin-3-rutinoside, 3-glucoside, 3-rutinoside, 3–7-diglucoside, and 3-rhamnoglucoside (Chothani and Vaghasiya, 2011).

Ojo et al. (2006) evaluated the hepatoprotective activity of stem bark extract of *B. aegyptiaca* against APAP-induced hepatotoxicity in Wistar albino rats. They reported that stem bark extract of *B. aegyptiaca* produced significant hepatoprotective effects by decreasing the activity of serum enzymes. Values recorded for AST, ALT, and ALP were significantly lower compared to those recorded for control rats. A higher inhibition of serum level elevation of ALP was observed with the extract.

Montasser et al. (2017) evaluated the ameliorative effects of Melatonin, ursodeoxycholic acid and *B. aegyptiaca* against hepatotoxicity-induced by methotrexate for one month. Methotrexate showed significant increase in ALT, AST, ALP, GGT, total and DBL, as well as TNF-α levels, oxidized glutathione (GSSG), MDA and NO. Whereas TP, ALB, TAC, reduced GSH, GPx, GR, glutathione *S*-transferase (GST), SOD, and CAT levels were significantly decreased in methotrexate treated group. These alterations were improved by melatanonin and *B. aegyptiaca* treatment, whereas no improvement was noticed in ursodeoxycholic treatment.

Baliospermum montanum (Willd.) Müll.-Arg. FAMILY: EUPHORBIACEAE

Wadekar et al. (2008) investigated the hepatoprotective activity of alcohol, chloroform, and aqueous extract of roots of *Baliospermum montanum* in PCM-induced liver damage model in rats. Liver damage in rats was produced by PCM (2 g/kg, p.o.) in tween 80. Alcohol, chloroform, and aqueous extracts of roots of the plant were administered to rats daily for 7 days. The biochemical parameters were investigated. Histopathological changes in liver were studied. Concurrently, Silymarin was used as standard hepatoprotective agent. The results indicated that biochemical changes produced by PCM were restored to normal by alcohol, chloroform and aqueous extracts. The alcohol and aqueous extract of roots of *B. montanum* showed significant hepatoprotective effect whereas chloroform extract (CE) showed moderate hepatoprotective activity against PCM-induced liver damage model in rats.

Kumar and Mishra (2009) evaluated the hepatoprotective activity of *B. montanum*. Rats and primary cultures of rat hepatocytes were used as the *in vivo* and *in vitro* models to evaluate the hepatoprotective activity of subfractions from total ME of *B. montanum*. CCl_4 was selected as hepatotoxin. Silymarin was the reference hepatoprotective agent. Among the ethyl methyl ketone and methanol subfractions tested (50, 100, and 150 mg/kg), methanol subfraction (150 mg/kg) of the bio-active total ME and Silymarin (100 mg/kg) enhanced liver cell recovery by restoring all the altered biochemical parameters back to normal. In the *in vitro* study, release of transaminases, TP together and hepatocyte viability were the criteria. Primary cultures of hepatocytes were treated with CCl_4 (10 μL/mL), and various concentrations (100, 500, and 1000 μg/mL) of ethyl methyl ketone and methanol subfractions of total ME and Silymarin (100 μg/mL). CCl_4 reduced the hepatocyte viability and also altered the biochemical parameters, which were restored significantly by ethyl methyl ketone (1000 μg/mL) and methanol (500 and 1000 μg/mL) subfractions. These results suggest that *B. montanum* possess the hepatoprotective activity against CCl_4-induced liver injury in both rats and primary cultures of rat hepatocytes.

Ballota glandulosissima Hub.-Mor & Patzak FAMILY: LAMIACEAE

Water extract of *Ballota glandulosissima* (BG) was investigated by Özbek et al. (2004) for hepatoprotective effect on CCl_4-induced hepatotoxicity in rats. Treatment of animals with BG (100 mg/kg, i.p.) +CCl_4 (0.8 mL/kg i.p.) for 7 days significantly ameliorated the levels of AST, ALT, and ALP elevated by the CCl_4 treatment alone. The results of biochemical tests were also confirmed by histopathological examination. BG together with CCl_4 treatment decreased the ballooning degeneration but did not produce apoptosis of hepatocytes, centrilobular and bridging necrosis observed in the CCl_4 treatment alone. BG, at 100 mg/kg per os, showed a significant reduction (34.22%) in rat paw oedema-induced by carrageenan.

Bambusa bambos (L.) Voss FAMILY: POACEAE

The hepatoprotective activity of methanolic shoot extract of *Bambusa bambos* was tested by Patil et al. (2018) against CCl_4-induced hepatotoxicity in female Wistar rats. Methanolic extract attenuated the increase in AST and ALT as well as ALP and TB that occur during liver injury after CCl_4 injection. Outcome of this study suggests that treatment with methanolic shoot extract of *B. bambos*-induced reduction in ALT, AST, ALP, and TB in rats indicating hepatoprotective potential of the extract.

Barleria prionitis L. FAMILY: ACANTHACEAE

Iridoid enriched fraction IF from the ethanol–water extract of aerial parts (leaves and stems) of *Barleria prionitis* was evaluated by Singh et al. (2005) for hepatoprotective activity in various acute and chronic animal test models of hepatotoxicity. It afforded significant hepatoprotection against CCl_4, galactosamine and PCM-induced hepatotoxicity. Silymarin was used as a reference hepatoprotective. In the safety evaluation study the oral LD_{50} was found to be more than 3000 mg/kg, with no signs of abnormalities or any mortality observed for 15 days period under observation after single dose of drug administration whereas intraperitoneal LD_{50} was found to be 2530 ± 87 mg/kg. SE (n=10) in mice. The studies revealed significant and concentration-dependent hepatoprotective potential of "IF" as it reversed the majority of the altered hepatic parameters in experimental liver damage in rodents.

Basella alba L. FAMILY: BASELLACEAE

Das et al. (2015) evaluated the possible hepatoprotective effects of aqueous leaf extracts of *Basella alba* in comparison with Silymarin in PCM-induced hepatotoxicity in albino rats. Aqueous leaf extracts of *B. alba* 100 mg/kg/day orally had significant hepatoprotective effect in PCM-induced hepatotoxicity in albino rats. The results were well comparable and even in some respects superior to standard drug Silymarin.

Bauhinia hookeri F. Muell. FAMILY: FABACEAE

The hepatoprotective and antioxidant activity of *Bauhinia hookeri* ethanol extract against CCl_4-induced liver injury was investigated in mice. Ethanol extract treatment significantly inhibited the CCl_4-induced increase in ALT (44% and 64%), AST (36 and 46%), ALP (28 and 42%), and MDA (39 and 51%) levels at the tested doses, respectively. Moreover, ethanol extract treatment markedly increased the activity of antioxidant parameters GSH, GPx, GR, GST, and SOD. Histological observations confirmed the strong hepatoprotective activity. These results suggest that a dietary supplement of extract could exert a beneficial effect against oxidative stress and various liver diseases by enhancing the antioxidant defense status, reducing LPO, and protecting against the pathological changes of the liver. The hepatoprotective activity of the extract is mediated, at least in part, by the antioxidant effect of its constituents (Sayed et al., 2014, 2015).

Bauhinia purpurea L. FAMILY: FABACEAE

Yahya et al. (2013) evaluated the hepatoprotective potential of ME of

Bauhinia purpurea leaves using the PCM-induced liver toxicity in rats. From the histological observation, lymphocyte infiltration and marked necrosis were observed in PCM-treated groups (negative control), whereas maintenance of the normal hepatic structure was observed in group pretreated with Silymarin and ME. Hepatotoxic rats pretreated with Silymarin or ME exhibited significant decrease in ALT and AST enzyme level. Moreover, the extract also exhibited antioxidant activity and contained high TPC. In conclusion, ME exerts potential hepatoprotective activity that could be partly attributed to its antioxidant activity and high phenolic content.

In an attempt to identify the hepatoprotective bioactive compounds in MEBP, the extract was prepared in different partitions and subjected to the PCM-induced liver injury model in rats by Zakaria et al. (2016). Dried ME was partitioned successively to obtain petroleum ether (PEBP), ethyl-acetate (EABP), and aqueous (AQBP) partitions, respectively. All partitions were subjected to *in-vitro* antioxidant (i.e., TPC, DPPH- and superoxide-radicals scavenging assay, and oxygen radical absorbance capacity (ORAC) assay) and anti-inflammatory (i.e., lipooxygenase (LOX) and xanthine oxidase (XO) assay) analysis. The partitions, prepared in the dose range of 50, 250, and 500 mg/kg, together with a vehicle (10% DMSO) and standard drug (200 mg/kg Silymarin) were administered orally for seven consecutive days prior to subjection to the 3 mg/kg PCM-induced liver injury model in rats. Of all partitions, EABP possessed high TPC value and demonstrated remarkable antioxidant activity when assessed using the DPPH- and superoxide-radical scavenging assay, as well as ORAC assay, which was followed by AQBP and PEBP. All partitions also showed low anti-inflammatory activity via the LOX and XO pathways. In the hepatoprotective study, the effectiveness of the partitions is in the order of EABP > AQBP > PEBP, which is supported by the microscopic analysis and histopathological scoring. In the biochemical analysis, EABP also exerted the most effective effect by reducing the serum level of ALT and AST at all doses tested in comparison to the other partitions. Phytochemical screening and HPLC analysis suggested the presence of: flavonoids, condensed tannins and triterpenes in EABP; flavonoids, condensed tannins, and saponins in PEBP and; only saponins in AQBP.

Bauhinia racemosa Lam. FAMILY: FABACEAE

The ME of *Bauhinia racemosa* stem bark was investigated by Gupta et al. (2004) for the antioxidant and hepatoprotective effects in Wistar albino rats. The ME and Silymarin produced significant hepatoprotective effect by decreasing the activity of serum enzymes, bilirubin and LPO and significantly increased the levels of GSH, SOD, CAT, and protein in a dose-dependent manner. Methanol extract also showed antioxidant effects on $FeCl_2$-ascorbate-induced LPO in rat liver homogenate and on superoxide scavenging activity. From these results, it was suggested that ME of *B. racemosa* could protect the liver cells from PCM and CCl_4-induced liver damages perhaps, by its antioxidative effect on hepatocytes; hence eliminating the deleterious effects of toxic metabolites from PCM or CCl_4.

Bauhinia variegata L. FAMILY: FABACEAE

Bauhinia variegata commonly known as Kachnar, is widely used in Ayurveda as tonic to the liver. Bodakhe and Ram (2007) assessed the potential of *B. variegata* bark as hepatoprotective agent. The hepatoprotective activity was investigated in CCl_4 intoxicated SD rats. *B. variegata* alcoholic stem bark extract (SBE) at different doses (100 and 200 mg/kg) were administered orally to male SD rats weighing between 100 and 120 g. The effect of SBE on the serum marker enzymes, namely, AST, ALT, ALP, and GGT and liver protein and lipids were assessed. The extract exhibited significant hepatoprotective activity. Hence, *B. variegata* appears to be a promising hepatoprotective agent.

Jha et al. (2009) have used liver slice culture model to demonstrate hepatoprotective activity of *B. variegata* leaves extract (BVLE) *in vitro*. CCl_4 (20 mM) has been used as a hepatotoxin and the cytotoxicity of CCl_4 is estimated by quantitating the release of lactate dehyrogenase (LDH) in the medium. CCl_4 induces twice the amount release of LDH from the liver as compared to the cells from untreated liver tissue and this was significantly reduced in presence of BVLE (10 μg/mL).

Al-Isawi and Al-Jumaily (2019) investigated the hepatoprotective activity and antioxidant enzymes of purified *B. variegata* leaves extract and purified flowers extract were administered (200 mg/kg, orally once daily) to reduce the effect of CCl_4-damage in rat's liver for three weeks. The purified *B. variegata* leaves and purified flowers significantly inhibited the CCl_4-induced increase in ALT, AST, ALP levels at the tested doses, respectively, after treatment. However, purified *B. variegata* leaves and purified flowers treatment noticeably improved the activity of antioxidant limitations: GSH, SOD, and CAT. The hepatoprotective activity was confirmed by histological finding. From these results, it can be concluded that the *B. variegata* leaves and flowers extracts contain remarkable flavonoid which can be used for reducing oxidative stress.

Berberis aristata DC. FAMILY: BERBERIDACEAE

Berberis aristata roots have been used in the treatment of jaundice in Ayurveda. Hepatoprotective and antioxidant activity of dried aerial part of *B. aristata* was investigated in aqueous and methanolic extract and berberine, against CCl_4-induced liver injury. Results obtained were comparable to standard drug Silymarin. Crude extract of *B. aristata* (shoot and fruit) showed PCM and CCl_4 protection against induced liver toxicity and it also indicates that hepatoprotective action of extract is partially through inhibition of MDME (Gilani and Janbaz, 1995d; Janbaz and Gilani, 2000). Butanolic extract of *B. aristata* shows effective action of hepatoprotection by selective inotropic activity (Gilani et al., 1999).

The hepatoprotective activity of *Berberis aristata* stem bark suspension was studied by Rathi et al. (2015) using CCl_4 overdose-induced liver damage in rats. When the vehicle treated group 1 was compared with control group 2 (treated with $CCl_{4)}$, there was a significant rise in the level of serum marker enzyme ALP, but no significant difference observed between the vehicle treated group and standard drug treated group. Increased levels of serum SGOT, SGPT, ALP, and bilirubin were observed in CCl_4 treated control group. But in different dose levels of EEBA (100 and 300 mg/kg of BW) treated groups, a significant reduction of serum SGOT, SGPT, ALP, and bilirubin (total and direct) were observed.

Berberis asiatica Roxb. ex DC. FAMILY: BERBERIDACEAE

Tiwari and Khosa (2010) investigated the hepatoprotective and antioxidant activities of dried aerial parts of *Berberis asiatica* aqueous (AqBA) and methanol (MeBA) extracts against CCl_4-induced hepatic injury. AqBA and MeBA at dose levels of 200 and 300 mg/kg offered significant hepatoprotective action by reducing the serum marker enzymes such as SGOT and SGPT. They also reduced the elevated level of serum ALP, serum acid ACP, and SB. Reduced enzymic and nonenzymic antioxidant levels and the elevated LPO level were restored to normal by administration of MeBA and AqBA. Histopathological studies further confirmed the hepatoprotective activity of these extracts when compared with CCl_4 treated control groups. MeBA extract at 200 mg/kg, p.o. showed significant hepatoprotective activity similar to that of the standard drug, Silymarin.

Berberis lycium Royle FAMILY: BERBERIDACEAE

Chand et al. (2011) evaluated the hepatoprotective role of feed-added *Berberis lycium* in broiler chicks. In this study, 320 1-day-old broiler chicks were randomly divided into four major groups A, B, C, and D and fed rations supplemented with 0, 15, 20, and 22.5 g *B. lycium* kg^{-1} ration, respectively. Each group was further divided into two subgroups, one vaccinated against Newcastle disease (ND) and infectious bursal disease (IBD), the other nonvaccinated. Antibody titre against IBD and ND, relative weight of lymphoid organs, post-challenge morbidity and mortality, serum hepatic enzymes, and total serum protein were observed. Group C had higher anti-IBD and anti-ND antibody titres. Relative bursa weight in groups C and D was higher until day 28, but birds in group C performed better at later stages of examination. Relative spleen weight was highest in group C. During initial stages, there was no effect on relative thymus weight, but at later stages the effect was significant. Groups C and D performed similarly in terms of relative thymus weight. The birds were challenged to field IBD through intramuscular injection at a dose rate of 0.5 mL per bird. Post-challenge morbidity was lowest in groups C and D, while treatment significantly affected mortality among affected (morbid) birds. Levels of serum ALT and ALP were lowest in group C. Serum protein was similar in all groups and in both vaccinated and nonvaccinated broiler chicks.

Berberis tinctoria Lesch. FAMILY: BERBERIDACEAE

The ME of *Berberis tinctoria* leaves was investigated for its hepatoprotective and antioxidant effects on PCM (750 mg/kg)-induced acute liver damage in Wistar albino rats by Murugesh et al. (2005). The methanol extract of *Berberis tinctoria* (MEBT) at the doses of (150 and 300 mg/kg) produced significant hepatoprotective effect by decreasing the activity of serum enzymes, bilirubin, and LPO while it significantly increased the levels of GSH, catalyse (CAT) and SOD in a dose-dependent manner. The effects of ME were comparable to that of the standard drug, Silymarin.

Berberis vulgaris L. FAMILY: BERBERIDACEAE

Fruit, leaves, and stem of *Berberis vulgaris* (barberry) have medical usages including hepatoprotection. *B. vulgaris* fruit extract contains various flavonoids that act as antioxidant (Hadaruga et al., 2010). Berberine, oxyacanthine, and other alkaloids such as berbamine, palmatine, columbamine, malic acid, jatrorrhizine, and berberrubine comprise some other compounds (Fatehi et al., 2005). Stigmasterol, terpenoids lupeol, oleanolic acid, stigmasterol glucoside, and polyphenols were also identified in this plant (Imanshahidi and Hosseinzadeh 2008). Berberine, an isoquinoline alkaloid with a long medicinal history, exists in roots, rhizomes, and stem bark of the plant. Berberine inhibits potassium and calcium currents in isolated rat hepatocytes. It has hepatoprotective effects, both preventive and curative, on $CC1_4$-induced liver injury through scavenging the peroxidative products. $CC1_4$ significantly increased the serum ALT, AST, and ALP levels in rats. Treatment with the methanolic extract of *B. vulgaris* fruit significantly helped these changes reach to an almost normal level. In addition, the extract could prevent $CC1_4$-induced liver oxidative damage in rats (Feng et al., 2010). Domitrović's study was indicative of berberine's effect on protecting the liver from $CC1_4$-induced injury. The hepatoprotective mechanisms of berberine could be attributed to the free-radical scavenging, decline in oxidative/nitrosative stress, and the inhibition of inflammatory response in the liver (Domitrović et al., 2011). In addition, *B. vulgaris* extract/β-cyclodextrin exhibited better hepatoprotective effects than free extract on oral administration possibly due to greater bioavailability. Formulated extract could be used as an economical phytotherapeutical supplement that is helpful for chronic or acute conditions or a support for routine therapies of serious hepatic disorders. In Hermenean's study, pretreatment with formulated or nonformulated extract prevented the increase in ALT, AST, and MDA levels, and helped the level of antioxidant enzymes return to normal values. According to histopathological and electron-microscopic examination, in both pretreated groups, more moderate damage in liver was observed with a more pronounced protective effect after administration of the formulated extract (Hermenean et al., 2012).

Berchemia lineata **(L.) DC.** **FAMILY: RHAMNACEAE**

Li et al. (2015) investigated the hepatoprotective effect of *Berchemia lineata* ethanol extract (BELE) on CCl_4-induced acute liver damage in mice. Compared with the model group, administration of 400 mg/kg BELE for 7 days in mice significantly decreased the serum ALT, ALP, and TBIL, along with the elevation of TP. In addition, BELE (100, 200, and 400 mg/kg, i.g.) treated mice recorded a dose-dependent increment of SOD and reduction of MDA levels. Histopathological examinations also confirmed that BELE can ameliorate CCl_4-induced liver injuries, characterized by extensive hepatocellular degeneration/necrosis, inflammatory cell infiltration, congestion, and sinusoidal dilatation.

Beta vulgaris L. FAMILY: AMARANTHACEAE

Ethanolic extract of *Beta vulgaris* roots given orally at doses of 1000, 2000, and 4000 mg/kg exhibited significant dose-dependent hepatoprotective activity against CCl_4-induced hepatotoxicity in rats. Hepatotoxicity and its prevention were assessed by serum markers, namely, cholesterol, TG, ALT and ALP (Agarwal et al., 2006). The hepatoprotective activity of *B. vulgaris* may be attributed to its antioxidant (Georgiev et al., 2010) and anti-inflammatory (Atta and Alkofahi, 1998) activities. The plant is safe to use even in large doses. Phytochemical studies on roots of *B. vulgaris* have shown the presence of betaine, betacyanins, betaxanthins, oxalic acid, and ascorbic acid (Chakole et al., 2011).

Betula utilis D. Don FAMILY: BETULACEAE

Betula utilis extracts was evaluated by Duraiswamy et al. (2012) for *in vitro* and *in vivo* hepatoprotective activity against D-GalN-induced hepatic damage. The phytochemical screening on *B. utilis* revealed the presence of primary metabolites such as, carbohydrates and protein, the secondary metabolites such as alkaloids, glycosides, saponins, sterols, flavonoids, phenolic compounds, and tannins. The hepatocytes challenged with 50 μL of 10 mM solution of D-GalN. The group to which toxicant was added showed a significant increase in the AST, ALT, and ALP enzyme level when compared to the normal. All the biochemical parameters were restored to normal significantly. Minimum restoration was observed at 62.5 μg/mL in both the plants extract. Moreover *in-vivo* treatment with the BUE and BUA extract showed significant decreases in serum levels of AST, ALT, ALP, LDH, and TB and a significant elevation in serum levels of the TP, TGL, and ALB in dose-dependent manner. However, the normal architecture of the liver completely lost and marked centrilobular necrosis, focal necrosis and bile duct proliferation were observed in liver of D-D-GalN treated group.

Bidens bipinnata L. FAMILY: ASTERACEAE

Zhong et al. (2007) evaluated the hepatoprotective activity of the total flavonoids of *Bidens bipinnata* against CCl_4-induced acute liver injury in mice and to determine its mechanism of action. Oral administration of total flavonoids at doses of 50, 100, and 200 mg/kg for 7 days significantly reduced the elevated relative values of liver weight, serum transaminases (ALT and AST) and the hepatic morphologic changes induced by CCl_4 in mice. In addition, total flavonoids markedly inhibited CCl_4-induced LPO and enhanced the activity of the antioxidant enzymes SOD and GPx. Moreover, pretreatment with flavonoids suppressed NO production and nuclear factor-kappaB activation in CCl_4-treated mice. The results suggest that TFB has significant hepatoprotective activity and its mechanism is related, at least in part, to its antioxidant properties.

Yuan et al. (2008a, b) evaluated the protective effects of total flavonoids of *B. bipinnata* against CCl_4-induced liver fibrosis in rats. The results showed that total flavonoids (80 and 160 mg/kg) treatment for 10 weeks significantly reduced the elevated liver index (liver weight/BW) and spleen index (spleen weight/BW), elevated levels of serum transaminases (ALT and AST), hyaluronic acid, type III procollagen and hepatic HYP. In addition, TFB markedly inhibited CCl_4-induced LPO and enhanced the activity of the antioxidant enzymes SOD and GPx. Moreover, TFB (80 and 160 mg/kg treatment improved the morphologic changes of hepatic fibrosis induced by CCl_4 and suppressed nuclear factor (NF)-kappaB, alpha-smooth muscle actin (SMA) protein expression and transforming growth factor (TGF)-beta1 gene expression in the liver of liver fibrosis of rats. In conclusion, TFB was able to ameliorate liver injury and protect rats from CCl_4-induced liver fibrosis by suppressing oxidative stress. This process may be related to inhibiting the induction of NF-kappaB on hepatic stellate cell activation and the expression of TGF-beta1.

Bidens pilosa **L. var.** ***minor*** **(Blume) Sherff,** ***B. pilosa*** **L.,** ***B. chilensis*** **DC. FAMILY: ASTERACEAE**

Bidens pilosa L. var. *minor* (Blume), *B. pilosa*, and *B. chilensis*, commonly known as "Ham-hong-chho" in Taiwan, have been traditionally used for medicinal purposes. To clarify and compare the hepatoprotective effects of these three plants, Chin et al. (1996) evaluated their potential effectiveness on CCl_4- and APAP-induced acute hepatic lesions in rats. The results indicated that the increase in SGOT and SGPT activities caused by CCl_4 (3.0 mL/kg, s.c.) and APAP administration (600 mg/kg, i.p.) could be significantly reduced by treating with the extracts of all the three kinds of "Ham-hong-chho" and the extract of *B. chilensis* exhibited the greatest hepatoprotective effects. These phenomena were also confirmed by histological observation. Liver damage induced by CCl_4 and APAP was markedly improved in the extract of *B. chilensis* treated groups, while groups treated with the extracts of *B. pilosa* var. *minor* and *B. pilosa* demonstrated only moderate protective effects. The pharmacological and pathological effects of these three crude groups were compared with *Bupleurum chinense*, which has been reported previously as a treatment criteria in the CCl_4 model, and with Silymarin as a standard reference medicine in the APAP model. The results suggest that *Bidens pilosa* var. *minor*, *B. pilosa*, and *B. chilensis* can protect liver injuries from various hepatotoxins and have potential as broad spectrum antihepatic agents.

Abdel-Ghany et al. (2016) investigated the protective effects of aqueous and MEs of *B. pilosa* using various *in vivo* and *in vitro* models of hepatic injury. One kilogram of the aerial parts of *B. pilosa* was used to prepare 80% methanol and aqueous extracts of the plant (500 g for each extract). The TPC, TFC, and antioxidant activity of both extracts were evaluated. The hepatoprotective activity of these extracts in CCl_4 (0.1%) and D-GalN (700 mg/kg)-induced liver injury, respectively, was investigated in mice. Paracetamol-induced liver injury was used as *in vitro* reference standard. TPC and TFC of ME were higher than those of the aqueous extract. The combination of ME and Silymarin showed the highest antioxidant activity. *In-vivo* administration of CCl_4 and D-GalN significantly increased the levels of ALT, AST, and ALP, but decreased the TP, ALB, and GSH contents of liver. Co-administration of the extracts (50 mg/kg) and Silymarin (100 mg/kg) effectively countered the effects of CCl_4 and D-GalN, while also exerting their antioxidant properties. Both methanol and aqueous extracts showed hepatoprotective activity in PCM-induced cytotoxicity in primary cultures of rat hepatocytes.

Bixa orellana L. FAMILY: BIXACEAE

In a study by Ahsan et al. (2009a, b), the MEs of *Bixa orellana*, *Cajanus cajan*, *Glycosmis pentaphylla*, and *Casuarina equisetifolia* were all shown to possess significant hepatoprotective activity. The four plants extracts at a dose of 500 mg/kg BW exhibited moderate protective effect by lowering the serum levels of ALT or SGPT, AST or SGOT, and cholesterol to a significant extent against liver damage induced by CCl_4. The result could also be expressed in the order of *Bixa orellana* > *Cajanus cajan* > *Glycosmis pentaphylla* > *Casuarina equisetifolia*.

Boehmeria nivea var. *nivea* and *B. nivea* var. *tenacissima* (= *B. frutescens*) FAMILY: URTICACEAE

Lin et al. (1998) investigated the relationship between liver protective effects and antioxidant activity of *Boehmeria nivea* var. *nivea* and *B. nivea* var. *tenacissima*. The water extracts of both plants exhibited a hepatoprotective activity against CCl_4-induced liver injury. Both the plants showed antioxidant effects in $FeCl_2$-ascorbate-induced LPO in rat liver homogenate. Moreover, the active oxygen species scavenging potencies were evaluated by an electron spin resonance (ESR) spin-trapping technique. *B. nivea* var. *tenacissima* displayed better superoxide radical scavenging activity than *B. nivea* var. *nivea*.

Boerhavia diffusa L. FAMILY: NYCTAGINACEAE

The roots of *Boerhavia diffusa* L., commonly known as "Punarnava," are used by a large number of tribes in India for the treatment of various hepatic disorders. An alcoholic extract of whole plant *B. diffusa* given orally exhibited hepatoprotective activity against experimentally induced CCl_4 hepatotoxicity in rats and mice (Chandan et al., 1991). The extract also produced an increase in normal bile flow in rats suggesting a strong choleretic activity. The extract does not show any signs of toxicity up to an oral dose of 2 g/kg in mice.

Rawat et al. (1997) investigated the effect of seasons, thickness of roots of *B. diffusa* and form of dose (either aqueous or powder) for their hepatoprotective action to prove the claims made by the different tribes of India. The hepatoprotective activity of roots of different diameters collected in three seasons, rainy, summer, and winter, was examined in thioacetamide intoxicated rats. The results showed that an aqueous extract (2 mL/kg) of roots of diameter 1–3 cm, collected in the month of May (Summer), exhibited marked protection of a majority of serum parameters, that is, GOT, GPT, ACP, and ALP, but not GLDH and bilirubin, thereby suggesting the proper size and time of collection of *B. diffusa* roots for the most desirable results. Further, the studies also proved that the aqueous form of drug (2 mL/kg) administration has more hepatoprotective activity than the powder form; this is probably due to the better absorption of the liquid form through the intestinal tract (Rawat et al., 1997).

Muthulingam (2008) investigated the antihepatotoxic effect of aqueous leaf extract of *B. diffusa* (BDEx) on rifampicin-induced liver injury. Significant elevation of serum hepatic marker enzymes (AST, ALT, ALP), bilirubin and cholesterol whereas protein level decreased in rats treated with rifampicin (1 g/kg BW orally one day only). Oral administration of BDEx (250 and 500 mg/kg BW once daily for 28 days) and Silymarin to rifampicin-induced liver injury rats caused significantly attenuated the aforementioned parameters. The maximum antihepatotoxic effect against rifampicin-induced liver injury was achieved with BDEx 500 mg/kg BW but doses higher than 500 mg/kg BW were less effective. These results are compared to the reference hepatoprotective agent Silymarin.

Aqueous and ethanolic extracts of *B. diffusa* significantly attenuated APAP (Olaleye et al., 2010) and ethanol (Devaki et al., 2004)-induced biochemical (rise of serum AST, ALT, APT, and bilirubin) and histopathological changes in liver suggesting its hepatoprotective action. The extract has been shown to possess significant antioxidant (Olaleye et al., 2010) and anti-inflammatory (Bhalla et al., 1968) activities which may contribute to its hepatoprotective activity. The aerial part of *B. diffusa* is a rich source of flavonoids, steroids, and alkaloids. Detailed phytochemical analysis showed the presence of campesterol, daucosterol, sitosterols, punarnavine, boeravinones A-F, borhavone, amino acids, lignans, and tetracosanoic, esacosanoic, stearic, and ursolic acids (Al-Asmari et al., 2014).

Jayavelu et al. (2013) evaluated the hepatoprotective activity of different parts of *B. diffusa* such as root and aerial parts against ibuprofen (IB)-induced hepatotoxicity in Wistar albino rats. The administration of ibuprofen (500 mg/kg BW) produced significant changes in the normal hepatic cells, resulting in the formation of gastric lesions, centrilobular necrosis, vacuolization, and hepatomegaly. The adverse effect of ibuprofen was reflected in the levels of biochemical parameters of liver marker enzymes such as ALT, AST, ALP, and bilirubin. The activities of natural antioxidant enzymes like SOD, CAT, GPx, and GST were decreased significantly. The ME (85%) of the root and aerial part of *B. diffusa* (500 mg/kg. BW) produced remarkable changes in affected hepatic cell architecture and restored nearly normal structure and functions of hepatic cells. Similarly, the different parts of the *B. diffusa* (500 mg/kg. BW) restored the altered biochemical parameters of liver marker enzymes close to normal control levels. The observed results show the root of *B. diffusa* possesses more hepatoprotective efficacy than the aerial part of the same plant. The results suggest that the hydro alcoholic (15:85%) extract of *B. diffusa* possesses significant potential effect as a hepatoprotective agent.

Beedimano and Jeevangi (2015) evaluated the hepatoprotective activity of aqueous extract of *B. diffusa* against CCl_4-induced toxicity in male albino rats. Administration of *B. diffusa* at doses 250 and 500 mg/kg orally demonstrated hepatoprotective activity by preventing the increase of ALT, AST, ALP, and SB and also confirmed by histopathology of the liver. The results were comparable to that of Silymarin.

Boesenbergia rotunda **(L.) Mansf. FAMILY: ZINGIBERACEAE**

Salama et al. (2012) investigated the effects of ethanol-based extract from *Boesenbergia rotunda* (BR) on liver cirrhosis. Data from the acute toxicity tests showed that the extract was safe to use. Histological analysis of the livers of the rats in cirrhosis control group revealed uniform coarse granules on their surfaces, hepatocytic necrosis, and lymphocytes infiltration. But, the surfaces morphologically looked much smoother and the cell damage was much lesser in those livers from the normal control, Silymarin and BR-treated groups. In the high-dose BR treatment group, the livers of the rats exhibited nearly normal looking lobular architecture, minimal inflammation, and minimal hepatocyte damage, the levels of the serum biomarkers and liver enzymes read nearly normal, and these results were all comparable to those observed or quantified from the normal and Silymarin-treated groups. The BR extract had the antioxidant activity about half of what was recorded for Silymarin. The progression of the liver cirrhosis can be intervened using the ethanol-based BR extract, and the liver's status quo of property, structure, and function can be preserved.

Bombax ceiba **L. FAMILY: MALVACEAE**

Hepatoprotective activity of methanolic extract of flowers of *Bombax ceiba* (MEBC) was investigated by Ravi et al. (2010) against hepatotoxicity produced by administering a combination of two antitubercular drugs Isoniazid and Rifampicin for 10 and 21 days, respectively, by intraperitoneal route in rats. MEBC was evident in all the doses as there was a significant decrease in AST, ALT, ALP, and total bilirubin levels, but increased the level of TP in comparison to control. MEBC significantly decreased the level of TBARS and elevated the level of GSH at all doses as compared to control. Histology of the liver section of the animals treated with MEBC improved the hepatotoxicity caused by antitubercular drugs. The results obtained from the analysis of biochemical parameters and histopathological studies, enabled to conclude that the MEBC were not able to revert completely the hepatic injury induced by INH + RIF, but it could limit the effect of INH + RIF to the extent of necrosis.

Borago officinalis L. FAMILY: BORAGINACEAE

Hamed and Wahid (2015) investigated the hepatoprotective effect of *Borago officinalis* aerial ethanolic extract (BAEE) against CCl_4-induced liver damage in comparison to Silymarin. The hepatoprotective potential of BAEE in rats was evaluated following oral administration of CCl_4, which enhanced hepatic LPO and notably depleted reduced GSH. CCl_4 administration caused over expression of the inflammatory markers TNF-α and NFκB protein levels, in addition to a significant increase in the release of liver serum biomarker levels. Administration of BAEE showed hepatic protection by significantly reducing elevated levels of serum enzyme levels. Notably, BAEE significantly reduced expression of the TNF-α and NFκB protein expression levels comparable to wild type and Silymarin. These findings were augmented with the histopathological results in which BAEE was able to show an improvement in the liver condition.

Boschniakia rossica (Chamisso & Schlecht.) B. Fedtschenko FAMILY: OROBANCHACEAE

Quan et al. (2011) investigated the hepatoprotective effect of *Boschniakia rossica* extract (BRE), rich in phenylpropanoid glycoside and iridoid glucoside, on CCl_4-induced liver damage. CCl_4 challenge not only elevated the serum marker enzyme activities and reduced ALB level but also increased liver oxidative stress, as evidenced by elevated lipid hydroperoxide (LOOH) and MDA concentrations, combined with suppressed potential of hepatic antioxidative defense system including SOD, GPx activities and reduced GSH content. Furthermore, serum TNF-α (TNF-α), hepatic nitrite level, iNOS, and cyclooxygenase-2 (COX-2) protein contents were elevated while cytochrome P450 2E1 (CYP2E1) expression and function were inhibited. Preadministration of BRE not only reversed the significant changes in serum toxicity markers, hepatic oxidative stress, xenobiotic metabolizing enzymes, and proinflammatory mediators induced by CCl_4 but also restored liver CYP2E1 level and function. Interestingly, the protein expression of HO-1 was further elevated by BRE treatment, which was markedly increased after CCl_4 challenge.

Boswellia sacra Fleuckiger FAMILY: BURSERACEAE

The water extract of oleo-gum-resin of *Boswellia sacra* is used in the treatment of liver problems in the Middle East and Arab–African countries.

Asad and Alhumoud (2015) evaluated its effect on liver injury induced by CCl_4 in acute and chronic liver injury models. The *Boswellia sacra* extract at tested doses of 2 and 5 mL/kg significantly reduced the elevated serum levels of biomarkers in liver damage. The hepatoprotective effect was supported by changes in histopathology. The GC-MS analysis of the extract revealed the presence of several phyto-constituents that included menthol, 3-cyclohexen-1-ol, and octanoic acid. The water extract oleo-gum-resin of *Boswellia sacra* possesses hepatoprotective activity, as claimed by traditional healers within the Middle East and Arab–African countries.

Brassica nigra (L.) K. Koch FAMILY: BRASSICACEAE

The protective effect of the ME of *Brassica nigra* leaves was investigated against D-GalN-induced hepatic and nephrotoxicity in Wistar rats. The D-GalN-induced toxicity was evident from a significant increase in the serum and tissue inflammatory markers in toxic rats, when compared with the control (saline alone treated animals). The *B. nigra* pretreated groups (200 and 400 mg/kg BW) showed significant reduction in the D-GalN-induced toxicity as obvious from biochemical parameters. Histopathological observations confirm the protective effect of *B. nigra* leaf extract by reduction in hepatic and renal tissue damage. Accordingly, the crude ME of *B. nigra* leaf lacks inherent toxicity and exhibits hepatic and nephroprotective (Rajamurugan et al., 2012).

Brassica rapa L. FAMILY: BRASSICACEAE

The pretreatment of rats with *Brassica rapa* juice protected the rats against CCl_4-induced hepatotoxicity. The treatment significantly reduced the SGOT, SGPT, ALP, and bilirubin level at a dose of 16 mL/kg BW. In addition, the juice was also replenished the lowered nonprotein sulfhydryl (NP-SH) concentration in the liver tissue after CCl_4 treatment (Al-Snafi et al., 2019). The protective effect of turnip root ethanolic extract (TREE) on early hepatic injuries was studied in alloxan-induced diabetic rats. TREE treatment groups received TREE (200 mg/kg) daily for 8 weeks through the gavage. TREE significantly decreased the levels of serum biomarkers of hepatic injury. Furthermore, it significantly decreased the LPO and elevated the decreased levels of antioxidant enzymes in diabetic rats. The study also showed that histopathological changes were in agreement with biochemical findings (Daryoush et al., 2011). The effect of aqueous extract of *B. rapa chinensis* (250, 500 mg/kg, p.o.) against the oxidative stress induced by *tert*-butyl hydroperoxide (*t*-BHP) in rats. The treatment with aqueous extract of *B. rapa chinensis* significantly combats the oxidative stress imposed by *t*-BHP in the hepatic tissues as evidenced by marked improvement in the antioxidant status and suppressing LPO levels. The results obtained were dose dependent with 500 mg/kg BW, dosage of *B. rapa chinensis* aqueous extract revealing more potential in curbing toxic insult of *t*-BHP (Kalava and Mayilsamy, 2014). The antifibrogenic and the therapeutic effect of turnip extracts was studied in thioacetamide (TAA)-induced liver fibrosis animal model. Antifibrogenic effect was demonstrated histopathologically and serologically after the animals fed with turnip extracts with synchronous TAA injections for 7 weeks. The animals fed with 20 mg/mL of turnip extracts showed the highest antifibrogenic effect (Shin et al., 2006). The level of hepatic fibrosis induced by TAA was compared among TAA-turnip group, TAA group, and vehicle control group. Nodules formed by TAA were observed; they were rarely shown in vehicle control group, observed in most area in TAA group, but only shown in periportal regions in TAA-turnip group. These results were confirmed through Masson's trichrom stain; fibrous structures increased in TAA group (fibrosis score: 4) but significantly decreased in TAA-turnip group (fibrosis score: 2–3) (Li et al., 2010).

The hepatoprotective effect of *B. rapa* seeds was evaluated by Fu et al. (2016). All extracts of *Brassica rapa* seeds exhibited high phenolic contents and strong antioxidant activities. The activities were in the order of ethanol extract≥50% ethanol extract > water extract. All the three extracts showed effective hepatoprotection. They could ameliorate liver damage and improve the antioxidant defense system. Sinapine thiocyanate was the main component and was hepatoprotective in a dose-dependent manner.

Bridelia ferruginea Benth. FAMILY: PHYLLANTHACEAE

Adetutu and Olorunnisola (2013) investigated the hepatoprotective potential of the aqueous extracts of leaves of *Bridelia ferruginea*. Extract of dose of 100 mg/kg BW was given to rats in five groups for seven consecutive days followed by a single dose of 2-AAF (0.5 mmol/kg BW). The rats were sacrificed after 24 h and their bone marrow smears were prepared on glass slides stained with Giesma. The micronucleated polychromatic erythrocyte cells (mPCEs) were thereafter recorded. The hepatoprotective effects of the plant extract against 2-AAF-induced liver toxicity in rats were evaluated by monitoring the levels of ALP, GGT, and histopathological analysis. The results of the 2-AAF induced liver toxicity experiments showed that rats treated with the plant extract (100 mg/kg) showed a significant decrease in mPCEs as compared with the positive control. The rats treated with the plant extract did not show any significant change in the concentration of ALP and GGT in comparison with the negative control group whereas the 2-AAF group showed a significant increase in these parameters. Leaf extract also showed a protective effect against histopathological alterations (Adetutu and Olorunnisola, 2013).

Bruguiera gymnorrhiza (L.) Lam. FAMILY: RHIZOPHORACEAE

Sur et al. (2016) evaluated the antioxidant and hepatoprotective actions of leaves extract of *Bruguiera gymnorrhiza*. Polyphenols such as GA, quercetin, and coumarin obtained from BR exhibited powerful antioxidant properties. Moreover, it produced dose-dependent protection against GalN-induced hepatitis in rats. It significantly reduced GalN-induced elevation of enzymes (ALT, AST, and ALP) in serum and resist oxidative stress marked by LPOs, GSH, and CAT in hepatic parenchyma.

Bryonia dioica Salisb. FAMILY. CUCURBITACEAE

The plant leaves extract was evaluated for its protective effect in hepatotoxicity induced in rats with CCl_4. Single oral dose of 250 mg/kg of different fractions extract was given to rats for 7 days. Serum activities of transaminases (ALT and AST) were used as the biochemical marker of hepatotoxicity. Histopathological changes in rat's liver section were also examined. The results indicated that pretreatment of rats with Bryonia extract prior to induction of hepatotoxicity offered a hepatoprotective action (Khadim, 2014).

Bryophyllum calycinum Salisb. FAMILY: CRASSULACEAE

The juice of the leaves and the ethanolic extract of the marc left after expressing were studied in rats against CCl_4-induced hepatotoxicity. It was found that they were effective hepatoprotective as evidenced by *in vitro*, *in vivo*, and histopathological studies. The juice was found to be more effective than the ethanolic extract (Devbhuti et al., 2008).

Radix Bupleuri-Bupleurum chinense DC. and *Bupleurum scorzonerifolium* Willd. FAMILY: APIACEAE

Radix Bupleuri, also called "Chaihu" in Chinese, is derived from the dried roots of *Bupleurum chinense* and *Bupleurum scorzonerifolium*. The liver protective effects against CCl_4-induced liver injury were investigated after treatment of mice with raw and vinegar-baked *Radix Bupleuri* (5 g/kg/day) for 14 days. The results showed that both raw and processed *Radix Bupleuri* showed liver protective effects against CCl_4-induced liver injury, and the vinegar-baked *Radix Bupleuri* exerted better effects than that of raw *Radix Bupleuri* (Li et al., 2015). Pretreated with saikosaponins, especially SSa or SSd, showed remarkable inhibition of D-GalN-induced hepatic injury through decreasing the activity of glucose-6-phosphatase and NADPH-cytochrome C reductase and increasing 5′-nucleotidase activity (Abe et al., 1980). Similarly, bupleurosides III, VI, IX, and XIII and saikosaponin b_3 isolated from *B. scorzonerifolium* were also found to exhibit protective effect on the D-GalN-induced cytotoxicity in primary cultured rat hepatocytes (Matsuda et al., 1997). Further studies also demonstrated that the protective effects of saikosaponins isolated from *Bupleurum chinense* could prevent hepatocyte injury through regulating intracellular calcium levels (Han et al., 2006). In a rat model with CCl_4-induced acute hepatic injury, the hepatic enzyme levels (GOT, GPT, and ALP) and the LPO in the liver were significantly reduced by the administration of SSd (Abe et al., 1982). Additionally, SSd significantly reduced collagen I deposition and ALT level on liver fibrosis rats and decreased the concentration of transforming growth factor $\beta 1$ (TGF-$\beta 1$). Moreover, SSd was able to alleviate hepatocyte injury from oxidative stress. The effect of SSd on liver fibrosis may be related to its ability to reduce LPO (Fan et al., 2007).

Bupleurum kaoi Liu et al. FAMILY: APIACEAE

A leaf infusion of *Bupleurum kaoi* was prepared and the antioxidant properties and *in vitro* hepatoprotective activity were demonstrated by Liu et al. (2006). The leaf infusion exerted DPPH free-radical scavenging activity, inhibitory capacity on superoxide anion formation and superoxide anion scavenging activity. The hepatotoxicity of APAP and CCl_4 on the rat liver cells were also decreased by the leaf infusion.

Fractionation with supercritical CO_2 is employed to divide ethanolic extract (E) of *B. kaoi* into four fractions (R, F1, F2, and F3).

To assess the selectivity of the fractionation, extracts of the four fractions were characterized in terms of the hepatoprotective capacity and activity of antioxidant enzymes against CCl_4 induced damage. The *in vitro* study revealed that pretreatment with *B. kaoi* extract or its fractions, except F3, significantly protected primary hepatocytes against damage by CCl_4. The R and F1 fractions had the highest saikosaponins content (175 and 200 mg/g dry weight, respectively) and most effectively protected the liver from damage by CCl_4. This study demonstrated that the oral pretreatment of *B. kaoi* (100 and 500 mg/kg), except F3, three days before a single dose of CCl_4 was administered significantly lowered the serum levels of hepatic enzyme markers (AST and ALT). A pathological examination showed that lesions, including ballooning degeneration, necrosis, hepatitis, and portal triaditis were partially healed by treatment with *B. kaoi* extract and fractions (Wang et al., 2004).

Yen et al. (2005) examined the hepatoprotective effect of three materials extracted or isolated from the roots of *B. kaoi* against DMN-induced hepatic fibrosis in rats. They were water extract (BKW), polysaccharide-enriched fractions (BKP), and saponin-enriched fractions (BKS). Treated with groups of BKW, BKP, BKS markedly reduced GOT, GPT levels in rats serum. In addition, treated with groups of BKW, BKP, BKS markedly raised TP levels in rats serum and liver homogenates. Furthermore, treated with groups of BKW, BKP markedly raised ALB levels in rats serum and liver homogenates. Treated with groups of BKW, BKP, and BKS markedly raised interferon-gamma (IFN-gamma) levels in rats serum, where only BKS and Silymarin markedly raised interkeukin-10 (IL-10) levels in rats serum compared to that of DMN-treated rats. None of test materials of *B. kaoi* except Silymarin reduced the MDA levels, but BKW, BKP markedly raised hepatic GSH levels to reveal the activity of anti-LPO. Otherwise, treated with groups of BKW, BKP, and BKS significantly reduced collagen contents in rats liver homogenates. *B. kaoi* demonstrated the anti-inflammatory and antifibrotic activities followed by antioxidant activity of enhanced GSH production, enhanced the liver cell regeneration and concerned with regulations of INF-gamma and IL-10. The ability of hepatoprotective and antifibrotic activities of *B. kaoi* are higher than *B. chinense*.

Butea monosperma (Lam.) Taub. FAMILY: FABACEAE

Oral administration of *Butea monosperma* flowers powder (100 mg/kg) effectively inhibited PCM-induced changes in the serum marker enzymes in rabbits. Increase in transaminases AST, ALT, and ALP was observed with PCM treated group. The results suggest that the BM flowers powder possessed significant potential as hepatoprotective agent. Isobutrin and butrin, the antihepatotoxic principles of flowers was reported and this activity was monitored by means of CCl_4 and GalN-induced liver lesion *in vitro*. The methanolic extract of *B. monosperma* possesses hepatoprotective effects and also it might suppress the promotion stage via inhibition of oxidative stress and polyamine biosynthetic pathway by significant reduction in TAA-induced serum AST, ALT (ALT/SGPT), LDH, and GGT activities (Wagner et al., 1986).

Sehrawat et al. (2006) investigated the effect of *B. monosperma*, a known liver acting drug on the tumor promotion related events of carcinogenesis in rat liver. TAA was used to induce tumor promotion response and oxidative stress and caused significant depletion in the detoxification and antioxidant enzyme armory with concomitant elevation in MDA formation, hydrogen peroxide (H_2O_2) generation, ornithine decarboxylase activity, and unscheduled DNA synthesis. However, B. *monosperma* pretreatment at two different doses restored the levels of the abovesaid parameters in a dose-dependent manner. The alcoholic extract of *B. monosperma* used in the present study seems to offer dose-dependent protection and maintain the structural integrity of hepatic cells. This was evident from the significant reduction in TAA-induced serum GOT, GPT, LDH, and GGT activities. Overall results indicate that the methanolic extract of *B. monosperma* possesses hepatoprotective effects and also it might suppress the promotion stage via inhibition of oxidative stress and polyamine biosynthetic pathway.

Hepatoprotective evaluation of *B. monosperma* flower extract against liver damage by PCM in rabbits was carried out by Maaz et al. (2010). Oral administration of *B. monosperma* flowers powder (100 mg/kg) effectively inhibited PCM-induced changes in the serum marker enzymes in rabbits. Increase in transaminases AST, ALT, and ALP was observed with PCM treated group. These values significantly decreased when PCM along with *B. monosperma* powder was given at a time and monitored after seven and fourteen days. However no significant difference was observed in *B. monosperma* powder treated group of rabbits. The results suggest that the flowers powder possessed significant potential as hepatoprotective agent.

Aqueous extract of flowers of *B. monosperma* was also evaluated by Sharma and Shukla (2011) at different dose levels (200, 400, 800 mg/kg, p.o.) for its protective efficacy against CCl_4 (1.5 mL/kg i.p.)-induced acute liver injury to validate its use in traditional medicines. The

CCl_4 administration altered various biochemical parameters, including serum transaminases, protein, ALB, hepatic LPO, reduced GSH and TP levels, which were restored toward control by therapy of *B. monosperma* Adenosine triphosphatase and glucose-6-phosphatase activity in the liver were decreased significantly in CCl_4 treated animals. Therapy of *B. monosperma* showed its protective effect on biochemical and histopathological alterations at all the three doses in dose-dependent manner. *B. monosperma* extract possesses modulatory effect on drug-metabolizing enzymes as it significantly decreased the hexobarbitone-induced sleep time and increased excretory capacity of liver which was measured by BSP retention. Histological studies also supported the biochemical finding and maximum improvement in the histoarchitecture was seen at a higher dose of *B. monosperma* extract. Gupta et al. (2012) evaluated the hepatoprotective potential ethanolic extract of *B. monosperma* in CCl_4-intoxicated rats.

Mathan et al. (2011) evaluated the hepatoprotective and antitumorigenic properties of the aqueous extract and butanol fractions of *B. monosperma* flowers in animal models. Dried flowers of *B. monosperma* were extracted with water and fractionated further using *n*-butanol. The hepatoprotective activity of the aqueous extract was initially confirmed in a CCl_4-induced liver damage model of rats. Oral administration of the aqueous extract produced a strong hepatoprotective effect similar to Silymarin and normalized the serum levels of ALT, AST, bilirubin, and TG in rats. However, it did not affect the levels of GSH and MDA which are oxidative stress markers in liver. Intraperitoneal administration of the aqueous extract in the X15-myc oncomice not only maintained liver architecture and nuclear morphometry but also downregulated the serum vascular endothelial growth factor (VEGF) levels. Immunohistochemical staining of liver sections with anti-ribosomal protein S27a antibody showed post-treatment abolition of this proliferation marker from the tumor tissue. The butanol fractions, however, did not show antitumorigenic activity. Thus, the aqueous extract of *B. monosperma* flowers is not only hepatoprotective but also antitumorigenic by preserving the nuclear morphometry of the liver.

Kaur et al. (2017) evaluated the hepatoprotective properties of EAF from *B. monosperma* bark in rat model. In preliminary antioxidant studies, the extract demonstrated pronounced superoxide scavenging (IC_{50} 88.85 μg/mL) and anti-LPO (IC_{50} 131.66 μg/mL) potential. In animal studies, the extract showed a protective effect against TAA-induced pathophysiology in liver of male Wistar rats. The levels of different parameters related to hepatic functions were altered by TAA treatment (300 mg/g BW) in rats. The pretreatment of rats with BEAC (50, 100, and 200 mg/kg BW) was able to normalize the biochemical markers, namely, SB, SGOT, SGPT, ALB, and ALP along with liver antioxidative molecules, namely, SOD, CAT, GSH and GR. Results of histopathological and colorimetric studies revealed that *Butea monosperma* bark extract treatment also restored the markers of fibrosis, that is, collagen and HYP toward a normal level. The bark extract considerably inhibited TAA-induced expression of p-PI3K, p-Akt, and p-mTOR in hepatocytes as revealed from immunohistochemical studies. This finding is the first evidence of inhibitory action of *B. monosperma* bark on these procarcinogenic proteins. HRMS analysis revealed the presence of quercetin, buteaspermin B, and ononin in the fraction of *B. monosperma*.

Byrsocarpus coccineus Schumach. & Thonn. FAMILY: CONNARACEAE

The leaf decoction of *Byrsocarpus coccineus* is drunk for the treatment of jaundice in West African traditional medicine. Akindele et al. (2010) investigated the hepatoprotective and *in vivo* antioxidant effects of *B. coccineus* in CCl_4-induced hepatotoxicity in rats. CCl_4 significantly increased the levels of ALT and AST and reduced TP. In CCl_4 treated animals, *B. coccineus* (200, 400, and 1000 mg/kg) dose dependently and significantly decreased ALT, AST, and ALP levels with peak effect produced at the highest dose. Conversely, *B. coccineus* produced significant increases in ALB and TP levels. The standard drug produced significant effects in respect of ALT (downward arrow), ALB (upward arrow), and TP (upward arrow). CCl_4 also produced significant reductions in the activity of CAT, SOD, peroxidase, and GSH, and conversely increased MDA level. *B. coccineus* produced significant and dose-dependent reversal of CCl_4-diminished activity of the antioxidant enzymes and reduced CCl_4-elevated level of MDA. The standard drug also significantly increased CCl_4-diminished antioxidant enzymes activity and reduced CCl_4-elevated MDA level. In general, the effects of the standard drug were comparable and not significantly different from those of *B. coccineus* (Akindele et al., 2010).

Caesalpinia bonduc (L.) Roxb. [Syn.: *C. bonducella* (L.) Fleming] FAMILY: FABACEAE

The hepatoprotective effect of ME of leaves of *Caesalpinia bonduc*

(Syn.: *C. bonducella*) was studied by Gupta et al. (2003) by means of PCM-induced liver damage in rats. The degree of protection was measured by using biochemical parameters such as serum transaminase (SGPT and SGOT), ALP, bilirubin, and TP. Further, the effects of the extract on LPO, GSH, SOD, and CAT were estimated. The methanol extract of *C. bonduc* (MECB) (50, 100, and 200 mg/kg) produced significant hepatoprotective effect by decreasing the activity of serum enzymes, bilirubin, and LPO, while it significantly increased the levels of GSH, SOD, CAT, and protein in a dose-dependent manner. The effects of MECB were comparable to that of the standard drug Silymarin. However, at a lower dose (25 mg/kg) it could not restore the deleterious effect produced by PCM.

Kumer et al. (2010) evaluated the hepatoprotective effect of the ME of *Caesalpinia bonduc* (Syn.: *C. bonducella*) (MECB) in Wistar albino rats administered with CCl_4. The MECB and Silymarin produced significant hepatoprotective effect by decreasing the activity of serum enzymes, bilirubin, uric acid, and LPO and significantly increased the levels of SOD, CAT, GSH, vitamin C, vitamin E, and protein in a dose-dependent manner. From these results, it was suggested that MECB possesses potent hepatoprotective and antioxidant properties.

Noorani et al. (2011) evaluated the protective effect of methanolic leaf extract of *C. bonduc* on gentamicin-induced hepatotoxicity and nephrotoxicity in rats. Administration of gentamicin resulted in damage to liver and kidney structures. Administration of methanolic extract of *C. bonduc* before gentamicin exposure prevented severe alterations of biochemical parameters and disruptions of liver structure. This study demonstrated that pretreatment with methanolic extract of *C. bonduc* significantly attenuated the physiological and histopathological alterations induced by gentamicin.

Ubhenin et al. (2016) evaluated the acute toxicity, hepatoprotective, and *in-vivo* antioxidant activities of ethanolic extract of *C. bonduc* leaf on CCl_4-induced liver damage using Swiss albino rats. The plant extracts at 250 and 500 mg/kg BW showed a remarkable hepatoprotective and *in vivo* antioxidant activities against CCl_4-induced hepatotoxity judged from the serum marker enzymes. The CCl_4-induced significant increase in AST, ALT, ALP, TB, and MDA with a reduction of TP, CAT, and GPx. Treatment of rats with different doses of plant extract (250 and 500 mg/kg BW) significantly altered serum maker enzymes and antioxidant levels to near normal levels.

Caesalpinia crista L. FAMILY: CAESALPINIACEAE

The ameliorating effect of *Caesalpinia crista* extract (CCME) on iron-overload-induced liver injury was investigated. CCME attenuated the percentage increase in liver iron and serum ferritin levels when compared to control group. CCME also showed a dose-dependent inhibition of LPO, protein oxidation, and liver fibrosis. The serum enzyme markers were found to be less, whereas enhanced levels of liver antioxidant enzymes were detected in CCME-treated group. In the presence of CCME, the reductive release of ferritin iron was increased significantly. Furthermore, CCME exhibited DPPH radical scavenging and protection against Fe^{2+}-mediated oxidative DNA damage (Sarkar et al., 2012).

Cajanus cajan L. FAMILY: FABACEAE

Cajanus cajan is a nontoxic edible herb, widely used in Indian folk medicine for the prevention of various liver disorders. Kundu et al. (2008) demonstrated that methanol-AF of *C. cajan* leaf extract could prevent the chronically treated alcohol-induced rat liver damage. Alcohol effected significant increase in liver marker enzyme activities and reduced the activities of antioxidant enzymes. Co-administration of methanol-AF reversed the liver damage due to alcohol; it decreased the activities of liver marker enzymes and augmented antioxidant enzyme activities. They also demonstrate significant decrease of the phase II detoxifying enzyme, UDP-glucuronosyl transferase (UGT) activity along with a three- and twofold decrease of UGT2B gene and protein expression, respectively. Methanol-AF co-administration normalized UGT activity and revived the expression of UGT2B with a concomitant expression and nuclear translocation of Nrf2, a transcription factor that regulates the expression of many cytoprotective genes.

Singh et al. (2011) evaluated the hepatoprotective activity of hydroalcoholic extract of the aerial parts of *C. cajan* against CCl_4-induced liver damage in Wistar rats. The extract was effective in protecting the liver injury as was significant reduction in serum enzyme AST, ALT, and increase in TP.

Cajanus scarabaeoides (L.) Thouars FAMILY: FABACEAE

Pattanayak et al. (2011) investigated the hepatoprotective activity of crude flavonoids extract of *Cajanus scarabaeoides* in PCM intoxicated albino rats. Paracetamol (640 mg/kg) has enhanced the ALT, AST, ALP, TP, and TB in liver. Treatment of crude flavonoids extract of *C. scarabaeoides* L (50 mg/kg) has brought back the altered levels of biochemical markers to the near normal levels in the dose-dependent manner and which is compared with standard Silymarin (100 mg/kg). Results of histopathological studies also provided supportive evidence for biochemical analysis.

Calendula officinalis L. Family: Asteraceae

The 80% methanolic extract of *Calendula officinalis* leaves was investigated against APAP-induced hepatic damage in 30 male albino rats. Acetaminophen produce 100% mortality at dose of 1 g/kg in mice, while pretreatment of mice with *C. officinalis* (1.0 g/kg) reduced the death to 30%. Pretreatment of mice with leaves extract (500 mg/kg orally, four doses at 12 hours interval) prevented the APAP (640 mg/kg induced rise in serum transaminases (GOT, GPT), SB, and serum ALP. Post-treatment with three successive doses of leaves extract (500 mg/kg, 6 hourly)) restricted the hepatic damage induce by APAP (Ali and Khan, 2006).

C. officinalis (marigold) is a medicinal plant and cosmetic herb popularly known in Europe and the USA. The dried flower heads or the dried ligulate flowers of this plant are used for pharmaceutical and/or cosmetic purposes (Hamburger et al., 2003). Antibacterial, anti-inflammatory, antiviral, and antioxidant activities have been already noted for *C. officinalis* (Preethi et al., 2006). It has been taken in order to treat fevers and jaundice and to promote menstruation. Extracts, tinctures, balms, and salves of *C. officinalis* have been applied directly to heal wounds and soothe inflamed and injured skin. *C. officinalis* compounds, which are potentially active chemical constituents, are monoterpenes, such as α-thujene and T-muurolol, sesquiterpene and flavonol glycosides, triterpene alcohols, triterpenoid saponins, flavonoids, carotenoides, xanthophylls, phenolic acids, mucilage, bitters, phytosterols, tocopherols, calendulin, resin, and volatile oil (Okoh et al., 2008; Singh et al., 2011). The anti-inflammatory features of *C. officinalis* flowers, according to *in vivo* pharmacological tests, have been associated with the triterpenoid fatty acid esters (Hamburger et al., 2003). In Singh's study, 80% effect of methanolic extract of leaves (500 mg/kg orally, four doses at 12 h interval) of *C. officinalis* was investigated against APAP-induced hepatic damage in albino rats. The potential hepatoprotective effects of *C. officinalis* extracts against CCl_4-induced oxidative stress and cytotoxicity in isolated primary rat hepatocytes were detected (Singh et al., 2011), confirmed by a significant improvement in cell viability and enzymes leakages (ALT, AST, and LDH). Also, the reduction of hepatocytolysis and steatosis, and return to normal values of various enzymes activity could be attributed to hepatoprotective effects (Khan et al., 2011). *C. officinalis* plant extracts significantly improved cell survival, contributing greatly to preserving the cellular membranes integrity against $CC1_4$. Moreover, plant extracts of *C. officinalis* protect the intracellular antioxidant defense system, indicated by preserving GST and inhibiting LPO (Preethi and Kuttan, 2009). Protective role of the flower extract of *C. officinalis* against $CC1_4$-induced acute hepatotoxicity and cisplatin-induced nephrotoxicity has been shown (Mishra et al., 2010). Possible mechanism of action of the flower extract may be due to its antioxidant activity and reduction of oxygen radicals (Braga et al., 2009).

Calendula flower hydroalcoholic extract caused 28.5% reduction in hepatocytolysis of CCl_4-intoxicated rat liver due to reduction in glutamo-pyruvatetransaminase and glutamo-oxalate-transaminase. Histo-enzymological studies showed steatosis reduction by succinate dehydrogenase, cytochromoxidase, LDH and Mg^{2+}-dependent ATPase (Rasu et al., 2005). Calendula flower hot water extract showed antihepatoma activity (25%–26% inhibition) against five human liver cancer cells: Hep3B, SK-HEP-1, HepG2/C3A, PLC/ PRF/5, and HA22T/VGH (Lin et al., 2002). Moreover, CCl_4 intoxicated rats pretreated with Calendula floral extract afford a protection against CCl_4-induced toxicity and showed an improvement in liver function due to significant antioxidant activity and free-radical scavenging activity of bioactive metabolites including flavonoids and terpenoids present in Calendula (Maysa et al., 2015). These bioactive metabolites have potent activities for scavenging the hydroxyl radicals (OH) and superoxide radicals (O.2) resulted from CCl_4 metabolites (Maysa et al., 2015).

Calotropis gigantea R. Br. FAMILY: APOCYNACEAE

Hepatoprotective activity of *Calotropis gigantea* root bark experimental liver damage induced by D-GalN in rats was evaluated by Deshmukh et al. (2008). Alcoholic extract of root bark of the plant *C. gigantea* at an oral dose of 200 mg/kg exhibited a significant protection effect by normalizing the levels of AST, ALT, SGOT, SGPT, ALP, total bilirubin, LDH, which were significantly increased in rats by treatment with D-GalN.

Ethanolic extract (50%) of stems of *C. gigantea* at doses of 250 and 500 mg/kg were studied by Lodhi et al. (2009) for hepatoprotective activity in male Wistar rats with liver damage induced using CCl_4. Various biochemical parameters such as AST, ALT, GSH, LPO, SOD, GPx, and CAT were evaluated. The results revealed that the *C. gigantea* extract significantly decreased AST, ALT, and LPO levels. The antioxidant parameters GSH, GPx, SOD, and CAT levels were increased considerably compared to their levels in groups not treated with *C. gigantea* extract.

Usmani and Kushwaha (2010) investigated the hepaotprotective activity of leaf extracts of *C. gigantea* in CCl_4-induced hepatotoxicity. Silymarin, CE, and ME caused very significant reduction in SGPT level. The increase in CCl_4-induced SGOT level was decreased significantly by methanolic extract and Silymarin. The ALP level was reduced significantly by Silymarin, chloroform, and ME. The methanolic extract and CE-treated animals showed liver histology improved compared to diseased animals with few necrotic patches here and there with very little change.

Calotropis procera (Aiton) Dryand FAMILY: APOCYNACEAE

Padhy et al. (2007) evaluated latex of *Calotropis procera* for its hepatoprotective effect against CCl_4-induced hepatotoxicity in rats. Subcutaneous injection of CCl_4, administered twice a week, produced a marked elevation in the serum levels of AST, ALT, and tumor necrosis factor alpha (TNF-α). Histological analysis of the liver of these rats revealed marked necro-inflammatory changes that were associated with increase in the levels of TBARS, PGE (2) and CAT and decrease in the levels of GSH, SOD, and GPx. Daily oral administration of aqueous suspension of dried latex (DL) of *C. procera* at 5, 50, and 100 mg/kg doses produced a dose-dependent reduction in the serum levels of liver enzymes and inflammatory mediators and attenuated the necro-inflammatory changes in the liver. The DL treatment also normalized various biochemical parameters of oxidative stress.

Hepatoprotctive activity of 70% ethanol extract of flowers of *C. procera* was evaluated by Qureshi et al. (2007) against CCl_4-induced hepatotoxicity in albino rats and mice. Pretreatment with 70% ethanolic extract reduced the biochemical markers of hepatic injury like SGPT, SGOT, ALP, bilirubin cholesterol, cholesterol, HDL, and tissue GSH levels. Similarly pretreatment with the ethanolic extract reduced the CCl_4-induced elevation in the pentobarbitone sleeping time. Histopathological observations also revealed that pretreatment with the extract protected the animals from CCl_4-induced liver damage. Ethanolic extract demonstrated dose-dependent reduction in the *in vitro* and *in vivo* LPO-induced by CCl_4. In addition, it showed dose-dependent free-radical scavenging activity.

Hydroethanolic extract (70%) of *C. procera* flowers was prepared and tested by Setty et al. (2007) for its hepatoprotective effect against

PCM-induced hepatitis in rats. Paracetamol (2 g/kg) has enhanced the SGPT, SGOT, ALP, bilirubin, and cholesterol levels and reduced the serum levels of HDL and tissue level of GSH. Treatment with hydro-ethanolic extract of *C. procera* flowers (200 and 400 mg/kg) has brought back the altered levels of biochemical markers to the near normal levels in the dose-dependent manner.

Chavda et al. (2010) evaluated possible hepatoprotective and antioxidant potential *C. procera*. Hepatoprotective activity of the ME (MCP) and phyto-constituents directed three subfractions hexane (HCP), ethylacetate (ECP), and chloroform (CCP) of the root bark was determined using CCl_4-induced liver injury in mice. First the MCP extracts and then three subfractions namely HCP and ECP and CCP from MCP extract evaluated, at an oral dose of 200 mg/kg. The MCP and its subfractions HCP and ECP exhibited a significant hepatoprotective effect by lowering the elevated serum levels of SGOT, SGPT, ALP, total and direct SB, cholesterol and significantly increasing HDL and moderately increasing TP and ALB. While, the CCP fraction does not show significant protective effect. These biochemical observations were supplemented by histopathological examination of liver sections. Further, the effects of the active fractions on antioxidant enzymes have also been investigated to elucidate the possible mechanism of its hepatoprotective activity. The fractions exhibited a significant effect by modifying the levels of reduced GSH, super oxide dimutase, CAT activity, and MDA equivalent, an index of LPO of the liver.

Administration of 150 and 300 mg/kg BW of the ethanolic extract of *C. procera* root *did not protect the liver* and kidney from CCl_4-induced toxicity. Pretreatment with the extract rather potentiated the toxicity induced by CCl_4. It is advised strongly that caution should be taken when ingesting alcoholic preparations of *C. procera* root (Dahiru et al., 2013).

Shehab et al. (2015) compared and evaluated the antioxidant and hepatoprotective activities of four deserts plants, *Fagonia indica* Burm. f., *Calotropis procera* R. Br., *Zygophylum hamiense* Schweinf., and *Salsola imbricata* Forssk. in correlation to their composition especially their phenolic content. The flavonol quercitrin and rosmarinic acid were major in the *F. indica*, *C. procera*, and *S. imbricata* samples, while rutin prevailed in that of *Z. hamiense*. The ethanolic and methanolic extracts showed noticeable DPPH radical-scavenging activity as compared to ascorbic acid. Assessment of liver enzymes revealed that oral administration of the extracts did not show any evidence of hepatotoxicity. Moreover, protection against

CCl_4-induced liver damage was evident upon administration of three plants extracts namely, *F. indica, C. procera,* and *S. imbricata*. Overall, hepatotoxicity induced by CCl_4 was effectively prevented by the three plants extracts through scavenging of free radicals and by boosting the antioxidant capacity of the liver. The protective effect of the plants could be attributed to their high quercitrin and rosmarinic acid contents.

Calpurnia aurea (Aiton) Benth. FAMILY: FABACEAE

The effect of hydroethanolic seed extract of *Calpurnia aurea* was evaluated by Nulata et al. (2015) against HAART-induced free-radical reactions in liver and liver cell damage in Wistar male albino rats. Increased free-radical reactions, ALP, amino transferases release, and decreased antioxidant profiles were detected in HAART-treated rats. The rats treated with the extract (300 mg/kg) reduce the HAART-induced liver toxicity but minimum dose of extract (100 mg/kg) did not show any significant change against HAART altered parameters.

Camellia oleifera Abel FAMILY: THEACEAE

The oil of tea seed (*Camellia oleifera*) is used extensively in China for cooking. Loee et al. (2007) evaluated the effects of tea-seed oil on CCl_4-induced acute hepatotoxicity in male SD rats. The results showed that a tea-seed oil diet significantly lowered the serum levels of hepatic enzyme markers (ALT, AST, and LDH), inhibited fatty degeneration, reduced the content of the peroxidation product MDA, and elevated the content of GSH. Pretreatment of animals with tea-seed oil (150 g/kg diet) could increase the activities of GPx, GR, and GSH S transferase in liver when compared with CCl_4-treated group.

Camellia sinensis (L.) Kuntze FAMILY: THEACEAE

Tamoxifen (TAM) citrate, is widely used for the treatment of breast cancer. It showed a degree of hepatic carcinogenesis. El-Beshbishy (2005) and Mahaboube (2016) investigated the antioxidant capacity of green tea (*Camellia sinensis*) extract against TAM-induced liver injury. The antioxidant flavonoid; epicatechin (a component of green tea) was not detectable in liver and blood of rats in either normal control or TAM-intoxicated group; however, TAM intoxication resulted in a significant decrease of its level in liver homogenate of TAM-intoxicated rats. The model of TAM-intoxication elicited significant declines in the antioxidant enzymes (GST, GPx, SOD, and

CAT) and reduced GSH concomitant with significant elevations in TBARS and liver transaminases; SGPT and SGOT levels. The oral administration of 1.5% green tea to TAM-intoxicated rats, produced significant increments in the antioxidant enzymes and reduced GSH concomitant with significant decrements in TBARS and liver transaminases levels.

The hepatoprotective activity of the aqueous extract of *C. sinensis* has been studied against experimentally-induced liver damage in rats. The extract significantly attenuated CCl_4-induced biochemical (serum ALT, AST, ALP, TP, and ALB) and histopathological changes in liver (Sengottivelu et al., 2008). Tea decoction has been shown to possess significant antioxidant, anti-inflammatory, and immunomodulatory activities (Ratnasooriya and Fernando, 2009; Chaudhuri et al., 2007), which may contribute to its hepatoprotective activity. The antioxidant and anti-inflammatory activity of tea has been attributed to saponin contents of *C. sinensis* (Sur et al., 2001). Some cases of green-tea-induced liver toxicity have been reported (Pedrós et al., 2003; Bonkovsky et al., 2006). Phytochemical studies on aerial parts of *C. sinensis* have shown the presence of saponins, flavonoids, quercetine, quercitrin, rutin, catechin, caffeine, theophylline, and theobromine (Al-Asmari et al., 2014).

The *in vitro* and *in vivo* protective effects of water extract of pu-erh tea (WEPT) on *tert*-butyl-hydroperoxide (*t*-BHP)-induced oxidative damage in hepatocytes of HepG2 cells and in rat livers were investigated by Duh et al. (2010). After treatment with 200 μg/mL of samples, the survival rate of HepG2 cells induced by *t*-BHP increased. WEPT concentration-dependently inhibited ROS generation in HepG2 cells in response to the oxidative challenge induced by *t*-BHP. Administration of WEPT (0.2, 0.5, and 1.0 g/kg of BW) to rats for 56 consecutive days before a single dose of *t*-BHP (0.5 mmol/kg, i.p.) exhibited a significant protective effect by lowering levels of SGOT and SGPT, as well as reducing the formation of MDA. Taken together, these results demonstrate that WEPT is able to protect against hepatic damage *in vitro* and *in vivo*, suggesting that the drinking of puerh tea may protect liver tissue from oxidative damage.

Wakawa and Ira (2015) appraised the hepatoprotective effects of aqueous leaf extract of *C. sinensis* against CCl_4-induced liver injury in rats. A significant decrease was observed in both the groups treated with 100 and 200 mg/kg BW of the leaf extract and silymarin on the levels of the enzymes and nonenzyme markers of tissue damage and LPO with no significant changes in the relative organ weight of the treated groups.

Canna indica L. FAMILY: CANNACEAE

The hepatoprotective activity of ME of aerial parts of *Canna indica* plant was evaluated against CCl_4-induced hepatotoxicity. Extract at doses (100 and 200 mg/kg) restored the levels of all serum parameters like SGPT, SGOT, TB which were elevated in CCl_4 administrated rats. A 10% liver homogenate was used for estimation of CAT, GSH content, LPO level for *in vivo* antioxidant status of liver. All LPOs, reduced GSH, and CAT levels were observed normal in extract-treated rats. Histopathology demonstrated profound necrosis, lymphocytic infiltration was observed in hepatic architecture of CCl_4-treated rats which were found to obtain near normalcy in extract plus CCl_4 administrated rats (Joshi et al., 2009).

C. indica is effective against hepatic necrosis and NAPI-mediated PCM induced hepatic damage. The plant rhizome extract exerts an inhibitory effect on hepatocytes necrosis by hepatocytes regeneration, decreased serum ALT and shows anti-inflammatory activity against NAPQI-mediated PCM poisoning (Longo et al., 2015).

Cannabis sativa L. FAMILY: CANNABINACEAE

Musa et al. (2012) elucidated the hepatoprotective and toxicological effect of *Cannabis sativa* oil on liver to explore the side effect of the oil for 4 weeks for toxicological parameter and 10 days for hepatoprotective activity. The results of study showed hepatorenal lesions but not causing death. Extracts of *C. sativa* showed no signs of abnormalities and no mortalities among of spring rats. No changes in the physiological behaviors were observed throughout the experiment. The oil was tested for the hepatoprotective activity, the hepatotoxicity produced by administration of CCl_4 in paraffin oil 1:9 at a dose of 0.2 mL/kg for 10 days, was found to be inhibited by simultaneous oral administration of oil of *C. sativa* seeds at a dose 1 and 0.5 mL/kg for 10 days, with evidence of decreased level of serum AST, ALT, ALP, and bilirubin. In addition, the concurrent administration of oil with CCl_4 for 10 days masked the liver changes induced by the hepatotoxic compound observed in the control rats and comparable with the hepatoprotective effect of the standard drug Silymarin.

Canscora decussata (Roxb.) Schult. & Schult.f. FAMILY: GENTIANACEAE

Aqueous and methanolic extract of *Canscora decussata* was investigated by Akhtar et al. (2013) to know the hepatoprotective effect against CCl_4-induced liver damage in rabbits. Liver marker enzymes AST, ALT, ALP, and bilirubin were evaluated. Significant hepatoprotective effect was observed in methanolic extract while moderate activity of aqueous extract was exhibited against CCl_4 treated animals. Akhtar et al. (2015) evaluated *in vivo* hepatoprotective properties of *Canscorra decussata* whole plant methanolic extract against PCM toxicity in rabbits. Hepatoprotective activities of methanolic extracts of *C. decussata* were examined against PCM-induced liver damage in rabbits using Silymarin as control. Enzyme activities of SGOT, SGPT, and ALP, bilirubin total and direct bilirubin were analyzed. Oral administration of methanolic extract exhibited significant hepatoprotective activity.

Capparis brevispina DC. FAMILY: CAPPARACEAE

Aniyathi et al. (2009) studied the effect of the ethanol extract of the stem bark of *Capparis brevispina* against PCM-induced hepatotoxicity in Wistar rats. Significant hepatoprotective effects were obtained against liver damage induced by over dose of PCM (Acetaminophen) as evident from the decreased serum level of SGPT, SGOT, ALP, and bilirubin in the extract treated groups. The hepatoprotective effect was further confirmed by histopathological studies of the liver, which showed improved architecture, absence of nuclear pycnosis, hepatocongestion and necrosis, when compared with the liver of the toxin group of animals.

Capparis decidua (Forssk.) Edgew. FAMILY: CAPPARACEAE

The hepatotoxicity produced by administration of CCl_4 in paraffin oil (1:9 v/v) at a dose of 0.2 mL/kg for 10 days was found to be inhibited by simultaneous oral administration of aqueous and methanolic extracts of *C. decidua* stems (200, 400 mg/kg BW) for 10 days, with evidence of decreased level of serum AST, ALT, ALP, and bilirubin (Ali et al., 2009, 2011).

Rehman et al. (2017) investigated the hepatoprotective effects of aqueous-ethanolic extract of *C. decidua* (stems) against PCM-induced liver injury in experimental animals. To observe the level of improvement, biochemical parameters such as SGPT, SGOT, ALP, and total bilirubin levels as well as histopathological changes in liver tissues were studied. Silymarin (50 mg/kg, p.o.) was used as reference drug. The levels of the biochemical parameters were increased in rabbits which were intoxicated by PCM. *C. decidua* extract (750 mg/kg, BW) treated rabbits showed maximum reduction of biochemical parameters in a significant manner. Histopathological examination of the liver tissues of control and treated groups also confirmed the hepatoprotective activity. The phytochemical screening of the extracts revealed the presence of tannins, alkaloids, saponins, and flavonoids. The results of this study therefore suggest that the different doses of *C. decidua* possess significant hepatoprotective effect and this effect might be due to the presence of flavonoids and tannins.

Capparis sepiaria L. FAMILY: CAPPARACEAE

The hepatoprotective effect of the alcohol extract of *Capparis sepiaria* stem against CCl_4-induced toxicity was studied by Satyanarayana et al. (2009) in albino rats. The rats were given daily pretreatment with alcohol extract of *C. sepiaria* (100 mg/kg) and the standard Silymarin (25 mg/kg) orally for 7 days. The toxicant used on 7th day was CCl_4 at a dose of 1.25 mL/kg as 1:1 mixture with olive oil. The extract produced significant reduction in the elevated levels of AST, ALT, TB, and rise of decreased TP level when compared with the toxic control. In the histopathological studies, the liver sections of rats treated with vehicle showed normal hepatic architecture, whereas that of CCl_4 treated group showed total loss of hepatic architecture.

Capparis spinosa L. FAMILY: CAPPARACEAE

The protective action of *Capparis spinosa* ethanolic root bark extract was evaluated by Aghel et al. (2007) in an animal model of hepatotoxicity, which was induced by CCl_4. Results of the biochemical studies of blood samples of CCl_4 treated animals showed a significant increase in the levels of serum enzyme activities, reflecting the liver injury caused by CCl_4. Whereas blood samples from the animals treated with ethanolic root bark extracts showed significant decrease in the levels of serum markers, indicating the protection of hepatic cells. The results revealed that ethanolic root bark extract of *C. spinosa* could afford a significant dose-dependent protection against CCl_4-induced hepatocellular injury (Aghel et al., 2007).

Treatment of the PCM-induced liver damage in rats with aqueous extract of *C. spinosa* (25, 50, 100, 200 mg/kg of BW) for 7, 14, 21 days decreased ALT, AST activity, TB and creatinine levels in comparison with nontreated group, as well as improving the damaged liver tissues with dose-dependent manner (Alnuaimy and Al-Khan, 2012).

Tlili et al. (2017) explored the antioxidant, nephroprotective, and hepatoprotective effects of methanolic extract of *C. spinosa* leaves (MECS) associated with its phytochemical content. The levels of total phenolics, flavonoids, and condensed tannins were 23.37 mg GAE/g, 9.05 mgQE/g and 9.35 mgTAE/g, respectively. HPLC analysis revealed nine compounds, namely rutin, resveratrol, coumarin, epicatechin, luteolin, catechin, kaempferol, vanillic acid, and GA. The MECS showed interesting antioxidant capacity. MECS-treatment significantly prevented the increase in serum ALT, AST, and LDH levels in acute liver damage induced by CCl_4, decreased the amount of hepatic MDA formation and elevated the activities of SOD, CAT, and GPx, and restored liver injury.

Kalantari et al. (2018) investigated the antioxidant and hepatoprotective effects of *C. spinosa* and quercetin in *tert*-butyl hydroperoxide (*t*-BHP)-induced acute liver damage. Different fractions of *C. spinosa* were examined for TPC and antioxidant property. Among these fractions, hydroalcoholic extract was used to assess the hepatoprotective effect in *t*-BHP-induced hepatotoxicity model by determining serum biochemical markers, sleeping time and antioxidant assay such as reduced GSH as well as histopathological examination of liver tissues. The total phenolic and quercetin contents of hydroalcoholic fraction were significantly higher than other fractions. It also showed high antioxidant activity. Pretreatment with hydroalcoholic fraction at the dose of 400 mg/kg and quercetin at the dose of 20 mg/kg showed liver protection against *t*-BHP-induced hepatic injury, as it was evident by a significant decrease in serum enzymes marker, sleeping time and MDA and an increase in the GSH, SOD, and CAT activities confirmed by pathology tests. The final results ascertained the hepatoprotective and antioxidant effects of *C. spinosa* and quercetin in a dose-dependent manner. Moreover, this study suggests that possible mechanism of this protection may be associated with its property of scavenging free radicals which may be due to the presence of phenolic compounds (Kalantari et al., 2018).

***Capsella bursa-pastoris* (L.) Medik. FAMILY: BRASSICACEAE**

Capsella bursa-pastoris showed hepatoprotective activity in toxicity-induced by CCl_4 in rats. The serum levels of SGOT and bilirubin in the group of *C. bursa-pastoris* (aerial parts) crude extract treated animals showed significant decreases by 26.9% and 31.7%, respectively, at the dose of 500 mg/kg BW. The smaller dose of the extract, although it lowered the levels of all parameters, did not do so by a significant amount (Al-Snafi et al., 2019).

Caralluma adscendens (Roxb.) R.Br. var. *attenuata* (Wight) Grav. FAMILY: APOCYNACEAE

Jyothi et al. (2018) carried out investigation on *in vitro* and *in vivo* antioxidant and hepatoprotective potential of *Caralluma adscendens* var. *attenuata* against ethanol toxicity.

In vitro antioxidant activity was evaluated by determination of total phenolics, flavonoids content, reductive ability, and free-radical scavenging activity assays. *In vivo* studies were performed against ethanol-induced liver toxicity in rats using Silymarin as standard. In this study serum CAT, SOD, and reduced GSH levels were quantified. The results obtained by this study revealed that the concentration of the liver protecting enzymes was not affected by ethanol when animals were co-administered with plant extract. Rats which were pretreated with methanolic extract of *C. adscendens* var. *attenuata* inhibited the stimulated serum levels of SGPT, SGOT, ALP and TB when compared with Silymarin. The research findings reveal that *C. adscendens* var. *attenuata* have significant antioxidant and hepatoprotective activities (Jyothi et al., 2018).

Cardiospermum halicacabum L. FAMILY: SAPINDACEAE

Arjuman et al. (2009) evaluated the hepatoprotective activity of *Cardiospermum halicacabum* against CCl_4-induced liver toxicity. In stem extract treated animals there was a decrease in serum levels of the markers and significant increase in TP, indicating the recovery of cells. These biochemical observations were supplemented by histological examination of liver section. The ethyl acetate stem extract (400 mg/kg) of *C. halicacbum* afforded significant protection against CCl_4-induced hepatocellular injury compared to all extracts.

Carduus acanthoides L. FAMILY: ASTERACEAE

Aktay et al. (2000) studied the hepatoprotective activity of ethanolic extracts of *Carduus acanthoides* using CCl_4-induced heptotoxicity model in rats. According to the results of biochemical tests, significant reductions were obtained in CCl_4-induced increases of plasma and liver tissue MDA levels and plasma enzyme activities by the treatment with extracts, which reflects functional improvement of hepatocytes. Necrosis that is a more severe form of injury was also markedly prevented by these extracts.

Carduus nutans L. FAMILY: ASTERACEAE

Aktay et al. (2000) studied the hepatoprotective activity of ethanolic extracts of *Carduus nutans* using CCl_4-induced heptotoxicity model in rats. According to the results of biochemical tests, significant reductions were obtained in CCl_4-induced increases of plasma and liver tissue MDA levels and plasma enzyme activities by the treatment with *C. nutans* extracts, which reflects functional improvement of hepatocytes. In spite of a significant inhibitory effect of *C. nutans* extract on biochemical parameters, this *was not supported* by histopathological examination. Moreover, extensive bleeding, mainly originating from endothelial damage of the vessels, was observed in the livers of experimental animals administered with *C. nutans* extract. *Therefore, this situation may suggest that some components of the plant induce such damage.*

Careya arborea Roxb. FAMILY: MYRTACEAE

Kumar et al. (2005) evaluated the hepatoprotective and antioxidant effect of the ME of *Careya arborea* stem bark in Wistar albino rats. The different groups of animals were administered with CCl_4. The ME of *C. arborea* and Silymarin produced significant hepatoprotective effect by decreasing the activity of serum enzymes, bilirubin, uric acid, and LPO and significantly increased the levels of SOD, CAT, GSH, vitamin C, vitamin E, and protein in a dose-dependent manner. From these results, it was suggested that MECA possess potent hepatoprotective and antioxidant properties.

Islam et al. (2018) evaluated the hepatoprotective effect of ethanolic extract of *C. arborea* bark on SD rats induced by PCM. The increased serum levels of hepatic marker enzymes were found in the PCM treated group, indicating the severity of hepatocellular damage induced by PCM. Treatment with CA as well as standard hepatoprotective agent Silymarin attenuated the increased levels of these hepatic enzymes. Body weight was improved insignificantly by CA, whereas liver weight was recovered significantly. ALT, AST, ALP, as well as bilirubin levels were improved very highly significantly by CA at 500 mg/kg dose. Also, the TP and ALB levels were increased significantly at the dose of 500 mg/kg. Silymarin produced very highly significant effect at 100 mg/kg dose.

Carica papaya L. FAMILY: CARICACEAE

Ethanol and aqueous extracts of *Carica papaya* has been evaluated by Rajkapoor et al. (2002) for its antihepatotoxic activity. The ethanol and aqueous extracts of *C. papaya* showed remarkable hepatoprotective activity against CCl_4-induced hepatotoxicity. The activity was evaluated by using biochemical parameters such as serum AST, ALT, ALP, TB and gamma GGTP. The histopathological changes of liver sample was compared with respect to control.

Adenye et al. (2009) reported the dose-dependent (100–400 mg/kg/day/oral route) and time-course protective effects of the 400 mg/kg/oral route of the aqueous seed extract of unripe and mature *C. papaya* fruit (CPE) in CCl_4 hepatotoxic rats. Results showed the extract to cause significant dose-related attenuation in the elevation of serum liver enzyme markers of acute hepatocellular injury (ALT, AST), serum lipids (TG, TC, HDL-c, LDL-c, and VLDL-c) and serum proteins (TP and ALB). Maximum hepatoprotection was offered at an oral dose of 400 mg/kg/day of the extract. The biochemical results obtained were corroborated by improvements in the CCl_4-induced hepatic histological changes. In addition, maximum hepatoprotection was offered at the 400 mg/kg of CPE for up to 3 hours post-CCl_4 induction.

Kantam (2009) reported that pretreatment with medium and high doses of *C. papaya* extracts such as 250 mg and 500 mg/kg p.o. significantly reversed the elevated serum enzyme markers in animals treated with TAA.

The aqueous and ethanol extract of dried fruit of *C. papaya* was evaluated for its hepatoprotective activity in rats against CCl_4-induced hepatotoxicity in rats. The aqueous (250 mg/kg, p.o.) and ethanol (250 mg/kg, p.o.) extracts of *C. papaya* showed significant hepatoprotection by lowering the biochemical parameters such as SGPT, SGOT, SB, and alkaline phosphatase (Sadeque et al., 2010).

A study was performed to evaluate the hepatoprotective effects of *C. papaya* against CCl_4-induced hepatotoxicity and compared it with that of vitamin E and results confirmed that *C. papaya* and vitamin E showed significant hepatoprotection against CCl_4-induced hepatotoxicity, but *C. papaya* showed more significant changes in ALP level than vitamin E (Sadeque et al., 2012). Pandit et al. (2013) examined the hepatoprotective effect of *C. papaya* leaves against ethanol, and antitubercular drug-induced liver damage and results revealed that hepatoprotective activity was evident by the significant reduction in the levels of all serum markers in both models.

Awodele et al. (2016) investigated the hepatomodulatory effects of aqueous extracts of *C. papaya* leaf (CPL) and unripe fruit (CPF) at doses of 100 and 300 mg/kg on CCl_4 and APAP (ACM)-induced liver toxicities in rats. There was a significant reduction in CCl_4 and ACM-induced increases in serum levels of ALT, AST, ALP, and DBL at 100 and 300 mg/kg, respectively. The levels of CAT, SOD, and reduced GSH were decreased in both models with corresponding significantly elevated level of MDA. However, these antioxidant enzymes were significantly increased in CPL and CPF-treated rats. Histopathological assessment of the liver confirmed the protective effects of CPL and CPF on CCl_4 and ACM-induced hepatic damage evidenced by the normal presentation of liver tissue architecture.

Carissa carandas L. FAMILY: APOCYNACEAE

Hegde and Joshi (2009) demonstrated significant hepatoprotective activity of ethanolic extract of the roots of *Carissa carandas* (ERCC; 100, 200, and 400 mg/kg, p.o.) against CCl_4 and PCM-induced hepatotoxicity by decreasing the activities of serum marker enzymes, bilirubin and LPO, and significant increase in the levels of uric acid, GSH, SOD, CAT, and protein in a dose-dependent manner that was confirmed by the decrease in the total weight of the liver and histopathological examination. While, Bhaskar and Balakrishnan (2009) reported hepatoprotective effects of the ethanol, and aqueous extracts of roots of *C. carandas* against ethanol-induced hepatotoxicity in rats. The ethanol and aqueous extracts at a dose level of 100 and 200 mg/kg produce significant hepatoprotection by decreasing serum transaminase (SGPT and SGOT), ALP, bilirubin and LPO, while significantly increased the levels of liver GSH, and serum protein.

The roots of *C. carandas* are used by the tribes of Western Ghats, Tamil Nadu, for the treatment of various liver disorders. Balakrishnan et al. (2011) studied the effect of EAF of the ethanol extract from roots of *Carissa carandas* against CCl_4-, PCM-, and ethanol-induced hepatotoxicity in rats. Significant hepatoprotective effects were obtained against liver damage induced by all the three toxins, as evident from changed biochemical parameters like serum transaminases (SGOT and SGPT), ALP, TB, TP, and TC. Parallel to these changes, the EAF prevented toxin-induced oxidative stress by significantly maintaining the levels of reduced GSH and MDA, and a normal architecture of the liver, compared to toxin controls.

Carissa opaca Stapf ex Haines FAMILY: APOCYNACEAE

The effect produced by methanolic extract of *Carissa opaca* leaves was investigated by Sahreen et al. (2011) on CCl_4-induced liver damages in rat. Hepatotoxicity induced with CCl_4 was evidenced by significant increase in LPO (TBARS) and H_2O_2 level, serum activities of AST, ALT, ALP, LDH, and γ-GT. Level of GSH determined in liver was significantly reduced, as were the activities of antioxidant enzymes; CAT, peroxidase (POD), SOD, GSH-Px, GSR, GST, and QR. On cirrhotic animals treated with CCl_4, histological studies showed centrilobular necrosis and infiltration of lymphocytes. Methanolic extract (200 mg/kg BW) and Silymarin (50 mg/kg BW) co-treatment prevented all the changes observed with CCl_4-treated rats. The phytochemical analysis of methanolic extract indicated the presence of flavonoids, tannins, alkaloids, phlobatannins, terpenoids, coumarins, anthraquinones, and cardiac glycosides. Isoquercetin, hyperoside, vitexin, myricetin, and kaempherol were determined in the methanolic extract.

Carissa spinarum L. FAMILY: APOCYNACEAE

Hegde and Joshi (2010) evaluated ethanolic extracts of *C. spinarum* roots in CCl_4-induced as well as PCM-induced hepatotoxicity. The significant elevation in the levels of serum marker enzymes in control group such as SGOT, SGPT, and SALP content of CCl_4/PCM shows degree of hepatotoxicity. Animals pretreated with extract (100, 200, and 400 mg/kg) as well as a standard drug (Silymarin) demonstrated significant hepatoprotection by decreasing serum marker enzymes in a dose-dependent manner (Hegde and Joshi, 2010).

Carthamus oxyacantha M. Bieb. FAMILY: ASTERACEAE

Bukhsh et al. (2014) investigated the protective and curative effects of alcoholic extract of *Carthamus oxyacantha* seeds against CCl_4-induced hepatic damage in rats (SD strain). In protective studies, plant extract (400 mg/kg BW) was given before hepatic damage while in curative studies; hepatic damage was induced before the application of plant extract. Results show that alcoholic extract of seeds of *C. oxycantha* possessed both hepatoprotective and hepatocurative activity. However, hepatoprotective activity was more pronounced as compared to hepatocurative activity. Histopathological studies also supported the biochemical parameters.

Carthamus tinctorious L. FAMILY: ASTERACEAE

Paramesha et al. (2011) evaluated the hepatoprotective and *in vitro* antioxidant effect of methanolic extract and its isolated constituent, dehydroabietylamine, in *Carthamus tinctorious*, var Annigeri-2-, an oil yielding crop. Both the methanolic extract (at 150 and 300 mg/kg BW) and dehydroabietylamine (at 50 mg/kg BW) showed significant liver protection against CCl_4-induced liver damage that was comparable with the standard drug, Silymarin (100 mg/kg BW), in reducing the elevated serum enzyme markers. The liver sections of the animals treated with dehydroabietylamine elicit a significant liver protection compared with the methanolic extract against CCl_4-induced liver damage. Further, both the methanolic extract and dehydroabietylamine exhibited a considerable and dose-dependent scavenging activity of DPPH, NO, and hydroxyl radical. Similarly, in the reducing power assay, the results were very persuasive. In addition, the Fe^{2+}chelating activity and the total antioxidant assay established the antioxidant property of the methanolic extract and its isolated constituent. Among the two experimental samples, dehydroabietylamine proved to be more effective for the said parameters.

The hepotaprotective activity of methanolic extract of flowers of *C. tinctorius* was investigated in CCl_4-induced liver injury in rats by Yar et al. (2012). CCl_4-treated rats showed a significant elevation in the serum level of AST, ALT, ALP, TB (TSB), and also elevation in tissue MDA content, serum TNF-α, and interleukin-6 (IL-6) with reduction of tissue GSH content, serum SOD, and CAT. Pretreatment of rats with *C. tinctorius* extract for 30 successive days prior to CCl_4 administration produced a significant reduction in the level of biochemical markers, aminotransferases, TSB, LPO product, and inflammatory cytokines with elevation in enzymatic and nonenzymatic antioxidants compared with CCl_4-treated rats. Histopathological changes in liver samples were compared with control and CCl_4-treated group.

Carum copticum Benth. FAMILY: APIACEAE

Gilani et al. (2005) described the hepatoprotective activity of the aqueous-methanolic extract of *Carum copticum* seeds (CSE) to rationalize some of its traditional uses. Pretreatment of rats with CSE (500 mg/kg orally for 2 days at 12 h intervals) prevented PCM (640 mg/kg) and CCl_4 (150 mL/kg)-induced rise in serum ALP, AST, and ALT. The same dose of CSE was able to prevent the CCl_4-induced prolongation in pentobarbital-induced sleeping time in mice confirming its hepatoprotectivity.

Cassia auriculata L. FAMILY: FABACEAE

MeOH extract from *Cassia auriculata* root has been evaluated by Jaydevkar et al. (2014) as a hepatoprotective agent against damage induced with EtOH 40% administered for 21 days or with a mixture of INH (27 mg/kg), RIF (54 mg/kg), and PZA (135 mg/kg) administered for 21 days in Wistar rats. This extract was administered at 300 and 600 mg/kg by oral via during 30 days in both assays. It was observed that the *C. auriculata* extract reduced AST, ALT, ALP, TB, and TC levels and stimulated the activity of the endogenous enzymatic systems, such as CAT, GPx, and SOD. The authors concluded that the hepatoprotective effect is due to maintaining the integrity of the hepatocyte cellular membrane, reducing the concentration of liver enzymes in serum and increasing the endogenous antioxidant effect of the organisms. These latter results were confirmed with histopathological studies (Jaydevkar et al., 2014).

Cassia fistula L. FAMILY: FABACEAE

Hepatoprotective activity of the *n*-heptane extract of *Cassia fistula* leaves was investigated by Bhakta et al. (1999, 2001) in rats by inducing hepatotoxicity with CCl_4:liquid paraffin (1:1) and PCM, respectively. The extract has been shown to possess significant protective effect by lowering the serum levels of transminases (SGOT and SGPT), bilirubin, and ALP. The extract of *C. fistula* at a dose of 400 mg/kg showed significant hepatoprotective activity which was comparable to that of a standard hepatoprotective agent. Ilavaran et al. (2001) also reported hepatoprotective activity of crude extract of *C. fistula* against CCl_4-induced hepatotoxicity.

Pradeep et al. (2005) evaluated the hepatoprotective activity of *C. fistula* leaf extract against subacute CCl_4 hepatotoxicity. Pretreatment of ethanolic leaf extract of *C. fistula* (500 mg/kg BW/d for 7 days) followed by CCl_4 treatment (0.1 mL/100 g BW from day 8 till day 14) completely reversed back LPO and the activities of CAT and GR in the liver tissue toward normalcy. This treatment also reversed the elevated levels of the enzymes in the serum. Ethanolic leaf extract alone treatment did not produce any change in all the parameters studied.

Chaudhari et al. (2009) evaluated the hepatoprotective activity of methanolic extract of *C. fistula* seeds against PCM-induced hepatic injury in rats. Alteration in the level of biochemical markers of hepatic damage like SGOT, SGPT, ALP, and billirubin were tested in both treated and untreated groups. Paracetamol (2 g/kg) has enhanced the SGPT, SGOT, ALP, and billirubin level reduced. Treatment with methanolic extract of *C. fistula* seeds (200 and 400 mg/kg) has brought back the altered level of biochemical markers to the near normal levels in the dose-dependent manner.

Jehangir et al. (2010) evaluated the hepatoprotective effect of *C. fistula* leaves in isoniazid and rifampicin-induced hepatotoxicity in rats. The antituberculosis group rats showed variable increase in serum ALT, AST, ALP, and TB levels. Groups treated with 400 mg/kg of BW *C. fistula* treatment decreased the level of these parameters in rats. However, group D rats treated with 500 mg/kg BW of extract of *C. fistula* dose significantly decreased levels of these biochemical parameters. The morphological examination of experimental group C rats showed slight recovery whereas the rats in experimental group D showed a significant recovery.

Hepatoprotective effect of *C. fistula* fruit extract was investigated in mice by Kalantari et al. (2011). Bromobenzene treatment alone elicited a significant increase in activities of AST, ALT, ALP (but not γGT), and it significantly elevated the levels of direct and TB. Co-treatment with *C. fistula* fruit extract, however, significantly and dose dependently decreased the above-mentioned enzyme activities (with exception of γGT) and bilirubin levels, producing a recovery to the naive state. The protective effect of *C. fistula* fruit extract against liver injury evoked by BB was confirmed by histological examination as well.

The effect of *C. fistula* seeds on subchronic hepatic toxicity in chicks fed with 30–50 mg/kg methanolic extract of the fruit seeds, for 7–21 days, was investigated by Iqbal et al. (2016). At 40 mg/kg dose of *C. fistula* to chicks showed a nonsignificant decrease in count of RBCs and other serological values, presumably through a provoked hemolysis of RBCs and/or toxicity of the plant. However, a significant reduction in liver function enzymes (ALP, ALT, and AST) urea, and creatinine levels with an increase in plasma protein reflected on its hepato-renal protection. Only nominal pathological lesions were observed in microscopic examination of the chick's hepatic tissues.

Chaerunnisa et al. (2018) evaluated the heptoprotective activity of the ethanol extract of *C. fistula* barks against hepatotoxicity induced by PCM on rats. Ethanol extract at doses of 150 and 300 mg/kg BW gave protective effect with SGPT levels 60.83 IU/L and 56.95 IU/L, respectively, and SGOT levels of 134.30 IU/L and 110.17 IU/L, respectively, significantly different from the controls. The extract had radical scavenging activity.

Cassia italica (Mill.) Spreng. FAMILY: FABACEAE

Aqueous and ethanolic extracts of *Cassia italica* leaf were tested by Modibbo and Nadro (2012) and Nadro and Onoagbe (2014) for hepatoprotective activity against CCl_4 hepatotoxicity. Pretreatment with 200 mg/kg BW of *C. italica* leaf extracts protected rats against CCl_4-induced liver damage. The serum and liver marker enzymes (AST, ALT, ALP, and GGT) were significantly lowered. Total bilirubin, cholesterol, TPs, and LPO when compared to CCl_4 control were significantly reduced. The data obtained suggests the beneficial role of *C. italica* leaf as a protective agent of liver damage or injury. These results suggest that the anti-jaundice effect of *C. italica* leaf may be attributed at least in part to the improved liver function.

Cassia obtusifolia L. FAMILY: FABACEAE

Xie et al. (2012) investigated the hepatoprotective effects of cassia seed ethanol (CSE) extract in CCl_4-induced liver injury in mice. Acute CCl_4 administration caused great lesion to the liver, shown by the elevation of the serum aminotransferase activities, mitochondria membrane permeability transition, and the ballooning degeneration of hepatocytes. However, these adverse effects were all significantly inhibited by CSE pretreatment. CCl_4-induced decrease of the CYP2E1 activity was dose dependently inhibited by CSE pretreatment. Furthermore, CSE dramatically decreased the hepatic and mitochondrial MDA levels, increased the hepatic and mitochondrial GSH levels, and restored the activities of SOD, GR, and GST. These results suggested that CSE could protect mice against CCl_4-induced liver injury via enhancement of the antioxidant capacity.

Meng et al. (2019) evaluated the hepatoprotective effects of *Cassia obtusifolia* seeds on nonalcoholic fatty liver disease (NAFLD). Twelve weeks of high fat diet administration significantly increased the levels of AST, ALT, TG, TC, TNF-α, IL-6, IL-8, and MDA, decreased SOD (199.42 vs. 137.70 U/mg protein) and GSH (9.76 vs. 4.55 mg/g protein) contents, compared to control group. *C. obtusfolia* administration group significantly decreased the elevated biomarkers with the ED_{50} = 1.2 g/kg for NAFLD rats. *C. obtusifolia* treatment also prevented the decreased expression of LDL-R mRNA, and improved the histopathological changes compared to model group.

Cassia occidentalis L. FAMILY: FABACEAE

The hepatoprotective effect of aqueous-ethanolic extract (50%, v/v) of leaves of *Cassia occidentalis* was studied by Jafri et al. (1999) on rat liver damage induced by PCM and ethyl alcohol by monitoring serum transaminase (AST and ALT), ALP, serum cholesterol, serum total lipids, and histopathological alterations. The extract of leaves of the plant produced significant hepatoprotection.

Uzzi and Grillo (2013) investigated the hepatoprotective potentials of aqueous leaf extract of *C. occidentalis* on PCM-induced hepatotoxicity in adult Wistar rats. Twenty adult rats weighing between 150 and 300 g were used for this study. They were randomly divided into four groups (A, B, C, and D), whereby group A served as the control, while groups B, C and D served as test groups. Hepatotoxicity was induced in the test groups via oral administration of PCM (800 mg/kg BW). However, while groups C and D were treated for 21 days with 250 and 500 mg/kg/BW of *C. occidentalis* leaf extract, respectively, group B was left untreated and served as the test control. Microscopy revealed normal histological hepatocytes in the control animals while those of test control were severe vascular congestion, periportal infiltrates of chronic inflammatory cells and periportal oedema. However, hepatic sections from groups C and D presented a dose-dependent healing actions compared to the features observed for group B (untreated hepatotoxic group).

Cassia roxburghii DC. FAMILY: FABACEAE

Seeds of *Cassia roxburghii* had been used in ethnomedicine for various liver disorders for its hepatoprotective activity. Arulkumaran et al. (2009) screened the hepatoprotective activity of seeds of *C. roxburghii*. The methanolic extract of seeds of *C. roxburghii* reversed the toxicity produced by ethanol and CCl_4 combination in dose-dependent manner in rats. The extract at the doses of 250 and 500 mg/kg are comparable to the effect produced by Liv. 52®, a well-established plants based hepatoprotective formulation against hepatotoxins.

Cassia singueana Delile FAMILY: FABACEAE

Ottu et al. (2013) evaluated *in vivo* antioxidant and hepatoprotective activities of the methanolic extract of the root of *Cassia singueana* in rats following acute and chronic CCl_4 intoxication. The liver, kidney, and heart showed significant reduction in the levels of MDA in the CCl_4 control protein in groups pretreated with the extract for three days at 5 mg/kg. Similarly, compared to the CCl_4 control, significant reduction in serum AST, ALT, and bilirubin as well as in level of TC and MDA with concomitant increase in

HDL cholesterol, SOD, and CAT levels when CCl_4-intoxicated rats were treated with *C. singueana* root extract for 2 weeks.

Cassia sieberiana DC. FAMILY: FABACEAE

Madubuike et al. (2015) evaluated the hepatoprotective and antioxidant activity of the methanolic extract of *Cassia sieberiana* leaves. Phytochemical analysis revealed the presence of saponins, alkaloids, flavonoids, terpene/sterol, glycosides, and tannins. The extract (100, 200, and 400 mg/kg) and Silymarin (100 mg/kg) produced a significant dose-dependent increase in ALT, AST, and ALP levels in serum of treated rats, when compared with the negative control group. The extract (25–400 μg/mL concentration) produced a concentration-dependent increase in antioxidant activity in DPPH photometric assay. The IC 50 of the extract in DPPH photometric assay was 50 μg/mL concentrations. The extract and Silymarin showed a significant dose dependent increase in catalase level in treated rats, when compared with the negative control group. Also, the extract (400 mg/kg) and Silymarin (100 mg/kg) produced a significant decrease in MDA level in treated rats, when compared with the negative control group.

Cassia sophera L. FAMILY: FABACEAE

Hepatoprotective activity of ethanolic extracts of *Cassia sophera* leaves was evaluated by Mondal et al. (2012) against CCl_4-induced hepatic damage in rats. The extracts at doses of 200 and 400 mg/kg were administered orally once daily. The hepatoprotection was assessed in terms of reduction in histological damage, changes in serum enzymes, SGOP, AST, ALT, serum ALP, TB, and TP levels. The substantially elevated serum enzymatic levels of AST, ALT, ALP, and TB were restored toward the normalization significantly by the extracts. The decreased serum TP level was significantly normalized. The biochemical observations were supplemented with histopathological examination of rat liver sections. The results of this study strongly indicate that *C. sophera* leaves have potent hepatoprotective action against CCl_4-induced hepatic damage in rats.

Cassia surattensis Burm. f. FAMILY: FABACEAE

The therapeutic potential of *Cassia surattensis* in reducing free-radical-induced oxidative stress and inflammation particularly in hepatic diseases was evaluated by Kumar et al. (2016). The polyphenol rich *C. surattensis* seed extract showed good *in vitro* antioxidant. *C. surattensis* seed extract contained TPC of 100.99 mg GAE/g dry weight and there was a positive correlation between TPC and the antioxidant activities of the seed extract. *C. surattensis* seed extract significantly reduced the elevated levels of serum liver enzymes (ALT, AST, and ALP) and relative liver weight in PCM-induced liver hepatotoxicity in mice. Moreover, the extract significantly enhanced the antioxidant enzymes and GSH contents in the liver tissues, which led to decrease of MDA level. The histopathological examination showed the liver protective effect of *C. surattensis* seed extract against PCM-induced histoarchitectural alterations by maximum recovery in the histoarchitecture of the liver tissue. Furthermore, histopathological observations correspondingly supported the biochemical assay outcome, that is, the significant reduction in elevated levels of serum liver enzymes.

Cassia tora L. FAMILY: FABACEAE

Rajan et al. (2009) determined the hepatoprotective effects of *Cassia tora* against CCl_4-induced liver damage in albino rats. The efficacy of the treatment was estimated by the serum level marker enzymes: SGOT, SGPT, and LDH. The treatment also includes the estimation of enzymatic antioxidants: SOD, GPx, GSH-S transferase, and CAT; nonenzymatic antioxidants: vitamin C and vitamin E. The results of this study revealed the remarkable increase of marker enzymes in induced rats and decreased level in *C. tora* treated ones. Furthermore, the level of enzymatic and nonenzymatic antioxidant level were elevated in treated rats compared to induced ones.

Saravanan and Malarvannan (2016) investigated the hepatoprotective activity of *C. tora* on CCl_4-induced hepatotoxicity in rats. CCl_4-treated rats showed a significant elevation in the serum activities of ALT, AST, and acid phosphatase while significantly decreasing the levels of TP and ALB as compared to the normal control rats, thereby indicating liver damage. Administration of *C. tora* leaf at doses of 500 mg/kg, significantly prevented the rise in the levels of the marker enzymes as well as it significantly prevented the decrease in the serum levels of TP and ALB. The diminished rise of serum enzymes, together with the diminished fall in the levels of TP and ALB in the extract treated groups, is a clear manifestation of the hepatoprotective effect of the extract.

Cassytha filiformis L. FAMILY: LAURACEAE

Raj et al. (2013) evaluated the hepatoprotective and antioxidant activity of *Cassytha filiformis* methanolic on the CCl_4-induced hepatotoxicity.

Phytochemical investigation showed that methanolic extract contains poly phenolic compounds, tannins, flavonoids, alkaloids and saponins. Acute toxicity study showed that methanolic extract was safe up to 5000 mg/kg BW. Compared to CCl_4 toxicant group wet liver weight, wet liver volume was markedly reduced in methanolic extract treated group. The toxicant induced a rise in the plasma enzyme levels of ALT, SGOT, SGPT, and TB level. This increased level was significantly lowered by the extract at 500 than 250 mg/kg BW. The histopathological changes, that is, fatty changes, necrosis, etc., were partly or fully prevented in animals treated with the extracts.

Castanea crenata Siebold & Zucc. FAMILY: FAGACEAE

The antioxidant effects of chestnut (*Castanea crenata*) inner shell extract (CISE) were investigated by Noh et al. (2010) in a *tert*-butylhydroperoxide (*t*-BHP)-treated HepG2 cells, and in mice that were administered CCl_4 and fed a high-fat diet. Preincubation with CISE significantly blocked the oxidative stress induced by *t*-BHP treatment in HepG2 cells and preserved the activities of CAT, SOD, GPx, and GR compared to group treated with *t*-BHP only. Similarly, the CCl_4- and HFD-induced reduction of antioxidant enzymes activities in liver was prevented by CISE treatment compared to control groups. Furthermore, hepatic LPO were remarkably lower in the CISE-treated groups with *t*-BHP or HFD. To determine the active compound of CISE, the fractionation of CISE has been conducted and scoparone and scopoletin were identified as main compounds. These compounds were also shown to inhibit the *t*-BHP-induced ROS generation and reduction in antioxidant enzyme activity in an *in vitro* model system. From these results, it was demonstrated that CISE has the ability to protect against damage from oxidative stressors such as *t*-BHP, CCl_4, and HFD in *in vitro* and *in vivo* models.

Castanopsis indica (Roxb. ex Lindl.) A.DC. FAMILY: FAGACEAE

Dolai et al. (2012) investigated the free-radical scavenging activity of ME of *Castanopsis indica* in mediating hepatoprotective activity of CCl_4 intoxicated rats. In case of hepatoprotective evaluation the levels of liver enzymatic, nonenzymatic systems (SGOT, SGPT, ALP), TB, TP, CAT, reduce GSH, SOD, and LPO were restored toward the normal value in *C. indica* treated CCl_4 intoxicated rats. The free-radical scavenging and antioxidant activities may be attributed to the presence of phenolic compound, which was (318.70±1.39) μg in 1000 μg of *C. indica* extract.

Catha edulis (Vahl) Endl. FAMILY: CELASTRACEAE

Al-Mehdar et al. (2012) evaluated the potential hepatic toxicity and the pro-oxidant effect of khat *(Catha edulis)* alone (2 g/kg BW) or in combination with CCl_4 (100 mg/kg twice a 4 weeks) in albino rats and to evaluate the possible protection effects of α-lipoic acid (100 mg/kg BW) or vitamin E (200 mg/kg BW). The results showed significant elevations in ALT, ALP, total and DBL and LPO (MDA) and significant reduction of food intake, BW, ALB, reduced GSH, CAT, and TAC in serum, liver, and kidney tissues of rats treated with khat compared to the control animals indicated the hepatotoxic and pro-oxidant effect of khat. The results also showed that co-administration of khat with CCl_4 exacerbate the toxic effects, complications of hepatic injury and oxidative stress. α-lipoic acid and vitamin-E-induced protective effects against khat and CCl_4-induced oxidative hepatotoxicity as indicated by significant improvement in BW, food intake, serum biochemical parameters, oxidative stress markers and the histological and histochemical pictures of liver. It could be concluded that both α-lipoicacid and vitamin E are important for people chewing khat to protect against the hepatotoxicity and oxidative stress.

***Catunaregam spinosa* (Thunb.) Tirveng. [Syn.: *Randia dumetorum* (Retz.) Lam.] FAMILY: RUBAICEAE**

Kandimalla et al. (2016) evaluated the antioxidant and hepatoprotective activity of leaf and bark of *Catunareguam spinosa* (Syn.: *Randia dumetorum*) against CCl_4-induced hepatic damage in male Wistar rats. CCl_4 administration-induced hepatic damage in rats resulted in increased levels of AST, ALT, ALP, LDH, TBARS, ALB, bilirubin, TNF-α, IL-1β, and decreased levels of TP and antioxidant enzymes like SOD, CAT, and GR. Leaf and bark MEs pretreatment exhibited protection against CCl_4-induced hepatotoxicity by reversing all the abnormal parameters to significant levels. Histopathological results revealed that *C. spinosa* leaf and bark extracts at 400 mg/kg protects the liver from damage induced by CCl_4.

***Cedrelopsis grevei* Baill. & Courchet FAMILY: MELIACEAE**

Mossa et al. (2015) evaluated the antioxidant and hepatoprotective effects of *Cedrelopsis grevei* leaves against cypermethrin (Cyp)-induced oxidative stress and liver damage in male mice. The antioxidant activity of *C. grevei* methanolic extract was the highest with an IC_{50} <225 μg/mL by DPPH assay. The high dose of methanolic extract (300 mg/kg BW) was effective to attenuate the perturbations in the tested enzymes. Histopathological examination in the liver tissue of those mice, demonstrated that a co-administration of methanolic extract (150 and 300 mg/kg/d) showed marked improvement in its histological structure in comparison to Cyp-treated group alone and represented by nil to moderate degree in inflammatory cells.

Ceiba pentandra (L.) Gaertn. FAMILY: MALVACEAE

Bairwa et al. (2010) reported protective activity of EAF of ME of stem bark of *Ceiba pentandra* against PCM-induced liver damage in rats. A significant reduction in serum enzymes SGOT, SGPT, ALT, AST, ALP, TB content, and histopathological screening in the rats treated gave indication that EAF of methanolic extract of *C. pentandra* possesses hepatoprotective potential against PCM-induced hepatotoxicity in rats.

Methanolic extract of *C. pentandra* was evaluated by Gandhare et al. (2012) for hepatoprotective and antioxidant activities in rats. The plant extract (200 and 400 mg/kg, p.o.) showed a remarkable hepatoprotective and antioxidant activity against TAA-induced hepatotoxicity as judged from the serum marker enzymes and antioxidant levels in liver tissues. Treatment of rats with different doses of plant extract (200 and 400 mg/kg) significantly altered serum marker enzymes and antioxidant levels to near normal against APAP-treated rats. Also the extract was effectively altered the drug metabolizing enzymes such as Cytochrome P450, NADPH Cytochrome C reductase, and GSH *S* transferase. The activity of the extract at dose of 400 mg/kg was comparable to the standard drug, Silymarin (50 mg/kg, p.o.). Histopathological changes of liver sample were compared with respective control.

Centella asiatica (L.) Urban FAMILY: APIACEAE

Antony et al. (2006) evaluated the hepatoprotective effects of the *Centella asiatica* extract in CCl_4-induced liver injury in SD rats. CCl_4-induced hepatotoxic effects were evident by a significant increase in the serum marker enzymes and a decrease in the total serum protein and ALB. Administration of extract of *C. asiatica* effectively inhibited these changes in a dose-dependent manner; maximum effect was with 40 mg/kg. Histopathological examination of liver tissue corroborated well with the biochemical changes. Hepatic steatosis, hydropic degeneration, and necrosis were observed in CCl_4-treated group, while these were completely absent in the treatment group. This effect is attributed to the presence of asiaticoside (14.5%) in the extract.

Choi et al. (2016) investigated the protective effects of *Centella asiatica* on DMN-induced liver injury in rats. *C. asiatica* significantly decreased the relative liver weights in the DMN-induced liver injury group, compared with the control. The assessment of liver histology showed that *C. asiatica* significantly alleviated mass periportal ± bridging necrosis, intralobular degeneration, and focal necrosis, with fibrosis of liver tissues. Additionally, *C. asiatica* significantly decreased the level of MDA, significantly increased the levels of antioxidant enzymes, including SOD, GPx, and CAT, and may have provided protection against the deleterious effects of ROS. In addition, *C. asiatica* significantly decreased inflammatory mediators, including IL-1β, IL-2, IL-6, IL-10, IL-12, TNF-α, interferon-γ, and granulocyte/macrophage colony-stimulating factor. These results suggested that *C. asiatica* had hepatoprotective effects through increasing the levels of antioxidant enzymes and reducing the levels of inflammatory mediators in rats with DMN-induced liver injury.

Sivakumar et al. (2018) evaluated the hepatoprotective potential of *C. asiatica* against PCM-induced hepatotoxicity in experimental rats. Treatment with *C. asiatica* significantly prevented the drug-induced increase in serum levels of hepatic enzymes. Furthermore, *C. asiatica* significantly reduced the LPO in the liver tissue and restored the activities of defense antioxidant enzymes SOD, CAT, and GSH toward normal. Histopathology of liver tissue showed that *C. asiatica* attenuated the hepatocellular necrosis, regeneration, and repair of cells toward normal.

Ceratonia siliqua L. Family: Fabaceae

Hsouna et al. (2011) evaluated the hepatoprotective effect of the leaf extract of *Ceratonia siliqua*. Among the tested extracts, the EAF exhibited the highest total phenolic and flavonoids content. The antioxidant activity *in vitro* systems showed a more significant potent free-radical scavenging activity of this extract than other analysis fractions. The HPLC finger print of EAFs active extract showed the presence of six phenolic compounds. The *in vivo* results showed that oral administration of CCl_4 enhanced levels of hepatic and renal markers (ALT, AST, ALP, LDH, γ-GT, urea, and creatinine) in the serum of experimental animals. It also increased the oxidative stress markers resulting in increased levels of the LPO with a concomitant decrease in the levels of enzymatic antioxidants (SOD, CAT,

GPx) in both liver and kidney. The pretreatment of experimental rats with 250 mg/kg (BW) of the EAF, by intraperitoneal injection for 8 days, prevented CCl_4 induced disorders in the levels of hepatic markers. The biochemical changes were in accordance with histopathological observations suggesting a marked hepatoprotective effect of the extract.

Ceriops decandra (Griff.) Ding Hou FAMILY: RHIZOPHORACEAE

Ceriops decandra is used to treat hepatitis. Gnanadesigan et al. (2017) carried out study to identify the hepatoprotective activity of plant parts (leaf, bark, collar, flower, and hypocotyls) of *C. decandra. In vitro* antioxidant studies were carried out with DPPH, HRSA, NO, FRAP, and LPO assays. The LD_{50} was calculated and *in vivo* hepatoprotective activity was carried out with the leaf extract, which was found to be the most potent. The *in vivo* hepatoprotective activity was performed as follows: Group 1, control animals; Group 2, CCl_4-treated animals; Group 3, Silymarin (100 mg/kg BW p.o.) treated animals; Groups 4, 5, and 6, *C. decandra* treatment groups (100, 200, and 400 mg/kg BW). Histopathological scores were calculated with standard protocols. Of the selected different plant parts, the leaf extract showed maximum antioxidant scavenging properties. A study of the oral acute toxicity found *C. decandra* extract to be nontoxic up to 2000 mg/kg BW. The *in vivo* hepatoprotective nature of the leaf extract was identified as dose dependent and the levels of SGOT, SGPT, ALP, bilirubin, CHL, and LDH were found to be significantly decreased compared with hepatotoxin groups. Histopathological scores did not show any significant variations between control and high dose (400 mg/kg BW) of leaf-extract treated animals. Preliminary phytochemical analysis of the leaf extract revealed the presence of phenolic groups, alkaloids, triterpenoids, flavonoids, catechin, and anthraquinone. In conclusion, the hepatoprotective nature of the *C. decandra* leaf extract might be due to the occurrence of unique secondary metabolites and their antioxidant scavenging properties (Gnanadesigna et al., 2017).

Cestrum nocturnum L. FAMILY: SOLANACEAE

The hepatoprotective activities of *Cestrum nocturnum* (Queen of Night) was evaluated against the PCM-induced hepatotoxicity in the mice by Qadir et al. (2014). Aqueous ethanol extract of *C. nocturnum* (250 and 500 mg/kg) produced significant hepatoprotective activities against PCM induced liver injury in Swiss albino mice. Histopathalogical studies of liver further supported the hepatoprotective effects of *C. noctrunum*. Phytochemical screening showed the presence of alkaloids, flavonoids, saponins, terpenes, phenolic compounds, carbohydrates, and volatile oils. Most of the flavonoids have hepatoprotective activity. Therefore, the hepatoprotective activity of *C. nocturnum* may be due to the presence of flavonoids and phenolic components.

Chenopodium album L. FAMILY: CHENOPODIACEAE

Nigam and Paarakh (2011) evaluated the hepatoprotective activity of aerial parts of *Chenopodium album* using alcohol as chronic hepatitis model. Alcoholic (ALCA) and aqueous (AQCA) extracts of the aerial parts of *C. album* were evaluated for hepatoprotective activity against alcohol-induced hepatotoxicity using biochemical markers and by histopathological method. The aqueous extract at a dose of 400 mg/kg was found to be more potent when compared to Silymarin. ALCA and AQCA (200 and 400 mg/kg) showed significant hepatoprotective activity against alcohol-induced hepatotoxicity as evident by restoration of serum transaminases, ALP, and bilirubin content. Histopathology of the liver tissue further confirmed the reversal of damage induced by hepatotoxin. Aqueous and alcoholic extract did not show any sign of toxicity up to oral dose of 5 g/kg in mice.

Pal et al. (2011) investigated the hepatoprotective activities of dried whole plant of *C. album* acetone and MEs in ratio of 50:50 against PCM-induced hepatic injury. Acetone and methanol extract at dose levels of 200 and 400 mg/kg offered significant hepatoprotective action by reducing the serum marker enzymes like SGOT and SGPT. They also reduced the elevated level of ALP, serum ACP and SB. Reduced enzymatic and nonenzymic antioxidant levels and elevated lipid peroxidase level were restored to normal by administration of methanol and acetone extract of *C. album*. Histopathological studies further confirmed the hepatoprotective activity of these extracts when compared with PCM-treated control groups.

The different extracts (Pet-Ether extract, ethyl acetate extract, and ME) of *C. album* aerial parts were studied by Nayak et al. (2012) for its hepatoprotective effect on CCl_4-induced hepatotoxic rats. The extracts were found to decrease significantly CCl_4-induced elevation of SGOT, SGPT, bilirubin, and TC. But it increased HDL cholesterol level and liver weight with respect to CCl_4 toxic rats. Histopathological profiles showed that out of all the extracts, ME had significant protective effect against CCl_4-induced liver injury which is comparable with the standard drug Silymarin (25 mg/kg, p.o.).

Chenopodium murale L. FAMILY: CHENOPODIACEAE

Saleem et al. (2014) evaluated the hepatoprotective activity of *Chenopodium murale*. The results showed that aqueous methanolic extract of *C. murale* (200 and 500 mg/kg) produced significant decrease in PCM induced increased levels of liver enzymes (ALT, AST, ALP) and TB. These findings were further supported by histopathological investigations by microscope and detection of phytoconstituents having hepatoprotective potential, for example, quercetin, kaempferol, and GA by HPLC.

Chrysanthemum indicum L. FAMILY: ASTERACEAE

Jeong et al. (2013) investigated the hepatoprotective effect of the hot water extract of *Chrysanthemum indicum* flower (HCIF) in *in vitro* and *in vivo* systems. Hepatoprotective activities were evaluated at 250 to 1000 μg/mL concentrations by an *in vitro* assay using normal human hepatocytes (Chang cell) and hepatocellular carcinoma cells (HepG2) against CCl_4-induced cytotoxicity. The hepatoprotective effects of HCIF significantly reduced the levels of SGOT and SGPT as compared with the vehicle control group (CCl_4 alone). The survival rates of HepG2 and Chang cells were significantly improved compared with the control group. HCIF [50 mg/kg BW] treatment significantly reduced the serum levels of SGOT, SGPT, ALP, and LDH as compared with the control group in this *in vivo* study. The expression level of cytochrome P450 2E1 (CYP2E1) protein was also significantly decreased at the same concentration (50 mg/kg BW).

Chrysophyllum albidum G.Don FAMILY: SAPOTACEAE

The leaf extract of *Chrysophyllum albidum* was studied for hepatoprotective activity by Adebayo et al. (2011) against rats with-induced liver damage by CCl_4. The results showed that the levels of AST, ALT, ALP, and TB were significantly higher in rats treated with CCl_4 indicating liver injury, while these parameters were reduced significantly after treatment of rats with the extract. The hepatoprotective activity of *C. albidum* was also supported by histopathological studies of liver tissue. The liver tissue of rats in the group treated with CCl_4 showed marked centrilobular fatty degeneration and necrosis while the groups treated with plant extract showed signs of protection against this toxicant as evidenced by the absence of necrosis.

Cicer arietinum L. FAMILY: FABACEAE

Mekky et al. (2016) evaluated the hepatoprotective activity of the seed extract on CCl_4-induced hepatotoxicity in rats and acute toxicity. Administration of the extract to rats in doses up to 2 g/kg) did not cause any mortalities or observable signs of toxicity. Further, the plant extract showed a strong hepatoprotective activity based on assessing serum ALT, AST, and ALP and levels of albumen, globulin, TP, TC, HDL, TGs, and LDL. The antioxidative activity was evaluated by assessing hepatic CAT and SOD activity as well as reduced GSH, and MDA levels. Additionally, anti-inflammatory activity was observed as the extract significantly lowered the hepatic TNF α content. Histopathological examination of liver tissues indicated that the extract-treated animals showed almost normal hepatic architecture with fewer pathological changes.

Cichorium endivia L. FAMILY: ASTERACEAE

Marzouk et al. (2011) evaluated the effect of hydroalcoholic extract of *Cichorium endivia* leaves (HCE) against APAP-induced oxidative stress and hepatotoxicity in male rats. Oral administration of APAP produced liver damage in rats as manifested by the significant increase in liver MDA and serum total lipids, TC, creatinine, TB and enzyme activities (AST, ALT, and ALP). While a significant decrease in the levels of liver GSH, GST, SOD, CAT, serum TP, and ALB was recorded. Pretreatment of rats with *C. endivia* leaves extract or Silymarin for 21 days succeeded to modulate these observed abnormalities resulting from APAP as indicated by the pronounced improvement of the investigated biochemical and antioxidant parameters. The hepatoprotective activity of HCE was found to be compatible with Silymarin.

Cichorium glandulosum Boiss. et Huet FAMILY: ASTERACEAE

In vivo hepatoprotective effect of *Cichorium glandulosum* root extract (CGRE) was evaluated by Upur et al. (2009) using two experimental models, CCl_4- and GalN-induced acute hepatotoxicity in mice. Pretreatment with CGRE (800 mg/kg/day, p.o.) for 7 days significantly reduced the impact of CCl_4 toxicity (10 mL/kg, i.p.) on the serum markers of liver damage, AST, ALT, and ALP. Protective effect was reconfirmed against GalN-induced injury (800 mg/kg BW, i.p.) and elevated serum enzymatic levels were significantly and dose dependently restored toward normalization by the extracts. The extract showed noticeable antioxidant activity, comparable with standard antioxidants, through its ability to scavenge several free radicals (DPPH, O(2)(−), NO) and efficiency against LPO.

C. glandulosum was used historically in Uyghur folk medicine. Its roots, seeds, and aerial parts are extensively used by Uyghur residents in Xinjiang to eliminate savda typhoid, dredge, and cure obstructive jaundice variety liver disorders. Tong et al. (2015) evaluated the hepatoprotective activity of total flavonoids (TFs) obtained from *C. glandulosum* seeds against CCl_4-induced liver damage *in vitro* and *in vivo*. The dried seeds of *C. glandulosum* were extracted with 70% aqueous ethanol, and the extract was chromatographed with D101 macroporous resin. *In vitro*, the antioxidant capacity against LPO was evaluated using ferrothiocyanate, thiobarbituric acid (TBA), β-carotene bleaching, and LPO inhibition assay. The cytotoxicity and hepatoprotective activity of TFs were evaluated in human liver hepatoma cells (HepG2). MTT assay, hepatic injury markers AST, ALT, LDH leakage, MDA, and GSH were performed. *In vivo*, the hepatoprotective activity of TFs against CCl_4-induced acute liver injury was evaluated in rats. A series of biochemical and antioxidant parameter levels were measured in liver homogenate. The suppressive effect on pancreatic lipase activity was determined. Results indicated that TFs showed antioxidant capacity against LPO. Administrating CCl_4 (1%, v/v) caused a significant decrease in HepG2 viability. Treatment with TFs at doses (62.5, 125, and 250 μg/mL) could significantly ameliorate the cytotoxicity and decline the levels of AST, ALT, and LDH induced by CCl_4. The markers including MDA and GSH, which were close to oxidative damage, were restored. Oral treatment with TFs *in vivo* at doses of 100, 200, and 400 mg/kg significantly reduced the levels of AST, ALT, ALP, TB, and TBARS in the serum compared with CCl_4-induced acute liver injury in rats. TFs showed dose-dependent suppressive effects on pancreatic lipase activity, and the IC50 was 1.318 ± 0.164 mg/mL. TFs from *C. glandulosum* seeds demonstrated significant hepatoprotection against CCl_4-induced hepatotoxicity. TFs exhibited significant suppression of LPO and pancreatic lipase capacity, which may be the mechanisms of hepatoprotective effects against CCl_4 (Tong et al., 2015).

Cichorium intybus **L. FAMILY: ASTERACEAE**

Gadgoli and Mishra (1997) evaluated the antihepatotoxic activity of aqueous extract and fractions of seeds of *Cichorium intybus* in CCl_4 and PCM-induced toxicity. The methanol soluble fraction exhibited maximum activity with significant reductions in SGOT, SGPT, ALP, and bilirubin levels. Similar results were also obtained with the isolated compound.

The natural root and root callus extracts of *C. intybus* were compared for their anti-hepatotoxic effects in Wistar strain of Albino rats against CCl_4-induced hepatic damage (Zafar and Ali, 1998). The increased levels of serum enzymes (AST and ALT) and bilirubin observed in rats treated with CCl_4 were very much reduced in the animals treated with natural root and root callus extracts and CCl_4. The decreased levels of ALB and proteins observed in rats after treatment with CCl_4 were found to increase in rats treated with natural root and root callus extracts and CCl_4. These biochemical observations were supplemented by histopathological examination of liver sections. Results of this study revealed that *C. intybus* root callus extract could afford a better protection against CCl_4-induced hepatocellular damage as compared to the natural root extract.

Aktay et al. (2000) studied the hepatoprotective activity of ethanolic extracts of *C. intybus* using CCl_4-induced hepatotoxicity model in rats. According to the results of biochemical tests, significant reductions were obtained in CCl_4-induced increases of plasma and liver tissue MDA levels and plasma enzyme activities by the treatment with *C. intybus* extracts, which reflects functional improvement of hepatocytes. In spite of a significant inhibitory effect of *C. intybus* extract on biochemical parameters, this was *not supported* by histopathological examination.

The different fractions of alcoholic extract and one phenolic compound AB-IV of seeds of *C. intybus* were screened for antihepatotoxic activity on CCl_4-induced liver damage in albino rats (Ahmed et al., 2003). The degree of protection was measured using biochemical parameters like AST, ALT, ALP, and TP. The methanol fraction and compound AB-IV were found to possess a potent antihepatotoxic activity comparable to the standard drug Silymarin (Silybon-70). The histopathological study of the liver was also carried out, wherein the methanolic fraction and compound AB-IV showed almost complete normalization of the tissues as neither fatty accumulation nor necrosis was observed.

The effects of different concentrations of the hydroalcoholic extract of dried powdered leaves of *C. intybus*, on CCl_4-induced hepatotoxicity *in vivo* in rats and CCl_4-induced cytotoxicity in isolated rat hepatocytes were investigated by Jamshidzadeh et al. (2006). The results showed that the *C. intybus* extract could protect the liver from CCl_4-induced damages with doses of 50 and 100 mg/kg, but concentrations higher than 200 mg/kg were less effective.

The extract with concentrations of 60 to 600 µg /mL protected the erthrocytes against CCl_4-induced cytotoxicity, but concentrations of ≥1.5 mg/mL and higher increased the CCl_4-induced cytotoxicity. The *C. intybus* extract itself was toxic toward isolated hepatocytes in concentrations above 3.6 mg/mL. The results of this study therefore supported the traditional believes on hepatoprotective effect of the *C. intybus* extract; however, high concentrations were hepatotoxic.

Heibatollah et al. (2008) evaluated the hepatoprotective activity of hydroalcholic extract of *C. intybus* using a CCl_4-induced liver injury in rats. The leaf extract at oral dosage of 200, 400, and 500 mg/kg exhibited significant protective effect against CCl_4-induced hepatoxocity. Level of serum markers such as AST, ALT, ALP, and TB were significantly increased in CCl_4-treated rats. Simultaneously, *C. intybus* extract significantly suppressed mainly the increase in plasma activities of AST, ALT, ALP, and TB concentration, which are considered as markers of liver functional state.

Subash et al. (2011) evaluated the hepatoprotective activity of *C. intybus* in CCl_4 intoxicated albino rats. In the groups where hepatic injury induced by CCl_4 and extract were given simultaneously, the toxic effect of CCl_4 was controlled significantly by maintenance of structural integrity of hepatocyte cell membrane and normalisation of functional status of liver. Histology of liver sections from *C. intybus* + CCl_4-treated rats revealed moderate centrilobular hepatocytes degeneration, few areas of congestion with mild fatty changes.

Acute liver inflammation in rats induced by CCl_4 was cured by treatment with *C. intybus* (Chicory) and *Taraxacum officinale* (Dandelion) water extract. Hepatoprotective activity of Chicory leaves alone or mixed with Dandelion leaves water extract was investigated against CCl_4 intoxication in Wister albino rats. A severe hepatic damage was noted as elevated levels of liver marker enzymes due to treatment with CCl_4. A significant reduction in liver marker enzymes was observed during treatment with Chicory/mixture of chicory/Dandelion water extract (Abdulrahman et al., 2013)

Hassan and Yousef (2010) elucidated the modulating effect of chicory (*Cichorium intybus*)-supplemented diet against nitrosamnine-induced oxidative stress and hepatotoxicity in male rats. The obtained results revealed that rats received nitrosamine precursors showed a significant increase in liver TBARS and total lipids, TC, bilirubin, and enzymes activity (AST, ALT, ALP, and GGT) in both serum and liver. While a significant decrease in the levels of GSH, GSH-Rx, SOD, CAT, TP, and ALB was recorded. Nevertheless, chicory-supplemented diet succeeded to modulate these observed abnormalities resulting from nitrosamine compounds as indicated by the reduction of TBARS and the pronounced improvement of the investigated biochemical and antioxidant parameters.

Saggu et al. (2014) assessed the modulating effect of chicory (*C. intybus*) fruit extract against 4-*tert*-OP induced oxidative stress and hepatotoxicity in male rats. The obtained results revealed that rats which received 4-*tert*-OP showed a significant increase in liver TBARS and bilirubin, AST, ALT, ALP, and GGT activities. While a significant decrease in the levels of GSH, SOD, CAT recorded. However, CFR extract succeeded to modulate these observed abnormalities resulting from 4-*tert*-OP as indicated by the reduction of TBARS and the pronounced improvement of the investigated biochemical and antioxidant parameters. Histopathological evidence, together with observed PCNA and DNA fragmentation, supported the detrimental effect of 4-*tert*-OP and the ameliorating effect of CFR extract on liver toxicity. Another study was conducted by Li et al. (2014) to investigate the hepatoprotective effect of *C. intybus* extract against CCl_4-induced liver damage in rats. A significant heatoprotective activity of *C. intybus* plant extract was observed on measurement of liver marker enzymes.

El-Gengaihi et al. (2016) assessed the phytochemical and hepatoprotective activity of different extracts of dried herb of *C. intybus* against CCl_4 intoxicated male albino rats. The hepatoprotective activity of different extracts at 500 mg/kg BW was compared with CCl_4-treated animals. There were significant changes in serum biochemical parameters such as ALT, AST, ALP, bilirubin, ALB, TP, and γ-glutamyl transferase (GGT) in CCl_4 intoxicated rats, which were restored toward normal values in *C. intybus*-treated animals. Histopathological examination of liver tissues further substantiated these findings.

Cineraria abyssinica Sch. Bip. ex A. Rich FAMILY: ASTERACEAE

Sintayehu et al. (2012) evaluated the hepatoprotective and antioxidant activities of the leaf extracts of *Cineraria abyssinica*. Hepatoprotective activities of the aqueous and 80% methanolic extracts as well as the methanol fraction of the leaves of *C. abyssinica* were investigated against CCl_4-induced liver damage in rats. Intraperitoneal administration of 2 mL/kg of CCl_4 (50% in liquid paraffin) significantly raised the plasma levels of ALP, ALT, and AST in the toxin group compared with the values in the control group. Pretreatment of rats with 200 mg/kg of the aqueous, 80% MEs and the methanol fraction reduced the toxin-induced rise in plasma ALP, ALT, and AST. The standard drug, Silymarin (100 mg/kg) reduced serum ALP, ALT, and AST. Bioactivity-guided fractionation of the methanol fraction resulted in the isolation of the flavonol glycoside rutin. The results of biochemical analysis were further verified by histopathological examination of the liver, which showed improved architecture, the absence of necrosis and a decrease in inflammation, compared with the findings in the toxin group of animals. Both the extracts and rutin showed potent DPPH free-radical scavenging activities. Acute toxicity studies showed that the total extracts of the plant are nontoxic up to a dose of 3 g/kg.

Cinnamomum zeylanicum L. FAMILY: LAURACEAE

The hepatoprotective activity of ethanolic extract of *C. zeylanicum* was investigated by Moselhy and Ali (2009) against CCl_4-induced liver damage in rats. It was indicated that ethanolic extract of *C. zeylanicum* had more hepatoprotective action than water extract against CCl_4 by lowering MDA level and increase the level of antioxidant enzymes like SOD and CAT. Eidi et al. (2012) evaluated the protective effect of cinnamon (*C. zeylanicum*) bark extract against CCl_4-induced liver damage in male Wistar rats. Administration with cinnamon extracts (0.01, 0.05, and 0.1 g/kg) for 28 days significantly reduced the impact of CCl_4 toxicity on the serum markers of liver damage, AST, ALT, and ALP. In addition, treatment of cinnamon extract resulted in markedly increased the levels of SOD and CAT enzymes in rats. The histopathological studies in the liver of rats also supported that cinnamon extract markedly reduced the toxicity of CCl_4 and preserved the histoarchitecture of the liver tissue to near normal.

Cirsium arisanense Kitamura FAMILY: COMPOSITAE

Cirsium arisanense has been used for hundreds of years in Taiwan as a folk medicine for hepatoprotection. Ku et al. (2008) extracted the phenol-containing aqueous components of *C. arisanense* roots (CaR) and leaves/stem (CaL), and then assessed their hepatoprotective activities in both human hepatocellular carcinoma Hep 3B cells and C57BL/6 mice strain. HPLC analysis revealed that the components of CaR and CaL differed from those of the positive control Silymarin. CaR exhibited a higher phenolic content and antioxidant capacity than CaL. Hep 3B cells treated with Silymarin (0–200 μg/mL) demonstrated a concentration-dependent decrease in viability;

however, both CaR and CaL did not exhibit any apparent cytotoxicity. Silymarin at 100 μg/mL, as well as CaR and CaL, not only protect Hep 3B cells from tacrine-induced hepatotoxicity but also decrease the expression of hepatitis B surface antigen (HBsAg). Moreover, an animal experiment demonstrated that CaR, CaL, and Silymarin have hepatoprotective effects in C57BL/6 mice injected with tacrine, and they significantly decrease the levels of plasma ALT and AST. These effects of CaR and Silymarin, but not of CaL, may occur via an increase in the hepatic GSH level and the elimination of the NO production.

Cirsium setidens Nakai FAMILY: ASTERACEAE

The antioxidant activity and hepatoprotective potential of *Cirsium setidens* were investigated by Lee et al. (2008). The *n*-butanol (*n*-BuOH) fraction of leaves and roots of *C. setidens* had a higher DPPH radical scavenging activity than the other soluble fractions. The *n*-BuOH fraction of roots of *C. setidens* had a significant hepatoprotective activity at a dose of 500 mg/kg compared to that of a standard agent. The biochemical results were confirmed by histological observations indicating that *C. setidens* extract decreased ballooning degeneration in response to CCl_4 treatment. The *n*-BuOH fraction reduced CCl_4-induced liver injury in rats, and transcript levels of genes encoding antioxidant enzymes such as GPx 1 (GPO1), GPx 3 (GPO3), and SOD were elevated in the livers of rats treated with this fraction (500 mg/kg).

Cissampelos pareira L. FAMILY: MENISPERMACEAE

Cissampelos pareira (EtOH extract) is another medicinal plant with hepatoprotective activity against damage liver induced by INH/RIF (50 mg/kg each) on Wistar rat; this extract was tested at 100, 200, and 400 mg/kg by intraperitoneal via administered for 28 days. The result showed that this extract reduces the levels of SGPT, SGOT, SALP, TP, ALB, TB; the effect was similar to that showed by Silymarin and was dose dependent (Balakrishnan et al., 2012). The EtOH extract of *C. pareira* was also active against CCl_4-induced hepatotoxicity damage (Surendran et al., 2011).

Cissus pteroclada Hayata FAMILY: VITACEAE

Polysaccharide of *Cissus pteroclada* (CPP) was extracted and purified by Li et al. (20115). Three major fractions (CPP I, CPP II-1, and CPP II-2) from the CPP were purified by column chromatography and investigated for their monosaccharide compositions, scavenging radical effects and hepatoprotective activities *in vitro*. The results showed that glucose and galactose were the main monosaccharides of three polysaccharide fractions, CPP II-1 and CPP II-2 were acidic polysaccharide fractions which contained glucuronic acid and galacturonic acid. Antioxidant activity determination suggested that CPP I and CPP II-1 had a higher scavenging effects on DPPH, superoxide radical, hydroxyl radical, and ABTS radical. Moreover, the results of antioxidant test *in vitro* showed that CPP II-2 could significantly increase the activities of SOD and GSH-Px and decreased MDA level in human hepatocyte cell line (HL7702 cell), which indicating that CPP II-2 possessed good hepatoprotective activity.

Citrullus colocynthis (L.) Schrad. FAMILY: CUCURBITACEAE

Vakiloddin et al. (2015) evaluated the hepatoprotective and antioxidant profile of *Citrullus colocynthis* fruits. Hepatoprotective profile of methanolic extract of *C. colocynthis* fruits was investigated on rats, which were made hepatotoxic using PCM. The antioxidant profile of the extract was evaluated by conducting CAT, SOD, LPO, and DPPH tests. During hepatoprotective investigation, the PCM treated group II showed significant increase in TB, SGOT, SGPT, and ALP level. The results so obtained showed that pretreatment of rats with the methanolic extract 300 mg/kg p.o. decreases the elevated TB, SGOT, SGPT, and ALP serum levels. Also, the inhibitory profile was found comparable with toxicant group (Paracetamol 2 g/kg, p.o.).

Citrullus lanatus (Thunb.) Matsum. & Nakai FAMILY: CUCURBITACEAE

C. lanatus is used in traditional herbal medicine. Its fruits are eaten as a febrifuge when fully ripe or almost putrid. The fruit is also diuretic and helpful for the treatment of dropsy and renal stones. The rind of the fruit is prescribed for alcoholic poisoning and diabetes. *C. lanatus* contains bioactive compounds including cucurbitacin, triterpenes, anthraquinones, sterols, alkaloids, flavanoids, saponins, tannins, flavones aglycone, and simple phenols. The aqueous extract of *C. lanatus* is believed to be a good source of glucose, fiber, vitamin C, lycopene, and beta carotene. Watermelon juice at 120 g/70 kg BW of rats decreased SOD activity and LDL-cholesterol, and increased CAT and HDL-cholesterol, which could indicate its antioxidant effects (Georgina et al., 2011). The majority of cucurbitacin has cytoprotective activity on HepG2 cells. Cucurbitacin was demonstrated to have high potential as liver antifibrosis agent (Bartalis, 2005). Studies have been done to investigate the effect of *C. lanatus* juice on LPO in rat's liver, kidney, and brain. *In vivo* administration of CCl_4 once a week for 28 days caused a significant increase in serum markers of liver damage, AST, ALT, and TB, and decline in ALB compared to the control group. However, administration of CCl_4 with watermelon juice or ursodeoxycolic acid attenuated these changes significantly. LPO level increased in the liver, kidney, and brain tissues after CCl_4 administration. However, watermelon juice and ursodeoxycolic acid treatment prevented increase in LPO. According to the results, watermelon juice protects the liver, kidney, and brain tissues from *in vitro* CCl_4 toxicity in rats probably thanks to the antioxidant activity and inhibition of LPO formation. Together, biological evidence supports watermelon juice usefulness in the treatment of chemical-induced hepatotoxicity (Altas et al., 2011).

Citrus hystrix DC. FAMILY: RUTACEAE

Putri et al. (2013) evaluated the hepatoprotective activity of *Citrus hystrix* peel ethanolic extract (ChEE) in doxorubicin-induced hepatotoxicity in SD rats. ChEE did not repair neither hepatohistopathology profile nor reduce serum activity of ALT and AST. Abirami et al. (2015) evaluated the hepatoprotective effects of *C. hystrix* methanolic leaf extracts on PCM-induced toxicity. The leaf extracts restored the liver function markers and hepatic antioxidants to the normal level than elevated levels noticed on PCM control. Reversal of hepatoarchitecture has also been registered.

Citrus limon (L.) Osbeck FAMILY: RUTACEAE

Bouzenna et al. (2016) investigated the protective effect of *Citrus limon* essential oil against a high dose of aspirin-induced acute liver and kidney damage in female Wistar albino rats. Aspirin induced an increase of serum biochemical parameters and it resulted in an oxidative stress in both liver and kidney. This was evidenced by significant increase in TBARS in liver and kidney by 108% and 55%, respectively, compared to control. However, a decrease in the activities of SOD by 78% and 53%, CAT by 53% and 78%, and GPx by 78% and 51% in liver and kidney, respectively. Administration of EOC to rats attenuated the induced an effect of the high dose of aspirin induced in the afore mentioned serum biochemical parameters.

Citrus macroptera Montrouz FAMILY: RUTACEAE

The protective effect of an ethanol extract of *Citrus macroptera* (EECM) against APAP-induced hepatotoxicity was investigated in rats by Paul et al. (2016). Pretreatment with EECM prior to APAP administration significantly improved all investigated biochemical parameters, that is, transaminase activities, ALP, LDH, γ-glutamyl transferase activities and TB, TC, TG and creatinine, urea, uric acid, sodium, potassium and chloride ions, and TBARS levels. These findings were confirmed by histopathological examinations. The improvement was prominent in the group that received 1000 mg/kg EECM.

Citrus maxima (Burm.) Merr. FAMILY: RUTACEAE

Abirami et al. (2015) evaluated the hepatoprotective effects of *Citrus maxima* (Red and White variety) methanolic leaf extracts on PCM-induced toxicity. The leaf extracts restored the liver function markers and hepatic antioxidants to the normal level than elevated levels noticed on PCM control. Reversal of hepatoarchitecture has also been registered.

Clematis hirsuta Guill. & Perr. FAMILY: RANUNCULACEAE

Alqasoumi et al. (2008) evaluated the hepatoprotective activity of the ethanolic extract of *Clematis hirsuta* using CCl_4 in Wistar albino rats. Treatment of animals with crude extract of *C. hirsuta* (aerial parts) at doses 250 and 500 mg/kg *showed little decreases in all the parameters. In addition, the decreases were not statistically significant.*

Cleome droserifolia (Forssk.) Delile FAMILY: CAPPARACEAE

The effect of ethanol extract from aerial parts of *Cleome droserifolia* was investigated by Abdel-Kader et al. (2009) against CCl_4-induced liver injury. The hepatoprotective activity was evaluated through the quantification of biochemical parameters and confirmed using histopathology analysis. Efficient hepatoprotective effect was achieved by crude extract, fractions and some pure compounds. The significant hepatoprotective activities of both extracts seemed to be strongly connected to their content of hydroxycinnamoyl quinic acids and flavonoids (El-Askary et al., 2019).

Cleome gynandra L. FAMILY: CAPPARACEAE

Kumar et al. (2014) investigated the hepatoprotective effect of methonolic extract of *Cleome gynandra* against CCl_4-induced hepatocellular injury in Wistar rats. Administration with methanolic extract (200 and 400 mg/kg, p.o.) for 14 days significantly reduced the impact of CCl_4 toxicity on the serum markers of liver damage, SGOT, SGPT, ALP, and TB. The histopathological studies in the liver of rats also supported that *C. gynandra* extract markedly reduced the toxicity of CCl_4 and preserved the histoarchitecture of the liver tissue to near normal.

Cleome viscosa L. FAMILY: CAPPARACEAE

Gupta and Dixit (2009a, b) evaluated the hepatoprotective activity of ethanolic extract of leaves of *Cleome viscosa* against CCl_4 and TAA-induced hepatotoxicity in experimental animal models. The test material was found effective as hepatoprotective, through *in vivo* and histopathological studies. Results presented in indicate that the elevated levels of SGOT, SGPT, ALP, and BLN (total and direct) due to CCl_4 and TAA intoxication were reduced significantly in rats, after treatment with ethanolic extract. The hepatoprotective effect of ethanolic extract was comparable to that of Silymarin, a standard hepatoprotective agent.

Yadav et al. (2010) investigated the *in vivo* hepatoprotective potential of coumarinolignoids (cleomiscosins A, B, and C) isolated from the seeds of *C. viscosa.* The study was performed against CCl_4-induced hepatotoxicity in albino rats. The coumarinolignoids were found to be effective as hepatoprotective against CCl_4-induced hepatotoxicity as evidenced by *in vivo* and histopathological studies in small animals. Safety evaluation studies also exhibit that coumarinolignoids are well tolerated by small animals in acute oral toxicity study except minor changes in RBC count and hepatic protein content at 5000 mg/kg BW as a single oral dose. Coumarinolignoids which is the mixture of three compounds (cleomiscosin A, B, and C) is showing the significant protective effects against CCl_4-induced hepatotoxicity in small animals and also coumarinolignoids are well tolerated by small animals in acute oral study.

Clerodendron glandulosum Coleb. FAMILY: LAMIACEAE

Jadeja et al. (2011) evaluated toxicological effects and hepatoprotective potential of *Clerodendron glandulosum* (CG) aqueous extract. Acute and subchronic toxicity tests revealed that CG extract is nontoxic and its median lethal dose (LD_{50}) value is >5000 mg/kg BW. Also, rats pretreated with CG extract followed by administration of CCl_4 recorded significant decrement in plasma marker enzymes of hepatic damage, TB content and hepatic LPO. While, hepatic reduced GSH, ascorbic acid content, activity levels of superoxide and CAT and plasma TP content were significantly increased. Microscopic examination of liver showed that pretreatment with CG extract prevented CCl_4-induced hepatic damage in CG + CCl_4 group. CG extract has hepatoprotective potential by modulating activity levels of enzymes and metabolites governing liver function and by helping in maintaining cellular integrity of hepatocytes that is comparable with that of standard drug Silymarin. CG extract exhibits potent hepatoprotective activity against CCl_4-induced hepatic damage but does not exhibit any toxic manifestations.

Clerodendrum inerme (L.) Gaertn. FAMILY: LAMIACEAE

The ethanolic extract of *Clerodendrum inerme* leaves was screened by Gopal and Sengottuvelu (2008) for its hepatoprotective activity in CCl_4 (0.5 mL/kg, i.p.)-induced liver damage in Swiss albino rats at a dose of 200 mg/kg BW. The ethanolic extract of *C. inerme* significantly decreased the serum enzymes ALT, AST, ALP, TGLs, TC, and significantly increased the GSH level. Silymarin (25 mg/kg), a known hepatoprotective drug used for comparison, exhibited significant activity. The extract did not show any mortality up to a dose of 2000 g/kg BW (Sengottuvelu, 2008).

The ethanolic extract of *C. inerme* leaves were screened by Haque et al. (2011) for its hepatoprotective activity in PCM-induced liver damage in Swiss albino rats at a dose of 200 mg/kg BW. The ethanolic extract exhibited a significant protective effect by lowering levels of SGOT, SGPT, ALP, and TB. Liv.52 was used as positive control. The effects of the drug were judged by changes in serum marker ALT, AST, ALP, protein, and bilirubin levels. The extract did not show any mortality up to a dose of 2000 g/kg BW.

Clerodendrum infortunatum L. FAMILY: LAMIACEAE

Hepatoprotective potential of methanolic extract of *Clerodendrum infortunatum* (MECI) was studied by Sannigrahi et al. (2009) against CCl_4 hepatotoxicity in rats. The substantially elevated serum enzymatic levels of AST, ALT, ALP, and TB were restored toward normalization significantly by the extract. Silymarin was used as standard reference and exhibited significant hepatoprotective activity against CCl_4 induced haptotoxicity in rats. MDA concentration was decreased, while the liver antioxidative enzyme activity was elevated in all the MECI-treated rats. All the results were compared with standard drug Silymarin. In addition, histopathology of liver tissue was investigated to observe the morphological changes, showed the reduction of fatty degeneration and liver necrosis.

Clerodendrum serratum (L.) Moon FAMILY: LAMIACEAE

The ethanol extract of *Clerodendrum serratum* roots and ursolic acid isolated from it were evaluated by Vidya et al. (2007) for hepatoprotective activity against CCl_4-induced toxicity in male Wistar strain rats. The parameters studied were estimation of liver function serum markers such as serum TB, TP, ALT, AST, and ALP activities. The ursolic acid showed more significant hepatoprotective activity than crude extract. The histological profile of the liver tissue of the root extract and ursolic acid treated animal showed the presence of normal hepatic cords, the absence of necrosis, and fatty infiltration as similar to the controls. The results when compared with the standard drug Silymarin revealed that the hepatoprotective activity of the constituent ursolic acid is significant as similar to the standard drug.

Clerodendrum volubile P. Beauv. FAMILY: LAMIACEAE

Molehin et al. (2017) investigated the protective effect of *Clerodendrum volubile* methanolic extracts against CCl_4-induced hepatotoxicity in rats. CCl_4 hepatotoxicity, characterized by a significant increase in the levels of ALT, AST, and ALP, hepatic degeneration, and inflammation was attenuated by *C. volubile* methanolic extracts. The serum lipid parameters which include HDL and LDL were significantly decreased, and increased, respectively, by CCl_4. Methanolic extracts of *C. volubile* significantly prevented the decrease in the level of HDL and the increase in LDL in a dose-dependent manner. Decrease in TP induced by CCl_4 was moderately increased following the administration of methanolic extracts of *C. volubile*. Lipid peroxidation was significantly reduced while the reduced GSH level and the activities of hepatic antioxidant enzymes (CAT, SOD, and GPx) were significantly elevated by *C. volubile* extract in the CCl_4-treated rats. Their findings indicate that *C. volubile* extract has a significant protective effect against CCl_4-induced hepatotoxicity in rats which may be due to its antioxidant properties which are comparable to the reference antioxidant, vitamin E, used in this study.

Clidemia hirta (L.) Don FAMILY: MELASTOMATACEAE

Amzar and Iqbal (2017) evaluated the hepatoprotective and antioxidative potential of *Clidemia hirta* against CCl_4-induced liver injuries and oxidative damage in a murine model. The development of oxidative stress was observed through the escalation of hepatic LPO, depletion of GSH, and reduced antioxidant enzymes (GPx, GR, CAT, GST, and QR). Hepatic damage was evaluated by measuring serum transaminase (ALT and AST). In addition, CCl_4-induced hepatic damage was further evaluated using histopathological assessments. However, most of these changes were dependently ameliorated by the pretreatment of mice with a *C. hirta* dose.

Clitoria ternatea L. FAMILY: FABACEAE

Methanolic extract of *Clitorea ternatea* (200 mg/kg) significantly attenuated CCl_4 (Solanki and Jain, 2011) and PCM (Nithianantham et al., 2011)-induced biochemical (serum ALT, AST, and bilirubin levels) and histopathological alterations in liver. "Ayush-Liv.04" a polyherbal formulation consisting of 20% *C. ternatea* leaves as one of its constituents also showed significant hepatoprotective activity against ethanol and CCl_4-induced liver damage in rats (Narayanasamy and Selvi, 2005). *C. ternatea* possess significant anti-inflammatory (Devi et al., 2003) and antioxidant (Zingare et al., 2012; Kamkaen and Wilkinson, 2009) activities which may contribute to its hepatoprotective effects. Phytochemical studies on leaves of *C. ternatea* showed the presence of flavonoids, saponins, tannins, glycosides, quercetin, steroids, taraxerol, taraxerone, ternatins, and taraxerol (Nithianantham et al., 2011; Zingare et al., 2012).

The leaf extracts of *C. ternatea* was tested for hepatoprotective and antioxidant study against PCM-induced damage in mice. This study also concluded that leaf extract was found associated with hepatoprotective activity which was related to its antioxidant potential (Nithianatham et al., 2011). Nithianatham et al. (2013) also evaluated the hepatoprotective and antioxidant activity of *Clitoria ternatea* flower extract against APAP-induced liver toxicity. The amount of total phenolics and flavonoids were estimated to be 105.40 ± 2.47 mg/g GAE and 72.21 ± 0.05 mg/g catechin equivalent, respectively. The antioxidant activity of *C. ternatea* flower extract was 68.9% at a concentration of 1 mg/mL and was also concentration dependent, with an IC_{50}value of 327.00 μg/mL. The results of APAP-induced liver toxicity experiment showed that mice treated with the extract (200 mg/kg) showed a significant decrease in ALT, AST, and bilirubin levels, which were all elevated in the PCM group. Meanwhile, the level of GSH was found to be restored in extract-treated animals compared to the groups treated with APAP alone. Therapy of extract also showed its protective effect on histopathological alterations and supported the biochemical finding.

Biomolecules extracted from *C. ternatea* were evaluated *in vivo* and *in vitro* for hepatoprotective effects against CCl_4 toxicity in rats. AST, ALP, and bilirubin reduced significantly in hepatotoxic rats against CCl_4 treatment. Results showed that *C. ternatea* extract dose (500 mg/kg) had best hepatoprotection against CCl_4 (Balaji et al., 2015).

Cnidoscolus aconitifolius (Mill.) I.M. Johnst. FAMILY: EUPHORBIACEAE

Adaramoye et al. (2011) evaluated the possible protective effect of *Cnidoscolus aconitifolius* leaf extract against hepatic damage induced by chronic ethanol administration in rats. Ethanol-treated rats had significantly elevated serum and liver post-mitochondrial MDA, an index of LPO. Ethanol toxicity lowered the antioxidant defense indices, such as reduced GSH, SOD, and CAT. Specifically, the activities of hepatic SOD and CAT decreased by 48% and 51%, respectively, while the level of GSH decreased by 56%. In addition, serum TC, TGs and LDLs-cholesterol levels were elevated in ethanol-treated rats. Also, significant elevation in serum ALT and AST, and γ-glutamyl transferase activities were observed in ethanol-treated rats. Supplementation with leaf extract significantly decreased the activities of liver marker enzymes, stabilized the lipid profiles and restored the antioxidants status of ethanol-treated rats. The activities of leaf extract were comparable with kolaaviron in the ethanol-treated rats. This observation was supported by histopathological examination of liver slides.

Cnidoscolus chayamansa McVaugh FAMILY: EUPHORBIACEAE

The ethanol (EtOH) extract from the leaves of *Cnidoscolus chayamansa* administered orally demonstrated a protective effect in Wistar rats against the hepatotoxicity induced by the mixture of RIF/INH (100 mg/kg each), this extract diminishing

AST, ALT, and ALP levels. The observed effect was similar to that of the positive control (Silymarin, 2.5 mg/kg); the authors attributing this protection to the flavonoids present in the plant extract (Kulagthuran et al., 2012).

Coccinia grandis (L.) Voigt (Syn.: *Coccinia indica* Wt. & Arn.) FAMILY: CUCURBITACEAE

The effect of the methanolic extract of *Coccinia grandis* fruits on CCl_4-induced liver damage in Wistar rats was studied by Swamy et al. (2007). Significant hepatoprotective effect was obtained against CCl_4-induced liver damage by oral administration of *C. grandis* methanolic extract as evident from decreased levels of serum enzymes in the treated groups, compared to the controls.

Alcoholic extract of the fruits of *C. grandis* was evaluated in CCl_4-induced hepatotoxicity in rats and levels of AST, ALT, ALP, TPs, total and DBL were evaluated. At a dose level of 250 mg/kg, the alcoholic extract significantly decreased the activities of serum enzymes (AST, ALT, and ALP) and bilirubin which was comparable to that of Silymarin revealing its hepatoprotective effect (Vadivu et al., 2008).

Umamaheswari and Chatteerjee (2008) evaluated the hepatoprotective and antioxidant activities of petroleum ether, chloroform, ethyl-acetate, and residual fractions of the hydromethanol extract of leaves of *C. grandis* against PCM-induced hepatic damage in albino mice. Oral administration of the fractions at a dose of 200 mg/kg BW for 7 days resulted in a significant reduction in serum biochemical parameters and liver MDA and an increase in the enzymatic and nonenzymatic antioxidants when compared with PCM-damage mice. Profound fatty degeneration, fibrosis, and necrosis observed in the hepatic architecture of PCM-treated mice were found to acquire near-normalcy in drug co-administered mice. The effect produced by the chloroform fraction was almost comparable with the Silymarin-treated group.

Sunilson et al. (2009) evaluated the hepatoprotective effect of crude ethanolic and aqueous extracts from the leaves of *C. grandis* against liver damage induced by CCl_4 in rats. The ethanolic extract at an oral dose of 200 mg/kg exhibited a significant protective effect as shown by lowering serum levels of glutamic oxaloacetic transaminase, GPT, ALP, TB, and TC and increasing levels of TP and ALB levels as compared to Silymarin, the positive control. These biochemical observations were supported by histopathological examination of liver sections. The activity may be due to the presence of flavonoid compounds. The extracts showed no signs of acute toxicity up to a dose level of 2000 mg/kg.

A diethyl ether extract of the leaves was studied by Kumar et al. (2010) for hepatoprotective activity against CCl_4-induced liver toxicity in rats. The results showed hepatoprotective activity of *C. grandis* leaves extract at dose 400 mg/kg BW was comparable with standard treatment 125 mg/kg BW of Silymarin, a known hepatoprotective drug. These data also supplement for histopathological study of rat liver section.

The hepatoprotective effect of leaves of *C. grandis* against liver damage induced by PCM and CCl_4 in rats has also been reported by Kundu et al. (2012). The activity exhibited/inhibited by the plant extract was found to be comparable to that of the standard hepatoprotective drug, Silymarin.

Cocculus hirsutus (L.) W. Theob. FAMILY: MENISPERMACEAE

Thakare et al. (2009) evaluated the hepatoprotective activity of ME of *Cocculus hirsutus* in Wistar rats. Different concentrations of methanolic extract of *C. hirsutus* were evaluated for *in vivo* GR activity. On bile duct ligation (BDL), the liver fibrosis was induced with significant rise in serum marker enzymes levels. The HYP accumulation caused by hydrophilic bile acids accompanied by elevated hepatic LPO and GSH levels. Treatment with *C. hirsutus* decreased the elevated levels of serum marker enzymes showing hepatoprotection, which was further confirmed by histopathological results.

Lavanya et al. (2017) estimated the protective or curative potency of an extract from *C. hirsutus* leaves against CCl_4 intoxication *via* its antioxidant property in rats. When CCl_4-induced rats were co-treated with *C. hirsutus* at doses of 250 and 500 mg/kg BW, all the altered levels of liver marker enzymes and oxidative stress markers were restored to near control values. Histopathological studies provided direct evidence of the hepatoprotective effect of *C. hirsutus*.

Cochlospermum planchonii Hook.f. ex Planch. FAMILY: COCHLOSPERMACEAE

Aqueous extracts of *Cochlospermum planchonii* rhizomes are used by native medical practitioners in northern Nigeria to treat jaundice. Aliyu et al. (1995) investigated the hepatoprotective activity of the aqueous extracts of this plant. An extract prepared by a laboratory adaptation of their method was hepatoprotective in CCl_4-treated rats and it inhibited cytochrome P-450 enzymes, which constitutes a plausible hepatoprotective mechanism. A crystalline inhibitor (0.3% of dry weight of rhizomes) was isolated using inhibition of two rat cytochrome P-450 enzymes, aminopyrine-*N*-demethylase and aniline hydroxylase, as bioassays to guide fractionation by solvent partitioning, polyamide column chromatography, preparative thin layer chromatography and fractional crystallization. The inhibitor was identified as zinc formate by inductively coupled plasma atomic emission spectroscopy, nuclear magnetic resonance spectroscopy, and comparison with synthetic material by power X-ray diffraction crystallography. Synthetic and plant-derived zinc formate were equally effective as inhibitors of cytochrome P-450 enzymes and as hepatoprotective agents in CCl_4-treated rats. *C. planchonii* rhizomes contain unusually high levels of manganese and zinc, although much higher levels have been observed in plants considered to be hyperaccumulators of these metals (Aliyu et al., 1995).

Aqueous extract of *C. planchonii* rhizome was investigated by Nafiu et al. (2011) for its toxic effects in albino rats using some liver and kidney functional indices as "markers." The extract significantly decreased ALP activities in the liver leading to 80.95% loss by the end of the experimental period. While there was no consistent pattern in the kidney ALP activity and SB level, the serum enzyme compared well with the control value. There was no effect on the acid phosphatase activity of the tissues and serum of the animals. The extract also reduced the urea, ALB, and creatinine content in the serum of the animals. The alterations in the biochemical parameters by the aqueous extract of *C. planchoni* may have consequential effects on the normal functioning of the liver and kidney of the animals. Therefore, the 50 mg/kg BW of the aqueous extract of *C. planchoni* rhizome may not be completely safe as an oral remedy.

***Cochlospermum tinctorium* Perrier ex A. Rich. FAMILY: COCHLOSPERMACEAE**

The hepatoprotective activity of the rhizome of *Cochlospermum tinctorium* was investigated by Diallo et al. (1992) using CCl_4 toxicity on mouse and *t*-BHP *in vitro* induction of LPO and hepatocyte lysis. Aqueous, hydroethanolic and ethanolic extracts showed significant dose-dependent hepatoprotective actions. The ethanolic extract showed a hepatoprotective activity at lower doses than Silymarin. The ethanolic and hydroethanolic extracts exhibited remarkable effects against the induction of LPO and hepatocyte lysis; the aqueous extract showed comparatively weaker effects. These differences were related to the chemical composition of the extracts. Among the identified constituents of the drug, phenolic and polyphenolic compounds (gallic and ellagic acids, ellagitannins, flavonoids), carotenoids, triterpenes could be related to the biological activity.

The hepatoprotective effect of aqueous root extract of *C. tinctorium* on CCl_4 on-induced hepatic damage in rats was reported by Etuk et al. (2009). Wistar rats were divided into normal control, induction control, extract, and prednisolone treated groups. Hepatotoxicity was induced in rats by intraperitoneal administration of CCl_4 (30% in olive oil) for 5 days. Treatment group received 200 mg/kg of extract post-hepatotoxicity induction orally for 7 days. The animals were sacrificed on the 8th day, blood and hepatic tissue collected for liver function test and histopathological analysis, respectively. Administration of CCl_4-induced hepatic damage in the rats was evidenced by a significant increase in the blood clotting time, serum levels of ALT, AST, ALP, and bilirubin as compared to the control. There was also a significant reduction in the serum TP, serum ALB, and reduced GSH levels. Treatment with the extract reversed the values of all the biochemical parameters to near normal values in control. The histopathological reports collaborate with the biochemical analysis results. Oral administration of aqueous root extract of *C. tinctorium* for 7 days has significantly reversed hepatic damage produced by CCl_4 in Wistar rats.

Akinloye et al. (2012) examined the potential hepatoprotective action of methanolic extract of *C. tinctorium* leaf on CCl_4-induced liver damaged in rats. Forty rats divided into four groups (group 1—control; group 2—CCl_4 induced; group 3—CCl_4 + *C. tinctorium* extract (200 mg/kg); group 4—CCl_4 + prednisolone (standard anti-inflammatory drug) of 10 rats in each were used. Administration of CCl_4-induced hepatic damage in rats, as evidenced by a significant increase in the levels of SGPT, SGOT, cholesterol, and bilirubin levels. The level of MDA (an index of oxidative stress) was also higher in the CCl_4 group, when compared to other groups. However, administration of methanolic extract of *C. tinctorium* (200 mg/kg) after CCl_4 administration (group 3) brought a significant reduction in values of these parameters compared to the CCl_4-treated group (group 1). The potential hepatoprotective activity of the extract was also demonstrated by its regenerative action on some damaged liver tissues, as evidenced by the histopathological studies of the representative liver sample of group 3 rats' liver section, which showed hepatic regeneration, with no visible pathology. This study therefore showed the potential of the *C. tinctorium* leaf extract to decrease the levels of serum markers enzymes, indicating the protection of hepatic cells, and to confer some levels of protection against CCl_4-induced hepatocellular injury.

Cochlospermum vitifolium (Willd.) Spreng. FAMILY: COCHLOSPERMACEAE

Cochlospermum vitifolium is a Mexican medicinal plant that is used in folk medicine for the treatment of hypertension, diabetes, hepatitis, and related diseases. Sanchez-Salgado et al. (2007) evaluated the hepatoprotective effect of ME on subchronic experimental assay. Methanol extract (at a dose of 100 mg/kg) was administered to bile duct-obstructed rats to determine its hepatoprotective activity, showing a statistically significant decrease of serum glutamic-pyruvic transaminase (GPT, 45%) and ALP (15%).

Cocos nucifera L. FAMILY: ARECACEAE

Hepatoprotective and antioxidant effects of tender coconut water (TCW) were investigated in CCl_4-intoxicated female rats by Loki and Rajamohan (2003). Liver damage was evidenced by the increased levels of SGOT, SGPT, and decreased levels of serum proteins and by histopathological studies in CCl_4-intoxicated rats. Increased LPO was evidenced by elevated levels of TBARS viz, MDA, HP, and conjugated dienes (CD), and also by significant decrease in antioxidant enzymes activities, such as SOD, CAT, GPx, and GR and also reduced GSH content in liver. Nevertheless, CCl_4-intoxicated rats treated with TCW retained almost normal levels of these constituents. Decreased activities of antioxidant enzymes in CCl_4-intoxicated rats and their reversal of antioxidant enzyme activities in TCW-treated rats, shows the effectiveness of TCW in combating CCl_4-induced oxidative stress. Hepatoprotective effect of TCW is also evidenced from the histopathological studies of liver, which did not show any fatty infiltration or necrosis, as observed in CCl_4-intoxicated rats.

Colchicum autumnale L. FAMILY: COLCHICACEAE

Colchicine, the major alkaloid in *Colchicum autumnale* protects the liver of experimental animals against several hepatotoxins, that is, D-GalN and PCM by its ability to bind microtubule protein. A colchicine derivative, trimethylcolchicinic acid (TMCA) that does not bind tubulin, that is, tested on chronic liver damage induced by CCl_4 and by BDL. So, both compounds were equally potent but that TMCA could be administered at larger doses than colchicines without side effects and better hepatoprotective actions (Muriel and Rivera Espinoza, 2008).

Coldenia procumbens L. FAMILY: BORAGINACEAE

An investigation has been carried out by Been et al. (2011) to evaluate the *in vitro* hepatoprotective effect of ethanolic extract of *Coldenia procumbens* using antitubercular drugs such as isoniazid, rifampicin, pyrazinamide (PZA), and GalN HCl as toxicants and Silymarin as standard drug. The CTC50 of antitubercular drugs and GalN HCl, which were used as hepatotoxicants to assess the hepatoprotective effect of the plant extracts were found to be 500 µg/mL and 40 µg/mL, respectively, against BRL-3 A cell lines. The plant extract showed over 80% protection for both the toxicants which is promising for further *in vivo* studies.

Ganesan et al. (2013) evaluated the hepatoprotective activity of *C. procumbens* whole plant CE against hepatotoxicity on rats induced with 200 mg/kg of D-GalN. The parameters assessed were serum levels of SGOT, SGPT, ALP, TP, ALB, globulin, TC, TB, and blood sugar changes in liver. There was significant reversal of biochemical changes induced by D-GalN treatment in rats by CE treatment, indicating promising hepatoprotective activity.

Combretum quadrangulare Kurz FAMILY: COMBRETACEAE

The MeOH extract of leaves of *Combretum quadrangulare* showed significant hepatoprotective effect on D-GalN/LPS-induced experimental liver injury in mice and on D-GalN/ TNF-α-induced cell death in primary cultured mouse hepatocytes (Banskota et al., 2000). Phytochemical investigation led to the isolation of 30 cycloartane-type triterpenes together with betulinic acid, beta-sitosterol, beta-sitosterol glucoside, 4 flavones, and 3 flavone C-glucosides. These compounds showed various potencies of hepatoprotective effect on D-GalN/TNF-α-induced cell death in primary cultured mouse hepatocytes. Quadrangularol B, methyl quadrangularate I, kamatakenin, 5,7,4′-trihydroxy-3,3′-dimethoxyflavone, 5,4′-dihydroxy-3,7,3′-trimethoxyflavone and isokaempferide showed strong inhibitory effect on TNF-α-induced cell death with IC50 values of 34.3, 33.7, 13.3, 22.4, 13.4, and 22.8 µM, respectively, whereas clinically used silibinin had an IC_{50} value of 39.6 µM and glycyrrhizin showed very weak inhibitory effect. Methyl quadrangularates A and N, norquadrangularic acid B and vitexin also showed potent inhibition on TNF-α-induced cell death with IC_{50} values of 45.7, 89.3, 67.6, and 40.1 μM, respectively. The flavonoids and some of the cycloartane-type triterpenes appeared to be the hepatoprotective principles of the leaves of *C. quadrangulare*.

Hepatoprotective effect of MeOH, MeOH-H_2O (1:1), and H_2O extracts of *C. quadrangulare* seeds were examined on D-GalN/tumor necrosis factor-α (TNF-α)-induced cell death in primary cultured mouse hepatocytes (Adnyana et al., 2000). The MeOH extract showed the strongest inhibitory effect on D-GalN/TNF-α-induced cell death (IC_{50}, 56.4 μg/mL). Moreover, the MeOH extract also significantly lowered the SGPT level on D-GalN/LPS-induced liver injury in mice. Bioguided separation of the MeOH extract led to the isolation of 38 compounds of various classes including triterpene glucosides, lignans and catechin derivatives. Among the isolated triterpene glucosides, lupane-type (IC50, 63.1, 59.8 and 76.2 μM, respectively) and ursane-type (IC50, 30.2 and 34.6 μM, respectively) compounds exhibited strong hepatoprotective activity. 1-*O*-Galloyl-6-*O*-(4-hydroxy-3,5-dimethoxy) benzoyl-beta-D-glucose (IC50, 7.2 μM), methyl gallate (IC50, 19.9 μM), and (−)-epicatechin (IC50, 71.2 μM) also had a potent hepatoprotective effect on D-GalN/TNF-α-induced cell death in primary cultured mouse hepatocytes.

Commiphora africana (A. Rich) Endl. FAMILY: BURSERACEAE

The effects of *Commiphora africana* ethanolic leaf extract on some biochemical markers of liver and kidney functions were investigated in rats by Aliyu et al. (2007). The results showed a significant change in ALP activity, while groups administered 100 and 150 mg/kg showed significant increase. The group administered 25 mg/kg showed significant increase in serum creatinine after 24 h of treatment. The results of the liver and kidney histology showed that there was no noticeable damage to the liver tissues of rats administered the extract. However, hydropic degeneration of the cortical-tubular epithelium and glomerulus was seen with the group administered 100 mg/kg. Similarly, the group treated with 150 mg/kg showed acute glomerulonephritis and proliferation of the mesangial cells. These results suggest that *C. africana* extract may enhance liver function at low doses and may cause adverse effects at high doses.

Commiphora berryi (Arn.) Engl. FAMILY: BURSERACEAE

The capacity of *Commiphora berryi* bark as an antioxidant to protect against CCl_4-induced oxidative stress and hepatotoxicity in Albino Wistar rats was investigated by Shankar et al. (2008). Intraperitoneal injection of CCl_4, administered twice a week, produced a marked elevation in the serum levels of AST, ALT, ALP, and bilirubin. Histopathological analysis of the liver of CCl_4-induced rats revealed marked liver cell necrosis with inflammatory collections that were conformed to increase in the levels of SOD, GPx, and CAT. Daily oral administration of methanolic extract of *C. berryi* bark at 100 and 200 mg/kg doses for 15 days produced a dose-dependent reduction in the serum levels of liver enzymes. Treatment with *C. berryi* normalized various biochemical parameters of oxidative stress and was compared with standard Silymarin.

Commiphora myrrha (Nees) Engl. FAMILY: BURSERACEAE

Ahmad et al. (2015) investigated the hepatoprotective activity of *Commiphora myrrha* ethanol extract against D-GalN/LPS-induced acute hepatic injury in an animal model. The administration of D-GalN/LPS increased plasma aminotransferases and TB levels, which were attenuated by *C. myrrha* treatment. Hepatic LPO activity and NO content also increased, while the antioxidant activity measured by GSH, SOD, and CAT was reduced. *C. myrrha* provided significant restoration of GSH, SOD, and CAT levels. Furthermore, the acute phase response elicited by D-GalN/LPS administration enhanced mRNA expressions of TNF-α, IL-6, IL-10, iNOS-2, and HO-1, which were ameliorated by *C. myrrha* treatment. These findings indicate that *C. myrrha* considerably reduces the oxidative stress of D-GalN/LPS-induced hepatic injury via multiple pathways including a down regulation of inflammatory mediators and cytokines. Such a property might be sufficient to combat cellular damage caused by various conditions that resemble fulminant hepatitis and could be of a potential clinical application.

Commiphora opobalsamum (L.) Engl. FAMILY: BURSERACEAE

The hepatoprotective activity of an ethanolic extract of *Commiphora opobalsamum* ("Balessan") was investigated by Al-Howiriny et al. (2004) in rats by inducing hepatotoxicity with CCl_4. This extract has been shown to possess significant protective effect by lowering serum transaminase levels (SGOT and SGPT transaminase), ALP, and bilirubin. Pretreatment with an extract of Balessan prevented the prolongation of the barbiturate sleeping time associated with CCl_4-induced liver damage in mice. Nevertheless, CCl_4-induced low-level nonprotein sulfhydryl concentration in the liver was replenished by the Balessan extract. Phytochemical studies on aerial part of *C. opobalsamum* showed the presence of saponins, volatile oil, sterol and/or triterpenes, friedelin, flavonoids, mearnsetin, and quercetin.

Convolvulus arvensis L. FAMILY: CONVOLVULACEAE

Convolvulus arvensis is traditionally used as laxative. Its decoction is used in cough, flu, jaundice, and in skin diseases. It is also used to treat the painful joints, inflammation, and swelling. The plant *C. arvesis* is traditionally used for the treatment of jaundice (Thakral et al., 2010). Ali et al. (2013) evaluated the hepatoprotective activity of *C. arvensis*. The results showed that the extract of *C. arvensis* (200 and 500 mg/kg) produced significant decrease in PCM-induced increased levels of liver enzymes and TB. Histopathological investigation and detection of active constituent, quercetin by HPLC also supported the results.

Convolvulus fatmensis Kuntze FAMILY: CONVOLVULACEAE

The ME *of Convolvulus fatmensis* was evaluated by Atta et al. (2007) for its potential antiulcerogenic, diuretic, and hepatoprotective effects as well as for its acute toxicity. The antiulcerogenic effect was evaluated against acute (ethanol model) and prolonged (aspirin model)-gastric ulceration. The results revealed that oral administration of the ME (400 mg/kg BW) significantly reduced the alcohol-induced gastric ulcers (curative ratio; 32.6%). It also produced high curative ratio (75%), decreased the number of gastric ulcers, total acidity, and TP in aspirin-induced gastric ulcer. Oral administration of 500 mg/kg BW MEs did not affect the urine output, or sodium excretion while potassium and chloride excretion were significantly increased. The higher dose (1000 mg/kg BW) significantly increased the sodium, potassium, and chloride excretion but did not affect urine volume. Moreover, it produced mild hepatoprotective effect against CCl_4-induced hepatotoxicity as indicated from serum biochemical and histopathological changes. No signs of acute toxicity were observed after oral administration of doses up to 2.75 g/kg BW. Two aglycones flavonoid compounds kaempferol and quercetin were isolated from *C. fatmensis*. Four coumarin compounds were isolated and identified as umbelliferone, scopoletin, asculetin, and scopoline for the first time. Two phenolic acids were also isolated in the same manner and identified by spectroscopic methods, they are ferulic acid and caffeic acid.

Conyza bonariensis (L.) Cronquist FAMILY: ASTERACEAE

Saleem et al. (2014) evaluated the hepatoprotective activity of *Conyza bonariensis* ethanolic extract against PCM-induced hepatotoxicity in mice. It showed that *C. bonariensis* extract (250 and 500 mg/kg) reduced the level of liver marker enzymes and TB which were raised on intoxication with PCM. It was supported by histopathological results of liver. HPLC analysis confirmed the presence of qurecetin as an active constituent which was present in ethanolic extract of *C. bonariensis.*

COPTIDIS RHIZOMA

Coptidis rhizoma (CR) is the dried rhizome of *Coptis chinensis* Franch., *C. deltoidea* C. Y. Cheng et Hsiao or *C. teeta* Wall. (Ranunculaceae). CR is traditionally used for heat clearing and toxic-scavenging and it belongs to liver meridian in Chinese medicine practice. Clinically, CR can be used for hepatic and biliary disorders. Ye et al. (2009) demonstrated that coptidis rhizoma aqueous extract has hepatoprotective effect against CCl_4-induced acute liver damage and this was related to antioxidant property. The aqueous extract inhibited significantly the activities of ALT and AST and increased the activity of SOD. Feng et al. (2011) explored the protection of Coptidis rhizome aqueous extract (CRAE) on chronic liver damage induced by CCl_4 in rats and its related mechanism. Serum AST and ALT activities were significantly decreased in rats treated with different doses of CRAE, indicating its protective effect against CCl_4-induced chronic liver damage. Observation on serum SOD activity revealed that CRAE might act as an antioxidant agent against CCl_4-induced chronic oxide stress. Histological study supported these observations. Erk1/2 inhibition may take part into CRAE's effect on preventing hepatocyte from apoptosis when exposed to oxidative stress.

Cordia macleodii Hook. f. & Thoms. FAMILY: BORAGINACEAE

Qureshi et al. (2009) evaluated the possible antioxidant and hepatoprotective potential of ethanolic extract of *Cordia macleodii* leaves. Antioxidant activity of the extracts was evaluated by four established, *in vitro* methods, namely, DPPH radical scavenging method, NO radical scavenging method, iron chelation method, and reducing power method. The extract demonstrated a significant dose-dependent antioxidant activity comparable with ascorbic acid. The extract was also evaluated for hepatoprotective activity by CCl_4-induced liver damage model in rats. CCl_4 produced a significant increase in levels of SGPT, SGOT, ALP, and TB. Pretreatment of the rats with ethanolic extract of *C. macleodii* (100, 200, and 400 mg/kg po) inhibited the increase in levels of GPT, SGOT, ALP, and TB and the inhibition was comparable with Silymarin (100 mg/kg po).

Coreopsis tinctoria **Nutt. FAMILY: ASTERACEAE**

Coreopsis tinctoria is a traditional remedy for the management of various diseases including hepatitis. Tsai et al. (2017) investigated the hepatoprotective potentials of the ethanol extract from *C. tinctoria* (CTEtOH) using an animal model of CCl_4-induced acute liver injury. The results revealed that the serum ALT and AST levels significantly decreased after treatment with the extract. Moreover, histological analyses indicated that extract (0.5 and 1.0 g/kg) and Silymarin reduced the extent of CCl_4-induced liver lesions. *C. tinctoria* ethanol extract (0.5 and 1.0 g/kg) reduced the levels of MDA, NO, and proinflammatory cytokines (TNF-α and IL-1β). Furthermore, the ethanol extract (1.0 g/kg) reduced the level of IL-6. The activities of antioxidant enzymes, namely SOD and GR, significantly increased after treatment with the extract (0.5 and 1.0 g/kg) and that of GPx increased after treatment with 1.0 g/kg of extract.

Coriandrum sativum **L. FAMILY: APIACEAE**

Sreelatha et al. (2009) investigated the antioxidant activity of *Coriandrum sativum* on CCl_4 treated oxidative stress in Wistar albino rats. CCl_4 injection induced oxidative stress by a significant rise in serum marker enzymes and TBARS along with the reduction of antioxidant enzymes. In serum, the activities of enzymes like ALP, ACP, and protein and bilirubin were evaluated. Pretreatment of rats with different doses of plant extract (100 and 200 mg/kg) significantly lowered SGOT, SGPT, and TBARS levels against CCl_4-treated rats. Hepatic enzymes like SOD, CAT, and GPx were significantly increased by treatment with plant extract, against CCl_4-treated rats. Histopathological examinations showed extensive liver injuries, characterized by extensive hepatocellular degeneration/necrosis, inflammatory cell infiltration, congestion, and sinusoidal dilatation. Oral administration of the leaf extract at a dose of 200 mg/kg BW significantly reduced the toxic effects of CCl_4. The activity of leaf extract at the dose of 200 mg/kg was comparable to the standard drug, Silymarin.

The protective effect of cold-pressed coriander (*C. sativum*) oil (CO) against the toxicity caused by CCl_4 in rats was studied by El-Hadary and Ramadan (2015). CO is characterized by its high levels of monounsaturated fatty acids and polyunsaturated fatty acids, tocopherols and phenolic compounds. Male Wistar rats were orally treated with two doses of CO (100 and 200 mg/kg) with administration of CCl_4 (1 mL/kg, CCl_4 in olive oil) for 8 weeks. Liver biochemical parameters were determined in animals treated with CO. The results clearly demonstrated that CO augments the antioxidants' defense mechanism against CCl_4-induced toxicity and provides evidence that CO may have a therapeutic role in free radical-mediated diseases. Treatment with CO significantly reduced the impact of CCl_4 toxicity on AST, ALT, and ALP, kidney function indicators, protein profile, lipid profile and antioxidant markers of CCl_4-induced liver injury rats. The overall potential of the antioxidant system was significantly enhanced by the CO supplements as the hepatic MDA levels were lowered, whereas reduced GSH levels were elevated. The hepatoprotective impact of CO was also supported by histopathological studies of liver tissue. Histopathological examination showed that CO reduced fatty degeneration, cytoplasmic vacuolization, and necrosis in CCl_4-treated rats.

Baghdadi et al. (2016) evaluated protective effect of *C. sativum* volatile oil on hepatotoxicity of ibuprofen in rats. The activity of ALT and AST was measured in the liver of different groups in addition to the histological examination of the sections of liver. The results showed that ibuprofen caused a significant decrease in the activity of ALT and AST in the liver. The histological examination of the liver showed many pathological changes. Administration of coriander volatile oil in the combination with ibuprofen was able to significantly increase the activity of both AST and ALT, in the liver and caused a significant decrease in the deleterious effect induced by IBU.

Coronopus didymus L. FAMILY: BRASSICACEAE

The aqueous extract of whole plant of *Coronopus didymus* was screened by Mantena et al. (2005) for hepatoprotective effects in rats. The extract showed significant hepatoprotective activity at 200 and 400 mg/kg doses on oral administration. Mechanistically, *C. didymus* acts as an antioxidant as evidenced by its ability to scavenge DPPH and superoxide radicals. The hepatoprotective activity may be due to the presence of flavonoids, and tannins as they are reported to possess a variety of biological activities.

Cosmos sulphureus Cav. and *Cosmos bipinnatus* Cav. FAMILY: ASTERACEAE

Saleem et al. (2019) evaluated the hepatoprotective activity of *Cosmos sulphureus* and *Cosmos bipinnatus*. Blood was collected for the evaluation of liver biochemical markers and livers were removed for histopathological evaluation 24 h post-PCM treatment. HPLC analysis revealed the presence of quercetin, GA, caffeic acid, and chlorogenic acid in both plant extracts. The extracts of both plants decreased the level of ALT and TB significantly, dose dependently and protected hepatocytes from PCM-induced hepatotoxicity. It was concluded that both plants may possess hepatoprotective activity possibly due to the presence of quercetin and phenolic compounds.

Costus afer Ker-Gawl. FAMILY: ZINGIBERACEAE

Phytochemical composition of *Costus afer* extract and its alleviation of CCl_4-induced hepatic oxidative stress and toxicity were evaluated by Chibueze et al. (2012). The qualitative and quantitative analyses showed the presence of alkaloids, saponins, flavonoids, tannins, and phenols in the aqueous stem extract while flavonoids, saponins and phenols were detected in the ethanol extract. The toxicological study showed serum elevation of ALT, AST, and ALP and increase in the levels of TBARS expressed as melondialdehyde in the liver of the rats in response to the oral administration of CCl_4. The rats fed with 400 mg/kg of extract of *C. afer* showed significant reduction in ALT and ALP levels close to the control than the rats fed with 200 mg/kg. Though there were reduction in AST and TBARS levels close to the control than the rats fed with 200 mg/kg, they were not significant. This indicated that biological active compounds of *C. afer* are more polar and could serve as source of bioactive compounds for nutrition and therapeutic purposes.

Anyasor et al. (2013) investigated the hepatoprotective and *in vivo* antioxidant activity of orally administered chloroform and ethanol *C. afer* leaf extracts (20–60 mg/kg BW) on 1 g/kg BW APAP-induced acute hepatic injury in albino rats for 7 days. Result showed no significant difference in plasma and hepatic superoxide activities. Plasma CAT activity was significantly reduced in 20 and 40 mg/kg CEs and 20 mg/kg ethanol extract treated groups. Hepatic CAT activity was reduced significantly at co-administered 20 mg/kg ethanol extract and 1 g/kg ACT treated group. Plasma and hepatic GSH levels were elevated significantly in chloroform and ethanol leaf extracts treated groups. The plasma and liver GST activity in the CE groups were significantly reduced while ethanol leaf extract showed no significant difference. The plasma and hepatic MDA concentrations in the chloroform and ethanol extracts treated groups were found reduced. Further studies showed that the reduction in AST and ALT elevated activities induced by ACT was more profound in ethanol treated groups than chloroform treated groups. Therefore, findings from this study indicate that the chloroform leaf extract possesses high antioxidant effect than ethanol leaf extracts while ethanol leaf extract possesses high hepatoprotective potentials than chloroform leaf extract.

Okoko (2018) evaluated the effect of the methanolic extract of *C. afer* stem on PCM-induced tissue injury in rats (in vivo) and its ability to reduce hydrogen peroxide-induced erythrocyte damage (in vitro). Pretreatment of rats with *C. afer* extract (p.o.) caused significant reduction in serum levels of ALT, AST, ALP, urea, creatinine and tissue TBARS closer to control values. Pretreatment of the rats with the plant extract also increased the activities of SOD and CAT in both liver and kidney. Apart from the levels of TBARS in liver and CAT in kidney, rats pretreated with 400 mg/kg BW extract showed greater response than pretreatment with 200 mg/kg BW The extract also reduced hydrogen peroxide-induced erythrocyte damage assessed as haemolysis and LPO which was concentration dependent.

Tchamgoue et al. (2018) evaluated the cardio-, reno-, and hepato-antioxidant status of hydroethanolic extract of *C. afer* on streptozotocin-intoxicated diabetic rats. Streptozotocin administration induced toxicity in the cardiac, hepatic, and renal tissues by stimulating significant increases in the levels of CAT and SOD, GSH, and MDA. Similarly, significant increases in the levels of ALT, AST, urea, and TP were observed in streptozotocin-treated rats, whereas decreases were observed in the levels of ALP, LDH, and creatinine. Following the treatments with *C. afer* hydroethanolic extract prevented the effect of streptozotocin by maintaining the tissue antioxidant status (CAT, SOD, GSH, and MDA) and the plasma biochemical parameters (AST, ALT, ALP, LDH, creatinine, and urea) toward the normal ranges. The histopathological examination revealed hepatovascular congestion and leucocyte infiltration as well as renovascular congestion, glomerulosclerosis, and tubular clarification in the untreated diabetic control and their absence in the group of animals treated with a high dose of *C. afer* extract. The findings of the investigation suggest that *C. afer* possesses antioxidant activities capable of regulating drug induced tissue damage.

Crassocephalum crepidioides (Benth.) S. Moore FAMILY: ASTERACEAE

Free-radical scavenging and protective actions against chemically-induced hepatotoxicity of *Crassocephalum crepidioides* were investigated by

Aniya et al. (2005). A water extract of *C. crepidioides* strongly scavenged superoxide anion, hydroxyl radical, and also stable radical DPPH. Galactosamine (GalN, 400 mg/kg) and LPS (LPS, 0.5 μg/kg)-induced hepatotoxicity of rats as seen by an elevation of serum ALT and AST and of LPO in liver homogenates was significantly depressed when the herbal extract was given intraperitoneally 1 and 15 h before GalN and LPS treatment. Similarly, CCl_4-induced liver injury as evidenced by an increase in AST and ALT activities in serum was also inhibited by the extract pretreatment. Isochlorogenic acids, quercetin and kaempferol glycosides were identified as active components of *C. crepidioides* with strong free-radical scavenging action. These results demonstrate that *C. crepidioides* is a potent antioxidant and protective against GalN plus LPS-or CCl_4-induced hepatotoxicity.

Crossostephium chinensis (L.) Makino FAMILY: ASTERACEAE

The hepatoprotective potential of *Crossostephium chinensis* water extract (CCW) on CCl_4-induced liver damage was evaluated by Chang et al. (2011) in preventive and curative rat models. The results showed that CCW (0.1, 0.5, and 1.0 g/kg) significantly reduced the elevated levels of GPT and GOT by CCl_4 administration. TBARS level was dramatically reduced, and SOD, CAT, GPx, and GSH activities were significantly increased. In addition, CCW decreased NO production and TNF-α activation in CCl_4-treated rats. Therefore, they speculated that CCW protects against acute liver damage through its radical scavenging ability. CCW inhibited the expression of MMP-9 protein, indicating that MMP-9 played an important role in the development of CCl_4-induced chronic liver damage in rats. In LC-MS-MS analysis, the chromatograms of CCW with good hepatoprotective activities were established. Scopoletin may be an important bioactive compound in CCW.

Crepis rueppellii Sch. Bipp. FAMILY: ASTERACEAE AND *Anisotes trisulcus* (Forrsk.) Nees FAMILY: ACANTHACEAE

Pharmacological investigations were carried out by Fleurentin et al. (1986) to evaluate the hepatoprotective effects of *Crepis rueppellii* and *Anisotes trisulcus*. Ethanolic extracts of these plants were investigated for their ability to reduce mortality of mice after ethanol intoxication and to lower the activities of plasma glutamic-pyruvic transaminase (GPT) after CCl_4-induced hepatitis in rats. *Crepis* and *Anisotes* extracts and a 50:50 mixture of both at 200 mg/kg presented significant hepatoprotective effects in both experimental situations. The traditional therapeutic indications of these plants have been largely confirmed.

Crocus sativus L. FAMILY: IRIDACEAE

Omidi et al. (2014) investigated the protective effects of hydroalcoholic extract, remaining from *Crocus sativus* petals (CSP) against APAP-induced hepatotoxicity by measuring the blood parameters and studying the histopathology of liver in male rats. The APAP treatment resulted in higher levels of ALT, AST, and bilirubin, along with lower TP and ALB concentration than the control group. The administration of CSP with a dose of 20 mg/kg was found to result in lower levels of AST, ALT, and bilirubin, with a significant higher concentration of TP and ALB. The histopathological results regarding liver pathology, revealed severe conditions including cell swelling, severe inflammation, and necrosis in APAP-exposed rats, which was quiet contrasting compared to the control group. The pretreated rats with low doses of showed hydropic degeneration with mild necrosis in centrilobular areas of the liver, while the same subjects with high doses of appeared to have only mild hepatocyte degeneration.

Riaz et al. (2016) investigated the hepatoprotective effect of aqueous and ethanolic extract of *C. sativus* in amiodarone-induced hepatotoxicity in rabbits. Both aqueous and ethanolic extract of *C. sativus* significantly decreased serum ALT and AST enzyme activities and significant results were obtained when compared to the amiodarone group. Based on the results it was concluded that addition of *C. sativus* to the treatment protocol of patients maintained on amiodarone for long time period can be recommended to prevent the liver injury.

Croton bonplandianus Baill. FAMILY: EUPHORBIACEAE

Dutta et al. (2018) evaluated hepatoprotective capacity of antioxidant rich extract of *Croton bonplandianus* (CBL) on CCl_4-induced acute hepatotoxicity in murine model. Hydromethanolic extract of *C. bonplandianus* leaf was used for evaluation of free-radical scavenging activity. Liver cells of experimental mice were damaged using CCl_4 and subsequently hepatoprotective potential of the plant extract was evaluated using series of *in vivo* and *in vitro* studies. In the hepatoprotective study, Silymarin was used as a positive control. Antioxidant enzymes, pro-inflammatory markers, liver enzymatic, and biochemical parameters were studied to evaluate hepatoprotective activity of *C. bonplandianus* leaf extract. Free-radical scavenging activity of CBL extract was also observed in WRL-68 cell line. The phytochemicals identified by GCMS analysis were scrutinized using in-silico molecular docking procedure. The results showed that CBL extract has potent free-radical scavenging capacity. The biochemical parameters were over expressed due to CCl_4 administration, which were significantly normalized by CBL extract treatment. This finding was also supported by histopathological evidences showing less hepatocellular necrosis, inflammation and fibrosis in CBL and Silymarin treated group, compared to CCl_4 group. ROS generated due to H_2O_2 in WRL-68 cell line were normalize in the highest group (200 μg/mL) when compared with control and negative control (CCl_4) group. After molecular docking analysis, it was observed that the compound α-amyrin present in the leaf extract of *C. bonplandianus* has better potentiality to protect hepatocellular damages than the standard drug Silymarin. The present study provided supportive evidence that CBL extract possesses potent hepatoprotective capacity by ameliorating haloalkane induced liver injury in the murine model. The antioxidant and anti-inflammatory activities also affirm the same. The synergistic effects of the phytochemicals present in CBL are to be credited for all the hepatoprotective activity claimed above.

Croton oblongifolius Roxb. FAMILY: EUPHORBIACEAE

The different extracts of *Croton oblongifolius* were tested by Ahmed et al. (2002) for their hepatoprotective activity against CCl_4-induced hepatotoxicity in albino rats. The degree of protection was measured by using biochemical parameters like serum transaminases (SGOT and SGPT), ALP, TP, and ALB. The methanolic extracts showed the most significant hepatoprotective activity comparable with standard drug Silymarin. Other extracts namely petroleum ether and acetone also exhibited a potent activity.

Cryptomeria japonica (Thunb. ex L.f.) D. Don FAMILY: CUPRESSACEAE

Shyur et al. (2008) characterize the anti-inflammatory and hepatoprotective activities of the phytocompounds from *Cryptomeria japonica* wood on LPS- or TPA-induced activation of proinflammatory mediators and CCl_4-induced acute liver injury in mice. A CJH7-2 fraction was purified from *C. japonica* extracts following bioactivity-guided fractionation, and it exhibited significant activities on inhibition of NO production and iNOS expression as well as upregulating HO-1 expression in LPS-stimulated macrophages. CJH7–2 also potently inhibits COX-2 enzymatic activity

(IC_{50} = 5 μg/mL) and TPA-induced COX-2 protein expression in mouse skin (1 mg/200 μL/site). CJH7-2 (10 mg/kg BW) can prevent CCl_4-induced liver injury and aminotransferases activities in mice. Chemical fingerprinting analysis showed that terpenes are the major bioactive compounds in the CJH7-2 fraction.

Cucumis dipsaceus Ehrenb. FAMILY: CUCURBITACEAE

Lata amd Mittal (2017) investigated the hepatoprotective activity of flavonoids rich extracts of *Cucumis dipsaceus* fruits against CCl_4-induced hepatic damage in liver of Wistar albino rats. The results of the biochemical parameter were found to be significantly normalized from reduced release of SGOT, SGPT, ALP, bilirubin, and total protein of CCl_4-treated rats. The results obtained from *in vivo* antioxidant enzyme assays for SOD, CAT, GSH and, TBARS/MDA were performed on homogenized part of rat liver after complete dosing were significant than CCl_4 treated group. The hepatoprotective activity was also supported by histopathological studies on liver or pretreated rat and CCl_4-treated rats.

Cucumis ficifolius A. Rich. FAMILY: CUCURBITACEAE

Araya et al. (2019) assessed the hepatoprotective and radical scavenging activity of 80% methanol crude extract and different fractions of *Cucumis ficifolius* root. *C. ficifolius* crude extract and its solvent fractions showed strong radical scavenging activity and the chloroform fraction had the highest activity. No sign of toxicity was shown in an acute toxicity test of the extract. Hepatoprotective activity evaluation on the crude extract by a pretreatment model with 125, 250, and 500 mg/kg doses revealed a significant reduction of the serum level of CCl_4-induced liver enzyme markers at the highest tested dose (500 mg/kg). The chloroform fraction that had the highest radical scavenging activity and the crude extract, both at 500 mg/kg, were again evaluated in a post-treatment model and the results revealed that both the extract and the chloroform fraction demonstrated significant hepatoprotective activities which support the results of the pretreatment model.

Cucumis prophetarum L. FAMILY: CUCURBITACEAE

Alqasoumi et al. (2008) evaluated the hepatoprotective activity of the ethanolic extract of *Cucumis prophetarum hirsuta* using CCl_4 in Wistar albino rats. The crude extracts at the lower (250 mg/kg BW) dose failed to protect against elevated levels of all biomarkers, whereas only SGPT and bilirubin levels were found to be reduced significantly at higher dose. Although there are some claims for the benefit of *C. prophetarum* for liver problems, the obtained results could not prove their effectiveness as hepatoprotective agent.

Cucumis trigonus Roxb. FAMILY: CUCURBITACEAE

The EtOH extract of *Cucumis trigonus* (fruit) was tested as hepatoprotective agent against liver damage induced by RIF/INH (at 50 mg/kg each) in Wistar rats. This extract was tested at 100, 250, and 500 mg/kg, administered for 21 days by i.p. route. The group that received 500 mg/kg of extract showed similar levels of SGOT, SGPT, SALP, and GGTP to the Silymarin group. Total bilirubin, conjugated bilirubin (CB), unconjugated bilirubin, TP, ALB, and globulin levels were better in the extract-treated group (500 mg/kg); in this group, the histological analyses revealed a normal liver architecture (Gopalakirshnan and Kalairasi, 2015).

Cucurbita maxima Duchesne FAMILY: CUCURBITACEAE

Saha et al. (2011) investigated the hepatoprotective activity of methanol extract of *Cucurbita maxima* (MECM) against CCl_4-induced hepatotoxicity in Wistar rats. In MECM-treated animals, the toxic effect of CCl_4 was controlled significantly by restoration of the biochemical parameters, such as, SGPT, SGOT, ALP, TP and total billirubin, as well as by the improvement of the antioxidant status to/toward near normal values. Histology of the liver sections of the animals treated with the extracts showed the presence of normal hepatic cords, absence of necrosis and fatty infiltration, which further evidenced the hepatoprotective activity of *C. maxima.*

Cudrania cochinchinensis (Lour.) Yakuro Kudo & Masam. var. *gerontogea* FAMILY: MORACEAE

Three components in the EtOAc and *n*-BuOH fractions, obtained from the ethanol extract of *Cudrania cochinchinensis* var. *gerontogea* were evaluated by Lin et al. (1996) for their hepatoprotective activities in rats on CCl_4 and D-GalN-induced hepatotoxicity. Three flavonoids (25 mg/kg), wighteone, naringenin, and populnin (kaempferol-7-glucoside), exhibited greater hepatoprotective effects on CCl_4-induced liver injury than on D-GalN-induced hepatotoxicity by reversing the altered serum enzymes (SGOT and SGPT) and preventing the development of hepatic lesions, including liver centrilobular inflammation, cell necrosis, fatty change, ballooning degeneration in CCl_4 intoxication and necrosis of the portal area in D-GalN intoxication. Wighteone and naringenin (25 mg/kg) isolated from the EtOAc fraction showed a better hepatoprotective effect against CCl_4-induced liver injury than that of populnin (25 mg/kg) obtained from the *n*-BuOH fraction. Furthermore, wighteone protected the liver, not only against CCl_4-induced hepatotoxicity, but also against D-GalN-induced liver injury. These results demonstrated that wighteone and naringenin are two active hepatoprotective principles from *C. cochinchinensis* var. *gerontogea.*

Cuminum cyminum L. FAMILY: APIACEAE

Mushtaq et al. (2014) carried out the study for evaluation of hepatoprotective activity of aqueous ethanolic extract of *Cuminum cyminum* seed against the hepatotoxicity induced by Nimesulide on albino rats. Liver marker enzymes SGPT, SGOT, ALP, and TB in rats when intoxicated with Nimesulide, were measured. When animals were treated with *C. cyminum* (100, 200, and 300 mg/kg) after intoxication, then liver marker enzymes restored to normal which were comparable to Silymarin (25 mg/kg) treated results. Histological examinations also confirmed the hepatoprotective activities.

Cupressus sempervirens L. FAMILY: CUPRESSACEAE

Three phenolic compounds cosmosiin, caffeic acid, and *p*-coumaric acid were isolated from the leaves of *Cupressus sempervirens*, together with cupressuflavone, amentoflavone, rutin, quercitrin, quercetin, and myricitrin (Ibrahim et al., 2007). The hepatoprotective activity of the MeOH extract was carried out in liver homogenate of normal and CCl_4-treated rats; a significant decrease in SGOT, SGPT, cholesterol level, and TGs, while a significant increase in the TP level, was observed after the oral administration of MeOH extract. The free-radical scavenging activity against stable DPPH was measured for MeOH extract and some of the isolated phenolic compounds showed high antioxidant activity for quercetin, rutin, caffeic acid, and *p*-coumaric acid (Ibrahim et al., 2007).

The methanolic extract *C. sempervirens* was investigated by Rizk et al. (2007) for its efficiency in reducing CCl_4-induced hepatotoxicity in rats. Rats injected with a single toxic dose of CCl_4 and sacrificed after 24 h induced remarkable disturbances in the levels of all tested parameters. However, rats injected with the toxic agent and left for one and a half month to self-recover showed moderate improvements in the serum and liver biochemical parameters. However, treatment with *C. sempervirens* extract ameliorated the levels of the disturbed biochemical parameters.

Ali et al. (2010) investigated the role of *C. sempervirens* extract as therapeutic effect against CCl_4 with biochemical, histopathological evaluations on female Wistar rats. Remarkable disturbances were observed in the levels of all tested parameters in rats treated with CCl_4. However, rats injected with the toxic agent and left for one and a half month to self-recover showed moderate improvements in the studied parameters while treatment with medicinal herbal extract ameliorated the levels of the disturbed biochemical parameters. The group treated with *C. sempervirens* extract showed a remarkable improvement in comparison to the CCl_4 treated group showing histopathological liver profiles close to those of the control group.

Curculigo orchioides Gaertn. FAMILY: HYPOXIDACEAE

Rao et al. (196a) suggested the anti-inflammatory and hepatoprotective activities of *Curculigo orchioides*. Rao et al. (1996b) showed a hepatoprotective activity against rifampicin-induced hepatotoxicities. Rao et al. (1997) isolated cucrculignin A and curculigol and screened for their antihepatoxic activity against TAA and galactomine-induced hepatotoxic. Venukumar et al. (2002a, b) showed antioxidant activity of ME in CCl_4-induced hepatopathy in rats.

Curcuma comosa Roxb. FAMILY: ZINGIBERACEAE

Suksen et al. (2016) assessed the protective effect of *Curcuma comosa* extracts and its isolated compounds against *t*-BHP-induced hepatotoxicity in isolated primary rat hepatocytes. *t*-BHP markedly caused the formation of MDA and ALT leakage from the hepatocytes. Pretreatment with the *C. comosa* ethanol extract showed greater protective effect than the hexane extract, and the effect was concentration related. Treating the hepatocytes with compound D-92 provided greater protective effect than compound D-91. IC_{50} values of compounds D-91, D-92, and Silymarin for the protection of ALT leakage at 30 min were 32.7 ± 1.1, 9.8 ± 0.7, and 160 ± 8 μM, respectively. Further investigation showed that compound D-92 was more effective in maintaining the intracellular GSH content in the *t*-BHP treated group, whereas the reduction in antioxidant enzymes, GPx, and GSH-*S*-transferase activities, were not improved.

Curcuma longa L. FAMILY: ZINGIBERACEAE

The potential efficacy of the turmeric antioxidant protein (TAP) isolated from aqueous extract of turmeric *(Curcuma longa)* in protecting tissues from peroxidative damage was studied in the CCl_4-treated rats by Subramanian and Selvam (1999). The increased basal as well as promoter induced LPO formation in the tissues of CCl_4-treated rats was significantly inhibited by 40%–70% by TAP administration. The observed decreased antioxidant enzyme activities of SOD, CAT, GPx, GST, and antioxidant concentrations of reduced GSH, total (TSH), protein (PSH) and nonprotein (NPSH) thiols and ascorbic acid in the liver of CCl_4-treated rats were nearly normalized on TAP pretreatment. The increased activities registered with glucose-6-phosphate dehydrogenase, LDH, ALT, and AST in the liver of CCl_4-treated rats were conferred protection by 50%–80% on TAP treatment. Glucose-6-phosphatase and the membrane bound ATPase activities were decreased in CCl_4 treated animals and were completely restored on TAP treatment; thus suggesting that TAP treatment prevents tissue injury by neutralizing the oxidative stress induced changes.

Somchit et al. (2005) investigated the hepatoprotective activity of the ethanol extract of *C. longa* against PCM-induced liver damage in rats. Paracetamol at 600 mg/kg,-induced liver damage in rats manifested by increased serum ALT, AST, and ALP. Histologically, livers from these rats revealed parenchymal necrosis and massive inflammation. Pretreatment of rats with the ethanolic extract of *C. longa* 100 mg/kg prior to PCM dosing lowered serum level enzyme activities. Livers of these rats showed normal histology.

Kalantari et al. (2007) evaluated the hepatoprotective effect of *C. longa* extract against APAP-induced hepatotoxicity in mice. The acute elevation of serum transaminases (ALT, AST) significantly reduced in the groups receiving *C. longa* and the difference with the positive group (A) was significant. Necrosis of liver showed appropriate decrease according to histopathologic observation.

Sengupta et al. (2011) evaluated the hepatoprotective and immunotherapeutic effects of aqueous extract of turmeric rhizome in CCl_4 intoxicated Swiss albino mice. CCl_4 administration increased the level of SGOT, SGPT, and bilirubin level in serum. However, the aqueous extract of turmeric reduced the level of SGOT, SGPT, and bilirubin in CCl_4 intoxicated mice. Apart from damaging the liver system, CCl_4 also reduced nonspecific host response parameters like morphological alteration, phagocytosis, NO release, MPO release, and intracellular killing capacity of peritoneal macrophages. Administration of aqueous extract of *C. longa* offered significant protection from these damaging actions of CCl_4 on the nonspecific host response in the peritoneal macrophages of CCl_4 intoxicated mice.

Thattakudian et al. (2011) evaluated the ability of *C. longa* and *Tinospora cordifolia* formulation to prevent antituberculosis treatment-induced hepatotoxicity. Patients with active TB diagnosis were randomized to a drug control group and a trial group on drugs plus an herbal formulation. Isoniazid, rifampicin, PZA, and ethambutol for first 2 months followed by continuation phase therapy excluding Pyrazinamide for 4 months comprised the anti-tuberculous treatment. Curcumin-enriched (25%) *Curcuma longa* and a hydroethanolic extract enriched (50%) *Tinospora cordifolia* 1 g each divided in two doses comprised the herbal adjuvant. Hemogram, bilirubin, and liver enzymes were tested initially and monthly till the end of study to evaluate the result. Incidence and severity of hepatotoxicity was significantly lower in trial group (incidence: 27/192 vs. 2/316. Mean AST (195.93 ± 108.74 vs 85 ± 4.24, ALT (75.74 ± 26.54 vs. 41 ± 1.41, and SB (5.4 ± 3.38 vs. 1.5 ± 0.42. A lesser sputum positivity ratio at the end of 4 weeks (10/67 vs. 4/137) and decreased incidence of poorly resolved parenchymal lesion at the end of the treatment (9/152 vs. 2/278) was observed. Improved patient compliance was indicated by nil drop-out in trial versus 10/192 in control group. The herbal formulation prevented hepatotoxicity significantly and improved the disease outcome as well as patient compliance without any toxicity or side effects (Thattakudian et al., 2011).

Salama et al. (2013) evaluated mechanisms of the hepatoprotective activity of *Curcuma longa* rhizome ethanolic extract on TAA-induced liver cirrhosis in rats. Histopathology, immunohistochemistry, and liver biochemistry were significantly lower in the *C. longa*-treated groups compared with controls. CLRE-induced apoptosis, inhibited hepatocytes proliferation but had no effect on hepatic CYP2E1 levels. The progression of liver cirrhosis could be inhibited by the antioxidant and anti-inflammatory activities of CLRE and the normal status of the liver could be preserved.

Moghadam et al. (2015) investigated hepatoprotective effects of turmeric in methotrexate-induced liver toxicity in male Wistar albino rats. Methotrexate significantly induced liver damage and decreased its antioxidant capacity, while turmeric was hepatoprotective. Liver tissue microscopic evaluation showed that methotrexate treatment induced severe centrilobular and periportal degeneration, hyperemia of portal vein, increased artery inflammatory cells infiltration and necrosis, while all of histopathological changes were attenuated by turmeric (200 mg/kg). Turmeric extract can successfully attenuate

MTX-hepatotoxicity. The effect is partly mediated through extract's antinflammatory activity.

Curcuma xanthorrhiza Roxb. FAMILY: ZINGIBERACEAE

Curcuma xanthorrhiza (CX) has been used for centuries in traditional system of medicine to treat several diseases such as hepatitis, liver complaints, and diabetes (Devaraj et al., 2014). It has been consumed as food supplement and "jamu" as a remedy for hepatitis. Devaraj et al. (2014) evaluated the antioxidant and hepatoprotective potential of CX rhizome. Hepatoprotective assay was conducted against CCl_4-induced hepatic damage in rats at doses of 125, 250, and 500 mg/kg of hexane fraction. The highest antioxidant activity was found in hexane fraction. In the case of hepatoprotective activity, CX hexane fraction showed significant improvement in terms of a biochemical liver function, antioxidative liver enzymes, and LPO activity. Good recovery was observed in the treated hepatic tissues histologically.

Cuscuta arvensis Beyr. FAMILY: CONVOVULACEAE

Cuscuta arvensis is a parasitic plant, and commonly known as "dodder" in Europe, in the United States, and "tu si zi shu" in China. It is one of the preferred spices used in sweet and savory dishes. Also, it is used as a folk medicine for the treatment particularly of liver problems, knee pains, and physiological hepatitis, which occur notably in newborns and their mothers in the southeastern part of Turkey. Koca-Caliskan et al. (2018) investigated the hepatoprotective effects and antioxidant activities of aqueous and methanolic extracts of *C. arvensis* on APAP-induced acute hepatotoxicity in rats. The hepatoprotective activity of both the aqueous and methanolic extracts at an oral dose of 125 and 250 mg/kg was investigated by observing the reduction levels or the activity of ALP, ALT, AST, blood urine nitrogen, and TB content. *In vivo* antioxidant activity was determined by analyzing the serum SOD, MDA, GSH, and CAT levels. Chromatographic methods were used to isolate biologically active compounds from the extract, and spectroscopic methods were used for structure elucidation. Both the methanolic and aqueous extracts exerted noticeable hepatoprotective and antioxidant effects supporting the folkloric usage of dodder. One of the bioactive compounds was kaempferol-3-*O*-rhamnoside, isolated and identified from the methanolic extract (Koca-Caliskan et al., 2018).

Cuscuta campestris Yunck. FAMILY: CONVOVULACEAE

Cuscuta seeds and whole plant have been used to nourish the liver and kidney. Peng et al. (2016) investigated the hepatoprotective activity of the ethanol extract of *Cuscuta campestris* whole plant (CC_{EtOH}). The hepatoprotective effect of ethanol extract (20, 100, and 500 mg/kg) was evaluated on CCl_4-induced chronic liver injury. Serum ALT, AST, TG, and cholesterol were measured and the fibrosis was histologically examined. Ethanol extract exhibited a significant inhibition of the increase of serum ALT, AST, TG, and cholesterol. Histological analyses showed that fibrosis of liver induced by CCl_4 was significantly reduced by the extract. In addition, 20, 100, and 500 mg/kg of the extract decreased the level of MDA and enhanced the activities of antioxidative enzymes including SOD, GPx, and GR in the liver. Peng et al. (2016) demonstrated that the hepatoprotective mechanisms of the ethanolic extract were likely to be associated to the decrease in MDA level by increasing the activities of antioxidant enzymes such as SOD, GPx, and GRd.

Cuscuta chinensis Lam. FAMILY: CONVOLVULACEAE

Kim et al. (2007) investigated the protective effect of seeds of *Cuscuta chinensis* (CS) on acute liver injury induced by DMN in SD rats. Liver injury caused by DMN injection was significantly inhibited in the CS-treated group compared to the Silymarin-treated group. The results of blood biological assay were significantly protected by CS in serum TP (T-protein), T-bilirubin (T-bili), Dbilirubin (D-bili), SGOT, SGPT, and ALP. The HYP content and amount of active a-SMA and PCNA were significantly decreased in the CS-treated group than in the Silymarin-treated group. CS exhibited an *in vivo* hepatoprotective effect and antifibrogenic effects against DMN-induced acute liver injury and inhibited the formation of HYP, which suggests that CS may be useful in preventing fibrogenesis after liver injury.

Tu-Si-Zi, the seeds of *C. chinensis*, is a traditional Chinese medicine that is commonly used to nourish and improve the liver and kidney conditions in China and other Asian countries. Yen et al. (2007) evaluated the hepatoprotective effect of the aqueous and ethanolic extracts of *C. chinensis* on APAP-induced hepatotoxicity in rats. The *C. chinensis* ethanolic extract at an oral dose of both 125 and 250 mg/kg showed a significant hepatoprotective effect relatively to the same extent by reducing levels of SGOT, SGPT, and ALP. In addition, the same ethanolic extract prevented the hepatotoxicity induced by APAP-intoxicated treatment as observed when assessing the liver histopathology. The data suggest that the ethanolic extract of *C. chinensis* can prevent hepatic injuries from APAP-induced hepatotoxicity in rats and this is likely mediated through its antioxidant activities (Yen et al., 2007).

Cuscuta epithymum (L.) L. FAMILY: CONVOLVULACEAE

The effect of a formulation, containing methanolic extracts of the whole plant of *Cuscuta epithymum*, roots of *Begonia laciniata* Roxb., and the whole plant of *Dendrobium ovatum* (L.) Kraenzl was evaluated in CCl_4-induced hepatotoxic albino rats. The treated rats showed reduction in the elevated serum level of AST/SGOT, ALT/SGPT, ALP/SALP, and TB. The treated rats with a BW of 100 mg/kg showed a slight decrease, whereas the decrease in those with BW of 200 and 400 mg/kg were significant. The effect of *Cuscuta epithymum* extract was close to the control drug Silymarin (50 mg/kg BW). The histopathological analysis showed improvement in shape and modeling of hepatocytes in treated groups when compared to the standard and control groups (Ganapaty et al., 2013).

Cuscuta reflexa Roxb. FAMILY: CONVOLVULACEAE

The *Cuscuta reflexa* (aerial parts) MeOH extract is another medicinal plant with significant hepatoprotective effect against hepatotoxicity induced by RIF (100 mg/kg) and INH (100 mg/kg). This extract was administered in Albino rats at 100, 200, and 400 mg/kg by i.p. via, and Silymarin was employed as a positive control. In this assay, the MeOH extract showed that ALT, ASP, ALP, g-GT, TB, and TPs levels were similar to those of the Silymarin group; these results were confirmed with histological analyses (Balakrishnana et al., 2010).

Richariya et al. (2012) evaluated the hepatoprotective and antioxidant activity of ethanolic extract of *C. reflexa* in albino rats by inducing liver damage by CCl_4. The ethanol extract at on oral doses of 100 mg/kg exhibited a highly significant protective effect by lowering serum levels of LPO, serum enzymes ALT, AST, ALP, and bilirubin. The highly significant increase was found in the serum levels of SOD, CAT, and TP. These biochemical observations were supplemented by histopathological examination.

Cyathea gigantea (Wall. ex Hook.) Holttum
FAMILY: CYATHEACEAE

Kiran et al. (2012) investigated the hepatoprotective activity of methanolic leaf extract of *Cyathea gigantea* against PCM-induced liver damage in rats. Wistar albino rats of either sex were divided into five groups of six animals each and are given orally the following treatment for seven days. The normal control group was given 1% Na.CMC 1 mL/kg BW, p.o. Paracetamol at a dose of 1 g/kg BW, p.o. was given as toxic dose for inducing hepatotoxicity. Silymarin (50 mg/kg, p.o.) was given as reference standard. Two doses of *C. gigantea* extract, that is, 100 and 200 mg/kg, p.o. were tested for hepatoprotective activity. The treatment was given for seven days and after 24 h of last treatment blood was collected from retro-orbital plexus and analyzed for various serum parameters like SGOT, SGPT, ALP, TB, and TP in different groups. The PCM intoxication leads to histological and biochemical deteriorations. The treatment with methanolic leaf extract of *C. gigantea* reduced the elevated levels of SGOT, SGPT, ALP, TB, and also reversed the hepatic damage toward normal which further supports the hepatoprotective activity of leaf extract of *C. gigantea*. Kiran et al. (2012) concluded that the methanolic extract of leaves of *C. gigantea* at doses of 100 mg/kg BW and 200 mg/kg BW have a significant effect on the liver of PCM-induced hepatotoxicity model in rats.

Cymbopogon citratus Stapf
FAMILY: POACEAE

The aqueous leaf extracts of *Cymbopogon citratus* showed antihepatotoxic action against cisplatin-induced hepatic toxicity in rats. Hence the extracts have the potential to be used for the management of hepatopathies and as a therapeutic adjuvant in cisplatin toxicity (Arhoghro, 2012).

Rahim et al. (2014) evaluated the protective effects of *C. citratus* aqueous extract against liver injury induced by hydrogen peroxide (H_2O_2), in male rats. *C. citratus* attenuated liver damage due to H_2O_2 administration as indicated by the significant reduction, in the elevated levels of ALT, AST, ALP, LDH, TB, and MDA in serum and liver homogenates; increase in TP and GSH levels in serum and liver homogenates; and improvement of liver histopathological changes. These effects of the extract were similar to that of vitamin C which used as antioxidant reference.

Saenthaweesuk et al. (2017) investigated the protective effect of *C. citratus* (*CS*) extract on PCM-induced hepatotoxicity in rats. Phytochemical screening of the extract indicates the presence of tannins, flavonoids, and phenolic compounds.

Elevation of serum AST, ALT, and MDA levels along with depletion GSH in the liver were observed in rats treated with PCM alone compared with control. Pretreatment of the animal with *CS* extract reduced the levels of hepatic markers (AST and ALT). Pretreatment with *CS* extract also significantly reduced oxidative stress induced by PCM as shown by an increase in GSH level and reduction of MDA compared to rats treated with PCM alone.

Cynara cornigera L. FAMILY: APIACEAE

The ethanol extract of *Cynara cornigera* was fractionated and the fractions were subjected to hepatoprotective assays using Wistar albino rats at a dose of 500 and 250 mg/kg. The liver injury was induced in rats using CCl_4. Biochemical parameters such as AST, ALT, ALP, and TB were estimated as reflections of the liver condition, with Silymarin as a positive control. Phytochemical investigation and chromatographic separation of the hepatoprotective fractions led to the isolation of a new sesquilignan namely cornigerin, along with eight known compounds: apigenin, luteolin, β-sitosterol glycoside, apigenin 7-*O*-β-D-glucopyranoside, luteolin-7-*O*-β-D-glucopyranoside, apigenin-7-*O*-rutinoside, cynarin 1,5-di-*O*-caffeoylquinic acid, and apigenin-7-*O*-β-D-glucuronide.

Cynara scolymus L. FAMILY: APIACEAE

Cynara scolymus (artichoke) is traditionally used for the treatment of digestive disorders, moderate hyperlipidemia, and liver and bile diseases. The leaf extract of *C. scolymus* has been used for its hepatoprotective effects (Gebhardt, 1997). In *C. scolymus* leaf extract, there are compounds such as cynarin, luteolin, chlorogenic acid, and caffeic acid, other flavonoids, and polyphenol compounds, some of which have antioxidant properties. *C. scolymus* leaf extract also positively affected the changes in rat serum liver enzyme induced by $CC1_4$ and histopathological damage to liver tissue (Fallah Huseini et al., 2011). In rats pretreated with artichoke extract, plasma transaminase activities significantly decreased and histopathological changes in the liver ameliorated (Mehmetçik et al., 2008). In the rabbits intoxicated with $CC1_4$, *C. scolymus* leaf extract counteracted the toxic effect of $CC1_4$, blood sugar, cholesterol, TGs, leukocytes, and a number of erythrocytes (Păunescu et al., 2009). In other studies, *C. scolymus* was significant in keeping the normal liver function parameters, maintained the hepatic redox status as it is manifested by a significant increase in antioxidant enzyme activities and reduction in GSH accompanied by inhibition of LPO and protein oxidation, decreased NO and TNF alpha, and stabilized membrane in the untreated PCM intoxicated rats (Ali et al., 2012).

Aktay et al. (2000) studied the hepatoprotective activity of ethanolic extracts of stems, bracts and receptaculum of *C. scolymus* using CCl_4-induced hepatotoxicity model in rats. According to the results of biochemical tests, significant reductions were obtained in CCl_4-induced increases of plasma and liver tissue MDA levels and plasma enzyme activities by the treatment with *C. scolymus* extracts, which reflects functional improvement of hepatocytes.

Hepatocurative effects of *C. scolymus* leaf extract on CCl_4-induced oxidative stress and hepatic injury in rats were investigated by Colak et al. (2016) by serum hepatic enzyme levels, oxidative stress indicator (MDA), endogenous antioxidants, DNA fragmentation, p53, caspase 3 and histopathology. Significant decrease of serum ALT and AST levels were determined in the curative group. MDA levels were significantly lower in the curative group. Significant increase of SOD and CAT activity in the curative group was determined. In the curative group, *C. scolymus* leaf extract application caused the DNA% fragmentation, p53 and caspase three levels of liver tissues toward the normal range. Their results indicated that *C. scolymus* leaf extract has hepatocurative effects on CCl_4-induced oxidative stress and hepatic injury by reducing LPO, providing affected antioxidant systems toward the normal range. It also had positive effects on the pathway of the regulatory mechanism allowing repair of DNA damage on CCl_4-induced hepatotoxicity.

Tang et al. (2017) evaluated the preventive effects of ethanolic extract from artichoke against acute alcohol-induced liver injury in mice. Artichoke extract significantly prevented elevated levels of AST, ALT, TG, TC, and MDA. Meanwhile, the decreased levels of SOD and GSH were elevated by artichoke administration. Histopathological examination showed that artichoke attenuated degeneration, inflammatory infiltration, and necrosis of hepatocytes. Immunohistochemical analysis revealed that expression levels of TLR 4 and nuclear factor-kappa B (NF-κB) in liver tissues were significantly suppressed by artichoke treatment. Results obtained demonstrated that artichoke extract exhibited significant preventive protective effect against acute alcohol-induced liver injury. This finding is mainly attributed to its ability to attenuate oxidative stress and suppress the TLR4/NF-κB inflammatory pathway.

Cynodon dactylon **(L.) Pers.**
FAMILY: POACEAE

Jain et al. (2013) evaluated the hepatoprotective activity of roots of

Cynodon dactylon in CCl_4-induced hepatotoxicity in albino rabbits. Hepatotoxicity was induced in rabbits by CCl_4 0.05 mg/kg, intraperitoneally. Alcoholic extracts of roots of *C. dactylon* were administered orally for 20 days from day 1 to day 20 in the doses of 100 mg/kg/day with the help of syringe. Group I: The rise of serum transaminase, serum ALP, SB and decrease in serum ALB due to hepatotoxic effect of CCl_4 when compared to zero day of same group. Group II: *C. dactylon* extract was able to bring down the level of serum transaminase, serum ALP, SB and increased in serum ALB in a highly significant amount, when compared with group I. *C. dactylon* extract found to be effective in reducing SGOT, SGPT, ALP, and bilirubin. *C. dactylon* extract helps in normalize serum ALB level. *C. dactylon* extract had shown protection in restoration of liver function and regeneration of liver cells as observed on histopathology.

Surendra et al. (2011) evaluated hepatoprotective activity of aerial parts of *C. dactylon* against CCl_4-induced hepatotoxicity in Wister rats. Various doses of ethanolic extract of aerial parts of *C. dactylon* such as 100, 250, and 500 mg/kg were administered to animals. Researchers assessed the SB, cholesterol, SGPT, SGOT, and ALP levels. It was found that the extract of *C. dactylon* significantly reversed the rise in SB and cholesterol levels. The ethanolic extract also prevented decrease in secretion of ascorbic acid in urine in CCl_4 intoxicated group. The hepatic damage in animals treated with ethanolic extract was minimal causing no damage to structure and architectural frame of hepatic cells. Researchers concluded that the activity of extract could be attributed to preservation of structural integrity of cell membrane of hepatocytes and thereby maintaining normal function of liver.

The hepatoprotective activity of roots of *C. dactylon* in CCl_4-induced hepatotoxicity was studied in albino rabbits. Alcoholic extracts of roots of *C. dactylon* was administered orally for 20 days in a doses of 100 mg/kg/day. *C. dactylon* extract was able to bring down the level of serum transaminase, serum ALP, SB and increased in serum ALB significantly, when compared with untreated group (Kowsalya et al., 2015).

Cyperus alternifolius L. FAMILY: CYPERACEAE

Awaad et al. (2012) explored the potential hepatoprotective activity of total ethanol and successive extracts of *Cyperus alternifolius* against CCl_4-induced hepatotoxicity in rats and to isolate their bioactive constituents. The plant proved to be safe for human use because it did not induce any signs of toxicity or mortality in mice when administered orally at doses up to 5000 mg/kg. The total alcoholic extract in doses of 100 and 200 mg/kg and the successive extracts (ether, chloroform, and ethyl acetate) in a dose of 10 mg/kg exhibited a significant protective effect by lowering the elevated serum levels of AST, ALT, ALP. These results concluded that *C. alternifolius* possesses a significant protective effect against hepatotoxicity induced by CCl_4. Eight phenolic compounds were isolated from *C. alternifolius* and identified as esculetin, umbelliferon, imperatorin, psoralen, xanthotoxin, quercetin, quercetin-3-*O*-rutinoside and GA.

***Cyperus articulatus* L. FAMILY: CYPERACEAE**

Datta et al. (2013) evaluated the hepatoprotective activity of the methanol extract of *Cyperus articulatus* (MECA) against PCM-induced liver damage in rats. In MECA-treated animals, the toxic effect of PCM was controlled significantly by restoration of the biochemical parameters, such as, SGPT, SGOT, ALP, TP, and total billirubin, as well as by the improvement of the antioxidant status to/toward near normal values. Histology of the liver sections of the animals treated with the extracts showed the presence of normal hepatic cords, the absence of necrosis and fatty infiltration, which further evidenced the hepatoprotective activity of MECA.

***Cyperus rotundus* L. FAMILY: CYPERACEAE**

Ethyl acetate extract and two crude fractions, solvent ether and ethyl acetate, of the rhizomes of *Cyperus rotundus* were evaluated by Kumar and Mishra (2004, 2005) for hepatoprotective activity in rats by inducing liver damage by CCl_4. The ethyl acetate extract at an oral dose of 100 mg/kg exhibited a significant protective effect by lowering serum levels of glutamic oxaloacetic transaminase, GPT, ALP, and TB. These biochemical observations were supplemented by histopathological examination of liver sections.

The effects of *C. rotundus* rhizome on cellular lipogenesis and nonalcoholic/diet-induced fatty liver disease, and the molecular mechanism of these actions were studied by Oh et al. (2015). It appeared that the hexane fraction of *C. rotundus* rhizome reduced the elevated transcription levels of sterol regulatory element binding protein-1c (SREBP-1c) in primary hepatocytes following exposure to the liver X receptor α (LXRα) agonist. The SREBP-1c gene was a master regulator of lipogenesis and a key target of LXRα.

CRHF inhibited not only the LXRα-dependent activation of the synthetic LXR response element (LXRE) promoter, but also the activation of the natural SREBP-1c promoter. Moreover, the hexane fraction of *C. rotundus* decreased (i) the recruitment of RNA polymerase II to the LXRE of the SREBP-1c gene; (ii) the LXRα-dependent upregulation of various lipogenic genes; and (iii) the LXRα-mediated accumulation of TGs in primary hepatocytes. Furthermore, the hexane fraction of *C. rotundus* ameliorated fatty liver disease and reduced the expression levels of hepatic lipogenic genes in high sucrose diet-fed mice. CRHF did not affect the expression of ATP-binding cassette transporter A1, another important LXR target gene that was required for reverse cholesterol transport and protected against atherosclerosis. Accordingly, these results suggested that the hexane fraction of *C. rotundus* might be a novel therapeutic remedy for fatty liver disease through the selective inhibition of the lipogenic pathway (Oh et al., 2015).

Cyperus scariosus R.Br. FAMILY: CYPERACEAE

The hepatoprotective activity of aqueous-methanolic extract of *Cyperus scariosus* was investigated against APAP and CCl_4-induced hepatic damage. Acetaminophen produced 100% mortality at a dose of 1 g/kg in mice while pretreatment of animals with plant extract (500 mg/kg) reduced the death rate to 30%. Acetaminophen at a dose of 640 mg/kg produced liver damage in rats as manifested by the rise in serum levels of ALP, SGOT, and SGPT compared to respective control values. Pretreatment of rats with plant extract (500 mg/kg) significantly lowered the respective serum ALP, SGOT, and SGPT levels. The hepatotoxic dose of CCl_4 (1.5 mL/kg; orally) raised serum ALP, SGOT, and SGPT levels compared to respective control values. The same dose of plant extract (500 mg/kg) was able to significantly prevent CCl_4-induced rise in serum enzymes and the estimated values of ALP, SGOT and SGPT. The plant extract also prevented CCl_4-induced prolongation in pentobarbital sleeping time confirming hepatoprotectivity.

Cyperus tegetum Roxb. FAMILY: CYPERACEAE

The methanol extract of *Cyperus tegetum* rhizome (MECT) was evaluated by Halder et al. (2011) for its effect on PCM-induced liver damage in Wistar rats. Serum biochemical parameters, namely, SGOT, SGPT, ALP, total serum protein, TB content and liver biochemical parameters such as TBARS and reduced GSH content were estimated. Biochemical and histopathological observations indicated that MECT had remarkable hepatoprotective effect against PCM-induced liver damage in rats.

Cytisus scoparius L. FAMILY: FABACEAE

Raja et al. (2007) investigated the antioxidant activity of *Cytisus scoparius* on CCl_4 treated oxidative stress in Wistar albino rats. Pretreatment of rats with different doses of plant extract (250 and 500 mg/kg) significantly lowered SGOT, SGPT, LDH, and TBARS levels against CCl_4-treated rats. GSH and hepatic enzymes like SOD, CAT, GPx, GRD, and GST were significantly increased by treatment with the plant extract, against CCl_4-treated rats. The activity of extract at the dose of 500 mg/kg was comparable to the standard drug, Silymarin (25 mg/kg).

Dalbergia spinosa Roxb. FAMILY: FABACEAE

Hepatoprotective effect of *Dalbergia spinosa* against PCM-induced toxicity in rats was investigated by Kumaresan et al. (2013). Levels of liver marker enzymes SGPT, SGOT, ALP, and SB were increased in rats treated with PCM prior to administration of methanolic and aqueous extracts. Values of those parameter were significantly reduced on treatment with methanolic and aqueous extract. These results were confirmed by histopathological examinations of liver of control and treated groups.

Daniella oliveri (Rolfe) Hutch. & Dalziel FAMILY: BIGNONIACEAE

Onoja et al. (2015) investigated the hepatoprotective and antioxidant activity of the methanolic extract of *Daniella oliveri* leaves. The hepatoprotective activity was investigated using CCl_4-induced hepatotoxicity in albino rats. The pretreatment with extract (100, 200, and 400 mg/kg) and Silymarin (100 mg/kg) produced a significant dose-dependent increase in hepatoprotective activity when compared with the negative control group. The extract (25–400 μg/mL concentration) produced a concentration-dependent increase in antioxidant activity in DPPH photometric assay. The IC_{50} of the extract in DPPH photometric assay was 400 μg/mL concentrations. The extract and Silymarin showed a significant dose-dependent increase in CAT level in treated rats when compared with the negative control group. Also, the extract and Silymarin produced a significant dose-dependent decrease in MDA level in treated rats when compared with the negative control group. The results of this study suggest that *D. oliveri* leaves have a potent hepatoprotective activity that may be linked to its antioxidant activities and validates its use in the traditional management of liver disorders.

Daucus carota L. FAMILY: APIACEAE

Bishayee et al. (1995) reported the hepatoprotective effect of aqueous extracts of fresh tuber roots of *Daucus carota* on CCl_4-induced acute liver damage. The increased serum enzyme levels by CCl_4 induction were lowered due to pretreatment with the extract. The extract also decreased the elevated SB and urea content due to CCl_4 administration. Results of this study revealed that *D. carota* could afford a significant protective action in the alleviation of CCl_4 induced hepatocellular injury.

Singh et al. (2012) assessed the *in vivo* antioxidant and hepatoprotective activity of methanolic extract of *D. carota* seeds in experimental animals. Methanolic extracts of *D. carota* seeds is used for hepatoprotection assessment. Oxidative stress was induced in rats by TAA 100 mg/kg s.c., in four groups of rats (two test, standard, and toxic control). Two test groups received *D. carota* seeds extract at doses of 200 and 400 mg/kg. Standard group received Silymarin (25 mg/kg) and toxic control received only TAA. Control group received only vehicle. On the 8th day, animals were sacrificed and liver enzyme like SGPT, SGOT, and ALP were estimated in blood serum and antioxidant enzyme like SOD, CAT, glutathione reductase (GRD), GPx, GST, and LPO were estimated in liver homogenate. A significant decrease in SGPT, SGOT, and ALP levels was observed in all drug treated groups as compared to TAA group and in case of antioxidant enzyme a significant increase in SOD, CAT, GRD, GPx, and GST was observed in all drug-treated groups as compared with TAA group. But in case of LPO, a significant reduction was observed as compared to toxic control group. Seed extract has contributed to the reduction of oxidative stress and the protection of liver in experimental rats.

Decalepis hamiltonii Wight & Arn. FAMILY: APOCYNACEAE

Hepatoprotective activity of the roots of *Decalepis hamiltonii* was studied by Harish and Shivanandappa (2010) using CCl_4 induced liver injury model in albino rats. The hepatotoxicity produced by acute CCl_4 administration was found to be inhibited by pretreating the rats with crude methanolic extract of the roots of *D. hamiltonii* prior to CCl_4 induction. Hepatotoxic inhibition was measured with the decreased levels of hepatic serum marker enzymes SGPT, SGOT, ALP, and LDH and LPO formation. Imbalance level of

GSH and antioxidant enzymes such as CAT, GPx, and GR were normalized in rats pretreated with Dh extract followed by CCl_4 administration. Pathological changes of hepatic lesions caused by CCl_4 were also improved by pretreatment with the *D. hamiltonii* root extract.

The hepatoprotective activity of the aqueous extract of the roots of *D. hamiltonii* with known antioxidant constituents was studied by Srivastava and Shivanandappa (2006, 2010) against CCl_4-induced oxidative stress and liver injury in rats. Pretreatment of rats with aqueous extract of the roots of *D. hamiltonii*, single (50, 100, and 200 mg/kg BW) and multiple doses (50 and 100 mg/kg BW for 7 days) significantly prevented the CCl_4 (1 mL/kg BW)-induced hepatic damage as indicated by the serum marker enzymes (AST, ALT, ALP, and LDH). Parallel to these changes, the root extract also prevented CCl_4-induced oxidative stress in the rat liver by inhibiting LPO and protein carbonylation, and restoring the levels of antioxidant enzymes (SOD, CAT, GPx, GR, and GST) and GSH. The biochemical changes were consistent with histopathological observations suggesting marked hepatoprotective effect of the root extract in a dose-dependent manner. Protective effect of the aqueous extract of the roots of *D. hamiltonii* against CCl_4-induced acute hepatotoxicity could be attributed to the antioxidant constituents (Srivastava and Shivanandappa, 2010).

Delonix regia (Bojer ex Hook.) Raf. FAMILY: FABACEAE

Ahmed et al. (2011) evaluated the possible beneficial effect of ME of aerial parts of *Delonix regia* against CCl_4-induced liver damage in rats. The ME of aerial parts of *D. regia* (400 mg/kg) was administered orally to the Wistar albino rats with hepatotoxicity induced by CCl_4 (2 mL/kg, p.o.). Silymarin (50 mg/kg, p.o.) was given as reference standard. The plant extract was effective in protecting the liver against the injury induced by CCl_4 in rats. This was evident from significant reduction in serum enzymes AST, ALT, ALP, ALB, TLP (total protein), DBIL, and TBIL. Histopathological observation showed hugely dilated central vein, disrupted cords of hepatocytes and few hepatocytes shows feathery change, mild inflammation and moderate degree of macro and micro vesicular steatosis. It can be concluded that the methanolic extract of aerial parts of *D. regia* possesses hepatoprotective activity against CCl_4-induced hepatotoxicity in rats.

Fractionation of the ethanolic extract of the flowers of *Delonix regia* led to the isolation of three sterols, namely, stigmasterol (1.54%), ß-sitosterol and its 3-*O*-glucoside (6.93%), a triterpene, namely, ursolic acid (3.61%) and four flavonoids: quercetin (2.92%), quercitrin (0.59%), isoquercitrin (3.87%) and rutin (5.12%) in addition to the amino acid L-azeditine-2-carboxylic acid (El-Sayed et al., 2011). The ethanolic extract and its two fractions were tested by El-Sayed et al. (2011) for hepatoprotective activity against CCl_4-induced hepatic cell damage in rats at two dose levels (50 and 100 mg/kg), and the flavonoid-rich fraction showed significant hepatoprotection at 100 mg/kg. The presence of the aforementioned flavonoids with their efficient free-radical scavenging properties may explain this liver protection ability.

Azab et al. (2013) evaluated the hepatopotective activity of *Delonix regia* using CCl_4-induced liver injury. The plant extract at dose of 50, 100, and 200 mg/kg BW reduced serum AST, ALT, and ALP as well as total and DBL in a dose-dependent manner. The treatment of rats with the extract at different doses significantly increased the liver tissue level of SOD, CAT, reduced GSH and TAC and decreased the level of MDA as compared to CCl_4-treated group.

Descurainia sophia (L.) Webb. ex Prntl. FAMILY: BRASSICACEAE

Moshaie-Nezhad et al. (2018) evaluated the protective effects of *Descurainia sophia* seed extract on PCM-induced oxidative stress and acute liver injury in mice. Pretreatment of mice with *D. sophia* seed extract, significantly prevented the PCM-induced elevation in the levels of serum ALT, AST, and ALP, TB and MDA. The results of histopathologic studies were consistent with the above findings.

Desmodium oojeinense (Roxb.). H. Ohashi FAMILY: FABACEAE

Jayadevaiah et al. (2017) investigated the hepatoprotective activity of ethanolic extract of stem bark of *Desmodium oojeinense* against CCl_4-induced hepatotoxicity in Wistar rats. The hepatoprotective activity of the ethanolic extract was assessed in CCl_4 induced hepatotoxic rats. Alteration in the levels of biochemical markers of hepatic damage like SGOT, SGPT, ALP, and TB were tested in both CCl_4 treated and untreated groups. CCl_4 has enhanced the SGOT, SGPT, ALP, and TB in liver. Treatment with ethanolic extract of *D. oojeinense* has brought back the altered levels of biochemical markers to the near normal levels in the dose-dependent manner. Results of histopathological studies also provided supportive evidence for biochemical analysis.

Desmostachya bipinnata (L.) Stapf FAMILY: POACEAE

The hepatoprotective effect of the polyphenolic fraction of *Desmostachya bipinnata* root (PFDB) was studied in liver damage induced in female SD rats. A dose-dependent increase in percentage viability was observed when ethanol-exposed BRL3A cells were treated with PFDB. Both the treatment groups upon pretreatment with PFDB exhibited a significant protective effect by lowering serum glutamic oxaloacetic transaminase, serum GPT, ALP, TGs, cholesterol, urea, uric acid, bilirubin, and creatinin levels and improving protein level in serum in dose-dependent manner, which was comparable to that of Silymarin group. In addition, PFDB prevented elevation of reduced GSH, GPx, superoxide dismutase (SOD), and CAT in the TAM-intoxicated rats in concentration-dependent manner and significantly reduced the LPO in the liver tissue. The biochemical observations were confirmed by histopathological studies, which showed the attenuation of hepatocellular necrosis (Rahate and Rajasekarn, 2015).

The hepatoprotective potential of dried powdered roots of *Desmostachya bipinnata* (100 and 200 mg/kg, orally for 7 days) was studied against PCM-induced liver damage in Wistar rats. Animals before treatment with aqueous extract of *D. bipinnata* showed significant reduction in the elevated level of serum marker enzymes, MDA, LH, bilirubin,and significant improvement in the antioxidant enzymes when compared to PCM damaged rats. *D. bipinnata* showed good hepatoprotective and antioxidant activity when compared to Silymarin (Gouri et al., 2014).

Dichrostachys cinerea (L.) Wight & Arn. FAMILY: FABACEAE

Suresh Babu et al. (2011) evaluated the hepatoprotective activity of the methanolic extract of *Dichrostachys cinerea* leaves using albino mice and rats. The methanolic extract did not show any mortality up to a dose of 3500 mg/kg BW. The methanolic extract showed significant hepatoprotectivity. CCl_4 treated animals showed significant increase in the levels of TB, AST, ALT, and ALP, as compared to the controls. In addition, the TP level was decreased, reflecting the liver injury due to the toxic effect of CCl_4. Animals treated with methanolic extract showed significant reduction in the levels of liver function serum markers. The effect was more pronounced in the animals treated with ME; thus, proving the significant hepatoprotective action of the extract. The histopathological profile of the drug-treated liver tissue demonstrated similar morphology as that of controls.

Dicranopteris linearis (Burm. f.) Underw. FAMILY: GLEICHENIACEAE

Ismail et al. (2014) studied hepatoprotective effects of *Dicranopteris linearis* aqueous leaf extract (DLAE) against CCl_4 or PCM-induced toxicity. Biochemical and histopathological assessment showed that high dose of DLAE significantly reduced the level of ALP, AST, and ALT against PCM or CCl_4. There was necrosis and inflammation in liver due to CCl_4 or PCM-induced toxicity. It showed that DLAE has strong hepatoprotective activities due to antioxidant properties and high contents of flavonoids. The mathanolic extract of leaves of the same plant was tested against CCl_4 induced liver toxicity in rats. It was observed that the serum level of AST, ALT, and ALP decreased in the mice which were pretreated with ME of *D. linearis*, which were increased when CCl_4 was administered (Kamisan et al., 2014). Similar study was conducted by Mamat et al. (2013) on the methanolic extract of *D. linearis* leaves for its hepatoprotective activity against PCM and CCl_4-induced liver damage in rats. The liver marker enzymes level was reduced to normal when treated with the extract. It was concluded that hepatoprotective activity of MEDL is attributed to its antioxidant activity and flavonoids contents.

Zakaria et al. (2017) investigated the potential of methanolic extract of *D. linearis* leaves to attenuate liver intoxication induced by APAP in rats. Pretreatment of rats with 10% DMSO failed to attenuate the toxic effect of APAP on the liver as seen under the microscopic examination. This observation was supported by the significant increase in the level of serum liver enzymes of ALT, AST, and ALP, and significantly decreased in the activity of endogenous antioxidant enzymes of CAT and SOD in comparison to Group 1. Pretreatment with MEDL, at all doses, significantly reduced the level of ALT and AST while the levels of CAT and SOD were significantly restored to their normal value. Histopathological studies showed remarkable improvement in the liver cells architecture with an increase in the dose of the extract. MEDL also demonstrated a low to none inhibitory activity against the respective LOX- and NO-mediated inflammatory activity. The HPLC and GCMS analyses of leaves extract demonstrated the presence of several nonvolatile (such as rutin, GA, etc.) and volatile (such as methyl palmitate, shikimic acid, etc.) bioactive compounds.

Digera muricata (L.) Mart. FAMILY: AMARANTHACEAE

Khan et al. (2011) evaluated the probable protective effects of *Digera muricata* methanol extract (DME) against acrylamide induced hepatocellular injuries in female SD rat. Treatment with ME (100, 150, and 200 mg/kg), dose dependently, ameliorated the toxicity of acrylamide and the studied parameters were reversed toward the control level. Hepatic lesions induced with acrylamide were reduced with DME treatment. Phytochemical screening indicates the presence of flavonoids, alkaloids, terpenoids, saponins, tannins, phlobatanin, coumarins, anthraquinones, and cardiac glycosides. Total phenolic and flavonoids contents were 205 ± 0.23 and 175.0 ± 0.65 mg/g as equivalent to GA and rutin, respectively, in ME.

Diospyros lotus L. FAMILY: EBENACEAE

Cho et al. (2015) investigated the protective effect of *Diospyros lotus* leaf extract (DLLE) against APAP-induced acute liver injury in mice. Administration of DLE significantly attenuated the levels of serum AST, ALT, and liver LPO in APAP-treated mice. Histopathological examination showed that DLE treatment decreased the incidence of liver lesions in APAP-treated mice. DLE treatment markedly increased SOD, CAT, GPx activity, and GSH levels in APAP-treated mice. Furthermore, DLE treatment significantly suppressed the production of proinflammatory factors such as the NO, IL-6, TNF-α, and iNOS in APAP-treated mice.

Dipteracanthus patulus (Jacq.) Nees FAMILY: ACANTHACEAE

The hepatoprotective activity as well as the possible underlying mechanism(s) of methanolic extract of *Dipteracanthus patulus* (MEDP) using CCl_4 and PCM rat models was evaluated by Shrinivas and Suresh (2012). In both CCl_4 and PCM rat models, MEDP treatment is able to restore depleted antioxidant liver enzymes (SOD, GSHPx, and CAT) and also depleted the elevated serum biomarkers significantly, respectively. In addition, histoarchitecture of liver was also improved with the treatment of MEDP. The protective activity of MEDP (500 mg/kg p.o.) was comparable to that of Silymarin (100 mg/kg p.o.). The hepatoprotective activity of MEDP seems to be related to its antioxidant activity possibly through free-radicals scavenging mechanism by activating antioxidant enzymes.

Dodonaea viscosa (L.) Jacq. FAMILY: SAPINDACEAE

Ahmad et al. (2012) determined the antihyperlipidemic and hepatoprotective activity of *Dodonaea viscosa* leaves extracts in the alloxan-induced diabetic rabbits. Serum levels of TGs, TC, LDL-cholesterol, HDL-cholesterol, ALT, and AST were estimated. The oral administration of aqueous:methanol (70:30) extract of the *D. viscosa* leaves significantly decreased the raised parameters (TG, TC, and LDL cholesterol) to normal values. But the extract has significantly increased HDL-cholesterol, ALT, and AST levels.

For the aqueous:methanol (70:30) extract given animals, the average serum level of TC was 60.00 ±1.30 mg/dL, LDL-cholesterol was 92.80 ±2.29 mg/dL, HDL-cholesterol was 31.80±1.0 mg/dL and TG was 15.40 ± 0.75 mg/dL while the average serum levels of ALT and AST were 45.60 ± 3.08 and 27.20 ± 1.36 IU/dL, respectively. It was concluded from the study that aqueous: methanolic (70:30) extract of *D. viscosa* leaves exerts antihyperlipidemic and hepatoprotective effects in the alloxan-induced diabetic rabbits.

Ali et al. (2014) investigated the hepatoprotective potential of the methanolic extract of the whole plant of *D. viscosa* and its ethyl acetate, aqueous, butanol, and *n*-hexane fractions against CCl_4-induced hepatoxicity in rats. Hepatoprotection was assessed in terms of reduction in serum enzymes (ALT, AST, and ALP) that occur after CCl_4 injury, and by histopathology and immunohistochemistry. The methanolic extract reduced the serum enzyme level (ALT, AST, and ALP) down to control levels despite CCl_4 treatment. It also reduced the CCl_4-induced damaged area to 0% as assessed by histopathology. The CD68+ macrophages were also reduced in number around the central vein area by the methanolic extract. These hepatoprotective effects were better than the positive control Silymarin. Similar hepatoprotective activities were found with the ethyl acetate, and AFs of the methanolic extract. The butanol and *n*-hexane fractions showed elevated levels of ALT, AST, and ALP as compared to the positive control Silymarin. Histopathology showed ~30% damage to the liver cells with the butanol and *n*-hexane fractions which still showed some protective activity compared to the CCl_4-treated control. HPLC fingerprinting suggested that hautriwaic acid present in the methanolic extract and its ethyl acetate, and AFs may be responsible for this hepatoprotective activity of *D. viscosa* which was confirmed by *in vivo* experiments.

Dolichos lablab L. FAMILY: FABACEAE

The hepatoprotective effects and underlying mechanism of *Dolichos lablab* water extract (DLL-Ex) were assessed using an *in vitro* cellular model in which NAFLD was simulated by inducing excessive FFA influx into hepatocytes. HepG2 cells were treated with DLL-Ex and FFAs for 24 h. DLL-Ex inhibited expression of CD36 in HepG2 cells, which regulates fatty acid uptake, as well as BODIPY-labeled fatty acid uptake. Additionally, DLL-Ex significantly attenuated FFA-mediated cellular energy depletion and mitochondrial membrane depolarization. Furthermore, DLL-Ex enhanced phosphorylation of AMPK, indicating that AMPK is a critical regulator of DLL-Ex-mediated inhibition of hepatic lipid accumulation, possibly through its antioxidative effect (Im et al., 2016).

Ecbolium viride (Forssk.) Alston FAMILY: ACANTHACEAE

Cheedella et al. (2013) investigated the possible antioxidant and protective role of *Ecbolium viride* against PCM-induced hepatotoxicity. The plant extract (100, 200, and 400 mg/kg, p.o.) showed a remarkable hepatoprotective activity against PCM-induced hepatotoxicity as judged from the serum marker enzymes in rats. Paracetamol induced a significant rise in AST, ALT, ALP, TB. Treatment of rats with different doses of plant extract (100, 200, and 400 mg/kg) significantly altered serum marker enzymes levels to near normal against PCM-treated rats. The activity of the extract at dose of 400 mg/kg was approximately comparable to the standard drug, Silymarin (25 mg/kg, p.o.). Antioxidant activity of the extract was evaluated by using DPPH-radical scavenging, NO radical scavenging, superoxide radical scavenging, H_2O_2 scavenging, hydroxy radical scavenging, and LPO inhibitory methods. Histopathological changes of liver samples were compared with respective control. This study revealed that ethanolic extract of *E. viride* roots has significant hepatoprotective activity against PCM-induced hepatotoxicity in rats and antioxidant potential.

Echinochloa colona (L.) Link FAMILY: POACEAE

Praneetha et al. (2017) evaluated the *in vitro* hepatoprotective activity of methanolic extract of caryopses of *Echinochloa colona* (ECME) against ethanol in HepG2 cell lines. The extract has shown a dose-dependent cytoprotective activity with maximum protection at 200 μg/mL. The percentage cell viability of the extract, ECME at 200 μg/mL was more, that is, 69.33% which was well comparable to that of standard drug, Silymarin (100 μg/mL).

Echinophora platyloba DC. FAMILY: APIACEAE

Heidarian et al. (2014) evaluated the effects of *Echinophora platyloba* extract on APAP-induced hepatotoxicity in rats. Treatment of rats with different doses of *E. platyloba* extract significantly reduced the elevated serum levels of ALT, AST, ALP, and MDA compared to the untreated group. The effects of 1000 mg/kg *E. platyloba* were similar to the control and treated Silymarin groups. Moreover, *E. platyloba* significantly increased the activity of liver CAT and serum antioxidant capacity at different doses. Also,

E. platyloba extract improved liver histopathological changes compared with the respective control group.

Echinops galalensis Schweinf. FAMILY: ASTERACEAE

The ME of the flowering aerial parts of *Echinops galalensis*, its fractions, and the isolated compounds have been reported for their hepatoprotective effects against CCl_4-induced cell damage in an *in vitro* assay on human hepatoma cell line (Huh7). The extract and isolated compounds at 100 μg/mL prior to CCl_4 challenge protected against cell injury by decreasing the level of AST, ALT, MDA, and increasing the activities of SOD (Abdallah et al., 2013). The protective effects of *E. galalensis* methanolic extract, its fractions as well as the isolated compounds are at least partly due to their antioxidant activities as evidenced by the reduction in MDA level and the increase in SOD activity.

Eclipta prostrata (L.) L. [Syn.: *Eclipta alba* (L.) Hassk.] FAMILY: ASTERACEAE

The protective effect of *Eclipta prostrata (Syn.: Eclipta alba)* on CCl_4-induced acute liver damage was studied by Ma-Ma et al. (1978) using 54 female guinea pigs as experimental animals. Pretreatment of the animals with *E. prostrata* gave significant protection from the hepatotoxic action of CCl_4. This was evidenced by studying the mortality rate, serum AST, serum ALT, and serum ALP activity in *E. prostrata*-protected and -unprotected groups of animals. The mortality rate at the end of 24 h was 77.7% in the unprotected group and 22.2% in the protected group. Serum enzyme activities were also significantly lower in the *E. prostrata*-protected group. The protective effect was also seen histologically, where centrilobular necrosis, hydropic degeneration, and fatty change of the hepatic parenchymal cells were markedly reduced in the animals receiving *E. prostrata* treatment before CCl_4 intoxication.

Alcoholic and CEs of *Eclipta prostrata, Tephrosia purpurea*, and *Boerhavia diffusa* were screened by Murthy et al. (1992) for antihepatotoxic activity. The extracts were given after the liver was damaged with CCl_4. Liver function was assessed based on liver-to-body-weight ratio, pentobarbitone sleep time, serum levels of transaminase (SGPT, SGOT), ALP, and bilirubin. Alcoholic extract of *Eclipta prostrata* was found to have good antihepatotoxic activity.

The hepatoprotective effect of the ethanol/water (1:1) extract of *E. prostrata* (Ep) has been studied by Saxena et al. (1993) at subcellular levels in rats against CCl_4-induced hepatotoxicity. Ep significantly counteracted CCl_4-induced inhibition of the hepatic microsomal drug metabolising enzyme amidopyrine *N*-demethylase and membrane bound glucose 6-phosphatase, but failed to reverse the very high degree of inhibition of another drug metabolising enzyme aniline hydroxylase. The loss of hepatic lysosomal acid phosphatase and ALP by CCl_4 was significantly restored by Ep. Its effect on mitochondrial succinate dehydrogenase and adenosine 5′-triphosphatase was not significant. The study showed that hepatoprotective activity of Ep is by regulating the levels of hepatic microsomal drug metabolising enzymes.

An alcoholic extract of freshly collected *E. prostrata* exhibited dose-dependent (62.5–500.0 mg/kg p.o.) significant hepatoprotective activity against CCl_4-induced liver injury in rats and mice (Singh et al., 1993). It indicated its protective role on parameters such as hexobarbitone-induced sleep, zoxazolamine-induced paralysis, BSP clearance, serum levels of transaminases, bilirubin, and protein. The extract did not show any signs of toxicity and the minimum lethal dose was greater than 2.0 g/kg when given orally and intraperitoneally in mice.

The hepatoprotective effects of *E. prostrata* were studied by Lin et al. (1996) on acute hepatitis induced in mice by a single dose of CCl_4 (31.25 µL/kg, i.p.) or APAP (600 mg/kg, i.p.) and in rats by a single dose of β-D-galactosamine (188 mg/kg, i.p.). The hepatoprotective activity was monitored by estimating the serum transaminases (SGOT and SGPT) levels and histopathological changes in the liver of experimental animals. The *E. prostrata* extracts significantly inhibited the acute elevation of serum transaminases induced by CCl_4 in mice and by β-D-GaLN in rats. However, in the experimental model of APAP, although an inhibiting tendency was noticed, no statistical significance was observed. Histopathologically, the crude *E. prostrata* extract significantly ameliorates either CCl_4 or GaLN-induced histopathological changes in the liver of experimental animals but no statistically significant improvement could be observed in APAP-induced liver damage (Lin et al., 1996).

The effect of *E. prostrata* (EP) extract was studied by Tabassum and Agrawal (2004) on PCM-induced hepatic damage in mice. Treatment with 50% ethanol extract of *E. prostrata* (100 and 250 mg/100 g BW) was found to protect the mice from hepatotoxic action of PCM as evidenced by significant reduction in the elevated serum transaminase levels. Histopathological studies showed marked reduction in fatty degeneration and centrizonal necrosis, in animals receiving different doses of *E. prostrata* along with PCM as compared to the control group. The mice administered Liv.

52 used for comparative evaluation, showed a significant reduction in serum enzyme activity and normal livers. It is stipulated that the extract treated groups were partially protected from hepatocellular damage caused by PCM.

Parmar et al. (2010) evaluated the hepatoprotective activity of leaves of *E. prostrata* against PCM-induced toxicity in male Wistar rats. Paracetamol-induced hepatotoxicity resulted in an increase in serum AST, ALT, ALP activity, and bilirubin level accompanied by significant decrease in albumin level. Co-administration of the aqueous extract protects the PCM-induced lipid peroxidatiom, restored latered serum marker enzymes and antioxidant level toward near normal. Prabu et al. (2011) also reported the protective effect of leaf extract of *E. prostrata* on PCM-induced toxicity in liver.

Thirumalai et al. (2011) elucidated the restorative effect of the aqueous leaf extract (85%) of *E. prostrata* against CCl_4-induced hepatotoxicity in male albino rats. CCl_4-induced oxidative stress was indicated by elevated levels of TBARS and HP, and augmented levels of serum AST, ALT, and ALP. The depleted activity levels of antioxidant enzymes such as SOD, CAT, GPx, and GST were found in CCl_4-induced animals. The aqueous leaf extract of *E. prostrata* (250 mg/kg BW) ameliorated the effects of CCl_4 and returned the altered levels of the biochemical markers near to the normal levels.

Anoopraj et al. (2014) studied the chemopreventive potential of *E. prostrata* in male Wistar rats. The study revealed a partial protective effect on DEN-induced hepatocarcinogenesis as indicated by histopathological examinations and values of liver function indices. This can be attributed to the protective effect on DEN-induced oxidative cell membrane damage.

Beedimani and Shetkar (2015) evaluated the hepatoprotective activity of aqueous extract of *E. prostrata* against CCl_4-induced toxicity in male albino rats. *E. prostrata* administration at doses 250 and 500 mg/kg orally demonstrated significant hepatoprotective activity by preventing the increase of ALT, AST, ALP, and SB and also confirmed by histopathology of the liver. The results were comparable to that of Silymarin. The results of the study confirmed the hepatoprotective activity of aqueous extracts of *E. alba* at doses of 250 and 500 mg/kg against CCl_4-induced hepatotoxicity in rats. However, the dose adjustments may be necessary to optimize the similar hepatoprotective efficacy in clinical settings.

Baranisrinivasan et al. (2009) evaluated the hepatoprotective effect of the aqueous leaf extracts of *Enicostema littorale* and *Eclipta prostrata* combine (1:1) at dose level of 250 mg/kg BW including ethanol-induced oxidative stress in liver tissue of Wistar male albino rats. The aqueous leaf extracts supplementation of *Enicostema littorale* and *Eclipta prostrata* combine (1:1) produced significant hepatoprotection and antioxidative effect during ethanol-induced hepatotoxicity.

Elephantopus scaber L. FAMILY: ASTERACEAE

The efficacy of the medicinal plant *Elephantopus scaber* to prevent CCl_4-induced chronic liver dysfunction in the rats was examined by Rajesh and Latha (2001) by determining different biochemical markers in serum and tissues. A marked elevation in the activities of enzymes like AST, ALT, ALP, and protein content was observed in CCl_4-treated rats. The concentrations of total lipid, cholesterol and phospholipids were studied in serum and the different tissues. The concentration of serum TGs was also studied. In rats, which received both *E. scaber* and CCl_4 the activities of the liver function marker enzymes and the concentrations of TP, total lipid, phospholipids, TGs, and cholesterol were maintained at near normal levels.

Ho et al. (2012) compared the *in vivo* hepatoprotective effect of *Elephantopus scaber* with *Phyllanthus niruri* on the ethanol-induced liver damage in mice. The total phenolic and total flavanoid content of *E. scaber* ethanol extract were determined in this study. Accelerating serum biochemical profiles (including AST, ALT, ALP, TG, and TB) associated with fat drop and necrotic body in the liver section were observed in the mice treated with ethanol. Low concentration of *E. scaber* was able to reduce serum biochemical profiles and the fat accumulation in the liver. Furthermore, high concentration of *E. scaber* and positive control *P. niruri* were able to revert the liver damage, which is comparable to the normal control. Added to this, *E. scaber* did not possess any oral acute toxicity on mice.

Elettaria cardamomum (L.) Maton FAMILY: ZINGIBERACEAE

Aboubakr (2016) evaluated the hepatoprotective activity of aqueous extract of cardamom in acute experimental liver injury induced by gentamicin. Twenty four male albino rats were randomly divided into four groups (six rats in each). Animals of the first group served as control and orally (p.o.) received (1 mL/kg saline). The second experimental group was given gentamicin (80 mg/kg i.p.) for 7 days. Third and fourth groups were given an aqueous extract of cardamom (100 and 200 mg/kg p.o.) + gentamicin for 7 days, respectively. The degree of hepatoprotection was measured using serum AST, ALT, bilirubin, ALB, and lipid profile levels. In the acute liver damage induced by gentamicin, cardamom aqueous extracts (100 and 200 mg/kg, p.o.) significantly reduced the elevated serum levels of AST, ALT, bilirubin, cholesterol, TGs, and LDL-cholesterol (LDL-chol) in gentamicin-induced hepatotoxicity. Also, cardamom aqueous extracts (100 and 200 mg/kg, p.o.) significantly increased the lowered serum levels of ALB and HDL cholesterol (HDL-chol) in gentamicin-induced hepatotoxicity rats. Histopathological examination of the liver tissues supported the hepatoprotection.

Elsholtzia densa Benth. FAMILY: LAMIACEAE

Zargar et al. (2018) evaluated the hepatoprotective activity of methanolic extract of *Elsholtzia densa* against experimentally induced acute (CCl_4) and chronic (PCM) liver injury in albino Wistar rats. Activity was measured by monitoring the serum levels of ALT, ALP AST and LDH, TP levels, bilirubin, and ALB. The results of the CCl_4 and PCM-induced liver toxicity experiments showed that the rats treated with the methanolic extract of *E. densa* exhibited a significant decrease in biochemical parameters as well as the proteins, which were all elevated in the CCl_4 and PCM group. The extract at a concentration of 300 mg/kg BW. showed a significant decline in the levels of AST, ALT, ALP, and LDH in CCl_4 injected animals and in PCM-treated animals when compared to the control group. The activities of tissue antioxidants GSH, GPx, GR, GST, and CAT was significantly restored in dose-dependent manner in animals treated with extracts as with acute and chronic hepatotoxic models.

Elytraria acaulis (L.f.) Lindau FAMILY: ACANTHACEAE

Reddy et al. (2015) evaluated the hepatoprotective effects of hydroalcoholic (EAEE) and aqueous extracts (EAAE) of whole plant of *Elytraria acaulis* on CCl_4-induced hepatotoxicity in albino Wistar rats. Serum parameters like SGPT values were increased in the CCl_4-induced rats. The decreased level of SGOT, SGPT, ALP were observed in the animals treated with Liv. 52 (standard drug) and in animals treated with the extracts of *E. acaulis.* The highest activity of observed for the *E. acaulis* whole plant hydroalcholic and aqueous extracts at a dose of 200 mg/kg BW. The study revealed significant decrease in the levels of SB in EAEE and EAAE-treated rats when compared to CCl_4 treated group rats. The hepatoprotective activity was also supported by histopathological studies of liver tissue.

Embelia ribes Burm. f. FAMILY: PRIMULACEAE

Tabassum and Agrawal (2003) evaluated the hepatoprotective activity of *Embelia ribes* on PCM-induced liver cell damage in mice. The mice treated with *E. ribes* extract (50, 100, and 200 mg/100 g/d) showed a dose-dependent fall of 41%, 47%, and 66%, respectively, of serum SGPT levels as compared to the elevated levels in the mice receiving PCM only. Histopathology of liver of mice revealed 67%, 70%, and 80% normal livers, respectively, in mice receiving the above doses of *E. ribes*.

Endostemon viscosus (Roth) M.R. Ashby [Syn.: *Orthosiphon diffusus* (Benth.) Benth.] FAMILY: LAMIACEAE

Ghaffari (2013) evaluated the mechanisms behind the antioxidant and hepatoprotective potential of *Endostemon viscosus* methanol active fraction (MAF) using *in vivo* (rat) and *in vitro* (cell culture) models using CCl_4-induced hepatotoxicity. Rats pretreated with *Endostemon viscosus* MAF demonstrated significantly reduced levels of serum LDH and ALP Similarly, multiple dose MAF administration demonstrated significantly enhanced levels of antioxidant enzymes in the liver homogenates. Histological analysis revealed complete neutralization of CCl_4-induced liver injury by the extract. The *in vitro* studies demonstrated that pretreatment of MAF effectively prevented H_2O_2-induced oxidative stress, genotoxicity, and significantly enhanced expression of genes for antioxidant enzymes.

Enicostema axillare (Poir. ex Lam.) A. Raynal FAMILY: GENTIANACEAE

The whole plant of *Enicostema axillare* is used in variety of diseases in traditional Indian system of medicine including hepatic ailments. Swertiamarin isolated from *E. axillare* was evaluated by Jaishree and Badamai (2010) for antioxidant and hepatoprotective activity. Swertiamarin, a secoiridoid glycoside, was found to contain a major constituent of the extract. D-GalN caused significant hepatotoxicity by alteration of several hepatic parameters. It also caused significant LPO and reduced the levels of antioxidant defense mechanisms. The treatment with swertiamarin at 100 and 200 mg/kg BW when administered orally for 8 days prior to D-GalN caused a significant restoration of all the altered biochemical parameters due to D-GalN toward the normal, indicating the potent antioxidant and hepatoprotective nature of swertiamarin (Jaishree and Badamai, 2010).

Enicostema littorale Blume
FAMILY: GENTIANACEAE

The alcohol extract of the whole plant of *Enicostema littorale* was evaluated by Senthilkumar et al. (2003) for its antihepatotoxic activity against CCl_4-induced hepatic damage in rats. Elevation of enzyme levels have been observed in the CCl_4-treated group. A significant reduction was observed in SGPT, SGOT, ALP, TB, GGTP, and increased protein levels in the groups treated with Silymarin and the alcohol extract of *E. littorale*. The enzyme levels were nearly restored to the normal level. The rats treated with Silymarin and the extract, along with toxicant, showed signs of protection against these toxicants to a considerable extent as evident from the formation of normal hepatic cards and the absence of necrosis and vacuoles.

Vishwakarma and Goyal (2004) evaluated the effect of *E. littorale* against CCl_4-induced acute liver damage in mice. Concomitant treatment with aqueous extract at 250, 500, and 1000 mg/kg showed marked reduction in the ALT and AST levels as compared to CCl_4 treated group. The reduction produced by aqueous extract on ALT and AST was maximum at 500 mg/kg. Serum ALP and bilirubin level was significantly raised by CCl_4 treatment. Concomitant treatment with aqueous extract at different doses showed marked reduction in the ALP and bilirubin levels as compared to CCl_4-treated group. The prolonged pentobarbitone sleeping time was significantly reduced in treatment group. The hepatic damage in animal pretreated with aqueous extract was minimal with distinct preservation of structures and architectural frame of the hepatic cells.

Epaltes divaricata (L.) Cass.
FAMILY: ASTERACEAE

Hewawasam et al. (2004) evaluated the hepatoprotective and antioxidative effects of an aqueous extract of *E. divaricata* plant against CCl_4-induced hepatocellular injury in mice.

The degree of hepatoprotection was measured using serum AST, ALT, ALP, and liver-reduced GSH level. The liver tissue was used for histopathological assessment of liver damage. Pretreatment of mice with the plant extract of *Epaltes* (0.9 g/kg) orally for 7 days significantly reduced serum levels of ALT, AST, and ALP enzymes by 21.40%, 47.36%, and 71.12%, respectively, and significantly increased the liver reduced GSH level by 42.32%, 24 h after the administration of CCl_4. A marked improvement in the enzyme activities and the liver-reduced GSH level was observed in the *Epaltes* pretreated mice 4 days after the administration of CCl_4. Histopathological studies provided supportive evidence for the biochemical analysis.

Ephedra foliata Boiss. ex C.A. Mey. FAMILY: EPHEDRACEAE

The hepatroprotective effect of *Ephedra foliata* was studied in Wistar albino rats by Alqasoumi et al. (2008b). Liver injury was induced in rats using CCl_4. The biochemical parameters; SGOT, SGPT, ALP, and TB were estimated as reflection of the liver condition. The hepatoprotective effect offered by *Ephedra foliata* (whole plant) crude extract at 500 mg/kg doses, was found to be significant in all parameters studied with 42.6%, 39.5%, 21.2%, and 46.2% reduction in SGOT, SGPT, ALP, and bilirubin, respectively. At the lower doses (250 mg/kg) the extract resulted in a significant reduction in SGOT, ALP, and bilirubin.

Eriocaulon quinquangulare L. FAMILY: ERIOCAULACEAE

Decoction prepared from the whole plant of *Eriocaulon quinquangulare* is prescribed to treat liver disorders. Fernando ad Soysa (2016) investigated the hepatoprotective activity and antioxidant capacity of the aqueous extract of *E. quinquangulare in vitro*. Hepatoprotective activity against ethanol-induced hepatotoxicity was carried out using porcine liver slices. The total phenolics and flavonoids were 10.3±1.6 w/w% GAEs and 45.6±3.8 w/w% (−)-epigallocatechin gallate equivalents, respectively. The values of EC_{50} for DPPH, hydroxyl radical and NO scavenging assays were 37.2±1.7 μg/mL, 170.5±6.6 μg/mL, and 31.8±2.2 μg/mL, respectively. The reducing capability of AEQ was 6.9±0.2 w/w% L-ascorbic acid equivalents (AAEs) in the FRAP assay. For hypotonic-solution-induced HEHA, the IC_{50} was 1.79±0.04 mg/mL. A significant decrease was observed in ALT, AST, and LDH release from the liver slices treated with AEQ compared to the ethanol treated liver slices. A significant reduction in LPO was also observed in liver slices treated with the plant extract compared to that of the ethanol-treated liver slices.

Eruca sativa Mill. FAMILY: BRASSICACEAE

The ethanolic extract of *Eruca sativa* leaves and seeds showed significant hepatoprotective activity against CCl_4 (Alqasoumi, 2010) and ethanol (Hussein et al., 2010) induced liver injury. The *E. sativa* extract also showed significant cytoprotective effect against liver cancer cells (Lamy et al., 2008). The hepatoprotective activity of *E. sativa* may be attributed to its antioxidant (Alam et al., 2007) and anti-inflammatory (Yehuda et al., 2009) activities. It is an edible plant with no reported toxicity. Phytochemical studies on leaves of *E. sativa* have shown the presence of large amount of polyphenols, flavonoids, erucin, erysolin, glucosinolates, quercetins, erucic acid, and phenylethyl isothiocyanate (Alqasoumi, 2010; Lamy et al., 2008).

Erycibe expansa Wall. ex G.Don FAMILY: CONVOLVULACEAE

The methanolic extract from the stems of *Erycibe expansa* was found to show a hepatoprotective effect on *D-GALN*-induced cytotoxicity in primary cultured mouse hepatocytes (Matsuda et al., 2004). By bioassay-guided separation, two new prenylisoflavones and a pterocarpane, erycibenins A, B, and C, were isolated from the active fraction (the EtOAc-soluble fraction) together with ten isoflavones and seven pterocarpanes. The isolated constituents, erycibenin A (IC_{50} = 79 μM), genistein (29 μM), orobol (36 μM), and 5,7,4′-trihydroxy-3′-methoxyisoflavone (55 μM) exhibited inhibitory activity on *D-GALN*-induced cytotoxicity in primary cultured mouse hepatocytes.

Erythrina indica Lam. FAMILY: FABACEAE

Mujahid et al. (2017) evaluated the antihepatotoxic potential of *Erythrina indica* against isoniazid (INH) and rifampicin (RIF)-induced hepatotoxicity in rats. Treatment with *E. indica* significantly prevented drug induced increase in serum levels of hepatic enzymes. Furthermore, *E. indica* significantly reduced the LPO in the liver tissue and restored activities of defense antioxidant enzymes GSH, SOD, and CAT toward normal. Histopathology of liver tissue showed that *E. indica* attenuated the hepatocellular necrosis, regeneration, and repair of cells toward normal.

Erythrina × neillii FAMILY: FABACEAE

The major phytoconstituents of *Erythrina × neillii,* and the hepatoprotective effect and underlying mechanisms were assessed by Baker et al. (2019). The dichloromethane extract revealed an abundance of alkaloids (25), in addition to tentatively identifying flavone, flavanone and three fatty acids. Additionally, 36 compounds belonging to different classes of phytoconstituents with a predominance of flavonoids, *O/C*-flavone and flavonol glycosides, followed by alkaloids (9), fatty acids (4) and (2), and phenolic glycoside were identified in the ethyl acetate extract. Compared with MTX, alcoholic leaf extract (500 mg/kg) ameliorated the MTX-induced alterations by improving several biochemical marker levels, fighting oxidative stress in serum and liver tissues, and decreasing inflammatory mediators; this finding was further confirmed by the histopathological study.

Erythrina senegalensis DC.
FAMILY: FABACEAE

Hepatoprotective effect of the ethanolic extract of *Erythrina senegalensis* stem bark was studied *in vivo* against CCl_4-induced liver damage as well as *in vitro* against rat liver slices intoxicated CCl_4. *E. senegalensis* extract at 100 mg/kg significantly attenuated hepatotoxin induced biochemical serum ALT, AST, and LPO in liver homogenate. Polyphenols including flavonoids have been characterized from this plant which could be implicated for its hepatoprotective potential (Njayou et al., 2008).

Wakawa and Hauwa (2013) evaluated the protective activity of aqueous leaf extract of *E. senegalensis* against CCl_4-induced liver injury. A significant decrease was observed in both the groups pretreated with the leaf extract on the levels of the enzymes and nonenzyme markers of tissue damage, LPO, and relative organ weights and this is shown to be dose dependent when compared to rats that were give CCl_4 only.

Etlingera elatior (Jack) R.M. Sm.
FAMILY: ZINGIBERACEAE

The efficacy of *Etlingera elatior* (torch ginger) to protect hepatotoxicity induced by lead acetate was evaluated experimentally in male SD rats by Haleagrahara et al. (2010). There was a significant decrease in TA and other antioxidant enzymes and increase in LPO and protein carbonyl content (PCC) with lead acetate ingestion. Concurrent treatment with *E. elatior* extract significantly reduced the LPO and PCC in serum and increased the antioxidant enzyme levels in the liver. Significant histopathological changes were seen in hepatic tissue with chronic lead ingestion. Treatment with *E. elatior* significantly reduced these lead-induced changes in hepatic architecture. *E. elatior* has also reduced the blood lead levels. Thus, there has been extensive biochemical and structural alterations indicative of liver toxicity with exposure to lead and *E. elatior* treatment significantly reduced these oxidative damages.

Eucalyptus maculata (Hook.) K.D. Hill & L.A.S. Johnson
FAMILY: MYRTACEAE

Mohamed et al. (2005) assessed the hepatoprotective and antioxidant activities of the chloroformic extract of the resinous exudate and its phenolic constituents obtained from the stems of *Eucalyptus maculata.* Acetaminophen at a dose of 1 g/kg BW produced 100% mortality in mice, while pretreatment of animals with the chloroformic extract (125 and 250 mg/kg) protected against the mortalities by 66%. Pretreatment of rats with either the chloroformic extract (250 mg/kg) or any of the pure isolates (20 mg/kg) significantly reduced the increase in serum level of AST, ALT, and ALP produced by APAP (640 mg/kg). Pretreatment of animals with the chloroformic extract or its isolates also protected against ascorbic acid depletion in serum and kidney tissues induced by oral administration of paraquat without modifying the serum level of GSH and glycogen content in liver tissue. The phenolic content of the chloroformic extract and the pure isolates produced an antioxidant activity which may be due to the formation of stable phenoxyl radical in addition to its effect through vitamin C.

Euclea natalensis A.DC. FAMILY: EBENACEAE

Lall et al. (2016) evaluated the hepatoprotective activity of the ethanolic extract of *Euclea natalensis* on isoniazid and rifampicin-induced hepatic damage in a rat model. The plant showed a hepatoprotective effect (50% at 12.5 μg/mL) and the ability to increase T-helper 1 cell cytokines; Interleukin 12, Interleukin 2, and Interferon α by up to 12-fold and the ability to decrease the T-helper 2 cell cytokine Interleukin 10 4-fold when compared to baseline cytokine production.

Eucommia ulmoides Oliv. FAMILY: EUCOMMIACEAE

The protective effects of water extract of Du-Zhong (*Eucommia ulmoides*) leaves and its active compound (protocatechuic acid) on liver damage were evaluated by Hung et al. (2006) by CCl_4-induced chronic hepatotoxicity in Wistar rats. It showed that CCl_4-treated rats increased the relative organ weights of liver and kidney. CCl_4-induced rats liver damage and significantly increased the GOT, GPT, LDH, and ALP levels in serum as compared with the control group. Treatment with extract or protocatechuic acid could decrease the GOT, GPT, LDH, and ALP levels in serum when compared with CCl_4-treated group. CCl_4-treated rats also significantly decreased the GSH content in liver and trolox equivalent antioxidant capacity in serum whereas increased MDA content in liver as compared with the control group. Treatment with extract or protocatechuic acid also significantly increased the GSH content and significantly decreased the MDA content in liver. Administration of extract or protocatechuic acid could increase the activities of GPx, GRd, and GST in liver. Liver histopathology showed that extract or protocatechuic acid reduced the incidence of liver lesions including hepatic cells cloudy swelling, lymphocytes infiltration, cytoplasmic vacuolization, hepatic necrosis, and fibrous connective tissue proliferated induced by CCl_4 in rats. The data suggest that oral administration with aqueous extract for 28 consecutive days significantly decrease the intensity of hepatic damage induced by CCl_4 in rats.

Eugenia jambolana Lam.
FAMILY: MYRTACEAE

Sisodia and Bhatnagar (2009) estimated the hepatoprotective effects of the methanolic seed extract of *Eugenia jambolana* in Wistar albino rats treated with CCl_4. Administration of *E. jambolana* (doses 100, 200, and 400 mg/kg p. o.) significantly prevented CCl_4-induced elevation of serum SGOT, SGPT, ALP, ACP, and bilirubin (total and direct) level. Histological examination of the liver section revealed hepatic regeneration, after administration of various doses of *E. jambolana*. The results were comparable to that of Liv. 52.

Eupatorium cannabinum L.
FAMILY: ASTERACEAE

Eupatorium cannabinum aqueous extract (125, 250, 500, 1000 mg/kg) possessed antinecrotic properties against CCl_4-induced hepatotoxicity. Pretreatment (30 min before CCl_4), with *E. cannabinum* showed a significant decrease of GPT levels at 250, 500, and 1000 mg/kg (Lexa et al., 1989). An aqueous extract of the plant exhibited antinecrotic activity against CCl_4-induced hepatotoxicity in rats. The effect is attributed to the presence of flavonoids, rutoside, hyperoside, and quercetin; phenolic acids, caffeic, and chlorogenic; and not due to the presence of eupatoriopicrin. Acrylic acid and the lactic, malic, and citric acids, present in the plant, also exhibited protective effect against acute toxicity induced by ethanol in mice (Khare et al., 2007).

Euphorbia antiquorum L.
FAMILY: EUPHORBIACEAE

Aqueous extract of the leaf of *Euphorbia antiquorum* has been reported to have hepatoprotective and antioxidant activity by Jyothi et al. (2008). *This report appears to be doubtful because E. antiquorum has no leaves perhaps the plant might have been misidentified.*

Euphorbia dracunculoides Lam.
FAMILY: EUPHORBIACEAE

Batool et al. (2017) evaluated the hepatoprotective effects of the ME of aerial parts of *Euphorbia dracunculoides* against CCl_4-induced toxicity in *SD* male rats. Analysis of serum indicated significant rise in the level of AST, ALT, ALP, and globulin whereas decrease was recorded for the TP and ALB in CCl_4-treated rats. In liver tissues the activity level of CAT, POD, SOD, GST, GSH was decreased while the level of LPOs; TBARS, nitrite, and hydrogen peroxide increased in CCl_4-treated rats as compared to the control group. Histopathological injuries and DNA damages were recorded in liver of rat with CCl_4 treatment. However, co-administration of ME, dose dependently, ameliorated the CCl_4-induced hepatic toxicity in these parameters.

Euphorbia fusiformis Buch.-Ham. ex D. Don FAMILY: EUPHORBIACEAE

The tubers of *Euphorbia fusiformis* are traditionally used in India by the Malayali tribes of Chitteri hills, Eastern Ghats, Tamil Nadu to treat liver disorders. The hepatoprotective potential of the ethanol extract of *E. fusiformis* tuber against rifampicin-induced hepatic damage was investigated in Wistar albino rats by Anusuya et al. (2010). The acute and subchronic toxicity was assessed in mice and rats, respectively. The ethanol extract of tubers (250 mg/kg p.o.) showed remarkable hepatoprotective effect against rifampicin-induced hepatic damage in Wistar albino rats. The degree of protection was measured using the biochemical parameters SGOT, SGPT, GGTP, ALP, TB, and TP. Treatment with ethanolic extract prior to the administration of rifampicin significantly restored the elevated levels of the said parameters on a par with the control group. The single dose LD (50) was found to be 10,000 mg/kg BW when administered orally in mice. Subchronic toxicity studies in rats with oral doses of 125, 250, and 500 mg/kg exhibited no significant changes in BW gain, general behavior, hematological, and biochemical parameters. The histological profile of liver and kidney also indicated the nontoxic nature of this drug (Anusuya et al., 2010).

Euphorbia hirta L. FAMILY: EUPHORBIACEAE

The antihepatotoxic effect of hydroalcoholic extract of whole *Euphorbia hirta* extracts was evaluated in experimental models of liver injury in rats induced by CCl_4 or PCM. *E. hirta* showed hepatoprotective activities at doses 125 and 250 mg/kg, since serum levels of ALT and AST in rats given the extracts were significantly lower compared to control CCl_4 or PCM-injured rats (Tiwari et al., 2011). The *in vivo* antimalarial activity of the *E. hirta* extract doses (200, 400, and 800 mg/kg BW) was studied against *P. berghei* infected mice. The results showed that the extract had significant suppressive activity of 51%–59% and prophylactic activity of 25%–50% when compared with chloroquine that gave 95% and 81% suppressive and prophylactic antiplasmodial activities, respectively. The antiplasmodial action of the extract was not related to the oxidation of RBC membrane lipids as increasing extract concentration results in the reduction of the enzymatic activities of SOD and GPx, and concentrations of GSH and TBARS (Jeje et al., 2016).

Euphorbia tirucalli L. FAMILY: EUPHORBIACEAE

A systemic and scientific investigation of aqueous extract of *Euphorbia tirucalli* for its antioxidant and hepatoprotective potential against carbon-tetrachloride-induced hepatic damage in rats was carried out by Jyothi et al. (2008). The aqueous extract has demonstrated dose-dependent *in vitro* antioxidant property (at 20, 40, 60, 80, 100 μg) in all the models of the study. Similarly, aqueous extract of *E. tirucalli* at the doses of 125 and 250 mg/kg produced significant hepatoprotective effect by decreasing the serum enzymes, bilirubin, cholesterol, TGs, and tissue LPO, while it significantly increased the levels of tissue GSH in a dose-dependent manner.

Fagonia schweinfurthii (Hadidi) Nabil & Hadidi FAMILY: ZYGOPHYLLACEAE

The whole plant of *Fagonia schweinfurthii* is used in a variety of diseases including hepatic ailments in deserts and dry areas of India. Pareek et al. (2013) evaluated the antioxidant and hepatoprotective activity of ethanolic extract from *F. schweinfurthii* in CCl_4-induced hepatotoxicity in HepG2 cell line and rats. *In vitro* cytotoxicity and hepatoprotective potential of the extracts were evaluated using HepG2 cells. Based on the cytotoxicity assay, ethanolic extract (50, 100, 200 μg/mL) was assessed for hepatoprotective potential against CCl_4 induced toxicity in HepG2 cell line by monitoring cell viability, AST, ALT, LDH leakage, LPO, and glutathione level (GSH). Further, *in vivo* hepatoprotective activity of the ethanolic extract was evaluated against CCl_4-induced hepatotoxicity in male Wistar albino rats. Rats were pretreated with the ethanolic extract (200 mg, 400 mg/kg/day p.o.) for 7 days followed by a single dose of CCl_4 (1.0 mL/kg, i.p.) on 8th day. Silymarin was used as a positive control. After 24 h of CCl_4 administration, various biochemical parameters like AST, ALT, ALP, TB, and TP levels were estimated in serum. The antioxidant parameters like SOD activity, CAT activity, GSH content, and MDA level in the liver homogenate were determined. Histopathological changes in the liver of different groups were also studied. The ethanolic extract possessed strong antioxidant activity *in vitro*. The results indicated that CCl_4 treatment caused a significant decrease in cell viability. The CCl_4-induced changes in the HepG2 cells were significantly ameliorated by treatment of the ethanolic extract. This ethanolic extract significantly prevented CCl_4-induced elevation of AST, ALT, ALP, TB, and CCl_4 induced a decrease in TP in rats. The ethanolic extract treated rat liver antioxidant parameters (SOD, CAT, MDA, and GSH,) were significantly antagonized for the pro-oxidant effect of CCl_4. Histopathological studies also supported the protective effect of ethanolic extract. The results of this study revealed that the ethanolic extract has significant hepatoprotective activity. This effect may be due to the ability of the extract to inhibit LPO and increase in the antioxidant enzymatic activity (Pareek et al., 2013).

Ferulago angulata (Schlecht.) Boiss. FAMILY: APIACEAE

Ferulago angulata is used to treat liver diseases and has been used both as food and therapeutics by many cultures for thousands of years because of the natural antioxidant compounds. Kiziltas et al. (2017) determined antioxidant properties of *F. angulata* flowers, evaluated the hepatoprotective effect of flowers against *N*-nitrosodimethylamine (NDMA) induced on liver tissue by assessing antioxidant enzymes and histopathological parameters in Wistar albino rats. The results of the study indicated that FASB flowers contain high levels of total antioxidant activity, phenolics and flavonoids. Due to the positive effect on significant changes in antioxidant enzymes of liver tissue and histopathological examination, it is thought that the plant could be used as a hepatoprotective.

Ficus benghalensis L. FAMILY: MORACEAE

Hepatoprotective effect of leucopelargonidin derivative, isolated from the bark of *Ficus benghalensis*, at a dose level of 100 mg/kg/day i.p. was evaluated in CCl_4-induced hepatotoxic rats, using vitamin E at a dose level 50 mg/kg/day i.p. as the reference standard. Result with regard to decrease in biochemical parameters like TC, HDL, LDL, FFA, TAG; decrease in the activities of glucose 6-phosphate dehydrogenase (G6PD), HMG-CoA reductase in the liver and enzymes like ALT, ALP, and AST in serum and liver; increase in the levels of antioxidant enzymes in liver; inhibition of fatty infiltration and fibrosis, was comparable to that of vitamin E (Augusti et al., 2005).

Methanolic extract of the aerial root of *F. benghalensis* (MEFB) was tested for hepatoprotective activity against isoniazid-rifampicin-induced liver injury in rats using Liv. 52 at a dose level of 10 mg/kg p.o. as the reference standard. Results of MEFB at a dose level of 100, 200, and 300 mg/kg p.o. about bilirubin level, TP level, ALB level, AST, and ALT were almost same as that of Liv. 52. Histopathological results about hepatocytic necrosis, inflammation, and neutrophil infiltration were also comparable to that of Liv. 52 (Parameswari et al., 2012).

The hepatoprotective potentials of the ethanolic extract of the *F. benghalensis* leaves were tested against CCl_4 and ethanol-induced liver damage in rats by Shinde et al. (2012). CCl_4, and ethanol elevated the levels of AST, ALT, and decreased levels of TP and TA. Treatment with the ethanolic extract of *F. benghalensis* 100, 200, and 400 mg/kg ameliorated the effects of the hepatoxins and significantly reduced the elevated levels of the biochemical marker enzymes.

Jyothilekshmi (2015) evaluated the protective role of *F. benghalensis* bark extract against ethanol-induced hepatotoxicity in rats. Alcohol intake increased the biochemical parameters like AST, ALT, ALP, TB and decreased the levels of ALB and TP along with changes in histological parameters (damage to hepatocytes). Treatment with water extract of *F. benghalensis* bark (at a dose of 400 mg/kg, p.o. daily for 28 days) significantly prevented the biochemical and histological changes induced by ethanol, indicating the recovery of hepatic cells. The activity of the extract was also comparable to that of Silymarin, a standard hepatoprotective drug.

Ficus carica L. FAMILY: MORACEAE

Oral application of methanolic extract of *Ficus carica* leaves was evaluated for its hepatoprotective activity in CCl_4-induced liver damaged rats by Mohan et al. (2007). An oral dose of 500 mg/kg exhibited a significant protective effect reflected by lowering the serum activities of AST, ALT, total SB concentrations, and MDA equivalent, an indicator of LPO of the liver.

The antioxidant, hepatoprotective, and kidney protective activities of MEs of *F. carica* leaves were evaluated by Singab et al. (2010). Liver and kidney damage were induced in rats by CCl_4. The extract was given intraperitoneally at doses of 50 mg/kg (leaf) and 150 mg/kg (fruit). The activity of the extracts was comparable to that of Silymarin, a known hepatoprotective agent. Hepatoprotective activity was evaluated by measuring serum levels of AST, ALT,

ALP, TB, and TP. These biochemical observations were supported by histopathological examination of liver sections. Kidney function was evaluated by measuring plasma urea and creatinine. Methanol extracts of *F. carica* showed potent antioxidant and hepatoprotective activities; in-depth chromatographic investigation of the most active extract (leaf extract) resulted in the identification of umbelliferone, caffeic acid, quercetin-3-*O*-β-d-glucopyranoside, quercetin-3-*O*-α-l-rhamnopyranoside, and kaempferol-3-*O*-α-l-rhamnopyranoside.

The protective action of *F. carica* leaf ethanolic extract (obtained by maceration) was evaluated by Aghel et al. (2011) in an animal model of hepatotoxicity induced by CCl_4. Liver marker enzymes were assayed in serum. Sections of livers were observed under a microscope for the histopathological changes. Levels of marker enzymes such as ALT and AST increased significantly in CCl_4 treated mice, pretreated with the plant extract and intoxicated with CCl_4, decreased activities of these two enzymes. Also, pretreatment with the extract in these groups resulted in less pronounced destruction of the liver architecture with no fibrosis and moderate inflammation. The observations suggested that the treatment with *F. carica* leaf extract in a dose of 200 mg/kg enhanced protection against CCl_4-induced hepatic damage.

Gond and Khadabadi (2008) extracted dried leaves of *F. carica* using petroleum ether and applied the extract to rats with rifampicin-induced toxic liver injury. Serum activities or concentrations of AST, ALT, bilirubin, and histological changes in liver were assessed. They observed significant reversal of rifampicin-dependent biochemical (AST, ALT, and TB), histological and functional changes in those rats receiving the extract.

The hepatoprotective activity of various extracts of *F. carica* leaves and fruits have been experimentally confirmed against CCl_4 (Mujeeb et al., 2011)-induced hepatotoxicity. The hepatoprotective activity of *F. carica* may be attributed to its marked anti-inflammatory (Ali et al., 2012) and antioxidant (Joseph and Justin Raj, 2011) activities. Phytochemical studies on leaves and fruits of *F. carica* have shown the presence of flavonoids, vitamins, nicotinic acid, tyrosine, ficusin, bergaptene, stigmasterol, furocoumarin, psoralen, taraxasterol, beta-sitosterol, rutin, and sapogenin (Al-Asmari et al., 2014).

***Ficus chlamydocarpa* Mildbr. & Burrett FAMILY: MORACEAE**

Hepatoprotective effect of *Ficus chlamydocarpa* was evaluated through the induction of acute hepatic damage in rats using CCl_4 (Donfack et al., 2010). In this study, the pretreatment with 50–200 mg/kg of methanolic extract of *F. chlamydocarpa* stem bark prevented serum increase of hepatic enzyme markers and LDH, enhanced hepatic reduced GSH level and decreased of hepatic MDA during CCl_4 intoxication. Previous phytochemical studies on stem bark of *F. chlamydocarpa* revealed the presence of the following flavonoids; alpinumisoflavone, genistein (4′,5,7-trihydroxyisoflavone) and luteolin (3′,4′, 5,7-tetrahydroxy flavones) with significant DPPH radical scavenging activities with IC50 (μg/mL of 6, 5.7, 5.0, respectively) (Donfack et al., 2010).

Ficus cordata Thunb. FAMILY: MORACEAE

Donfack et al. (2011) evaluated the hepatoprotective effect of crude extract and isolated compounds from *Ficus cordata* on the CCl_4-induced liver cell damage as well as the possible antioxidant mechanisms involved in the protective effect. The phytochemical investigation of this crude extract led to the isolation of five compounds identified as: β-amyrin acetate, lupeol, catechin, epiafzelechin, and stigmasterol. These compounds showed significant hepatoprotective activities as indicated by their ability to prevent liver cell death and LDH leakage during CCl_4 intoxication.

Ficus exasperata Vahl FAMILY: MORACEAE

The ethanol extracts of the leaves of *Ficus exasperata* showed significant hepatoprotective activities in APAP-induced hepatotoxic rats (Odutuga et al., 2014). The extract at 125–500 mg/kg significantly ameliorated toxin induced alterations in the liver ALT, AST, ALP, and bilirubin levels. The histological evaluation showed a partial prevention of inflammation, necrosis, and vacuolization induced by CCl_4 (Odutuga et al., 2014).

Ficus glomerata Roxb. FAMILY: MORACEAE

Channabasavaraj et al. (2008) evaluated *Ficus glomerata* extracts for antioxidant and hepatoprotective properties. The ME of the bark of *F. glomerata* showed potent *in vitro* antioxidant activity when compared to the root ME. In the *in vivo* studies, the CCl_4-treated control rats showed a significant alteration in the levels of antioxidant and hepatoprotective parameters. The ME of the bark when given orally along with CCl_4 at the doses of 250 and 500 mg/kg BW showed a significant reversal of these biochemical changes toward the normal when compared to CCl_4-treated control rats in serum, liver and kidney. The results were comparable to those observed for standard sylimarin. Histological studies also confirmed the same.

Ficus gnaphalocarpa (Miq.) Steud. ex A. Rich FAMILY: MORACEAE

The *in vitro* hepatoprotective effect of the methanolic extract from *Ficus gnaphalocarpa* on the CCl_4-induced liver cell damage as well as the possible antioxidant mechanisms involved in this protective effect, were investigated by Donfack et al. (2011). The phytochemical investigation of this methanolic extract led to the isolation of six compounds identified as: betulinic acid (1) 3-methoxyquercetin; (2) catechin; (3) epicatechin; (4) quercetin; (5) and quercitrin (6). The hepatoprotective activity of these compounds was tested *in vitro* against CCl_4-induced damage in rat hepatoma cells. In addition, radical-scavenging activity, β-carotene-linoleic acid model system, ferric-reducing antioxidant parameter, and microsomal LPO assays were used to measure antioxidant activity of crude extract and isolated compounds. Silymarin and trolox were used as standard references and, respectively, exhibited significant hepatoprotective and antioxidant activities. The compounds (5), (6), and (2) showed significant antioxidant and hepatoprotective activities as indicated by their ability to prevent liver cell death and LDH leakage during CCl_4 intoxication. These results suggest that the protective effects of crude extract of *F. gnaphalocarpa* against the CCl_4-induced hepatotoxicity possibly involve the antioxidant effect of these compounds.

Ficus hispida L. FAMILY: MORACEAE

The ME of the leaves of *Ficus hispida* was evaluated by Mondal et al. (2000) for hepatoprotective activity in rats by inducing acute liver damage by PCM (750 mg/kg, p.o.). The extract at an oral dose of 400 mg/kg exhibited a significant protective effect by lowering the serum levels of transaminase (SGOT and SGPT), bilirubin, and ALP. These biochemical observations were supplemented by histopathological examination of liver sections. The activity of extract was also comparable to that of Liv. 52 a known hepatoprotective formulation.

Ficus racemosa L. FAMILY: MORACEAE

An extract of the leaves of *Ficus racemosa* was evaluated by Mandal et al. (1999) for hepatoprotective activity in rats by inducing chronic liver damage by subcutaneous injection of 50% v/v CCl_4 in liquid paraffin at a dose of 3 mL/kg on alternate days for a period of 4 weeks. The biochemical parameters SGOT, SGPT, SB, and ALP were estimated to assess the liver function. The activity of extract was also comparable to a standard liver tonic (Neutrosec).

The hepatoprotective effects of petroleum ether (FRPE) and methanol (FRME) extract of *F. racemosa* stem bark were studied by Ahmed and Urooj (2010) using the model of hepatotoxicity induced by CCl_4 in rats. Pretreatment with the extracts restored TP and ALB to near normal levels. Both the extracts resulted in significant decreases in the activities of AST, ALT, and ALP, compared to CCl_4-treated rats. However, a greater degree of reduction was observed in ME pretreated group. The extracts improved the antioxidant status considerably as reflected by low TBARS and high GSH values. Methanol extract exhibited higher hepatoprotective activity than a standard liver tonic (Liv. 52), while the protective effect of petroleum ether extract (PEE) was similar to that of Liv. 52. The protective effect of *F. racemosa* was confirmed by histopathological profiles of the liver.

Ficus religiosa L. FAMILY: MORACEAE

The MeOH extract from *Ficus religiosa* leaves was tested against liver damage induced with isoniazid-rifampicin and PCM (100 mg/kg, each) in Wistar albino rats; the extract was tested at 100, 200, and 300 mg/kg and was administered during 21 days by oral. Liv. 52 used positive control. The result revealed that the MeOH extract reduced SGPT and SGOT levels at similar values to those of the Liv. 52 group but TP, ALB ad ALP levels are similar to the INH/RIF group. Histological analysis of the liver showed protection, because the animal group treated with MeOH extract shown a normal hepatic architecture (Parameswari et al., 2013). This extract showed a hepatoprotective effect against liver damage induced with PCM (Parameswari et al., 2013).

The hepatoprotective effects of *F. religiosa* latex in cisplatin induced liver injury were investigated in Wistar rats. Cisplatin-treated animals showed significant increase in serum ALT, AST, ALP, and hepatocytes cells degeneration, inflammatory infiltrate and necrosis, these changes were significantly alleviated by *F. religiosa* latex (Yadav, 2015).

Ficus sycomorus L. FAMILY: MORACEAE

The aqueous root-bark extract of *Ficus sycomorus* was tested for its chemical constituents, acute toxicity, and hepatoprotective effect against CCl_4-induced hepatotoxicity in rats by Garba et al. (2007). Phytochemical analysis of the extract revealed the presence of saponins, flavonoids, alkaloids, tannins and reducing sugar and LD_{50} was calculated as 3.20/0.6031 g/kg. Pretreatment of the rats with the extract was able to reduce though not significantly, changes in the biochemical parameters (decrease in ALB but increase in AST, ALT, ALP, and bilirubin) and preserved the liver parenchymal architecture against CCl_4-induced degenerative changes, fibroplasia, and cirrhosis. The plant extract had hepatoprotective effect on the parenchymal architecture of the liver against CCL_4-induced hepatotoxicity in rats.

Filipendula hexapetala Gilib. FAMILY: ROSACEAE

The influence of ME produced from the flowers of *Filipendula hexapetala* on some liver biochemical parameters in rats intoxicated with CCl_4 was evaluated by Ćebović and Maksimović (2012). Pretreatment with the extract inhibited CCl_4-induced liver injury by decreasing LPO and increasing the content of reduced GSH in a dosage-dependent manner, bringing the levels of all antioxidant enzymes close to control values. The administration of CCl_4 diminished hepatic antioxidant defense mechanisms by significant reduction of peroxidase and CAT activities. The CAT activity was significantly recovered in groups treated with the extract investigated and intoxicated with a single CCl_4 dose. A similar impact on hepatic peroxidase activity has also been observed, indicating a partial detoxication of hydrogen peroxide by both CAT and peroxidase.

Filipendula ulmaria (L.) Maxim. FAMILY: ROSACEAE

The extract of meadowsweet (*Filipendula ulmaria*) aerial parts and fractions of the extract exhibits hepatoprotective and antioxidant activity during experimental toxic CCl_4 hepatitis (Shilova et al., 2006, 2008). This extract improved liver function. Meadowsweet extract in 70% ethanol (100 mg/kg) was most potent and exhibited low toxicity. This extract produced a normalizing effect on activity of enzymes, markers of cytolysis, LPO, and antioxidant defense system in liver cells. Fractionation of the extract was accompanied by dissociation of the effect. These changes reflect specific action of a complex of bioactive substances. The ethyl acetate and chloroform fractions from this extract were most potent. The effectiveness of these fractions by several parameters surpassed that of Carsil.

Firmiana simplex (L.) W.Wight FAMILY: STERCULIACEAE

Kim et al. (2015) evaluated the hepatoprotective potential of the methanolic extract originating from *Firmiana simplex* stem bark against the ethanol-induced hepatotoxicity in rat primary hepatocytes. Two flavonoid glycosides, quercitrin (1) and tamarixetin 3-*O*-rhamnopyranoside (2), were isolated from the active EtOAc fraction. Compound 1 significantly protected rat primary hepatocytes against ethanol-induced oxidative stress by reducing the intracellular ROS level and preserving antioxidative defense systems such as GR, GSH-PX, and total GSH. The EtOAc fraction of *F. simplex* stem bark and its major constituent quercitrin (1) could function as hepatoprotective agents to attenuate the development of alcoholic liver disease.

Flacourtia indica (Burm. f.) Merr. FAMILY: FLACOURTIACEAE

The extracts of the aerial parts of *Flacourtia indica* were evaluated for hepatoprotective properties. In PCM-induced hepatic necrosis in rat models, all extracts were found to reduce serum AST, serum ALT, and serum ALP. The most significant reduction of the serum level of AST and ALT was exhibited by petroleum ether and ethyl acetate extracts at a single oral of dose of 1.5 g/kg of BW with a reduction of AST and ALT level compared to PCM (3 g/kg of BW) treated animals. Histopathological examination also showed good recovery of PCM-induced necrosis by petroleum ether and ethyl acetate extracts. On the other hand, the ME did not show any remarkable effect on PCM-induced hepatic necrosis (Nazneen et al., 2009).

Gnanaprakash et al. (2010) investigated the hepatoprotective activity of aqueous extract of leaves of *F. indica* against CCl_4-induced hepatotoxicity. CCl_4 has enhanced the AST, ALT, ALP, and the lipid peroxides (TBARS) in liver. Treatment of aqueous extract of *F. indica* leaves (250 and 500 mg/kg) exhibited a significant protective effect by altering the serum levels of AST, ALT, ALP, TP, TB, and liver TBARS. These biochemical observations were supported by histopathological study of liver sections.

Flacourtia montana J. Graham FAMILY: FLACOURTIACEAE

Flacourtia montana and its related species have been used traditionally for the treatment of various diseases. Joshy et al. (2016) evaluated the hepatoprotective, anti-inflammatory, and antioxidant activities of *F. montana* methanolic extract. The hepatoprotective effect of *F. montana* was evaluated against PCM-induced hepatotoxicity in Wistar rats. Administration of PCM (2 g/kg) showed a significant biochemical and histological deterioration in the liver of experimental animals. Pretreatment with *F. montana* (200 and 400 mg/kg BW p.o.) significantly reduced the elevated levels of serum enzymes like serum AST, serum ALT, ALP, and reversed the hepatic damage in the liver which evidenced the hepatoprotective activity. The anti-inflammatory activity of *F. montana* was evaluated by carrageenan-induced paw edema and cotton pellet-induced granuloma models. *F. montana* (200 and 400 mg/kg) showed a significant reduction in rat paw edema with 76.39% and 80.32%, respectively induced by carrageenan against the reference anti-inflammatory drug ibuprofen (10 mg/kg) (83.10%). Oral administration of *F. montana* (200 and 400 mg/kg) also significantly reduced the granuloma mass formation in cotton pellet granuloma method. The reducing power and hydrogen peroxide radical scavenging were increased at increasing doses of *F. montana* (Joshy et al., 2016).

Flagellaria indica L. FAMILY: FLAGELLARIACEAE

Gnanaraj et al. (2016b) evaluated the hepatoprotective mechanism of aqueous extract of leaves of *Flagellaria indica* (FI) against CCl_4-mediated liver damage in SD rats. Total phenolic content in the aqueous extract of FI leaves was 65.88 ± 1.84 mg GAE/g. IC_{50} value for free-radical scavenging activity of FI aqueous extract was reached at the concentration of 400 µg/mL. Biochemical studies show that the aqueous extract of FI was able to prevent the increase in levels of serum transaminases, ALT, and AST (38%–74% recovery), and MDA formation (25%–87% recovery) in a dose-dependent manner. Immunohistochemical results evidenced the suppression of oxidative stress markers (4-hydroxynonenal and 8-hydroxydeoxyguanosine) and pro-inflammatory markers (tumor necrosis factor-α, interleukin-6, and prostaglandin E2). Histopathological and hepatocyte ultrastructural alterations proved that there were protective effects in FI against CCl_4-mediated liver injury. Signs of toxicity were not present in rats treated with FI alone (500 mg/kg BW).

Flaveria trinervia (Spreng.) C. Mohr. FAMILY: ASTERACEAE

Umadevi et al. (2004) evaluated the hepatoprotective effect of methanolic extract leaves of *Flaveria trinervia* against CCl_4-induced liver damage in rats. The extract significantly reversed the elevated enzyme levels and altered biochemical parameters compared to control and Silymarin. Besides, a significant reduction of the CCl_4-induced changes in the liver histopathology was also observed. The extract showed a considerable potential for hepatoprotective activity possibly by the antioxidant property.

Flemingia macrophylla (Willd.) Kuntze FAMILY: FABACEAE

Hsieh et al. (2011) investigated the protective effect of the aqueous extract of *Flemingia macrophylla* against hepatic injury induced by CCl_4. After oral administration of the aqueous extract, 0.5 and 1.0 g/kg doses significantly decreased ALT and AST, attenuated the histopathology of hepatic injury, ameliorated oxidative stress in hepatic tissue, and increased the activities of CAT, SOD, and GSH-Px. The hepatoprotective effect of daidzein and genistein were consistent to that of aqueous extract.

Flemingia strobilifera R.Br. FAMILY: FABACEAE

Kumar and Gowda (2011) evaluated the hepatoprotective activity of the CE of *Flemingia strobilifera* leaf in Wistar albino rats induced by ethanol-CCl_4. Levels of AST, ALT, ALP, and TB were increased and the levels of TP were decreased in ethanol-CCl_4-treated rats. The CE at both doses *did not decrease the elevatd levels of all these biochemical parameters and did not restore the normalcy of total protein significantly*. LPO was increased significant in liver tissue in the ethanol-CCl_4-treated rats while the activities of the GSH, CAT, and SOD were decreased. Histopathology also showed similar results. From this study it was concluded that the *CE of F. strobilifera did not show significant hepatoprotective and antioxidant action.*

Foeniculum vulgare Mill. FAMILY: APIACEAE

Ghanem et al. (2012) evaluated the antioxidant and hepatoprotective activities of the 80% methanolic extract as well as the ethyl acetate (EtOAc) and butanol (BuOH) fractions of the wild fennel (*Foeniculum vulgare* subsp. *piperitum*) and cultivated fennel (*F. vulgare* var. *azoricum*). Two phenolic compounds, that is, 3,4-dihydroxyphenethylalchohol-6-*O*-caffeoyl-β-D-glucopyranoside and 3′,8′-binaringenin were isolated from the fennel wild herb. The EtOAc and BuOH fractions of wild fennel were found to exhibit a radical scavenging activity higher than those of cultivated fennel. An *in vitro* method of rat hepatocytes monolayer culture was used for the investigation of hepatotoxic effects of the 80% ME on the wild and cultivated fennel, which were >1000 and 1000 μg/mL, respectively. Moreover, their hepatoprotective effect against the toxic effect of PCM (25 mM) was exerted at 12.5 μg/mL concentration.

The potential protective effect of anise and fennel essential oils was studied against CCl_4-induced fibrosis in rats. Administration of

CCl_4 (1.5 mL/kg/kg BW) intraperitoneally in olive oil (1:7 dilution) for seven successive weeks resulted in liver damage manifested by significant increase in serum AST, ALT, ALP, decreased TP, and increased TGs, TC, LDL and decreased the HDL level. Rats treated orally with essential oil of *Foeniculum vulgare* (Fennel, 200 and 400 kg/BW) for seven successive weeks and intoxicated with CCl_4 showed a significant protection against induced increase in serum liver enzyme (AST, ALT, ALP), restored TP level and ameliorate the increased TGs, total, cholesterol, LDL, and decreased the HDL. These protective effects were further confirmed by histopathological examination (El-Sayed et al., 2015).

Fragaria ananassa **(Weston) Duchesne FAMILY: ROSACEAE**

Hamed et al. (2016) evaluated the hepatoprotective effect of strawberry (*Fragaria ananassa*) juice on experimentally induced liver injury in rats. Rats were intraperitoneally injected with CCl_4 with or without strawberry juice supplementation for 12 weeks and the hepatoprotective effect of strawberry was assessed by measuring serum liver enzyme markers, hepatic tissue redox status and apoptotic markers with various techniques including biochemistry, ELISA, quantitative PCR assays, and histochemistry. The hepatoprotective effect of the strawberry was evident by preventing CCl_4-induced increase in liver enzymes levels. Determination of oxidative balance showed that strawberry treatment significantly blunted CCl_4-induced increase in oxidative stress markers and decrease in enzymatic and nonenzymatic molecules in hepatic tissue. Furthermore, strawberry supplementation enhanced the antiapoptotic protein, Bcl-2, and restrained the pro-apoptotic proteins Bax and caspase-3 with a marked reduction in collagen areas in hepatic tissue.

Fraxinus rhynchophylla **Hance FAMILY: OLEACEAE**

Guo et al. (2017) investigated the hepatoprotective activity of EtOH–water extract from the seeds of *Fraxinus rhynchophylla* against CCl_4-induced liver injury in mice. The EtOH–water extract significantly alleviated liver damage as indicated by the decreased levels of serum ALT and AST, the MDA content, and increased the levels of SOD, GSH, and GSH peroxidise (GSH-Px), and reduced the pathological injury induced by CCl_4. This research indicates that the EWE from the seeds of *F. rhynchophylla* decreased liver index, inhibited the increase of serum aminotransferase induced by CCl_4, and decreased hepatic MDA content, SOD, and GSH-Px activities.

Fraxinus xanthoxyloides Wall. FAMILY: OLEACEAE

Younis et al. (2016) evaluated the ME of leaves of *Fraxinus xanthoxyloides* for its hepatoprotective potential against CCl_4-induced hepatic injuries in rat. HPLC-DAD analysis of methanpol extract revealed the existence of rutin and caffeic acid. In CCl_4-treated rats, the level of ALT, AST, TB was significantly increased while the ALB concentration in serum was decreased as compared to control group. The level of hepatic antioxidant enzymes, CAT, POD, SOD, GST, and GR was significantly decreased against the control group. Further, significant decrease in GSH while increase in lipid peroxides (TBARS), H_2O_2, DNA damages, and comet length was induced with CCl_4 in hepatic tissues of rat. In contrast, co-administration of ME and Silymarin restored the biochemical and histopathological status of the liver.

Fumaria asepala Boiss. FAMILY: FUMARIACEAE

Aktay et al. (2000) studied the hepatoprotective activity of ethanolic extracts of *Fumaria asepala* using CCl_4-induced heptotoxicity model in rats. According to the results of biochemical tests, significant reductions were obtained in CCl_4-induced increases of plasma and liver tissue MDA levels and plasma enzyme activities by the treatment with *F. asepala* extracts, which reflects functional improvement of hepatocytes. Liver sections were also studied histopathologically to confirm the biochemical results.

Fumaria indica (Hausskn.) Pugsley FAMILY: FUMARIACEAE

Fumaria indica showed hepatoprotective activity against CCl_4, PCM, and rifampicin-induced heptatotoxicity in albino rats. The PEE against $CCl_{4,}$ total aqueous extract against PCM, and methanolic extract against rifampicine-induced hepatotoxicities showed similar reductions in the elevated levels of some of the serum biochemical indicators in a manner similar to that of Silymarin, indicating its potential as a hepatoprotective agent (Rao and Mishra, 1997).

Further investigation revealed that an active compound monomethyl fumarate has been separated from methanolic extract of whole plant *F. indica* and the compound showed no hepatocytotoxicity up to the dose of 1 mg/mL *in vitro* and up to 50 mg/kg (p.o.) *in vivo* in albino rats. *In vivo*, monomethyl fumerate showed significant antihepatotoxic activity against CCl_4, PCM and rifampicine-induced hepatotoxicities to an extent almost similar to that of Silymarin, a known antihepatotoxic agent (Rao and Mishra, 1998). Nimbkar et al. (2003) have also reported that

F. indica have good hepatoprotective activity against hepatotoxicity caused by antitubercular drug.

Rathi et al. (2008) evaluated the hepatoprotective potential of 50% ethanolic water extract of whole plant of *F. indica* and its three fractions, viz., hexane, chloroform and butanol against D-galactosamine-induced hepatotoxicity in rats. The hepatoprotection was assessed in terms reduction in histological damage, changes in serum enzymes (SGOT, SGPT, and ALP) and metabolites bilirubin, reduced GSH and LPO (MDA content). Among fractions more than 90% protection was found with butanol fraction in which alkaloid protopine was quantified as highest, that is, about 0.2 mg/g by HPTLC. The isolated protopine in doses of 10–20 mg p.o. also proved equally effective hepatoprotectants as standard drug Silymarin (single dose 25 mg p.o.). In general, all treatments excluding hexane fraction proved hepatoprotective at par with Silymarin. Protopine present in *F. indica* at doses of 10–20 mg/kg (p.o.) also proved to be equally effective hepatoprotectants as standard drug Silymarin (single dose of 25 mg/kg, p.o.) (Rathi et al., 2008).

Hussain et al. (2012) investigated the chemopreventive potential of *F. indica* extract (FIE) on *N*-NDEA and CCl_4-induced hepatocarcinogenesis in Wistar rats. The co-treatment with FIE significantly prevented the decrease of the BW and also increased in relative liver weight caused by NDEA. The treatment with FIE significantly reduced the nodule incidence and nodule multiplicity in the rats after NDEA administration. The levels of liver cancer markers such as AST, ALT, ALP, γ-glutamyl transferase, TBL, AFP and, carcinoembryonic antigen were substantially increased by NDEA treatment. However, FIE treatment significantly reduced the liver injury and restored the entire liver cancer markers. Histological observations of liver tissues also correlated with the biochemical observations.

Fumaria officinalis L. FAMILY: FUMARIACEAE

Hepatoprotective activity of ethanolic extract of *Fumaria officinalis* was evaluated in CCl_4-induced liver damage in rats. The ethanolic extract at a dose of 200 and 500 mg/kg orally induced significant hepatoprotective effect by reducing the serum marker enzymes like SGPT, SGOT, and ALP. Extract also reduced the elevated levels of serum total and DBL, cholesterol, TGs. Ascorbic acid in rat's urine and histophathological studies further conform the hepatoprotective activity of *F. officinalis* when compared to the CCl_4 treated control groups (Sharma et al., 2012).

Fumaria parviflora Lam.
FAMILY: FUMARIACEAE

The hepatoprotective activity of an aqueous-methanolic extract of *Fumaria parviflora* was investigated by Gilani et al. (1996) against PCM- and CCl_4-induced hepatic damage. Pretreatment of rats with plant extract (500 mg/kg, orally twice daily for 2 days) prevented the PCM (640 mg/kg)-induced rise in serum enzymes ALP, SGOT and SGPT, whereas the same dose of the extract was unable to prevent the CCl_4-induced rise in serum enzyme levels. Post-treatment with three successive doses of the extract (500 mg/kg, 6 hourly) also restricted the PCM-induced hepatic damage. The plant extract (500 mg/kg; orally) caused significant prolongation in pentobarbital (75 mg/kg)-induced sleep as well as increased strychnine-induced lethality in mice, suggestive of an inhibitory effect on MDME.

Alqasoumi et al. (2009) evaluated the efficacy of *F. parviflora,* traditional Saudi plant used for liver problems. The ethanol extract of the aerial part of *F. parviflora* was subjected to hepatoprotective assays using Wistar albino rats. Liver injury induced in rats using CCl_4. The biochemical parameters; SGOT, SGPT, ALP, and TB were estimated as reflection of the liver condition. Based on the results of the biochemical parameters measurements, histopathological study was performed on the liver of rats treated with two extracts. The normal appearance of hepatocytes indicated a good protection of the extracts from CCl_4 hepatotoxicity. All the results were compared with Silymarin, the reference hepatoprotective drug.

Tripathi et al. (2010) elucidated the effect of *F. parviflora* (Fp) on nimesulide-induced cell death in primary rat hepatocyte cultures. Fp extract treated cells showed increased viability as compared to nimesulide stressed cells as assessed by MTT assay. LDH leakage increased significantly at 500 μM nimesulide, and the data suggested that apoptosis was the predominant mechanism responsible for cell death. Nimesulide-induced apoptosis was further confirmed by DNA fragmentation and chromatin condensation. Nimesulide exposure increased intracellular ROS, translocation of Bax and Bcl2 followed by mitochondrial depolarization and cytochrome c (Cyt c) release along with caspase-9/-3 activity confirming involvement of mitochondria in nimesulide-induced apoptosis. Events like membrane depolarization of mitochondria, expression of Bax, Bcl2, externalization of phosphatidyl serine are substantially reversed by the pretreatment of Fp extract. Thus, the study indicates that Fp extract modulates critical events regulating pro and antiapoptotic proteins in mitochondria-dependent apoptosis induced by nimesulide.

Fumaria vaillantii Lois. FAMILY: FUMARIACEAE

Aktay et al. (2000) studied the hepatoprotective activity of ethanolic extracts of *Fumaria vaillantii* using CCl_4-induced heptotoxicity model in rats. According to the results of biochemical tests, significant reductions were obtained in CCl_4-induced increases of plasma and liver tissue MDA levels and plasma enzyme activities by the treatment with *F. vaillantii* extracts, which reflects functional improvement of hepatocytes. Liver sections were also studied histopathologically to confirm the biochemical results. The extracts of *F. vailantii* significantly prevented the elevation of plasma and hepatic MDA formation (evidence of LPO) as well as enzyme levels (AST and ALT) in acute liver injury, which might be ascribed to their potent hepatoprotective activity. Liver sections were also studied histopathologically to confirm the biochemical results.

In another study methanolic extract of, *F. vaillantii* was investigated against CCl_4-induced hepatocellular injury in rat and the result revealed that CCl_4 administration caused severe acute liver damage in rats, demonstrated by significant elevation of serum AST, ALT levels and classic histopathological changes. It seems that post-treatment of ME of it significantly reduces the ALT, AST, and ALP levels in comparison with CCl_4 group. Histopathological studies also provided supportive evidence for the biochemical analysis (Seyed et al., 2007).

Galium verum L. FAMILY: RUBIACEAE

A hepatoprotective activity of *Galium verum* (I and II dry extracts, 25 mg/kg), was studied against CCl_4-induced acute hepatitis in rats. The hepatoprotective effect of I and II extracts at the dose of 25 mg/kg is characterized by a decreased activities of serum enzymes, Serum ALT decreased 2.7–3.5-fold, Serum AST decreased 2.4–3.4-fold and ALP decreased 1.2–1.3-fold; whereas the activity of cholinesterase increased 1.3–1.4. The administration of the I extract decreased TBARS levels 1.4-fold in the serum and 1.8-fold in the liver homogenate. The administration of the II extract decreased TBARS levels 1.6-fold in the serum and 2.0-fold in the liver homogenate. The histopathological analysis of the liver material of the experimental group showed neither degenerative-dystrophic changes nor significant hemodynamic changes in comparison with the control group. The hepatoprotective effect of the II extract was more pronounced than that of the I extract and was comparable to the hepatoprotective activity of the reference drug Silibor (Goryacha et al., 2017).

Garcinia dulcis **(Roxb.) Kurz** ***FAMILY: CLUSIACEAE***

Garcinia species have been traditionally used for the treatment of many ailments including the liver damage. Gogoi et al. (2017) investigated the antioxidant and hepatoprotective activity of fruit rind extract of this plant. Phytochemical investigation revealed the presence of both phenolic and flavonoid groups in the extract in a significant amount. Antioxidant activity of the plant extract was observed in all models and percentage of inhibition was dose dependent. Intoxicated with CCl_4, elevated the liver function enzymes, bilirubin, and suppressed the production of TP. Pretreatment with the extract decreased the SGOT, SGPT, ALP, and bilirubin level significantly and increased the production level of TP in a dose-dependent manner. The histopathological observation supported the hepatoprotective potentiality of the extract. The results indicate that fruit rind part of *G. dulcis* is nontoxic and the plant can utilize as an antioxidant source. The plant has a protective agent for liver damages and other diseases caused by free radicals.

Gardenia gummifera **L. f. FAMILY: RUBIACEAE**

Prabha et al. (2012) evaluated the antioxidant and antihepatotoxic effect of methanolic extract of *Gardenia gummifera* root (MEGG) on TAA-induced oxidative stress in male Wistar rats. MEGG significantly prevented the elevation of serum AST, ALT, ALP, LDH, and tissue MDA levels in both experimental groups, when compared to the TAA alone treated groups. The rats receiving TAA plus MEGG exhibited significant increases in hepatic and renal antioxidant activities including GSH, GST, GR, GPx and CAT levels. Quantification of histopathological changes also supported the dose-dependent protective effects of MEGG.

Gentiana asclepiadea **L. FAMILY: GENTINACEAE**

Mihailović et al. (2013) evaluated the hepatoprotective activity of MEs of aerial parts (GAA) and roots (GAR) *Gentiana asclepiadea* against CCl_4-induced liver injury in rats. In CCl_4 treated animals, GAA, and GAR significantly decreased levels of serum transaminases, ALP and TB, and increased the level of TP. Treatment with the extracts resulted in a significant increase in the levels of CAT, SOD, and reduced GSH, accompanied with a marked reduction in the levels of MDA, as compared to CCl_4 treated group. The histopathological studies confirmed protective effects of extracts against CCl_4-induced liver injuries. No genotoxicity was observed in liver cells after GAA treatment, while GAR showed only slight genotoxic effects by comet assay. Phytochemical analysis revealed the presence of sweroside, swertiamarin and gentiopicrin in high concentrations in both extracts.

Gentiana cruciata L. FAMILY: GENTIANACEAE

Mihailović et al. (2014) investigated the effects of the MEs of *Gentiana cruciata* aerial parts (GCA) and roots (GCR) against CCl_4-induced liver injury in rats. Pretreatment with GCA and GCR, containing sweroside, swertiamarin, and gentiopicrin in high concentrations, dose dependently and significantly decreased the levels of serum transaminases, ALP and TB, whereas an increase in the level of TP was found compared with the CCl_4-treated group. Moreover, oral administration of extracts significantly enhanced antioxidant enzyme activities (SOD and CAT), increased the content of GSH and decreased the content of TBARS. Microscopic evaluations of the liver revealed CCl_4-induced lesions and related toxic manifestations that were minimal in the liver of rats pretreated with extracts at the dose of 400 mg per kg BW. The results suggest that the use of *G. cruciata* extracts has a merit as a potent candidate in protecting the liver against chemical-induced toxicity.

Gentiana olivieri Griseb. FAMILY: GENTIANACEAE

Aktay et al. (2000) studied the hepatoprotective activity of ethanolic extracts of *Gentiana olivieri* using CCl_4-induced heptotoxicity model in rats. According to the results of biochemical tests, significant reductions were obtained in CCl_4-induced increases of plasma and liver tissue MDA levels and plasma enzyme activities by the treatment with *G. olivieri* extracts, which reflects functional improvement of hepatocytes. Liver sections were also studied histopathologically to confirm the biochemical results. The extracts of *G. olivieri* significantly prevented the elevation of plasma and hepatic MDA formation (evidence of LPO) as well as enzyme levels (AST and ALT) in acute liver injury, which might be ascribed to their potent hepatoprotective activity. Liver sections were also studied histopathologically to confirm the biochemical results.

Hepatoprotective effect of *G. olivieri* flowering herbs on subacute administration was studied by Orhan et al. (2003) using *in vivo* models in rats. For the activity assessment on CCl_4-induced hepatic damage following biochemical parameters were evaluated; plasma and hepatic tissue MDA formation, and liver tissue GSH level, as well as plasma transaminase enzyme levels (AST and ALT). Results of biochemical tests were also confirmed by histopathological examination. Through bioassay-guided fractionation procedures isoorientin, a known C-glycosylflavone, was isolated from the EAF as the active antihepatotoxic constituent by silica gel column chromatography. Isoorientin exhibited significant hepatoprotective effect at 15 mg/kg BW dose (Orhan et al., 2003).

Gentiana scabra Bunge FAMILY: GENTIANACEAE

In traditional Chinese medicine, the rhizomes of *Gentiana scabra* are used to treat hepatitis, stomatitis, and inflammatory diseases. Ko et al. (2011) examined the protective effects of *G. scabra* aqueous extract (GS) and polyphenols in liver of CCl_4-intoxicated mice. GS exhibited anti-LPO, DPPH, and superoxide radical scavenging activities with IC_{50} values of 45.84, 183.38, and 56.25 μ/mL, respectively. HPLC analysis revealed that the major polyphenolic constituents of GS were kaempferol, ellagic acid, and quercetin. Daily oral administration of 500 mg/kg GS (GS500) and 1000 mg/kg GS (GS1000) significantly prevented the elevation of GPT and TBA reactive substances levels, while enhancing the levels of antioxidant enzymes (SOD, CAT, and GPx) and serum total antioxidant activity in mice with hepatoxicity. At 1000 mg/kg, GS was as effective as 100 mg/kg Silymarin in reducing oxidative stress and preventing liver injury. Histopathological studies further confirmed the hepatoprotective activity of GS. Taken together, these results show that the antioxidant activities and polyphenolic compounds (kaempferol, ellagic acid, and quercetin) of GS may have contributed to its hepatoprotective activity in CCl_4-intoxicated mice, and its mechanism of action could be mediated through the reduction of oxidative stress in liver tissue.

Gentiana veitchiorum Hemsl. FAMILY: GENTIANACEAE

Zhang et al. (2014) evaluated the hepatoprotective effect of ME of *Gentiana veitchiorum* against CCl_4-induced oxidative stress and liver injury in mice. Oral administration of ME at 200 and 400 mg/kg for 15 days dose dependently inhibited the serum elevations of AST, ALT, and ALP, and recovered the reduction of SOD, CAT, and GPx in liver tissue. Hematoxylin and eosin staining examination performed in liver tissues suggested that the ME treatment ameliorated histopathological changes in CCl_4-induced mice. Western blotting analysis implied that ME increased HO^{-1} expression and recovered TNF-α alternation.

G. veitchiorum can protect the liver against CCl_4-induced damage in mice, and this hepatoprotective effect was due at least in part to its ability through scavenging CCl_4-associated free radical activities.

Gentianella turkestanorum (Gand.) Holub. FAMILY: GENTIANACEAE

Yang et al. (2017) investigated the contents of secoiridoid compounds (i.e., sweroside, swertiamarin and gentiopicrin) from *Gentianella turkestanorum* extracts, and the potential effects of *G. turkestanorum* extracts against CCl_4-induced liver injury in mice. Iridoid glycoside showed the highest content in the product extracted by butanol, but lower in the products extracted by ethyl acetate and water. All *G. turkestanorum* extracts showed protective effects against CCl_4-induced acute liver injury in mice, among which butanol extract showed the maximal protective effects. *G. turkestanorum* extracts induced significant decrease in the serum ALT, AST, ALP, and TB compared with those in the mice with acute lung injury. Obvious increase was noticed in serum TP. Moreover, such effects presented in a dose-dependent manner. Compared with the control group, the MDA was significantly elevated in the model group, while significant decrease was observed in the levels of SOD, GSH, and CAT in model group compared with the control group. Whereas, such phenomenon was completely reversed by *G. turkestanorum* extracts in a dose-dependent manner. In their further study, Yang et al. (2019) have shown that *G. turkestanorum* showed protective effects on hepatic injury by modulating the endoplasmic reticulum Stress and NF-κB signaling pathway. *G. turkestanorum* could inhibit the LPO and increase the antioxidant activity. Also, it could inhibit the cellular apoptosis through downregulating the transcriptional level of ERS-related genes and proteins. This process was associated with the nuclear translocation of NF-κB p65 protein.

Gentianopsis barbata (Froel) Ma FAMILY: GENTIANACEAE

The granulated dry extract from *Gentianopsis barbata* in a dose of 0.1 g/kg produced a pronounced therapeutic effect in rats with experimental toxic (tetrachloromethane) and drug-induced (tetracycline) hepatitis. The gentian extract improved the bile production and secretion functions of liver, normalized the protein, lipid, and pigment metabolism, and increased the antioxidant system activity in the test animals (Nikolaev et al., 2001).

***Ginkgo biloba* L. FAMILY: GINKGOACEAE**

Ozenirler et al. (1997) evaluated the protective effect of *Ginkgo biloba* extract on CCl_4-induced hepatic damage in rats. Hepatic MDA, GSH, and HYP levels and histopathologic alterations in liver specimens were assessed. Around 200 mg/kg/day *G. biloba* extracts were given orally to the animals for 10 days, then a single dose of 2 mL/kg BW CCl_4 was administered intraperitoneally. *G. biloba* extract treatment reduced hepatic MDA levels significantly, but did not alter GSH and HYP levels. The light and electron microscopic findings showed that *G. biloba* extract limited the CCl_4-induced hepatocyte necrosis and atrophy. These results suggest that this extract may protect the hepatocytes from CCl_4-induced liver injury.

Shenoy et al. (2001) assessed the protective activity of *G. biloba* against CCl_4-induced hepatotoxicity in rats and probe into its mechanism of action. In their study, elevation in the levels of end products of LPO in liver of rats treated with CCl_4 was observed. The increase in MDA levels in liver suggests enhanced LPO leading to tissue damage and failure of antioxidant defense mechanisms to prevent formation of excessive free radicals. Pretreatment with *G. biloba* significantly reversed these changes. Hence it is possible that the mechanism of hepatoprotection of *G. biloba* is due to its antioxidant effect. Histopathological studies showed that CCl_4 caused steatosis and hydropic degeneration of the liver tissue. *G. biloba* pretreatment exhibited protection, which confirmed the results of biochemical studies. All the effects of *G. biloba* were comparable with those of Silymarin, a proven hepatoprotective. The results of their study indicate that simultaneous treatment with *G. biloba* protects the liver against CCl_4-induced hepatotoxicity.

Luo et al. (2004) studied the reversing effect of *G. biloba* extract (GbE) on established CCl_4-induced liver fibrosis in rats. Compared with saline-treated group, liver fibrosis rats treated with GbE had decreased serum TB and aminotransferase levels and increased levels of serum ALB. Microscopic studies revealed that the livers of rats receiving GbE showed allieviation in fibrosis as well as expression of αSMA. The liver collagen and reticulum contents were lower in rats treated with GbE than saline-treated group. RT-PCR revealed that the level of TIMP-1 decreased while the level of MMP-1 increased in GbE group. Administration of GbE improved CCl_4-induced liver fibrosis. It is possibly attributed to its effect of inhibiting the expression of TIMP-1 and promoting the apoptosis of hepatic stellate cells.

Sener et al. (2006) investigated the possible beneficial effect of *G. biloba* (EGb), an antioxidant agent, against Acetaminophen toxicity in mice. ALT, AST levels, and TNF-α were increased significantly after AAP treatment, and reduced with EGb. Acetaminophen caused a significant decrease in GSH levels while MDA levels and MPO activity were increase in liver tissues. These changes were reversed by EGb treatment. Furthermore, luminal and lusigenin CL levels in the AAP group increased dramatically compared to control and reduced by EGb treatment. These results implicate that AAP causes oxidative damage in hepatic tissues and Ginkgo extract, by its antioxidant effects protects the tissues.

The protective effects of *G. biloba* phytosomes (GBP) on CCl_4-induced hepatotoxicity and the probable mechanism(s) involved in this protection were investigated in rats by Naik and Panda (2007). GBP (25 and 50 mg/kg) and Silymarin elicited significant hepatoprotective activity by decreasing the activities of serum marker enzymes and LPO and elevated the levels of GSH, SOD, CAT, GPx, GR, ALB, and TP in a dose-dependent manner.

Chávez-Morales et al. (2011) evaluated the protective effect of *Ginkgo biloba* extract on liver damage by a single dose of CCl_4 in male rats. The pretreatment with 4 mg/kg (p.o.) of GbE, for 5 days, prevented most of the damage caused by CCl_4: significantly decreased the serum activities of ALT and AST (54 and 65%, respectively), compared to CCl_4-treated rats; GbE partially prevented the increase of liver MDA and the decrease of ALB concentration. This pretreatment prevented the downregulation of TNF-α and upregulated the interleukine 6 (IL-6) mRNA steady-state level. Moreover, the GbE reduced the amount of necrotic areas in the central lobe area, compared to CCl_4-treated rats.

Chang et al. (2007) investigated the effects of the combined extracts of *Ginkgo biloba*, *Panax ginseng*, and *Schizandra chinensis* at different doses on hepatic antioxidant status and fibrosis in rats with CCl_4-induced liver injury in Male SD rats. The pathological results showed that herbal extract suppressed hepatic bile duct proliferation, and low-dose herbal extract inhibited liver fibrosis. Hepatic SOD activity was lower in the CCl_4 group, but there was no difference in the Silymarin or herbal extract treated groups compared to the control group. Hepatic CAT activity and the ratio of reduced to oxidized GSH were significantly higher in the high extract group than those in the CCl_4 group. Silymarin and herbal extract reversed the impaired hepatic total antioxidant status. Herbal extract partially reduced the elevated hepatic LPOs. Hepatic transforming growth factor-β1 (TGF-β1) level decreased significantly in the low extract group. Therefore, high-dose herbal extract improved hepatic antioxidant capacity through enhancing CAT activity and GSH redox status, whereas low-dose herbal extract inhibited liver fibrosis through decreasing hepatic TGF-β1 level in rats with CCl_4 induced liver injury.

Glossocardia bosallea (L.f.) DC. FAMILY: ASTERACEAE

Rajopadhye and Upadhye (2012) investigated the antioxidant and hepatoprotective activity of hexane (GH), ethanol (GE), and water extract (GW) of the whole plant *Glossocardia bosvallea*. GH and GE inhibited DPPH, NO and SOD radicals in a dose-dependent manner. The total antioxidant capacities of lipid-soluble substances in GH, GE and GW were 14.342, 12.656, and 9.890 nmol/g trolox equivalent, respectively. The trade of phenol content was GW < GH < GE. The hepatoprotective effect was assayed in CCl_4-induced cytotoxicity in a liver slice culture model. Depletion was observed in LDH, LPO, and antioxidative enzymes on administration of GH and GE or ascorbic acid as standard in CCl_4-induced cytotoxicity in the liver. GH and GE significantly prevented oxidative liver damage.

Glycine max (L.) Merr. FAMILY: FABACEAE

Nonalcoholic fatty liver disease (NAFLD), which is characterized by >5% deposition of TGs in hepatocytes, is often referred as a major risk factor for obesity, type 2 diabetes, and hypertension. Hong et al. (2016) investigated the hepatoprotective effect of whole soybean embryos containing bioactive substances such as isoflavones and soyasaponins. Hepatic SOD and GPx activity of the experimental groups increased during the period of the study. Hepatic mRNA expressions of TNF α, nuclear factor (erythroid-derived 2)-like 2, and Caspase 3 were decreased when soybean embryos were increased in the mice's diets. Both of the soybean embryo-treated groups showed significantly decreased serum and liver TG and TC. Adiponectin, AMP-activated protein kinase (AMPK) α, hydroxymethylglutaryl-CoA reductase, sterol regulatory element-binding protein-1c, fatty acid synthase, and apolipoprotein B mRNA expressions were decreased in the mice that were fed soybean embryos.

Glycosmis pentaphylla (Retz.) DC. FAMILY: RUTACEAE

Nayak et al. (2011) evaluated the hepatoprotective activity of *Glycosmis pentaphylla* against PCM-induced hepatotoxicity in Swiss albino mice. The effect of ME (200 and 400 mg/kg) and PEE (200 and 400 mg/kg) was studied on PCM-induced (250 mg/kg intraperitoneally) hepatic damage in mice for estimating the serum marker enzymes as ALT, AST, ALP, TB, and TP. Mice were protected from the hepatotoxic action of PCM as evidenced by significant reduction in the elevated serum level of ALT, AST, ALP, TB and an increased level of TP with a significant reduction in liver weight when compared with PCM treated group and Silymarin (50 mg/kg) was used as a positive control. These biochemical observations were supplemented by histopathological examination of liver sections from different experimental groups and corroborated the hepatoprotective efficacy of methanol and petroleum ether plant extract. The ME (400 mg/kg) of *G. pentaphylla* is able to alter the toxic condition of the hepatocytes so as to protect the membrane integrity against PCM-induced leakage of marker enzymes.

Glycyrrhiza glabra L. FAMILY: FABACEAE

The hepatoprotective activity of a water extract of *Glycyrrhiza glabra* was studied by Al-Qarawi et al. (2001) against CCl_4-induced hepatotoxicity in rats. Pre- and post-treatment with licorice extract showed a dose-dependent reducetion of CCl_4-induced elevated serum levels of enzyme activity with parallel increase in TP and ALB levels, indicating the licorice could preserve the normal functional status of the liver.

Rajesh and Latha (2004) evaluated the potential efficacy of *G. glabra* in protecting tissues from peroxidative damage in CCl_4-intoxicated rats. The increased LPO formation in the tissues of CCl_4-treated rats was significantly inhibited by *G. glabra*. The observed decreased antioxidant enzyme activities of SOD, CAT, GSH-Px, GST, and antioxidant concentration of GSH were nearly normalized by *G. glabra* treatment. CCl_4-induced damage produces alteration in the antioxidant status of the tissues, which is manifested by abnormal histopathology. *G. glabra* restored all these changes.

Huo et al. (2011) evaluated the antihepatotoxic effect of licorice aqueous extract CCl_4-induced liver injury in a rat model. Licorice extract significantly inhibited the elevated AST, ALP, and ALT activities and the decreased TP, ALB, and G levels caused by CCl_4intoxication. It also enhanced liver SOD, CAT, GSH-Px, GR, GST activities, and GSH level, reduced MDA level. Licorice extract still markedly reverses the increased liver HYP and serum TNF-α levels induced by CCl_4 intoxication.

Yin et al. (2011) evaluated the hepatoprotective and antioxidant effects of *G. glabra* extract (2.5, 5 and 10 μg/mL) on the CCl_4-induced carp hepatocyte damage *in vitro*. CCl_4 at 8 mM in the culture medium produced significantly elevated levels of LDH, SGOT, SGPT and MDA and significantly reduced levels of SOD and GSH-Px. Pretreatment (5 μg/mL) and pre- and post-treatment (5 and 10 μg/mL) of the hepatocytes with *G. glabra* extract significantly reduced the elevated levels of LDH, GOT, GPT, and MDA and increased the reduced levels of SOD and GSH-Px by CCl_4; post-treatment of the hepatocytes with *G. glabra* extract at 5 μg/mL reduced the GPT and GOT levels and increased the GSH-Px level, but had no effect on the other parameters at all the studied concentrations.

Al-Razzuqi et al. (2012) evaluated the hepatoprotective effect of aqueous extract of *G. glabra* roots in rabbit models with acute liver injury induced by CCl_4 at a dose of 1.25 mL/kg as a mixture with olive oil. Aqueous extract of *G. glabra* was administered in a dose of 2 g/kg/d orally for 7 days. Their results demonstrate that the aqueous extract of *G. glabra* had a significant effect in amiolerating liver functions as well as restoring hepatic tissue in acute liver diseases when it was given in a single dose per day of 2 g/kg BW. Therefore, the aqueous extract of *G. glabra* roots can be used for prevention and treatment of liver disorders.

Glycyrrhiza uralensis Fisch. FAMILY: FABACEAE

An inhibitor of beta-glucuronidase from the rhizomes of *Glycyrrhiza uralensis* was isolated and its hepatoprotective activity on CCl_4-induced hepatotoxicity of rats was investigated by Shim et al. (2000). From the water-soluble extract of *G. uralensis*, glycyrrhizin was isolated as a potent inhibitor of beta-glucuronidase. When glycyrrhizin was orally administered, it had a hepatoprotective activity. However, when glycyrrhizin was intraperitoneally administered, it did not have a hepatoprotective activity. 18 beta-glycyrrhetinic acid, which is a major metabolite of glycyrrhizin by human intestinal bacteria, was also a potent inhibitor of beta-glucuronidase. When 18 beta-glycyrrhetinic acid was intraperitoneally administered, it also had some hepatoprotective activity. These results suggest that glycyrrhizin may be a natural prodrug for the observed hepatoprotective effect in rats and that serum beta-glucuronidase levels have implications for the liver injury, as reductions of its activity by administration of inhibitors such as *G. uralensis* or its derived products and Silymarin correlate with reductions in biochemical indices of liver injury.

Jung et al. (2016) evaluated the effect of licorice on chronic alcohol-induced fatty liver injury mediated by inflammation and oxidative stress.

Alcohol consumption increased serum ALT and AST activities and the levels of TGs and TNF-α. Lipid accumulation in the liver was also markedly induced, whereas the GSH level was reduced. All these alcohol-induced changes were effectively inhibited by licorice treatment. In particular, the hepatic GSH level was restored and alcohol-induced TNF-α production was significantly inhibited by licorice. The phytoconstituents include glycyrrhizic acid, liquiritin, and liquiritigenin.

Gmelina asiatica L. FAMILY: VERBENACEAE

Merlin and Parthasarathy (2011) evaluated the hepatoprotective activity of chloroform and ethanol extracts of *Gmelina asiatica* aerial parts against CCl_4-induced hepatic damage in rats. The extracts at dose of 400 mg/kg were administered orally once daily. The substantially elevated serum enzyme levels of AST, ALT, serum ALP, and TB were restored toward normalization significantly by the extracts. The biochemical observations were supplemented with histopathological examination of rat liver sections. The results of this study strongly indicate that *G. asiatica* aerial parts have potent hepatoprotective action against CCl_4-induced hepatic damage in rats. Ethanol extract was found more potent hepatoprotective. *In vivo* antioxidant and free-radical scavenging activities were also screened which were positive for both chloroform and ethanol extracts. This study suggests that possible mechanism of this activity may be due to free-radical-scavenging and antioxidant activities which may be due to the presence of flavonoids in the extracts.

Gossweilerodendron balsamiferum (Vermoesen) Harms FAMILY: LEGUMINOSAE

Methanol extract of *Gossweilerodendron balsamifarum* leaves at 250 and 500 mg/kg was evaluated by Aloh et al. (2015) for hepatoprotective potentials using female albino rats. The plant extract (250 and 500 mg/kg), however showed no remarkable hepatoprotective and antioxidant activity against chloroform/ hexane-induced hepatotoxicity as judged from the serum marker enzymes and antioxidant levels in liver tissues. Chloroform/hexane induced a significant rise in AST, ALT, ALP, TB, LPO with a reduction of TP, SOD, CAT, GPx, and GST. Treatment of rats with different doses of the plant extract (250 and 500 mg/kg), however, could not reverse the condition. Also lipid profiles of the control group and that of the extract were found not to differ. This was ascertained by comparing the levels of (HDL, LDL, TC and triacylglycerol (TAG), in the two groups.

Gossypium hirsutum L. FAMILY: MALVACEAE

The protective effect of *Gossypium hirsutum* extracts was studied in acute experimental hepatic injury in rats. *G. hirsutum* extracts significantly decrease the serum transaminase activities, increased the SOD activities and decreased MDA content (Batur et al., 2008).

Graptopetalum paraguayense E. Walther FAMILY: CRASSULACEAE

Graptopetalum paraguayense, a vegetable consumed in Taiwan, has been used in folk medicine for protection against liver injury. Duh et al. (2011) investigated the protective effects of *G. paraguayense* against CCl_4-induced liver damage in rats. Water extracts ranging from 50 to 300 mg/kg BW administrations significantly lowered AST and ALT levels, and inhibited MDA generation in CCl_4-treated rats. Water extract increased cellular GSH level and antioxidant enzymes, including SOD, GR, and CAT. Serum TNF-α was decreased in the group treated with CCl_4 plus extract (150 and 300 mg/kg BW). Histopathological examination of livers showed that the water extract reduced fatty degeneration, cytoplasmic vacuolization and necrosis in CCl_4-treated rats. In contrary, 10 mg/kg BW of GA was administrated, this dose was related with water extract (300 mg/kg BW), and had significantly decreased the AST and ALT compared to the CCl_4 treated group. Aforesaid results suggested that GA from WGP offered antioxidative activity against CCl_4-induced oxidative liver damage.

Grewia mollis Juss. FAMILY: MALVACEAE

Methanolic extract of *Grewia mollis* leaves showed significant hepatoprotective activity against CCl_4-induced liver injury (Asuku et al., 2012). *G. mollis* extract possesses significant antioxidant (Asuku et al., 2012) and anti-inflammatory (Al-Youssef et al., 2012) activities which may contribute to its hepatoprotective effects. The pharmacological effect of *G. mollis* may be attributed to its steroidal and/or triterpenoidal constituent which have proven to be anti-inflammatory activity (Al-Youssef et al., 2012). Phytochemical studies on leaves of *G. mollis* has shown the presence of luteolin, tetrahydroxyflavone, 7β-hydroxy-23-enedeoxojessic acid, 7β-hydroxy-23-deoxojessic acid, β-sitosterol, and β-sitosterol-3-*O*-glucoside (Al-Youssef et al., 2012; Asuku et al., 2012).

Grewia tenax Forssk. FAMILY: MALVACEAE

The administration of ethanol extract of *Grewia tenax* significantly restored CCl_4-induced biochemical (serum AST, ALT, APT, TB, and gamma-glutamyl transferase) and histopathological changes in rats. Reversal of pentobarbital-induced prolongation of narcolepsy by the extract also suggested its hepatoprotective effect. The chronic administration of extract significantly reduced cholesterol, LDLs, and TGs level (Al-Said et al., 2011). The hepatoprotective effect of *G. tenax* is attributed to antioxidant and anti-inflammatory properties. Phytochemical studies on plant of *G. tenax* have shown the presence of triacontan1-ol, β-amyrin, β-Sitosterol, lupenne, erythrodiol, betulin, and tetratriacont-21-ol-12-one.

Grewia tiliifolia Vahl FAMILY: MALVACEAE

Bioassay-guided fractionation of methanolic extract of the *Grewia tiliifolia* bark by Ahamed et al. (2010) has resulted in the isolation of D-erythro-2-hexenoic acid gamma-lactone (EHGL) and gulonic acid gamma-lactone (GAGL). Hepatoprotective activity of the methanolic extract and the isolated constituents were evaluated against CCl_4-induced hepatotoxicity in rats. The treatment with methanolic extract, EHGL, and GAGL at oral doses of 100, 150, and 60 mg/kg, respectively, with concomitant CCl_4 intraperitoneal injection (1 mL/kg) significantly reduced the elevated plasma levels of aminotransferases, ALP, and the incidence of liver necrosis compared with the CCl_4-injected group without affecting the concentrations of SB and hepatic markers. EHGL and GAGL significantly inhibited the elevated levels of TBARS and GSH in liver homogenates. Histology of the liver tissues of the extract and isolated constituents treated groups showed the presence of normal hepatic cords, absence of necrosis and fatty infiltration as similar to the normal control. The results revealed that the hepatoprotective activity of EHGL is significant as similar to the standard drug Silymarin. To clarify the influence of the extract and isolated constituents on the protection of oxidative-hepatic damage, Ahamed et al. (2010) examined *in vitro* antioxidant properties of the test compounds. The extract and the constituents showed significant free-radical scavenging activity. These results suggest that the extract as well as the constituents could protect the hepatocytes from CCl_4-induced liver damage perhaps, by their antioxidative effect on hepatocytes, hence eliminating the deleterious effects of toxic metabolites from CCl_4.

Gundelia tournefortii L. FAMILY: ASTERACEAE

Gundelia tournefortii is used as an occasional food, and its extracts have been used for the prevention and treatment of liver diseases in Iran. The effects of different concentrations of the hydroalcoholic extract of dried powdered footstalks of *G. tournefortii* were investigated by Jamshidzadeh et al. (2005) on CCl_4-induced hepatotoxicity using *in vivo* model in rats and isolated rat hepatocytes. Rats received different concentrations of the *G. tournefortii* extract by i.p. injection for three consecutive days before the injection of CCl_4 (i.p.). The results showed that the *G. tournefortii* extract could protect the liver against CCl_4-induced damages with doses of 200 and 300 mg/kg, but concentrations higher than 300 mg/kg were less effective. For *in vitro* studies, the extract was added to the suspension of freshly isolated rat hepatocytes incubated in Krebs–Henseleit buffer under a flow of 95% O_2 and 5% CO_2, 20 min before the addition of 10 mM of CCl_4. The *G. tournefortii* extract with concentrations of 0.2–0.8 mg/mL protected the cells against CCl_4-induced cytotoxicity, and its maximum protective effect was about 0.5 mg/mL, but concentrations of 1.0 mg/mL and higher increased the CCl_4-induced cytotoxicity. The *G. tournefortii* extract itself was toxic toward isolated hepatocytes with concentrations above 1.0 mg/mL. Therefore, the results of this study support the traditional believes on hepatoprotective effects of *G. tournefortii* ; however, high concentrations were hepatotoxic (Jamshidzadeh et al., 2005).

Gymnosporia emarginata (Willd.) Ding Hou FAMILY: CELASTRACEAE

Rubab et al. (2013) evaluated the hepatoprotective activity of methanolic extracts of *Gymnosporia emarginata* against PCM-induced liver damage in rats. The methanolic extracts of *G. emarginata* (300 mg/kg) were administered orally to the animals with hepatotoxicity induced by PCM (3 g/kg). The plant extract was effective in protecting the liver against the injury induced by PCM in rats. This was evident from significant reduction in serum enzymes ALT, AST, ALP, and bilirubin.

Gymnosporia montana (Roth) Benth. FAMILY: CELASTRACEAE

Patel et al. (2010) explored the hepatoprotective activity of the ethanol extract of leaves of *Gymnosporia montana* against PCM-induced hepatotoxicity. Pre and post-treatment with ethanol extract of *G. montana* at doses of 50 and 100 mg/kg was studied by comparing the above-mentioned parameters with Silymarin (100 mg/kg) as standard. Both doses of ethanol extract of *G. montana* were found to be hepatoprotective. Extract at the dose of 100 mg/kg produced effects comparable to those of Silymarin.

Haldina cordifolia **(Roxb.) Ridsdale [Syn.:** ***Adina cordifolia*** **(Roxb.) Hook.f.] FAMILY: RUBIACEAE**

The acetone (AEAC) and aqueous extracts (AQEAC) of *Haldina cordifolia* were studied for hepatoprotective activity against Wister rats with liver damage induced by ethanol by Sharma et al. (2012). It was found that AEAC and AQEAC, at a dose of 500 mg/kg BW exhibited hepatoprotective effect by lowering the SGPT, SGOT, alkaline phosphate and TB to a significant extent and also significantly increased the levels of TP. The hepatoprotective activity was also supported by histopathological studies of liver tissue. Since results of biochemical studies of blood samples of ethanol-treated rats showed significant increase in the levels of serum enzyme activities, reflecting the liver injury caused by ethanol and blood samples from the animals treated with AEAC and AQEAC showed significant decrease in the levels of serum markers, indicating the protection of hepatic cells against ethanol-induced hepatocellular injury. The effects of AEAC and AQEAC were comparable with standard drug Silymarin.

Halenia elliptica **D.Don FAMILY: GENTIANACEAE**

Halenia elliptica, a medicinal herb of Tibetan origin, was commonly used in folk medicine to treat hepatitis. Huang et al. (2010) evaluated the hepatoprotective and antioxidant activity of *H. elliptica* against experimentally induced liver injury. The ME of *H. elliptica* was studied for its hepatoprotective effects against CCl_4-induced liver toxicity in rats. Activity was measured by monitoring the levels of ALT, AST, ALP, and TB. The ME possessed strong antioxidant activity *in vitro*. The results of CCl_4-induced liver toxicity experiment showed that rats treated with the ME of *H. elliptica* (100 and 200 mg/kg), and also the standard treatment, Silymarin (50 mg/kg), showed a significant decrease in ALT, AST, ALP, and TB levels, which were all elevated in the CCl_4 group. The results observed after administration of 100 mg/kg ME were comparable to those of Silymarin at 50 mg/kg. The ME did not show any mortality at doses up to 2000 g/kg BW. These results seem to support the traditional use of *H. elliptica* in pathologies involving hepatotoxicity, and the possible mechanism of this activity may be due to strong free-radical scavenging and antioxidant activities of ME (Huang et al., 2010).

Haloxylon salicornicum (Moq.) Bunge FAMILY: CHENOPODIACEAE

Ahmad and Eram (2011) investigated the *in-vivo* hepatoprotective effect of aerial parts of *Haloxylon salicornicum* in order to validate its traditional use in hepatobiliary disorders, by native people of Cholistan desert, Pakistan. Aerial parts (ethanolic extract) of *H. salicornicum*, (500 and 750 mg/kg/day, p.o. for 7 days) were evaluated on CCl_4 intoxicated rabbits (0.75 mL/kg., s/c.) by serum biochemical parameters and liver histopathological observations. Silymarin was used as a standard hepatoprotective drug. CCl_4 intoxicated group had elevated levels of SGOT, SGPT, and ALP significantly but TB level was normal as compared to control group. The extract (both doses of 500 and 750 mg/kg) showed hepatoprotective effect by significant restoration of SGOT, SGPT, ALP, and TB levels as compared to CCl_4 control. Around 500 mg/kg doses of *H. salicornicum* extract produced more significant results as compared to 750 mg/kg doses and Silymarin. Histopathological examination of liver tissues further substantiated these findings. Alqasoumi et al. (2012) reported antioxidant and anti-inflammatory activities of *H. salicornicum* which may contribute to its hepatoprotective activity.

Hammada scoparia (Pomel) Iljin FAMILY: AMARANTHACEAE

Bourogaa et al. (2013) evaluated the antioxidative activity and hepatoprotective effects of methanolic extract (ME) of *Hammada scoparia* leaves against ethanol-induced liver injury in male rats. The animals were treated daily with 35% ethanol solution (4 g/kg/day) during 4 weeks. This treatment led to an increase in the LPO, a decrease in antioxidative enzymes (CAT, SOD, and GPx) in liver, and a considerable increase in the serum levels of AST and ALT and ALP. However, treatment with ME protects efficiently the hepatic function of alcoholic rats by the considerable decrease in aminotransferase contents in serum of ethanol-treated rats. The glycogen synthase kinase-3 β was inhibited after ME administration, which leads to an enhancement of GPx activity in the liver and a decrease in LPO rate by 76%. These biochemical changes were consistent with histopathological observations, suggesting marked hepatoprotective effect of ME. These results strongly suggest that treatment with methanolic extract normalizes various biochemical parameters and protects the liver against ethanol-induced oxidative damage in rats.

Harungana madagascariensis Lam. ex Poir. FAMILY: HYPERICACEAE

The protective effects of 100–500 mg/kg/day of the aqueous root extract of *Harungana madagascariensis* were evaluated by Adeneye et al. (2008) on the average BW, relative liver-BW, ALT, and AST, ALP, TBL, and CB, TGs, TC, cholesterol fractions (HDL-c, LDL-c, VLDL-c), fasting blood glucose (FBG), TP, and ALB in the acute and repeated dose APAP hepatotoxic rats. Results showed that acute intraperitoneal injection of 800 mg/kg of APAP induced significant elevations in the serum concentrations of ALT, AST, ALP, and FBG but caused significant decreases in the serum concentrations of TP and ALB, while inducing nonsignificant alterations in the serum levels of lipids, TB and CB. However, pretreatments with 100–500 mg/kg of *H. madagascariensis* significantly attenuated elevations of ALT, AST, ALP, and FBG. In addition, the extract significantly attenuated reduction in the serum levels of TP and ALB while inducing nonsignificant alterations in the serum lipids. Repeated APAP hepatotoxicity caused similar effects in the measured parameters except that it was associated with significant reduction in the FBG while inducing significant increases in the serum TB and CB. Oral pretreatments with the extract significantly enhanced APAP-induced hypoglycemia while significantly attenuating significant elevations in the serum levels of TB and CB, in dose-related fashion. The associated histopathologic features of moderate-to-severe hepatic necrosis were also attenuated by the extract.

Heinsia crinita (Afzel) G. Taylor FAMILY: RUBIACEAE

Ebong et al. (2014) assessed the hepatoprotective activity of whole methanolic leaf extract of *Heinsia crinita* (HC) in alloxan-induced diabetic albino Wistar rats. Extract of *H. crinita* reduced serum glucose level by 15% within 14 days, while the diabetic control showed no reduction. The results further showed significant decrease in ALT, AST, and ALP activity in control group which gave high values of these enzymes. Animals treated with extract of *H. crinita* showed high values of K^+ and Cl^- but lower Na^+ and urea values when compared to diabetic control with low values of K^+ and Cl^- but high values of Na^+ and urea. Electrolyte modulation by the plant extract was shown in this study to be better than insulin. These findings showed that *H. crinita* may possess antidiabetic properties and beneficial in the management of type 1 diabetes and its attendant liver and kidney complications.

Helianthus annuus L. FAMILY: ASTERACEAE

The hepatoprotective activity of ethanolic and aqueous extracts of *Helianthus annuus* flowers was studied in CCl_4-induced hepatotoxicity in Wistar rats. Treatment with the *H. annuus* flower extracts significantly reduced elevated serum enzymatic level of SGOT, SGPT, ALP, and TB in CCl_4-induced rats treated with 200 mg/kg BW. The biochemical effects of the ethanolic and aqueous extracts of *H. annuus* flowers were further confirmed by histopathological examinations of liver (Vasavi et al., 2014).

Helianthus tuberosus L. FAMILY: ASTERACEAE

The effects of *Helianthus tuberosus* (HT) on hyperglycemia and hepatoprotection in streptozotocin (STZ)-induced diabetic rats were investigated by Kim and Han (2013). Water extract of HT reversed the increase in serum glucose levels in STZ-induced diabetic rats. Further HT treatment remarkably prevented elevation of serum AST, ALT, lactate dehydrogenase (LDH) and γ-glutanyl transpeptidase levels in STZ-induced diabetic rats. Treatment with HT also improved lipid profiles damaged by STZ in the liver by reducing serum levels of TGs and total cholesterol.

Helicanthes elastica (Desr.) Danser FAMILY: LORANTHACEAE

Aqueous and 95% ethanolic extract of whole plant of *Helicanthes elastica* were subjected to acute oral toxicity by Kumar et al. (2016). The aqueous and ethanolic extracts revealed no observable changes in the rats up to the dose level of 2000 mg/kg BW. The extracts were then screened for PCM-induced hepatic injury at dose levels of 200 and 400 mg/kg BW (1/10 and 1/5 LD_{50} based on toxicity study). The aqueous extract of whole plant of *H. elastica* was found to produce significant reversal of the PCM-induced changes in the measured biochemical and histopathological parameters at lower dose of 200 mg/kg which was found to be better than ethanol extract at the same dose level.

Helichrysum plicatum DC. subsp. plicatum FAMILY: ASTERACEAE

Tütüncü et al. (2010) evaluated the hepatoprotective acivity of diethyl ether extract *Helichrysum plicatum* subsp. *plicatum* against CCl_4-induced acute liver toxicity in rats. Some of the animals died during treatment. Serum ALT levels were higher in the treated groups than those in the CCl_4 groups. Histopathological findings were similar to the CCl_4 and extract treated groups. It was concluded that diethylether extract of *H. plicatum* subsp. *plicatum did not have a protective effect in CCl4-induced acute liver toxicity in rats and even exacerbated the toxicity*.

Helminthostachys zeylanica (L.) Hook. FAMILY: OPHIOGLOSSACEAE

Suja et al. (2004) reported the effect of the ME of *Helminthostachys zeylanica* rhizomes on CCl_4-induced liver damage in Wistar rats. The results showed that significant hepatoprotective effect was obtained against CCl_4-induced liver damage, by oral administration of *H. zeylanica* ME as evident from decreased levels of serum enzymes and almost normal architecture of the liver, in the treated groups, as compared to the controls.

Hemidesmus indicus R.Br. FAMILY: APOCYNACEAE

Oral treatment with the ethanol extract of *Hemidesmus indicus* roots (100 mg/kg, for 15 days) significantly prevented rifampicin and isoniazid-induced hepatotoxicity in rats (Prabakaran et al., 2000). Baheti et al. (2006) evaluated the hepatoprotective activity of methanolic extract of *H. indicus*. Treatment of rats with PCM and CCl_4 produced a significant increase in the levels of SGPT, SGOT, ALP, total and DBL. Rats pretreated with methanolic extract of roots of *H. indicus* (100–500 mg/kg BW, p.o.) exhibited rise in the levels of these enzymes but it was significantly less as compared to those treated with PCM or CCl_4 alone. The results of methanolic extract of *H. indicus* were comparable with the standard hepatoprotective agent Silymarin (100 mg/kg). Maximum hepatoprotective effect was found to be at the dose of 250 mg/kg BW in case of CCl_4-induced hepatic damage while 500 mg/kg BW in case of PCM-induced hepatic damage.

Herpetospermum caudigerum Wall. ex Chakrav. FAMILY: CUCURBITACEAE

Herpetospermum caudigerum is traditionally used for the treatment of liver diseases, cholic diseases, and dyspepsia as a well-known Tibetan medicine in China. Shen et al. (2015) investigated the hepatoprotective effect of extract of *H. cordigerum* and ascertain its active ingredients and possible mechanism. EAF and HPE significantly alleviated liver injury as indicated by the decreased levels of serum ALT and AST and reduce the pathological tissue damage induced by CCl_4. Moreover, they decreased the MDA content and increased the levels of SOD, GSH, and GSH-Px. Chemical analysis indicated that the EAF was rich in HPE. The lignans extract of *H. caudigerum* is effective for the prevention of CCl_4-induced hepatic damage in mice and HPE may be partially responsible for the pharmacological effect of hepatoprotection. The hepatoprotective effect may be related to its free-radical scavenging effect, inhibiting LPO and increasing antioxidant activity.

Heterotheca inuloides Cass. FAMILY: ASTERACEAE

A model of hepatotoxicity by CCl_4 in rats was used in order to evaluate the protective potential of the acetonic and methanolic extracts of *Heterotheca inuloides* (Coballase-Urrutia et al., 2011). Pretreatment with the two *H. inuloides* extracts attenuated the increase in the activity of serum AST and ALT observed in CCl_4-induced liver injury. The protective effect was confirmed by the analysis of tissue slides stained with hematoxylin-eosin and periodic acid/Schiff's reagent. Additionally, the two extracts are scavengers to the superoxide radical as was observed by electron paramagnetic resonance. Since the methanolic extract resulted in a better protective effect in the previous experiments, it was used to investigate in more detail the mechanism of hepatoprotection. Quercetin, one of the main components of the extract, with known hepatoprotective and antioxidant activity was used as a positive control. Pretreatment of animals with the methanolic extract or quercetin, was associated with the prevention of 4-hydroxynonenal and 3-nitrotyrosine increase in the liver, two markers of oxidative stress. Furthermore, the decrease in the activity of several antioxidant enzymes including SOD, CAT, and GPx in CCl_4-induced liver injury was alleviated by the pretreatment with *H. inuloides* methanolic extract or quercetin.

Hibiscus cannabinus L. FAMILY: MALVACEAE

Agbor et al. (2005) evaluated the hepatoprotective activity of aqueous leaf extract of *Hibiscus cannabinus* using CCl_4 and PCM liver toxicity. The aqueous leaf extract of *H. cannabinus* showed a significant hepatoprotective activity against this damage in lowering the plasma transaminases and bilirubin concentration significantly. The absence of necrosis in liver cells of rats pretreated with extract indicated a protective effect. The extract also inhibited LPO, suggesting a possible mechanism of action.

Hibiscus esculentus **(L.) Moench FAMILY: MALVACEAE**

The ethanol extract of *Hibiscus esculentus* roots inhibited the formation of ODFR *in vitro* for superoxide radical production, hydroxyl radical generation, NO radical formation and LPO formation, respectively. The oral administration of the extract (250 and 500 mg/kg BW), significantly reduced CCl_4-induced hepatotoxicity in rats, as judged from the serum and tissue activity of marker enzymes SGOT, SGPT, and ALP. These results were comparable with standard drug Silymarin (20 mg/kg, p.o.) (Sunilson et al., 2008).

Alqasoumi (2011) in his study made an attempt to validate the claimed uses of 'Okra' *H. esculentus* in liver diseases. The preventive action of ethanolic extract of okra against liver injury was evaluated in rodents using CCl_4-induced hepatotoxicity model. Ethanolic extract, at 250 and 500 mg/kg BW, exerted significant dose-dependent hepatoprotection by decreasing the CCl_4-induced elevation of serum SGOT, SGPT, ALP, GGT, cholesterol, TGs, and MDA nonprotein sulfhydryls (NP-SH) and TP levels in the liver tissue. A significant reduction was also observed in pentobarbital-induced sleeping time in mice. The hepatoprotective and antioxidant activities of the extract are being comparable to standard Silymarin. These findings were supported by histological assessment of the liver biopsy. The ability of okra extract to protect chemically induced liver damage may be attributed to its potent antioxidant property.

Hibiscus hispidissimus **Griff. FAMILY: MALVACEAE**

Hibiscus hispidissimus is used in tribal medicine of Kerala, the southern most state of India, to treat liver diseases. Krishnakumar et al. (2008) evaluated the effect of the ethanolic extract of *H. hispidissimus* whole plant on PCM-induced and CCl_4-induced liver damage in healthy Wistar albino rats. The results showed that significant hepatoprotective effects were obtained against liver damage induced by PCM and CCl_4 as evidenced by decreased levels of serum enzymes, SGOT, SGPT, serum ALP, SB, and an almost normal histological architecture of the liver of the treated groups compared to the toxin controls. The extract also showed significant antilipid peroxidant effects *in vitro*, besides exhibiting significant activity in quenching DPPH radical, indicating its potent antioxidant effects.

Hibiscus rosa-sinensis **L. FAMILY: MALVACEAE**

The ability of high (6.73) and low (5.02) pH solutions of lead acetate precipitate of *Hibiscus rosa-sinensis* petal anthocyanin extract to prevent CCl_4-induced LPO in rats has been investigated by Obi and Uneh (2012) using plasma MDA level as an index of the process. Relative to its value in the plasma of CCl_4 treated extract-free rats, the high pH form (pH 6.73; blue pigment) of the resolubilized precipitate significantly reduced the level of MDA in the plasma of rats to which it was administered prior to CCl_4 treatment. When compared with CCl_4-treated, extract-free rats, statistically significant reduction in the level of MDA was also observed in the plasma obtained from rats treated first with the low pH form (pH 5.02; red pigment) of the resolubilized precipitate before CCl_4 exposure. These data suggest that *H. rosa-sinensis* petal partially purified anthocyanin extract may be protective against CCl_4-induced lipoperoxidation.

Hibiscus sabdariffa L. FAMILY: MALVACEAE

Ali et al. (2003) investigated the effect of the water extract of the dried flowers of *Hibiscus sabdariffa* and Hibiscus anthocyanins (HAs) (which are a group of natural pigments occurring in the dried calyx of *H. sabdariffa*) on PCM-induced hepatotoxicity in rats. Given for 4 weeks (but not for 2 or 3 weeks) the extract significantly improved some of the liver function tests evaluated, but did not alter the histology of the PCM-treated rats or the pentobarbitone-induced sleeping time. At a dose of 200 mg/kg, the hepatic histology and the biochemical indices of liver damage were restored to normal. Lower dose were ineffective.

Amin and Hamza (2005) evaluated the hepatoprotective effect of water extract of *H. sabdariffa* in Azathioprine-induced hepatotoxicity in rats. Typically, administration of Azathioprine induces oxidative stress through depleting the activities of antioxidants and elevating the level of malonialdehyde in liver. This escalates levels of ALT and AST in serum. Pretreatment with the extract proved to have a protective effect against Azathioprine-induced hepatotoxicity. Animals pretreated with water extract not only failed to show necrosis of the liver after azathioprine administration, but also retained livers that for the most part, were histologically normal. In addition, the extract blocked the induced elevated levels of ALT and AST in serum. The Azathioprine-induced oxidative stress was relieved to varying degrees by the herbal extract. This effect was evident through reducing MDA levels and releasing the inhibitory effect of Azathioprine on the activities of GSH, CAT, and SOD.

Dried flower *H. sabdariffa* (HSE) extracts were studied by Liu et al. (2006) for their protective effects against liver fibrosis induced using

CCl_4 in male Wistar rats. HSE significantly reduced the liver damage including steatosis and fibrosis in a dose-dependent manner. Moreover, HSE significantly decreased the elevation in plasma AST and ALT. It also restored the decrease in GSH content and inhibited the formation of lipid peroxidative products during CCl_4 treatment. In the primary culture, HSE also significantly inhibited the activation of the hepatic stellate cells.

Olaleye and Rocha (2008) investigated the hepatoprotective activity of *H. sabdariffa* on APAP-induced liver damage in mice. Paracetamol caused liver damage as evident by significant increased in plasma activities of AST and ALT. There were general significant losses in the activities of SOD, GPx, CAT, and delta-ALA-D and an increase in TBARS in the liver of PCM-treated group compared with the control group. However, the extract of *H. sabdariffa* was able to counteract these effects.

Ubani et al. (2011) investigated the effect of *H. sabdariffa* leaf extract on liver marker enzymes such as AST, ALT, and ALP, for its hepatoprotective effect in phenobarbitone-induced rats. Phenobarbitone-treated rats showed a significant increase in the levels of circulatory AST, ALT, and ALP. These changes were significantly decreased in rats treated with HSEt and phenobarbitone. These results indicate that extract offers hepatoprotection by influencing the levels of liver markers in phenobarbitone-induced rats and this could be due to its free-radical scavenging property and the presence of natural antioxidants.

Yin et al. (2011) evaluated the hepatoprotective and antioxidant effects of *H. sabdariffa* extract on the CCl_4-induced hepatocyte damage in fish and provide evidence as to whether it can be potentially used as a medicine for liver diseases in aquaculture. CCl_4 at 8 mM in the culture medium produced significantly elevated levels of LDH, SGOT, SGPT, and MDA and significantly reduced levels of SOD and GSH-Px. Pretreatment and pre- and post-treatment of the hepatocytes with *H. sabdariffa* extract significantly reduced the elevated levels of LDH, GOT, GPT, and MDA and increased the reduced activities of SOD and GSH-Px in a dose-dependent manner; post-treatment did not show any protective effect. The results suggest that *H. sabdariffa* extract can be potentially used for preventing rather than curing liver diseases in fish.

The antioxidative and hepatoprotective activities of 200 and 300 mg/kg BW ethanolic extract of dried flower of *H. sabdariffa* were assessed by Usoh et al. (2012) in rats treated with 0.25 mL/kg BW (intraperitoneally) of CCl_4. The oral administration of the extracts showed a significant dose-dependent decrease in the

CCl_4-induced MDA formation in liver. The levels of vitamins A, C, and β-carotene were shown to be significantly decreased and increased, respectively, in CCl_4 and extract treated groups when compared with the control. The increase in the levels of these vitamins might not be unconnected with the antioxidant properties possessed by the extract. The extract also displayed a strong hepatoprotective effect as it significantly reduced CCl_4-induced hepatotoxicity in rats, as judged from the serum activities of ALT, AST, and ALP.

Adetutu and Owoade (2013) investigated the hepatoprotective potentials of *H. sabdariffa* polyphenolic rich extract (HPE), (a group of phenolic compounds occurring in the dried calyx of *Hibiscus sabdariffa*) against CCl_4-induced damaged in rats. The antioxidant investigation showed that HPE was able to scavenge the ABTS and DPPH radicals and these radicals scavenging abilities were found to be dose dependent. Pretreatment of rats with different doses of HPE (50 and 100 mg/kg) significantly lowered serum ALT, AST, ALP, LDH, and TBARS levels in CCl_4-treated rats. GSH, SOD, and CAT were significantly increased by pretreatment with the HPE, in CCl_4-treated rats. HPE was found to contain high level of TPC (140.78 mg/g in GAE/g dried weight).

Flavonoid-rich AF of methanolic extract of *H. sabdariffa* calyx was evaluated by Ghosh et al. (2015) for its antihepatotoxic activities in streptozotocin-induced diabetic Wistar rats. The ameliorative effects of the extract on STZ-diabetes-induced liver damage was evident from the histopathological analysis and the biochemical parameters evaluated in the serum and liver homogenates. Reduced levels of GSH, CAT, SOD, and GPx in the liver of diabetic rats were restored to a near normal level in the *H. sabdariffa*-treated rats. Elevated levels of AST, ALT, and ALP in the serum of diabetic rats were also restored in *H. sabdariffa*-treated rats. Examination of stained liver sections revealed hepatic fibrosis and excessive glycogen deposition in the diabetic rats. These pathological changes were ameliorated in the extract-treated rats.

Obouayeba et al. (2014) investigated the hepatoprotective potential of the aqueous extract of petals of *H. sabdariffa*. Three major results were obtained. The hepatotoxicity of DNPH expressed by the rats of Group 1 was significantly different from those of the other groups (control, 2–4) for both hepatotoxicity and oxidative stress markers. The hepatoprotective and antioxidant properties of the petal extracts and confirmation of those of Silymarin through the rats of Groups 2–4 were statistically identical to the control group for markers of hepatotoxicity and oxidative stress.

Famurewa et al. (2015) investigated ameliorative potential of methanolic extract *H. sabdariffa* against CCl_4-induced LPO, hepatic damage and oxidative stress in adult male Wistar rats. CCl_4-induced oxidative stress in experimental rats was evidenced by increase in MDA and reduction in SOD, CAT, and reduced GSH. *H. sabdariffa* extract treatment at 600 and 1000 mg/kg doses resulted in significant modulation of antioxidant indices and ALP, but failed to demonstrate significant effects in AST, ALT, and MDA. There were significant increases in the serum vitamins C and E at 600 and 1000 mg/kg doses of the extract.

Ezzat et al. (2016) evaluated the hepatoprotective effect and studied the metabolic profile of the anthocyanin-rich extract of *H. sabdariffa* calyces (HSARE). The UPLC-qTOF-PDA-MS analysis of HSARE enabled the identification of 25 compounds represented by delphinidin and its derivatives, cyanidin, kaempferol, quercetin, myricetin aglycones, and glycosides, together with hibiscus lactone, hibiscus acid, and caffeoylquinic acids. Compared to the TAA-intoxicated group, HSARE significantly reduced the serum levels of ALT, AST, and hepatic MDA by 37.96, 42.74 and 45.31%, respectively. It also decreased hepatic inflammatory markers, including tumor necrosis factor alpha, interleukin-6, and interferon gamma (INF-γ), by 85.39%, 14.96%, and 70.87%, respectively. Moreover, it decreased the immunopositivity of nuclear factor kappa-B and CYP2E1 in liver tissue, with an increase in the effector apoptotic marker (caspase-3 positive cells), restoration of the altered hepatic architecture and increases in the activities of SOD and GSH by 150.08 and 89.23%, respectively. HSARE revealed pronounced antioxidant and anti-inflammatory potential where SOD and INF-γ were significantly improved. HSARE possesses the added value of being more water soluble and of natural origin with fewer side effects expected compared to Silymarin.

Hibiscus vitifolius L. FAMILY: MALVACEAE

Samuel et al. (2012) evaluated four extracts [petroleum ether, chloroform ($CHCl_3$), methanol (MeOH) and the aqueous] obtained from *Hibiscus vitifolius* roots in a model of hepatotoxicity induced with the mixture of INH (7.5 mg/kg), RIF (10 mg/kg) and PZA (35 mg/kg) in Wistar rats. The MeOH extract, followed by the aqueous extract administered orally exhibited significant hepatoprotective activity on attenuating toxic effects in the liver produced by anti-TB drugs.

Biochemical parameters and reduction of hepatocellular necrosis were very similar to the positive control Silymarin (100 mg/kg); the authors attribute the antioxidant effect of the extract to the presence of flavonoids (Samuel et al., 2012).

Hippophae rhamnoides L. FAMILY: ELAEAGNACEAE

Geetha et al. (2008) evaluated the hepatoprotective activity of Seabuckthorn (*Hippophae rhamnoides*) on CCl_4-induced liver injury in SD male albino rats. Pretreatment of leaf extract at a concentration of 100 and 200 mg/kg BW significantly protected the animals from CCl_4-induced liver injury. The extract significantly restricted the CCl_4-induced increase of SGOT, SGPT, ALP, and bilirubin and better maintained protein levels in the serum. Further it also has enhanced GSH, and decrease MDA levels.

Hsu et al. (2009) examined the protective effects of *H. rhamnoides* seed oil on CCl_4-induced hepatic damage in male ICR mice. Oral administration of SBT seed oil at doses of significantly reduced the elevated levels of ALT, AST, ALP, TG, and cholesterol at least 13% in serum, and the level of MDA in liver at least 22%, that was induced by CCl_4 (1 mL/kg) in mice. Moreover, the treatment of SBT seed oil was also found to significantly increase the activities of SOD, CAT, GSH-Px, glutathione reductase (GSH-Rd), and GSH content in liver up to 134%. The optimal dose of SBT seed oil was 0.26 mg/kg, as the minimum amount exhibiting the greatest hepatoprotective effects on CCl_4-induced liver injury. Overall, the hepatoprotective effect of SBT seed oil at all tested doses was found to be comparable to that of Silymarin (200 mg/kg) and have been supported by the evaluation of the liver histopathology in mice.

Maheshwari et al. (2011) investigated antioxidant and hepatoprotective activities of phenolic rich fraction (PRF) of Seabuckthorn leaves on CCl_4-induced oxidative stress in SD rats. Total phenolic content was found to be 319.33 mg GAE/g PRF and some of its phenolic constituents, such as GA, myricetin, quercetin, kaempferol, and isorhamnetin were found to be in the range of 1.935–196.89 mg/g of PRF. Oral administration of PRF at dose of 25–75 mg/kg BW significantly protected from CCl_4-induced elevation in AST, ALT, γ-GGT and bilirubin in serum, elevation in hepatic LPO, HP, protein carbonyls, depletion of hepatic reduced GSH and decrease in the activities of hepatic antioxidant enzymes; SOD, CAT, GPx, GR, and GST. The PRF also protected against histopathological changes produced by CCl_4 such as hepatocytic necrosis, fatty changes, vacuolation, etc. The data obtained in the study suggests that PRF has potent antioxidant activity, prevent oxidative damage to major biomolecules and afford significant protection against CCl_4-induced oxidative damage in the liver.

Gayathiri et al. (2012) evaluated the hepatoprotective activity of aqueous extract of *H. rhamnoides* in CCl_4-induced hepatotoxicity in albino Wistar rats. There was a tremendous elevation in the liver enzymes such as ATP, ALT, and AST associated with CCl_4 administration. In addition to that administration of CCl_4 in rats caused liver tissue abnormalities such as centrizonal necrosis, sinusoidal dilation, and hepatic fatty degeneration. The aqueous extract of *H. rhamnoides* showed significant reduction in the liver tissue abnormalities and liver enzymes.

***Holostemma ada-kodien* Schult.**
FAMILY: APOCYNACEAE

Sunil et al. (2015) studied on hepatoprotective activity of *Holostemma ada-kodien* and they found that pretreatment of the rats with alcoholic extract prior to PCM administration caused a significant reduction in the values of AST, ALT, ALP, and SB approximately comparable to the hepatoprotective of standard drug Silymarin.

***Homalium letestui* Pellegr.**
FAMILY: FLACOURTIACEAE

Homalium letestui has been traditionally used by the Ibibios of Southern Nigeria to treat stomach ulcer, malaria, and other inflammatory diseases and Yorubas of western Nigeria as an antidote. Okokon et al. (2017) evaluated the hepatoprotective properties of the ethanol extract of the plant stem. The hepatoprotective effect of the extract of the stem of the plant (200–600 mg/kg) was evaluated by the assay of liver function parameters, namely total and DBL, serum protein and ALB, TC, ALT, AST, and ALP activities, antioxidant enzymes like SOD, CAT, and GPx, reduced GSH and histopathological study of the liver. Administration of the extract of the stem of the plant caused a significant dose-dependent reduction of high levels of liver enzymes (ALT, AST, and ALP), TC, direct and TB as well as elevation of serum levels of TP, ALB, and antioxidant enzymes (SOD, CAT, GPx, and GSH). Histology of the liver sections from extract and Silymarin-treated animals showed reductions in the pathological features compared to the PCM-treated animals. The chemical pathological changes were consistent with histopathological observations suggesting marked hepatoprotective effect of the extract of *H. letestui* stem. GCMS analysis of *n*-butanol fraction (NBF) revealed the presence of 16 bioactive compounds.

Hoslundia opposita Vahl FAMILY: LAMIACEAE

The hepatoprotective potentials of the stem solvent fractions of *Hoslundia opposita* were investigated by Akah and Odo (2010). The fractions were prepared and tested for hepatoprotective effect against CCl_4 and PCM-induced liver damage in rats. Changes in the levels of biochemical markers of hepatic injury, namely; AST, ALT, ALP, and bilirubin were determined in both treated and control groups of rats. The effects of the extracts were compared with that of silymarin (100 mg/kg). Phytochemical analysis and acute toxicity studies of the extract were also performed. The results showed that CCl_4 and PCM (2 g/kg) elevated the levels of AST, ALT, ALT, and bilirubin. Treatment with the ME and methanol and EAF of *H. opposita* (100 mg/kg) ameliorated the effects of the hepatotoxins and significantly reduced the elevated levels of the biochemical marker enzymes, while the chloroform and hexane fractions showed no significant hepatoprotective effect. The extracts showed good toxicity profile with an LD50 value above 5000 mg/kg for the ME. Phytochemical analysis showed the presence of resins, flavonoids, sterols/triterpenes, tannins, saponins, alkaloids, reducing sugars, cardiac glycosides, and proteins in the solvent fractions (Akah and Odo, 2010).

Hybanthus enneaspermus (L.) F.V. Muell. FAMILY: VIOLACEAE

Vuda et al. (2012) investigated the hepatoprotective, curative and antioxidant properties of aqueous extract of *Hybanthus enneaspermus* used against CCl_4-induced liver damage in rats. Liver damage was induced by CCl_4 (1 mL/kg i.p.), and Silymarin was used as a standard drug to compare hepatoprotective, curative and antioxidant effects of the extract. Rats were treated with aqueous extract of *H. enneaspermus* at a dose of either 200 or 400 mg/kg after division into pretreatment (once daily for 14 days before CCl_4 intoxication) and post-treatment (2, 6, 24 and 48 h after CCl_4 intoxication) groups. Pretreatment and post-treatment with aqueous extract of *H. enneaspermus* showed significant hepatoprotection by reducing the AST, ALT, and ALP enzymatic activities and TB levels which had been raised by CCl_4 administration. Pre-and post-treatment with aqueous extract significantly decreased hepatic LPO as well as producing a corresponding increase in tissue total thiols. Post-treatment with aqueous extract improved ceruloplasmin levels. The histopathological examination of rat liver sections treated with aqueous extract confirms the serum biochemical observations. This study results demonstrate the protective, curative, and antioxidant effects of *H.*

enneaspermus aqueous extract used against CCl_4-induced hepatotoxicity in rats, and suggest a potential therapeutic use of *H. enneaspermus* as an alternative for patients with acute liver disease.

Hygrophila schulli (Buch.-Ham.) M.R. Almeida & S.M. Almeida [Syn.: *H. auriculata* (K. Schu.) Heine; *Asteracantha longifolia* (L.) Nees] FAMILY: ACANTHACEAE

Seeds of *Hygrophila schulli* are used in Indian systems of medicine for the treatment of liver ailments. The antihepatotoxic effect of methanolic extracts of the seeds was studied by Singh and Hanada (1995) on rat liver damage induced by a single dose of PCM (3 g/kg, p.o.) or TAA (100 mg/kg, s.c.) by monitoring several liver function tests, namely, serum transaminases (SGOT and SGPT), ALP, sorbitol dehydrogenase, glutamate dehydrogenase and bilirubin in serum. Furthermore, hepatic tissues were processed for assay of TGs and histopathological alterations simultaneously. A significant hepatoprotective activity of the methanolic extract of the seeds was reported.

The hepatoprotective activity of the aqueous extract of the roots of *H. schulli* was studied by Shanmugasundaram and Venkataraman (2006) on CCl_4-induced liver toxicity in rats. The activity was assessed by monitoring the various liver function tests, namely, ALT, AST, ALP, TP, and TB. Furthermore, hepatic tissues were subjected to histopathological studies. The root extract was also studied for its *in vitro* antioxidant activity using ferric thiocyanate and TBA methods. The extract exhibited significant hepatoprotective and antioxidant activities.

The EtOH (95%) extract from the aerial part of *H. schulli* exhibited a hepatoprotective effect on male SD rats against the damage induced with the mixture of INH/RIF (50 mg/kg each), the EtOH extract was administered at dose of 500 mg/kg/d, by oral via for 28 days. The authors found that this extract restored ALT, AST, and ALP levels and accelerated the regeneration of hepatic cells (Lina et al., 2012). The hepatoprotective effect has also been described of the aqueous extract of *H. schulli* (complete plant and root) against the acute damage caused by CCl_4 and APAP (Hewawasam et al., 2003; Shailajan et al., 2005). The plant's MeOH extract also showed a hepatoprotective effect against damage caused by APAP (Shivashangari et al., 2004). The alkaloid-rich fraction from the MeOH extract of *H. schulli* demonstrated a hepatoprotective effect *in vitro* model [human hepatocellular carcinoma (HepG2)] and *in vivo* (in Wistar rats) on inducing injury with CCl_4; the protective effect observed similar to that of Silymarin (Raj et al., 2010).

Hygrophila spinosa **T. Anderson FAMILY: ACANTHACEAE**

Usha et al. (2007) analyzed the levels of some known antioxidant (both enzymic and nonenzymic) activities in the roots of *Hygrophila spinosa* to find out the hepatoprotective effect of the same in CCl_4-induced liver damage in albino rats. The roots were found to be rich in antioxidants. To find out the hepatoprotective activity, the aqueous extract of the plant root samples was administrated to rats for 15 days. The serum marker enzymes AST, ALT, and gama glutamyl were measured in experimental animals. The increased enzyme levels after liver damage with CCl_4 were nearing to normal value when treated with aqueous extract of the root samples. Histopathological observation also proved the hepatoprotectivity of the root samples.

Hypericum japonicum **Thunb. FAMILY: HYPERICACEAE**

Wang et al. (2008) investigated the hepatoprotective activity of different parts of *Hypericum japonicum* against CCl_4-induced hepatitis and α-naphthyl-isothiocyanate (ANIT)-induced cholestasis. Mice were divided into groups and then administrated orally with solutions extracted from herbs before they were modeled in the experiments. Levels of ALT, AST and TB (T-BIL) in serum were evaluated. HPLC fingerprint was used for phytochemical analysis of the extracts. The total aqueous extract of *H. japonicum* had an obvious effect on the decreasing of AST, ALT, and T-BIL levels in serum. The isolated fraction IV (F4) exhibited a preferable activity of ameliorating cholestasis, while Fraction V (F5) was more efficacious in protecting the liver from injury. Chemical fingerprint indicated that F5 contained several flavonoids which might be the active chemicals against hepatotoxicity (Wang et al., 2008).

Hypericum perforatum **L. FAMILY: HYPERICACEAE**

Ozturk et al. (1992) reported the hepatoprotective effect of alcoholic extract of aerial part of *Hypericum perforatum* extract. The extract significantly attenuated CCl_4 and ethanol (Roman et al., 2011; Ozturk et al., 1992)-induced hepatic toxicity. Experimental studies also showed significant choleretic activity of *H. perforatum* (Roman et al., 2011). The protective action of *H. perforatum* has been attributed to its anti-inflammatory, antioxidant, and immune-modulating activities. Phytochemical studies on plant of *H. perforatum* showed the presence of hypericin, pseudohypericin, hyperforin, adhyperforin, quercetin, hyperoside, rutin, campferol, myricetin, amentoflavone, kielcorin, and norathyriol (Al-Asmari et al., 2014).

Bayramoglu et al. (2014) investigated the effective role of *Hypericum perforatum* on hepatic ischemia–reperfusion (I/R) injury in rats. While the ALT, AST, LDH activities, and MDA levels were significantly increased, CAT, and GPx activities significantly decreased in only I/R-induced control rats compared to normal control rats. Treatment with HPE $_{50}$ significantly decreased the ALT, AST, LDH activities and MDA levels, and markedly increased activities of CAT and GPx in tissue homogenates compared to I/R-induced rats without treatment-control group. In oxidative stress generated by hepatic ischemia–reperfusion, *H. perforatum* as an antioxidant agent contributes an alteration in the delicate balance between the scavenging capacity of antioxidant defence systems and free radicals in favour of the antioxidant defence systems in the body.

Hypericum scabrum L. FAMILY: HYPERICACEAE

Dadkhah et al. (2014) examined the protective role of *Hypericum scabrum* oils (100 and 200 mg/kg BW, i.p) on APAP-induced liver damages in the rat. Increased levels of hepatic LP and FRAP and SOD activity were reversed in the rats treated with oils. In addition, the depleted GSH were compensated with the oil treatments. The protective effect of the oils was further confirmed by the histopathological examination carried out on liver biopsies. The data pointed out that *H. scabrum* oil could modulate the hepatic toxicity induced by the APAP through adjusting the oxidative stress/antioxidant parameters.

Hyptis suaveolens (L.) Poit. FAMILY: LAMIACEAE

Babalola et al. (2011) evaluated the possible hepatoprotective activity of the pretreatment with aqueous extract of the leaves of *Hyptis suaveolens* on APAP-induced hepatotoxicity in rabbits. The marker enzymes show significant elevation in APAP treated animals; these were significantly reduced toward an almost normal level in animals pretreated with aqueous extract of the leaves of *H. suaveolens*. The reduction was observed for the concentrations of TP and ALB.

Ghaffari et al. (2012) evaluated the heptoprotective property of ME *H. suaveolens* as an antioxidant to protect against CCl_4-induced oxidative stress, hepatotoxicity in Albino Wistar rats. Administration of CCl_4 markedly raised the serum level of enzymes such as AST, ALT, ALP, and LDH in rats. However, pretreatment of the rats with *H. suaveolens* methanolic extract at 50 and 100 mL/kg BW for 7 days prior to

CCl_4 administration resulted in a significant decrease in serum AST, ALT, ALP, and LDH activities. With respect to the histological examination, pretreatment with *H. suaveolens* suppressed the acute hepatic damage and was consistent with an improvement in the serum biological parameters of hepatotoxicity.

Ichnocarpus frutescens R.Br. FAMILY: APOCYNACEAE

Dash et al. (2007) evaluated the hepatoprotective effect of chloroform and ME (CEIF and MEIF) of whole plant of *Ichnocarpus frutescens* by PCM-induced liver damage in Wistar albino rats. Chloroform and MEs at a dose level of 250 and 500 mg/kg produce significant hepatoprotection by decreasing the activity of serum enzymes, bilirubin, and LPO, while they significantly increased the levels of GSH, SOD, and CAT in a dose-dependent manner. The effects of chloroform and MEs were comparable to that of standard drug, Silymarin.

Ilex latifolia Thunb. FAMILY: AQUIFOLIACEAE

Crude polysaccharides from the leaves of *Ilex latifolia* (ILPS) was fractionated by DEAE cellulose-52 chromatography, affording four fractions of ILPS-1, ILPS-2, ILPS-3, and ILPS-4 in the recovery rates of 32.3%, 20.6%, 18.4%, and 10.8%, respectively, based on the amount of crude ILPS used (Fan et al., 2014). The four fractions were mainly composed of arabinose and galactose in monosaccharide composition. Compared with ILPS-1 and ILPS-2, ILPS-3, and ILPS-4 had relative higher contents of sulfuric radical and uronic acid. The antioxidant activities *in vitro* of ILPS decreased in the order of crude ILPS>ILPS-4>ILPS-3>EPS-2>ILPS-1. Furthermore, the administration of crude ILPS significantly prevented the increase of serum ALT and AST levels, reduced the formation of MDA and enhanced the activities of SOD and GPx in CCl_4-induced liver injury mice.

Imperata cylindrica Beauv. var. major (Nees) C. E. Hubb. FAMILY: POACEAE

Three new C-methylated phenylpropanoid glycosides (1, 2), a new 8-4′-oxyneolignan (3), together with two known analogs (4, 5), were isolated by Ma et al. (2018) from the rhizomes of *Imperata cylindrica* var. *major*. Their structures were determined by spectroscopic and chemical methods. Compounds 1, 2, and 5 (10 μM) exhibited pronounced hepatoprotective activity against *N*-acetyl-*N*-aminophenol (APAP)-induced HepG2 cell damage *in vitro* assays. Furthermore, their antioxidant activities against Fe^{2+}-cysteine-induced rat liver microsomal LPO and the effects on the secretion of TNF-α in murine peritoneal macrophages (RAW264.7) induced by LPSs were evaluated.

Indigofera aspalathoides Vahl ex DC. FAMILY: FABACEAE

The alcoholic extract of stem of *Indigofera aspalathoides* was evaluated by Rajkapoor et al. (2006) for its antihepatotoxic activity against CCl_4-induced hepatic damage in rats. The activity was evaluated by using biochemical parameters, such as SGPT, SGOT, ALP, TB, and GGTP. The histopathological changes of liver sample were compared with respective control. The extract showed remarkable hepatoprotective effect.

Indigofera caerulea Roxb. FAMILY: FABACEAE

Ponmari et al. (2014) investigated the hepatoprotective effect of the methanolic extract of *I. caerulea* leaves (MIL) and elucidation of its mode of action against CCl_4-induced liver injury in rats. HPLC analysis of MIL when carried out showed peaks close to standard ferulic acid and quercetin. Intragastric administration of MIL up to 2000 mg/kg BW, did not show any toxicity and mortality in acute toxicity studies. During "*in-vivo*" study, hepatic injury was established by intraperitoneal administration of CCl_4 3 mL/kg BW twice a week for 4 weeks in SD rats. Further, hepatoprotective activity of MIL assessed using two different doses (100 and 200 mg/kg BW) showed that intragastric administration of MIL (200 mg/kg BW) significantly attenuates liver injury. Investigation of the underlying mechanism revealed that MIL treatment was capable of reducing inflammation by an antioxidant defense mechanism that blocks the activation of NF-κB as well as inhibits the release of proinflammatory cytokine TNF-α and IL-1β. The results suggest that MIL has a significant hepatoprotective activity which might be due to the presence of phytochemicals namely analogues of ferulic acid and other phytochemicals which together may suppress the inflammatory signaling pathways and promote hepatoprotective activity against CCl_4 intoxicated liver damage.

Indigofera oblongifolia Forssk. FAMILY: FABACEAE

Shahjahan et al. (2005) assessed the protective effect of *Indigofera oblongifolia* on CCl_4-induced hepatotoxicity in male Wistar rats. The activities of AST, ALT, ALP, and LDH were significantly increased in serum of CCl_4-induced animals when compared with control animals.

Antioxidant status was significantly lowered in CCl_4-treated animals with a significant increase in the levels of lipid peroxides (TBARS), significantly lower levels of GSH, and lowered activities of SOD, CAT, and GSH peroxidase (GPx). The protective effect of *I. oblongifolia* was evident from lowering of levels of marker enzymes in serum and maintenance of antioxidant status in the liver as seen from lowered levels of TBARS, increased levels of GSH, and increased activities of SOD, CAT, and GPx.

Indigofera suffruticosa Mill. FAMILY: FABACEAE

The aqueous extract of *Indigofera suffruticosa* (50 mg/kg, ip) showed the protective effect on liver tissue of mice bearing sarcoma 180 (Silva et al., 2014). Lima et al. (2014), using the purified indigo compound from the leaves of *I. suffruticosa*, did not observe a reduction in sarcoma 180 tumor but found that liver cells remained preserved, emphasizing its hepatoprotective effect. Lima et al. (2014) investigated the antitumor action of indica, a compound extracted from the leaves of *I. suffruticosa*, on mice bearing sarcoma 180 cell lines and found that such a compound did not promote a reduction in tumor growth; however, it is found that treatment with indica does not allow to alter the architecture of the liver cells, suggesting a hepatoprotective effect.

Indigofera tinctoria L. FAMILY: FABACEAE

The effect of pretreatment with *Indigofera tinctoria* (IT) extract against the toxicity of D-GalN and CCl_4 during "in situ" perfusion of the liver for 2 h was studied in rats by Sreepriya et al. (2001). Release of LDH and levels of urea in the liver effluent perfusate, was studied and the rate of bile flow was monitored. Perfusion with D-galactosamine (5 mM) or CCl_4 (0.5 mM) resulted in increased LDH leakage, decreased urea levels in the liver effluent and reduction in bile flow. IT pretreatment (500 mg/kg BW) *in vivo* ameliorated D-GalN and $CC1_4$-induced adverse changes toward near normalcy and thereby indicates its hepatoprotective effects in rats

Subhashini et al. (2017) evaluated the antihepatotoxic potential of methanolic extract of *I. tinctoria.* The preliminary phytochemical analysis showed the presence of major bioactive compounds. The plant extract also exhibited good antioxidant potential that showed 69.3% of free-radical scavenging at the highest concentration (5 mg/mL) tested. The plant extract showed good antihepatotoxic activity with inhibitory concentration at 1.8 mg/mL; thus preventing the damage to the liver by hepatotoxins.

Inula crithmoides L. FAMILY: ASTERACEAE

Fractionation directed by hepatoprotective activity of *Inula crithmoides* root resulted in the isolation of two new quinic acid derivatives, 3,5-di-*O*-caffeoylquinic acid 1-methyl ether (caffeoyl=(*E*)-3-(3,4-dihydroxyphenyl) prop-2-enoyl; quinic acid =1,3,4,5- tetrahydroxycyclohexanecarboxylic acid) and 4,5-di-*O*-caffeoylquinic acid 1-methyl ether, in addition to the well-known hepatoprotective compound, 1,5-di-*O*-caffeoylquinic acid. The hepatoprotective effect was indicated by the significant decrease in the level of four measured serum biochemical parameters (SGOT, SGPT, ALP, and bilirubin) in experimental rats.

Ipomoea aquatica Forssk. FAMILY: CONVOLVULACEAE

In the Indian system of traditional medicine (Ayurveda) it is recommended to consume *Ipomoea aquatica* to mitigate disorders like jaundice. The protective effects of ethanol extract of *I. aquatica* against liver damage were evaluated by Alkiyumi et al. (2012) in TAA-induced chronic hepatotoxicity in rats. There was no sign of toxicity in the acute toxicity study, in which SD rats were orally fed with *I. aquatica* (250 and 500 mg/kg) for two months along with administration of TAA. The results showed that the treatment of *I. aquatica* significantly lowered the TAA-induced serum levels of hepatic enzyme markers (ALP, ALT, AST, protein, ALB, bilirubin and prothrombin time). The hepatic content of activities and expressions SOD and CAT that were reduced by TAA were brought back to control levels by the plant extract supplement. Meanwhile, the rise in MDA level in the TAA receiving groups also were significantly reduced by *I. aquatica* treatment. Histopathology of hepatic tissues displayed that *I. aquatica* has reduced the incidence of liver lesions, including hepatic cells cloudy swelling, infiltration, hepatic necrosis, and fibrous connective tissue proliferation induced by TAA in rats.

Ipomoea staphylina Roem. & Schult. FAMILY: CONVOLVULACEAE

Bag et al. (2013) investigated the protective effect of hydroalcoholic extract of leaves of *Ipomoea staphylina* against CCl_4-induce hepatotoxicity in rats. The extract (200 mg/kg, p.o.) showed hepatoprotective activity against CCl_4-induced hepatotoxicity by significantly reducing the levels of AST, ALT, ALP, TB and significantly increasing the serum TP. The extract also improved the histology of the liver. The TPC and the TFC of the extract were 94.59 ± 2.54 mg GAE/g and 44.68 ± 2.00 mg rutin equivalent/g, respectively.

Jeyadevi et al. (2019) investigated the hepatoprotective and antioxidant potential of aqueous extract of *I. staphylina* leaves on Wistar rats against CCl_4-induced liver injury. Total phenolics and flavonoid content of AIS are 20.78 ± 0.86 mg of GAE/g and 17.78 ± 0.45 mg of RE/g of extract, respectively. ABTS and DPPH free-radical scavenging activity of AIS exhibits the IC_{50} value of 32.08 ± 0. 12 and 45.24 ± 0.65 µg/mL, respectively. HPLC analysis of AIS showed the presence of polyphenolic compounds such as protocatechuic acid, chlorogenic acid, caffeic acid, vanillin, *N*-coumaric acid, naringinin, qurecetin, rutin, and flavon. Treatment with AIS restored the biochemical parameters toward normal level, prevented the occurrence of LPO and significantly increased the levels of enzymatic (SOD and CAT) and nonenzymatic (reduced GSH) antioxidants in CCl_4 administered Wistar rats. Histopathology results confirmed the hepatoprotective effect of AIS.

Irvingia gabonensis (Aubry-Lecomte ex O'Rorke) Baill. FAMILY: IRVINGIACEAE

The ethanol extract of the leaves of *Irvingia gabonensis* has been investigated for its hepatoprotective activity in sodium arsenite-induced hepatotoxicity and clastogenicity in male Wistar rats (Gbadegesin et al., 2014). The extract at 250 or 500 mg/kg dose dependently attenuated sodium arsenite induced rise in liver enzymes including AST, ALT, and GGT and prevented histopathological alterations in the liver (Gbadegesin et al., 2014). Phytochemical studies on the ethanol extract of *I. gabonensis* showed the presence of tannins, saponins, alkaloids, terpenoids, flavonoids, and phenols (Gbadegesin et al., 2014).

Ixeris laevigata Sch-Bip. var. *oldhami* Kitam. FAMILY: ASTERACEAE

The hepatoprotective effects of *Ixeris laevigata* var. *oldhami* (IL) were studied on cholestatic hepatitis induced by α-naphthylisothiocyanate (ANIT) and acute hepatitis induced by CCl_4 in rats by Lu et al. (2002). Hepatoprotective activity was monitored by estimating the serum transaminases levels and the histopathological changes in the livers of experimental rats. The pretreatment of animals with IL, extract (0.3–2.0 g/kg orally) significantly inhibited the acute elevation of serum transaminases, as well as the hepatotoxin-induced histopathological changes in the livers of the experimental rats.

Ixora coccinea L. FAMILY: RUBIACEAE

Shyamal et al. (2010) reported on the protective effects of ethanolic extractof *Ixora coccinea* on the aflatoxin B1 (AFB1)-intoxicated livers of albino male Wistar rats. Pretreatment of the rats with oral administration of the ethanolic extract of *I. coccinea*, prior to AFB1 was found to provide significant protection against toxin-induced liver damage, determined 72 h after the AFB1 challenge (1.5 mg/kg, intraperitoneally) as evidenced by a significant lowering of the activity of the serum enzymes and enhanced hepatic reduced GSH status. Pathological examination of the liver tissues supported the biochemical findings. The plant extract showed significant antilipid peroxidant effects *in vitro*.

Juglans regia L. FAMILY: JUGLANDACEAE

Eidi et al. (2013) investigated the protective effect of walnut (*Juglans regia*) leaf extract against CCl_4-induced liver damage in rats. Administration of walnut leaf extract (ranging from 0.2 to 0.4 g/kg BW) significantly lowered serum ALT, AST, and ALP levels in CCl_4-treated rats. Walnut leaf extract increased antioxidant enzymes, including SOD and CAT. Histopathological examination of livers showed that walnut leaves extract reduced fatty degeneration, cytoplasmic vacuolization, and necrosis in CCl_4-treated rats.

Juncus subulatus Forssk. FAMILY: JUNCACEAE

The volatile oil, ethyl acetate, *n*-butanol, and total alcoholic extracts of *Juncus subulatus* were evaluated by Abdel-Razik et al. (2009) for their hepatoprotective and antioxidant activity in female rats against ethanol-induced hepatic injury. Serum liver enzymes (AST, ALT, and ALP), TP, ALB, cholesterol, TGs, NO, malondialdhyde (MDA) and TAC were measured colorimetrically. The results showed that all extracts of *J. subulatus* exhibited hepatoprotective activity in the following order: volatile oil extract > ethyl acetate extract > *n*-butanol extract > total alcoholic extract.

Juniperus communis L. FAMILY: CUPRESSACEAE

Ved et al. (2017) investigated the antioxidant, cytotoxic, and hepatoprotective activities of *Juniperus communis* leaves against various models. Total phenol content was found maximum 315.33 mg/GAE/g in EAF. Significant scavenging activities were found for EAF (IC_{50} = 177 μg/mL) as compared to standard BHT (IC_{50} = 138 μg/mL), while EAF showed good Fe^{2+} chelating ability having an IC_{50} value of 261 mg/ML compared to standard ethylenediaminetetraacetic acid (7.7 mg/mL). It was found that EAF treated group shows remarkable decrease in serum AST, serum ALT, TB, DBL, and ALP level in the treatment group as compared to the hepatotoxic group.

Juniperus phoenicea L. FAMILY: CUPRESSACEAE

The methanolic extract of *Juniperus phoenicea* was investigated by Rizk et al. (2007) for its efficiency in reducing CCl_4-induced hepatotoxicity in rats. Rats injected with a single toxic dose of CCl_4 and sacrificed after 24 h induced remarkable disturbances in the levels of all tested parameters. However, rats injected with the toxic agent and left for one and a half month to self-recover showed moderate improvements in the serum and liver biochemical parameters. However, treatment with the extract of *J. phoenicea* ameliorated the levels of the disturbed biochemical parameters.

Ali et al. (2010) investigated the role of *J. phoenicea* extract as therapeutic effect against CCl_4 with biochemical, histopathological evaluations on female Wistar rats. Remarkable disturbances were observed in the levels of all tested parameters in rats treated with CCl_4. However, rats injected with the toxic agent and left for one and a half month to self-recover showed moderate improvements in the studied parameters while, treatment with medicinal herbal extract ameliorated the levels of the disturbed biochemical parameters. The group treated with *J. phoenicea* extract showed a remarkable improvement in comparison to the CCl_4 treated group showing histopathological liver profiles close to those of the control group.

Alzergy and Elgharbawy (2017) evaluated the possible ameliorative effect of aqueous extract of *J. phoenicea* leaves on clinical, histological, ultra structural, and biochemical parameters against exposure to trichloroacetic acid (TCA)-induced oxidative stress and liver toxicity in Swiss albino mice. Mice treated with TCA showed loss of appetite, loss of body furs and decreased activity. These alterations decreased in mice administrated aqueous extract of *J. phoenicea* and TCA. Biochemical analysis revealed that intoxicated mice with TCA led to a significant increases in serum ALT and insignificant increase in AST. Significant amelioration in these parameters was found in mice treated with *J. phoenicea* and TCA. Serum TP in all treated groups were not found to be significantly different from the control. Histological and ultrastructure examinations also confirmed the protective efficacy of *J. phoenicea*. Administrated of TCA showed many severe pathological lesions include prominent vacuolated hepatocytes, dilatation and congestion of blood vessels with intravascular hemolysis of numerous red blood corpuscles, loss of normal histological architecture with stenosis of hepatic sinusoids and hyperplasia of Kupffer cells. Moreover, some hepatocytes exhibited abnormal division. Also, necrosis of some hepatocytes with pyknotic or karyolitic nuclei were noticed. However, focal necrotic areas associated with inflammatory cells infiltration were frequently observed. Furthermore, most hepatocytes revealed severe reactivity with periodic acid Schiff (PAS) technique. Mice treated with TCA and *J. phoenicea* showed marked tissue repair and disappearance of most pathological changes. Moderate reactivity of most hepatocytes with PAS stain were frequently noticed. Electron microscopic examination of liver of mice treated with TCA showed abnormal nuclear features with decrease and abnormal heterochromatin distribution and increase nucleoli. Crowded cytoplasm of hepatocytes with small electron dense granules represented lysosomes and mitochondria with indistinct details beside, few dilated rough endoplasmic reticulum with indistinct attached ribosomes and accumulated lipid droplets of variable size were also recorded. In addition, congested blood sinusoids and hypertrophied endothelial lining cells with few and poorly identifiable organelles were seen. Also, necrotic Kupffer cells with irregular fragmented nuclei were detected. No obvious ultrastructure changes were observed in hepatocytes of mice treated with *J. phoenicea*. However, few vacuoles, slight increase in glycogen content, and dispersed cytoplasm contained clumps of intact organelles in many hepatocytes were demonstrated. Generally, administration of *J. phoenicea* lessened most sever ultrastructure changes in hepatocytes of TCA intoxicated mice.

Laouar et al. (2017) investigated the antioxidant and hepatoprotective properties of *J. phoenicea* berries against CCl_4-induced oxidative damage in rats. Hepatotoxicity was induced in albino Wistar rats by a single dose of CCl_4 dissolved in olive oil (1 mL/kg BW, 1/1 in olive oil, i.p.). Aqueous extract of *J. phoenicea* berries (AEJP) was administered at the dose of 250 mg/kg/d by gavage for 12 days. Obtained results revealed that administration of CCl_4 caused a significant increase in plasma AST, ALT, ALP, and LDH activities and TB concentration, compared to the control group. While, ALB and TP concentration were significantly lower. Additionally, a significant decrease in the level of hepatic GSH, GPx and, GST activities associated with a significant increase of MDA content in CCl_4 group than those of the control. However, the treatment of experimental rats with AEJP prevented these alterations and maintained the antioxidant status. The histopathological observations supported the biochemical evidences of hepatoprotection. Authors concluded that the results of this investigation indicate that *J. phoenicea* possesses hepatoprotective activity and this effect may be due to its antioxidant properties. Phytochemical investigation of the berries of *J. phoenicea* led to the isolation of four compounds, namely; scutellarin, isoscutellarin, shikimic acid, and the new palmitoyl lactone derivative 16-hydroxy palmitic-1, 16-olide (Abou-Ela et al., 2005b).

Juniperus procera Hochst. ex Endl. FAMILY: CUPRESSACEAE

The ethanolic extracts of aerial part of *Juniperus procera* showed significant hepatoprotective activity against CCl_4-induced liver injury (Alqasoumi and Abdel-Kader, 2012). The hepatoprotective activity has been attributed to terpene contents of *J. procera (*Alqasoumi, 2007). *J. procera* possess significant antioxidant/free-radical scavenging and anti-inflammatory activities which may contribute to its hepatoprotective activity (Al-Asmari et al., 2014). Acute and chronic toxicity studies revealed that the extract of *J. procera* is free from toxicity even in high dose (Alqasoumi, 2007). Phytochemical studies on aerial part of *J. procera* showed the presence of terpenes, β-peltatin A, deoxypodophyllotoxin, and totarol (Alqasoumi and Abdel-Kader, 2012).

Juniperus sabina L. FAMILY: CUPRESSACEAE

Abdel-Kader et al. (2017) evaluated the hepatoprotective activity of *Juniperus sabina* against CCl_4-induced toxicity using male Wistar rats. Daily oral administration of two doses (200 and 400 mg/kg) of the extract of aerial parts of *J. sabina* for seven days followed by on dose of CCl_4 at day 6. The higher dose showed 47%, 50%, 38%, 17%, and 42% decrease in the levels of AST, ALT, GGT, ALP, and bilirubin, respectively. Animals received the total extract of *J. sabina* showed a significant dose-dependent recovery of the NP-SH contents, TPs nd reduction in the level of MDA in both liver and kidney tissues. Histopathological study revealed improvement of kidney function was less than liver function.

Jussiaea nervosa Poir. FAMILY: ONAGRACEAE

Hepatoprotective potentials of *Jussiaea nervosa* leaf extract against Cadmium-induced hepatotoxicity were investigated by Ibiam et al. (2013). Liver function parameters (ALT, AST, ALP, bilirubin) were significantly elevated in exposed rats in comparison to the controls, except for TP and ALB, which were significantly decreased. Histopathological assessment reveals renal pathology in exposed rats in sharp contrast with the controls. *J. nervosa* extract however lowered the values of liver function parameters with 100 mg/kg BW dose producing the highest ameliorative effects. Similarly, the serum ALB and TP significantly improved with normal liver architecture.

Justicia adhatoda L. (Syn.: *Adhatoda vasica* Nees) FAMILY: ACANTHACEAE

Pandit et al. (2004) evaluated the hepatoprotective effect of *Justicia adhtoda* (Syn.: *Adhatoda vasica*) in CCl_4-induced hepatotoxicity in rats. Pretreatment with the test drug *J. adhtoda* in both doses (100 and 200 mg/kg) as well as pretreatment with standard drug Silymarin significantly reduced these liver enzyme levels dose dependently, showing that *J. adhatoda* has hepatoprotective action. Histopathological findings indicated that pretreatment with *J. adhatoda* (100 and 200 mg/kg) offered protection to the hepatocytes from damage induced by CCl_4, with mild fatty changes in the hepatic parenchymal cells, which corroborated the changes observed in the hepatic enzymes.

J. adhatoda leaf showed significant hepatoprotective effect at doses of 50–100 mg/kg, p.o., on liver damage induced by D-GalN in rats (Bhattacharya et al., 2005). Pingale et al. (2009) investigated the potential hepatoprotective action of *J. adhatoda* whole plant powder against CCl_4-induced liver damaged Wister rat model. Blood and tissue biochemical parameters of liver have been examined for evaluating the hepatoprotection action. These biochemical markers are GOT, GPT, alkaline phosphate, glucose, bilirubin, triglycerides, γGT, cholesterol, DNA, RNA, TP etc. The effect of *J. adhatoda* whole plant powder is compared with Silymarin by standard protocol and is found to have better hepatoprotective action.

The hepatoprotective activity of ethyl acetate extract of *J. adhatoda* was investigated by Ahmad et al. (2013) against CCl_4-induced liver damage in Swiss albino rats. Pretreatment of rats with the ethyl acetate extract of *J. adhatoda* (100 and 200 mg/kg) prior to the CCl_4 dose at 1mL/kg lowered the three serum level enzymes and also bilirubin. Histopathological observations also coincided with the above results; however, 200 mg/kg dose was found to be more active.

Justicia gendarussa Burm.f. FAMILY: ACANTHACEAE

Krishna et al. (2009) evaluated the hepatoprotective activity of ME of leaves of *Justicia gendarussa* using CCl_4-induced hepatotoxicity in albino rats. The extract showed significant hepatoprotective activity at 300 mg/kg BW Interestingly, the hepatoprotective activity decreases as the dose increases.

Justicia schimperiana (Hochst. ex Nees) T. Anderson FAMILY: ACANTHACEAE

Umer et al. (2010) evaluated the *in vivo* hepatoprotective activity of *Justicia schimperiana* used in Ethiopian traditional medical practices for the treatment of liver diseases. The levels of hepatic marker enzymes were used to assess their hepatoprotective activity against CCl_4-induced hepatotoxicity in Swiss albino mice. The results revealed that pretreating mice with the hydroalcoholic extract significantly suppressed the plasma AST activity when compared with the CCl_4 intoxicated control. Among the Soxhlet extracts, the ME of *J. schimperiana* showed significant hepatoprotective activity. Further fractionation of this extract using solid phase extraction and testing them for bioactivity indicated that the fractions did not significantly reverse liver toxicity caused by CCl_4. However, the percentage hepatoprotection of the distilled water fraction was comparable with that of the standard drug Silymarin at the same dose (50 mg/kg) as evidenced by biochemical parameters. Histopathological studies also supported these results. *In vitro* DPPH assay conducted on the water fraction of *J. schimperiana* possess moderate radical scavenging activity which led to the conclusion that the hepatoprotective activity of the plants could be in part through their antioxidant action.

Justicia spicigera Schltdl. FAMILY: ACANTHACEAE

The antioxidant and hepatoprotective activities of EAF of the dried aerial part of *Justicia spicigera* were evaluated and the characterization of its anthocyanin content was done by Awad et al. (2015) against hepatic fibrosis, induced by CCl_4 in rats. The phenolic content in the EAF was 42.94 mg/g. Twelve anthocyanins were identified, the major of which are peonidin 3,5-diglucoside (64.30%), malvidin 3,5-diglucoside (10.59%) and petunidin 3,5-diglucoside (4.71%). Treatment of CCl_4 intoxicated rats with EAF recorded improvement in the liver function indices and oxidative stress markers. The histopathological observations confirmed the results.

Kalanchoe pinnata Pers. (Syn.: *Bryophyllum calycinum* Salisb., *B. pinnatum* Kurz) FAMILY: CRASSULACEAE

Juice of the fresh leaves of *Kalanchoe pinnata* is used very effectively for the treatment of jaundice in folk medicines of Bundelkhand region of India. The juice of the leaves and the ethanolic extract of the marc left after expressing the juice were studied in rats against CCl_4-induced hepatotoxicity. The test material was found effective as hepatoprotective as evidenced by *in vitro*, *in vivo* and histopathological studies. Results indicated that the level of SGOT, SGPT, SALP, and SB were reduced significantly after treatment of rats with the concentrate and ethanolic extract along with the toxicant (CCl_4). The juice was found more effective than ethanolic extract.

Khaya grandifoliola (Desr.) A. Juss. FAMILY: MELIACEAE

The hepatoprotective effect of *Khaya grandifoliola* has been studied against PCM (Njayou et al., 2013a), and CCl_4-induced hepatotoxicity (Njayou et al., 2013b) in rats. The methanol; methylchloride extract of the stem bark of this plant at 25 and 100 mg/kg dose dependently attenuated hepatotoxin-induced alterations in biochemical parameters (serum ALP, AST, ALT and TP and liver TBARS, SOD, GSH, and GR) and prevented toxin-induced alteration in liver histopathology. The extract also showed antioxidant and anti-inflammatory activities (Njayou et al., 2013b) which may be contributing to its hepatoprotective activity.

Khaya senegalensis (Desv.) A. Juss. FAMILY: MELIACEAE

Ojo et al. (2006) evaluated the hepatoprotective activity of stem bark extract of *Khaya senegalensis* against APAP-induced hepatotoxicity in Wistar albino rats. They reported that stem bark extract of *K. senegalensis* produced significant hepatoprotective effects by decreasing the activity of serum enzymes. Values recorded for AST, ALT, and ALP were significantly lower compared to those recorded for control rats. A higher inhibition of serum level elevation of ALP was observed with the extract.

Kigelia africana (Lam.) Benth. FAMILY: BIGNONIACEAE

Olaleye and Rocha (2008) investigated the hepatoprotective activity of *Kigelia africana* on APAP-induced liver damage in mice. Paracetamol caused liver damage as evident by significant increase in plasma activities of AST and ALT. There were general significant losses in the activities of SOD, GPx, CAT, and delta-ALA-D and an increase in TBARS in the liver of PCM-treated group compared with the control group. However, the extract of *K. aficana* was able to counteract these effects.

Kohautia grandiflora DC. FAMILY: RUBIACEAE

Garba et al. (2009) investigated the hepatoprotective effect of the aqueous extract of *Kohautia grandiflora* on PCM-induced hepatotoxicity in rats. Biochemical analysis of the serum obtained showed a significant increase in the levels of AST, ALT, ALP, and ALB in rats administered with 500 mg/kg of PCM and 300 mg/kg of the extract, respectively. Pretreatment of the animals with the extract caused a decrease in the levels of these enzymes. Histopathological assessments of the liver sections of rats administered with 500 mg/kg of PCM and 300 mg/kg of the extract showed congestion of the venous sinusoids, necrosis, edema, mononuclear infiltration and cloudy swellings with the severity higher in the PCM treated group. Pretreatment with 300 mg/kg of the extract revealed a slight hepatoprotection compared with the rats that were administered with PCM alone.

Kyllinga nemoralis L. FAMILY: CYPERACEAE

Somasundaram et al. (2010) evaluated the hepatoprotective activity of the rhizomes of *Kyllinga nemoralis* against CCl_4-induced hepatotoxicity in rats. Hepatotoxicity was induced in male Wistar rats by CCl_4 and olive oil (50%, v/v). i.p. ethanolic and PEEs of *K. nemoralis* rhizomes were administered to the experimental rats (100 and 200 mg/kg, p.o. for 7 days). The hepatoprotective effect of these extracts was evaluated by the assay of liver function biochemical parameters and histopathological studies of the liver compared with Silymarin. Both extracts showed significant hepatoprotection when compared to control, similar to standard Silymarin. Histology of liver sections also revealed that the extracts protected liver from injury.

Lagenaria breviflora (Benth.) Roberty FAMILY: CUCURBITACEAE

Saba et al. (2012) investigated the hepatoprotective and *in vivo* antioxidant effects of the ethanol extract of whole fruit of *Lagenaria breviflora* (LB) in experimental animals. Forty nine Wistar albino rats were divided into seven groups of seven. Group I served as the control group; rats in Group II were given i.p. CCl_4 (1.5 mL/kg) alone; Groups III–VI received different concentrations of plant extract (100, 250, and 500 mg/kg) with CCl_4 and Group VII received kolaviron at 200 mg/kg as a reference hepatoprotective agent. There was a significant increase in MDA and hydrogen peroxide (H_2O_2) generation in the serum of CCl_4-treated rats

(Group II) while the serum GSH level decreased significantly. Pretreatment with LB extract led to a significant increase in serum GSH and a ALT, AST, ALP, LDH, bilirubin, creatinine and BUN increased significantly in CCl_4-treated rats (Group II). This study suggested that treatment with LB extract enhances the recovery from CCl_4-induced hepatic damage and oxidative stress via its antioxidant and hepatoprotective properties.

Lagerstroemia speciosa (L.) Pers. FAMILY: LYTHRACEAE

Thambi et al. (2009) evaluated the hepatoprotective activity of *Lagerstroemia speciosa* leaf extract against hepatotoxicity induced by CCl_4, which was evaluated in terms of serum marker enzymes like SGPT, SGOT, ALP, serum TB, TP levels along with concomitant hepatic and antioxidants like SOD, CAT, GSH, GPx, and LPO enzymes were monitored. These biochemical parameters altered by the single dose level of CCl_4. Pretreatment with *L. speciosa* prior to the administration of CCl_4, at the doses of 50 and 250 mg/kg. BW/day, p.o. for 7 days, significantly restored all the serum and liver tissue parameters near to the normal levels, respectively. Silymarin was used as a reference standard, prior to the administration of CCl_4 to rats. These findings indicate the protective potential of *L. speciosa* against hepato toxicity which possibly involve mechanism related to its ability of selective inhibitors of (ROS-like antioxidants brought about significant inhibition of TBARS suggesting possible involvement of $O_2 \bullet^-$, $HO_2^\bullet$, and •OH. Authors opined that the amelioration may be attributed to the synergistic effects of its constituents rather than to any single factor as the leaves are rich in tannins, sterols, flavonoids, saponins, etc.

Prabhu et al. (2010) investigated the effect of alcoholic extract of *L. speciosa* against CCl_4-induced liver fibrosis in male albino Wistar rats. Treatment with alcoholic extract of *L. speciosa* (100 mg/kg BW) orally, reduced the HYP content of the liver, various serum enzymes level and total billirubin. The architecture of liver deranged by CCl_4 showed improvement following administration of the extract.

Radical scavenging activities of *L. speciosa* petal extracts and its hepato-protection in CCl_4-intoxicated mice have been investigated by Tiwary et al. (2017). The petal extracts can scavenge or neutralize free radicals of different origin and chelate ferrous ion. There was no toxic effect of the extract on murine spleenocytes and human MCF7 and HepG2 cell lines. The *in-vivo* tests have indicated that feeding of the petal extract has several manifestations, like reduction of MDA level, increase in GSH level, and restoration of CAT in CCl_4 intoxicated mice, to reverse liver damage to a considerable extent. Furthermore, GCMS analyses have confirmed the presence of various compounds reported as potential antioxidant. These compounds may have contributed toward protection against damages inflicted by free radicals.

Laggera pterodonta (DC.) Sch.Bip. ex Oliv. FAMILY: ASTERACEAE

Laggera pterodonta as a folk medicine has been widely used for several centuries to ameliorate some inflammatory ailments as hepatitis in China. Wu et al. (2006, 2007, 2009) evaluated the hepatoprotective effect of total flavonoids and total phenolics from *L. pterodonta* against CCl_4-, D-GalN-, TAA-, and *t*-BHP-induced injury was examined in primary cultured neonatal rat hepatocytes. The total flavonoids and total phenolics inhibited the cellular leakage of two enzymes, hepatocyte ASAT and ALAT, caused by these chemicals and improved cell viability. Moreover, it afforded much stronger protection than the reference drug silibinin. Neutralizing ROS by nonenzymatic mechanisms may be one of the main mechanisms of TPLP against chemical-induced hepatocyte injury (Wu et al., 2006, 2007, 2009).

Launaea intybacea (Jacq.) Beauverd FAMILY: ASTERACEAE

Hepatoprotective activity of ethyl acetate extract of aerial parts of *Launaea intybacea* were evaluated in PCM-induced hepatotoxicity in albino rats. Silymarin (200 mg/kg) was given as reference standard. The ethyl acetate extract of aerial parts of *L. intybacea* have shown very significant hepatoprotection against PCM-induced hepatotoxicity in albino rats in reducing serum TB, SALP, SGPT, SCOT levels and liver homogenates LPO, SOD, CAT, GPx, GST, and GSH levels (Takate et al., 2010).

Launaea procumbens (Roxb.) Ramayya & Rajagopal FAMILY: ASTERACEAE

Launaea procumbens is used as a folk medicine to treat hepatic disorders in Pakistan. The effect of CE of *L. procumbens* (LPCE) was evaluated by Khan et al. (2012) against CCl_4-induced liver damage in rats. Chloroform extract inhibited LPO, and reduced the activities of AST, ALT, ALP, and LDH in serum induced by CCl_4. GSH contents were increased as were the activities of antioxidant enzymes (CAT, SOD, GST, GSR, and GSH-Px) when altered due to CCl_4 hepatotoxicity. Similarly, absolute liver weight, relative liver weight and the number of hepatic lesions were reduced with co-administration of LPCE. Phyochemical analyses of CE indicated that it contained catechin, kaempferol, rutin, hyperoside, and myricetin.

Lavandula coronopifolia Poir. FAMILY: LAMIACEAE

Farshori et al. (2015) evaluated the protective potential of four extracts (namely petroleum ether extract (LCR), chloroform extract (LCM), ethyl acetate extract (LCE), and alcoholic extract (LCL)) of *Lavandula coronopifolia* on oxidative stress-mediated cell death induced by ethanol, a known hepatotoxin in human hapatocellular carcinoma (HepG2) cells. Cells were pretreated with LCR, LCM, LCE, and LCL extracts (10–50 μg/mL) of *L. coronopifolia* for 24 h and then ethanol was added and incubated further for 24 h. After the exposure, cell viability using (3-(4,5-dimethylthiazol-2-yl)-2,5-diphenyltetrazolium bromide) and neutral red uptake assays and morphological changes in HepG2 cells were studied. Pretreatment with various extracts of *L. coronopifolia* was found to be significantly effective in countering the cytotoxic responses of ethanol. Antioxidant properties of these *L. coronopifolia* extracts against ROS generation, LPO, and GSH levels induced by ethanol were investigated. Results show that pretreatment with these extracts for 24 h significantly inhibited ROS generation and LPO induced and increased the GSH levels reduced by ethanol. The data from the study suggests that LCR, LCM, LCE, and LCL extracts of *L. coronopifolia* showed hepatoprotective activity against ethanol-induced damage in HepG2 cells. However, a comparative study revealed that the LCE extract was found to be the most effective and LCL the least effective.

Lawsonia inermis L. (Syn.: *Lawsonia alba* Lam.) FAMILY: LYTHRACEAE

The hepatoprotective activity of the 50% ethanol extract of the bark of *Lawsonia inermis* was investigated against the CCl_4-induced oxidative stress by Ahmed et al. (2000). Pretreatment of rats with doses of 250 and 500 mg/kg of the plant extract significantly lowered SGOT, SGPT, and LDH levels, in a dose-dependent manner against the significant rise of these damage marker enzymes when challenged with CCl_4 (1 mL/kg, orally). Parallel to these changes, the plant extract prevented CCl_4-induced oxidative stress by significantly maintaining the levels of reduced GSH, its metabolizing enzymes and simultaneously inhibiting the production of free radicals. Pretreatment of rats with the extract also inhibited the peroxidation of microsomal lipids in a dose-dependent manner.

The oral administration of varying doses of aqueous suspension of extract of *L. inermis* bark extracts to rats for 10 days afforded good hepatoprotection against CCl_4-induced elevation of serum marker enzymes, SB, liver LPO and reduction in total serum protein, liver GSH, GSH peroxidise, glutathione-transferase, glycogen, SOD, and CAT activity (Bhandarkar and Khan, 2003).

Hepatoprotective activity of *L. inermis* was shown in a toxicity model by CCl_4 in rats. This research proved that animal pretreatment with a methanolic extract of *Lawsonia inermis* (100 and 200 mg/kg of weight) attenuated the increase in AST serum activity.

ALT, ALP, TB, and histological changes observed in the damage induced by CCl4 (Mohamed et al., 2016; Sanni et al., 2010). Previous reports have shown that *L. inermis* is rich in phenolic compounds such as phenolic acids, flavonoids, tannins, lignin, and others that possess antioxidant, anticarcinogenic, and antimutagenic effects as well as antiproliferative potentials (Uma and Aida, 2010), which may be responsible for its hepatoprotective activities.

Adetutu and Olorunnisola (2013) investigated the hepatoprotective potential of the aqueous extracts of leaves of *L. inermis*. Extract of dose of 100 mg/kg BW was given to rats in five groups for seven consecutive days followed by a single dose of 2-AAF (0.5 mmol/kg BW). The rats were sacrificed after 24 h and their bone marrow smears were prepared on glass slides stained with Giemsa. The mPCEs were thereafter recorded. The hepatoprotective effects of the plant extract against 2-AAF-induced liver toxicity in rats were evaluated by monitoring the levels of ALP, GGT, and histopathological analysis. The results of the 2-AAF-induced liver toxicity experiments showed that rats treated with the plant extract (100 mg/kg) showed a significant decrease in mPCEs as compared with the positive control. The rats treated with the plant extract did not show any significant change in the concentration of ALP and GGT in comparison with the negative control group whereas the 2-AAF group showed a significant increase in these parameters. Leaf extract also showed a protective effect against histopathological alterations. This study suggests that the leaf extract has hepatoprotective potential, thereby justifying their ethnopharmacological use (Adetutu and Olorunnisola, 2013).

Leea asiatica **(L.) Ridsdale**
FAMILY: VITACEAE

Sen et al. (2014) investigated the hepatoprotective and antioxidant activity of *Leea asiatica* leaves against APAP-induced hepatotoxicity. Higher dose exhibited significant hepatoprotective activity against APAP-induced toxicity. Level of SOD, CAT, GPx in liver tissue, and reduced GSH in liver and blood were also significantly increased in extract (300 mg/kg) treated animals compared to disease control group.

Leonotis nepetifolia (L.) R.Br. FAMILY: LAMIACEAE

Williams et al. (2016) evaluated the hepatoprotective activity of *Leonotis nepetifolia* against APAP-induced hepatotoxicity. Methanol and aqueous extracts as pretreatment and post-treatment protected against hepatic injury. Extracts abrogated the 14-fold and 4-fold APAP-induced increases in ALT and AST respectively, including histopathological damage. Extracts reversed APAP-induced 4-fold and 14-fold increases in GR and SOD activities, respectively. Additionally, extracts reversed APAP-induced decline in GPx activity; particularly the aqueous extract as pretreatment increased GPx activity up to 2.2-fold over saline-treated controls. Extracts, as pretreatment and post-treatment, prevented APAP-induced hepatic injury by modulating the activities of antioxidant enzymes. Extract-induced increase in GPx activity would facilitate the scavenging of hydroperoxide and peroxide reactive species generated by high dose APAP.

Lepidium sativum L. FAMILY: BRASSICACEAE

Lepidium sativum seed has been used in traditional medicine for the treatment of jaundice, liver problems, spleen diseases and gastrointestinal disorders. It was also reported to possess antihypertensive, diuretic, antiasthmatic, antioxidant, and anti-inflammatory activities. The role hepatoprotective of methanolic extract of *L. sativum* at a dose of 200 and 400 mg/kg was investigated in CCl_4-induced liver damage in rats. Significant reduction in all biochemical parameters were found in groups treated with *L. sativum.* The severe fatty changes in the livers of rats caused by CCl_4 were insignificant in the *L. sativum* treated groups (Abuelgasim et al., 2008).

CCl_4-treated rats receiving a methanolic extract of *L. sativum* showed a significant reduction in all biochemical parameters of liver injury and a striking attenuation of CCl_4-induced fatty degeneration of the liver (Afaf et al., 2008).

Al-Asmari et al. (2015) evaluated the hepatoprotective effect of ethanolic extracts of *L. sativum* seeds against CCl_4-induced acute liver injury in rats. The bioactive compounds responsible for this activity have been analyzed by GC-MS. To evaluate the hepatoprotective activity, six groups (n = 6) of rats were taken. First group was control, second was toxic and other groups received oral test solutions: 100 mg/kg Silymarin, or LSS (100, 200, and 400 mg/kg), once daily for seven consecutive days, followed by hepatotoxicity induction with CCl_4. Blood and liver tissues were collected for biochemical, antioxidant, and microscopic analyses. The bioactive constituents present in the extract were analyzed by GC-MS. Results showed that pretreatment with seed extract and Silymarin significantly reduced the level of serum ALT, AST, ALP, and bilirubin (BIL), which was increased significantly in toxic group treated with only CCl_4. Histological analysis of liver tissues in groups pretreated with seed extract and Silymarin showed mild necrosis and inflammation of the hepatocytes compared to the toxic group. GC-MS analysis of LSS showed the presence of 12 major fatty acids including alpha-linolenic acid as a major constituent. These results indicated that LSS exerts enhance hepatoprotective activity that could be attributed to its antioxidant activity, coupled together with the presence of anti-inflammatory compounds in LSS extract.

Hepatoprotective effect of the *L. sativum* ethanolic extract (LSEE) was assessed by D-GalN-induced/LPS (400 mg/kg and 30 μg/kg) liver damage model in rats by Raish et al. (2016). Marked amelioration of hepatic injuries by attenuation of serum and LPO has been observed as comparable with Silymarin (25 mg/kg p.o.). D-GalN/LPS induced significant decrease in oxidative stress markers protein level, and ALB. LSEE significantly downregulated the D-GalN/LPS-induced pro-inflammatory cytokines TNFα and IL-6 mRNA expression in dose-dependent fashion about 0.47 and 0.26 fold and upregulates the IL-10 by 1.9 and 2.8 fold, respectively, while encourages hepatoprotective activity by downregulating mRNA expression of iNOS and HO-1. MPO activity and NF-κB DNA-binding effect significantly increased and was mitigated by LSEE in a dose-dependent style as paralleled with Silymarin.

Leptadenia reticulata (Retz.) Wight & Arn. FAMILY: APOCYNACEAE

Nema et al. (2011) evaluated the hepatoprotective activity of ethanolic and aqueous extract of stems of *Leptadenia reticulata* in CCl_4-induced hepatotoxicity in rats. Treatment of animals with ethanolic and aqueous extracts significantly reduced the liver damage and the symptoms of liver injury by restoration of architecture of liver as indicated by lower levels of SB and protein as compared with the normal and Silymarin-treated groups. Histology of the liver sections confirmed that the extracts prevented hepatic damage induced by CCl_4 showing the presence of normal hepatic cords, absence of necrosis, and fatty infiltration. The ethanolic and aqueous extracts of stems of *L. reticulata* showed significant hepatoprotective activity. The ethanolic extract is more potent in hepatoprotection in CCl_4-indiced liver injury model as compared with aqueous extract.

Lespedeza cuneata G. Don FAMILY: FABACEAE

The aerial parts of *Lespedeza cuneata*, perennial legume native to Eastern Asia, have been used therapeutically in traditional Asian medicine to protect the function of liver, kidneys and lungs. The aerial parts of *L. cuneata* were used by Kim et al. (2011) to prepare an ethanol extract, which was then tested for hepatoprotective effects against injury by *t*-BHP. At a dose of 20 µg/mL, the ethanol extract significantly protected HepG2 cells against the cytotoxicity of *t*-BHP. Further fractionation of the extract with ethyl acetate allowed the isolation of five flavonoid compounds that were structurally identified by ^{1}H- and ^{13}C-NMR spectroscopy as isovitexin, hirsutrin, trifolin, avicularin, and quercetin. Hirsutrin, avicularin and quercetin (10 µM) showed clear hepatoprotective activity against injury by *t*-BHP in HepG2 cells, whereas isovitexin and trifolin showed no protective effects. The observed hepatoprotective effect of the investigated compounds showed a high correlation with radical scavenging activity, which followed the structure–activity relationships of the flavonoid aglycones.

Leucas aspera (Willd.) Link FAMILY: LAMIACEAE

The cold methanolic extract of the whole plant of *Leucas aspera* showed significant hepatoprotective activity in CCl_4-induced hepatotoxicity in rats (Mangathayaru et al., 2005). *L. aspera* leaves fresh juice was tested against CCl_4-induced liver damage. The evaluation markers used were GOT, GPT, ALP, glucose, bilirubin, cholesterol and TP. Silymarin was used as a standard for comparison. The fresh juice showed good result against liver disorders (Shirish and Pingale, 2010).

L. aspera was screened for its hepatoprotective activity by Radhika and Bridha (2011). Wistar strains of Albino rats were used as the experimental models. The Hepatic serum markers AST, ALT, ALP, GGT, serum protein, and SB were analyzed. The antioxidant status of the animals was also assessed by measuring the activity of GSH and SOD. The extent of lipid peroxidation was also measured. The presence of flavanoids in the extract was confirmed by TLC. Histopathology of the tissues was performed to provide a diagnostic support to the preclinical studies.

Banu et al. (2012) investigated the hepatoprotective, antioxidant, and protective effect of *L. aspera* on MDMEs in D-GalN-induced hepatotoxicity in rats. Pretreatment with *L. aspera* extract significantly protected the liver in D-GalN administered rats. *L. aspera* extract significantly elevated antioxidant enzymes like SOD, CAT, GPx, and decreased LPO levels in liver. The total phenolic and flavonoid content in aqueous extract was found to be 28.33 ± 0.19 GAEs mg/g of extract and 3.96 ± 0.57 rutin equivalent mg/g of extract, respectively. LA extract (200 and 400 mg/kg) treatment with CCl_4 decreased the hexobarbitone-induced sleeping time in mice by 56.67% and 71.30%, respectively, which indicated the protective effect of LA on hepatic MDMEs. Histological studies showed that LA at 400 mg/kg attenuated the hepatocellular necrosis in D-GalN intoxicated rats.

Irfan et al. (2012) evaluated the hepatoprotective activity of *L. aspera* leaves extract, against simvastatin-induced hepatotoxicity for 30 days. There was a marked elevation of biochemical parameters such as increases in SGPT, SGOT, ALP, SB, and decrease the TPs content in simvastatin-treated rats, which were restored toward normalization in *L. aspera* (200 and 400 mg/kg) treated animals. The 400 mg/kg dose of *L. aspera* shows the best results than 200 mg/kg, similar to Silymarin (20 mg/kg). Histopathological observations confirmed the beneficial roles of *L. aspera* and Silymarin against ethanol-induced liver injury in rats.

Hepatoprotective activity of hydroalcoholic leaf extract of *L. aspera* on male albino Wistar rats was investigated, hydroalcoholic leaf extract of *L. aspera* have shown hepatoprotective activity (Thenmozhi et al., 2013). A dose of 400 mg/kg hydroalcoholic leaf extract of *L. aspera* showed a significant reduction in the liver enzymes in the dose-dependent manner.

Leucas ciliata Wall. FAMILY: LAMIACEAE

Qureshi et al. (2010) evaluated ethanolic extract of *Leucas ciliata* leaves for possible antioxidant and hepatoprotective potential. Antioxidant activity of the extract was evaluated by using DPPH radical scavenging, NO radical scavenging, Iron chelation and Reducing power methods. Hepatoprotective activity of the extract was evaluated by CCl_4-induced liver damage model in rats. The extract demonstrated a significant dose-dependent antioxidant activity comparable with ascorbic acid. In hepatoprotective activity study, CCl_4 significantly increased the levels of SGPT, SGOT, ALP, and TB. Pretreatment of the rats with ethanolic extract of *L. ciliata* (100, 200, and 400 mg/kg po) inhibited the increase in serum levels of SGPT, SGOT, ALP, and TB and the inhibition was comparable with Silymarin (100 mg/kg po).

Leucas hirta (Heyne ex Roth) Spreng. FAMILY: LAMIACEAE

Methanol and aqueous leaf extracts of *Leucas hirta* demonstrated hepatoprotective activity against CCl_4-induced liver damage in rats (Manjunatha et al., 2005). The parameters studied were serum TB, TP, ALT, AST, and ALP activities. The hepatoprotective activity was also supported by histopathological studies of liver tissue. Results of the biochemical studies of blood samples of CCl_4 treated animals showed significant increase in the levels of serum markers and decrease in TP level reflecting the liver injury caused by CCl_4. Whereas blood samples from the animals treated with methanol and aqueous leaf extracts showed significant decrease in the levels of serum markers and increase in TP indicating the protection of hepatic cells. The results revealed that methanol leaf extract followed by aqueous extract of *L. hirta* could afford significant protection against CCl_4-induced hepatocellular injury.

Leucas lavandulaefolia Sm. FAMILY: LAMIACEAE

The hepatoprotective activity of *Leucas lavandulaefolia* aerial parts were tested against CCl_4-induced hepatic damage in rats. Histopathological examination of the liver section of the rats treated with toxicant showed intense centrilobular necrosis and vacuolization. Whereas, ethyl acetate extract given at a dose level of 400 mg/kg showed an significant reduction in SGPT, SGOT, ALP and TB levels and was comparable to that of Silymarin, used as a standard drug (Chandrashekar and Prasanna, 2010). Different extracts of the *L. lavendulaefolia* leaves were tested against D-GalN-induced liver toxicity in rats. SGOT, SGPT, ALP, GGT, of the serum and HTG of the rat's liver were estimated after 48 h of intoxication. Microscopic observation of the liver along with the BW, liver weight and also food intake were also studied during the experiments. The result indicated that ME (100 mg/kg, p.o.) exhibited significant hepatoprotective activity (Kotoky et al., 2008). Chloroform extract of aerial parts of *L. lavandulaefolia* at a dose of 200 and 400 mg/kg was administered orally as a fine suspension in 0.3% sodium carboxy methyl cellulose for 14 days. Liver damage was induced by administration of D (+) galactosamine. Treated group showed significant decrease in

AST, ALT, ALP, TB, LDH, TC levels in serum when compared with D (+) GalN administered group. It can be concluded that CE of this plant seems to possess hepatoprotective activity in rats. Further studies are needed to evaluate the potential usefulness of this extract in clinical conditions associated with liver damage (Chandrashekar et al., 2007).

Leucophyllum frutescens (Berland.) I. M. Johnst. FAMILY: SCROPHULARIACEAE

The protective action of *Leucophyllum frutescens* ME of aerial parts was evaluated by Balderas-Renteria et al. (2007) in an animal model of hepatotoxicity induced by CCl_4. Samples of livers were observed under microscope for the histopathological changes. Levels of marker enzymes such as ALT and AST were increased significantly in CCl_4-treated rats. Animals intoxicated with CCl_4 and treated with *L. frutescens* ME significantly decreased the activities of these two enzymes. Also these groups resulted in less pronounced destruction of the liver architecture, there is no fibrosis and have moderate inflammation.

Lippia nodiflora (L.) Michx. FAMILY: LAMIACEAE

Arumanayagam and Arunmani et al. (2015) evaluated the antibacterial and hepatoprotective effect of methanolic extracts of leaves of *Lippia nodiflora* (LN), on HepG2 cells. Authors have showed that LN reduced ROS production against LPS-induced toxicity on HepG2 cells, and thereby decreased the apoptotic gene expression and protect the liver cells against toxicity.

Litchi chinensis Sonn. FAMILY: SAPINDACEAE

The chloroform and MEs from *Litchi chinensis* leaf were evaluated by Basu et al. (2012) for their protective effects on PCM-induced liver damage in Wistar albino rats. All biochemical observations indicated that both the test extracts exerted significant hepatoprotective efficacy against PCM-induced hepatic damage in rats. The ME was found to be more effective than CE.

Limnophila repens L. FAMILY: SCROPHULARIACEAE

Venkateswarlu and Ganapathy (2019) assessed the hepatoprotective activity of ME of the whole plant of *Limnophila repens* against PCM toxicity in Wistar rats. The noxious effects of PCM had been considerably controlled in the extract treated groups that were demonstrated by the restoration of serum biochemical parameters to near normal levels. *L. repens* showed hepatoprotective effects in a dose-dependent manner. A high dose of *L. repens* showed a nonsignificant difference with the hepatoprotective effects of standard drug Silymarin. From the research, it had been figured that *L. repens* have significant hepatoprotective properties.

Limonia acidissima L. [Syn.: *Feronia elephantum* Correa; *Feronia limonia* (L.) Swingle] FAMILY: RUTACEAE

Kamat et al. (2030) evaluated the hepatoprotective activity of aqueous extract of leaves of *Limonia acidissima* using CCl_4-induced liver damage model in rats. The result indicated that physical, functional and biochemical changes produced by CCl_4 were restored to normal by aqueous extract of the leaves of *L. acidissima.* The hepatoprotective action of aqueous extract of leaves was confirmed by histopathological examination.

Jain et al. (2012a, b, c) evaluated the hepatoprotective potential of *Limonia acidissima* root, stem bark, and leaves extracts and fractions using experimental models. Activity levels of AST and ALT and cell viability were evaluated in HepG2 cells treated with CCl_4 in presence or absence of extracts or fractions. Also, plasma markers of hepatic damage, hepatic antioxidants, LPO, and histopathological alterations were assessed in rats treated with CCl_4 alone or in combination with 200 or 400 mg/kg BW of FSB-7 or 25 mg/kg BW of Silymarin. In vitro co-supplementation of extracts or fractions recorded varying degree of hepatoprotective potentials. Also, presupplementation of methanolic extract followed by CCl_4 treatment significantly prevented hepatic damage and depletion of cellular antioxidants. Also, CCl_4^+ extract showed minimal distortion in the histoarchitecture of liver and results were comparable to that of CCl_4+ Silymarin-treated rats.

Dar et al. (2012) evaluated the hepatoprotective effect of the ethanolic extract of fruit pulp of *L. acidissima* against PCM-induced hepatotoxicity in albino rats. Administration of 300 mg/kg BW of ethanolic extract of *L. acidissima* effectively reduced the pathological damages caused by PCM intoxication. In addition to serum parameters treatment of 300 mg/kg BW of ethanolic extract of *L. acidissima* also promotes the BW in albino rats.

Sharma et al. (2012) evaluated the hepatoprotective effects of leaves of *L. acidissima* against TA-induced liver necrosis in diabetic rats. LA significantly lowered the mortality rate and showed improvement in liver function parameters in TA-induced diabetic rats without change in liver weight, volume and serum glucose levels. Significant increase in serum ALT and AST was caused by single dose of TAA (300 mg/kg p.o.) compared to the normal rats. FE treatment (at

400 and 800 mg/kg) for 7 days caused significant reversal of AST and ALT values as compared to $DA_{+}TA$ group of rats.

Liquidamber styraciflua L. FAMILY: ALTINGIACEAE

The methanolic extract of the leaves of *Liquidamber styraciflua* (LSE) was evaluated by Eid et al. (2015) for hepatoprotective and antioxidant activities in CCl_4 liver-damaged rats. LSE exhibited a significant dose-dependent protective effect by lowering the serum levels of ALT, AST, ALP, MDA, and ameliorating the level of serum protein. In addition, LSE showed antioxidant activity through improving the levels of blood GSH, vitamin, vitamin E, and hepatic TP contents. The LSE revealed activity approached that of Silymarin, a known hepatoprotective agent. These biochemical observations were supported by examination of the histopathological features of the liver. Chromatographic fractionation of LSE afforded seven phenolic compounds. These were identified as: GA, isorugosin B, casuarictin, quercetin-3-*O*-β-D-$^{4}C_{1}$-glucopyranoside, myricetin-3-*O*-α-L-$^{1}C_{4}$-rhamnopyranoside (myricetrin), quercetin and nyricetin. The isolated phenolics probably account for the antioxidant and hepatoprotective effects exhibited by the parent extract.

Lonchocarpus sericeus (Poir.) Kunth ex DC. FAMILY: FABACEAE

Lonchocarpus sericeus is used in traditional medicine for the treatment of pathological conditions such as gastrointestinal and hepatic disorders. Agbonon and Gbeassor (2009) investigated the hepatoprotective effect of *L. sericeus* ethanolic extract on liver damage induced by CCl_4, test animals were administered orally 200 and 400 mg/kg of *L. sericeus* extract four times after CCl_4 intoxication. Results indicated *L. sericeus* extract reduced aminotransferase levels in the blood and MDA concentration in the liver. The antioxidant potential in the liver and blood was increased, an action that may be related to phenolics in the extract. The results demonstrated a hepatoprotective effect of *L. sericeus*.

Lonicera japonica Thunb. FAMILY: CAPRIFOLIACEAE

Lonicera japonica, a widely used traditional Chinese medicine, possessed antiviral, hepatoprotective, antitumor, and other activities. *L. japonica* could be used as healthy food, cosmetics, and so on. Four novel flavonoids, japoflavone A–D (2–3, 5, 12), together with 10 known flavonoids, were isolated and identified on the basis of spectroscopic evidence from *L. japonica* by Ge et al. (2018). In addition, all isolates were assayed for their antihepatoma and hepatoprotective activities in vitro. 5 and 8 showed significant antihepatoma activity in SMCC 7721 cell with IC_{50} values of 13.01±2.62 and 16.69±0.35 μg/mL; Compound 12 showed significant hepatoprotective activity against H_2O_2-induced injury in SMCC 7721 and HepG 2 cells. Drop in the levels of CAT and SOD caused by H_2O_2 were remarkably reversed in a dose-dependent manner after treatment with 12.

Lophatherum gracile **Brongn. FAMILY: POACEAE**

He et al. (2016) evaluated the heptoprotective activity of ethanol extract of *Lophatherum gracile* against CCl_4-induced hepatotoxicity in mice. Pretreatment with the ethanol extract significantly decrased the serum ALT, AST, TB, and total cholesterol in a dose-dependent manner. Two doses of the extract pretreatment also prevented the increase in MDA level and reduce in abilities of SOD, CAT, and GPx in the liver tissues of CCl_4 poisoned mice at different extent.

Lophira lanceolata **Tiegh. ex Keay FAMILY: OCHNACEAE**

Aba et al. (2014) investigated the acute toxicity effect of aqueous stem bark extract of *Lophira lanceolata* and the activities of liver enzymes and other markers of organ damage in rats pretreated with aqueous stem bark extract of extract and subsequently intoxicated with PCM. The serum activities of the AST, ALT, ALP, and other markers of organ damages (bilirubin and TP) were investigated. The extract did not cause any death in all the groups even at the highest dose (5000 mg/kg BW). The results also showed varying degrees in the activity of the enzymes in the serum in comparison with the negative control. The mean serum ALP, ALT, and AST activity of rats pretreated with 300 mg/kg of the extract and 1000 mg/kg PCM were significantly lower than that of rats intoxicated with 1000 mg/kg PCM. This study showed that the aqueous stem bark extract of *L. lanceolata* possesses some active constituents that have antihepatotoxic potentials.

Luffa acutangula **(L.) Roxb. FAMILY: CUCURBITACEAE**

Hepatoprotective activity of hydroalcoholic extract of *Luffa acutangula* against CCl_4 and rifampicin-induced hepatotoxicity in rats was evaluated by Jadhav et al. (2010) and probable mechanism(s) of action has been suggested. Administration of standard drug Silymarin and extract of fruits of *L. acutangula* showed significant hepatoprotection against CCl_4 and rifampicin-induced hepatotoxicity in rats. Hepatoprotective activity was due to the decreased levels of serum marker enzymes, namely (AST, ALT, ALP, and LDH) and increased TP including the improvement in histoarchitecture of liver cells of the treated groups as compared to the control group. Fruit extracts also showed a significant decrease in MDA formation, increased activity of nonenzymatic intracellular antioxidant, GSH and enzymatic antioxidants, CAT, and SOD. Results of this study demonstrated that endogenous antioxidants and inhibition of LPO of membrane contribute to hepatoprotective activity of hydroalcoholic extract of *L. acutangula*.

Furthermore, Ulanganathan et al. (2010) screened hepatoprotective activity of ethanolic extract of the leaves of *L. acutangula* var. *amara* against CCl_4. CCl_4-induced elevated levels of serum markers (SGPT, SGOT, and ALP) were brought to normal by oral administration of leaf extract. Tissue-specific antioxidant activity of extract has been observed with the help of improved levels of GPx, GST, reduced GSH, SOD, CAT, and LPO (Ulanganathan et al., 2010).

Ethanolic fruit extract showed significant hepatoprotective activity compared to pet ether extract in CCl_4-induced liver necrosis. It also significantly reduced SGPT, SGOT, serum ALP, SB, serum cholesterol, TG, serum high density lipoproteins, serum TPs, and serum ALB. Histopathological studies of liver showed early necrosis in PEE while no necrosis was observed in the ethanolic extract, indicating the hepatoprotective potential of the latter (Ibrahim et al., 2014).

Hepatoprotective activity of different fractions of alcoholic fruit extract of *L. acutangula* was evaluated by Mishra and Mukerjee (2017) against PCM-induced liver toxicity. Toluene, chloroform, and EACs of ethanolic extract were administered orally (100 mg/kg) and biochemical parameters were measured. Ethyl acetate fraction increased DBL level while ALT, AST, and ALP levels were restored to normal when compared with other fractions. Histopathological evaluation of liver cells indicated the absence of necrosis with less vacuole formation (Mishra and Mukerjee, 2017).

Luffa cylindrica (L.) M.J. Roem. FAMILY: CUCURBITACEAE

Methanolic extract of *Luffa cylindrica* leaves (MELC) was evaluated by Sharma et al. (2014) for its hepatoprotective potential against PCM intoxicated Wistar rats. Administration of MELC and standard drug Silymarin showed significant hepatoprotective protection in experimental animals. Elevated serum marker enzymes of SGOT, SGPT, ALP, and SB were significantly reduced to near normal level in MELC treated animals. Lipid peroxidation level was decreased significantly by MELC 250 and 500 mg/kg doses treatment groups. In case of antioxidant enzymes SOD, GSH, and CAT levels were increased significantly after treatment with MELC. The extract showed a dose-dependent reduction of PCM-induced elevated serum levels of enzyme activities indicating the extract could preserve the normal functional status of the liver.

Luffa echinata Roxb. FAMILY: CUCURBITACEAE

The different extracts of the fruits of *Luffa echinata* were tested by Ahmed et al. (2001) for their hepatoprotective activity against CCl_4-induced hepatotoxicity in albino rats. The degree of protection was measured by using biochemical parameters like SGOT, SGPT, ALP, TP, and TA. The petroleum ether, acetone, and methanolic extracts showed a significant hepatoprotective activity comparable with those of Silymarin.

Lumnitzera racemosa Willd. FAMILY: COMBRETACEAE

The hepatoprotective activity of *Lumnitzera racemosa* bark extract was evaluated by Gnanadesigan et al. (2011) while leaf extract was evaluated by Ravikummar and Gnanadesigan (2011) using Wistar albino rats against CCl_4-induced hepatotoxicity. In both these studies the level of SGOT, SGPT, ALP, bilurbin, cholesterol, sugar, and LDH were significantly increased in hepatotoxin-treated rats when compared with the control group. But, the maximum reduction of SGOT, SGPT, ALP, bilirubin, cholesterol, sugar, and LDH were observed with 300 mg/kg BW of bark and leaf extract-treated rats. Histopathological scores showed that no visible changes were observed with high dose (300 mg/kg BW) of bark extract-treated rats except mild fatty changes. The hepatoprotective and antioxidant activities of the bark extract might be to the presence of unique chemical classes such as flavonoids, alkaloids, and polyphenols.

Lysimachia clethroides Duby FAMILY: PRIMULACEAE

Wei et al. (2012) evaluated the antioxidant and hepatoprotective activity of the extracts of *Lysimachia clethroides* (LC) was assayed by the methods of DPPH and ABTS *in vitro*. The ethyl acetate (LCEA) and *n*-butanol extracts (LCBU) of *L. clethroides* were the higher antioxidant activity in DPPH and ABTS assay. LCEA had the highest antioxidant activity (DPPH: IC_{50} = 9.02 μg/mL, ABTS: IC50 = 7.43 μg/mL, respectively). Thus, hepatoprotective effect of the extracts of *L. clethroides* was evaluated on CCl_4-induced acute liver injury mice. Intragastric administration of LCPE, LCEA, and LCBU on CCl_4-induced acute liver injury in mice for 8 days, the level of SGOT and SGPT in each treatment group significantly decreased. The level of MDA in liver for each treatment group could significantly decrease, and the level of SOD in liver only in group of LCBU (150 mg/kg) had no significant increase, the other treatment groups had significant increase.

Lycium chinense Miller FAMILY: SOLANACEAE

Ha et al. (2005) investigated the protective effect of *Lycium chinense* fruit (LFE) against CCl_4-induced hepatotoxicity and the mechanism underlying these protective effects in rats. The pretreatment of LFE has shown to possess a significant protective effect by lowering the serum AST, ALT, and ALP. This hepatoprotective action was confirmed by histological observation. In addition, pretreatment of LFE prevented the elevation of hepatic MDA formation and the depletion of reduced GSH content and CAT activity in the liver of CCl_4-injected rats. The LFE also displayed hydroxide radical scavenging activity in a dose-dependent manner (IC_{50} = 83.6 μg/mL), as assayed by ESR spin-trapping technique. The expression level of cytochrome P450 2E1 (CYP2E1) mRNA and protein, as measured by reverse transcriptase-polymerase chain reaction (RT-PCR) and western blot analysis, was significantly decreased in the liver of LFE-pretreated rats when compared with that in the liver of control group. Based on these results, it was suggested that the hepatoprotective effects of the LFE might be related to antioxidative activity and expressional regulation of CYP2E1.

Lygodium flexuosum (L.) Sw. FAMILY: SCHIZAEACEAE

The preventive and curative effect of *Lygodium flexuosum* on experimentally induced liver damage by CCl_4 and D-GalN was evaluated in rats by Wills and Asha (2006a, b, c). Treatment with CCl_4 caused a significant decrease in body and liver weight. *L. flexuosum n*-hexane extract prevented or reversed the decline in body and liver weight. Treatment with the extract prevented or restored the elevation of serum AST, ALT, and LDH levels. *L. flexuosum* treatment remarkably prevented or reversed an increase in liver HYP content in chronically treated rats. Histopathological changes of hepatic lesions induced by CCl_4 were significantly improved by treatment with *L. flexuosum*. Wills and Asha (2007) also investigated the protective mechanism of *L. flexuosum* extract in treating and preventing CCl_4-induced hepatic fibrosis in rats. Treatment with the *n*-hexane extract (200 mg/kg) reduced the mRNA levels of proinflammatory cytokines, growth factors and other signaling molecules, which are involved in hepatic fibrosis. The expression levels of TNF-α, interleukin-1beta, TGF-β1, procollagen-I, procollagen-III, and tissue inhibitor of metalloproteinase-1 were elevated during CCl_4 administration and reduced the levels to normal by the treatment with the extract treatment. The increased levels of matrix metalloproteinase-13 in extract-treated rats were indicative of the protective action of *L. flexuosum n*-hexane extract.

The antiangiogenic effect of *L. flexuosum* extract was evaluated by Wills et al. (2006) in Wistar rats intoxicated with N-NDEA in preventive and curative models. Rats intoxicated with NDEA had elevated levels of serum gamma-GT, AST, ALT, LDH levels and hepatic MDA and decreased levels of hepatic GSH. When treated with *L. flexuosum* extract had normal levels of gamma-GT, AST, ALT, LDH levels, hepatic MDA and GSH. NDEA administered rat liver showed an overexpressed levels of angiopoietins 1 (Ang-1) and 2 (Ang-2) and its receptor Tie-2 mRNA. *L. flexuosum* extract treatment significantly reduced the levels of Ang-1, Ang-2, and Tie-2 in rat livers evidenced by RT-PCR. Immunohistochemical analysis showed that VEGF was overexpressed and localized around periportal area of liver sections intoxicated with NDEA and its overexpression was effectively reduced by the treatment with *L. flexuosum* extract. Histopathological observations also substantiated NDEA-induced hepatotoxicity and the effect was significantly reduced by *L. flexuosum* extract treatment.

***Lygodium microphyllum* (Cav.) R.Br. FAMILY: LYGODIACEAE**

Animal studies were carried out by Gnanaraj et al. (2017) to evaluate hepatoprotection of aqueous extract of *Lygodium microphyllum* at different doses (200, 400, and 600 mg/kg BW) against CCl_4-mediated liver injury and histopathological alterations. Total phenolic content in aqueous extract of *L. microphyllum* leaves was 206.38 ± 9.62 mg GAE/g. The inhibitory concentration (IC_{50}) for free-radical scavenging activity of *L. microphyllum* was reached at a concentration of 65 μg/mL. *L. microphyllum* was able to prevent the increase in levels of serum ALT, serum AST and hepatic MDA formation in a dose-dependent manner. Immunohistochemical results evidenced the suppression of oxidative stress markers (4-hydroxynonenal, 8-hydroxydeoxyguanosine) and proinflammatory cytokines (tumor necrosis factor-α, interleukin-6, prostaglandin E_2). Histopathological and hepatocyte ultrastructural alterations showed protective effects by *L. microphyllum* against CCl_4-mediated oxidative stress. Hepatoprotective mechanism of *L. microphyllum* can be attributed to its antioxidative effects through protection of ultrastructural organelles.

Lysimachia paridiformis Franch. var. *stenophylla* Franch. FAMILY: PRIMULACEAE

Wei et al. (2012) investigated the hepatoprotective effect of the extracts of *Lysimachia paridiformis* var. *stenophylla* petroleum ether extract (LPFPE), EtOAC extract (LPFEA) and *n*-BuOH extract (LPFBU), on CCl_4-induced acute liver injury mice. Intragastric administration of LPFPE (1000, 500, and 250 mg/kg, respectively), LPFEA (800, 400, and 200 mg/kg, respectively) and LPFBU (600, 300, and 150 mg/kg, respectively) were carried out on mice for 8 days. The level of glutamic-oxalacetic transaminase (GOT) and GPT in each treatment group significantly decreased. The level of MDA in liver for each treatment group significantly decreased, and the level of SOD in liver for each treatment group significantly increased. It was demonstrated that three extracts of LPF had good hepatoprotective effect for CCl_4-induced liver injury mice, which may be attributable to its antioxidant activity.

Macrothelypteris torresiana (Gaudich) Ching (Syn.: *Lastrea torresiana* Moore) FAMILY: THELYPTERIDACEAE

Mondal et al. (2017) evaluated the hepatoprotective potential of ethanol extract from *Macrothelypteris torresiana* aerial parts and detect the polyphenolic compounds present in the extract using high-performance thin layer chromatography (HPTLC). Hepatoprotective potential of the extract was tested at doses of 300 and 600 mg/kg, per os (p.o.), on Wistar albino rats. The extract and Silymarin treated animal groups showed a significant decrease in activities of different biochemical parameters like SGOT, SGPT, and ALP, which were elevated by CCl_4 intoxication. The levels of TB and TP along with the liver weight were also restored to normalcy by the extract and Silymarin treatment. After CCl_4 administration, the levels of hepatic antioxidant enzymes such as GSH and CAT were decreased whereas the level of hepatic LPO was elevated. The levels of these hepatic antioxidant enzymes were also brought to normalcy by the extract and Silymarin treatment. Histological studies supported the biochemical findings, and treatment with extract at doses of 300 and 600 mg/kg, p.o. was found to be effective in restoring CCl_4-induced hepatotoxicity in rats. A simple HPTLC analysis was conducted for the detection of polyphenolic compounds in the extract, and the result revealed the presence of caffeic acid as phenolic acid and quercetin as a flavonoid (Mondal et al., 2017).

Macrotyloma uniflorum (Lam.) Verdc. FAMILY: FABACEAE

Hepatoprotective activity of methanolic extract of *Macrotyloma uniflorum* seed was investigated against D-galactosamine and PCM-induced hepatotoxicity in Wistar albino rats by Parmar et al. (2012). The 95% methanolic extract of fruit of *Macrotyloma uniflorum* (MEMUS) at the dose of (200 and 400 mg/kg) produced a dose-dependent reduction in biochemical parameters like SGPT, SGOT, ALP, bilirubin (Direct and Total) as well as in morphological parameters in D-galctosamine and PCM-induced hepatotoxicity in rats. The histopathological study further supported the hepatoprotective activity of the test extract. Maximum protection was seen at 400 mg/kg MEMUS. *M. uniflorum* seed showed significant hepatoprotactive properties in Wistar albino rats.

The protective effect of a hydroalcoholic extract of the seeds of *M. uniflorum* (MUSE) in ethanol-induced hepatotoxicity and antitubercular drug-induced liver injury was investigated in rats by Panda et al. (2015a, b). MUSE elicited significant hepatoprotective and antioxidant activity by attenuating the alcohol and antitubercular drug-elevated levels of serum marker enzymes (AST, ALT, and ALP), bilirubin, hepatic and serum TGs and the LPO marker MDA, and by restoring the ethanol-depleted levels of ALB, TPs, reduced GSH and the antioxidant enzymes SOD, CAT, GPx, and GR.

Madhuca longifolia (J. Koenig) Macbr. FAMILY: SAPOTACEAE

The ethanolic extract of *Madhuca longifolia* bark was screened by Roy et al. (2012, 2015) for hepatoprotective activity against D-GalN-induced hepatotoxicity in Wistar rats. Pretreatment with the extract of *M. longifolia* bark (200 and 400 mg/kg BW) brought back the altered levels of biochemical markers to the near normal levels. Histopathological studies also confirmed the hepatoprotective activity of these extracts when compared with D-GalN treated groups.

Mahonia fortunei (Lindl.) Fedde. FAMILY: BERBERIDACEAE

Chao et al. (2009) investigated the hepatoprotective activity of ethanol extract from Shidagonglao (*Mahonia fortunei*) roots (SDGL (EtOH)) on CCl_4-induced acute liver injury. Rats pretreated orally with SDGL (EtOH) (100 and 500 mg/kg) and Silymarin (200 mg/kg) for three consecutive days prior to the administration of a single dose of 50% CCl_4 significantly prevented the increases in the activities of serum ALT and AST in CCl_4-treated rats. Histological analysis also showed that SDGL (EtOH) (100 and 500 mg/kg) and Silymarin reduced the incidence of liver lesions including vacuole formation, neutrophil infiltration and necrosis of hepatocytes induced by CCl_4 in rats. Moreover, the SDGL (EtOH) (100 and 500 mg/kg) increased the activities of antioxidative enzymes, SOD, GPx, and GSH reductase (GRd) and decreased MDA level in liver, as compared to those in the CCl_4-treated group. Furthermore, SDGL (EtOH) (100 and 500 mg/kg) and Silymarin attenuated the increased levels of TNF-α in serum and NO in liver as compared to the CCl_4-treated group. The hepatoprotective mechanisms of SDGL (EtOH) are likely related to inhibition of TNF-alpha, MDA, and NO productions via increasing the activities of antioxidant enzymes (SOD, GPx, and GRd). These experimental results suggest that SDGL (EtOH) can attenuate CCl_4-induced acute liver injury in rats.

Mallotus japonicus (L.f.) Müll.-Arg. FAMILY: EUPHORBIACEAE

The hepatoprotective effects of bergenin, a major constituent of *Mallotus japonicus,* were evaluated by Kim et al. (2000) and Lim et al. (2000) against CCl_4-induced liver damage in rats. Bergenin at a dose of 50, 100, or 200 mg/kg was administered orally once daily for successive 7 days and then a mixture of 0.5 mL/kg (ip) of CCl_4 in olive oil (1:1) was injected two times each at 12 and 36 h after the final administration of bergenin. The substantially elevated serum enzymatic activities of ALT/AST, sorbitol dehydrogenase and GGT due to CCl_4 treatment were dose dependently restored toward normalization. Meanwhile, the decreased activities of GST and GR were restored toward normalization. In addition, bergenin also significantly prevented the elevation of hepatic MDA formation and depletion of reduced GSH content in the liver of CCl_4-intoxicated rats in a dose-dependent fashion. The results of this study clearly indicate that bergenin has a potent hepatoprotective action against CCl_4-induced hepatic damage in rats (Kim et al., 2000; Lim et al., 2000).

The hepatoprotective effects of bergenin, a major constituent of *M. japonicus*, were evaluated by Lim et al. (2001) against GalN-induced liver damage in rats. Bergenin (50, 100, and 200 mg/kg) was given orally once daily for seven successive days and then GalN 400 mg/kg was injected intraperitoneally to rats at 24 and 96 h after the final administration of bergenin. Pretreatment with bergenin reduced the increased enzyme activities of ALT/AST, sorbitol dehydrogenase, GGT and the elevated level of MDA induced by GalN. Bergenin restored the decreased hepatic contents of GSH as well as the decreased activities of GST and GR by GalN toward normalization, suggesting that the hepatoprotective effects of bergenin may consist in maintaining adequate levels of hepatic GSH for the removal of xenobiotics.

Malva parviflora L. FAMILY: MALVACEAE

Mallhi et al. (2014) determined the hepatoprotective activity of aqueous methanolic extract of whole plant of *Malva parviflora*. Two doses of plant (250 and 500 mg/kg) were administered in PCM intoxicated mice and results were compared with Silymarin. Observational parameters were ALT, AST, ALP, and TB. The results showed that the extract of *M. parviflora* produced significant reduction in liver enzymes and TB. Results were supported by histopathological investigation, phytochemical screening, and detection of hepatoprotective constituents (kaempferol and apigenin) by HPLC.

Malva sylvestris L. FAMILY: MALVACEAE

Hussain et al. (2014) assessed the hepatoprotective effects of *Malva sylvestris* against PCM-induced hepatotoxicity in mice. Two different doses of *M. sylvestris* (300 and 600 mg/kg) were administered intraperitoneally for seven consecutive days followed by intraperitoneal administration of PCM (250 mg/kg). Paracetamol significantly induced oxidative stress in the liver, ultimately leading to increased serum levels of liver enzyme markers like ALT, AST, ALP, TB, and DBL. The extract of *M. sylvestris* significantly reduced the serum levels of these elevated liver enzyme markers in a dose-dependent manner. Histopathological examination of liver tissues also showed hepatoprotective effects of *M. sylvestris* in restoring normal functional ability of the liver.

Malvaviscus arboreus Cav. Family: MALVACEAE

Abdelhafez et al. (2018) explored the chemical composition and the hepatoprotective potential of *Malvaviscus arboreus* against CCl_4-induced hepatotoxicity. The total extract of the aerial parts and its derived fractions (petroleum ether, dichloromethane, ethyl acetate, and aqueous) were orally administered to rats for six consecutive days, followed by injection of CCl_4 on the next day. Results showed that the ethyl acetate and dichloromethane fractions significantly alleviated liver injury in rats as indicated by the reduced levels of ALT, AST, ALP, TB, and MDA, along with enhancement of the total antioxidant capacities of their livers, with the maximum effects were recorded by the EAF. Moreover, the protective actions of both fractions were comparable to those of Silymarin (100 mg/kg), and have been also substantiated by histopathological evaluations. Detailed chemical analysis of the most active fraction (*i.e.,* ethyl acetate) resulted in the isolation and identification of six compounds for the first time in the genus, comprising four phenolic acids; β-resorcylic, caffeic, protocatechuic, and 4-hydroxyphenylacetic acids, in addition to two flavonoids; trifolin and astragalin. Such phenolic principles, together with their probable synergistic antioxidant and liver-protecting properties, seem to contribute to the observed hepatoprotective potential of *M. arboreus*.

Mammea africana Sabine FAMILY: CLUSIACEAE

Okokon et al. (2016) evaluated for hepatoprotective potentials of *Mammea africana* against PCM-induced liver injury in rats. Administration of the stem bark extract caused a significant dose-dependent reduction of high levels of liver enzymes (ALT, AST, and ALP), TC, direct and TB as well as elevation of serum levels of TP, ALB, and antioxidant enzymes (SOD, CAT, GPx, and GSH). Histology of the liver sections of extract and Silymarin-treated animals showed reductions in the pathological features compared to the PCM-treated animals. The chemical pathological changes were consistent with histopathological observations suggesting marked hepatoprotective effect of the stem bark extract of *M. africana.*

Mangifera indica L. FAMILY: ANACARDIACEAE

Among Yoruba herbalists (Southwest Nigeria), hot water infusion of *Mangifera indica* stem bark is reputedly used for the treatment of fever, jaundice, and liver disorders. Adeneye et al. (2015) investigated the protective effects and mechanism(s) of chemopreventive and curative effects of 125–500 mg/kg/day of *M. indica* aqueous stem bark extract (MIASE) in acute CCl_4-induced liver damage in rats. Rats were treated intragastrically with 125, 250, and 500 mg/kg/day of MIASE for 7 days before and after the administration of CCl_4 (3 mL/

kg of 20% CCl_4, i.p.). The serum levels of ALT, AST, TP, ALB, TG, TC, HDL cholesterol (HDL-c), LDL cholesterol (LDL-c), TB, CB, and FBG levels were estimated. In addition, hepatic tissue reduced GSH and the MDA concentrations, CAT, SOD activities in the hepatic homogenate, and histopathological changes in the rat liver sections were determined. Preliminary qualitative phytochemical screening for bioactive compounds in MIASE was also conducted. Results showed that oral treatment with 125–500 mg/kg/day of MIASE significantly attenuated the increase in serum ALT, AST, ALP, FBG, TB, CB, and LDL-c levels in acute liver injury induced by CCl_4 treatment. Findings also revealed significant elevations in the serum TC, TG, HDL-c, TP, and ALB levels. There was marked architectural remodeling in the hepatic lesions of hepatocyte vacuolation and centrilobular necrosis induced by CCl_4 treatment, coupled with significant weight loss. MIASE also markedly enhanced SOD and CAT activities while reducing MAD formation; and increased GSH concentration in the hepatic homogenate compared with untreated CCl_4-intoxicated group, with more protection offered in the curative than the chemopreventive models of CCl_4 hepatotoxicity.

Omotayo et al. (2015) evaluated the hepatoprotective activity of aqueous and ethanol extract of *M. indica* stem bark in PCM-induced liver injury in Winstar albino rats. The ethanolic extract was found to be more effective than the aqueous extract (200 mg/kg) in PCM-induced liver damage by decreasing the activity of serum enzymes AST, ALT, and ALP while increasing significantly the levels of TPs and ALB. GSH, CAT, and SOD activity were significantly increased in the treated groups while MDA levels were reduced.

Marrubium vulgare L. FAMILY: LAMIACEAE

Marrubium vulgare, commonly known as "horehound," is traditionally used to treat various diseases. The plant possesses antihypertensive, analgesic, anti-inflammatory, and hypoglycemic effect, antidyslipidemic activity, and antioxidant properties (Elberry et al., 2010, 2011). Antimicrobial activity against Gram positive bacteria, analgesic properties, and antihypertensive, antidiabetic, and antioxidant properties were noted, particularly related to diterpenes, sterols, phenylpropanoids, and flavonoids. Totally 46 compounds, comprising 96.3% of the oil, were detected. The main components of the oil were (E)-caryophyllene, germacrene D, and bicyclogermacrene. Evaluation of the antihepatotoxic and antioxidant properties of the extract against $CC1_4$-induced hepatic damage in rats showed a significant antihepatotoxic effect via significantly reducing AST, ALT, and LDH (Elberry et al., 2010). In another study, aqueous extract of the whole plant was examined for antihepatotoxic activity against CCl_4-induced hepatic damage in male Wistar rats. The extract at 500 mg/kg BW dose for 7 days was compared with the standard drug Silymarin at 10 mg/kg BW dose. The aqueous extract had significant antihepatotoxic activity and reduced the elevated levels of serum enzymes such as SGOT, SGPT, and ALP and increasing TP. Some studies also reported the antioxidant effect of free-radical scavengers in the extract; thanks to flavonoid content (Elberry et al., 2010, 2011).

Akhter et al. (2013) evaluated the hepatoprotective activity of ME of whole plant of *M. vulgare* against PCM toxicity in Wistar rats. Methanol extract of *M. vulgare* was administered at the doses 100 and 200 mg/kg/day, p.o. for 7 days. Serum analysis was performed to estimate the levels of ALT, AST, ALP, ALB, TB, and TGs. Histopathology studies were also performed on the CAT liver samples. The toxic effects of PCM were significantly controlled in the extract-treated groups which was manifested by the restoration of serum biochemical parameters to near normal levels. From the study it was concluded that *M. vulgare* possess significant hepatoprotective properties.

Marsdenia volubilis (L.f.) T. Cooke FAMILY: APOCYNACEAE

Rubab et al. (2013) evaluated the hepatoprotective activity of methanolic extracts of *Marsedenia volubilis* against PCM-induced liver damage in rats. The methanolic extracts of *M. volubilis* (500 mg/kg) were administered orally to the animals with hepatotoxicity induced by PCM (3 g/kg). The plant extract was effective in protecting the liver against the injury induced by PCM in rats. This was evident from significant reduction in serum enzymes ALT, AST, ALP, and bilirubin. It was concluded from the result that the methanolic extract of *M. volubilis* possesses hepatoprotective activity against PCM-induced hepatotoxicity in rats.

Marsilea minuta L. FAMILY: MARSILEACEAE

Divya et al. (2013) carried out investigation to identify the most effective hapatoprotective fraction of methanolic extract of *Marsilea minuta* by fractionating and evaluating its fractions for hapatoprotective activity in three mechanistically devised models viz., CCl_4, PCM and ethanol-induced liver damage in rats. Pretreatment with fractions (toluene, 1-butanol, aqueous at 50, 100 mg/kg BW) significantly reversed the changes in serum biochemical parameters and histology of liver caused by the three hepatotoxins namely CCl_4, PCM and ethanol indicating their hepatoprotective activity. The hepatoprotective activity of butanol fraction was well supported in metabolizing enzymes, prothrombin time and DPPH studies. All the fractions exhibited significant hepatoprotective activity; out of all the fractions butanol fraction (50 mg/kg) was identified to be most effective hepatoprotective fraction.

MATRICARIA CHAMOMILLA L. [Syn.: *Chamomile recutita* (L.) Rauschert] FAMILY: ASTERACEAE

Hepatoprotective activity of aqueous ethanolic extract of *Matricaria chamomilla* capitula against PCM-induced hepatic damage in albino rats was evaluated by Gupta and Misra (2006). The effect of aqueous ethanolic extract of *M. chamomilla* capitula on blood and liver GSH, Na^+ K^+-ATPase activity, serum marker enzymes, SB, glycogen and thiobarbutiric acid reactive substances against PCM-induced damage in rats have been studied to find out the possible mechanism of hepatoprotection. It was observed that extract of chamomile has reversal effects on the levels of above-mentioned parameters in PCM hepatotoxicity. The extract of capitula of chamomile functions as a hepatoprotective agent and this hepatoprotective activity of chamomile may be due to normalization of impaired membrane function activity.

Gupta et al. (2006) evaluated the antioxidant and hepatoprotective activity of methanolic extract of capitula of *M. chamomilla* against CCl_4-induced hepatic damage in rats. The extract at dose level 300 mg/kg has shown a potent and prominent free-radical scavenging activity and hepatoprotective activity.

Al-Baroudi et al. (2014) investigated the effect of oral administration of aqueous extract of *M. chamomilla* capitula for 4 weeks on hepatotoxicity induced to rats by herbicide 2,4-dichlorophenoxy acetic acid (2,4-D). These effects could be explored by measuring BW gain, feed efficiency ratio and relative weight of the liver. Serum levels of liver enzymes; AST, ALT, and ALP; ALB, TP, TB, LDH enzyme, and antioxidant enzymes; GR and SOD were determined. Oral administration of *M. chamomilla* capitula extract to hepatotoxic rats for 28 days significantly decreased the elevated serum levels of liver enzymes (AST, ALT, and ALP), TB and LDH enzyme and positive groups. Levels of antioxidant enzymes GR and SOD were significantly increased as compared to the control positive groups. This study recommends that intake of *M. chamomilla* capitula extract as a herbal tea may be beneficial for patients who suffer from liver diseases and oxidative stress antioxidant enzymes.

Maytenus emarginata (Willd.) Ding Hou FAMILY: CELASTRACEAE

Parmar et al. (2009, 2010) evaluated the hepatoprotective activity of leaves of *Maytenus emarginata* against CCl_4 and PCM-induced toxicity in male Wistar rats. Toxin-induced hepatotoxicity resulted in an increase in serum AST, ALT, ALP activity and bilirubin level accompanied by significant decrease in ALB level. Co-administration of the aqueous extract protects the PCM-induced lipid peroxidation, restored altered serum marker enzymes and antioxidant level toward near normal.

Meconopsis integrifolia (Maxim.) Franch FAMILY: PAPAVERACEAE

Zhou et al. (2013) investigated the hepatoprotective and antioxidant effects of *Meconopsis integrifolia* ethanolic extract (MIE) *in vitro* and *in vivo* against $CC1_4$-induced hepatic damage in male Wistar rats. Ethanolic extract exhibited strong antioxidant ability *in vitro*. In the rats with CCl_4-induced liver injury, the groups treated with ME and Silymarin showed significantly lower levels of ALT, AST, ALP, and TB. Methanol extract demonstrated good antioxidant activities in both the liver and kidney of the rats *in vivo*. Methanol extract exhibited excellent hepatoprotective effects and antioxidant activities *in vitro* and *in vivo*, supporting the traditional use of *M. integrifolia* in the treatment of hepatitis.

Melastoma malabathricum L. FAMILY: MELASTOMATACEAE

Kamisan et al. (2013) investigated the hepatoprotective activity of a ME of *Melastoma malabathricum* leaves using two established rat models. Ten groups of rats (nZ6) were given a once-daily administration of 10% dimethyl sulfoxide (negative control), 200 mg/kg Silymarin (positive control), or MEMM (50, 250, or 500 mg/kg) for 7 days followed by induction of hepatotoxicity either using PCM or CCl_4. Based on the results obtained, MEMM exhibited a significant hepatoprotective activity against both inducers, as indicated by an improvement in the liver function test. These observations were supported by the histologic findings. Authors concluded that *M. malabathricum* leaves possessed hepatoprotective activity, which could be linked to their phytochemical constituents and antioxidant activity; this therefore requires further in-depth studies.

Mamat et al. (2013) assessed the hepatoprotective activity of methanol extract of *M. malabathricum* leaves (MEMM) against the PCM-induced liver toxicity in rats model. MEMM exerted significant and high antioxidant activity in which high TPC was recorded; while in the hepatotoxicity study, the extract exhibited significant hepatoprotective effects against the PCM-induced hepatotoxic model. The results observed for serum liver enzymes (ALT, ALP, and AST) as well as the microscopic observations and microscopic scoring supported the hepatoprotective potential of MEMM. The phytochemical and HPLC analysis of MEMM demonstrated the presence of flavonoids as its major constituents.

Melilotus officinalis (L.) Pall. FAMILY: FABACEAE

Alamgeer et al. (2017) investigated the hepatoprotective activity of methanolic extract of *Melilotus officinalis* against PCM and CCl_4-induced hepatic damage. *M. officinalis* at selected oral doses of 50 and 100 mg/kg showed significant hepatoprotective effects by decreasing the levels of serum marker enzymes such as TB, SGOT, SGPT, ALP, ALB, and TP, when compared with standard drug (Silymarin) and negative control. Similarly, histopathological studies also supported biochemical estimations. It was concluded that extract of *M. officinalis* has strong hepatoprotective activity against PCM and CCl_4-induced hepatotoxicity, which might be due to free-radical scavenging mechanisms exhibited by flavonoids and phenolics; thus affirming its traditional therapeutic role in liver injury.

Melothria heterophylla (Lour.) Cogn. FAMILY: CUCURBITACEAE

Mondal et al. (2011) investigated the hepatoprotective activity of ethanol extract of *Melothria heterophylla* against CCl_4-induced hepatic damage in rats. β-sitosterol was isolated by column chromatography and characterized spectroscopically. Two different doses (200 and 400 mg/kg BW) of ethanol extract were administered orally in alternate days. The hepatoprotective activity was studied in liver by measuring biochemical parameters such as serum AST, ALT, ALP, TP, and TB. Lipid peroxidation product and different antioxidant enzyme activities were assessed in liver homogenate. The ethanol extract reduced all biochemical parameters and LPO, as well as it increased the antioxidant enzyme activities in comparison with Silymarin. The protective effect of the extract on CCl_4-induced damage was confirmed by histopathological examination of the liver.

Melothria maderaspatana (L.) Cogn. FAMILY: CUCURBITACEAE

Investigations were carried out by Thabrew et al. (1988) and Jayathilaka et al. (1989) to evaluate the ability of *Melothria maderaspatana* to protect the livers of albino rats from CCl_4-mediated alterations in liver histopathology and serum levels of ALT, AST, and ALP. Treatment with an aqueous extract of *Melothria* aerial parts (either before or after CCl_4 administration) markedly decreased CCl_4-mediated alterations in liver histopathology as well as serum enzyme levels.

Melothria perpusilla (Blume) Cogn. FAMILY: CUCURBITACEAE

Yengkhom et al. (2017) assessed the hepatoprotective effect of aqueous extract of *Melothria perpusilla* (AEMP) against CCl_4-induced liver injury. ALT, AST, ALP, and bilirubin were significantly reduced in groups receiving both CCl_4 and AEMP when compared with CCl_4-treated group.

Mentha arvensis L. FAMILY: LAMIACEAE

Ethanol, chloroform and aqueous extracts of *Mentha arvensis* leaves were studied by Patil and Mall (2002) to evaluate its hepatoprotective effects against CCl_4-induced liver damage in rats. Histopathological changes in liver and biochemical parameters such as SGPT, SGOT, ALP and SB were analyzed. There was significant reduction in values of marker enzymes. The phytochemical investigations of *M. arvensis* leaves extract showed the presence of compounds of flavonoids, steroids, triterpenoids, alkaloids, glycosides, carbohydrates, tannins and phenols. Hepatoprotective and antioxidant activity of ethanol extract *M. arvensis* against CCl_4-induced hepatic damage in rats was investigated in another study. It was observed that the level of AST, ALT, ALP, and bilirubin was decreased while the level of antioxidant enzymes significantly increased on treatment of rats with *M. arvensis* leaves extract.

Mentha piperita L. FAMILY: LAMIACEAE

Sharma et al. (2007) studied the protective role of leaves of *Mentha piperita* (Mint) in adult Swiss albino mice against arsenic-induced hepatopathy. In the arsenic-treated group, there was a significant increase in ACP, ALP, SGOT, SGPT, and LPO content, whereas a significant decrease was recorded in BW, liver weight, GSH, and LDH activity in liver.

Pre- and post-treatment of Mentha with arsenic significantly alters the biochemical parameters in liver. A significant decline in ACP, ALP, SGOT, SGPT, and LPO content was observed. However, a significant increase in BW, liver weight, GSH content, and LDH activity in liver was estimated.

Bellassoued et al. (2018) investigated the composition and *in vitro* antioxidant activity of *M. piperita* leaf essential oil (MpEO) in CCl_4 –intoxicated Wistar rats. The *in vitro* antioxidant activity of MpEO was lower than that of Silymarin. Pretreatment of animals with MpEO at a dose of 5 mg/kg did not have a significant effect on ALT, AST, ALP, LDH, γGT, urea or creatinine levels in CCl_4-induced stress. Whereas pretreatment with MpEO at doses of 15 and 40 mg/kg prior to CCl_4, significantly reduced stress parameters (ALT, AST, ALP, LDH, γGT, urea, and creatinine) compared to the CCl_4-only group. Moreover, a significant reduction in hepatic and kidney LPO (TBARS) and an increase in antioxidant enzymes SOD, CAT, and GPx was also observed after treatment with MpEO (40 mg/kg) compared to CCl_4-treated rats. Furthermore, pretreatment with MpEO at 40 mg/kg can also markedly ameliorate the histopathological hepatic and kidney lesions induced by administration of CCl_4.

Mentha pulegium L. FAMILY: LAMIACEAE

Jain et al. (2012) evaluated the *in-vivo* antioxidant potential of ethanolic extract of *Mentha pulegium* against CCl_4-induced toxicity in rats. Pre-treatment with 600 mg/kg (p.o.) of ethanolic extract of *M. pulegium* improved the GSH, SOD, CAT, and peroxidase levels significantly as compared to control group. The studies revealed that *M. pulegium* has significant *in vivo* antioxidant activity and can be used to protect tissue from oxidative stress. The result showed that the activities of GSH, SOD, CAT, and peroxidase in group treated with CCl_4 declined significantly than that of normal group. Ethanolic extract of *M. pulegium* in the dose of 600 mg/kg, p.o., has improved the GSH, SOD, CAT, and peroxidase levels significantly, which were comparable with Liv. 52.

Mentha spicata L. FAMILY: LAMIACEAE

Ben Saad et al. (2018) investigated the protective effect of *Mentha spicata* supplementation against nicotine-induced oxidative damage in the liver and erythrocytes of Wistar rats. After 2 months of treatment, nicotine induced an increase in the level of WBCs and a marked decrease in erythrocytes, hemoglobin, and hematocrit. AST, ALT, ALP, and LDH activities were also found to be higher in nicotine-treated group than those of the control group. Furthermore, nicotine-treated rats exhibited oxidative stress, as evidenced by a decrease in antioxidant enzymes activities and an increase in LPO level in liver and erythrocytes. The oral administration of *M. spicata* extract by nicotine-treated rats alleviated such disturbances. *M. spicata* contained bioactive compounds that possess important antioxidant potential and protected liver and erythrocytes against nicotine-induced damage.

Merremia emarginata (Burm. f.) Hallier f. [Syn.: *Ipomoea reniformis* (Roxb.) Choisy] FAMILY: CONVOLVULACEAE

Neelima et al. (2017) evaluated the hepatoprotective activity of methanolic extract of *Merremia emarginata* (MEME) in experimentally induced hepatotoxic rats. The control group did not exhibit increase in serum parameters, but ethanol toxicant group showed significant increase in serum parameters such as SGOT, SGPT, bilirubin (total and direct), and LPO, whereas GSH and TP levels were markedly reduced. Silymarin, MEME low dose (200 mg/kg p.o.) and high dose (400 mg/kg, p.o.) treated groups showed significant decrease in SGOT, SGPT, total and DBL, TB, LPO, and increase in GSH and TP levels. Based on the study findings in serum marker enzyme levels and antioxidant parameters, it is concluded that MEME possesses hepatoprotective activity.

Micromeria croatica (Pers.) Schott FAMILY: LAMIACEAE

Vladimir-Knežević et al. (2015) investigated the hepatoprotective activity and possible underlying mechanisms of *Micromeria croatica* (MC) ethanolic extract using a model of CCl_4-induced liver injury in mice. CCl_4 intoxication resulted in liver cells damage and oxidative stress and triggered inflammatory response in mice livers. MC treatment decreased ALT activity and prevented liver necrosis. Improved hepatic antioxidant status was evident by increased Cu/Zn SOD activity and decreased 4-hydroxynonenal (4-HNE) formation in the livers. Concomitantly, nuclear factor erythroid 2-related factor 2 (Nrf2) and HO-1 were overexpressed. The hepatoprotective activity of MC was accompanied by the increase in NF-κB activation and TNF-α expression, indicating amelioration of hepatic inflammation. Additionally, MC prevented tumor growth factor-β1 (TGF-β1) and α-smooth muscle actin (α-SMA) expression, suggesting the potential for suppression of hepatic fibrogenesis.

Mikania scanden (L.) Willd. FAMILY: ASTERACEAE

The hepatoprotective effect of *M. scandens* was tested in animal model using rats, in which the injury was produced by ethyl alcohol and drugs such as PCM and sodium diclofenac, showing that ethanolic extract and fractions have a promising hepatoprotective effect (Maity et al., 2012a, b).

Millingtonia hortensis L. FAMILY: BIGNONIACEAE

Millingtonia hortensis is an abundant resource of flavonoids, which might be beneficial in protecting liver tissue from injury. The hepatoprotective and antioxidant potential of ethanolic extract of *M. hortensis* on CCl_4-induced hepatotoxicity and the possible mechanism involved therein were investigated by Babitha et al. (2012) in rats. CCl_4 treatment produced a profound increase in the levels of MDA, hepatic marker enzymes and bilirubin content compared with the control. Pretreatment with the flower extract of *M. hortensis* significantly enhanced the levels of endogenous antioxidants and reduced the levels of hepatic marker enzymes in relation to the CCl_4 treated group. Ballooning degeneration and fatty changes in hepatocytes were prevented by pretreatment with the flower extract.

Mimosa pudica L. FAMILY: FABACEAE

Leaves of *Mimosa pudica* have been reported for hepatoprotective activity against CCl_4-induced toxicity in albino rats at the dose of 200 mg/kg BW. Methanolic extract of leaves of *M. pudica* lowered the serum levels of various biochemical parameters like SGOT, SGPT, ALP, TB, and TC while there was increased level of TN and ALB (Rajendran et al., 2009).

Simultaneous administration of the leaf extract *M. pudica* along with the toxin ethanol in rats showed a considerable protection against the toxin-induced oxidative stress and liver damage as evidence by a significant increase in antioxidant activities. The study reveals that the co administration of *M. pudica* aqueous extract significantly lowered the level of LPO in alcohol fed mice (Nazeema and Brindha, 2009).

The ethanol extract of *M. pudica* leaves was evaluated for its hepatoprotective and antioxidant activities against CCl_4-induced liver damage, in Wistar albino rats. The substantially elevated levels of serum GOT, GPT, ALP, and TB, due to CCl_4 treatment, were restored toward near normal by *M. pudica* extract in a dose-dependent manner. Ethanol extract of *M. pudica* also increased the serum total proteins of CCl_4-intoxicated rats. Reduced enzymatic and nonenzymatic antioxidant levels and elevated LPO levels were restored toward near normal by administration of *M. pudica*.

Suneetha et al. (2011) screened the hepatoprotective activity of the methanolic root extract of *M. pudica*. Biochemical parameters like SGOT, SGPT, and SB were measured. The activity of tissue antioxidant enzymes namely LPO, CAT, reduced GSH, and histopathological evaluation of liver sections were also studied. CCl_4 administration in rats elevated the levels of SGPT, SGOT, cholesterol and bilirubin. Administration of the methanolic root extract of the *M. pudica* at a dose (400 mg/kg) significantly prevented this increase. The activity of antioxidant enzymes like CAT and reduced GSH was decreased and MDA content was increased in CCl_4-treated group. The enzyme levels of CAT and reduced GSH were significantly increased and MDA content significantly decreased in the group treated with *M. pudica* at a dose of 400 mg/kg. Histopathological studies revealed that the concurrent administration of CCl_4 with the *M. pudica* extract exhibited protection of the liver tissue. The study has confirmed the hepatoprotective activity of methanolic extract of *M. pudica*, which may be attributed to its antioxidant property.

Kumaresan et al. (2015) evaluated the hepatoprotective activity of *M. pudica* on experimentally induced hepatotoxic rats. Twelve healthy albino rats were taken for the study. These animals were segregated into three groups, namely, normal, untreated, and treated containing four animals in each group. Before the injection of hepatotoxic substance to the rats, they were fasted overnight. Then 0.3 mL of CCl_4 with paraffin in the ratio of 3:1 was injected for 10 days per animal. The crude powder of *M. pudica* was administered to the animals belonging to the treated group starting from the day of injection of CCl_4 and was continued for 10 days. The liver function of the three distinct groups of the animals was assessed by collecting the blood sample and liver homogenate. The elevation in the serum and tissue markers was found in the group of animals under the untreated group. The treated group of animals showed approximately equal levels of enzyme markers as found in the normal animals.

Purkayastha et al. (2016) investigated the hepatoprotective effects of ethanolic extract of leaves of *M. pudica* against CCl_4-induced hepatic injury in albino rats. Liver damage in rats treated with CCl_4 (1 mL/kg/BW), administered subcutaneously, on alternate days for one week) was studied by assessing parameters such SGOT, SGPT, ALP, and bilirubin (total and direct). Silymarin (100 mg/kg) was used as the standard drug. Administration of ethanolic extract of *M. pudica* significantly prevented CCl_4-induced elevation of serum SGOT, SGPT, ALP, and bilirubin (total and direct) level. Histological examination of the liver section revealed improved hepatic architecture, after administration of *M. pudica*. The results were comparable to that of Silymarin. No additive effect was observed.

Mitracarpus scaber Zucc. ex Schult. & Schult.f. FAMILY: RUBIACEAE

The effect of *Mitracarpus scaber* on CCl_4-induced acute liver damage in the rat has been evaluated by Germano et al. (1999). Results showed that treatment with *M. scaber* decoction resulted in significant hepatoprotection against CCl_4-induced liver injury both *in-vivo* and *in-vitro*. *In-vivo*, *M. scaber* pretreatment reduced levels of SGOT and SGPT previously increased by administration of CCl_4. *In-vitro* results indicated that addition to the culture medium of *M. scaber* extracts significantly reduced SGOT and LDH activity. *Mitracarpus* treatment also resulted in a good (> 93%) survival rate for the CCl_4-intoxicated hepatocytes as demonstrated by 3-(4,5-dimethylthiazol-2-yl)-2,5-diphenyl tetrazolium bromide assay. Moreover, as in the *in-vitro* assay, *M. scaber* had radical-scavenging properties, shown by its reaction with the DPPH radical (EC_{50}, the extract concentration resulting in a 50% reduction in the absorbance of DPPH blank solution, = 41.64 ± 1.5 μg/mL).

Mitragyna rotundifolia Kuntze FAMILY: RUBIACEAE

Gong et al. (2012) evaluated the protective role of *n*-butanol extract from *Mitragyna rotundifolia* barks (MRBBU) and leaves (MRLBU) against CCl_4-induced acute liver injury in mice. The results of the intragastric administration of MRBBU (300, 150, and 75 mg/kg BW per day, respectively) and MRLBU (300, 150, and 75 mg/kg BW per day, respectively) on CCl_4-induced acute liver injury in mice for 8 days demonstrated that the level of SGPT and SGOT in each treatment group decreased significantly. It was observed that the level of liver MDA in MRLBU (75 mg/kg) group did not significantly decreased, but that in the other treatment groups significantly decreased. With the exception of MRLBU group (150 and 75 mg/kg), the level of SOD in other treatment groups significantly increased. Results indicated that *M. rotundifolia* has hepatoprotective effect against CCl_4-induced oxidative damage in mice, and the hepatoprotective effect may be correlated with its antioxidant effects.

Mollugo pentaphylla L. FAMILY: AIZOACEAE

Valarmarthi et al. (2010) investigated the hepatoprotective activity of the alcoholic extract of *Mollugo pentaphylla* against CCl_4-induced hepatic toxicity in rats. Liver function was assessed by the determination of SGOT, SGOT, ALP, acyl carrier protein (ACP), and bilirubin. Histopathological studies revealed that concurrent administration of the extract with CCl_4 exhibited protective of the liver, which further evidenced its hepatoprotective activity. The result suggest that the use of alcoholic extract of *M. pentaphylla* exhibited significant protective from liver damage in $CC1_4$-induced liver damage model. Phytochemical screening of this plant revealed the presence flavanoids, saponins, terpenoids and tannis.

Momordica balsamina L. FAMILY: CUCURBITACEAE

Alqasoumi et al. (2009) evaluated the efficacy of *Momordica balsamina,* traditional Saudi plant used for liver problems. The ethanol extract of the leaves of *M. balsamina* were subjected to hepatoprotective assays using Wistar albino rats. Liver injury was induced in rats using CCl_4. The biochemical parameters; SGOT, SGPT, ALP, and TB were estimated as reflection of the liver condition. Based on the results of the biochemical parameters measurements, histopathological study was performed on the liver of rats treated with two extracts. The normal appearance of hepatocytes indicated a good protection of the extracts from CCl_4 hepatotoxicity. All the results were compared with Silymarin, the reference hepatoprotective drug.

Momordica charantia L. FAMILY: CUCURBITACEAE

Chaudhari et al. (2009) evaluated the hepatoprotective activity of hydroalcoholic extract of *Momordica charantia* leaves against CCl_4-induced hepatotoxicity in albino Wistar rats. In the hydroalcoholic extract of *M. charantia* leaves treated animals, the toxic effect of CCl_4 was controlled significantly by restoration of the increased levels of SGOT, SGPT, ALP and TB as compared to the toxicant control. The hydroalcoholic extract of leaves of *M. charantia* showed significant hepatoprotective activity.

Hossain et al. (2011) investigated the hypolipidemic and hepatoprotective effects of the different fractions (Petroleum ether, ethyl acetate and chloroform) of methanolic extract of *M. charantia.* The different fractions of extract were administered intraperitoneally as a single dose of

150 mg/kg BW to alloxan-induced diabetic rats and found to reduce blood lipid level (total cholesterol and triglycerides) significantly. The plant fractions also exhibited correction of altered biochemical parameters, namely, SGOT and SGPT levels in diabetic rats. The effect of plant fractions was compared with standard drug metformin. The phytochemical screening tests indicated that the different constituents such as saponins, tannins, triterpenes, alkaloids, and flavonoids, etc., were present in the plant which has hypolipidemic and hepatoprotective properties.

Thenmozhi and Subramanian (2011) investigated the effect of *M. charantia* fruit extract on circulatory liver markers, LPO, and antioxidants status in ammonium chloride (AC)-induced hyperammonemic in male albino Wistar rats. Administration of *M. charantia* fruit extract in hyperammonemic rats reduced the levels of ammonia and urea. The antioxidant property *M. charantia* fruit extract was studied by assessing the activities of TBARS, HP, and liver markers (ALT, AST, and ALP and the levels of GPx, SOD, CAT, reduced GSH, vitamins A, C, and E in AC-treated rats. Oxidative stress was effectively modulated by *M. charantia* fruit extract administration. *M. charantia* fruit extract significantly improved the status of antioxidants and decreased ammonia, urea, LPO, and liver markers enzymes as compared to AC treated group. However examination of the efficacy of the active constituents of the *M. charantia* on hyperammonemia is desirable.

Hepatoprotective and hepatocurative effects of *M. charantia* were investigated by Zahra et al. (2012) by analysis of different serum enzymes which included ALT, AST, and LDH. There were two phases of the study. In phase one, rabbits were induced toxicity with the administration of APAP and then extract of *M. charantia* was given to observe hepatocurative effects. It was observed that the elevated levels of enzymes were significantly decreased in APAP-induced rabbits. In second phase, prior oral administration of *M. charantia* extract was given for 15 days and then APAP was administrated. Reduced level of marker enzymes showed the hepatoprotective effect of *M. charantia.* Another study was conducted by Ajilore et al. (2012) to investigate the hepatoprotective effects of methanolic extract of *M. charantia* leaves. Hepatotoxicity in rats was induced by cadmium. Serum TP, ALB, ALT, and AST were evaluated. Total protein and ALB level was significantly reduced while ALT and AST level was raised. It was observed that toxic effect of cadmium was significantly controlled in the rats pretreated with methanolic extracts of *M. charantia.*

Mada et al. (2014) evaluated hepatoprotective effect of *M. charantia* leaves extract against CCl_4-induced hepatotoxicity in rats. Treatment with the extract significantly restored liver weight to near normal. The result showed a significant increase in hemoglobin (Hb) and packed cell volume (PCV) compared to toxin control group. Also treatment with the extract caused a significant decrease in the activities of ALT, AST, ALP, and the level of TB: and a significant increase in TP level compared to control group. Similarly, the extract caused a significant decrease in the level of reduced GSH and MDA and a significant elevation in the activities of SOD and CAT compared to toxin control group.

Deng et al. (2017) investigated the possible effects of *M. charantia* water extract against liver injury in restraint-stressed mice. *M. charantia* water extract reduced the serum AST and ALT, reduced the NO content and the protein expression level of iNOS in the liver; significantly reduced the mitochondrial ROS content, increased the MMP and the activities of mitochondrial respiratory chain complexes I and II in restraint-stressed mice.

Momordica dioica Roxb. FAMILY: CUCURBITACEAE

The hepatoprotective activity of ethanolic and aqueous extracts of *Momordica dioica* leaves were evaluated by Jain et al. (2008) against CCl_4-induced hepatic damage in rats. The extracts at a dose of 200 mg/kg were administered orally once daily. The substantially elevated serum enzymatic levels of AST, ALT, serum ALP, and TB were restored toward normalization significantly by the extracts. Silymarin was used as a standard reference and exhibited a significant hepatoprotective activity against CCl_4-induced hepatotoxicity in rats. The biochemical observations were supplemented with histopathological examination of rat liver sections. The results of this study strongly indicate that *Momordica dioica* Roxb. leaves have potent hepatoprotective action against CCl_4-induced hepatic damage in rats. Ethanolic extract was found more potent hepatoprotective. Meanwhile, *in-vivo* antioxidant and free-radical scavenging activities were also screened which were positive for both ethanolic and aqueous extracts. This study suggests that a possible mechanism of this activity may be due to free-radical scavenging and antioxidant activities which may be due to the presence of flavonoids in the extracts (Jain et al., 2008).

Momordica subangulata Blume FAMILY: CUCURBITACEAE

Asha (2001) evaluated the hepatoprotective activity of leaves of *Momordica subangulata* using PCM overdose-induced liver damage in rats. *Momordica subangulata* leaf suspension (500 mg/kg, fresh weight; 50 mg/kg, dry weight) protected rats from PCM-induced liver damage as judged from serum marker enzyme activities. It also stimulated bile flow in normal rats.

Momordica tuberosa Cogn. FAMILY: CUCURBITACEAE

Hydro alcoholic extract of tubers of *Momordica tuberosa* was subjected to preliminary phytochemical screening and evaluated by Kumar et al. (2008) for *in vitro* and *in vivo* antioxidant and hepatoprotective activity against CCl_4-induced liver damage in rats. Pretreatment with 70% ethanolic extract of *M. tuberosa* reversed CCl_4-induced elevation of levels of serum biomarkers to near normal levels, suggesting that the tubers of *M. tuberosa* possess hepatoprotective property and this property may be attributed to the antioxidant property of the plant.

Morinda citrifolia L. FAMILY: RUBIACEAE

Wang et al. (2008a, b) evaluated the protective effects of Noni (*Morinda citrifolia*) fruit juice on acute liver injury induced by CCl_4 in female SD rats. Liver damage (microcentrilobular necrosis) was observed in animals pretreated with 20% placebo (drinking water) + CCl_4. However, pretreatment with 20% Noni juice in drinking water + CCl_4 resulted in markedly decreased hepatotoxic lesions. Furthermore, serum ALP, ALT, and AST levels were significantly lower in the Noni group than the placebo group. In a correlative time-dependent study, one dose of CCl_4 (0.25 mL/kg in corn oil, p.o.) in female SD rats, pretreated with 10% placebo for 12 days, caused sequential progressive hepatotoxic lesions over a 24 h period, while a protective effect from 10% Noni juice pretreatment was observed. Histopathological examination revealed that liver sections from the Noni juice + CCl_4 appeared similar to controls, whereas typical hepatic steatosis was observed in the placebo + CCl_4 group. Serum ALP, AST, ALT, TC, TG, low-LDL, and VLDL levels were increased in the placebo group compared with the TNJ group. In contrast, HDL was increased in the TNJ group and decreased in the placebo group.

Morinda lucida Benth. FAMILY: RUBIACEAE

The propanol and aqueous leaves extracts of *Morinda lucida* were evaluated by Didunyemi et al. (2019) for their hepatoprotective potential on oxidative stress and acute liver damage induced by APAP in rats. Co-treatment of rats with Acetaminophen and *M. lucida* extracts, significantly ameliorated APAP-induced elevation of serum ALT, AST, ALP, TB, and TC and also alleviated the depletion of GSH, GPx, SOD, CAT, NO, MPO, and LPO levels in rats.

Morinda tinctoria Roxb. FAMILY: RUBIACEAE

Subramanian et al. (2013) investigated the hepatoprotective activity of aqueous and methanolic extracts of leaves of *Morinda tinctoria* against PCM-induced liver damage in rats. The phytochemical investigation of the both extracts showed the presence of alkaloids, flavonoids, glycosides, carbohydrates, saponin, tannin, and phenols. The PCM intoxication lead to histological and biochemical deterioration. The treatment with both aqueous and methanolic leaves extracts of *M. tinctoria* reduced the level of SGOT, SGPT, TB, DB and TC and also reversed the hepatic damage toward normal which further supports the hepatoprotective activity of leaf extracts of *M. tinctoria*.

Moringa oleifera Lam. (Syn.: *Moringa pterygosperma* Gaertn.*)* FAMILY: MORINGACEAE

The ethanolic extract of the leaves of *Moringa oleifera* (MO) was found to exhibit hepatoprotective effect against alcohol-induced hepatotoxicity in rats (Saalu et al., 2012). This research proved that animal pretreatment with ethanolic extract of *M. oleifera* (300 mg/kg of weight) significantly attenuated hepatotoxin-induced biochemical (serum AST, ALT, ALP, and GGT) and histopathological changes in the liver. Additionally, *M. oleifera* leaves also showed significant anti-inflammatory (Kumar and Mishra, 1998), and antioxidant potencies (Saalu et al., 2012; Kumar and Pari, 2003), which may be contributing to its hepatoprotective activity. A number of phytochemicals with antioxidant activities have been characterized from *M. oleifera* including; quercetin, rutin, kaempferol, and caffeoyqumic acids (Lawal et al., 2016).

The EtOH (95%) extract from *M. oleifera* leaves administered orally during 45 days at 150, 200, and 250 mg/kg also demonstrated a significant protective effect against the damage induced by INH, RIF, and PZA (at doses of 7.5, 10, and 35 mg/kg, respectively) on Wistar rats; in this study, the authors employed Silymarin as a positive control (200 mg/kg). The authors found that AST, ALT, ALP, TB, cholesterol, and TGs levels were similar in the extract group treated with 250 mg/kg and in the Silymarin group, also increase the level of antioxidants enzyme and showing a reduction of LPO in this group (Pari and Kumar, 2002; Kumar and Pari, 2003). Omodanisi et al. (2017) also reported hepatoprotective activity in diabetes-induced male Wistar rats on oral administration of methanolic extract of leaves of *M. oleifera* (250 mg/kg).

Studies on the hepatoprotective activity of different extracts of the stem bark of *M. oleifera* was carried out by Kurma and Mishra (2008) in albino rats. The total aqueous extract showed hepatoprotective action against CCl_4- and rifampicin-induced hepatotoxicities, while the PEE exhibited similar activity against PCM-induced hepatotoxicity. Caffeic and fumaric acids, isolated for the first time from the bioactive total aqueous extract, were characterized and assessed for their *in vitro* hepatoprotective activity. These showed significant hepatoprotective activity against galactosamine- and TAA-induced hepatic cytotoxicities.

Fakurazi et al. (2008) evaluated the hepatoprotective action of *M. oleifera* (MO) against a single high dose of APAP. The level of GSH was found to be restored in MO-treated animals compared to the groups treated with APAP alone. Fakurazi et al. (2012) reported that *M. oleifera* hydroethanolic extracts effectively alleviate APAP-induced hepatotoxicity in experimental rats through their antioxidant nature. Among MO edible parts the flower extracts contain the highest TPC and antioxidant capacity, followed by leaves extract. The oxidative marker MDA, as well as 4-HNE protein adduct levels were elevated and GSH, SOD, and CAT were significantly decreased in groups treated with hepatotoxin. The biochemical liver tissue oxidative markers measured in the rats treated with MO flowers and leaves hydroethanolic extracts showed a significant reduction in the severity of the liver damage.

Patel et al. (2010) evaluated the hepatoprotective activity of crude aqueous and ethanol (alcoholic) extract of leaves of *M. oleifera* against CCl_4-induced hepatocytes injury of rats *in vitro*. The extract was effective in reducing CCl_4-induced enhanced activities of GPT, GOT, LPO, and % viability. As per results obtained from the study, it is evident that low concentration of alcoholic extract has significant hepatoprotective activity at even 0.01 mg/L while that of aqueous extract was found at 0.1 mg/L and higher concentration of the drug.

Hepatoprotective effect of the ethanolic extract of *M. oleifera* leaves was studied against antitubercular drugs (isoniazid, rifampicin, and PZA) (Ajilore et al., 2012)-induced liver damage as well as against cadmium-induced hepatotoxicity in rats. Moringa extract significantly attenuated hepatotoxin-induced biochemical (serum AST, ALT, APT, and bilirubin) and histopathological changes in liver. Ezejindu et al. (2013) reported that extract of leaves of *M. oleifera* have protective effect on liver enzymes of

CCl_4-induced hepatotoxicity in adult Wister rats.

The antioxidant and hepatoprotective activities of the ethanol extract of *M. oleifera* leaves were investigated by Singh et al. (2014) and El-Bakry et al. (2016) against CCl_4-induced hepatotoxicity in rats. The activities were studied by assaying the serum marker enzymes like SGOT, SGPT, GGT, LDH, ALP, ACP, as well as TB, TP, and ALB in serum concomitantly with the activities of LPO, SOD, CAT, GSH, GR, and GPx in liver. The activities of all parameters registered a significant alteration in CCl_4-treated rats, which were significantly recovered toward an almost normal level in rats co-administered with *M. oleifera* extract in a dose-dependent manner. All the biochemical investigations were confirmed by the histopathological observations and compared with the standard drug Silymarin. Results suggest that the antioxidant and hepatoprotective activities of *M. oleifera* leaves are possibly related to the free-radical scavenging activity, which might be due to the presence of total phenolics and flavonoids in the extract and/or the purified compounds β-sitosterol, quercetin, and kaempferol, which were isolated from the ethanol extract of *M. oleifera* leaves.

Toppo et al. (2015) evaluated the hepatoprotective activity of *M. oleifera* against cadmium-induced toxicity in Wistar albino rats. Oral administration of cadmium chloride @ 200 ppm/kg for 28 days resulted in a significant increase in AST, ALT, ALP, significant increase of LPO and decrease in SOD, and increase in cadmium accumulation in liver. Treatment with *M. oleifera* @ 500 mg/kg significantly decreased the elevated ALP, AST, ALT, LPO levels, and increase in SOD levels, and as compared to cadmium chloride treated group. However, there was no significant difference in cadmium concentration in liver when compared with cadmium chloride treated group.

Shaik et al. (2017) compared the hepatoprotective effects of *M. oleifera* ethanolic extracts with other standard drug, Liv. 52, in Albino Wistar rats. Rats treated with ethanolic leaf extract of *M. oleifera* the SGOT values were significantly lower when compared to SGOT levels in control rats. Rats treated with ethanolic leaf extract of *M. oleifera* the SGPT values were significantly lower when compared to SGPT levels in control rats. They concluded that ethanolic extracts of *M. oleifera* leaves exhibited significant hepatoprotective activity in a dose-dependent manner.

Aladeyelu et al. (2018) investigated the hepatoprotective role of ethanolic extract of MO on liver function (biomarkers) in cadmium chloride ($CdCl_2$)-induced hepatotoxicity on the liver of albino Wistar rats. Significant increase in the serum levels of ALT, AST, and ALP of animals in group B, but some protective effect of MO in the treatment groups with significant decreases in ALT, AST, and ALP levels. Authors concluded that ethanolic extract of *MO* showed appreciable hepatoprotective values on liver functions (biomarkers) in $CdCl_2$hepatotoxicity.

Hepatoprotective and antioxidant activity of aerial parts of MO in prevention of NAFLD in Wistar rats was also investigated by Asgari-Kafrani et al. (2019). Treatment of rats with *MO* led to significant reduction in TC, TG, LDL, VLDL, ALT, and AST compared to the HFD group. On the contrary, increased levels of HDL-C. Histopathological examinations demonstrated steatosis and inflammatory in HFD group liver sections. Significant decreases were observed in steatosis and inflammatory in groups treated with *MO* compared to the HFD group.

Moringa peregrina (Forssk.) Fiori FAMILY: MORINGACEAE

Azim et al. (2017) assessed the ameliorative role of *Moringa peregrina* leaves extract against APAP toxicity in rats. Administration of *M. peregrina* leaves extract reversed APAP-related toxic effects through: powerful MDA suppression, GPx normalization, and stimulation of the cellular antioxidants synthesis represented by significant increase of GSH, CAT, and SOD in liver, blood and brain, besides, DNA fragmentation was significantly decreased in the liver tissue.

Morus alba L. FAMILY: MORACEAE

The hepatoprotective activities of MEs of root barks of *Morus alba* were evaluated by Singab et al. (2010). Liver damage was induced in rats by CCl_4 in a subcutaneous dose of 1 mL (40% v/v in corn oil)/kg. The root bark extract was given intraperitoneally at doses of 50 mg/kg. The activity of the extracts was comparable to that of Silymarin, a known hepatoprotective agent. Hepatoprotective activity was evaluated by measuring serum levels of AST, ALT, ALP, TB, and TP. These biochemical observations were supported by histopathological examination of liver sections. Methanol extracts of *M. alba* showed potent antioxidant and hepatoprotective activities.

Hogade et al. (2010) investigated the hepatoprotective activity of *M. alba* leaves extracts against CCl_4-induced hepatotoxicity. Leaves powder of *M. alba* was successively extracted with PEE, CE, ALE, and water extract (AQE) against CCl_4-induced hepatotoxicity using Standard drug Liv. 52. The ALE showed presence of alkaloids, flavonoides, carbohydrates, tannins and steroids, while carbohydrates, flavonoides, alkaloids were present with AQE. The PEE, CHE, ALE did not produce any mortality. CCl_4 produced significant changes in biochemical parameters (increases in SGPT, SGOT, ALP, and SB.) and histological (damage to hepatocytes) using standard drug Liv. 52. Pretreatment with ALE and AQE extracts significantly prevented the biochemical and histological changes induced by CCl_4 in the liver. This study shows that the ALE and AQE extracts possessed hepatoprotective activity.

Morus bombycis Koidzumi FAMILY: MORACEAE

The antioxidant activity and liver protective effect of *Morus bombycis* were investigated by Jin et al. (2005). Aqueous extracts of *M. bombycis* had higher superoxide radical scavenging activity than other types of extracts. The aqueous extract at a dose of 100 mg/kg showed significant hepatoprotective activity when compared with that of a standard agent. The biochemical results were confirmed by histological observations indicating that *M. bombycis* extract together with CCl_4 treatment decreased ballooning degeneration. The water extract recovered the CCl_4-induced liver injury and showed antioxidant effects in assays of $FeCl_2$-ascorbic-acid-induced LPO in rats. Based on these results, Jin et al. (2005) suggest that the hepatoprotective effect of the *M. bombycis* extract is related to its antioxidative activity.

Jin et al. (2006) investigated hepatoprotective activity and antioxidant effect of the 2,5-dihydroxy-4,3′-di(beta-D-glucopyranosyloxy)-*trans*-stilbene that purified from *M. bombycis* roots against CCl_4-induced liver damage in rats. The 2,5-dihydroxy-4,3′-di(beta-D-glucopyranosyloxy)-*trans*-stilbene displayed dose-dependent superoxide radical scavenging activity (IC50 = 430.2 μg/mL), as assayed by the ESR spin-trapping technique. The increase in AST activities in serum associated with CCl_4-induced liver injury was inhibited by 2,5-dihydroxy-4,3′-di(beta-D-glucopyranosyloxy)-*trans*-stilbene and at a dose of 400–600 mg/kg samples had hepatoprotective activity comparable to the standard agent, Silymarin. The biochemical assays were confirmed by histological observations showing that the 2,5-dihydroxy-4,3′-di(beta-D-glucopyranosyloxy)-*trans*-stilbene decreased cell ballooning in response to CCl_4 treatment. These results demonstrate that the 2,5-dihydroxy-4,3′-di(beta-D-glucopyranosyloxy)-*trans*-stilbene is a potent antioxidant with a liver protective action against CCl_4-induced hepatotoxicity.

Morus indica L. FAMILY: MORACEAE

Mous indica aqueous (MAq) and dechlorophyllised (MDc) extracts efficacy to protect against the CCl_4-induced hepatotoxicity was studied by Reddy and Urooj (2017) in rats in comparison with standard drug Liv. 52. The TP, ALB, urea, creatinine, and TB were within the normal level in all the groups whereas the hepatic enzymes AST, ALT, and ALP activity was less in Liv. 52 and *Morus* treated groups. Total cholesterol, TGs, were determined. The TC and TG levels in *Morus* treated groups were significantly lower than Liv. 52, CCl_4 and healthy control groups. Also pretreatment with *Morus* extracts restored the hepatic architecture near to the standard drug treatment which is showed in histopathologicl sections of liver.

Morus nigra L. FAMILY: MORACEAE

The possible hepatoprotective effect of the administration of *Morus nigra* ethanolic extract leaves against hepatotoxic effect of the antirheumatic drug, methotrexate was evaluated both *in vivo* (using animal models) and *in vitro* (human hepatoma HepG2 cells). A marked reduction in the viability of HepG2 cells was observed after 48 h with IC_{50} equal to 14.5 μg/mL of ethanolic extract administration. Treating the animals with the extract in combination with methotrexate mitigated liver injury, causing a significant reduction in activities of AST, ALT, ALP, and LDH as compared to the methotrexate group. The liver architecture revealed more or less normal appearance with the combined treatment when compared with methotrexate treatment alone. This study recommends that the co-administration of ethanolic extract with methotrexate that may have therapeutic benefits against methotrexate-hepato-cytotoxicity.

Mucuna pruriens (L.) DC. FAMILY: FABACEAE

Obogwu et al. (2014) investigated the hepatoprotective and *in vivo* antioxidant activities of the hydroethanolic extract of *Mucuna pruriens* leaves in antitubercular and alcohol-induced hepatotoxicity assays in rats. The hepatotoxicants significantly increased the levels of ALT, AST, ALP, bilirubin, and MDA; and reduced the levels of CAT, SOD, GPx, and reduced GSH compared to control. *M. pruriens* significantly reversed the elevation in the level of ALT, AST, ALP, and bilirubin caused by the hepatotoxicants. The extract (200 and 400 mg/kg) significantly reversed the diminution in the level of *in vivo* antioxidants and increased the level of MDA produced by

INH-RIF. *M. pruriens* (100–400 mg/kg) elicited significant reduction in the level of MDA compared to the alcohol group.

Murraya koenigii (L.) Spreng. FAMILY: RUTACEAE

Sathaye et al. (2011) reported hepatoprotective activity of crude aqueous extract of *Murraya koenigii* against ethanol-induced hepatotoxicity in experimental animals. Based on the inhibitory concentration (IC_{50}) obtained from the cell viability assay, graded concentrations of 100 and 500 μg/mL of aqueous extract (WE), isolated carbazole alkaloids and tannin (T) fraction were chosen to study the hepatoprotective activity against ethanol-induced hepatotoxicity using liver carcinoma cell lines (Hep G(2)). The tannins and the carbazole alkaloids from the aqueous extract exhibited excellent hepatoprotective activity with respect to the different parameters studied and maintained normal morphology even after ethanolic challenge to the cells as comparable to the protection offered by the standard drug L-ornithine L-aspartate. The modulating effect of the aqueous extract and isolates on liver metabolizing enzymes, reduction in LPO and decreased cellular damage were found to contribute to the hepatoprotective activity.

Desai et al. (2012) investigated hepatoprotective effects of polyphenol rich *M. koenigii* hydroethanolic leaf extract in CCl_4 treated hepatotoxic rats. *M. koenigii* pretreated rats with different doses (200, 400, and 600 mg/kg BW) showed significant decrement in activity levels of ALT, AST, ALP, TP, and bilirubin. Also, extract-treated rats recorded a dose-dependent increment in hepatic SOD, CAT, reduced GSH and ascorbic acid and, a decrement in LPO. Microscopic evaluations of liver revealed CCl_4-induced lesions and related toxic manifestations that were minimal in liver of rats pretreated with MK extract.

Musa paradisiaca L. FAMILY: MUSACEAE

Nirmala et al. (2012) investigated the hepatoprotective activity of stem of *Musa paradisiaca* in CCl_4 and PCM-induced hepatotoxicity models in rats. Administration of hepatotoxins (CCl_4 and PCM) showed significant biochemical and histological deteriorations in the liver of experimental animals. Pretreatment with ALE (500 mg/kg), more significantly and to a lesser extent the ALE (250 mg/kg) and aqueous extract (500 mg/kg), reduced the elevated levels of the serum enzymes like SGOT, SGPT, ALP, and bilirubin levels and alcoholic and aqueous extracts reversed the hepatic damage toward the normal, which further evidenced the hepatoprotective activity of stem of *M. paradisiaca.*

Musa sapientum L. FAMILY: MUSACEAE

Dikshit et al. (2011) evaluated the hepatoprotective activity of aqueous extract of central stem of *Musa sapientum* (AqMS) against CCl_4-induced hepatotoxicity in rats. There was significant rise in AST, ALT, and ALP in CCl_4 intoxicated control group II. Treatment with AqMS prevented rise in levels of these enzymes. There was significant rise in MDA and fall in GSH in blood and liver in group II, indicating increased LPO and oxidative stress upon CCl_4 administration. Treatment with AqMS prevented rise in MDA and increased GSH in treated group. SOD levels were decreased in group II while groups treated with AqMS showed significant rise. Maximum hepatoprotective effect was observed with 50 mg/kg dose. Hepatoprotective effect observed with this dose was comparable to standard hepatoprotective drug Silymarin. The results of pathological study also support the results of biochemical findings.

Musanga cecropioides R.Br. ex Tedlie FAMILY: URTICACEAE

The hepatoprotective activities and the mechanisms of actions of *Musanga cecropioides* stem bark aqueous extract (MCW) were investigated by Adeneye (2009) on acute hepatocellular injuries induced by CCl_4 and APAP in male Wistar rats. Among the Yorubas (southwest Nigeria), cold decoction of MCW is used as a natural antidote for oral gastric poisonings, infective hepatitis and other liver diseases. Pretreatment of rats with graded doses (125–500 mg/kg) of MCW significantly attenuated the acute elevation of the liver enzymes and the hepatotoxin-induced histopathological lesions in the rat livers. The presence of two active natural antioxidants (flavonoids and alkaloids) in high concentrations in MCW may account for the hepatoprotective activities observed in this study.

M. cecropioides leaf extract is used in ethnomedicine for the management of jaundice and other hepatic ailments in Ibibio, Nigeria. Nwidu et al. (2018) evaluated the hepatoprotective and antioxidant effects of *M. cecropioides* hydromethanolic leaf (MCHL) extract against CCl_4-induced hepatotoxicityin rats. The MCHL extracts significantly reduced the increase in AST, ALT, ALP, CB, and TB levels induced by CCl_4 intoxication. There was no significant alteration in hematological indices or weight following administration of the MCHL extracts. Histopathological examinations revealed mitotic bodies in the 141.4 mg/kg MCHL extract-treated rats, an indication of tissue repair processes.

Myrica rubra Sieb. et Zucc. FAMILY: MYRICACEAE

The relationship between the expression of mitochondrial voltage-dependent anion channels (VDACs) and the protective effects of *Myrica rubra* fruit extract (MCE) against CCl_4-induced liver damage was investigated. Pretreatment with 50 mg/kg, 150 mg/kg or 450 mg/kg MCE significantly blocked the CCl_4-induced increase in both AST and ALT levels in mice. Ultrastructural observations of decreased nuclear condensation, ameliorated mitochondrial fragmentation of the cristae and less lipid deposition by an electron microscope confirmed the hepatoprotection. The MMP dropped from −191.94 ± 8.84 mV to −132.06 ± 12.26 mV after the mice had been treated with CCl_4. MCE attenuated CCl_4-induced MMP dissipation in a dose-dependent manner. At a dose of 150 or 450 mg/kg of MCE, the MMPs were restored. Pretreatment with MCE also prevented the elevation of intra-mitochondrial free calcium as observed in the liver of the CCl_4-insulted mice. In addition, MCE treatment significantly increased both transcription and translation of VDAC inhibited by CCl_4. The above data suggest that MCE mitigates the damage to liver mitochondria induced by CCl_4, possibly through the regulation of mitochondrial VDAC, one of the most important proteins in the mitochondrial outer membrane.

Myristica fragrans Houtt. FAMILY: MYRISTICACEAE

Morita et al. (2003) evaluated the hepatoprotective activity of Nutmeg (*Myristica fragrans*) by feeding the extract in rats with liver damage caused by LPS plus D-GalN. As assessed by aminotransferase activities, nutmeg showed potent hepatoprotective activity. Bioassay-guided isolation of the active compound from nutmeg was carried out in mice by a single oral administration of the respective fractions. Myristicin, one of the major essential oils of nutmeg, was found to possess extraordinarily potent hepatoproteective activity. Myristicin markedly suppressed LPS/D-GalN-induced enhancement of serum TNF-α-concentrations and hepatic DNA fragmentation in mice. These findings suggest that the hepatoprotective activity of myristicin might be, at least in part, due to the inhibition of TNF-α release from macrophages.

Nardostachys jatamansi (D. Don) DC. FAMILY: CAPRIFOLIACEAE

Ali et al. (2000) reported that a 50% ethanolic extract of the rhizomes of *Nardostachys. jatamansi* is shown to possess hepatoprotective activity. Pretreatment of rats with the extract (800 mg/kg BW, orally) for three consecutive days significantly ameliorated the liver damage in rats exposed to the hepatotoxic compound TAA. Elevated levels of serum transaminases (aminotransferases) and ALP, observed in TAA alone treated group of animals, were significantly lowered in *N. jatamansi* pretreated rats. Pretreatment of the animals with the extract also resulted in an increase in survival in rats intoxicated with LD90 dose of the hepatotoxic drug.

Nasturtium officinale R.Br. FAMILY: BRASSICACEAE

Azarmehr et al. (20190 investigated the hepatoprotective and antioxidant activity of hydroALE of *Nasturtium officinale* (watercress, WC) in APAP-induced hepatotoxicity in rats. The results have shown that there was a significant increase in AST, ALT, FRAP, and PCO content in APAP group in comparison to control. Also, there was a significant reduction in T-SH levels and GPx activity in APAP group compared to control. However, administration of WC extract and Silymarin not only causes a significant decrease in AST activity, but also they markedly increased T-SH content and GPx activity compared to APAP group. GC-MS analysis showed the major compositions were found to be benzenepropanenitrile (48.30%), phytol (10.10%), α-cadinene (9.50%), and linolenic acid (8%).

Nauclea latifolia Sm. FAMILY: RUBIACEAE

Effiong et al. (2013) evaluated the hepatoprotective and antioxidant effect of the ethanol extract of *Nauclea latifolia* (NL) leaf in Wistar albino rats. Protective action of (NL) leaf extracts was evaluated using animal model of hepatotoxicity induced by Acetaminophen (Paracetamol) (2 g/kg, BW, p.o.). Levels of AST and ALT were increased and the levels of TP and ALB were decreased in Acetaminophen-treated rats. NL leaf at 400 mg/kg, BW doses decreased the elevated levels of the transaminases and restored the normalcy of TP and ALB significantly. The activities of CAT, GPx, and SOD were decreased in hepatotoxic rats but administration with NL leaf extract increased the GPx, CAT, and SOD. Histopathology showed restoration of the APAP damaged liver with NL administration.

Edagha et al. (2014) investigated the effect of ethanolic extracts of the leaves *Nauclea latifolia* (NL) on the hematological parameters and histomorphology of the liver of male Swiss albino mice infected with *Plasmodium berghei berghei*. The extract exhibited a hepatoprotective and reversibility effects at a dose-dependent level on the histological architecture of treated groups administered compared with the control, and also caused a significant reduction in the RBC parameters in a dose-dependent manner especially in nonparasitized mice. In conclusion, acute toxicity test of ethanolic extract of *Nauclea latifolia* up to 5000 mg/kg may be considered as relatively safe if mortality alone is the yardstick. However, at a dose of 1000 mg/kg, it is severely hepatoxic in nonparasitized mice, yet the extract at 500 mg/kg had beneficial effects in both the hematological indices and liver cytoarchitecture of parasitized host via a possible synergistic mechanism of its rich bioactive ingredients comparable with Coartem®.

Nelumbo nucifera Gaertn. FAMILY: NELUMBONACEAE

Ethanol extracts from *Nelumbo nucifera* (ENN) seeds were studied for possible antioxidative and hepatoprotective effects by Sohn et al. (2003). Antioxidative effects were measured spectrophotometrically by reduction of DPPH radicals. Hepatoprotective effects were tested using CCl_4 and aflatoxin B1 (AFB1)-induced hepatocyte toxicity models. ENN showed potent free-radical scavenging effects with a median inhibition concentration of 6.49 μg/mL. Treatment of hepatocytes with ENN inhibited both the production of serum enzymes and cytotoxicity by CCl_4. The genotoxic and cytotoxic effects of AFB1 were also inhibited by ENN in dose-dependent manners. These hepatoprotective effects of ENN against CCl_4 and AFB1 might result from its potent antioxidative properties.

Lotus (*N. nucifera*), an aquatic vegetable, is extensively cultivated in eastern Asia, particularly in China. An ethanolic extract of the leaves was studied by Huang et al. (2009) for its hepatoprotective activity against CCl_4-induced liver toxicity in rats. *In vitro* and *in vivo* antioxidant activity was also assessed. The results showed the hepatoprotective activity of LLE at doses of 300 and 500 mg/kg and *in vivo* antioxidant activity at 100 mg/kg that was comparable with that of a standard treatment comprising 100 mg/kg of Silymarin, a known hepatoprotective drug. These data were supplemented with histopathological studies of rat liver sections. The main flavonoids and phenolic compounds of LLE were analyzed by HPLC-DAD-ESI/MS methods. Six of the compounds detected were tentatively characterised, one as catechin glycoside and five as flavonoid glycoside derivatives: miricitrin-3-*O*-glucoside, hyperin, isoquercitrin, quercetin-3-*O*-rhamnoside and astragalin.

The protective effects of lotus germ oil on liver and kidney damage by CCl_4-induced chronic hepatotoxicity in mice, PC-12 cells, and DNA damage were investigated by Lv et al. (2012). The mice were treated orally with lotus germ oil or dl-α-tocopherol after administration CCl_4 for 49 consecutive days. The levels of key antioxidant enzymes, such as SOD,

CAT, and the concentration of GSH, as well as the concentration of MDA, an indicator of LPO, were determined in homogenates of the liver and the kidney. The pathological histology of the liver was also examined. The activities of SOD, CAT, and the concentration of GSH were increased significantly after treated with lotus germ oil in a concentration-dependent manner. Whereas, the content of the peroxidation product MDA were decreased significantly, similar to the serum levels of hepatic enzyme biomarkers (ALT and AST). Furthermore, lotus germ oil could inhibit the conversion of super-coiled pBR322 plasmid DNA to the open circular form and apoptosis of hydrogen peroxide-induced PC-12 cells. The result of this study suggested that the lotus germ oil could be recognized as powerful "functional oil" against oxidative stress.

Yuan et al. (2014) evaluated the antioxidant activity and heatoprotective activity of extracts of leaves of *N. nucifera.* The hepatoprotective effect of ethyl acetate and butanol extracts was evaluated on CCl_4-induced acute liver injury mice. The results showed that the levels of SGOT and SGPT in each treatment group significantly decreased (except for the group of ethyl acetate extract). The contents of MDA in liver had significant decrease and the level of SOD in liver for each treatment group could significantly decrease.

***Neoboutonia melleri* (Müll. Arg.) Prain var. *velutina* FAMILY: EUPHORBIACEAE**

Effa et al. (2018) evaluated the hepatoprotective properties of aqueous extract of *Neoboutonia melleri* var. *velutina,* a Cameroonian medicinal plant. Acute hepatitis models (CCl_4 and concanavalin A) were performed in mice receiving or not receiving, different extract doses by gavage. Liver injury was assessed using histology, transaminases and pro-inflammatory markers. Extract antioxidant and radical scavenging capacities were evaluated. The extract led to a significant decrease in pro-inflammatory cytokine expression *in vitro* and to a remarkable protection of mice from CCl_4-induced liver injury, as shown by a significant decrease in dose-dependent transaminases level. Upon extract treatment, inflammatory markers were significantly decreased and liver injuries were limited as well. In the concanavalin A model, the extract displayed weak effects.

***Nerium oleander* L. (Syn.: *Nerium indicum* Mill.) FAMILY: APOCYNACEAE**

Hepatoprotective capacity of *Nerium oleander* (Syn.: *N. indicum*) leaf extract (NILE) was evaluated by Dey et al. (2015) on CCl_4-induced acute hepatotoxicity in murine model. The results showed that the biochemical parameters were overexpressed due to CCl_4 administration, which were significantly normalized by NILE treatment. The findings were further supported by histopatholog ical evidences showing less hepatocellular necrosis, inflammation and fibrosis in NILE and Silymarin treated groups, compared to CCl_4 group. GC-MS analysis revealed the presence of different bioactive phytochemicals with hepatoprotective and antioxidant properties.

Singhal and Gupta (2012) investigated the antioxidant and hepatoprotective activity of methanolic flower extract of *N. oleander* against CCl_4-induced hepatotoxicity in rats. The methanolic extract at dose of 100, 200, and 400 mg/kg were administered orally once daily for seven days. Serum enzymatic levels of serum AST, ALT, ALP, and TB were estimated along with estimation of SOD and MDA levels in liver tissues. Further histopathological examination of the liver sections was carried out to support the induction of hepatotoxicity and hepatoprotective efficacy. The extract showed potent activities on reducing power, LPO, DPPH, ABTS, superoxide anion, hydroxyl radical, and metal chelation. The substantially elevated serum enzymatic levels of AST, ALT, ALP and TB were found to be restored toward normalization significantly by the methanolic extract in a dose-dependent manner with maximum hepatoprotection at 400 mg/kg dose level. The histopathological observations supported the biochemical evidences of hepatoprotection. Elevated level of SOD and decreased level of MDA further strengthen the hepatoprotective observations.

Newbouldia laevis (P. Beauv.) Seem. FAMILY: BIGNONIACEAE

Hepatoprotective effect and antioxidant properties of leaf extracts of *Newbouldia laevis* (NL) were investigated by Hassan et al. (2010) against CCl_4-induced acute hepatic injury in rats. Significantly increased levels of transaminases, ALP, ALB, bilirubin and cholesterol in CCl_4 intoxicated rats were brought to normal levels in rats treated with the extract and CCl_4 in a dose-dependent manner. Reduced enzymic and nonenzymic antioxidants levels and elevated LPOs levels were restored to normal by administration of the extract at 300 mg/kg. Histopathological examination further confirmed the hepatoprotective activity of NL when compared with CCl_4 treated group. Phenolics, glycosides, balsams, resins, and volatile oils were detected in the extract. The lethal dose (LD 50) of the plant was greater than 3000 mg/kg.

Nicotiana plumbaginifolia Viv. FAMILY: SOLANACEAE

Nicotiana plumbaginifolia is used as folk medicine in the treatment of liver dysfunction in Pakistan. Shah et al. (2016) investigated the hepatoprotective role of *N. plumbaginifolia* methanolic extract against CCl_4-induced oxidative damage in liver of chicks. Hepatoprotective activity was assessed by measuring the activities of the antioxidant enzymes: CAT, peroxidase, SOD, GPx, GR, and LPO (TBARS). Serum was analyzed for various biochemical parameters. The results revealed that CCl_4-induced oxidative stress as evidenced by the significant decrease in the activity levels of antioxidant enzymes, while an increase in the levels of TBARS in liver samples is compared with the control group. Serum levels LDH, TGs, TC, and LDL was elevated while reducing high-density lipoprotein compared to controls. Co-treatment of methanolic extract treatment reversed these alterations, which seems likely that the extract can protect the liver tissues against CCl_4-mediated oxidative damage.

Nigella sativa L. FAMILY: RANUNCULACEAE

Nigella sativa is an aromatic plant, traditionally used by the Middle East nations for asthma, cough, bronchitis, headache, rheumatism, fever, influenza, and eczema. Several biological activities, including antioxidant activity and resolution of hepatorenal toxicity have been reported for *N. sativa* seeds (Al-Ghamdi, 2001). Mollazadeh and Hosseinzadeh (2014) gave a detailed review on hepatoprotective studies of *N. sativa*. Aqueous suspension of seeds powder of *N. sativa* showed significant hepatoprotective activity against CCl_4 and ischemic-reperfusion-induced liver injury (Khan, 1999; Al-Suhaimi, 2012; Gani and John, 2013; Danladi et al., 2013).

Nehar and Kumari (2013) evaluated the ameliorating effect of *N. sativa* seed oil (5ml/kg BW, 10 mL/kg BW) on the liver damage caused by TAA (20 mg/kg BW) in albino rats for a period of eight weeks. A significant improvement in the altered levels of bilirubin, ALB, TP, ALT, ALP, and γ-glutamyl transferase was observed after treatment with 10 mL/kg BW of *N. sativa* oil. Antioxidant enzymes like CAT, SOD, GPx, TBARS and reduced GSH also showed significant improvement in their altered levels in *N. sativa* (10 mL/kg BW)-treated rats. The results confirm the ameliorating effect of *N. sativa* oil on liver injury caused by TAA and suggest the ability of *N. sativa* oil in scavenging the free radicals and protecting the liver cell against oxidative damage. The histopathological examination of liver section also confirm the ability of *N. sativa* oil in decreasing the severity of histopathological injury caused by TAA.

According to mechanism of PCM-induced hepatotoxicity Yesmin et al. (2013) described hepatoprotection of *N. sativa* with antioxidant and anti-inflammatory properties against PCM toxicity. *N. sativa* improved histopathological changes such as centrilobular necrosis, pyknosis of hepatocytes and neutrophils infiltration that were induced by oxidative stress in the liver in rats.

N. sativa contains more than 30 fixed oils. The volatile oil has been proved to contain thymoquinone and many monoterpenes such as *N*-cymene and α-pinene. The CCl_4 treatment increased the LPO and liver enzymes, and decreased the antioxidant enzyme levels. *N. sativa* treatment helped the elevated LPO and liver enzyme levels decrease and the reduced antioxidant enzyme levels increase (Kanter et al., 2005). The levels of liver enzymes and total oxidative status, oxidative stress index, and MPO in treated mice were significantly lower, and TAC in liver tissue was significantly higher compared to the controls (Yildiz et al., 2008). *N. sativa* is useful in the treatment of rheumatism and related inflammatory diseases and the anti-inflammatory effect was confirmed in rats (Alemi et al., 2013). Also, the aqueous extract of *N. sativa* has an anti-inflammatory effect demonstrated by its inhibitory effects on carrageenan-induced paw edema (Al-Ghamdi, 2001). Pretreatment of mice with 12.5 mg/kg thymoquinone (an *N. sativa* derived-compound) significantly reduced the elevated levels of serum enzymes as well as hepatic MDA content and significantly increased hepatic nonprotein sulfhydryl(-SH) (Paarakh, 2010). *N. sativa* contributes to inhibition of enzymes present in the neoglucogenesis pathway in the liver.

Bouasla et al. (2014) investigated the protective effects of *N. sativa* oil (NSO) supplementation against aluminium chloride ($AlCl_3$)-induced oxidative damage in liver and erythrocytes of rats. Simultaneously, a preliminary phytochemical study was affected in order to characterize the bioactive components containing in the NSO using chemical assays. The antioxidant capacities of NSO were evaluated by DPPH assay. The results showed that NSO was found to contain large amounts of total phenolics, flavonoids and tannins. Hepatoprotective investigation showed that $AlCl_3$ exhibited an increase in WBC counts and a marked decrease in erythrocyte counts and hemoglobin content. Plasma AST, ALT, ALP, and LDH activities and TB concentration were higher in $AlCl_3$ group than those of the control, while ALB and TP concentration were significantly lower. Compared to the control, a significant raise of hepatic and erythrocyte MDA level associated with a decrease in reduced GSH content, GPx, SOD, and CAT, activities of $AlCl_3$-treated rats. However, the administration of NSO alone or combined with $AlCl_3$ has improved the status of all parameters studied.

Thymoquinone, the active constituent of *N. sativa*, was tested by Daba and Abdel-Rahman (1998) in isolated rat hepatocytes as a hepatoprotective agent against *t*-BHP toxicity. TBHP (2 mM) was used to produce oxidative injury in isolated rat hepatocytes and caused progressive depletion of intracellular GSH, loss of cell viability as evidenced by trypan blue uptake and leakage of cytosolic enzymes, ALT, and AST. Preincubation of hepatocytes with 1 mM of either thymoquinone or silybin, which is a known hepatoprotective agent, resulted in the protection of isolated hepatocytes against TBHP-induced toxicity evidenced by decreased leakage of ALT and AST, and by decreased trypan blue uptake in comparison to TBHP treated hepatocytes. Both thymoquinone and silybin prevented TBHP-induced depletion of GSH to the same extent. Although thymoquinone protected the liver enzymes leakage, the degree of protection was less than that caused by silybin (Daba and Abdel-Rahman, 1998).

Suddek (2014) also reported hepatoprotective activity of thymoquinone against liver damage induced by TAM in female rats. Pretreatment with thymoquinone (50 mg·(kg body mass) (−1)·day (−1); orally, for 20 consecutive days, starting 10 days before TAM injection) significantly prevented the elevation in serum activity of the assessed enzymes. Thymoquinone significantly inhibited TAM-induced hepatic GSH depletion and LPO accumulation. Consistently, thymoquinone normalized the activity of SOD, inhibited the rise in TNF-α and ameliorated the histopathological changes.

Nili-Ahmadabadi et al. (2018) also explored the protective effects of thymoquinone on diazinon (DZN)-induced liver toxicity in the mouse model. Their findings showed that DZN caused a significant increase in ALT, AST, ALP serum levels, LPO and NO, the depletion of the TAC and TTM, and structural changes in the liver tissue. Following TQ administration, a significant improvement was observed in the oxidative stress biomarkers in the liver tissue. In addition, biochemical findings were correlated well to the histopathological examinations.

Hepatoprotective activity of ethanolic extract of *N. sativa* in PCM-induced acute hepatotoxicity was investigated in rats by Kushwah et al. (2014). Fasted male Wistar rats were orally treated with *N. sativa* extract in graded doses for 5 days followed by *N. sativa* extract and PCM 3 g/kg on 6 and 7th day. Circulatory liver markers and reduced GSH levels were estimated and histopathological study of liver performed. Paracetamol caused a significant increase in serum ALP, SGOT, SGPT, and total bilirubin and a significant decrease in GSH compared to control. *N. sativa* pretreatment significantly prevented the increase in liver enzymes and TB and decrease in GSH level as compared to PCM group. Liver histopathology showed marked reduction in sinusoidal dilatation, midzonal necrosis, portal triaditis and occasional apoptosis in *N. sativa* extract treated groups as compared to group receiving only PCM. *N. sativa* extract possesses hepatoprotective action against PCM-induced acute hepatoxicity.

Nilgirianthus ciliatus (Nees) Bremek. FAMILY: ACANTHACEAE

Hepatoprotective activity of bark extract of *Nilgirianthus ciliatus* was evaluated by Usharani et al. (2013) against PCM-induced toxicity in mice. Biochemical studies showed significant reduction in three levels of SGOT, SGPT, and ALP in the test groups when compared with the PCM control group and the treatment group also showed remarkable improvement in the total serum protein level. Histopathological examination of the liver tissue confirmed the hepatoprotective activity by the extract. The extract almost maintained the normal architecture of the liver when compared to the PCM control group which had hemorrhage and necrosis in extensive areas of the liver parenchyma and also the hepatocytes had vacuolated cytoplasm and a number of inflammatory cells and siderophages. The extract treated group also showed normal glomeruli.

Nymphaea lotus L. FAMILY: NYMPHAEACEAE

The protective effect of ME of *Nymphaea lotus* (NL) against CCl_4-induced chronic hepatotoxicity in rats was investigated by Oyeyemi et al. (2017). The activities of ALT, AST, and levels of TB in the serum, TBARS, SOD, CAT, GPx, and GSH in the liver, and histopathology of the liver were determined using standard procedures. NL significantly lowered the levels of ALT, AST, and TB and exhibited antioxidant potentials in rats exposed to CCl_4 relative to the control values. Specifically, NL at 100 and 200 mg/kg significantly increased CCl_4-induced decrease in hepatic GSH and GPx and also decreased the level of hepatic TBARS in CCl_4-intoxicated rats. Histopathological findings revealed cellular infiltration and fibrosis in rats that received CCl_4 only, which were ameliorated in rats that received NL+CCl_4.

Nymphaea stellata Willd. FAMILY: NYMPHAEACEAE

Nymphaea stellata is a medicinal plant mentioned in Ayurveda for the treatment of liver disorders. Bhandarkar and Khan (2004) investigated the hepatoprotective activity of extract of *N. stellata* flower against CCl_4-induced hepatic damage in albino rats. The oral administration of varying dosage of extract of *N. stellata* flower to rats for 10 days afforded the good hepatoprotection against CCl_4-induced elevation in serum marker enzymes, SB, liver LPO and reduction in liver GSH, liver GPx, glycogen, SOD, and CAT activity.

Nymphoides hydrophylla (Lour.) Kuntze FAMILY: MENYANTHACEAE

Bharathi et al. (2014) evaluated the *in vitro* antioxidant activity and *in vivo* hepatoprotective activity of *Nymphoides hydrophylla* in CCl_4-induced albino rats for acute liver injury. The hepatoprotective activity was assessed by estimating various biochemical parameters and histopathological studies. CCl_4 administration caused severe hepatic damage in rats as evidenced by elevated SGPT, SGOT, ALP, and TB levels. The ethanolic whole plant extract of *N. hydrophylla* significantly lowered the biological indicators and the results were compared with that of standard drug Silymarin. The histopathological study of liver was carried out and observed. They concluded that the ethanolic whole plant extract of *N. hydrophylla* has significant antioxidant activity and succeeded to restore the biochemical parameters and improved the histological alteration of the liver.

Ocimum americanum L. FAMILY: LAMIACEAE

The aqueous extract of *Ocimum americanum* (200 and 400 mg/kg) significantly attenuated PCM-induced biochemical (serum ALP, AST, ALT, and TBIL level) and histopathological alterations in the liver (Aluko et al., 2013). The hepatoprotective activity of *O. americanum* may be attributed to its antioxidant activities (Oboh, 2008).

Ocimum basilicum L. FAMILY: LAMIACEAE

Significant hepatoprotective effects were obtained by ethanolic extract of leaves of *Ocimum basilicum* against liver damage induced by H_2O_2 and CCl_4 as evidenced by decreased levels of antioxidant enzymes (enzymatic and nonenzymatic) (Meera et al., 2009). The extract also showed significant anti LPO effects *in vitro*, besides exhibiting significant activity in superoxide radical and NO radical scavenging, indicating their potent antioxidant effects.

Ocimum gratissimum L. FAMILY: LAMIACEAE

The effect of aqueous leaf extract of *Ocimum gratissimum* was investigated by Arhoghro et al. (2009) in rat models of liver injury induced by CCl_4. Treatment of separate groups of rats with 2.5 mL/kg BW of 5, 10 and 15% aqueous extracts of *O. gratissimum* for 3 weeks after establishment of CCl_4-induced liver damage, resulted in significantly less hepatotoxicity than with CCl_4 alone, as measured by ALP, ALT, and AST activities. For serum ALT activity decreased from 68.95 ± 21.38 U/L to 35.77 ± 1.48 U/L, while for AST, activity level decreased from 165.65 ± 17.75 to 110.10 ± 3.05 U/L and for ALP, activity level decreased from 364.65 ± 37.75 to 212.74 ± 15.27 U/L. The reduction though not statistically significant was dose dependent. Histopathological findings also suggest that treatment with aqueous extracts of *O. gratissimum* after establishment of CCl_4-induced liver damage significantly reduced and even reversed the liver damage in the rats.

Ujowundu et al. (2011) evaluated the effect of diesel petroleum intoxication in rats and the ability of phytochemicals and antioxidant content of *O. gratissimum* to ameliorate such toxicity. Diesel-induced hepatotoxicity was characterized by significant decrease in serum protein concentration and oxidative enzymes activities. Also increase in activities of liver function enzymes and cholesterol was observed. The group of rats which diet was supplemented with *O. gratissimum* showed significant improvement in the concentration of serum proteins and decrease in the activities of liver function enzymes. Similarly the activities of oxidative enzymes significantly increased, compared to the un-treated rats.

The ME of *O. gratissimum* was screened by Uhegbu et al. (2012) for hepatoprotective activity in albino rats intoxicated with CCl_4. It reduced significantly liver enzyme levels for animals treated with CCl_4 and the methanolic plant leaf extract concurrently compared to animals treated wwith CCl_4 only. Many histopathologcal changes in the liver such as marked dilation of the cental vein, blood vessel congestion and inflammatory leucophytic infiltrations which were observed in the CCl_4 treated animals were not observed in CCl_4 + plant extract treated animals. No apparent disruptions of the normal liver structure by histological and enzyme activities assessment were observed.

Adetutu and Olorunnisola (2013) investigated the hepatoprotective potential of the aqueous extracts of leaves of *O. gratissimum*. Extract of dose of 100 mg/kg BW was given to rats in five groups for seven consecutive days followed by a single dose of 2-AAF (0.5 mmol/kg BW). The rats were sacrificed after 24 hours and their bone marrow smears were prepared on glass slides stained with Giemsa. The mPCEs were thereafter recorded. The hepatoprotective effects of the plant extract against 2-AAF-induced liver toxicity in rats were evaluated by monitoring the levels of ALP, GGT, and histopathological analysis. The results of the 2-AAF-induced liver toxicity experiments showed that rats treated with the plant extract (100 mg/kg) showed a significant decrease in mPCEs as compared with the positive control. The rats treated with the plant extract did not show any significant change in the concentration of ALP and GGT in comparison with the negative control group whereas the 2-AAF group showed a significant increase in these parameters. Leaf extract also showed protective effect against histopathological alterations (Adetutu and Olorunnisola, 2013).

The *in vitro* antioxidant activities and hepatoprotective effects of methanol leaf extracts of *O. gratissimum* in rats intoxicated with DMN were investigated by Awogbindin et al. (2014). Results revealed a copious flavonoid content in the extracts. The extracts showed high DPPH radical scavenging activity and a significant inhibition of AAPH-induced LPO. The MEs also showed strong reductive potential. Pretreatments with the extracts, at 100 and 200 mg/kg, ameliorated the levels of ALT, AST and H_2O_2. In addition, the induction of GPx was also decreased by the two extracts at both doses. Moreover, the extracts significantly raised the level of nonenzymic antioxidant, GSH.

Ocimum tenuiflorum L. (Syn. *Ocimum sanctum* L.) FAMILY: LAMIACEAE

Leaves of sacred/holy basil, that is, green tulsi (*Ocimum tenuiflorum*) are used traditionally for their hepatoprotective effect. Lahon and Das (2011) evaluated the hepatoprotective activity of *O. tenuiflorum* and observed whether synergistic hepatoprotection exists with Silymarin. Albino rats (150–200 g) were divided into five groups. Groups A and B were normal and experimental controls, respectively. Groups C, D and E received the ALE of *O. tenuiflorum* leaves (OSE) 200 mg/kg BW/day, Silymarin 100 mg/kg BW/day and OSE 100 mg/kg BW/day + Silymarin 50 mg/kg BW/day p.o., respectively, for 10 days. Hepatotoxicity was induced in Groups B, C, D, and E on the eighth day with PCM 2 g/kg BW/day. The hepatoprotective effect was evaluated by performing an assay of the serum proteins, ALB globulin ratio, ALP, transaminases and liver histopathology. In groups

C, D and E, liver enzymes and ALB globulin ratio were significantly closer to normal than in group B. Reduction in sinusoidal congestion, cloudy swelling and fatty changes and regenerative areas of the liver were observed on histopathological examination in groups C, D and E, whereas group B showed only hepatic necrosis. They concluded that *O. tenuiflorum* alcoholic leaf extract shows significant hepatoprotective activity and synergism with Silymarin (Lahon and Das, 2011).

Chattopadhyay et al. (1992) evaluated the hepatoprotective activity of leaf extract of *O. tenuiflorum* on PCM-induced hepatic damage in rats. *O. tenuiflorum* was found to protect the rats from hepatotoxic action of PCM as evidenced by significant reduction in the elevated serum enzyme levels. Histopathological studies showed marked reduction in fatty degeneration in animals receiving *O. tenuiflorum* along with PCM as compared to the control group.

Oldenlandia corymbosa L. [Syn. *Hedyotis corymbosa* (L.) Lam.] FAMILY: RUBIACEAE

Oldenlandia corymbosa is used in traditional medicine of India and China to treat various hepatic disorders. Sadasivan et al. (2006) evaluated the hepatoprotective effect of the methanolic extract of the whole plant of *O. corymbosa* against PCM overdose-induced liver damage in Wistar rats. The methanolic extract of the plant produced significant hepatoprotective effects as evidenced by decreased serum enzyme activities, SGPT, SGOT, SAKP and SB and an almost normal histological architecture of the liver, in treated groups, compared to the controls. *H. corymbosa* shortened hexobarbitone-induced sleeping time in mice, besides showing significant antilipid peroxidant effect *in vitro*.

Chimkode et al. (2009) investigated the hepatoprotective activity of ethanolic extract of *O. corymbosa* which was separated into different fractions against CCl_4 intoxification. The results indicate that intoxification with CCl_4 increase the levels of SGOT and SGPT. The elevated levels of SGOT and SGPTwere significantly decreased by ether and butanol fractions and butanone and ethanol, whereas petroleum ether and ethyl acetate did not show any significant reduction in the level of SGOT and SGPT.

Rathi et al. (2009) evaluated the hepatoprotective effect of *O. corymbosa* on perchloroethylene-induced damage in rat liver. Perchloroethylene induction (1000 mg/kg BW) significantly increased the liver marker enzymes, LPO levels and significantly decreased the antioxidant status. Treatment of ethanolic extract of *O. corymbosa* at dose of (400 mg/kg BW) was administered orally once daily for 10 days decreased the liver marker enzyme levels (AST, ALT, and LDH) and LPO with increase in antioxidant enzyme levels.

Oldenlandia herbacea (L.) Roxb. FAMILY: RUBIACEAE

Pandian et al. (2013) evaluated the hepatoprotective effect of methanolic extract of the whole plant of *Oldenlandia herbacea* against D-galactosmine-induced rats. The extract showed significant reduction in the D-GalN-induced liver damage and symptoms of liver injury by restoration of the deviated levels of various biochemical parameters of liver which were observed in toxic control group. Histopathology of the liver sections confirmed that the extract prevented hepatic damage induced by D-GalN.

Oldenlandia umbellata L. FAMILY: RUBIACEAE

Gupta et al. (2007) evaluated the protective mechanisms of ME of *Oldenlandia umbellata* in CCl_4 intoxicated rats. Administration of the ME significantly prevented CCl_4-induced elevation level of serum SGPT, SGOT, ALP, and bilirubin. The TP level was decreased due to hepatic damage induced by CCl_4 and it was found to be increased in ME of *O. umbellata* treated group. Treatment of rats with CCl_4 led to a marked increase in LPO as measured by MDA. This was associated with a significant reduction of the hepatic antioxidant system, for example, GSH, and CAT. A comparative histopathological study of liver exhibited almost normal architecture, as compared to CCl_4-treated control group.

De et al. (2017) evaluated the *in vitro* hepatoprotective activity of *O. umbellata* on CCl_4-induced liver injury. Significant hepatoprotective effect was obtained against CCl_4-induced liver damage as judged from serum marker enzyme activities (SGOT, SGPT, ALT, and TB) and a normal architecture of liver compared to toxic control.

Olea oleaster Hoffmans & Link FAMILY: OLEACEAE AND *Juniperus procera* FAMILY: CUPRESSACEAE

The effect of *Olea oleaster* and *Juniperus procera* leaves extracts and their combination on TAA-induced hepatic cirrhosis were investigated by Al-Attar et al. (2016) in male albino mice. Administration of TAA for six and twelve weeks resulted in a decline in BW gain and increased the levels of ALP, AST, ALT and TB. Treatment of mice with these extracts showed a pronounced attenuation in TAA-induced hepatic cirrhosis associated with physiological and histopathological alterations.

Operculina turpethum (L.) Silva Manso FAMILY: CONVOLVULACEAE

The ethanolic extract obtained from roots of *Operculina turpethum* were evaluated for hepatoprotective activity in rats by inducing liver damage by PCM. The ethanol extract at an oral dose of 200 mg/kg exhibited a significant protective effect by lowering serum levels of SGOT, SGPT, ALP, and TB. These biochemical observations were supplemented by histopathological examination of liver sections. Silymarin was used as positive control.

Opuntia ficus-indica (L.) Mill. FAMILY: CACTACEAE

The protective effects of the juice of *Opuntia ficus-indica* fruit (prickly pear) against CCl_4-induced hepatotoxicity were examined in rats by Galati et al. (2005). The animals were treated orally with the juice (3 mL/rat) 2 h after administration of the hepatotoxic agent. Preventive effects were studied by giving the juice (3 mL/rat) for nine consecutive days. On day 9 the rats received the hepatotoxic agent. Morphological and biochemical evaluations were carried out for 24, 48, and 72 h after induction of the hepatic damage. Data show that *O. ficus-indica* fruit juice administration exerts protective and curative effects against the CCl_4-induced degenerative process in rat liver. Histology evaluation revealed a normal hepatic parenchyma at 48 h; the injury was fully restored after 72 h. Moreover, a significant reduction in CCl_4-induced increase of SGOT and SGPT plasma levels is evident; these data are in agreement with the functional improvement of hepatocytes. *O. ficus-indica* fruit juice contains many phenol compounds, ascorbic acid, betalains, betacyanins, and a flavonoid fraction, which consists mainly of rutin and isorhamnetin derivatives. Hepatoprotection may be related to the flavonoid fraction of the juice, but other compounds, such as vitamin C and betalains could, synergistically, counteract many degenerative processes by means of their antioxidant activity.

Ben Saad et al. (2017) determined the phytochemical composition of *O. ficus-indica* cactus cladode extract (CCE). It also investigates antioxidant activity and hepatoprotective potential of CCE against lithium carbonate (Li_2CO_3)-induced liver injury in rats. Treatment with Li_2CO_3 caused a significant change of some hematological parameters including RBCs, WBCs, hemoglobin content (Hb), hematocrit (Ht), and mean corpuscular volume (VCM) compared to the control group. Moreover, significant increases in the levels of glucose, cholesterol,

TGs and of AST, ALT, ALP, and LDH activities were observed in the blood of Li_2CO_3-treated rats. Furthermore, exposure to Li_2CO_3 significantly increased the LPO level and decreased SOD, CAT, and GPx activities in the hepatic tissues.

Opuntia microdasys **Lehm. FAMILY: CACTACEAE**

Chahdoura et al. (2017) The hepatoprotective activity of *Opuntia microdasys* aqueous flowers extract at post flowering stage (OFP) has been tested. The OFP extract showed inhibitory activity against α-glucosidase (IC_{50} = 0.17 ± 0.012 mg/mL) and α-amylase. The inhibitory potential of OFP extract on these enzymes suggests a positive and probable role of this extract in the management and treatment of diabetes mellitus, particularly, for type 2. Oral administration of the OFP at 200 mg/kg to diabetic male rats for 28 days demonstrated a significant protective effect by lowering the levels of glucose and hepatic marker enzymes (AST, ALT, LDH, γ-GT, BT, PAL, TC, LDL-C, HDL-C, and TG). OFP attenuated oxidative stress by decreasing the SOD, CAT, GPx activity and the levels of PC and MDA in the liver and restored the histological architecture of the rat liver. OFP has protective effects on the protection of liver, thereby reducing some of the causes of diabetes in experimental animals.

Opuntia monacantha **(Willd.) Haw. FAMILY: CACTACEAE**

The chloroform and MEs of *Opuntia monacantha* were studied by Saleem et al. (2015) for its hepatoprotective effect against PCM-induced liver damage in rabbits. The extracts at 200, 400, and 600 mg/kg BW in one week protocol showed significant hepatoprotective activity by reducing the magnitude of liver markers including ALT, AST, ALP, and TB levels. The results were supported by histopathological studies of liver tissue. Chemical analysis of *O. monacantha* indicated the presence of alkaloids, tannins, saponins, flavonoids, and polysaccharides and its hepatoprotective potential may be due to the presence of flavonoids. Its is concluded that 600 mg/kg is the potent dose of both extracts of *O. monacantha* as hepatoprotective plant.

Opuntia robusta **J.C. Wendl. *AND Opuntia streptacantha* Lem. FAMILY: CACTACEAE**

Opuntia spp. fruits are an important source of nutrients and contain high levels of bioactive compounds, including antioxidants. González-Ponce et al. (2016) evaluated the hepatoprotective effect of *Opuntia robusta* and *O. streptacantha* extracts against APAP-induced acute liver failure. Fruit extracts (800 mg/kg/day, orally) were given prophylactically to male Wistar rats before intoxication with APAP (500 mg/kg, intraperitoneally). Rat hepatocyte cultures were exposed to 20 mmol/L APAP, and necrosis was assessed by LDH leakage. *O. robusta* had significantly higher levels of antioxidants than *O. streptacantha*. Both extracts significantly attenuated APAP-induced injury markers AST, ALT, and ALP and improved liver histology. The *Opuntia* extracts reversed APAP-induced depletion of liver GSH and glycogen stores. In cultured hepatocytes, *Opuntia* extracts significantly reduced leakage of LDH and cell necrosis, both prophylactically and therapeutically. Both extracts appeared to be superior to *N*-acetyl-L-cystein when used therapeutically.

Origanum vulgare L. FAMILY: LAMIACEAE

The effect of an aqueous extract of *Origanum vulgare* (OV) leaves extract on CCl_4-induced hepatotoxicity was investigated by Sikander et al. (2013) in normal and hepatotoxic rats. OV administration led to significant protection against CCl_4-induced hepatotoxicity in dose-dependent manner, maximum activity was found in CCl_4+OV3 (150 mg/kg BW) groups and changes in the hepatocytes were confirmed through histopathological analysis of liver tissues. It was also associated with significantly lower serum ALT, ALP, and AST levels, higher GST, CAT, SOD, GPx, GR, and GSH level in liver tissue. The level of LPO also decreases significantly after the administration of OV leaves extract. The biochemical observations were supplemented with histopathological examination of rat liver sections.

Ornithogalum saundersiae Baker FAMILY: LILIACEAE

Ying-Wan et al. (2010) examined the protective effects of total saponins from *Ornithogalum saundersiae* on D-GalN and LPS-induced fulminant hepatic failure. D-GalN/LPS increased the serum aminotransferase levels and LPO, while decreased the reduced GSH level. The pretreatment with saponins from *O. saundersiae* attenuated these changes in a dose-dependent manner. Elevation of TNF-α level and activation of caspase-3, HIF-1α were observed in the D-GalN/LPS group, which was attenuated by OC. The survival rate of the OC groups was significantly higher than that of the D-GalN/LPS group.

Orthosiphon stamineus Benth. FAMILY: LAMIACEAE

The hepatoprotective activity of the ME of *Orthosiphon stamineus* was assessed in PCM-induced hepatotoxicity rat model. Change in the levels of biochemical markers such as AST, ALT, ALP, and LPOs were assayed in both PCM treated and control (untreated) groups. Treatment with the methanolic extract of *O. stamineus* leaves (200 mg/kg) has accelerated the return of the altered levels of biochemical markers to the near normal profile in the dose-dependent manner (Maheswari et al., 2008).

Chin et al. (2009) assessed the sub-chronic (14-day treatment) effect of oral administration of *O. stamineus* on hepatic drug metabolising enzymes and the protective effect against APAP-induced liver injury in SD rats. Isolated hepatocytes and liver cytosolic fraction were used to examine the effect of *O. stamineus* on the activity of aminopyrine *N*-demethylase and glutathione-*S*-transferase, respectively. Hepatoprotective effect of *O. stamineus* was judged by comparing the serum levels of hepatic markers, that is, AST, ALT, and ALP and the percentage of hepatocytes viability between *O. stamineus* treatment groups and APAP control group. *O. stamineus* extract at 125 and 500 mg/kg exhibited significant hepatoprotective effect by restoring the elevation of serum levels of AST and ALT and increased the percentage of hepatocytes viability in rats. The hepatoprotective effect exhibited by *O. stamineus* at dose 500 mg/kg was comparable to Silymarin at dose 20 mg/kg in APAP-induced liver injury rats. The activity of GST was 17% higher in the rats treated with 500 mg/kg of *O. stamineus* compared with the control group. There was no significant difference in BW gained, food consumption, water intake and relative organ weight between the treatment group and the control group. Methanol extract of *O. stamineus* protects against APAP-induced liver injury in rats by enhancing the activity of GST in liver (Chin et al., 2009).

O. stamineus as medicinal plant is commonly used in Malaysia for the treatment of hepatitis and jaundice. Alshawsh et al. (2011) evaluated the hepatoprotective effects in a TAA-induced hepatotoxic model in SD rats. The hepatotoxic animals showed a coarse granulation on the liver surface when compared to the smooth aspect observed on the liver surface of the other groups. Histopathological study confirmed the result; moreover, there was a significant increase in serum liver biochemical parameters (ALT, AST, ALP, and Bilirubin) and the level of liver MDA, accompanied by a significant decrease in the level of TP and ALB in the hepatotoxic control group when compared with that of the normal group. The high-dose treatment group (200 mg/kg) significantly restored the elevated liver function enzymes near to normal. This study revealed that 200 mg/kg extracts of *O. stamineus* exerted a hepatoprotective effect.

Osbeckia aspera Blume FAMILY: MELASTOMATACEAE

Identification of the active components of plants with hepatoprotective properties requires screening of large numbers of samples during fractionation and purification. A screening assay has been developed by Thabrew et al. (2011) based on protection of human liver-derived HepG2 cells against toxic damage. Various hepatotoxins were incubated with HepG2 cells in 96-well microtitre plates (30,000 cells $well^{-1}$) for 1 h and viability was determined by metabolism of the tetrazolium dye 3-(4,5-dimethylthiazol-2-yl)-5-(3-carboxymethoxy phenyl)-2-(4-sulphophenyl) -2*H*-tetrazolium (MTS). Bromobenzene (10 mm) and 2,6-dimethyl-*N*-acetyl-*p*-quinoneimine (2,6-diMeNAPQI, 200 mm) had greater toxic effects than *tert*-butyl hydroperoxide (1.8 mm) or galactosamine (10 mm), reducing mean viability to 44.6 ± 1.2% (s.e.m.) and 561 ± 21% of control, respectively. Protection against toxic damage by these agents was tested using a crude extract of a known hepatoprotective Sri Lankan plant, *Osbeckia aspera*, and two pure established hepatoprotective plant compounds, (+)-catechin and Silymarin (1 mg/mL) Viability was significantly improved by *Osbeckia* (by 37.7 ± 2.4%, and 36.5 ± 21%, for BB and 2,6-diMe-NAPQI toxicity, respectively). Comparable values for (+)-catechin were 68.6 ± 2.9% and 63.5 ±11%, and for Silymarin 24.9 ± 1.4% and 25.0 ± 1.6% (Thabrew et al., 2011).

Osbeckia octandra DC. FAMILY: MELASTOMATACEAE

Aqueous extracts of the leaves of *Osbeckia octandra* have been compared with (+)-3-cyanidanol with regard to their abilities to alleviate CCl_4-induced liver dysfunction in albino rats by comparing the abilities of these drugs to protect the liver against CCl_4-mediated alterations in the liver histopathology and SGOT, SGPT, and ALP (Jayathilaka et al., 1989). Although the protection offered by (+)-3-cyanidanol and *O. octandra* appears to be comparable in post-treatment, *Osbeckia* was significantly more effective in pretreatment.

Thabrew et al. (1995) investigated the effects of *O. octandra* leaf extract on PCM-induced liver injury, both *in vivo* in mice and in rat hepatocytes *in vitro*. Oral administration of *Osbeckia* extract (330 mg/kg) at the same time as PCM (450 mg/kg) to mice, resulted in a significant protection against liver damage, as assessed by improvements in the blood Normotest, total liver GSH, plasma AST level, and liver histopathology at 24 h after PCM administration. In experiments to assess the direct effects of *Osbeckia* extract, significant protection was also found in freshly isolated rat hepatocytes against damage induced by 185 µM 2,6-dimethyl *N*-acetyl *N*-quinoneimine (2,6-diMeNAPQI, an analogue of NAPQI, the toxic metabolite of PCM) *in vitro*. When *Osbeckia* extract (500 µg/mL) was added to the incubation medium at the same time as 2,6-diMeNAPQI significant changes in cell viability, cell reduced GSH level, and reduced release of LDH were demonstrated after 1 h incubation as compared with 2,6-diMeNAPQI alone. Significant protection was still obtained against 2,6-diMeNAPQI *in vitro* when addition of *Osbeckia* extract was delayed by 20 min. These results indicate that *Osbeckia* extract can protect against PCM-induced liver injury.

A comparison of the beneficial roles of *Pavetta indica* and *Osbeckia octandra* against CCl_4-induced liver damage has been studied in albino rats by Thabrew et al. (1987) assessing their ability to protect the liver against CCl_4-mediated alterations in the liver histopathology and the serum levels of AST, ALT, and ALP. Treatment with either *Pavetta* or *Osbeckia* leaf extract (before or after CCl_4 administration) markedly decreased the CCl_4-mediated alterations in the liver histopathology as well as the serum enzyme levels. However, on comparison of the two plant extracts, *Osbeckia* appears to be a better hepatotonic than *Pavetta*. Thus, in livers of rats pretreated with *Osbeckia* for 7 days, CCl_4 had hardly any effect on the serum enzymes or the liver cell architecture, while even after pretreatment with *Pavetta* for the same length of time, CCl_4 was still able to produce a 32%, 16%, and 26% increase in the activities of AST, ALT, and ALP, respectively. Post-treatment with *Osbeckia* also resulted in a faster recovery of the livers in comparison with those from animals post-treated with *Pavetta*.

Otostegia persica Boiss. FAMILY: LAMIACEAE

The hepatoprotective effect of the ME of aerial parts (shoot) from *Otostegia persica* was investigated by Bezenjani et al. (2012) against the CCl_4-induced acute hepatotoxicity in male rats. The ME of *O. persica* shoot is active at 300 mg/kg (per os) and it possesses remarkable antioxidant and hepatoprotective activities. Additionally, histopathological studies verified the effectiveness of this dose of extract in acute liver damage prevention.

Toori et al. (2013) evaluated the hepatoprotective properties of *O. persica* ethanol extract on CCl_4-induced liver damage in rats. The administration of CCl_4 increased AST, ALT, ALP, TB, and MDA in serum but it decreased TP, and ALB compared with normal control. Treatment with *O. persica* extract at three doses resulted in decreased enzyme markers, bilirubin levels, and LPO marker (MDA) and increased TP and ALB compared with CCl_4 group. The results of pathological study also support the hepatoprotective effects which were observed at doses of 80 and 120 mg/kg.

Oxalis corniculata L. FAMILY: OXALIDACEAE

Khan et al. (2012) characterized the chemical composition of *Oxalis corniculata* methanol extract (OCME) and its various fractions, and to determine the antioxidant potential by different *in vitro* assays. Methanol extract was also evaluated for its antioxidant capacity against hepatotoxicity induced with CCl_4 in rat. The results showed the presence of flavonoids, alkaloids, terpenoids, saponins, cardiac glycosides, phlobatannins and steroids while tannins were absent. Total amount of phenolic and flavonoids was affected by the solvents. Free radicals were scavenged by the extract/fraction in a dose response curve in all models. Biochemical parameters of serum; AST, ALT, ALP, LDH, gamma-glutamyl transpeptidase (γ-GTT), TB, cholesterol, and TGs were significantly increased while TP and ALB were decreased by CCl_4. Treatment of CCl_4 significantly decreased the liver contents of reduced GSH and activities of antioxidant enzymes; CAT (CAT), SOD, GSH-Px, GST, GSR, and QR whereas elevated the thiobarbituric acid reactive substances (TBARS) contents, and hepatic lesions. All the parameters were brought back to control levels by the supplement of ME.

Oxalis stricta L. FAMILY: OXALIDACEAE

Patel et al. (2016) determined the hepatoprotective effects of ethanolic extract of *Oxalis stricta* (EEOS) toward PCM intoxicated hepatic damage in albino rats. In PCM alone treated animals showed LPO was increased with decrease in SOD, CAT, reduced GSH levels which represents the hepatic antioxidant status. It was further confirmed by histopathological observations. It was concluded that the 70% ethanolic extract of *O. stricta* possesses hepatoprotective activity.

Panax ginseng C.A. Mey FAMILY: ARALIACEAE

Hafez et al. (2017) investigated the effect of ginseng extract on CCl_4-induced liver fibrosis in male Wistar rats. Treatment with ginseng extract decreased hepatic fat deposition and lowered hepatic reticular fiber accumulation compared with the CCl_4 group. The CCl_4 group showed a significant increase in hepatotoxicity biomarkers and upregulation of the expression of genes encoding TGF-β, TβR-I, TβR-II, MMP2, MMP9, Smad-2,-3,-4, and IL-8 compared with the control group. However, CCl_4 administration resulted in the significant downregulation of IL-10 mRNA expression compared with the control group. Interestingly, ginseng extract supplementation completely reversed the biochemical markers of hepatotoxicity and the gene expression alterations induced by CCl_4. Ginseng extract had an antifibrosis effect via the regulation of the TGF-β1/Smad signaling pathway in the CCl_4-induced liver fibrosis model. The major target was the inhibition of the expression of TGF-β1, Smad2, and Smad3.

Pandanus odoratissimus L.f. FAMILY: PANDANACEAE

Mishra et al. (2015) investigated the hepatoprotective activity of ethanolic extract of the root of *Pandanus odoratissimus* against PCM-induced hepatotoxicity in rats. Experimental findings revealed that the extract at dose level of 200 and 400 mg/kg of BW showed dose-dependent hepatoprotective effect against PCM-induced hepatotoxicity by significantly restoring the levels of serum enzymes to normal that was comparable to that of Silymarin, but the extract at dose level of 400 mg/kg was found to be more potent when compared to that of 200 mg/kg. Besides, the results obtained from histopathological study also support the study.

Parathelypteris nipponica (Franch. et Sav.) Ching. FAMILY: THELYPTERIDACEAE

Fu et al. (2010) evaluated the antioxidant, free-radical scavenging, hepatoprotective, and anti-inflammatory potential of *Parathelypteris nipponica*. The methanolic extract of *P. nipponica* (TMPN) exhibited strong antioxidant activity and strong free-radical scavenging activity. Hepatoprotective activity of TMPN was determined by the CCl_4-induced oxidative tissue injury in rat liver, the extract showed significant hepatoprotective activity that was evident by enzymatic examination and histopathological study. In assessing anti-inflammatory activity the carrageenan-induced rat paw oedema test was used, the extract reduced carrageenan-induced rat paw oedema in dose-dependent manner, achieving high degree of anti-inflammatory activity.

Parkia clappertoniana Keay FAMILY: FABACEAE

Patrick-Iwuanyanwu et al. (2010) evaluated the protective effects of *Parkia clappertoniana* against CCl_4-induced hepatotoxicity in rats. Serum ALT, AST, ALP and, LDH levels 24 h after CCl_4 administration decreased significantly in rats pretreated with *P. clappertoniana* than in CCl_4-treated rats only. Total SB also showed a remarkable decrease in rats pretreated with *P. clappertoniana* when compared to those administered CCl_4 alone. Lipid peroxidation expressed by MDA concentration was significantly decreased in rats pretreated with *P. clappertoniana* than in rats administered CCl_4 alone. Histopathological examinations of rats administered CCl_4 alone revealed severe hepatic damage to the liver. However, rats pretreated with *P. clappertoniana* showed significant improvements in the architecture of rat liver.

Parkinsonia aculeata L. FAMILY: FABACEAE

Hepatoprotective activity of 50% ethanolic leaf extracts of *Parkinsonia aculeata* was evaluated by Hassan et al. (2008) against CCl_4-induced hepatic damage in rats. Significantly elevated levels of AST, ALT, ALP, TB, and peroxide value in CCl_4-intoxicated rats were restored to normal levels in the animals treated with the extract at doses of 100, 200, and 300 mg kg^{-1} and CCl_4. The levels of TPs, ALB, vitamins C and E appreciated significantly in animals treated with different doses of the leaf extracts and CCl_4. The effects were dose-dependent.

Shah and Deval (2011) explored the *in vivo* hepatoprotective action of *P. aculeata* leaves extract induced by CCl_4 and PM. Extract was administered orally at a daily dose of 200 mg per kg and 300 mg per kg, for 7 days (*in vivo*). The levels of cytosolic enzymes, serum enzymes and endogenous antioxidant enzymes, used as a marker of oxidative damage to hepatocytes, was reversed to the same level as in normal group in dose-dependent manner. No obvious signs of toxicity were observed 300 mg per kg treatment dose.

Passiflora species FAMILY: PASSIFLORACEAE

The Passion Fruit (*Passiflora* sp.) that grows in the Indonesian region generally has three varieties, namely purple passion fruit (*Passiflora edulis* Sims.), red passion fruit (*Passiflora ligularis* Juss.), and yellow passion fruit (*Passiflora verrucifera* Lindl.).

The passion fruit peel is an economic waste that has not been utilised optimally, but has many efficacious phytochemical contents. Nerdy et al. (2019) examined hepatoprotective activity (with PCM-induced hepatotoxic) from three varieties of the passion fruit (purple passion fruit peel extract, red passion fruit peel extract and yellow passion fruit peel extract) in the albino rat. The hepatoprotective activity for positive control was similar to the 250 mg of purple passion fruit peel extract per kg of BW, 250 mg of red passion fruit peel extract per kg of BW, and 500 mg of yellow passion fruit peel extract per kg of BW.

Peganum harmala L. FAMILY: NITRARIACEAE

Ahmed et al. (2013) evaluated the protective effect of the purified protein from seeds of *Peganum harmala* against CCl_4-induced toxicity in male albino rats. Results of the dose-dependent experiment with purified protein prior to CCl_4 administration were higher at 4 mg/kg BW. The antioxidant activity of the purified protein was determined *in vitro* by DPPH radical scavenging test. Pretreatment with 4 mg/kg BW of the purified protein significantly altered the deteriorating damage induced by CCl_4 toxicity to a near normal range, which was similar to treatment with vitamin C. These results suggest that the purified protein possesses a protective effect against CCl_4-induced toxicity and probably acts as an antioxidative defense through free-radical scavenging activity.

Bourogaa et al. (2015) investigated the protective effects of *P. harmala* seeds extract (CPH) against chronic ethanol treatment. Ethanol treatment caused also a drastic alteration in antioxidant defence system; hepatic SOD, CAT, and GPx activities. A co-administration of CPH during ethanol treatment inhibited LPO and improved antioxidants activities. However, treatment with *P. harmala* extract protects efficiently the hepatic function of alcoholic rats by the considerable decrease of aminotransferase contents in serum of ethanol-treated rats.

Peltophorum pterocarpum (DC.) Backer FAMILY: FABACEAE

The 70% ethanolic extract of *Peltophorum pterocarpum* leaves were investigated for its hepatoprotective effect against PCM-induced acute liver damage on albino Wister rats (Biswas et al., 2010). Paracetamol (2 g/kg, po) significantly elevated the serum levels of biochemical markers like SGPT, SGOT, ALP, bilirubin (total and direct), TC, TGs, and depleted tissue GSH and increased the LPO upon administration of PCM (2 mg/kg, p.o.) to albino rats. This indicated that 70% ethanolic extract of leaves of *P. pterocarpum* at 100 and 200 mg/kg doses significantly reduced the elevated levels of biochemical markers mentioned above. Test extract treatment also increased the level of tissue GSH and significantly decreased tissue lipid preoxidation. The effect of 70% ethanolic extract (ELPP) was comparable with that of the standard Silymarin 100 mg/kg.

Hexane and ethanol leaves extracts of *P. pterocarpum* at doses of 75, 150, and 250 mg/kg BW and pure compound bergenin at doses of 25, 75, and 100 mg/kg BW were studied by Tasleem et al. (2014) for hepatoprotective activity in both sexes of albino rats. Liver injury was induced by CCl_4 using 1.5 mL/kg BW once during experimental period. The protective and therapeutic effects of test samples were compared with the standard drug containing betaine glucuronate + diethanolamine glucuronate + nicotinamide ascorbate. Different biochemical parameters such as ALT, ALP, γ-GT, DB (bilirubin direct), and TB (total bilirubin) were investigated and assessed to match the histopathological examination results. Among various biochemical tests, bergenin demonstrated maximum reduction in AST, ALT, γ-GT, DB, and TB serum levels at a dose of 100 mg/kg BW in prophylactic mode in comparison to toxicant group. The statistical significance was observed against ALT and TB serum levels when compared to toxicant group that were also noted in line with the histopathological examination of rat liver section where bile duct and central vein observed intact, however mild inflammation with absences of fibrosis and negative degeneration of lobular hepatocytes were recorded. Findings of the present study demonstrated bergenin as a potent hepatoprotective against CCl_4-induced hepatic toxicity.

Penthorum chinense Pursh FAMILY: PENTHORACEAE

Penthorum chinense, a well-known Miao ethnomedicine, has been traditionally used to treat several liver-related diseases, such as jaundice and viral hepatitis. Wang et al. (2017) evaluated the probable properties of the aqueous extract of *P. chinense* on CCl_4-induced acute liver injury in mice. C57BL/6 mice were orally administered an aqueous extract of *P. chinense* (5.15 and 10.3 g/kg BW) or Silymarin (100 mg/kg) once daily for 1 week prior to CCl_4 exposure. Silymarin was used as a positive drug to validate the effectiveness of PCP. A single dose of CCl_4 exposure caused severe acute liver injury in mice, as evidenced by the elevated serum levels of ALT, AST, and ALP, and the increased TUNEL-positive cells in liver, which were remarkably ameliorated by the pretreatment of PCP. Extract was also found to decrease the levels of MDA, restore the GSH and enhance the activities of SOD and CAT in the liver. In addition, the pretreatment of extract inhibited the degradation of hepatic cytochrome P450 2E1 (CYP2E1), upregulated the expression of nuclear factor erythroid 2-related factor 2 (Nrf2) and its target proteins in CCl_4-treated mice. Results indicated that the pretreatment of PCP (10.3 g/kg BW) effectively protected against CCl_4-induced acute liver injury, which was comparable to efficacy of Silymarin (100 mg/kg). This hepatoprotective effects might be attributed to amelioration of CCl_4-induced oxidative stress via activating Nrf2 signaling pathway (Wang et al., 2017).

Pergularia daemia (Forssk.) Chiov. *FAMILY: APOCYNACEAE*

The roots of *Pergularia daemia* are used by the tribes of Western Ghats, Tamil Nadu, for the treatment of various liver disorders. Kumar and Mishra (2006, 2008) evaluated the hepatoprotective effect of crude ethanolic and aqueous extracts and ethanolic extraction from the aerial parts of *P. daemia* in rats by inducing liver damage by CCl_4. The ethanolic extract at an oral dose of 200 mg/kg exhibited a significant protective effect by lowering serum levels of SGOT, SGPT, ALP, TB and TC and increasing the levels of TP and ALB levels as compared to Silymarin used as a positive control. These biochemical observations were supplemented by histopathological examination of liver sections. The activity may be a result of the presence of flavonoid compounds. Furthermore, the acute toxicity of the extracts showed no signs of toxicity up to a dose level of 2000 mg/kg. Thus it could be concluded that ethanolic extract of *P. daemia* possesses significant hepatoprotective properties. It was suggested that the presence of flavonoids in ethanol extract and its ethanol fraction may be responsible for hepatoprotective properties.

Bhaskar and Balakrishnan (2010) evaluated the hepatoprotective effect of aqueous and ethanol extracts of *P. daemia* roots by PCM and CCl_4-induced liver damage in Wistar albino rats. The degree of protection was measured by physical changes (liver weight), biochemical (SGPT, SGOT, ALP, DBL, TB, cholesterol, and decrease in protein), antioxidant enzymes (LPO and GSH levels), and histological changes. Pretreatment with the extracts significantly prevented the physical, biochemical, antioxidant enzyme levels and histological changes induced by PCM and CCl_4 in the liver. The effects of extracts were comparable to that of the standard drug Silymarin. The ethanol extract was found to exhibit greater hepatoprotective activity than the aqueous extract. These results indicate that *P. daemia* could be useful in preventing chemically induced acute liver injury.

The EAF of the ethanol extract from roots of *Carissa carandas* (CCF) and *P. daemia* were studied against CCl_4-, PCM-, and ethanol-induced hepatotoxicity in rats by Balakrishnan et al. (2011). Significant hepatoprotective effects were obtained against liver damage induced by all the three toxins, as evident from changed biochemical parameters like serum transaminases (SGOT and SGPT), ALP, TB, TP, and TC. Parallel to these changes, the EAF prevented toxin-induced oxidative stress by significantly maintaining the levels of reduced GSH and MDA, and a normal architecture of the liver, compared to toxin controls.

Periploca hydaspidis Falc. FAMILY: APOCYNACEAE

The antioxidant capacity of *Periploca hydaspidis* was assessed through various *in vitro* assays and by the hepatoprotective potential on CCl_4-induced toxicity in rat by Ali et al. (2018). Phytochemical analysis of ME indicated the existence of rutin, GA, and caffeic acid. Total phenolic (TPC) and TFC exhibited significant correlation with DPPH, NO, hydroxyl ion, inhibition of β-carotene oxidation, iron chelation, reducing power and TAC. In hepatic sample of rat, CCl_4 administration increased the level of nitrite, hydrogen peroxide (H_2O_2), TBARS, whereas a decline was recorded in antioxidant enzymes; SOD, POD, CAT and in reduced GSH. Concentration of ALT, ALP, AST, and globulin increased whereas level of TP and ALB decreased in serum of CCl_4-treated rats. Level of pro-inflammatory cytokines; TNF-α, tumor growth factor-β1 (TGF-β1) and resistin were increased in serum whereby anti-inflammatory markers; interleukin-10 (IL-10), adiponectin, and nuclear factor erythroid 2-related factor 2 (Nrf-2) decreased in hepatic tissues of CCl_4-treated rats. DNA damages and histopathological alterations were induced with administration of CCl_4 to rat. The altered levels of various parameters provoked by CCl_4 toxicity restored toward the control level by the ME of *P. hydaspidis* in a dose-dependent manner.

Periploca laevigata Aiton FAMILY: APOCYNACEAE

Tlili et al. (2018) evaluated the hepatoprotective activity of *Periploca laevigata* seeds extract. Pretreatment with *P. laevigata* extract ameliorated levels of serum parameters for liver together with oxidative stress indicators and antioxidant enzymes activities.

Persea americana Mill. FAMILY: LAURACEAE

The protective effects of aqueous extract of *Persea americana* against CCl_4-induced hepatotoxicity in male albino rats were investigated by Brai et al. (2014). Liver damage was induced in rats by administering a 1:1 (v/v) mixture of CCl_4 and olive oil [3 mL/kg, subcutaneously (sc)] after pretreatment for 7 days with AEPA. Hepatoprotective effects of AEPA were evaluated by estimating the activities of ALT, AST, ALP, and levels of TBL. The effects of AEPA on biomarkers of oxidative damage (LPO) and antioxidant enzymes namely, CAT, SOD, GPx, and GST were measured in liver post-mitochondrial fraction. AEPA and Reducdyn® showed significant hepatoprotective activity by decreasing the activities of ALT, AST, ALP, and reducing the levels of TBL. The activities of antioxidant enzymes, levels of MDA and protein carbonyls were also decreased dose dependently in the AEPA-treated rats. Pretreatment with AEPA also decreased the serum levels of GSH significantly. These data revealed that AEPA possesses significant hepatoprotective effects against CCl_4-induced toxicity attributable to its constituent phytochemicals. The mechanism of hepatoprotection seems to be through modulation of antioxidant enzyme system.

Petroselinum crispum Mill. FAMILY: APIACEAE

Ethanolic extract of *Petroselenium crispum* leaves has been pharmacologically investigated for its hepatoprotective activity (Al-Howiriny et al., 2003). The extract dose dependently attenuated CCl_4 induced increase in serum AST, ALT, ALP, and TB. The ethanolic extract of *P. crispum* leaves also showed significant anti-inflammatory (Al-Howiriny et al., 2003) and antioxidant activities which may contribute to its hepatoprotective action. Although perfectly safe in pharmacological doses, *P. crispum* may be toxic in excess, especially when used as essential oil.

Peumus boldus Molina FAMILY: MONIMIACEAE

Dried hydro-ALE of *Peumus boldus* has been evaluated by Lanhers et al. (1991) for hepatoprotective, choleretic and anti-inflammatory effects in mice and rats, in order to validate or to invalidate traditional therapeutic indications. This extract exerted a significant hepatoprotection of *t*-BHP-induced hepatotoxicity in isolated rat hepatocytes (*in vitro* technique) by reducing the LPO and the enzymatic leakage of LDH; this *in vitro* efficacy was reinforced by a significant hepatoprotection on CCl_4-induced hepatotoxicity in mice (*in vivo* technique), the plant extract reducing the enzymatic leakage of ALT. Boldine, the main alkaloid of *P. boldus* appears to be implicated in this hepatoprotective activity.

Phaseolus trilobus **Ait. FAMILY: FABACEAE**

Phaseolus trilobus is extensively used by tribal people of Nandurbar district (Maharashtra, India) in the treatment of jaundice and other liver disorders. Fursule and Patil (2010) assessed the medicinal claim of *P. trilobus* as hepatoprotective and antioxidant. Methanol and aqueous extracts of *P. trilobus* reduced elevated level of ALT, AST, ALP, LDH, bilirubin, and HYP significantly in bile duct ligated Wistar rats, proving hepatoprotective activity comparable with Silymarin. Both the extracts were found to reduce the elevated levels of serum TBARS and elevate superoxide scavenging radical activity proving antioxidant activity comparable with ascorbic acid. The reduced level of GSH was found to be elevated in liver proving antioxidant activity comparable with Silymarin.

Phoenix dactylifera **L. FAMILY: ARECACEAE**

The ameliorative activity of aqueous extracts of the flesh and pits of dates (*Phoenix dactylifera*) on CCl_4-induced hepatotoxicity was studied in rats by Al-Quarawi et al. (2004). Treatment with aqueous extract of date flesh or pits significantly reduced CCl_4-induced elevation in plasma enzyme and bilirubin concentration and ameliorated morphological and histological liver damage in rats. It is possible that β-sitosterol, a constituent of *P. dactylifera*, is at least partly responsible for the protective activity against CCl_4 hepatotoxicity. The data suggest that the daily oral consumption of an aqueous extract of the flesh and pits of dates, and as a part of the daily diet ad libitum, was prophylactic to CCl_4 poisoning, achieving about 80% protection with date palm flesh and 70% with pits. A similar percentage of protection was achieved when the aqueous extracts of the flesh and pits were used as a cure against CCl_4 poisoning after toxicity was induced.

The ameliorative activity of aqueous extract of the flesh of dates (*P. dactylifera*) and ascorbic acid on TAA-induced hepatotoxicity was studied in rats by Ahmed et al. (2008). Liver damage was assessed by estimation of plasma concentration of bilirubin and enzymes activities of AST, ALT, LDH, γ glutamyl transferase, and ALP and serum AFP and serum total testosterone. Treatment with aqueous extract of date flesh or by ascorbic acid significantly reduced TAA-induced elevation in plasma bilirubin concentration and enzymes. This study suggests that TAA-induced liver damage in rats can be ameliorated by administration of extract of date flesh and ascorbic acid.

Saafi et al. (2011) investigated the role of date palm fruit extract (*P. dactylifera*) in protection against oxidative damage and hepatotoxicity induced by subchronic exposure to dimethoate (20 mg/kg/day). Pretreatment with date palm fruit extract restored the liver damage induced by dimethoate, as revealed by inhibition of hepatic LPO, amelioration of SOD, GPx, and CAT activities and improvement of histopathology changes.

Okwuosa et al. (2014) evaluated the hepatoprotective activity of methanolic fruit extracts of *P. dactylifera* (date palm) against TAA-induced liver damage in male albino Wistar rats. The fruit of *P. dactylifera* had an oral LD_{50} >6000 mg/kg in rats. Preliminary phytochemical screening revealed the presence of carbohydrates, proteins, saponins, steroids, glycosides, flavonoids, tannins, and terpenoids. There was a significant rise in the level of biochemical makers of liver damage like ALT, AST, ALP, and TB and a fall in ALB in TA-treated groups when compared with the respective values for extract and Silymarin treated groups. Also, when compared with the normal control group for ALT, AST, ALP, bilirubin, and ALB, respectively, group B given low dose of extract had respective values that were significantly different from it. Histopathological findings in the test groups showed mild alteration in histoarchitecture when compared with TA group which showed extensive vacuolation, inflammatory cells and generalized necrosis.

Sangi et al. (2014) evaluated the effect of fruit of *P. dactylifera* in rats affected by hepatotoxicity and cirrhosis as a result of administration of CCl_4. The level of ALT, AST, and ALP was determined, regarding its relation to liver damage. Liver tissues were investigated to compare between healthy and infected ones. It was found that *P. dactylifera* significantly revised the changes produced by CCl_4 on hepatic cells and enzymes.

Antihyperlipidemic and hepatoprotective potential of native date variety Aseel has been assessed by Ahmed et al. (2016) in hyperlipidemia-induced albino rats. Fasting blood sugar, cholesterol, TGs, LDL, and VLDL along with cholesterol-HDL and LDL-HDL ratio were significantly decreased at 300 mg/kg without any increase in liver enzymes as observed in positive control group. Animals received 600 mg/kg also revealed significant decline in fasting blood sugar, TG, VLDL, and ALP.

Salem et al. (2018) sought to determine effects of antioxidant potential of aqueous and methanolic extracts of *P. dactylifera* leaves (PLAE and PLME) against the widely-used analgesic PCM (PCM)-induced hepatotoxicity. PCM significantly elevated serum liver markers, AST, ALT, ALP, GGT, and bilirubin compared to control (untreated) group. These PCM-induced effects were associated with oxidative stress as demonstrated by increased levels of MDA and reduced levels of hepatic antioxidant enzymes, GPx, CAT, and SOD. Pretreatment of PLME decreased ALT and AST by 78.2% and tissue MDA by 54.1%, and increased hepatic GPx (3.5-fold), CAT (7-fold) and SOD (2.5-fold) compared to PCM group. These PLME-mediated effects were comparable to NAC pretreatment. Histological analysis demonstrated that PLME conserved hepatic tissues against lesions such as inflammation, centrilobular necrosis, and hemorrhages induced by PCM. In contrast, PLAE-mediated effects were less effective in reducing levels of liver function enzymes, oxidative stress, and liver histopathological profiles, and restoring antioxidant defenses against PCM-induced intoxication.

Phyllanthus species FAMILY: PHYLLANTHACEAE

Phyllanthus species are traditionally well-known for their medicinal properties including hepatoprotective activity. Srirama et al. (2012) assessed the hepatoprotective and antioxidant activities of 11 *Phyllanthus* species, *P. amarus* Schumach., *P. urinaria* L., *P. debilis* Klein ex Willd, *P. tenellus* Roxb., *P. virgatus* G. Forst., *P. maderaspatensis* L., *P. reticulatus* Poir., *P. polyphyllus* Willd., *P. emblica* L., *P. indofischerii* Bennet. and *P. acidus* (L.) Skeels. The dried leaves and stems of each plant species were extracted in methanol and successively in water. The extracts were screened for hepatoprotective activity at a concentration of 50 µg/mL against *t*-BHP-induced toxicity in HepG2 cells. Seven extracts from five species that showed hepatoprotective activity were assessed for their 50% effective concentration (EC_{50}) values and their antioxidant activity using a DPPH assay. Phyllanthin and hypophyllanthin contents were also determined in these *Phyllanthus* species. The MEs of *P. polyphyllus*, *P. emblica* and *P. indofischeri* showed high levels of hepatoprotective activity with EC_{50} values of 12, 19, and 28 µg/mL and IC_{50} of 3.77, 3.38, and 5.8 µg/mL for DPPH scavenging activity respectively against an IC_{50} of 3.69 µg/mL for ascorbic acid. None of these activities could be attributed to phyllanthin and hypophyllanthin. The study also confirms the hepatoprotective and antioxidant activities of leaves of *P. emblica* and *P. polyphyllus*.

Phyllanthus species IN TAIWAN

Lee et al. (2006) examined the effect of oral administration of *Phyllanthus* methanolic extracts (PME) (i.e., *P. acidus, P. emblica, P. myrtifolius, P. multiflorus, P. amarus, P. debilis, P. embergeri, P. hookeri, P. tenellus, P. urinaria* L.s. *nudicarpus, P. urinaria* L.s. *urinaria*) or GA on the progression of acute liver damage induced by CCl_4 in rats by morphological and biochemical methods. *P. acidus, P. urinaria* L.s. *urinaria*, GA at a dose of 0.5 g/kg, and *P. emblica, P. urinaria* L.s. *nudicarpus* at a dose of 1.0 g/kg attenuated CCl_4-induced increase in serum glutamate-oxalate-transaminase (GOT). *P. acidus, P. urinaria* L.s. *nudicarpus, P. urinaria* L.s. *urinaria*, GA at a dose of 0.5 g/kg, and *P. emblica, P. amarus, P. hookeri, P. tenellus* at a dose of 1.0 g/kg attenuated CCl_4-induced increase in serum GPT. Concurrently, *P. acidus, P. multiflorus, P. embergeri, P. hookeri, P. tenellus*, and *P. urinaria* L.s. *urinaria* elevated the activity of liver reduced GSH-Px. Since the protective effects of *P. acidus, P. emblica, P. myrtifolius, P. embergeri, P. urinaria* L.s. *nudicarpus, P. urinaria* L.s. *urinaria*, and GA correlate with a reduction in liver infiltration and focal necrosis observed using histological methods, these data demonstrate that *P. acidus* and *P. urinaria* L.s. *urinaria* are hepatoprotective and antioxidant agents.

Phyllanthus acidus (L.) Skeels FAMILY: PHYLLANTHACEAE

Jain and Singhai (2011) investigated and compare the hepatoprotective effects of crude ethanolic and aqueous extracts of *Phyllanthus acidus* leaves on APAP and TAA-induced liver toxicity in Wistar rats. APAP or TAA administration caused severe hepatic damage in rats as evident from significant rise in serum AST, ALT, ALP, TB and concurrent depletion in total serum protein. The *P. acidus* extracts and Silymarin prevented the toxic effects of APAP or TAA on the above serum parameters indicating the hepatoprotective action. The aqueous extract was found to be more potent than the corresponding ethanolic extract against both toxicants. The phenolic and flavonoid content (175.02 ± 4.35 and 74.68 ± 1.28, respectively) and DPPH [$IC_{50} = (33.2 \pm 0.31)$ µg/mL] scavenging potential was found maximum with aqueous extract as compared to ethanolic extract.

Phyllanthus amarus Schum. & Thonn. FAMILY: PHYLLANTHACEAE

Pramyothin et al. (2007) investigated the protective effect and possible mechanism of aqueous extract from *Phyllanthus amarus* (PA) on ethanol-induced rat hepatic injury. In the *in vitro* study, PA (1–4 mg/mL) increased% MTT reduction assay and decreased the release of transaminases (AST and ALT) in rat primary cultured hepatocytes being treated with ethanol. Hepatotoxic parameters studied *in vivo* included serum transaminases (AST and ALT), serum triglyceride (STG), HTG, TNF-alpha, interleukin 1 beta (IL-1beta), together with histopathological examination. In acute toxicity study, single dose of PA (25, 50, and 75 mg/kg, p.o.) or SL (Silymarin, a reference hepatoprotective agent, 5 mg/kg), 24h before ethanol (5 g/kg, p.o.) lowered the ethanol-induced levels of transaminases (AST and/or ALT). The 75 mg/kg PA dose gave the best result similar to SL. Treatment of rats with PA (75 mg/(kg day), p.o.) or SL (5 g/(kg day), p.o.) for 7 days after 21 days with ethanol (4 g/(kg day), p.o.) enhanced liver cell recovery by bringing the levels of AST, ALT, HTG, and TNF-alpha back to normal. Histopathological observations confirmed the beneficial roles of PA and SL against ethanol-induced liver injury in rats. The possible mechanism may involve their antioxidant activity (Pramyothin et al., 2007).

Naaz et al. (2007) also evaluated the hepatoprotective effect of ethanolic extract of *P. amarus* on aflatoxin B(1)-induced liver damage in mice using different biochemical parameters and histopathological studies. Aflatoxin was administered orally (66.6 μg/kg BW 0.2 mL/day) to the mice of each group except control to which normal saline and ascorbic acid (0.1 g/kg BW 0.2 mL/day were given, respectively. Ethanolic extract of *P. amarus* (0.3 g/kg BW 0.2 mL/day was given to all groups except control groups (group I and group V) after 30 min of aflatoxin administration. The entire study was carried out for 3 months and animals were sacrificed after an interval of 30 days till the completion of study. *P. amarus* extract was found to show hepatoprotective effect by lowering down the content of TBARS and enhancing the reduced GSH level and the activities of antioxidant enzymes, GPx, GST, SOD, and CAT. Histopathological analyses of liver samples also confirmed the hepatoprotective value and antioxidant activity of the ethanolic extract of the herb, which was comparable to the standard antioxidant, ascorbic acid. The overall data indicated that *P. amarus* possesses a potent protective effect against aflatoxin B(1)-induced hepatic damage, and the main mechanism involved in the protection could be associated with its strong capability to reduce the intracellular level of ROS by enhancing the level of both enzymatic and nonenzymatic antioxidants (Naaz et al., 2007).

The hepatoprotective effect of methanolic extract of the leaf of *P. amarus* against ethanol-induced oxidative damage was investigated by Faremi et al. (2008) in adult male Wistar albino rats. Ethanol treatment markedly decreased the level of reduced GSH, SOD, and CAT in the liver, which was significantly enhanced by *P. amarus* treatment. GST, which was increased after chronic ethanol administration, was significantly reduced by *P. amarus* treatment in the liver. Also, *P. amarus* significantly increased the activities of hepatic ALT and AST as well as ALP, with a concomitant marked reduction in the plasma activity of the transaminases in the ethanol-challenged rats. Lipid peroxidation level, which was increased after chronic ethanol administration, was significantly reduced in the liver by *P. amarus* co-treatment.

The phytochemical screening and the effects of the methanolic extracts of the leaves of *P. amarus* on some biochemical parameters of male Guinea pigs were investigated by Obianime and Uche (2008). Phytochemical investigations revealed the presence of flavonoids, tannins, alkaloids, terpenoids, steroids, saponins, and cardiac glycosides. The ME of *P. amarus* leaves (50–800 mg/kg) caused a significant decrease in the levels of TC, AST, ALT, urea, uric acid, TP, prostatic, alkaline, and acid phosphatases. The highest reduction effect was obtained with uric acid at 400 mg/kg of *P. amarus* extract while the least effect was observed in TC. These effects were dose- and time-dependent. This shows that the leaves of *P. amarus* have hepatoprotective, nephroprotective, and cardio protective properties.

Enogieru et al. (2015) confirmed the hepatoprotective role of *P. amarus* experimently on Wistar rats. Liver protection was significantly shown by leaves extract of *P. amarus* and Silymarin against acetaminophen-induced toxicity. There was significant increase in ALT, AST, ALP, and reduction in TP in rats which were treated only by acetaminphen. It was further confirmed that level of ALT, AST, and ALP decreased when acetaminphen toxicated rats were treated with *P. amarus* leaves extract and Silymarin.

Phyllanthus emblica **L. (Syn.: *Emblica officinalis* Gaertn.) FAMILY: EUPHORBIACEAE**

Phyllanthus emblica is a constituent of many hepatoprotective formulations available in market. 50% ALE of *P. emblica* and quercetin isolated from it were studied by Gulati et al. (1995) for hepatoprotective effect against country made liquor and PCM challenge in albino rats and mice, respectively. The extract at the dose of 100 mg/100 g, p.o. and quercetin at the dose of 15 mg/100 g, p.o. produced significant hepatoprotection.

Phyllanthus emblica (amla) has been used in Ayurveda as a potent rasayan for the treatment of hepatic disorders. Sultana et al. (2004) reported that fruits of *P. emblica* inhibit TAA-induced oxidative stress and hyper-proliferation in rat liver. The administration of a single necrotic dose of TAA (6.6 mM/kg) resulted in a significant increase SGOT, SGPT, and GGT levels compared with saline-treated control values. Thioacetamide caused hepatic GSH depletion and a concomitant increase in MDA content. It also resulted in an increase in the activity of GST, GR, G6PD, and a decrease in GPx activity. Hepatic ornithine decarboxylase activity and thymidine incorporation in DNA were increased by TAA administration. Prophylactic treatment with *P. emblica* for seven consecutive days before TAA administration inhibited SGOT, SGPT and GGT release in serum compared with treated control values. It also modulated the hepatic GSH content and MDA formation. The plant extract caused a marked reduction in levels of GSH content and simultaneous inhibition of MDA formation. *P. emblica* also caused a reduction in the activity of GST, GR, and G6PD. GPx activity was increased after treatment with the plant extract at doses of 100 and 200 mg/kg. Prophylactic treatment with the plant caused a significant downregulation of ornithine decarboxylase activity and profound inhibition in the rate of DNA synthesis. In conclusion, the acute effects of TAA in rat liver can be prevented by pretreatment with *P. emblica* extract.

Tasduq et al. (2005) investigated the hepatoprotective property of a 50% hydroALE of the fruits of *P. emblica* against antituberculosis drugs-induced hepatic injury. The biochemical manifestations of hepatotoxicity induced by rifampicin (RIF), isoniazid (INH), and PZA, either given alone or in combination were evaluated. In vitro studies were done on suspension cultures of rat hepatocytes while subacute studies were carried out in rats. The hepatoprotective activity of fruit extract was found to be due to its membrane stabilizing, antioxidative and CYP 2E1 inhibitory effects. Mir et al. (2007) evaluated the hepatoprotective effect of ALE of fruits of *P. emblica* on CCl_4 and TAA-induced hepatotoxicity in rats. The extract reversed enzymatic alterations induced by CCl_4 and TAA.

The efficacy of *P. emblica* to prevent PCM-induced hepatotoxicity in rats was examined by the histopathological study of liver cells by Malar and Bai (2009). The total blood cell count in each group of animal was also calculated. Treatment with aqueous extract of fruits of *P. emblica* showed the appearance of normal hepatocytes, offset of necrosis and consequent appearance of leucocytes thus suggesting the hepatoprotective effect of this medicinal plant.

Chaphalkar et al. (2017) investigated the protective effect of the hydroalcoholic extract of *P. emblica* bark (PEE) in ethanol-induced hepatotoxicity model in rats. Total phenolic, flavonoid, and tannin content and *in vitro* antioxidant activities were determined by using H_2O_2 scavenging and ABTS decolorization assays. PEE was rich in total phenols (99.523 ± 1.91 mg GAE/g), total flavonoids (389.33 ± 1.25 mg quercetin hydrate/g), and total tannins (310 ± 0.21 mg catechin/g), which clearly support its strong antioxidant potential. HPTLC-based quantitative analysis revealed the presence of the potent antioxidants GA (25.05 mg/g) and ellagic acid (13.31 mg/g). Moreover, one-month PEE treatment (500 and 1000 mg/kg, p.o.) followed by 30-day 70% ethanol (10 mL/kg) administration showed hepatoprotection as evidenced by significant restoration of ALT, ALP, and TP and further confirmed by liver histopathology. PEE-mediated hepatoprotection could be due to its free-radical scavenging and antioxidant activity that may be ascribed to its antioxidant components, namely, ellagic acid, and GA.

Hepatoprotective activity of *P. emblica* (PE) and Chyavanaprash (CHY) extracts were studied using CCl_4-induced liver injury model in rats by Jose and Kuttan (2000). PE and CHY extracts were found to inhibit the hepatotoxicity produced by acute and chronic CCl_4 administration as seen from the decreased levels of serum and liver LPO, SGPT, ALP. Chronic CCl_4 administration was also found to produce liver fibrosis as seen from the increased levels of collagen-HYP and pathological analysis. PE and CHY extracts were found to reduce these elevated levels significantly, indicating that the extract could inhibit the induction of fibrosis in rats.

The effect of *P. emblica* fruit extract (PFE) against alcohol-induced hepatic damage in rats was investigated by Reddy et al. (2010). In vitro studies showed that PFE possesses antioxidant as well NO scavenging activity. In vivo administration of alcohol (5 g/kg BW/day) for 60 days resulted increased liver LPO, protein carbonyls, nitrite plus nitrate levels. Alcohol administration also significantly lowered the activities of SOD, CAT, GPx, GST, and reduced GSH as compared with control rats. Administration of PFE (250 mg/kg BW) to alcoholic rats significantly brought the plasma enzymes toward near normal level and also significantly reduced the levels of LPO, protein carbonyls, and restored the enzymic and nonenzymatic antioxidants level. This observation was supplemented by histopathological examination in liver.

Phyllanthus fraternus G.L. Webster FAMILY: PHYLLANTHACEAE

The effect of CCl_4 administration on liver mitochondrial function and the protective effect of an aqueous extract of *Phyllanthus fraternus* were studied in rats by Padma and Setty (1999). Administration of rats with an aqueous extract of *P. fraternus* prior to CCl_4 administration showed significant protection on the CCl_4-induced mitochondrial dysfunction on all the parameters studied.

Hepatoprotective and antioxidant properties of aqueous extract of *P. fraternus* (AEPF 200, 300, and 400 mg/kg BW, orally) were investigated by Lata et al. (2014) against cyclophosphamide (CPA 200 mg/kg, BW, intraperitoneally administered)-induced liver damage in mice. Histopathological studies of CPA administration cause liver injury, featuring substantial increase in SGOT, SGPT, LDH, ALP, acid phosphatase, and TB. Moreover, CPA intoxication also causes strong oxidative stress, which is evident from significant increase in LPO level. These changes were coupled with a decline in SOD and CAT as well as ALB, cholesterol level, RBC, and WBCs count. AEPF-treated mice displayed a significant inhibition of LPO and augmentation of endogenous antioxidants.

Phyllanthus maderaspatensis L. FAMILY: PHYLLANTHACEAE

Phyllanthus maderaspatensis has been used in folk medicine of many countries as a remedy against several pathological conditions including jaundice and hepatitis. The hexane extract of (200 and 100 mg/kg) of *P. maderaspatensis* showed significant hepatoprotection on CCl_4 and TAA-induced liver damage in rats (Asha et al., 2007). The protective effect was evident from serum biochemical parameters and histopathological analysis. Rats treated with *P. maderaspatensis* remarkably prevented the elevation of serum AST, ALT, and LDH and liver LPOs in CCl_4 and TAA-treated rats. Hepatic GSH levels significantly increased the treatment with the extracts. Histopathological changes induced by CCl_4 and TAA were also significantly reduced by the extract treatment. The activity of hexane extracts of *P. maderaspatensis* was comparable to that of Silymarin, the reference hepatoprotective drug.

Ilyas et al. (2015) evaluated hepatoprotective activity of *P. maderaspatensis* against galactosamine-induced toxicity and also investigation of polyphenols in each extract. The hydroalcoholic extract was found to contain comparatively high amount of kaempferol, quercetin, catechin, rutin, and ellagic acid which are responsible for hepatoprotection. Antioxidant parameters such as GSH, CAT, and SOD activity in liver tissues were restored toward the normalization more significantly by the hydroalcoholic extract when compared with other extracts. The biochemical observations were supplemented with histopathological examination.

Phyllanthus muellarianus (Kuntze) Exell. FAMILY: PHYLLANTHACEAE

Leaves of *Phyllanthus muellarianus* are widely used in the management of liver disorders in Nigeria. Hepatoprotective effect of *P. muellarianus* aqueous leaf extract was investigated by Ajiboye et al. (2017) in APAP-induced liver injury mice. Oral administration of *P. muellarianus* aqueous leaf extract significantly attenuated APAP-mediated alterations in serum ALP, ALT, AST, ALB and TB. Similarly, APAP-mediated decrease in activities of SOD, CAT, GPx, GR and G6PD were significantly attenuated in the liver of mice. Increased levels of CD, lipid HP, MDA, protein carbonyl, fragmented DNA, TNF-α, interleukin-6 and-8 were significantly lowered by *P. muellarianus* aqueous leaf extract.

Phyllanthus niruri L. FAMILY: PHYLLANTHACEAE

Antioxidant activity and hepatoprotective potential of *Phyllanthus niruri*, a widely used medicinal plant, were investigated by Harish and Shivanandappa (2006). Methanolic and aqueous extract of leaves and fruits of *P. niruri* showed inhibition of membrane LPO, scavenging of DPPH radical and inhibition of ROS *in vitro*. Antioxidant activity of the extracts was also demonstrable *in vivo* by the inhibition of the CCl_4-induced formation of LPOs in the liver of rats by pretreatment with the extracts. CCl_4-induced hepatotoxicity in rats, as judged by the raised serum enzymes, SGOT, and SGPT, was prevented by pretreatment with the extracts, demonstrating the hepatoprotective action of *P. niruri* (Harish and Shivanandappa, 2006).

The hepatoprotective action of the protein fraction of *P. niruri* was investigated against APAP (Bhattacharjee and Sil, 2006) and CCl_4 hepatotoxicity (Bhattacharjee and Sil, 2007). The partially purified protein fraction of *P. niruri* was injected intraperitoneally in mice either prior to (preventive) or after the induction of toxicity (curative). Levels of different liver marker enzymes in serum and different antioxidant enzymes, as well as LPO in total liver homogenates were measured in normal, control (toxicity induced) and *P. niruri* protein fraction-treated mice. *P. niruri* significantly reduced the elevated SGPT and ALP levels in the sera of toxicity-induced mice, compared with the control group. Lipid peroxidation levels were also reduced in mice treated with *P. niruri* protein fraction compared with the APAP and CCl_4 treated control groups. Among the antioxidant enzymes SOD, CAT, GST levels were restored to almost normal levels compared with the control group. *P. niruri* treatment also enhanced reduced hepatic GSH levels caused by APAP (Bhattacharjee and Sil, 2006) and CCl_4 administration (Bhattacharjee and Sil, 2007). The results demonstrated that the protein fraction of *P. niruri* protected liver tissues against oxidative stress in mice, probably acting by increasing antioxidative defense.

Chatterjee and Sil (2006) evaluated its effect on the prooxidant–antioxidant system of liver and the hepatoprotective potential of aqueous extract of the herb *P. niruri* (PN) on Nimesulide (NIM)-induced oxidative stress *in vivo* using a murine model, by determining the activities of hepatic antioxidant enzymes SOD and CAT, levels of reduced GSH and LPO (expressed as malonaldialdehyde, MDA). NIM administration (8 mg/kg BW) for 7 days caused significant depletion of the levels of SOD, CAT, and reduced GSH, along with the increased levels of LPO. Intraperitoneal administration of the extract at a dose of 50 mg/kg BW for 7 days, prior to NIM treatment, significantly restored most of the NIM-induced changes and the effect was comparable to that obtained by administering 100 mg/kg BW of the extract orally. Thus, results suggested that intraperitoneal administration of the extract could protect liver from NIM-induced hepatic damage more effectively than oral administration. Antioxidant property of the aqueous extract of PN was also compared with that of a known potent antioxidant, vitamin E. The PN extract at a dose of 100 mg/kg BW along with NIM was more effective in suppressing the oxidative damage than the PN extract at a dose of 50 mg/kg BW.

Iqbal et al. (2007) evaluated hepatoprotective activity of *P. niruri* (Bhui amla) in PCM administered rats. Propylene glycol appeared nontoxic to the liver while significant degrees of centrilobuler hepatotoxicity was observed in group PCM-treated rats. The ethanol extract + PCM treated group suggested significant improvements in the serum parameters but these parameters appeared better alleviated in the hexane + PCM group. Hepatic reduced GSH concentrations were replenished to the control level in both groups. Hepatic histology supported biochemical and other observations. Lesser degrees of alleviations were observed in the dichloromethane + PCM and butaene + PCM groups. However, the hexane extract-pretreated group appeared to provide the most significant hepatoprotection against PCM-induced hepatotoxicity in the rat. Titration of the dose following isolation of the active ingredient might offer complete alleviation.

The antioxidative and hepatoprotective potential of *P. niruri* were investigated by Sabir and Rocha (2008b). The aqueous extracts of leaves showed inhibition against TBARS, induced by different pro-oxidants (10 μM $FeSO_4$ and 5 μM sodium nitroprusside) in rat liver, brain and kidney homogenates. The extracts also lowered the formation of TBARS in phospholipids extracted from egg yolk. The plant exhibited strong antioxidant activity in the DPPH free radical and iron chelation assay. The hepatoprotective activity of the extracts were also demonstrated *in vivo* against PCM-induced liver damage, as evidenced by the decrease in SGOT, and SGPT, and increased CAT activity in the liver in treatment groups, compared to the control.

Parmar et al. (2009, 2010) evaluated the hepatoprotective activity of leaves of *P. niruri* against CCl_4 and PCM-induced toxicity in male Wistar rats. Toxin-induced hepatotoxicity resulted in an increase in serum AST, ALT, ALP activity and bilirubin level accompanied by significant decrease in ALB level. Co-administration of the aqueous extract protects the toxin-induced lipid peroxidatiom, restored latered serum marker enzymes and antioxidant level toward near normal.

To determine if the extract from PN plays a protective role against liver cirrhosis induced by TAA in rats Amin et al. (2012) carried out an investigation. Significant differences were observed between the TAA group and the other groups regarding body and liver weights, liver biochemical parameters, TAC, LPO, and oxidative stress enzyme levels. Gross visualization indicated coarse granules on the surface of the hepatotoxic rats' livers, in contrast to the smoother surface in the livers of the Silymarin and PN-treated rats. Histopathological analysis revealed necrosis, lymphocytes infiltration in the centrilobular region, and fibrous connective tissue proliferation in the livers of the hepatotoxic rats. But, the livers of the treated rats had comparatively minimal inflammation and normal lobular architecture. Silymarin and PN treatments effectively restored these measurements closer to their normal levels. Progression of liver cirrhosis induced by TAA in rats can be intervened using the PN extract and these effects are comparable to those of Silymarin.

The hepatoprotective effect of the aqueous extract of *P. niruri* was evaluated by Makoshi et al. (2013)

in an APAP-induced hepatotoxicity study using 24 male rabbits of the New Zealand White breed. The most significant healing or hepatoprotective effect of the extract of *P. niruri* was seen in the animals administered the extract at 25 mg/kg which showed no significant change in the liver, both grossly and histologically. In most groups, the liver enzyme assay and serum ALBs and globulins levels increased slightly, except the group administered 25 mg/kg extract of *P. niruri.*

Phyllanthus polyphyllus Willd. FAMILY: PHYLLANTHACEAE

Methanolic extract of *Phyllanthus polyphyllus* was evaluated by Rajkapoor et al. (2008) for hepatoprotective and antioxidant activities in rats. The plant extract (200 and 300 mg/kg, p.o.) showed a remarkable hepatoprotective and antioxidant activity against APAP-induced hepatotoxicity as judged from the serum marker enzymes and antioxidant levels in liver tissues. Acetaminophen induced a significant rise in AST, ALT, ALP, TB, GGTP, lipid peroxidase (LPO) with a reduction of TP, SOD, CAT, GPx, and GST. Treatment of rats with different doses of plant extract (200 and 300 mg/kg) significantly altered serum marker enzymes and antioxidant levels to near normal against APAP-treated rats. The activity of the extract at dose of 300 mg/kg was comparable to the standard drug, Silymarin (50 mg/kg, p.o.). Histopathological changes of liver sample were compared with respective control. Results indicate the hepatoprotective and antioxidant properties of *P. polyphyllus* against APAP-induced hepatotoxicity in rats.

Phyllanthus reticulatus Poir. FAMILY: PHYLLANTHACEAE

Two partially purified organic fractions designated by PR1 and PR2 of the fat free ethanol (95%) extract of aerial parts of *Phyllanthus reticulatus* were tested by Das et al. (2008) for the hepatoprotective activity in rats against CCl_4-induced liver damage. The rats receiving the fractions showed promising hepatoprotective activity as evident from significant changes of pentobarbital-induced sleeping time, changes in serum levels of SGPT, SGOT, ALP, and bilirubin and also from histopathological changes as compared to CCl_4-intoxicated rats.

Physalis peruviana L. FAMILY: SOLANACEAE

Physalis peruviana is a medicinal herb used by Muthuvan tribes and Tamilian native who reside in the shola forest regions of Kerala, India against jaundice. Arun and Asha

(2006) evaluated for its antihepatotoxic, phytochemical analysis and the acute toxicity of the most promising extract in rats. Water, ethanol and hexane extracts of *P. peruviana* (500 mg/kg BW) showed antihepatotoxic activities against CCl_4-induced hepatotoxicity. The ethanol and hexane extracts showed moderate activity compared to water extract, which showed activity at a low dose of 125 mg/kg. The results were judged from the serum marker enzymes. Histopathological changes induced by CCl_4 were also significantly reduced by the extract. Further, the extract administration to rats resulted in an increase in hepatic GSH and decrease in MDA. The extract was found to be devoid of any conspicuous acute toxicity in rats.

El-Gengaihi et al. (2012) investigated the potential of *P. peruviana* as hepato-renal protective agent against fibrosis. Two compounds (15-desacetylphysabubenolide and Betuline) were isolated and their structures were elucidated. The biological evaluation was conducted on different rats groups; control, control treated with the methanolic husk extract, CCl_4, CCl_4 treated with husk extract and CCl_4 treated with Silymarin. The evaluation was done through measuring oxidative stress markers; MDA, SOD and NO. Liver function indices; AST, ALT, ALP, GGT, bilitubin and total hepatic protein were estimated. Kidney disorder biomarkers; creatinine, urea, and serum protein were evaluated. The results revealed improvement of all the investigated parameters. Liver and kidney histopathological analysis confirmed our results.

Taj et al. (2014) investigated the antihepatotoxic effect of *P. peruviana* whole ripe fruit, water and ethanol extracts of fruit in normal as well as in CCl_4 intoxicated rats. Animal treated/fed with various preparations of *P. peruviana* showed significant lowering effect in the elevated levels of serum markers like ALAT, ASAT, ALP, LDH, creatinine, urea, and bilirubin indicating the protection against hepatic cell damage. The water extract of *P. peruviana* showed highest activity in both rat models while ripe fruit and ethanol extract showed moderate activity compared to standard drug.

The role of *P. peruviana* as functional food against hepato-renal fibrosis induced by CCl_4 was evaluated. The chemical composition of leaves referred the presence of with anolides and flavonoids. Two compounds, ursolic acid and lupeol, were isolated. The results revealed plant safety and decrease in NO, MDA, IgG, ALP, tissue protein, bilirubin, creatinine and urea levels. Increase in SOD, AST, ALT, GGT, and serum protein levels were observed. Improvement in liver and kidney histopathological architectures were also seen. *P. peruviana* recorded a significant protective role in liver and kidney against fibrosis.

Picralima nitida (Stapf) T. Durand & H. Durand FAMILY: APOCYNACEAE

MacDonal et al. (2016) investigated the *in vivo* CCl_4-mediated hepatotoxicity of methanolic seed extract of *Picralima nitida* using Wistar rats. Results show that treatment with *P. nitida* extract had no adverse effect on the BW of Wistar rats. Biochemical analysis showed an increase in CAT and GSH which are good antioxidant agents. Photomicrographs showed moderate amelioration from steatosis caused by CCl_4 in the treatment groups.

Picrorhiza kurroa Royle ex Benth. FAMILY: PLANTAGINACEAE

Anandan and Devaki (1999) evaluated the hepatoprotective effect of *Picrorrhiza kurroa* on tissue defence system in D-galactosamine-induced hepatitis in rats. In D-galactosamine-induced hepatitis in rats, a significant increase of LPO and a decrease in liver antioxidant enzymes levels were observed. Pretreatment with the ethanol extract of *P. kurroa* prevented these alterations.

The preventive hypolipemic effect of water extract of *Picrorhiza* rhizome was observed by Lee et al. (2006) in Poloxamer (PX)-407-induced hyperlipemic mice with their hepatoprotective effects. Doses of 50, 100, and 200 mg/kg of PR extracts were given orally once a day for 12 weeks initiated with intraperitoneal injection of PX-407 (0.5 g/kg), and changes in BW and gains, liver weight, serum AST and ALT levels were monitored with serum LDL, HDL, TG and TC levels. The efficacy of PX-407 was compared to that of 10 mg/kg of simvastatin (SIMVA). No meaningful changes in the BW were detected in all dosing groups compared to that of vehicle control group. Dramatic decrease of both absolute and relative liver weight was dose dependently observed in all PR extracts dosing groups compared to that of vehicle control group. The serum AST and ALT levels were significantly and dose dependently decreased in PR extracts dosing groups. However, slight increase of liver weight, serum AST and ALT levels were detected in SIMVA-dosing groups. The serum LDL, TG and TC levels were dose dependently decreased in PR extracts dosing groups and SIMVA-dosing group compared to that of vehicle control group, respectively. The serum HDL levels were slightly but dose dependently increased in PR extracts dosing groups compared to that in vehicle control group, respectively. However, the efficacy on the serum lipid levels of PR extracts was lower than that of SIMVA-about 200 mg/kg of PR extracts which showed similar effect compared to that of SIMVA 10 mg/kg. On the basis of these results, it was concluded that water extract of PR has a relatively good favorable preventive effect on the PX-407 inducing hyperlipemia with favorable hepatoprotective effect.

The hepatoprotective effect of the ethanol extract of *P. kurroa* rhizomes and roots (PK) on liver mitochondrial antioxidant defense system in isoniazid and rifampicin-induced hepatitis in rats was investigated by Jeyakumar et al. (2009). In liver mitochondria of antitubercular drugs administered rats, a significant elevation in the level of LPO with concomitant decline in the level of reduced GSH and in the activities of antioxidant enzymes was observed. Co-administration of the extract (50 mg/kg/day for 45 days) significantly prevented these antitubercular drugs-induced alterations and maintained the rats at near normal status.

Pimpinella anisum L. FAMILY: APIACEAE

Diethyl ether extract of *Pimpinella anisum* seed has been investigated for its hepatoprotective activity in rats. The extract dose dependently attenuated CCl_4-induced rise liver enzymes including AST and ALT (Cengiz et al., 2008). *P. anisum* possess significant antioxidant and anti-inflammatory activities which may contribute to its hepatoprotective efficacy. Phytochemical studies on plant of *P. anisum* have shown the presence of volatile oils (anethole, eugenol, methyl chavicol, and estragole), fatty acids (palmitic, petroselinic, vaccenic, and oleic acids), and coumarins (Al-Asmari et al., 2014).

El-Sayed et al. (2015) investigated the possible potential protective effect of Anise and Fennel essential oils, against CCl_4-induced fibrosis in rats. Rats treated orally with essential oil of *Pimpinella anisum* (Anise, 125 and 250 mg/kg) or *Foeniculum vulgare* (Fennel, 200 and 400 kg/BW) for seven successive weeks and intoxicated with CCl_4 showed a significant protection against induced increase in serum liver enzymes (AST, ALT, and ALP), restored TP level and ameliorate the increased TGs, total, cholesterol, LDL, and decreased the HDL. A significant corrective effect of either Anise or fennel oils on biochemical parameters were supported by histopathological examination of the rats.

Protective effects of different extracts and essential oil from *P. anisum* seeds were examined by Jamshidzadeh et al. (2015) against CCl_4-induced toxicity. The parameters such as serum transaminases, LDH activity, hepatic GSH content, liver LPO and histopathological changes of liver were assessed as toxicity markers. In the *in vitro* model of this study, markers such as cell viability, cellular reduced and oxidized GSH and LPO in HepG2 cells were evaluated. The data revealed that the *n*-hexane extract, effectively attenuated CCl_4-induced toxicity in both *in vitro* and *in vivo* models in current investigation.

Pinus roxburghii Sarg. FAMILY: PINACEAE

The hepatoprotective activity of wood oil of *Pinus roxburghii* at doses of 200, 300, and 400 mg/kg BW were studied by Khan et al. (2012) on rat liver damage induced by CCl_4 and ethanol. The substantially elevated serum enzymatic levels of SGPT, SGOT, ALP, TB, MDA and decreased level of GSH and TP induced by hepatotoxins were significantly restored toward normalization by the wood oil at doses of 200 and 300 mg/kg. Hepatoprotective effect of oil may be due to its inhibitory effect on free radical formation as evident by recovery of GSH contents and decreased LPO. Light microscopy of the liver tissue further confirmed the reversal of damage induced by hepatotoxins. Phytochemical analysis revealed presence of triterpenes and steroids, which have been known for their hepatoprotective activity.

Piper betle L. FAMILY: PIPERACEAE

Piper betle is a commonly used masticatory in Asia. Saravanan et al. (2002) carried out to investigate the hepatoprotective and antioxidant properties of *P. betle*, using ethanol intoxication as a model of hepatotoxic and oxidative damage. Ethanol-treated rats exhibited elevation of hepatic marker enzymes and disturbances in antioxidant defense when compared with normal rats. Oral administration of *P. betle* extract (100, 200, or 300 mg/kg BW) for 30 days significantly decreased AST, ALT, TBARS, and lipid HP in ethanol-treated rats. The extract also improved the tissue antioxidant status by increasing the levels of nonenzymatic antioxidants (reduced GSH, vitamin C, and vitamin E) and the activities of free radical-detoxifying enzymes such as SOD, CAT, and GPx in liver and kidney of ethanol-treated rats. The highest dose of *P. betle* extract (300 mg/kg BW) was most effective. The results were comparable with the known hepatoprotective drug, Silymarin.

Pushpavalli et al. (2008) investigated the *in vivo* antioxidant potential of *P. betle* leaf-extract against oxidative stress induced by D-GalN intoxication in male albino Wistar rats. Toxicity was induced by an intraperitoneal injection of D-GalN, 400 mg/kg BW for 21 days. Rats were treated with *P. betle* extract (200 mg/kg BW) via intragastric intubations. Pushpavalli et al. (2008) assessed the activities of liver marker enzymes (AST, ALT, ALP, GGTP) and levels of TBARS, lipid HP, SOD, CAT, GPx, vitamin C, vitamin E, and reduced GSH. The extract significantly improved the status of antioxidants and decreased TBARS, HP, and liver marker enzymes when compared with the D-GalN treated group, demonstrating its hepatoprotective and antioxidant properties.

Piper chaba Hunter FAMILY: PIPERACEAE

The 80% aqueous acetone extract from the fruit of *Piper chaba* was found to have hepatoprotective effects on D-GalN/LPS-induced liver injury in mice (Matsuda et al., 2009). From the ethyl acetate-soluble fraction, three new amides, piperchabamides E, G, and H, 33 amides, and four aromatic constituents were isolated. Among the isolates, several amide constituents inhibited D-GalN/TNF-α-induced death of hepatocytes, and the following structural requirements were suggested: (i) the amide moiety is essential for potent activity; and (ii) the 1,9-decadiene structure between the benzene ring and the amide moiety tended to enhance the activity. Moreover, a principal constituent, piperine, exhibited strong *in vivo* hepatoprotective effects at doses of 5 and 10 mg/kg, p.o., and its mode of action was suggested to depend on the reduced sensitivity of hepatocytes to TNF-alpha (Matsuda et al., 2009).

Piper cubeba L.f. FAMILY: PIPERACEAE

AlSaid et al. (2015) evaluated the effectiveness of *Piper cubeba* fruits ethanol extract (PCEE) in the amelioration of CCl_4-induced liver injuries and oxidative damage in the rodent model. Treatment with PCEE significantly and dose dependently prevented drug induced increase in serum levels of hepatic enzymes. Furthermore, PCEE significantly reduced the LPO in the liver tissue and restored activities of defense antioxidant enzymes NP-SH and CAT toward normal levels. The administration of PCEE significantly downregulated the CCl_4-induced proinflammatory cytokines TNFα and IL-6 mRNA expression in dose-dependent manner, while it upregulated the IL-10 and induced hepatoprotective effect by downregulating mRNA expression of iNOS and HO-1 gene.

Piper longum L. FAMILY: PIPERACEAE

Ethanol extract of *Piper longum* fruits and five different crude fractions, petroleum ether (40–60Â°), solvent ether, ethyl acetate, butanol and butanone were subjected to preliminary qualitative chemical investigations by Jalalpure et al. (2003). The ethanolic extract and all other fractions were screened orally for hepatoprotective activity in adult Wistar rats. The ethanolic extract and butanol fraction have shown significant activity, lowering the serum enzymes glutamic oxaloacetic transaminase and GPT in rats treated with CCl_4 when compared to control and Liv. 52-treated rats.

Christiana et al. (2006) evaluated the antifibrotic effect of ethanol extract of the fruits of *P. longum*. Liver fibrosis was induced in rats by CCl_4 administration. The extent of liver fibrosis was assessed by measuring the level of liver hydroxy proline (HP) and serum enzyme levels. Following CCl_4 administration HP was significantly increased and serum enzyme levels were elevated. Treatment with the ethanol extract of *P. longum* reduced the HP and also the serum enzymes. The liver weight that increased following CCl_4 administration due to the deposition of collagen was reduced by the ethanol extract. Hence, it is concluded that this extract inhibits liver fibrosis induced by CCl_4.

P. longum (fruits and roots powder) is given with boiled milk in the Indian traditional system of medicine for the treatment of liver ailments and jaundice. Patel and Shah (2009) investigated the hepatoprotective activity of *P. longum* milk extract. CCl_4 was used as a hepatotoxin at a dose of 0.5 mL/kg p.o. with olive oil (1:1) thrice a week for 21 days to produce the chronic reversible type of liver necrosis. Following treatment with *P. longum* milk extract (200 mg/day p.o. for 21 days), a significant hepatoprotective effect was observed in CCl_4-induced hepatic damage as evident from decreased level of serum enzymes, TB and DBL. The hepatoprotective effect of *P. longum* is comparable to the standard drug Silymarin (25 mg/kg/day p.o. for 21 days).

***Piper nigrum* L. FAMILY: PIPERACEAE**

Piper nigrum popularly known as 'Black Pepper' contains abundant amount of piperine alkaloids. Piperine alkaloids have been implicated in Hepatoprotective activity. Nirwane and Bapat (2012) evaluated effect of methanolic extract of *P. nigrum* fruits in Ethanol-CCl_4-induced hepatotoxicity Wistar rats. Ethanol-CCl_4 exhibited increase in the hepatic biomarkers (TG, AST, ALT, ALP, and Bilirubin), LPO which were significantly decreased after pretreatment with methanolic extract (100, 200 mg/kg) and piperine (50 mg/kg). Ethanol-CCl_4 significantly decreased levels of SOD, CAT, and GSH which were restored with MEPN and PPR. The results were similar to that of Liv. 52, which served as a reference standard. Histopathological studies were also in agreement of above. The study indicates that *P. nigrum* possess potential hepatoprotective activity which may be attributed to its piperine alkaloids, having therapeutic potential in treatment of liver disorders.

Piper trioicum Roxb. FAMILY: PIPERACEAE

The hepatoprotective activity of ethanolic extract of aerial parts of *Piper trioicum* (EEPT) against CCl_4-induced liver damage in albino rats was investigated by Lakshmi et al. (2017). There was a marked elevation of serum marker enzyme levels and LPO in CCl_4-treated rats, which were restored toward normalization in the extract (EEPT 100 and 200 mg/kg BW intraperitoneally (i.p), once daily for 14 days) treated animals. The levels of antioxidant enzymes like SOD, Catalase and Glutathione peroxidase (GPx) were markedly decreased in CCl_4-treated rats which were increased and brought back to normal level in the treated animals.

Pisonia aculeata L. FAMILY: NYCTAGINACEAE

The MeOH (95%) extract from *Pisonia aculeata* (250 and 500 mg/kg) suspended in acacia gum at 5% demonstrated a protective effect in Wistar rats against injury induced with RIF and INH (100 and 50 mg/kg, respectively), normalizing AST, ALT, ALP, and TB levels, inhibiting cytochrome P450, augmenting NADPH levels and diminishing LPO, the effect being close to that generated by the control of Silymarin (Anbarasu et al., 2011).

Ethanol extract of *P. aculeata* (EPA) was evaluated by Palanivel et al. (2008) for hepatoprotective and antioxidant activities in rats. The plant extract (250 and 500 mg/kg, p.o.) showed a remarkable hepatoprotective and antioxidant activity against CCl_4-induced hepatotoxicity as judged from the serum marker enzymes and antioxidant levels in liver tissues. CCl_4 induced a significant rise in AST, ALT, ALP, TB, GGTP, lipid peroxidase (LPO) with a reduction of TP, SOD, CAT, GPx, and GST. Treatment of rats with different doses of plant extract (250 and 500 mg/kg) significantly altered serum marker enzymes and antioxidant levels to near normal against CCl_4-treated rats. The activity of the extract at dose of 500 mg/kg was comparable to the standard drug, Silymarin (50 mg/kg, p.o.). Histopathological changes of liver sample were compared with respective control.

Pistacia lentiscus L. FAMILY: ANACARDIACEAE

Klibet et al. (2016) evaluated the protective effect of *Pistacia lentiscus* against sodium arsenite-induced hepatic dysfunction and oxidative stress in experimental Wistar rats. A significant decrease in the levels of RBCs, hemoglobin, hematocrit, reduced GSH and metallothionein associated with a significant increase of MDA were noticed in the arsenic-exposed group when compared to the control. The As-treated group also exhibited an increase in hepatic antioxidant enzymes namely SOD, GPx, and CAT. However, the co-administration of PLo has relatively reduced arsenic effect. The results showed that arsenic intoxication disturbed the liver pro-oxidant/antioxidant status. PLo co-administration mitigates arsenic-induced oxidative damage in rat.

Ljubuncic et al. (2005) evaluated the efficacy of an aqueous extract prepared from the dried leaves of *P. lentiscus* in a rat model of hepatic injury caused by the hepatotoxin, TAA. In TAA-treated rats, long-term administration of extract aggravated the inflammatory and fibrotic and GSH depleting responses without affecting the extent of LPO. Although their previous *in vitro* study established that extracts prepared from the leaves of *P. lentiscus* had antioxidant activity, this *in vivo* study establishes *these extracts also contains hepatotoxins* whose identity may be quite different from those compounds with antioxidant properties. The results of this study suggest complementing *in vitro* experiments with those involving animals are essential steps in establishing the safety of medicinal plants. Furthermore, these data confirm that complete reliance on data obtained using *in vitro* methodologies may lead to erroneous conclusions pertaining to the safety of phytopharmaceuticals.

Pistacia lentiscus L. FAMILY: ANACARDIACEAE; *Phillyrea latifolia* L. FAMILY: OLEACEAE, AND *Nicotiana glauca* Graham FAMILY: SOLANACEAE

The hepatoprotective effect of the boiled and nonboiled aqueous extracts of *Pistacia lentiscus, Phillyrea latifolia,* and *Nicotiana glauca*, that are alleged to be effective in the treatment of jaundice in Jordanian folk medicine, was evaluated *in vivo* using CCl_4 intoxicated rats as an experimental model by Janakat and Al-Merie (2002). Plant extracts were administrated orally at a dose of 4 mL/kg BW, containing various amounts of solid matter. Only total SB level was reduced by treatment with nonboiled aqueous extract of *N. glauca* leaves, while the boiled and nonboiled aqueous extracts of the *N. glauca* flowers were noneffective. Bilirubin level and the activity of ALP were both reduced upon treatment with boiled aqueous extract of *P. latifolia* without reducing the activity of ALT and AST.

Aqueous extract of *P. lentiscus* (both boiled and nonboiled) showed marked antihepatotoxic activity against CCl_4 by reducing the activity of the three enzymes and the level of bilirubin. The effect of the nonboiled aqueous extract was more pronounced than that of the boiled extract.

Pittosporum neelgherrense Wight & Arn. FAMILY: PITTOSPORACEAE

The stem bark of *Pittosporum neelgherrense* is used by the Kani and Malapandaram tribes of Kerala as an effective antidote to snake bite and for the treatment of various hepatic disorders. The effect of the methanolic extract of the stem bark of *P. neelgherrense* was studied by Shyamal et al. (2006) against CCl_4-, D-GalN- and APAP-induced acute hepatotoxicity in Wistar rats. Significant hepatoprotective effects were obtained against liver damage induced by all the three liver toxins, as evident from decreased levels of serum enzymes, SGOT, SGPT, and almost normal architecture of the liver in the treated groups, compared to the toxin controls.

Plantago lanceolata L. FAMILY: PLANTAGINACEAE

Aktay et al. (2000) studied the hepatoprotective activity of ethanolic extracts of *Plantago lanceolata* using CCl_4-induced heptotoxicity model in rats. According to the results of biochemical tests, significant reductions were obtained in CCl_4-induced increases of plasma and liver tissue MDA levels and plasma enzyme activities by the treatment with *P. lanceolata* extracts, which reflects functional improvement of hepatocytes. Liver sections were also studied histopathologically to confirm the biochemical results.

Plantago major L. FAMILY: PLANTAGINACEAE

Türel et al. (2009) investigated the hepatoprotective activities of *Plantago major* against CCl_4-induced hepatotoxicity. *P. major* (25 mg/kg) significantly reduced the serum ALT and AST levels when compared to the CCl_4 group. The histopathological findings showed a significant difference between the *P. major* (25 mg/kg) and CCl_4 groups. *P. major* also showed anti-inflammatory activity.

Platycodon grandiflorum A. DC FAMILY: CAMPANULACEAE

The protective effects of a *Platycodi radix* (Changkil: CK), the root of *Platycodon grandiflorum* on CCl_4-induced hepatotoxicity and the possible mechanisms involved in this protection were investigated by Lee and Jeong (2002) in mice. Pretreatment with CK prior to the administration of $CC1_4$ significantly prevented the increased serum enzymatic activities of ALT and AST in a dose-dependent manner. In addition, pretreatment with CK also significantly prevented the elevation of hepatic MDA formation and the depletion of reduced GSH content in the liver of $CC1_4$-intoxicated mice. However, hepatic reduced GSH levels and GST activities were not affected by treatment with CK alone. $CC1_4$-induced hepatotoxicity was also essentially prevented, as indicated by a liver histopathologic study. The effects of CK on the cytochrome P450 (P450) 2E1, the major isozyme involved in $CC1_4$ bioactivation were also investigated. Treatment of mice with CK resulted in a significant decrease of P450 2E1-dependent *p*-nitrophenol and aniline hydroxylation in a dose-dependent manner. CK showed antioxidant effects in $FeCl_2$-ascorbate-induced LPO in mice liver homogenate and in superoxide radical scavenging activity. The protective effects of CK against $CC1_4$-induced hepatotoxicity possibly involve mechanisms related to its ability to block P450-mediated $CC1_4$ bioactivation and free-radical scavenging effects.

Kim et al. (2012) evaluated the protective effect of the standardized aqueous extract of *P. grandiflorum* (BC703) on cholestasis-induced hepatic injury in mice. Serum ALT and serum AST increased to 395.2 ± 90.0 and 266.0 ± 45.6 unit/L in the BDL alone group and decreased with BC703 in a dose-dependent manner. Especially the 10 mg/kg of BC703-treated mice showed a 77% decrease of serum ALT and 56% of AST as compared with BDL alone. Decreased antioxidant enzyme levels in BDL alone group were elevated in BC703-treated groups ranging from 7 to 29% for GSH and from 13% to 25% for SOD. BC703 treatment also attenuated MDA (from 3 to 32%) and NO levels (from 32% to 50%) as compared with BDL alone. Histopathological studies further confirmed the hepatoprotective effect of BC703 in BDL-induced cholestesis.

Plectranthus amboinicus (Lour.) Spreng. (Syn.: *Coleus aromaticus* Benth.) FAMILY: LAMIACEAE

Vijayavel et al. (2013) investigated the protective effect of *Coleus aromaticus* leaf extract against naphthalene-induced hepatotoxicity in rats. Significant protective effect was observed against naphthalene-induced liver damage, which appeared evident from the response levels of marker enzymes (AST, ALT, acid phosphatase, ALP, and LDH). The biochemical components, namely, TGs, free fatty acids, cholesterol acyl transferase, high-density lipoprotein, LDL, cholesterol and bilirubin were found to be increased in liver and serum of naphthalene stressed rats when compared to control.

Plectranthus barbatus Andrews [Syn.: *Coleus barbatus* (Andrews) Benth. ex G.Don] FAMILY: LAMIACEAE

Battochio et al. (2013) tested the effects of water extract of *Plectranthus barbatus* on liver damage in biliary obstruction in young rats. Cholestasis increased the TS, ALT, AST, liver wet weight, MI, number of necrotic areas, intensity of fibrosis and intensity of ductal proliferation. The EAB decreased the TS and IM in the animals without cholestasis (sham operated animals). The EAB decreased the TS, ALT, AST, liver wet weight, MI, NN, and intensity of fibrosis of the cholestatic animals. In the LA group there was a positive correlation between the IPD and the intensity of fibrosis, a negative correlation between the intensity of ductal proliferation and the FM and a negative correlation between the intensity of fibrosis and the frequency of mitoses. In the LD group there was a negative correlation between the number of necrotic areas and the IPD.

Pluchea indica (L.) Less. FAMILY: ASTERACEAE

The methanol fraction of the extract of *Pluchea indica* roots exhibited significant hepatoprotective activity against experimentally-induced hepatotoxicity by CCl_4 in rats and mice (Sen et al., 1993). The extract caused significant reduction of the elevated serum enzyme levels (AST, ALT, LDH, and serum ALP) and SB content in acute liver injury. A significant increase of reduced serum TP, ALB and ALB/globulin ratio was also observed on extract treatment. The extract significantly reduced the prolonged pentobarbitone-induced sleeping time and also caused a significant reduction of plasma prothrombin time in comparison with CCl_4-treated animals. The extract caused significant reduction of the increased bromosulphalein retention by CCl_4 treatment.

Plumbago zeylanica L. FAMILY: PLUMBAGINACEAE

Kumar et al. (2009) evaluated the hepatoprotective activity of methanolic extract of aerial parts of *Plumbago zeylanica* in CCl_4-induced hepatotoxicity in Wistar rats. The extract of aerial parts of *P. zeylanica* have shown very significant hepatoprotection against CCl_4-induced hepatotoxicity in Wistar rats by reducing serum TB, SGPT, SGOT, and ALP levels. Histopathological studies also confirmed the hepatoprotective nature of the extract.

Petroleum ether extract of root of *P. zeylanica* was investigated by Kanchana and Sadiq (2011) for hepatoprotective activity against PCM-induced liver damage. The markers in the animals treated with PCM recorded elevated concentration indicating severe hepatic damage by parcetamol, whereas the blood samples from the animals treated with PEE of roots showed significant reduction in the serum markers indicating the effect of the plant extract in restoring the normal functional ability of the hepatocytes. The dosage of extract of plant roots used was 300 mg/kg BW of rat.

Polyalthia longifolia (Sonner) Thwaites FAMILY: ANNONACEAE

Jothy et al. (2012) evaluated the *in vitro* antioxidant, free-radical scavenging capacity, and hepatoprotective activity of MEs from *Polyalthia longifolia* using established *in vitro* models. Interestingly, all the extracts showed considerable *in vitro* antioxidant and free-radical scavenging activities in a dose-dependent manner when compared to the standard antioxidant which verified the presence of strong antioxidant compound in leaf extracts tested. Phenolic and flavonoid content of these extracts is significantly correlated with antioxidant capacity. *P. longifolia* extract was subjected for *in vivo* hepatoprotective activity in PCM-intoxicated mice. Therapy of *P. longifolia* showed the liver protective effect on biochemical and histopathological alterations. Moreover, histological studies also supported the biochemical finding, that is, the maximum improvement in the histoarchitecture of the liver. Results revealed that *P. longifolia* leaf extract could protect the liver against PCM-induced oxidative damage by possibly increasing the antioxidant protection mechanism in mice.

Polygala arvensis Willd. FAMILY: POLYGALACEAE

The suspensions of CE of leaves of *Polygala arvensis* in 0.3% carboxy methyl cellulose (CMC) was evaluated by Dhanabal et al. (2006) for hepatoprotective activity in Wistar albino rats by inducing hepatic injury with d-galactosamine (400 mg/kg). The CE of *P. arvensis* at an oral dose of 200 and 400 mg/kg exhibited a significant protection effect by normalizing the levels of AST, ALT, ALP, TB, LDH, TC, TGL, ALB, and TP which were significantly increased in rats by treatment with 400 mg/kg i.p. of d-galactosamine.

Polygala javana DC. FAMILY: POLYGALACEAE

The effect of ethanol extract of *Polygala javana* whole plant was evaluated in CCl_4-induced hepatotoxicity in rats. Ethanol extract of whole plant of *Polygala javana* pretreatment (100 and 200 mg/kg BW) significantly reduced CCl_4-induced elevation of SGOT, SGPT, ALP, total, conjugated and unconjugated bilirubin. While the reduced concentration of SOD, CAT, GPx, and GRD were reversed. Silymarin (100 mg/kg BW) a known hepatoprotective drug showed similar results.

Polygonum bistorta L. Family: POLYGONACEAE

Mittal et al. (2012) evaluated the efficacy of root extract of *Polygonum bistorta* (100 mg/kg), tannic acid (25 mg/kg, p.o.) and resveratrol (30 mg/kg, *p.o.*) against toxicants induced damage in liver and kidney. The activities of transaminases, ALP, and protein were increased in serum after 48 h days of toxicants administration. A significant rise was observed in LPO level however reduced GSH content was observed. A concomitant fall was observed in the enzymatic activities of adenosine triphosphatase, glucose-6-phosphatase. Administration of PB and TA and Resveratrol significantly brought the values of studied parameters toward normal and also reversed the histopathological alterations in liver and kidney. They concluded that *P. bistorta* and tannic acid can be used to reduce the hepatorenal damage and may serve as an alternative medicine.

Polygonum multiflorum Thunb. FAMILY: POLYGONACEAE

Lin et al. (2018) examined the hepatoprotective effects of the ethanolic extract of *P. multiflorum* (PME) in *in vitro* and *in vivo* models. The PME induced expression of antioxidant-response-element-(ARE-) related genes in HepG2 cells showed a dose-dependent manner. Pretreatment of HepG2 cell with PME suppressed H_2O_2- and APAP-(APAP-) induced cellular ROS generation and cytotoxicity. In APAP-induced mouse liver injury, pretreatment with PME also showed ability to increase the survival rate and reduce the severity of liver injury. Treatment with PME attenuated BDL-induced extrahepatic cholestatic liver injury and further increased multidrug resistance protein 4 (MRP4) and reduced organic anion-transporting polypeptide (OATP) expression. Furthermore, increased nuclear translocation of the nuclear factor erythroid 2-related factor 2 (Nrf2) was observed after treatment with PME in both *in vivo* models. This study showed the hepatoprotective activity of PME by regulating the redox state in liver injury through Nrf2 activation and controlling hepatic bile acid homeostasis in obstructive cholestasis, through bile acid transporter expression modulation.

Polygonum orientale L. FAMILY: POLYGONACEAE

Chiu et al. (2018) investigated the hepatoprotective activity of the ethanolic extract of *Polygonum orientale* (POE) fruits against CCl_4-induced acute liver injury (ALI). Acute toxicity testing indicated that the LD_{50} of POE exceeded 10 g/kg in mice. Mice pretreated with POE (0.5, 1.0 g/kg) experienced a significant reduction in their AST, ALT, and ALP levels and reduction in the extent of liver lesions. POE reduced the MDA, NO, TNF-α (TNF-α), interleukin-1β (IL-1β), and interleukin-6 (IL-6) levels, and increased the activity of SOD, GPx, and GR in liver. HPLC revealed peaks at 11.28, 19.55, and 39.40 min for protocatechuic acid, taxifolin, and quercetin, respectively. The hepatoprotective effect of POE against CCl_4-induced ALI was seemingly associated with its antioxidant and anti-proinflammatory activities.

Pongamia pinnata (L.) Pierre FAMILY: FABACEAE

Hepatoprotective effect of stem extracts of *Pongamia pinnata* in alloxan-induced diabetic Wistar albino rats was investigated by Morajhar et al. (2015). In diabetic rats the serum glucose, SGPT, SGOT, and bilirubin levels were significantly increased whereas TP and ALB levels were decreased in comparison with the control groups. Significant recovery was observed in all the parameters with aqueous and ALE of *P. pinnata*. Histopathological observations were also in coordination with these results.

Rajeshkumar and Kayalvizhi (2015) investigated the hepatoprotective activity of aqueous and ethanol extract of *P. pinnata* leaves against APAP-induced liver damage in albino rats. Silymarin as a standard drug for comparing the activity. Ethanolic extract treated group showed highly significant activity, whereas aqueous extract treated group has shown the significant action but less compared with ethanolic extract. Plant extracts restored biochemical enzymes and brings down to normal as compared to standard drug Silymarin.

Populus nigra L. FAMILY: SALICACEAE

Debbache-Benaida et al. (2013) evaluated antioxidant, anti-inflammatory, hepatoprotective and vasorelaxant activities of *Populus nigra* flower buds ethanolic extract. The results showed a moderate antioxidant activity (40%), but potent anti-inflammatory activity (49.9%) on carrageenan-induced mice paw edema, and also as revealed by histopathologic examination, complete protection against $AlCl_3$-induced hepatic toxicity. Relaxant effects of the same extract on vascular preparation from porcine aorta precontracted with high concentration of U46619 were considerable at 10^{-1} g/L, and comparable between endothelium-intact (67.74%, IC_{50}=0.04 mg/mL) and-rubbed (72.72%, IC_{50}= 0.075 mg/mL) aortic rings.

Portulaca oleracea L. FAMILY: PORTULACACEAE

The suspensions of methanol and PEEs of entire plant of *Portulaca oleracea* in CMC were evaluated by Prabhakaran et al. (2010) for hepatoprotective activity in Wister albino rats by inducing hepatic injury with D-galactosamine (400 mg/kg). D-GalN-induced hepatic damage was manifested by a significant increase in the activities of marker enzymes. Biochemical data exhibited significant hepatoprotective activity of ME of *P. oleracea* at oral dose of 200 and 400 mg/kg against D-GalN. Silymarin was used as reference standard also exhibited significant hepatoprotective activity against D-GalN. The biochemical observations were supplemented with histopathological examination of rat liver sections.

The hepatoprotective activity of the aqueous and ethanolic extract of *P. oleracea* whole plant has been investigated by several investigators (Anusha et al., 2011; Ahmad et al., 2013). The extract significantly attenuated CCl_4 induced rise in biochemical (serum AST, APT, TB, and TP) and histopathological changes in liver. It also antagonised CCl_4 and prolonged pentobarbitone-induced sleeping time clearly suggesting significant hepatoprotective activity. The extracts of *P. oleracea* also showed significant antioxidant and anti-inflammatory activities which may contribute to its hepatoprotective activity. *P. oleracea* contains several biologically active compounds that include, alkaloids, coumarins, flavonoids, cardiac glycosides, anthraquinone glycosides, alanine, saponins, tannins, and organic acids (free oxalic acids, cinnamic acids, caffeic acid, malic acids, and citric acids). Omega-3-acids, alpha-linolenic acid, vitamins, GSH, glutamic acid, and aspartic acid containing β-sitosterol have also been found in various parts of plants (Al-Asmari et al., 2014).

Ahmida (2010) investigated the effect of *P. oleracea* on the oxidative stress in PCM-induced hepatic toxicity in male rats. Paracetamol treatment resulted in an increase in the hepatic TBARS content and depletion in TAC, reduced GSH content, CAT, and SOD activities. Administration of *P. oleracea* with

PCM significantly ameliorated the indices of hepatotoxicity induced by PCM. In addition, *P. oleracea* alleviated PCM-induced oxidative changes in liver.

Eidi et al. (2015) investigated the protective effect of the ethanol extract of *P. oleracea* against CCl_4-induced hepatic toxicity in rats. Treatment with CCl_4 resulted in increased serum activities of marker enzymes with a concomitant decrease in SOD. Histological alterations were also observed in the liver tissue upon CCl_4 treatment. Administration of purslane extract (0.01, 0.05, 0.1, and 0.15 g/kg BW) significantly showed a marked tendency toward normalization of all measured biochemical parameters in CCl_4-treated rats. Histopathological changes also paralleled the detected alteration in markers of liver function. Farkhondeh and Samarghandian (2019) also reported protective effects of *P. oleracea* against hepato-gastric diseases.

Pouteria campechiana (Kunth) Baehni. FAMILY: SAPOTACEAE

Aseervatham et al. (2014) investigated the antioxidant and hepatoprotective effect of polyphenolic-rich *Pouteria campechiana* fruit extract against APAP-intoxicated rats. Treatment with *P. campechiana* fruit extract effectively scavenged the free radicals in a concentration-dependent manner within the range of the given concentrations in all antioxidant models. The animals were treated with APAP (250 mg/kg BW; p.o.) thrice at the interval of every 5 days after the administration of *P. campechiana* aqueous extract and Silymarin (50 mg/kg). Acetaminophen treatment was found to trigger an oxidative stress in liver, leading to an increase of serum marker enzymes. However, treatment with *P. campechiana* fruit extract significantly reduced the elevated liver marker enzymes (AST, ALT, and ALP) and increased the antioxidant enzymes (viz., SOD and CAT) and GSH indicating the effect of the extract in restoring the normal functional ability of hepatocytes.

Prangos ferulacea (L.) Lindl. FAMILY: APIACEAE

Prangos ferulacea grows in Southern Iran and is used in Iranian herbal medicine mainly for gastrointestinal disorders. *P. ferulacea* has been used in folk medicine as carminative, emollient, and tonic for gastrointestinal disorders, antiflatulent, sedative, anti-inflammatory, anti-viral, antihelminthic, antifungal, and antibacterial (Asadi-Samani et al., 2015). Monoterpenes, sesquiterpenes, coumarines, flavonoids, alkaloids, tannins, saponins, and terpenoids are some important compounds identified in this plant. *P. ferulacea* was shown that the oils (both fruit and leaf essential oils) were rich in monoterpenes, specially α-pinene, and β-pinene. Some of these components have an antioxidant effect against oxidative stress. In a study, the protective and antioxidant effects of *P. ferulacea* are reported to be higher compared to α-tocopherol (vitamin E) and the effect of GST has been demonstrated. The study of effects of *P. ferulacea* hydroalcoholic extract on changes in rats liver structure and serum activities of ALT and AST after alloxan injection indicated that in diabetic rats, the serum ALT and AST significantly increased. Moreover, necrosis of hepatocytes, cytoplasmic vacuolations, and lymphocytic inflammation were observed. Diabetic rats treated by root extract of *P. ferulacea* in contrast to the diabetic group exhibited a significant decrease in these enzymes. Also, root hydroalcoholic extract of *P. ferulacea* was shown to affect changes in aminotransferases and to prevent the histopathological changes of liver related to alloxan-induced diabetes in rats (Massumi et al., 2007; Akhgar et al., 2011; Kafash-Farkhad et al., 2012, 2013).

Premna corymbosa (Burm.f.) Rottl. & Willd. FAMILY: LAMIACEAE

Karthikeyan and Deepa (2010) investigated the hepetoprotective activity of *Premna corymbosa* leaves against CCl_4-induced hepatic damage. In treatment with the ethanolic extract, the toxic effect of CCl_4 was controlled significantly by restoration of the levels of biochemical parameters as compared to normal and standard drug Silymarin treated groups. The histopathology of liver sections evidenced the heptoprotective activity.

Premna esculenta Roxb. FAMILY: LAMIACEAE

Premna esculenta is a shrub used by the ethnic people of Chittagong Hill Tracts of Bangladesh for the treatment of hepatocellular jaundice. Mahmud et al. (2012, 2016) evaluated the hepatoprotective and the *in vivo* antioxidant activity of ethanolic extracts of leaves of the plant in CCl_4-induced liver damage in rats. The extract both at the doses of 200 and 400 mg/kg, p.o. significantly reduced the elevated levels of SGPT, SGOT, ALP, and increased the reduced levels of TP and ALB compared to the CCl_4-treated animals. The extracts also showed a significant increase in the reduced levels of SOD, CAT, and peroxidase. The effects of the extracts on these parameters were comparable with those of the standard, Silymarin. The findings of the study indicate that the leaf extract of *P. esculenta* showed a potential hepatoprotective activity and the protective action might have manifested by restoring the hepatic SOD, CAT, and peroxidase levels.

The results justify the traditional use of this plant in liver disorders.

Premna serratifolia L. FAMILY: LAMIACEAE

Vadivu et al. (2009) evaluated hepatoprotective activity of ALE of leaves of *Premna serratifolia* by CCl_4-induced hepato-toxicity in rats. The degree of protection in hepatoprotective activity has been measured by using biochemical parameters such as SGOT and SGPT, ALP, bilirubin and TP. The results suggest that the ALE at the dose level of 250 mg/kg has produced significant hepatoprotection by decreasing the activity of serum enzymes, bilirubin, and LPO which is comparable to that of standard drug Silymarin.

Prosopis africana (Guill. & Perr.) Taub. FAMILY: FABACEAE

Ojo et al. (2006) evaluated the hepatoprotective activity of stem bark extract of *Prosopis africana* against APAP-induced hepatotoxicity in Wistar albino rats. They reported that stem bark extract of *P. africana* produced significant hepatoprotective effects by decreasing the activity of serum enzymes. Values recorded for AST, ALT, and ALP were significantly lower compared to those recorded for control rats. A higher inhibition of serum level elevation of ALP was observed with the extract.

Prosthechea michuacana (Lindl.) W.E.Higgins FAMILY: ORCHIDACEAE

Methanol, hexane and CEs of *Prosthechea michuacana* were studied by Rosa et al. (2009) against CCl_4-induced hepatic injury in albino rats. Pretreatment with methanolic extract reduced biochemical markers of hepatic injury levels demonstrated dose-dependent reduction in the *in vivo* peroxidation induced by CCl_4. Likewise, pretreatment with extracts of *P. michuacana* on PCM-induced hepatotoxicity and the possible mechanism involved in this protection were also investigated in rats after administering the extracts of *P. michuacana* at 200, 400, and 600 mg/kg. The degree of protection was measured by monitoring the blood biochemical profiles. The methanolic extract of orchid produced significant hepatoprotective effect as reflected by reduction in the increased activity of serum enzymes, and bilirubin. These results suggested that methanolic extract of *P. michuacana* could protect PCM-induced lipid peroxidation thereby eliminating the deleterious effects of toxic metabolites of PCM. This hepatoprotective activity was comparable with Silymarin. Hexane and CEs did not show any apparent effect. The findings indicated that the methanolic extract of *P. michuacana* can be a potential source of natural hepatoprotective agent (Rosa et al., 2009).

Protium heptaphyllum (Aubl.) March. FAMILY: BURSERACEAE

In the search of hepatoprotective agents from natural sources, α- and β-amyrin, a triterpene mixture isolated from the trunk wood resin of folk medicinal plant, *Protium heptaphyllum* was tested by Oliveira et al. (2005) against APAP-induced liver injury in mice. Liver injury was analyzed by quantifying the serum enzyme activities and by histopathological observations. In mice, APAP (500 mg/kg, p.o.) caused fulminant liver damage characterized by centrilobular necrosis with inflammatory cell infiltration, an increase in serum ALT and AST activities, a decrease in hepatic GSH and 50% mortality. Pretreatment with α- and β-amyrin (50 and 100 mg/kg, i.p. at 48, 24, and 2 h before APAP) attenuated the APAP-induced acute increase in serum ALT and AST activities, replenished the depleted hepatic GSH, and considerably reduced the histopathological alterations in a manner similar to NAC, a sulfhydryls donor. Also, the APAP-associated mortality was completely suppressed by terpenoid pretreatment. Further, α- and β-amyrin could potentiate the pentobarbital (50 mg/kg, i.p.) sleeping time, suggesting the possible suppression of liver cytochrome-P450. These findings indicate the hepatoprotective potential of α- and β-amyrin against toxic liver injury and suggest that the diminution in oxidative stress and toxic metabolite formation as likely mechanisms involved in its hepatoprotection.

Prunella vulgaris L. FAMILY: LAMIACEAE

The hepatoprotective effect of ethanolic extract of leaves of *Prunella vulgaris* against CCl_4-induced hepatic damage was investigated by Mrudula et al. (2010). The degree of protection was determined measuring levels of serum marker enzymes like SGOT, SGPT, ALP, Total and DBL and liver weight of rat. Administration of ethanolic extract of *Prunella vulgaris* (50, 100 mg/kg, p.o.) markedly decreases CCl_4-induced elevation levels of Serum marker enzymes and liver weight in dose-dependent manner. The effects of extract were compared with standard, Silymarin at 100 mg/kg dose. In ethanolic extract treated animals, the toxic effect of CCl_4 was controlled significantly by restoration of the levels of enzymes as compared to the normal and standard drug Silymarin treated groups. Histology of the liver sections of the animals treated with extract showed the presence of liver cells, absence of necrosis and fatty infiltration, which further evidenced the hepato protective activity.

Prunus armeniaca L. FAMILY: ROSACEAE

The *Prunus armeniaca* (commonly called apricot) has many medicinal properties. It has organic acids, salicylic acid, tannins and potassium salts, *p*-coumaric acid, and protocatechuic, ferulic, and diferulic acids. The plant is used as antitussive and antiasthmatic, with hepatoprotective effects (Kshirsagar et al., 2011). The plant has two flavonoid glycosides, 4′,5,7-trihydroxy flavone-7-*O*-[β-D-mannopyranosyl (1‴→2″)]-β-D-allopyranoside and 3,4′,5,7-tetrahydroxy-3,5′-di-methoxy flavone 3-*O*-[α-Lrhamnopyranosyl (1‴→6″)]-β-D-galactopyranoside (Rashid et al., 2007). Fruit of the plant is rich in carotenoids, flavonoids, and phenols. Hepatoprotective effect of ground apricot kernel (GAK) (0.5, 1, and 1.5 mg/kg/BW/rat) was examined in rats injected with 10 mg/kg DMN, demonstrating that the GAK supplemented diet resulted in improving liver function, liver CAT, SOD, and GST and hence reducing AST, ALT, and MDA, which was confirmed by liver histology. Hierarchically high levels of GAK (1.5 mg/kg/BW/rat) yielded the best results compared to other tested levels (Abdel-Rahman, 2011). Animal studies have shown that *P. armeniaca* administration to rats with chronic ethanol feeding decreases the levels of ALT and AST in the serum, which reduces oxidative stress and LPO in the liver by increasing the levels of antioxidant enzymes. Studies showed that supplementation of β-carotene prevented ethanol-induced increase in the serum aminotransferases and inhibited the depletion of the antioxidant molecule GST in the liver.

Additionally, *in vitro* studies on the hepatocytes isolated from the ethanol-fed rats indicated that β-carotene enhanced the cell viability, and increased CAT activity and level of GST. In mechanistic studies performed on hepatocytes isolated from the rats fed with ethanol, β-carotene ameliorated the oxidative stress, enhanced antioxidant, and decreased the expression of CYP2E1 and apoptosis. Lutein and meso-zeaxanthin present in *P. armeniaca* in small quantities are effective on treatment of alcohol-induced damage. Administering lutein and meso-zeaxanthin, compared to alcohol, reduced the serum levels of aminotransferases, ALP, and bilirubin to decrease the levels of LPO, conjugated diene, and HP in rat liver. Based on histopathological studies, administering ethanol-treated rats with lutein and meso-zeaxanthin reversed the histopathological abnormalities and reduced HYP, which is an indicator of fibrosis (Shivashankara et al., 2012). *P. armeniaca* fed to Wistar rats decreased oxidative stress and reduced histological damage; its dietary intake can reduce the risk of liver steatosis caused by free radicals (Ahmed et al., 2010). Examining hepatoprotective effect and antioxidant role of sun, sulfite-dried *P. armeniaca*, and its kernel against ethanol-induced oxidative stress indicated that administration of sun, sulfite-dried apricot, but not its kernel, supplementation restored the ethanol-induced imbalance between MDA and antioxidant system toward nearly normal values particularly in tissues. Altogether, apricot has a hepatoprotective effect in rats fed with ethanol, probably through the antioxidative defense systems (Yurt and Celik, 2011). *P. armeniaca* feeding had beneficial effects on $CC1_4$-induced liver steatosis and damage probably due to its antioxidant nutrient (beta-carotene and vitamin) contents and high radical-scavenging activity (Ozturk et al., 2009).

Raj et al. (2016) evaluated the hepatoprotective effect of *P. armeniaca* (Apricot) leaf on PCM-induced liver toxicity in rats. The phytochemical investigation of the extracts showed the presence of Alkaloids, volatile oil, saponin glycosides, condensed tanins, terpenoids, steroids and flavonoids. Methanol and aqueous extract before the PCM administration caused a significant reduction in the values of SGOT, SGPT, ALP, TBARS, GGT, LDH, TP, Albumin and sB almost comparable to the ursodeoxycholic acid. The hepatoprotective activity was confirmed by histopathological examination of the liver tissue of control and treated animals.

Psidium guajava L. FAMILY: MYRTACEAE

Roy et al. (2006) and Roy and Das (2010) evaluated the hepatoprotective activity of different extracts (petroleum ether, chloroform, ethyl acetate, methanol and aqueous) of *P. guajava* in acute experimental liver injury induced by CCl_4 and PCM. The effects observed were compared with a known hepatoprotective agent, Silymarin (100 mg/kg, p.o.). In the acute liver damage induced by different hepatotoxins, *P. guajava* methanolic leaf extract (200 mg/kg, p.o.) significantly reduced the elevated serum levels of AST, ALT, ALP, and bilirubin in CCl_4 and PCM-induced hepatotoxicity. *P. guajava* ethyl acetate leaf extract (200 mg/kg, p.o.) significantly reduced the elevated serum levels of AST, ALT, and bilirubin in CCl_4-induced hepatotoxicity whereas *P. guajava* aqueous leaf extract (200 mg/kg, p.o.) significantly reduced the elevated serum levels of ALP, ALT, and bilirubin in CCl_4-induced hepatotoxicity. *P. guajava* ethyl acetate and aqueous leaf extracts (200 mg/kg, p.o.) significantly reduced the elevated serum levels of AST in PCM-induced hepatotoxicity. Histological examination of the liver tissues supported the hepatoprotection. The methanolic extract of leaves of *P. guajava* plant possesses better hepatoprotective activity compared to other extracts.

The effect of *P. guajava* extract on erythromycin-induced liver damage in albino rats was investigated by Sambo et al. (2009) using normal rats. Pretreatment with 150 mg/kg of *P. guajava* extract showed a slight degree of protection against the-induced hepatic injury caused by 100 mg/kg of erythromycin stearate. Biochemical analysis of the serum obtained revealed a significant increase in serum levels of hepatic enzymes measured in the groups administered with 100 mg/kg of erythromycin stearate and 300/450 mg/kg of *P. guajava* extract compared to the control groups and those pretreated with 150 mg/kg of *P. guajava* extract. This study has shown that the aqueous extract of *P. guajava* leaf possesses hepatoprotective property at lower dose and a hepatotoxic property at higher dose but further studies with prolonged duration is recommended.

D'Mello and Rana (2010) evaluated the hepatoprotective activity of ethanolic extract of *P. guajava* and the phospholipid complex of the extract with phosphatidylcholine against PCM-induced hepatic damage in albino rats. Significant hepatoprotective effects were observed against liver damage induced by PCM overdose as evident from decreased serum levels of SGOT, SGPT, ALP, and bilirubin in the extract treated groups (200, 400 mg/kg) and phospholipid complex (100 mg/kg) compared to the intoxicated controls. The hepatoprotective effect was further verified by histopathology of the liver. The phospholipid complex showed better activity than the plain extracts which was almost comparable to standard Silymarin. The aqueous extracts of *P. guajava* and the phospholipid complex exhibited protective effect against PCM-induced hepatotoxicity with the complex showing activity better than the plain extract.

Taju et al. (2010) evaluated the hepatoprotective activity of *P. guajava* leaf extract in PCM-induced liver damage. The levels of AST, ALT, protein and bilirubin were reduced by the effect of *P. guajava* leaf extracts. The liver weight increased when compared the toxin control by the increasing concentration of the *P. guajava* leaf extract (500 mg/kg, p.o.).

***Pterocarpus marsupium* Roxb.**
FAMILY: FABACEAE

Mankani et al. (2005) evaluated the hepatoprotective activity of *Pterocarpus marsupium* stem bark extracts against CCl_4-induced hepatotoxicity in male Wistar rats. Methanol and aqueous extracts of *P. marsupium* stem bark were administered to the experimental rats (25 mg/kg/day, p.o. for 14 days). The hepatoprotective effect of these extracts was evaluated by the assay of liver function biochemical parameters (TB, serum protein, ALT, AST, and ALP activities) and histopathological studies of the liver. In ME-treated animals, the toxic effect of CCl_4 was controlled significantly by restoration of the levels of SB, protein and enzymes as compared to the normal and the standard drug Silymarin-treated groups. Histology of the liver sections of the animals treated with the extracts showed the presence of normal hepatic cords, absence of necrosis and fatty infiltration, which further evidenced the hepatoprotective activity.

Devipriya et al. (2007) assessed the hepatoprotective effect of the *P. marsupium* against CCl_4-induced hepatotoxicity in Wistar albino rats. Levels of marker enzymes such as ALT, AST, ALP, and LDH and bilirubin were increased significantly in CCl_4 treated animals. These enzymes were significantly decreased in Group III treated with plant extracts.

Pterocarpus osun Craib FAMILY: FABACEAE

Ajiboye et al. (2010) investigated the *in vitro* antioxidant potentials and attenuation of APAP-induced redox imbalance by *Pterocarpus osun* leaf in Wistar rat liver. The extract (150 and 300 mg/kg BW) significantly attenuated the altered liver and serum enzymes of APAP treated animals. SOD, CAT, GPx, GR, and G6PDH activities as well as vitamins C and E, and GSH levels were significantly elevated by the extract. The activities of uridyl diphosphoglucuronosyl transferase (59%), quinone oxidoreductase (53%), and GST (73%) significantly increased. The extract of *P. osun* leaf extract at 1.0 mg/mL scavenged the DPPH, hydrogen peroxide, superoxide ion, and ABTS at 94%, 98%, 92%, and 86%, respectively, while ferric ion was significantly reduced. There was attenuation of MDA and lipid hydroperoxide. The results indicates that *P. osun* leaves attenuated APAP-induced redox imbalance, possibly acting as free-radical scavenger, inducer of antioxidant and drug-detoxifying enzymes, which prevented/reduced LPO.

Pterocarpus santalinus L.f. FAMILY: FABACEAE

Pterocarpus santalinus bark and heartwood are rich in flavonoids and protect liver against chemical-induced toxicity. Studies showed that aqueous (45 mg/mL) and ethanol (30 mg/mL) bark extracts of *P. santalinus* restored CCl_4-induced liver damage in rats (Manjunatha

2006). *P. santalinus* protects the liver from severe damage caused by D-galN and may serve as a useful adjuvant in several clinical conditions associated with liver damage. Studies also revealed that potential compounds present in *P. santalinus* heartwood, that is, pterocarpol and cryptomeridiol targeted against HBx proteins of hepatitis B-virus through *in silico* docking studies and reported as strong drug candidates (Manjunatha et al., 2010). Itoh et al. (2009, 2010) reported that *P. santalinus* suppresses hepatic fibrosis in chronic liver injury. Saradamma et al. (2016) demonstrated the protective effects of methanolic extract of *P. santalinus* heartwood against alcohol-induced hepatotoxicity. Also, researchers have to elucidate the underlying protective molecular mechanisms of isolated compounds against liver related diseases.

Pterospermum acerifolium (L.) Willd. FAMILY: STERCULIACEAE

The hepatoprotective activity of the ethanol extract of the leaf of *Ptrospermum acerifolium* was investigated by Kharpate et al. (2009) in rats for CCl_4-induced hepatotoxicity. Hepatotoxicity was induced in male Wistar rats by intraperitoneal injection of CCl_4 (0.1 mL/kg/d p.o. for 14 d). Ethanol extract of *P. acerifolium* leaves were administered to the experimental rats (25 mg/kg/d p.o. for 14d). The Hepatoprotective effect of these extracts was evaluated by liver function biochemical parameters (TB, serum protein, ALT, AST, and ALP activites) and histopathological studies of liver. In ethanol extract-treated animals, the toxicity effect of CCl_4 was controlled significantly by restoration of the levels of SB and enzymes as compared to the normal and standard drug Silymarin-treated groups. Histology of liver sections of the animals treated with the extracts showed the presence of normal hepatic cords, absence of necrosis and fatty infiltration which further evidence the hepatoprotective activity.

The petroleum ether and hydro ALEs of *P. acerifolium* were studied by George et al. (2016) for hepatoprotective activity against albino rats with liver damage induced by PCM. The PEE at 50 mg/kg was having best activity as it decreased the mean level of bilirubin from 2.3645 ± 0.07 to 0.2975 ± 0.13. Petroleum ether extract was found to highly protective at both the dose of 25 and 50 mg/kg for ALP, SGOT, and SGPT. For LPO level, SOD and, CAT level it was observed that both extract were having significantly protection at 25 and 50 mg/kg dose.

Punica granatum L. FAMILY: PUNICACEAE

Kaur et al. (2006) evaluated antioxidant and hepatoprotective activity of ethanolic extract of pomegranate flowers. The extract was found to contain a large amount of polyphenols and exhibit enormous reducing ability, both indicative of potent antioxidant ability. The results indicated pomegranate flower extract to exert a significant antioxidant activity *in vitro*. The efficacy of extract was tested *in vivo* and it was found to exhibit a potent protective activity in acute oxidative tissue injury animal model: ferric nitrilotriacetate (Fe-NTA)-induced hepatotoxicity in mice. Intraperitoneal administration of 9 mg/kg BW Fe-NTA to mice-induced oxidative stress and liver injury. Pretreatment with pomegranate flower extract at a dose regimen of 50–150 mg/kg BW. for a week significantly and dose dependently protected against Fe-NTA-induced oxidative stress as well as hepatic injury. The extract afforded up to 60% protection against hepatic LPO and preserved GSH levels and activities of antioxidant enzymes viz., CAT, GPx GR, and GST by up to 36%, 28.5%, 28.7%, 40.2% and 42.5%, respectively. A protection against Fe-NTA-induced liver injury was apparent as inhibition in the modulation of liver markers viz., AST, ALT, ALP, bilirubin and ALB in serum. The histopathological changes produced by Fe-NTA, such as ballooning degeneration, fatty changes, necrosis were also alleviated by the extract.

Celik et al. (2009) investigated the protective and antioxidant properties of *P. granatum* (PG) beverage against TCA-exposure in rats. The hepatopreventive and antioxidant potential of the plant's infusion was evaluated by measuring level of serum enzymes, antioxidant defense systems (ADS) and LPO content in various organs of rats. Three experimental groups: A (untreated=control), B (only TCA-treated), and C (TCA+PG treated). According to the results, while the levels of AST and ALT increased significantly in B groups they decreased significantly in the C groups. LDH and CK did not change significantly in B groups' whereas decreased significantly in the C groups. Liver, brain, kidney and heart tissues MDA content significantly increased in B groups, whereas no significant changes were observed in the C groups. On the other hand, SOD decreased significantly in liver of the B group but did not change significantly in the C groups. GST activity increased significantly in liver, brain and spleen of C group while significant decrease was observed for kidney as compared to those of control. Hence, the study reveals that constituents present in PG impart protection against carcinogenic chemical-induced oxidative injury that may result in development of cancer during the period of a 52-day protective exposure.

P. granatum is used as a medicinal plant, and its fruit concentrate has been used for the prevention and treatment of liver diseases in Iran. The effects of different concentrations of the hydroalcoholic, ethyl acetate and *n*-hexane extracts of the *P. granatum* (fruit juice and seed) were investigated against CCl_4-induced cytotoxicity in HepG2 cells. Concentrations (1–10000 μg/mL) of the extracts were added to the cells, 1 h before the addition of 100 mM of CCl_4. After 24 h, the cells were evaluated for toxicity, TBARs level and GSH content. The hydroalcoholic extracts of fruit juice and seeds with concentrations of 100 to 1000 μg/mL protected the cells against CCl_4-induced cytotoxicity, but the ethyl acetate extract of fruit juice with higher concentration (1000 μg/mL) protected the cells against CCl_4 cytotoxicity and the *n*-hexane extracts were less effective. The ethyl acetate and *n*-hexane extracts of seeds with different concentrations did not have any significant protective effect. The *P. granatum* extracts themselves were not toxic toward cells with concentrations up to 1 mg/mL.

The hepatoprotective effect is also reported by the acetonic (70%) extract from the fruit of *P. granatum* against the hepatotoxicity induced by the mixture of INH/RIF (50 mg/kg each, administered by intraperitoneal-i.p.-via). The INH/RIF was co-administrated with 400 mg/kg of the extract for 15 days on male Wistar rats by oral via; the authors found that the extract diminished OS by reducing lipoperoxidation. *P. granatum* inactivates FR and increasing the activity of the antioxidant enzymes SOD, CAT, GST, and GPx, these enzymes constituting the most important endogenous antioxidant defense systems which limit the toxicity associated with the FR formed during damage induced by anti-TB drugs (Yogeeta et al., 2007). Likewise, the EtOH extract of *P. granatum* (peel) and of *Vitis vinifera* (seeds) administered on male Wistar rats for 12 weeks confers protection from the hepatocellular injury caused by DEN (Kumar, 2015).

Pyrenacantha staudtii **(Engl.) Engl. FAMILY: ICACINACEAE**

The effect of ethanol extract of *Pyrenacantha staudtii* leaves on CCl_4-induced hepatotoxicity in rats was studied by Anosike et al. (2008). Phytochemical analysis of the extract revealed the presence of alkaloids, glycosides, saponnins, carbohydrates, tannins, flavonoids and resin. Result from the study showed that both concentrations of the extract (750 and 1500 mg/kg BW) significantly reduced CCl_4-induced elevations in the liver enzymes-ALT and AST in a dose-dependent manner. CCl_4-induced increases in total and CB were also significantly lowered by the extract. These results show that the extract of *P. staudtii* leaves has protective effect against CCl_4-induced liver toxicity and damage.

Quercus aliena Blume FAMILY: FAGACEAE

Jin et al. (2005) investigated the protective effect of *Quercus aliena* acorn extracts against CCl_4-induced hepatotoxicity in rats, and the mechanism underlying the protective effects. Pretreatment with *Q. aliena* acorn extracts reduced the increase in serum AST and serum ALT levels. The hepatoprotective action was confirmed by histological observation. The aqueous extracts reversed CCl_4-induced liver injury and had an antioxidant action in assays of $FeCl_2$-ascorbic acid-induced LPO in rats. Expression of cytochrome P450 2E1 (CYP2E1) mRNA, as measured by RT-PCR, was significantly decreased in the livers of *Q. aliena* acorn-pretreated rats compared with the livers of the control group. These results suggest that the hepatoprotective effects of *Q. aliena* acorn extract are related to its antioxidative activity and effect on the expression of CYP2E1 (Jin et al., 2005).

Quercus dilatata Lindl. ex Royle FAMILY: FAGACEAE

Quercus dilatata was evaluated by Kazmi et al. (2018) for *in vitro* polyphenol content and antioxidant potential as well as *in vivo* protective role against bisphenol A (BPA)-induced hepatotoxicity. The distilled water-acetone (QDDAE) and methanol-ethyl acetate (QDMEtE) extracts were standardized and administered in high (300 mg/kg BW) and low (150 mg/kg BW) doses to SD rats, injected with BPA (25 mg/kg BW). Silymarin (50 mg/kg BW) was used as positive control. Subsequently, blood and liver homogenates were collected after four weeks of treatment, and the defensive effects of both extracts against oxidative damage and genotoxicity were assessed via hematological and biochemical investigations, determination of endogenous expression of enzymes as well as levels of free radicals and comet assay. Between the two extracts, maximum phenolics (213 ± 0.15 μg GAE/mg dry extract (DE) and flavonoids (55.6 ± 0.16 μg quercetin equivalent/mg DE) content, DPPH scavenging activity (IC_{50}: 8.1 ± 0.5 μg/mL), antioxidant capacity (53.7 ± 0.98 μg AAE/mg DE) and reducing potential (228.4 ± 2.4 μg AAE/mg DE) were observed in QDMEtE. In *in vivo* analysis, a dose-dependent hepatoprotective activity was exhibited by both the extracts. QDDAE demonstrated maximum reduction in levels of ALT (49.77 ± 3.83 U/L), TBA reactant substances (33.46 ± 0.70 nM/min/mg protein), hydrogen peroxide (18.08 ± 0.01 ng/mg tissue) and nitrite (55.64 ± 1.79 μM/mL), along with decline in erythrocyte sedimentation rate (4.13 ± 0.072 mm/h), histopathological injuries and DNA damage in BPA intoxicated rats as compared with QDMEtE. Likewise, QDDAE also significantly restored activity levels of endogenous antioxidants, including SOD, CAT, GPx (POD) and GSH with values of 6.46 ± 0.15 U/mg protein, 6.87 ± 0.1 U/min, 11.94 ± 0.17 U/min and 16.86 ± 1.56 nM/min/mg protein, respectively. Comparative results were obtained for QDMEtE.

Quercus infectoria Oliv. FAMILY: FAGACEAE

Aqueous extracts from nutgalls of *Quercus infectoria* were investigated for their hepatoprotective potential by studying their antioxidant capacity using four different methods, by determining their *in vitro* anti–inflammatory activity against 5-lipoxygenase, and by evaluating their hepatoprotective potential against liver injury induced by CCl_4 in rats. *Q. infectoria* extract exhibited potent antioxidant and anti-inflammatory activities. Treatment of rats with *Q. infectoria* extracts reversed oxidative damage in hepatic tissues induced by CCl_4.

Raphanus sativus L. FAMILY: BRASSICACEAE

The ME of *Raphanus sativus* root extract showed a protective effect on PCM-induced hepatotoxicity in a dose-dependent manner (Chaturvedi et al., 2007; Chaturvedi and Machacha, 2007)). Degree of LPO caused by PCM was measured in terms of TBARS and protection was measured in reference to SGOT, SGPT, and blood and hepatic levels of antioxidants like GSH and CAT. Administration of extract along with PCM showed significant protection. Levels of TBARS were found to be low, activities of SGOT and SGPT were low, while hepatic GSH levels were significantly higher in experimental rats that received the mixture of PCM and the extract as compared to rats that received PCM only. Activities of CAT were also high in all experimental groups. Thus this study indicates the involvement of *R. sativus* root extract with antioxidants like GSH and CAT in rendering protection against PCM-induced LPO and hepatotoxicity.

The fresh juice obtained from the locally grown radish root was tested by Rafatullah et al. (2008) for possible hepatoprotective effect against CCl_4-induced hepatocellular damage in albino rats. The juice at two doses of 2 and 4 mL/kg/rat for five consecutive days, exhibited a significant dose-dependent protective effect. The protective effect was demonstrated in lowering the elevated serum levels of SGPT, SGOT, Bil, ALP, and increasing NP-SH level. Silymarin, a known hepatoprotective agent was used as a positive control. Biochemical data were further supported by the histopathological results. The phytochemical examination of the fresh juice revealed the presence of sulfurated, phenolic and terpenoid compounds in radish.

Elshazly et al. (2016) evaluated the impact of radish oil on the possible genotoxic and hepatotoxic effects of hexavalent chromium in male rats. The chromium exposure promoted oxidative stress with a consequently marked hepatic histopathological alterations, increased serum ALT and ALP activities, AFP levels, and micronucleated erythrocytes % in peripheral blood. Moreover, COMET assay of hepatic DNA revealed that SDD exposure significantly decreased the intact cells%, head diameter, and head DNA% compared to control, indicating DNA damage. However, radish oil co-administration with SDD resulted in marked amendment in the altered parameters as detected by improved liver function markers (ALT and ALP) and AFP level, decreased LPO, increased antioxidant markers, inhibited hepatic DNA damage and restored the hepatic histology by preventing the appearance of the altered hepatocytes' foci and decreasing chromium-induced histopathological lesions. It was concluded that radish oil was able to provide a convergent complete protection against the geno- and hepatotoxicity of chromium by its potent antioxidant effect.

Rauwolfia serpentina (L.) Benth. FAMILY: APOCYNACEAE

Hepatoprotective activity of aqueous ethanolic extract of rhizome of mature *Rauwolfia serpentina* against PCM-induced hepatic damage in rats was investigated by Gupta et al. (2006). The effect of ethanolic extract of rhizome of *R.serpentina* on blood and liver GSH, Na+K+-ARPase activity, serum marker enzymes, SB and TBARS, liver GSH peroxidise, glutathione-S-transferase, glutathione reductase, SOD, CAT acivity and glycogen against PCM-induced damage in rats have been studied. Aqueous ethanol extrat of rhizome has reversal effects on the levels of above-mentioned parameters in PCM hepatotoxicity.

Rhazya stricta Decne. FAMILY: APOCYNACEAE

Pretreatment with *Rhazya stricta* significantly protected mice against PCM-induced biochemical changes and prolongation of pentobarbitone-induced sleeping time. The hepatoprotective effect of *R. stricta* was comparable with Silymarin (Ali et al., 2001). The extract of *R. stricta* leaves also showed significant antioxidant and anti-inflammatory activities which may contribute to its hepatoprotective activity. Phytochemical studies on *R. stricta* showed the presence of alkaloids (rhazimine, stemmadenine, vincadine, and rhazimanine), carboline, and flavonoidal glycoside (Al-Asmari et al., 2014).

Rheum emodi Wall. FAMILY: POLYGONACEAE

Ibrahim et al. (2008) evaluated the hepatoprotective capacity of *Rheum emodi* extracts in CCl_4 treated male rats. The dried powder of *R. emodi* was extracted successively with petroleum ether, benzene, chloroform, and ethanol and concentrated in vacuum. Primary rat hepatocyte monolayer cultures were used for *in vitro* studies. *In vivo*, the hepatoprotective capacity of the extract of the rhizomes of *R. emodi* was analyzed in liver injured CCl_4-treated male rats. Primary hepatocytes monolayer cultures were treated with CCl_4 and extracts of *R. emodi*. A protective activity could be demonstrated in the CCl_4 damaged primary monolayer culture. Extracts of the fruit pericarp of rhizomes of *R. emodi* (3.0 mg/mL) were found to have protective properties in rats with CCl_4-induced liver damage as judged from serum marker enzyme activities.

Hepatoprotective effects of *R. emodi* roots and their aqueous and methanolic extracts were studied by Akhtar et al. (2009) against liver damage induced by PCM in albino rats. In addition, the effects of herbal preparation, Akseer-e-Jigar and a control drug, Silymarin were also studied. Pretreatment and post-treatment hepatoprotective effects of all these drugs were determined. The prevention of liver damage and curative effects of the drugs were judged by changes in serum ALT, AST, ALP, ALB, and bilirubin (total and direct) levels. Powdered *R. emodi* roots (1 and 1.5 g/kg) and their aqueous extract did not significantly affect serum enzymes, ALB, and bilirubin levels. However, treatment with powder (2 g/kg), methanolic extract (0.6 g/kg), Akseer-e-Jigar (1 g/kg) and Silymarin (50 mg/kg) in both pre and post-treatment studies significantly prevented the PCM-induced rise of serum enzymes and bilirubin levels whereas serum ALB was raised after treatment with these drugs. It is conceivable, therefore, that *R. emodi* roots and Akseer-e-Jigar possess hepatoprotective principles that can prevent and/or treat liver damage due to PCM. The study has supported empirical use of the plant and its compound preparation used in traditional medicine.

Rhinacanthus nasuta (L.) Kurz. FAMILY: ACANTHACEAE

Shyamal et al. (2010) evaluated the protective effects of *Rhinacanthus nasuta* ethanolic extracts on the aflatoxin B1 (AFB1)-intoxicated livers of albino male Wistar rats. Pretreatment of the rats with oral administration of the ethanolic extracts *R. nasuta*, prior to AFB1 was found to provide significant protection against toxin-induced liver damage, determined 72 hours after the AFB1 challenge (1.5 mg/kg, intraperitoneally) as evidenced by a significant lowering of the activity of the serum enzymes and enhanced hepatic reduced GSH status. Pathological examination of the liver tissues supported the biochemical findings. The plant extract showed significant antilipid peroxidant effects *in vitro*.

Rhizophora mucronata Lam. FAMILY: RHIZOPHORACEAE

Ravikumar and Gnanadesigan (2012) evaluated the hepatoprotective effect of bark, collar, hypocotyl and stilt root extracts of *Rhizophora mucronata*. Of the selected extracts, stilt root showed better hepatoprotective activity. The hepatoprotective activity of the *R. mucronata* stilt root extract was dose-dependent (75–300 mg/kg BW) showed that the level of SGOT, SGPT, ALP, bilurubin, cholesterol, sugar and LDH were significantly reduced by all the doses when compared with the levels in the hepatotoxin group rats. No histopathological alteration other than mild fatty changes was observed with the high dose (300 mg/kg) of stilt root extract. Phytochemcial analysis of the extracts showed the presence of various chemical constituents, including flavonoids, alkaloids, coumarins and polyphenols.

Rhodiola sachalinensis Borissova FAMILY: CRASSULACEAE

Nan et al. (2003) investigated the protective effect of an aqueous extract from the root of *Rhodiola sachalinensis* (RSE) on liver injury induced by repetitive administration of CCl_4 in rats. RSE was given orally to rats at doses of 50, 100 or 200 mg/kg throughout the CCl_4 treatment for 28 days. In rats treated with CCl_4, the levels of HYP and MDA in the liver, and serum enzyme activities were significantly increased. RSE treatment significantly reduced the levels of liver HYP and MDA, and serum enzyme activities, in accordance with improved histological findings. Immunohistological findings indicated RSE treatment inhibited hepatic stellate cell activation, which is a major step for collagen accumulation during liver injury. These data suggest that RSE protects the liver from repetitive injury induced by CCl_4 in rats.

Rhododendron arboreum Sm. FAMILY: ERICACEAE

Prakash et al. (2008) evaluated the hepatoprotective activity of pretreatment with ethanolic extract of leaves of *Rhododendron arboreum* against CCl_4-induced hepatotoxicity in Wistar rat model. The activities of all the marker enzymes registered a significant elevation in CCl_4-treated rats, which were significantly recovered toward an almost normal level in animals co-administered with ethanolic extract of leaves of *R. arboreum* at a dose of 60 and 100 mg/kg. Ethanolic extract of leaves of *R. arboreum* prevented decrease in the excretion of ascorbic acid in CCl_4-induced hepatotoxicity in rats. Histopathological analysis confirmed the biochemical investigations. This hepatoprotective property may be attributed to the quercetin related flavonoids, saponins and phenolic compounds present in the leaves of *R. arboreum.*

Rhoicissus tridentata (L.f.) Wild & R.B. Drumm. FAMILY: VITACEAE

Rhoicissus tridentata, a South African medicinal plant is commonly used for the treatment of ailments like epilepsy, kidney and bladder complaints (Opoku et al., 2007). The aqueous extract of the roots were shown to possess significant hepatoprotective effect against CCl_4-induced acute liver injury in rats. The variables investigated were the enzymes ALT, AST, and glucose-6-phospate (G-6-Pase). CCl_4 intoxication resulted in significant increases in all the variables investigated except G-6-Pase which was significantly decreased. The administration of *R. tridentata* extracts after CCl_4 intoxication resulted in significant decreases in all the variables investigated except G-6-Pase which was significantly increased (Opoku et al., 2007).

Rhus oxyacantha Schousb. ex Cav. FAMILY: ANACARDIACEAE

Miled et al. (2017) investigated the antioxidant activity and hepatoprotective effects of ethyl acetate extract of *Rhus oxyacantha* root cortex (RE) against DDT-induced liver injury in male rats. The RE exhibited high total phenolic, flavonoid and condensed tannins contents. The antioxidant activity *in vitro* systems showed a significant potent free-radical scavenging activity of the extract. The HPLC finger print of *R. oxyacantha* active extract showed the presence of five phenolic compounds with higher amounts of catechol and GA. Pretreatment of rats with RE at a dose of 150 and 300 mg/kg BW significantly lowered serum transaminases and LDH in treated rats. A significant reduction in hepatic TBARS and a decrease in antioxidant enzymes activities and hepatic MTs levels by treatment with plant extract against DDT were observed. These biochemical changes were consistent with histopathological observations, suggesting marked hepatoprotective effect of RE with the two doses used. These results strongly suggest that treatment with ethyl acetate extract normalizes various biochemical parameters and protects the liver against DDT-induced oxidative damage in rats and thus help in evaluation of traditional claim on this plant.

Ricinus communis L. FAMILY: EUPHORBIACEAE

In the Indian system of medicine, the leaves, roots and seed oil of this plant have been used for the treatment of inflammation and liver disorders for a long time. *Ricinus communis* leaves ethanolic extract 250/500 mg/kg BW possesses hepatoprotective activity due to their inhibitory activities of an increase in the activities of serum transaminases and the level of liver lipid per oxidation, protein, glycogen and the activities of acid and ALP in liver induced by CCl_4. The *R. communis* ethanolic extract 250/500 mg/kg BW also treated the depletion of GSH level and adenosine triphosphatase activity which was observed in the CCl_4-induced rat liver. The presence of flavonoids in ethanol extract of *R. communis* produces beneficial effect the flavonoids have the membrane stabilizing and antiperoxidative effects. Hence the *R. communis* increase the regenerative and reparative capacity of the liver due to the presence of flavonoids and tannins. The whole leaves of *R. communis* showed the protective effect against liver necrosis as well as fatty changes induced by CCl_4 while the glycoside and cold aqueous extract provide protection only against liver necrosis and fatty changes respectively (Natu et al., 1977; Shukla et al., 1992).

The protective effects of ethanol extract of *R. communis* leaves on CCl_4-induced liver damage were investigated in rats by Prince et al. (2011). An increase in the activities of serum transaminases and the level of liver LPO, protein, glycogen and the activities of acid and ALP in liver induced by CCl_4 were significantly inhibited by treatment with *R. communis* ethanolic extract (250/500 mg/kg BW). In addition, the depletion of GSH level and adenosine triphosphatase activity observed in the CCl_4-induced rat liver were effectively prevented by treatment with *R. communis* ethanolic extract (250/500 mg/kg BW). Histopathological examination further confirmed the hepatoprotective activity of *R. communis* ethanol extract when compared with the CCl_4-induced control rats.

Rosa canina L. FAMILY: ROSACEAE

Sadeghi et al. (2016) investigated the hepatoprotective activity of hydroethanolic fruit extract of *Rosa canina* against CCl_4-induced hepatotoxicity in rats. Hepatotoxicity was evidenced by considerable increase in serum levels of AST, ALT, ALP, and LPO (MDA) and decrease in levels of ALB and TP. Injection of CCl_4 also induced congestion in central vein, and lymphocyte infiltration. Treatment with hydro-alcoholic fruit extract of *R. canina* at doses of 500 and 750 mg/kg significantly reduced CCl_4-elevated levels of ALT, AST, ALP, and MDA. The extract also increased the serum levels of ALB and TP compared to CCl_4 group at the indicated dose. Histopathological studies supported the biochemical finding.

Taghizadeh et al. (2018) evaluated the hepatoprotective effects of *R. canina* extract on streptozotocin (STZ)-induced diabetes in rats by measuring the fasting blood glucose (FBG), TAC, and liver enzyme activity, including serum ALT and AST. In the untreated diabetic group, the results showed a significant increase in FBG, ALT, and AST levels compared to the other groups. The level of TAC decreased in this group, but not significantly compared to the other groups. In the treated groups, administration of *R. canina* extract significantly improved the mentioned parameters in a dose-dependent manner. Histological evaluations indicated that *R. canina* extract ameliorated defective liver caused by STZ.

Rosa damascena Mill. FAMILY: ROSACEAE

The fresh juice of *Rosa damascena* flower exhibited promising *in vitro* antioxidant potential. The partially purified acetone fraction (AF) from silica gel column chromatography was found to be the active fraction with antioxidant properties. The AF required for 50% inhibition of superoxide radical production, hydroxyl radical generation and LPO formation were 13.75, 135, and 410 m g/mL, respectively. Oral administration of AF at 50 mg/kg BW significantly reduced the serum ALP, SGPT and SGOT activity and LPO level in rats receiving an acute dose of CCl_4 (Achuthan et al., 2003).

The hepatoprotective activity of the ALE of *Rosa damascena* was studied by Alam et al. (2008) against PCM-induced acute hepatotoxicity in rats. Liver damage was assessed by estimating serum enzyme activities of AST, ALT, ALP, and histopathology of liver tissue. Pre- and post-treatment with ethanolic extracts showed a dose-dependent reduction of PCM-induced elevated serum levels of enzyme activity. The mechanism underlying the protective effects was assayed *in vitro* and the *R. damascena* extracts displayed dose-dependent free radical activity using DPPH and TBA method. The hepatoprotective action was confirmed by histopathological observation. The ethanolic extracts reversed PCM-induced liver injury. These results suggest that the hepatoprotective effects of *R. damascena* extracts are related to its antioxidative activity.

Rosmarinus officinalis **L.** **FAMILY: LAMIACEAE**

Lyophilised ethanol and aqueous extracts of *Rosmarinus officinalis* young sprouts and total plant have been evaluated for hepatoprotective activities in the rat (Hoefler et al., 1987). Aqueous extracts of young sprouts show a significant hepatoprotective effect on plasma GPT levels when given as pretreatment before CCl_4 intoxication while the whole plant extract was inactive. Both sprouts and whole plant aqueous extracts were ineffective when given after CCl_4 administration.

The effect of oral administration of *R. officinalis* on CCl_4-induced acute liver injury was investigated by Sotelo-Félix et al. (2002). Rats were daily treated with the plant extract at a dose of 200 mg/kg corresponding to 6.04 mg/kg of carnosol as determined by reverse phase HPLC. The treatment was initiated 1 h after CCl_4 administration and *R. officinalis* fully prevented CCl_4 effect on hepatic LPO after 24 h of CCl_4 administration. The increase in bilirubin level and ALT activity in plasma induced by CCl_4 was completely normalized by *R. officinalis*. The treatment also produced a significant recovery of CCl_4-induced decrease in liver glycogen content. CCl_4 did not modify the activity of liver cytosolic GST compared with that of control groups. However, *R. officinalis* increased liver cytosolic GST activity and produced an additional increment in plasma GST activity in rats treated with CCl_4. Histological evaluation showed that *R. officinalis* partially prevented CCl_4-induced inflammation, necrosis and vacuolation. *R. officinalis* might exert a dual effect on CCl_4-induced acute liver injury, acting as an antioxidant and improving GST-dependent detoxification systems.

Amin and Hamza (2005) evaluated the hepatoprotective effect of water extract of *R. offcinalis* in Azathioprine-induced hepatotoxicity in rats. Administration of azathioprine induces oxidative stress through depleting the activities of antioxidants and elevating the level of malonialdehyde in liver. This escalates levels of ALT and AST in serum. Pretreatment with the extract proved to have a protective effect against azathioprine-induced hepatotoxicity. Animals pretreated with water extract not only failed to show necrosis of the liver after azathioprine administration, but also retained livers that for the most part, were histologically normal. In addition, the extract blocked the induced elevated levels of ALT and AST in serum. The Azathioprine-induced oxidative stress was relieved to varying degrees by the herbal extract. This effect was evident through reducing MDA levels and releasing the inhibitory effect of Azathioprine on the activities of GSH, CAT, and SOD.

Soyal et al. (2007) carried out a study to observe the radioprotective effects of *R. officinalis* leaves extract (ROE) against radiation-induced histopathological alterations in liver of mice. Normal hepatocyte counts were found to be declined up to day 10th post-irradiation in both the groups but thereafter such cells increased reaching to near normal level at the last autopsy interval, only in experimental group. Contrary, frequency of abnormal hepatocytes increased up to day 10th after irradiation in both the groups. Binucleate hepatic cells showed a biphasic mode of elevation after irradiation, first at 12 hrs and second on day 10th in control group; whereas in experimental group, the elevation was comparatively less marked and even the second peak was not evident. Irradiation of animals resulted in an elevation in lipid peroxidation (LP) and a significant decrease in GSH concentration in liver as well as in blood. Conversely, experimental group showed a significant decline in LPx and an elevation in GSH concentration.

Fregozo et al. (2012) determined the effect of *R. officinalis* treatment on the expression of the glutamate transporter (GLT-1) and on neuronal damage in the frontal cortex of Wistar rats with hepatic damage that was induced by CCl_4. The morphological evaluation of the frontal cortex showed that *R. officinalis* exerted a protective effect that was correlated with increased GLT-1 expression. The protective effect of *R. officinalis* against CCl_4-induced hepatic damage may be due to improved hepatocellular function.

Ramadan et al. (2013) examined the effect of water extract (200 mg/kg BW) of *R. officinalis* in streptozotocin (STZ)-induced diabetic rats for 21 days. The hepatoprotective effects were investigated in the liver tissues sections. There was a significant increase in serum liver biochemical parameters (AST, ALT, and ALP), accompanied by a significant decrease in the level of TP and ALB in the STZ-induced rats when compared with that of the normal group. The high-dose treatment group (200 mg/kg BW) significantly restored the elevated liver function enzymes near to normal. This study revealed that rosemary extracts exerted a hepatoprotective effect.

El-Hadary et al. (2019) explored the hepatoprotective effects of cold-pressed *R. officinalis* oil (CPRO) against CCl_4-induced liver toxicity in the experimental rats. In CPRO the percentages of polyunsaturated, monounsaturated and saturated fatty acids were 42.3%, 41.7%, and 15.8%, respectively. CPRO contained high amounts of total phenolic compounds (7.20 mg GAE/g). α-, β-, γ- and δ-tocotrienols accounted for 18, 12, 29, and 158 mg/100 g CPRO, respectively, while α-, β-, γ- and δ-tocopherols accounted for 291, 22, 1145, and 41 mg/100 g CPRO, respectively. The LD_{50} at 24 h was 5780 mg/kg. Treatment with 200 mg/kg CPRO caused a decrease in creatine, urea, and uric acid levels to 0.66, 28.3 and 3.42 mg/dL, respectively. After 8 weeks of administration, levels of total lipids (TL), TC, total triglycerol (TAG), low-dentistry lipoprotein-cholesterol (LDL-C) and very low-dentistry lipoprotein-cholesterol (VLDL-C) were decreased to 565, 165, 192, 75.6 and 38.5 mg/L, respectively. MDA levels in liver were reduced and GSH levels were elevated in CPRO-treated rats. CPRO reduced the activity of ALT, AST, and ALP as well as kidney function markers, lipid and protein profiles. Histopathological examination of liver indicated that CPRO administration reduced fatty degenerations, cytoplasmic vacuolization and necrosis. CPRO possessed a hepatoprotective effect against CCl_4-induced toxicity, mediated possibly due to the antioxidant traits of CPRO. CPRO contains high levels of phenolics and tocols, which is a scavenger of reactive species making the oil a promising material for functional foods and pharmaceuticals.

Rosmarinus tomentosus Hub.-Mor. & Maire FAMILY: LAMIACEAE

A dried ethanol extract of the aerial parts of *Rosmarinus tomentosus* and its major fraction separated by column chromatography (fraction F19) were evaluated by Galisteo et al. (2000) for antihepatotoxic activity in rats with acute liver damage induced by a single oral dose of TAA. Silymarin was used as a reference antihepatotoxic substance. Pretreatment with *R. tomentosus* ethanol extract, fraction F19 or Silymarin significantly reduced the impact of TAA toxicity on plasma protein and urea levels as well as on plasma AST, ALT, LDH, and GGT activities compared with TAA-treated animals (group T). Pretreatment with *R. tomentosus* ethanol extract significantly reduced the impact of TAA damage on ALP and GGT activities compared with group T. Silymarin administration significantly reduced ALP and GGT activities compared with group T. Fraction F19 administration reduced only ALP activity compared with group T.

Rubia cordifolia L. FAMILY: RUBIACEAE

The hepatoprotective activity of an aqueous-ME of *Rubia cordifolia* was investigated by Gilani and Janbaz (1995c) against APAP and CCl_4-induced hepatic damage. Acetaminophen produced 100% mortality at a dose of 1 g/kg in mice while pretreatment of animals with plant extract (500 mg/kg) reduced the death rate to 30%. Acetaminophen at a dose of

640 mg/kg produced liver damage in rats as manifested by the rise in serum levels of GOT and GPT to 1447±182 and 899±201 IU/L (*n* =10) respectively, compared with respective control values of 97±10 and 36±11. Pretreatment of rats with plant extract (500 mg/kg) lowered significantly the respective serum GOT and GPT levels to 161±48 and 73±29.

Various extracts of roots of *R. cordifolia* were screened by Babita et al. (2007) for its hepatoprotative activity using TAA-induced hepatotoxicity in rats. The methanolic extract could protect the liver of the animals against TAA-induced hepatotoxicity by the restoration of the levels of SGPT and SGOT. Histology of the liver sections of animals treated with methanolic extract showed the normal hepatic architecture with absence of necrosis, which further evidence the hepatoprotective activity.

Rubus aleaefolius Poir. FAMILY: ROSACEAE

Various crude forms of *Rubus aleaefolius* have been evaluated by Hong et al. (2010) for their effects on CCl_4-induced acute liver injury in mice vivo experimental model. The low-dosage EAF was the most active when the fractions were compared. It was found to decrease AST, ALT; to prevent formation of hepatic MDA, NO, and intensify the activity of SOD. The histopathological changes induced by CCl_4 were also significantly reduced. The separation revealed the presence of six constituents by a bioassay-guided fractionation,-Sitosterol (1), 1-Hydroxyeuscaphic acid (2), Oleanolic acid (3), Myrianthic acid (4), Euscaphic acid (5), and Tomentic acid (6). 1-Hydroxyeuscaphic acid (major constituent) showed a tremendous activity and the results confirm the traditional uses of *R. aleaefolius* in treating hepatitis.

Rubus chingii Hu FAMILY: ROSACEAE

Different *in vitro* free radical generating systems were used to assess the antioxidative activity of aqueous extracts of the five herbal components of Wu-zi-yan-zong-wan, a traditional Chinese medicinal formula with a long history of use for tonic effects. *Rubus chingii* fruits was found to be the most potent. It was further investigated by Yau et al. (2002) using the primary rat hepatocyte system. tert-Butyl hydroperoxide (*t*-BHP) was used to induce oxidative stress. Being a short chain analog of lipid hydroperoxide, *t*-BHP is metabolized into free radical intermediates by the cytochrome P450 system in hepatocytes, which in turn, initiate LPO, GSH depletion and cell damage. Pretreatment of hepatocytes with *R. chingi* fruit extract (50 μg/mL to 200 μg/mL) for 24 h significantly reversed *t*-BHP-induced cell viability loss, LDH leakage and the associated GSH depletion and LPO. The amount of ROS formed was also decreased as visualized by the fluorescence probe 2′,7′-dichlorofluorescin diacetate.

Rumex alveolatus Losinsk. FAMILY: POLYGONACEAE

Naseri et al. (2019) explored the protective effects of *Rumex alveolatus* on the liver damage induced by CCl_4 in mice. All administrations were done intraperitonially. *R. alveolatus* extract (450 mg/kg) reduced the serum level of all three liver enzymes. At this dose, the volume of sinusoids, hepatocytes, and bile duct decreased significantly, but the volume of portal vein and central vein increased significantly. *R. alveolatus* extract has protective effects against CCl_4-induced damages in the liver.

Rumex dentatus L. FAMILY: POLYGONACEAE

Rumex dentatus at doses 250 and 500 mg/kg decreased the elevated level of ALT, AST, ALP, and bilirubin induced by PCM and results are comparable with Silymarin (Saleem et al., 2014). The results were supported by histopathological investigations, phytochemical screening and detection of hepatoprotective active constituents, for example, quercetin, kaempferol, and myricetin by HPLC.

Rumex vesicarius L. FAMILY: POLYGONACEAE

Tukappa et al. (2015) evaluated the hepatoprotective potential and in vitro cytotoxicity studies of whole plant ME of *Rumex vesicarius*. Methanol extract at a dose of 100 mg/kg BW and 200 mg/kg BW were assessed for its hepatoprotective potential against CCl_4-induced hepatotoxicity by monitoring activity levels of SGOT, SGPT, ALP, TP (Total protein), TB (Total bilirubin) and SOD, CAT (Catalase), MDA. The cytotoxicity of the same extract on HepG2 cell lines were also assessed using MTT assay method at the concentration of 62.5, 125, 250, and 500 μg/mL. Pretreatment of animals with whole plant MEs of *R. vesicarius* significantly reduced the liver damage and the symptoms of liver injury by restoration of architecture of liver. The biochemical parameters in serum also improved in treated groups compared to the control and standard (Silymarin) groups. Histopathological investigation further corroborated these biochemical observations. The cytotoxicity results indicated that the plant extract which were inhibitory to the proliferation of HepG2 cell line with IC_{50} value of 563.33±0.8 μg/

mL were not cytotoxic and appears to be safe (Tukappa et al., 2015).

Sabina przewalskii Kom. FAMILY: CUPRESSACEAE

Ip et al. (2004) evaluated the hepatoprotective activity of *Sabina przewalskii* in menadione-intoxicated female mice. Pretreatment with an ethyl acetate-soluble fraction of the plant suppressed the plasma ALT activity in menadione-intoxicated mice. *In vitro* studies showed that the *Sabina* extract could inhibit NADH-induced superoxide production in isolated hepatocytes as indicated by a decrease of lucigenin-amplified chemiluminescence. In addition, the extract inhibited *t*-BHP-induced LPO in isolated hepatocytes dose dependently. These results suggested that the protective action of the *Sabina* extract against menadione-induced hepatotoxicity is associated with its antioxidant activities including the scavenging of superoxide anion radicals and inhibition of LPO.

Saccharum officinarum L. FAMILY: POACEAE

Hepatoprotective activity of *Saccharum offcinarum* juice was evaluated against liver damage induced in Wistar rats by administration of PCM (Patel et al., 2010a), CCl_4 (Patel et al. (2010b) and 20% ethanol (Patel et al. (2010c). Treatment with *S. officinarum* juice significantly prevented the physical, biochemical, histological and functional changes induced by PCM, CCl_4 and ethanol in the liver, while *in vitro* model exhibited significant increase in% viability of cells with pre-exposure to *S. officinarum* and Silymarin as compared to toxin control.

INH is the drug of choice for the treatment of TB and it is a well-known cause of acute clinical liver injury which can be severe and sometimes fatal. Khan et al. (2015) investigated the effects of *S. officinarum* juice on oxidative liver injury due to INH in mice. Thirty mice were divided into three groups, containing 10 mice each. Group A being the control; group B and C were experimental and were treated orally with INH 100 mg/kg per day and INH 100 mg/kg per day plus *S. officinarum* L. juice 15 mL/kg per day respectively for a period of 30 days. Blood samples were taken at 30th day by cardiac puncture under anaesthesia and liver in each was taken out for microscopic examination. INH treated mice showed; rise in serum ALT, AST, ALP and TB levels, while group C mice treated with *S. officinarum* juice significantly decreased the levels of these biochemical parameters. The histopathological examination of group A showed normal liver structure which was deranged in (INH) group B, whereas group C showed significant recovery in histological structure. *S. officinarum* constituents, especially flavanoids and anthocyanins have strong antioxidant properties which provides hepatoprotection against oxidative liver injury produced by INH. INH-induced liver injury is associated with oxidative stress, and co-administration of *S. officinarum* juice (15 mL/kg BW) may reduce this damage effectively in mice (Khan et al., 2015).

Salacia reticulata Wight FAMILY: CELASTRACEAE

The hepatoprotective effects of the hot water and methanolic extracts from the roots and stems of *Salacia reticulata* were examined by Yoshikawa et al. (2002) using an oxidative stress-induced liver injury model. Both the extracts (400 mg/kg, p.o.) significantly suppressed the increase in SGOT and SGPT activities in CCl_4-treated mice. These extracts also inhibited CCl_4-induced TBARS formation, which indicates increased LPO in the liver. A good correlation was observed between the amount of phenolic compounds in the extracts and their inhibitions of TBARS formation. The IC50 values of the extracts on DPPH radical scavenging were less than 10 μg/mL and the antioxidative activities of six phenolic compounds from the roots of *S. reticulata* were examined. Against the CCl_4-induced SGOT and SGPT elevations and TBARS formation in mice, mangiferin and (−)-4′-*O*-methyl epigallocatechin showed potent activity at a dose of 100 mg/kg, but (−)-epicatechin-(4beta->8)-(−)-4′-*O*-methylepigallocatechin did not. These results suggest that the antioxidative activity of the principal phenolic compounds is involved in the hepatoprotective activity of *S. reticulata.*

Salix subserrata Willd. FAMILY: SALICACEAE

Wahid et al. (2016) investigated the hepatoprotective effects of *Salix subserrata* flower ethanolic extract against CCl_4-induced liver damage. The administration of flower extract showed hepatic protection at an oral dose of 150 mg/kg. Ethanolic extract of the flowers significantly reduced the elevated serum levels of intracellular liver enzymes as well as liver biomarkers in comparison to CCl_4-intoxicated group. Notably, the extract significantly reduced the expression levels of TNF-α and NFkB proteins compared to their levels in CCl_4 intoxicated group. These findings were confirmed with the histopathological observations, where the extract was capable of reversing the toxic effects of CCl_4 on liver cells compared to that observed in CCl_4-intoxicated animals. Their results show that the flower extract has potential hepatoprotective effects at 150 mg/kg. These effects can be regarded to the antioxidant and anti-inflammatory properties of the extract.

Salvia cryptantha Montbret & Aucher ex Benth. FAMILY: LAMIACEAE

Yalcin et al. (2017) evaluated hepatoprotective effects of *Salvia cryptantha* (black weed) plant extract against CCl_4-induced hepatic injury in rats. The antioxidant and antiapoptotic gene transcripts were decreased in all of the control and treatment groups, while Caspase 3 levels were not statistically different. The *S. cryptantha* plant extract treatment was also found to improve SOD, GPx, and CAT levels, while reducing the serum levels of MDA. The extract of *S. cryptantha* supplementation had a protective effect against CCl_4-induced liver damage.

Salvia officinalis L. FAMILY: LAMIACEAE

Amin and Hamza (2005) evaluated the hepatoprotective effect of water extract of *Salvia officinaalis* in Azathioprine-induced hepatotoxicity in rats. Administration of Azathioprine induces oxidative stress through depleting the activities of antioxidants and elevating the level of MDA in liver. This escalates levels of ALT and AST in serum. Pretreatment with the extract proved to have a protective effect against azathioprine-induced hepatotoxicity. Animals pretreated with water extract not only failed to show necrosis of the liver after azathioprine administration, but also retained livers that, for the most part, were histologically normal. In addition, the extract blocked the induced elevated levels of ALT and AST in serum. The azathioprine-induced oxidative stress was relieved to varying degrees by the herbal extract. This effect was evident through reducing MDA levels and releasing the inhibitory effect of azathioprine on the activities of GSH, CAT, and SOD.

Santalum album L. SANTALACEAE

Hydro-alcoholic extract of the leaves of *Santalum album* showed significant hepatoprotective activity against CCl_4 and PCM-induced hepatotoxicity by decreasing the activities of serum marker enzymes, bilirubin and LPO and significant increase in the levels of GSH, SOD, CAT, and protein in a dose-dependent manner, which was further confirmed by the decrease in the total weight of the liver and histopathological examinations (Hegde et al., 2014).

Santolina chamaecyparissus L.
FAMILY: ASTERACEAE

Santolina chamaecyparissus is used traditionally in Ayurvedic system of medicine in India for the treatment of liver diseases, and as a liver tonic. The hydroalcoholic extract of *S. chamaecyparissus* whole plant was tested for its hepatoprotective effect against D-GalN-induced hepatic damage in rats at a dose of 250 mg/kg. The substantially elevated levels of AST, ALT, ALP, TB, LDH, TC, TGL, TP, and ALB (TA) were restored significantly by the extract The hepatoprotective effect was comparable to that of standard Silymarin. Results of histopathological studies supported these findings.

Sapindus mukorossi Gaertn.
FAMILY: SAPINDACEAE

Ibrahim et al. (2008) evaluated the hepatoprotective capacity of *Sapindus mukorossi* extract in CCl_4 treated male rats. The dried powder of *S. mukorossi* was extracted successively with petroleum ether, benzene, chloroform, and ethanol and concentrated in vacuum. Primary rat hepatocyte monolayer cultures were used for *in vitro* studies. *In vivo*, the hepatoprotective capacity of the extract of the fruit pericarp of *S. mukorossi* was analyzed in liver injured CCl_4-treated male rats. Primary hepatocytes monolayer cultures were treated with CCl_4 and extracts of *S. mukorossi*. A protective activity could be demonstrated in the CCl_4 damaged primary monolayer culture. Extract of the fruit pericarp of *S. mukorossi* (2.5 mg/mL) was found to have protective properties in rats with CCl_4-induced liver damage as judged from serum marker enzyme activities.

Sapium ellipticum (Hochst.) Pax
FAMILY: EUPHORBIACEAE

Njouendou et al. (2014) asssesed the hepatoprotective activity of the extract of *Sapium ellipticum* at two different doses (100 and 200 mg/kg) on rat model of TAA-induced hepatotoxicity by investigating biochemical and histopatological markers after 29 days of treatment. Elevated SGOT, SGPT, ALP, TB and DBL observed in TAA toxic groups were restored toward the normal values, when animals received extract treatment at100 and 200 mg/kg. Administration of TAA lowered significantly the level of liver antioxidants markers SOD, CAT, GSH, and increased the level of LPO, which were moderated in group of animals treated with both extracts at 100 and 200 mg/kg. The activity of liver microsome CYP2E1 was significantly high when animals received TAA treatment; both extracts reduced the increase in enzyme activity. The histological and biochemical changes exhibited by *Sapium ellipticum* extract on toxic animals were comparable to those obtained with standard Silymarin. The hepatoprotective effects of AC extract observed in this study showed that these two medicinal plants are potential sources of antihepatotoxic drugs.

Sapium sebiferum (L.) Roxb. FAMILY: EUPHORBIACEAE

Sapium sebiferum leaves were used to determine its hepatoprotective effects against PCM-induced hepatotoxicity in mice (Hussain et al., 2015). A dose-dependent study was conducted using two different doses (200 and 400 mg/kg) of the extract of *S. sebiferum* against toxic effects of PCM (500 mg/kg) in experimental animal model. Paracetamol significantly increased the serum levels of liver enzyme markers like ALT, AST, ALP, TB, and DBL. The extract showed protective effects by normalizing the liver enzymes markers in a dose-dependent manner. Histopathological results confirmed the hepatoprotective effects of leaves of *S. sebiferum*.

Saponaria officinalis L. FAMILY: CARYOPHYLLACEAE

Talluri et al. (2018) carried out studies on root parts (rhizomes) of *S. officinalis* for phytochemical analysis and hepatoprotective activity on paracetmol-induced liver toxicity. The phytochemical analysis of *S. officinalis* roots' extracts showed presence of sterols, terpenoids, glycosides, carbohydrates, proteins, flavanoids, alkaloids, phenols, tannins and absence of saponins and oils. The methanolic extract showed more phenolic and alkaloid contents on their quantification. The *S. officinalis* roots extracts are found to be safe at 2000 mg/kg b. w. in acute toxicity study and showed dose-dependent percentage protection on liver toxicity. Methanol extract showed more activity at 500 mg/kg b. w. and is comparable with standard drug Liv. 52 on altered liver biomarker enzymes AST (SGOT), ALT (SGPT), ALP, TB and TP with percentage protection 66.67%, 60.63%, 65.93%, 64.24%, and 60.98%.

Saraca asoca (Roxb.) Willd. FAMILY: FABACEAE

Arora et al. (2015) investigated possible hepatoprotective activity of methanolic and hydroalcoholic extract of *Saraca asoca* stem bark. Administration of hepatoxin CCl_4 showed significant biochemical and histological deterioration in the liver of experimental animals. Pretreatment with methanolic extract more significantly and to a lesser extent hydroalcoholic extract reduced the elevated levels of serum enzymes like serum SGOT,

SGPT, and ALP and reversed the hepatic damage which evidenced the hepatoprotective activity of stem bark of *S. asoca*.

Sarcostemma brevistigma Wight FAMILY: APOCYNACEAE

Sethuraman et al. (2003) evaluated the hepatoprotective activity of ethyl acetate extract of stem of *Sarcostemma brevistigma* in male albino rats. A significant reduction was observed in SGPT, SGOT, ALP, TB, and GGTP levels in the groups treated with Silymarin and ethyl acetate extract of *S. brevistigma*. The enzyme levels were almost restored to normal. It was observed that the size of the liver was enlarged in CCl_4-intoxicated rats but it was normal in drug-treated groups. A significant reduction in liver weight supports this finding.

Saururus chinensis (Lour.) Baill. FAMILY: SAURURACEAE

Saururus chinensis has been used in Chinese folk medicine for the treatment of various diseases, such as edema, jaundice, gonorrhea, antipyretic, diuretic, and anti-inflammatory agents. Wang et al. (2009) evaluated the hepatoprotective and antifibrotic effects of *S. chinensis* extract in CCl_4-induced liver fibrosis rats. The extract (70 mg/kg) was administrated via gavage once a day starting from the onset of CCl_4 treatment (14 weeks) for subsequent 8 weeks. Evaluated with liver index, serum levels of ALT, AST, hyaluronic acid (HA), hepatic MDA content, and SOD activity, TC, TG, total lipoprotein (TP), ALB, HYP, total antioxidant capacity (T-AOC), laminin (LN), type III collagen terminal peptide (PC-IIINP), and type IV collagen (IV-C), as well as with histopathologic changes of liver. SC-E effectively reduced the elevated levels of liver index, serum ALT, AST, HA, and hepatic MDA contents, enhance the reduced hepatic SOD activity in CCl_4-treated rats. The histopathological analysis suggested that extract of *S. chinensis* obviously alleviated the degree of liver fibrosis induced by CCl_4.

Schisandra chinensis (Turcz.) Baill. FAMILY: SCHISANDRACEAE

Yan et al. (2009) investigated the synergistic hepatoprotective effect of lignans from Fructus *Schisandra chinensis* (LFS) with *Astragalus* polysaccharides (APS) on chronic liver injury in male SD rats. Subcutaneous injection of 10% CCl_4 twice a week for 3 months resulted in significantly elevated serum ALT, AST, ALP activities compared to controls. In the liver, significantly elevated levels of MDA, lowered levels of reduced GSH and CAT, SOD were observed following CCl_4 administration. "LFS+ASP" treatment of rats at doses of "LFS (45 mg/kg) +APS (150 mg/kg)" and "LFS (135 mg/kg)+APS (450 mg/kg)" displayed hepatoprotective and antioxidative effects than the administration of either LFS or APS, as evident by lower levels of serum ALT, AST, ALP, and hepatic MDA concentration, as well as higher SOD, CAT activities, GSH concentration compared to the toxin treated group. Histopathological examinations revealed severe fatty degeneration in the toxin group, and mild damage in groups treated with 'LFS+APS' were observed. The coefficients drug interaction between each individual drug and their combination (at the same dose of their single treatment) of these foregoing parameters were all less than 1, indicating that LFS and APS display hepatoprotective and antioxidant properties and act in a synergistic manner in CCl_4-induced liver injury in rats.

Cheng et al. (2013) investigated the antioxidant and hepatotective effects of *S. chinensis* pollen extract (SCPE) on CCl_4-induced acute liver damage in mice. Total phenolic content, TFC, individual phenolic compounds and antioxidant activities DPPH radical scavenging activity, chelating activity, and reducing power assay) were determined. In vivo study, SCPE (10, 20, and 40 g/kg) administered daily orally for 42 days prior to CCl_4-intoxicated. SCPE had high TPC (53.74±1.21 mg GAE/g), TFC (38.29±0.91 mg Rutin/g), quercetin and hesperetin may be the major contributor to strong antioxidant activities. Moreover, SCPE significantly prevented the increase in serum ALT and AST level in acute liver damage induced by CCl_4, decreased the extent of MDA formation in liver and elevated the activities of SOD and GSH-Px (GSH-Px) in liver. The results indicated that SCPE has strong antioxidant activities and significant protective effect against acute hepatotoxicity induced by CCl_4, and have been supported by the evaluation of liver histopathology in mice. The hepatoprotective effect may be related to its free-radical scavenging effect, increasing antioxidant activity and inhibiting LPO.

Schouwia thebaica Webb. FAMILY: BRASSICACEAE

Awaad et al. (2006) investigated the hepatoprotective activity of the total ALEs of *Schouwia thebaica*. The total extract was fractionated in turn with diethyl ether, chloroform, ethyl acetate, and *n*-butanol, respectively. These extracts were tested for possible hepatoprotective activity. It was found that the ethyl acetate and *n*-butanol extracts showed hepatoprotective activity. These extracts significantly reduced the increase in activities of ALT, AST, and GGT, and levels of glucose, TGs, and cholesterol in serum of CCl_4-treated rats. The extracts showing activity were found to contain flavonoids; one new compound, chrysoeriol-7-*O*-xylosoide-(1,2)-arabinofuranoside (2), in addition to another known four compound chrysoeriol (1), quercetin (3), quercetin-7-*O*-rhamnoside (4), and kaempferol-3-O-beta-D-glucoside (5). The isolated new compound was mainly found to be responsible for this activity when tested on animals in the laboratory.

Scoparia dulcis L. FAMILY: PLANTAGINACEAE

Praveen et al. (2009) assessed the hepatoprotective activity of 1:1:1 petroleum ether, diethyl ether, and methanol (PDM) extract of *Scoparia dulcis* against CCl_4-induced acute liver injury in mice. The extract at the dose of 800 mg/kg, p.o., significantly prevented CCl_4-induced changes in the serum and liver biochemistry and changes in liver histopathology. The above results were comparable to standard, Silymarin (100 mg/kg, p.o.). In the *in vitro* DPPH scavenging assay, the extract showed good free-radical scavenging potential.

Tsai et al. (2010) investigated the hepatoprotective activity and active constituents of the ethanol extract of *S. dulcis*. The hepatoprotective effect of ethanol extract (0.1, 0.5 and 1 g/kg) was evaluated on the CCl_4-induced acute liver injury. Mice pretreated orally with ethanol extract (0.5 and 1.0 g/kg) and Silymarin (200 mg/kg) for five consecutive days before the administering of a single dose of 0.2% CCl_4 (10 mL/kg of BW, ip) showed a significant inhibition of the increase of ALT and AST. Histological analyses also showed that ethanol extract (0.5 and 1.0 g/kg) and Silymarin reduced the extent of liver lesions induced by CCl_4, including vacuole formation, neutrophil infiltration and necrosis. Moreover, ethanol extract decreased the MDA level and elevated the content of reduced GSH in the liver as compared to those in the CCl_4 group. Furthermore, ethanol extract (0.5 and 1.0 g/kg) enhanced the activities of antioxidative enzymes including SOD, GPx, GR, and GST. The quantities of active constituents in ethanol extract were about 3.1 mg luteolin/g extract and 1.1 mg apigenin/g extract. The hepatoprotective mechanisms of ethanol extract were likely associated to the decrease in MDA level and increase in GSH level by increasing the activities of antioxidant enzymes such as SOD, GPx, GRd, and GST.

Scrophularia hypericifolia Wydler FAMILY: SCROPHULARIACEAE

The hepatoprotective and nephroprotective effects of the ethanol extract of the aerial parts of *Scrophularia hypericifolia* growing in Saudi Arabia were evaluated by Alqasoumi (2014) at 250 and 500 mg/kg doses using Wistar albino rats as experimental animal model. The ethanol extract of the aerial parts of *S. hypericifolia* showed dose-dependent moderate level of protection against PCM-induced hepatrotoxicity and nephrotoxicity as indicated from the obtained results. The reduction of the sodium and potassium levels by the higher dose of the extract exceeded that obtained by Silymarin.

Scrophularia koelzii Pennell FAMILY: SCROPHULARIACEAE

The ALE of the aerial parts of *Scrophularia koelzii* (100 mg/kg p.o. × 7 days) significantly protected the rat liver against TAA-induced toxicity. On its further fractionation, activity was localized in the chloroform fraction from which four iridoid glycosides, namely scropolioside-A, koelzioside, harpagoside and 6-0-(3″-0-*p*-methoxycinnamoyl)-α-L-rhamnopyranosyl catalpol were isolated. Among these, scropolioside-A showed maximum hepatoprotective activity. Immunostimulant response was observed with all the four iridoids with harpagoside exhibiting maximum induction of immune response (Garg et al., 1994).

Sechium edule (Jacq.) Sw. FAMILY: CUCURBITACEAE

Ethanolic extract of fruits of *Sechium edule* and its different fractions (100, 200 mg/kg, p.o.) showed significant hepatoprotective activity against CCl_4-induced hepatotoxicity in rats by reducing the levels of AST, ALT, ALP, total bilirubin and hepatic LPO and increasing the levels of antioxidants markers like hepatic GSH, CAT, SOD and TP in a dose-dependent manner, which was confirmed by histopathological examination. Thus, the ethanolic extract of fruits of *S. edule* could protect the liver cells from CCl_4-induced liver damages, by its antioxidative effect on hepatocytes.

Sedum sarmentosum Bunge FAMILY: CRASSULACEAE

The known *δ*-amyrone, 3-epi-*δ*-amyrin, *δ*-amyrin as well as a new hydroperoxy triterpene, 18*β*- hydroperoxy-olean-12-en-3-one (named sarmentolin) were isolated as hepatoprotective agents from *Sedum sarmentosum* (Amin et al., 1998).

Selaginella lepidophylla (Hook. & Grev.) Spring. FAMILY: SELAGINELLACEAE

Alcoholic and aqueous extracts of *Selaginella lepidophylla* were evaluated by Tiwari et al. (2014) for their hepatoprotective activity using CCl_4 and PCM-induced acute hepatic injury model. Treatment with CCl_4 and PCM significantly increased liver weight and volume compared to the normal group. Pretreatment with Silymarin, alcoholic and aqueous extracts significantly prevent increase in liver weight and volume. The alcoholic and aqueous extracts of *S. lepidophylla* exhibited significant hepatoprotective activity against CCl_4 and PCM-induced hepatotoxicity in rats, as result showed in physical, biochemical and histopathological parameter.

Senna alata (L.) Roxb. FAMILY: FABACEAE

The leaves of *Senna alata* are well known for their medicinal uses for various diseases of the liver. The hepatoprotective effect of the plant has been shown in Wistar albino rat intoxicated with CCl_4. This study reported that ME and fractions (ethanol and butanol) of *S. alata* leaves administered orally at 400 mg/kg decreased hepatic enzyme levels (serum ALT, AST, ALP,) total and DBL, liver TBARS induced by CCl_4 damage. Administration of the ME of this plant showed maintenance of the hepatocytes membrane's structural integrity (Patrick-Iwuanyanwu et al., 2011). The extract also showed strong antioxidant and anti-inflammatory (Hennebelle et al., 2009), activities which may contribute to its hepatoprotective property.

Senna occidentalis (L.) Link (Syn.: *Cassia occidentalis* L.) FAMILY: FABACEAE

Senna occidentalis is used in Unani medicine for liver ailments and is an important ingredient of several polyherbal formulations marketed for liver diseases. The hepatoprotective effect of aqueous-ethanolic extract (50%, v/v) of leaves of *S. occidentalis* was studied by Jafri *et al.* (1999) on rat liver damage induced by PCM and ethyl alcohol by monitoring serum transaminase (AST and serum ALT), ALP, serum cholesterol, serum total lipids and histopathological alterations. The extract of leaves of the plant produced significant hepatoprotection. An ethanol extract of leaves of *S. occidentalis* was evaluated by Saraf et al. (1994) for its antihepatotoxic activity using CCl_4 and TAA as hepatotoxins.

Usha et al. (2007) analyzed the levels of some known antioxidant (both enzymic and nonenzymic) activities in the roots of *S. occidentalis* to find out the hepatoprotective effect of the same in CCl_4-induced liver damage in albino rats. The roots were found to be rich in antioxidants. To find out the hepatoprotective activity, the aqueous extract of the plant root samples were administrated to rats for 15 days. The serum marker enzymes AST, ALT and, Gama Glutamyl were measured in experimental animals. The increased enzyme levels after liver damage with CCl_4 were nearing to normal value when treated with aqueous extract of the root samples. Histopathological observation also proved the hepatoprotectivity of the root samples.

Chinaka et al. (2011) evaluated the hepatoprotective effects and antioxidant activities of methanol leaf extract of *S. occidentalis* against APAP-induced hepatic injury in rats. The extract at 150 and 300 mg/kg doses significantly reduced APAP-mediated increase in serum transaminases (AST, ALT, and ALP) and bilirubin with increase in total serum protein values. The extract exerted maximal effect at 300 mg/kg. Acetaminophen-induced prolonged pentobarbitone sleeping time was significantly reduced in the groups treated with 150 and 300 mg/kg of the extract. Using the DPPH assay, the extract showed 60% antioxidant activity when compared with ascorbic acid (79%) at 400 μg/mL. FRAP assay gave 1.7 FRAP value compared to 2.0 from ascorbic acid at 400 μg/mL. The extract markedly stabilized rat erythrocyte cell membranes.

Senna singueana (Delile) Lock FAMILY: FABACEAE

Sobeh et al. (2017) profiled the chemical constituents of a ME from *Senna singueana* bark and 36 secondary metabolites were identified. Proanthocyanidins dominated the extract. Monomers, dimers, trimers of (epi) catechin, (epi) gallocatechin, (epi) guibourtinidol, (ent) cassiaflavan, and (epi) afzelechin represented the major constituents. The extract demonstrated notable antioxidant activities *in vitro*: In DPPH (EC_{50} of 20.8 μg/mL), FRAP (18.16 mM $FeSO_4$/mg extract) assays, and TPC amounted 474 mg GAE/g extract determined with the Folin-Ciocalteu method. The extract showed a remarkable hepatoprotective activity against D-GalN-induced hepatic injury in rats. It significantly reduced elevated AST, and TB. Moreover, the extract induced a strong cytoplasmic Bcl-2 expression indicating suppression of apoptosis. In conclusion, the bark extract of *S. sengueana* represents an interesting candidate for further research in antioxidants and liver protection.

Senna surattensis (Burm.f.) Irwin & Barneby FAMILY: FABACEAE

The flavonoidal content, as well as the antioxidant and hepatoprotective activities of leaves of *Senna surattensis* were investigated by El-Sawi and Sleem (2010). Quercetin, quercetin 3-O-glucoside 7-*O*-rahmnoside and rutin were isolated from the ethyl acetate extract. The antioxidant activity was determined by radical scavenging activity of DPPH free radical using ESR spectroscopy. 0.5 mg/mL of the 80% ethanol extract resulted in 99% inhibition of the peak area of DPPH. For the investigation of the hepatoprotective activity, liver damage was induced to male albino rats by intraperitoneal injection of CCl_4. The relative potencies of the extract with respect to Silymarin in reducing serum AST, ALT, and ALP after 30 days of administration were 90%, 85% and 87%, respectively. The significant reduction in AST, ALT, and ALP in plasma indicates the efficacy of *S. surattensis* leaf extract as a hepatoprotective.

Sesamum indicum L. FAMILY: PEDALIACEAE

The hepatoprotective activity of ME of seeds of *Sesamum indicum* was investigated in CCl_4-induced liver injury in rats by Nwachukwu et al. (2011). Results showed significant increase in the levels of biochemical markers of hepatic damage like ALT, AST, and ALP in CCl_4 control group (D) when compared with the baseline control group F. Treatment with ME (200, 400, and 800 mg/kg) prior to CCl_4 administration, significantly protected rats from injury as evident by moderate changes in liver histoarchitecture in groups A-C and significant reduction in levels of ALT, AST, and ALP when compared to the CCl_4 control group. The CCl_4 control group showed marked vacuolar degeneration, inflammatory cell infiltration, and necrosis of hepatic tissue.

Cengiz et al. (2013) investigated sesame (*S. indicum*) for its hepatoprotective effect in CCl_4-induced experimental liver damage. To this end, 0.8 mg/kg of sesame fixed oil was provided intraperitoneally to rats whose livers were damaged by CCl_4. Tissue and blood samples were taken at the end of the experiments and evaluated histologically and biochemically. Ballooning degenerations and an increase in lipid droplets in liver parenchyma and increases in serum ALT, AST, and bilirubin were found in the CCl_4 group. Biochemical and histopathological findings in the sesame fixed oil treated group were not significantly different from the CCl_4 group. *Sesame did not show* a hepatoprotective effect in CCl_4-induced liver toxicity.

Sesbania grandiflora (L.) Pers. FAMILY: FABACEAE

Sesbania grandiflora is widely used in Indian folk medicine for the treatment of liver disorders. Oral administration of an ethanolic extract of *S. grandiflora* leaves (200 mg/kg/day) for 15 days produced significant hepatoprotection against erythromycin estolate (800 mg/kg/day)-induced hepatotoxicity in rats (Pari and Uma, 2003). The increased level of serum enzymes (AST, ALT, ALP), bilirubin, cholesterol, TGs, phospholipids, free fatty acids, plasma TBARS and HP observed in rats treated with erythromycin estolate were significantly decreased in rats treated concomitantly with sesbania extract and erythromycin estolate. The sesbania extract also restored the depressed levels of antioxidants to near normal. The results of the study reveal that sesbania could afford a significant protective effect against erythromycin estolate-induced hepatotoxicity. The effect of sesbania was compared with that of Silymarin, a reference hepatoprotective drug.

Kale et al. (2012) investigated the hepatoprotective activity of ethanolic and aqueous extract of *S. grandiflora* flower in CCl_4-induced hepatotoxicity models in rats. The ethanolic and aqueous extract of *S. grandiflora* flower significantly decreased the biochemical parameters (SGOT, SGPT, ALP, TP, and TB). The extract did not show any mortality up to a dose of 2000 g/kg BW. These findings suggest that the ethanolic and aqueous extract of *S. grandiflora* flower (500 mg/kg) was effective in bringing about functional improvement of hepatocytes. The healing effect of this extract was also confirmed by histological observations.

Bhoumik et al. (2016) evaluated the hepatoprotective activity of aqueous extract of leaves of *S. grandiflora* (AESG) at the dose of 250 and 500 mg/kg BW per oral using CCl_4-induced liver damage in Wistar albino rats. Aqueous extract showed significant hepatoprotective effect by lowering the serum levels of various biochemical parameters such as SGOT, SGPT, ALP, TBIL, total cholesterol (CHL) and by increasing the levels of TN and ALB, in the selected model. These biochemical observations were inturn confirmed by histopathological examinations of liver sections and were comparable with the standard hepatoprotective drug Silymarin (100 mg/kg BW i.p.) which served as a reference control.

Sida acuta Burm. f. FAMILY: MALVACEAE

Sreedevi et al. (2009) evaluated the hepatoprotective properties of the methanolic extract of the root of *Sida acuta* (SA) and the phytochemical analysis of SA. Significant hepatoprotective effects were obtained against liver damage induced by PCM overdose as evident from decreased serum levels of SGPT, SGOT, ALP, and bilirubin in the SA treated groups (50, 100, 200 mg/kg) compared to the intoxicated controls. The hepatoprotective effect was further verified by histopathology of the liver. Pretreatment with *S. acuta* extract significantly shortened the duration of hexobarbitone-induced narcosis in mice indicating its hepatoprotective potential. Phytochemical studies confirmed the presence of the phenolic compound, ferulic acid in the root of *S. acuta*, which accounts for the significant hepatoprotective effects.

Sida cordata (Burm.f.) Borssum FAMILY: MALVACEAE

Gnanasekaran et al. (2012) evaluated the hepatoprotective activity of ethanolic extract of the whole plant *Sida cordata* on the Chang cell line (normal human liver cells). The percentage viability of the cell line was carried out. The cytotoxicity of *S. cordata* on normal human liver cell was evaluated by the SRB assay [Sulphorhodamine B assay] and MTT assay [(3-(4,5-dimethylthiazole-2 yl)-2,5 diphenyl tetrazolium bromide) assay]. The principle involved is the cleavage of tetrazolium salt MTT into a blue coloured derivative by living cells which contains mitochondrial enzyme succinate dehydrogenase. Cells which are exposed only with toxicant CCl_4 showed a percentage viability of 42% while cells which are pretreated with extract showed an increase in percentage viability and the results were highly significant (when compared to CCl_4 intoxicated cells). The percentage viability ranged from 72% to 78% cell viability, when pretreated with the extracts.

Mistry et al. (2013) investigated the hepatoprotective potential of *S. cordata* in Wistar albino rats. Treatment with leaf extract significantly and dose dependently reduced CCl_4 induced elevated serum level of hepatic enzymes. Furthermore, the extract significantly reduced the LPO in the liver tissue and restored activities of defence antioxidant enzymes GSH, SOD, and CAT toward normal levels, which was confirmed by the histopathological studies.

Sida rhombifolia L. FAMILY: MALVACEAE

Rao and Mishra (1997) evaluated the hepatoprotective effects of different parts of *Sida rhombifolia* on chemical and drug-induced hepatotoxicants and on carrageenan-induced paw oedema in rats. The powdered roots aerial parts and their aqueous extract showed significant hepatoprotective activity.

The ethanolic extract of *S. rhombifolia* whole plants were investigated for its hepatoprotective effect on PCM (2 g/kg/BW/p.o. suspended in 0.5% CMC) induced acute liver damage in Wistar albino rats by Ramadoss et al. (2012). The ethanolic extract of *S. rhombifolia* (SRE) at the dose of (100 and 200 mg/kg/p.o.) produced significant hepatoprotective effect by decreasing the activity of serum enzymes, bilirubin and proteins. The effects of ethanolic extract of *S. rhombifolia* were comparable to that of standard drug Silymarin.

Sideroxylon mascatense (A.DC.) T.D. Penn. [Syn.: *Monothcca buxifolia* (Falc.) A.DC.] FAMILY: SAPOTACEAE

Jan and Khan (2016) evaluated the protective potential of the ME of *Sideroxylon mascatense* (MSM) in rat exposed to CCl_4 toxicity. HPLC-DAD analysis of MBM indicated the existence of GA, catechin, caffeic acid and rutin. MBM administration significantly alleviated the toxic effect of CCl_4 in rat and decreased the elevated level of RBCs, pus, and epithelial cells, specific gravity, creatinine, urobilinogen, urea and ALB while increased the pH and urinary protein. Increase in the level of urobilinogen, BUN, urea and TB while decrease of ALB and TP in serum was restored by the administration of MSM to CCl_4 fed rat. Administration of MSM to CCl_4 exposed rats significantly increased the activity level of phase I and phase II enzymes and GSH while decreased the level of TBARS, H_2O_2, nitrite and DNA damages in renal tissues of rat. Furthermore, histopathological alterations induced with CCl_4 in renal tissues of rat were also diminished with the administration of MSM.

Ullah et al. (2016) investigated the hepatoprotective potential of *S. mascatense* fruit hydroethanolic extract, against isoniazid- and rifampicin-induced hepatotoxicity in rats. Phytochemical investigations lead to the isolation of oleanolic acid and isoquercetin. Pretreatment with *S. mascatense* extract at doses of 150 and 300 mg/kg for 21 days, restored the isoniazid- and rifampicin-induced elevation of serum levels of ALT, AST, ALP, bilirubin and TPs as well as afforded significant protection against histopathological changes in the liver.

Silybum marianum (L.) Gaertn. FAMILY: ASTERACEAE

Madani et al. (2008) evaluated the hepatoprotective effect of polyphenolic extracts of seeds of *Silybum marianum* on TAA-induced hepatotoxicity in rat. The extracts were injected to the rats at a dose of 25 mg/kg BW together with TAA at a dose of 50 mg/kg BW. Significant decrease in the activity of aminotransferases, ALP, and bilirubin was observed in the groups treated with extracts and TAA compared with the group that was treated only with TAA.

Ethyl acetate (100 mg/kg BW) and ethanol seed extracts for *S. marianum* (100 mg/kg BW) were tested against the injection (i.p.) by CCl_4 (2 mL/kg BW) the inducer of liver damage by Shaker et al. (2010).

Their activity was compared with standard hepatic drug hepaticum (100 mg/kg BW) for 10 days. Ethanolic extract showed the most significantly decrease in the liver enzymes. For the oxidative experiments, ethyl acetate showed the most increase for GSH level and the risk factor HDL/LDL significantly. Hepaticum was the most powerful group for the significant decreasing for MDA and fucosidase activity. Some equal improvements were noticed in the histopathological studies for the protective groups.

The hepatoprotective and antioxidant effects of ethanolic extract of *S. marianum* plant were evaluated by Ramadan et al. (2011) by measuring liver function, tissue antioxidant enzymes and histological examination of liver. Antioxidant enzymes as SOD, CAT, and GSH were measured in liver homogenate. In vitro determination of the extract activity was compared to DPPH and measured spectrophotometrically. Oral administration of *S. marianum* extract significantly decreased liver enzyme activity when given in repeated doses. The small and large doses increased the activity of antioxidant enzyme. The obtained results proved the protective effect of ethanolic extract on liver cells. The protective effect of this extract may be attributed to presence of flavonoids compounds and their antioxidant effects.

Simarouba amara Aublet FAMILY: SIMAROUBACEEAE

Maranhão et al. (2014) investigated the hepatoprotective effects of the aqueous extract of *S. amara* stem bark on CCl_4-induced hepatic damage in rats. The extract was evaluated by HPLC. The aqueous extract of stem bark decreased the levels of liver markers and LPO in all doses and increased the CAT levels at doses 250 and 500 mg/kg. Immunohistochemical results suggested hepatocyte proliferation in all doses. These results may be related to catechins present in the stem bark extract.

Simarouba glauca DC. FAMILY: SIMAROUBACEAE

The ethanolic and CE of *Simarouba glauca* were tested for their hepatoprotective activity against Para-cetamol-induced hepatic damage by John et al. (2005). The degree of protection was measured by using biochemical parameters like SGPT, SGOT, and ALP. Pretreatment with Silymarin (100 mg/kg, p.o.), Chloroform extract (200 and 400 mg/kg) and ethanol extract (200 and 400 mg/kg) of leaves of *S. glauca* for 5 days has significantly reduced the elevated serum enzyme level when compared with that of the hepatoxic control group. Both chloroform and ethanol extracts of leaves of *S. glauca* reduced the histological changes caused by PCM.

Smilax china L. FAMILY: SMILACACEAE

The Antihepatotoxic activity of the ethanolic extract of *Smilax china* roots was evaluated by Raju et al. (2012) using CCl_4-induced hepatotoxicity in albino rats. Administration of CCl_4 resulted in a significant rise in the levels of SGPT, SGOT, ALP, ACP, and bilirubin (total and direct) when compared to the vehicle treated group (Group-I). Pretreatment with test extract significantly reduced the elevated levels of biochemical parameters in dose-dependent manner. The results indicated that the effect of test extract on biochemical markers was found to be less potent than the reference standard, Silymarin.

Smilax regelii Killip & C.V. Morton FAMILY: SMILACACEAE

The hepatoprotective effect of the ethanol extract of roots of *Smilax regelii* has been studied in rats. Ethanolic extract of sarsaparilla significantly inhibited CCl_4 induced rise in AST, ALT, and bilirubin, in serum in rats. The extract showed strong antioxidant, anti-inflammatory, and immunomodulating activities which may contribute to its hepatoprotective property. No known toxicity or side effects have been documented for sarsaparilla; however ingestion of large doses may cause gastric irritation. Phytochemical studies on plant of *S. regelii* showed the presence of cetyl-parigenin, astilbin, beta-sitosterol, caffeoyl-shikimic acids, dihydroquercetin, diosgenin, engeletin, essential oils, epsilon-sitosterol, eucryphin, eurryphin, ferulic acid, glucopyranosides, isoastilbin, isoengetitin, kaempferol, parigenin, parillin, pollinastanol, resveratrol, rhamnose, saponin, sarasaponin, sarsaparilloside, sarsaponin, sarsasapogenin, shikimic acid, sitosterol-d-glucoside, smilagenin, smilasaponin, smilax saponins A-C, smiglaside A-E, smitilbin, stigmasterol and taxifolin, and titogenin (Al-Asmari et al., 2014).

Smilax zeylanica L. FAMILY: SMILACACEAE

The protective effects of the ME (200 and 400 mg/kg) of root and rhizome of *Smilax zeylanica* were studied by Murali et al. (2012) on PCM-induced (1 g/kg) hepatic damage in Wistar rats by estimating the serum levels of AST, ALT, ALP, TPs, TB, and ALB. Sections of liver were observed for histopathological changes in liver architecture. Rats were protected from the hepatotoxic action of PCM as evidenced by the significant reduction in the elevated serum levels of ALT, AST, ALP, TB, and an increased level of TP and ALB with a significant reduction in liver weight, when compared with the PCM control. Silymarin (100 mg/kg) was used as the standard. The biochemical observations were supplemented by the histopathological studies on the liver sections of different groups. The ME of *S. zeylanica* was found to alter the damage caused to hepatocytes by PCM and prevent the leakage of vital serum markers, which confirmed the hepatoprotective effect of this plant.

Solanum alatum Moench. FAMILY: SOLANACEAE

Solanum alatum aqueous extract was investigated by Lin et al. (1995) on carrageenin-induced edema and on CCl_4-induced liver injury. The extract (100 or 200 mg/kg BW) exhibited both anti-inflammatory and hepatoprotective activities. The effects were more prominent at the dose of 200 mg/kg. Histological changes such as necrosis, fatty changes, ballooning degeneration, and inflammatory infiltration of lymphocytes and Kupffer cells around the central veins were concurrently improved by treatment with the *S. alatum* aqueous extract.

Solanum fastigiatum Willd. FAMILY: SOLANACEAE

Sabir and Rocha (2008a) evaluated the potential antioxidant and hepatoprotective activity of aqueous extracts of leaves using *in vitro* and *in vivo* models to validate the folkloric use of the plant. The extract showed inhibition against TBARS, induced by 10 μM $FeSO_4$ and 5 μM sodium nitroprusside in rat liver, brain and phospholipid homogenates from egg yolk. The plant exhibited strong antioxidant activity in the DPPH assay. The aqueous extract also showed significant hepatoprotective activity that was evident by enzymatic examination and brought back the altered levels of TBARS, nonprotein thiol and ascorbic acid to near the normal levels in a dose-dependent manner. Acute toxicity studies revealed that the LD (50) value of the extract is more than the dose 4 g/kg BW of mice.

Solanum indicum L. FAMILY: SOLANACEAE

Parmar et al. (2009, 2010) evaluated the hepatoprotective activity of leaves of *Solanum indicum* against CCl_4 and PCM-induced toxicity in male Wistar rats. Toxin-induced hepatotoxicity resulted in an increase in serum AST, ALT, ALP activity, and bilirubin level accompanied by significant decrease in albumn level. Co-administration of the aqueous extract protects the toxin-induced lipid peroxidatiom, restored altered serum marker enzymes, and antioxidant level toward near normal.

Solanum melongena L. FAMILY: SOLANACEAE

Five varieties of eggplant (*Solanum melongena*) were evaluated by

Akanitapichat et al. (2010) for total phenolic and flavonoid content, antioxidant activity and hepatoprotection against cytotoxicity of *t*-BHP (*t*-BuOOH) in human hepatoma cell lines, HepG2. Pretreatment of HepG2 cells with 50 and 100 μg/mL of SM1-SM5 significantly increased the viability of *t*-BuOOH-exposed HepG2 cells by 14.49 ± 1.14% to 44.95 ± 2.72%. The antioxidant activities of the eggplant were correlated with the total amounts of phenolic and flavonoid. Significant correlation was found between hepatoprotective activities and total phenolic/ flavonoid content and antioxidant activities, indicating the contribution of the phenolic antioxidant present in eggplant to its hepatoprotective effect on *t*-BuOOH-induced toxicity.

Hamzah et al. (2016) evaluated the effect of ME of *S. melongena* on CCl_4-induced liver damage in albino rats. Phytochemical screening revealed the presence of bioactive compounds such as tannins, saponins, flavonoids and alkaloids. The activities of ALT, AST, and ALP significantly increased in CCl_4-induced groups when compared to all treated groups. The administration of ME of *S. melongena* at 500 mg/kg BW and 1500 mg/kg BW decreased the activity of ALT in the treated groups. The activity of SOD and CAT in the CCl_4-induced group was decreased and this was significantly increased on treatment with the extracts at 500 and 1500 mg/kg BW. However, methanolic extract of *S. melongena* at 1500 mg/kg BW showed a more significant increase in the activity of these enzymes. Also, increase in the level of MDA in CCl_4 treated group was observed when compared with the normal group and this was decreased on administration with ME of *S. melongena* at 500 and 1500 mg/kg BW.

Solanum nigrum L. FAMILY: SOLANACEAE

Ethanol extract of *Solanum nigrum* was investigated for its hepatoprotective activity against CCl_4-induced hepatic damage in rats. The ethanol extract showed remarkable hepatoprotective activity. The activity was evaluated using biochemical parameters such as serum AST, ALT, ALP and TB. The histopathological changes of liver sample in treated animals were compared with respect to control (Kuppuswamy et al., 2003).

The effects of *S. nigrum* extract were evaluated on TAA-induced liver fibrosis in mice. Mice in the three TAA groups were treated daily with distilled water and SNE (0.2 or 1.0 g/kg) via gastrogavage throughout the experimental period. SNE reduced the hepatic HYP and L-smooth muscle actin protein levels in TAA treated mice. SNE inhibited TAA-induced collagen (L1) (I), transforming growth factor-M1

(TGF-M1) and mRNA levels in the liver. Histological examination also confirmed that SNE reduced the degree of fibrosis caused by TAA treatment. Oral administration of SNE significantly reduced TAA-induced hepatic fibrosis in mice, probably through the reduction of TGF-M1 secretion (Lin et al., 2008).

Mir et al. (2010) evaluated the efficacy of *S. nigrum* on the liver functions in CCl_4-induced injuries. Post treatment of rats with aqueous extracts of plant (500 mg/orally, two doses with 24 h interval) prevented CCl_4-induced rise in activity of serum Transaminases (ALT and AST) and ALP and ALE did not prevent the rise of same enzymes compared to the sham control group in which liver was damaged by CCl_4 no treatment given. Histological examination of the liver of treated animals with aqueous extract of plant showed that fatty acids change was less in comparison to the sham control group. In treated group reduction in BW was minimal and liver enlargement was also less compared to the animals in sham control group.

S. nigrum aqueous and methanolic extracts were studied for hepatoprotective activity in rats injected with 0.2 mL/kg CCl_4 for 10 consecutive days. The water extracts showed a hepatoprotective effect against CCl_4-induced liver damage, which was evident by the decrease in serum AST, ALT, and ALP activities bilirubin concentration and by mild histopathological lesions when compared with the group of rats injected with CCl_4 alone. The methanolic extracts of *S. nigrum* (250 to 500 mg/kg) also had hepatoprotective effects with levels of serum AST, ALT, ALP, and bilirubin decreasing significantly in animals treated with *S. nigrum* methanolic extract compared to an untreated group (Elhag et al., 2011).

Subash et al. (2011) evaluated the hepatoprotective activity of *S. nigrum* in CCl_4 intoxicated albino rats. In the groups where hepatic injury induced by CCl_4 and extract were given simultaneously, the toxic effect of CCl_4 was controlled significantly by maintenance of structural integrity of hepatocyte cell membrane and normalisation of functional status of liver. Histology of liver sections from *S. nigrum* + CCl_4-treated rats revealed moderate centrilobular hepatocytes degeneration, few areas of congestion with mild fatty changes.

Kumar et al. (2012) evaluated the hepatoprotective activity of aqueous extract of *S. nigrum* on CCl_4-induced hepatic damage in Albino Wistar rats. ALT, ALP, and total bilirubin were estimated and liver was subjected to histopathological examination. ALT, ALP, and total bilirubin levels were significantly increased in CCl_4 treated group but *S. nigrum* displayed significant limitation of rise in these parameters. This hepatoprotection was also reflected in histologal changes.

The extracts of whole plant of *S. nigrum* significantly attenuated CCl_4 (Raju et al., 2003) and TAA (Hsieh et al., 2008)-induced biochemical (serum AST, ALT, APT, and TB) and histopathological changes in liver. The hepatoprotective action of *S. nigrum* may be attributed to its antioxidant and anti-inflammatory constituents. Phytochemical studies on *S. nigrum* showed the presence of glycoalkaloids, glycoproteins, polysaccharides, GA, catechin, protocatechuic acid, caffeic acid, epicatechin, rutin, and naringenin (Al-Asmari et al., 2014).

Solanum paniculatum L. FAMILY: SOLANACEAE

de Souza et al. (2019) evaluated the chemical composition and the hepatoprotective and analgesic activities of *S. paniculatum* leaf extracts. HPLC-ESIMS analysis of the SPOE fraction tentatively identified 35 flavonoids, esters of hydroxycinnamic acid and isomers of chlorogenic acid. SPAE (600 and 1200 mg/kg BW) and SPOE (300 mg/kg BW) antagonized the rise in ALT and AST, and the depletion of GSH, and elevation of TBARs levels in the liver caused by AP. The liver protective effects of SPOE (300 mg/kg BW) against AP-induced liver toxicity mimicked those of *N*-acetyl-cysteine (NAC 300 or 600 mg/kg BW ip).

Solanum pseudocapsicum L. FAMILY: SOLANACEAE

The total alkaloid fraction of the ME of leaves of *Solanum pseudocapsicum* was tested by Vijayan et al. (2008) for its hepatoprotective activity against CCl_4-induced toxicity in freshly isolated rat hepatocytes, HepG2 cells and animal models. The total alkaloid fraction was able to normalise the levels of AST, ALT, ALP, TGL, TPs, ALB, TB and DBL, which were altered due to CCl_4 intoxication in freshly isolated rat hepatocytes and also in animal models. The antihepatotoxic effect of the total alkaloid fraction was observed at very low concentrations (6–10 μg/mL) and was found to be superior to that of the standard used. A dose dependent increase in the percentage viability was observed when CCl_4 exposed HepG2 cells were treated with different concentrations of the total alkaloid fraction. The highest percentage viability of HepG2 was observed at a concentration of 10 μg/mL. Its *in vivo* hepatoprotective effect at 20 mg/kg BW was comparable with that of the standard at 250 mg/kg BW. The total alkaloid fraction merits further investigation to identify the active principles responsible for the hepatoprotective properties (Vijayan et al., 2008). The results from this investigation also indicate well correlation between the *in vivo* and *in vitro* studies.

Solanum pubescens Willd. FAMILY: SOLANACEAE

Hemamalini et al. (2012) evaluated the hepatoprotective activity of methanolic extracts of *Solanum pubescens* against PCM-induced liver damage in rats. The methanolic extracts of *S. pubescens* (300 mg/kg) were administered orally to the animals with hepatotoxicity induced by PCM (3 g/kg). Silymarin (25 mg/kg) was given as reference standard. The plant extract was effective in protecting the liver against the injury induced by PCM in rats. This was evident from significant reduction in serum enzymes ALT, AST, ALP, and total bilirubin (TB).

Ethanol extract of *S. pubescens* was evaluated by Pushpalatha and Ananthi (2010) for hepatoprotective and antioxidant activities in rats. The plant extract (500 mg/kg/day) showed a remarkable hepatoprotective and antioxidant activity against CCl_4-induced hepatotoxicity as judged from the serum marker enzymes and antioxidant levels in liver tissues. CCl_4 induced a significant rise in AST, ALT, ALP, TB, LPO with a reduction of TP, SOD, CAT, and reduced GSH. Treatment of rats with plant extract (500 mg/kg) significantly altered serum marker enzymes and antioxidant levels to near normal against CCl_4-treated rats. The activity of the extract at dose of 500 mg/kg was comparable to the standard drug, Silymarin (50 mg/kg, p.o.). Histopathological examination of the liver tissues supported the hepatoprotective activity of plant.

Solanum torvum Sw. FAMILY: SOLANACEAE

Mohan et al. (2011) examined the protective effect of *Solanum torvum* in Doxorubicin (DOX)-induced hepatotoxicity in Wistar rats. Treatment with *S. torvum* (100 and 300 mg/kg) significantly decreased the levels of ALT, AST, and increased the antioxidant defence enzyme levels of SOD and CAT. Histopathological changes showed that DOX caused significant structural damages to liver like inflammation, congestion and necrosis which was reversed with *S. torvum*.

Solanum trilobatum L. FAMILY: SOLANACEAE

Protective action of *Solanum trilobatum* extract (STE) was evaluated by Shahjahan et al. (2004) in an animal model of hepatotoxicity induced by CCl_4. Levels of marker enzymes such as ALT, AST, ALP, and LDH were increased significantly in CCl_4-treated rats. STE brought about a significant decrease in the activities of all these enzymes. Lipid peroxidation (LP) was increased significant in liver tissue in the CCl_4-treated rats while the activities of GPx, CAT, and SOD were decreased. STE treatment led to the recovery of these levels to near normal.

Solanum tuberosum L. FAMILY: SOLANACEAE

Han et al. (2006) investigated the hepatoprotective effect of purple potato extract (PPE) against GalN-induced liver injury in rats. PPE (400 mg) was administered once daily for 8 d, and then GalN (250 mg/kg of BW) was injected at 22 h before the rats were killed. Serum TNF-α, LDH, ALT, and AST levels increased significantly after injection of GalN, but PPE inhibited GalN-induced alterations in serum TNF-α, LDH, ALT, and AST levels. Hepatic LPO and GSH levels in the control + GalN group were higher and lower respectively than those in the control group, and those in the PPE + GalN group did not differ from that in the control group. The LPO level in hepatic microsomes treated with 2,2′-azobis (2-amidinopropane) dihydrochloride in the PPE group was significantly lower than that in the control group. This suggests that PPE has hepatoprotective effects against GalN-induced hepatotoxicity via inhibition LPO and/or inflammation in rats.

Singh et al. (2008) investigated the potential of potato peel extract to offer protection against acute liver injury in rats. Rats pretreated with potato peel extract (oral, 100 mg/kg BW/day for 7 days) were administered a single oral dose CCl_4 (3mL/kg BW, 1:1 in groundnut oil) and sacrificed 8 h of post-treatment. Hepatic damage was assessed by employing biochemical parameters (transaminase enzyme levels in plasma and liver [AST; ALT, LDH]). Further, markers of hepatic oxidative damage were measured in terms of MDA, enzymic antioxidants (CAT, SOT, GST, GPx) and GSH (reduced GSH) levels. In addition, the CCl_4-induced pathological changes in liver were evaluated by histopathological studies. Pretreatment of rats with potato peel extract significantly prevented the increased activities of AST and ALT in serum, prevented the elevation of hepatic MDA formation as well as protected the liver from GSH depletion. Potato peel extract pretreatment also restored CCl_4-induced altered antioxidant enzyme activities to control levels. The protective effect of the extract was further evident through the decreased histological alterations in liver. These findings provide evidences to demonstrate that potato peel extract pretreatment significantly offsets CCl_4-induced liver injury in rats, which may be attributable to its strong antioxidant propensity.

Solanum xanthocarpum Schrad. & W. Wendl. FAMILY: SOLANACEAE

The EtOH extract from whole *Solanum xanthocarpum* was tested on Wistar rats. This extract was administered at doses of 125 and 250 mg/kg during 28 days and liver damage was induced with INH/RIF (50 mg/kg of each); Silymarin at 100 mg/kg was utilized as positive control. The EtOH extract showed a hepatoprotective effect through regulation to normal values of liver enzyme (ALP, AST, and ALT) levels in serum; in addition, it increased antioxidant activity in liver (SOD, CAT, and GSH). The authors associated this beneficial effect with the content of secondary metabolites such as alkaloids and flavonoids, which decrease the oxidative damage on liver cells (Verma et al., 2015). On the other hand, the EtOH extract from *S. xanthocarpum* (fresh and matured fruits) at 100 200 and 400 mg/kg was administered by i.g. via in the Wistar rats during 35 d. Hepatotoxicity damage was induced with INH/RIF/PZA (7.5, 10, and 35 mg/kg each) and Silymarin was used as positive control (100 mg/kg). This extract reduce the LPO levels, restored the endogenous antioxidant system (GSH, SOD, and CAT), reduced hepatocellular necrosis and inflammatory cell infiltration; this effect was dose dependent (Hussain et al., 2012).

The hepatoprotective and antioxidant effects of total extracts and steroidal saponins of *S. xanthocarpum* (Sx) and *S. nigrum* (Sn) on PCM-induced hepatotoxicity and the possible mechanism involved therein in rats were investigated by Gupta et al. (2009). The total extracts and steroidal saponins of Sx and Sn at two dose levels (100 and 200 mg/kg, p.o.) and a standard hepatoprotective herbal drug Silymarin (25 mg/kg BW/day p.o. for 7 days) were administered to the PCM (500 mg/kg p.o. × 7 days) intoxicated rats. The degree of protection was measured by using biochemical parameters, such as, serum transaminase (SGOT and SGPT), ALP, TB and TP estimation. Further, the effects of total extracts and steroidal saponins of Sx and Sn on LPO, reduced GSH, SOD, and CAT status were estimated. The total extracts and steroidal saponins of Sx and Sn at the above doses produced significant hepatoprotective effects by attenuating the activity of serum enzymes, bilirubin and LPO, while they significantly increased the levels of GSH, SOD, and CAT in a dose-dependent manner. The effects of the total extracts and steroidal saponins of Sx and Sn were comparable to those of the standard drug, Silymarin. The antioxidant effects of the compounds may be responsible, at least partly, for the beneficial effects. These compounds have provided remedial measures against the deleterious effects of toxic metabolites of PCM. The hepatoprotective and antioxidant effects of steroidal saponins of Sx and Sn were found to be greater than those of the total extracts.

Gupta et al. (2011) investigated the hepatoprotective potential of *S. xanthocarpum* in experimental rats. Obtained results demonstrated that the treatment with the extract significantly (and dose dependently) prevented chemically induced increase in serum levels of hepatic enzymes. Furthermore, the extract significantly reduced the LPO in the liver tissue and restored activities of defence antioxidant enzymes GSH, SOD, and CAT toward normal levels. Histopathology of the liver tissue showed that the extract attenuated the hepatocellular necrosis and led to reduction of inflammatory cells infiltration.

Singh et al. (2016) investigated the combination of ethanolic fruits extract of *S. xanthocarpum* (SX) and *Juniperus communis* (JC) against PCM and azithromycin-induced liver toxicity in rats. Liver toxicity was induced by combine oral administration of PCM and azithromycin for 7 days in Wistar rats. Fruit extract of *S. xanthocarpum* (200 and 400 mg/kg) and *Juniperus communis* (200 and 400 mg/kg) were administered daily for 14 days. Chronic treatment of SX and JC extract significantly and dose dependently attenuated the liver toxicity by normalizing the biochemical factors and no gross histopathological changes were observed in liver of rats. Furthermore, combined administration of lower dose of SX and JC significantly potentiated their hepatoprotective effect which was significant as compared to their effect per se.

Solidago microglossa DC. FAMILY: ASTERACEAE

The antioxidative and hepatoprotective potential of *Solidago microglossa* was investigated by Sabir et al. (2012). The leaf extract showed inhibition against TBARS induced by different prooxidants in rat liver, brain and phospholipid homogenates from egg yolk. Moreover, the free-radical scavenging activities of the extract were evaluated by the scavenging of DPPH and hydroxyl radical on benzoic acid hydroxylation and deoxyribose assays. The ethanolic extract showed significant hepatoprotective activity against PCM (250 mg/kg)-induced liver damage in mice in a dose-dependent manner. The phenolic composition and their quantification by HPLC resulted in the identification of GA and flavonoids: quercetrin (quercetin-3-*O*-rhamnoside), rutin (quercetin-3-*O*-rutinoside), and quercetin.

Sonchus arvensis L. FAMILY: ASTERACEAE

Sonchus arvensis is traditionally reported in various human ailments including hepatotoxicity in Pakistan. Alkreathy et al. (2014) assessed the protective effects of methanolic extract of *S. arvensis* against CCl_4-induced genotoxicity and DNA oxidative damages in hepatic tissues of male SD rats. Results of this study showed that treatment with methanolic extract reversed the activities of serum marker enzymes and cholesterol profile as depleted with CCl_4 treatment. Activities of endogenous antioxidant enzymes of liver tissue homogenate; CAT, SOD, GSH-Px, GST, and GR were reduced with administration of CCl_4, which were returned to the control level with SME treatment. CCl_4-induced hepatic cirrhosis decreased hepatic GSH and increased lipid peroxidative products (TBARS), were normalized by treatment with SME. Moreover, administration of CCl_4 caused genotoxicity and DNA fragmentation which were significantly restored toward the normal level with the extract.

Sonchus asper (L.) Hill FAMILY: ASTERACEAE

Hepatoprotective activity of ME of *Sonchus asper* against CCl_4-induced injuries in male rats was investigated by Khan et al. (2012). The administration of ME of *S. asper* and Silymarin significantly lowered the CCl_4-induced serum levels of hepatic marker enzymes (AST, ALT, and LDH), cholesterol, LDL, and TGs while elevating high-density lipoprotein levels. The hepatic contents of GSH and activities of CAT, SOD, GPx, GST, and GR were reduced. The levels of TBARS that were increased by CCl_4 were brought back to control levels by the administration of SAME and Silymarin. Liver histopathology showed that the extract reduced the incidence of hepatic lesions induced by CCl_4 in rats.

Aftab-Ullah et al. (2015) conducted a study to confirm the hepatoprotective activity of aqueous methanolic extract of *S. asper* in rabbits against PCM-induced hepatotoxicity. The whole plant was evaluated for its *in vivo* activity. Rabbits were divided into 5 groups (2 per group) for acute toxicity study. Biochemical parameters had depicted those alterations. Histopathological studies also supported them. When PCM was administered in lethal doses it blocked the GSH production and it's binding to *N*-acetyl-*p*-benzoquinone imine, then extra free PCM metabolite binds to cellular proteins and caused hepatocellular damage and elevation in serum enzymes. Extract in 500 and 750 mg/kg had restored these levels significantly due to the presence of phyto constituents like saponins, tannic acid, flavonoids and proanthocyanidins. Their studies showed that the aqueous methanolic extract of *S. asper* has potent hepatoprotective activity upon PCM hepatic damage more appropriately at a dose of 750 mg/kg.

Sonneratia apetala **Buch.-Ham. FAMILY: LYTHRACEAE**

Liu et al. (2019) evaluated the antioxidant activity and hepatoprotective effect of a mangrove plant *Sonneratia apetala* fruit extract (SAFE) on APAP-induced liver injury in mice. The results manifested that SAFE significantly improved survival rates, attenuated hepatic histological damage, and decreased the ALT and AST levels in serum in APAP-exposed mice. SAFE treatment also increased GSH level and GSH-Px activity, enhanced CAT, and total antioxidant capacity (T-AOC), as well as reducing MDA level in liver. In addition, the formation of TNF-α, interleukin 6 (IL-6), and elevation of MPO in APAP-exposed mice were inhibited after SAFE treatment. And SAFE also displayed high DPPH radical scavenging activity and reducing power *in vitro*. The main bioactive components of SAFE such as total phenol, flavonoid, condensed tannin, and carbohydrate were determined.

Spathodea campanulata **P. Beauv. FAMILY: BIGNONIACEAE**

The extract of the stem bark of *Spathodea campanulata* produced significant hepatoprotection (Dadzeasah, 2012). In this study it was reported that the methanolic extracts of the stem bark of *S. campanulata*, at doses of 100, 300, and 625 mg/kg significantly attenuated CCl_4 induced rise in biochemical (serum AST, ALT, and GGT) and histopathological changes in the liver (Dadzeasah, 2012). Phytochemical studies on *S. campanulata* showed the presence of flavonoids, tannins, spathoside, *n*-alkanes, linear aliphatic alcohols, beta-sitosterol-3-*O*-beta-D-glucopyranoside, oleanolic acid, pomolic acid, *N*-hydroxybenzoic acid, phenylethanol esters, reducing sugars. The *in vitro* testing which gave positive results for reducing power and TPC (Dadzeasah, 2012) also support the activity of the plant extract with reference to its hepatoprotection.

Spermacoce hispida **L. FAMILY: RUBIACEAE**

The serum biochemical analysis results suggest that the use of ethanolic extract of *Spermacoce hispida* exhibited significant protective effect from hepatic damage in CCl_4-induced hepatotoxicity model. Histopathological studies revealed that concurrent administration of the extract with CCl_4 exhibited protective effect on the liver, which further evidenced its hepatoprotective activity (Karthikeyan et al., 2011).

Sphaeranthus amaranthoides **Burm.f. FAMILY: ASTERACEAE**

Ethanol extract from *Sphaeranthus amaranthoides* was studied by Swarnalatha and Neelakanta Reddy

(2012) against the D-galactosamine hepatotoxicity. Significant hepatoprotective effect was obtained against liver damage induced by the D-galactosamine as evident from changed antioxidant enzymes like CAT, SOD, GPx, GST, GSH, G6PD, and GR and a normal architecture of liver and mitochondria compared to toxin controls. De et al. (2017) evaluated the *in vitro* hepatoprotective activity of *S. amaranthoides* on CCl_4-induced liver injury. Significant hepatoprotective effect was obtained against CCl_4-induced liver damage as judged from serum marker enzyme activities (SGOT, SGPT, ALT, and TB) and a normal architecture of liver compare to toxic control.

Sphaeranthus indicus L. FAMILY: ASTERACEAE

Tiwari and Khosa (2008) investigated the protective effect of *Sphaeranthus indicus* against CCl_4-induced hepatotoxicity and the mechanism underlying these protective effects in rats. The animals receiving the extracts of *S. indicus* has shown to possess a significant protective effect by lowering the serum AST, ALT, and ALP. This hepatoprotective action was confirmed by hexobarbitone-induced sleeping time in mice, which was increased by CCl_4 treatment and in addition the extract-stimulated bile flow (choleretic activity) in anaesthetized normal rats.

Sphenocentrum jollyanum Pierre FAMILY: MENISPERMACEAE

The methanolic extract of *Sphenocentrum jollyanum* stem bark showed significant hepatoprotective activity against CCl_4-induced liver injury. At a concentration of 50, 100, and 200 mg/kg, the extract showed a remarkable hepatoprotective and antioxidant activity against CCl_4-induced liver injury in a concentration dependent manner. This was evident from significant reduction in serum marker enzymes, AST, ALT, ALP, TB and LPO. It was also observed that the extract of *S. jollyanum* stem bark conferred a significant protection against CCl_4-induced TP, SOD, CAT, GPx, and GST depletion in the liver. The plant extract restored the activities of the marker enzymes to near normal. In addition, this extract possesses significant antioxidant activities with IC_{50} values of 13.11 and 30.04 μg/mL in superoxide and hydrogen radical scavenging activity, respectively and anti-inflammatory, activities which may be contributing to its hepatoprotective effects (Olorunnisola et al., 2011).

Spilanthes ciliata Kunth FAMILY: ASTERACEAE

Shyamal et al. (2010) evaluated the protective effects of ethanolic extracts whole plant of *Spilanthes ciliata* on the aflatoxin B1 (AFB1)-intoxicated livers of albino male Wistar rats. Pretreatment of the rats with oral administration of the plant ethanolic extract prior to AFB1 was found to provide significant protection against toxin-induced liver damage, determined 72 hours after the AFB1 challenge (1.5 mg/kg, intraperitoneally) as evidenced by a significant lowering of the activity of the serum enzymes and enhanced hepatic reduced GSH status. Pathological examination of the liver tissues supported the biochemical findings. The plant extract showed significant antilipid peroxidant effects *in vitro*.

Spondias mombin L. FAMILY: ANACARDIACEAE

Hameno (2010) evaluated the hepatoprotctive activity of leaf extract of *Spondias mombin* on CCl_4-induced hepatotoxicity. The extract was found to possess profound therapeutic ability as it decreased bilirubin levels in the group treated with CCl_4. Additionally, ALP levels decreased from 219.7±30.02 U/L in CCl_4 only treated group to 40.25 ± 2.39 U/L in groups treated with CCl_4 and 1000 mg/kg, respectively. This effect was clearly evident in histopathological studies of livers of rats. The therapeutic ability of the aqueous extract was comparable to Silymarin.

The *in vitro* antioxidant activities and hepatoprotective effects of methanol leaf extract of *S. mombin* in rats intoxicated with DMN were investigated by Awogbindin et al. (2014). Results revealed a copious flavonoid content in the extracts. The extracts showed high DPPH radical scavenging activity and a significant inhibition of AAPH-induced LPO. The MEs also showed strong reductive potential. Acute oral DMN administration led to hepatotoxicity as evident by elevated levels of ALT and AST. The antioxidant status and oxidative stress were monitored by determining the levels of hepatic reduced GSH, hydrogen peroxide (H_2O_2), and GPx activity. H_2O_2 generation was significantly enhanced, GSH level was significantly reduced and GPx activity was significantly induced in DMN intoxicated group. However, pretreatments with the extracts, at 100 and 200 mg/kg, ameliorated the levels of ALT, AST and H_2O_2. In addition, the induction of GPx was also decreased by the two extracts at both doses. Moreover, the extracts significantly raised the level of nonenzymic antioxidant, GSH.

S. mombin is used in folk medicine in Nigeria for the treatment of hepatitis. Nwidu et al. (2017, 2018) evaluated the *in vivo* hepatoprotective and antioxidant effects of *S. mombin* leaf (SML) and *S. mombin* stem (SMS) methanolic extracts in a rat model of hepatotoxicity. CCl_4 treatment-induced liver injury, with significantly increased levels of markers of hepatocellular injury

ALT, AST, ALP, TBIL, and CB, as well as a significant reduction of total circulatory protein. SML or SMS plant extracts at 500 and 1000 mg/kg prior to CCl_4 treatment significantly ameliorated liver injury, and decreased the levels of ALT, AST, TBIL, and CB. SML or SMS extracts significantly increased cellular levels of GSH, the activities of CAT and SOD, and significantly decreased TBARS.

Spondias pinnata **(L.f.) Kurz. FAMILY: ANACARDIACEAE**

The ethyl acetate and methanolic extracts of *S. pinnata* stem heartwood possess a marked *in vivo* hepatoprotective effect on CCl_4 intoxicated rats. The ethyl acetate and methanolic extracts were administered at doses of 100, 200, and 400 mg/kg, p.o., and the results showed a protective activity in a dose-dependent manner as evidenced by the significant decreases in ALT and AST to their normal levels, which was comparable to Silymarin. The hepatoprotective effect in this study was attributed to the presence of flavonoids. Histopathological examination was also carried out on CCl_4 intoxicated rats and revealed that normal hepatic architecture was retained in rats treated with *S. pinnata* extracts (Rao and Raju, 2010).

Hazra et al. (2013) evaluated the effect of *S. pinnata* stem bark ME on iron-induced liver injury in mice. Intraperitoneal administration of iron dextran induced an iron overload and led to liver damage along with a significant increase in serum hepatic markers (ALT, AST, ALP, and bilirubin). The administration of *S. pinnata* ME in doses of 50, 100, and 200 mg/kg induced a marked increase in antioxidant enzymes, along with dose-dependent inhibition of LPO, protein oxidation, and liver fibrosis. Meanwhile, the levels of serum enzyme markers and ferritin were also reduced, suggesting that the extract is potentially useful as an iron chelating agent for iron overload diseases.

Stachys pilifera **Benth. FAMILY: LAMIACEAE**

Stachys pilifera has long been used to treat infectious diseases, respiratory and rheumatoid disorders in Iranian folk medicine. Antitumor and antioxidant activity of the plant have been reported. Kokhdan et al. (2017) assessed the hepatoprotective activity of ethanol extract of *S. pilifera* in CCl_4-induced hepatotoxicity in rats. The rats were randomly divided into six equal groups (n = 7). Group I was treated with normal saline; Group II received CCl_4 (1 mL/kg. i.p., twice a week) for 60 consecutive days; Groups III, IV, and V were given CCl_4 plus *S. pilifera* (100, 200, and 400 mg/kg/d, p.o.); Group VI received the extract (400 mg/kg/d, p.o.). Histopathological analysis and measurement of serum AST, ALT, ALP, MDA, TP, and ALB were performed. CCl_4 caused a significant increase in the serum levels of AST, ALT, ALP, and MDA as well as decreased ALB, and TP serum levels. The extract (200 and 400 mg/kg/d) significantly normalized the CCl_4-elevated levels of ALT, AST, ALP, and MDA. The extract (200 and 400 mg/kg/d) also increased the serum levels of TP compared to CCl_4 group. The extract (200 and 400 mg/kg/d) also decreased the histological injuries (inflammation and fatty degeneration) by CCl_4.

Stachytarpheta indica **(L.) Vahl FAMILY: VERBENACEAE**

Hepatoprotective activity of ethanolic extract of *Stachytarpheta indica* was evaluated by Joshi et al. (2010) by CCl_4-induced toxicity with standard hepatoprotective drug as Liv. 52. Ethanolic extract of *S. indica* produced reduction in CCl_4-induced elevated levels of SGPT, SGOT, ALP, and SB and reversed TP in rats indicating hepatoprotective activity at the dose of 200 mg/kg BW and was comparable to that of standard drug Liv. 52 (1 mL/kg BW). Further histopathological studies indicated that the animals pretreated with ethanolic extract of *S. indica* was minimal with distinct preservation of structures and architectural frame of the hepatic cells, and is comparable to standard Liv. 52.

Sterculia setigera **Delile FAMILY: MALVACEAE**

Garba et al. (2018) investigated the potency of stem bark extract of *Sterculia setigera* as a hepatoprotective agent against acute administration (overdose) of APAP in experimental animals. Experimental animals were grouped into six treatments with each group containing five rats. Group 1 was the placebo, Group II was the standard treatment orally administered APAP at a dosage of 250 mg/kg BW and thereafter treated with the standard drug Silymarin at 100 g/kg BW after 6 h, to Group III (negative treatment) was orally administered APAP only, at a dosage of 250 mg/kg BW without follow up treatment with standard drug (Silymarin). Groups IV, V and VI were orally administered 70% methanol stem bark extract at a dosage of 200, 400, and 600 mg/kg BW six hours after being orally administered with the hepatotoxic APAP. The trial treatment was carried out for a period of three weeks. From the result obtained it is clear that the extract substantially protects the liver from the oxidative damage that usually characterise the continuous administration of APAP.

Strychnos potatorum L. FAMILY: LOGANIACEAE

Sanmugapriya and Venkataraman (2006) described the hepatoprotective and antioxidant activities of the seed powder (SPP) and aqueous extract (SPE) of *Strychnos potatorum* seeds against CCl_4-induced acute hepatic injury. Hepatic injury was achieved by injecting 3 mL/kg, s.c. of CCl_4 in equal proportion with olive oil. Both SPP and SPE at the doses 100 and 200 mg/kg, p.o. offered significant hepatoprotective action by reducing the serum marker enzymes like SGOT and SGPT. They also reduced the elevated levels of ALP and SB. Reduced enzymic and nonenzymic antioxidant levels and elevated LPO levels were restored to normal by administration of SPP and SPE. Histopathological studies further confirmed the hepatoprotective activity of SPP and SPE when compared with the CCl_4 treated control groups. The results obtained were compared with Silymarin (50 mg/kg, p.o.), the standard drug. In conclusion, SPE (200 mg/kg, p.o.) showed significant hepatoprotective activity similar to that of the standard drug, Silymarin (50 mg/kg, p.o.).

Suaeda fruticosa Forssk. FAMILY: AMARANTHACEAE

Rehman et al. (2013) evaluated the protective effects of the aqueous-methanolic extract of *Suaeda fruticosa* (500 and 750 mg/kg BW, p.o.) against PCM-induced hepatotoxicity in rabbits. Silymarin 100 mg/kg, p.o. was served as standard. The degree of protection was determined by measuring the levels of biochemical marker such as SGOT, SGPT, ALP, and bilirubin (total). The results showed that extract of *S. fruticosa* significantly decreased the PCM-induced increased levels of liver enzymes and TB in a dose-dependent manner. Histopathological studies also revealed the hepatoprotective effects of *S. fruticosa.*

Suaeda maritima L. FAMILY: AMARANTHACEAE

The ethanolic extracts of *Suaeda maritima* leaves significantly attenuated concanavalin (a hepatotoxin)-induced biochemical (serum AST, ALT, APT, and bilirubin) and histopathological changes in liver (Ravikumar et al., 2011). The extract of plant also showed significant antioxidant, anti-inflammatory, antiviral, and antibacterial activities which may contribute to its hepatoprotective activity. Phytochemical studies on plant of *S. maritima* showed the presence of alkaloid, flavonoid, sterols, phenolic compounds, and tannins.

Suaeda monoica Forsk. ex Gmel FAMILY: AMARANTHACEAE

Ravikumar et al. (2010) investigated the hepatoprotective activity of *Suaeda monoica* on Con A induced male Wistar albino rats. In acute toxicity study, ethanolic extract of *S. monoica* leaves 250–3500 mg/kg BW was administered through oral gavages and observed for the physical signs of toxicity for 14 days. Pretreatment of rats with the plant extract (300 mg/kg BW) showed significant reduction of protective effect by reducing the serum level of AST, ALT, ALP, cholesterol and bilurubin parameters and increased level of PTN and ALB. Histopathological examination of the liver sections revealed that the normal liver architecture was disturbed by hepatotoxin intoxication. The section obtained from the extract group and Silymarin group were retained the cell architecture, although less visible changes observed in the sections of the rats treated with ethanolic extract and Con-A. LD50 was calculated as 3.0 g/kg BW.

Swertia chirata Buch.-Ham. ex Wall. FAMILY: GENTIANACEAE

Mukherjee et al. (1997) investigated the hepatoprotective effect of *Swertia chirata* in albino rats. Intraperitoneal injection of CCl_4 (1 mL/kg BW on every 72 hr. for 16 days) significantly increased AST, ALT, and ALP activities and bilirubin level in rat, but liver glycogen and serum cholesterol levels were decreased. Histologically it produced hepatocytic necrosis especially in the centrilobular region. Simultaneous treatments with *S. chirata* (in different doses, viz., 20, 50, and 100 mg/kg BW daily) and CCl_4 caused improvement at both biochemical and histopathological parameters compared to that of CCl_4 treatment alone but it was most effective when *S. chirata* was administered in a moderate dose (50 mg/kg BW).

The extracts of *S. chirata* were evaluated by Karan et al. (1999a, b) for antihepatotoxic activity using CCl_4, PCM and galactosamine models. The ME of the whole plant was found active at a dose of 100 mg/kg i.p. On fractionating this extract into chloroform soluble and butanol soluble fractions, the activity was retained in the chloroform soluble fraction which was most active at a dose level of 25 mg/kg i.p. with overall protection of 81% and 78% against PCM and galactosamine, respectively. The butanol soluble fraction, rich in bitter secoiridoids, showed marginal or was devoid of significant activity. The protective effect observed against these three hepatotoxins which are different in their mechanisms of inducing hepatotoxicity, suggests broader and nonspecific protection of the liver against these three toxins by nonbitter components of *S. chirata*.

Nagalekshmi et al. (2011) investigated the ability of the extract of *S. chirata* to offer protection against acute hepatotoxicity induced by PCM (150 mg/kg) in Swiss albino mice. Oral administration of *S. chirata* extract (100–200 mg/kg) offered a significant dose-dependent protection against PCM-induced hepatotoxicity as assessed in terms of biochemical and histopathological parameters. The PCM-induced elevated levels of serum marker enzymes such as SGPT, SGOT, ALP, and bilirubin in peripheral blood serum and distorted hepatic tissue architecture along with increased levels of LPOs and reduction of SOD, CAT, reduced GSH, and GPx in liver tissue. Administration of the plant extract after PCM insult restored the levels of these parameters to control (untreated) levels.

Swietenia mahagani (L.) Jacq. FAMILY: MELIACEAE

Study of an aqueous leaf extract of *Swietenia mahagani* in chronic alcohol-induced liver injury in rats exhibited hepatoprotective activity (Udem et al., 2011). Study of a defatted ME of SM bark exerted remarkable hepatoprotective activity against PCM-induced hepatic damage in rats, probably attributable to its augmenting endogenous antioxidant mechanisms (Halder et al., 2011).

Symplocos racemosa Roxb. FAMILY: SYMPLOCACEAE

Wakchaure et al. (2011) evaluated the hepatoprotective activity of ethanol extract of *Symplocos racemosa* bark on CCl_4-induced hepatic damage in rats. CCl_4 with olive oil (1 : 1) (0.2 mL/kg, *i.p.*) was administered for 10 days to induce hepatotoxicity. Bark extract (200 and 400 mg/kg, *p.o.*) and Silymarin (100 mg/kg *p.o.*) were administered concomitantly for fourteen days. The degree of hepatoprotection was measured using serum transaminases (AST and ALT), ALP, bilirubin, ALB, and TP levels. Metabolic function of the liver was evaluated by thiopentone-induced sleeping time. Antioxidant activity was assessed by measuring liver MDA, GSH, CAT, and SOD levels. Histopathological changes of liver sample were also observed. Significant hepatotoxicity was induced by CCl_4 in experimental animals. EESR treatment showed significant dose-dependent restoration of serum enzymes, bilirubin, ALB, TPs, and antioxidant levels. Improvements in hepatoprotection and morphological and histopathological changes were also observed in the EESR-treated rats.

Synedrella nodiflora (L.) Gaertn. FAMILY: ASTERACEAE

Gnanaraj et al. (2016a) evaluated the ability of the crude aqueous extract of *S. nodiflora* leaves to protect against CCl_4-mediated hepatic injury in SD rats. Biochemical, immunohistochemical, histological, and ultrastructural findings were in agreement to support the hepatoprotective effect of *S. nodiflora* against CCl_4-mediated oxidative hepatic damage.

Syringa oblata Lindl. FAMILY: OLEACEAE

Li et al. (2018) evaluated the hepatoprotective activity of *S. oblata* leaves ethanol extract against CCl_4-induced hepatotoxicity in primary hepatocytes and mice with the indicator of GST alpha 1. CCl_4 negatively modulated biochemical parameters and liver antioxidant activities. However, the use of *S. oblata* leaves ethanol extract restored altered-serum biochemical parameters and liver antioxidant activities in a dose-dependent manner. Importantly, the trends in *S*-transferase alpha 1 were similar to ALT and AST level, and *S*-transferase alpha 1 was suggested to be a marker for the evaluation of hepatoprotective activity of *S. oblata* leaves ethanol extract. Histopathological examination showed that CCl_4 causes significant hepatic relative to control group. The above findings suggested that *S. oblata* leaves ethanol extract has hepatoprotective effects against CCl_4-induced hepatic injury and *S*-transferase alpha 1 may be an indicator to evaluate the protective effects of *S. oblata* leaves ethanol extract.

Syzygium aromaticum (L.) Merr. & L.M. Perry FAMILY: MYRTACEAE

El-Hadary and Hassanien (2016) investigated the protective effects of cold-pressed *Syzygium aromaticum* oil (CO) against CCl_4-induced liver toxicity in rats. High levels of monounsaturated fatty acids (39.7%) and polyunsaturated fatty acids (42.1%) were detected in CO. The oil contained high amounts of tocols and phenolics. The LD_{50} value at 24 h was approximately 5950 mg/kg. Treatment with 200 mg/kg CO resulted in a decrease of creatinine, urea, and uric acid levels to 0.86, 32.6, and 2.99 mg/dL, respectively. Levels of TL, TC, TAG, LDL-C, and VLDL-C were decreased to 167, 195.3, 584.5, 74.6, and 39.0 mg/L, respectively, after 8 weeks of treatment. Hepatic MDA levels were reduced and GSH levels were elevated in CO-treated rats. CO reduced the activities of AST, ALT, and ALP as well as kidney function markers, protein, and lipid profiles, respectively. Histopathological examination of liver indicated that CO treatment reduced fatty degenerations, cytoplasmic vacuolization, and necrosis.

Hepatoprotective activity of aqueous extract of clove (*Syzygium aromaticum*) in albino rats was observed against controversial hepatotoxicity effects of

PCM-induced liver toxicity in rats by Thuwaini et al. (2016). Analysis of the treated rats with PCM (600 mg/kg) showed PCM-induced male rat hepatotoxicity represented by significant decline in the serum total ALB. Conversely, the study declared significantly increment, bilirubin, ALT, AST, and ALP as shown in group 2 (induction group) in compared with group1 (control group). While, simultaneous administration of clove aqueous extract (100 and 200 mg/kg) with PCM, was displayed significantly attenuated the adverse changes in the serum total ALB, bilirubine, ALT IU/L, AST, and ALP. The histopathological examination in the liver of rats also encouraging that clove extract markedly diminished the toxicity of PCM and keeps the histoarchitecture of the liver tissue to near normal.

Syzygium cumini L. FAMILY: MYRTACEAE

Moresco et al. (2007) evaluated the effect of the aqueous *Syzygium cumini* leaf extract on hepatotoxicity induced by CCl_4 in rats. A significant increase in the AST and ALT activities occurred after CCl_4 administration alone, which was significantly lowered by preadministration with the aqueous extract of *S. cumini*, but not by a single dose. This suggests that the extract may be useful for liver protection but needs to be given over a significant period and prior to liver injury.

Islam et al. (2015) investigated the hepatoprotective effects of ME of seeds of *S. cumini* in CCl_4-induced stress SD rats. The values of liver function markers were found to be significantly lower while serum protein level was significantly higher in control and treated groups as compared to that of the CCl_4 treated group. Histological examination of liver tissues also indicated that the extract of *S. cumini* seeds in both the doses, and Silymarin protected the liver from CCl_4-induced stress.

Tabernaemontana divaricata (L.) R.Br. ex Roem. & Schult. [Syn.: *Ervatamia coronaria* (Jacq.) Stapf] FAMILY: APOCYNACEAE

Gupta et al. (2004) found, in an *in vivo* study, that ME from leaves of *Tabernaemontana divaricata* produced a significant hepatoprotective effect by decreasing LPO and significantly increasing the level of antioxidant agents such as GSH, SOD, and CAT in a dose-dependent manner.

Poornima et al. (2014) evaluated the hepatoprotective activity of *T. divaricata* against DEN and Fe NTA-induced liver necrosis in rats. In simultaneously treated animals, the plant extract significantly decreased the levels of uric acid, bilirubin, AST, ALT, and ALP in serum and increased the levels of liver marker enzymes in liver. Treatment with the extracts resulted in a significant increase in the levels of antioxidants accompanied by a marked reduction in the levels of MDA when compared to DEN and Fe NTA treated group. When compared with 200 mg/kg BW rats, 400 mg/kg BW rats and 5-fluorouracil treated rats showed better results in all the parameters. The histopathological studies confirmed the protective effects of extract against DEN and Fe NTA-induced liver necrosis.

Tabebuia rosea (Bertol) Bertero ex A.DC. FAMILY: BIGNONIACEAE

Hemamalini et al. (2012) evaluated the hepatoprotective activity of methanolic extracts of *Tabebuia rosea* against PCM-induced liver damage in rats. The methanolic extracts of *T. rosea* (500 mg/kg) were administered orally to the animals with hepatotoxicity-induced by PCM (3 g/kg). Silymarin (25 mg/kg) was given as reference standard. All the test drugs were administered orally by suspending in 1% Tween-80 solution. The plant extract was effective in protecting the liver against the injury induced by PCM in rats. This was evident from significant reduction in serum enzymes ALT, AST, ALP, and total bilirubin (TB).

Tagetes erecta L. FAMILY: ASTERACEAE

Hepatoprotective activity of *Tagets erecta* against CCl_4-induced hepatic damage in rats was investigated by Giri et al. (2011). There was a significant increase in serum ALT, AST, ALP, and bilirubin levels in CCl_4-intoxicated group compared to the normal control group. Ethyl acetate fraction of *T. erecta* (EATE) at the dose of 400 mg/kg orally significantly decreased the elevated serum marker enzymes and level of bilirubin almost to the normal level compared to CCl_4-intoxicated group. Histological changes in the liver of rats treated with 400 mg/kg of EATE extract and CCl_4 showed a significant recovery except cytoplasmic vascular degenerations around portal tracts, mild inflammation and foci of lobular inflammation.

Hepatoprotective activity of *T. erecta* was investigated by Karwani and Sisodia (2015) in CCl_4 intoxicated Albino Wistar rats and the results were compared with, Silymarin. Subcutaneous injection of CCl_4, produced a marked elevation in the level of biochemical markers. Oral administration of *T. erecta* at dose 200 mg/kg, 400 and 600 mg/kg in CCl_4 intoxicated rats showed marked decrease in the level of biochemical markers and results were at par with the effect shown by Silymarin. The results of histopathological analysis were in compliance with the findings of blood biochemical parameter analysis.

Tamarindus indica L. FAMILY: FABACEAE

Protective effect of *Tamarindus indica* was evaluated by Pimple et al. (2007) by intoxicating the rats with PCM. The aqueous extracts of different parts of *T. indica* such as fruits, leaves (350 mg/kg p.o.) and unroasted seeds (700 mg/kg p.o.) were administered for 9 days after the third dose of PCM. Biochemical estimations such as AST, ALT, ALP, TB and TP were recorded on 4th and 13th day. Liver weight variation, thiopentone-induced sleeping time, and histopathology were studied on 13th day. Silymarin (100 mg/kg p.o.) was used as a standard. A significant hepatoregenerative effect was observed for the aqueous extracts of tamarind leaves, fruits and unroasted seeds as judged from the parameters studied.

The hepatoprotective effect of the aqueous extract from *T. indica* fruit (250 and 500 mg/kg) was evaluated against liver damage induced with INH (50 mg/kg) and RIF (100 mg/kg) in Wistar rats; this extract decreased the liver enzyme (AST, ALT, and ALP) level as well as bilirubin and TBARS in serum. Additionally, the extract at 500 mg/kg increased the activity of the antioxidant systems (GSH, SOD, and CAT) with simultaneously lowered values of LPO. The effect observed was better at 500 mg/kg than at lower doses (250 mg/kg); finally, the microscopic structure of hepatocytes was similar to that shown by Silymarin after 14 days of administration (Amir et al., 2015). In addition, tablets with decoction of *T. indica* leaves showed a hepatoprotective effect against CCl_4 liver damage (Rodríguez-Amado et al., 2016).

Tamarix nilotica (Ehrenb.) Bunge FAMILY: TAMARICACEAE

The hydroalcoholic extract of *Tamarix nilotica* flower showed marked hepatoprotective activity against CCl_4-induced liver injury (Abouzid and Sleem, 2011). Experimental studies also showed highly significant antioxidant and anti-inflammatory activities of *T. nilotica* which may contribute its hepatoprotective activity. Phytochemical studies on *T. nilotica* showed the presence of flavonoids, tannins, syringaresinol, isoferulic acid, niloticol, 3-hydroxy-4-methoxycinnamaldehyde, methyl and ethyl esters of GA, para-methoxygallic acid, kaempferol, quercetin 3-*o*-glucuronides, 3-*o*-sulfated kaempferol, 7,4-dimethyl ether, and free flavonols.

Tanacetum balsamita L. subsp. *balsamitoides* (Sch.-Bip.) Griesrson FAMILY: ASTERACEAE

Tütüncü et al. (2010) evaluated the hepatoprotective acivity of diethylether extract *Tanacetum balsamita* subsp. *balsamitoides* against CCl_4-induced acute liver toxicity in rats. Some of the animals died during treatment. Serum ALT levels were higher in three treated groups than those in the CCl_4 groups. Histopathological findings were similar in the CCl_4 and extract treated groups. It was concluded that diethylether extract of *T. balsamita* subsp. *balsamitoides* did not have a protective effect in CCl_4-induced acute liver toxicity in rats and even exacerbated the toxicity.

Tanacetum parthenium (L.) Sch. Bip. FAMILY: ASTERACEAE

Mahmoodzadeh et al. (2017) investigated the effects of *Tanacetum parthenium* extract (TPE) on LPO, antioxidant enzymes, biochemical factors, and liver enzymes in the rats damaged by CCl_4. Pre- and post-treatment with TPE could significantly decrease ALT, AST, ALP, TG, LDL, TC, and glucose levels and increase HDL, and ALB levels and CAT, SOD, and GPx activities compared to the CCl_4-damaged control group.

Tapinanthus bangwensis (Eng. & K.Krause) *Danser* FAMILY: LORANTHACEAE

Ihegboro et al. (2018) determined the cytotoxicological and hepatocurative effect of AF of *Tapinanthus bangwensis* in APAP (PCM)-induced Wistar rats. Data obtained were analyzed by 2-way analysis of variance at 95% confidence level and reported as mean ± standard deviation. The concentrated AF of *T. bangwensis* was found to be 23.3 g (58.25%). Quantitative determination of some vital phytochemicals revealed the following: flavonoid (84.6 ± 0.41 mg/100 g), phenol (147.5 ± 1.07 mg/100 g), tannin (31 ± 0.85 mg/ 100 g), alkaloid (23.45 ± 0.09 mg/100 g), and saponin (0.146 ± 0.0 mg/100 g). Treatment of rats with the aqueous extract of *T. bangwensis* significantly decreased PCM-induced elevation of activities of liver function indices, ALP, ALT, AST, TG, TC level and increased the ALB, TP, and high-density lipoprotein levels. The plant extract also attenuated the PCM elevated LPO product, MDA. The research findings suggest that aqueous extract of *T. bangwensis* is slightly cytotoxic, possesses appreciable antioxidant property and exhibited hepatocurative effect against PCM-induced hepatoxicity.

Taraphochlamys affinis (Griff.) Bremekhu FAMILY: APOCYNACEAE

Huang et al. (2012) evaluated the hepatoprotective activity of *Taraphochlamys affinis* in CCl_4-induced liver toxicity. Rats were orally treated with the total saponins of *T. affinis* (TSTA) daily with administration of CCl_4 twice a week for

8 weeks. Compared to the normal control, CCl_4-induced liver damage significantly increased the activities of ALT, AST, and ALP in serum and decreased the activities of SOD, CAT, GSH-Px, and GSH reductase (GSH-Rd) in liver. Meanwhile content of MDA, which was oxidative stress marker, was increased. Histological finding also confirmed the hepatotoxic characterization in rats. Furthermore, proinflammatory mediators including TNF-α in serum, prostaglandin E2 (PGE2), iNOS, and cyclooxygenase-2 (COX-2) in liver were detected with elevated contents, while expression of xenobiotic metabolizing enzyme—cytochrome P4502E1 (CYP2E1) was inhibited. The results revealed that TSTA not only significantly reversed CCl_4 originated changes in serum toxicity and hepatotoxic characterization, but also altered expression of hepatic oxidative stress markers and proinflammatory mediators, combined with restoring liver CYP2E1 level.

Taraxacum officinale (L.) Weber ex F.H.Wigg. FAMILY: ASTERACEAE

Taraxacum officinale, commonly known as dandelion, has a long history of traditional use in the treatment of hepatobiliary problems. Its root has been shown to have sesquiterpene lactones, triterpenes, carbohydrates, fatty acids (myristic), carotenoids (lutein), flavonoids (apigenin and luteolin), minerals, taraxalisin, coumarins, and cichoriin. Aesculin has been reported from the leaf. Germacrane and guaiane-type sesquiterpene lactones including taraxinic acid derivatives were obtained from the roots of this plant (Tabassum et al., 2010). Also, several flavonoids, for example, caffeic acid, chlorogenic acid, luteolin, and luteolin 7-glucoside, have been isolated from the dandelion (Choi et al., 2010). Ethanolic extract of *T. officinale* was effective on decrease in serum ALT levels (Tabassum et al., 2010). Hydroalcoholic acid extract of the root enhanced levels of SOD, CAT, GST, and LPO (Das and Mukherjee, 2012). Oral administration of extracts of the *T. officinale* roots has proved to increase bile flow (Kumar et al., 2011). Another study distinguished that treatment with root extract of *T. officinale* was effective on reduction of serum ALT and ALP levels in rats (Fallah et al., 2010). Root extract reduces serum AST, ALT, ALP, and LDH activities and increases hepatic antioxidant activities such as CAT, GST, GR, GPx, and GSH. Thus, aqueous extract of *T. officinale* root protects against alcohol-induced toxicity in the liver by elevating antioxidant activity and decreasing LPO. Sesquiterpene lactones in the plant have a protective effect against acute hepatotoxicity induced by the administration of CCl_4 in mice, which was indicated by reduced levels of hepatic enzyme markers, such as serum transaminase (ALT, AST), ALP, and TB (Mahesh et al., 2010).

The protective effects of *T. officinale* root against alcoholic liver damage were investigated in HepG2/2E1 cells and ICR mice by You et al. (2010). When an increase in the production of ROS was induced by 300 mM ethanol *in vitro*, cell viability was drastically decreased by 39%. However, in the presence of hot water extract (TOH) from *T. officinale* root, no hepatocytic damage was observed in the cells treated with ethanol, while ethanol-extract (TOE) did not show potent hepatoprotective activity. Mice, which received TOH (1 g/kg BW/d) with ethanol revealed complete prevention of alcohol-induced hepatotoxicity as evidenced by the significant reductions of serum AST, ALT, ALP, and LDH activities compared to ethanol-alone administered mice. When compared to the ethanol-alone treated group, the mice receiving ethanol plus TOH exhibited significant increases in hepatic antioxidant activities, including CAT, glutathione-S-transferase, GPx, GR, and glutathione. Furthermore, the amelioration of MDA levels indicated TOH's protective effects against liver damage mediated by alcohol *in vivo*. These results suggest that the aqueous extract of *T. officinale* root has protective action against alcohol-induced toxicity in the liver by elevating antioxidative potentials and decreasing LPO.

Colle et al. (2012) evaluated the hepatoprotective activity of *T. officinale* leaf extract against APAP-induced hepatotoxicity. *T. officinale* was able to decrease TBARS levels induced by 200 mg/kg APAP (p.o.), as well as prevent the decrease in sulfhydryl levels caused by APAP treatment. Furthermore, histopathological alterations, as well as the increased levels of serum AST and ALT caused by APAP, were prevented by *T. officinale* (0.1 and 0.5 mg/mL). In addition, *T. officinale* extract also demonstrated antioxidant activity *in vitro*, as well as scavenger activity against DPPH and NO radicals.

Al-Malki et al. (2013) investigated the hepatoprotective activity of *T. officinale* leaves water extract against CCl_4-induced liver toxicity in mice by evaluation of biochemicals parameters. Genomic DNA integrity and histopathological studies of mice liver further supported the hepatoprotective activity. CCl_4 had severe damage on liver which was confirmed by elevated levels of marker enzyme of liver. Normal function of hepatocytes was confirmed by reduced levels of marker enzymes with the treatment of Dandelion leaf water extract.

Hfaiedh et al. (2016) assessed the efficiency of *T. officinale* leaf extract (TOE) in treating sodium dichromate hazards; it is a major environmental pollutant known for its wide toxic manifestations with-induced liver injury. Their results using Wistar rats showed that sodium dichromate significantly increased serum biochemical parameters. In the liver, it was found to induce an oxidative stress, evidenced from increase in LPO and changes in antioxidative activities. In addition, histopathological observation revealed that sodium dichromate causes acute liver damage, necrosis of hepatocytes, as well as DNA fragmentation. Interestingly, animals that were pretreated with TOE, prior to sodium dichromate administration, showed a significant hepatoprotection, revealed by a significant reduction of sodium dichromate-induced oxidative damage for all tested markers.

Tecoma stans (L.) Juss. ex Kunth FAMILY: BIGNONIACEAE

Kameshwaran et al. (2013) screened hepatoprotective activity with ethanolic extract of flowers of *Tecoma stans* (EETS). The extract significantly reduced the elevated serum levels of AST, ALT, ALP, and bilirubin. EETS at the dose of 500 mg/kg, p.o. prevented the increase in liver weight when compared to hepatotoxin treated control while the extract at the dose 250 mg/kg was ineffective except in the PCM-induced liver damage. In the chronic liver injury induced by CCl_4, EEPG at the dose of 500 mg/kg, p.o. was found to be more effective than the extract at the dose of 250 mg/kg, p.o. Histological examination of the liver tissues supported the hepatoprotection.

Tecomella undulata Seem. FAMILY: BIGNONIACEAE

Crude methanolic extract from the bark parts of *Tecomella undulata* was evaluated by Rana et al. (2008) for hepatoprotective activity in rats by inducing liver damage by CCl_4. The methanolic extract at an oral dose of 200 mg/kg exhibited a significant protective effect by lowering serum levels of SGOT, SGPT, ALP, TB and TC and increasing the levels of TP and ALB levels as compared to Silymarin used as a positive control. These biochemical observations were supplemented by histopathological examination of liver sections. The activity may be a result of the presence of flavonoid compounds. Furthermore, the acute toxicity of the extracts showed no signs of toxicity up to a dose level of 2000 mg/kg.

Khatri et al. (2009) evaluated the hepatoprotective activity of stem bark of *T. undulata* against TAA-induced hepatotoxicity. Hepatotoxicity was induced in albino rats by subcutaneous injection of TAA. Ethanolic extract of stem bark of *T.*

undulata (200, 500, and 1000 mg/kg/day was evaluated. Oral administration of *T. undulata* at 1000 mg/kg resulted in a significant reduction in serum AST (31%), ALT (42%), GGTP (49%), ALP (37%), TB (48%) and liver MDA levels (50%), and significant improvement in liver GSH (68%) when compared with TAA damaged rats. Histology of the liver sections of the animals treated with the extracts also showed dose-dependent reduction of necrosis.

Singh and Gupta (2011) investigated the hepatoprotective activity of the ME of *T. undulata* leaves tested against liver damage of albino rats. 30% alcohol and PCM-induced significant increase in TBARS alongwith the alterations in the activities of enzymatic and nonenzymatic antioxidants and serum markers in the liver and serum of treated rats. Simultaneously, oral treatment with *T. undulata* reversed to all the serum and liver parameters, dose dependently, in 30% alcohol and PCM-treated rats. The biochemical results were also compared with the standard drug, Silymarin. These findings indicate the hepatoprotective potential of *T. undulata* leaves against liver damage might be due to the presence of flavonoids, quinones, and other bioactive constituents.

Jain et al. (2012) evaluatied possible hepatoprotective role of isolated compounds from *T. undulata* stem bark (TSB) using *in vitro* and *in vivo* experimental models. TSB-2 and MS-1 accounted for significant cell death whereas; TSB-1, TBS-7, TSB-9, TSB-10, and MS-2 did not register significant cytotoxicity. Further, noncytotoxic components exhibited ascending grade of hepatoprotection *in vitro* (TSB-10<TSB-1<TSB-7<TSB-9<MS-2). Pretreatment of TSB-7 or MS-2 to CCl_4-treated rats prevented hepatocyte damage as evidenced by biochemical and histopathological observations.

Tectona grandis L.f. FAMILY: LAMIACEAE

Sachan et al. (2014) evaluated the hepatoprotective activity of hydroalcoholic extract of *Tectona grandis* leaf in CCl_4-induced hepatotoxic model in Wistar rats. The biochemical parameters (SGOT, SGPT, and ALP) in rat serum were found to be increased in CCl_4-induced hepatotoxicity model. Hydroalcoholic extract of *T. grandis* leaf was found to decrease the biochemical parameters (SGOT, SGPT, and ALP) of rat serum at the doses of 300 and 600 mg/kg. Hydroalcoholic extract of *T. grandis* leaf significantly and dose dependently produced hepatoprotective activity at the doses of 300 and 600 mg/kg.

Telfairia occidentalis Hook.f. FAMILY: CUCURBITACEAE

Fluted pumpkin (*Telfairia occidentalis*) leaf is a darkish-green leafy vegetable popularly used in soup and in herbal preparations for the management of many diseases in Nigeria. Oboh (2005) investigated the hepatoprotective property of ethanolic and aqueous extracts of *T. occidentalis* leaf (earlier confirmed to have a high level of antioxidant activity) against garlic induced-oxidative stress in rat hepatocytes. Oxidative stress was induced in Wistar strain albino rats by over-dosing them with raw garlic (4%) for 14 days, and this caused a significant increase in serum ALP, SGOT, and SGPT, while there was no significant change in SB, ALB, globulin, and TPs. However, intubation of some of the rats fed raw garlic with 5 mg or 10 mg/0.5 mL of *T. occidentalis* leaf extract (ethanolic or aqueous) caused a significant decrease in serum ALP, GOT, and GPT when compared with rats fed raw garlic without intubation with the *T. occidentalis* leaf extract. Moreover, 10 mg/0.5 mL of extract was more effective than 5 mg/0.5 mL of extract, while the aqueous extracts appeared to be more effective than the ethanolic extracts in protecting hepatocytes. It could be inferred that both aqueous and ethanolic extracts of *T. occidentalis* leaf have hepatoprotective properties, although the aqueous extract is more effective than the ethanolic extract, which could be attributed to the higher antioxidant activity of the aqueous extract than the ethanolic extracts of *T. occidentalis* leaves. Nwanna and Oboh (2007) reported antioxidant and hepatoprotective properties of polyphenol extracts from *Telfairia occidentalis* leaves on APAP-induced liver damage. They reported that free soluble polyphenols had higher protective effect on the liver than the bound polyphenols; however their actions were not dose dependent.

The effect of *T. occidentalis* leaves and Silymarin on PCM-induced liver damage was investigated by Danladi et al. (2012) in rats. This study showed that ethanolic extract of *T. occidentalis* leaves possess hepatoprotective property at a dose 500 mg/kg.

Tephrosia purpurea (L.) Pers. FAMILY: FABACEAE

Ramamurthy and Srinivasam (1993) evaluated the hepatoprotective action of *Tephrosia purpurea* in rats by inducing hepatotoxicity with D-GalN HCl (acute) and CCl_4 (chronic). *T. purpurea* (aerial parts) powder was administered orally at a dose of 500 mg/kg. Serum levels of transaminases (SGOT and SGPT) and bilirubin were used as the biochemical markers of hepatotoxicity. Histopathological changes in the liver were also studied. The results of the study indicated that the administration of *T. purpurea* along with the hepatotoxins offered a protective action in both acute (D-galactosamine) and chronic (CCl_4) models.

Khatri et al. (2009) evaluated the hepatoprotective activity of aerial parts of *T. purpurea* against TAA-induced hepatotoxicity. Hepatotoxicity was induced in albino rats by subcutaneous injection of TAA. Aqueous-ethanolic extract of aerial parts of *T. purpurea* (100, 300, and 500 mg/kg/day were evaluated. Oral administration of *T. purpurea* at 500 mg/kg resulted in a significant reduction in serum AST (35%), ALT (50%), GGTP (56%), ALP (46%), TB (61%) and liver MDA levels (65%), and significant improvement in liver GSH (73%) when compared with TAA damaged rats. Histology of the liver sections of the animals treated with the extracts also showed dose-dependent reduction of necrosis.

Hepatoprotective action of EAF of ethanol extract of *Tephrosia purpurea* (EETP) root was evaluated by Shah et al. (2011) on commonly used model of experimental hepatic damage in rats against CCl_4-induced hepatotoxicity. The results showed that oral administration of EETP resulted in a significant reduction in AST, ALT, ALP, and TB when compared with CCl_4 damaged rats. A comparative histopathological study of liver from test group exhibited almost normal architecture, as compared to CCl_4-treated group. The results are comparable to that of Silymarin. Hepatoprotective activity of EETP exhibited better effectiveness than Silymarin in certain parameters, concluded its hepatoprotective potential.

Terminalia arjuna Roxb. FAMILY: COMBRETACEAE

Manna et al. (2006) evaluated the protective role of the aqueous extract of the bark of *Termnalia arjuna* (TA) on CCl_4-induced oxidative stress and resultant dysfunction in the livers and kidneys of mice. CCl_4 caused a marked rise in serum levels of GPT and ALP. TBARS level was also increased significantly whereas GSH, SOD, CAT, and GST levels were decreased in the liver and kidney tissue homogenates of CCl_4 treated mice. Aqueous extract of TA successfully prevented the alterations of these effects in the experimental animals. Data also showed that the extract possessed strong free-radical scavenging activity comparable to that of vitamin C.

Sinha et al. (2007) investigated the antioxidative properties of aqueous extract of the bark of *T. arjuna* (TA) on sodium fluoride (NaF)-induced oxidative damages in the livers and kidneys of Swiss albino mice. The mice were treated with 600 ppm NaF for one week in drinking water and the activities of antioxidant enzymes, like SOD, CAT, and GST, and the levels of nonprotein thiol, reduced GSH, along with

LPO in the liver and kidney, were determined. Fluoride administration significantly altered the levels of all of the factors compared to that of normal mice. Dose- and time-dependent studies suggest that the aqueous extract of the bark of TA showed optimum protective activity against NaF-induced oxidative damages at a dose of 40 mg/kg BW for 10 days. Oral administration of the extract for the specified dose and time followed by NaF treatment (600 ppm) normalized the levels of the hepatic and renal antioxidant enzymes, GSH, and LPO significantly to almost normal levels. The effects of a known antioxidant, vitamin E, and a nonrelevant agent, bovine serum albumin (BSA), have also been included in the study. In addition, TA extract has been found to possess radical scavenging activity.

Ghosh et al. (2010) investigated the protective role of the fruit extract of *T. arjuna* (AE) against cadmium-induced oxidative liver impairment using a murine model. Cadmium reduced hepatocytes viability, activated MAPKs, disturbed Bcl-2 family protein balance, increased ROS production and induced apoptotic cell death by mitochondria dependent caspases-3 activation. AE treatment, however, suppressed all the apoptotic actions of cadmium. Similarly, mice treated with cadmium altered a number biomarkers related to hepatic oxidative stress and other apoptotic indices. Oral administration of AE both pre- and post-prevented all the Cd-induced hepatic damages.

The methanolic extract of *T. arjuna* was evaluated by Biswas et al. (2011) for its protective effect on PCM-induced liver damage in Wistar rats. Serum and liver biochemical observations indicated that methanolic extract of *T. arjuna* exerted remarkable hepatoprotective efficacy against PCM-induced hepatic damage in Wistar rats that is may be due to its augmenting endogenous antioxidant mechanisms.

The aqueous extract of *T. arjuna* bark was investigated by Doorika and Ananthi (2012) for its hepatoprotective effect against isoniazid-induced acute liver damage on albino rats. Isoniazid (100 mg/kg) significantly elevatd serum levels of biochemical markers like SGPT, SGOT, ALP, ACP, bilirubin, protein, and depleted antioxidant enzymes GSH and SOD upon administration of isoniazid to albino rats. The aqueous extract of bark of *T. arjuna* at 200 mg/kg dose significantly reduced the elevated levels of biochemical markers mentioned above. Test extract treatment also increases the level of SOD and GSH.

Ghosh et al. (2013) investigated the protective effect of *T. arjuna* (TA) aqueous bark extract against ROS-induced liver tissue damage using Cu-ascorbate system as an *in vitro* model of oxidative stress. Incubation of goat liver homogenate with

Cu ascorbate showed an increased level of LPO, PCC, DNA smearing in agarose gel indicating direct tissue damage. The sharp increase in the activity of pro-oxidant enzymes like XO is also an indication of tissue damage as this enzyme is involved in formation of uric acid and O_2−. The increase in the level of O_2− is also evident from increased Cu–Zn SOD activity in cytosol and Mn-SOD activity in mitochondrial fraction. Catalase activity of hepatic tissue is concomitantly decreased, which is an indication of excess H_2O_2 accumulation in liver tissue. The TA extract dose dependently prevented the changes of oxidative stress biomarkers, pro-oxidant and antioxidant enzymes in liver tissue when co-incubated with the above-mentioned system.

Screening hepatoprotective potential of ethanolic and aqueous extract of *T. arjuna* bark against PCM/CCl_4-induced liver damage in Wistar albino rats was carried out by Vishwakarma et al. (2013). Hepatotoxicity was induced by PCM/CCl_4 and the biochemical parameters such as SGPT, SGOT, and serum ALP, SB and histopathological changes in liver were studied along with Silymarin as standard hepatoprotective agents. The phytochemical investigation of the extracts showed the presence of alkaloids, glycosides, steroids, and flavonoids, saponin, tanins. Pretreatment of the rats with ethanol and aqueous extract prior to PCM/CCl_4 administration caused a significant reduction in the values of SGOT, SGPT, ALP, and sB almost comparable to the Silymarin. The hepatoprotective was confirmed by histopathological examination of the liver tissue of control and treated animals.

The ME of *T. arjuna* leaves was evaluated as hepatoprotective against Swiss albino rats by Said et al. (2014). The extract at dose (500 mg/kg) showed significant reduction in ALT serum level by 19.9% while at dose (1000 mg/kg) it reduced significantly serum ALT, AST, and ALP levels by −26.15%, −25.46%, and −23.69%, respectively, as compared with PCM treated group.

Haidry and Malik (2014) investigated the hepatoprotective effect of *T. arjuna* stem bark on cadmium provoked toxicity. Treatment of rats with the extract significantly reversed the effects of cadmium toxicity. Chaudhari and Mahajan (2016) evaluated the hepatoprotective activity of methanolic extract of *T. arjuna* stem bark (MeOH-TASB) and its extracted flavonoids baicalein (Bai) and quercetin (Que) by using a simple *in vitro* goat liver slice culture model. CCl_4 was used to induce hepatotoxicity in liver slice of goat. The treatment of liver cells with CCl_4 caused twice increase in LPO of cells besides release of ALT, AST, ALP, and LDH was 4.26, 4.88,

2.89, and 3.66 times, respectively, as compared to untreated liver cells. Thus, a toxic effect of CCl_4 was significantly reduced by the treatment of MeOH-TASB, Bai, and Que. Moreover, the protective effect was a dose-dependent manner, not only for MeOH-TASB extract but also for its phyto-ingredients. These results indicate that all MeOH-TASB and its extracted flavonoids protect the liver cells from CCl_4-induced oxidative/free radical mediated damage *in vitro*. Que shows more protective effect than Bai.

Terminalia bellirica (Gaertn.) Roxb. FAMILY: COMBRETACEAE

Jadon et al. (2007) evaluated the protective effect of *Terminalia bellirica* fruit extract and its active principle, GA (3,4,5-trihydroxybenzoic acid) at different doses against CCl_4 intoxication. Toxicant caused significant increase in the activities of serum transaminases and serum ALP. Hepatic LPO level increased significantly whereas significant depletion was observed in reduced GSH level after CCl_4 administration. A minimum elevation was found in protein content on the contrary a significant fall was observed in glycogen content of liver and kidney after toxicant exposure. Activities of adenosine triphosphatase and succinic dehydrogenase inhibited significantly in both the organs after toxicity. Treatment with TB extract (200, 400, and 800 mg/kg, p.o.) and GA (50, 100, and 200 mg/kg, p.o.) showed dose-dependent recovery in all these biochemical parameters but the effect was more pronounced with GA. Thus it may be concluded that 200 mg/kg dose of GA was found to be most effective against CCl_4-induced liver and kidney damage.

Hepatoprotective activity of the ethanolic and aqueous extracts of *T. bellirica* was studied by Jain et al. (2008) in Wistar rats using the ethanol-induced liver hepatotoxicty. Ethanol administration resulted in significant elevation of physical parameters (viz. rat liver weight and liver volume), biochemical parameters like AST, ALT, ALP, DBL, and TB levels, while ALB and TP were found to be decreased compared to normal group. Pretreatment with Silymarin, ethanolic extract of *T. bellirica* (ALTB) and aqueous extract of *T. bellirica* (AQTB) significantly prevented the physical and biochemical changes induced by these hepatotoxins. Histopathology of liver confirmed their finding as the treatment with the extracts resulted in minor liver cell damage compared to toxic control group. The hepatoprotective effect offered by ALTB (400 mg/kg, p.o.) was found to be significantly greater than AQTB (400 mg/kg, p.o.) and standard (Silymarin 50 mg/kg, p.o.) group.

Terminalia catappa L. FAMILY: COMBRETACEAE

Terminalia catappa, distributed throughout tropical and subtropical beaches, is used for the treatment of dermatitis and hepatitis. Punicalagin and Punicalin isolated from the leaves of *T. catappa* were evaluated by Lin et al. (1998, 2001) for antihepatotoxic activity in the rat liver. After evaluating the changes of several biochemical functions in serum, the levels of AST and ALT were increased by APAP administration and reduced by punicalagin and punicalin. Histological changes around the hepatic central vein and oxidative damage induced by APAP were also recovered by both compounds. The data show that both punicalagin and punicalin exert antihepatotoxic activity, but treatment with larger doses enhanced liver damage. These results suggest that even if punicalagin and punicalin have antioxidant activity at small doses, treatment with larger doses will possibly induce some cell toxicities (Lin et al., 1998, 2001).

Punicalagin and punicalin isolated from the leaves of *T. catappa* reduced hepatitis by reducing levels of AST and ALT which increased by APAP administration in rat (Lin et al., 2001).

Gao et al. (2004) evaluated the effect of the CE of *T. catappa* leaves (TCCE) on CCl_4-induced acute liver damage and D-GalN-induced hepatocyte injury. Moreover, the effects of ursolic acid and asiatic acid, two isolated components of TCCE, on mitochondria, and free radicals were investigated to determine the mechanism underlying the action of TCCE on hepatotoxicity. In the acute hepatic damage test, remarkable rises in the activity of serum ALT and AST (5.7- and 2.0-fold) induced by CCl_4 were reversed and significant morphological changes were lessened with pretreatment with 50 and 100 mg/kg TCCE. In the hepatocyte injury experiment, the increases in ALT and AST levels (1.9- and 2.1-fold) in the medium of primary cultured hepatocytes induced by D-GalN were blocked by pretreatment with 0.05, 0.1, 0.5 g/L TCCE. In addition, Ca^{2+}-induced mitochondrial swelling was dose dependently inhibited by 50–500 μM ursolic acid and asiatic acid. Both ursolic acid and asiatic acid, at concentrations ranging from 50 to 500 μM, showed dose-dependent superoxide anion and hydroxyl radical scavenging activity.

The protective effects of CEs of TCCE on CCl_4-induced liver damage and the possible mechanisms involved in the protection were investigated in mice by Tang et al. (2006). They found that increases in the activity of serum AST and ALT and the level of liver LPO (2.0-fold, 5.7-fold, and 2.8-fold) induced by CCl_4 were significantly inhibited by oral pretreatment with 20, 50, or 100 mg/kg of TCCE. Morphological observation further confirmed the hepatoprotective effects of TCCE. In addition, the disruption of MMP (14.8%), intramitochondrial Ca^{2+} overload (2.1-fold), and suppression of mitochondrial Ca^{2+}-ATPase activity (42.0%) in the liver of CCl_4-insulted mice were effectively prevented by pretreatment with TCCE.

The antioxidant and hepatoprotective actions of *T. catappa* collected from Okinawa Island were evaluated *in vitro* and *in vivo* by Kinoshita et al. (2007) using leaves extract and isolated antioxidants. A water extract of the leaves of *T. catappa* showed a strong radical scavenging action for DPPH and superoxide (O(2)(.−)) anion. Chebulagic acid and corilagin were isolated as the active components from *T. catappa.* Both antioxidants showed a strong scavenging action for O(2)(.−) and peroxyl radicals and also inhibited ROS production from leukocytes stimulated by phorbol-12-myristate acetate. Galactosamine (GalN, 600 mg/kg, s.c.,) and LPS (0.5 μg/kg, i.p.)-induced hepatotoxicity of rats as seen by an elevation of serum ALT, AST, and GST activities was significantly reduced when the herb extract or corilagin was given intraperitoneally to rats prior to GalN/LPS treatment. Increase of free-radical formation and LPO in mitochondria caused by GalN/LPS treatment were also decreased by pretreatment with the herb/corilagin. In addition, apoptotic events such as DNA fragmentation and the increase in caspase-3 activity in the liver observed with GalN/LPS treatment were prevented by the pretreatment with the herb/corilagin.

Aqueous and alcoholic bark extracts of *Terminalia catappa* were evaluated by Vahab and Harindran (2016) for the hepatoprotective activity. Preliminary photochemical analysis revealed the presence of carbohydrate, tannins, phenols, flavonoids and saponins in all extracts. LCMS and GCMS studies showed the presence of steroid in ALE. Chemical analysis showed the trace elements within the limit. Acute toxicity studies of both extracts, selected the effective dose as 400 mg/kg. Isoniazid-induced hepatotoxicity study was conducted for 21 days. At the end of 21st day the animals were killed. SGPT, SGOT, ALP bilirubin and TP were measured. After the treatment with the extracts Bilirubin, SGPT, SGOT, and ALP were decreased and total proteins were increased. Histopathologic examination assured the hepatoprotective activity of the extracts.

Terminalia chebula Gaertn. FAMILY: COMBRETACEAE

Terminalia chebula is an important herbal drug in Ayurvedic pharmacopea. A 95% ethanolic extract of *T.*

chebula fruit, which was chemically characterized on the basis of chebuloside II as a marker, was investigated by Tasduq et al. (2006) for hepatoprotective activity against antituberculosis drug-induced toxicity. The ethanolic extract was found to prevent the hepatotoxicity caused by the administration of RIF, INH, and PZA (in combination) in a subchronic mode (12 weeks). The hepatoprotective effect of the extract could be attributed to its prominent antioxidative and membrane stabilizing activities. The changes in biochemical observations were supported by histological profile.

Ahmadi-Najo et al. (2017) evaluated the effects of the hydroalcoholic extract of *T. chebula*, on some biochemical and histopathological parameters of liver tissue in DZN-administered rats. In DZN-treated group, the levels of serum urea, HDL, and liver SOD, CAT, and vitamin C significantly decreased compared to control. Also, in this group, serum TG, TC, VLDL, protein carbonyl (PC), MDA, TNF-α, and *TNF-α* gene expression significantly increased as compared to the control (vehicle-treated rats). Treatment with *T. chebula* resulted in a significant increase in CAT, SOD, vitamin C, HDL and a significant decrease in the level of urea, MDA, PC, TG, TC, VLDL, TNF-α protein, and the gene expression of *TNF-α* compared with test without treatment group. Histopathological evidence demonstrated that treatment with *T. chebula* extract could decrease liver lymphocyte infiltration. This study suggests that *T. chebula* fruit extract has protective effects against DZN-induced oxidative stress.

Terminalia paniculata Roth FAMILY: COMBRETCEAE

Eesha et al. (2011) evaluated the hepatoprotective activity of *Terminalia paniculata* against PCM-induced hepatic damage in rats. Paracetamol (2 g/kg) increased the serum levels of ALT, AST, ALP, and the LPOs. Treatment of Liv. 52, Silymarin, and ethanolic extract of *T. paniculata* (200 mg/kg) altered levels of biochemical marker and showed significant hepatoprotective activity. Ethanolic extract revealed the presence of phenolic compound and flavanoids. Their findings suggested that ethanolic bark extract of *T. paniculata* possessed hepatoprotective activity in a dose-dependent manner.

Tetracarpidium conophorum Müll.Arg. FAMILY: EUPHORBIACEAE

Tetracarpidium conophorum is used in ethno medicine for treating various diseases including hepatic ailments. Oriakhi et al. (2018)

investigated the effect of ME of *T. conophorum* seeds in rats intoxicated with CCl_4 24 and 48 h after intoxication, respectively. There were significant increases in serum hepatic enzyme markers (ALT, AST, ALP, and γ-GT) activities, as well as bilirubin and significant reduction in antioxidant enzymes in rats intoxicated with CCl_4 when compared to control group, but administration (pretreatment and post-treatment) of ME of *T. conophorum* seeds at doses of 250 and 500 mg/kg BW and standard sylimarin drug attenuated the toxic insult of CCl_4 in a dose-dependent manner at 24 and 48 h after intoxication, respectively.

Tetracera loureiri (Finet & Gagnep.) Pierre ex W. G. Craib FAMILY: DILLENIACEAE

Kukongviriyapan et al. (2003) evaluated the *in vitro* and *in vivo* antioxidant and hepatoprotective activities of *Tetraptera loureiri*. The ethanol extract of *T. loureiri* possessed potent antioxidant and strong free-radical scavenging properties assayed using FRAP and DPPH, respectively. The cytoprotective effects of *T. loureiri* were demonstrated in ethanolic extracts of freshly isolated rat hepatocytes against the chemical toxicants PCM and tertiary-butylhydroperoxide. The cells pretreated with the extract maintained the GSH/GSSG ratio and suppressed LPO in a dose-dependent manner. Pretreating rats with the ethanol extract orally, one hour prior to intraperitoneal injection of toxic doses of PCM, significantly prevented elevations of plasma ALT and AST.

Tetrapleura tetraptera (Schum. & Thonn.) Taub. FAMILY: FABACEAE

The antioxidant and hepatoprotective activity of *Tetrapleura tetraptera* extracts in CCl_4-induced liver injury in rats was investigated. *T. tetraptera* extracts showed varying levels of protective action against CCl_4-induced liver damage as evidenced through the significant reduction in the activities of serum marker enzymes for liver damage (ALT, AST, and ALP), and bilirubin levels when compared with CCl_4-intoxicated control rats. The extracts decreased the elevation in the activities of the enzymes in the liver. They also protected against CCl_4-induced LPO. The extracts reduced CCl_4-liver-induced necrosis in dose-dependent manner. These results indicated that fruit extracts of *T. tetraptera* possess a hepatoprotective property against CCl_4-induced liver damage which was mediated through its antioxidative defenses.

Teucrium polium L. FAMILY: LAMIACEAE

The hydroalcoholic extract of aerial part of *Teucrium polium* dose dependently attenuated CCl_4 (Panovska et al., 2007) and APAP (Kalantari et al., 2013)-induced biochemical (serum ALT, AST, APT, and TB) and histological changes in liver. Experimental studies of *T. polium* on cultured hepatocytes also confirmed its strong antioxidant (Panovska et al., 2007; Bomzon et al., 2010) and anti-inflammatory activities which may contribute to its hepatoprotective activity. Phytochemical studies on *T. polium* showed the presence of flavonoids, terpenes including syrapoline, thujene, caryophyllene, cedrol, epi-cadinol, and bisabolene (Al-Asmari et al., 2014).

Protective effect of *T. polium* extract on APAP-induced hepatotoxicity was investigated in mice by Forouzandeh et al. (2013). The results of this study showed the protective effect in all doses but the most significant protection was observed in doses of 250 and 500 mg/kg. Also these findings were supported and confirmed by histological examination.

Thaumatococcus daniellii (Benn.) Benth. FAMILY: MARANTACEAE

Chinedu et al. (2018) evaluated the *in vivo* hepatoprotective activity of ethanolic extract of *Thaumatococcus daniellii* leaves (ETD) in male Wistar rats at an oral dose of 500–1500 mg/kg daily for 14 days. Oral treatment with ETD increased organ SOD activity, renal reduced GSH and plasma HDL concentrations while reducing plasma ALT activity, plasma cholesterol (CHOL), bilirubin (DBIL) and organ MDA concentrations. Data was supported by histological report showing no pathologic abnormality. This data indicate ethanolic extract of *T. daniellii* leaves shows antioxidant, hypolipidemic, and hepatoprotective potential.

Thespesia lampas (Cav.) Dalz. & Gibs. FAMILY: MALVACEAE

Ambrose et al. (2012) evaluated the hepatoprotective activity of *Thespesia lampas* against PCM-induced liver damage. *n*-Butanol fraction (200 mg/kg) exhibited maximum hepatoprotective activity which was concentration dependent. *T. lampas* contains three known hepatoprotective compounds (tannic acid, ellagic acid, and rutin) and a major unidentified hepatoprotective phenolic compound.

Study in CCl_4-induced hepatocellular damage, *T. lampas* exhibited a statistically significant reduction of elevated enzymes. Hepatoprotective effect was confirmed by histopathological liver exam. Results provide pharmacologic support for its folkloric use as a hepatoprotective agent (Sangameswaran et al., 2008).

Stem extracts of *T. lampas* was evaluated for hepatoprotective and antioxidant activity against CCl_4-induced hepatic damage in rats. The extracts showed significant hepatoprotective and antioxidant effect by lowering the serum levels of transaminases (SGOT and SGPT), ALP, bilirubin, protein, cholesterol and TG as compared to Silymarin as a standard hepatoprotective agent. The extracts showing increased levels of SOD, CAT, and reduced GSH and decreased level of LPO. The biochemical observations were supplemented with histopathological examination of rat liver sections.

Thespesia populnea (L.) Sol. ex Correa FAMILY: MALVACEAE

Yuvaraj and Subramoniam (2009) evaluated the possible hepatoprotective activities of the *Thespesia populnea* in CCl_4-induced liver injury in rats. The water suspension (500 mg/kg BW) of leaf, flower and stem bark of *T. populnea* showed varying levels of protective action against CCl_4-induced liver damage as evidenced from significant reduction in the activities of serum marker enzymes for liver damage (ALT, AST, and ALP), and bilirubin levels when compared with CCl_4-intoxicated control rats. The stem bark suspension showed maximum hepatoprotection compared with leaf and flower. An ethanol extract of the stem bark was more active than *n*-hexane and water extracts, showing remarkable protection at a dose of 60 mg/kg BW. The hepatoprotective effect of this extract was almost comparable to that of Silymarin (100 mg/kg), a reference herbal drug.

Thuja occidentalis L. FAMILY: CUPRESSACEAE

Dubey and Batra (2008) evaluated the possible hepatoprotective activity of ethanolic fraction of *Thuja occidentalis* aerial part assessed against CCl_4-induced liver damage in rats. A dose of EFTO 400 mg/kg p.o. exhibited significant protection from liver damage in acute and chronic CCl_4-induced liver damage model. Histopathological examination was carried out after the treatment to evaluate hepato protection. The fraction was found to possess good hepatoprotective property.

Thonningia sanguinea Vahl FAMILY: BALANOPHORACEAE

The hepatoprotective effect of *Thonningia sanguinea* was studied by Gyamfi et al. (1999) on acute hepatitis induced in rats by a single dose of galactosamine (GalN, 400 mg/kg, IP) and in mice by CCl_4 (25 μL/kg, IP). GalN-induced hepatotoxicity in rats as evidenced by an increase in ALT and GSH-*S*-transferase activities in serum was significantly inhibited when *T. sanguinea* extract (5 mL/kg, IP) was given to rats 12 and 1 h before GalN treatment.

The activity of liver microsomal GSH S-transferase, which is known to be activated by oxidative stress, was increased by the GalN treatment and this increase was blocked by *T. sanguinea* pretreatment. Similarly, *T. sanguinea* pretreatment also inhibited CCl_4-induced hepatotoxicity in mice.

Thunbergia laurifolia Lindl. FAMILY: ACANTHACEAE

Pramyothin et al. (2005) investigated the hepatoprotective activity of *Thunbergia laurifolia*. Primary cultures of rat hepatocyte and rats were used as the *in vitro* and *in vivo* models to evaluate the hepatoprotective activity of aqueous extract. Ethanol was selected as hepatotoxin. Silymarin was the reference hepatoprotective agent. In the *in vitro* study, MTT reduction assay and release of transaminases (ALT and AST) were the criteria for cell viability. Primary cultures of rat hepatocyte (24 h culturing) were treated with ethanol (96 microl/mL) and various concentrations of TLE (2.5, 5.0, 7.5, and 10.0 mg/mL) or SL (1, 2 and 3 mg/mL) for 2 h. Ethanol decreased MTT (%) nearly by half. Both TLE and SL increased MTT reduction and brought MTT (%) back to normal. Ethanol induced release of ALT and AST was also reduced by TLE (2.5 and 5.0 mg/mL) and SL (1 mg/mL). In the *in vivo* study, serum transaminases, STG together with HTG and histopathological examination were the criteria for evidences of liver injury. Ethanol (4 g/(kg day), p.o. for 14 days) caused the increase in ALT, AST, HTg, and centrilobular hydropic degeneration of hepatocytes. TLE at 25 mg/(kg day), p.o., or SL at 5 mg/(kg day), p.o., for 7 days after ethanol enhanced liver cell recovery by bringing HTg, ALT, and/or AST back to normal. These results suggest that TLE and SL possess the hepatoprotective activity against ethanol-induced liver injury in both primary cultures of rat hepatocyte and rats (Pramyothin et al., 2005).

Thymus linearis Benth. FAMILY: LAMIACEAE

Alamgeer et al. (2014) evaluated the hepatoprotective activity of aqueous and ether extracts of *Thymus linearis* (250 and 500 mg/kg orally) against CCl_4- and PCM-induced hepatic damage in mice. Serum levels of ALT, AST, ALP were assessed. Antioxidant activity of both the extracts was also determined using DPPH scavenging method. The results indicated that both the extracts significantly produce a dose-dependent reduction in serum levels of ALT, AST, ALP when compared to CCl_4- and PCM-treated groups. The maximum effect in all the parameters was observed at a dose of 500 mg/kg. The extracts also demonstrated a significant antioxidant activity. LD_{50} of both extracts was found to be a 1050 and 900 mg/kg, respectively. It is conceivable that the hepatoprotective activity of *T. linearis* might be due to the presence of certain pharmacologically active compounds.

Tinospora cordifolia (Willd.) Miers ex Hook. f. & Thoms. FAMILY: MENISPERMACEAE

Kumar et al. (2013) explored the hepatoprotective activity of *Tinospora cordifolia* against experimentally-induced hepatotoxicity in albino rats. ALT, ALP, and TB levels were significantly increased in CCl_4 treated group while *T. cordifolia* displayed significant reduction in rise in these parameters. This hepatoprotection was also reflected in histological changes. It is assumed that this hepatoprotective effect of *T. cordifolia* may be due to several reasons such as antioxidant and/or free-radical scavenger property and ability to induce hepatic regeneration.

The effect of *T. cordifolia* extract on modulation of hepatoprotective and immunostimulatory functions in CCl_4 intoxicated mature rats is reported by Bishayi et al. (2002). Treatment with *T. cordifolia* extract (100 mg/kg BW for 15 days) in CCl_4 intoxicated rats was found to protect the liver, as indicated by enzyme level in serum. A significant reduction in serum levels of SGOT, SGPT, ALP, and bilirubin were observed following *T. cordifolia* treatment during CCl_4 intoxication.

Goyal and Kumar (2016) explored the hepatoprotective activity of *T. cordifolia* in Albino Wistar rats against hepatotoxicity induced by CCl_4. ALT, ALP, and total bilirubin levels were significantly increased in CCl_4 treated group while *T. cordifolia* displayed significant reduction in rise in these parameters.

Tinospora crispa (L.) Hook.f. & Thoms. FAMILY: MENISPERMACEAE

Kadir et al. (2011) determined the effect of ethanolic extract of the dried stems of *Tinospora crispa* in a male rat model of hepatic fibrosis caused by the hepatotoxin, TAA. A significant increase in the activity of liver enzymes, bilirubin and G-glutamyl transferase and gross and histopathological changes were determined. Although previous *in vitro* study established that this extract had strong antioxidant activity, this *in vivo* study establishes that this extract contains hepatotoxins whose identity may be quite different from those compounds with antioxidant properties. The study confirms that complete reliance on data obtained using *in vitro* methodologies may lead to erroneous conclusions pertaining to the safety of phytopharmaceuticals.

***Toona sinensis* (A.Juss.) Roem.**
FAMILY: MELIACEAE

Two polysaccharide fractions (TSP-1 and TSP-2) with molecular weights of 833.6 and 81.6 kDa were isolated from *Toona sinensis* leaves by hot water extraction, DEAE Cellulose-52 chromatography and Sephacryl S-400 gel permeation chromatography. Structural analysis indicated that TSP-1 and TSP-2 consisted of Manp, GlcpA, Glcp, Galp, Xylp, and Araf with different molar ratios. Methylation and NMR analysis revealed that the backbone of TSP-1 might consist of 1,6-linked-Glcp, 1,3,6-linked-Manp and 1,6-linked-Galp, while TSP-2 was mainly composed of 1,3,5-linked-Araf, 1,6-linked-Glcp, 1,4-linked-Xylp and 1,6-linked-Galp. Congo red assay indicated that TSP-1 and TSP-2 had no triple-helix structure, which was consistent with the results of AFM. In vivo hepatoprotective activity showed that TSP-1 and TSP-2 could improve CCl_4-induced mice liver injury by reducing the activities of AST, ALT, and the level of MDA, increasing the activities of SOD, GSH-Px, and CAT and the level of GSH in liver and decreasing the expression levels of TNF-α and IL-6 in liver. These results suggest that TSP-1 and TSP-2 have promising potential to serve as hepatoprotective agents (Cao et al., 2019).

***Tournefortia sarmentosa* Lam.**
FAMILY: BORAGINACEAE

Teng et al. (2012) investigated the effect of the aqueous extract of the *Tournefortia sarmentosa* on APAP induced hepatotoxicity *in vivo* and *in vitro*. *T. sarmentosa* significantly reduced the elevated liver function (SGOT, SGPT, and ALP) and inflammatory markers (TNF-α, IL-1β, and IL-6) in serum of APAP-intoxicated rats. MDA level and antioxidant enzyme levels (CAT, SOD, and GPx) were also reduced in APAP-intoxicated rats treated with *T. sarmentosa*. Incubation of rat hepatocyte cell line clone-9 cells with APAP reduced cell viability and increased the extent of LPO. APAP stimulation also reduced the level of GSH and caused reduction in the activities of the antioxidant enzymes, CAT, SOD, and GPx. Pretreatment of hepatocytes with *T. sarmentosa* aqueous extract before and during APAP stimulation attenuated the extent of LPO, increased cell viability and GSH level, and enhanced the activities of antioxidant enzymes.

***Tragopogon porrifolius* L.**
FAMILY: ASTERACEAE

Tragopogon porrifolius, known as purple salsify, is grown for its edible root and shoot (Mroueh et al., 2011). It has bioactive compounds which prevent cancer or other free-radical associated illnesses. The nutritional value of this plant is derived from monounsaturated and essential fatty acids, polyphenols, vitamins, and fructo-oligosaccharides, having probiotic effects on the intestinal microflora. The most abundant compounds of this plant include 4-vinyl guaiacol (19.0%), hexadecanoic acid (17.9%), hexahydrofarnesylacetone (15.8%), and hentriacontane (10.7%) (Formisano et al., 2010; Konopiński, 2009). *T. porrifolius* has apparently yielded the hepatogenic/hepatoprotective effects against liver diseases or hepatotoxicity induced by a variety of hepatotoxic agents such as chemicals, drugs, pollutants, and infection with parasites, bacteria, or viruses (hepatitis A, B, and C). These beneficial effects of plants are related to the polyphenolic compounds. The study of antioxidant activity of the methanolic extract of the aerial part of *T. porrifolius* as well as its protection against $CC1_4$-induced hepatotoxicity in rats showed a dose-dependent increase in the activity of liver antioxidant enzymes. About 250 mg/kg BW dose increased the activity of CAT, SOD, and GST. Also, substantial hepatoprotective capacity against $CC1_4$-induced hepatic injury has been shown, attributable to restoring the activity of AST, ALT, and LDH to normal levels (Mroueh et al., 2011; Govind, 2011). Investigation of the effects of water extract of *T. porrifolius* shoot on lipemia, glycemia, inflammation, oxidative stress, hepatotoxicity, and gastric ulcer using a rat model showed that after one month of *T. porrifolius* water extract intake, a significant decrease in the levels of serum cholesterol, triglyceride, glucose, and liver enzyme (ALP, ALT, and LDH) was observed. Pretreating rats with *T. porrifolius* extract demonstrated considerable anti-inflammatory effects in both acute and chronic inflammation caused by carrageenan and formalin. In addition, *T. porrifolius* revealed effective antioxidant capability owing to its remarkable scavenging activity (Zeeni et al., 2014).

Trianthema decandra L. FAMILY: AIZOACEAE

Balamurugan and Muthusamy (2008) investigated hepatoprotective effect and antioxidant activity of the ethanol extract of the roots of *Trianthema decandra* (200 and 400 mg/kg) in rats treated with CCl_4 for 8 weeks. Extract at the tested doses restored the levels of all serum (AST, ALT, ALP, TB, and TP) and liver homogenate enzymes (GPx, GR, SOD, and CAT) significantly. Histology demonstrated profound steatosis degeneration and nodule formation were observed in the hepatic architecture of CCl_4-treated rats which were found to acquire near-normalcy in extract plus CCl_4

administrated rats, and supported the biochemical observations.

Sengottuvelu et al. (2012) appraised the hepatoprotective activity of aqueous extract of *T. decandra* roots against CCl_4-induced liver damage in male Wistar rats. *T. decandra* (100 and 200 mg/kg) results in a significant reduction in serum hepatic enzymes when compared to rats treated with CCl_4 alone. There was a significant increase in the serum TP and ALB when compared to rats treated with CCl_4 alone.

***Trianthema portulacastrum* L.** **FAMILY: AIZOACEAE**

The antihepatotoxic potential of an ethanolic extract of the whole plant of *Trianthema portulacastrum* (excluding the roots) was evaluated by Bishayee et al. (1996) against alcohol-CCl_4-induced acute liver damage in mice. The extract at a dose of 50, 100, or 150 mg/kg was administered *per* as once daily for successive three days concomitant with alcohol-CCl_4, treatment. The substantially elevated serum enzymatic activities of SGOT, SGPT, LDH, ALP, sorbitol, and glutamate dehydrogenase due to alcohol-CCl_4, treatment were dose dependently restored toward normalization following the extract therapy. There was a marked inhibition of SB and urea levels in the plant extract-treated groups which were otherwise drastically increased in alcohol-CCl_4 control animals. The extract also significantly prevented the elevation of hepatic MDA formation (evidence of LPO) and depletion of reduced GSH content in liver of mice intoxicated with alcohol-CCl_4, in a dose-responsive fashion. The results of this study clearly indicate that the plant possesses a potent hepatoprotective action against alcohol-CCl_4-induced hepatocellular injury which corroborates its use in hepatic disorders as well as alcohol-evoked liver ailments in traditional oriental medicine.

Kumar et al. (2005) evaluated the antioxidant and hepatoprotective activity of *T. portulacastrum*. The results show that pretreatment of rats with 100 mg or 200 mg/kg, p.o., of ethanolic extracts of *T. portulacastrum* prevented significantly the PCM- and TAA-induced reduction of blood and liver GSH, liver Na-K-ATPase level. TBARS of toxicants-treated animals were significantly higher than the control animals. Administration of the ethanolic extract markedly decreased the level of TBARS. The degree of protection was more with the higher dose of the extract. Pretreatment of rats with *T. portulacastrum* extract significantly reduced the elevated levels of TBARS and increased the concentration of hepatic and blood GSH.

The effect of an ethanolic extract of the plant *T. portulacastrum* on the CCl_4-induced chronic hepatocellular damage of Swiss albino mice has been investigated by Mandal et al. (1997a, b). The CCl_4 administration alone caused hepatocellular necrosis, severe anemialeucopaenia, lymphocypaenia, neutrophilia, eosinophilia, and hemoglobinaemia along with the alterations of plasma ALB and globulin. The administration of plant extract (at 100 or 150 mg/kg) restored the CCl_4-induced alterations of the hematological parameters to the normal level. The extract elicited a marked protection against CCl_4-induced hepatotoxicity as indicated by the several hematological parameters, related indices of formed elements, and different fractions of plasma protein. They also observed the dose-dependent antihepatotoxic effect of the extraction on these mice. The 150 mg/kg of extract was found to be more effective in normalizing the toxic effects of CCl_4 on the above parameters of mice. These results suggest that the hepatoprotective effect of *T. portulacastrum* could be caused by its critical involvement in modulating several factors associated with erythropoiesis and the boosting of general immunity of the host.

Mehta et al. (2003) evaluated the hepatoprotective effect of aerial parts of *T. portulacastrum* using PCM and rifampicin as liver toxicants in rats. The potency of alcoholic and aqueous extracts of aerial parts was compared with that of Silymarin at a dose of 100 mg/kg, p.o. Alcoholic extract of aerial parts caused significant fall in the enzyme levels in serum in rats.

The ethanolic extract of *T. portulacastrum* showed a significant dose-dependent (100 mg, 200 mg/kg p.o. 10×) protective effect against PCM and TAA-induced hepatotoxicity in albino rats (Kumar et al., 2004). The degree of protection was measured by using biochemical parameters like SGOT, SGPT, ALP, BRN, and TP. The plant extract completely prevented the toxic effects of PCM (APAP) and TAA on the above serum parameters. A significant hepatoprotective activity of the ethanolic extracts of *T. portulacastrum* was reported.

Aflatoxins are potent hepatotoxic and hepatocarcinogenic agents. Reactive oxygen species and consequent peroxidative damage caused by aflatoxin are considered to be the main mechanisms leading to hepatotoxicity. Banu et al. (2009) assessed the hepatoprotective effect of ethanolic leaves extract of *T. portulacastrum* on aflatoxin B1 (AFB1)-induced hepatotoxicity in a rat model. The hepatoprotection of *T. portulacastrum* is compared with Silymarin, a well known standard hepatoprotectant. LDH, ALP, ALT, and AST were found to be significantly increased in the serum and decreased in the liver of AFB1 administered (1 mg/kg BW, orally) rats, suggesting hepatic damage. Marked increase in the LPO levels and a concomitant decrease in the enzymic (SOD, CAT, GPx, GR, glucose-6-phosphate dehydrogenase, and glutathione-S-transferase) and nonenzymic (reduced GSH, vitamin C, and vitamin E) antioxidants in the hepatic tissue were observed in AFB1 administered rats. Pretreatment with *T. portulacastrum* (100 mg/kg/p.o.) and Silymarin (100 mg/kg/p.o.) for 7 days reverted the condition to near normal.

The ethanolic leaves extract of *T. portulacastrum* showed a significant dose-dependent (100 mg and 200 mg/kg p.o.) hepatoprotective effect against two well-known hepatotoxins namely PCM and TAA-induced hepatotoxicity in albino rats (Bhattacharya and Chatterjee, 1998; Shyam et al., 2010).

Tribulus terrestris L. FAMILY: ZYGOPHYLLACEAE

The aqueous and hydroalcoholic extracts of fruit of *Tribulus terrestris* dose dependently attenuated PCM (Kavitha et al., 2011) and ferrous sulfate (Sambasivam et al., 2013)-induced liver damage.Two compounds (tribulusamides A and B), isolated from the fruits of *T. terrestris* significantly, protected cultured hepatocytes against D- GalN-induced toxicity (Li et al., 1998). *T. terrestris* has also been reported to possess antioxidant and anti-inflammatory activities which may contribute to its hepatoprotection. Phytochemical studies on *T. terrestris* showed the presence of Tribulusamides A and B, tigogenin, neotigogenin, terrestrosid F, and gitonin.

Abdel-Kader et al. (2016) evaluated the hepatoprotective and nephroprotective activities of the ethanolic plant extract and petroleum ether, dichloromethane and aqueous methanol fractions against CCl_4-induced toxicity in adult Wistar rats. The effect of the total 95% ethanol extract at 400 mg/kg on serum and tissue liver parameters was weak. However, protective effect on kidney was promising. The best effect was observed on the urea and creatinine levels. Both malondialdehyde and nonprotein sulfhydryl groups in kidney tissues were improved to levels comparable with those obtained by Silymarin.

Trichilia roka Chiov. FAMILY: MELIACEAE

Trichilia roka is a tree widely distributed in tropical Africa. It has been used in Mali folk medicine for the treatment of various illnesses. A decoction of the roots is taken as a remedy for colds and pneumonia, and it is used as a diuretic and in hepatic disorders. Germanò et al. (2001) evaluated the hepatoprotective effects of a decoction of *T. roka* root on CCl_4-induced acute liver damage in rats. Treatment with the decoction showed a significant protective action made evident by its effect on the levels of SGOT and SGPT in the serum, on the protein content and LPO levels in the liver homogenate. Histopathological changes produced by CCl_4, such as necrosis, fatty change, ballooning degeneration and inflammatory infiltration of lymphocytes around the central veins, were clearly recovered by the treatment with Trichilia root decoction. On fractionating this extract into diethyl ether-soluble and water-soluble fractions, the activity was retained in the diethyl ether-soluble fraction. Moreover, the administration of decoction prevented a preferential deposition of collagen around the sinusoidal cell layer, which is responsible for the perisinusoidal fibrosis in the early stage of CCl_4 damage. This study showed that treatment with *T. roka* extracts or Silymarin (as reference) appeared to enhance the recovery from CCl_4-induced hepatotoxicity. The hepatoprotective properties of *T. roka* may be correlated to polyphenol content of the decoction and its diethyl ether-soluble fraction (Germanò et al., 2001).

Trichodesma sedgwickianum S.P. Banerjee FAMILY: BORAGINACEAE

Saboo et al. (2019) investigated the antioxidant and hepatoprotective potential of *Trichodesma sedgewickianum* assessed by CCl_4-induced oxidative stress in rats. Among the extracts, successive ethanol extract showed higher concentration of polyphenols and *in vitro* antioxidant property. The *in vivo* antioxidant efficiency was confirmed by comparing the enzymatic level, SOD, CAT, reduced GSH, and MDA in test group with the standard and control. Hepatoprotective effect was observed by changes in the serum enzyme level which were further supported by histological examination. Phytochemically ethanol extract contains GA and catechin along with other constituents.

Trichopus zeylanicus Gaertn. FAMILY: DIOSCOREACEAE

Hepatoprotection of *Trichopus zeylanicus* extract has been evaluated by Subramoniam et al. (1998). The plant leaf suspension (1000 mg/kg; wet weight) as well as its ME showed a remarkable hepatoprotective activity against paracetamol-induced hepatotoxicity as judged from the serum marker enzymes, liver histology, and levels of LPOs in liver. The effect of the ME was found to be concentration dependent. They also showed that the water and hexane extracts did not showed any hepatoprotective activity. Palanisami et al. (2013) showed that

T. zeylanicus leaf extract has the ability to attenuate the liver damage caused by $HgCl_2$ in rats.

Trichosanthes cucumerina L. var. cucumerina FAMILY: CUCURBITACEAE

Different parts of the plant *Trichosanthes cucumerina* var. *cucumerina* are used to treat liver disorders, traditionally. It is one among the constituents in various Ayurvedic formulations used for the treatment of liver disorders and other diseases. Kumar et al. (2009) evaluated the protective effect of *T. cucumerina* against experimentally-induced liver injury. The methanolic extract of whole plant of *T. cucumerina* (TCME) was evaluated for the hepatoprotective activity against CCl_4-induced hepatotoxicity in rats. Various biochemical parameters like ALT, AST, ALP, TB, TP, and ALB levels were estimated in serum as well as the GSH and MDA levels in the liver were determined. Histopathological changes in the liver of different groups were also studied. The pretreatment of TCME at dose levels of 250 and 500 mg/kg BW p.o. had controlled the raise of AST, ALT, ALP, TB, and MDA levels and the effects were comparable with standard drug (Silymarin 100 mg/kg BW p.o.). The GSH, TP, and ALB levels were significantly increased in the animals received pretreatment of the extract. The animals received pretreatment of the extract showed decreased necrotic zones and hepatocellular degeneration when compared to the liver exposed to CCl_4 intoxication alone. Thus the histopathological studies also supported the protective effect of the extract (Kumar et al., 2009).

Trichosanthes lobata Roxb. FAMILY: CUCURBITACEAE

Rajasekaran and Periasamy (2012) investigated possible hepatoprotective activities of ethanolic extract of *Trichosanthes lobata* against PCM-induced hepatotoxicity. Blood samples from rats treated with ethanolic extract of *T. lobata* (200 and 400 mg/kg BW) had significant reductions in serum markers in PCM administered animals, indicating the effect of the extract in restoring the normal functional ability of hepatocytes. Silymarin (100 mg/kg, p.o.) was used as a reference drug.

Trichuriella monsoniae (L.f.) Bennet [Syn.: *Aerva monsoniae* (Retz.) Mart.] FAMILY: AMARANTHACEAE

Palati and Vanapatla (2018) investigated the protective effect of methanolic extract of the whole plant of *Trichuriella monsoniae* (METM) and selenium on cadmium (Cd)-induced oxidative liver damage in albino Wistar rats. Oral administration of Cd significantly elevated the levels of hepatic markers such as ALT, AST, ALP, γ-glutamyl transferase, cholesterol, TB, DBL, and decreased levels of TPs and ALB. Co-administration of METM and selenium in Cd-intoxicated rats, the altered biochemical parameters, and pathological changes were recovered significantly than the individual effects of METM and selenium.

Tridax procumbens L. FAMILY: ASTERACEAE

The hepatoprotective activity of aerial parts of *Tridax procumbens* was investigated by Ravikumar et al. (2005) against d-galactosamine/LPS (D-GalN/LPS)-induced hepatitis in rats. D-GalN/LPS (300 mg/kg BW/30 μg/kg BW)-induced hepatic damage was manifested by a significant increase in the activities of marker enzymes (AST, ALT, ALP, LDH, and GGT) and bilirubin level in serum and lipids both in serum and liver. Pretreatment of rats with a chloroform insoluble fraction from ethanolic extract of *T. procumbens* reversed these altered parameters to normal values. The biochemical observations were supplemented by histopathological examination of liver sections.

Hemalatha (2010) investigated the hepatoprotective activity of *T. procumbens*. In vitro studies on *T. procumbens* revealed the antioxidant potential of the herb with chloroform fraction of the ethanolic extract showing maximum activity. It is also reported to possess antioxidant minerals such as iron, magnesium, copper, and zinc. In vivo studies on rodents on the antioxidant potential of the plant induced through LPS, $CCl_{4,}$ alloxan, and PCM intoxication-induced hepatitis confirmed the results from *in vitro* studies as a potential anti-hepatotoxic herb.

Hepatoprotective activity of *T. procumbens* was evaluated by Wagh and Shinde (2010) against PCM (APAP)-induced hepatic damage in male albino rats. Paracetamol (2 g/kg BW)-induced hepatic damage was well manifested by significant increase in the activities of ALT, AST, ALP in serum and enhanced LPO. However, the activities of SOD and CAT in liver tissue were lowered. Consequent to PCM-induced hepatic injury, the SB level was increased. Paracetamol toxicity, also resulted in, significant reduction in total serum protein and the hepatic GSH and glycogen contents. The oral administration of varying doses of ethanolic extract of *T. procumbens* (100, 200, 300, and 400 mg/kg BW) for the period of 7 days reversed these altered parameters to normal levels indicating the antioxidative and hepatoprotective efficacy of *T. procumbens* against PCM-induced liver injury.

Adetutu and Olorunnisola (2013) investigated the hepatoprotective potential of the aqueous extracts of leaves of *T. procumbens*. Extract of dose of 100 mg/kg BW was given to rats in five groups for seven consecutive days followed by a single dose of 2-AAF (0.5 mmol/kg BW). The rats were sacrificed after 24 hours and their bone marrow smears were prepared on glass slides stained with Giemsa. The mPCEs were thereafter recorded. The hepatoprotective effects of the plant extract against 2-AAF-induced liver toxicity in rats were evaluated by monitoring the levels of ALP, GGT, and histopathological analysis. The results of the 2-AAF-induced liver toxicity experiments showed that rats treated with the plant extract (100 mg/kg) showed a significant decrease in mPCEs as compared with the positive control. The rats treated with the plant extract did not show any significant change in the concentration of ALP and GGT in comparison with the negative control group whereas the 2-AAF group showed a significant increase in these parameters. Leaf extract also showed protective effect against histopathological alterations.

Wagh and Shinde (2018) investigated the efficacy of *T. procumbens* (TP) in normalizing rifampicin-induced liver dysfunction as judge on the basis of alteration in different indices of hepatotoxicity in male albino rats. Administration of toxic dose of RIF (50 mg/kg/day) to albino rats resulted in increased activities of AST, ALT, and ALP in serum, and decreased activities of SOD and CAT in liver as compared with normal controls that confirms the development of hepatotoxicity in animals. The levels of hepatic TBARS as a marker of LP were elevated in RIF toxicity. Nevertheless, hepatic GSH, total serum protein and tissue glycogen were found to be decreased in RIF toxicity as against normal controls. However, co-administration of TP extract to RIF intoxicated rats reversed all these RIF induced changes in toxicity parameters. RIF toxicity resulted in altered values of various hepatotoxicity markers indicating severe liver injury and hepatic dysfunction. Treatment with TP extract normalized the altered parameters significantly over a period of 10 days in a dose-dependent manner. TP extract at a dose level of 400 mg/kg/day was proved to be most effective as the values of the marker parameters were comparable well with and approached to those of the normal counterparts thereby confirming the protective efficacy of TP extract against RIF-induced hepatotoxicity in male albino rats.

***Trigonella foenum-graecum* L.**
FAMILY: FABACEAE

Öner et al. (2008) investigated the anti-inflammatory and

hepatoprotective activities of *Trigonella foenum-graecum* (TFG). TFG significantly reduced the serum ALT and AST levels when compared to CCl_4 group. The histopathological findings showed a significant difference between the TFG and CCl_4 groups.

Significant hepatoprotective effects were obtained by ethanolic extract of leaves of *T. foenum-graecum* against liver damage induced by H_2O_2 and CCl_4 as evidenced by decreased levels of antioxidant enzymes (enzymatic and nonenzymatic) (Meera et al., 2009). The extract also showed significant anti LPO effects *in vitro*, besides exhibiting significant activity in superoxide radical and NO radical scavenging, indicating their potent antioxidant effects.

Sakr and Abo-El-Yazid (2011) evaluated the effect of aqueous extract of fenugreek (*T. foenum-graecum*) seeds against hepatotoxicity induced in albino rats by the anticancer drug adriamycin (ADR). Animals were given single dose of ADR (10 mg/kg BW) and were killed after 2 and 4 weeks. Liver of ADR-treated animals showed histopathological and biochemical alterations. The histopathological changes include hepatic tissue impairment, cytoplasmic vacuolization of the hepatocytes, congestion of blood vessels, leucocytic infiltrations, and fatty infiltration. Moreover, the expression of proliferating cell nuclear antigen was increased in ADR-treated rats. The liver enzymes, ALT and AST were increased in the sera of treated rats. Moreover, ADR significantly increased the concentration of MDA and decreased the activities of SOD and CAT in hepatic tissue. Treating animals with ADR and aqueous extract of fenugreek (0.4 g/kg BW) seeds led to an improvement in histological and biochemical alterations induced by ADR. The biochemical results showed that AST and ALT appeared normal together with reduction in the level of MDA (LPO marker) and increase in SOD and CAT activities.

Sakr and Abo-El-Yazid (2012) evaluated the effect of aqueous extract of fenugreek (*T. foenum-graecum)* seeds against hepatotoxicity induced in albino rats by the ADR. Liver of ADR-treated animals showed histopathological and biochemical alterations. The histopathological changes include hepatic tissue impairment, cytoplasmic vacuolization of the hepatocytes, congestion of blood vessels, leucocytic infiltrations, and fatty infiltration. Moreover, the expression of proliferating cell nuclear antigen was increased in ADR-treated rats. The liver enzymes, ALT, and AST were increased in the sera of treated rats. Moreover, ADR significantly increased the concentration of MDA and decreased the activities of SOD and CAT in hepatic tissue. Treating animals with ADR and aqueous extract of fenugreek (0.4

g/kg BW) seeds led to an improvement in histological and biochemical alterations induced by ADR. The biochemical results showed that AST and ALT appeared normal together with reduction in the level of MDA (LPO marker) and increase in SOD and CAT activities.

Trigonella foenum-graecum L. FAMILY: FABACEAE +*Rosmarinus officinalis* FAMILY: LAMIACEAE+ *Cinnamomum zeylanicum* FAMILY: LAURACEAE

Albasha and Azab (2014) evaluated effectiveness of fenugreek seeds, rosemary and cinnamon against cadmium-induced hepatotoxicity in guinea pigs from the histological and biochemical aspects. In cadmium-treated animals, there were severe structural damages in the liver. Most of hepatocytes appeared fused together forming eosinophilic syncytial masses. The hepatocytes appeared irregularly arranged with disorganization of hepatic architecture. The hepatocytes appeared large with light and foamy cytoplasm filled with numerous vacuole-like spaces. The nuclei appeared with pyknotic nuclei. The central vein appeared dilated and congested with massive hemorrhage extending to the nearby cells. Mild periductal fibrosis around bile duct in the portal area was observed. Also, there were focal degenerative and necrotic changes along with inflammatory cell infiltration. Decrease in BW and increase in liver weight were observed. Biochemically, the serum ALT, AST, ALP), and γ-glutamyltransferase activities, serum total and DBL were elevated. Co-adminstration of fenugreek, rosemary and cinnamon significantly improved the structural changes in the liver and also all the above-mentioned biochemical parameters were significantly declined.

Tulbaghia violacea Harv. FAMILY: ALLIACEAE

The rhizomes extract of *Tulbaghia violacea* dose dependently attenuated atherosclerogenic-induced alteration in markers of endothelial dysfunction, lipid profile, liver enzymes and histological changes. The antioxidant and cytotoxicity activities of *T. violacea* as well as its phytochemical components such flavonoids and saponins (Olorunnisola et al., 2011) may be responsible for its hepatoprotective properties.

Tylophora indica (Burm.f.) Merr. FAMILY: ASCLEPIADACEAE

Gujrati et al. (2007) investigated the hepatoprotective activity of alcoholic and aqueous extracts of leaves of *Tylophora indica* against ethanol-induced hepatotoxicity. The hepatoprotective activity was assessed in ethanol-induced hepatotoxic rats.

The alcoholic extract did not produce any mortality even at 5000 mg/kg while LD50 of aqueous extract was found to be 3162 mg/kg. Ethanol produced significant changes in physical (increased liver weight and volume), biochemical (increase in serum ALT, AST, ALP, DBL, TB, cholesterol, TGs, and decrease in TP and ALB level), histological (damage to hepatocytes) and functional (thiopentone-induced sleeping time) liver parameters. Pretreatment with alcoholic extract and aqueous extracts significantly prevented the physical, biochemical, histological and functional changes induced by ethanol in the liver.

The methanolic extract of *T. indica* leaves was screened by Mujeeb et al. (2009) for hepatoprotective activity in CCl_4-induced hepatotoxicity in albino rats. Hepatoprotective activity of methanolic extract at a dose of 200 and 300 mg/kg BW, i.p., was compared with Silymarin (25 mg/kg, i.p.) treated animals. *T. indica* leaves (200 and 300 mg/kg) exhibited significant reduction in serum hepatic enzymes when compared to rats treated with CCl_4 alone. Furthermore, histopathological studies were also done to support the study.

Uraria picta (Jacq.) DC. FAMILY: FABACEAE

Hem et al. (2017) investigated the hepatoprotective activity of ME of the roots of *Uraria picta* (UPME) at a dose of 100, 200, and 400 mg/kg, p.o. in experimental models of rats in PCM-induced liver injury. Methanolic roots extract showed significant activity. Administration of PCM 2000 mg/kg-induced liver injury in rats, and therefore, increased the level of enzymes ALT, ALP, and AST in the blood. Administration of UPME 400, 200, and 100 decreased the level of enzymes ALT, ALP, and AST significantly which were found comparable with the standard drug Silymarin 100 mg/kg.

Urtica dioica L. FAMILY: URTICACEAE

Joshi et al. (2015) isolated hepatoprotective component from *Urtica dioica* (whole plant) against CCl_4-induced hepatotoxicity *in-vitro* (HepG2 cells) and *in-vivo* (rats) model. Antioxidant activity of hydroalcoholic extract and its fractions petroleum ether fraction, EAF, NBF, and AF were determined by DPPH and NO radicals scavenging assay. Fractions were subjected to *in-vitro* HepG2 cell line study. Further, the most potent fraction (EAF) was subjected to *in-vivo* hepatoprotective potential against CCl_4 challenged rats. The *in-vivo* hepatoprotective active fraction was chromatographed on silica column to isolate the bioactive constituent(s). EAF of hydroalcoholic extract of *U. dioica* possessed the potent antioxidant activity. The *in-vitro*

HepG2 cell line study showed that the EAF prevented the cell damage. The EAF significantly attenuated the increased liver enzymes activities in serum and oxidative parameters in tissue of CCl_4-induced rats, suggesting hepatoprotective and antioxidant action, respectively. Column chromatography of most potent antioxidant fraction (EAF) lead to the isolation of 4-hydroxy-3-methoxy cinnamic acid (ferulic acid) which is responsible for its hepatoprotective potential.

Uvaria afzelii SC Elliot FAMILY: ANNONACEAE

The hepatoprotective activity of *Uvaria afzelii* was evaluated in the experimental acute hepatic damage induced by CCl_4 in rat (Ofeimun et al., 2013). In this study, it was reported that the methanolic extracts of the root of *U. afzelii*, at doses of 125, 250, and 500 mg/kg, significantly reduced the serum hepatic enzymes, total and unconjugated bilirubin. Phytochemical studies of this plant have shown the presence of syncarpic acid, dimethoxym atteucinol, emorydone, 2-hydroxydemethoxymatteucinol, uvafzelic acid, syncarpurea, afzeliindanone, flavonoids, triterpenoids, and phenols (Lawal et al., 2016).

Uvaria chamae P. Beauv. FAMILY: ANNONACEAE

The hepatoprotective activity of *Uvaria chamae* methanol root bark extracts were tested *in vivo* and *in vitro* by Madubunyi (2012). An oral administration of the ME (60 mg/kg) significantly reduced pentobarbitone-induced sleep in rats poisoned with APAP. In this model, a protection of 92% against the cytotoxicity of APAP is obtained by pretreatment with the ME as compared to a protection of 89.6% when the animals were pretreated with silibinin. The *n*-hexane extract was without a significant hepatoprotective effect in this model. Intraperitoneal injection of the ME into rats showed no significant effect on pentobarbitone-induced hypnosis. The elevation of SGOT, SGPT, ALP, and urea induced by PCM intoxication in rats was also significantly attenuated by the ME. The ME did not influence the concentration of microsomal proteins in the serum. This *in vivo* efficacy was substantiated by significant hepatoprotection on APAP (AA)-induced hepatotoxicity in isolated rat hepatocytes. The ME, at a dose of 1 mg/mL, remarkably reduced the leakage of LDH in primary cultured rat hepatocytes and showed a significant effect on LPO. The AA-induced elevation of the LPO in rats was significantly decreased in the presence of *U. chamae* root bark ME. A protection of 56% against the *t*-BHP-induced LPO in rats was obtained by pretreatment with the ME. The ME also showed a significant antioxidant effect.

Vaccinium oxycoccos L. FAMILY: ERICACEAE

Hussain et al. (2017) investigated the hepatoprotective efficacy of cranberry (*Vaccinium oxycoccos*) extract (CBE) against CCl_4-induced hepatic injury using *in-vivo* animal model. The exposure of experimental animals to CCl_4 did induce significant hepatotoxicity. The oral administration of CBE demonstrated a significant dose-dependent alleviation in the liver enzymes (AST, ALT, and ALP), increased antioxidant defense (GSH, SOD, and CAT), and reduced MDA levels in the serum of treated animals compared to the animals without treatment. The resulting data showed that the administration of CBE decreased the serum levels of ALT, AST, and ALP compared to the CCl_4-induced group.

Vaccaria pyramidata Medik. FAMILY: CARYOPHYLLACEAE

Biswas et al. (2017) determined the hepatoprotective effects of ethanolic root extract of *Vaccaria pyramidata* against CCl_4-induced hepatic injury in Wistar albino rats. Pretreatment with Silymarin, ethanolic root extract of *V. pyramidata* showed better protection against CCl_4-induced hepatotoxicity. Test indicated a reduction in elevated serum enzyme levels, that is, SGOT, SGPT, and ALP significantly in test groups when compared with toxic control. Pretreatment with standard Silymarin, ethanolic extract significantly reduced the levels of direct and TB when compared to toxic control group. The rats pretreated with Silymarin and ethanolic root extract of *V. pyramidata* exhibited significant decreased in TG levels when compared to the toxic group. This was also confirmed by the result of histopathological examination, which revealed dose dependent decrease in incidence and severity of histopathological changes.

Vaccinium vacillans Kalm ex Torr. FAMILY: ERICACEAE

Sadik et al. (2008) studied the efficacy of dietary supplementation with blueberries (*Vaccinium vacillans*) on DEN-initiated hepatocarcinogenesis in male Wistar rats. BB caused significant decrease in the elevated serum levels of AFP, homocysteine along with levels of GSH, DNA, ribonucleic acid (RNA) and activity of GR in liver. Normalization of elevated 2-macroglobulin (2M) and TAC levels in serum, hepatic GST, GPx activities and liver weight was achieved whereas BW was significantly decreased. Moreover, no significant change was observed in elevated relative liver weight, hepatic glucose-6-P-dehydrogenase (G6PD), LDH along with serum aminotransferases, ALP and -glutamyltransferase (-GT) activities. Significant increase in reduced hepatic activity of XO was achieved and histopathological damage was minimized in BB-treated group.

Valeriana wallichii DC. FAMILY: VALERIANACEAE

Syed et al. (2014) evaluated the hepatoprotective activity of aqueous extract of the roots of *Valeriana wallichii* in albino rats induced by CCl_4. The extracts in the dose of 500 mg/kg were shown a significant decrease in the levels of AST and ALT and CAT activity. Around 300 mg/kg dose of extract showed minimal hepatoprotection. The findings were confirmatory to histopathology.

Vanilla planifolia Jacks. ex Andrews FAMILY: ORCHIDACEAE

Geegi et al. (2011) evaluated the hepatoprotective activity of ethanolic extract of *Vanilla planifolia* against PCM-induced liver damage in rats. The plant extract was effective in protecting the liver against the injury induced by PCM in rats. This was evident from significant reduction in serum enzymes ALT, AST, ALP, and bilirubin.

Ventilago leiocarpa Benth. FAMILY: RHAMNACEAE

Two animal experiments models including carrageenin-induced oedema and CCl_4-induced liver injury were investigated by Lin et al. (1995) to compare and elucidate anti-inflammatory and hepatoprotective effects of four fractions ($CHCl_3$, EtOAc, *n*-BuOH, H_2O) from the stem bark of *Ventilago leiocarpa*. The results showed that each fraction displayed both anti-inflammatory and hepatoprotective activities. The H_2O fraction (50, 100 mg/kg) was even more effective than indomethacin (10 mg/kg) in reducing carrageenin-induced oedema, and also had the greatest protection against CCl_4-induced liver injury. It significantly lowered the acute increase in SGOT and SGPT levels caused by CCl_4. Histopathological changes such as necrosis, fatty change, ballooning degeneration and inflammatory infiltration of lymphocytes and Kupffer cells around the central veins were also concurrently improved by the treatment with each fraction (50 mg/kg) or the H_2O fraction (100 mg/kg).

Verbascum sinaiticum Benth. FAMILY: SCROPHULARIACEAE

Umer et al. (2010) evaluated the *in vivo* hepatoprotective activity of *Verbascum sinaiticum* used in Ethiopian traditional medical practices for the treatment of liver diseases. The levels of hepatic marker enzymes were used to assess their hepatoprotective activity against CCl_4-induced hepatotoxicity in Swiss albino mice. The results revealed that pretreating mice with the hydroalcoholic extract significantly suppressed the plasma AST, ALT activity when compared with the CCl_4 intoxicated control. Histopathological studies also supported these results.

Vernonia ambigua Kotschy & Peyr. FAMILY: ASTERACEAE

The hepatoprotective activity of leaf extract of *Vernonia ambigua* has been investigated using CCl_4-induced hepatotoxicity in albino rats. The extract significantly attenuated CCl_4-induced biochemical (ALT, AST, and ALP, TB, CHOL, TGA, TP and ALB (Orji et al., 2015)). The hepatoprotective properties of plants from genus *Vernonia* may be attributed to presence of mainly flavonoids, steroids and polysaccharides (Leonard et al., 2002), that has been characterized previously from this genus.

Vernonia amygdalina Delile FAMILY: ASTERACEAE

Iwalokun et al. (2006) and Adetutu and Olorunnisola (2013) investigated the hepatoprotective potential of the aqueous extracts of leaves of *Vernonia amygdalina*. Extract of dose of 100 mg/kg BW was given to rats in five groups for seven consecutive days followed by a single dose of 2-AAF (0.5 mmol/kg BW). The hepatoprotective effects of the plant extract against 2-AAF-induced liver toxicity in rats were evaluated by monitoring the levels of ALP, GGT, and histopathological analysis. The results of the 2-AAF-induced liver toxicity experiments showed that rats treated with the plant extract (100 mg/kg) showed a significant decrease in mPCEs as compared with the positive control. The rats treated with the plant extract did not show any significant change in the concentration of ALP and GGT in comparison with the negative control group whereas the 2-AAF group showed a significant increase in these parameters. Leaf extract also showed protective effect against histopathological alterations. This study suggests that the leaf extract has hepatoprotective potential, thereby justifying their ethnopharmacological use (Adetutu and Olorunnisola, 2013).

The effect of aqueous extract of the leaves of *V. amygdalina* on CCl_4-induced liver damage was investigated by Arhoghro et al. (2009) in experimental rats. Treatment of separate groups of rats with 2.5 mL/kg BW of 5%, 10%, and 15% aqueous leaf extracts of *V. amygdalina* for 3 weeks after establishment of CCl_4-induced liver damage, resulted in significantly less hepatotoxicity than CCl_4 alone, as measured by serum ALP, ALT, and AST activities. The effect of extract though not statistically significant was close dependent. Histopathological study also showed significant reduction and even reversal of liver damage in the rats. The results of this study showed that aqueous leaf extract of *V. amygdalina* has a potent antihepatotoxic action against CCl_4-induced liver damage in rats.

The possible modulatory effect of methanolic extract of *V. amygdalina* on the toxicity of CCl_4, in male rats was investigated by Adesanoye and Farombi (2010). Oral administration of CCl_4 at a dose of 1.2 g/kg BW 3 times a week for 3 weeks significantly induced marked hepatic injury as revealed by increased activity of the serum enzymes ALT, AST, SALP and gamma-GT. Methanolic extract of *V. amygdalina* administered 5 times a week for 2 weeks before CCl_4 treatment at 250 and 500 mg/kg doses of the extract ameliorated the increase in the activities of these enzymes. Similarly the methanolic extract of *V. amygdalina* reduced the CCl_4-induced increase in the concentrations of cholesterol, TG and phospholipid by 37.8%, 30.6%, and 8.5%, respectively, and a reduction in the cholesterol/phospholipids ratio. These parameters were however increased at 750 mg/kg extract pretreatment. CCl_4-induced LPO was likewise attenuated by 57.2% at 500 mg/kg dose of the methanolic extract of *V. amygdalina.* Similarly, administration of the extract increased the activities of the antioxidant enzymes: SOD, GST, and reduced GSH concentration significantly at 500 mg/kg and CAT activity at 500–1000 mg/kg doses.

Minari (2012) investigated the hepatoprotective properties of *V. amygdalina* leaves in albino rats using CCl_4. Administration of CCl_4 alone to rats significantly reduced the activities of liver ALT and AST by 71% and 60%, respectively. Simultaneous treatment of CCl_4 injection with oral administration of 20 and 60 mg/kg BW of the methanolic extract significantly reversed these changes in the liver. The activities of liver and kidney GGT were considerably reduced by CCl_4 administration and were also reversed by the plant extract.

Johnson et al. (2015) investigated the hepatoprotective effect of ethanolic leaf extract of *V. amygdalina* against APAP-induced hepatotoxicity in *SD* male albino rats. Animals treated with Silymarin, *V. amygdalina* extracts significantly have reduced WBC count compared to PCM control group. HGB, RBC, and HCT values in all the groups administered with Silymarin, vitamin C and *V. amygdalina* extracts were significantly increased when compared to the PCM-intoxicated animals without treatment. Treatment with *V. amygdalina* extract showed effective hepatoprotective effect as evidence in the decrease in the plasma levels of liver biomarker enzymes and reduction in oxidative stress parameters. Histopathological evaluation of the liver architecture also revealed that all the treated animals have reduced the incidence of PCM-induced liver lesions.

Usunomena et al. (2015) evaluated the protective role of *V. amygdalina* pretreatment on some biochemical indices in DMN-induced liver damage in male albino rats. In rats administered 20 mg/kg DMN, liver damage was clearly shown by increased activities of serum hepatic marker enzymes namely AST, ALT, ALP, and GGT, increased lipid profile parameters such as TC and TGs as well as increased level of LPO indices, MDA in liver. The toxic effect of DMN was also indicated by significantly decreased levels of antioxidants such as SOD, CAT, and reduced GSH. However, in rats pretreated with 400 mg/kg *V. amygdalina* and dosed thereafter with DMN, there were significant reversal in the activities of serum hepatic marker enzymes, lipid profiles, LPO, and significant restoration of antioxidant levels in the liver when compared to DMN-alone-treated rats. Histopathological studies in the liver of rats also showed that *V. amygdalina* pretreatment markedly reduced the toxicity of DMN and significantly preserved the normal histological architecture of the tissue.

Chemoprotective effect of methanolic extract of *V. amygdalina* against 2-acetylaminofluorene-induced hepatotoxicity in rats was studied by Adesanoye et al. (2016). The extract at 250 and 500 mg/kg significantly ameliorated the oxidative damage, functional impairments, and histopathological changes associated with 2-AAF toxicity by reducing the activities of serum enzymes, upregulating the antioxidant defense enzymes and GSH with decrease in MDA level.

Vernonia condensata Baker FAMILY: ASTERACEAE

Vernonia condensata is traditionally used in South American Countries as an anti-inflammatory, analgesic, and hepatoprotective. da Silva et al. (2017) investigated the *in vivo* hepatoprotective and antioxidant, and the *in vitro* anti-inflammatory activities of the ethyl acetate partition (EAP) from the ethanolic extract of this medicinal plant leaves. EAP was able to inhibit all the acute biochemical alterations caused by APAP overdose. EAP inhibited MDA formation, maintained the CAT and increased the GR activities. Also, EAP decreased NO, IL-6, and TNF-α levels at concentrations from 10 to 20 μg/mL. 1,5-dicaffeoylquinic acid was isolated and identified as the major compound in EAP. Apigenin, luteolin, and chlorogenic acid were also identified. EAP anti-inflammatory action may be due to its antioxidant activity or its capacity to inhibit the pro-inflammatory cytokines.

Viburnum punctatum Buch.-Ham. ex D.Don FAMILY: CAPRIFOLIACEAE

Alex et al. (2014) evaluated *in vitro* hepatoprotective activity of chloroform and MEs of *Viburnum punctatum* (200 and 400 μg/mL) against CCl_4 induced toxicity. The Chang liver cells were treated with different concentrations of chloroform and MEs of *V. punctatum*, showed a dose dependent increase in percentage viability and the results were highly significant when compared with CCl_4-induced group. The percentage viability ranged between 62% and 84% at 200–400 μg/mL concentrations. The methanolic extract exhibited more hepatoprotective activity.

Vicia calcarata Desf. FAMILY: FABACEAE

The hepatoprotective activity of flavonol glycosides rich fraction, prepared from 70% alcohol extract of the aerial parts of *Vicia calcarata*, was evaluated by Singab et al. (2005) in a rat model with a liver injury induced by daily oral administration of CCl_4. Treatment of the animals with rich fraction using a dose of 25 mg/kg, BW during the induction of hepatic damage by CCl_4 significantly reduced the indices of liver injuries. The hepatoprotective effects of rich fraction significantly reduced the elevated levels of the following serum enzymes: ALT, AST, ALP, and LDH. The antioxidant activity of rich fraction markedly ameliorated the antioxidant parameters including GSH content, GSH-Px, SOD, plasma CAT and packed erythrocytes G6PDH to be comparable with normal control levels. In addition, it normalized liver MDA levels and creatinine concentration. Chromatographic purification of F-2 resulted in the isolation of two flavonol glycosides that rarely occur in the plant kingdom, identified as quercetin-3, 5-di-*O*-β-D-diglucoside and kaempferol-3,5-di-*O*-β-D-diglucoside in addition to the three known compounds identified as quercetin-3-*O*-α-L-rhamnosyl-(1→6)-β-D-glucoside [rutin], quercetin-3-O-β-D-glucoside [isoquercitrin] and kaempferol-3-*O*-β-D-glucoside [astragalin]. Moreover, the spectrophotometric estimation of the flavonoids content revealed that the aerial parts of the plant contain an appreciable amount of flavonoids (0.89%) calculated as rutin. The data obtained from this study revealed that the flavonol glycosides of rich fraction protect the rat liver from hepatic damage induced by CCl_4 through inhibition of LPO caused by CCl_4 reactive free radicals.

Vigna radiata (L.) R.Wilczek FAMILY: FABACEAE

Mung bean (*Vigna radiata*) is a hepatoprotective agent in dietary supplements. Fermentation and germination processes are well recognized to enhance the nutritional values especially the concentration of active compounds such as amino acids and GABA of various foods. Ali et al. (2013) compared the antioxidant and hepatoprotective effects of freeze-dried mung bean and amino-acid- and GABA-enriched germinated and fermented mung bean aqueous extracts. Liver SOD, MDA, FRAP, NO levels, and serum biochemical profile such as AST, ALT, TG, and cholesterol and histopathological changes were examined for the antioxidant and hepatoprotective effects of these treatments. Germinated and fermented mung bean have recorded an increase of 27.9 and 7.3 times of GABA and 8.7 and 13.2 times of amino acid improvement, respectively, as compared to normal mung bean. Besides, improvement of antioxidant levels, serum markers, and NO level associated with better histopathological evaluation indicated that these extracts could promote effective recovery from hepatocyte damage. These results suggested that freeze dried, germinated, and fermented mung bean aqueous extracts enriched with amino acids and GABA possessed better hepatoprotective effect as compared to normal mung bean.

Wu et al. (2001) evaluated the effects of various water extract concentrations (100, 500, and 1000 mg/kg BW.) of mung bean (*Vigna radiata*), adzuki bean (*Vigna angularis*), black bean (*Phaseolus vulgaris*) and rice bean (*Vigna umbellata*) and Silymarin (25 mg/kg BW. on APAP-induced liver injury by measuring SGOT and SGPT activities in rats. The results showed that the SGOT and the SGPT activities, increased by APAP, were decreased significantly through treatment with increasing amounts up to 1000 mg/kg BW. of the exracts. In particular, the mung bean aqueous extract showed the best hepatoprotective effect on APAP-induced hepatotoxicity. The pathological changes of liver injury caused by APAP improved by the treatment with all the legume extracts, which were compared to Silymarin as a standardized drug. In addition to these results, the extract of mung bean acted as a potential hepatoprotective agent in dietary supply.

Viola canescens Wall. ex. Roxb. FAMILY: VIOLACEAE

Abdullah et al. (2017) investigated the hepatoprotective activity of solvent extracts of whole plant of *Viola canescens*. Phytochemical screening ME showed the presence of alkaloids, phenols, flavonoids, saponins, carbohydrates, tannins, and triterpenes. Methanol extract and EAF (at 200 and 400 mg/kg BW) and partially purified EAF at 50 mg/kg BW significantly reduced the level of ALT, ALP, TB and restored the level of serum protein in comparison to CCl_4 treated group. A significant reduction in MDA and elevation in CAT and SOD level was observed in extract treated animals as compared to CCl_4. The liver biopsy of mice treated with the solvent extracts showed remarkable restoration of normal histological architecture.

Viola odorata L. FAMILY: VIOLACEAE

Traditionally *Viola odorata* is used for liver protection. To provide scientific support to its traditional use, aqueous methanolic extract of *V. odorata* (250 and 500 mg/kg) was given to mice intoxicated with PCM (Qadir et al., 2014). The extract significantly reduced PCM-induced increase levels of serum hepatic enzymes and TB. Histopathological studies showed that the plant attenuated the hepatocellular necrosis and inflammation. HPLC results showed the presence of hepatoprotective flavonoids (isorhamnetin and luteolin) in the extract.

Jamshed et al. (2019) evaluated the hepatoprotective activity of *V. odorata* extract in PCM-induced hepatic dysfunction in mice. Methanol extract of *V. odorata* leaves has the potential of limiting hepatic damage by lowering AST, ALT, and GGT.

Vitellaria paradoxa Gaertn.f. FAMILY: SAPOTACEAE

Ojo et al. (2006) evaluated the hepatoprotective activity of stem bark extract of *Vitellaria paradoxa* against APAP-induced hepatotoxicity in Wistar albino rats. They reported that stem bark extract of *V. paradoxa* produced significant hepatoprotective effects by decreasing the activity of serum enzymes. Values recorded for AST, ALT, and ALP were significantly lower compared to those recorded for control rats. A higher inhibition of serum level elevation of ALP was observed with the extract.

Vitex doniana Sweet FAMILY: LAMIACEAE

The effect of methanolic extract of *Vitex doniana* fruits on APAP-induced protein oxidation, LPO, and DNA fragmentation in mice was investigated by Ajiboye (2015). Antioxidant activity of the extract (0.2–1.0 mg/mL) was investigated *in vitro* using DPPH radical, superoxide ion, hydrogen peroxide, hydroxyl radical, and ferric ion reducing system. *V. doniana* extract at 1.0 mg/mL scavenged DPPH, superoxide ion, hydrogen peroxide, and hydroxyl radical by 86%, 78%, 80%, and 72%, respectively, it also reduced ferric ion significantly. Hepatoprotective effect of *V. doniana* fruits was monitored in APAP-induced hepatotoxicity in mice. Acetaminophen-mediated alterations in serum ALP, ALT, AST, ALB, and TB levels in mice were significantly attenuated by the extract.

Similarly, APAP-mediated decrease in activities of SOD, CAT, GPx, GR, and G6PD was significantly attenuated in the liver of mice. Increased levels of CD, lipid HP, MDA, protein carbonyl, and fragmented DNA were significantly lowered by methanolic extract of *V. doniana* fruits.

Vitex negundo L. FAMILY: LAMIACEAE

Hepatoprotective activity of *Vitex negundo* leaf ethanolic extract was investigated by Tandon et al. (2008) against hepatotoxicity produced by administering a combination of three antitubercular drugs isoniazid −7.5 mg/kg, rifampin −10 mg/kg and PZA-35 mg/kg for 35 35 days by oral route in rats. *V. negundo* leaf ethanolic extract was administered in three graded doses of 100, 250, and 500 mg/kg orally, 45 min prior to antitubercular challenge for 35 days. Hepatoprotective effect of *V. negundo* leaf ethanolic extract was evident in the doses of 250 and 500 mg/kg as there was a significant decrease in TB, AST, ALT, and ALP levels in comparison to control. Histology of the liver section of the animals treated with the *V. negundo* leaf ethanolic extract in the doses of 250 and 500 mg/kg further confirms the HP activity.

Mahalakshmi et al. (2010) evaluated the hepatoprotective activity on ethanolic extract of leaves of *V. negundo* was determined by using male Wistar Albino rats. The *V. negundo* is a natural plant product, its leaves are used with the added advantage to revert Ibuprofen-induced hepatotoxicity. Oral administration of ethanol extract of *V. negundo* (100 and 300 mg) produced a significant and dose-dependent inhibition to the acute hepatotoxic-induced rats and various parameters were analyzed, when compared with negative control *V. negundo* showed a significant activity in 300 mg/kg/BW. They exhibited a significant inhibition of hepatic toxicity by using various marker enzymes and the histopathological analysis. The inhibitory effect of the *V. negundo* on hepatotoxicity was compared to that of positive control group. The various parameters such glucose, protein, TGs, bilirubin, urea, creatinine, ALP, ACP, SGPT, SGOT, and histopathological parameters were measured by dissection the rats. A significant index and values were observed in the acute assays; an effective significant alteration in all biochemical and histopathological sections weres observed. From these results, it was concluded that the *V. negundo* having the potential effectiveness at the dose of 300 mg/kg/BW, significance in a dose-dependent manner.

The hepatoprotective activity of ethanolic extract from the leaves of *Vitex negundo* (VN) was assessed by Kadir et al. (2013) against

TAA-induced hepatic injury in SD rats. After 12 weeks, the rats administered with VN showed a significantly lower liver to BW ratio. Their abnormal levels of biochemical parameters and liver MDA were restored closer to the normal levels and were comparable to the levels in animals treated with the standard drug, Silymarin. Gross necropsy and histopathological examination further confirmed the results. Progression of liver fibrosis induced by TAA in rats was intervened by VN extract administration, and these effects were similar to those administered with Silymarin.

Vitex trifolia L. FAMILY: LAMIACEAE

Aqueous and ethanol extracts of leaf of *Vitex trifolia* was investigated by Manjunatha and Vaidya (2008) for hepatoprotective activity against CCl_4-induced liver damage. To assess the hepatoprotective activity of the extracts, various biochemical parameters viz., TB, TP, ALT, AST, and ALP activities were determined. Results of the serum biochemical estimations revealed significant reduction in TB and serum marker enzymes and increase in TP in the animals treated with ethanol and aqueous extracts. However significant rise in these serum enzymes and decrease in TP level was noticed in CCl_4 treated group indicating the hepatic damage. The hepatoprotective activity was also supported by histological studies of liver tissue. Histology of the liver tissue treated with ethanol and aqueous extracts showed normal hepatic architecture with few fatty lobules.

Vitis thunbergii Siebold & Zucc. FAMILY: VITACEAE

Deng et al. (2012) investigated the *in vitro* and *in vivo* antioxidant effects and hepatoprotective activity of the ethanol extract of *Vitis thunbergii*. Ethanol extracts exhibited strong antioxidant ability *in vitro*. After oral administration of extract significantly decreased ALT and AST, and ameliorated the oxidative stress in hepatic tissue and increased the activity of CAT, SOD, GPx, and GSH. Serum TNF-α, interleukin-1β (IL-1β), and NO were decreased in the group treated with CCl_4 plus ethanol extract. Western blotting revealed that ethanol extract blocked protein expression of iNOS and cyclooxygenase-2 (COX-2) in CCl_4-treated rats, significantly. Histopathological examination of livers showed that ethanol extract of *V.thunbergii* reduced fatty degeneration, cytoplasmic vacuolization, and necrosis in CCl_4-treated rats.

Vitis vinifera L. FAMILY: VITACEAE

Maheswari and Rao (2005) studied the effect of oral administration of grape

(*Vitis vinifera*) seed oil (GSO) against CCl_4-induced hepatotoxicity in rats. Oral administration of GSO (3.7 g/kg, BW orally) for 7 days resulted in a significant reduction in serum AST, ALT, and ALP levels and liver MDA and HP and significant improvement in GSH, SOD, CAT, and TP, when compared with CCl_4 damaged rats. The antioxidant effect of GSO at 3.7 g/kg for 7 days was found to be comparable with vitamin E (100 mg/kg, orally) in CCl_4-treated rats. Profound fatty degeneration, fibrosis, and necrosis observed in the hepatic architecture of CCl_4-treated rats were found to acquire near-normalcy in drug co-administered rats.

The hepatoprotective effect of ethanolic extract and its four different fractions ($CHCl_3$, EtOAc, *n*-BuOH, and remaining water fraction) of *V. vinifera* leaves was investigated by Orhan et al. (2007) against CCl_4-induced acute hepatotoxicity in rats. The ethanolic extract was found active at 125 mg/kg dose (per os). The ethanolic extract was fractionated through successive solvent–solvent extractions and the *n*-BuOH fraction in 83 mg/kg dose possessed remarkable antioxidant and hepatoprotective activities. Liver damage was assessed by using biochemical parameters (plasma and liver tissue MDA, transaminase enzyme levels in plasma [AST, ALT] and liver GSH [glutathione] levels). Additionally, the pathological changes in liver were evaluated by histopathological studies. Legalon 70 Protect was used as standard natural originated drug (Orhan et al., 2007).

Grape juice is a source of polyphenols, as catechin, anthocyanidins, resveratrol, and others. Some health benefits have been attributed to these compounds (e.g., antioxidant and antitumorigenic properties). Dani et al. (2008) investigated the possible antioxidant activity of two different grape juices: organic purple grape juice and conventional purple grape juice. The group treated with organic grape juice showed the highest SOD and CAT activities in both plasma and liver when compared with the conventional and control groups. In plasma, authors observed a positive correlation among SOD and CAT activities, resveratrol, and all anthocyanin contents, suggesting that these polyphenols may be, at least in part, responsible for this increased antioxidant defense. The grape juices were capable of reducing carbonyl and LPO levels in plasma and liver. However, in plasma, the organic group showed lower carbonyl and TBARS levels when compared to the conventional grape juice group. Their findings suggest that the intake of purple grape juice, especially of organic juice, induces a better antioxidant capacity when compared to conventional juice.

Amelioration of TAM-induced liver injury in rats by grape seed extract, black seed extract and curcumin was reported by

El-Beshbishy et al. (2010). Liver injury was induced in female rats using TAM. Grape seeds (*V. vinifera*) extract (GSE), black seed (*Nigella sativa*) extract (NSE), curcumin (CUR) or Silymarin (SYL) were orally administered to TAM-intoxicated rats. Liver histopathology of TAM-intoxicated rats showed pathological changes. TAM-intoxication elicited declines in liver antioxidant enzymes levels (GPx, GR, SOD, and CAT), reduced GSH and GSH/GSSG ratio plus the hepatic elevations in LPOs, oxidized glutathione (GSSG), TNF-α and serum liver enzymes; ALT, AST, ALP, LDH, and GGT levels. Oral intake of NSE, GSE, CUR, or SYL to TAM-intoxicated rats, attenuated histopathological changes and corrected all parameters mentioned above. Improvements were prominent in case of NSE (similarly SYL) > CUR > GSE. Data indicated that NSE, GSE, or CUR act as free radicals scavengers and protect TAM-induced liver injury in rats.

The ethanolic extract of the root of *V. vinifera* was evaluated by Sharma et al. (2012) for hepatoprotective activity in rats with liver damage induced by CCl_4. The extract at an oral dose of 200 mg/kg exhibited a significant protective effect by lowering the serum levels of SGPT, SGOT, ALP, and TB. The extract at this dose also increased the level of TP. These biochemical observations were supplemented by histopathological examination of liver sections. The activity of extract was also comparable to that of Silymarin, a known hepatoprotective drug.

Suosuo grape (the fruits of *V vinifera*) has been used for the prevention and treatment of liver diseases in Uighur folk medicine in China besides its edible value. The hepatoprotective effects of total triterpenoids (VTT) and total flavonoids (VTF) from Suosuo grape were evaluated by Liu et al. (2012) in BCG-plus-LPS-induced immunological liver injury (ILI) in mice. Various dose groups (50, 150, and 300 mg/kg) of VTT and VTF alleviated the degree of liver injury of ILI mice, effectively reduced the BCG/LPS-induced elevated liver index and spleen index, hepatic NO, and MDA content, increased liver homogenate ALT and AST levels, and restored hepatic SOD activity in ILI mice. VTT and VTF also significantly inhibited intrahepatic expression of Th1 cytokines (IFN-γ and IL-2) in ILI mice and increased intrahepatic expression of Th2 cytokines (IL-4 and IL-10). Moreover, the increased Bax/Bcl-2 ratio was significantly downregulated by VTT and VTF in liver tissue of ILI mice. These results are comparable to those of biphenyl dicarboxylate (DDB, the reference hepatoprotective agent) and suggest that VTT and VTF play a protective role against ILI, which may have important implications for our understanding of the immunoregulatory mechanisms of this plant.

Almajwal and Elsadek (2015) evaluated the protective effect of red grape dried seeds (RGDS) on antioxidant properties, lipid metabolism, and liver and kidney functions of rats with PCM (750 mg/kg)-induced hepatotoxicity. A significant decrease in levels of serum cholesterol, TGs, LDL-C, and VLDL-C, with a significant increase in level of HDL-C for RGDS groups compared to induced control. Rats administered a diet containing RGDS levels produced significant hepatoprotection by decreasing the activities of liver enzymes, kidney parameters, and LPO, while levels of GSH, SOD, and CAT were increased significantly to near the normal levels.

Wedelia calendulacea L. FAMILY: ASTERACEAE

The ALE of whole plant *Wedelia calendulacea* exhibited protective activity against CCl_4-induced liver injury *in vivo*. The extract also increased the bile flow in rats suggesting a stimulation of liver secretory capacity. The minimum lethal dose was greater than 200 mg/kg p.o. in mice (Sharma et al., 1989).

The ME of *W. calendulacea* was tested by Halder et al. (2007) for its hepatoprotective activity against TAA-induced hepatotoxicity in Wister albino rats. The oral administration of the extract (100, 200, and 400 mg/kg BW) significantly reduced TAA-induced hepatotoxicity in rats, as judged from the estimation of various biochemical parameters *viz*, SGOT, SGPT, ALP, LP, GSH, TP, and bilirubin. These results were comparable with standard drug Silymarin (20 mg/kg BW, p.o.).

The hepatoprotective activity of ethanolic extract of *W. calendulacea* was studied against CCl_4-induced, acute hepatotoxicity in rats by Murugaian et al. (2008). The treatment with ethanolic extract of *W. calendulacea* showed a dose-dependent reduction of CCl_4-induced elevated serum levels of enzyme activities with parallel increase in TP and bilirubin, indicating the extract could preserve the normal functional status of the liver. The weight of the organs such as liver, heart, lung, spleen, and kidney in CCl_4-induced experimental animals administered with the extract showed an increase over CCl_4 control group.

Wedelia chinensis (Osbeck) Merr. FAMILY: ASTERACEAE

Lin (1994) evaluated the hepatoprotective activity of *Wedelia chinensis* against acute hepatitis induced by three hepatotoxins: CCl_4, APAP and D(+)-glactosamine in rats. After treatment with *W. chinensis* (300 mg/kg, p.o.) a reduction in the elevation of SGOT, SGPT levels was observed 24 h after hepatotoxins were administered. These serological observations were confirmed by histopathological examinations.

Woodfordia fruticosa Kurz. FAMILY: LYTHRACEAE

Dried flowers of *Woodfordia fruticosa* are used in variety of diseases in traditional Indian system of medicine including hepatic ailments. Chandan et al. (2008) evaluated the hepatoprotective activity of petroleum ether (WF1), chloroform (WF2), ethyl alcohol (WF3), and aqueous (WF4) extracts of the flowers of *W. fruticosa* against CCl_4-induced hepatotoxicity using biochemical markers, hexobarbitone sleep time, BSP clearance test and effect on bile flow and bile solids. The aqueous extract (WF4) was most potent among the four extracts studied in detail. WF4 showed significant hepatoprotective activity against CCl_4-induced hepatotoxicity as evident by restoration of serum transaminases, ALP, bilirubin and TGs. The restoration of microsomal aniline hydroxylase and amidopyrine-*N*-demethylase activities indicated the improvement in functional status of endoplasmic reticulum. Restoration of LPO and GSH contents suggests the antioxidant property of WF4. The recovery in bromosulphalein clearance and stimulation of bile flow suggested the improved excretory and secretory capacity of hepatocytes. Light microscopy of the liver tissue further confirmed the reversal of damage induced by hepatotoxin (Chandan et al., 2008).

Wrightia tinctoria R.Br. FAMILY: APOCYNACEAE

Wrightia tinctoria was reported to have hepatoprotective activity (Chandrashekhar et al., 2004). Further, Bigoniya and Rana (2010) investigated the hepatoprotective effect of triterpene fraction isolated from the stem bark of *W. tinctoria* (containing lupeol, β-amyrin and β-sitosterol) against CCl_4-induced hepatotoxicity in comparison with known standard Silymarin in the rat. Pretreatment with triterpene fraction 125, 250, and 400 mg/kg, p.o. once a day for 4 days before CCl_4 and continued further 3 d, attenuated the CCl_4-induced acute increase in SGPT, SGOT, and ALP activities and considerably reduced the histopathological alterations. Further, triterpene fraction reduced thiopentone-induced sleeping time, suggesting the protection of liver metabolizing enzymes. The authors claim triterpenes administration changed the tissue redox system by scavenging the free radicals and by improving the antioxidant status of the liver replenished the depleted hepatic GSH and SOD. Triterpene pretreatment improved bromsulphalein clearance of the CCl_4-intoxicated liver and increased the cellular viability. These effects substantiate protection of cellular phospholipid from peroxidative damage induced by highly reactive toxic intermediate radicals formed during biotransformation of CCl_4. Triterpene fraction afforded protection against the hepatic abnormalities due to the presence of lupeol and β-amyrin.

Recently, Jamshed et al. (2019) evaluated the hepatoprotective activity of *W. tinctoria* seeds extract in PCM-induced hepatic dysfunction in mice. Methanol extract of *W. tinctoria* seed has the potential of limiting hepatic damage by lowering AST, ALT, and GGT.

Ximenia americana L. var. *caffra* FAMILY: OLACACEAE

Sobeh et al. (2017) investigated the hepatoprotective activities of a root extract from *Ximenia americana* var. *caffra* in rat models with D-GaIN-induced hepatotoxicity. The root extract is rich in tannins with 20 compounds including a series of stereoisomers of (epi) catechin, (epi) catechin-(epi) catechin, (epi) catechin-(epi) catechin-(epi) catechin, and their galloyl esters. Promising antioxidant potential was observed *in vitro*. Significant reduction of serologic enzymatic markers and hepatic oxidative stress markers such as ALT, AST, MDA, GGT, and TB, as well as elevation of GSH and ALB were observed in rats with D-galactosamine-induced liver damage treated with the extract. These findings agree with a histopathological examination suggesting a hepatoprotective potential for the root extract.

Xylopia aethiopica (Dunal) A. Rich FAMILY: ANNONACEAE

Patrick-Iwuanyanwu et al. (2010) evaluated the protective effects of *Xylopia aethiopica* against CCl_4-induced hepatotoxicity in rats. Serum ALT, AST, ALP, and LDH levels 24 h after CCl_4 administration decreased significantly in rats pretreated with *X. aethiopica* than in CCl_4-treated rats only. Total SB also showed a remarkable decrease in rats pretreated with *X. aethiopica* when compared to those administered CCl_4 alone. Lipid peroxidation expressed by MDA concentration was significantly decreased in rats pretreated with *X. aethiopica* than in rats administered CCl_4 alone. Histopathological examinations of rats administered CCl_4 alone revealed severe hepatic damage to the liver. However, rats pretreated with *X. aethiopica* showed significant improvements in the architecture of rat liver.

Adewale et al. (2014) evaluated the hepatoprotective effects of aqueous extract of *X. aethiopica* stem bark on CCl_4-induced liver damage in SD rats. Serum ALT, AST, and ALP levels 24 hrs after CCl_4 administration decreased significantly in rats pretreated with *X. aethiopica* than in CCl_4-treated rat only. Total SB also showed a remarkable decrease in rats pretreated with *X. aethiopica* when compared to those administered with CCl_4 alone. The activities of GST and CAT in liver tissues were increased in the rats pretreated with *X. aethiopica* compared with CCl_4 alone. Lipid peroxidation expressed by MDA concentration was significantly decreased in rats pretreated with *X. aethiopica* compared with CCl_4 treated rat. The rats pretreated with *X. aethiopica* showed significant improvements in the cytoarchitecture of rat liver. The results suggested that aqueous extract of *X. aethopica* could palliate the liver injuries perhaps by its antioxidative effect, hence eliminating the deleterious effect of toxic metabolites from the CCl_4.

Adewale and Orhue (2015) examined the hepatoprotective effect of aqueous extract of the fruits of *X. aethiopica* using the CCl_4 model. Whereas CCl_4 administration resulted in significant elevations in plasma ALT, AST, and ALP, there was a significant reduction in both plasma TP and ALB. In addition, histopathological changes were observed with CCl_4. Analysis of the data obtained for MDA, SOD, and CAT suggest that the plant extract exerts its protective effect probably by inhibiting CCl_4-induced LPO in liver tissue. It can be suggested that *X. aethiopica* fruits has the ability to offer a significant degree of protection to liver cells against CCl_4-induced hepatotoxicity in Wistar albino rats by antioxidant mechanism of action.

Zanthoxylum armatum **DC. FAMILY: RUTACEAE**

Zanthoxylum armatum is described as a hepatoprotective in Ayurveda, the Indian system of medicine. Ranawat et al. (2010) carried out study to evaluate the hepatoprotective activity of ethanolic extract of bark of *Z. armatum* in CCl_4-induced hepatotoxicity in male Wistar rats. Ethanolic extracts at doses of 100, 200, and 400 mg/kg were administered orally once daily for 7 days. The hepatoprotective activity was assessed using various biochemical parameters like ALT, AST, ALP, SB, TP, and serum antioxidant enzymes along with histopathological studies of liver tissue. The substantially elevated serum enzymatic levels of serum transaminases, ALP and TB were significantly restored toward normalization by the extracts. Bark extracts significantly increased the levels of antioxidant enzymes: SOD, CAT and, GSH. Phytochemical analysis revealed the presence of isoquinoline alkaloid, berberine, as well as flavonoids and phenolic compounds, which have been known for their hepatoprotective activities (Ranawat et al., 2010).

The hepatoprotective activity of the leaves ethanolic extract of *Z. armatum* was evaluated in CCl_4-induced hepatotoxicity in rats by Verma and Khosa (2010). The extract at a dose of 500 mg/kg registered a significant decrease in the levels of SGOT, SGPT, ALP, and SB and liver inflammation, which was supported by histopathological studies on liver, thus exhibited a significant hepatoprotective activity. The phytochemical screening of defatted ethanolic extract showed the presence of sterols, alkaloids, flavonoids, and reducing sugars.

Zataria multiflora Boiss FAMILY: LAMIACEAE

Ahmadipour et al. (2015) evaluated the protective effects of methanolic extract of *Zataria multiflora* against hepatic damage induced by cisplatin in male Wistar rats. Rats treated with cisplatin resulted in a significant increase in serum activity, AST, ALT, and ALP in treated mice. Management with *Z. multiflora* reduced the business of these enzymes to nearly normal levels. In parallel with these changes, this extract reduced cisplatin-induced oxidative stress by inhibiting LPO and protein carbonylation, and restoring the antioxidant enzyme (SOD, CAT, and GSH-Px) and elevation of the GSH level. Biochemical and histological observations showed the hepatoprotective effect was found in a dose-dependent manner in *Z. multiflora* methanolic extract. This protective effect can be attributed to the antioxidant compounds.

Atayi et al. (2018) evaluated the protective effects of *Zataria multiflora* aromatic water (AW) against the hepatic injury induced by long-term albendazole treatment in mice with CE. A decrease in serum liver enzyme activity in the both *Z. multiflora*+albendazole groups was observed when compared to control, *Z. multiflora* and albendazole groups; however, the results for *Z. multiflora*+albendazole 100 were significant and superior compared to those for *Z. multiflora*+albendazole 200. No significant differences for oxidative stress markers were observed between the different groups. A combined therapy with *Z. multiflora* AW and albendazole is effective against hepatic injury induced by CE and/or long term albendazole administration in mice with cystic echinococcosis.

Zingiber officinale Roscoe FAMILY: ZINGIBERACEAE

The effect of the ethanol extract of the rhizome of *Zingiber officinale* was tested against CCl_4 and APAP-induced liver toxicities in rats by

Yemitan and Izegbu (2006). Increases in serum and liver marker enzymes such as ALT, AST, LDH, ALP as well as sorbitol and glutamate dehydrogenases were produced in normal rats that were not pretreated with the extract. However, extract-pretreated rats attenuated in a dose-dependent manner, CCl_4 and APAP-induced increases in the activities of ALT, AST, ALP, LDH, and SDH in the blood serum. The protective effect of the extract on CCl_4 and APAP-induced damage was confirmed by histopathological examination of the liver.

Hepatoprotective activity of aqueous ethanol extract of *Z. officinale* was evaluated by Ajith et al. (2007) against single dose of APAP-induced (3 g/kg, p.o.) acute hepatotoxicity in rat. Aqueous extract of *Z. officinale* significantly protected the hepatotoxicity as evident from the activities of serum transaminase and ALP. SGPT, SGOT, and ALP activities were significantly elevated in the APAP alone treated animals. Antioxidant status in liver such as activities of SOD, CAT, GPx, and GST, a phase II enzyme, and levels of reduced GSH were declined significantly in the APAP alone treated animals (control group). Hepatic LPO was enhanced significantly in the control group. Administration of single dose of aqueous extract of *Z. officinale* (200 and 400 mg/kg, p.o.) prior to APAP significantly declines the activities of serum transaminases and ALP. Further the hepatic antioxidant status was enhanced in the *Z. officinale* plus APAP treated group than the control group. The hepatoprotective effect of aqueous ethanol extract of *Z. officinale* against APAP-induced acute toxicity is mediated either by preventing the decline of hepatic antioxidant status or due to its direct radical scavenging capacity.

The effect of ginger (*Z. officinale*) upon hepatotoxicity induced in albino rats by the anticancer drug, ADR was studied by Sakr et al. (2011). Injecting animals with ADR induced various histological changes in the liver. These changes include congestion of blood vessels, leucocytic infiltration, cytoplasmic vacuolization of the hepatocytes and fatty infiltration. Adriamycin caused significant elevation in serum ALT and AST enzymes after 4 and 6 weeks of treatment. It also caused an increase in MDA (LPO marker) and depletion of the antioxidant enzyme, SOD. Treating animals with water extract of ginger and ADR led to an improvement in the histological changes induced by ADR together with significant decrease in ALT and AST activity. Moreover, ginger reduced the level of MDA and increased the activity of SOD.

Abdel-Azeem et al. (2013) investigated efficacy of ginger pretreatment in alleviating APAP-induced acute hepatotoxicity in rats. Administration of APAP elicited significant liver injury that was manifested by remarkable increase in plasma ALT, AST, ALP, arginase activities, and TB concentration. Meanwhile, APAP significantly decreased plasma TPs and ALB levels. Ginger or vitamin E treatment prior to APAP showed significant hepatoprotective effect by lowering the hepatic marker enzymes (AST, ALT, ALP, and arginase) and TB in plasma. In addition, they remarkably ameliorated the APAP-induced oxidative stress by inhibiting LPO (MDA). Pretreatment by ginger or vitamin E significantly restored TAGs, and TP levels. Histopathological examination of APAP-treated rats showed alterations in normal hepatic histoarchitecture, with necrosis and vacuolization of cells. These alterations were substantially decreased by ginger or vitamin E. These results demonstrated that ginger can prevent hepatic injuries, alleviating oxidative stress in a manner comparable to that of vitamin E. Combination therapy of ginger and APAP is recommended especially in cases with hepatic disorders or when high doses of APAP are required.

Bardi et al. (2013) assessed the hepatoprotective activity of the ethanolic extract of rhizomes of *Z. officinale* (ERZO) against TAA-induced hepatotoxicity in rats. Results confirmed the induction of liver cirrhosis in group treated with TAA whilst administration of Silymarin or ERZO significantly reduced the impact of TAA toxicity. These groups decreased fibrosis of the liver tissues. Immunohistochemistry assessment against proliferating cell nuclear antigen did not show remarkable proliferation in the ERZO-treated rats when compared with toxicity-induced animals. Moreover, factions of the ERZO extract were tested on Hep-G2 cells and showed antiproliferative activity (IC50 38–60 μg/mL). This study showed hepatoprotective effect of ERZO.

Ziziphus jujuba Mill. FAMILY: RHAMNACEAE

Kumar et al. (2008) investigated the hepatoprotective effect of methanolic extract of *Ziziphus jujuba* fruits (MEZJ), in rat models of PCM and TAA-induced hepatic damage. The low- and medium-doses of MEZJ significantly inhibited the acute elevation of biomarkers in serum and elevated the fall of biomarkers in liver tissue homogenate (LTH). The activities of antioxidants enzymes were significantly increased in LTH of rats pretreated with low and medium doses of MEZJ.

Shen et al. (2009) investigated the protective effect against hepatic injury induced by CCl_4 for the ethanolic extract of *Z. jujuba.* After administration of the ethanolic extract, at the dose of 200 mg/kg, significantly decreased ALT and

AST, and attenuated histopathology of hepatic injury, and ameliorated the oxidative stress in hepatic tissue. Partly assayed indexes were ameliorated after administrated extract at the dose of 100 mg/kg.

Ziziphus mauritiana Lam. FAMILY: RHAMNACEAE

Dahiru et al. (2005, 2010) reported the protective effect of the ethanol extract of the leaves of *Ziziphus mauritiana* on CCl_4-induced liver damage. Pretreatment of rats with 200 and 300 mg/kg BW of *Z. mauritiana* leaf extract protected rats against CCl_4 liver injury by significantly lowering AST, ALT, ALP, TB, and LPO levels compared to control.

The aqueous extract of *Z. mauritiana* fruit (Zm) was evaluated by Dahiru et al. (2010) for its protective activity against CCl_4-induced liver damage. 250, 500 mg/kg BW of Zm fruit extract or 100 mg/kg Silymarin (standard) were administered to different groups of rats prior to CCl_4 administration. Both 250 and 500 mg/kg BW of Zm fruit extract significantly reduced (dose dependently) the levels of enzymes and nonenzymes markers of tissue damage when compared to rats given CCl_4 only. These findings were supported by liver histology and suggest that Zm fruit possessed reach hepatoprotective principles that inhibited the toxicity of CCl_4 against the liver.

Ziziphus oenoplia (L.) Mill. FAMILY: RHAMNACEAE

Rao et al. (2012) evaluated the hepatoprotective potential of ethanolic extract of *Ziziphus oenoplia* roots against antitubercular drugs. The study was performed on Wistar rats treated with the EtOH (50%) extract from *Z. oenoplia* roots in a hepatotoxicity model induced by INH/RIF (50 mg/kg each); the extract was administered orally for 21 days. The results at doses of 300 mg/kg were very similar to those of the control group (Silymarin, 100 mg/kg) in restoring the serum levels of AST, ALT, ALP, and bilirubin (Rao et al., 2012).

Ziziphus rotundifolia L. FAMILY: RHAMNACEAE

Parameshwar et al. (2012) evaluated the hepatoprotective acgtivity of ethanolic extract of leaves of *Ziziphus rotundifolia* against CCl_4-induced liver damage in rats. Liver damage was evidence by elevated levels of biochemical parameters such as SGOT, SGPT, and serum ALP. Treatment with ethanolic extract of *Z. rotundifolia* (300 mg/kg p.o.) produced a significant reversal in the above biochemical parameters and reducing power, superoxide anion scavenging activity and reduced histopathological scores.

Ziziphus spina-christi (L.) Willd. FAMILY: RHAMNACEAE

Amin and Ghoneim (2009) examined the effects of the water extract of *Ziziphus spina-christi* (ZSC) on CCl_4-induced hepatic fibrosis. Histopathological, biochemical and histology texture analyses revealed that ZSC significantly impede the progression of hepatic fibrosis. Extract resulted in a significant amelioration of liver injury judged by the reduced activities of serum ALT and AST. Oral administration of ZSC has also restored normal levels of MDA and retained control activities of endogenous antioxidants such as SOD, CAT, and GSH. Furthermore, ZSC reduced the expression of alpha-smooth muscle actin, the deposition of types I and III collagen in CCl_4-injured rats. Texture analysis of microscopic images along with fibrosis index calculation showed improvement in the quality of type I collagen distribution and its quantity after administration of ZSC extract.

Yossef et al. (2011) evaluated the protective effect of the *Z. spina-christi* fruits as an antioxidant against CCl_4-induced oxidative stress and hepatotoxicity in Albino Wistar rats. Subcutaneous injection of CCl_4 produced a marked elevation in the serum levels of AST, ALT, and ALP. Daily dietary containing powder of ZSCF at 2.5, 5, 10, and 15% of basal diet for 6 weeks produced a reduction in the serum levels of liver enzymes. ZSCF has also restored normal levels of MDA and retained control activities of endogenous antioxidants such as SOD, and GSH.

Mohamed (2012) evaluated the hepatoprotective effect of aqueous leaves extract of *Psidium guajava* and *Ziziphus spina-christi* on PCM-induced hepatic damage. The hepatoprotective activities of the two extracts were compared with a known hepatoprotective drug, Silymarin. Oral administration of PCM-induced liver damage in rats as manifested by a significant rise in serum levels of AST, ALT, ALP, LDH, gamma glutamyl transferase (γ-GT), and TB. Oral administration of *Psidium guajava* and *Ziziphus spina-christi* aqueous leaf extracts, significantly attenuated hepatotoxicity induced by PCM resulting in a significant decrease in serum levels of hepatic enzymes marker and TB. The results also showed that oral administration of PCM-induced oxidative stress in liver reflected in a significant decrease in hepatic SOD and CAT activities. Treatment with aqueous leaves extract of either *Psidium guajava* or *Ziziphus spina-christi* ameliorated the oxidative stress in PCM hepatotoxic rats and restored normal activities of hepatic antioxidant enzymes.

Ziziphus vulgaris Lam. FAMILY: RHAMNACEAE

Ebrahimi et al. (2013) investigated the hepatoprotective potential of ethanolic extract of *Ziziphus vulgaris* in laboratory rats against CCl_4-induced hepatic injury. Levels of AST, ALT, ALP, ALB, TP, and bilirubin were measured as well as pathological assessment of the liver samples for scoring of portal inflammation and hepatocellular necrosis. Results revealed that although there was a significant decrease in liver enzymes of the *Z. vulgaris* treated groups there were insignificant differences in protein and ALB concentrations between the experimental animal groups. The results showed hepatoprotective impact against CCl_4-induced liver injury according to both serological and pathological investigation.

CAMBODIAN MEDICINAL PLANTS

Lee et al. (2017) investigated the hepatoprotective activities of 64 crude ethanol extracts of Cambodian medicinal plants against *t*-BHP-induced cytotoxicity in human liver-derived HepG2 cells, and assessed their cytoprotective mechanism pertaining to the expression of heme oxygenase (HO)-1 and nuclear factor E2-related factor 2 (Nrf2). Of the 64 extracts, 19 extracts exhibited high hepatoprotective activities: *Ampelocissus martini, Bauhinia bracteata, Bombax ceiba, Borassus flabellifer, Cardiospermum halicacabum, Cayratia trifolia, Cinnamomum caryophyllus, Cyperus rotundus, Dasymaschalon lomentaceum, Ficus benjamina, Mangifera duperreana, Morinda citrifolia, Pandanus humilis, Peliosanthes weberi, Phyllanthus emblica, Quisqualis indica, Smilax glabra, Tinospora crispa*, and *Willughbeia cochinchinensis*, with half maximal effective concentrations ranging between 59.23 and 157.80 μg/mL. Further investigations revealed that, of these 19 extracts, HO-1 and Nrf2 were expressed in *P. weberi* and *T. crispa* expressed in a dose-dependent manner. In addition, the activities of ROS were suppressed following treatment of these two extracts in *t*-BHP-induced HepG2 cells. These results indicated that of the 64 Cambodian plants, *P. weberi* and *T. crispa* exhibited hepatoprotective effects on *t*-BHP-induced cytotoxicity in HepG2 cells, possibly by the induction of Nrf2-mediated expression of HO-1. Taken together, these results suggested that *T. crispa* or *P. weberi* may offer potential for therapeutic applications in liver disease characterized by oxidative stress.

KEYWORDS

- **hepatoprotective plants**
- **medicinal plants for liver protection**
- **traditional medicine**
- **herbal medicine**
- **Ayurvedic medicine**

REFERENCES

Aba, P.E.; Ozioko, I.E.; Udem, N.D.; Udem, S.C. Some biochemical and haematological changes in rats pretreated with aqueous stem bark extract of *Lophira lanceolata* and intoxicated with paracetamol (acetaminophen). *J. Complement. Integr. Med.* **2014,** *11*(4), 273–277.

Abdallah, H.M.; Ezzat, S.M.; El Dine, R.S.; Abdel-Sattar, E.; Abdel-Naim, A.B. Protective effect of *Echinops galalensis* against CCl_4-induced injury on the human hepatoma cell line (Huh7). *Phytochem Lett.* **2013,** *6,* 73–78.

Abdel-Azeem, A.S.; Hegazy, A.M.; Ibrahim, K.S.; Farrag, A.R.; El-Sayed, E.M. Hepatoprotective, antioxidant, and ameliorative effects of ginger (*Zingiber officinale* Roscoe) and vitamin E in acetaminophen treated rats. *J. Diet. Suppl.* **2013**, *10*(3), 195–209.

Abdel-Ghany, R.H.; Barakat, W.M.; Shahat, A.A.; Abd-Allah, WE-S.; Ali, E.A. *In vitro* and *in vivo* hepatoprotective activity of extracts of aerial parts of *Bidens pilosa* L. (Asteraceae). *Trop. J. Pharm. Res.* **2016,** *15,* 2371–2381.

Abd El-Ghffar, E.A.; El-Nashar, H.A.; Eldahshan, O.A.; Singab, A.N.B. GC-MS analysis and hepatoprotective activity of the *n*-hexane extract of *Acrocarpus fraxinifolius* leaves against paracetamol-induced hepatotoxicity in male albino rats. *Pharm. Biol.* **2017,** *55,* 441–449.

Abdelhafez, O.H.; Fawzy, M.A.; Fahim, J.R.; Desoukey, S.Y.; Krischke, M.; Mueller, M.J.; Abdelmohsen, U.R. Hepatoprotective potential of *Malvaviscus arboreus* against carbon tetrachloride-induced liver injury in rats. *PLos One* **2018,** *13*(8), e0202362. doi.10.1371/journal.pone.0202362

Abdel-Kader, M.S.; Alqasoumi, S.I.; Al-Taweel, A.M. Hepatoprotective constituents from *Cleome droserifolia. Chem. Pharm. Bull.* (Tokyo). **2009,** *57*(6), 620–624.

Abdel-Kader, M. S.; Al-Qutaym, A.; Saeedan, A. S. B.; Hamad, A. M.; Alkharfy, K. M. Nephroprotective and hepatoprotective effects of *Tribulus terrestris* L. growing in Saudi Arabia. *J. Pharm. Pharmacogn. Res.*, **2016**, *4*(4), 144–152.

Abdellatif, A.G.; Gargoum, H.M.; Debani, A.A.; Bengleil, M.; Alshalmani, S,; El Zuki, N.; El Fitouri, O. Camel thorn has hepatoprotective activity against carbon tetrachloride or acetaminophen induced hepatotoxicity, but enhances the cardiac toxicity of adriamycin in rodents. *Int. J. Med. Health, Pharm. Biomed. Eng.* **2014**; *8*(2), 118–122.

Abdel-Rahman, M.K. Can apricot kernels fatty acids delay the atrophied hepatocytes from progression to fibrosis in dimethylnitrosamine (DMN)-induced liver injury in rats? *Lipids Health Dis.* **2011,** *10*, 114.

Abdel-Rahman, M. K.; El-Megeid, A. A. A. Hepatoprotective effect of soapworts (*Saponaria officinalis*), pomegranate peel (*Punica granatum* L) and cloves (*Syzygium aromaticum* Linn) on mice with CCl hepatic intoxication. *World J. Chem*, **2006,** *1*(1), 41–46.

Abdel-Razika, A.F.; Elshamya, A.I.; Nassara, M.I.; El-Kousyb, S.M.; Hamdyc, H. Chemical constituents and hepatoprotective activity of *Juncus subulatus. Rev. Latinoamer. Quím.* **2009,** *37*(1), 70–84.

Abdullah; Khan, M.A.; Ahmad, W.; Ahmad, M.; Nisar, M. Hepatoprotective effect of the solvent extracts of *Viola canescens* Wall. ex Roxb. against CCl_4 induced toxicity through antioxidant and membrane stabilizing activity. *BMC Complement. Altern. Med.* **2017**. 17, 10. doi: 10.1186/s12906-016-1537-7

Abdulrahman, L.; Malki, A.; Kamel, M.; Golayel, A. Hepatoprotective efficacy of chicory alone or combined with Dandelion leaves against induced liver damage. *Life Sci. J.* **2013**, *10*(4), 140–157.

Abe, H.; Sakaguchi, M.; Odashima, S.; Arichi, S. Protective effect of saikosaponin-d isolated from *Bupleurum falcatum* L. on CCl_4-induced liver injury in the rat. *Naunyn-Schmiedeberg's Arch. Pharmacol.* **1982**, *320*(3), 266–271.

Abe, H.; Sakaguchi, M.; Yamada, M.; Arichi, S.; Odashima, S. Pharmacological actions of saikosaponins isolated from *Bupleurum falcatum*. I. Effects of saikosaponings on liver function. *Planta Medica.* **1980**, *40*(4), 366–372.

Abirami, A.; Nagarani, G.; Siddhuraju, P. Hepatoprotective effect of leaf extracts from *Citrus hystrix* and *C. maxima* against paracetamol induced liver injury in rats. *Food Sci. Human Wellness* **2015**, *4*, 35–41.

Aboubakr, M.; Abdelazem, A.M. Hepatoprotective effect of aqueous extract cardamom against gentamicin-induced hepatic damage in rats. *Int. J. Basic Appl. Sci.* **2016,** *5,* 1–4.

Aboul-Ela, M.; El-Shaer, N.; Abd El-Azim, T. Chemical constituents and hepatotoxic effect of the berries of *Juniperus phoenicea* Part 2. *Nat. Prod. Sci.* **2005,** *11*(4), 240–247.

Abouzid, S.; Sleem, A. Hepatoprotective and antioxidant activities of *Tamarix nilotica* flowers. *Pharm. Biol.* **2011**, *49*(4), 392–395.

Abuelgasim, A.A.; Nuha, H.S.; Mohammed, A.H. Hepatoprotective effect of *Lepidium sativum* against carbon tetrachloride induced damage in rats. *Res. J. Animal Vet. Sci.* **2008,** *3*, 20–28.

Achuthan, C.R.; Babu, B.H.; Padikkala, J. Antioxidant and hepatoprotective effects of *Rosa damascena*. *Pharm. Biol.* **2003**, *41*, 357–361.

Adaramoye, O.A.; Aluko, A.; Oyagbemi, A.A. *Cnidoscolus aconitifolius* leaf extract protects against hepatic damage induced by chronic ethanol administration in Wistar rats. *Alcohol Alcohol.* **2011**, *46*(4), 451–458.

Adebayo, A.H.; Abolaji, A.O.; Kela, R. Hepatoprotective activity of *Chrysophyllum albidum* against carbon tetrachloride induced hepatic damage in rats. SENRA *Acad. Pub. Bur. Brit. Col.* **2011**, *5*, 1597–1602.

Adeneye, A.A. Protective activity of the stem bark aqueous extract of *Musanga cecropioides* in carbon tetrachloride- and acetaminophen-induced acute hepatotoxicity in rats. *Afr. J. Tradit. Complement. Altern. Med.* **2009**, *6*, 131–138.

Adeneye, A.A.; Awodele, O.; Aiyeola, S.A.; Benebo, A.S. Modulatory potentials of the aqueous stem bark extract of *Mangifera indica* on carbon tetrachloride-induced hepatotoxicity in rats. *J. Tradit. Complement. Med.* **2015**, *5*, 106–115.

Adeneye, A.A.; Olagunju, J.A.; Banju, A.A.F.; Abdul, S.F.; Sanusi, O.A.; Sanni, O.O.; Osarodion, B.A.; Shoniki, O.E. The aqueous seed extract of *Carica papaya* Linn. prevents carbon tetrachloride induced hepatotoxicity in rats. *Int. J. Appl. Res. Nat. Prod.* **2009**, *2*(2), 19–32.

Adeneye, A.A.; Olagunju, J.A.; Elias, S.O.; Olatunbosun, D.O.; Mustafa, A.O.; Adeshile, O.I.; et al., Protective activities of the aqueous root extract of *Harungana madagascariensis* in acute and repeated acetaminophen hepatotoxic rats. *Int. J. Appl. Res. Nat. Prod.* **2008,** *1*, 29–42.

Adesanoye, O.A.; Adekunle, A.E.; Adewale, O.B.; Mbagwu, A.E.; Delima, A.A.; Adefegha, S.A.; Molehin, O.R.; Farombi, E.O. Chemoprotective effect of *Vernonia amygdalina* Del. (Astereaceae) against 2-acetylaminofluorene-induced hepatotoxicity in rats. *Toxicol. Ind. Health.* **2016**, *32*(1), 47–58.

Adesanoye, O.A.; Farombi, E.O. Hepatoprotective effects of *Vernonia amygdalina* (Astereaceae) in rats treated with carbon tetrachloride. *Exp. Toxicol. Pathol.* **2010**, *62*, 197–206.

Adetutu, A.; Olorunnisola, O.S. Hepatoprotective potential of some local medicinal plants against 2-acetylaminoflourene-induced damage in rat. *J. Toxicol.* **2013**. doi.: 10.1155/2013/272097

Adetutu, A.; Owoade, A.O. Hepatoprotective and antioxidant effect of *Hibiscus* polyphenol rich extract (HPE) against carbon tetrachloride (CCl_4) induced damage in rats. *British J. Med. Medical Res.* **2013**, *3*, 1574–1586.

Adewale, O.B.; Adekeye, A.O.; Akintayo, C.O.; Onikanni, A.; Sabiu, S. Carbon tetrachloride (CCl_4)-induced hepatic damage in experimental SD rats: Antioxidant potential of *Xylopia aethiopica*. *J. Phytopharmacol.* **2014**, *3*, 118–123.

Adewale, S.A.; Orhue, N.E.J. Aqueous extract of the fruits of *Xylopia aethiopica* (Dunal) A. Rich. protects against carbon tetrachloride-induced hepatotoxicity in rats. *European J. Med. Plants* **2015**, *9*(4), 1–10.

Adewole, S.A.; Ojewole, J.A. Protective effect of *Annona muricata* Linn. (Annonaceae) leaf aqueous extract on serum lipid profiles and oxidative stress in hepatocytes of streptozotocin-treated diabetic rats. *Afr. J. Tradit. Complement. Altern. Med.* **2009**; *6*, 30–41.

Adewusi, E.A.; Afolayan, A.J. A review of natural products with hepatoprotective activity. *J. Med. Plants Res.* **2010**, *4*, 1318–1334.

Adeyemi, D.O.; Ukwenya, V.O.; Obuotor, E.M.; Adewole, S.O. Anti-hepatotoxic activities of *Hibiscus sabdariffa* L. in animal model of streptozotocin diabetes-induced liver damage. *BMC Complement. Altern. Med.* **2014**, 14, 277. doi: 10.1186/1472-6882-14-277.

Adhvaryu, M.R.; Reddy, N.M., Vakharia, B.C. Prevention of hepatotoxicity due to anti-tuberculosis treatment: A novel integrative approach. *World J. Gastroenterol.* **2008**. *14*(30), 4753–4762.

Adnyana, I,; Tezuka, Y.; Banskota, A.H.; Tran, K.Q.; Kadot, S. Hepatoprotective constituents of the seeds of *Combretum quadrangulare*. *Biol. Pharm. Bull.* **2000, 23**(11), 1328–32.

Afaf, I.; Abuelgasim; Nuha, H.S.; Mohammed, A.H. Hepatoprotective effect of *Lepidium sativum* against carbontetrachloride induced damage in rats. *Res. J. Anim. Vet. Sci.* **2008**, *3*, 20–23.

Aftab-Ullah, Ahmad, M.; Ahmad, T.; Naseer, F.N. Hepatoprotective activity of *Sonchus asper* in PCM-induced hepatic damage in rabbits. *Bangladesh J. Pharmacol.* **2015**, *10*, 115.

Agarwal, M.; Srivastava, V.K.; Saxena, K.K.; Kumar, A. Hepatoprotective activity of *Beta vulgaris* against CCl_4-induced hepatic injury in rats. Fitoterapia. **2006**. *77*(2), 91–93.

Agbonon, A.; Gbeassor, M. (2009). Hepatoprotective effect of *Lonchocarpus sericeus* leaves in CCl_4-induced liver damage. *J. Herbs Spices Med. Plants.*, *15*(2), 216–226.

Agbor, G.A.; Oben, J.E.; Nkegoum, B.; Takala, J.P.; Ngogang, J.Y. Hepatoprotective activity of *Hibiscus cannabinus* (Linn.) against carbon tetrachloride and paracetamol induced liver damage in rats. *Pak. J. Biol. Sci.* **2005**. *8*, 1397–1401.

Aghel, N.; Kalantari, H.; Rezazadeh, S. Hepatoprotective effect of *Ficus carica* leaf extract on mice intoxicated with carbon tetrachloride. *Iran J. Pharm. Sci.* **2011**, *10*, 63–68.

Aghel, N.; Rashidi, I.; Mombeini, A. Hepatoprotective activity of *Capparis spinosa* root bark against CCl_4 induced hepatic damage in mice. *Iranian J. Pharm. Res.* **2007**, *6*(4), 285–290.

Ahamed, M.B.K.; Krishna, V.; Dandin, C.J. In vitro antioxidant and *in vivo* prophylactic effects of two gamma-lactones isolated from *Grewia tiliaefolia* against hepatotoxicity in carbon tetrachloride intoxicated rats. *Eur. J. Pharmacol.* **2010,** *631*(1–3), 42–52.

Ahangarpour, A.; Heidari, H.; Oroojan, A. A.; Mirzavandi, F.; Esfehani, K. N.; Mohammadi, Z. D. Antidiabetic, hypolipidemic and hepatoprotective effects of *Arctium lappa* root's hydro-alcoholic extract on nicotinamide-streptozotocin induced type 2 model of diabetes in male mice. *Avicenna J. Phytomed.*, **2017**, *7*(2), 169.

Ahmad, A.; Raish, M.; Ganaie, M.A.; Ahmad, S.R.; Mohsin, K.; Al-Jenoobi, F.I.; Al-Mohizea, A.M.; Alkharfy, K.M. Hepatoprotective effect of *Commiphora myrrha* against d-GalN/LPS-induced hepatic injury in a rat model through attenuation of pro inflammatory cytokines and related genes. *Pharm. Biol.* **2015**, *53*(12), 1759–1767.

Ahmad, M.; Erum, S. Hepatoprotective studies on *Haloxylon salicornicum*: A plant from Cholistan desert. *Pak. J. Pharm. Sci.* **2011**, *24*, 377–382.

Ahmad, M.; Itoo, A.; Baba, I.; Jain, S.M.; Saxena, R.C. Hepatoprotective activity of *Portulaca oleracea* Linn. on experimental animal model. *Int. J. Pharm. Pharm. Sci.* **2013**, *5*(3), 267–269.

Ahmad, M.; Mahmood, Q.; Gulzar, K.; Akhtar, M.S.; Saleem, M.; Qadir, M.I. Antihyperlipidemic and hepatoprotective activity of *Dodonaea viscosa* leaves extracts in alloxan-induced diabetic rabbits (*Oryctolagus cuniculus*). *Pak. Vet. J.* **2011**, *32*, 50–54.

Ahmad, R.; Raja, V.; Sharma, M. Hepatoprotective activity of ethyl acetate extract of *Adhatoda vasica* in Swiss albino rats. *Int. J. Cur. Res. Rev.* **2013**, 5, 16–21.

Ahmadi-Naji, R.; Heidarian, E.; Ghatreh-Samani, K. Evaluation of the effects of the hydroalcoholic extract of *Terminalia chebula* fruits on diazinon-induced liver toxicity and oxidative stress in rats. *Avicenna J. Phytomed.* **2017**, *7*(5), 454–466.

Ahmadipour, A.; Sharififar, F.; Nakhaipour, F.; Samanian, M.; Karami-Mohajeri, S. Hepatoprotective effect of *Zataria multiflora* Boisson cisplatin-induced oxidative stress in male rat. *J. Med. Life.*, **2015**, *8*(Special Issue 4), 275.

Ahmed, A.; Alam, T.; Khan, S.A. Hepatoprotective activity of *Luffa echinata* fruits. *J. Ethnopharmacol.* **2001**, *76*, 187–189.

Ahmed, B.; Alam, T.; Varshney, M.; Khan, S.A. Hepatoprotective activity of two plants belonging to the Apiaceae and the Euphorbiaceae family. *J. Ethnopharmacol.* **2002**, *79*, 313–316.

Ahmed, B.; Al-Howiriny, T.A.; Siddiqui, A.B. Antihepatotoxic activity of seeds of *Cichorium intybus*. *J. Ethnopharmacol.* **2003**, *87*, 237–240.

Ahmed, B.M.; Khater, R.M. Evaluation of the protective potential of *Ambrosia maritima* extract on acetaminophen-induced liver damage. *J. Ethnopharmacol.* **2001**, *75*, 169–171.

Ahmed, F.; Urooj, A. Hepatoprotective effects of *Ficus racemosa* stem bark against carbon tetrachloride-induced hepatic damage in albino rats. *Pharm. Biol.* **2010**, *48*(2), 210–216. doi: 10.3109/13880200903081788

Ahmed, H.; Elzahab, H.A.; Alswiai, G. Purification of antioxidant protein isolated from *Peganum harmala* and its protective effect against CCl_4 toxicity in rats. *Turk. J. Biol.* **2013**, *37*, 39–48.

Ahmed, J.; Nirmal, S.; Dhasade, V.; Patil, A.; Kadam, S.; Pal, S.; Mandal, S.; Pattan, S. Hepatoprotective activity of methanol extract of aerial parts of *Delonix regia*. *Phytopharmacology* **2011**, 5, 118–122.

Ahmed, M.B.; Hasona, N.A.; Selemain, H.A. Protective effects of extract from dates (*Phoenix dactylifera L.)* and ascorbic acid on thioacetamide-induced hepatotoxicity in rats. *Iranian J. Pharm. Res.* **2008**, *7*(3), 193–201.

Ahmed, S.; Alam Khan, R.; Jamil, S.Anti hyperlipidemic and hepatoprotective effects of native date fruit variety Aseel" (*Phoenix dactylifera*). *Pak. J. Pharm. Sci.*, **2016**, *29*(6), 1945–1950.

Ahmed, S.; Rahman, A.; Alam, A.; Saleem, M.; Athar, M.; Sultana, S. Evaluation of the efficacy of *Lawsonia alba* in the alleviation of carbon tetrachloride induced oxidative stress. *J. Ethnopharmacol.* **2000**, *69*(2), 157–164.

Ahmed, T., Sadia, H.; Khalid, A.; Batool, S.; Janjua, A. Report: Prunes and liver function: A clinical trial. *Pak. J. Pharm. Sci.* **2010**, *23*, 463–466.

Ahmida, M.H. Evaluation of *in vivo* antioxidant and hepatoprotective activity of *Portulaca oleracea* L. against paracetamol-induced liver toxicity in male rats. *Am. J. Pharmacol. Toxicol.* **2010,** *5*(4), 167–176.

Ahsan, H.; Khan, K. U.; Khan, I. U.; Saleem, M.; Khadija, H. A. Sharif, A.; Naz, H. Evaluation of hepatoprotective activity of *Melilotus officinalis* L. against PCM and carbon tetrachloride induced hepatic injury in mice. *Acta Pol. Pharm.* **2017**, *74*(3), 903–909.

Ahsan, R.; Islam, K.M.; Musaddik, A.; Haque, E. Hepatoprotective activity of methanol extract of some medicinal plants against carbon tetrachloride induced hepatotoxicity in albino rats. *Global J. Pharmacol.* **2009a**, *3*(3), 116–122.

Ahsan, R.; Islam, K.M.; Bulbul, I.J.; Musaddik, M.A.; Haque, E. Hepatoprotective activity of methanol extract of some medicinal plants against carbon tetrachloride-induced hepatotoxicity in rats. *Eur. J. Sci. Res.* **2009b,** *37*(2), 302–310.

Ai, G.; Liu, Q.; Hua, W.; Huang, Z.; Wang, D. Hepatoprotective evaluation of the total flavonoids extracted from flowers of *Abelmoschus manihot* (L.) Medic.: *In vitro* and *in vivo* studies. *J. Ethnopharmacol.* **2013**, *146*, 794–802.

Ajayi, G. O.; Adeniyi, T. T.; Babayemi, D. O. Hepatoprotective and some haematological effects of *Allium sativum* and vitamin C in lead-exposed Wistar rats. *Int. J. Med. Med. Sci.* **2009**, *1*(3), 064–067.

Ajiboye, T.O.; Ahmad, F.M.; Daisi, A.O.; et al., Hepatoprotective potential of *Phyllanthus muellarianus* leaf extract: Studies on hepatic, oxidative stress and inflammatory biomarkers. *Pharm. Biol.* **2017**, *55*(1), 1662–1670.

Ajiboye, T.O.; Salau, A.K.; Yakubu, M.T.; Oladiji, A.T.; Akanji, M.A.; Okogun, J.I.; *et al.,* Acetaminophen perturbed redox homeostasis in Wistar rat liver: Protective role of aqueous *Pterocarpus osun* leaf extract. *Drug Chem. Toxicol.* **2010,** *33,* 77–87.

Ajiboye, T.P. Standardized extract of *Vitex doniana* Sweet stalls protein oxidation, lipid peroxidation and DNA fragmention in acetaminophen-induced hepatotoxicity. *J. Ethnopharmacol.* **2015**, *164*, 273–82. doi: 10.1016/j.jep.2015.01.026.

Ajilore, B.S.; Atere, T.G.; Oluogun, W.A.; Aderemi, V.A. Protective effects of *Moringa oleifera* Lam. on cadmium-induced liver and kidney damage in male Wistar rats. *Int. J. Phytother. Res.* **2012**, *2*(3), 42–50.

Ajilore; Yannuga, B.S.A.; Olugbenga. Hepatoprotective potentials of methanolic extract of the leaf of *Momordica charantia* Linn. on cadmium-induced hepatotoxicity in rats. *J. Nat. Sci. Res.* **2012**, *2*(7), 41–47.

Ajith, T.A.; Hema, U.; Aswathy, M.S. *Zingiber officinale* Roscoe prevents APAP induced acute hepatotoxicity by enhancing hepatic antioxidant status. *Food Chem. Toxicol.* **2007**, *45*, 2267–2272.

Akah, P.A.; Odo, G.L. Hepatoprotective effect of the solvent fractions of the stem of *Hoslundia opposita* Vahl (Lamiaceae) against carbon tetrachloride and paracetamol induced liver damage in rats. *Int. J. Green Pharm.,* **2010**, *4*, 54–58. doi: 10.4103/0973-8258.62159

Akanitapichat, P.; Phraibung, K.; Nuchklang, K.; Prompitakkul, S. Antioxidant and hepato-protective activities of five eggplant varieties. *Food Chem. Toxicol.* **2010** *48*(10), 3017–21.

Akhgar, M.R.; Pahlavanzadeh-Iran, S.; Lotfi-Anari, P.; Faghihi-Zarandi, A. Composition of essential oils of fruits and leaves of *Prangos ferulacea* (L.) Lindl. growing wild in Iran. *Trends Mod. Chem.* **2011**, *1*, 1–4.

Akhtar, M.S., Amin, M., Maqsood, M.; Alamgeer. Hepatoprotective effect of *Rheum emodi* roots (Revand chini) and Akseer-e-Jigar against paracetamol-induced hepatotoxicity in rats. *Ethnobotanical Leaflets*. **2009**, *13*, 310-315.

Akhtar, S.M.; Qayyum, I.M.; Irshad, N.; Hussain, R.; Hussain, A.; Yaseen, M.; et al., Studies on hepatoprotective properties of different extracts of *Canscora decussata* Schult. against carbon tetrachloride induced hepatotoxicity. *Int. J. Curr. Pharm. Res.* **2013**, *5*(1), 36–37.

Akhtar, M.S.; Qayyum, M.I.; Irshad, N.; Yaseen, M.; Hussain, A.; Altaf, H.; Saif-ur-Rehman, Suleman, N. Hepatoprotective properties of methanolic extract of *Canscora decussata* Schult. against PCM induced liver toxicity in rabbits. *Int. J. Innov. Appl. Stud.* **2015**, *10*, 701–706.

Akhter, N.; Shawl, A.S.; Sultana, S.; Chandan, B.K.; Akhter, M. Hepatoprotective activity of *Marrubium vulgare* against PCM induced toxicity. *J. Pharm. Res.* **2013**, *7*, 565–570.

Akindele, A.J.; Ezenwanebe, K.O.; Anunobi, C.C.; Adeyemi, O.O. Hepatoprotective and *in vivo* antioxidant effects of *Byrsocarpus coccineus* Schum. & Thonn. (Connaraceae). *J. Ethnopharmacol.* **2010**, *129*, 46–52.

Akinloye, O.A.; Ayankojo, A.G.; Olaniyi, M.O. Hepatoprotective activity of *Cochlospermum tinctorium* against carbon tetrachloride induced hepatotoxicity in rats. *Rom. J. Biochem.* **2012**, *49*(1), 3-12.

Aktaya, G.; Deliorman, D.; Ergun, E.; Ergun, F.; Yeşilada, E.; Cęvik, C. Hepatoprotective effects of Turkish folk remedies on experimental liver injury. *J. Ethnopharmacol.* **2000**, *73*, 121–129.

Aladeyelu, S. O.; Onyejike, D. N.; Ogundairo, O. A.; Adeoye, O. A.; Gbadamosi, M. T.; Ogunlade, B.; Ojewale, A. O. Hepatoprotective role of *Moringa oleifera* ethanolic leaf extract on Liver functions (Biomarker) in cadmium chloride induced hepatotoxicity in Albino Wistar rats. *Int. J. Basic, Appl. Inn. Res.* **2018**, *7*(1), 12–17.

Alam, J.; Mujahid, M.; Jahan, Y.; Bagga, P.; Rahman, M.A. Hepatoprotective potential of ethanolic extract of *Aquilaria agallocha* leaves against paracetamol induced hepatotoxicity in SD rats. *J. Tradit. Complement. Med.* **2017**, *7*, 9–13.

Alam, M. A.; Nyeem, M. A. B.; Awal, M. A.; Mostofa, M.; Alam, M. S.; Subhan, N.; Rahman, M. M. Antioxidant and hepatoprotective action of the crude ethanolic extract of the flowering top of *Rosa damascena. Orient Pharm. Exp. Med.*, **2008**, *8*(2), 164–170.

Alam, M.S.; Kaur, G.; Jabbar, Z.; Javed, K.; Athar, M. *Eruca sativa* seeds possess antioxidant activity and exert a protective effect on mercuric chloride induced renal toxicity. *Food Chem. Toxicol.* **2007**, *45*(6), 910–920.

Alamgeer; Nawaz, M.; Ahmad, T.; Mushtaq, M.N.; Batool, A. Hepatoprotective activity of *Thymus linearis* against paracetamol and carbon tetrachloride-induced hepatotoxicity in albino mice. *Bangladesh J. Pharmacol.* **2014**, *9*, 230–234.

Alamgeer; Nasir, Z.; Qaisar, M.N.; Uttra, A.M.; Ahsan, H.; Khan, K.U.; Khan, I.U.; Saleem, M.; Khadijai; Asif, H.; Sharif, A.; Younis, W.; Naz, H. Evaluation of hepatoprotective activity of *Melilotus officinalis* L. against paracetamol and carbon tetrachloride induced hepatic injury in mice. *Acta Pol. Pharm.* **2017**, 74(3), 903–909.

Al-Asmari, A.K.; Athar, M.T.; Al-Shahrani, H.M.; Al-Dakheel, S.I.; Al-Ghamdi, M.A. Efficacy of *Lepidium sativum* against carbon tetra chloride induced hepatotoxicity and determination of its bioactive compounds by GC-MS. *Toxicol Rep.* **2015**, *2*, 1319–1326. doi: 10.1016/j.toxrep.2015.09.006

Al-Asmari, A.K.; Al-Elaiwi, A.M.; Athar, M.T.; Tariq, M.; Al Eid, A.; Al-Asmary, S.M. A Review of hepatoprotective plants used in Saudi traditional medicine. *Evid. Based Compl. Altern. Med.* **2014**, *2014*, 1–22.

Al-Attar, A.M.; Alrobai, A.A.; Almalki, D.A. Effect of *Olea oleaster* and *Juniperus procera* leaves extracts on thioacetamide induced hepatic cirrhosis in male albino mice. *Saudi J. Biol. Sci.*, **2016**, *23*(3), 363–371.

Al-Baroudi, D.A.A.; Arafat, R.A.; El-kholy, T.A. Hepatoprotective effect of chamomile capitula extract against 2,

4-dichlorophenoxyacetic acid-induced hepatotoxicity in rats. *Life Sci. J.* **2014**, *11*(8), 34–40.

Albasha, M.O.; Azab, A.E. Effect of cadmium on the liver and amelioration by aqueous extracts of fenugreek seeds, rosemary, and cinnamon in guinea pigs: Histological and biochemical study. *Cell Biol.,* **2014**, *2*(2): 34–44.

Al-Dosari, M. S. The effectiveness of ethanolic extract of *Amaranthus tricolor* L.: A natural hepatoprotective agent. *Am. J. Chin. Med.*, **2010**, *38*(06), 1051–1064.

Alemi, M.; Sabouni, F.; Sanjarian, F.; Haghbeen, K.; Ansari, S. Antiinflammatory effect of seeds and callus of *Nigella sativa* L. extracts on mix glial cells with regard to their thymoquinone content. *AAPS Pharm. Sci. Tech.* **2013**, *14*, 160–167.

Alex, A.R.; Ilango, K.; Boguda, V.A.; Ganeshan, S. In vitro hepatoprotectvive activity of extracts of *Viburnum punctatum* Buch.-Ham. ex D.Don against carbon tetrachloride induced toxicity. *Int. J. Pharm. Pharm. Sci.* 2014, 6(7), 392–394.

Al-Ghamdi, M.S. The anti-inflammatory, analgesic and antipyretic activity of *Nigella sativa. J. Ethnopharmacol.* **2001**, *76*, 45–48.

Al-Howiriny, T. A.; Al-Sohaibani, M. O.; Al-Said, M. S.; Al-Yahya, M. A.; El-Tahir, K. H.; Rafatullah, S. Hepatoprotective properties of *Commiphora opobalsamum* (Balessan), A traditional medicinal plant of Saudi Arabia. *Drugs Under Exp. Clin. Res.* **2004**, *30*(5/6), 213–220.

Al-Howiriny, T.A.; Al-Sohaibani, M.O.; El-Tahir, K.H.; Rafatullah, S. Preliminary evaluation of the anti-inflammatory and anti-hepatotoxic activities of 'Parsley' *Petroselinum crispum* in rats. *J. Nat. Remedies*, **2003**, *3*(1), 54–62.

Ali, B.; Mujeeb, M.; Aeri, V.; Mir, S.R.; Faiyazuddin, M.; Shakeel, F. Anti-inflammatory and antioxidant activity of *Ficus carica* Linn. leaves. *Nat. Prod. Res.* **2012**, *26*(5), 460–465.

Ali, B.H.; Bashir, A.K.; Rasheed, R.A. Effect of the traditional medicinal plants *Rhazya stricta, Balanitis aegyptiaca* and *Haplophylum tuberculatum* on paracetamol induced hepatotoxicity in mice. *Phytother. Res.* **2001**, *15*(7), 598– 603.

Ali, B.H.; Mousa, H.M.; El-Mougy, S. The effect of a water extract and anthocyanins of *Hibiscus sabdariffa* L. on paracetamol-induced hepatoxicity in rats. *Phytother. Res.* **2003**, *17*, 56–59.

Ali, H.; Kabir, N.; Muhammad, A.; Shah, M.R.; Musharraf, S.G.; Iqbal, N.; et al., Hautriwaic acid as one of the hepatoprotective constituent of *Dodonaea viscosa. Phytomedicine* **2014**, *21*, 131–140.

Ali, J.; Khan, A. Preventive and curative effects of *Calendula oficinalis* leaves extract on APAP-induced hepatotoxicity. *JPMI.* **2006**, *20*(4), 370–373.

Ali, M.; Qadir, M.I.; Saleem, M.; Janbaz, K.H.; Gul, H.; Hussain, L.; Ahmad, B. Hepatoprotective potential of *Convolvulus arvensis* against paracetamol-induced hepatotoxicity. *Bangladesh J. Pharmacol.* **2013**, *8*, 300–304.

Ali, N.M.; Yusof, H.M.; Long, K.; Yeap. S.K.; Ho, W.Y.; Beh, B.K., Koh, S.P.; Abdullah, M.P.; Alitheen, N.B. Antioxidant and hepatoprotective effect of aqueous extract of germinated and fermented mung bean on ethanol mediated liver damage. *BioMed Res. Int.* **2013**; *2013*, 693613 doi:10.1155/2013/693613.

Ali, S.; Ansari, K.A.; Jafry, M.A.; Kabeer, H.; Diwakar, G. *Nardostachys jatamansi* protects against liver damage induced by thioacetamide in rats. *J. Ethnopharmacol.* **2000**, *71*(3), 359–363.

Ali, S.; Khan, M. R.; Shah, S. A.; Batool, R.; Maryam, S.; Majid, M.; Zahra, Z. Protective aptitude of *Periploca hydaspidis* Falc against CCl_4 induced hepatotoxicity in experimental rats. *Biomed. Pharmacother.*, **2018**, *105*, 1117–1132.

Ali, S.A., Al-Amin, T.H.; Mohamed, A.H.; Gameel, A.A. Hepatoprotective activity of aqueous and methanolic extracts of *Capparis decidua* stems against carbon tetrachloride induced liver damage in rats. *J. Pharmacol. Toxicol.*, **2009**, *4*, 167–172.

Ali, A.S., Gameel, A.A., Mohamed, A.H., Hassan, T. Hepatoprotective activity of *Capparis decidua* aqueous and methanolic stems extracts against carbon tetrachloride induced liver histological damage in rats. *J. Pharmacol. Toxicol.* **2010**, *6*(1), 62–68.

Ali, S.A.; Rizk., M.Z.; Ibrahim, N.A.; Abdallah, M.S.; Sharra, H.M.; Moustafa, M.M. Protective role of *Juniperus phoenicea* and *Cupressus sempervirens* against CCl_4. *World J. Gastrointest. Pharmacol. Ther.* **2010**, *1*(6), 123–131.

Ali, Z.Y. Biochemical evaluation of some natural products against toxicity induced by anti-tubercular drugs in rats. *N. Y. Sci. J.* **2012**, *5*(10), 69–80.

Ali, Z.Y.; Atia, H.A.; Ibrahim, N.H. Possible hepatoprotective potential of *Cynara scolymus, Cupressus sempervirens* and *Eugenia jambolana* against paracetamol-induced liver injury: *In-vitro* and *in-vivo* evidence. *Nat. Sci.* **2012**, *10*, 75–86.

Al-Isawi, J.K.T.; Al-Jumaily, E.F. Antioxidant and hepatoprotectivee study of a purified *Bauhinia variegata* leaves and flowers against carbon tetrachloride-induced toxicity in experimental rats. *Biomed. Pharmacol.* **2019**, *12*(1). doi.10.13005/bpj/1655

Aliyu, R.; Adebayo, A.H.; Gatsing, D.; Garba, I.H. The effects of ethanolic leaf extract of *Commiphora africana* (Burseraceae) on rat liver and kidney functions. *J. Pharmacol. Toxicol.* **2007**, *2*, 373–379.

Aliyu, R.; Okoye Z.S.; Shier, W.T. The hepatoprotective cytochrome P-450 enzyme inhibitor isolated from the Nigerian medicinal plant *Cochlospermum planchonii* is a zinc salt. *J. Ethnopharmacol.* **1995**, *48*, 89–97.

Alkiyumi, S.S.; Abdullah, M.A.; Alrashdi, A.S.; Salama, S.M.; Abdelwahab, S.I.; Hadi, A.H. *Ipomoea aquatica* extract shows protective action against thioacetamide-induced hepatotoxicity. *Molecules* **2012**, *17*(5), 6146–6155.

Alkreathy, H.M.; Khan, R.A.; Khan, M.R.; Sahreen, S. CCl_4 induced genotoxicity and DNA oxidative damages in rats: Hepatoprotective effect of *Sonchus arvensis*. *BMC Complement Altern. Med.* **2014**, *14*, 452. doi: 10.1186/1472–6882-14-452.

Allam, R. M.; Selim, D. A.; Ghoneim, A. I.; Radwan, M. M.; Nofal, S. M.; Khalifa, A. E.; El-Sebakhy, N. A. Hepatoprotective effects of *Astragalus kahiricus* root extract against ethanol-induced liver apoptosis in rats. *Chin. J. Nat. Med.*, **2013**, *11*(4), 354–361.

Allan, J.J.; Damodaran, A.; Deshmukh, N.S.; Goudar, K.S.; Amit, A. Safety evaluation of a standardized phytochemical composition extracted from *Bacopa monniera* in SpragueDawley rats. *Food Chem. Toxicol.* **2007**, *45*(10), 1928–1937.

Almajwal, A.M.; Elsadek, N.F. Lipid-lowering and hepatoprotective effects of *Vitis vinifera* dried seeds on paracetamol-induced hepatotoxicity in rats. *Nutr. Res. Pract.* **2015**. *9*(1), 37–42.

Al-Malki, A. L.; Abo-Golayel, M. K.; Abo-Elnaga, G.; Al-Beshri, H. Hepatoprotective effect of dandelion (*Taraxacum officinale*) against induced chronic liver cirrhosis. *J. Med. Plant Res.*, **2013**, *7*(20), 1494–1505.

Al-Mehdar, A.A.; El-Denshary, E.S.; Abdel-Wahhab, M. Alpha lipoic acid and alpha-tocopherol counteract the oxidative stress and liver damage in rats sub-chronically treated with Khat (*Catha edulis*) extract. *Global J. Pharmacol.* **2012**, *6*(2), 94–105.

Alnuaimy, R.J.M.; Al-Khan, H.I.A. Effect of aqueous extract of *Capparis spinosa* on biochemical and histological changes in paracetamol-induced liver damage in rats. *Iraqi J. Vet. Sci.* **2012**, *26*(1), 1–10.

Aloh, G.S.; Obeagu, E.I.; Odo, C.E.; Kanu, S.N.; Okpara, K.E.; Udezuluigbo, C.N.; Ugwu, G.U. Hepatoprotective potentials of methanol extracts of *Gossweilerodendron balsamifarum* and lipid profile of albino rats. *Eur. J. Pharm. Med. Res.* **2015**, *2*, 124–129.

Al-Qarawi, A.A.; Abdel-Rahman, H.A.; ElMougy, S.A. Hepatoprotective activity of licorice in rat liver injury models. *J. Herbs Spices Med. Plants.* **2001**, *8*, 7–14.

Al-Qarawi, A.A.; Al-Damegh, M.A.; El-Mougy, S.A. Hepatoprotective influence of *Adansonia digitata* pulp. *J. Herbs Spices Med. Plants* **2003**, *10*, 3.

Al-Quarawi, A.A.; Mousa, H.M.; Hamed Ali, B.E.; Abdel-Rahman, H.; El-Mougy, S.A. Protective effects of extracts from dates (*Phoenix dactylifera* L) on carbon tetrachloride-induced hepatotoxicity in rats. *Int. J. Appl. Res. Vet. Med.,* **2004**, *2*, 176–180.

Alqasoumi, S.I. Isolation and chemical structure elucidation of hepatoprotective constituents from plants used in traditional medicine in Saudi Arabia. College of Pharmacy, King Saud University, **2007.**

Alqasoumi, S.I. Carbon tetrachloride-induced hepatotoxicity: Protective effect of "Rocket" *Eruca sativa* L. in rats. *Am. J. Chin. Med.* **2010**, *38*(1), 75–88.

Alqasoumi, S.I. 'Okra' *Hibiscus esculentus* L. A study of its hepatoprotective activity. *Saudi Pharm. J.* **2012**, *20*(2), 135–141.

Alqasoumi, S.I. Evaluation of the hepatroprotective and nephroprotective activities of *Scrophularia hypericifolia* growing in Saudi Arabia. *Saudi Pharm J.* **2014**, *22*(3), 258–263.

Alqasoumi, S.I.; Abdel-Kader, M.S. Terpenoids from *Juniperus procera* with hepatoprotective activity. *Pak. J. Pharm. Sci.* **2012a**, *25*(2), 315–322.

Alqasoumi, S.I.; Abdel-Kader, M.S. Screening of some traditionally used plants for their hepatoprotective effect. In: Phytochemicals as nutraceuticals-global approaches to their role in nutrition and health. Rao, V. (ed.), InTech., Rijeka, Coratia. **2012b.** http://www.intechopen.com/books.

Alqasoumi, S. I.; Al-Dosari, M. S.; AlSheikh, A. M.; Abdel-Kader, M. S. Evaluation of the hepatoprotective effect of *Fumaria parviflora* and *Momordica balsamina* from Saudi Folk medicine against experimentally induced liver injury in rats. *Res. J. Med. Plant.* **2009a**, *3*(1), 9-15.

Alqasoumi, S. I.; Al-Howiriny, T. A.; Abdel-Kader, M. S. Evaluation of the hepatoprotective effect of *Aloe vera, Clematis hirsuta, Cucumis prophetarum* and Bee Propolis against experimentally induced liver injury in rats. *Int. J. Pharmacol.* **2008a**, *4*(3), 213-217.

Alqasoumi, S. I.; Al-Rehaily, A. J.; Abdulmalik M.; AlSheikh, A.M.; Abdel-Kader, M. S. Evaluation of the hepatoprotective effect of *Ephedra foliata, Alhagi maurorum, Capsella bursa-pastoris* and *Hibiscus sabdariffa* against experimentally induced liver injury in rats. *Nat. Prod. Sci.* **2008b**, *14*(2), 95–99.

Alqasoumi, S.I.; Soliman, G. A. E. H.; Awaad, A. S.; Donia, A. E. R. M. Anti-inflammatory activity, safety and protective effects of *Leptadenia pyrotechnica, Haloxylon salicornicum* and *Ochradenus baccatus* in ulcerative colitis. *Phytopharmacol.* **2012**, *2*(1), 58–71.

Al-Razzuqi, R.A.M.; Al-Jawad, F.; Al-Hussaini, J.; Al-Jeboori, A. Hepatoprotective effect of *Glycyrrhiza glabra* in carbon tetrachloride-induced model of acute liver injury. *J. Phys. Pharm. Adv.* **2012**, *2*, 259–263.

Al-Said, M.S.; Mothana, R.A.; Al-Sohaibani, M.O.; Rafatullah, S. Ameliorative effect of *Grewia tenax* (Forssk) Fiori fruit extract on CCl_4-induced oxidative stress and hepatotoxicity in rats. *J. Food Sci.* **2011**, *76*(9), T200–T206.

Al-Said, M.S.; Mothana, R.A.; Al-Yahya, M.M.; Rafatullah, S.; Al-Sohaibani, M.O.; Khaled, J.M.; Alatar, A.; Alharbim N.S.; Kurkcuoglu, M.; Baser, H.C. GC-MS Analysis: In vivo hepatoprotective and antioxidant activities of the essential oil of *Achillea biebersteinii* Afan. growing in Saudi Arabia. *Evid. Based Compl. Altern. Med,* **2016**, Article ID 1867048, doi.10.1155/2016/1867048

AlSaid, M.; Mothana, R.; Raish, M.; et al., Evaluation of the effectiveness of *Piper cubeba* extract in the amelioration of CCl_4-induced liver injuries and oxidative damage in the rodent model. *BioMed Res. Int.* **2015**, Article ID 359358, 11 pages.

Al-Sayed, E.; Martiskainen, O.; Seif el-Din, S.H.; Sabra, A.A.; Hammam, O.A.; El-Lakkany, N.M.; Abdel-Daim, M.M. Hepatoprotective and antioxidant effect of *Bauhinia hookeri* extract against carbon tetrachloride-induced hepatotoxicity in mice and characterization of its bioactive compounds by HPLC-PDA-ESI-MS/MS. *Biomed. Res. Int.* **2014**, *2014*, 245171.

Alshawsh, M.A.; Abdulla, M.A.; Ismail, S.; Amin, Z.A. Hepatoprotective effects of *Orthosiphon stamineus* extract on thioacetamide-induced liver cirrhosis in rats. *Evid. Based Compl. Altern. Med.* **2011**, *2011*, 6. doi.10.1155/2011/103039

Al-Snafi, A.E. Pharmacological effects of *Allium* species grown in Iraq. An overview. *Int. J. Pharm. Health Care Res.* **2013**, *1*(4), 132–147.

Al-Snafi, A.E.; Mousa, H.N.; Majid, W.J. Medicinal plants possessed hepatoprotective activity. *IOSR J. Pharm.* **2019**, *9*(8), 26–56.

Al-Suhaimi, E.A. Hepatoprotective and immunological functions of *Nigella sativa* seed oil against hypervitaminosis A in adult male rats. *Int. J. Vitam. Nutr. Res.* **2012**, *82*(4), 288–297.

Altas, S.; Kizil. G.; Kizil. M.; Ketani. A.; Haris, P.I. Protective effect of DiyarbakIr watermelon juice on carbon tetrachloride-induced toxicity in rats. *Food Chem. Toxicol.* **2011**, *49*, 2433–2438.

Aluko, B.T.; Oloyede, O.I.; Afolayan, A.J. Hepatoprotective activity of *Ocimum americanum* L. leaves against paracetamol-induced liver damage in rats. *Am. J. Life Sci.* **2013**, *1*, 37–42.

Alwan, A.H.; Al-Gaillany, K.A.S.; Naji, A. Inhibition of the binding of 3H-Benzo (a) pyrene to rat liver microsomal protein by plant extracts. *Pharm. Biol.* **1989**, *27*(1), 33–37.

Al-Youssef, H.M.; Amina, M.; El-shafae, A.M. Biological evaluation of constituents from *Grewia mollis*. *J. Chem. Pharm. Res.* **2012**, 4(1), 508–518.

Alzergy, A.A.; Saad M.S. Elgharbawy, S.M.S. Hepatoprotective effects of *Juniperus phoenicea* L. on trichloroacetic acid induced toxicity in mice: Histological, ultrastructure and biochemical studies. *J. Am. Sci.* **2017**, *13*(12), 41–61.

Amagase, H.; Petesch, B.L.; Matsuura, H.; Kasuga, S.; Itakura, Y. Intake of garlic and its bioactive components. *J. Nutr.* **2001**, *131*, 955S-962S.

Amat, N.; Upur, H.; Blazeković, B. *In vivo* hepatoprotective activity of the aqueous extract of *Artemisia absinthium* L. against chemically and immunologically induced liver injuries in mice. *J. Ethnopharmacol.* **2010**, *131*, 478–484.

Ambrose, S.S.; Solairaj, P.; Subramoniam, A. Hepatoprotective activity of active fractions of *Thespesia lampas* Dalz. & Gibs. (Malvaceae). *J. Pharmacol. Pharmacother.* **2012**, *3*(4), 326–328.

Amer, O.S.O.; Dkhil, M.A.; Al-Quraishy, S. Antischistosomal and hepatoprotective activity of *Morus alba* leaves extract. *Pak. J. Zool.,* **2013**, *45*(2), 387–393.

Amin, A.; Ghoneim, M.D. *Zizyphus spina-christi* protects against carbon tetrachloride-induced liver fibrosis in rats. *Food Chem. Toxicol.* **2009**, *47*, 2111–2119.

Amin, A.; Hamza, A. Hepatoprotective effects of *Hibiscus*, *Rosmarinus* and *Salvia* on azathioprine-induced toxicity in rats. *Life Sci.* **2005**, *77*(3): 266–278.

Amin, H.; Mingshi, W.; Hong, Y.H.; Decheng, Z.; Lee, K.H. Hepatoprotective triterpenes from *Sedum sarmentosum*. *Phytochem.*, **1998**, *49*(8), 2607–2610.

Amin, M.; Pipelzadeh, M.H.; Mehdinejad, M.; Rashidi, I. An *in vivo* toxicological study upon shallomin, the active antimicrobial constitute of Persian shallot (*Allium hirtifolium* Boiss) extract. *Jundushapur J. Nat. Pharm. Prod.* **2012**. *7*, 17–21.

Amin, Z.A.; Bilgen, M.; Alshawsh, M.A.; Ali, H.M.; Hadi, A.H.; Abdulla, M.A. Protective role of *Phyllanthus niruri* extract against thioacetamide-induced liver cirrhosis in rat model. *Evid. Based Complement. Alternat. Med.* **2012**, *2012*, 241583. doi.10.1155/2012/241583

Amir, M.; Khan, M.A.; Ahmad, S.; Akhtar, M.; Mujeeb, M.; Ahmad, A.; et al., Ameliorating effects of *Tamarindus indica* fruit extract on anti-tubercular drugs induced liver toxicity in rats. *Nat. Prod. Res.* **2015**, *30*(6), 715–719.

Amzar, N.; Iqbal, M. The hepatoprotective effect of *Clidemia hirta* against carbon tetrachloride (CCl_4)-induced oxidative stress and hepatic damage in Mice. *J. Environ. Pathol. Toxicol. Oncol.*, **2017**, *36*(4), 1–6.

Anandan, R., Devaki, T. Hepatoprotective effect of *Picrorrhiza kurroa* on tissue defence system in dgalactosamine-induced hepatitis in rats. *Fitoterapia*, **1999**, *70*(1), 54–57.

Anbarasu, C.; Rajkapoor, B.; Bhat, K.S.; John, G.; Arul Amuthan, A.; Satish, K. Protective effect of *Pisonia aculeata* on thioacetamide induced hepatotoxicity in rats. *Asian Pac. J. Trop. Biomed.* **2012**, *2*(7), 511–515.

Aneja, S.; Vats, M.; Aggarwal, S.; Sardana, S. Phytochemistry and hepatoprotective activity of aqueous extract of *Amaranthus tricolor* Linn. roots. *J. Ayurveda Integr. Med.* **2013**, *4*(4), 211–215. doi: 10.4103/0975-9476.123693.

Aniya, Y.; Koyama, T.; Miyagi, C.; Miyahira, M.; Inomata, C.; Kinoshita, S.; Ichiba, T. Free radical scavenging and hepatoprotective actions of the medicinal herb, *Crassocephalum crepidioides* from the Okinawa Islands. *Biol. Pharm. Bull.*, **2005**. *28*(1), 19–23.

Aniya, Y.; Shimabukuro, M.; Shimoji, M.; Kohatsu, M.; Gyamfi, M.A.; Miyagi, C.; Kunii, D.; Takayama, F.; Egashira, T. Antioxidant and hepatoprotective actions of the medicinal herb *Artemisia campestris* from the Okinawa Islands. *Biol. Pharm. Bull.* **2000**, *23*(3), 309–312.

Aniyathi, M.J.A.; Latha, P.G.; Manikili, P.; Suja, S.R.; Syamal, S.; Shine, V.J.; Anuja, G.I.; Shikha, P.; Vidyadharan, M.K.; Rajasekharan, S. 2009. Evaluation of hepatoprotective activity of *Capparis brevispina* DC. stem bark. *Nat. Prod. Radiance* **2009**, *8*(5), 514–519.

Anonymous, *Artemisia scoparia*, http://www.naturalmedicinalherbs.net/herbs/a/artemisia-scoparia.php

Anoopraj, R.; Hemalatha, S., Balachandran, C. A preliminary study on serum liver function indices of diethylnitrosamine induced hepatocarcinogenesis and chemoprotective potential of *Eclipta alba* in male Wistar rats. *Vet. World,* **2014**, *7*, 6, 439–442.

Anosike CA, Ugwu UB, Nwakanma O. Effect of ethanol extract of *Pyrenacantha staudtii* leaves on carbontetrachloride induced hepatotoxicity in rats. *Biokemistri* **2008**, *20*, 17–22.

Antony, B.; Santhakumari, G.; Merina, B.; Sheeba, V.; Mukkadan. J. Hepatoprotective effect of *Centella asiatica* (L) in carbon tetrachloride-induced liver injury in rats. *Indian J. Pharm. Sci.* **2006**, *68*(6), 772–776.

Anusha, M.; Venkateswarlu, M.; Prabhakaran, V.; Taj, S.S.; Kumari, B.P.; Ranganayakulu, D. Hepatoprotective activity of aqueous extract of *Portulaca oleracea* in combination with lycopene in rats. *Indian J. Pharmacol.* **2011**, *43*(5), 563–567.

Anusuya, N. Raju, K.; Manian, S. Hepatoprotective and toxicological assessment of an ethnomedicinal plant *Euphorbia fusiformis* Buch.-Ham. ex D. Don. *J. Ethnopharmacol.* **2010**. *127*, 463–467.

Anyasor, G.N.; Odunsanya, K.; Ibeneme, A. Hepatoprotective and *in vivo* antioxidant activity of *Costus afer* leaf extract against acetaminophen induced hepatotoxicity in rats. *J. Investig. Biochem.* **2013**, *2*, 53–61.

Arbab, A.H.; Parvez, M.K.; Dosari, M.S.; Rehaily, A.J.; Ibrahim, K.E.; Alam, P. et al., Therapeutic efficacy of ethanolic extract of *Aerva javanica* aerial parts in the amelioration of CCl_4-induced hepatotoxicity and oxidative damage in rats. *Food Nutr Res.* **2016**, *60*, 30864.

Araya, E. M.; Adamu, B. A.; Periasamy, G.; Sintayehu, B.; Hiben, M. G. (2019). In vivo hepatoprotective and *in vitro* radical scavenging activities of *Cucumis ficifolius* A. Rich root extract. *J. Ethnopharmacol.* **2019**, *242*, doi.10.1016/j.jep.2019.112031

Arbab, A. H.; Parvez, M. K.; Al-Dosari, M. S.; Al-Rehaily, A. J.; Al-Sohaibani, M.; Zaroug, E. E.; Rafatullah, S. Hepatoprotective and antiviral efficacy of *Acacia mellifera* leaves fractions against hepatitis B virus. *Biomed. Res. Int.*, **2015**, *2015*. doi.10.1155/2015/929131

Arhoghro, E.M.; Ekpo, K.E.; Anosike, E.O.; Ibeh, G.O. Effect of aqueous extract of bitter leaf (*Vernonia amygdalina* Del.) on carbon tetrachloride (CCl_4)-induced liver damage in albino Wistar rats. *Eur. J. Sci. Res.* **2009**. *26*, 122–130.

Arhoghro, E.M.; Ikeh, C.H.; Eboh, A.S.L.; Angalabiri-owei, B. Liver function of Wistar rats fed the combined ethanolic leaf extract of *Alchornea cordifolia* and *Costus afer* in paracetamol-induced toxicity. *World J. Pharm. Res.* **2015**, *4*(5), 1–12.

Arhoghro, E.M.; Ekpo, K.E.; Ibeh, G.O. Effects of aqueous extracts of scent leaf (*Ocimum gratissimum*) on CCl_4-induced liver damage in Wistar albino rat. *Afr. J. Pharm. Pharmacol.*, **2009**, *3*(11), 562–567.

Arhoghro, E.M.; Kpomah, E.D.; Uwakwe, A.A. Curative potential of aqueous extract of lemon grass *Cymbopogon citratus* on cisplatin induced hepatotoxicity in albino Wistar Rats. *J. Physiol. Pharmacol. Adv.*, **2012**, *28*, 282-294.

Arjumand, A.; Srinivas Reddy, K.; Reddy, C.S. Hepatoprotective activity of *Cardiospermum halicacabum* stem extracts against carbon tetrachloride-induced hepatotoxicity in Wistar rats. *Int. J. Pharm. Sci. Nanotech.* **2009**, *2*(1), 488–492.

Arora, B.; Choudhary, M.; Arya, P.; Kumar, S.; Choudhary, N.; Singh, S. Hepatoprotective potential of *Saraca asoka* (Roxb.) De Wilde bark by carbon tetrachloride-induced liver damage in rats. *Bull. Faculty Pharm., Cairo Univ.*, **2015**, *53*(1), 23–28.

Arthur, F.K.N.; Woode, E.; Terlabi, E.O.; Larbie, C. Evaluation of hepatoprotective effect of aqueous extract of *Annona muricata* (Linn.) leaf against carbon tetrachloride and APAP-induced liver damage. *J. Nat. Pharm.* **2012**, *3*(1), 25–30.

Arulkumaran, K.S.; Rajasekaran, A.; Ramasamy, A.; Jegadeesan, M.; Kavimani, S.; Somasundaram, A. *Cassia roxburghii* seeds protect liver against toxic effects of ethanol and carbon tetrachloride in rats. *Int. J. Pharm. Tech. Res.* **2009**, *1*(2), 273–246.

Arumanayagam, S.; Arunmani, M. Hepatoprotective and antibacterial activity of *Lippia nodiflora* Linn. against lipopolysaccharides on HepG2 cells. *Pharmacogn Mag.*, **2015**, *11*(41), 24–31.

Arun, M.; Asha, V.V. Preliminary studies on antihepatotoxic effect of *Physalis peruviana* Linn. (Solanaceae) against carbon tetrachloride induced acute liver injury in rats. *J. Ethnopharmacol.* **2007**, 111(1), 110–114.

Asad, M.; Alhumoud, M. Hepatoprotective effect and GC-MS analysis of traditionally used *Boswellia sacra* oleo gum resin (Frankincense) extract in. *Afr. J. Tradit. Complement. Altern. Med.*, **2015**, *12*(2), 1–5.

Asadi-Samani, M.; Bahmani, M.; Rafieian-Kopaei, M. The chemical composition, botanical characteristic and biological activities of *Borago officinalis*: A review. *Asian Pac. J. Trop. Med.* **2014**, *7*, S22-S28.

Asadi-Samani, M.; Kafash-Farkhad, N.; Azimi, N.; Fasihi, A.; Alinia-Ahandani, E.; Rafieian-Kopaei, M. Medicinal plants with hepatoprotective activity in Iranian folk medicine. *Asian Pac. J. Tropical Biomed.* **2015**, *5*(2), 146–157. doi: 10.1016/S2221-1691(15)30159–3.

Asadi-Samani, M.; Rafieian-Kopaei, M.; Azimi, N. *Gundelia*: A systematic review of medicinal and molecular perspective. *Pak. J. Biol. Sci.* **2013**, *16*, 1238–1247.

Aseervatham, G.S.; Sivasudha, T.; Sasikumar, J.M.; Christabel, P.H.; Jeyadevi, R.; Ananth, D.A.; et al., Antioxidant and hepatoprotective potential of *Pouteria campechiana* on APAP-induced hepatic toxicity in rats. *J. Physiol. Biochem.* **2014**, *70*, 1–4.

Asgari, S.; Setorki, M.; Rafieian-Kopaei, M.; Heidarian, E.; Shahinfard, N.; Ansari, R.; et al., Postprandial hypolipidemic and hypoglycemic effects of *Allium hertifolium* and *Sesamum indicum* on hypercholesterolemic rabbits. *Afr. J. Pharm. Pharmacol.* **2012**, *6*, 1131–1135.

Asgari-Kafrani, A.; Fazilati, M.; Nazem, H. Hepatoprotective and antioxidant activity of aerial parts of *Moringa oleifera* in prevention of non-alcoholic fatty liver disease in Wistar rats. *S. Afr. J. Bot.*, **2019**, *2019*, 1–9.

Asha, V.V. Preliminary studies on the hepatoprotective activity of *Momordica subangulata* and *Naragamia alata. Indian J. Pharmacol.* **2001**, *33*, 276–279.

Asha, V.V.; Sheeba, M.S.; Suresh, V.; Wills, P.J. Hepatoprotection of *Phyllantus maderaspatensis* against experimentally induced liver injury in rats. *Fitoterapia* **2007**, *78*(2), 134–141.

Asif, H.M.; Akram, M.; Usmanghani, K.; Akhtar, N.; Shah, P.A.; Uzair, M.; et al., Monograph of *Apium graveolens* Linn. *J. Med. Plants Res.* **2011**, *5*, 1494–1496.

Asuku, O.; Atawodi, S.E.; Onyike, E. Antioxidant, hepatoprotective, and ameliorative effects of methanolic extract of leaves of *Grewia mollis* Juss. on carbon tetrachloride-treated albino rats. *J. Med. Food,* **2012**, *15*(1), 83–88.

Atayi, Z.; Borji, H.; Moazeni, M.; Darbandi, M. S.; Heidarpour, M. *Zataria multiflora* would attenuate the hepatotoxicity of long-term albendazole treatment in mice with cystic echinococcosis. *Parasitol. Int.*, **2018**, *67*(2), 184–187.

Atta, A.H.; Alkofahi, A. Anti-nociceptive and anti-inflammatory effects of some Jordanian medicinal plant extracts. *J. Ethnopharmacol.* **1998**, 60(2), 117–124.

Atta, A.H.; Mohamed, N.H.; Naser, S.M.; Mouneir, S.M. Phytochemical and pharmacological studies on *Convolvulus fatmensis* Ktze. *J. Nat. Rem.* **2007**, *7*, 109–119.

Augusti, K.; Anuradha, P.S.; Smitha, K.; Sudheesh, M.; George, A.; Joseph, M. Nutraceutical effects of garlic oil, its nonpolar fraction and a *Ficus* flavonoid as compared to vitamin E in CCl_4-induced liver damage in rats. *Indian. J. Exp. Biol.* **2005**, 43, 437–444.

Awaad, A.S.; Maitland, D. J.; Soliman, G. A. Hepatoprotective activity of *Schouwia thebica* Webb. *Bioorg. Med. Chem. Lett.* **2006**, *16*, 4624–4628.

Awaad, A.S.; Soliman, G.A.; El-Sayed, D.F.; El-Gindi, O.D.; Alqasoum, S.I. Hepatoprotective activity of *Cyperus alternifolius* on carbon tetrachloride-induced hepatotoxicity in rats. *Pharm. Biol.* **2012**, *50*(2), 155–161.

Awad, N.E.; Abdelkawy, M.A.; Hamed, M.A.; Souleman, A.M.A.; Abdelrahman, E.H.; Ramadan, N.S. Antioxidant and hepatoprotective effects of *Justicia spicigera* ethyl acetate fraction and characterization of its anthocyanin content. *Int. J. Pharm. Pharm. Sci.*, **2015**, 7(8), 91–96.

Awodele, O.; Yemitan, O.; Ise, P.U.; Ikumawoyi, V.O. Modulatory potentials of aqueous leaf and unripe fruit extracts of *Carica papaya* Linn. (Caricaceae) against carbon tetrachloride and APAP induced hepatotoxicity in rats. *J. Intercult. Ethnopharmacol.* **2016**, *5*, 27–35.

Awogbindin, I.O., Tade, O.G., Metibemu, S. D., Olorunsogo, O.O., Farombi, E.O. 2014. Assessment of flavonoid content, free-radical scavenging and hepatoprotective activities of *Ocimum gratissimum* and *Spondias mombin* in rats treated with dimethylnitrosamine. *Arch. Bas. App. Med.* **2014**, *2*, 45–54.

Azab, S.S.; Daim, M.A.; Eldahshan, O.A. Phytochemical, cytotoxic, hepatoprotective and antioxidant properties of *Delonix regia* leaves extract. *Med. Chem. Res.* **2013**, *22*(9), doi: 10.1007/s0004-012-0420-4.

Azadi, H.G.; Riazi, G.H.; Ghaffari, S.M.; Ahmadian, S.; Khalife, T.J. Effects of *Allium hirtifolium* (Iranian shallot) and its allicin on microtubule and cancer cell lines. *Afr. J. Biotechnol.* **2009**, *8*, 5030–5037.

Azarmehr, N.; Afshar, P.; Moradi, M.; Sadeghi, H.; Sadeghi, H.; Alipoor, B.; Doustimotlagh, A. H. Hepatoprotective and antioxidant activity of watercress extract on acetaminophen induced hepatotoxicity in rats. *Heliyon*, **2019**, *5*(7), e02072.

Azebaze, A.G.; Ouahouo, B.M.; Vardamides, J.C.; Valentin, A.; Kuete, V.; Acebey, L.; et al., Antimicrobial and antileishmanial xanthones from the stem bark of *Allanblackia gabonensis* (Guttiferae). *Nat. Product Res.* **2008**, *22*, 333–341.

Azim, S.A.A.; Abdelrahem, M.T.; Said, M.M.; Khattab, A. Protective effect of *Moringa peregrina* leaves extract on acetaminophen-induced liver toxicity in albino rats. *African J. Tradit. Complement. Altern. Med.* **2017**, *14*, 206–16.

Babalola, O.; Ojo, O.E.; Oloyede, F.A. Hepatoprotective activity of aqueous extract of the leaves of *Hyptis suaveolens* on acetaminophen induced hepatotoxicity in rabbits. *Res. J. Chem. Sci.* **2011**, *1*, 85–88.

Babitha S., Banji D., Banji O. J. F. Antioxidant and hepatoprotective effects of flower extract of *Millingtonia hortensis* Linn. on carbon tetrachloride induced hepatotoxicity. *J. Pharm. Bioallied Sci.* **2012**, 4(4), 307–312.

Babita, M.H.; Chhaya, G.; Goldee, P. Hepatoprotective activity of *Rubia cordifolia. Pharmacologyonline*. **2007**, *3*, 73–79.

Babu, B.H.; Shylesh, B.S.; Padikkala, J. Antioxidant and hepatoprotective effect of *Acanthus ilicifolius. Fitoterapia* **2001**, *72*, 272–277.

Bag, A.K.; Mumtaz, S.M.F. Hepatoprotective and nephroprotective activity of hydroalcoholic extract of *Ipomoea staphylina* leaves. *Bangladesh J. Pharmacol.* **2013**, *8*, 263–68.

Baghdadi, H.H.; El-Demerdash, F.M.; Radwan, E.H.; Hussein, S. The protective effect of *Coriandrum sativum* L. oil against liver toxicity induced by ibuprofen in rats. *J. Biosci. Appl. Res.* **2016**, *2*, 197–202.

Baheti, J.R.; Goyal, R.K.; Shaw, G.B. Hepatoprotective activity of *Hemidesmus indicus* in rats. *J. Exp. Biol.* **2006**, *44*, 399–402.

Bairwa, N.K.; Sethiya, N.K.; Mishra, S.H. Protective effect of stem bark of *Ceiba pentandra* Linn. against paracetamol-induced hepatotoxicity in rats. *Pharmacog. Res.* **2010**, **2**(1), 28–30.

Bais, S.S.; Mali, P.Y. Protective effect of *Amorphophallus campanulatus* tuber extracts against H_2O_2 induced oxidative damage in human erythrocytes and leucocytes. *Int. J. Green Pharm.* **2013**, *7*, 111–116.

Bajpai, V.K.; Kim, J.-E.; Kang, S.C. Protective effect of heat-treated cucumber (*Cucumis sativus* L.) juice against lead-induced detoxification in rat model. *Indian J. Pharm. Edu. Res.* **2017**, *51*, 59–69.

Bakr, R. O.; Fayed, M. A.; Fayez, A. M.; Gabr, S. K.; El-Fishawy, A. M.; El-Alfy, T. S. Hepatoprotective activity of *Erythrina* × *neillii* leaf extract and characterization of its phytoconstituents. *Phytomedicine*, **2019**, *53*, 9–17.

Balaji, B.; Bookya, K.; Bai, G.; Mangilal, T. Hepatoprotective agent present in pods of *Clitoria ternatea* with evidence of histopathological analysis. *Int. J. Pharm. Res.* **2015**, *4*(4), 32–38.

Balakrishman, B.; Sagameswaran, B.; Bhaskar, V.H. Effect of methanol extract of *Cuscuta reflexa* aerial parts on hepatotoxicity induced by antitubercular drugs in rats. *Int. J. App. Res. Nat. Prod.* **2010**, *3*(1), 18–22.

Balakrishnan, N.; Balasubaramaniam, A.; Sangameswaran, B.; Bhaskar, V. H. Hepatoprotective activity of two Indian medicinal plants from Western Ghats, Tamil Nadu. *J. Nat. Pharm.* **2011**, *2*(2), 92–98.

Balakrishnan, S.; Khurana, B.S.; Singh, A.; Kaliappan, I.; Dubey, G. Hepatoprotective effect of hydroalcoholic extract of *Cissampelos pareira* against rifampicin and isoniazid-induced hepatotoxicity. *Cont. J. Food Sci. Technol.* **2012,** *6*(1), 30–35.

Balamurugan, G.; Muthusamy, P. Observation of the hepatoprotective and antioxidant activities of *Trianthema decandra* Linn. (Vallai sharunnai) roots on carbon tetrachloride-treated rats. *Bangladesh J. Pharmacol.* **2008**, *3*, 83–89.

Balanehru, S.; Nagarajan, B. Protective effect of oleanolic acid and ursolic acid against lipid peroxidation. *Biochem. Int.* **1991**, *24*(5), 981–990.

Balderas-Renteria, I.; Camacho-Corona Mdel, R.; Carranza-Rosales, P.; Lozano-Garza, H.G.; Castillo-Nava, D.; Alvarez-Mendoza, F.J.; Tamez-Cantú, E.M. Hepatoprotective effect of *Leucophyllum frutescens* on Wistar albino rats intoxicated with carbon tetrachloride. *Ann. Hepatol.* **2007**, *6*(4), 251–254.

Banskota, A.H.; Tezuka, Y.; Adnyana, I.K.; Xiong, Q.; Hase, K.; Tran, K.Q.; Tanaka, K.; Saiki, I.; Kadota, S. Hepatoprotective effect of *Combretum quadrangulare* and its constituents. *Biol. Pharm. Bull.* **2003**, *23*(4), 456–460.

Banu, G. S.; Kumar, G.; Murugesan, A. G. Effect of ethanolic leaf extract of *Trianthema portulacastrum* L. on aflatoxin-induced hepatic damage in rats. *Indian J. Clin. Biochem.* **2009**, *24*(4), 414–418.

Banu, S.; Bhaskar, B.; Balasekar P. Hepatoprotective and antioxidant activity of *Leucas aspera* against D-galactosamine-induced liver damage in rats. *Pharm. Biol.* **2012**, *50*(12), 1592–1595.

Baranisrinivasan, P.; Elumalai, E. K.; Sivakumar, C.; Therasa, S. V.; David, E. Hepatoprotective effect of *Enicostemma littorale* Blume and *Eclipta alba* during ethanol induced oxidative stress in albino rats. *Int. J. Pharmacol.* **2009**, *5*(4), 268–272.

Bardi, D.A.; Halabi, M.F.; Abdullah, N.A.; Rouhollahi, E.; Hajrezaie, M.; Abdulla, M.A. *In vivo* evaluation of ethanolic extract of *Zingiber officinale* rhizomes for its protective effect against liver cirrhosis. *BioMed Res. Int.* **2013**, *2013*: 918460. doi: 10.1155/2013/918460

Bardi, D.A.; Halabi, M.F.; Hassandarvish, P.; Rouhollahi, E.; Paydar, M.; Moghadamtousi, S.Z.; et al., *Andrographis paniculata* leaf extract prevents thioacetamide-induced liver cirrhosis in rats. *PLoS One* **2014**, *9*(10): e109424. doi.10.1371/journal.pone.0109424

Bartalis J. Hepatoprotective activity of cucurbitacin [dissertation]. Brookings: South Dakota State University. **2005.**

Basu, A., T. Sen, R.N. Ray and A. Chaudhuri, 1992. Hepatoprotective effects of *Calotropis procera* root extract on experimental liver damage in animals. *Fitoterapia*, **1992**, *63*, 507–514.

Batool, R.; Khan, M.R.; Majid, M. *Euphorbia dracunculoides* L. abrogates carbon tetrachloride induced liver and DNA damage in rats. *BMC Complement Altern Med.* **2017**, *17*(1), 223. doi: 10.1186/s12906-017-1744-x.

Battochio, A.P.; Coelho, K.L.; Sartoli, M.S.; Coelho, C.A. Hepatoprotective effect of water soluble extract of *Coleus barbatus* on cholestasis on young rats. *Acta Cir. Bras.* **2008**, *23*, 220-229.

Batur, M.; Cheng, L.F.; Yan, D.; Parhat, K. Hepatoprotective effect of *Gossipium hirsutum* extract on acute experimental hepatitis on rat liver injury. *Zhongguo Zhong Yao Za Zhi* **2008**, *33*(15), 1873–1876.

Bayramoglu, G.; Bayramoglu, A.; Engur, S.; Senturk, H.; Ozturk, N.; Colak, S. The hepatoprotective effects of *Hypericum perforatum* L. on hepatic ischemia/reperfusion injury in rats. *Cytotechnology*, **2014**, *66*(3), 443–448.

Beckier, B. The contribution of wild plants to human nutrition in Ferlo (Northern Senegal). *Agroforest. Syst.,* **1983**, *1*(3), 257–267.

Beedimani, R.S.; Jeevangi, S.K. Evaluation of hepatoprotective activity of *Boerhaavia diffusa* against carbon tetrachloride induced liver toxicity in albino rats. *Int. J. Basic Clin. Pharmacol.* **2015**, *4*, 153–58.

Beedimani, R.S.; Shetkar, S. Hepatoprotective activity of *Eclipta alba* against carbon tetrachloride-induced hepatotoxicity in albino rats. *Int. J. Basic Clin. Pharmacol.* **2015**, *4*, 404–409.

Beena, P.; Purnima, S.; Kokilavani, R. In vitro hepatoprotective activity of ethanolic extract of *Coldenia procumbens* Linn. *J. Chem. Pharm. Res*., **2011**, *3*(2), 144–149.

Bellassoued, K.; Ben Hsouna, A.; Athmouni, K.; van Pelt, J.; Makni Ayadi, F.; Rebai, T.; Elfeki, A. Protective effects of *Mentha piperita* L. leaf essential oil against CCl_4-induced hepatic oxidative damage and renal failure in rats. *Lipids Health Dis.* **2018**, *17*(1), 9. doi: 10.1186/s12944-017-0645-9.

Ben Saad, A.; Dalel, B.; Rjeibi, I.; Smida, A.; Ncib, S.; Zouari, N.; Zourgui, L. Phytochemical, antioxidant and protective effect of *cactus cladodes* extract against lithium-induced liver injury in rats. *Pharm. Biol.*, **2017**, *55*(1), 516–525.

Ben Saad, A.; Rjeibi, I.; Alimi, H.; Ncib, S.; Bouhamda, T.; Zouari, N. Protective effects of *Mentha spicata* against nicotine-induced toxicity in liver and erythrocytes of Wistar rats. *Appl. Physiol. Nutr. Metab.*, **2018**, *43*(1), 77–83.

Bezenjani, S.N.; Pouraboli, I.; Malekpour Afshar, R.; Mohammadi, Gh. Hepatoprotective effect of *Otostegia persica* Boiss. shoot extract on carbon tetrachloride-induced acute liver damage in rats. *Iran J. Pharm. Res.,* **2012**, *11*, 1235–1241.

Bhalla, T.N.; Gupta, M.B.; Sheth, P.K.; Bhargava, K.P. Antiinflammatory activity of *Boerhaavia diffusa. Indian J. Physiol. Pharmacol.* **1968**, *6*(1), 11–16.

Bhakta, T.;, Banerjee S.; Mandal S.C.; Maity, T.K.; Saha, B.P.; Pal, M. Hepatoprotective activity of *Cassia fistula* leaf extract. *Phytomedicine* **2001**, *8*(3), 220–224.

Bhakta, T., Mukherjee, P.K., Mukherjee, K., Banerjee, Mandal, S.C., Maity, T.K., Pal, M., Saha, B.P. Evaluation of hepatoprotective activity of *Cassia fistula* leaf extract. *J. Ethnopharmacol.* **1999**, *66*, 277–282.

Bhandarkar, M.; Khan, A. Protective effect of *Lawsonia alba* Lam., against CCl_4-induced hepatic damage in albino rats. *Ind. J. Expt. Biol.* 2003, *41*, 85–87.

Bhandarkar, M.R.; Khan, A. Antihepatotoxic effect of *Nymphaea stellata* Willd. against Carbon tetrachloride hepatic damage in albino rats. *J. Ethnopharmacol*. **2004**, *91*, 61–64.

Bharathi, R.; Ravi Shankar, K.; Geetha, K. In vitro antioxidant activity and *in vivo* hepatoprotective activity of ethanolic whole plant extract of *Nymphoides hydrophylla* in CCl_4-induced liver damage in albino rats. *Int. J. Res. Ayurved Pharm.* **2014**, *5*(6), 667–672.

Bhaskar, V.H.; Balakrishnan, N. Hepatoprotective activity of laticiferous plant species (*Pergularia daemia* and *Carissa carandas*) from Western Ghats, Tamil Nadu, India. *Der Pharm. Lett.* **2009**, *1*, 130–142.

Bhaskar, V.H.; Balakrishnan, N. Protectivee effects of *Pergularia daemia* roots against paracetamol and carbon tetrachloride-induced hepatotoxicity in rats. *Pharm. Biol.* **2010**, *48*(1), 1265-1272.

Bhatt, S.; Kumar, H.; Sharma, M.; Saxena, K.K.; Garg, G.; Singh, G. Evaluation of hepatoprotective activity of *Aloe vera* in drug induced hepatitis. *World J. Pharm. Pharm. Sci.* **2015**; *4*, 935–944.

Bhatt, S.; Virani, S.; Sharma, M.; Kumar, H.; Saxena, K.K. Evaluation of hepatoprotective activity of *Aloe vera* in acute viral hepatitis. *Int. J. Pharm. Sci. Res.* **2014**, *5*(6): 2479–2485.

Bhattacharjee, R.; Sil, P.C. The protein fraction of *Phyllanthus niruri* plays a protective role against acetaminophen-induced hepatic disorder via its antioxidant properties. *Phytother. Res.* **2006**, *20*(7), 595–601.

Bhattacharjee, R.; Sil, P.C. Protein isolate from the herb, *Phyllanthus niruri* L. (Euphorbiaceae), plays hepatoprotective role against carbon tetrachloride-induced liver damage via its antioxidant properties. *Food Chem. Toxicol.* **2007**, *45*(5), 817–826.

Bhattacharyya, D.; Pandit, S.; Jana, U.; Sen, S.; Sur, T. S. Hepatoprotective activity of *Adhatoda vasica* aqueous leaf extract on d-galactosamine-induced liver damage in rats. *Fitoterapia* **2005**, *76*(2), 223– 225.

Bhattacharya, S.; Chatterjee, M. Protective role of *Trianthema portulacastrum* against diethylnitrosamine induced experimental hepatocarcinogenesis. *Cancer Lett.* **1998**, *129*, 7–13.

Bhoumik, D.; Mallik, A.; Berhe, A.H. Hepatoprotective activity of aqueous extract of *Sesbania grandiflora* Linn. leaves against carbon tetrachloride induced hepatotoxicity in albino rats. *Int. J. Phytomed.* **2016**, *8*(2), 294–299.

Bigoniya, P., Rana, A. C. Protective effect of *Wrightia tinctoria* bark triterpenoidal fraction on carbon tetrachloride-induced acute rat liver toxicity. *Int. J. Pharm. Tech.*, **2010**, *9*(2), 55–62.

Bijesh, V.; Pramod, C. Protective effect of *Annona reticulata* Linn. against simvastatin induced toxicity in Chang liver cells. *J. Sci. Innov. Res.* **2014**, *3*(5): 495–499.

Bishayee, A.; Mandal A.; Chatterjee, M. Prevention of alcohol-carbon tetrachloride induced signs of early hepatotoxicity in mice by *Trianthema portulacastrum* L. *Phytomedicine* **1996**, *3,* 155–161.

Bishayee, A.; Sarkar, A.; Chatterjee, M. Hepatoprotective activity of carrot (*Daucus carota* L.) against carbon tetrachloride intoxication in mouse liver. *J. Ethnopharmacol.* **1995**. *47*(2), 69–74.

Bishayi, B.; Roychowdhury, S., Ghosh, S.; Sengupta, M. Hepatoprotective and immunomodulatory properties of *Tinospora cordifolia* in CCl_4 intoxicated mature albino rats. *J. Toxicol. Sci.* **2002**, *27*, 139–46.

Biswas, K.; Kumar, A.; Babaria, B.; Prabhu, K.; Setty, R. Hepatoprotective effect of leaves of *Peltophorum pterocarpum* against paracetamol induced acute liver damage in rats. *J. Basic Clin. Pharm.*, **2010**, *1*(1), 10–15.

Biswas M, Karan TK, Kar B, Bhattacharya S, Ghosh AK, Bhattacharya, S.; Ghosh, A.K.; Kumar, R.B.S.; Haldar, P.K. Hepatoprotective activity of *Terminalia arjuna* leaf against paracetamol-induced liver damage in rats. *Asian J. Chem.* **2011**, *23*, 1739.

Biswas, P.; Trivedi, N.; Singh, B. K.; Jha K.K. Evaluation of hepatoprotective activity of ethanolic root extract of *Vaccaria pyramidata* against CCl_4-induced hepatotoxicity in Wister rats. *European J. Pharm. Med. Res.* **2017**, *4*(9), 532–538.

Bodakhe, S.H.; Ram, A. Hepatoprotective properties of *Bauhinia variegata* bark extract. *Yakugaku Zasshi.*, **2007**, *127*, 1503–1507.

Bomzon, A.; Shtukmaster, S.; Ljubuncic, P. The effect of an aqueous extract of *Teucrium polium* on glutathione homeostasis *in vitro*: A possible mechanism of its hepatoprotectant action. *Adv. Pharm. Sci.*, **2010**, *2010*, Article ID 938324, 7 pages.

Bonkovsky, H.L.Md. Hepatotoxicity associated with supplements containing Chinese green tea (*Camellia sinensis*). *Ann. Internal Med.* **2006**, *144*(1), 68–71.

Bouasla, I.; Bouasla, A.; Boumendjel, A.; Messarah, M.; Abdennour, C.; Boulakoud, M.S.; Feki, A. *Nigella sativa* oil reduces aluminium chloride-induced oxidative injury in liver and erythrocytes of rats. *Biol. Trace Elem. Res.* **2014**, *162*(1–3), 252–261.

Bourogaa, E.; Jarraya, R.M.; Damak, M.; Abdelfattah Elfeki, A. Hepatoprotective activity of *Peganum harmala* against ethanol-induced liver damages in rats. *Arch. Physiol. Biochem.* **2015,** *121*, 2, 62–67.

Bourogaa, E.; Nciri, R.; Mezghani-Jarraya, R.; Racaud-Sultan, C.; Damak, M.; El

Feki, A. Antioxidant activity and hepatoprotective potential of *Hammada scoparia* against ethanol-induced liver injury in rats. *J. Physiol. Biochem.*, **2013**, *69*(2), 227–237.

Bouzenna, H.; Dhibi, S.; Samout, N.; Rjeibi, I.; Talarmin, H.; Elfeki, A.; Hfaiedh, N. The protective effect of *Citrus limon* essential oil on hepatotoxicity and nephrotoxicity induced by aspirin in rats. *Biomed. Pharmacother.*, **2016**, *83*, 1327–1334.

Braga, P.C.; Dal Sasso, M.; Culici, M.; Spallino, A.; Falchi, M.; Bertelli, A.; et al., Antioxidant activity of *Calendula officinalis* extract: Inhibitory effects on chemiluminescence of human neutrophil bursts and electron paramagnetic resonance spectroscopy. *Pharmacology* **2009**, *83*, 348-355.

Brai, B.I.; Adisa, R.A.; Odetola, A.A. Hepatoprotective properties of aqueous leaf extract of *Persea americana* Mill. (Lauraceae) 'avocado' against CCL_4-induced damage in rats. *Afr. J. Tradit. Complement. Altern. Med.* 2014, *11*(2), 237–244.

Bukhsh, E.; Malik, S.A.; Ahmad, S.S.; Erum, S. Hepatoprotective and hepatocurative properties of alcoholic extract of *Carthamus oxyacantha* seeds. *African J. Plant Sci.* **2014**, *8*(1), 34–41.

Cao, J. J.; Lv, Q. Q.; Zhang, B.; Chen, H. Q. Structural characterization and hepatoprotective activities of polysaccharides from the leaves of *Toona sinensis* (A. Juss) Roem. *Carbohydr Polym.*, **2019**, *212*, 89–101.

Cao, L.; Du, J.; Ding, W. Hepatoprotective and antioxidant effects of dietary *Angelica sinensis* extract against carbon tetrachloride-induced hepatic injury in Jian Carp (*Cyprinus carpio* var. Jian). *Aquac Res.* **2016**, *47*(6), 1852–1863.

Ćebović, T.; Maksimović, Z. Hepatoprotective effect of *Filipendula hexapetala* Gilib. (Rosaceae) in carbon tetrachloride-induced hepatotoxicity in rats. *Phytother. Res.* **2012**, *26*, 1088–1091.

Celik, I.; Temur, A.; Isik, I. Hepatoprotective role and antioxidant capacity of pomegranate (*Punica granatum*) flowers infusion against trichloroacetic acid-exposed in rats. *Food Chem Toxicol.* **2009**, *47*, 145–151.

Cengiz, N.; Kavak, S.; Guzel, A.; Ozbek, H.; Bektas, H.; Him, A.; Erdogan, E.; Balahoroglu, R. Investigation of the hepatoprotective effects of Sesame (*Sesamum indicum* L.) in carbon tetrachloride-induced liver toxicity. *J. Membr. Biol.* **2013**, *246*, 1–6. doi: 10.1007/s00232-012-9494-7.

Cengiz, N.; Ozbek, H.; Him, A. Hepatoprotective effects of *Pimpinella anisum* seed extract in rats. *Pharmacologyonline*, **2008**, *3*, 870–874.

Chaerunisa, A. Y.; Ramadhani, F. N.; Nurani, T. D.; Najihudin, A.; Susilawati, Y.; Subarnas, A. Hepatoprotective and antioxidant activity of the ethanol extract of *Cassia fistula* L. barks. *J. Pharm. Sci. Res.* **2018**, *10*(6), 1415–1417.

Chahdoura, H.; Adouni, K.; Khlifi, A.; Dridi, I.; Haouas, Z.; Neffati, F.; Achour, L. Hepatoprotective effect of *Opuntia microdasys* (Lehm.) Pfeiff flowers against diabetes type II induced in rats. *Biomed. Pharmacother.* **2017**, *94*, 79–87.

Chakole, R.; Zade, S.; Charde, M. Antioxidant and anti-inflammatory activity of ethanolic extract of *Beta vulgaris* Linn. roots. *Int. J. Biomed. Adv. Res.* **2011**, *2*(4), 124–130.

Chand, N.; Durrani, F.R.; Ahmad, S.; Khan, A. Immunomodulatory and hepatoprotective role of feed-added *Berberis lycium* in broiler chicks. *J. Sci. Food Agric.* **2011,** *91*, 1737–1745.

Chandan, B.K.; Saxena, A.K.; Shukla, S.; Sharma, N.; Gupta, D.K.; Suri, K.A.; Suri, J.; Bhadauria, M.; Singh, B. Hepatoprotective potential of *Aloe barbadensis* Mill. against carbon tetrachloride induced hepatotoxicity. *J. Ethnopharmacol.* **2007**, *111*, 560–566.

Chandan, B.K.; Sharma, A.K.; Anand, K.K. *Boerhavia diffusa*: A study of its hepatoprotective activity. *J. Ethnopharmacol.* **1991**, *31*(3), 299–307.

Chandan, K.; Saxena, A.K.; Shukla, S.; Sharma, N.; Gupta, D.K.; Singh, K.; Suri, J.; Bhadauria, M.; Qazi, G.N. Hepatoprotective activity of *Woodfordia fruticosa* Kurz flowers against carbontetrachloride induced hepatotoxicity. *J. Ethnopharmacol.* **2008**. *119*, 218–224.

Chander, R.; Srivastava, V.; Tandon, J.S.; Kapoor, N.K. Antihepatotoxic activity of diterpenes of *Andrographis paniculata* (Kalmegh) against *Plasmodium berghei*-induced hepatic damage in *Mastomys natalensis*. *Int. J. Pharmacogn.*, **1995**, *33*(2), 135–138.

Chandrashekar, K.S.; Prasanna, K.S. Hepatoprotective activity of *Leucas lavandulaefolia* against carbon tetrachloride-induced hepatic damage in rats. *Int. J. Pharm. Sci. Res.*, **2010**, *2*, 101–103.

Chandrashekar, K.S.; Prasanna, K.S.; Joshi, A.B. Hepatoprotective activity of the *Leucas lavandulaefolia* on D(+) galactosamine-induced hepatic injury in rats. *Fitoterapia.* 2007, *78*(6), 440–442.

Chandrashekhar, V.M.; Abdul Haseeb, T.S.; Habbum, P.V.; Nagappa, A.N. Hepatoprotective activity of *Wrightia tinctoria* Roxb. in rats. *Indian Drugs* **2004**, *41*, 366–370.

Chang, H.-F.; Lin, Y.-H.; Chu, C.-C.; Wu, S.-J.; Tsai, Y.-H.; Chao, J. C.-J. Protective Effects of *Ginkgo biloba*, *Panax ginseng*, and *Schizandra chinensis* extract on liver injury in rats. *Am. J. Chin. Med.* 2007, 35(6), 995–1009.

Chang, T. N.; Ho, Y. L.; Huang, G. J.; Huang, S. S.; Chen, C. J.; Hsieh, P. C.; Chang, Y. S. Hepatoprotective effect of *Crossostephium chinensis* (L.) Makino in rats. *Am. J. Chin. Med.*, **2011**, *39*(03), 503–521.

Channa, S.; Dar, A.; Anjum, S.; Yaqoob, M. Anti-inflammatory activity of *Bacopa monniera* in rodents. *J. Ethnopharmacol.* **2006**, *104*(1–2), 286–289.

Channabasavaraj, K.P.; Badami, S.; Bhojraj, S. Hepatoprotective and antioxidant activity of methanol extract of *Ficus glomerata*. *J. Nat. Med.* **2008**. *62*(3), 379–383.

Chao, J.; Lee, M.-S.; Amagaya, S.; Liao, J.-W.; Wu, J.-B.; Ho, L.-K.; Peng, W.-H. Hepatoprotective effect of shidagonglao on acute liver injury induced by carbon tetrachloride. *Am. J. Chin. Med.* **2009**, *37*(6), 1085–1097.

Chaphalkar, R.; Apte, K.G.; Talekar, Y.; Ojha, S.K.; Nandave, M. Anti-oxidants of *Phyllanthus emblica* L. bark extract provide hepatoprotection against ethanol-induced hepatic damage: A comparison with Silymarin. *Oxid. Med. Cell Longev.* **2017**, *2017*. 3876040. doi: 10.1155/2017/3876040.

Chatterjee, M.; Sil, P.C. Hepatoprotective effect of aqueous extract of *Phyllanthus niruri* on nimesulide-induced oxidative stress *in vivo*. *Indian J. Biochem. Biophy.*, **2006**, *43*, 299–305.

Chattopadhyay, R.R. Possible mechanism of hepatoprotective activity of *Azadirachta indica* leaf extract. Part II. *J. Ethnopharmacol.* **2003**, *89*, 217–219.

Chattopadhyay, R.R.; Bandyopadhyay, M. Possible mechanism of hepatoprotective activity of *Azadirachta indica* leaf extract against paracetamol-induced hepatic damage in rats: Part III. *Indian J. Pharmacol.* **2005**, *37*, 184–185.

Chattopadhyay, R.R.; Sarkar, S.K.; Ganguly, S.; Banerjee, R.N.; Basu, T.K.; Mukherjee, A. Hepatoprotective activity of *Azadirachta indica* leaves on paracetamol-induced hepatic damage in rats. *Indian J. Exp. Biol.* **1992,** *30*(8), 738–740.

Chattopadhyay, R.R.; Sarkar, S.K.; Ganguly, S.; Medda, C.; Basu, T.K. Hepatoprotective activity of *Ocimum sanctum* leaf extract against paracetamol-induced hepatic damage in rats. *Indian J. Pharmacol.* **1992**, *24,* 163–165.

Chaturvedi, P.; George, S.; Machacha, C.N. Protective role of *Raphanus sativus* root extract on paracetamol-induced hepatotoxicity in albino rats. *Int. J. Vitam. Nutr. Res.* **2007**, *77*, 41–45.

Chaturvedi, P.; Machacha, C.N. Efficacy of *Raphanus sativus* in the treatment of paracetamol-induced hepatotoxicity in albino rats. *Brit. J. Biomed. Sci.* **2007**, *64* (3), 105–108.

Chaudhuri, A.K.N.; Karmakar, S.; Roy, D.; Pal, S.; Pal, M.; Sen, T. Anti-inflammatory activity of Indian black tea (Sikkim variety). *Pharmacol. Res.* **2005**, *51*(2), 169–175.

Chaudhari, B.P.; Chaware, V.J.; Joshi, Y. R.; Biyani, K.R. Hepatoprotective activity of Hydroalcoholic extract of *Momordica charantia* Linn. leaves against carbon tetrachloride induced hepatopathy in rats. *Int. J. ChemTech. Res.* **2009**, *1*(2), 355–358.

Chaudhari, G.M.; Mahajan, R.T. In vitro hepatoprotective activity of *Terminalia arjuna* stem bark and its flavonoids against CCl_4 induced hepatotoxicity in goat liver slice culture. *Asian J. Plant Sci. Res.* **2016**, *6*(6), 10–17.

Chaudhari N. B.; Chittam K. P.; Patil V. R. Hepatoprotective activity of *Cassia fistula* seeds against paracetamol-induced hepatic injury in rats. *Arch. Pharm. Sci. Res.* **2009**, *1*(2), 218–221.

Chavda, R.; Vadalia, K.R.; Gokani, R. Hepatoprotective and antioxidant activity of root bark of *Calotropis procera* R.Br (Asclepediaceae). *Int. J. Pharmacol.* **2010**, *6*(6), 937–943.

Chávez-Morales, R.M.; Jaramillo-Juárez, F.; Posadas del Río, F.A. Reyes-Romero, M.A.; Rodríguez-Vázquez, M.L.; Martínez-Saldaña, M.C. Protective effect of *Ginkgo biloba* extract on liver damage by a single dose of CCl_4 in male rats. *Human Exp. Toxicol.* 2011, *30*(3), 209–216.

Cheedella, H.K.; Alluri, R.; Ghanta, K.M. Hepatoprotective and antioxidant effect of *Ecbolium viride* (Forssk.) Alston roots against paracetamol-induced hepatotoxicity in Albino Wistar rats. *J. Pharmacy Res.* **2013**, *7*, 496–501.

Chellappan, D.K.; Ganasen, S.; Btunalai, S.; Canasamy, M.; Krishnappa, P.; Dua, K.; Chellian, J.; Gupta, G. The protective action of the aqueous extract of *Auricularia polytricha* in paracetamol-induced hepatotoxicity in rats. *Recent Pat. Drug Deliv. Formul.* **2016**, *10*, 72–76.

Cheng, N.; Ren, N.; Gao, H.; Lei, X.; Zheng, J.; Cao, W. Antioxidant and hepatoprotective effects of *Schisandra chinensis* pollen extract on CCl_4-induced acute liver damage in mice. *Food Chem. Toxicol.* **2013**, *55*, 234–240.

Chibuez, U.F.; Kingssley, N.A.; Okorie, O.N.; Akugwu, A.; Sucess, N.N. Phytochemical composition of *Costus afer* extract and its alleviation of carbon tetrachloride-induced hepatic oxidative stress and toxicity. *Int. J. Mod. Bot.* **2012**, *2*, 120–126.

Chimkode, R.; Patil, M.B., Jalalpure, S.; Pasha, T.Y.; Sarkar, S. A study of hepatoprotective activity of *Hedyotis corymbosa* Linn, in albino rats. *Anc. Sci. Life*, **2009**, *28*(4), 32–35.

Chih, H.W.; Lin, C.C.; Tang, K.S. The hepatoprotective effects of Taiwan folk medicine Ham-hong-chho in rats. *Am. J. Chin. Med.* **1996**, *24*, 231–240.

Chin, J.H., Hussin A.H., Ismai, S. Anti-hepatotoxicity effect of *Orthosiphon stamineus* Benth. against acetaminophen-induced liver injury in rats by enhancing hepatic GST activity. *Pharmacogn. Res.* **2009**, *1*, 53–58.

Chinaka, O.N., Okwoche, J.O.; Dozie, N.O. The hepatoprotective effect of *Senna occidentalis* methanol leaf extract against acetaminophen-induced hepatic damage in rats. *J. Pharmacol. Toxicol.* **2011**, *6*(7), 637–646.

Chinedu, S. N.; Iheagwam, F. N.; Makinde, B. T.; Thorpe, B. O.; Emiloju, O. C. Data on *in vivo* antioxidant, hypolipidemic and hepatoprotective potential of *Thaumatococcus daniellii* (Benn.) Benth leaves. *Data Brief*, **2018**, *20*, 364–370.

Ching, L.C.; Denen, S.; Hong, Y.M.; Lin, C.C.; Shieh, D.E.; Yen, M.H. Hepatoprotctive effect of the fractions of Ban-zhi-lian on experimental liver injuries in rats. *J. Ethnopharmacol.* **1997**, *56*, 193–200.

Chiu, Y-J.; Chou, S-C.; Chiu, C-S.; Kao, C-P.; Wu, K-C.; Chen, C-J.; et al., Hepatoprotective effect of the ethanol extract of *Polygonum orientale* on carbon tetrachloride-induced acute liver injury in mice. *J. Food Drug Anal.* **2018,** *26,* 369–379.

Cho, B. O.; Yin H. H.; Fang C. Z.; Kim S. J.; Jeong S. I.; Jang S. I. Hepatoprotective effect of *Diospyros lotus* leaf extract against acetaminophen-induced acute liver injury in mice. *Food Sci. Biotech.* **2015**, *24*(6):2205–2212.

Cho, E. K.; Jung, K. I.; Choi, Y. J. Antidiabetic, alcohol metabolizing enzyme, and hepatoprotective activity of *Acer tegmentosum* Maxim. stem extracts. *J. Korean Soc. Food Sci. Nutr.*, **2015**, *44*(12), 1785–1792.

Choi, M.J.; Zheng, H.M.; Kim, J.M.; Lee, K.W.; Park, Y.H.; Lee, D.H. Protective effects of *Centella asiatica* leaf extract on dimethylnitrosamine-induced liver injury in rats. *Mol. Med. Rep.* 2016, *14*(5), 4521–4528.

Choi, M.K.; Han, J.M.; Kim, H.G.; Lee, J.S.; Lee, J.S.; Wang, J.H.; Son, S.W.; Park, H.J.; Son, C.G. Aqueous extract of *Artemisia capillaris* exerts hepatoprotective action in alcohol-pyrazole-fed rat model. *J. Ethnopharmacol.* 2013, *147*(3), 662–670.

Choi, U.K.; Lee, O.H.; Yim, J.H.; Cho, C.W.; Rhee, Y.K.; Lim, S.I.; et al., Hypolipidemic and antioxidant effects of dandelion (*Taraxacum officinale*) root and leaf on cholesterol-fed rabbits. *Int. J. Mol. Sci.* **2010**, *11*, 67–78.

Chothani D.L.; Vaghasiya, H.U. A review on *Balanites aegyptiaca* Del (desert date): Phytochemical constituents, traditional uses, and pharmacological activity. *Pharmacogn. Rev.* **2011**, 5(9), 55–62.

Christiana, A. J.; Saraswathy, G. R.; Robert, S. J.; Kothai, R.; Chidambaranathan, N.; Nalini, G.; Therasal R. L. Inhibition of CCl_4 induced liver fibrosis by *Piper longum* Linn. *Phytomedicine*, **2006**, *13*, 196–198.

Chumbale, D.S.; Upasani, C.D. Hepatoprotective and antioxidant activity of *Thespesia lampas* (Cav.) Dalz & Gibs. *Phytopharmacology* **2012**, *2*(1), 114–122.

Coballase-Urrutia, E.; Pedraza-Chaverri, J.; Cardenas-Rodriguez, N.; Huerta-Gertrudis, B.; Garcia-Cruz, M.E.; Ramirez-Morales, A.; et al., Hepatoprotective effect of acetonic and methanolic extracts of *Heterotheca inuloides* against CCl_4-induced toxicity in rats. *Exp. Toxicol. Pathol.* **2011**, *63*, 363–370.

Colak, E.; Ustuner, M.C.; Tekin, N.; Colark, E.; Burukoglu, D.;, Degirmenci, I.; Gunes, H.V. The hepatocurative effects of *Cynara scolymus* L. leaf extract on carbon tetrachloride-induced oxidative stress and hepatic injury in rats. *Springer Plus* **2016**, *5*, 1–9.

Colle, D.; Arantes, L.P.; Gubert, P.; da Luz, S.C.; Athayde, M.L.; Teixeira Rocha, J.B.; Soares, F.A. Antioxidant properties of *Taraxacum officinale* leaf extract are involved in the protective effect against hepatoxicity induced by acetaminophen in mice. *J. Med. Food.* 2012 *15*(6), 549–556.

Correia, H.S.; Batista, M.T.; Dinis, T.C. The activity of an extract and fraction of *Agrimonia eupatoria* L. against reactive species. *Biofactors* **2007**, *29*, 91–104.

Daba, M.H.; Abdel-Rahman, M.S. Hepatoprotective activity of Thymoquinone in isolated rat hepatocytes. *Toxicol. Lett.* **1998**, *95*, 23–29.

Dadkhah, A.; Fatemi, F.; Farsani, M. E.; Roshanaei, K.; Alipour, M.; Aligolzadeh, H. Hepatoprotective effects of Iranian *Hypericum scabrum* essential oils against oxidative stress induced by acetaminophen in rats. *Braz. Arch. Biol. Technol.*, **2014**, *57*(3), 340–348.

Dadzeasah, P.E.A. Safety evaluation and hepatoprotective activity of the aqueous stem bark extract of *Spathodea campanulata* Kumasi: A thesis submitted Kwame Nkrumah University of Science and Technology. **2012**.

Dahanukar SA, Kulkarni A, Rege NN. Pharmacology of medicinal plants and natural products. *Indian J. Pharmacol.* **2000,** *32*, 81–118.

Dahiru, D.; Amos, D.; Sambo, S.H. Effect of ethanol extract of *Calotropis procera* root bark on carbon tetrachloride-induced hepatonephrotoxicity in female rats. *Jordan J. Biol. Sci.* **2013**, *6*(3), 227–230.

Dahiru, D.; Mamman, D.N.; Wakawa, H.Y. *Ziziphus mauritiana* fruit extract inhibits carbon tetrachloride-induced hepatotoxicity in male rats. *Pak. J. Nutr.* **2010**, *9*, 990–993.

Dahiru, D.; William, E.T.; Nadro, M.S. Protective effect of *Ziziphus mauritiana* leaf extract on carbon tetrachloride-induced liver injury. *Afr. J. Biotechnol.* **2005**, *4*(10), 1177–1179.

Dani, C.; Oliboni, L. S.; Pasquali, M. A.; Oliveira, M. R.; Umezu, F. M.; Salvador, M.; Henriques, J.A. Intake of purple grape juice as a hepatoprotective agent in Wistar rats. *J. Med. Food*, **2008**, *11*(1), 127–132.

Danladi, J.; Abayomi, K.B.; Mairiga, A.A.; Dahiru, A.U. Comparative study of the hepatoprotective effect of ethanolic extract of *Telfairia occidentalis* (Ugu) leaves and Silymarin on paracetamol-induced liver damage in Wistar rats. *Int. J. Anim. Veter. Adva.* **2012**, *4*, 235–239.

Danladi, J.; Abdulsalam, A.; Timbuak, J.A.; Ahmed, S.A.; Aa, M.; Dahiru, A.U. Hepatoprotective effect of black seed (*Nigella sativa*) oil on carbon tetrachloride (CCl_4) induced liver toxicity in adult Wistar rats. *IOSR J. Dent. Med. Sci.* **2013**, *4*(3), 56–62.

Dar, A.I.; Saxena, R.C.; Bansal, S.K. Assessment of hepatoprotective activity of fruit pulp of *Feronia limonia* (Linn.) against paracetamol induced hepatotoxicity in albino rats. *J. Nat. Prod. Plant Res.* **2012**, *2*(2): 226–233.

Daryoush, M.; Bahram, A.T.; Yousef, D.; Mehrdad, N. Protective effect of turnip root (*Brassica rapa* L.) ethanolic extract on early hepatic injury in alloxanized diabetic rats. *Austral. J. Basic Appl. Sci.* **2011**, *5*(7), 748–756.

Das, S.; Bandyopadhyay, S.; Ramasamy, A.; Mondal, S. Evaluation of hepatoprotective activity of aqueous extracts of leaves of *Basella alba* in albino rats. *Nat. Prod. Res.* **2015**, *29*(11), 1059–1064.

Das, B.K.; Bepary, S.; Datta, B.K.; Chowdhury, A.A.; Ali, M.S.; Rouf, A.S. Hepatoprotective activity of *Phyllanthus reticulatus*. *Pak. J. Pharm. Sci.* **2008**, *21*(4), 333–337.

Das, S.K.; Mukherjee, S. Biochemical and immunological basis of Silymarin effect, a milk thistle (*Silybum marianum*) against ethanol-induced oxidative damage. *Toxicol. Mech. Meth. 2012*, **22**(5), 409–413.

Dash, D.K.; Yeligar, V.C.; Nayak, S.S.; Ghosh, T.; Rajalingam, D.; Sengupta, P.; et al., Evaluation of hepatoprotective and antioxidant activity of *Ichnocarpus frutescens* (Linn.) R. Br. on paracetamol-induced hepatotoxicity in rats. *Trop. J. Pharm. Res.* **2007,** *6*, 755–765.

da Silva, J. B.; de Freitas Mendes, R.; Tomasco, V.; Pinto, N. D. C. C.; de Oliveira, L. G.; Rodrigues, M. N.; Ribeiro, A. New aspects on the hepatoprotective potential associated with the antioxidant, hypocholesterolemic and anti-inflammatory activities of *Vernonia condensata* Baker. *J. Ethnopharmacol.,* **2017**, *198*, 399–406.

Datta, S.; Dhar, S.; Nayak, S.S.; Dinda, S.C. Hepatoprotective activity of *Cyperus articulatus* Linn. against paracetamol induced hepatotoxicity in rats. *J. Chem. Pharm. Res.* **2013**, *5*(1), 314–319.

De, S.; Suresh, R.; Babu, A.M.S.S.; Aneela, S. *In-vivo* hepatoprotective activity of methanolic extracts of *Sphaeranthus amaranthoides* and *Oldenlandia umbellata. Pharmacogn. J.* **2017**, *9*, 98–101.

Debbache-Benaida, N.; Atmani-Kilani, D.; Schini-Keirth, V. B.; Djebbli, N.; Atmani, D.

Pharmacological potential of *Populus nigra* extract as antioxidant, anti-inflammatory, cardiovascular and hepatoprotective agent. *Asian Pac. J. Trop. Biomed.*, **2013**, *3*(9), 697–704.

Deng, J.S.; Chang, Y.C.; Wen, C.L.; Liao, J.C.; Hou, W.C.; Amagaya, S.; Huang, S.S.; Huang, G.J. Hepatoprotective effect of the ethanol extract of *Vitis thunbergii* on carbon tetrachloride-induced acute hepatotoxicity in rats through anti-oxidative activities. *J. Ethnopharmacol.* **2012**, *142*(3), 795–803.

Deng, Y.; Tang, Q.; Zhang, Y.; Zhang, R.; Wei, Z.; Tang, X.; Zhang, M. Protective effect of *Momordica charantia* water extract against liver injury in restraint-stressed mice and the underlying mechanism. *Food Nutr. Res.* 2017, *61*(1), 1348864. doi: 10.1080/16546628.2017.1348864

Desai, S.N.; Patel, D.K.; Devkar, R.V.; Patel, P.V.; Ramachandran, A.V. Hepatoprotective potential of polyphenol rich extract of *Murraya koenigii* L.: An *in vivo* study. *Food Chem. Toxicol.* **2012**, *50*, 310–314.

Deshmukh, P.; Nandgude, T.; Rathode, M.S.; Midha, A.; Jaiswal, N. Hepatoprotective activity of *Calotropis gigantea* root bark experimental liver damage induced by D-galactosamine in rats. *Int. J. Pharm. Sci. Nanotechn.* **2008**, *1*(3), 281–286.

de Souza, G.R.; De-Oliveira, A.C.A.; Soares, V.; Chagas, L.F.; Barbi, N. S.; Paumgartten, F.J.R.; da Silva, A.J.R. Chemical profile, liver protective effects and analgesic properties of a *Solanum paniculatum* leaf extract. *Biomed. Pharmacother.*, **2019**, *110*, 129–138.

Devaki, T.; Shivashangari, K.S.; Ravikumar, V.; Govindaraju, P. Hepatoprotective activity *Boerhaavia diffusa* on ethanol-induced liver damage in rats. *J. Nat. Remedies*, **2004**, *4*(2), 109–115.

Devaraj, S.; Ismail, S.; Ramanathan, S.; Yam, M.F. Investigation of antioxidant and hepatoprotective activity of standardized *Curcuma xanthorrhiza* rhizome in carbon tetrachloride-induced hepatic damaged rats. *ScientificWorld J.*, **2014**, *2014*, 1–8.

Devi, B.P.; Boominathan, R.; Mandal, S.C. Anti-inflammatory, analgesic and antipyretic properties of *Clitoria ternatea* root. *Fitoterapia*, **2003**, *74*(4), 345–349.

Devipriya, D.; Gowri, S.; Nideesh, T.R. Hepatoprotective effect of *Pterocarpus marsupium* against carbontetrachloride induced damage in albino rats. *Ancient Sci. Life* **2007**, *27*(1), 19–25.

Dey, P.; Dutta, S.; Sarkar, M.P.; Chaudhuri, T.K. Assessment of hepatoprotective potential of *N. indicum* leaf on haloalkane xenobiotic-induced hepatic injury in Swiss albino mice. *Chem. Biol. Interact.* 2015, *235*, 37–46.

Dey, P.; Saha, M.R.; Sen, A. Hepatotoxicity and the present herbal hepatoprotective scenario. *Int. J. Green Pharm.* **2013**, *7*, 265–273.

Dhanabal, S.P.; Jain, R.; Priyanka, D.L.; Muruganantham, N.; Raghu, P.S. Hepatoprotective activity of *Santolina chamaecyparissus* Linn. against D-galactosamine induced hepatotoxicity in rats. *Pharmacogn. Commun.* **2012**, *2*(2), 67–70.

Dhanabal, S.P.; Syamala, G.; Satish Kumar, M.N.; Suresh, B. Hepatoprotective activity of the Indian medicinal plant *Polygala arvensis* on d-galactosamine-induced hepatic injury in rats. *Fitoterapia* **2006**, *77*(6), 472–474.

Diallo, B.; Vanhaelen-Fastre, R.; Vanhaelen, M.; Fiegel, C.; Joyeux, M.; Roland, A.; Fleurentin, J. Further studies on the hepatoprotective effects of *Cochlospermum tinctorium* rhizomes. *J. Ethnopharmacol.* **1992**, *36*(2), 137–142.

Didibe, M., Scheuring, J.F., Tembely, D., Sidibe, M.M., Hofman, P., Frigg, M. Baobab-homegrown vitamin C for Africa. *Agrofor. Today* **1996**, *8*, 13–15.

Didunyemi, M.O.; Adetuyi, B.O.; Oyebanjo, O.O. *Morinda lucida* attenuates acetaminophen-induced oxidative damage and hepatotoxicity in rats. *J. Biomed. Sci.* *8*(2), 1–7.

Dikshit, P.; Tyagi, M.K.; Shukla, K.; Sharma, S.; Gambhir, J.K.; Shukla, R.; et al., Hepatoprotective effect of stem of *Musa sapientum* Linn in rats intoxicated with carbon tetrachloride. *Ann. Hepatol.* **2011**, *10*, 333–339.

Divya, B.; Praneetha, P.; Swaroopa Rani, V.; Ravi Kumar, B. Hepatoprotective effect of whole plant extract fractions of *Marsilea minuta* Linn. *Asian J. Pharm. Clin. Res.* **2013**, *6*(3), 100–107.

D'Mello, P.; Rana, M. Hepatoprotective activity of *Psidium guajava* extract and its phospholipid complex in paracetamol-induced hepatic damage in rats. *Int. J. Phytomed.* **2010**, 2, 85–93.

Dolai, N.; Karmakar, I.; Suresh Kumar, R.B.; Kar, B.; Bala, A.; Haldar, P.K. Free radical scavenging activity of *Castanopsis indica* in mediating hepatoprotective activity of carbaon tetrachloride intoxicated rats. *Asian Pac. J. Trop. Biomed.* **2012**, *2*(1), Suppl. S243-S-251.

Domitrović, R.; Jakovac, H.; Blagojević, G. Hepatoprotective activity of berberine is mediated by inhibition of TNF-α, COX-2, and iNOS expression in CCl_4-intoxicated mice. *Toxicology* **2011**, *280*, 33–43.

Donfack, H.J.; Amadou, D.; Florence, T.N.; et al., In vitro hepatoprotective and antioxidant activities of crude extract and isolated compounds from *Ficus gnaphalocarpa. Inflammopharmacology.* **2011**, *19*, 35–43.

Donfack, H.J.; Kengap, R.T.; Ngameni, B.; Chuisseu, P.; Tchana, A.N.; Buonocore, D.; Ngadjui, B.T.; Moundipa, P.F.; Marzatico, F. *Ficus cordata* Thunb (Moraceae) is a potential source of some hepatoprotective and antioxidant compounds. *Pharmacologia* **2011**, *2*, 137–145.

Donfack, J.H., Simob, C.F.F., Ngamenic, B., Tchanaa, A.N., Kerrd, P.G., Finzie, P.V., et al., Antihepatotoxic and antioxidant activities of methanol extract and isolated compounds from *Ficus chlamydocarpa. Nat. Prod. Comm.* **2010**, *5*, 1607–1612.

Doorika, P.; Ananthi, T. Antioxidant and hepatoprotective properties of *Terminalia arjuna* bark on isoniazid induced toxicity in albino rats. *Asian J. Pharm. Tech.*, **2012**, *2*(1), 15-18.

Dubey, S.K., Batra, A. Hepatoprotective activity from ethanol fraction of *Thuja occidentalis. Asian J. Res. Chem.* **2008**, *1*, 32–38.

Duh, P.-D.; Lin, S.L.; Wu, S.C. Hepatoprotection of *Graptopetalum paraguayense* E. Walther on CCl_4-induced liver damage and inflammation. *J. Ethnopharmacol.* **2011**, *134*(2), 379–385.

Duh, P.-D.; Wang, B.-S.; Liou, S.-J.; Lin, C.-J. Cytoprotective effects of pu-erh tea on hepatotoxicity *in vitro* and *in vivo* induced by tert-butyl-hydroperoxide. *Food Chem.* **2010**, *119*(2), 580–585.

Duraiswamy, B.; Satishkumar, M.N.; Gupta, S.; Rawat, M., Murugan, O.P. Hepatoprotective activity of *Betula utilis* bark on D-galactosamne-induced hepatic insult. *World J. Pharm. Pharm. Sci.* **2012**, *1*(1), 456–471.

Dutta, S.; Chakraborty, A.K.; Dey, P.; Kar, P.; Guha, P.; Sen, S.; Kumar, A.; Sen, A.; Chaudhuri, T.K. Amelioration of CCl_4 induced liver injury in swiss albino mice by antioxidant rich leaf extract of *Croton bonplandianus* Baill. *PLoS One* **2018,** *13,* e0196411.

Dwijayanti, A.; Frethernety, A.; Hardiany, N.S.; Purwaningsih, E.H. Hepatoprotective effects of *Acalypha indica* and *Centella asiatica* in rat's liver against hypoxia. *Procedia Chem.* **2015**, *14*, 11–14.

Ebong, P.E.; Igile, G.O.; Mgbeje, B.I.A.; Iwara, I.A.; Odongo, A.E.; Onofiok, U.L.; Oso, E.A. Hypoglycemic, hepatoprotective and nephroprotective effects of methanolic leaf extract of *Heinsia crinita* (Rubiaceae) in alloxan-induced diabetic albino Wistar rats. *IOSR J. Pharm.* 2014, *4*, 37–43.

Ebrahimi, S.; Ashkani-Esfahani, S.; Emami, Y.; Riazifar, S. Hepatoprotective effect of *Ziziphus vulgaris* on carbon tetrachloride (CCl_4)-induced liver damage in rats as animal model. *Galen Med. J.* **2013**, *2*(3), 88–94.

Edagha, I. A.; Atting, I. A.; Bassey, R. B.; Bassey, E. I.; Ukpe, S.J. Erythropoietic and hepatoprotective potential of ethanolic extract of *Nauclea latifolia* in mice infected with *Plasmodium berghei berghei. Am. J. Med. Sci.*, **2014**, *2*(1), 7–12.

Eesha, B.R., Mohanbabu Amberkar, V., Meena Kumari, K., Sarath, B., Vijay, M., Lalit, M., Rajput, R. Hepatoprotective activity of *Terminalia paniculata* against paracetamol induced hepatocellular damage in Wistar albino rats. *Asian Pac. J. Trop. Med.* 2011, *4*(6), 466–469.

Effa, A.M.; Gantier, E.; Hennebelle, T.; Roumy, V.; Rivière, C.; Dimo, T.; Kamtchouing, P.; Desreumaux, P.; Dubuquoy, L. *Neoboutonia melleri* var. *velutina* Prain: *In vitro* and *in vivo* hepatoprotective effects of the aqueous stem bark extract on acute hepatitis models. *BMC Compl. Altern. Med.* 2018, *22*, 18(1):24. doi: 10.1186/s12906-018-2091-2.

Effiong, G.S.; Udoh, I.E.; Udo, N.M.; Asuquo, E.N.; Wilson, L.A.; Ntukidem, I.U.; Nwoke, I.B. Assessment of hepatoprotective and antioxidant activity of *Nauclea latifolia* leaf extract against acetaminophen induced hepatotoxicity in rats. *Int. Res. J. Plant Sci.* **2013**, *4*, 55–63.

Eid, H.H.; Labib, R.M.; Hamid, N.S.A.; Hamed, M.A.; Ross, S. A. Hepatoprotective and antioxidant polyphenols from a standardized methanolic extract of the leaves of *Liquidambar styraciflua* L. *Bull. Fac. Pharm., Cairo University*, **2015**, *53*(2), 117–127.

Eidi, A.; Moghadam, J. Z.; Mortazavi, P.; Rezazadeh, S.; Olamafar, S. Hepatoprotective effects of *Juglans regia* extract against CCl_4-induced oxidative damage in rats. *Pharm. Biol.*, **2013**, *51*(5), 558–565.

Eidi, A., Mortazavi, P., Bazargan, M., Zaringhalam, J. Hepatoprotective activity of *Cinnamon* ethanolic extract against CCl_4-induced liver injury in rats. *EXCLI J.*, **2012**, *11*, 495–507.

Eidi, A.; Mortazavi, P.; Moghadam, J.Z.; Mardani, P.M. Hepatoprotective effects of *Portulaca oleracea* extract against CCl_4-induced damage in rats. *Pharm. Biol.*, **2015**, *53*, 7, 1042–1051.

Ela, M. A. A.; El-Lakany, A. M.; Abdel-Kader, M. S.; Alqasoumi, S. I.; Shams-El-Din, S. M.; Hammoda, H. M. New quinic acid derivatives from hepatoprotective *Inula crithmoides* root extract. *Helv. Chim. Acta*, **2012**, *95*(1), 61–66.

El-Askary, H.; Handoussa, H.; Badria, F.; El-Khatib, A. H.; Alsayari, A.; Linscheid, M. W.; Motaal, A.A. Characterization of hepatoprotective metabolites from *Artemisia annua* and *Cleome droserifolia* using HPLC/PDA/ESI/MS–MS. *Rev Bras. Farmacogn.*, **2019**, *29*(2), 213–220.

El-Bakry, K.; Toson, E. S.; Serag, M.; Aboser, M. Hepatoprotective effect of *Moringa oleifera* leaves extract against carbon tetrachloride-induced liver damage in rats. *World J. Pharm. Res.*, **2016**, *5*(5), 76–89.

Elberry, A.A.; Harraz, F.M.; Ghareib, S.A.; Gabr, S.A.; Nagy, A.A.; Abdel-Sattar, E. Methanolic extract of *Marrubium vulgare* ameliorates hyperglycemia and dyslipidemia in streptozotocin-induced diabetic rats. *Int. J. Diabetes Mellit.* **2011**, *11*, 1877–1878.

Elberry, A.A.; Harraz, F.M.; Ghareib, S.A.; Nagy, A.A.; Gabr, S.A.; Suliaman, M.I.; et al., Antihepatotoxic effect of *Marrubium vulgare* and *Withania somnifera* extracts on carbon tetrachloride-induced hepatotoxicity in rats. *J. Basic Clin. Pharm.* **2010**, *1*, 247–254.

El-Beshbishy, H.A. Hepatoprotective effect of green tea (*Camellia sinensis*) extract against tamoxifen-induced liver injury in rats. *J. Biochem. Mol. Biol.* **2005**, *38*, 563–570.

El-Beshbishy, H.A. Aqueous garlic extract attenuates hepatitis andoxidative stress induced by galactosamine/

lipoploysaccharide in rats. *Phytother. Res.* **2008**, *22*, 1372–1379.

El-Beshbishy, H.A.; Mohamadin, A.M.; Nagy, A.A.; Abdel-Naim, A.B. Amelioration of tamoxifen-induced liver injury in rats by grape seed extract, black seed extract and curcumin. *Indian J. Exp. Biol.* **2010**, *48*, 280 288.

El-Gengaihi, S.; Hamed, M.; Hassan, E.; Zahran, H.; Arafa, M. Chemical composition and nutritional effect of *Physalis peruviana* husk as hepato-renal protective agent. *Int. J. Phytomed.* **2012**, *4*, 229–236.

El-Gengaihi, S.; Mossa, A.T.; Refaie, A.A.; Aboubaker, D. Hepatoprotective efficacy of *Cichorium intybus* L. extract against carbon tetrachloride-induced liver damage in rats. *J. Diet Suppl.* 2016, *13*(5), 570–584.

El-Hadary, A.E.; Elsanhoty, R.M.; Ramadan, M.F. In vivo protective effect of *Rosmarinus officinalis* oil against carbon tetrachloride (CCl_4)-induced hepatotoxicity in rats, *PharmaNutrition*, **2019**, *9*. doi: 10.1016/j.phanu.2019.100151

El-Hadary, A. E.; Hassanien, M.F.H. Hepatoprotective effect of cold-pressed *Syzygium aromaticum* oil against carbon tetrachloride (CCl_4)-induced hepatotoxicity in rats. *Pharm. Biol.*, **2016**, *54*(8), 1364–1372.

El-Hadary, A.E., Ramadan, M.F. Potential protective effect of cold-pressed *Coriandrum sativum* oil against carbon tetrachloride-induced hepatotoxicity in rats. *J. Food Biochem.* **2015**, *40*, 190–200.

Elhag, R.A.M.; Badwi, M.A.E.; Bakhiet, A.O.; Galal, M. Hepatoprotective activity of *Solanum nigrum* extracts on chemically-induced liver damage in rats. *J. Vet. Med. Animal Health,* **2011**, *3*(4), 45-50.

El-Sawi, S.A.; Sleem, A.A. Flavonoids and hepatoprotective activity of leaves of *Senna surattensis* (Burm.f.) in CCl_4 induced hepatotoxicity in rats. *Aust. J. Basic Appl. Sci.* **2010**, *4*, 1326–1334.

El-Sayed, M.G.A.; Elkomy, A.; Samer, S.; ElBanna, A.H. Hepatoprotective effect of *Pimpinella anisum* and *Foeniculum vulgare* against carbon tetrachloride induced fibrosis in rats. *World J. Pharm. Pharm. Sci.* **2015**, *4*, 78–88.

El-Sayed, A.M.; Ezzat, S.M.; Salam, M.M.; Sleem, A.A. Hepatoprotective and cytotoxic activities of *Delonix regia* flower extracts. *Phcog. J.* **2011**, *3*(19), 49–56.

Elshazly, M. O.; Morgan, A. M.; Ali, M. E.; Abdel-mawla, E.; El-Rahman, S. S. A. The mitigative effect of *Raphanus sativus* oil on chromium-induced geno- and hepatotoxicity in male rats. *J. Adv. Res.* **2016**, *7*(3), 413–421.

El Sohafy, S.M.; Metwally, A.M.; Omar, A.A.; Amer, M.E.; Radwan, M.M.; Abdel-Kader, M.S.; ElSohly, M.A. Cornigerin, a new sesqui-lignan from the hepatoprotective fractions of *Cynara cornigera* L. *Fitoterapia*, **2016**, *115*, 101–105.

Enogieru, A.B.; Charles, Y.O.; Omoruyi, S.I.; Momodu, O.I. *Phyllanthus amarus*: A hepatoprotective agent in acetaminophen induced liver toxicity in adult wistar rats. *SMU Med. J.* **2015**, *2*(1), 150–165.

Esmaeili, M.A.; Alilou, M. Naringenin attenuates CCl_4-induced hepatic inflammation by the activation of an Nrf2-mediated pathway in rats. *Clin. Exp. Pharmacol. Physiol.* **2014**, *41*(6), 416–422.

Etim, O.E.; Farombi, E.O.; Usoh, I.F.; Akpan, E.J. The protective effect of *Aloe vera* juice on lindane induced hepatotoxicity and genotoxicity. *Pak. J. Pharm. Sci.* **2006**, *19*, 333–337.

Etuk, E.U.; Francis, U.U.; Garba, I. Regenerative action of *Cochlospermum tinctorium* aqueous root extract on experimentally-induced hepatic damage in rats. *African J. Biochem. Res.* **2009**, *1*, 1– 4.

Evans, W.C. 1996. An overview of drugs having antihepatotoxic and oral hypoglycaemic activities. In: Trease and Evans, Pharmacognosy. 14th ed. U.K: W.D. Sanders Company Ltd.

Ezejindu, D.; Chinweife, K.; Ihentuge, C. The effects of moringa extract on liver enzymes of carbon tetrachloride induced hepatotoxicity in adult Wister rats. *Int. J. Eng. Sci.* **2013,** *2*, 54–59.

Ezzat, S.M.; Salama, M.M.; Seif el-Din, S.H.; Saleh, S.; El-Lakkany, N.M.; Hammam, O.A.; Salem, M.B.; Botros, S.S. Metabolic profile and hepatoprotective activity of the anthocyanin-rich extract of *Hibiscus sabdariffa* calyces. *Pharm. Biol.* **2016**. *54*, 3172–3181.

Fakurazi, S.; Hairuszah, I.; Nanthini, U. *Moringa oleifera* Lam. prevents acetaminophen induced liver injury through restoration of glutathione level. *Food Chem. Toxicol.*, **2008**, *46*(8): 2611–2615.

Fakurazi, S.; Sharifudin, S.A.; Arulselvan, P. *Moringa oleifera* hydroethanolic extracts effectively alleviate acetaminophen-induced hepatotoxicity in experimental rats through their antioxidant nature. *Molecules*. **2012**, *17*, 8334–8350.

Fallah, H.H.; Zareei Mahmoudabady, A.; Ziai, S.A.; Mehrazma, M.; Alavian, S.M.; Mehdizadeh, M.; et al., 2011. The effects of *Cynara scolymus* L. leaf and *Cichorium intybus* L. root extracts on carbon tetrachloride induced liver toxicity in rats. *J. Med. Plants* **2011**, *10*, 33–40.

Fallah, H.H.; Zarrei, M.; Ziai, M.; Mehrazma, M.; Alavian, S.M.; Kianbakht, S.; et al., The effects of *Taraxacum officinale* L. and *Berberis vulgaris* L. root extracts on carbon tetrachloride induced liver toxicity in rats. *J. Med. Plants* **2010**, *9*, 45–52.

Famurewa, A.C.; Kanu, S.C.; Uzoegwu, P.N.; Ogugua, V.N. Ameliorative effects of *Hibiscus sabdariffa* extract against carbon tetrachloride-induced lipid peroxidation, oxidative stress and hepatic damage in rats. *J. Pharm. Biomed. Sci.* **2015**, *5*, 877–883.

Fan, J.; Li X.; Li P.; et al., Saikosaponin-d attenuates the development of liver fibrosis by preventing hepatocyte injury. *Biochem. Cell Biol.* **2007**, *85*(2), 189–195.

Fan J. L.; Wu Z. W.; Zhao T. H., Sun, Y.; Ye, H.; Xu, R.; Zeng, X. Characterization, antioxidant and hepatoprotective activities of polysaccharides from *Ilex latifolia* Thunb. *Carbohydr. Polym.* **2014**, *101*(1), 990–997.

Faremi, T. Y.; Suru, S. M.; Fafunso, M. A.; Obioha, U. E. Hepatoprotective potentials of *Phyllanthus amarus* against ethanol-induced oxidative stress in rats. *Food Chem. Toxicol.*, **2008**, *46*(8), 2658–2664.

Farghali, H.; Canová, N.K.; Zakhari, S. Hepatoprotective properties of extensively studied medicinal plant active constituents: Possible common mechanisms. *Pharm. Biol.* 2015, *53*(6), 781–791.

Farkhondeh, T.; Samarghandian, S. The therapeutic effects of *Portulaca oleracea* L. in hepatogastric disorders. *Gastroenterol Hepatol.*, **2019**, *42*(2), 127–132.

Farombi, E.O.; Shrotriya, S.; Na, H.K.; Kim, S.H.; Surh, Y.J. Curcumin attenuates dimethylnitrosamine-induced liver injury in rats through Nrf2-mediated induction of heme oxygenase-1. *Food Chem. Toxicol.* **2008**, *46*(4), 1279–1287.

Farshori, N.N.; Al-Sheddi, E.S.; Al-Oqail, M.M.; Hassan, W.H.; Al-Khedhairy, A.A.; Musarrat, J.; Siddiqui, M.A. Hepatoprotective potential of *Lavandula coronopifolia* extracts against ethanol induced oxidative stress-mediated cytotoxicity in HepG2 cells. *Toxicol Ind. Health*, **2015**, *31*(8), 727–737.

Fatehi, M.; Saleh, T.M.; Fatehi-Hassanabad, Z.; Farrokhfal, K.; Jafarzadeh, M.; Davodi, S.; A pharmacological study on *Berberis vulgaris* fruit extract. *J. Ethnopharmacol.* **2005**, *102*, 46–52.

Fattorusso, E.; Iorizzi, M.; Lanzotti, V.; Taglialatela-Scafati, O. Chemical composition of shallot (*Allium ascalonicum* Hort.). *J. Agric. Food Chem.* **2002**, *50*, 5686–5690.

Fazal, S.S.; Singla, R.K. 2012. Review on the pharmacognostical and pharmacological characterization of *Apium graveolens* Linn. *Indo Glob. J. Pharm. Sci.* **2012**, 2, 36–42.

Feng, Y.; Wang, N.; Ye, X. Hepatoprotective effect and its possible mechanism of coptidis rhizoma aqueous extract on carbon tetrachloride-induced chronic liver hepatotoxicity in rats. *J. Ethnopharmacol.* **2011**, *138*(3), 683–690.

Fernando, C.D.; Soysa, P. Total phenolic, flavonoid contents, *in-vitro* antioxidant activities and hepatoprotective effect of aqueous leaf extract of *Atalantia ceylanica. BMC Complement Altern Med.* **2016**, *14*, 395. doi: 10.1186/1472-6882-14-395.

Fernando, C.D., Soysa, P. Evaluation of hepatoprotective activity of *Eriocaulon quinquangulare in vitro* using porcine liver slices against ethanol induced liver toxicity and free radical scavenging capacity. *BMC Complement. Altern. Med.* **2016**, *16*, 74. doi: 10.1186/s12906-016-1044-x.

Firdous, S.M.; Sravanthi, K.; Debnath, R.; Neeraja, K. Protective effect of ethanolic extract and its ethyl acetate and *n*-butanol fractions of *Sechium edule* fruits against carbon tetrachloride induced hepatic injury in rats. *Int. J. Pharm. Pharm. Sci.* **2012**, *4*, 354–359.

Fleurentin, J.; Hoefler, C.; Lexa, A.; Mortier, F.; Pelt, J.M. Hepatoprotective properties of *Crepis rueppellii* and *Anisotes trisulcus*: Two traditional medicinal plants of Yemen. *J. Ethnopharmacol.* **1986**, *16*(1), 105–111.

Formisano, C.; Rigano, D.; Senatore, F.; Bruno, M.; Rosselli, S. Volatile constituents of the aerial parts of white salsify (*Tragopogon porrifolius* L., Asteraceae). *Nat. Prod. Res.* **2010**, *24*, 663–668.

Forouzandeh, H.; Azemi, M.E.; Rashidi, I.; Goudarzi, M.; Kalantari, H. Study of the protective effect of *Teucrium polium* L. extract on acetaminophen-induced hepatotoxicity in mice. *Iran. J. Pharm. Res.* **2013**, *12*(1), 123–129.

Fregozo C.S.; Beltrán, M.M.; Soto, M.E.F.; Vega, M.I.P.; Rodríguez, R.Y.R.; López-Velázquez, A.L.L.; et al., Protective effect of *Rosmarinus officinalis* L. on the expression of the glutamate transporter (GLT-1) and neuronal damage in the frontal cortex of CCl_4-induced hepatic damage. *J. Med. Plants Res.* **2012**, *6*(49), 5886–5894.

Fu, R.; Zhang, Y.; Guo, Y.; Peng, T.; Chen, F. Hepatoprotection using *Brassica rapa* var. *rapa* L. seeds and its bioactive compound, sinapine thiocyanate, for CCl_4-induced liver injury. *J. Funct. Foods.*, **2016**, *22*, 73–81.

Fu, W.; Chen, J.; Cai, Y.; Lei, Y.; Chen, L.; Pei, L.; Zhou, D.; Ling, X.; Ruan, J. Antioxidant, free radical scavenging, anti-inflammatory and hepatoprotective potential of the extract from *Parathelypteris nipponica* (Franch. et Sav.) Ching. *J. Ethnopharmacol.* **2010**, *130*, 521–528.

Fursule, R. A.; Patil, S. D. Hepatoprotective and antioxidant activity of *Phaseolus trilobus* Ait. on bile duct ligation induced liver fibrosis in rats. *J. Ethnopharmacol.* **2010**, *129*(3), 416–419.

Gadgoli, C., Mishra, S.H., Antihepatotoxic activity of *Cichorium intybus*. *J. Ethnopharmacol.* **1997**, *58*, 131–134.

Galati, E.M.; Mondello, M.R.; Lauriano, E.R.; Taviano, M.F.; Galluzzo, M.; Miceli, N. *Opuntia ficus-indica* (L.) Mill. fruit juice protects liver from carbon tetrachloride induced injury. *Phytother. Res.* **2005**, *19*, 796–800.

Galisteo, M.; Suarez, A.; del Pilar Montilla, M.; del Pilar Utrilla, M.; Jiménez, J.; et al., Antihepatotoxic activity of *Rosmarinus tomentosus* in a model of acute hepatic damage induced by thioacetamid. *Phytother. Res. J.* **2000**, *14*, 522–526.

Ganapaty, S.; Ramaiah, M.; Yasaswini, K.; Kumar, C.R. Determination of total phenolic, flavonoid, alkaloidal contents and *in vitro* screening for hepatoprotective activity of *Cuscuta epithymum* (L.) whole plant against CCl_4-induced liver damage animal model. *Int. J. Pharm. Pharm. Sci.* **2013,** *5*(4), 738–742.

Gandhare, B.; Kavimani, S,; Rajkapoor, B. Protective effect of *C. pentandra* on thioacetamide-induced hepatotoxicity in rats. *Int. J. Biol. Pharm. Res.* **2012**, *3*(1), 23–29.

Ganesan, R.; Venkatanarasimhan, M.; Sharad, P.; Pramod reddy, G.; Anandan, T.; Masilamani, G. Hepatoprotective effect of *Coldenia procumbens* Linn against D-galactosamine induced acute liver damage in rats. *Int. J. Integr. Sci. Innovation Tech.* **2013**, *2*(2), 9–11.

Gani, S.M.; John, S.A. Evalution of hepatoprotective effect of *Nigella sativa* L. *Int. J. Pharm. Pharm. Sci.* **2013**, *5*, 12–19.

Gao, J.; Tang, X.; Dou, H.; Fan, Y.; Zhao, X.; Xu, Q. Hepatoprotective activity of *Terminalia catappa* L. leaves and its two triterpenoids. *J. Pharm. Pharmacol.* **2004**, *56*(11), 1449–1455.

Garba, S. H.; Prasad, J.; Sandabe, U. K. Hepatoprotective effect of the aqueous root bark extract of *Ficus sycomorus* (Linn) on carbon tetrachloride induced hepatotoxicity in rats. *J. Biol. Sci.* **2007**, *7*(2), 276–281.

Garba, H.S.; Sambo, N.; Bala, U. The effect of the aqueous extract of *Kohautia grandiflora* on paracetamol induced liver damage in albino rats. *Nigerian J. Physiol. Sci.* **2009**, *24*, 1723.

Garba M. H.; Sherifat M. L.; Abdul-Majeed A. O.; Hafsa L. M.; Awal A.B.; Sa'adu A. A.; Lekene B. J. Hepato-protective potentials of *Sterculia setigera* stem bark extract on acetominophen induced hepatotoxicity in Wistar albino rats. *J. Med. Plants Res.* **2018**, *12*(29), 557–562.

Garg, H.S.; Bhandari, S.P.S.; Tripathi, S.C.; Patnaik, G.K.; Puri, A.; Saxena, R.; Saxena, R.P. Antihepatotoxic and immunostimulant properties of iridoid glycosides of *Scrophularia koelzii. Phytother. Res.* **1994**, *8*(4), 224–228.

Gargoum, H.M.; Muftah, S.S.; Al Shalmani, S.; Mohammed, H.A.; Alzoki, A.; Debani; Al Fituri, A.H.O.; El Shari, F.; El Barassi, I.; Meghil, S.E. Abdellatif, A.G. Phytochemical screening and investigation of the effect of *Alhagi maurorum* (camel thorn) on carbon tetrachloride, acetaminophen and adriamycin induced toxicity in experimental animals. *J. Scient. Innov. Res.* **2013**, *2*(6), 1023–1033.

Gaur, K.; Nema, R.K.; Kori, M.L.; Sharma, C.S.; Singh, V. Anti-inflammatory and analgesic activity of *Balanites aegyptiaca* in experimental animal models. *Int. J. Green Pharm.* **2008**, *2*(4), 214–217.

Gayathiri, S.; Vetriselvan, S.; Jothi, S.; Ishwin, S.; Devi, H.; Kaur, S.; Yaashini, A. Hepatoprotective activity of aqueous extract of *Hippophae rhamnoides* L. in carbon tetrachloride induced hepatotoxicity in albino Wistar rats. *Int. J. Biol. Pharm. Res.* **2012**, *3*(4), 531–537.

Gbadegesin, M.A.; Adegoke, A.M.; Ewere, E.G.; Odunola, O.A. Hepatoprotective and anticlastogenic effects of ethanol extract of *Irvingia gabonensis* (IG) leaves in sodium arsenite-induced toxicity in male Wistar rats. *Niger. J. Physiol. Sci.* **2014**, *2*, 29–36.

Ge, L.; Li, J.; Wan, H.; Zhang, K.; Wu, W.; Zou, X.; Zeng, X. Novel flavonoids from *Lonicera japonica* flower buds and validation of their anti-hepatoma and hepatoprotective activity *in vitro* studies. *Ind. Crops Prod.*, **2018**, *125*, 114–122.

Gebhardt, R. Antioxidative and protective properties of extracts from leaves of the artichoke (*Cynara scolymus* L.) against hydroperoxide-induced oxidative stress in cultured rat hepatocytes. *Toxicol. Appl. Pharmacol.* **1997**, *144*, 279–286.

Geegi, P.G.; Anitha, P.; Anthoni Samy, A.; Kanimozhi, R. Hepatoprotective activity of *Vanilla planifolia* against paracetamol induced hepatotoxicity in albino rats. *Int. J. Instit. Pharm. Life Sci.* **2011**, *1*(3), 70–74.

Geetha, S.; Jaymurthy, P.; Pal, K.; Pandey, S.; Kumar, R.; Sawhney, R.C. Hepatoprotective effects of sea buckthorn (*Hippophae rhamnoides* L.) against carbon tetrachloride induced liver injury in rats. *J. Sci. Food Agric.* **2008**, *88*, 1592–1597.

George, M.; Joseph, L.; Deshwal, N.; Joseph, J. Hepatoprotective activity of different extracts of *Pterospermum acerifolium* against paracetamol induced hepatotoxicity in albino rats. *Pharm. Innov. J.* **2016**, *5*, 32–36.

Georgiev, V.G.; Weber, J.; Kneschke, E.-M.; Denev, P. N.; Bley, T.; Pavlov, A. I. Antioxidant activity and phenolic content of betalain extracts from intact plants and hairy root cultures of the red beetroot *Beta vulgaris* cv. Detroit Dark Red. *Plant Foods Hum. Nutr.* **2010**, *65*(2), 105–111.

Georgina, E.O.; Kingsley, O.; Esosa, U.S.; Helen, N.K.; Frank, A.O.; Anthony, O.C. Comparative evaluation of antioxidant effects of watermelon and orange, and their effects on some serum lipid profile of Wister albino rats. *Int. J. Nutr. Metab.* **2011**, *3*, 97–102.

Germanò, M.P.; D'Angelo, V.; Sanogo, R.; Morabito, A.; Pergolizzi, S.; De Pasquale, R. Hepatoprotective activity of *Trichilia roka* on carbon tetrachloride-induced liver damage in rats. *J. Pharm. Pharmacol.* **2001**, *53*(11), 1569–1574.

Germano, M.P.; Sanogo, R.; Costa, C.; Fulco, R.; D'angelo, V.; Torre, E.A.; Viscomi, M.G.; De Pasquale, R. Hepatoprotective properties in the rat of *Mitracarpus scaber*. *J. Pharm. Pharmacol.* **1999**, *51*(6), 729–734.

Ghaffari, H.; Ghassam, B.J.; Prakash, H.S. Hepatoprotective and cytoprotective properties of *Hyptis suaveolens* against oxidative stress-induced damage by CCl_4 and H_2O_2. *Asian Pac. J. Trop. Med.* **2012**, *2012*, 868–874.

Ghaffari, H.; Venkataramana, M.; Nayaka, S.C.; Ghassam, B.J.; Angaswamy, N.; Shekar, S.; Sampath Kumara, K.K.; Prakash, H.S. Hepatoprotective action of *Orthosiphon diffusus* (Benth.) methanol active fraction through antioxidant mechanisms: An *in vivo* and *in vitro* evaluation. *J. Ethnopharmacol.* **2013**, *149*(3), 737–744.

Ghalehkandi, J.G.; Sis, N.M.; Nobar, R.S. Anti-hepatotoxic activity of garlic (*Allium sativum*) aqueous extract compared with chromium chloride in male rats. *Austral. J. Basic Appl. Sci.* **2012**, *6*(7), 80–84.

Ghanem, M.T.; Radwan, H.M.; Mahdy, EL-SM.; Elkholy, Y.M.; Hassanein, H.D.; Shahat, A.; et al., Phenolic compounds from *Foeniculum vulgare* (subsp. *piperitum*) (Apiaceae) herb and evaluation of hepatoprotective antioxidant activity. *Pharmacogn. Res.* **2012**, *4*, 104–108.

Ghorbel, H,; Feki, I.; Friha, I.; Khabir, A.M.; Boudawara, T.; Boudawara, M.; et al., Biochemical and histological liver changes occurred after iron supplementation and possible remediation by garlic consumption. *Endocrine* **2011**, *40*, 462–471.

Ghosh, A.K.; Mitra, E.; Das, N.; Dutta, M.; Bhattacharjee, S.; et al., Protective effect of aqueous bark extract of *Terminalia arjuna* against copper-ascorbate induced oxidative stress *in vitro* in goat liver. *J. Cell Tissue Res.* **2013**, *13*, 3729–3737.

Ghosh, J.; Das, J.; Manna, P.; Sil, P.C. Protective effect of the fruits of *Terminalia arjuna* against cadmium-induced oxidant stress and hepatic cell injury via MAPK activation and mitochondria dependent pathway. *Food Chem.* **2010**, *123*, 1062–1075.

Gião MS, Pereira CI, Fonseca SC, Pintado ME, Malcata FX. 2009. Effect of particle size upon the extent of extraction of antioxidant power from the plants *Agrimonia eupatoria, Salvia* sp. and *Satureja montana*. *Food Chem.* **2009**, *117*, 412–416.

Gilani, A.H.; Jabeen, Q.; Ghayur M.N.; Janbaz, K.H.; Akhtar, M.S. Studies on the antihypertensive, antispasmodic, bronchodilator and hepatoprotective activities of the *Carum copticum* seed extract. *J. Ethnopharmacol.* **2005**. *98*, 127–135.

Gilani, A.H.; Janbaz, K.H. Protective effect of *Artemisia scoparia* extract against acetaminophen induced hepatotoxicity. *General Pharmacol.* **1993**, *24*(6), 1455–1458.

Gilani, A. H.; Janbaz, K. H. Hepatoprotective effects of *Artemisia scoparia* against carbon tetrachloride: An environmental contaminant. *J. Pak. Med. Assoc.* **1994**, *44*(3), 65–68.

Gilani, A.H.; Janbaz, K.H. Preventive and curative effects of *Artemisia absinthium* on acetaminophen and carbontetrachloride-induced hepatotoxicity. *Gen. Pharmacol.* **1995a**, *26*(2), 309–315.

Gilani, A.H.; Janbaz, K.H. Studies on protective effect of *Cyperus scariosus* extract on acetaminophen and CCl_4 induced hepatotoxicity. *Gen. Pharmacol.*, **1995b**, *26*(3), 627–631.

Gilani, A. H.; Janbaz, K. H. Effect of *Rubia cordifolia* extract on acetaminophen and CCl_4 induced hepatotoxicity. *Phytother. Res.* **1995c**, *9*(5), 372–375. doi.: 10.1002/ptr.2650090513

Gilani, A.H.; Janbaz, K.H. Preventive and curative effects of *Berberis aristata* fruit extract on paracetamol- and CCl_4-induced hepatotoxicity. *Phytother. Res* **1995d**; *9*, 489–494.

Gilani A.H.; Janbaz K.H.; Aziz N.; Herzig M.J.U.; Kazmi M.M.; Choudhary M.I.; Herzig J.W. Possible mechanism of selective inotropic activity of the n-butanolic fraction from *Berberis aristata* fruit-distinguishing hype from hope. *General Pharmacol.* **1999**, *33*(5), 407–414.

Gilani, A.H.; Janbaz, K.H.; Akhtar, M.S. Selective protective effect of an extract from *Fumaria parviflora* on paracetamol-induced hepatotoxicity. *Gen. Pharmacol.* **1996**. *27*(6), 979–983.

Gilani, A.H.; Yaeesh, S.; Jamal, Q.; Ghayur, M. Hepatoprotective activity of aqueous-methanol extract of *Artemisia vulgaris*. *Phytother. Res.* **2005**, *19*(2), 170–172.

Giri, R.K.; Bose, A.; Mishra, S.K. Hepatoprotective activity of *Tagets erecta* against carbon tetrachloride-induced hepatic damage in rats. *Acta Pol. Pharm.* **2011**, *68*(6), 999–1003.

Girish, C.; Pradhan, S.C. Indian herbal medicines in the treatment of liver diseases: Problems and promises. *Fundam. Clin. Pharmacol.* **2012**, *26*(2), 180–189.

Gnanadesigan, M.; Ravikumar, S.; Anand, M. Hepatoprotective activity of *Ceriops decandra* (Griff.) Ding Hou mangrove plant against CCl_4 induced liver damage. *J. Taibah Univ. Sci.* **2017**, *11*(3), 450–457.

Gnanadesigan, M.; Ravikumar, S.; Inbanesan, S.J. Hepatoprotective and antioxidant properties of marine halophyte *Lumnitzera racemosa* bark extract in CCl_4 induced hepatotoxicity, *Asian Pac. J. Trop. Med.* **2011**. *4*(6), 462–465.

Gnanaprakash, K.; Madhusudhana Chetty, C.; Ramkanth, S.; Alagusundaram, M.; Tiruvengadarajan, V.S.; Arameswari, S.A.; Saleem, T.S.M. Aqueous extract of *Flacourtia indica* prevents carbon tetrachloride induced hepatotoxicity in rat. *Int. J. Biol. Life Sci.* **2010**, *6*(1), 51–55.

Gnanaraj, C.; Shah, M. D.; Haque, A. E.; Makki, J. S.; Iqbal, M. Hepatoprotective and immunosuppressive effect of *Synedrella nodiflora* L. on carbon tetrachloride (CCl_4)-intoxicated rats. *J. Environ. Pathol. Toxicol. Oncol.*, **2016a**, *35*(1), 29–42.

Gnanaraj, C.; Shah, M. D.; Makki, J. S.; Iqbal, M. Hepatoprotective effects of *Flagellaria indica* are mediated through the suppression of pro-inflammatory cytokines and oxidative stress markers in rats. *Pharm. Biol.*, **2016b**, *54*(8), 1420–1433.

Gnanaraj, C.; Shah, M. D.; Song, T. T.; Iqbal, M. Hepatoprotective mechanism of *Lygodium microphyllum* (Cav.) R. Br. through ultrastructural signaling prevention against carbon tetrachloride (CCl_4)-mediated oxidative stress. *Biomed Pharmacother.*, **2017**, *92*, 1010–1022.

Gnanasekaran, D.; Umamaheswara Reddy, C.; Jayaprakash, B.; Narayanan, N.; Ravikiran, Y. In vitro hepatoprotective activity against CCl_4 induced toxicity of some selected Siddha medicinal plants. *Am. J. Pharm. Tech. Res.* **2012**, *2*(1), 466–473.

Gogoi, N.; Gogoi, A.; Neog, B.; Baruah, D.; Singh, K.D. Evaluation of antioxidant and hepatoprotective activity of fruit rind extract of *Garcinia dulcis* (Roxburgh) Kurz. *Pharmacogn. Res.* **2017**, *9*(3), 266–272.

Gole M.K.; Das Gupta, S. Role of plant metabolites in toxic liver injury. *Asia Pacific J. Clin. Nutr.*, **2002**, *11*(1), 48–50.

Gomase, P. V.; Rangari V. D.; Verma P. R. Phytochemical evaluation and hepatoprotective activity of fresh juice of young stem (tender) bark of *Azadirachta indica* A. Juss. *Int. J. Pharm. Pharm. Sci.* **2011**, *3*(Suppl 2), 55-59.

Gond, N.Y.; Khadabadi, S.S. Hepatoprotective activity of *Ficus carica* leaf extract on rifampicin-induced hepatic damage in rats. *Indian J. Pharm. Sci.* **2008**, *70*, 364–366

Gong, F.; Yin, Z.; Xu, Q.; Kang, W. Hepatoprotective effect of *Mitragyna rotundifolia* Kuntze on CCl_4 induced acute liver injury in mice. *Afr. J. Pharm. Pharmacol.* **2012**, *6*, 330–335.

González-Ponce, H.A.; Martínez-Saldaña, M.C.; Rincón-Sánchez, A.R.; Sumaya-Martínez, M.T.; Buist-Homan, M.; Faber, K.N.; Moshage, H.; Jaramillo-Juárez, F. Hepatoprotective effect of *Opuntia robusta* and *Opuntia streptacantha* fruits against acetaminophen-induced acute liver damage. *Nutrients* **2016**, *8*(10), 607, https://doi.org/10.3390/nu8100607

Gopal, N.; Sengottuvelu, S. Hepatoprotective activity of *Clerodendrum inerme* against CCl_4-induced hepatic injury in rats. *Fitoterapia* **2008**, *79*, 24–26.

Gopalakrishnan, S.B.; Kalaiarasi, T. Hepatoprotective activity studies of *Cucumis trigonus* Roxb. against rifampicin-isoniazid-induced toxicity in rats. *Eur. J. Pharm. Med. Res.* **2015**, *2*(6), 141–146.

Goryacha, O.V.; Ilyina, T.V.; Kovalyova, A.M.; Koshovyi, O.M.; Krivoruchko, O.V.; Vladimirova, I.M.; Komisarenko, A.M. A hepatoprotective activity of *Galium verum* L. extracts against carbon tetrachloride-induced injury in rats. *Der Pharm. Chemica*, **2017**, *9*(7), 80–83.

Gouri, N., Umamaheswari, M.; Asokkumar, K.; Sivashanmugam, T.; Subeesh, V. Protective effect of *Desmostachya bipinnata* against pracetamol induced hepatotoxicity in rats. *UJP* **2014**, *3*(5), 20–24.

Govind, P. Medicinal plants against liver diseases. *Int. Res. J. Pharm.* **2011**, *2*, 115–121.

Goyal, Y.K., Kumar, V. 2016. Exploration and comparison of hepatoprotective activity of aqueous extract of *Tinospora cordifolia*: An experimental study. *Int. J. Sci. Res.* **2016**, *5*(3), 707–709.

Gressner, O.A. Chocolate shake and blueberry pie or why your liver would love it. *J. Gastroenterol. Hepatol. Res.* **2012**, *1*, 171–195.

Gu, H.; Gu, X.; Xu, Q.; Kang, W. Y. Antioxidant activity in vitro and hepatoprotective effect of *Phlomis maximowiczii in vivo*. *Afr. J. Tradit. Complement. Altern. Med.* **2014**, *11*(3), 46–52.

Gujrati, V.; Patel, N.; Rao, V.N.; Nandakumar, K.; Gouda, T.S.; Shalam, M.D.; Shanta Kumar, S.M. Hepatoprotective activity of alcoholic and aqueous extracts of leaves of *Tylophora indica* (Linn.) in rats. *Indian J. Pharmacol.* **2007**, *39*, 43–47.

Gulati, R.K.; Agarwal, S.; Agarwal, S.S. Hepatoprotective studies on *Phyllanthus niruri* and quercetin. *J. Exptl. Biol.***1995**, *33*(4), 261–268.

Guo, S.; Guo, T.; Cheng, N.; Liu, Q.; Zhang, Y.; Bai, L.; Bai, N. Hepatoprotective standardized EtOH–water extract from the seeds of *Fraxinus rhynchophylla* Hance. *J. Tradit. Complement. Med.* **2017**, *7*(2), 158–164.

Gupta, A.; Sheth, N. R.; Pandey, S.; Yadav, J.S.; Shah, D.R.; Vyas, B.; Joshi, S. Evaluation of protective effect of *Butea monosperma* (Lam.) Taub in experimental hepatotoxicity in rats. *J. Pharmacol. Pharmacother.* **2012**, *3*(2), 183–185.

Gupta, A.K.; Chitme, H.; Dass, S.K.; Misra, N. Hepatoprotective activity of *Rauwolfia serpentina* rhizome in paracetamol intoxicated rats. *J. Pharmacol. Toxicol. Methods* **2006**, *1*, 82–88.

Gupta, A.K.; Chitme, H.; Dass, S.K.; Misra, N. Antioxidant activity of *Chamomile recutita* capitula methanolic extracts against CCl_4-induced liver injury in rats. *J. Pharmacol. Toxicol.* **2006**, *1*, 101–107.

Gupta, A.K.; Ganguly, P.; Majumder, U.K.; Gosal, S. Hepatoprotective and anti oxidant effect of total extracts and steroidal saponins of *Solanum xanthocarpum* and *Solanum nigrum* in paracetamol induced hepatotoxicity in rats. *Pharmacologyonline* **2009**, *1*, 757–768.

Gupta, A.K.; Misra, N. Hepatoprotective activity of aqueous ethanolic extract of *Chamomile capitula* in paracetamol intoxicated albino rats. *Am. J. Pharmacol. Toxicol.* **2006**, *1*(1): 17–20.

Gupta, M.; Mazumder, U.K.; Sambath Kumar, R. Hepatoprotective and antioxidant role of *Caesalpinia bonducella* on paracetamol induced liver damage in rats. *Nat. Prod. Sci.* **2003**, *9*, 186–191.

Gupta, M.; Mazumder, K.U.; Kumar, S.T.; Gomathi, P.; Kumar, R.S. Antioxidant and hepatoprotective effects of *Bauhinia racemosa* against paracetamol and carbontetrachloride induced liver damage in rats. *Iranian J. Pharmacol. Ther.*, **2004**, *3*, 12–20.

Gupta, M.; Mazumder, U.K.; Kumar, R.S.; Sivakumar, T.; Gomathi, P. Antioxidant and protective effects of *Ervatamia coronaria* Stapf., leaves against carbon tetrachloride-induced liver injury. *Eur. Bull. Drug Res.* **2004**, *12*, 13–22.

Gupta, M.; Mazumder, U.K.; Thamil selvan, V.; Manikandan, L.; Senthil kumar, G.P.; Suresh, K.; Kakoti, B.K. Potential hepatoprotective effect and antioxidant role of methanol extract of *Oldenlandia umbellata* in carbontetrachloride induced hepatotoxicity in Wistar rats. *Iranian J. Pharamacol. Ther.* **2007**, *6*, 5–9.

Gupta, N.K.; Dixit, V.K. Evaluation of hepatoprotective activity of *Cleome viscosa* Linn. extract. *Indian J. Pharmacol.*, **2009a**, *41*, 36–40.

Gupta, N.K., Dixit, V.K. Hepatoprotective activity of *Cleome viscosa* Linn. extract against thioacetamide-induced hepatotoxicity in rats. *Nat. Prod. Res.* **2009b**, *23*(14), 1289–1297.

Gupta, R.; Gupta, A.; Singh, R.L. Hepatoprotective activities of Triphala and its constituents. *Int. J. Pharm. Res. Rev.* **2015**, *4*, 34–55.

Gupta, R.K.; Hussain, T.; Panigrahi, G.; Das, A.; Singh, G.N.; Sweety, K.; Faiyazuddin, Md.; Rao, C.V. Hepatoprotective effect of *Solanum xanthocarpum* fruit extract against CCl_4-induced acute liver toxicity in experimental animals. *Asian Pac. J. Trop. Med.* **2011**, *4*(12), 964–968.

Gutiérrez-Rebolledo, G.A.; Siordia-Reyes, G.A.; Meckes-Fischer, M.; Jiménez-Arellanes, A. Hepatoprotective properties of oleanolic and ursolic acids in anti-tubercular drug-induced liver damage. *Asian Pac. J. Trop. Med.* **2016**, *9*(7), 644–651.

Gutiérrez R.M.P.; Solís, R.V. Hepatoprotective and inhibition of oxidative stress in liver of *Prostechea michuacana*. *Rec. Nat. Prod.* **2009**, *3*, 46–51.

Gyamfi, M.A.; Yonamine, M.; Aniya, Y. Free-radical scavenging action of medicinal herbs from Ghana: *Thonningia sanguinea* on experimentally-induced liver injuries. *Gen. Pharmacol.*, **1999**, *32*(6), 661–667.

Ha, K.T.; Yoon, S.J.; Choi, D.Y.; Kim, D.W.; Kim, J.K.; Kim, C.H. Protective effect of *Lycium chinese* fruit on carbon tetrachloride-induced hepatotoxicity. *J. Ethnopharmacol.* **2005**, *96*, 529–535.

Habib, M.; Waheed, I. Evaluation of anti-nociceptive, anti-inflammatory and antipyretic activities of *Artemisia scoparia* hydromethanolic extract. *J. Ethnopharmacol.* **2013**, *145*(1), 18–24.

Hadaruga, D.I.; Hadaruga, N.G.; Bandur, G.N.; Rivis, A.; Costescu, C.; Ordodi, V.L.; et al., *Berberis vulgaris* extract/β cyclodextrin nanoparticles synthesis and characterization. *Rev. Chim.* (Bucharest) **2010**, *61*, 669–675.

Hafez, M.M.; Hamed, S.S.; El-Khadragy, M.F.; Hassan, Z.K.; Al Rejaie, S.S.; Sayed-Ahmed, M.M. Effect of ginseng extract on the TGF-β1 signaling pathway in CCl_4-induced liver fibrosis in rats. *BMC Complement. Altern. Med.* **2017**, *17*, 45. doi: 10.1186/s12906-016-1507-0.

Haidry, M.T.; Malik, A. Hepatoprotective and antioxidative effects of *Terminalia arjuna* against cadmium provoked toxicity in Albino rats (*Ratus norvigicus*). *Biochem Pharmacol,* **2014**, *3*, 1–4.

Halder, P.K.; Bera, S.; Panda, S.P.; Bhattacharya, S.; Chaulya, N.C.; Mukherjee, A. Hepatoprotective activity of *Cyperus tegetum* rhizome against paracetamol-induced liver damage in rats. *J. Complement. Integr. Med.* **2011**. *8*. doi: 10.2202/1553-3840.1398.

Halder, P.K.; Gupta, M.; Mazumder, U.K.; Kander, C.C.; Mankandan, L. Hepatoprotective effect of *Wedelia calendulacea* against thioacetamide induced liver damage in rats. *Pharmacologyonline* **2007**, *3*, 414–421.

Haldar, P.K.; Adhikari, S.; Bera, S.; Bhattacharya, S.; Panda, S.P.; Kandar, C.C. Hepatoprotective efficacy of *Swietenia mahagoni* (L.) Jacq. (Meliaceae) bark against paracetamol-induced hepatic damage in rats. *Ind. J. Pharm. Edu. Res.,* **2011**, *45*(2).

Haleagrahara, N.; Jackie, T.; Chakravarthi, S.; Rao, M.; Kulur, A. Protective effect of *Etlingera elatior* (torch ginger) extract on lead acetate-induced hepatotoxicity in rats. *J. Toxicol. Sci.* **2010**, *35*, 663–671.

Hamburger, M.; Adler, S.; Baumann, D.; Förg, A.; Weinreich, B. Preparative purification of the major anti-inflammatory triterpenoid esters from Marigold (*Calendula officinalis*). *Fitoterapia* **2003**, *74*, 328–338.

Hamed, A. N.; Wahid, A. (2015). Hepatoprotective activity of *Borago officinalis* extract against CCl_4-induced hepatotoxicity in rats. *J. Nat. Prod.* **2015**, *8*, 113–122.

Hamed, S.S.; Al-Yhya, N.A.; El-Khadragy, M.F.; Al-Olayan, E.M.; Alajmi, R.A.; Hassan, Z.K. The protective properties of the strawberry (*Fragaria ananassa*) against carbon tetrachloride-induced hepatotoxicity in rats mediated by anti-apoptotic and up-regulation of antioxidant genes expression effects. *Front. Physiol.* **2016**, 7.

Hamenoo, N.A. Hepatoprotective and toxicological assessment of *Spondias mombin* L. (Anacardiaceae) in rodents. MSc. Thesis submitted to Kwame Nkrumah University of Sci & Tech **2010**, pp. 15–17.

Hamzah, R.; Agboola, A.; Busari, M.; Omogu, E.; Umar, M.; Abubakar, A. Evaluation of hepatoprotective effect of methanol extract of *Solanum melongena* on carbon tetrachloride induced hepatotoxic rats. *Eur. J. Med. Plants.* **2016**, *13*, 1–12.

Han, K.-H.; Hashimoto, N.; Shimada, K.; Sekikawa, M.; Noda, T.; Yamauchi, H.; Hashimoto, M.; Chiji, H., Topping, D.L.; Fukushima, M. Hepatoprotective effect of purple potato extract against D-galactosamine-induced liver injury in rats. *Biosci. Biotechnol. Biochem.* **2006**, *70*(6), 1432–1437.

Han, W.; Wu, D.; Lu, Y.; Wang, L.; Hong, G.; Qiu, Q.; et al., Curcumin alleviated liver oxidative stress injury of rat induced by paraquat. *Zhonghua Lao Dong Wei Sheng Zhi Ye Bing Za Zhi* **2014**, *32*(5), 352–356.

Han X. H., Gai X. D., Xue Y. J., Chen M. Effects of the extracts from *Bupleurum chinese* DC. on intracelluar free calcium concentration and vincristine accumulation in human hepatoma BEL-7402 cells. *Tumor*. **2006**, 26, 314–317.

Hanafy, A.; Aldawsari, H.M.; Badr, J.M.; Ibrahim, A.K.; Abdel-Hady, S.E.-S. 2016. Evaluation of hepatoprotective activity of *Adansonia digitata* extract on acetaminophen-induced hepatotoxicity in rats. *Evid. Based Complement. Altern. Med.* **2016**. doi.: 10.1155/2016/4579149

Haque, R.; Mohdal, S.; Sinha, S.; Roy, M.G.; Sinha, D.; Gupta, S. Hepatoprotective activity of *Clerodendron inerme* against paracetamol induced hepatic injury in rats for pharmaceutical product. *Int. J. Drug Dev. Res.* **2011**, *3*(1), 118–126.

Harish, R.; Shivanandappa, H.T. Antioxidant activity and hepatoprotective potential of *Phyllanthus niruri. Food Chem.* **2006,** *95*, 180–185.

Harish, R.; Shivanandappa, H.T. Hepatoprotective potential of *Decalepis hamiltonii* (Wight & Arn) against carbon tetrachloride-induced hepatic damage in rats. *Pharm. Bioallied Sci.* **2010,** *2*(4), 341–345.

Hase, K.; Li, J.; Basnet, P.; Xiong, Q.; Takamura, S.; Namba, T.; Kadota, S. Hepatoprotective principles of *Swertia japonica* on galactosamine/lipopolysaccharide-induced liver injury in mice. *Chem. Pharm. Bull.*, **1997**, *45*(11), 1823–1827.

Hashem, M.M.; Salama, M.M.; Mohammed, F.F.; Tohamy, A.F.; El Deeb, K.S. Metabolic profile and hepatoprotective effect of *Aeschynomene elaphroxylon* (Guill. & Perr.). *PLoS ONE* **2019**, *14*(1), e0210576. https://doi.org/10.1371/journal.pone.0210576

Hassan, H.A.; Yousef, M.I. Ameliorating effect of chicory (*Cichorium intybus* L.)-supplemented diet against nitrosamine precursors-induced liver injury and oxidative stress in male rats. *Food Chem. Toxicol.* **2010**, *48*, 2163–2169.

Hassan, S.W.; Salawu, K.; Ladan, M.J.; Hassan, L.G.; Umar, R.A.; Fatihu, M.Y. Hepatoprotective, antioxidant and phytochemical properties of leaf extracts of *Newbouldia leavies*. *Int. J. Pharm. Tech. Res.* **2010**, *2*, 573–584.

Hassan, S.W.; Umar, R.A.; Ebbo, A.A.; Akpeji, A.J.; Matazu, I.K. Hepatoprotective effect of leaf extracts of *Parkinsonia aculeata* L. against CCl_4 intoxication in albino rats. *Int. J. Biol. Chem.* **2008**, *2*(2), 42–48.

Hazra, B.; Sarkar R.; Mandal, N. *Spondias pinnata* stem bark extract lessens iron overloaded liver toxicity due to hemosiderosis in Swiss albino mice. *Ann. Hepatol.* **2013**, *12*(1), 123–129.

He, Q.; Li, Y.; Liu, J.; Zhang, P.; Yan, S.; He, X.; Zhang, A. Hepatoprotective activity of *Lophatherum gracile* leaves ethanol extract against carbon tetrachloride-induced liver damage in mice. *Int. J. Pharmacol.* **2016,** *12*(4), 387–393.

Hegde, K.; Deepak, T.K.; Kabitha, K.K. Hepatoprotective potential of hydroalcoholic extract of *Santalum album* Linn. leaves. *Int. J. Pharm. Sci. Drug Res.* **2014**, *6*(3), 224–228.

Hegde, K.; Joshi, A.B. Hepatoprotective effect of *Carissa carandas* Linn. root extract against CCl_4 and paracetamol induced hepatic oxidative stress. *Indian J. Exp. Biol.* **2009**, *47*(8), 660–667.

Hegde, K.; Joshi, A.B. Hepatoprotective and antioxidant effect of *Carissa spinarum* root extract against CCl_4 and paracetamol induced hepatic damage in rats. *Bangladesh J. Pharmacol.* **2010**, *5*, 73–76.

Heibatollah, S.; Reza, N.M.; Izadpanah, G.; Sohailla, S. Hepatoprotective effect of *Cichorium intybus* on CCl_4 induced liver damage in rats. *Afr. J. Biochem. Res.,* **2008**, *2*(6), 141–144.

Heidarian, E.; Saffari, J.; Jafari-Dehkordi, E. Hepatoprotective action of *Echinophora platyloba* DC. leaves against acute toxicity of acetaminophen in rats. *J. Diet Suppl.* **2014,** *11*(1), 53–63.

Hem, K.; Singh, N.K.; Singh, M.K. Anti-inflammatory and hepatoprotective activities of the roots of *Uraria picta. Int. J. Green Pharm.* **2017** (Suppl), *11*(1), S166-S173.

Hemabarathy, B.; Budin, S.B.; Feizal, V. Paracetamol hepatoxicity in rats treated with crude extract of *Alpinia galanga. J. Biol. Sci.* **2009**, *9*(1), 57–62.

Hemalatha, R. Anti-hepatotoxic and anti-oxidant defense potential of *Tridax procumbens. Int. J. Green Pharm.* **2010**, *2*(3), 164–169.

Hemamalini, K,; Ramya Krishna, V.; Anurag Bhargav, D.R.; Uma Vasireddy, D.R. Hepatoprotective activity of *Tabebuia rosea* and *Solanum pubescens* against paracetamol induced hepatotoxicity in rats. *Asian J. Pharm. Clin. Res.* **2012**, *5*(4), 153–156.

Hena; Tiwari, P.; Srivastava, M.; Ghoshal, S. Hepatoprotective and histopathological activity of ethanol and aqueous extracts of stem of *Aloe vera* Linn. (Ghee gangwar) against paracetamol-induced liver damage in rats. *Int. J. Pharm. Bio Sci.* **2016**, 1, 1–7.

Hennebelle, T.; Weniger, B.; Joseph, H.; Sahpaz, S.; Bailleul, F. *Senna alata*. *Fitoterapia*. **2009**, *80*, 385–393.

Hermenean, A.; Ardelean, A.; Stan, M.; Hadaruga, N.; Mihali, C.V.; Costache, M.; et al., Antioxidant and hepatoprotective effects of naringenin and its b-cyclodextrin formulation in mice intoxicated with carbon tetrachloride: A comparative study. *J. Med. Food* **2014**, *17*(6): 670–677.

Hermenean, A.; Popescu, C.; Ardelean, A.; Stan, M.; Hadaruga, N.; Mihali, C.V.; et al., Hepatoprotective effects of *Berberis vulgaris* L. extract/β cyclodextrin on carbon tetrachloride-induced acute toxicity in mice. *Int. J. Mol. Sci.* **2012**, *13*, 9014–9034.

Hewawasam, R.P.; Jayatilaka, K.A.; Pathirana, C.; Mudduwa, L.K. Protective effect of *Asteracantha longifolia* extract in mouse liver injury induced by carbon tetrachloride and paracetamol. *J. Pharm. Pharmacol.* **2003**, *55*(10), 1413–1418.

Hewawasam, R.P.; Jayatilaka, K.A.P.W.; Pathirana, C.; Mudduwa, L.K.B. Hepatoprotective effect of *Epaltes divaricata* extract on carbon tetrachloride-induced hepatotoxicity in mice. *Indian J. Med. Res.*, **2004**, *120*, 30–34.

Hfaiedh, M.; Brahmi, D.; Zourgui, L. Hepatoprotective effect of *Taraxacum officinale* leaf extract on sodium dichromate-induced liver injury in rats. *Environ. Toxicol.* **2016**, *31*(3), 339–349.

Himaja, N. Comparative study of hepatoprotective activity of *Acanthospermum hispidum* plant extract and herbal niosomal suspension against anti-tubercular drug induced hepatotoxicity in rats. *Asian J. Pharm. Clin. Res.* **2015**, *8*(5), 256–259.

Ho, W.Y.; Yeap, S.K.; Ho, C.L.; Abdul Rahim, R.; Alitheen, N.B. Hepatoprotective activity of *Elephantopus scaber* on alcohol-induced liver damage in mice. *Evid. Based Complement. Altern. Med.* **2012**, *2012*, 8.

Hoefler, C.; Fleurentin, J.; Mortier, F.; Pelt, J.; Guillemain, J. Comparative choleretic and hepatoprotective properties of young sprouts and total plant extracts of *Rosmarinus officinalis* in rats. *J. Ethnopharmacol.* **1987**, *19*(2), 133–143.

Hogade, M.G.; Patil, K.S.; Wadkar, G.H., et al., Hepatoprotective activity of *Morus alba* (Linn.) leaves extract against carbon tetrachloride induced hepatotoxicity in rats. *Afr. J. Pharm. Pharmacol.* **2010**, *4*, 731–734.

Hong, J.; Kim, S.; Kim, H-S. Hepatoprotective effects of soybean embryo by enhancing adiponectin-mediated AMP-activated protein kinase α pathway in high-fat and high-cholesterol diet-induced nonalcoholic fatty liver disease. *J. Med. Food* **2016**, *19*, 549–559.

Hong, Z.; Chen, W.; Zhao, J.; Wu, Z.; Zhou, J. H.; Li, T.; Hu J. Hepatoprotective effects of *Rubus aleaefolius* Poir. and identification of its active constituents. *J. Ethnopharmacol.* **2010**, *129*(2), 267– 272.

Honmore, V.; Kandhare, A.; Zanwar, A.A.; Rojatkar, S.; Bodhankar, S.; Natu, A. *Artemisia pallens* alleviates acetaminophen induced toxicity via modulation of endogenous biomarkers. *Pharm Biol.* **2015**, *53*(4), 571–581.

Hosbas, S.; Hartevioglu, A.; Pekcan, M.; Orhan, D. D. Assessment of hepatoprotective activity of *Achillea biebersteinii* ethanol extract on carbon tetrachloride-induced liver damage in rats. *FABAD J. Pharm. Sci.* **2011**, *36*(1), 33–39.

Hossain, M. S.; Ahmed, M.; Islam, A. Hypolipidemic and hepatoprotective effects of different fractions of methanolic extract of *Momordica charantia* Linn. in alloxan induced diabetic rats. *Int. J. Pharm.. Sci. Res.*, **2011**, *2*(3), 601.

Houcher, Z.; Boudiaf, K.; Benboubetra, M.; Houcher. B. Effects of methanolic extract and commercial oil of *Nigella sativa* L. on blood glucose and antioxidant capacity in alloxan-induced diabetic rats. *Pteridines* **2007**, *18*, 8–18.

Hsieh, C.-C.; Fang, H.-L.; Lina, W.-C. Inhibitory effect of *Solanum nigrum* on thioacetamide-induced liver fibrosis in mice. *J. Ethnopharmacol.* **2008**, 119(1), 117–121.

Hsieh, C.W.; Ko, W.C.; Ho, W.J.; Chang, C.K.; Chen, G.J.; Tsai, J.C. Antioxidant and hepatoprotective effects of *Ajuga nipponensis* extract by ultrasonic-assisted extraction. *Asian Pac. J. Trop. Biomed.*, **2016**, *9*(5), 420–425.

Hsieh, P.C.; Ho, Y.L.;Huang, G.J.; Huang, M.H.; Chiang, Y.C.; Huang, S.S.; Hou. W.C.; Chang, Y.S. Hepatoprotective effect of the aqueous extract of *Flemingia macrophylla* on carbon tetrachloride-induced acute hepatotoxicity in rats through antioxidative activities. *Am. J. Chin. Med.* **2011**, *39*(2), 349–365.

Hsouna, A.B.; Saoudi, M.; Trigui, M.; Jamoussi, K.; Boudawara, T.; Jaoua, S.; et al., Characterization of bioactive compounds and ameliorative effects of *Ceratonia siliqua* leaf extract against CCl_4 induced hepatic oxidative damage and renal failure in rats. *Food Chem. Toxicol.* **2011**, *49*(12), 3183–3191.

Hsu, W.; Tsai, C.F.; Chen, W.K.; Lu, F.J. Protective effects of sea buckthorn (*Hippophae rhamnoides* L.) seed oil against carbon tetrachloride induced hepatotoxicity in mice. *Food Chem. Toxicol.* **2009**, *47*, 2281–2288.

Hu, X.-P.; Shin, J.-W.; Wang, J.-H.; Cho, J.-H.; Son, J.-Y.; Cho, C.-K.; Son, C.-G. Antioxidative and hepatoprotective effect of Cgx, An herbal medicine, against toxic acute injury in mice. *J. Ethnopharmacol.* **2008**. *120*, 51–55.

Huang, B.; Ban, X.; He, J.; Tong, J.; Tian, J.; Wang, Y. Hepatoprotective and antioxidant activity of ethanolic extracts of edible lotus (*Nelumbo nucifera* Gaertn.) leaves. *Food Chem.* **2010**. *120*, 873–878.

Huang, B.; Ban, X.; He J.; Zeng, H.; Zhang, P.; Wang, Y. Hepatoprotective and antioxidant effects of the methanolic extract from *Halenia elliptica. J. Ethnopharmacol.* **2010**, *131*(2), 276–281.

Huang, Q.; Zhang, S.; Zheng, L.; He, M.; Huang, R.; Lin, X. Hepatoprotective effects of total saponins isolated from *Taraphochlamys affinis* against carbon tetrachloride induced liver injury in rats. *Food Chem Toxicol.* **2012**, *50*, 713–718.

Hübner, S.; Hehmann, M.; Schreiner, S.; Martens, S.; Lukacin, R.; Matern, U. Functional expression of cinnamate 4-hydroxylase from *Ammi majus* L. *Phytochemistry* **2003**, *64*, 445–452.

Huda, M.; Gargoum; Muftah, S.S.; Shalmani, A.S.; Hamdoon, A.; Mohammed; et al., Phytochemical screening and investigation of the effect of *Alhagi maurorum* (Camel thorn) on carbon tetrachloride, acetaminophen and adriamycin induced toxicity in experimental animals. *J. Sci. Inno Res.* **2013**, *2*(6) 1023–33.

Hung, M.Y.; Fu, T.Y.; Shih, P.H.; Lee, C.P.; Yen, G.C. Du-Zhong (*Eucommia ulmoides* Oliv.) leaves inhibits CCl_4-induced hepatic damage in rats. *Food Chem. Toxicol.* **2006**, *44*, 1424–1431.

Huo, H.Z,; Wang, B.; Liang, Y.K.; Bao, Y.Y.; Gu, Y. Hepatoprotective and antioxidant effects of licorice extract against CCl_4-induced oxidative damage in rats. *Int. J. Mol. Sci.* **2011**, *12*(10), 6529–6543.

Hussain, F.; Malik, A.; Ayyaz, U.; Shafique, H.; Rana, Z.; Hussain, Z. Efficient hepatoprotective activity of cranberry extract against CCl_4-induced hepatotoxicity in Wistar albino rat model: Down-regulation of liver enzymes and strong antioxidant activity. *Asian Pac. J. Trop. Med.*, **2017**, *10*(11), 1054–1058.

Hussain, L.; Ikram, J.; Rehman, K.; Tariq, M.; Ibrahim, M.; Akash, M.S.H.

Hepatoprotective effects of *Malva sylvestris* L. against paracetamol-induced hepatotoxicity. *Turkish J. Biol.* **2014**, *38*, 891–896.

Hussain, L.; Akash, M.S.; Tahir, M.; Rehman, K. Hepatoprotective effects of *Sapium sebiferum* in paracetamol-induced liver injury. *Bangladesh J. Pharmacol.* **2015**, *10*, 393–98.

Hussain, L.; Akash, M.S.H.; Tahir, M.; Rehman, K.; Ahmed, K.Z. Hepatoprotective effects of methanolic extract of *Alcea rosea* against acetaminophen-induced hepatotoxicity in mice. *Bangladesh J. Pharmacol.* **2014**, *9*, 322–27.

Hussain, T.; Gupta, R.K.; Khan, M.S.; Hussain, M.D.; Arif, M.D.; Hussain, A.; et al., Evaluation of antihepatotoxic potential of *Solanum xanthocarpum* fruit extract against antitubercular drugs induced hepatopathy in experimental rodents. *Asian Pac. J. Trop. Biomed.* **2012**, *2*(6), 454–460.

Hussain, T.; Siddiqui, H.H.; Fareed, S.; Vijayakumar, M.; Rao, C.V. Evaluation of chemopreventive effect of *Fumaria indica* against N-nitrosodiethylamine and CCl_4-induced hepatocellular carcinoma in Wistar rats. *Asian Pac. J. Trop. Med.* **2012**, *5*(8), 623–629.

Hussein, J.; Salah, A.; Oraby, F.; El-Deen, A.N.; El-Khayat, Z. Antihepatotoxic effect of *Eruca sativa* extracts on alcohol induced liver injury in rats. *J. Am. Sci.* **2010**, *6*(11), 381–389.

Ibiam, A. U.; Ugwuja, E.I.; Ejeogo, C.; Ugwu, O. Cadmium-induced toxicity and the hepatoprotective potentials of aqueous extract of *Jussiaea nervosa* leaf. *Adv. Pharm. Bull.* *3*(2), 309–313.

Ibrahim, M.; Abid, M.; Abid, Z.; Ahamed, M.F.; Aara A. Phytochemical and hepatoprotective activity of fruit extracts of *Luffa acutangula* Roxb.var. *amara*. *J. Med. Pharm. Inn.* **2014**, *1*, 49–56.

Ibrahim, M.; Khaja, M.N.; Aara, A.; Khan, A.A.; Habeeb, M.A.; Devi, Y.P.; Narasu, M.L.; Habibullah, M. Hepatoprotective activity of *Sapindus mukorossi* and *Rheum emodi* extracts: *In vitro* and *in vivo* studies. *World J. Gastroenterol.* **2008**, *14*, 256671.

Ibrahim, N.A.; El-Seedi, H.R.; Mohammed, M.M. Phytochemical investigation and hepatoprotective activity of *Cupressus sempervirens* L. leaves growing in Egypt. *Nat. Prod. Res.* **2007**, *21*, 857–866.

Ihegboro, G. O.; Ononamadu, C. J.; Afor, E.; Odogiyan, G. D. Cytotoxic and Hepatocurative Effect of aqueous fraction of *Tapinanthus bangwensis* against paracetamol-induced hepatotoxicity. *J. Evid. Based Integr. Med.*, **2018**, *23*, 2515690X18801577.

Ikewuchi, J.C.; Uwakwe, A.A.; Onyeike, E.N.; Ikewuchi, C.C. Hepatoprotective effect of an aqueous extract of the leaves of *Acalypha wilkesiana* 'Godseffiana' Muell.-Arg. (Euphorbiaceae) against carbon tetrachloride induced liver injury in rats. *EXCLI J.* **2011**, *10*, 280–289.

Ikyembe, D.; Pwavodi, C.; Agbon, A.N. Hepatoprotective effect of methanolic leaf extract of *Anacardium occidentale* (cashew) on carbon-tetrachloride-induced liver toxicity in Wistar rats. *Sub-Saharan Afr. J. Med.* **2014,** *1*, 124–131.

Ilaiyaraja, N.; Khanum, F. Amelioration of alcohol-induced hepatotoxicity and oxidative stress in rats by *Acorus calamus*. *J. Diet Suppl.* **2011**, *8*, 331–345.

Ilavararan, R.; Mohideen, S.; Vijayalakshmi M.; Manonmani, G. Hepatoprotective effect of *Casia angustifolia* Vahl. *Indian J. Pharm. Sci.* **2001**, *63*, 540–547.

Ilyas, N.; Sadiq, M.; Jehangir, A. Hepatoprotective effect of garlic (*Allium sativum*) and milk thistle (Silymarin) in isoniazid induced hepatotoxicity in rats. *Biomedica* **2011**, *27*, 166–170.

Ilyas, U.K.;, Katare, D.P.; Aeri, V. Comparative evaluation of standardized alcoholic, hydroalcoholic, and aqueous extracts of *Phyllanthus maderaspatensis* Linn. against galactosamine-induced hepatopathy in albino rats. *Pharmacogn. Mag.* **2015,** *11*(42), 277–282. doi: 10.4103/0973-1296.153079.

Im, A.R.; Kim, Y.H.; Lee, H.W.; Song, K.H. Water extract of *Dolichos lablab* attenuates hepatic lipid accumulation in a cellular nonalcoholic fatty liver disease Model. *J. Med. Food* **2016**, *19*(5), 495–503.

Imanshahidi, M.; Hosseinzadeh, H. Pharmacological and therapeutic effects of *Berberis vulgaris* and its active constituent, berberine. *Phytother. Res.* **2008**, *22*, 999–1012.

Iniaghe, O.M.; Malomo, S.O.; Adebayo, J.O. Hepatoprotective effect of the aqueous extract of leaves of *Acalypha racemosa* in carbon tetrachloride treated rats. *J. Med. Plants Res.* **2008**, *2*, 301–305.

Ip, S.-P.; Woo, K.-Y.; Lau, O.-W.; Che1, C.-T. Hepatoprotective effect of *Sabina przewalskii* against menadione-induced toxicity. *Phytother. Res.* **2004**, *18*(4), 329–331.

Iqbal, I.; Afzal, M.; Aftab, M.N.; Manzoor, F.; Kaleem, A.; Kaleem, A. Hepatotoxicity of *Cassia fistula* extracts in experimental chicks and assessment of clinical parameters. *Kuwait J. Sci.* **2016**, *43*, 135–141.

Iqbal, M.J.; Dewan, Z.F.; Choudhury, S.A.R.; Mamun, M.I.R.; Mashiuzzaman, M.; Begum M. Pretreatment by hexane extract of *Phyllanthus niruri* can alleviate paracetamol-induced damage of the rat liver. *Bangladesh J. Pharmacol.* **2007**, *2*, 43–48.

Irfan, M.; Mohammed, F.A.; Syed, A.B.; Ibrahim, M. Hepatoprotective activity of *Leucas aspera* Spreng against simvastatin induced hepatotoxicity in rats. *Ann. Phytomed.* **2012**, *1*(2), 88–92.

Iroanya, O.; Okpuzor, J.; Adebesin, O. Hepatoprotective and antioxidant properties of a triherbal formulation against carbon tetrachloride induced hepatotoxicity. *IOSR J. Pharm.* **2012**, *2*, 130–136.

Islam, M.; Hussain, K.; Latif, A.; Hashmi, F.; Saeed, H.; Bukhari, N.; Hassan, S.S.; Danish, M.Z.; Ahmad, B. Evaluation of extracts of seeds of *Syzygium cumini* L. for hepatoprotective activity using CCl_4-induced stressed rats. *Pak. Vet. J.* **2015**, *35*, 197–200.

Islam, M.R.; Alam, M.J.; Khan, M.A.S.; Douti, K. M. N. Investigation of hepatoprotective properties of the ethanolic extract of *Careya arborea* Roxb. bark in paracetamol induced hepatotoxicity in rats. *J. Pharm. Res. Int.*, **2018**, *22*(4), 1–9.

Ismail, A.N.; Shamsahal, S.N.; Mamat, S.S.; Zabidi, Z.; Noraziemah; Zainulddin, W.; et al., Effect of aqueous extract of *Dicranopteris linearis* leaves against paracetamol and carbon tetrachloride-induced liver toxicity in rats. *Pak. J. Pharm/ Sci.* **2014**, *27*(4), 831–835.

Itoh, A.; Isoda, K.; Kondoh, M.; Kawase, M.; Kobayashi, M.; Tamesada, M.; Yagi, K. Hepatoprotective effect of syringic acid and vanillic acid on concanavalin A-induced liver injury. *Biol. Pharm. Bull.* **2009**, *32*, 1215–1219

Itoh, A.; Isoda, K.; Kondoh, M.; Kawase, M.; Watari, A.; Kobayashi, M.; Tamesada, M.; Yagi, K. Hepatoprotective effect of syringic acid and vanillic acid on CCl_4-induced liver injury. *Biol. Pharm. Bull.* **2010**, *33*, 983–987.

Iwalokun, B.A.; Efedede, B.U.; Alabi-Sofunde, J.A.; Oduala, T.; Magbagbeola, O.A.; Akinwande, A.I. Hepatoprotective and antioxidant activities of *Vernonia amygdalina* on acetaminophen-induced hepatic damage in mice. *J. Med. Food. 9*, **2006**, 524–530.

Jadeja, R.N.; M.C. Thounaojam, Ansarullah, S.V. Jadav, M.D. Patel, D.K. Patel, S.P. Salunke, G.S. Padate, R.V. Devkar, A.V.Ramachandran. Toxicological evaluation and hepatoprotective potential of *Clerodendron glandulosum* Coleb leaf extract. *Hum. Exp. Toxicol.* **2011**, *30*(1), 63–70.

Jadhav, V.B.; Thakare, V.N.; Suralkar, A.A.; Deshpande, A.D.; Naik, S.R. Hepatoprotective activity of *Luffa acutangula* against CCl_4 and rifampicin-induced liver toxicity in rats: A biochemical and histopathologicl evaluation. *Indian J. Exp. Biol.* **2010**, *48*, 822–829.

Jadon, A.; Bhadauria, M.; Shukla, S. Protective effect of *Terminalia belerica* Roxb. and gallic acid against carbon tetrachloride induced damage in albino rats. *J. Ethnopharmacol.*, **2007**, *109*(2), 214–218.

Jafri, A.; Jalis Subhani, M.; Javed, K.; Singh, S. Hepatoprotective activity of leaves of *Cassia occidentalis* against paracetamol and ethyl alcohol intoxication in rats. *J. Ethnopharmacol.* **1999**, *66*, 355–361.

Jahan, S.; Khan, M.; Imran, S.; Sair, M. The hepatoprotective role of Silymarin in isoniazid-induced liver damage of rabbits. *J. Pak. Med. Assoc.* **2015**. *65*(6), 620–623.

Jain, A.; Jain, I.P.; Singh, S.P.; Asha, A. Evaluation of hepatoprotective activity of roots of *Cynodon dactylon*: An experimental studies. *Asian J. Pharm. Clin. Res.* **2013**, *6*(2), 109–112.

Jain, A.; Soni, M.; Deb, L.; Jain, A.; Roult, A.P.; Gupta, V. B.; Krishna, K.L. Antioxidant and hepatoprotective activity of ethanolic and aqueous extracts of *Momordica dioica* leaves. *J. Ethnopharmacol.* **2008**, *115*, 61–66.

Jain, M., Kapadia, R.; Jadeja, R.N.; Thounaojam, M.C.; Devkar, R.V.; Mishra, S.H. Hepatoprotective activity of *Feronia limonia* root. *J. Pharm. Pharmacol.* **2012a**, *64*, 888–896.

Jain, M.; Kapadia, R.; Jadeja, R.N.; Thounaojam, M.C.; Devkar, R.V.; Mishra, S.H. Protective role of standardized *Feronia limonia* stembark methanolic extract against carbon tetrachloride induced hepatotoxicity. *Ann. Hepatol.* **2012b**, *11*(6), 935–943.

Jain, M.; Kapadia, R.; Jadeja, R.N.; Thounaojam, M.C.; Devkar, R.V.; Mishra, S.H. Amelioration of carbon tetrachloride induced hepatotoxicity in rats by standardized *Feronia limonia* Linn. leaf extracts. *EXCLI J.* **2012c**, *11*, 250–259.

Jain, N.K.; Singhai, A.K. Protective effects of *Phyllanthus acidus* (L.) Skeels leaf extracts on acetaminophen and thioacetamide induced hepatic injuries in Wistar rats. *Asian Pac. J. Trop. Med.* **2011**, 4(6), 470–474.

Jain, R.; Nandakumar, K.; Srivastava, V,; Vaidya, S.K., Patel, S., Kumar, P. Hepatoprotective activity of ethanolic and aqueous extract of *Terminalia belerica* in rats. *Pharmacologyonline* **2008**, *2*, 411–427.

Jain, S.; Dixit, V.K.; Malviya, N.; Ambawatia, V. Antioxidant and hepatoprotective activity of ethanolic and aqueous extracts of *Amorphophallus campanulatus* Roxb. tubers. *Acta Pol. Pharm.* **2009**, *66*(4), 423–428.

Jain, S.; Jain, D.K.; Balekar, N. *In-vivo* antioxidant activity of ethanolic extract of *Mentha pulegium* leaf against CCl_4 induced toxicity in rats. *Asian Pac. J. Trop. Biomed.* **2012**, *2*(2), S737-S740.

Jaiprakash, B.; Aland, R.; Karadi, R.V.; Savadi, R.V.; Hukkeri, V.I. Hepatoprotective activity of fruit pulp of *Balanites aegyptiaca. Indian Drugs* **2003**, *40*, 296–297.

Jaishree, V.; Badami, S. Antioxidant and hepatoprotective effect of Swertiamarin from *Enicostema axillare* against D-galactosamine induced acute liver damage in rats. *J. Ethnopharmacol.* **2010,** *130*, 103–106.

Jalalpure, S.S.; Patil, M.B.; Prakash, N.S.; Hemalatha, K.; Manvi, F.V. Hepatoprotective activity of fruits of *Piper longum* L. *Indian J. Pharm. Sci.* **2003**. *65*, 360–366.

Jameel, A.; Sunil, N.; Vipul, D.; Anuja, P.; Sagar, K.; Pal, S.; Subhash, M.; Shashikant, P. Hepatoprotective activity of methanol extract of aerial parts of *Delonix regia. Phytopharmacol.* **2011**, *1*(5), 118–122.

Jamshed, H.; Siddiqi, H.S.; Gilani, A.; Arslan, J.; Qasim, M.; Gul, B. Studies on antioxidant, hepatoprotective, and vasculoprotective potential of *Viola odorata* and *Wrightia tinctoria. Phytother. Res.* **2019**, 33, 1–9.

Jamshidzadeh, A.; Fereidooni, F.; Salehi, Z.; Niknahad, H. Hepatoprotective activity of

Gundelia tournefortii. *J. Ethnopharmacol.* **2005**, *101*, 233–237.

Jamshidzadeh, A.; Heidari, R.; Razmjou, M.; Karimi, F.; Moein, M. R.; Farshad, O.; Shayesteh, M. R. H. An *in vivo* and *in vitro* investigation on hepatoprotective effects of *Pimpinella anisum* seed essential oil and extracts against carbon tetrachloride-induced toxicity. *Iran J. Basic Med. Sci.*, **2015**, *18*(2), 205.

Jamshidzadeh, A.; Khoshnoud, M.J.; Zahra, D. Hepatoprotective activity of *Cichorium intybus* L. leaves extract against carbon tetrachloride induced toxicity. *Iranian J. Pharm. Res.* **2006**, *1*, 41–46.

Jan, S.; Khan, M.R. Protective effects of *Monotheca buxifolia* fruit on renal toxicity induced by CCl_4 in rats. *BMC Complement. Altern. Med.* **2016**, *16*, 289. doi: 10.1186/s12906-016-1256–0

Janakat, S.; Al-Merie, H. Evaluation of hepatoprotective effect of *Pistacia lentiscus, Phillyrea latifolia* and *Nicotiana glauca*. *J. Ethnopharmacol.* **2002**, *83*, 135–138.

Janbaz, K.H.; Gilani, A.H. Evaluation of the protective potential of *Artemisia maritima* extract on acetaminophen- and CCl_4-induced liver damage. *J. Ethnopharmacol.* **1995**, *47*(1), 43–47.

Janbaz, K.H.; Gilani, A.H. Studies on preventive and curative effects of berberine on chemical-induced hepatotoxicity in rodents. *Fitoterapia* **2000**; *71*, 25–33.

Janbaz, K.H.; Saeed, S.A.; Gilani, A.H. Protective effect of rutin on paracetamol and CCl_4-Induced hepatotoxicity in rodents. *Fitoterapia* **2002**, *73*, 557–563.

Jayathilaka, K.A.; Thabrew, M.I.; Pathirana, C.; de Silva, D.G.; Perera, D.J. An evaluation of the potency of *Osbeckia octandra* and *Melothria maderaspantana* as anti-hepatotoxic agents. *Planta Med.* **1989**, *55*(2), 137–139.

Jaydeokar, A.V.; Bandawane, D.D.; Bibave, K.H.; Patil, T.V. Hepatoprotective potential of *Cassia auriculata* roots on ethanol and antitubercular drug-induced hepatotoxicity in experimental models. *Pharm. Biol.* **2014**, *52*(3), 344–355.

Jayadevaiah, K.V.; Ishwar Bhat, K.; Joshi, A.B.; Vijayakumar, M.M.J.; Pinkey, R. Hepatoprotective activity of *Desmodium oojeinense* (Roxb.) H. Ohashi against paracetamol induced hepatotoxicity. *Asian J. Pharm. Health Sci.* **2012**, *2*(2), 312–315.

Jayasekhar, P.; Mohanan, V.; Rathinam, K. Hepatoprotective activity of ethyl acetate extract of *Acacia catechu*. *Indian J. Pharmacol.* **1997**, *29*, 426–428.

Jayavelu, A.; Natarajan, A.; Sundaresan, S.; Devi, K.; Senthilkumar, B. Hepatoprotective activity of *Boerhavia diffusa* Linn. (Nyctaginaceae) against Ibuprofen induced hepatotoxicity in Wistar albino rats. *Int. J.Pharma Res. Rev.* **2013**, *2*(4), 1–8.

Jehangir, A.; Nagi, A.H.; Shahzad, M.; Zia, A. The hepatoprotective effect of *Cassia fistula* (Amaltas) leaves in isoniazid and rifampicin induced hepatotoxicity in rodents, *Biomedica* **2010**, *26*, 25–29.

Jeje, T.O.; Ibraheem, O.; Brai, B.C.; Ibukun, E.O. Pharmacological potential of asthma weed (*Euphorbia hirta*) extract toward eradication of *Plasmodium berghei* in infected albino mice. *Int. J. Toxicol. Pharmacol. Res.* **2016**, *8*(3), 130–137.

Jeong, S.C.; Kim, S.M.; Jeong, Y.T.; Song, C.H. Hepatoprotective effect of water extract from *Chrysanthemum indicum* L. flower. *Chin. Med.*, **2013**, *8*(1), 7.

Jeyadevi, R.; Ananth, D.A.; Sivasudha, T. Hepatoprotective and antioxidant activity of *Ipomoea staphylina* Linn. *Clin. Phytosci.*, **2019**, *5*, Artilce 18.

Jeyakumar, R.; Rajesh, R.; Rajaprabhu, D.; Ganesan, B.; Buddhan, S.; Anandan, R. Hepatoprotective effect of *Picrorrhiza kurroa* on antioxidant defense system in antitubercular drugs induced hepatotoxicity in rats. *Afr. J. Biotech.* **2009**, *8*(7), 1314-1315.

Jha, M.; Nema, N.; Shakya, K.; Ganesh, N.; Sharma, V. In vitro hepatoprotective activity of *Bauhinia variegata*. *Pharmacologyonline*, **2009**, *3*, 114–118.

Jiménez-Arellanes, M.A.; Gutiérrez-Rebolledo, G. A.; Meckes-Fischer, M.; León-Díaz, R. Medical plant extracts and natural compounds with a hepatoprotective effect against damage caused by antitubercular drugs: A review. *Asian Pac. J. Trop. Med.* **2016**, *9*(12), 1141–1149.

Jin, Y.S.; Heo, S.I.; Lee, M.J.; Rhee, H.I.; Wang, M.H. Free radical scavenging and hepatoprotective actions of *Quercus aliena* acorn extract against CCl_4-induced liver. *Free Radic Res.* **2005**, *39*(12), 1351–1358.

Jin, Y.S.; Lee, M.J.; Han, W.; Heo, S.I.; Sohn, S.I.; Wang, M.H. Antioxidant effects and hepatoprotective activity of 2,5-dihydroxy-4,3′-di(beta-d-glucopyranosyloxy)-trans-stilbene from *Morus bombycis* Koidzumi roots on CCl_4-induced liver damage. *Free Radic. Res.* **2006**, 40(9), 986–992.

Jin, Y.S., Sa, J.H., Shim, T.H., Rhee, H.I., Wang, M.H. 2005. Hepatoprotective and antioxidant effects of *Morus bombycis* Koidzumi on CCl_4-induced liver damage. *Biochem. Biophys. Res. Commun.* **2005**, *329*(3), 991–995.

John, P.P.; Jose, N.; Carla, Sr. B. Preliminary pharmacological screening of *Simarouba glauca* DC. leaf extracts for hepatoprotective activity. *World J. Pharm. Pharm. Sci.* **2016**, *5*, 1714–1724.

Johnson, M.; Olufunmilayo, L.A.; Anthony, D.O.; Olusoji, E.O. Hepatoprotective effect of ethanolic leaf extract of *Vernonia amygdalina* and *Azadirachta indica* against acetaminophen-induced hepatotoxicity in SD male albino rats. *Am. J. Pharmacol. Sci.* **2015**, *3*, 79–86.

Jose, J.K.; Kuttan, R. Hepatoprotective activity of *Emblica officinalis* and Chyavanprash. *J. Ethnopharmacol.* **2000**, *72*, 129–35.

Joseph, B.; Justin Raj, S. Pharmacognostic and phytochemical properties of *Ficus carica* Linn.—an overview. *Int. J. PharmTech Res.* **2011**, 3(1), 8–12.

Joshi, V.; Sutar, P.; Karigar, A.; Patil, S.; Gopalakrishna, B.; Sureban, R. Screening of ethanolic extract of *Stachytarpheta indica* L. (Vahl) leaves for hepatoprotective activity. *Int. J. Res. Ayurveda Pharm.* **2010**, *1*, 174–179.

Joshi, Y.M.; Kadam, V.J.; Patil, Y.V.; Kaldhone, P.R. Investigation of hepatoprotective activity of aerial parts of *Canna indica* L. on carbon tetrachloride treated rats. *J Pharm. Res.* **2009**, *2*(12), 1879–1882.

Joshy, C., P.A. Thahimon, R. A. Kumar, B. Carla, C. Sunil. Hepatoprotective, anti-inflammatory and antioxidant activities of *Flacourtia montana* J. Grah. leaf extract in male Wistar rats. *Bull. Faculty Pharm. Cairo Univ.* **2016**, *54*, 209–217.

Jothy, S. L.; Aziz, A.; Chen, Y.; Sasidharan, S. Antioxidant activity and hepatoprotective potential of *Polyalthia longifolia* and *Cassia spectabilis* leaves against paracetamol-induced liver injury. *Evid. Based Complement. Alternat. Med.*, **2012**, *2012*, 1–10.

Jung, J.-C.; Lee, Y.-H.; Kim, S.H.; Kim, K.-J; Kim K.-M.; Oh, S.; Jung, Y.-S. Hepatoprotective effect of licorice, the root of *Glycyrrhiza uralensis* Fischer, in alcohol-induced fatty liver disease. *BMC Complement. Alternat. Med.* **2016**, *16*, Article number: 19.

Jyoti, T.M.; Prabhu, K.; Lakshminarasu, S.; Ramachandra, S.S. Hepatoprotective and antioxidant activity of *Euphorbia antiquorum*. *Pharmacogn. Mag.* **2008**, *4*, 127–133.

Jyothi, T.M.; Shankaraiah, M.M.; Prabhu, K.; Lakshminarasu, S.; Srinivasa, G.M.; Ramachandra, S.S. Hepatoprotective and antioxidant activity of *Euphorbia tirucalli*. *Iranian J. Pharmacol. Therapeut.* **2008**, *7*, 25–30.

Jyothi, V.M.; Sudheer, A; Bhargav, E.; Rao, R.P.; Naresh babu, C.; Naznen, P.V.; Kumar, B.P.; Salamma, S. In vitro & in vivo antioxidant and hepatoprotective potential of *Caralluma adscendens* var.

attenuata against ethanol toxicity. *Indo-Am. J. Pharm. Sci.* **2018**, *5*(4), 2541–2549.

Jyothilekshmi, S.; Protective role of *Ficus benghalensis* bark extract against ethanol-induced hepatotoxicity in rats. *Int. J. Curr. Res.* **2015**, *7*(11), 22285–22288.

Kadarian, C.; Broussalis, A.M.; Mino, J.; Lopez, P.; Gorzalczany, S.; Ferraro, G.; Acevedo, C. Hepatoprotective activity of *Achyrocline satureioides* (Lam.) D.C. *Pharmacol. Res.*, **2002**, *45*(1), 57–61.

Kadhim, E.J. Phytochemical investigation and hepatoprotective studies of Iraqi *Bryonia dioica* (Family Cucurbitaceae). *Int. J. Pharm. Pharm. Sci.* **2014**, *6*(4), 187–190.

Kadir, F.A.; Oathman, F.; Abdulla, M.A.; Hussam, F.; Hassandarvish, P. Effect of *Tinospora crispa* on thioacetamide-induced liver cirrhosis in rats. *Indian J. Pharmacol.* **2011,** *43*(1), 64–68.

Kadir, F.A.; Kassim, N.M.; Abdulla, M.A.; Yehye, W.A. Hepatoprotective role of ethanolic extract of *Vitex negundo* in thioacetamide-induced liver fibrosis in male rats. *Evid. Based Complement. Alternat. Med.* **2013** doi: 10.1155/2013/739850

Kadir, F.A.; Othman, F.; Abdulla, M.A.; Hussan, F.; Hassandarvish, P. Effect of *Tinospora crispa* on thioacetamide-induced liver cirrhosis in rats. *Indian J. Pharmacol.* **2011**, *43*, 64–68.

Kafash-Farkhad, N.; Asadi-Samani, M.; Rafieian-Kopaei, M. A review on phytochemistry and pharmacological effects of *Prangos ferulacea* (L.) Lindl. *Life Sci. J.* **2013**. *10*, 360–367.

Kafash Farkhad, N.; Farokhi, F.; Tukmacki, A.; Soltani Band, K. Hydroalcoholic extract of the root of *Prangos ferulacea* (L.) Lindl can improve serum glucose and lipids in alloxan-induced diabetic rats. *Avicenna J. Phytomed.* **2012**, *2*, 179–187.

Kalantari, H.; Forouzandeh, H.; Azemi, M.E.; Rashidi, I.; Goudarzi, M. Study of the protective effect of *Teucrium polium* L. extract on acetaminophen-induced hepatotoxicity in mice. *Iranian J. Pharm. Res.* **2013**, *12*(1), 123–129.

Kalantari, H., Foruozandeh, H., Khodayar, M.J., Siahpoosh, A., Saki, N., Kheradmand, P. Antioxidant and hepatoprotective effects of *Capparis spinosa* L. fractions and Quercetin on tert-butyl hydroperoxide-induced acute liver damage in mice. *J. Tradit. Complement. Med.* **2018**, *8*(1), 120–127.

Kalantari, H.; Jalali, M.; Jalali, A.; Mahdavinia, M.; Salimi, A.; Juhasz, B.; Tosaki, A.; Gesztelyi, R. Protective effect of *Cassia fistula* fruit extract against bromobenzene-induced liver injury in mice. *Hum. Exp. Toxicol.* **2011**, *30*(8), 1039–1044.

Kalantari, H.; Khorsandi, L.S.; Taherimobarekeh, M. The protective effect of the *Curcuma longa* extract on acetaminophen induced hepatotoxicity in mice. *Jundishapur J. Nat. Pharm. Prod.*, **2007**, *2*(1), 7–12.

Kalava, S.; Mayilsamy, D. Aqueous extract of *Brassica rapa chinensis* ameliorates tert-butyl hydroperoxide induced oxitative stress in rats. *Int. J. Curr. Pharm. Res.* **2014**, *6*(3), 58–61.

Kale, B.P.; Kothekar, M.A.; Tayade, H.P.; Jaju, J.B.; Mateenuddin, M. Effect of aqueous extract of *Azadirachta indica* leaves on hepatotoxicity induced by antitubercular drugs in rats. *Indian J. Pharmacol.* **2003**, *35*(3), 177–180.

Kale, I.; Khan, M.A.; Irfan, Y.; Goud, V.A. Hepatoprotective potential of ethanolic and aqueous extract of flowers of *Sesbania grandiflora* (Linn) induced by CCl_4. *Asian Pac. J. Trop. Biomed.* **2012**, *2*(2), S670-S679.

Kamat, C.D.; Khandelwal, K.R.; Bodhankar, S.L.; Ambawade, S.D.; Mhetre, N.A. Hepatoprotective activity of leaves of *Feronia elephantum* Correa (Rutaceae) against carbon tetrachloride-induced liver damage in rats. *J. Nat. Remed.* **2003**, *3*, 148–54.

Kameshwaran, S.; Kothai, A.R.; Jothimanivannan, C.; Senthilkumar, R. Evaluation of hepatoprotective activity of *Tecoma stans* flowers. *Pharmacologia.* **2013**, *4*, 236–242.

Kamisan, F.H., Yahya, F., Ismail, N.A., Din, S.S., Mamat, S.S., Zabidi, Z., Zainulddin, W.N.W., Mohtarrudin, N., Husain, H., Ahmad, Z., Zakaria, Z.A. Hepatoprotective activity of methanol extract of *Melastoma malabathricum* leaf in rats. *J. Acupunct. Meridian. Stud.* **2013**, *6*(1), 52–55.

Kamisan, F.H.; Yahya, Y.; Mamat, S.S.; Kamarolzaman, M.F.F.; Mohtarrudin, N.; Kek, T.L.; et al., Effect of methanol extract of *Dicranopteris linearis* against carbon tetrachloride-induced acute liver injury in rats. *BMC Complement. Alternat. Med.* **2014**, *14*, 123.

Kamkaen, N.; Wilkinson, J.M. The antioxidant activity of *Clitoria ternatea* flower petal extracts and eye gel. *Phytother. Res.* **2009**, 23(11), 1624–1625.

Kanchana, N.; Sadiq, A.M. Hepatoprotective effect of *Plumbago zeylanica* on paracetamol induced liver toxicity in rats. *Int. J. Pharm. Pharmac. Sci.* **2011**, *3*, 32–39.

Kandimalla, R.; Kalita, S.; Saikia, B.; Choudhury, B.; Singh, Y.P.; Kalita, K.; Dash, S.; Kotoky, J. Anti-oxidant and hepatoprotective potentiality of *Randia dumetorum* Lam. leaf and bark via inhibition of oxidative stress and inflammatory cytokines. *Front Pharmacol.* **2016**, *7*, 205.

Kang, S.; Lee, C.; Koo, H.; Ahn, D. Choi, H.; Lee, J.; et al., Hepatoprotective effects of aqueous extract from aerial part of agrimony. *Korean J. Pharmacogn.* **2006**, *37*, 28–32.

Kanter M, Coskun O, Budancamanak M. Hepatoprotective effects of *Nigella sativa* L and *Urtica dioica* L. on lipid peroxidation, antioxidant enzyme systems and liver enzymes in carbon tetrachloride-treated rats. *World J. Gastroenterol.* **2005**, *11*, 6684–6688.

Kantham, S. Influence of *Carica papaya* Linn. extracts on paracetamol and thioacetamide-induced hepatic damage in rats. *Internet J. Pharmacol.* **2009**, *9*, 1.

Karan, M.; Vasisht, K.; Handa, S.S. Antihepatotoxic activity of *Swertia chirata* on carbon tetrachloride induced hepatotoxicity in rats. *Phytother. Res* **1999a**, *13*(1), 24–30.

Karan, M.; Vasisht, K.; Handa, S.S. Antihepatotoxic activity of *Swertia chirata* on paracetamol and galactosamine induced hepatotoxicity in rats. *Phytother. Res* **1999b**, *13*(2), 95–101.

Karthikeyan, M.; Deepa, K. Hepatoprotective effect of *Premna corymbosa* (Burm. f.) Rottl. & Willd. Leaves extract on CCl_4-induced hepatic damage in Wistar albino rats. *Asian Pac. J. Trop. Med.* **2010**, *3*(1), 17-20.

Karthikeyan, M.; Wadhwane, S.S.; Kannan, M.; Rajasekar, S. Hepatoprotective activity of ethanolic extract of *Spermacoce hispida* Linn. against carbon tetrachloride (CCl_4) induced hepatotoxicity on albino Wistar rats. *Int. J. Pharma Res. Dev.* **2001**, *2*(11), 45–52.

Karwani, G.; Sisodia, S.S. Hepatoprotctive activity of *Tagetes erecta* Linn. in carbontetrachloride induced hepatotoxicity in rats. *World J. Pharm. Sci.* **2015**, *3*(6), 1191–1197.

Kaur, G.; Jabbar, Z.; Athar, M.; Alam, M.S. *Punica granatum* (pomegranate) flower extract possesses potent antioxidant activity and abrogates Fe-NTA induced hepatotoxicity in mice. *Food Chem. Toxicol.* **2006**, *44*(7), 984–993.

Kaur, V.; Kumar, M.; Kaur, P.; Kaur, S.; Singh, A.P.; Kaur, S. Hepatoprotective activity of *Butea monosperma* bark against thioacetamide-induced liver injury in rats. *Biomed. Pharmacother.* **2017**, *89*, 332–41.

Kavitha, P.; Ramesh, R.; Bupesh, G.; Stalin, A.; Subramanian, P. Hepatoprotective activity of *Tribulus terrestris* extract against acetaminophen-induced toxicity in a freshwater fish (*Oreochromis mossambicus*). *In Vitro Cell. Dev. Biol.-Animal*, **2011**, 47(10), 698–706.

Kazemi, S.; Asgary, S.; Moshtaghian, J.; Rafieian, M.; Adelnia, A.; Shamsi, F. Liver-protective effects of hydroalcoholic extract of *Allium hirtifolium* Boiss. in rats with alloxan-induced diabetes mellitus. *ARYA Atheroscler.* **2010**, *6*, 11–15.

Kazemifar, A.M.; Hajaghamohammadi, A.A.; Samimi, R.; Alavi, Z.; Abassi, E.; Asl, M.N. Hepatoprotective property of oral Silymarin is comparable to N-Acetyl cysteine in acetaminophen poisoning. *Gastroenterol. Res.* **2012**, *5*(5), 190–194.

Kazmi, S. T. B., Majid, M., Maryam, S., Rahat, A., Ahmed, M., Khan, M. R., & ul Haq, I. (2018). *Quercus dilatata* Lindl. ex Royle ameliorates BPA induced hepatotoxicity in SD rats. *Biomed. Pharmacother.*, **2018**, *102*, 728–738.

Khalaf-Allah, Ael-R.; El-Gengaihi, S.E.; Hamed, M.A.;, Zahran, H.G.; Mohammed, M.A. Chemical composition of golden berry leaves against hepato-renal fibrosis. *J. Diet Suppl.* **2016**, *13*(4), 378–392.

Khan, I.; Singh, V.; Chaudhary, A.K. Hepatoprotective activity of *Pinus roxburghii* Sarg. wood oil against carbon tetrachloride- and ethanol-induced hepatotoxicity. *Bangladesh J. Pharmacol.* **2012**, *7*, 94–99.

Khan, M.L.A. ibb-Al-Nabvi: *Nigella sativa*. *Islamic Voice*, **1999**, *13–18*(152), 1–2.

Khan, M.R.; Afzaal, M.; Saeed, N.; Shabbir, M. Protective potential of methanol extract of *Digera muricata* on acrylamide induced hepatotoxicity in rats. *Afr. J. Biotechnol.* **2011**, *10*, 8456–8464.

Khan, M. R.; Marium, A.; Shabbir, M.; Saeed, N.; Bokhari, J. Antioxidant and hepatoprotective effects of *Oxalis corniculata* against carbon tetrachloride (CCl_4) induced injuries in rat. *Afr J. Pharm. Pharmacol.*, **2012**, *6*(30), 2255–2267.

Khan, M.U., Rohilla, A., Bhatt, D., Afrin, S., Rohilla, S., Ansari, S.H. Diverse belongings of *Calendula officinalis*: An overview. *Int. J. Pharm. Sci. Drug Res.* **2011**, *3*, 173–177.

Khan, M.Z.; Jogeza, N.; Tareen, J.K.; Shabbir, M.I.; Malik, M.A.; Khan, A.U.R. Compilation on medicinal plants with hepatoprotective activity. *Isra Med. J.* **2016**, *8*, 196–202.

Khan, R.A.; Khan, M.A.; Ahmed, M.; Sahreen, S.; Shah, N.A.; Shah, M.S.; Bokhari, J.; Rashid, U.; Ahmad, B.; Jan. S. Hepatoprotection with a chloroform extract of *Launaea procumbens* against CCl_4-induced injuries in rats. *BMC Complement. Alternat. Med.* **2012**, *12*, 114–119.

Khan, R.A., Khan, M.R., Sahreen, S., Shah, N.A. Hepatoprotective activity of *Sonchus asper* against carbon tetrachloride-induced injuries in male rats: A randomized controlled trial. *BMC Complement. Alternat. Med.* **2012**, *12*, 90.

Khan, S.W., Tahir, M., Lone, K.P., Munir, B., Latif, W. Protective effect of *Saccharum officinarum* L. (sugar cane) juice on isoniazid induced hepatotoxicity in male albino mice. *J. Ayub Med. Coll. Abbottabad* **2015**, *27*(2), 346–350.

Khan, T.H., Sultana, S. Effect of *Aegle marmelos* on DEN initiated and 2-AAF promoted hepatocarcinogenesis: A chemopreventive study. *Toxicol. Mech. Meth.*, **2011**, *21*(6), 453–462.

Khare, C.P. Indian Medicinal Plants: An Illustrated Dictionary. Springer-Verlag, New York, USA, **2007**, pp. 249–250.

Kharpate, S.; Vadnerkar, G.; Jain, D.; Jain, S. Evaluation of hepatoprotective activity of ethanol extract of *Pterospermum acerifolium* Ster leaves. *Indian J. Pharm. Sci.* **2007**, *69*, 850–852.

Khatri, A., Garg, A., Agrawal, S.S. Evaluation of hepatoprotective activity of aerial parts of *Tephrosia purpurea* and stem bark of *Tecomella undulata*. *J. Ethnopharmacol.* **2009**, *122*, 1–5.

Kim, E.Y.; Kim, E.K.; Lee, H.S.; Sohn, Y.; Soh, Y.; Jung, H.S.; Sohn, N.W. Protective effects of *Cuscustae semen* against dimethylnitrosamine-induced acute liver injury in SD rats. *Biol. Pharm. Bull.*, **2007**, *30*(8), 1427–1431.

Kim, H. S.; Han, G. D. Hypoglycemic and hepatoprotective effects of Jerusalem artichoke extracts on streptozotocin-induced diabetic rats. *Food Sci. Biotechnol.* **2013**, *22*(4), 1121–1124.

Kim, H.S.; Lim, H.K.; Chung, M.W.; Kim, Y.C. Antihepatotoxic activity of bergenin, the major constituent of *Mallotus japonicus*, on carbon tetrachloride-intoxicated hepatocytes. *J. Ethnopharmacol.* **2000**, *69* (1), 79 83.

Kim, J. W.; Yang, H.; Cho, N.; Kim, B.; Kim, Y. C.; Sung, S. H. Hepatoprotective constituents of *Firmiana simplex* stem bark against ethanol insult to primary rat hepatocytes. *Pharmacogn. Mag.*, **2015**, *11*(41), 55.

Kim, S.M.; Kang, K.; Jeon, J.S.; Jho, E.H.; Kim, C.Y.; Nho, C.W.; Um, B.H. Isolation of phlorotannins from *Eisenia bicyclis* and their hepatoprotective effect against oxidative stress induced by tert-butyl hyperoxide. *Appl. Biochem. Biotechnol.* **2011**, *165*(5–6), 1296–1307.

Kim, S.M.; Kang, K.; Jho, E.H.; Jung, Y.J.; Nho, C.W.; Um, B.H.; Pan, C.H. Hepatoprotective effect of flavonoid glycosides from *Lespedeza cuneata* against oxidative stress induced by tert-butyl hyperoxide. *Phytother. Res.* **2011**, 25(7), 1011–1017.

Kim, T.W.; Lee, H.K.; Song, I.B.; Kim, M.S.; Hwang, Y.H.; Lim, J.H.; Yun, H.I. Protective effect of the aqueous extract from the root of *Platycodon grandiflorum* on cholestasis-induced hepatic injury in mice. *Pharm. Biol.*, **2012**, *50*(12), 1473–1478.

Kinoshita, S.; Inoue, Y.; Nakama, S.; Ichiba, T.; Aniya, Y. Antioxidant and hepatoprotective actions of medicinal herb, *Terminalia catappa* L. from Okinawa Island and its tannin corilagin. *Phytomedicine*, **2007**, *14*(11), 755–762.

Kiran, P.M.; Raju, A.V.; Rao, B.G. Investigation of hepatoprotective activity of *Cyathea gigantea* (Wall. ex. Hook.) leaves against paracetamol induced hepatotoxicity in rats. *Asian Pac. J. Trop. Biomed.* **2012**. *2*(5), 352–356.

Kiziltas, H.; Ekin, S.; Bayramoglu, M.; Akbas, E.; Oto, G.; Yildirim, S.; Ozgokce, F. Anti-oxidant properties of *Ferulago angulata* and its hepatoprotective effect against N-nitrosodimethylamine-induced oxidative stress in rats. *Pharm. Biol.* **2017**, *55*, 888–897.

Klibet, F.; Boumendjel, A.; Khiari, M.; El Feki, A.; Abdennour, C.; Messarah, M. Oxidative stress-related liver dysfunction by sodium arsenite: Alleviation by *Pistacia lentiscus* oil. *Pharm. Biol.* **2016**, *54*(2), 354–363.

Ko, H.J.; Chen, J.H.; Ng, L.T. Hepatoprotection of *Gentiana scabra* extract and polyphenols in liver of carbon tetrachloride-intoxicated mice. *J. Environ. Pathol. Toxicol. Oncol.* **2011**, *30*(3), 179–187.

Koca-Caliskan, U.; Yilmaz, I.; Taslidere, A.; Yalcin, F.N.; Aka, C.; Sekeroglu, N. *Cuscuta arvensis* Beyr "Dodder": In vivo hepatoprotective effects against acetaminophen induced hepatotoxicity in rats. *J. Med. Food* **2018**. *21*(6), doi: 10.1089/jmf.2017.0139

Kokhdan, E.P.; Ahmadi, K.; Sadeghi, H.; Sadeghi, H.; Dadgary, F.; Danaei, N.; Aghamaali, M.R. Hepatoprotective effect of *Stachys pilifera* ethanol extract in carbon tetrachloride-induce hepatotoxicity in rats. *Pharm. Biol.* **2017**, *55*(1), 1389–1393.

Kondeva-Burdina, M.; Shkondrov, A.; Simeonova, R.; Vitcheva, V.; Krasteva, I.; Ionkova, I. In vitro/in vivo antioxidant and hepatoprotective potential of defatted extract and flavonoids isolated from *Astragalus spruneri* Boiss. (Fabaceae). *Food Chem. Toxicol.* **2018**, *111*, 631–640.

Konopiński, M. Influence of intercrop plants and varied tillage on yields and nutritional value of salsify (*Tragopogon porrifolius* L.) roots. *Acta Sci. Pol. Hortorum Cultus* **2009**, *8*, 27–36.

Kotoky, J.; Dasgupta, B.; Sarma, G.K. Protective properties of *Leucas lavandulaefolia* extracts against D-galactosamine induced hepatotoxicity in rats. *Fitoterapia* **2008**, *79*(4), 290–292.

Kowsalya, R.; Kaliaperumal, J.; Vaishnavi, M.; Namasivayam, E. Anticancer activity of *Cynodon dactylon* L. root extract against diethyl nitrosamine-induced hepatic carcinoma. *South Asian J. Cancer* **2015**, *4*(2), 83–87.

Krishna, B.K.L.; Mruthunjaya, K.; Patel, J.A. Antioxidant and hepatoprotective activity of leaf extract of *Justicia gendarussa. Int. J. Biol. Chem.* **2009**, *3*, 99–110.

Krishnakumar, N.M.; Latha, P.G.; Suja, S.R.; Shine, V.J.; Shyamal, S.; Anuja, G.I.; Sini, S.; Pradeep, S.; Shikha, P.; Unni, P.K.; Rajasekharan, S. Hepatoprotective effect of *Hibiscus hispidissimus* Griffith, ethanolic extract in paracetamol and CCl_4-induced hepatotoxicity in Wistar rats. *Indian J. Exp. Biol.* **2008**, *46*(9): 653–659.

Kshirsagar, A.D.; Mohite, R.; Aggrawal, A.S.; Suralkar, U.R. Hepatoprotective medicinal plants of ayurveda—a review. *Asian J. Pharm. Clin. Res.* **2011**, *4*, 1–8.

Ku, K.L.; Tsai, C.T.; Chang, W.M.; Shen, M.L.; Wu, C.T.; Liao, H.F. Hepatoprotective effect of *Cirsium arisanense* Kitamura in tacrine-treated hepatoma Hep 3B cells and C57BL mice. *Am. J. Chin. Med.* **2008**, *36*(02), 355–368.

Kukongviriyapan, V.; Janyacharoen, T.; Kukongviriyapan, U.; Laupattarakasaem, P.; Kanokmedhakul, S.; Chantaranothai, P. Hepatoprotective and antioxidant activities of *Tetracera loureiri. Phytother. Res.* **2003**, *17*(7), 717–721.

Kulathuran, P.K.; Chidambaranathan, N.; Mohamed, H.; Jayaprakash, S.; Narayanan, N. Hepatoprotective activity of *Cnidoscolus chayamansa* against rifampicin and isoniazid induced toxicity in Wistar rats. *Res. J. Pharm. Biol. Chem. Sci.* **2012**, *3*(2), 577–585.

Kumar, A.K. Protective effect of *Punica granatum* peel and *Vitis vinifera* seeds on DEN-induced oxidative stress and hepatocellular damage in rats. *Appl. Biochem. Biotechnol.* **2015,** *175*(1), 410–420.

Kumar, A.S.; Samanta, K.C.; Chipa, R.C. Hepatoprotective activity of alcoholic and aqueous extracts of leaves of *Anisochilus carnosus* Wall. *Int. J. Pharm. Res. Dev.* **2010**, *2*(8), 17.

Kumar, B.S.A.; Lakshman, K.; Kumar, P.A.; Viswantha, G.L.; Veerapur, V.P.; Thippeswamy, B.S.; Manoj, B. Hepatoprotective activity of methanol extract of *Amaranthus caudatus* Linn. against paracetamol-induced hepatic injury in rats. *J. Chin. Integr. Med.* **2011**, *9*, 194–200.

Kumar, B.S.A.; Lakshman, K.; Swamy, V.; Kumar, P.; Shekar, D.; Manoj, B.; Vishwantha, G. Hepatoprotective and antioxidant activities of *Amaranthus viridis* Linn. *Macedonian J. Med. Sci.* **2011**, *4*, 125–130.

Kumar, B.S.; Gnanasekaran, D.; Jaishree, V.; Channabasavaraj, K.P. Hepatoprotective activity of *Coccinia indica* leaves extract. *Int. J. Pharm. Biomed. Res.* **2010**, *1*(4), 154–156.

Kumar, C.H.; Ramesh, A.; Kumar, J.N.S.; Ishaq, B.M. A review on hepatoprotective activity of medicinal plants. *Int. J. Pharm. Sci. Res.* **2011**, *2*, 501–515.

Kumar, G.; Banu G.S.; Pandian M.R. Evaluation of the antioxidant activity of *Trianthema portulacastrum. Indian J. Pharmacol.* **2005**, *37*, 331–34.

Kumar, G.; Banu, G.S.; Vanitha, P.P.; Sundararajan, M.; Rajasekara, P.M. Hepatoprotective activity of *Trianthema portulacastrum* L. against paracetamol and thioacetamide intoxication in albino rats. *J. Ethnopharmacol.*, **2004**, *92*(1), 37–40.

Kumar, K.E.; Harsha, K.N.; Sudheer, V.; Giri babu, N. In vitro antioxidant activity and *in vivo* hepatoprotective activity of aqueous extract of *Allium cepa* bulb in ethanol induced liver damage in Wistar rats. *Food Sci. Human Wellness.* **2013**, *2*(3), 132–138.

Kumar, K.S.; Rajakrishnan, R.; Thomas, J.; Reddy, G.A. Hepatoprotective effect of *Helicanthus elastica. Bangladesh J. Pharmacol.* **2016,** *11,* 525–530.

Kumar, K.V.A.; Gowda, K.V. Evaluation of hepatoprotective and antioxidant activity of *Flemingia strobilifera* R. Br. against experimentally induced liver injury in rats Int *J. Pharm. Pharm. Sci,* **2011**, *3*(3), 115–119.

Kumar, N.A.; Pari, L. Antioxidant action of *Moringa oleifera* Lam. (drumstick) against antitubercular drugs induced lipid peroxidation in rats. *J. Med. Food* **2003,** *6*(3), 255–259.

Kumar, P.; Deval Rao, G.; Ramachandra Setty, S. Antioxidant and hepatoprotective activity of tubers of *Momordica tuberosa* Cogn. against CCl_4 induced liver injury in rats. *Indian J. Exp. Biol.* **2008**, *46*, 510–513.

Kumar, R.; Kumar, S.; Patra, A.; Jayalakshmi, S. Hepatoprotective activity of aerial parts of *Plumbago zeylanica* Linn against carbon tetrachloride-induced hepatotoxicity in rats. *Int. J. Pharm. Pharm. Sci.* **2009**, *1*, Suppl, 171–175.

Kumar, R.S.; Kumar, K.A.; Murthy, N.V. Hepatoprotective and antioxidant effects of *Caesalpinia bonducella* on carbon tetrachloride-induced liver injury in rats. *Int. Res. J. Plant Sci.* **2010**, *1*(3), 62–68.

Kumar, R.S.; Sivakumar, T.; Sivakumar, P.; Nethaji, R.; Vijayabasker, M.; Perumal, Gupta, M.P.; Mazumder, U.K. Hepatoprotective and *in vivo* antioxidant effects of *Careya arborea* against carbon tetrachloride-induced liver damage in rats. *Intl. J. Mol. Med. Adv. Sci.* **2005**, *1*, 418–424.

Kumar, S.N.; Reddy, K.R.; Chandra Sekhar, K.B.; Rakhal Chandra Das, R.C. Hepatoprotective effect of *Cleome gynandra* on carbon tetrachloride-induced hepatotoxicity in Wistar rats. *Am. J. Adv. Drug Deliv.* **2014,** *2*(6), 734–740.

Kumar, S.P.; Asdaq, S.M.B.; Kumar, N.P.; Asad, M.; Khajuria, D.K. Protective effect of *Zizyphus jujuba* fruit extract against paracetamol and thioacetamide-induced hepatic damage in rats. *Internet J. Pharmacol.* **2009**, 7(1).

Kumar, S.V.S.; Mishra, S.H. Protective effect of rhizome extract of *Cyperus rotundus* Linn. against thioacetamide induced hepatotoxicity in rats. *Ind. Drugs* **2004**, *41*(3), 165–169.

Kumar, S.V.S.; Mishra, S.H. Hepatoprotective activity of rhizomes of *Cyperus rotundus* Linn. against carbon tetrachloride induced hepatotoxicity. *Ind. J. Pharm. Sci.*, **2005**, 67, 84–88.

Kumar, S.V.S.; Mishra, S.H. Hepatoprotective effect of *Pergularia daemia* Forsk. ethanol extract and its fraction. *Indian J. Exp. Biol.* **2006**, 46(6), 447–452.

Kumar, S.V.; Mishra, S.H. Hepatoprotective activity of extracts from *Pergularia daemia* Forsk. against carbon tetrachloride induced toxicity in rats. *Pharmacogn. Mag.* **2008**, *3*(11), 187–191.

Kumar, S.V.S.; Mishra, S.H. Hepatoprotective effect of *Baliospermum montanum* (Willd.) Muell.-Arg. against thioacetamide induced toxicity. *Int. J. Compreh. Pharm.* **2012**, 9, 1–4.

Kumar, S.V.S.; Kumar, C.V.; Vardhan, A.V. Hepatoprotective activity of *Acalypha indica* Linn against thioacetamide induced toxicity. *Int. J. Pharm. Pharm. Sci.*; **2013**, *5*, 3–7.

Kumar, S.V.S.; Sujatha, C.; Syamala, J.; Nagasudha, B.; Mishra, S.H. Protective effect of root extract of *Operculina turpethum* Linn. against paracetamol induced hepatotoxicity in rats. *Indian J. Pharm. Sci.* **2006**, *68*, 32–35.

Kumar, U.S.U.; Chen, Y.; Kanwar, J.R.; Sasidharan, S. Redox control of antioxidant and antihepatotoxic activities of *Cassia surattensis* seed extract against paracetamol intoxication in mice: In vitro and *in vivo* studies of herbal green antioxidant. *Oxid. Med. Cell Longev.* **2016**, *2016*: 6841348. doi: 10.1155/2016/6841348

Kumar, V.; Modi, P.K.; Saxena, K.K. Exploration of hepatoprotective activity of aqueous extract of *Tinospora cordifolia*-an experimental study. *Asian J. Pharm. Clin. Res.* **2013**, *6*(1), 87–91.

Kumar, V.; Sharma, S.; Midi, P.K. Exploration of hepatoprotective activity of aqueous extract of *Solanum nigrum*-an experimental study. *Int. J. Pharm. Sci. Res.* **2013**, *4*(1): 464–470.

Kumaresan, P.T.; Pandae, V. Hepatoprotective activity of *Aeschynomene aspera* Linn. *Pharmacologyonline* **2011,** *3*, 297–304.

Kumaresan, R.; Veerakumar, S.; Elango, V. A study on hepatoprotective activity of *Mimosa pudica* in albino rats. *Int. J. Pharm. Phytochem. Res.* **2015**, *7*, 337–339.

Kumaresan, T.P.; Arunjosph; Gururajesh, U.; Renugadevi; Kasula, S.C. Hepatoprotective activity of *Dalbergia spinosa* (Roxb) against paracetamol induced hepatotoxicity In rats. *Int. J. Pharm.Biol. Sci.* **2013**, 355–358

Kundu, M.; Mazumder, R.; Kushwaha, M.D. Evaluation of hepatoprotective activity of ethanol extract of *Coccinia grandis* (L.) Voigt. leaves on experimental rats by acute and chronic models. *Orient Pharm. Exp. Med.*, **2012**, *12*, 93–97.

Kundu, R.; Dasgupta, S.; Biswas, A.; Bhattacharya, A.; Pal, B.C.; Bandyopadhyay, D.; Bhattacharya, S.; Bhattacharya, S. *Cajanus cajan* Linn. (Leguminosae) prevents alcohol-induced rat liver damage and auguments cytoprotective function. *J. Ethnopharmacol.* **2008**, *118*, 440-447.

Kuppuswamy, R.; Govindaraju, A.; Velusamy, G.; Balasubramanian, R.; Balasundarm, J.; Sellamuthu, M. Effect of dried fruits of *Solanum nigrum* L. against CCl_4-induced hepatic damage in rats. *Biol. Pharm. Bull.* **2003**, *26*(11), 1618–1619.

Kuriakose G. C.; Kurup M. G. Antioxidant and hepatoprotective activity of *Aphanizomenon flos-aquae* Linn. against paracetamol intoxication in rats. *Indian J. Exp. Biol.* **2010**, *48*, 1123–1130.

Kurma, S.R.; Mishra, S.H. Hepatoprotective principles from the stem bark of *Moringa pterygosperma*. *Pharm. Biol.* **1998**, *36*(4), 295-300.

Kushwah, D.S.; Salman, M.T.; Singh, P.; Verma, V.K.; Ahmad, A. Protective effects of ethanolic extract of *Nigella sativa* seed in paracetamol induced acute hepatotoxicity *in vivo*. *Pak. J. Biol. Sci.* **2014**, *17*(4), 517–522.

Lahon, K.; Das, S. Hepatoprotective activity of *Ocimum sanctum* alcoholic leaf extract against paracetamol-induced liver damage in Albino rats. *Pharmacogn. Res.* **2011**, *3*(1), 13–18. doi: 10.4103/0974-8490.79110.

Lakshmi, T.; Sri Renukadevi, B.; Senthilkumar, S.; et al., Seed and bark extracts of *Acacia catechu* protects liver from acetaminophen induced hepatotoxicity by modulating oxidative stress, antioxidant enzymes and liver function enzymes in Wistar rat model. *Biomed. Pharmacother.* **2018**, *108*, 838–844.

Lalee, A.; Bhattacharya, B.; Das, M.; Mitra, D.; Kaity, S.; Bera, S.; Samanta, A. Hepatoprotective activity of ethanolic extract of *Aerva sanguinolenta* (Amaranthaceae) against paracetamol induced liver toxicity on Wistar Rats. NSHM *J. Pharm. Healthcare Manag.* **2012**, *03*, 57–65.

Lall, N.; Kumar, V.; Meyer, D.; Gasa, N.; Hamilton, C.; Matsabisa, M.; Oosthuizen, C. In vitro and In vivo antimycobacterial, hepatoprotective and immunomodulatory activity of *Euclea natalensis* and its mode of action. *J. Ethnopharmacol.*, **2016**, *194*, 740–748.

Lamy, E.; Schroder, J.; Paulus, S.; Brenk, P.; Stahl, T.; Mersch-Sundermann, V. Antigenotoxic properties of *Eruca sativa* (rocket plant), erucin and erysolin in human hepatoma (HepG2) cells towards benzo(a) pyrene and their mode of action. *Food Chem. Toxicol.*, **2008**, *46*(7), 2415–2421.

Lanhers, M.C.; Joyeux, M.; Soulimani, R.; Fleurentin, J.; Sayag, M.; Mortier, F.; Younos, C.; Pelt, J. Hepatoprotective and anti-inflammatory effects of a traditional medicinal plant of Chile, *Peumus boldus*. *Planta Medica* **1991**, *57*, 110-115.

Laouar A.; Klibet, F.; Bourogaa, E.; Benamara A.; Boumendjell, A.; Chefrour, A.; Messarah M. Potential antioxidant properties and hepatoprotective effects of *Juniperus phoenicea* berries against

CCl_4-induced hepatic damage in rats. *Asian Pac. J. Trop. Med.* **2017**, *10*(3), 263–269.

Lakshmi, A.I.; Gobinath, M.; Bhattacharya, J.; Priyanka, P. Hepatoprotective activity of ethanolic extract of aerial parts of *Piper trioicum* Roxb. against CCl_4-induced liver damage in rats. *World J. Pharm. Pharm. Sci.* **2016**, *6*, 949–961.

Lashgari, A.P.; Gazor, R.; Asgari, M.; Mohammadghasemi, F.; Nasiri, E.; Roushan, Z.A. Hepatoprotective effect of *Acantholimon gilliati* Turril, on formaldehyde induced liver injury in adult male mice. *Iranian J. Pharm. Res.* **2017**, *16*, S135–141.

Lata, S.; Mittal, S.K. In vitro and *in vivo* hepatoprotectivity of flavonoids rich extracts on *Cucumis dipsaceus* Ehrenb. (fruit). *Int. J. Pharmacol.* **2017**, *13*(6), 563–572.

Lata, S.; Singh, S.; Tiwari, K.N. and Upadhyay R. Evaluation of the antioxidant and hepatoprotective effect of *Phyllanthus fraternus* against a chemotherapeutic drug cyclophosphamide. *Appl. Biochem. Biotech.*, **2014**, *173*, 2163–2173.

Lavanya, G.; Manjunath, M.; Kavitha, P.; Parthasarathy, R.P. Preventive and curative effects of *Cocculus hirsutus* (Linn.) Diels leaves extract on CCl_4 provoked hepatic injury in rats. *Egyptian J. Basic Appl. Sci.* **2017**, *4*(4), 264–269.

Lawal, B., Shittu, O.K., Oibiokpa, F.I., Berinyuy, E.B., Mohammed, H. 2016. African natural products with potential antioxidants and hepatoprotectives properties: A review. *Clin. Phytosci.* **2016**, *2*, 23. doi. 10.1186/s40816-016-0037-0

Lazarova, I.; Simeonova, R.; Vitcheva, V.; Kondeva-Burdina, M.; Gevrenova, R.; Zheleva-Dimitrova, D.; Danchev, N. D. Hepatoprotective and antioxidant potential of *Asphodeline lutea* (L.) Rchb. roots extract in experimental models *in vitro/in vivo*. *Biomed. Pharmacother.*, **2016**, *83*, 70–78.

Lee, C.P., Shih, P.H., Hsu, C.L., Yen, G.C. Hepatoprotection of tea seed oil (*Camellia oleifera* Abel.) against CCl_4-induced oxidative damage in rats. *Food Chem. Toxicol.* **2007**, *45*, 888–895.

Lee, C.Y.; Peng, W.H.; Cheng, H.Y.; Chen, F.N.; Lai, M.T.; Chiu, T.H. Hepatoprotective effect of *Phyllanthus* in Taiwan on acute liver damage induced by carbon tetrachloride. *Am. J. Chin. Med.* **2006**, *34*(03), 471–482.

Lee, D.S.; Keo, S.; Cheng, S.K.; Oh, H.; Kim, Y.C. Protective effects of Cambodian medicinal plants on tert-butyl hydroperoxide induced hepatotoxicity via Nrf2-mediated heme oxygenase-1. *Mol. Med. Rep.*, **2017**, *15*(1), 451–459.

Lee, H., McGregor, R.A., Choi, M.S., Seo, K.I., Jung, U.J., Yeo, J., et al., Low doses of curcumin protect alcohol-induced liver damage by modulation of the alcohol metabolic pathway, CYP2E1 and AMPK. *Life Sci.* **2013**, *93*(18–19), 693–699.

Lee, H.S.; Ahn, H.C.; Ku, S.K. Hypolipidemic effect of water extracts of *Picrorrhiza rhizoma* in PX-407 induced hyperlipemic ICR mouse model with hepatoprotective effects: A prevention study. *J. Ethnopharmacol.* **2006**, *105*, 380–386.

Lee, K.J.; Jeong, H.G. Protective effect of Platycodi radix on carbon tetrachloride-induced hepatotoxicity. *Food Chem. Toxicol.* **2002**, *40*, 517–524.

Lee, M.H., Yoon, S., Moon, J.O. The flavonoid naringenin inhibits dimethylnitrosamine-induced liver damage in rats. *Biol. Pharm. Bull.* **2004**, *27*(1), 72–76.

Lee, S.H.; Heo, S.I.; Li, L.; Lee, M.J.; Wang, M.H. Antioxidant and hepatoprotective activities of *Cirsium setidens* Nakai against CCl_4-induced liver damage. *Am. J. Chin. Med.* **2008**, *36*(1), 107–114.

Leonard, S., Karen, L., Bruce, B., Thomas, K., Jay, H. Complementary and Alternative medicine in chronic liver disease. *Hepatol.* **2002**, *34*, 595–603.

Levy, C., Seeff, L.D., Lindor, K.D. 2004. Use of herbal supplements for chronic liver disease. *Clin. Gastroenterol. Hepatol.* **2004**, *2*, 947–956.

Lexa, A.; Fleurentins, J.; Lehr, P.R.; Mortier, F.; Pruvost, M.; Pelt, J.M. Choleretic and hepatoprotective properties of *Eupatorium cannabinum* in the rat. *Planta Med.* **1989**, *55*, 127–132.

Li, C.; Yi, L.T.; Geng, D.; Han, Y.Y.; Weng, L.J. Hepatoprotective effect of ethanol extract from *Berchemia lineata* against CCl_4-induced acute hepatotoxicity in mice. *Pharm. Biol.* **2015**, *53*, 767–772.

Li, G.Y.; Gao, H.Y.; Huang, J.; Lu, J.; Gu, J.K.; Wang, J.H. Hepatoprotective effect of *Cichorium intybus* traditional Uighur medicine, against carbontetrachloride-induced hepatic fibrosis in rats. *World J. Gastroenterol.* **2014**, *20*(16), 4753–60.

Li, J.-X.; Shi, Q.; Xiong Q.-B.; et al., Tribulusamide A and B, new hepatoprotective lignanamides from the fruits of *Tribulus terrestris*: Indications of cytoprotective activity in murine hepatocyte culture. *Planta Med.* **1998**, *64*(7), 628–631.

Li, L.; Park, D.H.; Li, Y.C.; Park, S.K.; Lee, Y.L.; Choi, H.M.; Han, D.S.; Yang, H.J.; et al., Anti-hepatofibrogenic effect of turnip water extract on thioacetamide-induced liver fibrosis. *Lab Anim. Res.* **2010**, *26*(1), 1–6.

Li, Y.; Li, Z.; Li, C.; Ma, X.; Chang, Y.; Shi, C.; Liu, F. Evaluation of hepatoprotective activity of *Syringa oblata* leaves ethanol extract with the indicator of glutathione S-transferase A1. *Rev. Bras. Farmacogn.* **2018**, *28*(4), 489–494.

Li, Y.J.; Lin, D.D.; Jiao, B.; Xu, C.T.; Qin, J.K.; Ye, G.J.; Su, G.F. Purification, antioxidant and hepatoprotective activities of polysaccharide from *Cissus pteroclada* Hayata. *Int. J. Biol. Macromol.* **2015**, *77*, 307–313.

Li, Z.-Y., Sun H.-M., Xing J., Qin X.-M., Du G.-H. Chemical and biological comparison of raw and vinegar-baked Radix Bupleuri. *J. Ethnopharmacol.* **2015**, *165*, 20–28.

Lim, H.K.; Kim, H.S.; Choi, H.S.; Choi, J.; Kim, S.H.; Chang, M.J. Effects of bergenin, the major constituent of *Mallotus japonicus* against D-galactosamine-induced hepatotoxicity in rats. *Pharmacology.* **2001**, *63*(2), 71–75.

Lim, H.-K.; Kim, H.-S.; Choi, H.-S.; Oh, S.; Choi, J. Hepatoprotective effects of Bergenin, a major constituent of *Mallotus japonicus*, on carbon tetrachloride-intoxicated rats. *J. Ethnopharmacol.* **2000**, *72*, 469–474.

Lima, I.R.; Vieira, J. R. C.; Silva, I. B.; Leite, R. M. P.; Maia, M. B.; Leite, S. P. Indican from Añil (*Indigofera suffruticosa* Miller) as an herbal protective agent to the liver. *Anal. Quantit. Cytol. Histol.*, **2014**, *36*(1)0, 41–45.

Lin, C-C.; Hsu, Y-F.; Lin, T-C.; Hsu, H-Y. Antioxidant and hepatoprotective effects of punicalagin and punicalin on acetaminophen-induced liver damage in rats. *Phytother. Res.* **2006**, *15*, 206–212.

Lin, C.-C.; Hsu, Y.-F.; Lin, T.-C.; Hsu, F.L.; Hsu, H.-Y. Antioxidant and hepatoprotective activity of punicalagin and punicalin on carbon tetrachloride-induced liver damage in rats. *Pharm Pharmacol.* **1998**, *50*(7), 789-94.

Lin, C.C.; Huang, P.C. Antioxidant and hepatoprotective effects of *Acathopanax senticosus. Phytother. Res.* **2002**, *14*(7), 489–494.

Lin, C.C.; Lee, H.Y.; Chang, C.H.; Namba, T.; Hattori, M. Evaluation of the liver protective principles from the root of *Cudrania cochinchinensis* var. *gerontogea. Phytother. Res.*, **1996**. *10*(1), 13–17.

Lin, C.C.; Lin, W.C.; Chang, C.H.; Namba, T. Antiinflammatory and hepatoprotective effects of *Ventilago leiocarpa. Phytother. Res.*, **1995**. *9*(1), 11–15.

Lin, C.C.; Lin, W.C.; Yang, S.R.; Shieh, D.E. Anti-inflammatory and hepatoprotective effects of *Solanum alatum. Am. J. Chin. Med.* **1995**, *23*(1), 65–69.

Lin, C.C.; Yen, M.H.; Lo, T.S.; Lin, J.M. Evaluation of the hepatoprotective and antioxidant activity of *Boehmeria nivea* var. *nivea* and *B. nivea* var. *tenacissima*. *J. Ethnopharmacol.* **1998**, *60*, 9–17.

Lin, E.-Y.; Chagnaadorj, A.; Huang S.-J.; Wang C.-C.; Chiang Y.-H.; Cheng C.-W. Hepatoprotective activity of the ethanolic extract of *Polygonum multiflorum* Thunb. against oxidative stress-induced liver injury. *Evid. Based Complement. Alternat. Med.* **2018**, *2018*, 9. doi: 10.1155/2018/4130307.4130307

Lin, H.-M.; Tseng, H.-C.; Wang, C.-J.; Lin, J.-J.; Lo, C.-W.; Chou, F.-P. Hepatoprotective effects of *Solanum nigrum* Linn. extract against CCl_4 induced oxidative damage in rats. *Chem. Biol. Interact.* **2008**, *171*, 283–293.

Lin, L.T.; Liu, L.T.; Chiang, L.C.; Lin, C.C. In vitro antihepatoma activity of fifteen natural medicines from Canada. *Phytother. Res.* **2002**, *16*, 440–444.

Lin, S.C.; Chung, T.C.; Lin, C.C.; et al., Hepatoprotective effects of *Arctium lappa* on carbon tetrachloride- and acetaminophen-induced liver damage. *Am. J. Chin. Med.* **2000**, *28*, 163–173.

Lin, S.C.; Lin, C.H.; Lin, C.C.; et al., Hepatoprotective effects of *Arctium lappa* Linne on liver injuries induced by chronic ethanol consumption and potentiated by carbon tetrachloride. *J. Biomed. Sci.* **2002**, 9, 401–409.

Lin, S.C.; Lin, C.C.; Lin, Y.H.; Shyuu, S.J. Hepatoprotective effects of Taiwan folk medicine: *Wedelia chinensis* on three hepatotoxin-induced hepatotoxicity. *Am. J. Chin. Med.* **1994**, *22*(2), 155–168.

Lin, S.C.; Lin, Y.H.; Shyuu, S.J.; Lin, C.C. Hepatoprotective effects of Taiwan folk medicine, *Alternanthera sessilis* on liver damage induced by various hepatotoxins. *Phytother. Res.* **1994**, *8*(7), 391–398

Lin, S.C., Yao, C.J., Lin, C.C., Lin, Y.H. Hepatoprotective activity of Taiwan folk medicine: *Eclipta prostrata* Linn. against various hepatotoxins induced acute hepatotoxicity. *Phytother. Res.* **1996**, *10*, 483–490.

Lina, S.M.M.; Ashab, I.; Ahmed, M.I.; Al-Amin, M.; Shahriar, M. Hepatoprotective activity of *Asteracantha longifolia* (Nees.) extract against anti-tuberculosis drugs induced hepatic damage in SD rats. *PhOL* **2012**, *3*(1), 13–19.

Liu, C.T.; Chuang, P.T.; Wu, C.Y.; Weng, Y.M.; Wenlung, C.; Tseng, C.Y. Antioxidative and *in vitro* hepatoprotective activity of *Bupleurum kaoi* leaf infusion. *Phytother. Res.*, **2006**, *20*(11), 1003–1008.

Liu, J.; Luo, D.; Wu, Y.; Gao, C.; Lin, G.; Chen, J.; Wu, X.; Zhang, Q.; Cai, J.; Su, Z. The Protective effect of *Sonneratia apetala* fruit extract on acetaminophen-induced liver injury in mice. *Evid. Based Complement. Alternat. Med.* **2019**; *2019*: 6919834. doi: 10.1155/2019/6919834

Liu, J.; Liu, Y.; Parkinson, A.; Klaassen, C.D. Effect of oleanolic acid on hepatic toxicant-activating and detoxifying systems in mice. *J. Pharmacol. Exp. Ther.* **1995**, *275*(2), 768–774.

Liu, J-Y.; Chen, C.-C.; Wang, W.-H.; Hsu, J.-D.; Yang, M.-Y.; Wang, C.-J. The protective effects of *Hibiscus sabdariffa* extract on CCl_4-induced liver fibrosis in rats. *Food Chem. Toxicol.*, **2006**, *44*, 336–43.

Liu, T.; Zhao, J.; Ma, L.; Ding, Y.; Su, D. Hepatoprotective effects of total triterpenoids and total flavonoids from *Vitis vinifera* L.against immunological liver injury in mice. *Evid. Based Complement. Alternat. Med.* **2012**, *2012*, 969386.

Ljubuncic, P.; Song, H.; Cogan U, Azaizeh H, Bomzon A. The effects of aqueous extracts prepared from the leaves of *Pistacia lentiscus* in experimental liver disease. *J. Ethnopharmacol.* **2005**, *100*, 198–204.

Lodhi, G.; Singh, H.K.; Pant, K.K.; Hussain, Z. Hepatoprotective effects of *Calotropis gigantea* extract against carbon tetrachloride induced liver injury in rats. *Acta Pharm.* **2009**, *59*(1), 89–96.

Loki, A.L.; Rajamohan, T. Hepatoprotective and antioxidant effect of tender coconut water on carbon tetrachloride-induced liver injury in rats. *Indian J. Biochem. Biophys.* **2003**, *40*, 354–357.

Longo, F.; Teuwa, A.; Fogue, S.K.; Spiteller, M.; Ngoa, L.E. Hepatoprotective effects of *Canna indica* L. rhizome against acetaminophen (paracetamol). *World J. Pharm. Pharm. Sci.* **2015**, *4*, 1609–24.

Lu, K.-L.; Tsai, C.-C.; Ho, L.-K.; Lin, C.-C.; Chang, Y.S. Preventive effect of the Taiwan folk medicine *Ixeris laevigata* var. *oldhami* on a-naphthyl-isothiocyanate and carbon tetrachloride-induced acute liver injury in rats. *Phytother. Res.* **2002**. *16*, S45–50.

Luangchosiri, C.; Thakkinstian, A.; Chitphuk, S.; Stitchantrakul, W.; Petraksa, S.; Sobhonslidsuk, A. A double-blinded randomized controlled trial of Silymarin for the prevention of antituberculosis drug-induced liver injury. *BMC Complement. Altern. Med.* **2015**, *15*(1), 334.

Luo, Y.-J.; Yu, J-P.; Shi, Z.-H.; Wang, L. *Gingkgo biloba* extract reverses CCl_4-induced liver fibrosis in rats. *World. J. Gastroenterol.* **2004**, *10*, 1037-1042.

Lv, L.; Jiang, C.; Li, J.; Zheng, T. Protective effects of lotus (*Nelumbo nucifera* Gaertn.) germ oil against carbon tetrachloride-induced injury in mice and cultured PC-12 cells. *Food Chem. Toxicol.*, **2012**, *50*(5), 1447–1453.

Lv, Y., Zhang, B., Xing, G., Wang, F., Hu, Z. Protective effect of naringenin against acetaminophen-induced acute liver injury in metallothionein (MT)-null mice. *Food Funct.* **2013**, *4*(2), 297–302.

Ma, J.; Sun, H.; Liu, H.; Shi, G.N.; Zang, Y.D.; Li, C.J.; Yang, J.Z.; Chen, F.Y.; Huang, J.W.; Zhang, D.; Zhang, D.M. Hepatoprotective glycosides from the rhizomes of *Imperata cylindrica*. *J. Asian Nat. Prod/Res.* **2018**, *20*(5), 451–459.

Maaz, A.; Bhatti, A.S.A.; Maryam, S.; Afzal, S.; Ahmad, M.; Gilani, A.N. Hepatoprotective evaluation of *Butea monosperma* against liver damage by paracetamol in rabbits. *Special Ed. Annals* **2010**, *16*(1), 73–76.

MacDonald, I.; Oghale, O-U.; Ikechi, E.G.; Orji, O.A. Hepatoprotective potentials of *Picralima nitida* against *in vivo* carbon tetrachloride-mediated hepatotoxicity. *J. Phytopharmacol.* **2016**, 5, 6–9.

Mada, S.B.; Inuwa, H.M.; Abarshi, M.M.; Mohammed, H.A.; Aliyu, A. Hepatoprotective effect of *Momordica charantia* extract against CCl_4-induced liver damage in rats. *Brit. J. Pharm. Res.* **2014**, *4*(3), 368–380.

Madani, H.; Talebolhosseini, M.; Asgary, S.; Naderi, G.H. Hepatoprotective activity of *Silybum marianum* and *Cichorium intybus* against thioacetamide in rats. *Pak. J. Nutr.* **2008**, *7*(1), 172–176.

Madubunyi, I. I. Hepatoprotective activity of *Uvaria chamae* root bark methanol extract against acetaminophen-induced liver lesions in rats. *Comp. Clin. Pathol.*, **2010**, *21*, 137–145.

Madubuike, G.K.; Onoja, S.O.; Ezeja, M.I. Antioxidant and hepatoprotective activity of methanolic extract of *Cassia sieberiana* leaves in carbon tetrachloride-induced hepatotoxicity in rats. *J. Adv. Med. Pharm. Sci.* **2015**, *2*(1), 1–9.

Mahalakshmi, R.; Rajesh, P.; Ramesh, N.; Balasubramanian, V.; Rajesh Kannan, V. Hepatoprotective activity on *Vitex negundo* Linn. (Verbenaceae) by using Wistar albino rats in Ibuprofen induced model. *Int. J. Pharmacol.* **2010**, *6*(5), 658–663.

Mahboub, F.A. The effect of green tea (*Camellia sinensis*) extract against hepatotoxicity induced by tamoxifen in rats. *J. Food Process. Tech.* **2016**, *7*(10), doi: 10.4172/2157-7110.1000625

Mahesh, A.; Jeyachandran, R.; Cindrella, L.; Thangadurai, D.; Veerapur, V.; Muralidhara Rao, D. Hepatocurative potential of sesquiterpene lactones of *Taraxacum officinale* on carbon tetrachloride induced liver toxicity in mice. *Acta Biol. Hung.* **2010**, *61*, 175–190.

Maheswari, C.; Maryammal, R.; Venkatanarayanan, R. Hepatoprotective activity of *Orthosiphon stamineus* on liver damage caused by paracetamol in rats. *Jordan J. Biol. Sci.* **2008**, *1*, 105-108.

Maheshwari, D.T.; Yogendra Kumar, M.S.; Verma, S.K.; Singh, V.K.; Singh, S.N. Antioxidant and hepatoprotective activities of phenolic rich fraction of Seabuckthorn (*Hippophae rhamnoides* L.) leaves. *Food Chem. Toxicol.* **2011**, *49*(9), 2422-2428.

Maheswari, M. U.; Rao, P. G. M. Anti hepatotoxic effect of grape seed oil. *Indian J. Pharmacol.* **2005**, *37*(3), 179-182.

Mahmoed, M.Y.; Rezq, A.A. Hepatoprotective effect of avocado fruits against carbon tetrachloride-induced liver damage in male rats. *World Appl. Sci. J.*, **2013**, *21*, 1445–1452.

Mahmoodzadeh, Y.; Mazani, M.; Rezagholizadeh, L. Hepatoprotective effect of methanolic *Tanacetum parthenium* extract on CCl_4-induced liver damage in rats. *Toxicol. Rep.* **2017,** *4,* 455–462.

Mahmud, Z., Bachar, S., Qais, N. Antioxidant and hepatoprotective activities of ethanolic extracts of leaves of *Premna esculenta* Roxb. against carbon tetrachloride-induced liver damage in rats. *J. Young Pharm.* **2012**, *4*(4), 228–234.

Mahmud, Z.A., Emran, T.B., Qais, N., Bachar, S.C., Sarker, M., Uddin, M.M. Evaluation of analgesic, anti-inflammatory, thrombolytic and hepatoprotective activities of roots of *Premna esculenta* (Roxb). *J. Basic Clin. Physiol. Pharmacol.* **2016**, *27*(1), 63–70.

Maity, T.; Ahmad, A.; Pahari, N. Evaluation of hepatotherapeutic effects of *Mikania sandens* (L.) Willd. on alcohol induced hepatotoxicity in rats. *Int. J. Pharm. Pharm. Sci.* **2012a**, *4,* 490–494.

Maity, T.; Ahmad, A.; Pahari, N. Hepatoprotective activity of *Mikania scandens* (L.) Willd. against diclofenac sodium induced liver toxicity in rats. *Acad. Sci.* **2012b**, *5*, 185–189.

Makoshi, M.S.; Adanyeguh I. M.; Nwatu L.I. Hepatoprotective effect of *Phyllanthus niruri* aqueous extract in acetaminophen sub-acutec exposure rabbits. *J. Vet. Med. Anim. Health.* **2013**, *5*(1), 8–15.

Malar, H.L.V.; Bai, S.M.M. Hepatoprotective activity of *Phyllanthus emblica* against paracetamol induced hepatic damage in Wister albino rats. *Afri. J. Basic Applied Sci.* **2009**, 1, 21–25.

Mallhi, T.H.; Abbas, K.; Ali, M.; Qadir, M.I.; Saleem, M.; Khan, Y.H. Hepatoprotective activity of methanolic extract of *Malva parviflora* against paracetamol-induced hepatotoxicity in mice. *Bangladesh J. Pharmacol.* **2014**, *9*, 342–46.

Ma-Ma, K.; Nyunt, N.; Tin, K.M. The protective effect of *Eclipta alba* on carbon tetrachloride-induced acute liver damage. *Toxicol. Appl. Pharmacol.* **1978**, *45*(3), 723–728.

Mamat, S.S.; Kamarolzaman, M.F.; Yahya, F.; Mahmood, N.D.; Shahril, M.S.; Jakius, K.F.; et al., Methanol extract of *Melastoma malabathricum* leaves exerted antioxidant and liver protective activity in rats. *BMC Complement. Altern. Med.* **2013**, *13*, 326

Mamat, S.S.; Kamisan, F.H.; Zainulddin, W.N.W.; Ismail, N.A.; Yahya, F.; Din, S.S.; et al., Effect of methanol extract of *Dicranopteris linearis* leaves against paracetamol and carbon tetrachloride (CCl_4)-induced liver toxicity in rats. *J. Med. Plants Res.* **2013**, *7*(19), 1305-1309.

Mandal, A.; Bishayee, A.; Chatterjee, M. *Trianthema portulacastrum* affords antihepatotoxic activity against carbon tetrachloride-induced chronic liver damage in mice: Reflection in subcellular levels. *Phytother Res.* **1997a**; *11*, 216–221.

Mandal, A.; Bandyopadhyay, S.; Chatterjee, M. *Trianthema portulacastrum* L. reverses hepatic lipid peroxidation, glutathione status and activities of related antioxidant enzymes in carbon tetrachloride-induced chronic liver damage in mice. *Phytomedicine* **1997b**; 239–244.

Mandal, S.C., Maity, T.K., Das, J., Pal, M., Saha, B.P. Hepatoprotective activity of *Ficus racemosa* leaf extract on liver damage caused by carbon tetrachloride in rats. *Phytother. Res.* **1999**, *13*(5), 430–432.

Mandal, S.C., Saraswathi, B., Ashok Kumar, C.K., Mohana Lakshmi, S., Maiti, B.C. Protective effect of leaf extract of *Ficus hispida* Linn. against paracetamol-induced hepatotoxicity in rats. *Phytother. Res.* **2000**, *14*(6), 457–459.

Mangathayaru, K.; Grace, X.F.; Bhavani, M.; Meignanam, E.; Karna, S.L.; Kumar, D.P. Effect of *Leucas aspera* on hepatotoxicity in rats. *Indian J. Pharmacol.*, **2005**, *37*(5), 329–330.

Manjunatha, B.K. Hepatoprotective activity of *Pterocarpus santalinus* an endangered medicinal plant. *Indian J. Pharmacol.* **2006**, *38*(1), 25–28.

Manjunatha, B.K.; Amit, R.; Priyadarshini, P.; Paul, K. Lead findings from *Pterocarpus santalinus* with hepatoprotective potentials through in silico methods. *Int. J. Pharm. Sci. Res.* **2010**, *7*, 265–270.

Manjunatha, B.K.; Bidya, S.M.; Dhiman, P.; et al., Hepatoprotective activity of *Leucas hirta* against CCl_4 induced hepatic damage in rats. *Indian J. Exp. Biol.* **2005**, *43*, 722–727.

Manjunatha, B.K.; Vidya, S.M. Hepatoprotective activity of *Vitex trifolia* against CCl_4 induced hepatic damage. *Indian J. Pharm. Sci.* **2008**, *70*(2), 241–245.

Mankani, K.L.; Krishna, V.; Manjunatha, B.K.; Vidya, S.M.; Jagadeesh Singh, S.D.; Manohara, Y. N.; Raheman, A.; Avinash, K. R. Evaluation of hepatoprotective activity of stem bark of *Pterocarpus marsupium* Roxb. *Indian J. Pharmacol.* **2005**, *37*, 165–168.

Manna, P.; Sinha, M.; Sil, P.C. Aqueous extract of *Terminalia arjuna* prevents carbon tetrachloride-induced hepatic and renal disorders. *BMC Comlement. Alt. Med.* **2006**, *6*, 33–43.

Manokaran, S.; Jaswanth, A.; Sengottuvelu, S.; Nandhakumar, J.; Duraisamy, R.; Karthikeyan, D.; Mallegaswari. R. Hepatoprotective activity of *Aerva lanata* against paracetamol induced hepatotoxicity in rats. *Res. J. Pharm. Tech.* **2008**, 1(4), 398–400.

Mansi, K.; Abushoffa, A.M.; Disi, A.; Aburjai, T. Hypolipidemic effects of seed extract of celery (*Apium graveolens*) in rats. *Pharmacogn. Mag.* **2009**, *5*, 301–305.

Mantena, S.K.; Mutalik, S.; Srinivasa, H.; Subramanian, G.S.; Prabhakar, R.K.R.; Srinivasan, K.K.; Unnikrishnan, M.K. Antiallergic, antipyretic, hypoglycemic and hepatoprotective effects of aqueous extract of *Coronopus didymus* Linn. *Biol. Pharm. Bull.*, **2005**, *28*(3), 468.

Maranhão, H.M.L.; Vasconcelos, C.F.B.; Rolim L.A.; Neto, P.J.R.; Neto, J.C.S.; Filho, R.C.S.; Fernandes, M.P.; Costa-Silva, J.H.; Araújo, A.V.; Wanderley, A.G. Hepatoprotective effect of the aqueous extract of *Simarouba amara* Aublet (Simaroubaceae) stem bark against carbon tetrachloride (CCl_4)-induced hepatic damage in rats. *Molecues* **2014**, *19*(11), 17735–17746.

Maruthappan, V.; Sakthi, K.S. Hepatoprotective effect of *Azadirachta indica* (Neem) leaves against alcohol induced liver injury in albino rats. *J. Pharm. Res.*, **2009**, *2*, 655–659.

Marzouk, M.; Sayed, A.A.; Soliman, A.M. Hepatoprotective and antioxidant effects of *Cichorium endivia* L. leaves extract against acetaminophen toxicity on rats. *J. Med. Med. Sci.*, **2011**, 2(12), 1273–1279.

Massumi, M.A.; Fazeli, M.R.; Alavi, S.H.R.; Ajani, Y. Chemical constituents and antibacterial activity of essential oil of *Prangos ferulacea* (L.) Lindl. fruits. *Iran J. Pharm. Sci.* **2007**, *3*, 171–176.

Mathan, G.; Fatima, G.; Saxena, A.K.; Chandan, B.K.; Jaggi, B.S.; Gupta, B.D.; Qazi, G.N.; Balasundaram, C.; Anand Rajan, K.D.; Kumar, V.L.; Kumar, V. Chemoprevention with aqueous extract of *Butea monosperma* flowers results in normalization of nuclear morphometry and inhibition of a proliferation marker in liver tumors. *Phytother. Res.* **2011**, *25*(3), 324–328.

Matsuda, H.; Morikawa, T.; Fengming, X.; Ninomiya, K.; Yoshikawa, M. New isoflavones and pterocarpane with hepatoprotective activity from the stems of *Erycibe expansa*. *Planta Med.*, **2004**, *70*(12), 1201–1209.

Matsuda, H.; Murakami, T.; Ninomiya, K.; Inadzuki, M.; Yoshikawa, M. New hepatoprotective saponins, bupleurosides III, VI, IX, and XIII, from Chinese Bupleuri Radix: Structure-requirements for the cytoprotective activity in primary cultured rat hepatocytes. *Bioorgan. Med. Chem. Lett.* **1997**, *28*(52), 2193–2198.

Matsuda, H.; Ninomiya, K.; Morikawa, T.; Yasuda, D.; Yamaguchi, I.; Yoshikawa, M. Hepatoprotective amide constituents from the fruit of *Piper chaba* : Structural requirements, mode of action, and new amides. *Bioorgan. Med. Chem.* **2009**, *17*, 7313–7323.

Maysa, M.; El Mallah; Mohamed, R.A. Hepatoprotective effect of *Calendula officinalis* Linn (Asteraceae) flowers against CCL_4 induced hepatotoxicity in rats. *World Appl. Sci. J.* **2015**, *33*, 949–959.

Mayuresh, J.; Sunita, S. Hepatoprotective and antioxidative Activity of *Avicennia marina* Forsk. *Gastroenterol. Hepatol. Int. J.* **2016**, *2*(1), 000114.

McBride, A.; Augustin, K.M.; Nobbe, J.; Westervelt, P. *Silybum marianum* (milk thistle) in the management and prevention of hepatotoxicity in a patient undergoing reinduction therapy for acute myelogenous leukemia. *J. Oncol. Pharm. Pract.* **2012**, *18*, 360–365.

Meera, R.; Devi, P.; Kameswari, B.; Madhumitha, B.; Merlin, N.J. Antioxidant and hepatoprotective activities of *Ocimum basilicum* Linn. and *Trigonella foenum-graecum* Linn. against H_2O_2 and CCl_4 induced hepatotoxicity in goat liver. *Indian J. Exp. Biol.* **2009**, *47*, 584–590.

Mehmetçik, G.; Ozdemirler, G.; Kocak-Toker, N.; Çevikbaş, U.; Uysal, M. Effect of pretreatment with artichoke extract on carbon tetrachloride-induced liver injury and oxidative stress. *Exp. Toxicol. Pathol.* **2008**, *60*, 475–480.

Mehta, R.S.; Shankar, M.B.; Geetha, M.; Saluja, A.K. Preliminary evaluation of hepatoprotective activity of *Trianthema portulacastrum* Linn. *J. Nat. Remedies* **2003**, *3*(2), 180–184.

Mekky, R.H.; Fayed, M.R.; El-Gindi, M.R.; Abdel-Monem, A.R.; Contreras, M.D.; Segura-Carretero, A.; Abdel-Sattar, E. Hepatoprotective effect and chemical assessment of a selected Egyptian chickpea cultivar. *Front. Pharmacol.* **2016**, *7*, 344. doi: 10.3389/fphar.2016.00344

Meng, Y.; Liu, Y.; Fang, N.; Guo, Y. Hepatoprotective effects of *Cassia* semen ethanol extract on non-alcoholic fatty liver disease in experimental rat. *Pharm. Biol.* **2019**, *57*(1), 98–104.

Menon, B.R.; Rathi, M.A.; Thirumoorthi, L.; and Gopalakrishnan, V.K. Potential effect of *Bacopa monniera* on nitrobenzene-induced liver damage in rats. *Indian J. Clin. Biochem.* **2010**, *25*(4), 401–404.

Merlin, N.J.; Parthasarathy, V. Antioxidant and hepatoprotective activity of chloroform and ethanol extracts of *Gmelina asiatica* aerial parts. *J. Med. Plant. Res.*, **2011**, *5*, 533–538.

Mihailović, V.; Katanić, J.; Mišić, D.; Stanković, V.; Mihailović, M.; Uskoković, A.; Arambašić, J.; Solujić, S.; Mladenović, M.; Stanković, N. Hepatoprotective effects of secoiridoid-rich extracts from *Gentiana cruciata* L. against carbon tetrachloride-induced liver damage in rats. *Food Funct.* **2014**, *5*, 1795–1803.

Mihailovic, V.; Mihailovic, M.; Uskokovic, A.; Arambasic, J.; Misic, D.; Stankovic, V.; et al., Hepatoprotective effects of *Gentiana asclepiadea* L. extracts against carbon tetrachloride induced liver injury in rats. *Food Chem. Toxicol.* **2012**, *52*, 83–90.

Miled, H.B.; Barka, Z.B.; Hallègue, D.; Lahbib, K.; Ladjimi, M.; Tlili, M.; Sakly, M.; Rhouma, K.B.; Ksouri, R.; Tebourbi,

O. Hepatoprotective activity of *Rhus oxyacantha* root cortex extract against DDT-induced liver injury in rats. *Biomed. Pharmacother.* **2017**, *90*, 203–215.

Minari, J.B. Hepatoprotective effect of methanolic extract of *Vernonia amygdalina* leaf. *J. Nat. Prod.* **2012**, *5*, 188–192.

Mir, A.; Anjum, F.; Riaz, N.; Iqbal, H.; Wahedi, H.M.; Zaman, J.; Khattak, K.; Khan, M.A.; Malik, S. Carbon tetrachloride (CCl_4)-induced hepatotoxicity in rats: Curative role of *Solanum nigrum*. *J. Med. Plants Res.* **2010**, 4(23), 525–532.

Mir, A.I.; Kumar, B.; Tasduq, S. A.; Gupta, D. K.; Bhardwaj S.; Johri R. K. Reversal of hepatotoxin-induced pre-fibrogenic events by *Emblica officinalis*—a histological study. *Indian J. Exp. Biol.*, **2007**, *45*, 626–629.

Mishra, A.K.; Mishra, A.; Chattopadhyay, P. *Calendula officinalis*: An important herb with valuable therapeutic dimensions-an overview. *J. Glob. Pharm. Technol.* **2010**, *2*, 14–23.

Mishra, B.; Mukerjee, A. In vivo and ex vivo evaluation of *Luffa acutangula* fruit extract and its fractions for hepatoprotective activity in Wistar rats. *Int. J. Pharm. Sci. Res.* **2017**, *8*, 5227–5233.

Mishra, G.; Khosa, R.; Singh, P.; Jha, K. Hepatoprotective potential of ethanolic extract of *Pandanus odoratissimus* root against paracetamol-induced hepatotoxicity in rats. *J. Pharm. Bioallied Sci.* **2015**, *7*, 45–48.

Mishra, S.; Dutta, M.; Mondal, S.K.; Dey, M.; Paul, S.; et al., Aqueous bark extract of *Terminalia arjuna* protects against adrenaline-induced hepatic damage in male albino rats through antioxidant mechanism(s): A dose response study. *J. Pharm. Res.* **2014**, *8*, 1264–1273.

Mistry, S.; Dutt, K.R.; Jena, J. Protective effect of *Sida cordata* leaf extract against CCl_4 induced acute liver toxicity in rats. *Asian Pac. J. Trop. Med.* **2013**, *6*(4), 280–284.

Mitra, P.; Ghosh, T.; Mitra, P.K. Seasonal variation in hepatoprotective activity of Titeypati (*Artemisia vulgaris* L.) leaves on antitubercular drugs induced hepatotoxicity in rats. *SMU Med. J.* **2016**, *3*(1), 763–774.

Mittal, D.K.; Joshi, D.; Shukla, S. Protective effects of *Polygonum bistorta* (Linn.) and its active principle against acetaminophen-induced toxicity in rats. *Asian J. Exp. Biol. Sci.* **2010**, *1*, 951–58.

Modibbo, A.A.; Nadro, M.S. Hepatoprotective effects of *Cassia italica* leaf extracts on carbon tetrachloride-induced jaundice. *Asian J. Biochem. Pharm. Res.* **2012**, *2*, 382–387.

Moghadam, A.R.; Tutunchi, S.; Namvaran-Abbas-Abad, A.; Yazdi, M.; Bonyadi, F.; Mohajeri, D.; Mazani, M.; Marzban, H.; Łos, M.J.; Ghavami, S. Pre-administration of turmeric prevents methotrexate-induced liver toxicity and oxidative stress. *BMC Complement. Altern. Med.* **2015**, *15*, 246.

Mohamad, N.E.; Yeap, S.K.; Lim, K.L.; Yusof, H.M.; Beh, B.K.; Tan, S.W. Antioxidant effects of pineapple vinegar in reversing of paracetamol-induced liver damage in mice. *Chin. Med.* **2015**, *10*, 3.

Mohamed, A.F.; Ali Hasan, A.G.; Hamamy, M.I.; Abdel-Sattar, E. Antioxidant and hepatoprotective effects of *Eucalyptus maculata*. *Med. Sci. Monit.* **2005**, *11*, 426–431.

Mohamed, E.A.K. Hepatoprotective effect of aqueous leaves extract of *Psidium guajava* and *Zizyphus spina-christi* against paracetamol induced hepatotoxicity in rats. *J. Appl. Sci. Res.* **2012**, *8*, 2800–2806.

Mohamed, M.A.; Eldin, I.M.T.; Mohammed, A.-E.H.; Hassan, H.M. Hepatoprotective effect of *Adansonia digitata* L. (baobab) fruits pulp extract on CCl_4-induced hepatotoxicity in rats. *World J. Pharm. Res.* 2015, *4*(8), 368–377.

Mohamed, M.A.; Eldin, I.M.T.; Mohammed, A-EH.; Hassan, H.M. Effects of *Lawsonia inermis* L. (Henna) leaves methanolic extract on carbon tetrachloride-induced hepatotoxicity in rats. *J. Intercult. Ethnopharmacol.* **2016**, *5*, 22–26.

Mohammadian, A.; Moradkhani, S.; Ataei, S.; Shayesteh, T.H.; Sedaghat, M.; Kheiripour, N.; Ranjbar, A. Antioxidative and hepatoprotective effects of hydroalcoholic extract of *Artemisia absinthium* L. in rat. *J. Herb. Med. Pharmacol.* **2016**, *5*, 29–32.

Mohan G.K.; Pallavi, E.; Kumar B.R.; Ramesh, M.; Venkatesh, S. Hepatoprotective activity of *Ficus carica* Linn. leaf extract against carbon tetrachloride-induced hepatotoxicity in rats. *DARU J. Pharm. Sci.* **2007**, *15*(3), 162–166.

Mohan, M.; Kamble, S.; Satyanarayana, J.; Nageshwar, M.; Reddy, N. Protective effect of *Solanum torvum* on Doxorubicin induced hepatotoxicity in rats. *Int. J. Drug Dev. Res.*, **2011**, *3*(3), 131–138

Molehin, O.R.; Oloyede, O.I.; Idowu, K.A.; Adeyanju, A.A.; Olowoyeye, A.O.; Tubi, O.I.; Komolafe, O.E.; Gold, A.S. White butterfly (*Clerodendrum volubile*) leaf extract protects against carbon tetrachloride-induced hepatotoxicity in rats. *Biomed. Pharmacother.* **2017**. *96*, 924–929.

Mollazadeh, H.; Hosseinzadeh, H. The protective effect of *Nigella sativa* against liver injury: A review. *Iran J. Basic Med. Sci.* **2014**, **17**(12), 958–966.

Mondal, A.; Kara, S.K.; Singha, T.; Rajalingam, D.; Maity, T.K. Evaluation of hepatoprotective effect of leaves of *Cassia sophera* Linn. *Evid. Based Complement. Alternat. Med* **2012**, *2012*, 1–5.

Mondal, A.; Maity, T.K.; Pal, D.; Sannigrahi, S.; Singh, J. Isolation and *in vivo* hepatoprotective activity of *Melothria heterophylla* (Lour.) Cogn. against chemically induced liver injuries in rats. *Asian Pac. J. Trop. Med.* **2011**, *4*(8), 619–623.

Mondal, S.; Ghosh, D.; Ganapaty, S.; Chekuboyina, S.V.G.; Samala, M. Hepatoprotective activity of *Macrothelypteris torresiana* (Gaudich.) aerial parts against CCl_4-induced hepatotoxicity in rodents and analysis of polyphenolic compounds by HPTLC. *J. Pharm. Anal.* **2017**, *7*, 181–189.

Montasser, A.O.S.; Saleh, H.; Ahmed-Farid, O.A.; Saad, A.; Marie, M.S. Protective effects of *Balanites aegyptiaca* extract, melatonin and ursodeoxycholic acid against hepatotoxicity induced by Methotrexate in male rats. *Asian Pac. J. Trop. Med.* **2017**, *10*(6), 557–565.

Morajhar, A.S.; Hardikar, B.; Sharma, B. Hepatoprotective effects of extracts of *Pongamia pinnata* in Alloxan-induced diabetic rats. *Int. J. Zool. Res.* **2015**, *11*(2), 37–47.

Moresco, R.N.; Sperotto, R.L.; Bernardi, A.S.; Cardoso, R.F.; Gomes, P. Effect of the aqueous extract of *Syzygium cumini* on carbon tetrachloride-induced hepatotoxicity in rats. *Phytother. Res.* **2007**, *21*, 793–795.

Morita, T.; Jinno, K.; Kawagishi, H.; Arimoto, Y.; Suganuma, H.; Inakuma, T.; Sugiyama, K. Hepatoprotective effect of myristin from nutmeg (*Myristica fragrans*) on lipopolysaccharide/d-galactosamine-induced liver injury. *J. Agric. Food Chem.*, **2003**, *51*, 1560–1565.

Moselhy, S.S.; Ali, H.K. Hepatoprotective effect of Cinnamon extracts against carbon tetrachloride-induced oxidative stress and liver injury in rats. *Biol. Res.* **2009**, 42, 93–98.

Moshaie-Nezhad, P.; Iman, M.; Faed, M.F.; Khamesipour, A. Hepatoprotective effect of *Descurainia sophia* seed extract against paracetamol-induced oxidative stress and hepatic damage in mice. *J. Herbmed. Pharmacol.* **2018**, *7*, 267–272.

Mossa, A.-T.H.; Heikal, T.M.; Belaiba, M.; Raoelison, E.G.; Ferhout, H.; Bouajila, J. Antioxidant activity and hepatoprotective potential of *Cedrelopsis grevei* on cypermethrin induced oxidative stress and liver damage in male mice. *BMC Complement. Altern. Med.* **2015**, *15*, 251. doi: 10.1186/s12906-015-0740-2.

Mroueh, M.; Daher, C.; El Sibai, M.; Tenkerian; C. Antioxidant and hepatoprotective activity of *Tragopogon porrifolius* methanolic extract. *Planta Med.* **2011**. doi: 10.1055/s-0031-1282460.

Mrudula, G.; Rai, P.M.; Jayaveera, K.N.; Pasha, S.Y.; Basha, K.M.; Fatima, H.; Swaditha, E.; et al., Evaluation of hepatoprotective activity of *Prunella vulgaris* by carbon tetrachloride induced hepatotoxicity. *Pharmacologyonline* **2010**, *2*, 1045–1053.

Mubashir, H.M.; Bahar, A.; Suroor, A.K.; Shah, M.Y.; Shamshir, K. Antihepatotoxic activity of aqueous extract of *Marrubium vulgare* whole plant in CCl_4 induced toxicity. *Indian J. Nat. Prod.* **2009**, *25*, 3–8.

Mujahid, M.; Siddiqui, H.H.; Hussain, A.; Hussain, M. S. Hepatoprotective effects of *Adenanthera pavonina* (Linn.) against antitubercular drugs-induced hepatotoxicity in rats. *Pharmacog. J.*, **2013**, *5*, 286–90.

Mujahid, M.; Hussain, T.; Siddiqui, H.H.; Hussain, A. Evaluation of hepatoprotective potential of *Erythrina indica* leaves against antitubercular drugs induced hepatotoxicity in experimental rats. *J. Ayurveda Integr. Med.* **2017**, *8*(1), 7–12.

Mujeeb, M.; Khan, S.A.; Aeri, V.; Ali, B. Hepatoprotective activity of the ethanolic extract of *Ficus carica* Linn. leaves in carbon tetrachloride-induced hepatotoxicity in rats. *Iranian J. Pharm. Res.* **2011**, *10*(2), 301–306.

Mujeeb, M.; Aeri, V.; Bagri, P.; Khan, S.A. Hepatoprotective activity of methanolic extract of *Tylophora indica* (Burm.f.) Merill. leaves. *Int. J. Green Pharm.*, **2010**, *3*(2), 125–127.

Mujumdar, A.M.; Upadhey, A.S.; Pradhan, A.M. Effect of *Azadirachta indica* leaf extract on carbon tetrachloride-induced hepatic damage in albino rats. *Indian J. Pharm. Sci.* **1998**, *60*, 363–367.

Mukherjee, S.; Sur, A.; Maiti, B.R. Hepatoprotective effect of *Swertia chirata* on rat. *Indian J. Exp. Biol.* **1997**, *35*, 384–388.

Mukul, K.G.; Dasgupta, S. Role of plant metabolites in toxic liver injury. *Asian Pac. J. Clin. Nutr.* **2002,** *11*(1), 48–50.

Mulata, H.N.; Daniel, S.; Melaku, U.; Ergete, W.; Gnanasekaran, N. Protective Effects of *Calpurnia aurea* seed extract on HAART hepatotoxicity. *Eur. J. Med. Plants.* **2015**, *9*, 1–12.

Mulla, W.A.; Salunkhe, V.R.; Bhise, S.B. Hepatoprotective activity of hydroalcoholic extract of leaves of *Alocasia indica* Linn. *Indian J. Exp. Biol.* **2009**, *47*, 816–821.

Murali, A.; Ashok, P.; Madhavan, V. Effect of *Smilax zeylanica* roots and rhizomes in paracetamol induced hepatotoxicity. *J. Complement. Integr. Med.* **2012**, *9*, Article 29. doi: 10.1515/1553-3840.1639.

Muriel, P., Rivera Espinoza, Y. Beneficial drugs for liver diseases. *J. Appl. Toxicol.* **2008**, *28*, 93–103.

Murthy, V.N.; Reddy, B.P.; Venkateshwarlu, V.; Kokate, C.K. Antihepatotoxic activity of *Eclipta alba, Tephrosia purpurea* and *Boerhaavia diffusa. Anc. Sci. Life* **1992**, *11*(3–4), 182–186.

Murugaian, P.; Ramamurthy, V.; Karmegam, N. Hepatoprotective activity of *Wedelia calendulacea* L. against acute hepatotoxicity in rats. *Res. J. Agric. Biol. Sci.* **2008**, *4*, 685–87.

Murugesh, K.S.; Yelgar, V.C.; Maiti, B.C.; Maity, T.K. Hepatoprotective and antioxidant role of *Berberis tinctoria* lesch leaves on paracetamol induced hepatic damage. *Iran. J. Pharmacol. Ther.* **2005**, 4, 64–69.

Musa, E.M.; El-Badwi, S. M.; Jah Elnabi, M.A.; Osman, E.A.; Dahab, M. Hepatoprotective and toxicity assessment of *Cannabis sativa* seed oil in Albino rat. *Int. J. Chem. Biol. Sci.* **2012**, *1*, 69–76.

Mushtaq, A.; Ahmad, M.; Jabeen, Q.; Saqib, A.; Wajid, M.; Akram, A.M. Hepatoprotective investigations of *Cuminum cyminum* dried seeds in nimesulide intoxicated albino rats by phytochemical and biochemical methods. *Int. J. Pharm. Sci.* **2014**, *6*(4), 506–510.

Muthukumaran, P.; Pattabiraman, K.; Kalaiyarasan, P. Hepatoprotective and antioxidant activity of *Mimosa pudica* on carbon tetra chloride-induced hepatic damage in rats. *Int. J. Curr. Res.* **2010**, *10*, 46–53.

Muthulingam, M. Antihepatotoxic effects of *Boerhaavia diffusa* L. on antituberculosis drug, rifampicin-induced liver injury in rats. *J. Pharmacol. Toxicol.* **2008**, *3*, 75–83.

Mutlag, S.H.; Ismael, D.K., Al-Shawi, N.N. Study the possible hepatoprotective effect of different doses of *Ammi majus* seeds'extract against CCl_4 induced liver damage in rats. *Pharm. Glob. (IJCP)* **2011**, *9*, 1–5.

Naaz, F.; Javed, S.; Abdin, M.Z. Hepatoprotective effect of ethanolic extract of *Phyllanthus amarus* Schumm. et Thonn. on Aflatoxin B1-induced liver damage in mice. *J. Ethnopharmacol.* **2007**, *113*, 503–509.

Nadro, M.S.; Onoagbe, I.O. Protective effects of aqueous and ethanolic extracts of the leaf of *Cassia italica* in CCl_4-induced liver damage in rats. *Am. J. Res. Comm.* **2014**, *2*, 122–130.

Nafiu, M.O.; Akanji, M.A.; Yakubu, M.T. Effect of aqueous extract of *Cochlospermum planchonii* rhizome on some kidney and liver functional indicies of albino rats. *Afr. J. Tradit. Complement. Altern. Med.* **2011**, *8*, 22–26.

Nagalekshmi, R.; Menon, A.; Chandrasekharan, D.K.; Nair, C.K.K. Hepatoprotective activity of *Andrographis paniculata* and *Swertia chirayita*. *Food Chem. Toxicol.* **2011**, *49*, 3367–3373.

Nagella, P.; Ahmad, A.; Kim, S.J.; Chung, I.M. Chemical composition, antioxidant activity and larvicidal effects of essential oil from leaves of *Apium graveolens*. *Immunopharmacol. Immunotoxicol.* **2012**, *34*, 205–209.

Naik, R.S.; Mujumdar, A.M.; Ghaskadbi, S. Protection of liver cells from ethanol cytotoxicity by curcumin in liver slice culture *in vitro*. *J. Ethnopharmacol.* **2004**, *95*(1), 31–37.

Naik, S.R.; Panda V.S. Antioxidant and hepatoprotective effects of *Ginkgo biloba* phytosomes in carbon tetrachloride-induced liver injury in rodents. *Liver Int.* **2007**, *27*, 393–399.

Nan, J.X.; Jiang, Y.Z.; Park, E.J.; Ko, G.; Kim, Y.C.; Sohn, D.H. Protective effect of *Rhodiola sachalinensis* extract on carbon tetrachloride-induced liver injury in rats. *J. Ethnopharmacol.*, **2003**, *84*, 143–148.

Narayanasamy, K.; Selvi, V. Hepatoprotective effect of a polyherbal formulation (ayush-liv. 04) against ethanol and CCl_4 induced liver damage in rats. *Anc. Sci. Life,* **2005**, *25*(1), 28–33.

Naseri, L.; Khazaeri, M.; Ghanbari, E.; Bazm. M.A. *Rumex alveollatus* hydroalcoholic extract protects CCl_4-induced hepatotoxicity in mice. *Comp. Clin. Pathol.* **2019**, *28*(2), 557–565.

Nasim, I.; Sadiq, M.; Jehangir, A. Hepatoprotective effect of garlic (*Allium sativum*) and milk thistle (Silymarin) in isoniazid induced hepatotoxicity in rats. *Biomedica* **2011**, *27*(1), 166–170.

Nasir, A.; Abubakar, M.G.; Shehu, R.A.; Aliyu, U.; Toge, B.K. Hepatoprotective effect of the aqueous leaf extract of *Andrographis paniculata* Nees against carbon tetrachloride induced hepatotoxicity in rats. *Nig. J. Basic Appl. Sci.* **2013**, *21*, 45–54.

Natu, V.; Agarwal, S.; Agarwal, S. L.; Agarwal, S. Protective effect of *Ricinus communis* leaves in experimental liver injury. *Indian J. Pharmacol.* **1977**, *9*(4), 265–268.

Nayak, D.P.; Dinda, S.C.;, Swain, P.K.; Kar, B.; Patro, V.J. Hepatoprotective activity against CCl_4-induced hepatotoxicity in rats of *Chenopodium album* aerial parts. *J. Phytother. Pharmacol.* **2012**, *1*(2), 33–41.

Nayak, S.S.; Jain, R.; Sahoo, A.K. Hepatoprotective activity of *Glycosmis pentaphylla* against paracetamol induced hepatotoxicity in Swiss albino mice. *Pharm. Biol.* **2011**, *49*, 111–117.

Nayak, V.; Gincy, T.B.; Prakash, M.; Joshi, C.; Rao, S.; Somayaji, S.N. et al.,

Hepatoprotective activity of *Aloe vera* gel against paracetamol induced hepatotoxicity in Albino rats. *Asian J. Phar. Biol. Res.* **2011**, *1*(2), 94–98.

Nazeema, T.H.; Brindha, V. Antihepatotoxic and antioxidant defense potential of *Mimosa pudica. Int. J. Drug Disc.* **2009**, *1*(2), 1–4.

Nazneen, M.; Mazid, M.A.; Kundu, J.K., Bachar, S.C.; Begum, F.; Datta, B.K. Protective effects of *Flacourtia indica* aerial parts extracts against paracetamol induced hepatotoxiciy in rats. *J. Taibah Univ. Sci.* **2009**, *2*, 1–6.

Neelima, S.; Pradeep Kumar, M.; Hari Kumar, C. An investigation of hepatoprotective activity of methanolic extract of *Ipomoea reniformis* on experimentally induced ethanol hepatotoxicity in rats. *Clin. Exp. Pharmacol.* **2017**, *7*, 227. doi:10.4172/2161-1459.1000227

Negahban, M.; Moharramipour, S.; Yousefelahi, M. Efficacy of essential oil from *Artemisia scoparia* Waldst. & Kit. against *Tribolium castaneum* (Herbst) (Coleoptera: Tenebrionidae). In: Proceedings of the 4th International Iran & Russia Conference, Agricultural and Natural Resources, Shahrekord, Iran, September **2004**.

Nehar, S.; Kumari, M. Ameliorating effect of *Nigella sativa* oil in thioacetamide-induced liver cirrhosis in albino rats. *Indian J. Pharm. Education Res.* **2013**, *47*(2), 135–139.

Nema, A.K.; Agarwal, A.; Kashaw, V. Hepatoprotective activity of *Leptadenia reticulata* stems against carbon tetrachloride-induced hepatotoxicity in rats. *Indian J. Pharmacol.* **2011**, *43*(3), 254–257.

Nerdy, N.; Ritarwan, K. Hepatoprotective activity and nephroprotective activity of peel extract from three varieties of the passion fruit (*Passiflora sp.)* in the albino rat. *Open Access Maced. J. Med. Sci.* **2019**, *7*(4), 536–542.

Nicoletti, N.F.; Rodrigues-Junior, V.; Santos, A.A.; Leite, C.E.; Dias, A.C.; Batista, E.L.; et al., Protective effects of resveratrol on hepatotoxicity induced by isoniazid and rifampicin via SIRT1 modulation. *J. Nat. Prod.* **2014**. *77*(10), 2190–2195.

Nigam, V.; Paarakh, P.M. Hepatoprotective activity of *Chenopodium album* Linn. against paracetamol induced liver damage. *Pharmacol. Online*. **2011**, *3*, 312–328.

Nikolaev, S.M.; Tsyrenzhapov, A.V.; Sambueva, Z.G.; Nikolaeva, G.G.; Ratnikova, G.V.; Tankhaeva, L.M. [Hepatoprotective effect of granules of the dry extract obtained from *Gentianopsis barbata* (Froel) Ma]. *Eksp. Klin. Farmakol.* **2001**, *64*(1), 49–52. (in Russian)

Nili-Ahmadabadi, A.; Alibolandi, P.; Ranjbar, A.; Mousavi, L.; Nili-Ahmadabadi, H.; Larki-Harchegani. A.; Ahmadimoghaddam, D.; Omidifar, N. Thymoquinone attenuates hepatotoxicity and oxidative damage caused by diazinon: An *in vivo* study. *Res. Pharm. Sci.* **2018,** *13*(6), 500–508.

Nimbkar, S.R.; Juvekar, A.R.; Jogalekar, S.N. Hepatoprotective activity of *Fumaria indica* in hepatotoxicity induced by antitubercular drugs treatment. *Indian Drugs* **2000**, *37*, 537–542.

Nirmala, M.; Girija, K.; Lakshman, K.; Divya, T. Hepatoprotective activity of *Musa paradisiaca* on experimental animal models. *Asian Pac. J. Trop. Biomed.* **2012**, *2*, 11–15.

Nirwane, A.M.; Bapat, A.R. Effect of methanolic extract of *Piper nigrum* fruits in ethanol-CCl_4 induced hepatotoxicity in Wistar rats. *Der Pharm. Lett.* **2012**, *4*(3), 795–802.

Nithianantham, K.; Ping, K.Y.; Latha, L.Y.; Jothy, S.L.; Darah, I.; Chen, Y.; Chew, A.-L.; Sasidharan, S. Evaluation of hepatoprotective effect of methanolic extract of *Clitoria ternatea* (Linn.) flower against acetaminophen induced liver damage. *Asian Pac. J. Trop. Dis*. **2013**, *3*(4): 314–319.

Nithianantham, K.; Shyamala, M.; Chen, Y.; Latha, L.Y.; Jothy, S.L.; Sasidharan, S. Hepatoprotective potential of *Clitoria ternatea* leaf extract against paracetamol induced damage in mice. *Molecules* **2011**, *16*, 10134–10145.

Njayou, F.N.; Ngoungoure, F.P.; Tchana, A.; Moundipa, P.F. Protective effect of *Khaya grandifoliola* C. DC. stem bark extract on carbon tetrachloride-induced hepatotoxicity in rats. *Int. J. Indige. Med. Plants.* **2013a,** *29*, 11–16.

Njayou, F.N.; Aboudi, E.C.E.; Tandjang, M.K.; Tchana, A.K.; Ngadjui, B.T.; Moundipa, P.F. Hepatoprotective and antioxidant activities of stem bark extract of *Khaya randifoliola* (Welw.) C.DC and *Entada africana* Guill. et Perr. *J. Nat. Prod.* **2013b**. *6*, 73–80.

Njayou, N.I., Moundipa, P.F., Donfack, J.H., Chuisseu, P.D., Tchana, A.N., Ngadjui, B.T., et al., Hepato-protective, antioxidant activities and acute toxicity of a stem bark extract of *Erythrina senegalensis* DC. *Int. J. Biol. Chem. Sci.* **2010**, *3*, 738–747.

Njouendou, A. J.; Nkeng-Efouet, A. P.; AssobNguedia, J. C.; Chouna, J. R.; Veerapur, V. P.; Thippeswamy, B. S.; Badami, S.; Wanji, S. Protective effect of *Autranella congolensis* and *Sapium ellipticum* stem bark extracts against hepatotoxicity-induced by thioacetamide. *Pharmcologyonline* **2014**, *2*, 38–47.

Nkosi, C.Z.; Opoku, A.R.; Terblanche, S.E. *In vitro* antioxidative activity of pumpkin seed (*Cucurbita pepo*) protein isolate and its *in vivo* effect on alanine transaminase and aspartate transaminase in acetaminophen-induced liver injury in low protein fed rats. *Phytother. Res*., **2006**, *20*, 9, 780–783.

Noguchi, T.; Lai, E. K.; Alexander, S. S.; et al., Specificity of a phenobarbital-induced cytochrome P-450 for metabolism of carbon tetrachloride to the trichloromethyl radical. *Biochem. Pharmacol.* **1982**, *31*(5), 615–624.

Noh, J.-R.; Gang, G.-T.; Kim, Y.-H. Antioxidant effects of the chestnut (*Castanea crenata*) inner shell extract in *t*-BHP-treated HepG2 cells, and CCl_4- and high-fat diet-treated mice. *Food Chem. Toxicol.* **2010**, *48*(11), 3177–3183.

Noorani, A.; Gupta, K.; Bhadada, K.; Kale, M.K. Protective effect of methanolic leaf extract of *Caesalpinia bonduc* L. on gentamicin-induced hepatotoxicity and nephrotoxicity in rats. *Iranian J. Pharmacol. Therapeut.*, **2011**, *10*, 21–25.

Nwachukwu, D.C.; Okwuosa, C.N.; Chukwu, P.U.; Nkiru, A.N., Udeani, T. Hepatoprotective activity of methanol extract of the seeds of *Sesamum indicum* in carbon tetrachloride induced hepatotoxicity in rats. *Indian J. Novel Drug Delivery.* **2011**, *3*(1), 36–42.

Nwaehujor, C.O.; Eban, L.K.; Ode, J.O.; Ejiofor, C.E.; Igile, G.O. Hepatotoxicity of methanol seed extract of *Aframomum melegueta* (Roscoe) K. Schum. (grains of paradise) in Sprague Dawley rats. *Am. J. Biomed. Res.* **2014**, *2*(4), 61–66.

Nwanna, E.E.;, Oboh, G. Antioxidant and hepatoprotective properties of polyphenol extracts from *Telfairia occidentalis* (fluted pumpkin) leaves on acetaminophen-induced liver damage. *Pak. J. Biol. Sci.* **2007**, *10*(16), 2682–2687.

Nwidu, L.L.; Oboma, Y.I.; Elmorsy, E.; Carter, W.G. Hepatoprotective effect of hydromethanolic leaf extract of *Musanga cecropioides* (Urticaceae) on carbon tetrachloride-induced liver injury and oxidative stress. *J. Taibah Univ. Med. Sci.* **2018**, *13*(4), 344–354.

Nwidu, L.L.; Elmorsy, E.; Yibala, O.I.; Carter, W.G. Hepato-protective and antioxidant effects of *Spondias mombin* leaf and stem extracts upon carbon tetrachloride-induced hepatotoxicity and oxidative stress. *J. Basic Clin. Pharm.* **2017**, *8*, S11-S19.

Nwidu, L.L.; Elmorsy, E.; Yibala, O.I.; Carter, W.G. Hepatoprotective and antioxidant activities of *Spondias mombin* leaf and stem extracts against carbon tetrachloride-induced hepatotoxicity. *J. Taibah Univ. Med. Sci.* **2018**, *13*(3), 262–271.

Nwobodo, E.I.; Nwosu, D.C.; Nwanjo, H.U.; Ihim, A.C.; Nnodim, J.K.; Nwobodo, C.I.; Edwrad, U.C. Vitamins C and E levels are enhanced by *Azadirachta indica* leaves aqueous extract in paracetamol induced hepatotoxicity in Wistar rats. *J. Med. Plants Res.* **2016**, *10*, 338–43.

Nwozo, S.O.; Oyinloye, B.E. Hepatoprotective effect of aqueous extract of *Aframomum melegueta* on ethanol-induced toxicity in rats. *Acta Biochem. Pol.* **2011**, *58*, 355–358.

Obi, F.O.; Uneh, E. pH dependent prevention of carbon tetrachloride-induced lipoperoxidation in rats by ethanolic extract of *Hibiscus rosa-sinensis* petal. *Biokemistri.* **2003**, *13*, 42–50.

Obianime, A.W.; Uche, F.I. The Phytochemical screening and the effects of methanolic extract of *Phyllanthus amarus* leaf on the biochemical parameters of male guinea pigs. *J. Appl. Sci.Environ. Manage.* **2008**, *12*(4), 73-77.

Obioha, U.E.; Suru, S.M.; Ola-Mudathir, K.F.; Faremi, T.Y. Hepatoprotective potentials of onion and garlic extracts on cadmium-induced oxidative damage in rats. *Biol. Trace Elem. Res.* **2009**, *129*, 143–156.

Obogwu M. B.; Akindele A. J.; Adeyemi O. O. Hepatoprotective and *in vivo* antioxidant activities of the hydroethanolic leaf extract of *Mucuna pruriens* (Fabaceae) in antitubercular drugs and alcohol models. *Chin. J. Nat. Med.* **2014**, *12*(4), 273–283.

Oboh, G. Hepatoprotective property of ethanolic and aqueous extracts of fluted pumpkin (*Telfairia occidentalis*) leaves against garlic-induced oxidative stress. *J. Med. Food* **2005**, *8*(4). 560–563.

Oboh, G. Antioxidative potential of *Ocimum gratissimum* and *Ocimum canum* leaf polyphenols and protective effects on some pro-oxidants induced lipid peroxidation in rat brain: An *in vitro* study. *Am. J. Food Technol.* **2008**, *3*, 325–334.

Obouayeba, A.P.; Boyvin, L.; M'Boh, G.M.; Diabete, S., Kouakou, T.H.; Djaman, A.J.; N'Guessan, J.D. Hepatoprotective and antioxidant activities of *Hibiscus sabdariffa* petal extracts in Wistar rats. *Int. J. Basic Clin. Pharmacol.* **2014**, *3*(5), 774–780.

Odutuga, A.A.; Dairo, J.O.; Ukpanukpong, R.U.; Eze, F.N. Hepatoprotective activity of ethanol extracts of *Ficus exasperata* leaves on acetaminophen-induced hepatotoxic rats. *Merit Res. J. Biochem. Bioinform.* **2014**, *2*, 28–33.

Ofeimun, J.O.; Eze, G.I.; Okirika, O.M.; Uanseoje, S.O. Evaluation of the Hepatoprotective effect of the methanol extract of the root of *Uvaria afzelii* (Annonaceae). *J. Appl. Pharma Sci.* **2013**, *3*, 125–129.

Oh, G.S.; Yoon, J.; Lee, G.G.; Kwak, J.H.; Kim, S.W. The Hexane fraction of *Cyperus rotundus* prevents non-alcoholic fatty liver disease through the inhibition of liver X receptor α-mediated activation of sterol regulatory element binding protein-1c. *Am. J. Chin. Med.* **2015**, *43*(3), 477–494.

Ojo, O.O.; Nadro, M.S.; Tella, I.O. Protection of rats by extracts of some common Nigerian trees against acetaminophen-induced Hepatotoxicity. *Afr. J. Biotechnol.* **2006**, *5*, 755–760.

Okoh, O.O.; Sadimenko, A.P.; Asekun, O.T.; Afolayan, A.J. The effects of drying on the chemical components of essential oils of *Calendula officinalis* L. *Afr. J. Biotechnol.* **2008**, *7*(10), 1500–1502.

Okoko, T. Paracetamol and Hydrogen peroxide-induced tissue alterations are reduced by *Costus afer* extract. *Am. J. Res. Med. Sci.* **2018**; *2*(2), 66–72

Okokon, J.E.; Bawo, M.B.; Mbagwu, H.O. Hepatoprotective activity of *Mammea africana* ethanol stem bark extract. *Avicenna J. Phytomed.* **2016**, *6*, 248–59.

Okokon, J.E.; Simeon, J.O.; Umoh, E.E. Hepatoprotective activity of the extract of *Homalium letestui* stem against paracetamol-induced liver injury. *Avicenna J. Phytomed.* **2017**, *7*, 27–36.

Okoli, J.T.N.; Agbo, M.O.; Ukekwe, I.F. Antioxidant and hepatoprotective activity of fruit extracts of *Tetrapleura tetraptera* (Schum. & Thonn.) Taubert. *Jordan J. Biol. Sci.* **2014**, *7*(4), 251–254.

Okoye, J.O.; Nwachukwu, D.A.; Nnatuanya, I.N.; Nwakulite, A.; Alozie, I.; Obi, P.E.; Faloye, G.T. Anticholestasis and antisinusoidal congestion properties of aqueous extract of *Annona muricata* stem bark following acetaminophen induced toxicity. *European J. Exp. Biol.* **2016**, *6*, 1–8.

Okwuosa, C.N.; Udeani, T.K.; Umeifekwem, J.E.; Conuba, E.; Anioke, I.E.; Madubueze, R.E. Hepatoprotective effect of methanolic fruit extracts of *Phoenix dactylifera* (Arecaceae) on thioacetamide-induced liver damage in rats. *Am. J. Phytomed. Clin. Ther.* **2014**, *2*, 290–300.

Olaleye, M.T.; Adegboye, O.O.; Akindahunsi, A.A. *Alchornea cordifolia* extract protects Wistar albino rats against acetaminophen induced liver damage. *Afr. J. Biotechnol.*, **2006**, *5*(24), 2439–2445.

Olaleye, M.T.; Akinmoladun, A.C.; Ogunboye, A.A.;, Akindahunsi, A.A. Antioxidant activity and hepatoprotective property of leaf extracts of *Boerhaavia diffusa* Linn against acetaminophen-induced liver damage in rats. *Food Chem. Toxicol.*, **2010**, *48*(8–9), 2200–2205.

Olaleyea, M.T.; Rochaa, B.T. Acetaminophen-induced liver damage in mice: Effects of some medicinal plants on the oxidative defence system. *Exper. Toxicol. Pathol.* **2008**, *59*, 319–327.

Oliveira, F.A.; Chaves, M.H.; Almeida, F.R.C.; Lima, R.C.P.; Silva, R.M.; Maia, J.L.; Brito, G.A.C.; Santos, F.A.; Rao, V.S. Protective effect of α- and β-amyrin, a triterpene mixture from *Protium heptaphyllum* (Aubl.) March. trunk wood resin, against acetaminophen-induced liver injury in mice. *J. Ethnopharmacol.*, **2005**, *98*, 103–108.

Olorunnisola, O.S.; Akintola, A.O.; Afolayan, A.J. Hepatoprotective and antioxidant effect of *Sphenocentrum jollyanum* (Menispermaceae) stem bark extract against CCl_4-induced oxidative stress in rats. *African J. Pharm. Pharmacol.* **2011**, *5*(9), 1241–1246.

Omeodu, S.I.; Ezeonwumelu, E.C.; Aleme, B.M.; Oretan, H.O. Hepatoprotective effect of aqueous leaf extract of soursop on some biochemical parameters of paracetamol-induced liver damage of Wistar rats. *Scientia Africana* **2017**, *16*(1).

Omidi, A.; Riahinia, N.; Montazer Torbati, M., Behdani, M. Hepatoprotective effect of *Crocus sativus*(saffron) petals extract against acetaminophen toxicity in male Wistar rats. *Avicenna J. Phytome*. **2014**, *4*, 330–336.

Omodanisi, E.I.;, Aboua, Y.G.;, Chegou, N.N.;, Oguntibeju, O.O. Hepatoprotective, antihyperlipidemic, and anti-inflammatory activity of *Moringa oleifera* in diabetic-induced damage in male Wistar rats. *Pharmacogn. Res.* **2017**, *9*(2), 182–187.

Omotayo, M.A.; Ogundare, O.C.; Longe, A.O.; Adenekan, S. Hepatoprotective effect of *Mangifera indica* stem bark extracts on paracetamol-induced oxidative stress in albino rats. *Eur. Sci. J.* **2015**, *11*, 1857–7431.

Öner, A.C.; Mercan, U.; Öntürk, H.; Cengiz, N.; et al., Anti-inflammatory and hepatoprotective activities of *Trigonella foenum-graecum* L. *Pharmacologyonline* **2008**, *2*, 126–132.

Onoja, S.O.; Madubuike, G.K.; Ezeja, M.I. Hepatoprotective and antioxidant activity of hydromethanolic extract of *Daniella oliveri* leaves in carbon tetrachloride-induced hepatotoxicity in rats. *J. Basic Clin. Physiol. Pharmacol.* **2015**, *26*(5), 465–470.

Opoku, A.R.; Ndlovu, I.M.; Terblanche, S.E.; Hutchings, A.H. *In vivo* hepatoprotective effects of *Rhoicissus tridentata* subsp. *cuneifolia*, a traditional Zulu medicinal plant, against CCl_4-induced acute liver injury in rats. *South Afr. J. Bot.* **2007**, *73*, 372–377.

Orhan, D.D.; Aslan, M.; Aktay, G.; Ergun, E.; Yesilada, E.; Ergun, F. Evaluation of hepatoprotective effect of *Gentiana olivieri* herbs on subacute administration and isolation of active principle. *Life Sci.* **2003**, *72*, 2273–2283.

Orhan, D.D.; Orhan, N.; Ergun, E.; Ergun, F. Hepatoprotective effect of *Vitis vinifera* leaves on carbontetrachloride-induced acute liver damage in rats. *J. Ethnopharmacol.* **2007**, *112*, 145–151.

Oriakhi, K.; Uadia, P.O.; Eze, I.G. Hepatoprotective potentials of methanol extract of *T. conophorum* seeds of carbon tetrachloride induced liver damage in Wistar rats. *Clin. Phytosci.* **2018**, *4*, 25, doi.: 10.1186/s40816-018-0085–8

Orji, O.U.; Ibiam, U.A.; Aja, P.M.; Uraku, A.J.; Inya-Agha, O.R.; Ugwu Okechukwu, P.C. Hepatoprotective activity of ethanol extract of *Vernonia ambigua* against carbon tetrachloride induced hepatotoxicity in albino rats. *IOSR J. Dental Med. Sci.* **2015,** *14*, 22–29.

Osadebe, P.O.; Festus, B.C.; Philip, F.U.; Nneka, R.N.; Ijeoma, E.A.; Nkemakonam, C.O. Phytochemical analysis, hepatoprotective and antioxidant activity of *Alchornea cordifolia* methanol leaf extract on carbon tetrachloride-induced hepatic damage in rats. *Asian Pac. J. Trop. Med.* **2012**, *5*, 289–293.

Ottu, O.J.; Atawodi, S.E.; Onyike, E. Antioxidant, hepatoprotective and hypolipidemic effects of methanolic root extract of *Cassia singueana* in rats following acute and chronic carbon tetrachloride intoxication. *Asian Pac. J. Trop. Med.* **2013**, *6*, 609–615.

Oyeyemi, I.T.; Akanni, O.O.; Adaramoye, O.A.; Bakare, A.A. Methanol extract of *Nymphaea lotus* ameliorates carbon tetrachloride-induced chronic liver injury in rats via inhibition of oxidative stress. *J. Basic Clin. Physiol. Pharmacol.* **2017**, *28*, 43–50.

Özbek, H.; Çitoğlu, G. S.; Dülger, H.; Uğraş, S.; Sever, B. Hepatoprotective and anti-inflammatory activities of *Ballota glandulosissima. J. Ethnopharmacol.* **2004**, *95*(2–3), 143–149.

Ozenirler, S.; Dinçer, S.; Akyol, G.; Ozoğul, C.; Oz, E._ The protective effect of *Ginkgo biloba* extract on CCl_4-induced hepatic damage. *Acta Physiol. Hungarica* **1997**, *85*(3), 277–285.

Ozturk, F.; Gul, M.; Ates, B.; Ozturk, I.C.; Cetin, A.; Vardi, N.; et al., Protective effect of apricot (*Prunus armeniaca* L.) on hepatic steatosis and damage induced by carbon tetrachloride in Wistar rats. *Br. J. Nutr.* **2009**, *102*: 1767–1775.

Öztürk, Y.; Aydin, S.; Baser, K. H. C.; Kirimer, N.; KurtarOzturk, N. Hepatoprotective activity of *Hypericum perforatum* L. alcoholic extract in rodents. *Phytother. Res.* **1992**, 6(1), 44–46.

Paarakh, P.M. *Nigella sativa* Linn.-A comprehensive review. *Indian J. Nat. Prod. Resour.* **2010**, *1*, 409–429.

Padhy, B.M.; Srivastava, A.; Kumar, V.L. *Calotropis procera* latex affords protection against carbon tetrachloride induced hepatotoxicity in rats. *J. Ethnopharmacol.* **2007**, *113*, 498–502.

Padma, P.; Setty, O.H. Protective effect of *Phyllanthus fraternus* against carbon tetrachloride-induced mitochondrial dysfunction. *Life Sci.* **1999**, *64*, 2411–2417.

Pal, A.; Banerjee, B.; Banerjee, T.; Masih, M.; Pal, K. Hepatoprotective activity of *Chenopodium album* Linn. plant against paracetamol-induced hepatic injury in rats. *Int. J. Pharm. Pharm. Sci.* **2011**, *3*, 55–57.

Palanisamy, N.; Dass, S.M. Beneficial effects of *Trichopus zeylanicus* extract on mercuric chloride-induced hepatotoxicity in rats. *J. Basic Clin. Physiol. Pharmacol.* **2013**, *24*, 51–57.

Palanivel, M.G.; Rajkapoor, B.; Senthil, K.R.; Einstein, J.W.; Kumar, E.P.; Rupesh, K.M.; et al., Hepatoprotective and antioxidant effect of *Pisonia aculeata* L. against CCl_4 induced hepatic damage in rats. *Sci. Pharm.* **2008**, *76*, 203–215.

Palati, D.J.; Vanapatla, S.R. Protective role of *Aerva monsoniae* and selenium on cadmium-induced oxidative liver damage in rats. *Asian J. Pharm. Clin. Res.* **2018**, *11*(6),

Pal, R.; Vaiphei, K.; Sikander, A.; Singh, K.; Rana, S.V. Effect of garlic on isoniazid and rifampicin-induced hepatic injury in rats. *World J. Gastroenterol.* **2006**, *12*(4), 636–639.

Panda, V.; Kharat, P.; Sudhamani, S. Hepatoprotective effect of the *Macrotyloma uniflorum* seed (Horse gram) in ethanol-induced hepatic damage in rats. *J. Biol. Active Prod. Nat.* **2015a**, *5*(3), 178–191.

Panda, V.S.; Kharat, P.; Sudhamani, S. Antioxidant and hepatoprotective effect of *Macrotyloma uniflorum* seed in antitubercular drug induced liver injury in rats. *J. Phytopharmacol.* **2015b**, *4*(1), 22–29.

Pandian, S.; Badami, S.; Shankar, M. Hepatoprotective activity of methanolic extract of *Oldenlandia herbacea* against D-galactosamine induced rats. *Int. J. Appl. Res. Nat. Prod.* **2008**, *6*(1), 16–19.

Pandit, A.; Sachdeva, T.; Bafna, P. Ameliorative effect of leaves of *Carica papaya* in ethanol and antitubercular drug induced hepatotoxicity. *Br. J. Pharm. Res.* **2013**, *3*, 648–61.

Pandit, S.; Sur, T.K.; Jana, U.; Debnath, P.K.; Sen, S.; Bhattacharyya, D. Prevention of carbon tetrachloride-induced hepatotoxicity in rats by *Adhatoda vasica* leaves. *Indian J. Pharmacol.* **2004**, *36*, 312–313.

Panovska, T.K.; Kulevanova, S.; Gjorgoski, I.; Bogdanova, M.; Petrushevska, G. Hepatoprotective effect of the ethyl acetate extract of *Teucrium polium* L. against carbontetrachloride-induced hepatic injury in rats. *Acta Pharm.*, **2007**, *57*(2), 241–248.

Paramesha, M.; Ramesh, C.K.; Krishna, V.; Ravi Kumar, Y.S.; Parvathi, K.M.M. Hepatoprotective and *in vitro* antioxidant effect of *Carthamus tinctorious* L., var Annigeri-2-, an oil-yielding crop, against CCl_4-induced liver injury in rats. *Pharmacogn. Mag.* **2011**, *7*(28), 289–297.

Parameshwar, P.; Reddy, Y.N.; Aruna Devi, M. Hepatoprotective and antioxidant activities of *Ziziphus rotundifolia* (Linn.) against carbon tetrachloride-induced hepatic damage in rats. *Int. J. Pharm. Sci. Nanotech.* **2012**, *5*(3), 1775–1779.

Parameswari, S.A.; Saleem, T.; Chandrasekar, K.; Chetty, C.M. Protective role of *Ficus benghalensis* against isoniazid-rifampicin-induced oxidative liver injury in the rat. *Rev. Brasil. Farmacognosia* **2012**, *22*, 604–610.

Parameswari, S.A.; Chetty, C.M.; Chandrasekhar, K.B. Hepatoprotective activity of *Ficus religiosa* leaves against isoniazid+ rifampicin and paracetamol induced hepatotoxicity. *Pharmacogn. Res.* **2013**, *5*(4), 271–276.

Pareek, A.; Godavarthi, A.; Issarani, R.; Nagori, B.P. Antioxidant and hepatoprotective activity of *Fagonia schweinfurthii* (Hadidi) Hadidi extract in carbon tetrachloride induced hepatotoxicity in HepG2 cell line and rats. *J. Ethnopharmacol.* **2013**, *150*(3), 973–981.

Pari, L.; Kumar, N.A. Hepatoprotective activity of *Moringa oleifera* on antitubercular drug-induced liver damage in rats. *J. Med. Food.* **2002**, *5*(3), 171–177.

Pari, L.; Uma, A. Protective effect of *Sesbania grandiflora* against erythromycin estolate-induced hepatotoxicity. *Therapie* **2003**, *58*, 439–443.

Park, E.J.; Jeon, C.H.; Ko, G.; Kim, J.; Sohn, D.H. Protective effect of curcumin in rat liver injury induced by carbon tetrachloride. *J. Pharm. Pharmacol.* **2000**, *52*(4), 437–440.

Park, E.Y.; Ki, S.H.; Ko, M.S.; Kim, C.W.; Lee, M.H.; Lee, Y.S.; et al., Garlic oil and DDB, comprised in a pharmaceutical composition for the treatment of patients with viral hepatitis, prevents acute liver injuries potentiated by glutathione deficiency in rats. *Chem. Biol. Interact.* **2005**, *155*, 82–96.

Parmar, H.B.; Das, S.K.; Gohil, K.J. Hepatoprotective activity of *Macrotyloma uniflorum* seed extract on paracetamol and d-galactosamine induced liver toxicity in albino rats. *Int. J. Pharmacol. Res.* **2012**, *2*(2), 86–91.

Parmar, S.; Dave, G.; Patel, H.; Kalia, K. Hepatoprotective value of some plants extract against carbon tetrachloride toxicity in male rats. *J. Cell Tissue Res.* **2009**, *9*, 1737–1743.

Parmar, S.R.; Vashrambhai, P.H.; Kalia, K. Hepatoprotective activity of some plants extract against paracetamol-induced hepatotoxicity in rats. *J. Herbal. Med. Toxicol.* **2010**, *4*, 101–06.

Patel, P.B.; Patel, T.K.; Shah, P.; Baxi, S.N.; Sharma, H.O.; Tripathi, C. B. Protective effect of ethanol extract of *Gymnosporia montana* (Roth) Benth. in paracetamol-induced hepatotoxicity in rats. *Indian J. Pharm. Sci.* **2010**, *72*(3), 392–396.

Patel, B.A.; Patel, J.D.; Rajval, B.P. Hepatoprotective activity of *Saccharum officinarum* L. against paracetamol induced hepatotoxicity in rats. *Int. J. Pharm. Sci. Res.* **2010a**, *4*(1), 102–108.

Patel, B.A.; Patel, J.D.; Raval, B.P.; Gandhi, T.R. The protective activity of *Saccharum officinarum* L. against CCl_4 induced hepatotoxicity in rats. *Int. J. Pharm. Res. 2010b.* *2*(3), 5–8.

Patel, B.A.; Patel, J.D.; Rajval, B.P.; Gandhi, T.R.; Patel, K.; Patel, P.U. Hepatoprotective activity of *Saccharum officinarum* L. against ethyl alcohol induced hepatotoxicity in rats. Scholar's research library. *Der Pharm. Lett.* **2010c**, *2*(1), 94–101.

Patel, J.A.; Shah, U.S. Hepatoprotective activity of *Piper longum* traditional milk extract on carbon tetrachloride induced liver toxicity in Wistar rats. *Boletín Latinoamericano y del Caribe de Plantas Medicinales y Aromáticas* **2009**, *8*, 121–129.

Patel, P.B.; Patel, T.K.; Shah, P.; Baxi, S.N.; Sharma, H.O.; Tripathi, C.B. Protective effect of ethanol extract of *Gymnosporia montana* (Roth) Benth. in paracetamol induced hepatotoxicity in rats. *Indian J. Pharmaceut. Sci.* **2010**. *72*(3), 392–396.

Patel, R.K., Patel, M.M.; Kanzariya, N.R.; Vaghela, K.R.; Patel, R.K.; Patel, N.J. In vitro hepatoprotective activity of *Moringa oleifera* Lam. leaves on isolated rats hepatocytes. *Int. J. Pharm. Sci.* **2010**, *2*(1), 457–463.

Patel, S.A.; Rajendra, S.; Setty, S.R. Hepatoprotective activity of *Oxalis stricta* Linn. on paracetamol-induced hepatotoxicity in albino rats. *Algerian J. Nat. Prod.* **2016**, *4*, 233–240.

Patel, T.; Shirode, D.; Roy, S.P.; Kumar, S.; Setty, S.R. Evaluation of antioxidant and hepatoprotective effects of 70% ethanolic bark extract of *Albizzia lebbeck* in rats. *Int. J. Res. Pharm. Sci.* **2010**, 1(3), 270–276.

Patil, S,; Shetti, A.; Hoskeri, H.J.; Rajeev, P.; Kulkarni, B.; Kamble, G.R.; Hiremath, G.B.; Kalebar, V.; Hiremath, S.V. Hepatoprotective activity of methanolic shoot extract of *Bambusa bambos* against carbon tetrachloride induce acute liver toxicity in Wistar rats. *J. App/ Biol. Biotech.* **2018**, *6*(04), 37–40.

Patrick-Iwuanyanwu, K.C.; Iwuanyanwu, P.; Matthew, O.; Makhmoor, T. Hepatoprotective effect of crude methanolic extract and fractions of Ring worm plant *Senna alata* (L.) Roxb. leaves from Nigeria against carbon tetrachloride induced hepatic damage in rats. *Eur. J. Exp. Biol.* **2011,** *1*, 128–138.

Patrick-Iwuanyanwu, K.C.; Wegwu, M.O.; Ayalogu, E.O. Prevention of CCl_4-induced liver damage by ginger, garlic and vitamin E. *Pak. J. Biol. Sci.* **2007**, *10*, 617–621.

Patrick-Iwuanyanwu, K.C.; Wegwu, M.O.; Okiyi, J.K. Hepatoprotective effects of African locust bean (*Parkia clappertoniana*) and Negro Pepper (*Xylopia aethiopica*) in CCl_4-induced liver damaged Wistar Albino rats. *Int. J. Pharmacol.* **2010**, *6*, 744–749.

Pattanayak, S.; Nayak, S.S.; Panda, D.P.; Dinda, S.C.; Shende, V.; Jadav, A.

Hepatoprotective activity of crude flavonoids extract of *Cajanus scarabaeoides* in paracetamol intoxicated albino rats. *Asian J. Pharm. Biol. Res.* **2011**, *1*, 22–27.

Paul, S.; Islam, M.A.; Tanvir, E.; Ahmed, R.; Das, S.; Rumpa, N-E.; Hossen, M.S.; Parvez, M.; Gan, S.H.; Khalil, M.I. Satkara (*Citrus macroptera*) fruit protects against acetaminophen-induced hepatorenal toxicity in rats. *Evid. Based Complement. Altern. Med.* **2016**, *2016*. doi: 10.1155/2016/9470954

Păunescu, A.; Ponepal, C.M.; Drăghici, O.; Marinescu, A.G. The CCl_4 action upon physiological indices in *Lepus timidus* and the protective role of some substances. *An. UO Fasc. Biol.* **2009**, *16*, 104–107.

Pedrós, C.; Cereza, G.; Garcia, N.; Laporte, J.-R. Liver toxicity of *Camellia sinensis* dried ethanolic extract. *Med. Clin.*, **2003**, *121*(15), 598–599.

Peng, W.H.; Chen, Y.W.; Lee, M.S.; Chang, W.T.; Tsai, J.C.; Lin, Y.C.; Lin, M.K. Hepatoprotective effect of *Cuscuta campestris* Yunck. whole plant on carbon tetrachloride induced chronic liver injury in mice. *Int. J. Mol. Sci.* **2016**, *17*(12), doi: 10.3390/ijms17122056

Pimple B.P.; Kadam P.V.; Badgujar N.S.; Bafna A.R.; Patil M.J. Protective effect of *Tamarindus indica* Linn. against paracetamol-induced hepatotoxicity in rats. *Indian J. Pharm. Sci.* **2007**, 69(6), 827–831.

Pingale, S.S. Hepatosuppression by *Adhatoda vasica* against CCl_4-induced liver toxicity in rat. *Pharmacologyonline*. **2009**, *3*, 633–639.

Pithayanukul, P.; Nithitanakool, S.; Bavovada, R. Hepatoprotective potential of extracts from seeds of *Areca catechu* and nutgalls of *Quercus infectoria*. *Molecules* **2009**. *14*, 4987–5000.

Ponmari, G.; Annamalai, A.; Gopalakrishnan, V.K.; Lakshmi, P.T.V.; Guruvayoorappan, C. NF-kB activation and proinflammatory cytokines mediated protective effect of *Indigofera caerulea* Roxb. on CCl_4-induced liver damage in rats. *Int. Immunopharm.* **2014**, *23*(2), 672–680.

Poornima, K.; Perumal, P.C.; Gopalakrishnan, V.K. Protective effect of ethanolic extract of *Tabernaemontana divaricata* (L.) R. Br. against DEN and Fe NTA induced liver necrosis in wistar albino rats. *BioMed. Res. Int.* **2014**, doi.org/10.1155/2014/240243

Porchezhian, E.; Ansari, S. H. Hepatoprotective activity of *Abutilon indicum* on experimental liver damage in rats. *Phytomedicine* **2005**, *12*(1–2), 62–64.

Prabha SP, Ansil PN, Nitha A, Wills PJ, Latha MS: Preventive and curative effect of methanolic extract of *Gardenia gummifera* Linn. on thioacetamide induced oxidative stress in rats. *Asian Pac. J. Trop. Dis.* **2012**, *2012*, 90–98.

Prabakaran, M.; Rangasamy, D.T. Protective effect of *Hemidesmus indicus* against rifampicin and isoniazid induced hepatotoxicity in rats. *Fitoterapia* **2000**, *71*, 55–59.

Prabhakaran, V.; Ashok Kumar B.S.; Shekar D.S.; Nandeesh, R.; Subramanyam, P.; Ranganayakulu, D. Evaluation of the hepatoprotective activity of *Portulaca oleracea* L. on D-galactosmaine-induced hepatic injury in rats. *Boletín Latinoamericano y del Caribe de Plantas Medicinales y Aromáticas* **2010**, *9*(3), 199–205.

Prabhu, K.; Kanchana, N.; Sadique, A.M. Hepatoprotective effect of *Eclipta alba* on paracetamol induced liver toxicity in rats. *J. Microbiol. Biotechnol. Res.* **2011**, *1*, 75–79.

Prabhu, V.V.; Chidambaranathan, N.; Nalini, G.; Venkataraman, S.; Jayaprakash, S.; Nagarajan, M. Evaluation of anti-fibrotic effect of *Lagerstroemia speciosa* (L) Pers. on carbon tetrachloride induced liver fibrosis. *Curr. Pharm. Res.* **2010**, *1*(1), 7–12.

Pradeep, H.A.; Khan, S.; Ravikumar, K.; Ahmed, M.F.; Rao, M.S.; Kiranmai, M.; Reddy, D.S.; Ahamed, S.R.; Ibrahim, M. Hepatoprotective evaluation of *Anogeissus latifolia: In vitro* and *in vivo* studies. *World J. Gastroenterol.* **2009**, *15*, 4816–4822.

Pradeep, K.; Raj Mohan, C.V.; Gobianand, K.; Karthikeyan, S. Protective effect of *Cassia fistula* Linn. on diethylnitrosamine induced hepatocellular damage and oxidative stress in ethanol pre-treated rats. *Biol Res.* **2010**, *43*(1), 113–25. doi: 0716-97602010000100013.

Prakash, T.; Fadadu, S.D.; Sharma, U.R.; Surendra, V.; Goli, D.; Stamina, P.; Kotresha, D. Hepatoprotective activity of leaves of *Rhododendron arboreum* in CCl_4 induced hepatotoxicity in rats. *J. Med. Plant Res.* **2008**, 2, 315–320.

Pramyothin, P.; Chirdchupunsare, H.; Rungsipipat, A.; Chaichantipyuth C. Hepatoprotective activity of *Thunbergia laurifolia* Linn. extract in rats treated with ethanol: *In Vitro* and *in vivo* studies. *J. Ethnopharmacol.* **2005**. *102*, 408–411.

Pramyothin, P.; Ngamtin, C.; Poungshompoo, S.; Chaichantipyuth, C. Hepatoprotective activity of *Phyllanthus amarus* Schumm. et Thonn. extract in ethanol treated rats: *In vitro* and *in vivo* studies. *J. Ethnopharmacol.* **2007**, *114*, 169–173.

Praneetha, P.; Durgaih, G.; Narsimha Reddy, Y.; Ravi Kumar, B. In vitro hepatoprotective effect of *Echinochloa colona* on ethanol-induced oxidative damage in HepG2 cells. *Asian J. Pharm. Clin. Res.* 2017, 10(9), 259–261.

Praveen, T.K.; Dharmaraj, S.D.; Bajaj, J.; Dhanabai, S.P.; Manimaran, S.; Nanjan, M.J.; Razdan, R. Hepatoprotective activity of petroleum ether, diethyl ether and methanol extract of *Scoparia dulcis* L. against CCl_4-induced acute liver injury in mice. *Indian J. Pharmacol.* **2009**, *41*, 110–114.

Preethi, K.C.; Kuttan, R. Hepato and reno protective action of *Calendula officinalis* L. flower extract. *Indian J. Exp. Biol.* **2009**, *47*, 163–168.

Preethi, K.C.; Kuttan, G.; Kuttan, R. Antioxidant potential of an extract of *Calendula officinalis* flowers *in vitro* and *in vivo*. *Pharm. Biol.* **2006**, *44*, 691–697.

Prince, E.S.; Parameswari, P.; Khan, R.M. Protective effect of *Ricinus communis* leaves extract on carbon tetrachloride induced hepatotoxicity in albino rats. *Iran. J. Pharm.* **2011**, *7*(4), 269–278.

Prochezian, E.; Ansari, S.H. Hepatoprotective activity of *Abutilon indicum* on experimental liver damage in rats. *Phytomedicine* **2005**, *12*, 62–64.

Purkayastha, A., Chakravarty, P., Dewan, B. Evaluation of hepatoprotective activity of the ethanolic extract of leaves of *Mimosa pudica* Linn in carbon tetrachloride induced hepatic injury in albino rats. *Int. J. Basic Clin. Pharmacol.* **2016**, *5*(2), 496–501.

Pushpalatha, M.; Ananthi, T. Protective effect of *Solanum pubescens* Linn. on CCl_4 induced hepatotoxicity in albino rats. *Mintage J. Pharm. Med. Sci.* **2012**, *1*(1), 11–13.

Pushpavalli, G.; Veeramani, C.; Pugalendi, K.V. Influence of *Piper betle* on hepatic marker enzymes and tissue antioxidant status in D-galactosamine-induced hepatotoxic rats. *J. Basic Clin. Physiol. Pharmacol.* **2008**, *19*(2), 131–150.

Putri, H.; Nagadi, S.; Larasati, Y.A.; Wulandari, N.; Hermawan, A.; Nugroho, A.E.; Cardioprotective and hepatoprotective effects of *Citrus hystrix* peels extract on rats model. *Asian Pac. J. Trop. Biomed.* **2013**, 3:371–375.

Qadir, M.I.; Ahmed, B.; Ali, M.; Saleem, M.; Ali, M. Natural hepatoprotectives: Alternative medicines for hepatitis. *RGUHS J. Pharm. Sci.* **2013**, *3*(4), 26–27.

Qadir, M.I.; Ali, M.; Ali, M.; Saleem, M.; Hanif, M. Hepatoprotective activity of aqueous methanolic extract of *Viola odorata* against paracetamol-induced liver injury in mice. *Bangladesh J. Pharmacol.* **2014**, *9*, 198–202.

Qadir, M.I.; Murad, M.S.A.; Ali, M.; Saleem, M.; Farooqi, A.A. Hepatoprotective effect of leaves of aqueous ethanol extract of

Cestrum nocturnum against paracetamol-induced hepatotoxicity. *Bangladesh J. Pharmacol.* **2014**, *9*, 167–70.

Quan, J.; Yin, X.; Xu, H. *Boschniakia rossica* prevents the carbon tetrachloride-induced hepatotoxicity in rat. *Exp Toxicol Pathol.* **2011**, *63*(1–2), 53–59.

Qureshi, A.A.; Prakash, T.; Patil, T.; Swamy, A.H.M.V.; Gouda, A.V.; Prabhu, K.; Setty, S.R. Hepatoprotective and antioxidant activities of flowers of *Calotropis procera* (Ait.) R.Br. in CCl_4 induced hepatic damage. *Indian J. Exp. Biol.* **2007**, 304–310.

Qureshi, M.N.; Kuchekar, B.S.; Logade, N.A.; Haleem, M.A. In vitro antioxidant and *In-vivo* hepatoprotective activity of *Leucas ciliata* leaves. *Rec. Nat. Prod.*, **2010**, *4*(2), 124–130.

Qureshi, N.N.; Kuchekar, B.S.; Logade N.A.; Haleem M.A. Antioxidant and hepatoprotective activity of *Cordia macleodii* leaves. *Saudi Pharm. J.*, **2009**, *17*, 299– 302.

Qureshi, S.A.; Rais, S.; Usmani, R.; Zaidi, S.S.M.; Jehan, M.; Lateef, T.; Azmi, M.B. *Centratherum anthelminticum* seeds reverse the carbon tetrachloride-induced hepatotoxicity in rats. *Afr. J. Pharm. Pharmacol.* **2016**, *10*, 533–39.

Radhika, J.; Brinda, P. Free radical scavenging and hepatoprotective activity of *Leucas aspera* Willd. against carbon tetrachloride induced hepatotoxcity in albino rats. *Int. J. Pharm. Tech.* **2011**, *3*(1), 1456–1469.

Rahman, H.; Vakati, K.; Eswaraiah, M.C.; Dutta, A.M. Evaluation of hepatoprotective activity of ethanolic extract of *Aquilaria agallocha* leaves (EEAA) against CCl_4 induced hepatic damage in rat. *Sch. J. Appl. Med. Sci.* **2013**, *1*, 9–12.

Rafatullah, S.; Al-Sheikh, A.; Alqasoumi, S.; Al-Yahya, M.; El-Tahir, K.; Galal, A. Protective effect of fresh radish juice (*Raphanus sativus* L.) against carbon tetrachloride-induced hepatotoxicity. *Int. J. Pharmacol.* **2008**, *4*(2), 130-134.

Rafatullah, S.; Mossa, J.S.; Ageel, A.M.; Al-Yahya, M.A.; Tariq, M. Hepatoprotective and safety evaluation studies on sarsaparilla. *Int. J. Pharmacogn.* **1991,** *29*(4), 296–301.

Rahaman, S.; Chaudhry, A.M. Evaluation of antioxidant and hepatoprotective effect of *Acacia modesta* Wall. against paracetamol induced hepatotoxicity. *Brit. J. Pharm. Res.* **2015**, *5(*5), 336–43.

Rahate, K.P.; Rajasekaran, A. Hepatoprotection by active fractions from *Desmostachya bipinnata* (L.) Stapf against tamoxifen-induced hepatotoxicity. *Indian J. Pharmacol.* **2015**, *47*(3), 311–315.

Rahim, S.M.; Taha, E.M.; Al-janabi, M.S.; Al-douri, B.I.; Simon, K.D.; Mazlan, A.G. Hepatoprotective effect of *Cymbopogon citratus* aqueous extract against hydrogen peroxide-induced liver injury in male rats. *Afr. J. Tradit. Complement. Altern. Med.* **2014**, *11*(2), 447–451.

Rahiman, O.M.F.; Kumar M.R.; Mani T.T.; Niyas K.M.; Kumar B, S.; Phaneendra, P.; Surendra, B. Hepatoprotective activity of "*Asparagus racemosus* root" on liver damage caused by paracetamol in rats. *Indian J. Novel Drug Deliv.* **2011**, 3(2), 112–117.

Raish, M.; Ahmad, A.; Alkharfy, K.M.; Ahamad, S.R.; Mohsin, K.; Al-Jenoobi, F.I.; Al-Mohizea, A.M.; Ansari, M.A. Hepatoprotective activity of *Lepidium sativum* seeds against D-galactosamine/lipopolysaccharide induced hepatotoxicity in animal model. *BMC Complement. Altern. Med.* **2016**. *16*, 501.

Raj, B.; Singh, S.D.J.; Samual, V.J.; John, S.; Siddiqua, A. Hepatoprotective and antioxidant activity of *Cassytha filiformis* against CCl_4 induced hepatic damage in rats. *J. Pharm. Res.* **2013**, *7*, 15–19.

Raj, D.S.; Vennil, J.J.; Aiyavu, C.; Pannerselvam, K. Hepatoprotective effect of alcoholic extracts of *Annona squamosa* leaves on experimentally induced liver injury in swiss albino mice. *Int. J. Integr. Biol.* **2009**, *5*(3), 182-187.

Raj, V.; Mishra, A.K.; Mishra, A.; Khan, N.A. Hepatoprotective effect of *Prunus armeniaca* L. (Apricot) leaf extracts on paracetamol induced liver damage in Wistar rats. *Pharmacog. J.* **2016**, *8*(2), 154–158.

Raj, V.P.; Chandrasekhar, R.H.; Rao, M.C.; Rao, V.J.; Nitesh, K. In vitro and *in vivo* hepatoprotective effects of the total alkaloid fraction of *Hygrophila auriculata* leaves. *Indian J. Pharmacol.* **2010**, *42*(2): 99–104.

Raja, S.; Ahamed, K.F.; Kumar, V.; Mukherjee, K.; Bandyopadhyay, A.; Mukherjee, P.K. Antioxidant effect of *Cytisus scoparius* against carbon tetrachloride treated liver injury in rats. *J. Ethnopharmacol.* **2007**, *109*, 41–47.

Rajamurugan, R.; Suyavaran, A.; Selvaganabathy, N.; Ramamurthy, C.H.; Reddy, G.P.; Sujatha, V.; Thirunavukkarasu, C. *Brassica nigra* plays a remedy role in hepatic and renal damage. *Pharm. Biol.* **2012**, *50*(12), 1488–1497.

Rajan, A.V.; Shanmugavalli, C.; Sunitha, C.G.; Umashanka, V. Hepatoprotective effect of *Cassia tora* on carbon tetrachloride-induce liver damage in albino rats. *Indian J. Sci. Tech.*, **2009**, *2*, 41–44.

Rajasekaran, A.; Periyasamy, M. Hepatoprotective effect of ethanolic extract of *Trichosanthes lobata* on paracetamol-induced liver toxicity in rats. *Chin. Med.* **2012** *7*(1), 12. doi: 10.1186/1749-8546-7-12

Rajendran, R.; Hemalatha, S.; Akasakalai, K.; MadhuKrishna, C.H.; Sohil, B.; Sundaram, M. Hepatoprotective activity of *Mimosa pudica* leaves against carbon tetrachloride induced toxicity. *J. Nat. Prod.* **2009**, *2*, 116–122.

Rajesh, K.; Swamy, V.; Shivakumar, S.; Inamdar; Joshi, V.; Kurnool, N.A. Hepatoprotective and antioxidant activity of ethanol extract of *Mentha arvensis* leaves against carbon tetrachloride-induced hepatic damage in rat. *Int J. Pharm. Tech Res.* **2013**, *5*(2), 426.

Rajesh, M.G.; Latha, M.S. Hepatoprotection by *Elephantopus scaber* Linn. in CCl_4 induced liver injury. *Indian J. Physiol. Pharmacol.* **2001**, *45*, 481–486.

Rajesh, G.; Latha, M.S. Protective activity of *Glycyrrhiza glabra* Linn. on carbontetrachloride-induced peroxidative damage. *Indian J. Pharmacol.* **2004**, *38*, 284–287.

Rajesh, S.V.; Rajkapoor, B.; Kumar, R.S.; Raju, K. Effect of *Clausena dentata* (Willd.) M. Roem. against paracetamol induced hepatotoxicity in rats. *Pak. J. Pharm. Sci.* **2009,** *22*(1), 90–93.

Rajeshkumar, S.; Kayalvizhi, D. Anti-oxidant and hepatoprotective effect of aqueous and ethanolic extracts of important medicinal plant *Pongamia pinnata* (Family: Leguminoseae). *Asian J. Pharm. Clin. Res.* **2015**, *8*, 1–4.

Rajeshkumar, S.; Tamilarasan, B.; Sivakumar, V. Phytochemical screening and hepatoprotective efficacy of leaves extracts of *Annona squamosa* against paracetamol induced liver toxicity in rats. *Int. J. Pharmacogn.,* **2015**, 2, 178–185.

Rajkapoor, B.; Jayakar, B.; Kavimani, S.; Murugesh, N. Effect of dried fruits of *Carica papaya* Linn on hepatotoxicity. *Biol. Pharm. Bull.* **2002**, *25*(12), 1645–1646.

Rajkapoor, B.; Jayakar, B.; Kavimani, S.; Murugesh, N. Protective effect of *Indigofera aspalathoides* against CCl_4-induced hepatic damage in rats. *J. Herb. Pharmacother.* **2006**, *6*(1), 49–54.

Rajkapoor, B.; Venugopal, Y.; Anbu, J.; Harikrishnan, N.; Gobinath, M.; Ravichandran, V. Protective effect of *Phyllanthus polyphyllus* on acetaminophen induced hepatotoxicity in rats. *Pak J. Pharm. Sci.* **2008**, *21*(1), 57–62.

Rajopadhye, A.A.; Upadhye, A.S. In vitro antioxidant and hepatoprotective effect of the whole plant of *Glossocardia bosvallea* (L. f.) DC. against CCl_4-induced oxidative stress in liver slice culture model. *J. Herbs Spices Med. Plants*, **2012**, *18*, 274–286.

Raju, B.G.S.; Ganga Rao, B.; Manju Latha, Y.B.; Srinivas, K. Antihepatotoxic activity of *Smilax china* roots on CCl_4-induced hepatic damage in rats. *Int. J. Pharm. Pharm. Sci.* **2012**, *4*(1), 494–496.

Raju, K.; Anbuganapathi, G.; Gokulakrishnan, V., Rajkapoor, B.; Jayakar, B.; Manian, S. Effect of dried fruits of *Solanum nigrum* Linn. against CCl_4-induced hepatic damage in rats. *Biol. Pharm. Bull.* **2003**, *26*, 1618–1619.

Ramadan, K.S.; Khalil, O.A.; Danial, E.N.; Alnahdi, H.S.; Ayaz, N.O. Hypoglycemic and hepatoprotective activity of *Rosmarinus officinalis* extract in diabetic rats. *J. Physiol. Biochem.* **2013**, *69*, 779–783.

Ramadan, S.I.; Shalaby, M.A.; Afifi, N.; El-Banna, H.A. Hepatoprotective and antioxidant effects of *Silybum marianum* plant in rats. *Int. J. Agro Veter. Med. Sci.* **2011**, *5*, 541–47.

Ramadoss, S.; Kannan, K.; Balamurugan, K.; Jeganathan, N.S.; Manavalan, R. Evaluation of hepatoprotective activity in the ethanolic extract of *Sida rhombifolia* Linn. against paracetamol-induced hepatic injury in albino rats. *Res. J. Pharm. Biol. Chem. Sci.* **2012**, *3*(1), 497–502.

Ramamurthy M.S.; Srinivasan, M. Hepatoprotective effect of *Tephrosia purpurea* in experimental animals. *Indian J. Pharmacol.* **1993**, *25*(1), 34–36.

Rana, A.C.; Avadhoot, Y. Hepatoprotective effects of *Andrographis paniculata* against carbontetrachloride-induced liver damage. *Arch. Pharm. Res.*, **1991**, *14*, 93–95.

Rana, M.G.; Katbamna, R.V.; Dudhrejiya A.V.; Sheth, N.R. Hepatoprotection of *Tecomella undulata* against experimentally induced liver injury in rats. *Pharmacologyonline*, **2008**, *3*, 674–682.

Ranawat, L.S., Bhatt, J., Patel, J. 2010. Hepatoprotective activity of ethanolic extracts of bark of *Zanthoxylum armatum* DC in CCl_4-induced hepatic damage in rats. *J. Ethnopharmacol. 127*, 777–780.

Rao, B.G.;, Raju, N.J. Investigation of hepatoprotective activity of *Spondias pinnata*. Int. *J. Pharma Sci. Res.* **2010**, *1*(3), 193–198.

Rao, C.V.; Rawat, A.K.; Singh, A.P.; Singh, A.; Verma, N. Hepatoprotective potential of ethanolic extract of *Ziziphus oenoplia* (L.) Mill roots against antitubercular drugs induced hepatotoxicity in experimental models. *Asian Pac. J. Trop. Med.* **2012**, *5*, 283–288.

Rao K. S.; Mishra S. H. Studies on *Curculigo orchioides* Gaertn. for anti-inflammatory and hepatoprotective activities. *Indian Drugs*. **1996a**, 33(1):, 20–25.

Rao K. S.; Mishra S. H.Effect of rhizomes of *Curculigo orchioides* Gaertn. on drug induced hepatotoxicity. *Indian Drugs*. **1996b**, 33(9), 458–461.

Rao, K.S.; Mishra, S.H. Hepatoprotective activity of the whole plants of *Fumaria indica*. *Indian J. Pharm. Sci*., **1997**, *59*(4), 165–170.

Rao, K.S.; Mishra, S.H. Anti-inflammatory and hepatoprotective activities of *Sida rhombifolia* Linn. *Indian J. Pharmacol.* **1997**, *29*, 110–116.

Rao, K.S.; Mishra, S.H. Antihepatotoxic activity of monomethyl fumarate isolated from *Fumaria indica*. *J. Ethnopharmacol*., **1998**, *60*(3), 207–213.

Rashid, F.; Ahmed, R. Mahmood, A.; Ahmad, Z.; Bibi, N.; Kazmi, S.U. Flavonoid glycosides from *Prunus armeniaca* and the antibacterial activity of a crude extract. *Arch. Pharm. Res*. **2007**, *30*, 932–937.

Rasu, M.A.; Tamas, M.; Puica, C.; Roman, I.; Sabadas, M. The hepatoprotective action of ten herbal extracts in CCl_4 intoxicated liver. *Phytother. Res.* **2005**, *19*, 744–749.

Rathi, A.; Srivastava, A.K.; Shirwaikar, A.; Rawat, A.K.S.; Mehrotra, S. Hepatoprotective potential of *Fumaria indica* Pugsley whole plant extracts, fractions and an isolated alkaloid protopine. *Phytomedicine*. **2008**, *15*(6–7):470–7.

Rathi, B.; Sahu, J.; Koul, S.; Khosa, R.; Raghav, P. Hepatoprotective activity of ethanolic extract of stem bark of *Berberis aristata* against carbon tetrachloride (CCl_4). *Int. J. Pharm. Sci.* **2015**, *1*, 43–45.

Rathi, M.A.; Baffila Pearl, D.L.; Sasikumar, J.M.; Gopalakrishnan, V.K. Hepatoprotective activity of ethanolic extract of *Hedyotis corymbosa* on perchloroethylene induced rats. *Pharmacologyonline*, **2009**, *3*, 230–239.

Rathi, M.A.; Meenakshi, P.; Gopalakrishnan, V.K. Hepatoprotective activity of ethanolic extract of *Alysicarpus vaginalis* against nitrobenzene-induced hepatic damage in rats. *South Indian J. Biol. Sci.* **2015**, *1*(2), 60–65.

Ratnasooriya, W.D.; Fernando, T.S.P. Gastric ulcer healing activity of Sri Lankan black tea (*Camellia sinensis* L.) in rats. *Pharmacog. Mag.* **2009**, *1*(1), 11–20.

Ravi, V.; Patel, S.S.; Verma, N.K.; Dutta, D.; Saleem, T.S. Hepatoprotective activity of *Bombax ceiba* Linn against isoniazid and rifampicin-induced toxicity in experimental rats. *Int. J. Appl. Res. Nat. Prod.* **2010**, *3*, 19–26.

Ravikumar, S.; Gnanadesigan, M. Hepatoprotective and antioxidant activity of mangrove plant *Lumnitzera racemosa. Asian Pac. J. Trop. Biomed.* **2011**, 1(4), 348–352.

Ravikumar, S.; Gnanadesigan, M. Hepatoprotective and antioxidant properties of *Rhizophora mucronata* mangrove plant in CCl_4 intoxicated rats. *J. Exp. Clin. Med.* **2012**, *4,* 66–72.

Ravikumar, S.; Gnanadesigan, M.; Jacob Inbanesan, S.; Kalaiarasi. A. Hepatoprotective and antioxidant properties of *Suaeda maritima* (L.) Dumort. ethanolic extract on concanavalin-A induced hepatotoxicity in rats. *Indian J. Exp. Biol.* **2011,** *49*, 455–460.

Ravikumar, S.; Gnanadesigan, M.; Serebiah, J.S.; Jacob Inbanesan, S.; Hepatoprotective effect of an Indian salt marsh herb *Suaeda monoica* Forsk. ex Gmel. against concanavalin-A induced toxicity in tats. *Life Sci. Med. Res.* **2010**, *2*, 1–9.

Ravikumar, V.; Shivashangari, K.S.; Devaki, T. Hepatoprotective activity of *Tridax procumbens* against D-galatosamine/lipopolysaccharide-induced hepatitis in rats. *J. Ethnopharmacol.* **2005**, *101*, 55–60.

Rawat, A.K.S.; Mehrotra, S.; Tripathi S.C.; Shome, U. Hepatoprotective activity of *Boerhavia diffusa* roots—A popular Indian ethnomedicine. *J. Ethnopharmacol.* **1997,** *56*, 61-66.

Ray, D.; Sharatchandra, K.; Thokchom, I.S. Antipyretic, antidiarrhoeal, hypoglycaemic and hepatoprotective activities of ethyl acetate extract of *Acacia catechu* Willd. in albino rats. *Indian J. Pharmacol.*, **2006**, *38*(6), 408–413.

Reddy, G.V.; Kumar, R.V.; Rama, V.; Reddy, M.K.; Reddy, Y.N. Preliminary hepatoprotective activity of medicinal plant extracts against carbon tetrachloride induced hepatotoxicity in albino rats. *Int. J. Rec. Sci. Res.* **2015**, 6, 4946–4951.

Reddy, V.D.; Padmavathi, P.; Gopi, S.; Paramahamsa, M.; and Varadacharyulu, N.Ch. Protective effect of *Emblica officinalis* against alcohol-induced hepatic injury by ameliorating oxidative stress in rats. *Indian. J. Clin. Biochem.*, **2010**, *25*, 419–424

Reddy, P.V., Urooj, A. Evaluation of hepatoprotective activity of *Morus indica* Linn. against toxicity induced by carbon tetrachloride in rats. *Int. J. Pharm. Sci. Res.* **2017**, *8*, 845–851.

Rehman, J., Akhtar, N., Asif, H.M., Sultana, S., Ahmad, M. Hepatoprotective evaluation of aqueous-ethanolic extract of *Capparis decidua* (Stems) in paracetamol induced hepatotoxicity in experimental rabbits. *Pak. J. Pharm. Sci.* **2017**, *30*(2), 507–511.

Rehman, J.; Akhtar, N.; Khan, Y.M.; Ahmad, K.; Ahmad, M.; Sultana, S.; et al., Phytochemical screening and hepatoprotective effect of *Alhagi maurorum* Boiss (Leguminosae) against paracetamol-induced hepatotoxicity in rabbits. *Trop. J. Pharm. Res.* **2015**, *14*(6), 1029–34.

Rehman, J.U.; Saqib, N.U.; Akhtar, N.; Jamshaid, M.; Asif, H.M.; Sultana, S.;

Rehman, R.U. Hepatoprotective activity of aqueous-methanolic extract of *Suaeda fruticosa* in paracetamol-induced hepatotoxicity in rabbits. *Bangladesh J. Pharmacol.* **2013**, *8*, 378–381.

Reshi, N.A.; Shankarsshingh, S.M.; Vasanaika, G.H. Evaluation of hepatoprotective potential of leaf and leaf callus extracts of *Anisochilus carnosus* (L.) Wall. *Int. J. Phytomed.* **2018**, *10*(3), 156–161.

Rezaei, A.; Foroush, S.S.; Ashtiyani, S.C.; Agababa, H.; Zaei, A.; Azizi, M.; Yarmahmodi, H. The effects of *Artemisia aucheri* extract on hepatotoxicity induced by thioacetamide in male rats. *Avicenna J. Phytomedicine* **2013**, *3*(4), 293–301.

Rezende, T.P.; do A Corrêa, J.O.; Aarestrup, B.J.; Aarestrup, F.M.; de Sousa, O.V.; da Silva Filho, A.A. Protective effects of *Baccharis dracunculifolia* leaves extract against carbon tetrachloride- and acetaminophen-induced hepatotoxicity in experimental animals. *Molecules* **2014**, *19*, 9257–9272.

Riaz, H.; Saleem, N., Ahmad, M., Mehmood, Y., Raza, S.A.; Khan, S.; Anwar, R.; Kamran, S.H. Hepatoprotective effect of *Crocus sativus* on amiodarone-induced liver toxicity. *Br. J. Pharm. Res.* **2016**, *12*(4), 1–11.

Richaria, A.; Singh, A.K.; Singh, S.; Singh, S.K. Hepatoprotective and antioxidants activity of ethanolic extract of *Cuscuta reflexa* roxb. *IOSR J. Pharmacy* **2012**, *2*(2), 142–147.

Rizk, M.Z., Abdallah, M.S., Sharara, H.M., Ali, S.A., Ibrahim, N.A., Mousstafa, M.M. Efficiency of *Cupressus sempiverens* L. and *Juniperus phoenicea* against carbon tetrachloride hepatotoxicity in rats. *Trends Med. Res.* **2007**, *2*, 83–94.

Rodríguez-Amado, J.R.; Lafourcade-Prada, A.; Escalona-Arranz, J.C.; Pérez-Rosés, R.; Morris-Quevedo, H.; Keita, H.; et al., Antioxidant and hepatoprotective activity of a new tablets formulation from *Tamarindus indica* L. *Evid. Based Complement. Alternat. Med.* **2016**. *2016*, 3918219.

Roger, D.; Pamplona, R. Liver toxicity. Encyclopaedia of Medicinal Plants. Editorial Safeliz, London, **2001**. pp. 392–395.

Roman, I.; Cristescu, M.; Puica, C. Effects of *Hypericum perforatum* and *Hypericum maculatum* extracts administration on some morphological and biochemical parameters in rat liver intoxicated with alcohol. *Studia Universitatis "Vasile Goldis," Seria Stiintele Viet,* ii, **2011**, *21*(2), 361–370.

Roome, T.; Dar, A.; Ali, S.; Naqvi, S. A study on antioxidant, free-radical scavenging, anti-inflammatory and hepatoprotective actions of *Aegiceras corniculatum* stem extracts. *J. Ethnopharmacol.* **2008**, *118*(3), 514–521.

Rosa, M.P.; Gutierrez, Rosario V.S. Hepatoprotective and inhibition of oxidative stress of *Prostechea michuacana. Rec. Nat. Prod.* **2009**, *3*(1), 46–51.

Rose, P.; Whiteman, M.; Moore, P.K.; Zhu, Y.Z. Bioactive S-alk(en)yl cysteine sulfoxide metabolites in the genus *Allium*: The chemistry of potential therapeutic agents. *Nat. Prod. Rep.* **2005**, *22*, 351–368.

Roy, C.K.; Das, A.K. Comparative evaluation of different extracts of leaves of *Psidium guajava* Linn. for hepatoprotective activity. *Pak. J. Pharm. Sci.* **2010**, *23*(1), 15–20.

Roy, C.K.; Kamath, J.V.; Asad, M. Hepatoprotective activity of *Psidium guajava* Linn. leaf extract. *Indian J. Exp. Biol.* **2006**, *44*(4), 305–311.

Roy, S.P.; Gupta, R.; Kannadasan, T. Hepatoprotective activity of ethanolic extract of *Madhuca longifolia* leaves on D-galactosamine-induced liver damage in rats. *J. Chem. Pharm. Sci.* **2012**, *4*, 205–209.

Roy, S.P.; Kannadasan, T.; Gupta, R. Screening of hepatoprotective activity of *Madhuca longifolia* bark on D-Galactosamine induced hepatotoxicity in rats. *Biomed. Res.* **2015**, 26(2), 365–369.

Rubab, A.; Hemamalini, K.; Shashi Priya, G.; Uma, V. Hepatoprotective activity of *Gymnosporia emarginata* (Willd) and

Marsedenia volubillis (Linn.f.) Stapf against paracetamol induced hepatotoxicity in rats. *Int. J. Current Pharm. Rev.* Res. **2013**, *4*(2), 36–41.

Saafi, E.B.; Louedi, M.; Elfeki, A.; Zakhama, A.and Najjar, M.F. Protective effect of date palm fruit extract (*Phoenix dactylifera* L.) on dimethoate-induced oxidative stress in rat liver. *Exp. Toxicol. Pathol.*, **2011**, *63*, 433–441.

Saalu, L.C.; Ogunlade, B., Ajayi, G.O., Oyewopo, A.O., Akunna, G.G., Ogunmodede, O.S. The hepatoprotective potentials of *Moringa oleifera* leaf extract on alcohol induced hepato-toxicity in Wistar rat. *Am. J. Biotechnol. Mol. Sci.* **2012**, *2*, 6–14.

Saba, A.B.; Onakoya, O.M.; Oyagbemi, A.A. Hepatoprotective and *in vivo* antioxidant activities of ethanolic extract of whole fruit of *Lagenaria breviflora*. *J. Basic Clin. Physiol. Pharmacol.* **2012**, *23*(1), 27–32.

Sabir, S.M.; Ahmad, S.D.; Hamid, A.; Khan, M.Q.; Athayde, M.L.; Santos, D.B.; Boligon, A.A.; Rocha, J.B.T. Antioxidant and hepatoprotective activity of ethanolic extract of leaves of *Solidago microglossa* containing polyphenolic compounds. *Food Chem.* **2012,** *131*(3), 741–747.

Sabir, S. M.; Rocha, J.B.T. Antioxidant and hepatoprotective activity of aqueous extract of *Solanum fastigiatum* (false "Jurubeba") against paracetamol-induced liver damage in mice. *J. Ethnopharmacol.* **2008a**, *120*(2), 226– 232.

Sabir, S.M.; Rocha, T.B.J. Water-extractable phytochemicals from *Phyllanthus niruri* exhibit distinct *in vitro* antioxidant and *in vivo* hepatoprotective activity against paracetamol-induced liver damage in mice. *Food Chem.* **2008b**, *111*, 845–851.

Saboo, S.S.; Tapadiya, G.; Farooqui, I.A.; Khadabadi, S.S. Free radical scavenging, *in vivo* antioxidant and hepatoprotective activity of folk medicine *Trichodesma sedgwickianum*. *Bangladesh J. Pharmacol.* **2013**, *8*, 58–64.

Sachan, K.; Singh, P.Kr.; Ashwlayan, V.D.; Singh, R. Evaluation of hepatoprotective activity of *Tectona grandis* Linn. *Int. J. Pharm. Med. Res.* **2014**, *2*(3), 105–108.

Sadasivan, S.; Latha, P.G.; Sasikumar, J.M.; Rajashekaran, S.; Shyamal, S.; Shine, V.J. Hepatoprotective studies on *Hedyotis corymbosa* (L.) Lam. *J. Ethnopharmacol.* **2006**, *106*, 245–249.

Sadeghi, H.; Hosseinzadeh, S.; Touri, M.A.; Ghavamzadeh, M.; Barmak, M.J.; Sayahi, M.; Sadeghi, H. Hepatoprotective effect of *Rosa canina* fruit extract against carbon tetrachloride-induced hepatotoxicity in rat. *Avicenna J. Phytomed.* **2016**, *6*, 181–88.

Sadeque, M.Z.; Begum, Z.A. Protective effect of dried fruits of *Carica papaya* on hepatotoxicity in rat. *Bangladesh J. Pharmacol.* **2010**, *5*, 48–50.

Sadeque, M.Z.; Begum, Z.A.; Umar, B.U.; Ferdous, A.H.; Sultana, S.; Uddin, M.K. Comparative efficacy of dried fruits of *Carica papaya* Linn. and Vitamin-E on preventing hepatotoxicity in rats. *Faridpur Med. College J.* **2012**, *7*, 29

Sadik, N.A.H.; Maraghy, S.A.; Ismail, M.F. Diethylnitrosamine-induced hepatocarcinogenesis in rats: Possible chemoprevention by blueberries. *Afr. J. Biochem. Res.* **2008**, *2*, 81–87.

Saenthaweesuk, S.; Munkong, N.; Parklak, W.; Thaeomor, A.; Chaisakul, J.; Somparn, N. Hepatoprotective and antioxidant effects of *Cymbopogon citratus* Stapf (lemon grass) extract in paracetamol-induced hepatotoxicity in rats. *Trop. J. Pharm. Res.* **2017**, *16*, 101–07.

Saggu, S.; Sakeran, M.I.; Zidan, N.; Tousson, E.; Mohan, A.; Rehman, H. Ameliorating effect of chicory (*Chichorium intybus* L.) fruit extract against 4-tert-octylphenol induced liver injury and oxidative stress in male rats. *Food Chem. Toxicol.* **2014**, *72*, 138–146.

Saha, P.; Mazumder, U.K.; Haldar, P.K.; Bala, A.; Kar, B.; Naskar, S. Evaluation of hepatoprotective activity of *Cucurbita maxima* aerial parts. *J. Herbal Med. Toxicol.*, **2011**, *5*, 7–22.

Sahreen, S.; Muhammad, R.K.; Rahma, A.K. Hepatoprotective effects of methanol extract of *Carissa opaca* leaves on CCl_4-induced damage in rat. *BMC Complem. Altern. Med.* **2011**, *11*, 48–56.

Said, A.; El-Fishawy, A.M.; El-Shenawy, S.; Hawas, U.W.; Aboelmagd, M. Hepatoprotective and gastro protective studies of *Terminalia arjuna* leaves extract and phytochemical profile. *Banat's J. Biotechnol.* **2014**, *5*, 30–36.

Saka, W.A.; Aknigbe, R.E.; Ishola, O.S.; Ashmu, E.A.; Ojayemi, O.T.; Adeleke, G.E. Hepatotherapeutic effect of *Aloe vera* in alcohol-induced hepatic damage. *Pak. J. Biol. Sci.* **2011**, *14*, 742–746.

Sakr, S.; Mahran, H., Lamfon, H. Protective effect of ginger (*Zingiber officinale*) on adriamycin-induced hepatotoxicity in albino rats. *J. Med. Plants Res.* **2011,** *5*(1), 133–140.

Sakr, S.A.; Abo-El-Yazid, S.M. Effect of fenugreek seed extract on adriamycin-induced hepatotoxicity and oxidative stress in albino rats. *Toxicol. Ind. Health* **2011,** *28*(10), 876–885.

Sakthidevi, G.; Alagammal, M.; Mohan, V.R. Evaluation of hepatoprotective and antioxidant activity of *Polygala javana* DC. whole plant-CCl_4 induced hepatotoxicity in rats. *Int. J. Pharm. Chem. Sci.* **2013**, *2*(2), 764–770.

Salama, S.M.; Abdulla, M.A.; AlRashdi, A.S.; Ismail, S.; Alkiyumi, S.S.; Golbabapour, S. Hepatoprotective effect of ethanolic extract of *Curcuma longa* on thioacetamide induced liver cirrhosis in rats. *BMC Complement. Alternat. Med.* **2013**, *13*(1), 56–73.

Salama, S.M.; Bilgen, M.; Al Rashdi, A.S.; Abdulla, M.A. Efficacy of *Boesenbergia rotunda* treatment against thioacetamide-induced liver cirrhosis in a rat model. *Evid. Based Complement. Alternat. Med* **2012**, *2012*: 137083.

Salem, G.A.; Shaban, A.; Diab, H.A.; et al., *Phoenix dactylifera* protects against oxidative stress and hepatic injury induced by paracetamol intoxication in rats. *Biomed. Pharmacother*. **2018**, *104*, 366–374.

Saleem, M.; Ahmed, B.; Karim, M.; Ahmed, S.; Ahmad, M.; Qadir, M.I.; Syed, N.I.H. Hepatoprotective effect of aqeous methanolic extract of *Rumex dentatus* in paracetamol induced hepatotoxicity in mice. *Bangladesh J. Pharmacol.* **2014**, *9*, 284–89.

Saleem, M.; Ali, H.A.; Akhtar, M.F.; Saleem, U.; Saleem A.; Irshad, I. Chemical characterisation and hepatoprotective potential of *Cosmos sulphureus* Cav. and *Cosmos bipinnatus* Cav. *Nat. Prod. Res.* **2019**, *33*(6), 897–900.

Saleem, M.; Ahmed, B.; Qadir, M.I.; Rafiq, M.; Ahmad, M.; Ahmad, B. Hepatoprotective effect of *Chenopodium murale* in mice. *Bangladesh J. Pharmacol.* **2014**, *9*, 124–28.

Saleem, M.; Irshad, I.; Baig, M.K.; Naseer, F. Evaluation of hepatoprotective effect of chloroform and methanol extracts of *Opuntia monacantha* in paracetamol-induced hepatotoxicity in rabbits. *Bangladesh J. Pharmacol.* **2015**, *10*, 16–20.

Saleem, M.; Naseer, F.; Ahmad, S.; Nazish, A.; Bukhari; Rehman, A.; et al., Hepatoprotective activity of ethanol extract of *Conyza bonariensis* against paracetamol induced hepatotoxicity in Swiss albino mice. *Am. J. Med. Biol. Res.* **2014**, *2*(6), 124–27.

Saleem, M.T.S.; Christina, A.J.M.; Chidambaranathan, N.; Ravi, V.; Gauthaman, K. Hepatoprotective activity of *Annona squamosa* Linn. on experimental animal model *Int. J. Appl. Res. Nat. Prod.*, **2008**, *1*(3), 1–7.

Samal, P.K. Hepatoprotective activity of *Ardisia solanacea* in carbon tetrachloride induced hepatotoxic albino rats. *Asian J. Res. Pharm. Sci.* **2013**, *3*(2), 79–82.

Sambasivam, M.; Ravikumar, R.; Thinagarbabu, R.; Davidraj, C.; Arvind, S. Hepatoprotective potential of *Azima tetracantha* and *Tribulus terrestris* on ferrous sulfate-induced toxicity in rat. *Bangladesh J. Pharmacol.* **2013**, *8*(3), 357– 360.

Sambo, N.; Garba, S.H.; Timothy, H. Effect of the aqueous extract of *Psidium guajava* on erythromycin-induced liver damage in rats. *Niger. J. Physiol. Sci.* **2009**, *24*, 171–176.

Samuel, A.J.S.J.; Mohan, S.; Chellappan, D.K.; Kalusalingam, A.; Ariamuthu, S. *Hibiscus vitifolius* (Linn.) root extracts shows potent protective action against anti-tubercular drug induced hepatotoxicity. *J. Ethnopharmacol.* **2012**, *141*, 396–402.

Sánchez-Salgado, J.C.; Ortiz-Andrade, R.R.; Aguirre-Crespo, F.; Vergara-Galicia, J.; Leon-Rivera, I.; Montes, S.; Villalobos-Molina, R.; Estrada-Soto, S. Hypoglycemic, vasorelaxant and hepatoprotective effects of *Cochlospermum vitifolium* (Willd.) Sprengel: A potential agent for the treatment of metabolic syndrome. *J. Ethnopharmacol.* **2007**, *109*, 400–405.

Sangameswaran, B.; Chumbhale, D.; Balakrishnan, B.R.; Jayakar, B. Hepatoprotective effects of *Thespesia lampas* Dalz. and Gibson in CCl_4 induced liver injury in rats. *Dhaka University J. Pharm. Sci.* **2008**, *7*, 11–13.

Sangameswaran, B.; Reddy, T.C.; Jayakar, B. Hepatoprotective effects of leaf extracts of *Andrographis lineata* Nees on liver damage caused by carbon tetrachloride in rats. *Phytother. Res.*, **2008**, *22*(1), 124-126.

Sangi, S.; El-feky, S.A.; Ali, S.S.; Ahmedani, E.I.; Tashtoush, M.H. Hepatoprotective effects of oleuropein, thymoquinone and fruit of *Phoenix dactylifera* on CCl_4 induced hepatotoxicity in rats. *World J. Pharmaceut. Res.* **2014**, *3*(2), 3475–3486.

Sanmugapriya, E.; Venkataraman, S. Studies on hepatoprotective and antioxidant actions of *Strychnos potatorum* Linn. seeds on CCl_4-induced acute hepatic injury in experimental rats. *J. Ethnopharmacol.* **2006**, *105*(1–2), 154–160.

Sanni, S.; Thilza, I.B.; Ahmed, M.T.; Sanni, F.S.; Talle, M.; Okwor, G.O. The effect of aqueous leaves extract of henna (*Lawsonia inermis*) in carbon tetrachloride induced hepato-toxicity in swiss albino mice. *Acad. Arena* **2010**, *2*, 87–89.

Sannigrahi, S.; Mazumder, U.K.; Pal, D.; Mishra, S.L. Hepatoprotective potential of methanol extract of *Clerodendrum infortunatum* Linn. against CCl_4 induced hepatotoxicity in rats. *Pharmacogn. Mag.* **2009**, *5*, 394–399

Saradamma, B.; Reddy, V.D.; Padmavathi, P.; Paramahamsa, M.; Varadacharyulu, N. Modulatory role of *Pterocarpus santalinus* against alcohol-induced liver oxidative/ nitrosative damage in rats. *Biomed. Pharmacother.* **2016**, *83*, 1057–1063.

Saraf, S.; Dixit, V. K.; Tripathi, S. C.; Patnaik, G. K. Antihepatotoxic activity of *Cassia occidentalis*. *Int. J. Pharmacogn.*, **1994**, *32*(2), 38–183

Saravanan, G.; Malarvannan, L. Hepatoprotective activity of *Cassia tora* on carbon tetrachloride induced hepatotoxicity. *Asian J. Innov. Res.* **2016**, *1*, 1–5.

Saravanan, R.; Prakasam, A.; Ramesh, B.; Pugalendi, K.V. Influence of *Piper betle* on hepatic marker enzymes and tissue antioxidant status in ethanol-treated Wistar rats. *J. Med. Food.* **2002**, *5*(4), 197–204.

Saravanan, R.; Viswanathan, P.; Viswanathan-Pugalendi, K. Protective effect of ursolic acid on ethanol-mediated experimental liver damage in rats. *Life Sci*. **2006**, 78(1), 713–718.

Sarkar, R.; Hazra, B.; Mandal, N. Hepatoprotective potential of *Caesalpinia crista* against iron-overload-induced liver toxicity in mice. *Evid. Based Complement. Alternat. Med* **2012**. doi.10.1155/2012/896341

Sasidharan, S.; Aravindran, S.; Latha, L.Y.; Vijenthi, R.; Saravanan, D.; Amutha, S. *In vitro* antioxidant activity and hepatoprotective effects of *Lentinula edodes* against paracetamol induced hepatotoxicity. *Molecules* (Basel, Switzerland). **2010,** *15*(6), 4478–4489.

Sathaye, S.; Bagul, Y.; Gupta, S.; Kaur, H.; Redkar, R. Hepatoprotective effects of aqueous leaf extract and crude isolates of *Murraya koenigii* against *in vitro* ethanol-induced hepatotoxicity model. *Exp. Toxicol. Pathol.* **2011**, *63*(6), 587–591.

Sathesh Kumar, B.; Ravi Kumar, G.; Krishna Mohan, G. Hepatoprotective effect Of *Trichosanthes cucumerina* var. *cucumerina* L. on carbontetrachloride induced liver damage in rats. *J. Ethnopharmacol.* **2009**, *123*, 347–350.

Satyanarayana, T.; Kshama Devi, K.; Mathews, A.A. Hepatoprotective activity of *Capparis sepiaria* stem against carbon tetrachloride-induced hepatotoxicty in rats. *JPRHC* **2009**, *1*(1), 34–45.

Saxena, A.K.; Singh, B.; Anand, K.K. Hepatoprotective effects of *Eclipta alba* on sub cellular levels in rats. *J. Ethnopharmacol.* **1993**, *40*, 155–161.

Sayed, A.E.; Martiskainen, O.; Sayed, H.; Seif el-Din; Nasser, A.; Sabra; et al., Hepatoprotective and antioxidant effect of *Bauhinia hookeri* extract against carbon tetrachloride-induced hepatotoxicity in mice and characterization of its bioactive compounds by HPLC-PDA-ESI MS/MS. *BioMed. Res. Int.* **2014**; 1–9.

Sayed, A.E.; Mohamed, M.; Daim, A.; Omnia, E.; Kilany; Karonen, M.; et al., Protective role of polyphenols from *Bauhinia hookeri* against carbon tetrachloride-induced hepato and nephrotoxicity in mice. *Ren. Fail.* **2015**, *37*, 7.

Sehrawat, A.; Khan, T.H.; Prasad, L.; Sultana, S. *Butea monosperma* and chemomodulation: Protective role against thioacetamide-mediated hepatic alterations in Wistar rats. *Phytomed. Int.* **2006**, *13*(3), 157–163.

Selim, Y.A., Ouf, N.H. Anti-inflammatory new coumarin from the *Ammi majus* L. *Org Med. Chem. Lett.* **2012**, 2, 1.

Sen S.; De B.; Devanna N.; Chakraborty R. Hepatoprotective and antioxidant activity of *Leea asiatica* leaves against acetaminophen-induced hepatotoxicity in rats. *Tang [Humanitas Medicine]* **2014**, *4*(3), 18. doi: 10.5667/tang.2014.0005.

Sen, T.; Basu, A.; Ray, R. N.; Nag Shaudhuri, A. K. Hepatoprotective effects of *Pluchea indica* (Lees) extract in experimental acute liver damage in rodents. *Phytother. Res.* **1993**, *7*(5), 352–355.

Sener, G.; Omurtag, G.Z.; Sehirli, O.; Tozan A.; Yuksel, M.; et al., Protective effects of *Ginkgo biloba* against acetaminophen-induced toxicity in mice. *Mol. Cell. Biochem.* **2006**, *283*, 39–45.

Sengottuvelu, S.; Duraisami, S.; Nandhakumar, J.; Duraisami, R.; Vasudevan, M. Hepatoprotective activity of *Camellia sinensis* and its possible mechanism of action. *Iran. J. Pharmacol. Therapeut.* **2008**, *7*(1), 9–14.

Sengottuvelu, S.; Srinivasan, D.; Duraisami, R.; Nandhakumar, J.; Vasudevan, M.; Sivakumar, T. Hepatoprotective activity of *Trianthema decandra* on CCl_4 induced hepatotoxocity on rats. *Int. J. Green Pharm.* **2012**, pp. 122–125.

Sengupta, M.; Sharma, G.D.; Chakraborty, B. Hepatoprotective and immunomodulatory properties of aqueous extract of *Curcuma longa* in carbon tetrachloride intoxicated Swiss albino mice. *Asian Pac. J. Trop. Biomed.* **2011**, *1*(3), 193–199.

Senthilkumar, K.T.M.; Rajkapoor, B.; Kavimani, S. Protective effect of *Enicostema littorale* against carbontetrachloride-induced hepatic damage in rats. *Pharm. Biol.* **2005**, *43*(5), 485–487.

Serairi-Beji, R.; Aidi Wannes, W.; Hamdi, A.; et al., Antioxidant and hepatoprotective effects of *Asparagus albus* leaves in carbon tetrachloride-induced liver injury rats. *J. Food Biochem.* **2017**, *42*(1), e12433.

Sethuraman, M.G.; Lalitha, K.G.; Kapoor, B.R. Hepatoprotective activity of *Sarcostemma brevistigma* against carbon tetrachloride-induced hepatic damage in rats. *Curr. Sci.*, **2003**, *84*(9), 1186–1187.

Setty, R.S.; Quereshi, A.A.; Swamy, A.H.; Patil, T.; Prakash, T.; Prabhu, K.; et al., Hepatoprotective activity of *Calotropis procera* flowers against paracetamol-induced hepatic injury in rats. *Fitoterapia* **2007**, *78*, 451–454.

Seyed, R.M.; Nassiri-Asl, M.; Farahani-Nick, Z.; Savad, S.; Seyed, K.F. Protective effects of *Fumaria vaillantii* extract on carbon tetrachloride induced heaptotoxicity in rats; *Pharmacologyonline* **2007**, *3*, 385–393.

Shah, A.S.; Khan, R.A.; Ahmed, M.; Muhammad, N. Hepatoprotective role of *Nicotiana plumbaginifolia* Linn. against carbon tetrachloride-induced injuries. *Toxicol. Ind. Health.* **2016**, *32*(2), 292–298.

Shah, R.; Parmar, S.; Bhatt, P.; Chanda, S. Evaluation of hepatoprotective activity of ethyl acetate fraction of *Tephrosia purpurea*. *Pharmacologyonline* **2011**, *3*, 188–194.

Shah, V.N.; Deval, K. Hepatoprotective activity of leaves of *Parkinsonia aculeata* Linn. against paracetamol induced hepatotoxicity in rats. *Int. J. Pharm.* **2011**, *1*, 59–63.

Shahjahan, M.; Sabitha, K.E.; Jainu, M.; et al., Effect of *Solanum trilobatum* against carbon tetrachloride-induced hepatic damage in albino rats. *Indian J. Med. Res.* **2004**, *120*, 194–198.

Shahjahan, M.; Vani, G.; Devi, C.S. Protective effect of *Indigofera oblongifolia* in CCl_4-induced hepatotoxicity. *J. Med. Food* **2005**, *8*, 261–265.

Shaikh, H., Bhosle, D., Shaikh, A., Bhagat, A., Khan, S., Quazi, Z. To evaluate the hepatoprotective activity of ethanolic leaf extract of *Moringa oleifera* plant in albino Wistar rats. *European J. Pharm. Med. Res.* **2017**, *4*(9), 601–604.

Shailajan S, Chandra N, Sane RT, Menon S. Effect of *Asteracantha longifolia* Nees against CCl_4 induced liver dysfunction in rat. *Indian J. Exp. Biol.* **2005**, *43*(1), 68–75.

Shaker, E.; Mahmoud, H.; Mnaa, S. Silymarin, the antioxidant component and *Silybum marianum* extracts prevent liver damage. *Food Chem. Toxicol.* **2010**, *46*, 803–806.

Shankar, N. L. G.; Manavalan, R.; Venkappayya, D.; Raj, C. D. Hepatoprotective and antioxidant effects of *Commiphora berryi* (Arn.) Engl. bark extract against CCl_4-induced oxidative damage in rats. *Food Chem. Toxicol.* **2008**, *46*(9), 3182– 3185.

Shanmugasundaram, S.; Venkataraman, S. Hepatoprotective and antioxidant effects of *Hygrophila auriculata* (K. Schum) Heine Acanthaceae root extract. *J. Ethnopharmacol.* **2006**, *104*, 124–128.

Sharma, A.; Chakraborti, K.K.; Handa, S.S. Anti-hepatotoxic activity of some Indian herbal formulations as compared to Silymarin. *Fitoterapia* **1991**, *62*, 229–235.

Sharma, A.; Sangameswaran, B.; Jain, V.; Saluja, M.S.Hepatoprotective activity of *Adina cordifolia* against ethanol induced hepatotoxicity in rats. *Int. Curr. Pharm. J.* **2012**, *1*(9), 279–284.

Sharma, A.; Sharma, M.K.; Kumar, M. Protective effect of *Mentha piperita* against arsenic-induced toxicity in liver of Swiss albino mice. *Basic Clin. Pharmacol. Toxicol.* **2007**, *100*(4), 249–257.

Sharma, A.K.; Anand, K.K.; Pushpangadan, P.; Chandan, B.K.; Chopra, C.L.; Prabhakar, Y.S.; Damodaran, N.P. Hepatoprotective effects of *Wedelia calendulacea*. *J. Ethnopharmacol.* **1989**, *25*(1), 93–102.

Sharma, A.; Shing, R.T.; Sehgal, V.; Handa, S.S. Anti-hepatotoxic activity of some plants used in herbal formulations. *Fitoterapia* **1991**, *62*, 131–138.

Sharma, N.; Shukla, S. Hepatoprotective potential of aqueous extract of *Butea monosperma* against CCl_4 induced damage in rats. *Exp. Toxicol. Pathol.* **2011**, *63*, 671–676.

Sharma, N.K.; Priyanka; Jha, K.K.; Singh, H.K.; Shrivastava, A.K. Hepatoprotective activity of *Luffa cylindrica* (L.) M.J.Roem.

leaf extract in paracetamol intoxicated rats. *Indian J. Nat. Prod. Res.* **2014**, *5*(2), 143–148.

Sharma, P.; Bodhankar, S.L.; Thakurdesai, P.A. Protective effect of *Feronia elephantum correa* leaves on thioacetamide induced liver necrosis in diabetic rats. *Asian Pac. J. Trop. Biomed.* **2012,** *2*(9), 691–695.

Sharma, S.K.; Vasudeva, N. Hepatoprotective activity of *Vitis vinifera* root extract against carbon tetrachloride-induced liver damage in rats. *Acta Pol. Pharm.* **2012,** *69*(5), 933–937.

Sharma, S.K.; Suman, N.; Vasudeva, N. Hepatoprotective activity of *Vitis vinifera* root extract against carbon tetrachloride-induced liver damage in rats. *Acta Pol. Pharm.* **2012**, *69*, 933–937.

Sharma, U.R.; Prakash, T.; Surendra, V.; Roopakarki, N.; Goli, D. Hepatoprotective activity of *Fumaria officinalis* against CCl_4-induced liver damage in rats. *Pharmacologia* **2012**, *3*(1), 9–14.

Sharma, V.; Pandey, D. Protective role of *Tinospora cordifolia* against lead-induced hepatotoxicity. *Toxicol. Int.* **2010**, *17*, 12–17.

Shauka, S.; Perveen, S.; Ahmed, M.; Jabeen, Q. Hepatoprotective effects of *Aerva javanica* against paracetamol induced liver toxicity. *JSZMC.* **2015**, *6*(3), 839–844.

Shehab, N.G.; Abu-Gharbieh, E.; Bayoumi, F.A. Impact of phenolic composition on hepatoprotective and antioxidant effects of four desert medicinal plants. *BMC Complement. Altern. Med.* **2015**, *15*, 401.

Shen, B.; Chen, H.; Shen, C.; Xu, P.; Li, J.; Shen, G., Yuan, H., Han, J. Hepatoprotective effects of lignans extract from *Herpetospermum caudigerum* against CCl_4-induced acute liver injury in mice. *J. Ethnopharmacol.* **2015**, *164*, 46–52.

Shen, X.; Tang, Y.; Yang, R.; Yu, L.; Fang, T.; Duan, J.A. The protective effect of *Zizyphus jujuba* fruit on carbon tetrachloride-induced hepatic injury in mice by anti-oxidative activities. *J. Ethnopharmacol.* **2009**, *122*(3), 555–560.

Shenoy, K.A.; Somayaji, S.N.; Bairy, K.L. Hepatoprotective effects of *Ginkgo biloba* against carbon tetrachloride induced hepatic injury in rats. *Indian J. Pharmacol.* **2001**, *33*, 260–266.

Sheshidhar, G.B.; Yasmeen, A.M.; Arati, C.; Sangappa, V.K.; Pundarikaksha, H.P.; Vijay, D.; Manjula, R. Evaluation of hepatoprotective activity of ethanolic extract of *Acacia catechu* Wlld in paracetamol induced hepatotoxicity in albino rats. *Int. J. Pharm. Biol. Sci.* **2013**, *3*(2), 264–270.

Shilova, I.V.; Zhavoronok, T.V.; Souslov, N.I.; Novozheeva, T.P.; Mustafin, R.N.; Losseva, A.M. Hepatoprotective properties of fractions from Meadowsweet extract during experimental toxic hepatitis. *Bull. Exp. Biol. Med. Pharmacol. Toxicol.*, **2008**, *146*(1), 49-51.

Shilova, I.V.; Zhavoronok, T.V.; Suslov, N.I.; Krasnov, E.A.; Novozheeva, T.P.; Veremeev, A.V.; Nagaev, M.G.;, Petina, G.V. Hepatoprotective and antioxidant activity of meadowsweet extract during experimental toxic hepatitis. *Bull. Exp. Biol. Med.* **2006**, *142*(2), 216–218.

Shim, S.B.; Kim, N.J.; Kim, D.H. β- Glucuronidase inhibitory activity and hepatoprotective effect of 18β- zglycyrrhetinic acid from the rhizomes of *Glycyrrhiza uralensis*. *Planta Med.*, **2000**, *66*(1), 40–43.

Shin, H.W.; Park, S.Y.; Lee, K.B.; Shin, E.; Lee, M.J.; Kim, Y.J.; Jang, J.J. Inhibitory effect of turnip extract on thioacetamide induced rat hepatic fibrogenesis. *Cancer Prev. Res.* **2006**, *11*, 265–272.

Shinde, M.; Shete, R.V.; Kore, K.J.; Attal, A.R. Hepatoprotective activity of *Ficus bengalensis* Linn leaves. *Curr. Pharm. Res.* **2012**, *2*(2), 503–507.

Shirish, S.; Pingale. Heoatoprotection by fresh juice of *Leucas aspera* leaves. *Der Pharm. Sin.*, **2010**, *1*(2), 136–140

Shirode, D.S.; Jain, B.B.; Mahendra kumar, C.B.; Setty, S.R. Hepatoprotective and antioxidant effects of *Albizzia lebbeck* against thioacetamide induced hepatotoxicity in rats. *J. Chem. Pharm. Sci.* **2012**, *5*, 199–204.

Shivashangari, K.S.; Ravikumar, V.; Devaki, T. Evaluation of the protective efficacy of *Asteracantha longifolia* on acetaminophen-induced liver damage in rats. *J. Med. Food,* **2004**, *7*(2), 245–251.

Shivashangari, K.S.; Ravikumar, V.; Devaki, T. Evaluation of the protective efficacy of *Asteracantha longifolia* on acetaminophen-induced liver damage in rats. *J. Med. Food* **2004**, *7*(2), 245–251.

Shivashankara, A.R.; Azmidah, A.; Haniadka, R.; Rai, M.P.; Arora, R.; Baliga MS. Dietary agents in the prevention of alcohol-induced hepatotoxicty: Preclinical observations. *Food Funct.* **2012**, *3*, 101–109.

Shrinivas, B.; Suresh, R.N. Hepatoprotective activity of methanolic extract of *Dipteracanthus patulus* (Jacq.) Nees: Possible involvement of antioxidant and membrane stabilization property. *Int. J. Pharm. Pharm. Sci.* **2012**, *4*(2), 685–690.

Shukla, B.; Visen, P.K.S.; Patnaik, G.K.; Dhawan, B.N. Choleretic effect of picroliv, the hepatoprotective principle of *Picrorhiza kurroa. Planta Med.* **1991**, *57*, 29–33.

Shukla, B.; Visen, P. K. S.; Patnaik, G.K.; Kapoor N.K.; Dhawan B.N. Hepatoprotective effect of an active constituent isolated from the leaves of *Ricinus communis* Linn. *Drug Dev. Res.* **1992**, *26*(2), 183–193.

Shyam, A.S.; Reddy, A.R.N.; Rajeshwar, Y.; Kira, N.G.; Prasad, K.D.; Baburao, B. et al., Protective effect of methanolic extract of *Trianthema portulacastrum* in atherosclerotic diet induced renal and hepatic changes in rats. *Der Pharm. Lett.* **2010**, *2*, 540–545.

Shyamal, S.; Latha, P.G.; Shine, V.J.; Suja, S.R.; Rajasekharan, S.; Ganga Devi, T. Hepatoprotective effects of *Pittosporum neelgherrense* Wight & Arn., a popular Indian ethnomedicine. *J. Ethnopharmacol.* **2006**, *107*, 151–155.

Shyamal, S.; Latha, P.G.; Suja, S.R.; Shine, V.J.; Anuja, G.I.; Sini, S.; et al., Hepatoprotective effect of three herbal extracts on aflatoxin B1-intoxicated rat Liver. *Singapore Med. J.* **2010**, *51*, 326–331.

Shyur, L.F.; Huang, C.C.; Lo, C.P.; Chiu, C.Y.; Chen, Y.P., Wang, S.Y., Chang, S.T. Hepatoprotective phytocompounds from *Cryptomeria japonica* are potent modulators of inflammatory mediators. *Phytochemistry* **2008**, *69*, 1348–1358.

Sikander, M.; Malik, S.; Parveen, K.; Ahmad, M.; Yadav, D.; Hafeez, Z.B.; Bansal, M. Hepatoprotective effect of *Origanum vulgare* in Wistar rats against carbon tetrachloride-induced hepatotoxicity. *Protoplasma.* **2013**, *250*(2), 483–493.

Silva, I. B.; Lima, I. R.; Santana, M. A. N.; Leite, R. M. P.; Leite, S. P. *Indigofera suffruticosa* Mill. (Fabaceae): Hepatic responses in mice bearing sarcoma 180. *Int. J. Morphol.* **2014**, 32(4), 1228–1233.

Simeonova, R.; Bratkov, V.M.; Kondeva-Burdina, M.; Vitcheva, V.; Manov, V.; Krasteva, I. Experimental liver protection of n-butanolic extract of *Astragalus monspessulanus* L. on carbon tetrachloride model of toxicity in rat. *Redox Rep.* **2015**, *20*(4), 145–153.

Singab, A.N.; Ayoub, N.A.; Ali, E.N.; Mostafa, N.M. Antioxidant and hepatoprotective activities of Egyptian moraceous plants against carbon tetrachloride-induced oxidative stress and liver damage in rats. *Pharm. Biol.* **2010**, 48(11), 1255–64.

Singab, A.N.; Youssef, D.T.; Noaman, E.; Kotb, S. Hepatoprotective effect of flavonol glycosides and rich fraction from Egyptian *Vicia calcarata* Desf. Against CCl_4-induced liver damage in rats. *Arch. Pharm. Res.* **2005**, *28*, 791–797.

Singanan, V.; Singanan, M.; Begum, H. The hepatoprotective effect of Bael leaves (*Aegle marmelos*) in alcohol induced liver injury in albino rats. *Int. J. Sci. Technol.*, **2007**, *2*(2), 83–92.

Singh, A.; Handa, S.S. Hepatoprotective activity of *Apium graveolens* and *Hygrophila auriculata* against paracetamol and thioacetamide intoxication in rats. *J. Ethnopharmacol.* **1995**, *49*, 119–126.

Singh, A.K.; Singh, S.; Chandel, H.S. Evaluation of hepatoprotective activity of *Abelmoschus moschatus* seed in paracetamol induced hepatotoxicity on rat. *IOSR J. Pharm.* **2012**, *2*(5), 43–50.

Singh, B.; Chandan, B.K.; Prabhakar, A.; Taneja, J.; Singh, H.; Qazi, N. Chemistry and hepatoprotective activity of an active fraction from *Barleria prionitis* in experimental animals. *Phytother. Res.*, **2005**, *19*, 391–404.

Singh, B.; Chandan, B.K.; Sharma, N.; Bhardwaj, V.; Satti, N.K.; Gupta, V.N.; et al., Isolation, structure elucidation and *in vivo* hepatoprotective potential of trans-tetracos-15 enoic acid from *Indigofera tinctoria* Linn. *Phytother. Res.* **2006**, *20*, 831–839.

Singh, B.; Saxena, A.K.; Chandan, B.K.; Agarwal, S.G.; Bhatia, M.S.; Anand, K.K. Hepatoprotective effect of ethanolic extract of *Eclipta alba* on experimental liver damage in rats and mice. *Phytother. Res.,* **1993**, *7*(2), 154–158.

Singh, D.; Arya, P.V.; Aggarwal, V.P.; Gupta, R.S. Evaluation of antioxidant a hepatoprotective activities of *Moringa oleifera* Lam. leaves in carbon tetrachloride-intoxicated rats. *Antioxidants* **2014**, *3*, 569–591.

Singh, D.; Gupta, R.S. Hepatoprotective activity of methanol extract of *Tecomella undulata* against alcohol and paracetamol induced hepatotoxicity in rats. *Life Sci. Med. Res.* **2011**, *26*, 1–6.

Singh, H.; Prakash, A.; Kalia, A.N.; Majeed, A.B.A. Synergistic hepatoprotective potential of ethanolic extract of *Solanum xanthocarpum* and *Juniperus communis* against paracetamol and azithromycin induced liver injury in rats. *J. Trad. Complem. Med.* **2016**, *6*(4), 370–376.

Singh, H. P.; Mittal, S.; Kaur, S.; Batish, D. R.; Kohli, R. K. Chemical composition and antioxidant activity of essential oil from residues of *Artemisia scoparia. Food Chem.* **2009**, *114*(2), 642–645.

Singh, K.; Singh, N.; Chandy, A.; Manigauha, A. In vivo antioxidant and hepatoprotective activity of methanolic extracts of *Daucus carota* seeds in experimental animals. *Asian Pac. J. Trop. Biomed.* **2012**, *2*(5), 385–388.

Singh, M.; Sasi, P.; Gupta, V.H.; Rai, G.; Amarapurkar, D.N.; Wangikar, P.P. Protective effect of curcumin, Silymarin and *N*-acetylcysteine on antitubercular drug-induced hepatotoxicity assessed in an *in vitro* model. *Hum. Exp. Toxicol.* **2012**. *31*(8), 788-797.

Singh, M.K.; Sahu, P.; Nagori, K.; Dewangan, D.; Kumar, T.; Alexander, A.; et al., Organoleptic properties *in-vitro* and *in-vivo* pharmacological activities of *Calendula officinalis* Linn: An over review. *J. Chem. Pharm. Res.* **2011**, *3*, 655–663.

Singh, N.; Kamath, V.; Narasimhamurthy, K.; Rajini, P.S. Protective effects of potato peel extract against carbon tetrachloride-induced liver injury in rats. *Environ. Toxicol. Pharmacol.* **2008**, *26*, 241–246.

Singh, R.; Rao, H. S. Hepatoprotective effect of the pulp/seed of *Aegle marmelos* Correa ex Roxb. against carbon tetrachloride-induced liver damage in rats. *Int. J. Green Pharm.* **2010**, *2*(4), 232–234.

Singh, S.; Lata, S.; Tiwari, K. *Aegle marmelos* leaves protect liver against toxic effects of cyclophosphamide in mice. *New York Sci. J.* **2014**, *7*(9), 43–53.

Singh, S.; Mehta, A.; Mehta, P. Hepatoprotective activity of *Cajanus cajan* against carbon tetrachloride-induced liver damage. *Int. J. Pharm. Pharmac. Sci.* **2011**, *3*, 1–7.

Singh S.K.; Rajasekar, N.; Raj, N.A.V.; Paramaguru, R. Hepatoprotective and antioxidant effect of *Amorphophallus campanulatus* against acetaminophen induced hepatotoxicity in rats. *Int. J. Pharm. Pharm. Sci.* **2011**, *3*(2), 202–205.

Singhal, K.G.; Gupta, G.D. Hepatoprotective and antioxidant activity of methanolic extract of flowers of *Nerium oleander* against CCl_4-induced liver injury in rats. *Asian Pac. J. Trop. Med.* **2012**, *5*(9), 677–685.

Sinha, M.; Manna, P.; Sil, P.C. Aqueous extract of the bark of *Terminalia arjuna* plays a protective role against sodium-fluoride-induced hepatic and renal oxidative stress. *J. Nat. Med.* **2007**, *61*, 251–260

Sinha, S.; Bhat, J.; Joshi, M.; Sinkar, V.; Ghaskadbi, S. Hepatoprotective activity of *Picrorhiza kurroa* Royle ex. Benth extract against alcohol cytotoxicity in mouse liver slice culture. *Int. J. Green Pharm.* **2011**, *5*, 244–53.

Sintayehu, B.; Bucar, F.; Veeresham, C.; Asres, K. Hepatoprotective and free-radical scavenging activities of extracts and a major compound isolated from the leaves of *Cineraria abyssinica* Sch-Bip. ex A. Rich. *Pharmacog. J.* **2012,** *4*, 40–46.

Sisodia, S.S.; Bhatnagar, M. Hepatoprotective activity of *Eugenia jambolana* Lam. in carbon tetrachloride treated rats. *Indian J. Pharmacol.* **2009,** *41*(1), 23–27. doi: 10.4103/0253-7613.48888

Sivaraj, A.; Vinothkumar, P.; Sathiyaraj, K.; Sundaresan, S.; Devi, K.; et al., Hepatoprotective potential of *Andrographis paniculata* aqueous leaf extract on ethanol. *J. Appl. Pharm. Sci.* **2011**, *1*, 204–208.

Sivakrishnan, S.S.; Kottaimuthu, A. Hepatoprotective activity of ethanolic extract of aerial parts of *Albizia procera* Roxb. *Int. J. Pharm. Pharm. Sci.* **2014**, *6*(1), 233–238.

Sobeh, M.; Mahmoud, M.F.; Abdelfattah, M.A.O.; El-Beshbishy, H.A.; El-Shazly, A.M.; Wink, M. Hepatoprotective and hypoglycemic effects of a tannin rich extract from *Ximenia americana* var. *caffra* root. *Phytomedicine*. **2017**, *33*, 36–42.

Sobeh, M.; Mahmoud, M.F.; Abdelfattah, M.A.O.; El-Beshbishy, H.A.; El-Shazly, A.M.; Wink, M. *Albizia harveyi*: Phytochemical profiling, antioxidant, antidiabetic and hepatoprotective activities of the bark extract. *Med. Chem. Res.* **2017**, *26*(12), 3091–3105.

Sobeh, M.; Mahmoud, M.F.; Hasan, R.A.; Cheng, H.; El-Shazly, A.M.; Wink, M. *Senna singueana*: Antioxidant, hepatoprotective, antiapoptotic properties and phytochemical profiling of a methanol bark extract. *Molecules*. **2017**, *8*, 22(9). pii: E1502. doi: 10.3390/molecules22091502.

Sohn, D.H.; Kim, Y.C.; Oh, S.H.; Park, E.J.; Li, X.; Lee, B.H. Hepatoprotective and free-radical scavenging effects of *Nelumbo nucifera. Phytomedicine* **2003**, *10*, 165–169.

Solanki, Y.B.; Jain, S.M. Hepatoprotective effects of *Clitoria ternatea* and *Vigna mungo* against acetaminophen and carbon tetrachloride-induced hepatotoxicity in rats. *J. Pharmacol. Toxicol.* **2011**, *6*, 1, 30–48.

Soliman, M.M.; Abdo-Nassan, M.; Ismail, T.A. Immunohistochemical and molecular study on the protective effect of curcumin against hepatic toxicity induced by paracetamol in Wistar rats. *BMC Complement. Alternat. Med.* **2014**, 14(1), 457–468.

Somasundaram, A.; Karthikeyan, R.; Velmurugan, V.; Dhandapani, B.; Raja, M. Evaluation of hepatoprotective activity of *Kyllinga nemoralis* Hutch & Dalz. rhizomes. *J. Ethnopharmacol.* **2009**, *127* (2), 555–557.

Somchit, M.N.; Zuraini, A.; Ahmad Bustaman, A.; Somchit, N.; Sulaiman, M.R.; Noratunlina, R. Protective activity of turmeric (*Curcuma longa*) in paracetamol-induced hepatotoxicity in rats. *Int. J. Pharmacol.* **2005**, *1*, 252–256.

Sotelo-Félix, J.I.; Martinez-Fong, D.; Muriel, P.; Santilla, R.L.; Castillo, D.; Yahuaca, P. Evaluation of the effectiveness of *Rosmarinus officinalis* (Lamiaceae) in the alleviation of carbon tetrachloride-induced acute hepatotoxicity in the rat. *J. Ethnopharmacol.* **2002**, *81*(2), 145–154.

Soyal, D.; Jindal, A.; Singh, I.; Goyal, P.K. Protective capacity of Rosemary extract against radiation-induced hepatic injury in mice. *Iran. J. Radiat. Res.* **2007**, *4*(4), 161–168.

Sreedevi, C. D.; Latha, P. G.; Ancy, P.; Suja, S. R.; Shyamal, S.; Shine, V. J.; Sini, S.; Anuja, G.I.; Rajasekharan, S. Hepatoprotective studies on *Sida acuta* Burm. f. *J. Ethnopharmacol.* **2009**, *124*(2), 171– 175.

Sreelatha, S.; Padma, P.R.; Umadevi. M. Protective effects of *Coriandrum sativum* extracts on carbon tetrachloride-induced hepatotoxicity in rats. *Food Chem. Toxicol.* **2009**, *47*, 702–708.

Sreepriya, M.; Devaki, T.; Nayeem, M. Protective effects of *Indigofera tinctoria* L. against D-galactosamine and carbon tetrachloride challenge on in situ perfused rat liver. *Indian J. Physiol. Pharmacol.* **2001**, *45*(4), 428–434.

Srirama, R.; Deepak, H.B.; Senthilkumar, U.; Ravikanth, G.; Gurumurthy, B.R.; Shivanna, M.B.; Chandrasekaran, C.V.; Agarwal, A.; Shaanker, R.U. Hepatoprotective activity of Indian *Phyllanthus*. *Pharm. Biol.* **2012**, *50*, 948–953.

Srivastava, A.; Shivanandappa, T. Hepatoprotective effect of the aqueous extract of the roots of *Decalepis hamiltonii* against ethanol-induced oxidative stress in rats. *Hepatol. Res.* **2006,** *35*(4), 267–275.

Srivastava, A.; Shivanandappa, T. Hepatoprotective effect of the root extract of *Decalepis hamiltonii* against carbon tetrachloride-induced oxidative stress in rats. *Food Chem.* **2010**, *118*, 411–417.

Subash, K.R.; Ramesh, K.S.; Chariaan, B.V.; Britto, F.; Jagan Rao, N.; Vijayakumar, S. Study of hepatoprotective activity of *Solanum nigrum* and *Cichorium intybus*. *Int. J. Pharmacol.* **2011**, *7*(4), 504–509.

Subhashini, S.; Narayanan, S.; Rejani, K.; Kamath, A.T.; Kamak, D.H.; Aravind, A. Studies on the *in vitro* antihepatotoxic activity of *Indigofera tinctoria* (Linn.) against Hep G2 Human liver carcinoma cell lines. *J. Pharm. Res.* **2017**, *11*(9), 1086–1094.

Subasini, U.; Rajamanickam, G.V.; Dubey, G.P.; Prabu, P.C.; Sahayam, C.S.; et al., Hydroalcoholic extract of *Terminalia arjuna*: A potential hepatoprotective herb. *J. Biol. Sci.* **2007**, *7*, 255–262.

Subramoniam, A.; Evans, D.A.; Rajasakhran, S.P. Hepatoprotective activity of *Trichopus zeylanicus* extracts against paracetmol induced damage in rats. *Ind. J. Expt. Biol.* **1998**, *36*, 385–389.

Subramanian, L.; Selvam, R. Prevention of CCl_4 induced hepatotoxicity by aqueous extract of turmeric. *Nutr. Res.* **1999**, *19*, 429–441.

Subramanian M.; Balakrishnan S.; Chinnaiyan S. K.; Sekar V. K.; Chandu A. N. Hepatoprotective effect of leaves of *Morinda tinctoria* Roxb. against paracetamol-induced liver damage in rats. *Drug Inv. Today*. **2013**, *5*(3), 223–228.

Suddek, G.M. Protective role of thymoquinone against liver damage induced by tamoxifen in female rats. *Can. J. Physiol. Pharmacol.* **2014**, *92*(8), 640–644. doi: 10.1139/cjpp-2014-0148.

Sudharani, D.; Krishna, K.L.; Deval, K.; Safia, A.K. Pharmacological profiles of *Bacopa monnieri*: A review. Int. J. Pharm. **2011**, *1*(1), 15–23.

Suja, S.R.; Latha, P.G.; Pushpangadan, P.; Rajasekharan, S. Evaluation of hepatoprotective effects of *Helminthostachys zeylanica* (L.) Hook. against carbon tetrachloride-induced liver damage in Wistar rats. *J. Ethnopharmacol.*, **2004**, *92*, 61–66.

Suksen, K.; Charaslertrangsi, T.; Noonin, C. Protective effect of diarylheptanoids from *Curcuma comosa* on primary rat hepatocytes against *t*-butyl hydroperoxide-induced toxicity. *Pharm. Biol.* **2016**, *54*(5), 853–862.

Suky, T.M.G.; Parthipan, B.; Kingston, C.; Mohanand, P.V.R.; Soris, T. Hepatoprotective and antioxidant effect of *Balanites aegyptiaca* (L.) Del against CCl_4 induced hepatotoxicity in rats. *Int. J. Pharm. Sci. Res.* **2011**, *2*(4), 887–892,

Sultana, S.; Ahmed, S.; Jahangir, T.; Sharma, S. Inhibitory effect of celery seeds extract on chemically induced hepatocarcinogenesis: Modulation of cell proliferation, metabolism and altered hepatic foci development. *Cancer Lett.* **2005**, *221*, 11–20.

Sumathy, T.; Subramanian, S.; Govindasamy, S.; Balakrishna, K.; Veluchamy, G. Protective role of *Bacopa monniera* on morphine induced hepatotoxicity in rats. *Phytother. Res.*, **2001**, *15*(7), 643-645.

Sumitha, P.; Thirunalasundari, T. Hepatoprotective activity of *Aegle marmelos* in CCl_4 induced toxicity: An *in vivo* study. *J. Phytol.* **2011**, *3*, 5–9.

Suneetha, B.; Pavan Kumar, P.; Prasad, K.V.S.R.G.; Vidyadhara, S.; Sambasiva Rao, K.R.S. Hepatoprotective and antioxidant activities of methanolic extract of *Mimosa pudica* roots against carbon tetrachloride induced hepatotoxicity in albino rats. *Int. J. Pharm.* **2011**, *1*(1), 46–53.

Sunilson, J.; Anbu, J.; Jayaraj, P.; Mohan, S.; Kumari A.G.; Varatharajan, R. Antioxidant and hepatoprotective effect of the roots of *Hibiscus esculentus* Linn. *Int. J. Green Pharm.*, **2008**, *2*, 200–203.

Sunil, J.; Krishna, J.Y.; Bramhachari, P.V. Hepatoprotective activity of *Holostemma adakodien* Schult., extract against paracetamol-induced hepatic damage in rats. *European J. Med. Plants.* **2015**. *6*, 45.

Sunilson, J.A.J.; Jayaraj, P.; Mohan, M.S.; Kumari, A.A.G.; Varatharajan, R. Antioxidant and hepatoprotective effect of the roots of *Hibiscus esculentus* Linn. *Int. J. Green Pharm.* **2010**, *2*, 200–203.

Sunilson, J.; Muthappan, M.; Das, A.; Suraj, R.; Varatharajan, R.; Promwichit, P. Hepatoprotective activity of *Coccinia grandis* leaves against carbon tetrachloride-induced hepatic injury in rats. *Int. J. Pharmacol.* **2009**, *5*, 222–227.

Sur, P.; Chaudhuri, T.; Vedasiromoni, J.R.; Gomes, A.; Ganguly, D.K. Antiinflammatory and antioxidant property of saponins of tea [*Camellia sinensis* (L.) O. Kuntze] root extract. *Phytother. Res.* **2001**, *15*(2), pp. 174–176.

Sur, T.K.; Hazra, A.; Hazra, A.K.; Bhattacharyya, D. Antioxidant and hepatoprotective properties of Indian Sunderban mangrove *Bruguiera gymnorrhiza* L. leaves. *J. Basic Clin. Pharm.* **2016**; *7*, 75–79.

Surendra, V,; Prakash, T.; Sharma, U.R.; Goli, D.; Dayalal, S.; Kotresha, F. Hepatoprotective activity of aerial plants of *C. dactylon* against $CC1_4$ induced hepatotoxicity in rats. *J. Pharmacogn. Mag.* **2008,** 4, 195–201.

Surendran, S.; Eswaran, M.B.; Vijayakumar, M.; Rao, C.V. *In vitro* and *in vivo* hepatoprotective activity of *Cissampelos pareira* against carbon-tetrachloride-induced hepatic damage. *Indian J. Exp. Biol.* **2011**, *49*, 939–945.

Suresh Babu, P.; Krishna, V.; Maruthi, K.R.; Shankaramurthy, K.; Babu, R.K. Evaluation of acute toxicity and hepatoprotectve activity of the methanolic extract of *Dichrostachys cinerea* (Wight & Arn.) leaves. *Pharmacognosy Res.* **2011**, *3*(1), 40–43.

Sureshkumar, S.V., Mishra, S.H. Hepatoprotective effect of extracts from *Pergularia daemia* Forsk. *J. Ethnopharmacol.* **2006**, *107*(2), 164–168.

Swamy, B.M.V.; Kumar, G.S.; Shiva kumar, S.I.; Suresh, H.M.; Rajashekar, U.; Sreedhar, C. Evaluation of hepatoprotective effects of *Coccinia grandis* Linn. against carbon tetrachloride-induced liver damage in Wistar rat. *Asian J. Chem.* **2007**, *19*, 2550–2554.

Swarnalatha, L.; Neelakanta Reddy, P. Hepatoprotective activity of *Sphaeranthus amaranthoides* on D-galactosamine induced hepatitis in albino rats. *Asian Pac. J. Trop. Biomed.* **2012**, S1900-S1905.

Syed, S.N.; Rizvi, W.; Kumar, A.; Khan, A.A.; Moin, S.; Ahsan, A. A study to evaluate antioxidant and hepatoprotective activity of aqueous extract of roots of *Valeriana wallichii* in CCl_4 induced hepatotoxicity in rats. *Int. J. Basic Clin. Pharmacol*, **2017**, *3*, 354–358.

Szolnoki, T.W. Food and Fruit Trees of the Gambia. Published in conjunction with the Bundesforschungsanstalt fur Forst-und Holzwirtschaft, Stiftung Walderhaltung in Afrika, Hamburg. **1985**. p. 132.

Tabassum, N.; Agrawal, S.S. Hepatoprotective activity of *Emebelia ribes* against paracetamol induced acute hepatocellular damage in mice. *Exp. Med.*, **2003**, *10*, 43–44.

Tabassum, N.; Agrawal, S. Hepatoprotective activity of *Eclipta alba* Hassk. against paracetamol induced hepatocellular damage in mice. *Exp. Med.* **2004**, *11*, 278–280.

Tabassum, N.; Shah, M.Y.; Qazi, M.A.; Shah, A. Prophylactic activity of extract of *Taraxacum officiale* Weber. against hepatocellular injury induced in mice. *Pharmacologyonline* **2010**, *2*, 344–352.

Tag, H.M. Hepatoprotective effect of mulberry (*Morus nigra*) leaves extract against methotrexate induced hepatotoxicity in male albino rat. *BMC Complement. Altern. Med.* **2015**, *15*, 252. doi: 10.1186/s12906-015-0744-y

Taghizadeh, M.; Rashid, A.A.; Taherian, A.A.; Vakili, A.; Mehran, M.. The protective effect of hydroalcoholic extract of *Rosa canina* (Dog rose) fruit on liver function and structure in streptozotocin-induced diabetes in rats. *J. Diet. Suppl.* **2018**, *15*(5), 624–635.

Taji, F.; Shirzad, H.; Ashrafi, K.; Parvin, N.; Kheiri, S.; Namjoo, A.; et al., A comparison between the antioxidant strength of the fresh and stale *Allium sativum* (garlic) extracts. *Zahedan J. Res. Med. Sci.* **2012**, *14*, 25–29.

Taju, G.; Jayanthi, M.; Nazeer basha, M.; Nathiga, K.; Sivaraj, A. Hepatoprotective effect of Indian medicinal plant *Psidium guajava* Linn. leaf extract on paracetamol induced liver toxicity in Albino rats. *J. Pharm. Res.* **2010**, *3*(8), 1759–1763.

Takate, S.B.; Pokharkar, R.D.; Chopade, V.V.; Gite, V.N. Hepato-protective activity of the aqueous extract of *Launaea intybacea* (Jacq.) Beauv. against carbon tetrachloride-induced hepatic injury in albino rats. *J. Pharm. Sci. Tech.* **2010**, *2*(7), 247–251.

Talluri, M.R.; Gummadi, V.P.; Battu, G.R. Chemical composition and hepatoprotective activity of *Saponaria officinalis* on paracetamol-induced liver toxicity in rats. *Pharmacogn. J.* **2018**, *10*(6), 1196–1201.

Tandon, V. R.; Khajuria, V.; Kapoor, B.; Kour, D.; Gupta, S. Hepatoprotective activity of *Vitex negundo* leaf extract against anti-tubercular drugs induced hepatotoxicity. *Fitoterapia* **2008**, *79*, 533– 538.

Tang, X.; Gao, J.; Wang, Y.; Fan, Y.M.; Xu, L.Z.; Zhao, X.N.; Xu, Q.; Qian, Z.M. Effective protection of *Terminalia catappa* L. leaves from damage induced by carbon tetrachloride in liver mitochondria. *J. Nutr. Biochem.* **2006**, *17*, 177–182.

Tang, X.; Wei. R.; Deng, A.; Lei, T. Protective effects of ethanolic extracts from artichoke, an edible herbal medicine, against acute alcohol-induced liver injury in mice. *Nutrients* **2017**, *9*(9), 1000. doi: 10.3390/nu9091000

Tasduq, S.A.; Kaisar, P.; Gupta, D.K.; Kapahi, B.K.; Maheshwari, H.S.; Jyotsna, S.; Johri, R.K. Protective effect of a 50% hydroalcoholic fruit extract of *Emblica officinalis* against anti-tuberculosis drugs induced liver toxicity. *Phytother. Res.* **2005**, *19*(3), 193–197.

Tasduq, S.A.; Singh, K.; Satti, N.K.; Gupta, D.K.; Suri, K.A.; Johri, R.K. *Terminalia chebula* (fruit) prevents liver toxicity caused by sub-chronic administration of rifampicin, isoniazid and pyrazinamide in combination. *Hum. Exp. Toxicol.* **2006**, *25*, 111–118.

Tasleem, F.; Mahmood, S.B.Z.; Imam, S.; Hameed, N.; Jafrey, R.; Azhar, I.; Gulzar, R.; Mahmoo, Z.A. Hepatoprotective effect of *Peltophorum pterocarpum* leaves extracts and pure compound against carbon tetra chloride induced liver injury in rats. *Med. Res. Arch.* **2017**, *5*(4), 1–14.

Tatiya, A.U.; Surana, S.J.; Sutar, M.P.; Gamit, N.H. Hepatoprotective effect of poly herbal formulation against various hepatotoxic agents in rats. *Pharmacogn. Res.* **2012**, *4*, 50–56.

Thamgoue, A.D.; Tchokouaha, L.R.Y.; Tsabang, N.; Tarkan, P.A.; Kuiate, J.-R.; Agbor, G.A. *Costus afer* protects cardio-, hepato-, and reno-antioxidant status in streptozotocin-intoxicated Wistar rats. *BioMed. Res. Int.* **2018**, doi.org/10.1155/2018/4907648

Teng, C.Y.;, Lai, Y.L.; Huang, H.I.; Hsu, W.H.; Yang, C.C.; Kuo, W.H. *Tournefortia sarmentosa* extract attenuates acetaminophen-induced hepatotoxicity. *Pharm. Biol.* **2012**, *50*(3), 291–396. doi: 10.3109/13880209.2011.602695.

Thabrew, M.; Joice, P.A. Comparative study of the efficacy of *Pavetta indica* and *Osbeckia octanda* in the treatment of liver dysfunction. *Planta Med.*, **1987**, *53*, 239–241.

Thabrew, M.I.; Hughes, R.D.; McFarlane, I.G. Screening of hepatoprotective plant components using a HepG2 cell cytotoxicity assay. *J. Pharm. Pharmacol.* **2011**. https://doi.org/10.1111/j.2042-7158.1997.tb06055.x

Thabrew, M.I., Jayatilaka, K.A., Perera, D.J. Evaluation of the efficacy of *Melothria maderaspatana* in the alleviation of carbon tetrachloride-induced liver dysfunction. *J. Ethnopharmacol.* **1988**, *23*(2–3), 305–312.

Thakare, S.P.; Jain, H.N.; Patil, S.D.; Upadhyay, U.M. Hepatoprotective effect of *Cocculus hirsutus* on bile duct ligation-induced liver fibrosis in albino Wistar rats. *Bangladesh J. Pharmacol.* **2009**, *4*, 126–30.

Thakral, J.; Borar, S.; Kalia, J.A.R. Antioxidant potential fractionation from methanol extract of aerial parts of *Convolvulus arvensis* Linn. (Convolvulaceae). *Int. J. Pharm. Sci. Drug Res.* **2010**, *2*, 219–23.

Thambi, P.; Sabu, M.C.; Chungath, J. Hepatoprotective and free-radical scavenging activities of *Lagerstroemia speciosa* Linn. leaf extract. *Orient Pharm. Exp. Med.* **2009**, *9*, 225–231.

Thattakudian, S.; Uduman, M.S.; Sundarapandian, R.; Muthumanikkam, A.; Kalimuthu, G.; Parameswari, S.; et al., Protective effect of methanolic extract of *Annona squamosa* Linn. in isoniazid-rifampicin induced hepatotoxicity in rats. *Pak. J. Pharm. Sci.* **2011**, *24*(2), 129–134.

Thenmozhi, A. J.; Subramanian, P. Hepatoprotective effect of *Momordica charantia* in ammonium chloride induced hyperammonemic rats. *J. Pharm. Res.* **2011**, *4*(3), 700–702.

Thenmozhi, M.; Dhanalakshmi, M.; Manjula devi, K.; Sushila, K.; Thenmozhi, S. Evaluation of hepatoprotective activity of *Leucas aspera* hydroalcoholic leaf extract during exposure to lead acetate in male albino Wistar rats. *Asian J. Pharm. Clin. Res.* **2013**, *6*(1), 78–81.

Thirumalai, T.; David, E.; Therasa, S.V.; Elumalai, E.K. Restorative effect of *Eclipta alba* in CCl_4 induced hepatotoxicity in male albino rats. *Asian Pac. J. Trop. Dis.* **2011**, *1*(4), 304–307.

Thuwaini, M.M.; Abdul-Mounther, M.; Kadhem, H.S. Hepatoprotective effects of the aqueous extract of clove (*Syzygium aromaticum*) against paracetamol induced hepatotoxicity and oxidative stress in rats. *European J. Pharm. Med. Res.* **2016**, *3*(8), 36–42.

Thyagarajan, S.P.; Jayaram, S.; Gopalakrishnan, V.; Hari, R.; Jeyakumar, P.; Sripathi, M.S. Herbal medicines for liver diseases in India. *J. Gastroenterol. Hepatol.* **2002**, *17*, S370-S376.

Tiwari, P.; Kumar, K.; Pandey, A.K.; Pandey, A.; Sahu, P.K. Antihepatotoxic activity of *Euphorbia hirta* and by using the combination of *Euphorbia hirta* and *Boerhaavia diffusa* extracts on some experimental models of liver injury in rats. *Int. J. Innov. Pharm. Res.* **2011**, *2*(2), 126–130.

Tlili, N.; Feriani, A.; Saadoui, E.; Nasri, N.; Khaldi, A. *Capparis spinosa* leaves extract: Source of bioantioxidants with nephroprotective and hepatoprotective effects. *Biomed Pharmacother.* **2017**, *87*, 171–179.

Tlili, N.; Tir, M.; Feriani, A.; Yahia, Y.; Allagui, M.S.; Saadaoui, E.; El Cafsi, M.; Nasri, N. Potential health advantages of *Periploca laevigata*: Preliminary phytochemical analysis and evaluation of *in vitro* antioxidant capacity and assessment of hepatoprotective, anti-inflammatory and analgesic effects. *J. Funct. Foods,* **2018**, *48*, 234–242.

Tiwari, B.K.; Khosa, R.L. Evaluation of the hepatoprotective activity of *Sphaeranthus indicus* flower heads extract. *J. Nat. Rem.* **2008**, *8*, 173–178.

Tiwari, B.K.; Khosha, R.L. Evalution of the hepatoprotective and antioxidant effect of *Berberis asiatica* against exeperimentally induced liver injury in rats. *Int. J. Pharm. Pharm. Sci.* **2010**, *2*(1), 92–97.

Tiwari, P.; Ahirwae, D.; Chandy, A.; Ahirwar, B. Evaluation of hepatoprotective activity of alcoholic and aqueous extracts of *Selaginella lepidophylla. Asian Pac. J. Trop. Dis.* **2014**, *4*(1), 81–86.

Tiwary, B.K., Dutta, S., Dey, P., Hossain, M., Kumar, A., Bihani, S., Nanda, A.K., Chaudhuri, T.K., Chakraborty, R. Radical scavenging activities of *Lagerstroemia speciosa* (L.) Pers. petal extracts and its hepatoprotection in CCl_4-intoxicated mice. *BMC Complement. Altern. Med.* **2017**, *17*, 55.

Tong, J.; Yao, X.; Zeng, H.; Zhou, G.; Chen, Y.; Ma, B.; Wang, Y. Hepatoprotective activity of flavonoids from *Cichorium glandulosum* seeds *in vitro* and *in vivo* carbon tetrachloride-induced hepatotoxicity. *J. Ethnopharmacol.* **2015,** *174*, 355–363. doi: 10.1016/j.jep.2015.08.045.

Toori, M.A.; Joodi, B.; Sadeghi, H.; Sadeghi, H.; Jafari, M.; Talebianpoor, M.S.; Mehraban, F.; Mostafazadeh, M.; Ghavamizadeh, M. Hepatoprotective activity of aerial parts of *Otostegia persica* against carbon tetrachloride-induced liver damage in rats. *Avicenna J. Phytomed.* **2015**, *5*, 238–46.

Toppo, R.; Roy, B.K.; Gora, R.H.; Baxla, S.L.; Kumar, P. Hepatoprotective activity of *Moringa oleifera* against cadmium toxicity in rats. *Vet. World.* **2015**, *8*, 537–540.

Touloupakis, E,; Ghanotakis, D.F. Nutraceutical use of garlic sulfur containing compounds. *Adv. Exp. Med. Biol.* **2010**, *698*, 110–121.

Tripathi, M.; Singh, B.K.; Mishra, C.; Raisuddin, S.; Kakkar, P. Involvement of mitochondria mediated pathways in hepatoprotection conferred by *Fumaria parviflora* Lam. extract against nimesulide induced apoptosis *in vitro*. *Toxicol. In Vitro* **2010**, *24*, 495–508.

Trivedi, N.P.; Rawal, U.M. Hepatoprotective and antioxidant property of *Andrographis paniculata* (Nees) in BHC-induced liver damage in mice. *Indian J. Exp. Biol.* **2001**, *39*, 41–46.

Tsai, J.-C.; Chiu, C.-S.; Chen, Y.-C.; Lee, M.-S.; Hao, X.-Y.; Hsieh, M.-T.; Cao, C.-P.; Peng, W.-H. Hepatoprotective effect of *Coreopsis tinctoria* flowers against carbon tetrachloride-induced liver damage in mice. *BMC Complement. Altern. Med.* **2017**, *17*, 139.

Tsai, J.C.; Peng, W.H.; Chiu, T.H.; Huang, S.C.; Huan, T.H.; Lai, S.C.; Lai, Z.R.; Lee, C.Y. Hepatoprotective effect of *Scoparia dulcis* on carbon tetrachloride induced acute liver injury in mice. *Am. J. Chin. Med.* **2010**, *38*, 761–775.

Tukappa, N.K.A., Londonkar, R.L., Nayaka, H.B., Kumar, C.B.S. Cytotoxicity and hepatoprotective attributes of methanolic extract of *Rumex vesicarius* L. *Biol. Res.* **2015**, *48*, 19. doi: 10.1186/s40659-015-0009-8.

Tütüncü, M.; Özbek, H.; Bayram, I.; Cengiz, N.; Özgökce, F. and Him, A. The effects of diethylether extract of *Helichrysum plicatum* DC. subsp. *plicatum* and *Tanacetum balsamita* L. subsp. *balsamitoides* (Sch. Bip.) Grierson (Asteraceae) on the acute liver toxicity in rats. *Asian. J. Anim. Vet. Adv.*, **2010**, 5(7), 465–471.

Tung, Y.T.; Wu, J.-H.; Huang, C.-C.; Peng, H.-C.; Chen, Y.-L.; Yang, S.-C.; Chang, S.-T. Protective effect of *Acacia confusa* bark extract and its active compound gallic acid against carbon tetrachloride-induced chronic liver injury in rats. *Food Chem. Toxicol.* **2009**, *47*(6), 1385–1392.

Türel, I.; Ozbek, H.; Erten, R.; Oner, A.C.; Cengiz, N.; Yilmaz, O. Hepatoprotective and anti inflammatory activities of *Plantago major*. *Indian J. Pharmacol.* **2009**, *41*, 120–124.

Ubani, C.S.; Joshua, P.E.; Anieke, U.C. Effects of aqueous extract of *Hibiscus sabdariffa* L. calyces on liver marker enzymes of phenobarbitone-induced adult wistar albino rats. *J. Chem. Pharm. Res.*, **2011**, *3*(4): 528–537.

Ubhenin, A.; Igbe, I.; Adamude, F.; Falodun, A. Hepatoprotective effects of ethanol extract of *Caesalpinia bonduc* against carbon tetrachloride induced hepatotoxicity in albino rats. *J. Appl. Sci. Environ. Manag.* **2016**, *20*, 396–401.

Udem, S.; Nwaogu, I.; Onyejekwe, O. Evaluation of hepatoprotective activity of aqeous leaf extract of *Swietenia mahogani* (Meliaceae) in chronic alcohol-induced liver injury in rats. *Maced. J. Med. Sci.* **2011**, *15*, 31–36.

Uhegbu, F.O.; Elekwa, I.; Emmanuel, Akubugwo, E.I.; Chinyere, G.; Iweala, E.E.J. Analgesic and hepatoprotective activity of methanoilc leaf extract of *Ocimum gratissimum*. *Res. J. Med. Plant.* **2012**, *6*, 108–119.

Ujowundu, C.O.; Nwokedinobi, N.; Kalu, F.N.; Nwaoguikpe, R.N.; Okechukwu, R.I. Chemopreventive potentials of *Ocimum gratissimum* in diesel petroleum induced hepatotoxicity in albino Wistar rats. *J. Appl. Pharm. Sci.* **2011**, *1*, 56–61.

Ulanganathan, I.; Divya, D.; Radha, K.; Vijayakumar, T.M.; Dhanaraju, M.P. Protective effect of *Luffa acutangula* var. *amara* against carbon tetrachloride-induced hepatotoxicity in experimental rats. *Res. J. Biol. Sci.* **2010**, *5*(9), 615–624.

Ulicná, O.; Greksak, M.; Vancova, O.; Zlatos, L.; Galbavy, S.; Bozek, P.; Nakano, M. Hepatoprotective effect of Rooibos tea (*Aspalathus linearis*) on CCl_4 induced liver damage in rats. *Physiol. Res.*, **2003**, *52*, 461-466.

Ullah, I., Khan, J.A.; Adhikari, A.; Shahid, M. Hepatoprotective effect of *Monotheca buxifolia* fruit against antitubercular drugs-induced hepatotoxicity in rats. *Bangladesh J. Pharmacol.* **2016**, *11*, 248–56.

Uma, D., Aida, W. Optimization of extraction parameters of total phenolic compounds from Henna (*Lawsonia inermis*) leaves. *Sains Malays.* **2010**, *39*, 119–128.

Umadevi, S.; Mohanta, G.P.; Kalaiselvan, R.; Manna, P.K.; Manavalan, R.; Sethupathi, S.; Shantha, K. Studies on hepatoprotective effect of *Flaveria trinervia*. *J. Nat. Rem.*, **2004**, *4*(2), 168–173.

Umamaheswari, M.; Kumar, C.T. Hepatoprotective and antioxidant activities of the fractions of *Coccinia grandis* against paracetamol-induced hepatic damage in mice. *Int. J. Biomed. Pharm. Sci.*, **2008**, *2*, 103–107.

Umer, S.; Asres, K.; Veeresham, C. Hepatoprotective activities of two Ethiopian medicinal plants. *Pharm. Biol.* **2010**, *48*(4), 461–468, doi: 10.3109/13880200903173593

Upur, H.; Amat, N.; Blazekovic, B.; Talip, A. Protective effect of *Cichorium glandulosum* root extract on carbon tetrachloride-induced and galactosamine-induced hepatotoxicity in mice. *Food Chem.Toxicol.* **2009**, 47, 2022–30.

Usha, K.; Kasturi, M.; Hemalatha, P. Hepatoprotective effect of *Hygrophila* spinosa and Cassia occidentalis on Carbon tetrachloride-induced liver damage in experimental rats. *Indian J. Clin. Biochem.* **2007**, 22(2), 132–135.

Usharani, K.; Amirtham, D.; Nataraja, T.S. Hepatoprotective activity of *Nilgirianthus ciliatus* (Nees) Bremek in paracetamol induced toxicity in Wistar albino rats. *Afr. J. Int. Med.* **2013**, *1*(4), 026–030.

Usmani, S.; Kushwaha, P. A study on hepatoprotective activity of *Calotropis gigantea* leaves extract. *Int. J. Pharm. Pharm. Sci.* **2010**, 2, 101–103.

Usoh, I.F.; Itemobong, S.; Ekaidem, O.E.; Etim, E.; Akpan, H.D.; Akpan, E.J.; et al., Antioxidant and hepatoprotective effects of dried flower extracts of *Hibiscus sabdariffa* L. on rats treated with carbon tetrachloride. *J. Appl. Pharm. Sci.* **2012,** *2*, 186–189.

Usunomena, U. Protective effects of *Annona muricata* ethanolic leaf extract against dimethylnitrosamine (DMN)-induced hepatotoxicity. *IOSR J. Pharm. Biol. Sci.* **2014**, *9*(4), 1–6.

Usunomena, U.; Ngozi, O.; Ikechi, E.G. Effect of *Vernonia amygdalina* on some biochemical indices in dimethylnitrosamine (DMN)-induced liver injury in rats. *Int. J. Anim. Biol.* **2015**, *1*, 99–105.

Uzzi, H.O.; Grillo, D.B. The hepatoprotective potentials of aqueous leaf extract of *Cassia occidentalis* against paracetamol induced hepatotoxicity in adult Wistar rats. *Int. J. Herbs Pharmacol. Res.* **2013**, *2*, 6–13.

Vadivu, R.; Krithika, A., Biplab, C., Dedeepya, P., Shoeb, N., Lakshmi, K.S. Evaluation of hepatoprotective activity of the fruits of *Coccinia grandis* Linn. *Int. J. Health Res.* **2008**, *1*(3), 163–168.

Vadivu, R.; Suresh, J.; Girinath, K.; Boopathikannan, P.; Vimala, R.; Kumar, N.M.S. Evaluation of hepatoprotective and *in-vitro* cutotoxic activities of leaves of *Premna serratifolia* Linn. *J. Sci. Res.,* **2009**, *1*(1), 145–152.

Vahab, A.; Harindran, J. Hepatoprotective activity of bark extracts of *Terminalia catappa* Linn. in albino rats. *World J. Pharm. Pharm. Sci.* **2016**, *5*, 1002–1016.

Vakiloddin, S.; Fuloria, N.; Fuloria, S.; Dhanaraj, S.A.; Balaji, K.; Karupiah, S. Evidences of hepatoprotective and antioxidant effect of *Citrullus colocynthis* fruits in paracetamol induced hepatotoxicity. *Pak. J. Pharm. Sci.* **2015**, *28*(3), 951–957.

Valarmathi, R.; Rajendran, A.; Gopal, V.; Senthamarai, R.; Akilandeswari, S.; Srileka, B. Protective effcet of the whole plant of *Mollugo pentaphylla* Linn. against carbon tetrachloride induced hepatotoxicity in rats. *Int. J. PharmTech Res.* **2010**, *2*(3), 1658–1661.

Valcheva-Kuzmanova, S.; Borisova, P.; Galunska, B.; Belcheva, A. Hepatoprotective effect of the natural fruit juice from *Aronia melanocarpa* on carbon tetrachloride-induced acute liver damage in rats. *Exp. Toxicol. Pathol.* **2004**, *156*(3), 195–201.

Vasavi, A.K.E.; Satapathy, D.K.; Tripathy, S.; Srinivas, K. Evaluation of hepatoprotective and antioxidant activity of *Helianthus annuus* flowers against carbon tetrachloride (CCl_4)-induced toxicity. *Int. J. Pharmacol. Toxicol.* **2014**, *4*(2), 132-.

Ved, A.; Gupta, A.; Rawat, A.K.S. Antioxidant and hepatoprotective potential of phenol-rich fraction of *Juniperus communis* Linn. leaves. *Pharmacog. Mag.* **2017,** *13*, 108–113.

Venkatesh, P.; Dinakar, A.; Senthilkumar, N. Hepatoptotective activity of an ethanolic extract of stems of *Anisochilus carnosus* against carbon tetrachloride induced hepatotoxicity in rats. *Int. J. Pharm. Pharm. Sci.* **2011**, *3*(1), 243-245.

Venkateswarlu, G.; Ganapathy, S. Hepatoprotective activity of *Limnophila repens* against paracetamol-induced hepatotoxicity in rats. *Int. J. Green Pharm.* **2019**, 13(3), 268–274.

Venukumar, M.R., Latha, M.S. Antioxidant activity of *Curculigo orchioides* in carbon tetrachloride-induced hepatopathy in rats.

Indian J. Clin. Biochem. **2002**, *17*(2), 80–87.

Venukumar, M.R., Latha, M.S. Hepatoprotective effect of the methanolic extract of *Curculigo orchioides* in CCl_4-treatred male rats. *Int. J. Pharmacol.* **2002**, *34*, 269–275.

Verma, N.; Khosa, R.L. Hepatoprotective activity of leaves of *Zanthoxylum armatum* DC in CCl_4 induced hepatotoxicity in rats. *Indian J. Biochem. Biophys.* **2010**, *47*, 124–127.

Verma, P.; Paswan, S.; Singh, S.P.; Shrivastva, S.; Rao, C.V. Assessment of hepatoprotective potential of *Solanum xanthocarpum* (whole plant) Linn. against isoniazid & rifampicin-induced hepatic toxicity in Wistar rats. *Elixir Appl. Bot.* **2015**, *87*(1), 35578–35583.

Vetriselva, S.; Subasini, U. Hepatoprotctive activity of *Andrographis paniculata. Int. J. Res. Pharm. Nano Sci.* **2012**, *1*(2), 307–316.

Vetriselvan, S.; Rajamanickam, V.; Muthappan, M.; Gnanasekaran, D.; Chellappann, D.K. Hepatoprotective effects of aqueous extract of *Andrographis paniculata* against CCl_4 induced hepatotoxicity in albino Wistar rats. *Asian J. Pharm. Clin. Res.* **2011**, *4*, 99–100.

Vidya, S.M.; Krishna, V.; Manjunatha, B.K.; Mankani, K.L.; Ahmed, M.; Jagadeesh Singh, S.D. Evaluation of hepatoprotective activity of *Clerodendrum serratum* L. *Indian J. Exp. Biol.* **2007**, *45*, 538–542.

Vijayan, P.; Prahant, H.C.; Dhanaraj, S.A.; Badami, S.; Suresh, B. Hepatoprotective effect of total alkaloid fraction of *Solanum pseudocapsicum* leaves. *Pharm. Biol.* **2003**, *41*, 443–448.

Vijayavel, K.; Anbuselvam, C.; Ashokkumar, B. Protective effect of *Coleus aromaticus* Benth. (Lamiaceae) against naphthalene-induced hepatotoxicity. *Biomed. Environ. Sci.* **2013**, *26*(4), 295–302.

Vishwanath, J.; Preetham, G.B.; Anil Kumar, B.; Satwik Reddy, J. A review on hepatoprotective Plants. *Int. J. Drug Dev. Res.*, **2012**, *4*(3), 1–8.

Vishwakarma, A.P.; Vishwe, A.; Sahu, P.; Chaurasiya, A. Screening of hepatoprotective potential of ethanolic and aqueous extract of *Terminalia arjuna* bark against paracetamol/CCl_4-induced liver damage in Wistar albino rats. *Int. J. Pharm. Arch.* **2013**, *2*, 243–250.

Vishwakarma, S.L.; Goyal, R.K. Hepatoprotective activity in *Enicostemma littorale* in CCl_4-induced liver damage. *J. Nat. Remedies,* **2004**, *4*, 120–126.

Vladimir-Knezevic, S.; Cvijanovic, O.; Blazekovic, B.; Kindl, M.; Stefan, M.B.; Domitrovic, R. Hepatoprotective effects of *Micromeria croatica* ethanolic extract against CCl_4-induced liver injury in mice. *BMC Complement. Altern. Med.* **2015**, *15*, 233. doi: 10.1186/s12906–015-0763-8

Vouffo, E.Y.; Donfack, F.M.; Temdie, R.J.; Ngueguim, F.T.; Donfack, J.H.; Dzeufiet, D.S.; et al. Hepato-nephroprotective and antioxidant effect of stem bark of *Allanblackia gabonensis* aqueous extract against acetaminophen-induced liver and kidney disorders in rats. *J. Exp. Integr. Med.* **2012**, *2*(4), 337–344.

Vuda, M.; D'Souza, R.; Upadhya, S.; Kumar, V.; Rao, N.; Kumar, V.; Boillat, C.; Mungli, P. Hepatoprotective and antioxidant activity of aqueous extract of *Hybanthus enneaspermus* against CCl_4-induced liver injury in rats. *Exp. Toxicol. Pathol.* **2012**, *64*(7–8), 855–859.

Wadekar, R.R.; Supale, R.S.; Tewari, K.M.; Patil, K.S.; Jalalpure, S.S. Screening of roots of *Baliospermum montanum* for hepatoprotective activity against paracetamol-induced liver damage in albino rats. *Int. J. Green Pharm.* **2010**, 220–223.

Wagh, S.S.; Shinde, G.B. 2010. Antioxidant and hepatoprotective activity of *Tridax procumbens* Linn. against paracetamol induced hepatotoxicity in male albino rat. *Adv. Stud. Biol.*, **2010**, *2*(3), 105–112.

Wagner H.; Geyer B.; Fiebig M.; Kiso Y.; Hikino H. Isobutrin and Butrin, the antihepatotoxic principles of *Butea monosperma* flowers. *Planta. Med.* **1986**. *52*(2), 77–79.

Wahid, A.; Hamed, A.N.; Eltahir, H.M.; Abouzied, M.M. Hepatoprotective activity of ethanolic extract of *Salix subserrata* against CCl_4-induced chronic hepatotoxicity in rats. *BMC Complement. Altern. Med.* **2016**, *16*, 263. doi: 10.1186/s12906-016-1238-2

Wakawa, H.; Ira, M. Protective effects of *Camellia sinensis* leaf extract against carbon tetrachloride induced liver injury in rats. *Asian J. Biochem.* **2015**, *10*, 86–92

Wakawa, H.Y.; Franklyne, E.A. Protective effects of *Abrus precatorius* leaf extract against carbon tetrachloride-induced liver injury in rats. *J. Nat. Sci. Res.* **2015**, *5*, 15–19.

Wakawa, H.Y.; Musa H. Protective effect of *Erythrina senegalensis* (DC) leaf extract on CCl_4-induced liver damage in rats. *Asian J. Biol. Sci.* **2013**, *6*, 234–238.

Wakchaure, D.; Jain, D.; Singhai, A.K.; SomanI, R. Hepatoprotective activity of *Symplocos racemosa* Roxb. bark extract in carbon tetrachloride-induced liver damage in rats. *J. Ayurveda Integr. Med.* **2011**, *2*, 137–143.

Wang, B.-J.; Liu, C.-T.; Tseng, C.-Y.; Wu, C.-P.; Yu, Z.-R. Hepatoprotective and antioxidant effects of *Bupleurum kaoi* Liu (Chao et Chuang) extract and its fractions fractionated using supercritical CO_2 on CCl_4-Induced liver damage. *Food Chem. Toxicol.* **2004**, *42*, 609–617.

Wang, L.; Cheng, D.; Wang, H.; Di, L.; Zhou, X.; Xu, T.; Yang, X.; Liu, Y. The hepatoprotective and antifibrotic effects of *Saururus chinensis* against carbontetrachloride-induced hepatic fibrosis in rats. *J. Ethnopharmacol.* **2009**, *126*, 487–491.

Wang, M.; Zhang, X.J.; Feng, R.; Jiang, Y.; Zhang, D.Y.; He, C.; Li, P.; Wan, J.B. Hepatoprotective properties of *Penthorum chinense* Pursh against carbon tetrachloride-induced acute liver injury in mice. *Chin. Med.* **2017**, *12*, 32. doi: 10.1186/s13020-017-0153-x.

Wang, M.Y.; Anderson, G.; Nowicki, D.; Jensen, J. Hepatic protection by *Morinda citrifolia* (noni) fruit juice against CCl_4-induced chronic liver damage in female SD rats. *Plant Foods Hum. Nutr.*, **2008a**, *63*(3), 141-145.

Wang, M.Y., Nowicki, D., Anderson, G., Jensen, J.; West, B. Liver protective effect of *Morinda citrifolia* (Noni). *Plant Foods Human Nutr.* **2008b**, *63*(2), 59–63.

Wang, N., Li, P., Wang, Y., Peng, W., Wu, Z., Tan, S., Liang, S., Shen, X. and Su, W. Hepatoprotective effect of *Hypericum japonicum* extract and its fractions. *J. Ethnopharmacol.* **2008**, *116*, 1–6.

Wei, J.-F.; Li, Y.-Y.; Yin, Z.-H.; Gong, F.; Shang, F.-D.; Hepatoprotective effect of *Lysimachia paridiformis* Franch. var. *stenophylla* Franch. on CCl_4-induced acute liver injury in mice. *Afr. J. Pharm. Pharmacol.* **2012**, *6*(13), 956–960.

Wei, J.F.; Li, Y.Y.; Yin, Z.H.; Gong, F.; Shang, F.D. Antioxidant activities *in vitro* and hepatoprotective effects of *Lysimachia clethroides* Duby on CCl_4-induced acute liver injury in mice. *Afr. J. Pharm. Pharmacol.* **2012**, *6*, 743–750.

Weissner, W. Brahmi and Cognition: Nature's Brainpower Enhancer," http://ayurveda-nama.org/pdf/resources/NAMA Brahmi Weissner.pdf.

Werawatganon, D.; Linlawan S.; Thanapirom K.; et al., *Aloe vera* attenuated liver injury in mice with acetaminophen-induced hepatitis. *BMC Complement. Alternat. Med.*, **2014**, *14*, 229.

Willett, K.L.; Roth, R.A.; Walker, L. Workshop overview: Hepatotoxicity assessment for botanical dietary supplements. *Toxicol. Sci.* **2004**, *79*, 4–9.

Williams, A.F.; Clement, Y.N.; Nayak, S.B.; Rao, A.V.C. *Leonotis nepetifolia* protects against acetaminophen induced hepatotoxicity: Histological studies and the role of antioxidant enzymes. *Nat. Prod. Chem. Res.* **2016**, *4*, 222.

Wills, P. J.; Asha, V. V. Preventive and curative effect of *Lygodium flexuosum* (L.) Sw. on carbon tetrachloride-induced hepatic fibrosis in rats. *J. Ethnopharmacol.* **2006a**, *107*(1), 7–11.

Wills, P. J.; Asha, V. V. Protective effect of *Lygodium flexuosum* (L.) Sw. (Lygodiaceae) against d-galactosamine induced liver injury in rats. *J. Ethnopharmacol.* **2006b**, *108*(1), 116–123.

Wills, P. J.; Asha, V. V. Protective effect of *Lygodium flexuosum* (L.) Sw. (Lygodiaceae) against crbon tetrachloride induced liver injury in rats. *J. Ethnopharmacol.* **2006c**, *108*(3), 320–326.

Wills, P. J.; Asha, V. V. Protective mechanism of *Lygodium flexuosum* extract in treating and preventing carbon tetrachloride-induced hepatic fibrosis in rats. *Chem. Biol. Interact.* **2007**, *165*(1), 76–85.

Wills, P. J.; Suresh, V.; Arun, M.; Asha, V. V. Antiangiogenic effect of *Lygodium flexuosum* against N-nitrosodiethylamine-induced hepatotoxicity in rats. *Chem Biol Interact.* **2006**, *164*(1–2), 25–38.

Wu, J.-B.; Lin, W.-L.; Hsieh C.-C.; Ho, H.-Y.; Tsay, H.-S.; Lin, W.-C. The hepatoprotective activity of kinsenoside from *Anoectochilus formosanus*. *Phytother. Res.* **2007**, *21*, 58–61.

Wu, L., Chen, Y. Protective effect of celery root against acute liver injury by CCl_4. West *China J. Pharm. Sci.* **2008**. doi: 10.3969/j.issn.1006-0103.2008.04.013.

Wu, S.J., Wang, J.S., Lin, C.C., Chang, C.H. Evaluation of hepatoprotective activity of Legumes. *Phytomedicine* **2001**, *8*(3), 213–219.

Wu Y.; Yang L., B.; Wang F.; Wu, X.; Zhou, C.; Shi, S.; Mo, J.; Zhao, Y. Hepatoprotective and antioxidative effects of total phenolics from *Laggera pterodonta* on chemical-induced injury in primary cultured neonatal rat hepatocytes. *Food Chem. Toxicol.* **2007**, 45: 1349–1355.

Wu, Y.; Wang, F.; Zheng, Q.; Lu, L.; Yao, H.; Zhou, C.; Wu, X.; Zhao, Y. Hepatoprotective effect of total flavonoids from *Laggera alata* against carbon tetrachloride-induced injury in primary cultured neonatal rat hepatocytes and in rats with hepatic damage. *Biomed. Sci.* **2006,** *13*(4), 569–578.

Wu, Y.H.; Zhang, X.M.; Hu, M.H.; Wu, X.M.; Zhao, Y. Effect of *Laggera alata* on hepatocyte damage induced by carbon tetrachloride *in vitro* and *in vivo*. *J. Ethnopharmacol.* **2009**, *126*(1), 50–6.

Wu, Z.R.; Bai, Z.T.; Sun, Y.; Chen, P.; Yang, Z.G.; Zhi, D.J.; et al., Protective effects of the bioactive natural product N-trans-Caffeoyldopamine on hepatotoxicity induced by isoniazid and rifampicin. *Bioorg. Med. Chem. Lett.* **2015**, *25*(22), 5424–5426.

Xie, Q.; Guo, F.F.; Zhou, W. Protective effects of *Cassia* seed ethanol extract against carbon tetrachloride-induced liver injury in mice. *Acta Biochim Pol.* **2012**, *59*(2), 265–70.

Xiong, Q.; Fan, W.; Tezuka, Y.; Ketut, A.I.; Stampoulis, P.; Hattori, M.; Namba, T.; Kadota, S. Hepatoprotective effect of *Apocynum venetum* and its active constituents. *Planta. Med.*, **2000**, *66*(2), 127–133.

Xu, L.; Gao, J.; Wang, Y.; Yu, W.; Zhao, X.; Yang, X.; Zhong, Z.; Qian, Z. *Myrica rubra* extracts protect the liver from CCl_4-induced damage. *Evid. Based Compl. Altern. Med,* **2011**, 518302. doi.: 10.1093/ecam/nep196

Yadav, N.P.; Chanda, D.; Chattopadhyay, S.K.; Gupta, A.K.; Pal, A. Hepatoprotective effects and safety evaluation of coumarinolignoids isolated from *Cleome viscosa* seeds. *Indian J. Pharm. Sci.* **2010,** *72*(6), 759–765. doi: 10.4103/0250-474X.84589.

Yadav, N.P.; Dixit, V.K. Hepatoprotective activity of leaves of *Kalanchoe pinnata* Pers. *J. Ethnopharmacol.*, **2003**, *86*, 197–202.

Yadav, Y.C. Hepatoprotective effect of *Ficus religiosa* latex on cisplatin induced liver injury in Wistar rats. *Rev. Bras. Farmacog.* **2015**, *25*, 278–283.

Yahya, F., Mamat, S.S., Kamarolzaman, M.F.F., Seyedan, A.A., Jakius, K.F., Mahmood, N.D., Shahril, M.S., Suhaili, Z., Mohtarrudin, N., Susanti, D., Somchit, M.N., Teh, L.K., Salleh, M.Z., and Zakaria, Z.A. 2013. Hepatoprotective activity of methanolic extract of *Bauhinia purpurea* leaves against paracetamol-induced hepatic damage in rats. *Evid. Based Complement. Alternat. Med.* **2013**, Article ID 636580, 10 pages doi.10.1155/2013/636580

Yalcin, A.; Yumrutas, O, Kuloglu T.; Elibol, E.; Parlar A.; Yilmiz, I.; Pehlivan M.; Dogukan, M.; Uckardes, F.; Aydin, H.; Turk, A.; Uludag, O.; Sahin, I.; Ugur, K.; Aydin, S. Hepatoprotective properties for *Salvia cryptantha* extract on carbon tetrachloride-induced liver injury. *Cell Mol.Biol.* **2017**, *63*(12), 56–62.

Yan, F., Zhang, Q-Y., Jiao, L., Han, T., Zhang, H., Qin, L.P., Khalid, R. Synergistic hepatoprotective effect of *Schisandrae lignans* with *Astragalus polysaccharides* on chronic liver injury in rats. *Phytomedicine* **2009**, *16*, 805–813.

Yan, J.Y.; Ai, G.;, Zhang, X.J.; Xu, H.J.; Huang, Z.M. Investigations of the total flavonoids extracted from flowers of *Abelmoschus manihot* (L.) Medic. against α-naphthylisothiocyanate-induced cholestatic liver injury in rats. *J. Ethnopharmacol.* **2015,** *172*, 202–213.

Yang, J.; Zhu, D.; Ju, B.; Jiang, X.; Hu, J. Hepatoprotective effects of *Gentianella turkestanerum* extracts on acute liver injury induced by carbon tetrachloride in mice. *Am. J. Transl. Res.* **2017**, *9*, 569–579.

Yang, J.; Zhu, D.; Wen, L.; Xiang, X.; Hu, J. *Gentianella turkestanerum* showed protective effects on hepatic injury by modulating the endoplasmic reticulum stress and NF-κB signaling pathway. *Curr. Mol. Med.* **2019,** *19*(6), 452–460.

Yantih, N.; Harahp, Y.; Sumaryono, W.; Setiabudy, R.; Rahayu, L. Hepatoprotective activity of pineapple (*Ananas comosus*) juice on isoniazid-induced rats. *J. Biol. Sci.* **2017**, *17*(8), 388–393.

Yar, H.S.; Ismail, D.K.; Alhamed, M.N. Hepatoprotective effect of *Carthamus tinctorius* L. against carbon tetrachloride induced hepatotoxicity in rats. *Pharmacie Globale Int. J. Compr. Pharm.* **2012**, *9*(2), 1–5.

Yasmin, S.; Kashmiri, A.M.; Anwar, K. Screening of aerial parts of *Abutilon bidentatum* for hepatoprotective activity in rabbits. *J. Med. Plants Res.* **2011**, 5, 349–353.

Yau, M.-H.; Che, C.-T.; Liang, S.-M.; Kong, Y.-C.; Fong, W.-P. An aqueous extract of *Rubus chingii* fruits protects primary rat hepatocytes against tert-butyl hydroperoxide induced oxidative stress. *Life Sci.* **2002**, *72*(3), 329–338.

Yaeesh, S.; Jamal, Q.; Khan, A.U.; Gilani, A.H. Studies on hepatoprotective, antispasmodic and calcium antagonistic activities of the aqueous methanol extract of *Achillea millefolium*. *Phyto. Res.* **2006**, 20, 546–551

Ye, X.; Feng, Y.; Tong, Y.; Ng, K.-M.; Tsao, S.; Lau, G. K. K.; Sze, C.; Zhang, Y.; Tang, J.; Shen, J.; Kobayashi, S. Hepatoprotective effects of Coptidis rhizoma aqueous extract on carbon tetrachloride-induced acute liver hepatotoxicity in rats. *J. Ethnopharmacol.* **2009**, *124*(1), 130–136.

Yehuda, H.; Khatib, S.; Sussan, I.; Musa, R.; Vaya, J.; Tamir, S. Potential skin antiinflammatory effects of 4-methylthiobutylisothiocyanate (MTBI) isolated from rocket (*Eruca sativa*) seeds. *BioFactors*, **2009**, *35*(3), 295–305.

Yemitan, O.K.; Izegbu, M.C. Protective effects of *Zingiber officinale* (Zingiberaceae) against carbon tetrachloride and acetaminophen induced hepatotoxicity in rats. *Phytotherap. Res.*, **2006**, *20*: 997–1002.

Yen, F.-L.; Wu, T.-H.; Lin, L.-T.; Lin, C.-C. Hepatoprotective and antioxidant effects of *Cuscuta chinensis* against acetaminophen-induced hepatotoxicity in rats. *J. Ethnopharmacol.* **2007**. *111*, 123–128.

Yen, M.H.; Weng, T.C.; Liu, S.Y.; Chai, C.Y.; Lin, C.C. The hepatoprotective effect of *Bupleurum kaoi*, an endemic plant to Taiwan, against dimethylnitrosamine-induced hepatic fibrosis in rats. *Biol. Pharm. Bull.* **2005**, *28*, 442–448.

Yengkhom, N.S.; Gunindro, N.; Kholi, S.M.; Moirangthem, R.S.; Rajkumari, B.D. Hepatoprotective effect of aqueous extract of *Melothria perpusilla* against carbon tetrachloride induced liver injury in albino rats. *Int. J. Res. Med. Sci.* **2017**, *5*, 806–810.

Yesmin, F.; Rahman, Z.; Dewan, J.F.; Helali, A.M.; Islam, Z.; Rahman, A.M.; et al., Hepatoprotective effect of aqueous and N-hexane extract of *Nigella sativa* in paracetamol (acetaminophen) induced liver disease of rats: A histopathological evaluation. *J. Pharm.* **2013**, *4*, 90–94.

Yildiz, F.; Coban, S.; Terzi, A.; Ates, M.; Aksoy, N.; Cakir, H.; et al., *Nigella sativa* relieves the deleterious effects of ischemia reperfusion injury on liver. *World J. Gastroenterol.* **2008**, *14*, 5204–5209.

Yin, G.; Cao, L.; Xu, P.; Jeney, G.; Nakao, M. Hepatoprotective and antioxidant effects of *Hibiscus sabdariffa* extract against carbon tetrachloride-induced hepatocyte damage in *Cyprinus carpio*. *In Vitro Cell Dev. Biol. Anim.* **2011**, *47*(1), 10–15.

Yin, G.; Cao, L; Xu, P.; Jeney, G.; Nakao, M.; Lu, C. Hepatoprotective and antioxidant effects of *Glycyrrhiza glabra* extract against carbon tetrachloride CCl_4-induced hepatocyte damage in common carp (*Cyprinus carpio*). *Fish Physiol. Biochem.* **2011**, *37*(1), 209–16. doi: 10.1007/s10695-010-9436–1.

Ying-Wan; Wu, Y.L.; Feng, X.C.; Lian, L.H.; Jiang, Y.Z.; Nan, J.X. The protective effects of total saponins from *Ornithogalum saundersiae* (Liliaceae) on acute hepatic failure induced by lipopolysaccharide and D-galactosamine in mice. *J. Ethnopharmacol.* **2010**, *132*, 450–455.

Ymele, V.E.; Dongmo, A.B.; Dimo, T. Analgesic and anti-inflammatory effect of the aqueous extract of the stem bark of *Allanblackia gabonensis* (Guttiferae). *Inflammopharmacology.* **2011**. doi:10.1007/s10787-011-0096–2.

Yogeeta, S.; Ragavender, H.R.B.; Devaki, T. Antihepatotoxic effect of *Punica granatum* acetone extract against isoniazid- and rifampicin-induced hepatotoxicity. *Pharm. Biol.* **2007**, *45*(8), 631-637.

Yoon, S.J.; Koh, E.J.; Kim, C.S.; Jee, O.P.; Kwak, J.H.; Jeong, W.J.; et al., *Agrimonia eupatoria* protects against chronic ethanol-induced liver injury in rats. *Food Chem. Toxicol.* **2012**, *50*, 2335–2341.

Yoshikawa, M.; Ninomiya, K.; Shimoda, H.; Nishida, N.; Matsuda, H. Hepatoprotective and antioxidative properties of *Salacia reticulata*: Preventive effects of phenolic constituents on carbontetrachloride-induced liver injury in mice. *Biol. Pharm. Bull.*, **2002**, *25*(1), 72–76.

Yossef, H.E.; Khedr, A.A.; Mahran, M.Z. Hepatoprotective activity and antioxidant effects of El Nabka (*Zizyphus spina-christi*) fruits on rats hepatotoxicity induced by carbon tetrachloride. *Nat. Sci.* **2011**, *9*(2), 1–7.

You, Y.; Yoo, S.; Yoon, H.G.; Park, J.; Lee, Y.H; Kim, S.; Oh, K.T.; Lee, J.; Cho, H.Y.; Jun, W. *In vitro* and *in vivo* hepatoprotective effects of the aqueous extract from *Taraxacum officinale* (dandelion) root against alcohol-induced oxidative stress. *Food Chem. Toxicol.* **2010**, *48*, 1632–1637.

Younis, T.; Khan, M.R.; Sajid, M. Protective effects of *Fraxinus xanthoxyloides* (Wall.) leaves against CCl_4 induced hepatic toxicity in rat. *BMC Complement. Altern. Med.* **2016**, *16*(1), 407.

Yuan, H.-D.; Jin, G.-Z.; Piao, G.-C. Hepatoprotective effects of an active part from *Artemisia sacrorum* Ledeb. against acetaminophen-induced toxicity in mice. *J. Ethnopharmacol.* **2010**, *127*(2), 528– 533.

Yuan, L.;, Gu, X.; Yin, Z.; Kang, W. Antioxidantactivities *in vitro* and hepatoprotective effects of *Nelumbo nucifera* leaves *in vivo* complement altern med. *Afr. J. Tradit.* **2014**, *11*, 85–91

Yuan, L.P.; Chen, F.H.; Ling, L.; Dou, P.F.; Bo, H.; Zhong, M.M.; Xia, L.J. Protective effects of total flavonoids of *Bidens pilosa* L. (TFB) on animal liver injury and liver fibrosis. *J. Ethnopharmacol.* **2008a**, *116*(3), 539–546.

Yuan, L.P.;, Chen, F.H.; Ling, L.; Bo, H.; Chen, Z.W.; Li, F.; Zhong, M.M.; Xia, L.J. Protective effects of total flavonoids of *Bidens bipinnata* L. against carbon tetrachloride-induced liver fibrosis in rats. *J. Pharm. Pharmacol.* **2008b**, *60*(10), 1393–1402.

Yurt, B.; Celik, I. Hepatoprotective effect and antioxidant role of sun, sulphited-dried apricot (*Prunus armeniaca* L.) and its kernel against ethanol-induced oxidative stress in rats. *Food Chem. Toxicol.* **2011**, *49*, 508–513.

Yuvaraj, P.; Subramoniam, A. Hepatoprotective property of *Thespesia populnea* against carbon tetrachloride-induced liver damage in rats. *J. Basic Clin. Physiol. Pharmacol.* **2009**, *20*, 169–177.

Zafar, R.; Ali S.M. Anti-hepatotoxic effects of root and root callus extracts of *Cichorium intybus* L. *J. Ethnopharmacol.* **1998**, *63*, 227–231.

Zahra, K.; Malik, A.M.; Mughal, S.M.; Arshad, M.; Sohail, I.M. Hepatoprotective role of extracts of *Momordica charantia* in acetaminophen-induced toxicity in rabbits. *J. Anim. Plant Sci.* **2012**, *22*(2), 273–77.

Zakaria, Z.A.; Kamisan, F.H.; Omar, M.H.; Mahmood, N.D.; Othman, F.; Abdul Hamid, S.S.; Abdullah, M.N.H. Methanol extract of *Dicranopteris linearis* L. leaves impedes acetaminophen-induced liver intoxication partly by enhancing the endogenous antioxidant system. *BMC Complement. Altern. Med.* **2017**, *17*(1), 271. doi: 10.1186/s12906-017-1781-5.

Zakaria, Z.A.; Yahya, F.; Mamat, S.S.; Mahmood, N.D.; Mohtarrudin, N.; Taher, M.; Hamid, S.S.A.; Teh, L.K.; Salleh, M.Z. Hepatoprotective action of various partitions of methanol extract of *Bauhinia purpurea* leaves against paracetamol-induced liver toxicity: Involvement of the antioxidant mechanisms. *BMC Complement. Altern. Med.* **2016**, *16*, 175.

Zarezade, V.; Moludi, J.; Mostafazadeh, M.; Mohammadi, M.; Veisi, A. Antioxidant and hepatoprotective effects of *Artemisia dracunculus* against CCl_4-induced hepatotoxicity in rats. *Avicenna J. Phytomed.* **2018**, *8*(1), 51–62.

Zargar, O. A.; Bashir R.; Ganie S. A.; Masood A.; Zargar M. A.; Hamid R. Hepatoprotective potential of *Elsholtzia densa* against acute and chronic models of liver damage in Wistar rats. *Drug Research.* **2018**, *68*(10), 567–575.

Zeashan, H.; Amresh, G.; Singh, S.; Rao, C.V. Hepatoprotective activity of *Amaranthus spinosus* in experimental animals. *Food Chem. Toxicol.* **2008**, *46*, 3417–3421.

Zeashan, H.; Amresh, G.; Singh, S.; Rao, C.V. Hepatoprotective and antioxidant activity of *Amaranthus spinosus* against CCl_4 induced toxicity. *J. Ethnopharmacol.* **2009**, *125*(2), 364–366.

Zeeni, N.; Daher, C.F.; Saab, L.; Mroueh M. *Tragopogon porrifolius* improves serum lipid profile and increases short-term satiety in rats. *Appetite* **2014**, *72*, 1–7.

Zhang, W.; Dong, Z.; Chang, X.; Zhang, C.; Rong, G.; Gao, X.; Zeng, Z.; Wang, C.; Chen, Y.; Rong, Y.; Qu, J.; Liu, Z.; Lu, Y. Protective effect of the total flavonoids from *Apocynum venetum* L. on carbon tetrachloride-induced hepatotoxicity *in vitro* and *in vivo*. *J. Physiol. Biochem.* **2018**, *74*(2), 301–312.

Zhang, Z.F.; Liu, Y.; Lu, L.Y.; Luo, P. Hepatoprotective activity of *Gentiana veitchiorum* Hemsl. against carbon tetrachloride-induced hepatotoxicity in mice. *Chin. J. Nat. Med.*, **2014**, *12*, 488-494.

Zhong, M.M.; Chen, F.H.; Yuan, L.P.; Wang, X.H.; Wu, F.R.; Yuan, F.L.; Cheng, W.M. Protective effect of total flavonoids from *Bidens bipinnata* L, against tetrachloride-induced liver injury in mice. *J. Pharm. Pharmacol.* **2007**, *59*, 1017–1025.

Zhou, D.; B, Ruan, J.; Cai, Y.; Xiong, Z.; Fu, W.; Wei, A. Anti-antioxidant and hepatoprotective activity of ethanol extract of *Arachniodes exilis* (Hance) Ching. *J. Ethanopharmacol.* **2010**, *129*, 232–237.

Zhou, G.; Chen, Y.; Liu, S.; Yao, X.; Wang, Y. In vitro and *in vivo* hepatoprotective and antioxidant activity of ethanolic extract from *Meconopsis integrifolia* (Maxim.) Franch. *J. Ethnopharmacol.* **2013**, *148*, 664–670.

Zingare, M.L.; Zingare, P.L.; Dubey, A.K.; Ansari, M.A. *Clitoria ternatea* (Aparajita): A review of the antioxidant, antidiabetic and hepatoprotective potentials. *Int. J. Pharm. Biol. Sci.* **2012**, *3*(1), 203–213.

CHAPTER 4

Phytoconstituents with Hepatoprotective Activity

ABSTRACT

The chemical compounds obtained from plant species is known as phytoconstituents. These are responsible for morphological and biological properties of plants and used in the treatment of variety of human diseases and disorders. This chapter concentrated on phytoconstituents with hepatoprotective activity.

INTRODUCTION

For millennia, medicinal plants have been a valuable source of therapeutic agents, and still many of today's drugs are plant-derived natural products or their derivatives. Recent trends in the study of hepatoprotective herbal source have turned toward isolation and purification of pure compounds from plants and assessment of their hepatoprotective activity. Various bioactive compounds from plant sources possessing antioxidant, anticancer, and immunostimulatory effect are also being tested for their possible hepatoprotective potential. A wide variety of flavonoids such as quercetin (*Helichrysum arenarium*), myricitoside C (*Cercis siliquastrum*), stachyrin (*Stachys recta*), and eupatolin (*Artemisia capillaris*); alkaloids such as atropine (*Datura metel*), pilocarpine (*Aristoclochia clementis*), and berberine (BBR) (*Beriberis vulgaris*); organic acids and lipids such as glycolic acid (*Cynara scolymus*), dihydrocholic acid (*Curcuma longa*) have shown potent antihepatotoxic activity (Valan et al., 2010). Farghali et al. (2014) reviewed the possible common mechanisms of hepatoprotective properties of extensively studied medicinal plant active constituents.

Adewusi and Afolayan (2010) encountered 58 chemically defined natural molecules reported in the literature, which have been evaluated for hepatoprotective activity. These compounds can serve as important leads for the discovery of new drugs in the treatment of liver diseases. However, for many of these compounds, the clinical data are very limited. Clinical efficacy and potential toxicity of active compounds in larger trials require further assessment, before recommendations concerning their routine use can be identified.

HEPATOPROTECTIVE PHYTOCONSTITUENTS

ACANTHOIC ACID

Park et al. (2004) investigated the protective effect of acanthoic acid, a diterpene isolated from the root bark of *Acanthopanax koreanus* (Family: Araliaceae) on liver injury induced by either *tert*-butyl hydroperoxide (*t*-BHP) or CCl_4 *in vitro* and *in vivo*. *In vitro*, the cellular leakage of lactic dehydrogenase (LDH) following treatment with 1.5 mM tBH for 1 h was significantly inhibited by cotreatment with acanthoic acid (25 and 5 μg/mL) and the ED_{50} of acanthoic acid was 2.58 μg/mL (8.5 μM). The cellular leakage of LDH following 1 h of treatment with 2.5 mM CCl_4 was significantly inhibited by co-treatment with acanthoic acid (25 μg/mL) and the ED_{50} of acanthoic acid was 4.25 μg/mL (14.1 μM). Co-treatment with acanthoic acid significantly inhibited the generation of intracellular reactive oxygen species (ROS) and intracellular glutathione (GSH) depletion induced by tBH or CCl_4. Acanthoic acid pretreatment (100 mg/kg/d for four consecutive days, *p.o.*) significantly reduced levels of AST and ALT in acute liver injury models induced by either tBH or CCl_4. Treatment with acanthoic acid (100 mg/kg, *p.o.*) at 6, 24, and 48 h after CCl_4 subcutaneous injection significantly reduced the levels of AST and ALT in serum. Histological observations revealed that fatty acid changes, hepatocyte necrosis, and inflammatory cell infiltration in CCl_4-injured liver were improved upon treatment with acanthoic acid. *In vivo* treatment with acanthoic acid was not able to modify CYP2E1 activity and protein expression in liver microsomes at the dose used, showing that the hepatoprotective effect of acanthoic acid was not mediated through inhibition of CCl_4 bioactivation. From the results above, acanthoic acid had a protective effect against tBH- or CCl_4-induced hepatotoxicity (INH) *in vitro* and *in vivo*.

AGATHISFLAVONE

Canarium manii (Burseraceae) was chemically investigated and the presence of the biflavonoid agathisflavone is reported from this plant by Anand et al. (1992). Pharmacologically, this biflavonoid in doses 50.0 and 100.0 mg given orally exhibited dose-dependent hepatoprotective activity against experimentally induced CCl_4-hepatotoxicity in rats and mice.

ALLICIN

Allicin (diallythiosulfinate) is the main biologically active component of freshly crushed garlic (*Allium sativum* L.) cloves. In D-galactosamine/lipopolysaccharide (D-GalN/LPS)-induced hepatitis rats, a significant increase of lipid peroxidation (LPO) and decreased liver antioxidant enzyme levels are observed. Pretreatment with allicin, the active component of freshly crushed garlic cloves prevented these alterations (Vimal and Devaki, 2004).

ALMOND OIL

Jia et al. (2011) evaluated the protective effects of almond oil against acute hepatic injury induced by CCl_4 in rats. Animals received almond oil prior to the administration of CCl_4 significantly decreased serum ALT, AST, ALP, and LDH activities, and total cholesterol (TC), triglyceride (TG), and low-density lipoprotein (LDL) content, and increased serum high-density lipoprotein (HDL) content,whereas, pretreatment with almond oil markedly increased rat hepatic SOD, catalase (CAT) and glutathione peroxidase (GPx) levels, and decreased the malonaldehyde (MDA) level. These results combined with liver histopathology demonstrated that almond oil has potent hepatoprotective effects.

AMENTOFLAVONE

Amentoflavone (AF) isolated from *Selaginella tamariscina* was screened for antioxidant activities *in vitro*, and hepatoprotective activity *in vivo* by Yue and Kang (2011). AF had no antioxidant activity in 2,2-diphenyl-1-picrylhydrazyl (DPPH) and 2,2'-azino-bis(3-ethylbenzothiazoline-6-sulfonic acid (ABTS) assay and poor reducing power in ferric reducing antioxidant power (FRAP) assay. Intragastric administration of AF (200 mg/kg/BW/d), AF (100 mg/kg), and AF (50 mg/kg) to mice injected with CCl_4 to induce acute hepatic injury for 8 days, the level of serum glutamic oxaloacetic transaminase (SGOT) and serum glutamic-pyruvic transaminase (SGPT) in group of AF (200 mg/kg) significantly decreased. The level of hepatic malondialdehyde (MDA) in groups of AF (200, 100, and 50 mg/kg, respectively) significantly decreased and the level of hepatic superoxide dismutase (SOD) only in group of AF (200 mg/kg) significantly increased.

δ-AMYRONE, 3-EPI-δ-AMYRIN, δ-AMYRIN, AND 18β-HYDROPEROXY-OLEAN-12-EN-3-ONE (NAMED SARMENTOLIN)

The known *δ*-amyrone, 3-epi-*δ*-amyrin, *δ*-amyrin as well as a new hydroperoxy triterpene, 18β- hydroperoxy-olean-12-en-3-one (named sarmentolin) were isolated as hepatoprotective agents from *Sedum sarmentosum*. Their structures were elucidated by spectral analysis and chemical reactions (Aimin et al., 1998).

ANDROGRAPHOLIDE

The diterpenes andrographolide (I), andrographiside (II), and neoandrographolide (III) isolated from *Andrographis paniculata* (AP) were investigated by Kapil et al. (1993) for their protective effects on hepatotoxicity induced in mice by CCl_4 or t-BHP intoxication. Pretreatment of mice with the diterpenes (I, II, and III; 100 mg/kg, i.p.) for three consecutive days produced a significant reduction in MDA formation, reduced GSH depletion, and enzymatic leakage of glutamic-pyruvate transaminase (GPT) and ALP in either group of the toxin-treated animals. A comparison with the known hepatoprotective agent silymarin (SM) revealed that I exhibited a lower protective potential than II and III, which were as effective as SM with respect to their effects on the formation of the degradation products of LPO and release of GPT and ALP in the serum. GSH status was returned to normal only by III. The greater protective activity of II and III could be due to their glucoside groups which may act as strong antioxidants.

Andrographolide showed a significant dose dependent (0.75–12 mg/kg p.o. × 7) protective activity against paracetamol-induced toxicity on *ex vivo* preparation of isolated rat hepatocytes (Visen et al., 1993). It significantly increased the percent viability of the hepatocytes as tested by trypan blue exclusion and oxygen uptake tests. It completely antagonized the toxic effects of paracetamol on certain enzymes (SGOT, SGPT, and ALP) in serum as well as in isolated hepatic cells. Andrographolide was found to be more potent than SM, a standard hepatoprotective agent.

Enhancing bioavailability and hepatoprotective activity of andrographolide from AP through its herbosome was investigated by Maiti et al. (2010). Andrographolide herbosome equivalent to 25 and 50 mg/kg andrographolide significantly protected the liver of rats, restoring hepatic enzyme activities with respect to CCl_4-treated animals. The rat plasma concentration of andrographolide obtained from the complex equivalent to 25 mg/kg andrographolide was higher than that obtained from 25 mg/kg andrographolide, and the complex maintained its effective plasma concentration for a longer period.

Chao and Lin (2012) reviewed the antihepatotoxic effects of *Andrographis paniculata* extract and derivative compounds, such as andrographolide, the major active compound, most studied for its bioactivities. Neoandrographolide shows anti-inflammatory and antihepatoxic properties, 14-deoxy-11,12-didehydroandrographolide, and 14-deoxyabdrographolide have immunostimulatory, antiatherosclerotic, and antihepatotoxic activities. The hepatoprotective activities include (1) inhibiting CCl_4, t-BHP-induced hepatic toxicity; (2) acting as cytochrome P450 (CYP) enzymes inducers; (3) modulating GSH content; (4) influence glutathione-*S*-transferase (GST) activity and phosphatidylinositol-3-kinase/Akt (PI3k/Akt) pathway; (5) synergistic effect with anticancer drug-induced apoptosis contributing to the bioactivities of AP extracts and isolated bioactive compounds.

Rajalakshmi et al. (2012) evaluated the efficacy of AP, especially its major bioactive compound—andrographolide on paracetamol-induced liver damage in rats with the help of serum biochemical markers and histological studies. Liver necrosis was produced by administering paracetamol (2 g/kg BW, p.o.) for continuous 7 days. Then, the hepatic damage was evidenced by the elevated levels of markers, such as AST, ALT, GGT, ALP, LDH, and bilirubin and decreased level of protein. But, after the treatment with AP powder extract (500 mg/kg BW, p.o.) the result turned reverse. Also, for further clarification, treatment was compared using a polyherbal syrup called Aeroliv (5 mL/kg BW, p.o.). Due to its multiherbal formulation, Aeroliv was found to be less effective than Andrographolide (200 mg/kg BW, p.o.). The dose selection was made based on previous investigations.

ANTHOCYANIN-RICH EXTRACT FROM BILBERRIES (*Vaccinium myrtillus*) AND BLACKCURRANTS (*Ribes nigrum*)

Cristani et al. (2015) investigated the protective effect of an anthocyanin-rich extract from bilberries and blackcurrants (AE) against acetaminophen (APAP)-induced acute hepatic damage in rats. The treatment with AE normalized blood activities of SGLT and SGPT and prevented APAP-induced plasmatic and tissutal alterations in biomarkers of oxidative stress, probably due to various bioproperties of the components of the extract.

APIGENIN-7-GLUCOSIDE

The effects of apigenin-7-glucoside (APIG) isolated from *Ixeris chinesis* (Thunb.) Nakai against liver injury caused by CCl_4 were investigated by Zheng et al. (2005). CCl_4 significantly increased the enzyme activities of SGPT and SGOT in blood serum, as well as the level of

MDA and 8-OHdG in liver tissue and decreased the levels of GSH. Pretreatment with APIG was able not only to suppress the elevation of SGPT, SGOT, MDA, and 8-OHdG, and inhibit the reduction of GSH in a dose-dependent manner *in vivo*, but also to reduce the damage of hepatocytes *in vitro*. On the other hand, the authors also found that APIG had strong antioxidant activity against ROS *in vitro* in a concentration-dependent manner.

ARJUNOLIC ACID

Manna et al. (2007) investigated the hepatoprotective role of arjunolic acid, a triterpenoid saponin, against arsenic-induced oxidative damages in murine livers. Administration of sodium arsenite at a dose of 10 mg/kg BW for 2 days significantly reduced the activities of antioxidant enzymes, SOD, CAT, GSH-*S*-transferase, GSH reductase, and GPx peroxidase as well as depleted the level of reduced GSH and total thiols. In addition, sodium arsenite also increased the activities of serum marker enzymes, ALT and ALP, enhanced DNA fragmentation, protein carbonyl (PCO) content, LPO end-products, and the level of oxidized GSH. Studies with arjunolic acid show that *in vitro*, it possesses free-radical scavenging and *in vivo* antioxidant activities. Treatment with arjunolic acid at a dose of 20 mg/kg BW for 4 days prior to arsenic administration prevents the alterations of the activities of all antioxidant indices and levels of the other parameters studied. Histological studies revealed less centrilobular necrosis in the liver treated with arjunolic acid prior to arsenic intoxication compared to the liver treated with the toxin alone.

BACOSIDE-A

Bacoside-A (B-A), a major constituent of *Bacopa monnieri,* was evaluated for its hepatoprotective activity against D-GalN-induced liver injury in rats. B-A (10 mg/kg of b.w.) was administered orally once daily for 21 days and then D-GalN (300 mg/kg of BW) was injected on 21st day after final administration of B-A. B-A reduces the elevated levels of serum ALT, AST, ALP, gamma-glutamyl transferase (γ-GT, LDH, 5¢nucleotidase (5¢ND). In addition, B-A also significantly restored toward normalization of the decreased levels of vitamin C, and vitamin E induced by D-GalN in both liver and plasma (Sumathi and Nongbri, 2008).

BAKUCHIOL, BAKUCHICIN, AND PSORALEN

Bioassay-guided fractionation of the water extract of the seeds of

Psoralea corylifolia furnished one hepatoprotective compound, bakuchiol, together with two moderately active compounds, bakuchicin and psoralen, on tacrine-induced cytotoxicity in human liver-derived Hep G2 cells. The EC_{50} values of compounds are 1.0, 47.0, 50.0 μg/mL, respectively. SM as positive control showed the EC_{50} value with 5.0 μg/mL (Hyun et al., 2001).

BERBERINE

BBR is a natural isoquinoline alkaloid with multiple pharmacological activities such as antioxidant, antiapoptotic, and anti-inflammatory effects. BBR has an efficacy in nonalcoholic fatty liver patients. A significant reduction of hepatic fat content and better improvement in body weight (BW), HOMA-IR, and serum lipid profiles were displayed in patients treated with BBR and lifestyle intervention (Yan et al., 2015). This efficacy may be related to the lipid metabolism regulatory effect of BBR. Pretreatment of BBR L02 hepatic cell lines exposed to hydrogen peroxide could increase cell viability and reduce apoptosis via the upregulation of sirtuin 1 and downregulation of apoptosis-related proteins (Zhu et al., 2013).

Feng et al. (2010) determined the hepatoprotective effects of BBR on serum and tissue SOD levels, the histology in CCl_4-induced liver injury in Sprague Dawley rats. Serum ALT and AST activities significantly decreased in a dose-dependent manner in both pretreatment and post-treatment groups with BBR. BBR increased the SOD activity in liver. Histological examination showed lowered liver damage in BBR-treated groups. This study demonstrated that BBR possesses hepatoprotective effects against CCl_4-INH and that the effects are both preventive and curative.

Domitrović et al. (2011) investigated the protective effects of isoquinoline alkaloid BBR on the CCl_4-INH in mice. The rise in serum levels of ALT, AST, and ALP in CCl_4-intoxicated mice was markedly suppressed by BBR in a concentration-dependent manner. The decrease in hepatic activity of (Cu/Zn SOD) and an increase in LPO were significantly prevented by BBR. Histopathological changes were reduced and the expression of tumor necrosis factor-α (TNF-α), cyclooxygenase-2 (COX-2), and inducible nitric oxide synthase (iNOS) was markedly attenuated by BBR 10 mg/mg.

Hepatotoxicity is one of the major side effects of methotrexate (MTX), which restricts the clinical use of this drug. Mehrzadi et al. (2018) studied the effect of BBR on MTX-induced hepatotoxicity. Results showed that MTX administration significantly increases AST, ALT, and ALP levels. It also increases MDA, PC, and NO levels and MPO activity. Moreover, MTX decreases hepatic GSH level, SOD, GPx, and CAT activities. Pretreatment with BBR for 10 days prevented some of these changes. Serum levels of AST and ALT decreased. Hepatic MDA level decreased and GSH level as well as GPx activity increased. Their results indicated that BBR might be useful for the prevention of the hepatotoxicity induced by MTX via ameliorative effects on biochemical and oxidative stress indices.

Mahmoud et al. (2017) investigated the protective effect of BBR against MTX hepatotoxicity, focusing on its ability to attenuate oxidative stress and apoptosis and to activate nuclear factor (erythroid-derived 2)-like 2 (Nrf2)/heme oxygenase-1 (HO-1) signaling and peroxisome proliferator activated receptor gamma (PPARγ). MTX-induced rats showed significant BW loss, increased serum liver function marker enzymes, bilirubin, and TNF-α. Liver LPO, nitric oxide (NO), and caspase-3 were significantly increased following MTX administration. BBR supplemented either before or after MTX significantly ameliorated BW, liver function markers, TNF-α, LPO, NO, and caspase-3. BBR increased serum albumin and liver antioxidant defenses in MTX-induced rats. Histological and immunohistochemical examination showed improved histological structure and decreased expression of Bax in liver of MTX-induced rats treated with BBR. In addition, BBR upregulated Nrf2, HO-1, and PPARγ expression in the liver of MTX-induced rats. In conclusion, BBR attenuated MTX-induced oxidative stress and apoptosis, possibly through upregulating Nrf2/HO-1 pathway and PPARγ. Therefore, BBR can protect against MTX-induced liver injury.

BERGENIN

The hepatoprotective effects of bergenin, a major constituent of *Mallotus japonicus*, were evaluated by Lim et al. (2000) against CCl_4-induced liver damage in rats. Bergenin at a dose of 50, 100, or 200 mg/kg was administered orally once daily for successive 7 days and then a mixture of 0.5 mL/kg (i.p.) of CCl_4 in olive oil (1:1) was injected two times each at 12 and 36 h after the final administration of bergenin. The substantially elevated serum enzymatic activities of ALT/AST, sorbitol dehydrogenase, and γ-GT due to CCl_4treatment were dose dependently restored toward normalization. Meanwhile, the decreased activities of GST and GSH reductase were restored toward normalization. In addition, bergenin also significantly prevented the elevation of hepatic MDA formation and depletion of reduced GSH content in the liver of CCl_4-intoxicated rats in a dose dependent fashion.

BETAINE

Reduction of CCl_4-induced hepatotoxic effects by oral administration of betaine in male Han–Wistar rats through a morphometric histological study was investigated by Junnila et al. (2000). Oral betaine, after the acclimation period of a week, increased the number of mitochondria but not mitochondria size (day 0), compared with the case in control rats. Exposure to CCl_4 resulted in centrilobular hepatic steatosis, and the administration of betaine significantly reduced this. Morphometric analyses also revealed that the addition of betaine increased the volume density of rough endoplasmic reticulum in the perinuclear areas of liver cell cytoplasm (day 7). Additionally, the administration of betaine prevented the reduction of Golgi complexes and mitochondrial figures in the cytoplasm observed after the exposures to CCl_4. Also, the volume density of mitochondria was smallest in the CCl_4-group, but the difference was not statistically significant.

BIOCHANIN A

Biochanin A (BCA) is an isoflavone found in red clover possessing multiple pharmacological activities including antimicrobial, antioxidant, and anticancer ones. Breikaa et al. (2013) assessed its hepatoprotective potential at different doses in a CCl_4-INH model in rats. The effects on hepatic injury were explored by measuring serum levels of ALT, AST, and ALP. Furthermore, the serum levels of glucose (GLU), urea, creatinine, total bilirubin, total proteins, TGs, and TC were determined. The metabolic capacity of the liver was assessed by measuring changes in CYP 2E1 (CYP2E1) activity. The underlying mechanisms were substantiated by measuring oxidative stress markers as CAT, SOD, GSH peroxidase, GSH transferase, GSH reductase, reduced GSH, total antioxidant capacity, and LPO, as well as inflammation markers such as NO, iNOS, COX-2, TNF-α, and leukocyte-common antigen. The results were confirmed by histopathological examination, and the median lethal dose was determined to confirm the safety of the drug. BCA successively protected against CCl_4-induced damage, normalizing many parameters to that of the control group. The study indicates that BCA possesses multimechanistic hepatoprotective activity that can be attributed to its antioxidant, anti-inflammatory, and immunomodulatory actions.

BIXIN

The carotenoid bixin present in annatto is an antioxidant that can protect cells and tissues against the deleterious effects of free radicals.

Moreira et al. (2014) evaluated the protective effect of bixin on liver damage induced by CCl_4 in rats. The animals were divided into four groups with six rats in each group. CCl_4 (0.125 mL/kg/BW) was injected intraperitoneally (i.p.), and bixin (5.0 mg/kg/BW) was given by gavage 7 days before the CCl_4 injection. Bixin prevented the liver damage caused by CCl_4, as noted by the significant decrease in serum aminotransferases release. Bixin protected the liver against the oxidizing effects of CCl_4 by preventing a decrease in GSH reductase activity and the levels of reduced GSH and nicotinamide adenine dinucleotide phosphate hydrogen (NADPH). The peroxidation of membrane lipids and histopathological damage of the liver was significantly prevented by bixin treatment. Therefore, they concluded that the protective effect of bixin against hepatotoxicity induced by CCl_4 is related to the antioxidant activity of the compound.

CAFFEIC ACID AND QUERCETIN

Caffeic acid (CA) and quercetin, the well-known phenolic compounds widely present in the plant kingdom, were investigated by Janbaz et al. (2004) for their possible protective effects against paracetamol and CCl_4-induced hepatic damage. Paracetamol at the oral dose of 1 g/kg produced 100% mortality in mice while pretreatment of separate groups of animals with CA (6 mg/kg) and quercetin (10 mg/kg) reduced the death rate to 20% and 30%, respectively. Oral administration of sublethal dose of paracetamol (640 mg/kg) produced liver damage in rats as manifested by the significant rise in serum levels of aminotransferases (AST and ALT) compared to respective control values. The serum enzyme values were significantly lowered on pretreatment of rats with either CA (6 mg/kg) or quercetin (10 mg/kg). Similarly, the hepatotoxic dose of CCl_4 (1.5 mL/kg; orally) also raised significantly the serum AST and ALT levels as compared to control values. The same dose of the CA and quercetin was able to prevent CCl_4-induced rise in serum enzymes. CA and quercetin prevented the CCl_4-induced prolongation in pentobarbital sleeping time confirming their hepatoprotectivity.

CARNOSOL

Sotelo-Félix et al. (2002) evaluated the effectiveness of carnosol to normalize biochemical and histological parameters of CCl_4-induced acute liver injury in male Sprague Dawley rats. Carnosol normalized bilirubin plasma levels, reduced MDA content in the liver by 69%, reduced ALT activity in plasma by 50%, and partially prevented the fall of liver glycogen content and distortion of the liver parenchyma. Carnosol prevents acute liver damage, possibly by improving the structural integrity of the hepatocytes. To achieve this, carnosol could scavenge free radicals induced by CCl_4, consequently avoiding the propagation of lipid peroxides. It was suggested that at least some of the beneficial properties of *Rosmarinus officinalis* are due to carnosol.

CHLOROGENIC ACID

Chlorogenic acid (CGA) isolated from *Anthocephalus cadamba* was screened by Kapil et al. (1995) for hepatoprotective activity by *in vitro* and *in vivo* assay methods using CCl_4 as a model of liver injury. Intraperitoneal administration of CGA to mice at a dose of 100 mg/kg BW for 8 days caused significant reversal in LPO, enzymatic leakage, Cyt P450 inactivation, and produced enhancement of cellular antioxidant defense in CCl_4-intoxicated mice, revealing that the antioxidative action of CGA is responsible for its liver protective activity. CGA exhibited a better therapeutic protective action than SM, in CCl_4-administered mice.

CGA is one of the most abundant dietary polyphenols, possessing well-known antioxidant capacity. Ji et al. (2013) evaluated the protection provided by CGA against APAP-induced liver injury in mice *in vivo* and the underlying mechanisms engaged in this process. Serum transaminases analysis and liver histological evaluation demonstrated the protection of CGA against APAP-induced liver injury. CGA treatment decreased the increased number of liver apoptotic cells induced by APAP in a dose-dependent manner. CGA also inhibited APAP-induced cleaved activation of caspase-3, 7. Moreover, CGA reversed APAP-decreased liver reduced GSH levels, glutamate-cysteine ligase (GCL), and GSH reductase activity. Further, results showed that CGA increased messenger RNA (mRNA) and protein expression of the catalytic subunit of GCL (GCLC), thioredoxin (Trx) 1/2, and thioredoxin reductase (TrxR) 1. Furthermore, CGA abrogated APAP-induced phosphorylated activation of ERK1/2, c-Jun N-terminal kinase (JNK), p38 kinases, and molecular signals upstream. The results of this study demonstrate that CGA counteracts AP-induced liver injury at various levels by preventing apoptosis and oxidative stress damage, and more specifically, both the GSH and Trx antioxidant systems and the mitogen-activated protein kinase (MAPK) signaling cascade appear to be engaged in this protective mechanism.

CHRYSIN

Chrysin, a natural flavonoid has been reported to possess potent anti-inflammatory, anticancer, and antioxidation properties. Anand et al. (2011) evaluated the protective effect of chrysin, an isoflavone, on CCl_4-induced toxicity in male Wistar rats. I.P. administration of CCl_4 (2 mL/kg) to rats for 4 days resulted in significantly elevated serum levels of SGOT, SGPT, ALP, and LDH, when compared to normal rats. In addition, the tissues (liver, kidney, and brain) and hemolysate samples showed considerable increase in levels of MDA and lowered levels of reduced GSH, vitamin C and vitamin E when compared to values in normal rats. Quantitative analysis of CAT, SOD, and GPx exhibited lower activities of these antioxidant enzymes in the tissues and haemolysate of CCl_4-administered rats. The protective action of chrysin on CCl_4-induced rat was demonstrated with SGPT, SGOT, ALP, and LDH resuming to near normal levels, while the mean levels of GSH and of vitamin C and vitamin E were elevated, the mean activities of CAT, SOD, and GPx were enhanced and the mean level of MDA was lowered in the tissue and hemolysate samples when compared to the CCl_4-exposed untreated rats. The expression of the iNOS gene appeared to be upregulated in the liver and kidney samples of CCl_4-exposed untreated rats, whereas in CCl_4-exposed chrysin-treated rats, the mRNA transcript levels of iNOS approximated normal levels. These results strongly suggest that chrysin is able to prevent the oxidative damage induced by CCl_4 in the liver, brain, kidney, and hemolysate of male Wistar rats.

CITRAL

Uchida et al. (2017) evaluated the effects of citral in a murine model of hepatotoxicity induced by APAP. Citral pretreatment significantly decreased the levels of ALT, AST, ALP, and γGT, MPO activity, and NO production. The histopathological analysis showed an improvement of hepatic lesions in mice after citral pretreatment. Citral inhibited neutrophil migration and exhibited antioxidant activity.

COLCHICINE

Colchicine is a well-established drug for its anti-inflammatory actions and is used as a medication for several diseases. It is effective clinically with ursodiol in a subset of patients with primary biliary cirrhosis and was found to be an effective and safe antifibrotic drug for long-term treatment of chronic liver disease in which fibrosis progresses toward Cirrhosis (Leung et al., 2011; Muntoni et al., 2010).

Antioxidant and antifibrotic properties of colchicine were investigated by Das et al. (2000) in the CCl_4 rat model. Urinary

8-isoprostane, kidney and liver MDA, and kidney glutathione levels increased following CCl_4 treatment, but only the rise in kidney MDA was significantly inhibited by colchicine pretreatment. Serum total antioxidant levels were significantly higher in the colchicine pretreatment group. The long-term effects of colchicine treatment on CCl_4-induced liver damage were investigated using liver histology and biochemical markers (hydroxyproline and type III procollagen peptide). Coadministration of colchicine with sublethal doses of CCl_4 over 10 weeks did not prevent progression to cirrhosis. However, rats made cirrhotic with repeated CCl_4 challenge and subsequently treated with colchicine for 12 months, all showed histological regression of cirrhosis. The antioxidant effect of colchicine *in vitro* was evident only at very high concentrations compared to other plasma antioxidants. Colchicine has only weak antioxidant properties but does afford some protection against oxidative stress; more importantly, long-term treatment with this drug may be of value in producing regression of established cirrhosis.

COUMARINOLIGNOIDS

Yadav et al. (2012) investigated the *in vivo* hepatoprotective potential of coumarinolignoids (cleomiscosins A, B, and C) isolated from the seeds of *Cleome viscosa* against CCl_4-INH in albino rats. Rats were divided into four groups. The coumarinolignoids were found to be effective as hepatoprotective against CCl_4-INH as evidenced by *in vivo* and histopathological studies in small animals. Safety evaluation studies also exhibit that coumarinolignoids are well-tolerated by small animals in acute oral toxicity study except minor changes in red blood cell count and hepatic protein content at 5000 mg/kg BW as a single oral dose. Coumarinolignoids, which is the mixture of three compounds (cleomiscosin A, B, and C), showed the significant protective effects against CCl_4-INH in small animals and also coumarinolignoids are well-tolerated by small animals in acute oral study.

CURCUMIN

Curcumin, as well-known dietary pigment derived from the rhizome of *C.longa* L., is a potent anti-inflammatory, antioxidant, and anticarcinogenic agent. Curcumin is another natural compound in which the *in vivo* hepatoprotective effect has been demonstrated (Wistar rats) against liver damage caused by APAP; the authors reported that it diminishes the expression of matrix metalloproteinase 8 (MMP-8) and increases the expression of genes encoding antioxidant enzymes in the liver (Soliman et al., 2014). This compound also demonstrated a hepatoprotective effect against acute and subacute liver damage caused by CCl_4 (Girish and Pradhan 2012; Park et al., 2000), paraquat (Han et al., 2014), EtOH *in vitro* and *in vivo* (Lee et al., 2013; Naik et al., 2004), and dimethylnitrosamine (Farombi et al., 2008).

Park et al. (2000) investigated the protective effects of curcumin on acute or subacute CCl_4-induced liver damage in rats. In rats with acute liver injury, curcumin (100 and 200 mg/kg) lowered the activity of serum ALT and AST. In rats with subacute liver injury, curcumin (100 mg/kg) lowered the activity of serum ALT and ALP of control rats. The liver hydroxyproline content in the curcumin (100 mg/kg)-treated group was reduced to 48% of the CCl_4 control group. MDA levels in curcumin (100 mg/kg)-treated rat liver was decreased to 67% of the control rat liver in subacute injury.

Farghaly and Hussein (2010) examined the protective effect of curcumin against paracetamol-INH in an attempt to understand its mechanism of action. Paracetamol (500 mg/kg BW) administration to rats resulted in massive elevation in serum and hepatic LDH activity and thiobarbituric acid reactive substances (TBARS) as well as in serum TNF-α levels, with a significant decrease in serum protein thiols (Pr-SHs), blood glutathione (GSH) levels, blood SOD, and GPx activities. However, paracetamol hepatotoxicity resulted in an increase in hepatic TBARS and depletion of hepatic GSH and Pr-SHs levels as well as in hepatic SOD, GPx, GR, glutathione-*S*-transferase (GST) and CAT activities. Oral administration of curcumin at a concentration of 50 mg/kg BW daily for 15 days to rats treated with paracetamol produced a significant protection against-induced increase in serum and hepatic LDH activities as well as TBARS and TNF-α levels. Also, curcumin (50 mg/kg BW) could inhibit reduce in serum Pr-SHs, blood GSH levels, and enhance increase in blood SOD and GPx activities. Hepatic TBARS level was suppressed by administration of curcumin to paracetamol-treated rats. In addition, curcumin enhance increase in hepatic GSH and Pr-SHs levels as well as in hepatic SOD, GPx, GR, GST, and CAT activities. These data indicate that curcumin is a natural antioxidant hepatoprotective agent against hepatotoxicity induced by paracetamol model.

According to a recent report, more than 65 human clinical trials of curcumin, which included more than 1000 patients, have been completed, and as many as 35 clinical trials are underway (Gupta et al., 2013). Curcumin was suggested to counteract the adverse hepatotoxic side effects of antituberculosis drug combinations, a possible valuable hepatoprotective application (Singh et al., 2012).

DIALLYL DISULFIDE

Lee et al. (2015) investigated the potential effects of diallyl disulfide (DADS) on CCl_4-induced acute hepatotoxicity and to determine the molecular mechanisms of protection offered by DADS in rats. DADS was administered orally at 50 and 100 mg/kg/d once daily for five consecutive days prior to CCl_4 administration. The single oral dose of CCl_4 (2 mL/kg) caused a significant elevation in serum aspartate and ALT activities, which decreased upon pretreatment with DADS. Histopathological examinations showed extensive liver injury, characterized by extensive hepatocellular degeneration/necrosis, fatty changes, inflammatory cell infiltration, and congestion, which were reversed following pretreatment with DADS. The effects of DADS on CYP2E1, the major isozyme involved in CCl_4 bioactivation, were also investigated. DADS pretreatment resulted in a significant decrease in CYP2E1 protein levels in dose-dependent manner. In addition, CCl_4 caused a decrease in protein level of cytoplasmic nuclear factor E2-related factor 2 (Nrf2) and suppression of nuclear translocation of Nrf2 concurrent with downregulation of detoxifying phase II enzymes and a decrease in antioxidant enzyme activities. In contrast, DADS prevented the depletion of cytoplasmic Nrf2 and enhanced nuclear translocation of Nrf2, which, in turn, upregulated antioxidant and/or phase II enzymes. These results indicate that the protective effects of DADS against CCl_4-INH possibly involve mechanisms related to its ability to induce antioxidant or detoxifying enzymes by activating Nrf2 and block metabolic activation of CCl_4 by suppressing CYP2E1.

DIHYDROQUERCETIN

Dihydroquercetin (DHQ) is a flavonoid in the Chinese traditional herbal medicine, Ramulus Euonymi, which has anti-inflammatory, antioxidant, and anticancer bioactivity. Chen et al. (2017) investigated the protective effects of DHQ on APAP-induced liver injury in a mouse model for the first time. DHQ treatment significantly attenuated serum ALT and AST levels as well as rescued hepatomegaly. It also downregulated TNF-α and IL-6, increased Nrf2 and SOD2 mRNA expression, downregulated Bax, and overexpressed Bcl-2 and Pro-caspase-3. Their data suggest that DHQ treatment can effectively attenuate APAP-induced liver injury by downregulating inflammatory factors, improving antioxidant capacity, and inhibiting hepatocyte apoptosis.

DIPRENYLATED ISOFLAVONOIDS

About three diprenylated isoflavonoids were isolated from 40%

hydroethanolic stem bark extract of *Erythrina senegalensis* and identified as 2,3-dihydro-2¢-hydroxyosajin (1), osajin (2), and 6,8-diprenylgenistein (3) (Donfack et al., 2008). These compounds were tested for hepatoprotective activities against *in vitro* CCl_4-induced hepatitis in rat liver slices. The obtained hepatoprotective percentages of compounds (1), (2), and (3) were, respectively 71.8 ± 1.45, 67.54 ± 3.56, and 69.41 ± 2.56. By comparison to compounds (2) and (3), compound (1) showed significant antioxidant effect with EC50 values of 41.28 ± 1.2, 31.27 ± 2.14, 19.17 ± 1.2, and 15.99 ± 0.49 μg/mL, respectively for the radical-scavenging action, inhibition of microsomal LPO, ß-CLAMS, and FRAP assays. The hepatoprotective activity of SM used as a reference compound was lower than the activities of isolated compounds. The results obtained provide promising baseline information for the potential use of this crude extract as well as some of the isolated compounds for their hepatoprotective and antioxidant activities.

ECHINACOSIDE

Wu et al. (2006) investigated the possible protective effects of echinacoside, one of the phenylethanoids isolated from the stems of *Cistanche salsa*, a Chinese herbal medicine, on the free-radical damage of liver caused by CCl_4 in rats. Treatment of rats with CCl_4 produced severe liver injury, as demonstrated by dramatic elevation of serum ALT, AST levels, and typical histopathological changes including hepatocyte necrosis or apoptosis, hemorrhage, fatty degeneration, etc. In addition, CCl_4 administration caused oxidative stress in rats, as evidenced by increased ROS production and MDA concentrations in the liver of rats, along with a remarkable reduction in hepatic SOD activity and GSH content. However, simultaneous treatment with echinacoside (50 mg/kg, i.p.) significantly attenuated CCl_4-INH. The results showed that serum ALT, AST levels and hepatic MDA content as well as ROS production were reduced dramatically, and hepatic SOD activity and GSH content were restored remarkably by echinacoside administration, as compared to the CCl_4-treated rats. Moreover, the histopathological damage of liver and the number of apoptotic hepatocytes were also significantly ameliorated by echinacoside treatment.

EMODIN (1,3,8-TRIHYDROXY-6-METHYL ANTHRAQUINONE)

The curative effect of emodin (1,3,8-trihydroxy-6-methyl anthraquinone), an active compound of the plant species *Ventilago maderaspatana* Gaertn, was evaluated by

Bhadauria et al. (2009) against CCl_4-induced hepatic CYP enzymatic and ultrastructural alterations in rats. The CCl_4-induced-toxic effects were observed with sharp elevation in the release of serum transaminases, ALP, LDH, and γ-glutamyl transpeptidase. An initial study for an optimum dose of emodin among different dose levels revealed that a 30 mg/kg dose was effective in restoring all the enzymatic variables and the liver histoarchitecture in a dose-dependent manner. Exposure to CCl_4 diminished the activities of CYP enzymes (i.e., aniline hydroxylase and amidopyrine-*N*-demethylase and microsomal protein contents with a concomitant increase in microsomal LPO). Emodin at 30 mg/kg effectively reversed the CCl_4-induced hepatotoxic events, which was consistent with ultrastructural observations. Hexobarbitone-induced sleep time and plasma bromosulphalein retention improved liver functions after emodin therapy. By reversal CYP activity and ultrastructural changes, emodin shows strong hepatoprotective abilities.

ESSENTIAL OILS FROM FAMILY LAMIACEAE; *Ajuga iva* (L.) SCHREBER, *Marrubium vulgare* L., *R. officinalis* L., AND *Thymus capitatus* HOFF. ET LINK

Following an ethnobotanical survey for hepatoprotective remedies, four Libyan medicinal plants from family Lamiaceae; *Ajuga iva, Marrubium vulgare, R.officinalis,* and *T. capitatus*, were studied for phytochemistry, antioxidant, and hepatoprotective activities. The major constituents of the essential oils (EOs)of *A. iva* and *R. officinalis* were carvacrol (35.07%) and 1,8-cincol (35.21%), respectively, thymol was the major constituent of the EO of *M. vulgare* (20.11%) and *T. capitatus* (90.15%). The oil of *M. vulgare* had the most powerful antioxidant activity by restoring GSH levels in the blood of alloxan-induced diabetic rats. The rats treated with the EOs of *M. vulgare, R. officinalis* and *T. capitatus* (50 mg/kg) showed a significant decrease in liver enzymes, which were elevated by CCl_4. The EO of *M. vulgare* had the most potent hepatoprotective activity.

EUGENOL

The chemoprotection extended by eugenol against CCl_4 intoxication was established by studies on drug-metabolizing phase I and phase II enzymes. An overall decrease in drug-metabolizing enzymes, namely NADPH-cytochrome *c* reductase, NADH-cytochrome reductase, coumarin hydroxylase, 7-ethoxy coumarin-*O*-deethylase, UDP-glucuronyltransferase, and GST, was observed with CCl_4 intoxication, with a subsequent decrease in

CYP and cytochrome b_5 content. CCl_4 caused a significant decrease in microsomal phospholipids and the marker enzymes glucose-6-phosphatase and 5-ND, and an increase in TBARS. Simultaneous administration of eugenol with CCl_4 inhibited the accumulation of TBARS and the decrease in the microsomal phospholipids and marker enzymes. Further, the chemical onslaught imposed by CCl_4 on the drug-metabolizing system was removed successfully by eugenol. Eugenol appears to act as an *in vivo* antioxidant and as a better inducer of phase II enzymes than phase I enzymes.

FLAVONE (5-HYDROXY, 7,8,2′TRIMETHOXY FLAVONE)

Nagaraja and Krishna (2016) evaluated the hepatoprotective activity of aqueous extract and isolated flavone (5-hydroxy, 7,8,2′-trimethoxy flavone) compound of *Andrographis alata* against CCl_4 INH. The hepatotoxicity was induced in albino rats CCl_4 (i.p.). Analysis of serum ALT, AST, and ALP activities with the concentrations of albumin, total protein, and bilirubin was carried out. The activities of all the marker enzymes reported a significant elevation in CCl_4-treated rats, which were significantly recovered toward an almost normal level in animals simultaneously administered with aqueous extract and flavone compound.

GALLIC ACID AND METHYL GALLATE

Rasool et al. (2010) investigated the hepatoprotective and antioxidant effects of gallic acid in paracetamol-induced liver damage in mice. Increased activities of liver marker enzymes and elevated TNF-α and LPO levels were observed in mice exposed to paracetamol, whereas the antioxidant status was found to be depleted when compared with the control group. However, gallic acid treatment (100 mg/kg BW i.p.) significantly reverses the above changes by its antioxidant action compared to the control group as observed in the paracetamol-challenged mice.

Chaudhuri et al. (2016) evaluated the activity of the methanolic extract of *Spondias pinnata* bark against iron-induced liver fibrosis and hepatocellular damage. In an iron-overloaded liver, iron reacts with cellular hydrogen peroxide to generate hydroxyl radicals, which, in turn, initiate the propagation of various free radicals; this situation leads to oxidative stress. Two compounds (gallic acid and methyl gallate) were isolated from the ethyl acetate fraction of this extract; an *in vivo* study showed that methyl gallate exhibited better iron chelation properties than gallic acid. It was proved that methyl gallate overcomes hepatic fibrosis by ameliorating oxidative stress and sequestrating the stored iron in cells. These results were in accordance with previous studies of Nabavi et al. (2013) which indicated the *in vivo* protective effect of gallic acid isolated from *Peltiphyllum peltatum* against sodium fluoride INH and oxidative stress. The results showed that gallic acid (10 and 20 mg/kg) prevented the sodium fluoride-induced abnormalities in the hepatic biochemical markers; these effects were comparable to the reference drug SM (10 mg/kg).

GENISTIN

Saleh et al. (2014) investigated the possible protective effects of genistein (GEN), a phytoestrogen, on the liver injury induced in rats by thioacetamide (TTA; 200.0 mg/kg body mass); administered three times a week by i.p. injection). GEN (0.5, 1.0, or 2.0 mg/kg body mass); by subcutaneous injection) was concurrently administered on a daily basis for 8 weeks, and its effects were evaluated 24 h after the administration of the last dose. The results from this study revealed that TTA-induced liver injury was associated with massive changes in the serum levels of liver biomarkers, oxidative stress markers, and liver inflammatory cytokines. Treatment of TAA-induced liver injury in rats with GEN decreased the elevated serum levels of AST, ALT, and total and direct bilirubin, and increased the serum level of albumin. GEN also restored the liver levels of MDA and reduced GSH, as well as TNF-α, interleukin (IL)-6, and their modulator nuclear factor kappa-light-chain-enhancer of activated B cells. From our results, it can be concluded that GEN attenuates the liver injury-induced in rats with TAA, and this hepatoprotective role is attributed to its anti-inflammatory and antioxidant properties.

GINSENOSIDE

Ginsenoside Ro, an oleanane-type saponin has been screened for activity in experimental models of acute and chronic hepatitis by Matsuda et al. (1991). Ginsenoside Ro (50 and 200 mg/kg, p.o.) inhibited the increase of SGOT and SGPT levels in D-GalN- and CCl_4-induced acute hepatitic rats. Ginsenoside Ro inhibited the increase of connective tissue in the liver of CCl_4-induced chronic hepatitic rats. Ginsenoside Ro showed a stronger inhibitory effect on the GalN-induced acute hepatitic model than those of the aglycone of ginsenoside Ro, oleanolic acid, or glycyrrhizic acid and its aglycone, glycyrrhetinic acid.

Ginsenoside Rg1 (Rg1) was used to investigate the potential roles of adenosine monophosphate (AMP)-activated protein kinase (AMPK) in CCl_4-INH. The experimental data indicated that treatment with Rg1 significantly decreased the elevation of plasma aminotransferases and alleviated hepatic histological abnormalities in CCl_4-exposed mice. Treatment with Rg1 also inhibited the increase of myeloperoxidase (MPO) and MDA, the induction of TNF-α, IL-6, iNOS, NO, and the upregulation of matrix metalloproteinase 2 (MMP-2), MMP-3, and MMP-9 in mice exposed to CCl_4. These effects were associated with suppressed nuclear accumulation of NF-κB p65.

GLYCYRRHIZIN

Glycyrrhizin (GLZ) is the major active component extracted from licorice (*Glycyrrhiza glabra*) roots, one of the most widely used herbal preparations for the treatment of liver disorders. Lee et al. (2007) evaluated the potential beneficial effect of GLZ in a mouse model of CCl_4-induced liver injury. The mice were treated i.p. with CCl_4 (0.5 mL/kg). They received GLZ (50, 100, 200, 400 mg/kg) 24 h and 0.5 h before and 4 h after administering CCl_4. The serum activities of aminotransferase and the hepatic level of MDA were significantly higher 24 h after the CCl_4 treatment, while the concentration of reduced GSH was lower. These changes were attenuated by GLZ. CCl_4 increased the level of circulating TNF-α markedly, which was reduced by GLZ. The levels of hepatic iNOS, COX-2, and HO-1 protein expression were markedly higher after the CCl_4 treatment. GLZ diminished these alterations for inducible nitric oxide and COX-2 but the protein expression of HO-1 was further elevated by the treatment of GLZ. CCl_4 increased the level of TNF-α, iNOS, COX-2, and HO-1 mRNA expressions. The mRNA expression of HO-1 was augmented by the GLZ treatment, while GLZ attenuated the increase in TNF-α, iNOS, and COX-2 mRNA expressions.

The protective effects of 18-beta-glycyrrhetinic acid (GA), the aglycone of GLZ derived from licorice, on CCl_4-INH and the possible mechanisms involved in this protection were investigated in mice by Jeong et al. (2002). Pretreatment with GA prior to the administration of CCl_4 significantly prevented an increase in serum alanine transaminase (ALT), aspartate aminotransferase (AST) activity, and hepatic LPO in a dose-dependent manner. In addition, pretreatment with GA also significantly prevented the depletion of GSH content in the livers of CCl_4-intoxicated mice. However, reduced hepatic GSH levels and GST activities were unaffected by treatment with GA alone. CCl_4-INH was also prevented, as indicated by a liver histopathologic study. The effects of GA on the CYP2E1, the major isozyme involved in CCl_4 bioactivation, were also investigated. Treatment of mice with GA resulted in a significant decrease of the P450 2E1-dependent hydroxylation of *p*-nitrophenol and aniline in a dose-dependent manner. Consistent with these observations, the P450 2E1 expressions were also decreased, as determined by immunoblot analysis. GA also showed antioxidant effects upon $FeCl_2$-ascorbate-induced LPO in mice liver homogenate and upon superoxide radical scavenging activity. These results show that protective effects of GA against the CCl_4-INH may be due to its ability to block the bioactivation of CCl_4, primarily by inhibiting the expression and activity of P450 2E1, and its free-radical scavenging effects.

GOODYEROSIDE A AND GOODYEROSIDE B

The antihepatotoxic activity of Goodyeroside A and Gooyeroside B isolated from *Goodyera* species were studied by Du et al. (2007) on injury induced by CCl_4 in primary culturedrat hepatocytes by measuring the levels of LDH, SGOT, and SGPT. In the CCl_4-treated control group, there were marked increases in LDH, SGOT, and SGPT activities compared with the normal group. In contrast, these levels were suppressed in Goodyeroside A and Goodyeroside B-treated groups. Goodyerin, a typical flavones glycoside, exhibited a significant and dose-dependent sedative and anticonvulsant effect.

3,4,5-TRIHYDROXY BENZOIC ACID (GALLIC ACID)

3,4,5-Trihydroxy benzoic acid (gallic acid) isolated from fraction TB5 of *Terminalia belerica* was evaluated by Anand et al. (1997) for its hepatoprotective activity against CCl_4-induced physiological and biochemical alterations in the liver. The main parameters studied were hexobarbitone-induced sleep, zoxazolamine-induced paralysis, serum levels of transaminases, and bilirubin. The hepatic markers assessed were LPO, drug metabolizing enzymes, glucose-6-phosphatase, and TGs. Administration of gallic acid led to significant reversal of majority of the altered parameters.

HEPATOPOIETIN (HPP) CN

Cui et al. (2009) evaluated the protective effect of rhHPPCn on liver injury and fibrosis induced by CCl_4 injection. The results showed that exogenous rhHPPCn could alleviate hepatocyte necrosis and protect the liver from the development of fibrotic lesions by proliferation stimulation. Additionally, HPPCn could reduce ALT/AST levels in rat serum following single and repeated CCl_4 injection.

HESPERIDIN

Tirkey et al. (2005) evaluated the protective effect of hesperidin (HD), a citrus bioflavonoid, on CCl_4-induced oxidative stress and resultant dysfunction of rat liver and kidney. CCl_4 caused a marked rise in serum levels of ALT and AST. TBARS levels were significantly increased, whereas GSH, SOD, and CAT levels decreased in the liver and kidney homogenates of CCl_4-treated rats. HD (200 mg/kg) successfully attenuated these effects of CCl_4.

Ahmad et al. (2012) evaluated the hepatoprotective and nephroprotective activity of HD, a naturally occurring bioflavonoid, against APAP-induced toxicity. APAP induces hepatotoxicity and nephrotoxicity as was evident by abnormal deviation in the levels of antioxidant enzymes. Moreover, APAP induced renal damage by inducing apoptotic death and inflammation in renal tubular cells, manifested by an increase in the expression of caspase-3, caspase-9, NFkB, iNOS, and Kim-1 and decrease in Bcl-2 expression. These results were further supported by the histopathological examination of the kidney. All these features of APAP toxicity were reversed by the co-administration of HD. Therefore, our study favors the view that HD may be a useful modulator in alleviating APAP-induced oxidative stress and toxicity.

Çetin et al. (2016) investigated the protective effect of HP on CCl_4-INH in rats. It was found that CCl_4-induced oxidative stress via a significant increase in the formation of TBARS and caused a significant decline in the levels of GSH, CAT, and SOD in rats. In contrast, HP blocked these toxic effects induced by CCl_4, causing an increase in GSH, CAT, and SOD levels and decreased formation of TBARS. In addition, histopathological damage increased with CCl_4 treatment. In contrast, HP treatment eliminated the effects of CCl_4 and stimulated antiapoptotic events, as characterized by reduced caspase-3 activation.

HYDROXYCHALCONES

Polyphenolics form a major part of the dietary antioxidant capacity of fruits and vegetables, identified as chemopreventive or anticancer agents. Hydroxychalcones are polyphenols abundantly distributed throughout the plant kingdom and are compounds with two aromatic rings (benzene or phenol) and an unsaturated side chain. Natanzi et al. (2011) investigated effect of phloretin (apple major flavonoid), 4-hydroxychalcone, and 4'-hydroxychalcone against APAP-induced acute liver damage. APAP produced 100% mortality at the dose of 1 g/kg in mice, while pretreatment and posttreatment (i.p., twice daily for 48 h) of animals with phloretin and 4-hydroxychalcone (50 mg/kg) and 4'-hydroxychalcone (25 mg/kg) significantly reduced the mortality rate. APAP produced acute toxicity at the dose of 640 mg/kg in mice, while pre and posttreatments of animals with phloretin and hydroxychalcones significantly lowered the rise in SGOT and SGPT. Liver sections collected for histological examination showed cellular changes including centrilobular necrosis, extensive portal inflammation, and micro and macro vesicular structures in the APAP group. These cellular changes were reduced following treatment of mice with Phloretin and hydroxychalcones. Authors concluded that phloretin and hydroxychalcones have hepatoprotective activity against APAP liver injury in mice.

3β-HYDROXY-2,3-DIHYDROWITHANOLIDE F

Protective effect of 3β-hydroxy-2,3-dihydrowithanolide F against CCl_4-INH has been assessed and the compound was found to possess marked protective effect by Budhiraja et al. (1986). A comparison of the protective properties showed that it is more active than hydrocortisone on a weight basis.

HYPEROSIDE

The hepatoprotective effects of hyperoside, a flavonoid glycoside isolated from *Artemisia capillaris*, have been examined against CCl_4-induced liver injury by Choi et al. (2011). Mice were treated i.p. with vehicle or 1 (50, 100, and 200 mg/kg) 30 min before and 2 h after CCl_4 (20 μL/kg) injection. Levels of serum aminotransferases were increased 24 h after CCl_4 injection, and these increases were attenuated by hyperoside. Histological analysis showed that hyperoside prevented portal inflammation, centrizonal necrosis, and Kupffer cell hyperplasia. LPO was increased, and hepatic GSH content was decreased significantly after CCl_4 treatment, and these changes were reduced by administration of hyperoside. Protein and mRNA expression of TNF-α, iNOS, COX-2, and HO-1) and nuclear protein expression of nuclear factor erythroid 2-related factor 2 (Nrf2) significantly increased after CCl_4 injection. Hyperoside suppressed TNF-α, iNOS, and COX-2 protein and mRNA expression and augmented HO-1 protein and mRNA expression and Nrf2 nuclear protein expression.

INDIGTONE

A bioactive fraction, indigtone (12.5–100 mg/kg, p.o.) characterized as trans-tetracos-15-enoic acid, obtained by fractionation of a petroleum ether extract of the aerial parts of

Indigofera tinctoria, showed a significant dose dependent hepatoprotective activity against paracetamol (200 mg/kg i.p.) and CCl_4 (0.5 mL/kg p.o. mixed with liquid paraffin 1:1)-induced liver injury in rats and mice. Pretreatment reduced Hexobarbitone-induced sleep time, and zoxazolamine induced paralysis time. Pre and posttreatment reduced levels of transaminases, bilirubin, TG, LPO, and restored the depleted GSH in serum (Singh et al., 2006).

IRIDOID GLUCOSIDES

Quan et al. (2009) examined the protective effect of iridoid glucosides from *Boschniakia rossica* (BRI) against CCl_4-induced liver injury. The CCl_4 challenge caused a marked increase in the levels of serum animotransferases, TNF-α and of hepatic iNOS protein, depleted reduced GSH, and propagated LPO. The liver antioxidative defense system, including SOD, GPx, and glutathione reductase (GR), as well as the CYP2E1 expressions were suppressed. However, preadministration of BRI reversed the significant changes of all liver function parameters induced by CCl_4 and restored the liver CYP2E1 content and function. These results demonstrated that BRI produced a protective action on CCl_4-induced acute hepatic injury via reduced oxidative stress, suppressed inflammatory response, and improved CYP2E1 function in the liver.

KAEMPFEROL

Shakya et al. (2014) analyzed the effect of kaempferol on oxidative stress induced by alcohol and thermally oxidized polyunsaturated fatty acid (ΔPUFA) in male albino Wistar rats. The levels of GGT, TBARS, and LH were significantly increased in liver of the alcohol + ΔPUFA group and were found to be reduced on treatment with kaempferol. The levels of both enzymatic and nonenzymatic antioxidants were decreased in liver of the alcohol + ΔPUFA group and were found to be restored on treatment with kaempferol.

KOLAVIRON

Farombi (2000) examined the protective mechanisms of kolaviron, a biflavonoid fraction from *Garcinia kola* (Heckel) seeds in rats treated with CCl_4. When administered at a dose of 1.2 g/kg, three times a week for 2 weeks, it significantly depressed the activities of microsomal aniline hydroxylase, aminopyrine N-demethylase, ethoxyresorufin O-demethylase, and p-nitroanisole O-demethylase. Kolaviron (200 mg/kg), administered for

14 days consecutively, inhibited the CCl_4-mediated decrease in the activities of these enzymes by 60%, 65%, 55%, and 63%, respectively. Kolaviron exerted its protective action by acting as an *in vivo* natural antioxidant and by enhancement of drug-detoxifying enzymes.

Kolaviron, a biflavonoid complex from *G. kola* seeds, possesses a variety of biological activities, including antioxidant. Adaramoye et al. (2009) investigated *in vivo* whether kolaviron may attenuate oxidative stress in liver of Wistar albino rats following chronic ethanol administration, while Adaramoye and Adeyemi (2006) investigated the effect of Kolaviron on D-GalN-induced toxicity. Experimentally, chronic ethanol or D-GalN administration led to hepatotoxicity as evidenced by the increase in levels of serum ALT, AST, ALP, and enhanced the formation of MDA in the liver. Levels of hepatic SOD, CAT, GST, and GSH were significantly reduced by ethanol and D-GalN treatment. Coadministration of kolaviron during ethanol and D-GalN treatment inhibited hepatic LPO and ameliorated SOD and GST activities. These findings demonstrated that kolaviron could have a beneficial effect by inhibiting the oxidative damage in liver of Wistar rats caused by chronic ethanol and D-GalN administration.

The effect of kolaviron, a mixture of Garcinia biflavonoid 1 (GB1), GB2, and kolaflavanone, used in the treatment of various ailments in southern Nigeria on hepatotoxicity and LPO induced by 2-acetylaminofluorene (2-AAF) in rats was investigated by Farombi et al. (2012). The ability of butylated hydroxyanisole (BHA) to attenuate the toxic effect of 2-AAF was also examined. Kolaviron administered orally to rats at a dose of 100 mg/kg BW twice a day for 1 week before challenge with 2-AAF (200 mg/kg feed) and continuously for 3 weeks at a single dose of 200 mg/kg BW reversed the 2-AAF-mediated decrease in final BW and relative organ weights, especially the liver. BHA was administered at a dose of 7.5 g/kg feed to the animals for 4 weeks. The extract significantly decreased the 2-AAF-mediated increase in the activity of AST, ALT, γ-GT, and ornithine carbamyl transferase by 58%, 62%, 60%, and 67%, respectively. BHA elicited, respectively, 55%, 63%, 57%, and 65% reduction in the 2-AAF induced-increase in the activities of these enzymes. Histological examination of the liver slices correlated with the changes in serum enzyme alterations. Similarly, kolaviron decreased the 2-AAF reduction of 5'-ND and glucose-6-phosphatase activities by 63% and 60%, respectively, while BHA elicited 59% and 61% decrease in the activities of these enzymes. Simultaneous administration of kolaviron with 2-AAF inhibited microsomal LPO as assessed by the TBARS formation by 66%. BHA produced a 64% reduction in TBARS formation. In the present study, kolaviron appears to act as an *in vivo* natural antioxidant and an effective hepatoprotective agent and is as effective as BHA.

LANTADENE A

Grace-Lynn et al. (2012) evaluated the hepatoprotective activity of lantadene A against APAP-induced liver toxicity in mice was studied. Activity was measured by monitoring the levels of AST, ALT, ALP, and bilirubin, along with histo-pathological analysis. SM was used as a positive control. A bimodal pattern of behavioral toxicity was exhibited by the lantadene A-treated group at the beginning of the treatment. However, treatment with lantadene A and SM resulted in an increase in the liver weight compared with the APAP-treated group. The results of the APAP-induced liver toxicity experiments showed that mice treated with lantadene A (500 mg/kg) showed a significant decrease in the activity of ALT, AST, and ALP and the level of bilirubin, which were all elevated in the APAP-treated group. Histological studies supported the biochemical findings and a maximum improvement in the histoarchitecture was seen. The lantadene A-treated group showed remarkable protective effects against histopathological alterations, with comparable results to the SM-treated group.

LIMONIN

Mahmoud et al. (2014) have investigated the effects of limonin isolated from the dichloromethane fraction of the seeds of bittersweet orange (*Citrus aurantium* var. *bigaradia*) in two dose levels (50 and 100 mg/kg) against D-GalN-induced liver toxicity in comparison with standard SM treatment on Toll-like receptors expression and hepatic injury, using a well-established rat model of acute hepatic inflammation. Oral administration of limonin before D-GalN-injection, significantly attenuated markers of hepatic damage (elevated liver enzyme activities and total bilirubin) and hepatic inflammation (TNF-α, infiltration of neutrophils), oxidative stress, and expression of TLR-4, but not TLR-2 in D-GalN-treated rats. Limonin effects were similar in most aspects to that of the lignan SM. The higher dose of limonin (100 mg/kg) performed numerically better for AST and bilirubin, and both the doses yielded similar results for ALT and GGT. While the lower dose of limonin (50 mg/kg) performed better against oxidative stress and liver structural damage as compared to the higher dose. Limonin exerts protective effects on liver toxicity associated with inflammation and tissue injury via attenuation of inflammation and reduction of oxidative stress.

LUCIDONE

Kumar et al. (2012) characterized the mechanisms underlying the hepatoprotective effect of lucidone against alcohol-induced oxidative stress *in vitro*. Human hepatoma (HepG2) cells were pretreated with lucidone (1–10 μg/mL) and then hepatotoxicity was stimulated by the addition ethanol (100 mM). With response to ethanol-challenge, increased amount of ALT and AST release were observed, whereas lucidone pretreatment significantly inhibited the leakage of AST and ALT in HepG2 cells without appreciable cytotoxic effects. Kumar et al. (2012) also found that lucidone pretreatment significantly decreased ethanol-induced NO, TNF-α, MDA, ROS, and GSH depletion in HepG2 cells. Furthermore, Western blot and quantitative-PCR analyses showed that ethanol-exposure apparently down-regulated endogenous antioxidant hemoxygenase-1 (HO-1) expression, whereas pretreatment with lucidone significantly upregulates HO-1 expression followed by the transcriptional activation of NF-E2 related factor-2 (Nrf-2). Interestingly, the profound up-regulation of HO-1 and Nrf-2 were observed in only ethanol-challenged cells, which evidenced that lucidone-induced induction of HO-/Nrf-2 were specific with oxidative stress.

LUPEOL

Lupeol, a pentacyclic triterpene and its ester derivative, lupeol linoleate, were investigated by Sunitha et al. (2001) for their possible hepatoprotective effect against cadmium (Cd)-induced toxicity in rats. Cd intoxicated rats showed elevated levels of MDA (basal and induced), and decreased levels of antioxidants and antioxidizing enzymes in the liver. The oral administration of triterpenes (150 mg/kg, once a day for 3 days before the injection of Cd chloride) changed the tissue redox system by scavenging the free radicals and by improving the antioxidant status of the liver. Lupeol linoleate had a better effect on the antioxidant status of the liver, when compared to lupeol.

Lupeol, a triterpene present in mango and other fruits, is known to exhibit a number of pharmacological properties including antioxidant, antilithiatic, and antidiabetic effects. Chemopreventive properties of lupeol and mango pulp extract (MPE) were evaluated by Prasad et al. (2007) against 7,12-dimethylbenz(a) anthracene (DMBA)-induced alteration in liver of Swiss albino mice.

Lupeol (25 mg/kg BW) or 1 mL of 20% w/v aqueous MPE/mouse were daily given once for 1 week after a single dose of DMBA (50 mg/kg BW). Lupeol/MPE supplementation effectively influenced the DMBA-induced oxidative stress, characterized by restored antioxidant enzyme activities and decrease in LPO. A reduction of apoptotic cell population in the hypodiploid region was observed in lupeol and MPE supplemented animals. The inhibition of apoptosis was preceded by decrease in ROS level and restoration of mitochondrial transmembrane potential followed by decreased DNA fragmentation. In DMBA-treated animals, downregulation of anti-apoptotic Bcl-2 and upregulation of proapoptotic Bax and Caspase 3 in mouse liver was observed. These alterations were restored by lupeol/MPE, indicating inhibition of apoptosis. Thus, lupeol/MPE was found to be effective in combating oxidative stress-induced cellular injury of mouse liver by modulating cell-growth regulators.

LUTEOLIN-7-GLUCOSIDE

Qiusheng et al. (2004) investigated the effects of luteolin-7-glucoside (LUTG) isolated from *Ixeris chinensis* against liver injury caused by CCl_4. CCl_4 significantly increased the enzyme activities of glutamic pyruvic transaminase (GPT) and SGOT in blood serum, as well as the level of MDA and 8-hydroxydeoxyguanosine (8-OHdG) in liver tissue, and decreased the levels of reduced GSH. Pretreatment with LUTG was not only able to suppress the elevation of GPT, GOT, MDA, and 8-OHdG, and inhibit the reduction of GSH in a dose-dependent manner *in vivo*, but also reduce the damage of hepatocytes *in vitro*. On the other hand, authors also found LUTG has a strong antioxidant activity against ROS *in vitro* in a concentration-dependent manner. The hepatoprotective activity of LUTG was possibly due to its antioxidant properties, acting as scavengers of ROS.

LYCOPENE

Jiang et al. (2016) evaluated the hepatoprotective effect of lycopene (Ly) on nonalcoholic fatty liver disease (NAFLD) in rat. A significant decrease was observed in the levels of serum AST, ALT, and the blood lipid TG and TC in the dose of 20 mg/kg Ly-treated rats, compared to the model group. Pretreatment with 5, 10, and 20 mg/kg of Ly significantly raised the levels of antioxidant enzyme SOD in a dose-dependent manner, as compared with the model group. Similarly, the levels of GSH were significantly increased after the Ly treatment. Meanwhile, pretreatment with 5, 10, and 20 mg/kg of Ly significantly reduced MDA amount by 30.87%, 45.51%, and 54.49% in the liver homogenates, respectively. The Ly treatment group showed significantly decreased levels of lipid products LDL-C, improved HDL-C level, and significantly decreased content of FFA, compared to the model group. Furthermore, the Ly-treated group also exhibited a down-regulated TNF-α and CYP2E1 expression, decreased infiltration of liver fats and reversed histopathological changes, all in a dose-dependent manner.

LYCORINE

Çitoğlu et al. (2012) reported the potential antinociceptive, anti-inflammatory, and hepatoprotective activities of lycorine from *Sternbergia fischeriana* (Herbert) Rupr. (Amaryllidaceae). Lycorine was evaluated on mice by using acetic-acid-induced writhing and tail-flick tests. The anti-inflammatory activity of lycorine was not found to be significant at a dose of 0.5 mg/kg. However, at doses of 1.0 and 1.5 mg/kg, i.p. showed a significant reduction with 53.45% and 36.42%, respectively, in rat paw oedema induced by carrageenan against the reference anti-inflammatory drug indomethacin (3 mg/kg, i.p.) (95.70%). The ED(50) of lycorine was determined as 0.514 mg/kg. Hepatoprotective activity of lycorine on CCl_4 induced acute liver toxicity following biochemical parameters were also evaluated. Rats were treated with lycorine at doses of 1.0 and 2.0 mg/kg, i.p. Results of biochemical tests were confirmed by histopathological examination. Lycorine exhibited significant hepatoprotective effect at a dose of 2.0 mg/kg i.p. dose.

MAGNOLOL

Magnolol, one major phenolic constituent of *Magnolia officinalis*, have been known to exhibit potent antioxidative activity. The anti-hepatotoxic activity of magnolol on APAP-induced toxicity in the Sprague Dawley rat liver was examined by Chen et al. (2009). After evaluating the changes of several biochemical parameters in serum, the levels of AST, ALT, and LDH were elevated by APAP (500 mg/kg) i.p. administration (8 and 24 h) and reduced by treatment with magnolol (0.5 h after APAP administration; 0.01, 0.1, and 1 mug/kg). Histological changes around the hepatic central vein, LPO (TBARS), and GSH depletion in liver tissue induced by APAP were also recovered by magnolol treatment. The data show that oxidative stress followed by LPO may play a very important role in the pathogenesis of APAP-induced hepatic injury; treatment with lipid-soluble antioxidant, magnolol, exerts antihepatotoxic activity.

MAJONOSIDE R2

The hepatoprotective effect of majonoside R2 (MR2) the major saponin constituent of Vietnamese ginseng (*Panax vientnamensis*, Araliaceae) was evaluated *in vivo* on D--GalN/LPS-induced hepatic apoptosis and subsequent liver failure in mice. Pretreatment of mice with MR2 (50 or 10 mg/kg, i.p.) at 12 and 1 h before D-GalN/LPS injection significantly inhibited apoptosis and suppressed following hepatic necrosis. Importantly, the elevation of TNF-α level, an important mediator for apoptosis in this model, was significantly inhibited by MR2 at a dose of 50 mg/kg. On the other hand, MR2 was found to protect primary cultured mouse hepatocytes from cell death by inhibiting apoptosis induced by D-GalN/TNF-α *in vitro*, as evidenced by DNA fragmentation analysis. These findings suggested that MR2 may have protected the hepatocytes from apoptosis via an inhibition of TNF-α production by activated macrophages and a direct inhibition of apoptosis induced by TNF-α.

MANGIFERIN

Mangiferin, a xanthone glucoside, is well known to exhibit antioxidant, antiviral, antitumor, anti-inflammatory, and gene-regulatory effects. Das et al. (2012) isolated mangiferin from the bark of *Mangifera indica* and assessed its beneficial role in GalN-induced hepatic pathophysiology. GalN (400 mg/kg BW) exposed hepatotoxic rats showed elevation in the activities of serum ALP, ALT, levels of TGs, TC, lipid-peroxidation, and reduction in the levels of serum total proteins, albumin, and cellular GSH. Besides, GalN exposure (5 mM) in hepatocytes induced apoptosis and necrosis, increased ROS and NO production. Signal transduction studies showed that GalN exposure significantly increased the nuclear translocation of NFκB and elevated iNOS protein expression. The same exposure also elevated TNF-α, IFN-γ, IL-1β, IL-6, IL-12, IL-18, and decreased IL-10 mRNA expressions. Furthermore, GalN also decreased the protein expression of Nrf2, NADPH:quinine oxidoreductase-1, HO-1, and GSTα. However, mangiferin administration in GalN intoxicated rats or coincubation of hepatocytes with mangiferin significantly altered all these GalN-induced adverse effects.

MELATONIN AND CRANBERRY FLAVONOIDS

Cheshchevik et al. (2012) evaluated the hepatoprotective potential of the antioxidant melatonin, as well as succinate and cranberry flavonoids in rats. Acute intoxication resulted in considerable impairment of mitochondrial respiratory parameters in the liver. The activity of mitochondrial succinate dehydrogenase (complex

II) decreased (by 25%). Short-term melatonin treatment (10 mg/kg, three times) of rats did not reduce the degree of toxic mitochondrial dysfunction but decreased the enhanced NO production. After 30-day chronic intoxication, no significant change in the respiratory activity of liver mitochondria was observed, despite marked changes in the redox-balance of mitochondria. The activities of the mitochondrial enzymes, succinate dehydrogenase, and GSH peroxidase, as well as that of cytoplasmic CAT in liver cells were significantly inhibited. Mitochondria isolated from the livers of the rats chronically treated with CCl_4 displayed obvious irreversible impairments. Long-term melatonin administration (10 mg/kg, 30 days, daily) to chronically intoxicated rats diminished the toxic effects of CCl_4, reducing elevated plasma activities of ALT, AST, and bilirubin concentration, prevented accumulation of membrane LPO products in rat liver and resulted in apparent preservation of the mitochondrial ultrastructure. The treatment of the animals by the complex of melatonin (10 mg/kg) plus succinate (50 mg/kg) plus cranberry flavonoids (7 mg/kg) was even more effective in prevention of toxic liver injury and liver mitochondria damage.

METADOXINE (PYRIDOXINE-PYRROLIDONE CARBOXYLATE)

Metadoxine has been reported to improve liver function tests in alcoholic patients. Gutiérrez-Ruiz et al. (2001) investigated the effect of metadoxine on some parameters of cellular damage in hepatocytes and hepatic stellate cells (HSCs) in culture treated with ethanol and acetaldehyde. HepG2 and CFSC-2G cells were treated with 50 mM ethanol or 175 μM acetaldehyde as initial concentration in the presence or absence of 10 μg/mL of metadoxine. About 24 h later reduced and oxidized GSH content, LPO damage, collagen secretion, and IL-6, IL-8, and TNF-α secretion were determined. The results suggest that metadoxine prevents GSH depletion and the increase in LPO damage caused by ethanol and acetaldehyde in HepG2 cells. In hepatic stellate cells, metadoxine prevents the increase in collagen and attenuated TNF-α secretion caused by acetaldehyde. Thus, metadoxine could be useful in preventing the damage produced in early stages of alcoholic liver disease as it prevents the redox imbalance of the hepatocytes and prevents TNF-α induction, one of the earliest events in hepatic damage.

p-METHOXY BENZOIC ACID

p-Methoxy benzoic acid isolated from the methanolic soluble fraction of the aqueous extract of *Capparis spinosa* L. (Capparaceae) was found to possess significant antihepatotoxic activity against CCl_4 and paracetamol INH *in vivo* and TTA and GalN INH in isolated rat hepatocytes, using *in vitro* technique. HPTLC analysis of methanol soluble fraction indicated that the compound constitutes 33% w/w of the active fraction.

METHYL HELICTERATE

Huang et al. (2012) investigated the effect of methyl helicterate isolated from *Helicteres angustifolia* on liver fibrosis in Sprague Dawley rats induced by CCl_4 and to explore its underlying mechanism. Methyl helicate (33.45 and 66.90 mg/kg) treatment significantly inhibited the loss of BW and the increase of liver index in rats induced by CCl_4. Methyl helicate also improved the liver functioning as indicated by decreasing serum enzymatic activities of ALT, AST, TP, and Alb. Histological results indicated that methyl helicate alleviated liver damage and reduced the formation of fibrous septa. Moreover, methyl helicate significantly decreased liver Hyp, HA, LN, and PCIII. Research on mechanism showed that methyl helicate could markedly reduce liver MDA concentration, increase activities of liver SOD, GPx, and inhibit the expression of TGF-β1 mRNA and Smad3 protein. Their findings indicated that MH can inhibit CCl_4-induced hepatic fibrosis, which may be ascribed to its radical scavenging action, antioxidant activity, and modulation of TGF-β-Smad3 signaling pathway.

MONO METHYL FUMARATE

Mono Methyl Fumarate (MMF) is an active component of *Fumaria* sp. methanolic extract. In Iranian traditional medicine, Fumitory is used widely as a remedy for several diseases including liver dysfunction. The effect of monomethyl fumarate and *Fumaria* extract was investigated by Moghaddam et al. (2012) against APAP-induced acute liver damage and compared to a known hepatoprotective plant, *Silybum marianum* and its active ingredient, SM. APAP produced acute toxicity at the dose of 640 mg/kg in mice while pretreatment and posttreatment of animals (i.p., twice daily for 48 h) with monomethyl fumarate, SM (25, 50, and 100 mg/kg) or the plant extracts (300 and 500 mg/kg) significantly lowered the rise of ALT, AST, LDH, ALKP, and bilirubin (total and direct) in serum. In addition, both the compounds and the plant extracts significantly increased GSH content and reduced MDA level of the liver.

MORIN

Lee et al. (2008) investigated the possible beneficial effects of morin on CCl_4-induced acute hepatotoxicity in rats. Rats received a single dose of CCl_4 (150 μL/100 g 1:1 in corn oil). Morin treatment (20 mg/kg) was given at 48, 24, and 2 h before CCl_4 administration. CCl_4 challenge elevated serum ALT, AST, and ALP levels, but these effects were prevented by the pretreatment of rats with morin. To identify the mechanism of protective activity of morin in CCl_4-INH in rats, Lee et al. (2002) investigated expressions of TNF-α, IL-1β, IL-6, and iNOS. The expressions of TNF-α, IL-1β, IL-6, and iNOS were increased by CCl_4 treatment and increased expressions of those were decreased by morin. These findings suggest that morin prevents acute liver damage by inhibiting the production of TNF-α, IL-1β, IL-6, and iNOS.

NARINGENIN

Another natural compound that shows a hepatoprotective effect *in vivo* is the flavonoid naringenin; in the case of damage caused by CCl_4 in mice, the authors observed that naringenin restores ALT and AST levels and improves the activity of SOD and GPx, avoiding LPO and, in turn, avoiding hepatocyte necrosis, steatosis, and fibrosis (Esmaeli and Alilou, 2014; Hermenean et al., 2014). These flavones also provide protection from the liver damage caused by dimethylnitrosamine in rats, against the damage generated by APAP in mice and against the chronic damage (60 days) caused by EtOH. The authors reported that naringenin restores ALT, AST, ALP, bilirubin, albumin, and total proteins serum levels and, reducing the lipid oxidation in the liver (Lee et al., 2004; Lv et al. 2013). It is noteworthy that naringenin possesses significant antioxidant and hepatoprotective activity.

Renugadevi and Prabhu (2010) evaluated the protective role of naringenin against Cd-induced oxidative stress in the liver of rats. Administration of naringenin at a dose of (50 mg/kg) significantly reversed the activities of serum hepatic marker enzymes to their near-normal levels when compared to Cd-treated rats. In addition, naringenin significantly reduced LPO and restored the levels of antioxidant defense in the liver. The histopathological studies in the liver of rats also showed that naringenin (50 mg/kg) markedly reduced the toxicity of Cd and preserved the normal histological architecture of the tissue.

NICOTINAMIDE

Shi et al. (2012) investigated the potential protective effects of Nicotinamide (NAM) on APAP-induced acute liver injury in mice. Pretreatment with NAM for 30 min significantly decreased plasma levels of ALT, AST, and MDA, and diminished histopathologic evidence of hepatic toxicity in mice following APAP administration. Similarly, post-treatment with NAM also decreased plasma ALT and AST levels in APAP-administrated mice. Furthermore, both pretreatment and posttreatment with NAM prolonged the survival rate of acute liver injury mice, accompanied by a significant reduction in the plasma levels of proinflammatory cytokines TNF-α, interferon-γ (INF-γ), and IL-6.

N-TRANS-CAFFEOYLDOPAMINE

N-trans-Caffeoyldopamine isolated from *Capsicum annuum, Theobroma cacao,* and *Lycium chinense* was tested at 2.5 mg/kg against oxidative damage induced by anti-TB drugs (INH/RIF 100 mg/kg each) on male Wistar rats. This natural compound showed a protective effect due to its good antioxidant activity, increasing SOD and GSH levels in hepatic tissue, and significantly decreasing the liver-enzyme levels (AST, ALT, and ALP), as well as the LPO in liver. It is, therefore, suggested that *N*-trans-Caffeoyldopamine can provide a definite protective effect against acute hepatic injury caused by INH/RIF in rats, which may mainly be associated with its antioxidative effect. This compound inhibited LPO through the CYP4502E1 downregulation (Wu et al., 2015).

OCTACOSANOL

Ohta et al. (2008) examined whether octacosanol, the main component of policosanol, attenuates disrupted hepatic ROS metabolism associated with acute liver injury progression in rats intoxicated with CCl_4. Octacosanol (10, 50, or 100 mg/kg) administered orally to CCl_4-intoxicated rats for 6 h after intoxication attenuated the increased activities of serum transaminases and the increased hepatic MPO and xanthine oxidase activities and LPO concentration and the decreased hepatic SOD and CAT activities and GSH concentration found at 24 h after intoxication dose-dependently. Octacosanol (50 or 100 mg/kg) administered to untreated rats decreased the hepatic LPO concentration and increased the hepatic GSH concentration. These results indicate that octacosanol attenuates disrupted hepatic ROS metabolism associated with acute liver injury progression in CCl_4-intoxicated rats.

OLEUROPEIN

Sangi et al. (2014) evaluated the effect of Oleuropein in rats affected by hepatotoxicity and cirrhosis as a result of administration of CCl_4. The levels of ALT, AST, and ALP were determined, regarding their relation to liver damage. Liver tissues were investigated to compare between healthy and infected ones. It was found that Oleuropein significantly revised the changes produced by CCl_4 on hepatic cells and enzymes.

ONONITOL MONOHYDRATE

Ononitol monohydrate, structurally similar to glycoside was isolated from *Cassia tora* leaves. Fifty Male rats were divided into five groups. Group I served as normal control. Groups II, III, and IV rats were INH by CCl_4 administering single dose of CCl_4 on 8th day only. Group III was treated with ononitol monohydrate (20 mg/kg BW) and group IV was treated with reference drug SM (20 mg/kg BW) both dissolved in corn oil and administering for 8 days. Ononitol monohydrate with corn oil alone was given for 8 days (group V). At the end of the experimental period, all the animals were sacrificed and analyzed for biochemical parameters to assess the effect of ononitol monohydrate treatment in CCl_4-induced hepatotoxicity. In *in vivo* study, ononitol monohydrate decreased the levels of serum transaminase, LPO, and TNF-α but increased the levels of antioxidant and hepatic GSH enzyme activities. Compared with reference drug SM, ononitol monohydrate possessed high hepatoprotective activity. Histopathological results also suggested the hepatoprotective activity of ononitol monohydrate with no adverse effect.

OSTHOLE

Osthole, a natural coumarin found in traditional Chinese medicinal herbs, has therapeutic potential in the treatment of various diseases. Cai et al. (2018) investigated the effects of osthole against APAP-INH in mice. Pretreatment with osthole significantly attenuated APAP-induced hepatocyte necrosis and the increases in ALT and AST activities. Compared with the mice treated with APAP alone, osthole pretreatment significantly reduced serum MDA levels and hepatic H_2O_2 levels, and improved liver GSH levels and the GSSG-to-GSH ratio. Meanwhile, osthole pretreatment markedly alleviated the APAP-induced up-regulation of inflammatory cytokines in the livers, and inhibited the expression of hepatic CYP enzymes, but it increased the expression of hepatic UDP-glucuronosyltransferases (UGTs) and sulfotransferases (SULTs). Furthermore, osthole pretreatment reversed APAP-induced reduction of hepatic cAMP levels, but pretreatment with H89, a potent selective PKA inhibitor, failed to abolish the beneficial effect of osthole, whereas pretreatment with L-buthionine sulfoximine, a GSH synthesis inhibitor, abrogated the protective effects of osthole on

APAP-induced liver injury, and abolished osthole-caused alterations in APAP-metabolizing enzymes. In cultured murine primary hepatocytes and Raw264.7 cells, however, osthole (40 μmol/L) did not alleviate APAP-induced cell death, but it significantly suppressed APAP-caused elevation of inflammatory cytokines.

PICROLIV

Administration of picroliv, a standardized fraction of alcoholic extent of *Picrorhiza kurroa* (Pk) (3–12 mg/kg/d for 2 weeks) simultaneously with *Plasmodium berghei* infection showed significant protection against hepatic damage in *Mastomys natalensis* (Chander et al., 1990). The increased levels of SGOT, SGPT, ALP, lipoprotein-X (LP-X), and bilirubin in the infected animals were reduced by different doses of picroliv. In the liver, picroliv decreased the levels of lipid peroxides and hydroperoxides and facilitated the recovery of SOD and glycogen. Picroliv had no effect on the degree of parasitaemia.

Hepatoprotective activity of picroliv against CCl_4-induced liver damage in rats was assessed by Dwivedi et al. (1990). Administration of CCl_4 to normal rats increased activities of hepatic 5(1)-¢ND, acid phosphatase, and acid ribonuclease, while the activities of succinate dehydrogenase, glucose 6-phosphatase, SOD, and CYP were decreased. Levels of lipid peroxides, total lipids, and cholesterol of liver were also increased. The activities of SGLT, SGPT, and ALP were increased. Other serum parameters showing changes after CCl_4 were: bilirubin, proteins, cholesterol, TGs, and lipoprotein-X. Picroliv in doses of 6 and 12 mg/kg provided a significant protection against most of the biochemical alterations produced by CCl_4. The degree of protection afforded by picroliv, when administered simultaneously or as a pretreatment was almost equal.

Tripathi et al. (1991) evaluated the hepatoprotective activity of picroliv, the active standardized fraction of *Picrorhiza kurroa*, against alcohol -$CC1_4$-induced liver damage in rat at the doses of 3, 6, and 12 mg/kg/d, p.o. for 7 days. It showed a significant dose-dependent hepatoprotective activity as evidenced by lowering of the elevated levels of SGLT, SGPT, acid phosphatase, ALP, glutamate dehydrogenase, and bilirubin in the serum of alcohol-$CC1_4$-treated rats. The depleted glycogen content of liver was also restored significantly. The activity of picroliv was compared with that of SM, a known hepatoprotective agent.

Picroliv, an iridoid glycoside isolated from *P. kurroa* demonstrated a dose-dependent protective activity on isolated hepatocytes against paracetamol-induced hepatic damage in rats. It also restored the normal values of enzymes (glutamic oxaloacetic transaminase, glutamic-pyruvic transaminase, and ALP) both in the isolated hepatocyte suspension as well as in the serum (Visen et al., 1991). Picroliv was found to be more potent than SM, a known hepatoprotective agent.

Picroliv, the active principle of *P. kurroa*, and its main components, which are a mixture of the iridoid glycosides, picroside-I, and kutkoside, were studied *in vitro* as potential scavengers of oxygen free radicals by Chander et al. (1992a). The superoxide (O_2-) anions generated in a xanthine-xanthine oxidase system, as measured in terms of uric acid formed and the reduction of nitroblue tetrazolium were shown to be suppressed by picroliv, picroside-I and kutkoside. Picroliv as well as both glycosides inhibited the nonenzymic generation of O_2- anions in a phenazine methosulphate NADH system. MDA generation in rat liver microsomes as stimulated by both the ascorbate-Fe^{2+} and NADPH-ADP-Fe^{2+} systems was shown to be inhibited by the Picroliv glycosides. Some known antioxidants tocopherol (vitamin E) and BHA were also compared with regard to their antioxidant actions in the above system. It was found that BHA afforded protection against ascorbate-Fe(2+)-induced MDA formation in microsomes but did not interfere with enzymic or nonenzymic O_2- anion generation; and tocopherol inhibited LPO in microsomes by both prooxidant systems and the generation of O_2- anions in the nonenzymic system but did not interfere with xanthine oxidase activity. This study shows that picroliv, picroside-I, and kutkoside possess the properties of antioxidants, which appear to be mediated through activity like that of SOD, metal ion chelators, and xanthine oxidase inhibitors.

Administration of picroliv, the active principle from *P. kurroa*, at a dose of 6 mg/kg, p.o. for 2 weeks showed a significant protection against changes in liver and brain GSH metabolism of *P. berghei* infected *M. natalensis* (Chander et al., 1992b). The depletion of reduced GSH level and inhibition of GST, GSH reductase, and GPx activities due to *P. berghei* infection were markedly recovered by picroliv. The increased levels of LPO products in damaged tissues were also reduced along with the recovery of GSH metabolism. Picroliv at the dose of 6 mg/kg, p.o. for 2 weeks provided significant protection against depletion of reduced GSH levels in liver and brain of *P. berghei* infected *M. natalensis* (Chander et al., 1994). The activation of γ-GT enzyme and decreased levels of cysteine, sulphydryl groups as well as GSH synthesis in both tissues due to *P. berghei* infection were reversed by picroliv. Enzymatic and nonenzymatic LPO in microsomes *in vitro* was significantly reduced by picroliv along with the recovery of reduced GSH.

Chander et al. (1998) also reported that Picroliv at the dose of 6 mg/kg p.o. for 2 weeks provided significant protection against the generation of LPO products in serum beta-lipoproteins of *P. berghei* infected *M. coucha.* Incubation of normal rat hepatocytes with very low density lipoprotein (LDL) isolated from infected animals caused significant generation of lipid peroxides followed by a decrease in the viability of these cells; however, these effects were partially reversed with the lipoproteins from infected and picroliv-treated groups. HDL from infected animals was not toxic to hepatocytes *in vitro*.

Picroliv showed a dose-dependent (1.5–12 mg/kg, p.o. for 7 days) hepatoprotective activity against oxytetracycline-induced hepatic damage in rat (Saraswat et al., 1997). It increased the number of viable hepatocytes (*ex vivo*) significantly. Increase in bile volume and its content in conscious rat suggests potent anticholestatic property. Picroliv also antagonised alterations in enzyme levels (GOT, GPT, and ALP) in isolated hepatocytes and serum, induced by oxytetracycline (200 mg/kg, i.p.) feeding. Picroliv was more potent than SM, a known hepatoprotective drug.

Picroliv, the standardized active principle from the plant *P. kurroa* showed significant curative activity *in vitro* in primary cultured rat hepatocytes against toxicity induced by TTA (200 μg/mL), GalN (400 μg/mL), and CCl_4 (3 μL/mL) (Visen et al., 1998). Activity was assessed by determining the change in hepatocyte viability and rate of oxygen uptake and other biochemical parameters (SGOT, SGPT, and AP). The toxic agents alone produced a 40%–62% inhibition of cell viability and a reduction of biochemical parameters after 24 h of incubation at 37 °C which (on removal of the toxic agents) was reversed after further incubation for 48 h. Incubation of damaged hepatocytes with picroliv exhibited a concentration (1–100 μg/mL)-dependent curative effect in restoring altered viability parameters.

PINORESINOL

Forsythiae Fructus is known to have diuretic, antibacterial, and anti-inflammatory activities. Kim et al. (2010) examined the hepatoprotective effects of pinoresinol, a lignan isolated from *F. Fructus*, against CCl_4-induced liver injury. Mice were treated i.p. with vehicle or pinoresinol (25, 50, 100, and 200 mg/kg) 30 min before and 2 h after CCl_4 (20 μL/kg) injection. In the vehicle-treated CCl_4 group, serum aminotransferase activities were significantly increased 24 h after CCl_4 injection and these increases were attenuated by pinoresinol at all doses. Hepatic GSH contents were significantly decreased and LPO was increased after CCl_4 treatment. These changes were attenuated by 50 and 100 mg/kg of pinoresinol. The levels of protein and mRNA expression of inflammatory mediators, including TNF-α, iNOS, and COX-2, were significantly increased after CCl_4 injection; and these increases were attenuated by pinoresinol. Nuclear translocation of nuclear factor-κB (NF-κB) and phosphorylation of c-Jun, one of the components of activating protein 1 (AP-1), were inhibited by pinoresinol. Pinoresinol ameliorates CCl_4 -induced acute liver injury, and this protection is likely due to antioxidative activity and downregulation of inflammatory mediators through inhibition of NF-κB and AP-1.

PTEROSTILBENE

El-Sayed el-SM et al. (2015) evaluated the protective effect of pterostilbene against APAP-induced hepatotoxicity. SM was used as a standard hepatoprotective agent. A single dose of APAP (800 mg/kg i.p.), injected to male rats, caused significant increases in serum levels of ALT, AST, ALP, bilirubin, TC, TGs, TNF-α, and hepatic contents of MDA, NO, caspase-3, hydroxyproline, with significant decreases in serum HDL-cholesterol, total proteins, albumin, and hepatic activities of reduced GSH, SOD and CAT as compared with the control group. On the other hand, administration of each of pterostilbene (50 mg/kg, p.o.) and SM (100 mg/kg, p.o.) for 15 days before APAP ameliorated liver function and oxidative stress parameters. Histopathological evidence confirmed the protection offered by pterostilbene from the tissue damage caused by APAP. In conclusion, pterostilbene possesses multimechanistic hepatoprotective activity that can be attributed to its antioxidant, anti-inflammatory, and antiapoptotic actions.

PUERARIN

Puerarin, the main isoflavone glycoside found in the root of *Pueraria lobata*. Hwang et al. (2007) investigated the protective effects of puerarin against hepatotoxicity induced by CCl_4 and the mechanism of its hepatoprotective effect. In mice, pretreatment with puerarin prior to the administration of CCl_4 significantly prevented the increased serum enzymatic activity of ALT, AST and hepatic MDA formation in a dose-dependent manner. In addition, pretreatment with puerarin significantly prevented both the depletion of reduced GSH content and the decrease in GST activity in the liver of CCl_4-intoxicated mice. Hepatic

GSH levels and GST activity were increased by treatment with puerarin alone. CCl_4-INH was also prevented, as indicated by liver histopathology. The effects of puerarin on CYP2E1, the major isozyme involved in CCl_4 bioactivation, were also investigated. Treatment of the mice with puerarin resulted in a significant decrease in the CYP2E1-dependent aniline hydroxylation in a dose-dependent manner. Consistent with these observations, the CYP2EI protein levels were also lowered. Puerarin exhibited antioxidant effects on $FeCl_2$-ascorbate-induced LPO in mouse liver homogenates, and on Superoxide radical scavenging activity.

QUERCETIN

Pavanato et al. (2003) investigated the protective effects of chronic administration of the flavonoid quercetin (150 μmol/kg BW/d i.p.) in rats with CCl_4-induced fibrosis. In animals rendered cirrhotic by administration of CCl_4 for 16 weeks, cell necrosis, fibrosis, and inflammatory infiltration were found. Histological abnormalities were accompanied by a higher hepatic content of collagen and TBARS. Expression of iNOS was significantly increased in the liver. Treatment with quercetin for 3 weeks improved liver histology and reduced collagen content, iNOS expression, and LPO. These effects were associated with an increased total peroxyl radical-trapping antioxidant capacity of liver.

RESVERATROL

Studies on resveratrol pretreatment effects on the synergy between D-GalN and LPS-induced liver failure in rats, and its effects on chemical prooxidants in immobilized perfused hepatocytes were investigated by Cerny et al. (2009) and Farghali et al. (2009). Parameters studied included liver function; plasma nitrite as a measure of NO; nonenzymatic and enzymatic antioxidants in plasma and liver homogenate; and morphological examinations were performed using light and electron microscopy. Observations related to pharmacological increases of iNOS-2/NO and inducible hemeoxygenase (HO-1)/carbon monoxide (CO) in fulminant hepatic failure and modulation by resveratrol were followed up by real-time reverse transcription PCR in liver tissue. Reduction in NO production, downregulation of NOS-2 expression, modification of oxidative stress parameters, and modulation of HO-1 are among the mechanisms responsible for the cytoprotective effect of resveratrol in the LPS/D-GalN liver toxicity, and in t-BHP-induced hepatocyte toxicity models. Resveratrol pretreatment led to the overall improvement in hepatotoxic markers and morphology after the hepatic insult. Several studies have also highlighted the hepatoprotective properties of resveratrol (Bishayee et al., 2010).

Resveratrol is a natural compound that has been evaluated against the acute liver damage caused by the mixture of INH (50 mg/kg) and RIF (100 mg/kg) in male Balb/C mice (Nicoletti et al., 2014). This compound, at a dose of 100 mg/kg, was administered 30 min prior to the RIF/INH mixture during 3 days. The results showed that resveratrol reduced AST and ALT serum levels by 36% and 58%, respectively, with regard to animals treated only with anti-TB drugs alone; the authors also observed that GSH and CAT levels were higher in the Resveratrol/RIF/INH-treated group with respect to the RIF/INH group; likewise, the group that received resveratrol-RIF/INH exhibited lower MPO values (19%) than the RIF/INH group. Through histological analysis, the authors observed less microvesicular steatosis and apoptosis in the livers of animals that received resveratrol-RIF/INH with respect to the group that was administered anti-TB drugs. The authors recommend the use of this substance to revert the damage caused by anti-TB drugs. This compound also protects against liver damage caused by APAP and EtOH (Nicoletti et al., 2014).

RUBIADIN

The hepatoprotective effects of rubiadin, a major constituent isolated from *Rubia cordifolia*, were evaluated by Rao et al. (2006) against CCl_4-induced hepatic damage in rats. Rubiadin at a dose of 50, 100, and 200 mg/kg was administered orally once daily for 14 days. The substantially elevated serum enzymatic activities of SGOT, SGPT, serum ALP, and γ-GT due to CCl_4 treatment were dose dependently restored toward normalization. Meanwhile, the decreased activities of GST and GSH reductase were also restored toward normalization. In addition, rubiadin also significantly prevented the elevation of hepatic MDA formation and depletion of reduced GSH content in the liver of CCl_4 intoxicated rats in a dose-dependent manner. SM used as a standard reference also exhibited significant hepatoprotective activity on post-treatment against CCl_4 INH in rats. The biochemical observations were supplemented with histopathological examination of rat liver sections. The results of this study strongly indicate that rubiadin has a potent hepatoprotective action against CCl_4-induced hepatic damage in rats.

RUTIN

Rutin, a well-known flavonoid was investigated by Janbaz et al. (2002) for its possible protective effect against paracetamol- and CCl_4-induced hepatic damage. Paracetamol produced 100% mortality at the dose of 1 g/kg in mice, while pretreatment of animals with rutin (20 mg/kg) reduced the death rate to 40%. Oral administration of a sublethal dose of paracetamol (640 mg/kg) produced liver damage in rats as manifested by the rise in serum level of transaminases (AST and ALT). Pretreatment of rats with rutin (20 mg/kg) prevented the paracetamol-induced rise in serum enzymes. The hepatotoxic dose of CCl_4 (1.5 mL/kg; orally) also raised the serum AST and ALT levels. The same dose of rutin (20 mg/kg) was able to prevent the CCl_4-induced rise in serum enzymes. Rutin also prevented the CCl_4-induced prolongation in pentobarbital sleeping time confirming its hepatoprotectivity.

Khan et al. (2012) evaluated the hepatoprotective effect of rutin against CCl_4-induced liver injuries in Sprague Dawley male rats. Rutin showed significant protection with the depletion of ALT, AST, ALP, and γ-GT in serum as was raised by the induction of CCl_4. Concentration of serum TGs, TC, and LDL was increased while HDL was decreased with rutin in a dose-dependent manner. Activity level of endogenous liver antioxidant enzymes; CAT, SOD, GPx, GST and glutathione reductase (GSR), and GSH contents were increased while LPO (TBARS) was decreased dose dependently with rutin. Moreover, increase in DNA fragmentation and oxo8dG damages while decrease in p53 and CYP 2E1 expression-induced with CCl_4 was restored with the treatment of rutin.

Hafez et al. (2015) investigated the hepato-protective effect of rutin on CCl_4-induced INH in male Wistar rats. Expression of the following genes was monitored with real-time PCR: IL-6, dual-specificity protein kinase 5 (MEK5), Fas-associated death domain protein (FADD), epidermal growth factor (EGF), signal transducer and activator of transcription 3 (STAT3), Janus kinase (JAK), B-cell lymphoma 2 (Bcl2), and B-cell lymphoma-extra-large (Bcl-XL). The CCl_4 groups showed significant increases in biochemical markers of hepatotoxicity and up-regulation of expression levels of IL-6, Bcl-XL, MEK5, FADD, EGF, STAT3, and JAK compared with the control group. However, CCl_4 administration resulted in significant down-regulation of Bcl2 expression compared with the control group. Interestingly, rutin supplementation completely reversed the biochemical markers of hepatotoxicity and the gene expression alterations induced by CCl_4. CCl_4 administration causes alteration in expression of IL-6/STAT3 pathway genes, resulting in hepatotoxicity. Rutin protects against CCl_4-INH by reversing these expression changes.

SAPONIN FRACTION

Rao et al. (2012) have reported that the saponin fraction from the dried pericarp of *Sapindus mukorossi* has a protective capability both *in vitro* on primary hepatocyte cultures and *in vivo* in a rat model of CCl_4-mediated liver injury. Saponin fraction pretreatment improves bromsulphalein clearance and also increases cellular viability. Saponin administration replenished depleted hepatic GSH and SOD by improving the antioxidant status of the liver and liver function enzymes. These effects substantiate protection of cellular phospholipids from peroxidative damage induced by highly reactive toxic intermediate radicals formed during biotransformation of CCl_4.

SILYMARIN

SM is a standardized mixture with high antioxidant power, a mechanism by which it is thought to exert its hepatoprotective effect on damage produced by FR generated by anti-TB drugs, EtOH, APAP, CCl_4, and others. SM is obtained from *Silybum marianum* and is frequently used for the treatment of liver diseases worldwide; it is endowed with antioxidant, anti-inflammatory, immunomodulatory, antiproliferative, antiviral, and antifibrotic activity (Eminzade et al., 2008). It contains at least seven flavolignans, the most important of which comprise silybin A, silybin B, silydianin, silycristins and isosilybins A and B, among others. Silybin A is the most important compound and represents 50%–70% of the extract of SM, being absorbed 20%–50%. A study was carried out on Wistar rats with liver damage caused by the administration of RIF/INH (100 and 50 mg/kg, respectively) and with the mixture of RIF/INH/PZA (100, 50, and 350 mg/kg, respectively), coadministered for 14 days with SM (200 mg/kg). The results showed that SM protects the liver damage caused by the mixture of anti-TB drugs and it regenerates in liver, the biochemical changes induced by the mixture of RIF/INH or RIF/INH/PZA. This effect is due to that SM reduced ALT and AST levels to normal values and also reduced serum albumin, total protein, and bilirubin values. In addition, the liver of the animals under study that received anti-TB drugs and SM did not induce steatosis, necrosis, or fibrosis. The authors suggest that SM can be employed as a nutritional supplement in patients treated with anti-TB drugs (Eminzade et al., 2008). Furthermore, it has been demonstrated that SM (150 mg/kg) administered orally possesses a hepatoprotective effect similar to that of *N*-acetyl-cysteine in the case of acute injury caused by APAP (Kazemifar et al., 2012). It is noteworthy that SM is frequently employed as a positive control in the search for substances with a hepatoprotective effect. SM has also exhibited a beneficial effect on anti-TB-INH in rabbits (INH 50 mg/kg). At a 50 mg/kg dose administered during 6 months, bilirubin and ALT serum levels were reduced with respect to anti-TB group (Jahan et al., 2015). The *in vivo* hepatoprotective effect of SM has been reported in a double-blind study of patients with TB and its treatment (INH, 5 mg/kg/d; RIF, 10 mg/kg/d; PZA, 25 mg/kg/ d, and EMB, 15 mg/kg/d). The treated group received SM (140 mg tablet, three times a daily) during 2 months, while the placebo group received tablets similar in appearance to those of SM. The author found that SM protects the liver damage caused by anti-TB drugs without demonstrating an adverse effect. It was observed that serum ALP, GGT, ALT, and AST values were higher than those shown by the placebo group. Also, it was described that SOD restoration can comprise a mechanism, which explains the beneficial effect exhibited by SM (Luangchosiri et al., 2015).

SM, a flavonolignan from the seeds of "milk thistle" (*S. marianum*), has been widely used from ancient times because of its excellent hepatoprotective action. SM has been a standard hepatoprotective agent for numerous studies and accounts for 180 million US dollars business in Germany alone (Pradhan and Girish, 2006). Fraschini et al. (2002) reviewed the pharmacology of SM including hepatoprotective properties. SM is a mixture of flavonolignans—silybin, silydianin, and silychristine—extracted from the seeds of milk thistle (*S. marianum* Gaertner) and has a history as a medical plant for almost two millennia. The main component of SM is silibinin (in a 50:50 mixture of Silybin A and Silybin B); the remaining components are silydianin, silycristin, isosilybin A, isosilybin B, isosilycristin, and taxifolin. Silybin, also known as silibinin, is the major constituent (70%–80%) as well as the most active biological component of milk thistle. Laboratory studies revealed that there is no real difference in activity between SM and silibinin in some experimental models.SM has been used medicinally to treat liver disorders, including acute and chronic viral hepatitis, toxin/drug-induced hepatitis, and cirrhosis and alcoholic liver diseases. It has also been reported to be effective in certain cancers. Its mechanism of action includes inhibition of hepatotoxin binding to receptor sites on the hepatocyte membrane; reduction of GSH oxidation to enhance its level in the liver and intestine; antioxidant activity; and stimulation of ribosomal RNA polymerase and subsequent protein synthesis, leading to enhanced hepatocyte regeneration. It is orally absorbed but has very poor bioavailability due to its poor water solubility (Dixit et al., 2007).

SM is not soluble in water and is usually administered in an encapsulated form. SM is absorbed when given orally. Peak plasma concentration is achieved in 6–8 h. The oral absorption of SM is only about 23%–47%, leading to low bioavailability of the compound; it is administered as a standard extract (70%–80% SM). After oral administration, the recovery in bile ranges from 2% to 3%. Silybin and the other components of SM are rapidly conjugated with sulfate and glucuronic acid in the liver and excreted through the bile. The poor water solubility and bioavailability of SM led to the development of enhanced formulations; for example, silipide (Siliphos$^{\beta}$), a complex of SM and phosphatidylcholine that is ten times more bioavailable than SM; an inclusion complex formed between SM and b-cyclodextrin, which is approximately 18 times more soluble than SM. There have been reports of silybin glycosides that have better solubility and stronger hepatoprotective activity (Dixit et al., 2007).

Silibinin has strong antioxidative and antifibrotic properties (Boigk et al., 1987; Dehmlow et al., 1996), which makes it a potentially useful drug for treatment of chronic liver diseases. The protective effect of SM is brought by competitively blocking the binding of phalloidin to receptors on the hepatocyte membrane surface and hindering α-amanitin to penetrate through the membrane into the cell nucleus. *In vitro* studies conducted with nuclei and nucleoli from rat liver point to another mechanism for the protective action of SM. Recently, it was demonstrated that high-dosage silibinin infusion treatment could significantly decrease the number of hepatitis C viruses after 4-weeks application.

Insulin resistance and oxidative stress are the major pathogenetic mechanisms leading the hepatic cell injury in these patients. The SM exerts membrane-stabilizing and antioxidant activity, it promotes hepatocyte regeneration; furthermore, it reduces the inflammatory reaction, and inhibits the fibrogenesis in the liver. These results have been established by experimental and clinical trials. According to open studies, the long-term administration of SM significantly increased survival time of patients with alcohol-induced liver cirrhosis. Based on the results of studies using methods of molecular biology, SM can significantly reduce tumor cell proliferation, angiogenesis as well as insulin resistance (Féher and Lengyel, 2012).

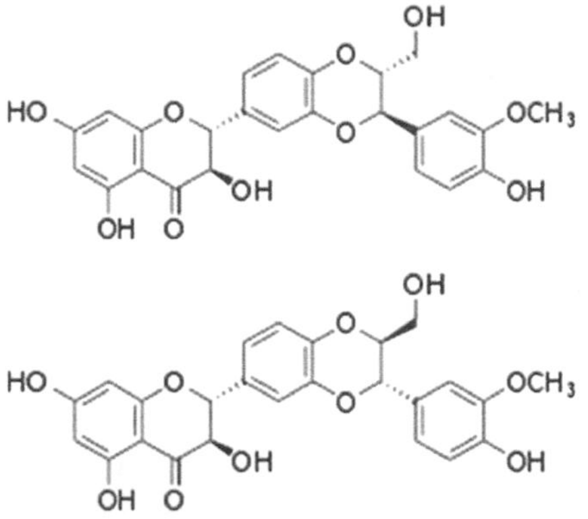

Silibinin A and Silibinin B

Pharmacological and clinical studies showed that SM is relatively safe at 100–150 mg daily in divided doses studied clinically (Ferenci et al., 1989). Silymarin's protective effect against oxidative stress was partially attributed to a reduction in the intracellular calcium in a model of perfused immobilized hepatocytes, resulting in better functioning hepatocytes (Farghali et al., 2000). The possible mechanism(s) that contribute to the hepatoprotective effect of SM and the role played by intracellular calcium was investigated using t-BHP and D-GalN toxicity in the isolated immobilized and perfused hepatocytes model. Silibinin was also found to be hepato-protective against steatosis and insulin resistance both *in vivo* and *in vitro*, partly through regulating the IRS-1/PI3K/Akt pathway which is involved in the pathogenesis of NAFLD (Zhang et al., 2013). Silibinin was also reported to improve hepatic oxidative stress and inflammation through the c-Jun NH2-terminal kinase (Salamone et al., 2012a) and nuclear factor kappa B (NF-kB) (Salamone et al., 2012b) pathways. In a clinical trial, silibinin given for 12 months in combination with vitamin E and phosphatidylcholine improved hepatic enzymes and insulin resistance in NAFLD patients (Loguercio et al., 2012). The anti-inflammatory, immunemodulating activity, antifibrotic, and antioxidant effects of SM that contribute to its hepatoprotective effect are summarized in a review by Abenavoli et al. (2010).

Various experimental studies using compounds that directly or indirectly cause liver damage have been carried out to demonstrate the hepatoprotective action of SM in xenobiotic intoxication and fungal intoxication.

Carbon Tetrachloride: SM has been shown to prevent CCl_4-induced LPO and hepatotoxicity (Lettiron et al., 1990; Muriel et al., 1990). This effect of SM is attributed to its ability to normalize the levels of the transaminases that are elevated in hepatotoxicity (Sharma et al., 1991). SM has been shown to protect harmful increase in the membrane ratios of cholesterol: phospholipids and sphingomyelin: phosphatidylcholine, thus providing protection from CCl_4-induced cirrhosis in rats (Mourelle and Franco, 1991). SM has also been found to reduce the increased collagen content in the CCl_4-induced chronic liver damage (Lettiron et al., 11990; Mourelle et al., 1989; Favari and Perez-Alvarez, 1997.

Liver fibrosis, a common condition occurring during the evolution of almost all chronic liver diseases, is the consequence of hepatocyte injury that leads to the activation of Kupffer cells and HSCs. Clichici et al. (2015) investigated the effects of two different doses of SM on a CCl_4-induced model of liver fibrosis with a focus on the early stages of liver injury. One month after the experiment began, authors explored hepato-cytolysis (aminotransferases and LDH), oxidative stress, fibrosis (histological score, hyaluronic acid), markers of HSC activation (transforming growth factor β1 [TGF-β1], and α-smooth muscle actin [α-SMA] expression by western blot) and activation of Kupffer cells by immunohistochemistry. SM 50 mg/BW had the capacity of reducing oxidative stress, hepato-cytolysis, fibrosis, activation of Kupffer cells, and the expression of α-SMA and TGF-β1 with better results than SM 200 mg/BW Thus, the usual therapeutic dose of SM, administered in the early stages of fibrotic changes is capable of inhibiting the fibrogenetic mechanism and the progression of initial liver fibrosis.

Wafay et al. (2012) investigated the modulatory effects of SM and garlic oil on CCl_4-INH in male albino rats. Intoxication of rats with CCl_4-induced significant elevation in serum liver enzymes as well as oxidative stress through the increase in oxidation markers over the antioxidant one as compared to that of the controls. Oral administration of SM or garlic oil improved these adverse effects.

Phenylhydrazine: Valenzuela et al. (1985) using a rat erythrocytic model, showed that SM inhibited the hemolysis and LPO caused by phenylhydrazine. The authors, in a separate study using rat liver, showed that SM provided protection from phenylhydrazine-induced liver GSH depletion and lipid superoxidation (Valenzula et al (1987).

tert-Butyl Hydroperoxide: t-BHP has been found to induce microsomal LPO and has been used as the model in different studies demonstrating the protective effect of SM. Valenzuela and Guerra (1986) demonstrated that SM inhibited oxygen consumption by rat microsomes, while Davila et al. (2000) showed that SM reduced enzyme loss and morphological alterations in neonatal rat hepatocytes. Farghali et al. (2000) demonstrated the inhibition of LPO by SM-perfused rat hepatocytes.

Ethanol: Administration of ethanol produces a decrease in the hepatic content of GSH, which is an important biomolecule that affords protection against chemically induced cytotoxicity (Thakur, 2002). The administration of ethanol enhances the ALT,

AST, and γ-GT levels and also makes the reduced GSH/oxidized GSH ratio abnormal. Wang et al. (1996) showed the protective effect of SM against ethanol-induced changes in these parameters. Valenzuela et al. (1996) showed the neutralization of the LPO, using acute intoxication in rats as the experimental model, while Valenzuela and Garrido (1994) showed reduction in the liver alterations, using chronic intoxication in rats as the model.

Halothane: Siegers et al. (1983) used acute intoxication caused by halothane in hypoxic rat models and demonstrated that SM provided protection from the hepatotoxic effects of halothane.

Thioacetamide: Antihepatotoxic effects and hepatoprotection were observed with SM in acute and chronic intoxication induced by TTA in rats (Schriewer and Weinhold, 1973; Hahn et al., 1968).

Galactosamine: GalN produces liver damage and administration in rats produces cholestatis due to inhibition of bile acid synthesis. An anticholestatic effect of SM has been reported by Saraswat et al. (1995). Datta et al. (1999) showed the effect of SM in normalizing elevated levels of serum transaminases and ALP in isolated rat hepatocytes with GalN-induced damage. Barbarino et al. (1981) showed the protective effect of SM in acute hepatitis in rats. Inhibition of GalN-induced LPO was also observed in perfused rat hepatocytes (Farghali et al., 2000). When experimental hepatitis in rats was used as the model, inhibition of the toxic effects on protein synthesis was also observed (Tyutyulkova et al., 1983).

Paracetamol: Paracetamol is known to cause centrilobular hepatic necrosis in mice/rats. SM, by its stabilizing action on the plasma membrane, has been shown to normalize the paracetamol-induced elevated biochemical parameters in the liver and serum (Ramellini and Meldolesi, 1976). Muriel et al. (1992) and Campos et al. (1989) used this model to show the protective effects of SM in paracetamol-induced LPO and GSH depletion. Renganathan studied the hepatoprotective effect of SM in mice and showed that SM prevented the hepatic cell necrosis induced by paracetamol in 87.5% of the animals; however, it provided protection in only 16% of the animals with hepatic necrosis induced by CCl_4 (Renganathan, 1999). He concluded that SM shows its hepatoprotective action either by preventing hepatic cell necrosis or by inducing hepatic cell regeneration. Silybin has also been shown to lessen the paracetamol-induced injury of kidney cells (Deak et al., 1990).

Erythromycin Estolate: Davila et al., using neonatal rat hepatocytes as the model, showed that SM reduced the enzyme loss and morphological alterations induced by erythromycin estolate (Davila et al., 1989).

Microcystin: This compound produces acute hepatotoxicity in mice/rats. Using this model, Mereish et al. demonstrated the neutralization of microcystin's lethal effects and pathological alterations by SM (Mereish et al., 1991).

Nitrosodiethylamine: The efficacy of SM on the antioxidant status of *N*-nitrosodiethylamine (NDEA)-induced hepatocarcinogenesis in Wistar albino male rats were assessed by Ramakrishnan et al. (2006). In hepatocellular carcinoma-induced animals, there was an increase in the number of nodules, relative liver weight. The levels of lipid peroxides were elevated with subsequent decrease in the BW, GSH, SOD, CAT, GPx, glutathione reductase (GR), and glucose-6-phosphate dehydrogenase (G6PD). In contrast, SM + NDEA-treated groups 4 and 5 animals showed a significant decrease in the number of nodules with concomitant decrease in the LPO status. The levels of GSH and the activities of antioxidant enzymes in both hemolysate and liver were improved when compared with hepatocellular carcinoma-induced group 2 animals. The electron microscopy studies were also carried out which supports the chemopreventive action of the SM against NDEA administration during liver cancer progression.

Amanita Phalloids Toxin: Administration of this compound produces acute intoxication in mice/rats/rabbits/dogs. SM at 50–150 mg/kg intravenous dose provided protection and cure (Desplaces et al., 1975). The increase in liver enzymes and reduction in coagulation factor seen with sublethal doses of A. Phalloids can be prevented with SM (Floershiem et al., 1978). This is evident at 5 h with SM given intravenously at a dose of 50 mg/kg and at 24 h with a dose of 30 mg/kg.

Antituberculosis Drugs: Eminzade et al. (2008) investigated the protective actions of SM against hepatotoxicity caused by different combinations of antituberculosis drugs in male Wistar albino rats. Animals were treated with intraperitoneal injection of isoniazid (INH, 50 mg/kg) and rifampicin (RIF, 100 mg/kg); and intragastric administration of pyrazinamid (PZA, 350 mg/kg) and SM (200 mg/kg). Hepatotoxicity was induced by a combination of drugs with INH+RIF and INH+RIF+PZA. Hepatoprotective effect of SM was investigated by co-administration of SM together with the drugs. Serum biochemical tests for liver functions and histopathological examination of livers were carried out to demonstrate the protection of liver against antituberculosis drugs by SM. Treatment of rats with INH+RIF or INH+RIF+PZA INH as evidenced by biochemical measurements: ALT, AST, and ALP activities and the levels of total bilirubin (T-Bil) were elevated, and the levels of albumin and total protein were decreased in drugs-treated animals. Histopathological changes were also observed in livers of animals that received drugs. Simultaneous administration of SM significantly decreased the biochemical and histological changes induced by the drugs. The active components of SM had protective effects against hepatotoxic actions of drugs used in the chemotherapy of tuberculosis in animal models. Since, no significant toxicity of SM is reported in human studies, this plant extract can be used as a dietary supplement by patients taking antituberculosis medications.

Freitag et al. (2015) investigated the effect of SM on the hypertension state and the liver function changes induced by APAP in spontaneously hypertensive rat (SHR). Animals normotensive (N) or hypertensive (SHR) were treated or not with APAP (3 g/kg, oral) or previously treated with SM. About 12 h after APAP administration, plasmatic levels of liver function markers: ALT, AST, GLU, γ-GT, and ALP of all groups, were determined. Liver injury was assessed using histological studies. Samples of their livers were then used to determine the MPO activity and NO production and were also sectioned for histological analysis. No differences were observed for ALT, γ-GT, and GLU levels between SHR and normotensive rats groups. However, AST and ALP levels were increased in hypertensive animals. APAP treatment promoted an increase in ALT and AST in both SHR and N. However, only for SHR, γ-GT levels were increased. The inflammatory response evaluated by MPO activity and NO production showed that SHR was more susceptible to APAP effect, by increasing leucocyte infiltration. SM treatment (Legalon) restored the hepatocyte functional and histopathological alterations induced by APAP in normotensive and hypertensive animals.

Clinical Trials: Effect of SM on chemical, functional, and morphological alterations of the liver has been investigated in patients with liver disease by Salmi and Sarna (1982). The patients were randomly allocated into a group treated with SM (treated) and a group receiving placebo (controls). Ninety-seven patients complete the 4-weeks trial-47 treated and 50 controls. In general, the series represented a relatively slight acute and subacute liver disease, mostly induced by alcohol abuse. There was a statistically highly significantly greater decrease of S-SGPT (S-ALT) and S-SGOT (S-AST) in the treated group than in controls. Serum total and conjugated bilirubin decreased more in the treated than in controls, but the differences were not statistically significant. BSP retention returned to normal significantly more often in the treated group. The mean percentage decrease of BSP was also markedly higher in the treated. Normalization of histological changes occurred significantly more often in the treated than in controls.

THEASINENSIN A

Theasinensins have been identified as a major group of unique catechin dimers mainly found in oolong tea and black tea. Among several types of theasinensins, theasinensin A (TSA), an epigallocatechin gallate dimer with an R-biphenyl bond, is the most abundant theasinensin prevalent in oolong tea. Hung et al. (2017) investigated the inhibitory effect of TSA on CCl_4-induced hepatic fibrosis in mice. After i.p. injection of CCl_4 for 8 weeks, histological lesions in the liver tissue and elevated serum levels of ALT and ALP were found in mice. Conversely, oral administration of TSA relieved CCl_4-induced liver injury as well as ameliorated liver functions. Immunohistochemical staining results revealed that collagen deposition was profoundly reduced due to supplementation with TSA. Hepatic α-SMA and matrix metallopeptidase 9 (MMP-9) expressions were suppressed through the inhibition of transforming growth factor β (TGF-β).

THYMOQUINONE

The effects of thymoquinone (TQ) and desferrioxamine (DFO) against CCl_4-INH were investigated by Mansour (2000). A single dose of CCl_4 (20 μL/kg, i.p.)-induced hepatotoxicity, manifested biochemically by significant elevation of activities of serum enzymes, such as ALT, AST, and LDH. Hepatotoxicity was further evidenced by significant decrease of total sulfhydryl (–SH) content, and CAT activity in hepatic tissues and significant increase in hepatic LPO measured as MDA. Pretreatment of mice with DFO (200 mg/kg i.p.) 1 h before CCl_4 injection or administration of TQ (16 mg/kg/d, p.o.) in drinking water, starting 5 days before CCl_4 injection and continuing during the experimental period, ameliorated the hepatotoxicity induced by CCl_4, as evidenced by a significant reduction in the elevated levels of serum enzymes as well as a significant decrease in the hepatic MDA content and a significant increase in the total sulfhydryl content 24 h after CCl_4 administration. In a separate *in vitro* assay, TQ and DFO inhibited the nonenzymatic LPO of normal mice liver homogenate induced by Fe^{3+}/ascorbate in a dose-dependent manner. These results indicate that TQ and DFO are efficient cytoprotective agents against CCl_4-induced hepotoxicity, possibly through inhibition of the production of oxygen free radicals that cause LPO.

Sangi et al. (2014) investigated the effect of TQ in rats affected by hepatotoxicity and cirrhosis as a result of administration of CCl_4. The levels of ALT, AST, and ALP were determined, regarding their relation to liver damage. Liver tissues were investigated to compare between healthy and infected ones. It was found that TQ significantly revised the changes produced by CCl_4 on hepatic cells and enzymes.

TQ is the main active constituent of *Nigella sativa* seeds. Nili-Ahmadabadi et al. (2018) explored the protective effects of TQ on diazinon (DZN)-induced liver toxicity in the mouse model. DZN caused a significant increase in ALT, AST, ALP serum levels, LPO and NO, the depletion of the total antioxidant capacity, total thiol molecule, and structural changes in the liver tissue. Following TQ administration, a significant improvement was observed in the oxidative stress biomarkers in the liver tissue. In addition, biochemical findings were correlated well to the histopathological examinations.

TORMENTIC ACID

The hepatoprotective effect of tormentic acid (TA) on APAP-induced liver damage was investigated in mice by Jiang et al. (2017). TA was i.p. administered for 6 days prior to APAP administration. Pretreatment with TA prevented the elevation of serum AST, ALT, T-Bil, TC, triacylglycerol (TG), and liver lipid peroxide levels in APAP-treated mice and markedly reduced APAP-induced histological alterations in liver tissues. Additionally, TA attenuated the APAP-induced production of NO, ROS, TNF-α, IL-1β, and IL-6. Furthermore, the Western blot analysis showed that TA blocked the protein expression of iNOS and COX-2, as well as the inhibition of nuclear factor-kappa B (NF-κB) and MAPKs activation in APAP-injured livertissues. TA also retained the SOD, GPx, and CAT in the liver. These results suggest that the hepatoprotective effects of TA may be related to its anti-inflammatory effect by decreasing TBARS, iNOS, COX-2, TNF-α, IL-1β, and IL-6, and inhibiting NF-κB and MAPK activation. Antioxidative properties were also observed, as shown by HO-1 induction in the liver and decreases in lipid peroxides and ROS.

TRITERPENES

Triterpenes Oleanolic Acid (OA, 3-bhydroxy-olea-12-en-28-oic acid) and ursolic acid (UA, 3-bhydroxyurs-12-en-28-oic acid) possess an important hepatoprotective effect against the liver damage caused by EtOH, CCl_4, D-GalN, APAP, Cd, bromobenzene, phalloidin, TTA, and other hepatotoxic substances (Balanehru and Nagarajan 1991; Liu et al., 1995; Saravanan et al., 2006). In China, clinical-phase experiments have been conducted on patients with acute and chronic hepatitis, who were treated with OA administered orally, for 3 months or more. The authors found that the compound demonstrated a beneficial effect with respect to the placebo group; this compound diminished AST and ALP serum levels, as well as cirrhosis in cases of chronic hepatitis. Recently published results on the hepatoprotective effect of the mixture of (OA/UA, obtained from *Bouvardia ternifolia*) against liver injury induced with INH (10 mg/kg), RIF (10 mg/kg), and PZ (30 mg/kg) in BALC/c mice. The animals received by subcutaneous via this OA/UA mixture at 100 and 200 mg/mouse daily for 11 weeks. The results showed that this mixture generates an increase on BW gain compared to the anti-TB-drugs group, also reduced the level of ALT and AST. In addition, histological analysis of livers from anti-TB group showed a steatosis and increased apoptosis, these effects were not detected in anti-TB + OA/UA group (Guti´errez-Rebolledo et al., 2016).

URSOLIC ACID

Shukla et al. (1992) evaluated the choleretic, anticholestatic, and hepatoprotective activities of ursolic acid, isolated from *Eucalyptus* hybrid, in rats. It produced a dose-dependent (5–20 mg/kg) choleretic effect. A significant anticholestatic activity (27.9%–100%) was observed against paracetamol (2.0 g/kg)-induced cholestasis. The compound also showed a marked hepatoprotective activity against paracetamol and GalN (800 mg/kg) INH by reversing the altered values in viability of the isolated hepatocytes and the altered biochemical liver and serum parameters. The activity of UA compared well with the known hepatoprotective drug, SM.

UA isolated from the leaves of *Eucalyptus* hybrid *E. tereticornis* showed a dose-dependent (5–20 mg/kg) hepatoprotective activity (21%–100%) in rats against TTA, GalN, and CCl_4 INH in rats (Saraswat et al., 1996). These hepatotoxins decreased the viability of hepatocytes as assessed by trypan blue exclusion and rate of oxygen uptake tests and decreased the volume of bile as well as the level of its contents. Pretreatment with UA increased the viability of rat hepatocytes significantly. A potent dose dependent anticholestatic activity was observed in conscious rat by noticing an increase in bile flow and its contents. UA had comparable activity to that of SM (Saraswat et al., 1996).

Saraswat et al. (2000) reported that UA has shown a significant preventive effect *in vitro* against ethanol-induced toxicity in isolated rat hepatocytes. Compared with the incubation of isolated hepatocytes with ethanol only, the simultaneous presence of UA in the cell suspension preserved the viability of hepatocytes and reversed the ethanol-induced loss in the level of all the marker enzymes (AST, ALT, and AP) studied. Ethanol alone resulted in a 48%–54% decrease in the viability and a 42%–54% reduction in the biochemical parameters of the hepatocytes. UA showed a concentration dependent (1–100 μg/mL) preventive effect (12%–76%) on alcohol-induced hepatocyte toxicity by restoring the altered parameters. The results, thus, suggest the effective use of an *in vitro* test system as an alternative for *in vivo* assessment of hepatoprotective activity of purified material (Saraswat et al. 2000).

UA is a triterpenoid that exists in nature and is the major component of some traditional medicinal herbs. UA has been evaluated for its hepatoprotective effect against chronic ethanol-mediated toxicity in rats by Saravanan et al. (2006). Ethanol administration (7.9 g/kg/d) for 60 days resulted in increased oxidative stress, decreased antioxidant defense and liver injury. It also negatively affected the serum total protein, albumin and A/G ratio. Subsequent to the experimental induction of toxicity (i.e., after the initial period of 30 days) UA treatment performed by coadministering UA (10, 20, and 40 mg/kg BW) for 30 days along with the daily dose of ethanol. While this treatment causing a significant improvement in BW, food intake and serum protein levels, it decreases serum aminotransferase activities (AST and ALT) and T-Bil levels. UA improved the antioxidant status of alcoholic rats, which is evaluated by the decreased levels of LPO markers in plasma (TBARS and lipid hydroperoxides) and increased levels of circulatory antioxidants such as reduced GSH, ascorbic acid, and alpha-tocopherol. Histopathological observations were also in correlation with the biochemical parameters. The activity of UA (20 mg/kg) compares well with SM, a known hepatoprotective drug, and seems to be better in certain parameters.

VANILLIN

Makni et al. (2011) investigated the protective effects of vanillin against CCl_4-INH in rat. Pretreatment with vanillin prior the administration of CCl_4 significantly prevented the decrease of protein synthesis and the increase in plasma ALT and AST. Furthermore, it inhibited hepatic LPO (MDA) and PCO formation and attenuated the CCl_4-mediated depletion of antioxidant enzyme CAT and SOD activities and GSH level in the liver. In addition, vanillin markedly attenuated the expression levels of proinflammatory cytokines such as TNF-α, IL-1β, and IL-6 and prevented CCl_4-induced hepatic cell alteration and necrosis, as indicated by liver histopathology.

VASICINONE

Vasicinone was isolated from leaves of *Justicia adhatoda*, evaluated for hepatoprotective activity using CCl_4 induced acute hepatotoxicity model in mice. CCl_4 treatments lead to significant increase in SGOT, SGPT, and ALP levels. Pretreatment with vasicinone and SM (25 mg/kg/d for 7 days) significantly decreased these enzyme levels. Histopathology of the livers from vasicinone and SM-pretreated animals showed normal hepatic cords and absence of necrotic changes suggesting pronounced recovery from CCl_4-induced liver damage. Both vasicinone and SM significantly decrease the CCl_4 mediated increase in pentobarbital indiced sleeping time in experimental animals, thus indicating recovery of liver function.

WEDELOLACTONE

Wedelolactone is isolated from the dried leaves of *Eclipta alba* and reported to be effective as a potential hepatoprotective, antibacterial, and antihemorrhagic. Pharmacokinetic studies of wedelolactone reveal its poor absorption through the intestine. Upadhyay et al. (2012) attempted to enhance the bioavailability of wedelolactone by its complexation with phosphatidyl choline and then formulating it as phytovesicles for hepatoprotective activity. The complex of wedelolactone rich fraction was prepared with phosphatidyl choline and characterized on the basis of solubility, melting point, thin layer chromatography (TLC), UV, IR, and NMR spectroscopy. The complex was further converted into phytovesicles and characterized. The hepatoprotective potential of phytovesicles was compared with complex, wedelolactone rich fraction and physical mixture of wedelolactone rich fraction and phosphatidyl choline by *in vitro* method. The results revealed that hepatoprotective activity is better in case of phytovesicles as compared to the complex, physical mixture and the wedelolactone itself. Enhanced bioavailability of the wedelolactone complex may be due to the amphiphilic nature of the complex, which greatly enhance the water and lipid solubility of the compound. The study by Upadhyay et al. (2012) clearly indicates the superiority of phytovesicles over the complex and wedelolactone, in terms of better absorption and improved hepatoprotective activity.

MIXTURE OF SM, CURCUMIN, AND N-ACETYLCYSTEINE

In a study, *in vitro* with the HepG2 cell line, Singh et al. (2012) tested the hepatoprotective effect of the mixture of SM, curcumin, and *N*-acetylcysteine against the liver damage caused with the INH/RIF/PZA mixture. The results revealed that the mixture of these natural compounds increases cell viability maintains cell morphology and does not alter mitochondrial activity (Singh et al., 2012).

ESSENTIAL OILS

Following an ethnobotanical survey for hepatoprotective remedies, four Libyan medicinal plants from family Lamiaceae; *Ajuga iva (*L.*)* Schreber, *Marrubium vulgare* L., *Rosmarinus officinalis* L., and *Thymus capitatus* Hoff. et Link growing widely in Beida, Libya, were selected by El-Hawary for hepatoprotective study. The chemical composition of EOs hydrodistilled from the dried aerial parts, were analyzed by GC-MS. The major constituents of the EOs of *A. iva* and *R. officinalis* were carvacrol (35.07%) and 1,8-cineol (35.21%), respectively, thymol was the major constituent of the EO of *M. vulgare* (20.11%) and *T. capitatus* (90.15%). The oil of *M. vulgare* had the most powerful antioxidant activity by restoring GSH levels in the blood of alloxan-induced diabetic rats. The rats treated with the EOs of *M. vulgare, R. officinalis,* and *T. capitatus* (50 mg/kg) showed a significant decrease in liver enzymes which were elevated by CCl_4. The EO of *M. vulgare* had the most potent hepatoprotective activity.

1-O-GALLOYL-6-O-(4-HYDROXY-3,5-DIMETHOXY) BENZOYL-BETA-D-GLUCOSE

A new gallic acid derivative, 1-*O*-galloyl-6-*O*-(4-hydroxy-3,5-dimethoxy)benzoyl-beta-D-glucose has been isolated from an H_2O-fraction of MeOH extract of *Combretum quadrangulare* seeds. This compound exhibited potent hepatoprotective activity against D-GalN/TNF-alpha-induced cell death in primary cultured mouse hepatocytes with an IC50 of 3.3 μM (Adnyana et al., 2001).

ESSENTIAL OIL OF Achillea wilhelmsii C. KOCH

The EO of *Achillea wilhelmsii* (100 and 200 mg/kg BW, i.p.) was evaluated against APAP-induced hepatic injuries in rats by Dadkhan et al. (2014). For this purpose, the activities of CYP, GST, and markers of liver injuries (ALT, AST, and ALP) together with level of GSH measured analytically in time intervals (2, 4, 8, 16, and 24 h) after treatments confirmed by histophatological consideration in rat livers. Administration of APAP (500 mg/kg BW, i.p.) significantly increased the activity of CYP450 concomitant with increasing the release of ALT and

AST. Whereas, GSH level and GST activity were decreased significantly after APAP treatment. Treatment of rats with *A. wilhelmsii* EOs significantly modulate these parameters to normal values. In addition, histophatological analysis of liver biopsies was consistent with the biochemical findings.

FLAVONOID C-GLYCOSIDE FROM *Solanum elaeagnifolium*

A new flavonoid C-glycoside, kaempferol 8-C-beta-galactoside, along with 12 known glycosidic flavonoids was isolated from the aqueous methanolic extract of *Solanum elaeagnifolium* by Hawas et al. (2013). Groups of 6 mice were administered *S. elaeagnifolium* extracts at 25, 50, and 75 mg/kg BW prior to or post administration of a single dose of paracetamol (500 mg/kg BW). The extract showed significant hepatoprotective and curative effects against histopathological and histochemical damage induced by paracetamol in liver. The extract also ameliorated the elevation in SGOT, SGPT, and ALP levels. These findings were accompanied by a nearly normal architecture of the liver in the treated groups, compared to the paracetamol control group. As a positive control, SM was used, an established hepatoprotective drug against paracetamol-induced liver injury. This study provides the first validation of the hepatoprotective activity of *S. elaeagnifolium*.

FLAVONOIDS FROM *Anastatica hierochuntica*

New skeletal flavonoids, anastatins A and B, were isolated by Yoshikawa et al. (2003) from the methanolic extract of an Egyptian medicinal herb, the whole plants of *Anastatica hierochuntica*. Their flavanone structures having a benzofuran moiety were determined on the basis of chemical and physicochemical evidence. Anastatins A and B were found to show hepatoprotective effects on D-GalN-induced cytotoxicity in primary cultured mouse hepatocytes and their activities were stronger than those of related flavonoids and commercial silybin.

PHYTOCONSTITUENTS FROM *Artemisia capillaris* Thunb.

Since an extract of the crude drug "inchinko," the buds of *A. capillaris*, showed significant antihepatotoxic activity by means of CCl_4-induced liver lesion *in vivo* and *in vitro*, the extract was fractionated by monitoring the activity to yield a number of flavonoids and a coumarin (6,7-dimethylesculetin). Antihepatotoxic effects of these constituents were determined by *in vitro* assay methods using CCl_4- and GalN-induced cytotoxicity in primary cultured rat hepatocytes. Some analogues of dimethylesculetin were also assayed for liver-protective activity. Both the antihepatotoxic activity and the dimethylesculetin content in this plant were found to vary markedly with the date of harvesting, which was assumed to be a reason for remarkable variations of the antihepatotoxic activity in commercially available preparations of this crude drug (Kiso et al., 1984).

Gao et al. (2016) evaluated the hepatoprotective effect of the EO of *A. capillaris* on CCl_4-induced liver injury in mice, the levels of serum AST and ALT, hepatic levels of reduced GSH, activity of GSH peroxidase, and the activities of SOD and MDA were assayed. Administration of the EO of *A. capillaris* at 100 and 50 mg/kg to mice prior to CCl_4 injection was shown to confer stronger *in vivo* protective effects and could observably antagonize the CCl_4-induced increase in the serum ALT and AST activities and MDA levels as well as prevent CCl_4-induced decrease in the antioxidant SOD activity, GSH level and GPx activity. The oil mainly contained β-citronellol, 1,8-cineole, camphor, linalool, α-pinene, β-pinene, thymol, and myrcene.

FLAVONOID RHAMNOCITRIN 4′-β-D-GALACTOPYRANOSIDE FROM Astragalus hamosus L.

The hepatoprotective activity of flavonoid rhamnocitrin 4′-β-D- galactopyranoside obtained from leaves of *Astragalus hamosus* was documented against *N*-diethylnitrosamine (DENA)-induced hepatic cancer in Wistar albino rats (Saleem et al., 2013; Al-Snafi, 2015).

ANTIHEPATOTOXIC AGENT FROM THE ROOT OF Astragalus membranaceus

Baek et al. (1996) isolated the components of *Astragalus membranaceus* and their structures were characterized as 3,4-methylenedioxypyrrole-aldehyde, 7-*O*-β-D-glucopyranosyl 7,3′-dihydroxy-4′-methoxy isoflavone and a naphthalene derivative on the basis of spectral and physical methods. 7-*O*-β-D-glucopyranosyl 7,3′-dihydroxy-4′-methoxy isoflavone and a naphthalene compound showed protective effect on CCl_4-induced cytotoxicity in primary cultured rat hepatocytes.

Bixa orellana SEED OIL

The effects of pretreatment with *Bixa orellana* seed oil on CCl_4-induced liver damage were determined in male Wister rats by Obidah et al. (2011). Rats were pretreated with *B. orellana* seed oil at 0%, 1%,

5%, or 10% (w/w) through dietary exposure for 4 weeks before a single intrapretoneal injection of CCl_4. Serum biochemical parameters, liver LPO, and relative organ weights were determined. Pretreatment with *B. orellana* seed oil (10%) resulted in significant reduction in serum liver marker enzymes activities, T-Bilconcentration, LPO, and relative liver weights induced by CCl_4 administration. These results show that dietary exposure to *B. orellana* seed oil exhibited moderate protection against CCl_4-INH in rats.

ISORHAMNETIN 3-*O*-GLUCOSIDE FROM *Brassica campestris*

Isorhamnetin 3-*O*-glucoside, which was combined together with isorhamnetin 3,7-di-*O*-glucoside in the plant leaves, suppressed increases in the plasma ALT and AST activities of mice with liver injury induced by the injection of CCl_4, but no suppression by isorhamnetin 3,7-di-*O*-glucoside was apparent. This result indicates that the release of glucose at the 7-position in isorhamnetin 3,7-di-*O*-glucoside was very important for mitigating liver injury (Igarashi et al., 2008).

ESSENTIAL OIL FROM WHITE CABBAGE [Brassica oleracea L. var. *capitata* (L.) Alef. f. *alba* DC.]

Chemical composition, antioxidant activity, and hepatoprotective effects EO from white cabbage (*Brassica oleracea* L. var. *capitata* f. *alba*(Bocfal) were investigated by Morales-López et al. (2017). Bocfal EO contained organic polysulphides, such as dimethyl trisulphide (DMTS) 65.43±4.92% and dimethyl disulphide 19.29±2.16% as major constituents. Bocfal EO and DMTS were found to be potent TBARS inhibitors with IC_{50} values of 0.51 and 3 mg/L, respectively. Bocfal EO demonstrated better hepatoprotective properties than DADS did, although both slightly affected the hepatic parenchyma per se, as observed using histopathology.

OCTADECENOIC ACID FROM Caesalpinia gilliesii

The hepatoprotective activity of dichloromethane fraction and isolated compounds from *Caesalpinia gilliesii* flowers dichloromethane fraction were studied in CCl_4-intoxicated rat liver slices by Osman et al. (2016) by measuring liver injury markers (ALT, AST, and GSH). A new 12,13,16-trihydroxy-14(Z)-octadecenoic acid was identified in addition to the known β-sitosterol-3-*O*-butyl, daucosterol, isorhamnetin, isorhamnetin-3-*O*-rhamnoside, luteolin-7,4'-dimethyl ether, genistein-5-methyl ether, luteolin-7-*O*-rhamnoside, isovanillic acid, and *p*-methoxybenzoic acid. Dichloromethane fraction and isorhamnetin were able to significantly protect the liver against intoxication. Moreover, the dichloromethane fraction and the isolated phytosterols-induced GSH were above the normal level. The hepatoprotective activity of *C. gilliesii* may be attributed to its high content of phytosterols and phenolic compounds.

43 kD HEPATOPROTECTIVE PROTEIN FROM *Cajanus indicus*

Sarkar et al. (2006) isolated, purified, and characterized the active principle(s) responsible for the hepatoprotective activity. The protein purified is composed of a single polypeptide chain having an apparent molecular mass of 43 kD as determined by SDS-PAGE and gel filtration through sephadex G-75 column. The isoelectric point of the protein determined was 4.8. Loss of biological activity after heat and protease treatment confirmed that the active molecule is a protein. Peptide fragments of the protein generated by trypsin cleavage were subjected to MALDI-TOF as well as LC-MS analyses and among the various fragments, four were very prominent and used for the determination of the amino acid sequence of the hepatoprotective protein. While one of the peptide fragments revealed strong sequence homology with plastocyanin, another fragment showed some similarity with a tomato protein present in the NCBI nonredundant database. The third peptide, on the other hand, is unique as it did not show any sequence homology with any known protein in the database. The protein showed maximum hepatoprotective activity when administered at a dose of 2 mg/kg BW for 5 days after CCl_4 administration. Histopathological studies also supported the hepatoprotective nature of the protein. Along with its curative property, the protein also possesses preventive role against a number of toxin-induced hepatic damages.

Ghosh and Sil (2007) investigated the protective effect of a 43 kDa protein isolated from the herb *C. indicus*, against APAP-induced hepatic and renal toxicity. Male albino mice were treated with the protein for 4 days (i.p., 2 mg/kg body wt) prior or post to oral administration of APAP (300 mg/kg body wt) for 2 days. Levels of different marker enzymes (namely, GPT and ALP), creatinine and blood urea nitrogen were measured in the experimental sera. Intracellular ROS production and total antioxidant activity were also determined from APAP and protein-treated hepatocytes. Indices of different antioxidant enzymes (namely, SOD, CAT, and GST) as well as LPO end-products and GSH were determined in both liver and kidney homogenates. In addition, CYP activity was also measured from liver microsomes. Finally, histopathological studies were performed from liver sections of control, APAP-treated and protein pre- and post-treated (along with APAP) mice. Administration of APAP increased all the serum markers and creatinine levels in mice sera along with the enhancement of hepatic and renal LPO. Besides, application of APAP to hepatocytes increased ROS production and reduced the total antioxidant activity of the treated hepatocytes. It also reduced the levels of antioxidant enzymes and cellular reserves of GSH in liver and kidney. In addition, APAP enhanced the CYP activity of liver microsomes. Treatment with the protein significantly reversed these changes to almost normal. Apart from these, histopathological changes also revealed the protective nature of the protein against APAP-induced necrotic damage of the liver tissues. The results suggest that the protein protects hepatic and renal tissues against oxidative damages and could be used as an effective protector against APAP-induced hepato-nephrotoxicity.

CHRYSOPHANOL FROM *Cassia occidentalis* L.

Rani et al. (2010) evaluated the protective role of methanol fraction and its pure compound chrysophanol from *Cassia occidentalis* against paracetamol-INH in adult male albino Wistar rats. Liv-52 was used as a standard reference. Chrysophanol (50 mg/kg BW) and methanol fraction (*COLMF*) (200 mg/kg BW) were administered to the paracetamol treated rats for 7 days. Oral administration of chrysophanol and *COLMF* significantly normalized the values of SOD, CAT, GPx, GSH, and vitamin C and vitamin E. And, the elevated serum enzymatic levels of AST, ALT, ACP, and ALP were significantly restored toward normalization by pretreatment with chrysophanol and *COLMF*. The histopathological studies also confirmed the hepatoprotective nature of the extracts.

FLAVONOID RUTINOSIDES FROM *Cinnamomum parthenoxylon*

Pardede et al. (2017) evaluated the hepatoprotective and antioxidant activity of flavonoid rutinosides from *Cinnamomum parthenoxylon.* The EtOAc fraction of *C. parthenoxylon* leaves showed potent hepatoprotective activity on *t*-BHP-induced cytotoxicity in HepG2 cells and higher antioxidant activity. Phytochemical analysis revealed that flavonoid rutinosides; rutin, nicotiflorin, and isorhofolin are major constituents in the etOAc fraction. The catechol group B ring in the structure of rutin holds potential for hepatoprotective and antioxidant activity.

PHENYLETHANOIDS ISOLATED FROM THE STEMS OF *Cistanche deserticola*

Four phenylethanoids isolated from the stems of *Cistanche deserticola*, acteoside (1), 2′-acetylacteoside

(2), isoacteoside (3), and tubuloside B (4), significantly suppressed NADPH/CCI4-induced LPO in rat liver microsomes (Xiong et al., 1998). Additionlly, primary cultured rat hepatocytes efficiently prevented cell damage induced by exposure to Cd4 or *o*-D-GalN. Acteoside (1) further showed pronounced antihepatotoxic activity against Cd4 *in vivo*.

ACYLATED PHENYLETHANOID OLIGOGLYCOSIDES FROM Cistanche tubulosa

The methanolic extract from fresh stems of *Cistanche tubulosa* (Orobanchaceae) was found to show hepatoprotective effects against D-GalN/LPS-induced liver injury in mice (Morikawa et al., 2010). From the extract, 3 new phenylethanoid oligoglycosides, kankanosides H(1) (1), H(2) (2), and I(3), were isolated together with 16 phenylethanoid glycosides (4–19) and two acylated oligosugars (20 and 21). Among the isolates, echinacoside [IC(50)=10.2 μM], acteoside (4.6 μM), isoacteoside (5.3 μM), 2′-acetylacteoside (4.8 μM), and tubuloside A (10, 8.6 μM) inhibited D-GalN-induced death of hepatocytes. These five isolates, 4 (31.1 μM), 5 (17.8 μM), 6 (22.7 μM), 8 (25.7 μM), and 10 (23.2 μM), and cistantubuloside B(1) (11, 21.4 μM) also reduced TNF-alpha-induced cytotoxicity in L929 cells. Moreover, principal constituents (4–6) exhibited *in vivo* hepatoprotective effects at doses of 25–100 mg/kg, p.o.

Citrus depressa FLAVONOIDS

Akachi et al. (2010) attempted to isolate the constituent(s) responsible for the suppressive effect of the juice of shekwasha, a citrus produced in Okinawa Prefecture, on D-GalN-induced liver injury in rats. Liver injury-suppressive activity, as assessed by plasma ALT and AST activities, was found only in the fraction that was extracted with *n*-hexane when three fractions were added to the diet and fed to rats. Of five compounds isolated from the *n*-hexane-soluble fraction by silica gel column chromatography, three compounds had liver injury-suppressive effects when five compounds were singly force fed to rats at a level of 300 mg/kg BW 4 h before the injection with GalN. The structures of the three active compounds were determined as 3′,4′,5,6,7,8-hexamethoxyflavanone (citromitin), 4′,5,6,7,8-pentamethoxyflavone (tangeretin), and 3′,4′,5,6,7,8-hexamethoxyflavone (nobiletin), which are known flavonoids mainly existing in citrus. Nobiletin, the most important compound in the *n*-hexane-soluble fraction, also had suppressive effects on liver injuries induced by CCl_4, APAP, and GalN/LPS in addition to liver injury-induced GalN. Nobiletin suppressed

GalN/LPS-induced increases in plasma TNF-α and NO concentrations and hepatic mRNA levels for inducible NO synthase and DNA fragmentation. These results suggest that nobiletin suppressed GalN/LPS-induced liver injury at least by suppressing the production of both TNF-alpha and NO. The results obtained here indicate that the hepatoprotective effect of shekwasha juice is mainly ascribed to several polymethoxy flavonoids included in the juice.

PHYTOCONSTITUENTS FROM Cnidium monnieri (L.) Cusson *FAMILY: APIACEAE*

Bioassay-guided fractionation of the EtOH extract of *Cnidium monnieri* furnished two hepatoprotective sesquiterpenes, torilin and torilolone, together with a new derivative, 1-hydroxytorilin (Oh et al., 2002). Torilin and torilolone showed hepatoprotective effects on tacrine-induced cytotoxicity in human liver-derived Hep G2 cells. The EC_{50} values of torilin and torilolone were 20.6 ± 1.86 and 3.6 ± 0.1 μM, respectively. Silybin as a positive control showed an EC_{50} value of 69.0 ± 3.4 μM.

COCONUT (Cocos nucifera) OIL

Zakaria et al. (2011) determined the hepatoprotective effect of MARDI-produced virgin coconut oils, prepared by dried- or fermented-processed methods, using the paracetamol-induced liver damage in rats. Liver injury induced by 3 g/kg paracetamol increased the liver weight 100 g/BW indicating liver damage. Histological observation also confirms liver damage indicated by the presence of inflammations and necrosis on the respective liver section. Interestingly, pretreatment of the rats with 10, but not 1 and 5 mL/kg of both VCOs significantly reduced the liver damage caused by the administration of paracetamol, which is further confirmed by the histological findings.

COFFEE-SPECIFIC DITERPENES

The hepatoprotective effects of kahweol and cafestol, coffee-specific diterpenes, on the CCl_4-induced liver damage as well as the possible mechanisms involved in these protections were investigated by Lee et al. (2007). Pretreatment with kahweol and cafestol prior to the administration of CCl_4 significantly prevented the increase in the serum levels of hepatic enzyme markers (ALT and AST) and reduced oxidative stress, such as reduced GSH content and LPO, in the liver in a dose-dependent manner. The histopathological evaluation of the livers also revealed that kahweol and cafestol reduced the incidence of liver lesions induced by CCl_4. Treatment of the mice with kahweol and cafestol also resulted in a significant decrease in the CYP2E1, the major isozyme involved in CCl_4 bioactivation, specific enzyme activities, such as *p*-nitrophenol and aniline hydroxylation. Kahweol and cafestol exhibited antioxidant effects on $FeCl_2$-ascorbate-induced LPO in a mouse liver homogenate, and on superoxide radical scavenging activity. These results suggest that the protective effects of kahweol and cafestol against the CCl_4-INH possibly involve mechanisms related to their ability to block the CYP2E1-mediated CCl_4 bioactivation and free-radical scavenging effects.

ESSENTIAL OILS OF Coriandrum sativum **L.** *AND Carum carvi* **L.** *FRUITS*

EOs of *Coriandrum sativum* and *Carum carvi* fruits were assayed for their *in vitro* and *in vivo* antioxidant activity and hepatoprotective effect against CCl_4 damage. The *in vitro* antioxidant activity was evaluated as a free-radical scavenging capacity, measured as scavenging activity of the EOs on DPPH and OH radicals and effects on LPO in two systems of induction. Some liver biochemical parameters were determined in animals pretreated with EOs and later intoxicated with CCl_4 to assess *in vivo* hepatoprotective effect. Tested EOs were able to reduce the stable DPPH in a dose-dependent manner and to neutralize H_2O_2, reaching 50% neutralization with IC_{50} values of <2.5 μL/mL for Carvi aetheroleum and 4.05 μL/mL for Coriandri aetheroleum. Caraway EO strongly inhibited LP in both the systems of induction, whereas coriander EO exhibited prooxidant activity. *In vivo* investigation conferred leak of antioxidative capacity of coriander EO, whereas the EO of caraway appeared promising for safe use in folk medicine and the pharmaceutical and food industries.

Cymbopogon citratus OR LEMONGRASS ESSENTIAL OIL

Uchida et al. (2017) investigated the hepatoprotective effect of *Cymbopogon citratus* or lemongrass essential oil (LGO) in an animal model of acute liver injury induced by APAP. LGO pretreatment decreased significantly the levels of ALT, AST, and ALP compared with APAP group. MPO activity and NO production were decreased. The histopathological analysis showed an improved of hepatic lesions in mice after LGO pretreatment. LGO inhibited neutrophil migration and exhibited antioxidant activity.

PHENOLIC COMPONENTS FROM ARTICHOKE (Cynara scolymus)

The main phenolic constituents of the artichoke leaf extract are a variety of mono and dicaffeoylquinic acids (e.g.,

CGA, cynarin [(1,3-dicaffeoylquinic acid; (1R,3R,4S,5R)-1,3-bis[[3-(3,4-dihydroxyphenyl)propenoyl]oxy]-4,5-dihydroxycyclohexanecarboxylic acid)]), caffeic acid, and flavonoids (e.g., luteolin-7-*O*-glucoside) for which several pharmacodynamic effects have been observed *in vitro* and *in vivo*. However, *in vivo* is not only the genuine extract constituents, but also their metabolites may contribute to efficacy. In addition, all parts of the plant contain sesquiterpene lactone cynaropecrin and inulin (Lattanzio and van Sumere, 1987).

Cynarin and, to a lesser extent, CA exhibited hepatoprotective activity in CCl_4-treated rats; however, a minimum of 1% polyphenols and 0.2% flavonoids in the dried leaves was required for the activity (Adzet et al., 1987). A protective effect of cynarin was also observed for D-GalN-pretreated primary-cultured mouse and rat hepatocytes (Kiso et al., 1983).

STEROIDS FROM *Cynanchum otophyllum*

About seven undescribed C21 steroids, namely cynanchin A–G, together with thirteen known analogues, were isolated by Dong et al. (2019) from the roots of *Cynanchum otophyllum*. All of the isolates were tested for their antihepatic fibrosis activity. Among them, compounds 4–6, 10–12, and 14–17 showed moderate or significant inhibitory effects for the proliferation of HSCs induced by transforming growth factor-β1 (TGF-β1) *in vitro*.

POLYSACCHARIDES ISOLATED FROM *Dendrobium officinale*

The effect of polysaccharides isolated from *Dendrobium officinale* (DOP) on APAP-INH and the underlying mechanisms involved were investigated by Lin et al. (2018). The results showed that DOP treatment significantly alleviated the hepatic injury. The decrease in ALT and AST levels in the serum and ROS, MDA, and MPO contents in the liver, as well as the increases in GSH, CAT, and T-AOC in the liver, were observed after DOP treatment. DOP treatment significantly induced the dissociation of Nrf2 from the Nrf2-Keap1 complex and promoted the Nrf2 nuclear translocation. Subsequently, DOP-mediated Nrf2 activation triggered the transcription and expressions of the GCLC) subunit, GCL regulatory subunit, HO-1, and NAD(P)H dehydrogenase quinone 1 (NQO1) in APAP-treated mice. Further investigation about mechanisms indicated that DOP exerted the hepatoprotective effect by suppressing the oxidative stress and activating the Nrf2-Keap1 signaling pathway.

TOTAL SAPONINS FROM *Dioscorea nipponica MAKINO*

Yu et al. (2014) investigated the effects and possible mechanisms of the total saponins from *Dioscorea nipponica* (TSDN) against CCl_4-INH in mice. The mice were orally administrated with TSDN for 7 days and then given CCl_4 (0.3%, 10 mL/kg i.p.). The results showed that TSDN significantly attenuated the activities of ALT and AST, consistent with hematoxylin–eosin staining. The ALP levels and relative liver weight were significantly decreased by TSDN compared with model group. Moreover, TSDN dramatically decreased MDA, iNOS, and NO levels, while the levels of GSH, GSH-Px, and SOD were increased. Further investigations showed that TSDN inhibited CCl_4-induced metabolic activation and CYP2E1 expression, downregulated the levels of MAPKs phosphorylation, NF-κB, HMGB1, COX-2 as well as effectively suppressed the expressions of Caspase-3, Caspase-9, PARP, and Bak. Quantitative real-time PCR assay demonstrated that TSDN obviously decreased the gene expressions of TNF-a, IL-1β, IL-6, IL-10, Fas, FasL, Bax as well as modulated Bcl-2 mRNA level.

PHYTOCONSTITUENTS FROM *Eclipta prostata* (SYN.: *E.alba*)

The alcoholic extract of fresh leaves of the plant *Eclipta prostata* (Syn.: *E. alba*), previously reported for is hepatoprotective activity was fractionated into three parts to chemically identify the most potent bioactive fraction. The hepatoprotective potential of the fraction prepared from extract was studied *in vivo* in rats and mice against CCl_4-induced hepatotoxicity. The hepatoprotective activity was determined on the basis of their effects on parameters like hexobarbitone sleep time, zoxazolamine paralysis time, bromosulphaline clearance, serum transaminases, and serum bilirubin. Fraction EaII (10–80 mg/kg, p.o.) containing coumestan wedelolactone and desmethylwedelolactone as major components with apigenin, luteolin, 4-hydroxybenzoic acid, and protocateuic acid as minor constituents exhibited maximum hepatoprotective activity and is the active fraction for hepatoprotective activity of *E. alba* leaves (Singh et al. 2001). The acute toxicity studies have shown that like Ea, Fraction EaII also has high safety margin.

PHENOLIC PETROSINS AND FLAVONOIDS FROM *Equisetum arvense*

Hepatoprotective activity-guided fractionation of the MeOH extract of *Equisetum arvense* by Oh et al. (2004) resulted in the isolation of two phenolic petrosins, onitin (1) and onitin-9-*O*-glucoside (2), along with four flavonoids, apigenin (3), luteolin (4), kaempferol-3-*O*-glucoside (5), and quercetin-3-*O*-glucoside (6). Among these, compounds

(1) and (4) exhibited hepatoprotective activities on tacrine-induced cytotoxicity in human liver-derived Hep G2 cells, displaying EC50 values of 85.8 + 9.3 μM and 20.2 + 1.4 μM, respectively. Silybin, used as a positive control, showed the EC50 value of 69.0 + 3.3 μM. Compounds (1) and (4) also showed superoxide scavenging effects (IC50 = 35.3 + 0.2 μM and 5.9 + 0.3 μM, respectively) and DPPH free-radical scavenging effect [IC(50) of 35.8 + 0.4 μM and 22.7 + 2.8 μM, respectively]. These results support the use of this plant for the treatment of hepatitis in oriental traditional medicine (Oh et al., 2004).

OA ISOLATED FROM CHLOROFORM EXTRACT OF *Flaveria trinervia*

Hoskeri et al. (2011) investigated the prophylactic effects of OA isolated from chloroform extract (CE) of *Flaveria trinervia* against ethanol-induced liver toxicity using rats. CE and OA at three different doses were tested by administering orally to the ethanol-treated animals during the last week of the 7 weeks study. SM was used as the standard reference. The substantially elevated serum enzymatic levels of serum glutamate oxaloacetate transaminase, GPT, alkaline phosphatase, and bilirubin in ethanol-treated animals were restored toward normalcy by treatment of CE and OA. *In vivo* antioxidant and *in vitro* free-radical scavenging activities were also positive for all the three concentrations of CE and OA. However, OA at 150 mg/kg showed significant activity when compared to the other two doses.

Biochemical observations in support with histopathological examinations revealed that CE and OA possess hepatoprotective action against ethanol INH in rats.

VOLATILE COMPONENTS OF FENNEL (*Foeniculum vulgare* Mill.) SEED EXTRACTS

The leaves, stalks, and seeds (fruits) of *Foeniculum vulgare* (fennel) are edible. Volatile components of fennel seed extracts include trans-anethole, fenchone, methylchavicol, limonene, L-pinene, camphene, M-pinene, M-myrcene, L-phellandrene, 3-carene, camphor, and cisanethole (Simandi et al., 1999). Hepatoprotective activity of *F. vulgare* (fennel) EO was studied using a CCl_4-induced liver fibrosis model in rats. The hepatotoxicity produced by chronic CCl_4 administration was found to be inhibited by *F. vulgare* EO with evidence of decreased levels of serum AST, ALT, ALP, and bilirubin.

Hepatoprotective activity of *F. vulgare* (fennel) EO was studied by Özbek et al. (2003, 2004) using a CCl_4-induced liver fibrosis model in rats. The hepatotoxicity produced by chronic CCl_4 administration was found to be inhibited by *F. vulgare* EO with evidence of decreased levels of serum AST, ALT, ALP, and bilirubin. Histopathological findings also suggest that *F. vulgare* EO prevents the development of chronic liver damage. The changes in BWs in the rats assigned to the study groups supported these biochemical and histopathological findings.

POLYPRENOLS FROM Ginkgo biloba L.

The hepatoprotective effects of polyprenols from *Ginkgo biloba* leaves were evaluated by Yang et al. (2011) against CCl_4-induced hepatic damage in Sprague Dawley rats. The elevated levels of serum ALT, AST, ALP, ALB, TP, HA, LN, TG, and CHO were restored toward normalization significantly by GBP in a dose dependent manner. The biochemical observations were supplemented with histopathological examination of rat liver sections. Meanwhile, GBP also produced a significant and dose-dependent reversal of CCl_4-diminished activity of the antioxidant enzymes and reduced CCl_4-elevated level of MDA. In general, the effects of GBP were not significantly different from those of the standard drug Essentiale.

SOYA (Glycine max) SAPONINS

Lijie et al. (2016) investigated the effects of soyasaponin Bb (Ss-Bb) on oxidative stress in alcohol-induced rat hepatocyte injury. It has been shown that the administration of Ss-Bb could significantly restore antioxidant activity in BRL 3A cells. Moreover, the impaired liver function and morphology changes resulting from ethanol exposure were improved by Ss-Bb treatment. Treatment with a pharmacological inhibitor of hemoxygenase-1 (HO-1) indicated a critical role of HO-1 in mediating the protective role. The authors found that pretreatment with Ss-Bb to ethanol exposure cells increased the expression level of HO-1. It was suggested that Ss-Bb may protect against alcohol-induced hepatocyte injury through ameliorating oxidative stress, and the induction of HO-1 was an important protective mechanism.

TRITERPENE SAPONINS FROM THE ROOTS OF Glycyrrhiza inflata

About two novel oleanane-type triterpene saponins, licorice-saponin P2 (1) and licorice-saponin Q2 (3), together with nine known compounds 2, 4-11, have been isolated from the water extract of the roots of *Glycyrrhiza inflata* by Zheng et al. (2015).

In *in vitro* assays, compounds 2-4, 6, and 11 showed significant hepatoprotective activities by lowering the ALT and AST levels in primary rat hepatocytes injured by D-GalN. In addition, compounds 2-4, 6, 7, and 11 were found to inhibit the activity of PLA2 with IC50 values of 6.9, 3.6, 16.9 27.1, 32.2, and 9.3 μM, respectively, which might be involved in the regulation of the hepatoprotective activities observed by Zheng et al. (2015).

HIBISCUS ANTHOCYANINS

Hibiscus anthocyanins (HAs), a group of natural pigments occurring in the dried flowers of *Hibiscus sabdariffa* L., which is a local soft drink material and medical herb, were studied by Wang et al. (2000) for antioxidant bioactivity. The preliminary study showed that HAs were able to quench the free radicals of 1,1-diphenyl-2-picrylhydrazyl. This antioxidant bioactivity was further evaluated using the model of t-BHP-induced cytotoxicity in rat primary hepatocytes and hepatotoxicity in rats. The results demonstrated that HAs, at the concentrations of 0.10 and 0.20 mg/mL, significantly decreased the leakage of LDH and the formation of MDA induced by a 30-min treatment of t-BHP (1.5 mM). The *in vivo* investigation showed that the oral pretreatment of HAs (100 and 200 mg/kg) for 5 days before a single dose of t-BHP (0.2 mmol/kg, i.p.) significantly lowered the serum levels of hepatic enzyme markers (alanine and AST) and reduced oxidative liver damage. The histopathological evaluation of the liver revealed that Hibiscus pigments reduced the incidence of liver lesions including inflammatory, leucocyte infiltration, and necrosis induced by t-BHP in rats.

SEABUCKTHORN (*Hippophae rhamnoides* L.) BERRY POLYSACCHARIDES

Wang et al. (2017) investigated the protective effects and mechanisms of Seabuckthorn polysaccharide (SP) against APAP-induced hepatotoxicity. Pretreatment with SP led to decreased levels of ALT and AST in APAP mice, without affecting APAP metabolism. This was accompanied by diminished liver injuries, increased levels of GSH and GSH-Px, reduced NO and iNOS expression. SP increased the activity of SOD as well as SOD-2 expression in APAP mice. SP suppressed APAP-induced JNK phosphorylation and increased the ratio of Bcl-2/Bax. Furthermore, SP decreased the expression of Keap-1 and increased the nuclear expression of Nrf-2. The expression of Nrf-2 target gene HO-1 was increased by SP pretreatment in APAP mice.

BENZOIC ACID FROM Hypoestes triflora

Hypoestes triflora is frequently used in Rwandese native medicine to treat hepatic diseases. Van Puyvelde et al. (1989) evaluated the hepatoprotctive activity of *H. triflora*. Premedication with a water extract of the leaves prevented the prolongation of the barbiturate sleeping time associated with CCl_4-induced liver damage in mice. The compound responsible for this protective activity was benzoic acid. Mice previously treated with benzoic acid also showed a significant diminution of the increased GOT and GPT levels seen after CCl_4 administration.

***CRUDE POLYSACCHARIDES FROM THE LEAVES OF Ilex latifolia* Thunb.**

Crude polysaccharides from the leaves of *Ilex latifolia* (ILPS) was fractionated by DEAE cellulose-52 chromatography, affording four fractions of ILPS-1, ILPS-2, ILPS-3, and ILPS-4 in the recovery rates of 32.3%, 20.6%, 18.4%, and 10.8%, respectively, based on the amount of crude ILPS used (Fan et al., 2014). The four fractions were mainly composed of arabinose and galactose in monosaccharide composition. Compared with ILPS-1 and ILPS-2, ILPS-3 and ILPS-4 had relative higher contents of sulfuric radical and uronic acid. The antioxidant activities *in vitro* of ILPS decreased in the order of crude ILPS>ILPS-4>ILPS-3>EPS-2>ILPS-1. Furthermore, the administration of crude ILPS significantly prevented the increase of serum ALT and AST levels, reduced the formation of MDA, and enhanced the activities of SOD and GPx in CCl_4-induced liver injury mice (Fan et al., 2014).

TOTAL FLAVONOIDS OF Indocalamus latifolius

The total flavonoids of *Indocalamus latifolius* were evaluated in term of their antioxidant and hepatoprotective activities by Tan et al. (2015). The results showed that *in vitro* hepatoprotective and antioxidant activities of total flavonoids at doses of 200 and 400 and 100 mg/kg, respectively, were comparable to those of the known hepatoprotective drug SM at 100 mg/kg. These data were supplemented with histopathological studies of rat liver sections. About seven of the main flavonoid compounds purified by column chromatography using silica gel, sephadex LH-20 and develosil ODS, and determined to be vitexin, orientin, isovitexin, homoorientin, tricin, tricin-7-*O*-β-D-glucopyranoside, and quercetin-3-*O*-glucopyranoside.

ANTHOCYANIN FRACTION FROM Ipomoea batatas

Choi et al. (2009) examined the protective effects of an anthocyanin fraction (AF) obtained from purple-fleshed sweet potato on APAP (paraceptamol [APAP])-INH in mice and to determine the mechanism involved. Mice pretreated with AF prior to APAP administration showed significantly lower increases in serum ALP and AST activities and hepatic MDA formation than APAP-treated animals without AF. In addition, AF prevented hepatic GSH depletion by APAP, and hepatic GSH levels and GST activities were upregulated by AF. APAP-INH was also prevented by AF, as indicated by liver histopathology findings. In addition, the effects of AF were examined on CYP2E1, the major isozyme involved in APAP bioactivation. Treatment of mice with AF significantly and dose-dependently reduced CYP2E1-dependent aniline hydroxylation and CYP2E1 protein levels. Furthermore, AF had an antioxidant effect on $FeCl_2$/ascorbate-induced LPO in mouse liver homogenates and had superoxide radical scavenging activity. These results suggest that AF protects against APAP-INH by blocking CYP2E1-mediated APAP bioactivation, by upregulating hepatic GSH levels, and by acting as a free radical scavenger.

Zhang et al. (2015) evaluated the antioxidant activities of anthocyanins obtained from purple sweet potato (PSPAs) and their protective effect on hepatic fibrosis induced by CCl_4 administration in mice. PSPAs showed high scavenging effects against DPPH and HO. Tested mice were orally treated with PSPAs every day for 3 weeks and in combination with CCl_4 at last. The results indicated that treatment with CCl_4 caused hepatotoxicity, which was assessed by an increase in the levels of relative liver weight, serum AST and ALT, as well as in hepatic lipid peroxide MDA. Moreover, the activities of SOD and GPx were reduced. However, these changes were inhibited by the treatment with PSPAs prior to the administration of CCl_4. The biochemical alternations were accompanied by histopathological changes, such as vacuolization, necrosis, and congestion.

Juglans regia (WALNUT) POLYPHENOLS

The polyphenol-rich fraction (WP, 45% polyphenol) prepared from the kernel pellicles of walnuts was assessed for its hepatoprotective effect in mice by Shimoda et al. (2008). A single oral administration of WP (200 mg/kg) significantly suppressed SGOT and SGPT elevation in liver injury induced by CCl_4, while it did not suppress DGalN-induced liver injury. In order to identify the active principles in WP, Shimoda et al. (2008) examined individual constituents for the protective effect on cell damage induced by CCl_4 and D-GalN in primary cultured rat hepatocytes. WP was effective against both CCl_4- and D-GalN-induced hepatocyte damages. Among the constituents, only ellagitannins with a galloylated glucopyranose core, such as tellimagrandins I, II, and rugosin C, suppressed CCl_4-induced hepatocyte damage significantly. Most of the ellagitannins including tellimagrandin I and 2,3-*O*-hexahydroxydiphenoylglucose exhibited remarkable inhibitory effect against D-GalN-induced damage. Telliamgrandin I especially completely suppressed both CCl_4- and D-GalN-induced cell damage, and thus is likely the principal constituent for the hepatoprotective effect of WP.

Juniperus CONSTITUENTS

Alqasoumi et al. (2013) investigated the hepatoprotective effect of *Juniperus phoenicea* constituents. Different fraction obtained from the aerial parts of *J. phoenicea* showed significant activity as hepatoprotective when investigated against CCl_4-induced liver injury. The hepatoprotective activity was evaluated through the quantification of biochemical parameters and confirmed using histopathology study. Phytochemical investigation of the petroleum ether, chloroform, and methanol fractions utilizing different chromatographic techniques resulted in the isolation of known diterpenoids namely: 13-epicupressic acid (1), imbricatolic acid (2), 7α-hydroxysandaracopimaric acid (3), 3β-hydroxysandaracopimaric acid (4), isopimaric acid (5), four flavonoid derivatives: cupressuflavone (6), hinokiflavone (7), hypolaetin-7-*O*-β-xylopyranoside (9), (−) catechin (10), inaddition to sucrose (8). Both the physical and spectral data were used for structure determination and all the isolates were evaluated for their hepatoprotective activity. Compounds 2 and 6 were effective; however, 7 was the most active. Hepatoprotective activity of 7 is comparable with the standard drug SM in reducing the elevated liver enzymes and restoring the normal appearance of hepatocytes. Hepatoprotective effect of combination of 6, 7, and SM with the diterpene sugiol was also explored.

TOTAL PHENOLICS FROM *Laggera pterodonta*

Laggera pterodonta as a folk medicine has been widely used for several centuries to ameliorate some inflammatory ailments as hepatitis in China. The hepatoprotective effect of total phenolics from *L. pterodonta* (TPLP) against CCl_4-, D-GalN-, TAA-, and *t*-BHP-induced injury was examined in primary cultured neonatal rat hepatocytes by Wu et al. (2007). TPLP inhibited the cellular leakage of two enzymes, hepatocyte AST and ALT, caused by these chemicals and improved cell viability. Moreover, TPLP afforded much stronger protection than the reference drug silibinin. Meanwhile, DPPH and superoxide radicals scavenging activities of TPLP were also determined. The total phenolic content of *L. pterodonta* and its main component type were quantified, and its principle components isochlorogenic acids were isolated and authenticated.

APPLE (Malus pumila) POLYPHENOLS

The hepatoprotective effects of apple polyphenols (AP, Appjfnol) against CCl_4-induced acute liver damage in Kunming mice as well as the possible mechanisms were investigated by Yang et al. (2010). Mice were treated with AP (200, 400, and 800 mg/kg, ig) for seven consecutive days prior to the administration of CCl_4. AP significantly prevented the increase in serum ALT and AST levels in acute liver injury induced by CCl_4 and produced a marked amelioration in the histopathological hepatic lesions coupled to weight loss. The extent of MDA formation was reduced; the SOD activity was enhanced, and the GSH concentration was increased in the hepatic homogenate in AP-treated groups compared with the CCl_4-intoxicated group. AP also exhibited antioxidant effects on FeSO(4)-L-Cys-induced LPO in rat liver homogenate and DPPH free-radical scavenging activity *in vitro*.

Manihot esculenta CONSTITUENTS

The expression levels of phase I (CYP1A1 and CYP1A2) and phase II (GST and UGT) enzyme-coded genes were measured in liver microsomes of 50 Wistar rats fed sweet cassava polysaccharides (SCP), an arabinogalactan type root mucilage (Charles and Huaang, 2009). The antioxidant properties of the SCP were investigated *in vitro* screened and investigated for its hepatoprotective activity in rat. There was significant induction of GSTYa1 and inhibition of CYP1A2. Moreover, an SCP diet was found to significantly increase UGT1A6 mRNA levels and to decrease CYP1A1 mRNA levels in chemically injured rat liver. SCP ethanol extracts exhibited hydroxyl radical and superoxide scavenging activities in a dose-dependent manner. *In vitro* and intracellular antioxidative enzyme activity assays demonstrated and confirmed the potential of cassava root extracts as a natural source of mucopolysaccharide substances with potential use in chemoprevention medicine.

PHYTOCONSTITUENTS FROM *Melaleuca styphelioides*

Al-Sayed and Esmat (2016) determined the *in vitro* hepatoprotective activity of the isolated pure compounds from *Melaleuca styphelioides* leaves using the CCl_4-challenged HepG2 cell model. Some compounds showed marked hepatoprotection, including tellimagrandin I, which produced 42%, 36%, and 31% decrease in ALT and 47%, 43%, and 37% decrease in AST, at the tested concentrations, respectively, pedunculagin (32%, 32%, and 30% decrease for ALT and 48%, 48%, and 45% for AST), tellimagrandin II (38%, 32%, and 26% decrease for ALT and 45%, 40% and 34% for AST) and pentagalloyl glucose (30%, 28%, and 26% decrease for ALT and 45%, 38%, and 36% for AST). Tellimagrandin I and II showed the highest increase in GSH (113%, 105%, and 81% and 110%, 103%, and 79%, respectively), which was comparable to Sil. Pedunculagin produced the highest increase in SOD (497%, 350%, and 258%).

PRENYLFLAVONOIDS, COUMARIN, AND STILBENE FROM *Morus alba*

Chemical investigation of the EtOH extract of *Morus alba* L., as guided by free-radical scavenging activity, furnished 5,7-dihydroxycoumarin 7-methyl ether (1), two prenylflavones, cudraflavone B (2) and cudraflavone C (3), and oxyresveratrol (4). Compounds 1 and 4 showed superoxide scavenging effects with the IC_{50} values of 19.1 ± 3.6 and 3.81 ± 0.5 μM, respectively. Compound 4 exhibited a DPPH free-radical scavenging effect (IC_{50} = 23.4 ± 1.5 μM). Compounds 2 and 4 showed hepatoprotective effects with EC_{50} values of 10.3 ± 0.42 and 32.3 ± 2.62 μM, respectively, on tacrine-induced cytotoxicity in human liver-derived Hep G2 cells (Oh et al., 2002).

ALKALOID FRACTIONS FROM THE ETHANOLIC EXTRACT OF *Murraya koenigii*

Sangale and Patil (2017) evaluated the protective effect of alkaloid fractions from the ethanolic extract of *Murraya koenigii* leaves (AMK) in acute experimental liver injury induced by CCl_4 and paracetamol. AMK (30 and 100 mg/kg, p.o.) significantly reduced the serum levels of AST, ALT, and ALP suggesting a hepatoprotective effect and also prevented the increases in liver weight compared to hepatotoxin-treated controls. AMK significantly increased the levels of SOD, CAT and reduced GSH, while levels of LPO were significantly reduced. Histological examination of liver tissue supported the idea that AMK was acting via hepatoprotective mechanism.

Myrtus communis ESSENTIAL OIL

Hsouna et al. (2019) evaluated the protective effect of *Myrtus communis* essential oil (*Mc*EO) on CCl_4-INH in rat. The major components of *Mc*EO are α-pinene (35.20%), 1,8-cineole (17%), linalool (6.17%), and limonene (8.94%) which accounted for 67.31% of the whole oil. The antioxidant activity of *Mc*EO was evaluated using DPPH scavenging ability, β-carotene bleaching inhibition, and hydroxyl radical-scavenging activity. Moreover, the effect of *Mc*EO (250 mg/kg BW) administrated for 14 consecutive days was evaluated in Wistar rat. Administration of a single dose of CCl_4 caused hepatotoxicity as monitored by an increase in LPO (TBARS) as well in PCO level but decreased in antioxidant markers in the liver tissue. The *Mc*EO pretreatment significantly prevented the increased plasma levels of hepatic markers and lipid levels induced by CCl_4 in rats. Furthermore, this fraction improved biochemical and histological parameters as compared to CCl_4-treated group.

TRITERPENE SAPONINS FROM *Panax vietnamensis*

The methanol extract of Vietnamese ginseng (*Panax vietnamensis*) was found to possess hepatocytoprotective effects on D-GalN/TNF-α (TNF-alpha)-induced cell death in primary cultured mouse hepatocytes. Further chemical investigation of the extract afforded two new dammarane-type triterpene saponins, ginsenoside Rh(5) (1) and vina-ginsenoside R(25) (2), as well as 8 known dammarane-type triterpene saponins, majonoside R(2) (3), pseudoginsenoside RT(4) (4), vina-ginsenosides R(1) (5), R(2) (6), and R(10) (7), ginsenosides Rg(1) (8), Rh(1) (9), and Rh(4) (10), and a known sapogenin protopanaxatriol oxide II (11). Among the compounds isolated, majonoside R(2) (3), the main saponin in Vietnamese ginseng, showed strong protective activity against D-GalN/TNF-alpha-induced cell death in primary cultured mouse hepatocytes. This demonstrates that the hepatocytoprotective effect of Vietnamese ginseng is due to dammarane-type triterpene saponins that have an ocotillol-type side chain, a characteristic constituent of Vietnamese ginseng (Tran et al., 2001).

CAFFEIC ACID AND ROSMARINIC ACID, MAJOR COMPOUNDS OF *Perilla frutescens*

Perilla frutescens leaves are often used in East Asian gourmet food. Yang et al. (2013) investigated the hepatoprotective effects of CA, rosmarinic acid (RA), and their combination. *P. frutescens* contains

1.32 μg CA/mg dry material (DM) and 26.84 μg RA/mg DM analyzed by HPLC-DAD and HPLC-MS. CA remarkably reduced the oxidative damage than RA in an *in vitro* study. Oral intubation with CA or RA alone for 5 days was conducted prior to treatment with a single dose of t-BHP (0.5 mmol/kg BW, i.p.), which led to a significant reduction of indicators of hepatic toxicity, such as AST, ALT, oxidized GSH, LPO and enzyme activities related to antioxidant such as CAT, GPx, and SOD. Interestingly, compared to treatment with CA or RA alone, a combination of both compounds more increased the endogenous antioxidant enzymes and GSH and decreased LPO in livers.

COMPOUND FROM *Phyllanthus niruri*

Among phyllanthin, hypophyllanthin, triacontanal, and tricontanol isolated from a hexane extract of *Phyllanthus niruri*, phyllanthin, and hypophyllanthin protected against CCl_4- and GalN-induced cytotoxicity in primary cultured rat hepatocytes, while triacontanal was protective only against GalN-induced toxicity (Syamasundar et al., 1985).

PICROSIDES FROM *Picrorhiza kurroa*

Picrorhiza kurroa (Pk), a known hepatoprotective plant, was studied by Vaidya et al. (1996) in experimental and clinical situations. The standardization of active principles–Picroside 1 and 2 was done with high-performance liquid chromatography. Picroside 1 ranged 2.72–2.88 mg/capsule and picroside 2 5.50–6.00 mg/capsule. In the GalN-induced liver injury in rats, Pk at a dose of 200 mg/kg, p.o. showed a significant reduction in liver lipid content, GOT and GPT. In a randomized, double-blind placebo controlled trial in patients diagnosed to have acute viral hepatitis (HBsAg negative), Pk root powder 375 mg three times a day was given for 2 weeks (n = 15) or a matching placebo (n = 18) was given. Difference in values of bilirubin, SGOT, and SGPT was significant between placebo and Pk groups. The time in days required for total serum bilirubin to drop to average value of 2.5 mg% was 75.9 days in placebo as against 27.44 days in Pk group. The study has shown a biological plausability of efficacy of Pk as supported by clinical trial in viral hepatitis, hepatoprotection in animal model and an approach for standardizing extracts based on picroside content.

POLYSACCHARIDES FROM *Pinus massoniana* POLLEN

Zhou et al. (2018) characterized the chemical composition, antioxidant activity, and hepatoprotective effect of the polysaccharides from Taishan *Pinus massoniana* pollen (TPPPS). In *in vitro* assays, TPPPS exhibited different degrees of dose-dependent antioxidant activities, and this was further verified by suppression of CCl_4-induced oxidative stress in the liver with three tested doses of TPPPS (100, 200, and 400 mg/kg BW) in rats. Pretreatment with TPPPS significantly decreased the levels of AST, ALT, ALP, LDH, and MDA against CCl_4 injuries, and elevated the activities of SOD as well as GPx. Histopathological observation further confirmed that TPPPS could protect the liver tissues from CCl_4-induced histological alternation. These results suggest that TPPPS has strong antioxidant activities and significant protective effect against acute hepatotoxicity induced by CCl_4. The hepatoprotective effect may partly be related to its free-radical scavenging effect, increasing antioxidant activity, and inhibiting LPO.

COMPOUNDS FROM *Pistacia integerrima*

In vitro antioxidant and hepatoprotective activity of isolated compounds from *Pistacia integerrima* was investigated by Joshi and Mishra (2010). The compounds isolated from the ethyl acetate fraction of methanol extract of *P. integerrima* were subjected to determination of antioxidant activity by DPPH free-radical activity, reducing power assay, scavenging of hydroxyl radicals, etc. There is a close relationship between antioxidant and hepatoprotective activity and, therefore, the isolated compounds were subjected to *in vitro* hepatoprotective studies using paracetamol INH in primary rat hepatocytes. *In vitro* hepato protective activity was assessed by determining the change in hepatocytes viability and other parameters like glutamic transaminase, GPT, and total protein. The fractions showed significant protective effect by restoring altered parameters in the selected *in vitro* model. The fractions isolated for ethyl acetate fraction of methanol extract of *P. integerrima* showed presence of phenolics and flavonoids, which are potent antioxidants.

GLUCOSYLOXYBENZYL SUCCINATE DERIVATIVES FROM THE *Pleione bulbocodioides*

About ten previously undescribed glucosyloxybenzyl succinate derivatives, pleionosides A-J, and 15 known compounds were isolated from the pseudobulbs of *Pleione bulbocodioides* (Franch.) Rolfe. Furthermore, three compounds exhibited potent hepatoprotective activity against *N*-acetyl-*p*-aminophenol (APAP)-induced HepG2 cell damage in *in vitro* assays, with cell survival rates of 31.89%, 31.52%, and 31.97% at 10 μM. About four compounds exhibited moderate antioxidant activity with increasing viability at 10 μM of 36.1%, 45.0%, 25.5%, and 20.7%.

FLAVONOIDS FROM Prosthechea michuacana

Gutierrez et al. (2011) investigated the effect of flavonoids isolated from methanol extract of *Prosthechea michuacana* on CCl_4-induced liver damage in mice. From the bulbs of *P. michuacana,* four known flavonoids were isolated (scutellarein 6-methyl ether, DHQ, apigenin 7-*O*-glucoside, and apigenin-7-neohesperidoside), together with the new flavonol glycoside apigenin-6-*O*-β-D-glucopyranosil-3-*O*-α-L-rhamnopyranoside. Treatment with flavonoids significantly prevented the biochemical measurable changes induced by CCl_4 in the liver. Compounds 1, 4, and 5 were found to exhibit good hepatoprotective effect. These effects were comparable to that of the standard drug SM, a well-known hepatoprotective agent. These results demonstrate that flavonoids contained in the bulbs of *P. michuacana* contribute to the hepatoprotective activity attributed to the plant.

TWO TRITERPENES FROM Protium heptaphyllum **(Aubl.) March**

Two triterpenes α- and β-amyrin isolated from *Protium heptaphyllum* were tested against APAP-induced liver injury in mice. Liver injury was analyzed by quantifying the serum enzyme activities and by histopathological observation. Pretreatment with α- and β-amyrin attenuated the acetaminophen-induced acute increase in serum ALT and AST activities replenished the depleted hepatic GSH, and considerably reduced the histopathological alterations. These findings demonstrated the hepatoprotective potential of α- and β-amyrin against toxic liver injury and suggest that the diminution in oxidative stress and toxic metabolite formation as likely mechanisms involved in its hepatoprotection (Oliveira et al., 2005).

PHENOLIC COMPOUNDS FROM Rhodiola sachalinensis

In total, two hepatoprotective phenolic compounds, kaempferol and salidroside, were isolated by Song et al. (2003) from the roots of *Rhodiola sachalinensis* together with two inactive compounds cinnamyl alcohol and daucosterol based on the hepatoprotective activity against tacrine-induced cytotoxicity in human liver-derived Hep G2 cells. The EC_{50} values of kaempferol and salidroside were 33.5 and 51.3 μm, respectively. Silybin as a positive control showed an EC_{50} value of 68.4 μm.

RICININE AND N-DEMETHYL-RICININE FROM *Ricinus communis*

Ricinus communis (leaf extract) was evaluated by Visen et al. (2008) for hepatoprotective, choleretic, and anticholestatic activity. In a preliminary test with albino rats, an ethanol extract showed significant protection against GalN-induced hepatic damage. It also showed dose-dependent choleretic and anticholestatic activity, and hepatoprotective activity as judged by hepatocytes isolated from paracetamol-treated rats. On fractionation of the ethanol extract, maximum activity was localized in the butanol fraction. Subsequent chromatographic fractionation and testing in the GalN model led to the isolation of two active fractions which, in turn, yielded two pure compounds: ricinine and *N*-demethyl-ricinine. *N*-Demethyl-ricinine was found to be more active and it reversed the biochemical changes produced by GalN at a dose of 6 mg/kg × 7 days. It possessed marked choleretic activity and demonstrated an anticholestatic effect against paracetamol-induced cholestasis.

ALKALOID EXTRACT OF THE ROOTS OF *Rubus alceifolius* Poir.

Lin et al. (2011) examined the effect of an alkaloid extract of the roots of *Rubus alceifolius* on liver damage and cytochrome enzymes, and underlying mechanism. Hepatotoxicity was induced in rats by treatment with CCl_4. Rats were then treated with the hepatoprotective drug bifendate, or with low, medium, and high doses of an alkaloid extract from the roots of *R. alceifolius*. Both bifendate and alkaloid treatment decreased the increase in liver enzymes and cell damage caused by CCl_4. CCl_4 treatment alone caused a decrease in total CYP content, an increase in CYP2E1 and CYP3A1 mRNA levels, and an increase in CYP2E1 and a decrease in CYP3A1 enzymatic activity. Alkaloid treatment brought these concentrations and activities back toward normal.

CONSTITUENTS OF *Sambucus formosana* Nakai

A new triterpene ester, named sambuculin A, β-amyrin, and OA were isolated from *Sambucus formosana* by Lin and Tome (1988) and their structures elucidated by physical and chemical methods. Sambuculin A and mixture of α-amyrin and β-amyrin palmitate exhibited strong activity against liver damage induced by CCl_4. Benzoylation of β-sitosterol and the mixture of α-amyrin enhanced the protective activity against CCl_4-produced liver damage.

CONSTITUENTS OF *Scutellaria rivularis* Benth. *FAMILY: LAMIACEAE*

The hepatoprotective effect of various fractions (*n*-hexane, $CHCl_3$,

EtOAc, *n*-BuOH, and H_2O) of Ban-zhi-lian derived from *Scutellaria rivularis* was studied by Lin et al. (1997) CCl_4, D-GalN and APAP-induced acute hepatotoxicity in rats. Liver damage was assessed by quantifying serum activities of SGOT and SGPT, as well as by histopathological examination. The results indicated that the $CHCl_3$ fraction and EtOAc fractions exhibited the greatest hepatoprotective effects on CCl_4-induced liver injuries, the $CHCl_3$ fraction and *n*-hexane fraction are most potent against D-GalN-induced intoxication, and the $CHCl_3$ fraction represented the most liver-protective effect on APAP-induced hepatotoxicity. The pathological changes of hepatic lesions caused by these three hepatotoxicants were improved by treatment with the fractions mentioned above, which were compared to GLZ and SM as standard reference medicines.

Baicalein, Baicalin, and Wogonin are three major components isolated from the entire plant of *S. rivularis.* Wogonin (5 mg/kg i.p.), exhibit best effect in CCl_4 and D-GalN-treated rats. Baicalein and Baicalin at the dose 20 mg/kg i.p. in D-GalN and APAP; at dose 10 mg/kg i.p. in CCl_4-treated rats exhibit best effect. Protective effects were seen by comparing the serum GOT, GPT, and histopathologic examination (hepatic lesions) (Lin and Shieh, 1996).

SESAME OIL

Azab (2014) investigated the possible hepatoprotective role of sesame oil against lead acetate INH in albino mice from the histological and biochemical aspects. In lead-treated animals, there were severe structural damages in the liver. The hepatocytes appeared irregularly arranged with disorganization of hepatic architecture. The hepatocytes appeared large with light and foamy cytoplasm filled with numerous vacuole-like spaces. The nuclei appeared with pyknotic nuclei. The central vein appeared dilated and congested with massive hemorrhage extending to the nearby cells. Also, there were focal degenerative and necrotic changes along with inflammatory cell infiltration. Decrease in BW and increase in liver weight were observed. Biochemically, the serum ALT, AST, ALP, and γ-glutamyl transferase activities were increased and serum total proteins and albumin were decreased. Co-administration of sesame oil significantly improved the structural changes in the liver and also the serum ALT, AST, ALP, and γ-glutamyl transferase activities were significantly declined and serum total proteins and albumin were elevated.

Sophora tonkinensis *POLYSACCHARIDES*

Cai et al. (2018) isolated two polysaccharides, STRP1 and STRP2, purified from *Sophorae tonkinensis* Radix via column chromatography. Structural analyses indicated that STRP1 and STRP2 were consisted of mannose, rhamnose, glucuronic acid, GLU, galactose, and arabinose in a similar molar ratio with main backbones of (1 → 3)-linked-α-D-Gal and (1 → 4)-linked-α-D-Glc, while average molecular weights were 1.30×10^4 and 1.98×10^5 Da, respectively. The authors observed a strong chelating ability on ferrous ions; substantial radical scavenging activities on DPPH, hydroxyl, and superoxide anion radicals *in vitro*; and significant attenuation on APAP-induced hepatic oxidative damage in mice for STRP1 and STRP2.

COMPOUNDS FROM *Suaeda glauca*

Bioassay-guided fractionation of the MeOH extract of *Suaeda glauca* yielded four phenolic compounds, methyl 3,5-di-*O*-caffeoyl quinate (1) and 3,5-di-*O*-caffeoyl quinic acid (2), isorhamnetin 3-*O*-beta-D-galactoside (3), and quercetin 3-*O*-beta-D-galactoside (4). Compounds 1 and 2 were hepatoprotective against tacrine-induced cytotoxicity in human liver-derived Hep G2 cells with the EC_{50} values of 72.7 ± 6.2 and 117.2 ± 10.5 μM, respectively. Silybin, as a positive control showed an EC_{50} value of 82.4 ± 4.1 μM.

Several chemically defined molecules have been isolated from crude plant extracts with proven hepatoprotective activity.

New skeletal flavonoids, anastatins A and B, were isolated from the methanol extract of *A. hierochuntica* L. Anastatins A and B were found to show hepatoprotective effects on D-GalN-induced cytotoxicity in primary cultured mouse hepatocytes and their activities were stronger than those of related flavonoids and commercial silybin—a known hepatoprotective compound (Yoshikawa et al., 2003).

TOTAL SAPONINS OF *Taraphochlamys affinis*

Hepatoprotective effects of TS isolated from *Taraphochlamys affinis* (TSTA) against CCl_4-induced liver injury in rats were investigated by Huang et al. (2012). In this study, rats were orally treated with the TS of TSTA daily with administration of CCl_4 twice a week for 8 weeks. Compared to the normal control, CCl_4-induced liver damage significantly increased the activities of ALT, AST, and ALP in serum and decreased the activities of SOD, CAT, GPx, and GSH reductase (GSH-Rd) in liver. Meanwhile, content of hepatic MDA, which was an oxidative stress marker, was increased. Histological finding also confirmed the hepatotoxic characterization in rats. Furthermore, proinflammatory mediators including TNF-α in serum, prostaglandin E2 (PGE2), iNOS, and COX-2 in liver were detected with elevated contents, while expression of xenobiotic metabolizing enzyme—CYP2E1 was inhibited. The results revealed that TSTA not only significantly reversed CCl_4 originated changes in serum toxicity and hepatotoxic characterization, but also altered expression of hepatic oxidative stress markers and proinflammatory mediators, combined with restoring liver CYP2E1 level. The results indicated that protective effect of TSTA against CCl_4-induced hepatic injury may rely on its effect on reducing oxidative stress, suppressing inflammatory responses, and improving drug-metabolizing enzyme activity in liver.

POLYPRENOLS FROM Taxus chinensis var. *mairei*

Yu et al. (2012) investigated the antifibrotic effects and the possible underlying mechanisms of taxus polyprenols (TPs) isolated from the needles of *Taxus chinensis* var. *mairei*. TPs successfully attenuated liver injury induced by CCl_4 shown by histopathological sections of livers and improved liver function as indicated by decreased ALT, AST, and ALP levels and increased ALB levels in serum of the rats. TPs significantly increased the hepatic Cu/Zn SOD and GSH-Px activities along with GSH content while a remarkable decrease in MDA content. Both immunohistochemical staining and mRNA expression levels of α-SMA indicated a profound suppression of HSCs activation. Furthermore, it significantly inhibited the mRNA expression of the profibrotic cytokines Col α1(I), Col α1(III), MMP-2, TIMP-1, TIMP-2, PDGF-β, TGF-β1, CTGF, and TNF-α and restored the hepatoprotective factor HGF.

COMPOUNDS FROM Tecomella undulata

Jain et al. (2012) evaluated the possible hepatoprotective role of isolated compounds from *Tecomella undulata* stem bark (TSB) using *in vitro* and *in vivo* experimental models. *In vitro* cytotoxicity and hepatoprotective potential of various extracts, fractions, and isolated compounds from TU stem bark were evaluated using HepG2 cells. Rats were pretreated with TU methanolic extract (TSB-7) or betulinic acid (MS-2) or SM for 7 days followed by a single dose of CCl_4 (0.5 mL/kg, i.p.). Plasma markers of hepatic damage, hepatic antioxidants, and indices of LPO along with microscopic evaluation of liver were assessed in control and treatment groups. TSB-2 and MS-1 accounted for significant cell death whereas; TSB-1, TBS-7, TSB-9, TSB-10, and

MS-2 did not register significant cytotoxicity. Further, noncytotoxic components exhibited ascending grade of hepatoprotection *in vitro* (TSB-10<TSB-1<TSB-7<TSB-9<MS-2). Pretreatment of TSB-7 or MS-2 to CCl_4-treated rats prevented hepatocyte damage as evidenced by biochemical and histopathological observations. It can be concluded that hepatoprotective potential of TU stem bark is partially due to the presence of betulinic acid.

ESSENTIAL OILS FROM *Thymus capitatus* AND *Salvia officinalis* FAMILY: LAMIACEAE

El-Banna et al. (2013) investigated the possible potential protective effect of EOs of *Thymus capitatus* and *Salvia officinalis* against paracetamol-induced hepatotoxicity. Administration of paracetamol (500 mg/kg b.w.) resulted in liver damage as manifested by significant increase in serum and hepatic LDH activity with a significant decrease in blood and hepatic GSH levels, as well as blood and hepatic SOD, and GPx activities. Rats pretreated orally with EO of *T. capitatus* and *S. officinalis* (50 mg/kg b.w. daily) for 15 days then intoxicated with paracetamol showed a significant protection against-induced increase in serum and hepatic LDH activities and inhibit reduce GSH levels and enhance increase SOD and GPx activities in blood and liver.

Thymus vulgaris ESSENTIAL OIL

Grespan et al. (2014) investigated the hepatoprotective effect of *Thymus vulgaris* essential oil (TEO). TEO reduced the levels of the serum marker enzymes AST, ALT, ALP, and MPO activity. The histopathological analysis indicated that TEO prevented APAP-induced necrosis. The EO also exhibited antioxidant activity, reflected by its DPPH radical-scavenging effects and in the LPO assay.

IRIDOID GLYCOSIDES FRACTION FROM *Veronica ciliata* Fisch.

Tan et al. (2017) analyzed the chemical composition of the iridoid glycosides fraction (IGF) isolated from *Veronica ciliata* and evaluated the antioxidant and hepatoprotective properties. Tan et al. (2017) determined the *in vitro* antioxidant ability of the IGF through radical scavenging assays and assessed the *in vivo* hepatoprotective potential in an APAP-induced acute liver injury murine model. The IGF was separated by HSCCC and three major iridoid glycosides (verproside, catalposide, and amphicoside) were identified as potent antioxidants and hepatoprotective compounds. Treatment with the IGF significantly suppressed the APAP-induced elevation in serum ALT, AST, and TNF-α; improved serum total antioxidant capacity; decreased MDA formation; elevated SOD and GSH activity; and decreased expression of proinflammatory factors (TNF-α, nuclear factor kappa B) in the liver. Authors examined the histopathology of resected livers for the evidence of hepatoprotection. The protection conferred by the IGF may be related to the reinforcement of antioxidant defense systems.

COMPOUNDS FROM *Viburnum tinus* L. FAMILY: ADOXACEAE

Mohamed et al. (2005) studied the hepatoprotective activity of leaf extract of *Viburnum tinus*. From the leaves of *V. tinus,* two acylated iridoid glucosides (viburtinosides A and B), a coumarin diglucoside scopoletin 7-*O*-beta-D-sophoroside and a natural occurred dinicotinic acid ester 2,6-di-C-methyl-nicotinic acid 3,5-diethyl ester were isolated. In addition to these, 10 known compounds were isolated, namely two bidesmosidic saponins, a hexamethoxy-flavone, and five flavonol glycosides, as well as suspensolide A and OA were isolated for the first time in this genus and species, respectively. Toxicity of the investigated extract was determined (LD50 = 500 mg/kg). CCl_4-INH has been evaluated in terms of the determination of ALT, AST, lipid peroxide, and NO levels in serum and compared using adult male rats. Their highly elevated levels were significantly reduced by treatment with the investigated aqueous methanol extract in a dose-dependent manner.

KAURENE GLYCOSIDES FROM *Xanthium strumarium* L.

Wang et al. (2011) evaluated the toxic effects on acute liver injury in mice of the two kaurene glycosides [atractyloside (ATR) and carboxyatractyloside (CATR)], which are main toxic constituents isolated from Fructus Xanthii on acute liver injury in mice. Histopathological examinations revealed that there were not obviously visible injury in lungs, heart, spleen, and the central nervous system in the mice by i.p. injection of ATR, at the doses 50, 125, and 200 mg/kg and CATR, at the doses 50, 100, and 150 mg/kg for 5 days. However, it revealed extensive liver injuries compared with the normal group. In the determination of enzyme levels in serum, i.p. injection of ATR and CATR resulted in significantly elevated serum ALT, AST, and ALP activities compared to controls. In the hepatic oxidative stress level, antioxidant-related enzyme activity assays showed that ATR and CATR administration significantly increased hepatic MDA concentration, as well as decreased SOD, CAT activities and GSH concentration, and this was in good agreement with the results of serum aminotransferase activity and histopathological examinations.

POLYSACCHARIDES EXTRACTED FROM Ziziphus jujuba cv. Huanghetanzao

Liu et al. (2015) evaluated the chemical composition and hepatoprotective function of the polysaccharides extracted from *Ziziphus jujuba cv. Huanghetanzao* (HJP). The composition of HJP was determined as heteropolysaccharides with galactose and arabinose being the main components. The pretreatment of mice with HJP significantly reduced the activities of serum hepatic AST, ALT, and LDH induced by CCl_4 or APAP, while the commercial liver-injury treatment drug silybin did not show any prevention effects. The mechanistic study results indicate that the administration of the CCl_4- or APAP-injured mice with HJP-enhanced SOD and GSH-Px and decreased MDA, indicating that antioxidation and detoxification could be the pathways for the liver protection observed. In addition, the liver prevention and treatment effects of HJP on the liver damage induced by CCl_4 or APAP obtained from the liver enzyme analyses were confirmed by the hepatic histopathology studies in mice. Therefore, HJP could be used as a prevention and treatment agent for liver injury induced by liver toxic chemicals and drugs.

KEYWORDS

- **phytoconstituents**
- **chemical constituents**
- **hepatoprotective activity**
- **hepatoprotective compounds**
- **hepatoprotective chemicals**

REFERENCES

Adaramoye, O.A.; Adeyemi, E.O. Hepatoprotection on D-galactosamine-induced toxicity in mice by purified fractions from *Garcinia kola* seeds. *Basic Clin. Pharmacol. Toxicol.*, **2006**, *98*(2), 135–141.

Abenavoli, L.; Capasso, R.; Milic, N.; Capasso, F. Milk thistle in liver disease: Past, present, future. *Phytother. Res.*, **2010**, *24*, 423–432.

Adaramoye, O.A.; Awogbindin, I.; Okusaga, J.O. Effect of kolaviron, a biflavonoid complex from *Garcinia kola* seeds, on ethanol-induced oxidative stress in liver of adult Wistar rats. *J. Med. Food,* **2009**, *12*(3), 584–590.

Adewusi, E.A.; Afolayan, A.J. A review of natural products with hepatoprotective activity. *J. Med. Plants Res.,* **2010**, *4*, 1318–1334.

Adnyana, K.; Tezuka, Y.; Awale, S.; Banskota, A.H.; Kim, Q.T.; Kadota, S. 1-0-galloyl-6-0-(4-hydroxy-3, 5-dimethoxy) benzoyl-β-dglucose, a new hepatoprotective constituent from *Combretum quadrangulare. Planta Med.,* **2001**, *67*(4), 370–371.

Adzet, T.; Camarasa, J.; Laguna, J.C. Hepatoprotective activity of polyphenolic compounds from *Cynara scolymus* against CCl_4 toxicity in isolated rat hepatocytes. *J. Nat. Prod.,* **1987**, *50*, 612–617.

Ahmad, S.T.; Arjumand, W.; Nafees, S.; Seth, A.; Ali, N.; Rashid, S.; Sultana, S. Hesperidin alleviates acetaminophen induced toxicity in Wistar rats by abrogation of oxidative stress, apoptosis and inflammation. *Toxicol. Lett.,* **2012**, *208*, 149–161.

Aimin, H.; Mingshi, W.; Hong, Y.H.; Decheng, Z.; Lee, K.H. Hepatoprotective triterpenes from *Sedum sarmentosum*. *Phytochemistry*, **1998**, *49*(8), 2607–2610.

Akachi, T.; Shiina, Y.; Ohishi, Y.; Kawaguchi, T.; Kawagishi, H.; Morita, T.; et al. Hepatoprotective effects of flavonoids from shekwasha (*Citrus depressa*) against d-galactosamine-induced liver injury in rats. *J. Nutr. Sci. Vitaminol.* (Tokyo), **2010**, *56*(1), 60–67.

Alqasoumi, S.I.; Farraj, A.I.; Abdel-Kader, M.S. Study of the hepatoprotective effect of *Juniperus phoenicea* constituents. *Pak. J. Pharm. Sci.,* **2013**, *26*(5), 999-1008.

Al-Sayed, E.; Esmat, A. Hepatoprotective and antioxidant effect of ellagitannins and galloyl esters isolated from *Melaleuca styphelioides* on carbon tetrachloride-induced hepatotoxicity in HepG2 cells. *Pharm. Biol.*, **2016**, *54*(9), 1727–1735.

Al-Snafi, A.E. Chemical constituents and pharmacological effects of *Astragalus hamosus* and *Astragalus tribuloides* grown in Iraq. *Asian J. Pharm. Sci. Tech.,* **2015**, *5*(4), 321–328.

An, R.B.; Sohn, D.H.; Jeong, G.S.; Kim, Y.C. In vitro hepatoprotective compounds from *Suaeda glauca*. *Arch. Pharm. Res.,* **2008**, *31*(5), 594–597.

Anand, K.K.; Gupta, V.N.; Rangari, V.; Singh, B.; Chandan, B.K. Structure and hepatoprotective activity of a biflavonoid from *Canarium manii*. *Planta Med.,* **1992**, *58*(6), 493–495.

Anand, K.K.; Singh, B.; Saxena, A.K.; Chandan, B.K.; Gupta, V.N.; Bhardwaj, V. 3, 4, 5-Trihydroxybenzoic acid (gallic acid), the hepatoprotective principle in the fruits of *Termmalia belerica*—bioassay guided activity. *Pharmacol. Res.,* **1997**, *36*, 315–321.

Anand, K.V.; Anandhi, R.; Pakkiyaraj, M.; Geraldine, P. Protective effect of chrysin on carbon tetrachloride (CCl_4)-induced tissue injury in male Wistar rats. *Toxicol. Ind. Health,* **2011**, *27*, 923–933.

Azab, A.E. Hepatoprotective effect of sesame oil against lead induced liver damage in albino mice: Histological and biochemical studies. *Am. J. BioSci.,* **2014**, *2*(6–2), 1–11.

Baek, N.L.; Kim, Y.S.; Kyung, J.S.; Park, K.H. Isolation of anti-hepatotoxic agent from the roots of *Astragalus membranaceous*. *Korean J. Pharmacogn.,* **1996**, *27*, 111–116.

Barbarino, F.; Neumann, E.; Deaciuc, I.; Ghelberg, N.W.; Suciu, A.; Cotuiu, C.; et al. Effect of silymarin on experimental liver lesions. *Med. Interne.,* **1981**, *19*, 347–357.

Bhadauria, M.; Nirala, S.K.; Shrivastava, S.; Sharma, A.; Johri, S.; Chandan, B. K.; Singh, B.; Saxena, A.K.; Shukla, S. Emodin reverses CCl_4 induced hepatic cytochrome P450 (CYP) enzymatic and ultra structural changes: The in vivo evidence. *Hepatol. Res.,* **2009**, *39*(3), 290–300.

Bishayee, A.; Darvesh, A.S.; Politis, T.; McGory, R. Resveratrol and liver disease: From bench to bedside and community. *Liver Int.,* **2010**, *30*, 1103–1114.

Boigk, G.; Stroedter, L.; Herbst, H.; Waldschmidt, J.; Riecken, E.O.; Schuppan, D. Silymarin retards collagen accumulation in early and advanced biliary fibrosis secondary to complete bile duct obliteration in rats. *J. Hepatol.,* **1987,** *26*, 643–649.

Breikaa, R.M.; Algandaby, M.M.; El-Demerdash, E.; Abdel-Naim, A.B. Biochanin A protects against acute carbon tetrachloride-induced hepatotoxicity in rats. *Biosci. Biotechnol. Biochem.,* **2013**, *77*(5), 909–916.

Budhiraja, R.D.; Garg, K.N.; Sudhir, S.; Arora, B. Protective effect of 3-hydroxy, 2,3-dihydrowithanolide F against CCl_4-inducd hepatotoxicity. *Planta Med.*, **1986**, *52*, 28–29.

Cai, L.L.; Zou, S.S.; Liang, D.P.; Luan, L.B. Structural characterization, antioxidant and hepatoprotective activities of polysaccharides from *Sophorae tonkinensis* Radix. *Carbohyd. Polym.*, **2018**, *184*, 354–365.

Cai, Y.; Sun, W.; Zhang, X.-X.; Lin, Y.-D.; Chen, H.; Li, H. Osthole prevents acetaminophen-induced liver injury in mice. *Acta Pharmacol. Sin.*, **2018**, 39(1), 74–84.

Campos, R.; Garrido, A.; Guerra, R.; Valenzuela, A. Silybin dihemisuccinate protects against glutathione depletion and lipid peroxidation induced by acetaminophen on rat liver. *Planta Med.*, **1989**, *55*, 417–419.

Cerny, D.; Canova, N.K.; Martinek, J.; et al. Effects of resveratrol pretreatment on tert-butylhydroperoxide induced hepatocyte toxicity in immobilized perifused hepatocytes: Involvement of inducible nitric oxide synthase and hemoxygenase-1. *Nitric Oxide-Biol. Ch.*, **2009**, *20*, 1–8.

Çetin, A.; Çiftçi, O.; Otlu, A. Protective effect of hesperidin on oxidative and histological liver damage following carbon tetrachloride administration in Wistar rats. *Arch. Med. Sci.*, **2016**, *12*, 486–493.

Chander, R.; Diwedi, Y.; Rastogi, R.; Sharma, S.K.; Garg, N.K.; Kapoor, N.K.; Dhawan, B.N. Evaluation of hepatoprotective activity of picroliv in *Mastomys natalensis* infected with *Plasmodium berghei*. *Ind. J. Med Res.* [B], **1990**, *92*, 34–37.

Chander, R.; Kapoor, N.K.; Dhawan, B.N. Picrolive, Picroside-1 and Kutkoside from *Picrorrhiza kurroa* an scavenger of super oxide anion. *Biochem. Pharmacol.*, **1992a**, *44*(1), 180–183.

Chander, R.; Kapoor, N.K.; Dhawan, B.N. Effect of picroliv on glutathione metabolism in liver and brain of *Mastomys natalensis* injected with *Plasmodium berghei*. *Indian J. Exp. Biol.*, **1992b**, *30*, 711–714.

Chander, R.; Kapoor, N.K.; Dhawan, B.N. Picroliv affects gamma-glutamyl cycle of liver and brain of *Mastomys natalensis* infected with *Plasmodium berghei*. *Indian J. Exp. Biol.*, **1994**, *32*, 324–327.

Chander, R.; Singh, K.; Visen, P.K.S.; Kapoor, N.K.; Dhawan, B.N. Picroliv prevents oxidation in serum lipoprotein lipids of *Mastomys coucha* infected with *Plasmodium berghei*. *Indian J. Exp. Biol.*, **1998**, *36*, 371–374.

Chander, R.; Srivastava, V.; Tandon, J.S.; Kapoor, N.K. Antihepatotoxic activity of diterpenes of *Andrographis paniculata* (Kalmegh) against *Plasmodium berghei* induced hepatic damage in *Mastomys natalensis*. *Int. J. Pharmacogn.*, **1995**, *33*(2), 135–138.

Chao, W.W.; Lin, B.F. Hepatoprotective diterpenoids isolated from *Andrographis paniculata*. *Chin. Med.*, **2012**, *3*, 136–143.

Charles, A.L.; Huang, T.C. Sweet cassava polysaccharide extracts protects against CCl_4 liver injury in Wistar rats. *Food Hydrocolloids*, **2009,** *23*, 1494–1500.

Chaudhuri, D.; Ghate N. B.; Panja S.; Mandal N. Role of phenolics from *Spondias pinnata* bark in amelioration of iron overload induced hepatic damage in Swiss albino mice. *BMC Pharmacol. Toxicol.*, **2016**, 17(1), Art. no. 34, doi: 10.1186/s40360-016-0077-6

Chen, X.; Huang, J.; Hu, Z.; Zhang, Q.; Li, X.; Huang, D. Protective effects of dihydroquercetin on an APAP-induced acute liver injury mouse model. *Int. J. Clin. Exp. Pathol.*, **2017**, *10*(10), 10223–10232.

Chen, Y.-H.; Lin, F.-Y.; Liu, P.-L.; et al. Antioxidative and hepatoprotective effects of magnolol on acetaminophen-induced liver damage in rats. *Arch. Pharmacol. Res.*, **2009**, 32(2), 221–228.

Cheshchevik, V.; Lapshina, E.; Zabrodskaya, S.; Reiter, R.; Prokopchik, N.; Zavodnik, I. Rat liver mitochondrial damage under acute or chronic carbon tetrachloride-induced intoxication: protection by melatonin and cranberry flavonoids. *Toxicol. Appl. Pharmacol.,* **2012**, 261, 271–279.

Choi, J.H.; Choi, C.Y.; Lee, K.J.; Hwang, Y.P.; Chung, Y.C.; Jeong, H.G. et al. Hepatoprotective effects of an anthocyanin fraction from purple-fleshed sweet potato against acetaminophen-induced liver damage in mice. *J. Med. Food,* **2009**, *12*, 320–326.

Choi, J.H.; Kim, D.W.; Yun, N.; Choi, J.S.; Islam, M.N.; Kim, Y.S.; Lee, S.M. Protective effects of hyperoside against carbon tetrachloride-induced liver damage in mice. *J. Nat. Prod.,* **2011**, *74*, 1055–1060.

Çitoğlu, G. S.; Acıkara, Ö. B.; Yılmaz, B. S.; Özbek, H. Evaluation of analgesic, anti-inflammatory and hepatoprotective effects of lycorine from *Sternbergia fisheriana* (Herbert) Rupr. *Fitoterapia,* **2012**, *83*(1), 81–87.

Clichici, S.; Olteanu, D.; Nagy, A-L.; Oros, A.; Filip, A.; Mircea, P.A. Silymarin inhibits the progression of fibrosis in the early stages of liver injury in CCl_4-treated rats. *J. Med. Food,* **2015**, *18*, 290–298.

Cristani, M.; Speciale, A.; Mancari, F.; Arcoraci, T.; Ferrari, D.; et al. Protective activity of an anthocyanin rich extract from bilberries and blackcurrants on acute acetaminophen-induced hepatotoxicity in rats. *Nat. Prod. Res.,* **2016**, *16*, 2845–2849.

Cui, C.P.; Wei, P.; Liu, Y.; Zhang, D.J.; Wang, L.S.; Wu, C.T. The protective role of hepatopoietin Cn on liver injury induced by carbon tetrachloride in rats. *Hepatol. Res.*, **2009**, *39*(2), 200–206.

Curreli, F.; Friedman, K.; Alvin, E.; Flore, O. Protective mechanism of glycyrrhizin on acute liver injury induced by carbon tetrachloride in mice. *Biol. Pharm. Bull.*, **2007**, *30*(10), 1898–1904.

Daba, M.H.; Abdel-Rahman, M.S. Hepatoprotective activity of Thymoquinone in isolated rat hepatocytes. *Toxicol. Lett.,* **1998,** *95*, 23–29.

Dadkhah, A.; Fatemi, F.; Ababzadeh, S.; Roshanaei, K.; Alipour, M.; Tabrizi, B. S. Potential preventive role of Iranian *Achillea wilhelmsii* C. Koch essential oils in acetaminophen-induced hepatotoxicity. *Bot. Stud.*, **2014**, *55*(1), 1–10.

Das, D.; Pemberton, P.W.; Burrows, P.C.; Gordon, C.; Smith, A.; McMahon, R.F.; Warnes, T.W. Antioxidant properties of colchicines in acute carbon tetrachloride induced rat liver injury and its role in the resolution of established cirrhosis. *Biochim. Biophys. Acta,* **2000**, *1502*, 351–362.

Das, J.; Ghosh, J.; Roy, A.; Sil, P.C. Mangiferin exerts hepatoprotective activity against D-galactosamine induced acute toxicity and oxidative/nitrosative stress via Nrf2-NFkB pathways. *Toxicol. Appl. Pharmacol.,* **2012**, *260*, 35–47.

Das, S.K.; Vasudevan, D.M. Protective effects of silymarin, a milk thistle (*Silybum marianum*) derivative on ethanol-induced oxidative stress in liver. *Indian J. Biochem. Biophys.,* **2006**, *43*(5), 306–311.

Datta, S.; Sinha, S.; Bhattacharya, P. Hepatoprotective activity of a herbal protein CI-1, purified from *Cajanus indicus* against α-galactosamine HCl toxicity in isolated rat hepatocytes. *Phytother. Res.,* **1999**, *13*, 508–512.

Davila, J.C.; Lenherr, A.; Acosta, D. Protective effect of flavonoids on the drug-induced hepatotoxicity *in vitro. Toxicology,* **1989**, 57, 2576–2586.

Deak, G.; Muzes, G.; Lang, I.; Nekam, K.; Gonzalez-Cabello, R.; Gergely, P.; et al. Effects of two bioflavanoids on certain cellular immune reactions *in vitro. Acta Physiol. Hung.,* **1990**, *76*, 113–121.

Dehmlow, C.; Erhard, J.; Groot, H. Inhibition of Kupffer cell functions as an explanation for the hepatoprotective properties of silibin. *Hepatology,* **1996**, *23*, 749–754.

Desplaces, A.; Choppin, J.; Vogel, G. The effects of silymarin on experimental phalloidine poisoning. *Arzneimittelforschung,* **1975**, *25*, 89–96.

Dhanasekaran, M.; Ignacimuthu, S.; Agastian, P. Potential hepatoprotective activity of ononitol monohydrate isolated from *Cassia tora* L. on carbon tetrachloride induced hepatotoxicity in Wistar rats. *Phytomedicine,* **2009,** *16*, 891–895.

Dwivedi, Y.; Rastogi, R.; Ramesh, C.; Sharma, S.K.; Kapoor, N.K.; Garg, N.K.; Dhawan, B.N. Hepatoprotective activity of picroliv against carbon tetrachloride induced liver damage in rats. *Ind. J. Med. Res.* [B], **1990,** *92*, 195–200.

Dixit, N.; Baboota, S.; Kohli, K.; Ahmad, S.; Ali, J. Silymarin: A review of pharmacological aspects and bioavailability enhancement approaches. *Indian J. Pharmacol.,* **2007**, *39*, 172–179.

Domitrovic, R.; Jakovacb, H.; Blagojevic, G. Hepatoprotective activity of berberine is mediated by inhibition of TNF-α, COX-2 and iNOS expression in CCl_4-intoxicated mice and iNOS expression in CCl_4-intoxicated mice. *Toxicology*, **2011**, 280, 33–43.

Donfack, J.H.; Nico, F.N.; Ngameni, B.; Tchana, A.; Chuisseu, P.D.; Finzi, P.V.; Ngadjui, B.T.; Moundipa, P. In vitro hepatoprotective and antioxidant activities of diprenylated isoflavonoids from *Erythrina senegalensis* (Fabaceae). *Asian J. Trad. Med.,* **2008**, *5*, 172–178.

Dong, J.; Peng, X.; Lu, S.; Zhou, L.; Qiu, M. Hepatoprotective steroids from roots of *Cynanchum otophyllum*. *Fitoterapia*, **2019**, *136*, 104–171.

Du, X.-M.; Irino, N.; Furusho, N.; Hayashi, J.; Shoyama, Y. Pharmacologically active compounds in the *Anoectochilus* and *Goodyera* species. *J. Nat. Med.*, **2007**, *62*, 132–148, doi: 10.1007/s11418-007-0169-0

Du, X.M.; Sun, N.Y.; Chen, Y.; Irino, N.; Shoyama, Y. Hepatoprotective aliphatic glycosides from three *Goodyera* species. *Biol. Pharm. Bull.*, **2000**, *23*(6), 731–734.

El-Banna, H.; Soliman, M.; Al-wabel, N. Hepatoprotective effects of *Thymus* and *Salvia* essential oils on paracetamol-induced toxicity in rats. *J. Physiol. Pharmacol. Adv.*, **2013**, *3*(2), 41–47.

El-Hawary, S.; El-Shabrawy, A.; Ezzat, S.; El-Shibany, F. Gas chromatography– mass spectrometry analysis, hepatoprotective and antioxidant activities of the essential oils of four Libyan herbs. *J. Med. Plants Res.*, **2013**, *7*, 1746–1753.

El-Sayed el-SM; Mansour, A.M.; Nady, M.E. Protective effects of pterostilbene against acetaminophen-induced hepatotoxicity in rats. *J. Biochem. Mol. Toxicol.*, **2015**, *29* (1), 35–42.

Eminzade, S.; Uraz, F.; Izzettin, F.V. Silymarin protects liver against toxic effects of anti-tuberculosis drugs in experimental animals. *Nutr. Metab.*, **2008**, *5*(18), 1–8.

Fan, J.L.; Wu, Z.W.; Zhao, T.H.; et al. Characterization, antioxidant and hepatoprotective activities of polysaccharides from *Ilex latifolia* Thunb. *Carbohydr. Polym.*, **2014**, *101*(1), 990–997.

Farghali, H.; Canová, N.K.; Zakhari, S. Hepatoprotective properties of extensively studied medicinal plant active constituents: Possible common mechanisms. *Pharm. Biol.*, **2015**, *53*(6), 781–791.

Farghali, H.; Cerny, D.; Kamenikova, L.; et al. Resveratrol attenuates lipopolysaccharide-induced hepatitis in D-galactosamine sensitized rats: Role of nitric oxide synthase 2 and heme oxygenase-1. *Nitric Oxide-Biol. Ch.*, **2009**, *21*, 216–225.

Farghali, H.; Kamenikova, L.; Hynie, S. Silymarin effects of intracellular calcium and cytotoxicity: A study in perfused rat hepatocytes after oxidative stress injury. *Pharmacol. Res.*, **2000**, *41*, 231–237.

Farghali, H.; Kamenikova, L.; Hynie, S.; Kmonickova, E. Silymarin effects of intracellular calcium and cytotoxicity: A study in perfused rat hepatocytes after oxidative stress injury. *Pharmacol. Res.*, **2000**, *41*, 231–237.

Farghali, H.; Kamenickova, E.; Lotkova, H.; Martinek, J. Evaluation of calcium channel blockers as potential hepatoprotective agents in oxidative stress injury of perfused hepatocytes. *Physiol. Res.*, **2000**, *49*, 261–268.

Farghaly, H.S.; Hussein, M.A. Protective effect of curcumin against paracetamol induced liver damage. *Aust. J. Basic Appl. Sci.*, **2010**, *4*(9), 4266–4274.

Farombi, E.O. Mechanisms for the hepatoprotective action of kolaviron: Studies on hepatic enzymes, microsomal lipids and lipid peroxidation in carbon tetrachloride-treated rats. *Pharmacol. Res.*, **2000**, *42*(1), 75–80.

Farombi, E.O.; Tahnteng, J.G.; Agboola, A.O.; Nwankwo, J.O.; Emerole, G.O. Chemoprevention of 2-acetylaminofluorine-induced hepatotoxicity and lipid peroxidation in rats by kolaviron—a *Garcinia kola* seed extract. *Food Chem. Toxicol.*, **2012**, *38*, 535–541.

Favari, L.; Perez-Alvarez, V. Comparative effects of colchicines and silymarin on carbon tetrachloride chronic liver damage in rats. *Arch. Med. Res.*, **1997**, *28*, 11–17

Féher, J.; Lengyel, G. Silymarin in the prevention and treatment of liver diseases and primary liver cancer. *Curr. Pharm. Biotechnol.*, **2012**, *13*(1), 210–217.

Feng, Y.; Siu, K.Y.; Ye, X.; Wang, N.; Yuen, M.F.; Leung, C.H.; et al. Hepatoprotective effects of berberine on carbon tetrachloride-induced acute hepatotoxicity in rats. *Chin. Med.*, **2010**, *5*, 33.doi: 10.1186/1749-8546-5-33.

Ferenci, P.; Dragosics, B.; Dittrich, H.; et al. Randomized controlled trial of silymarin treatment in patients with cirrhosis of the liver. *J. Hepatol.*, **1989**, *9*, 105–113.

Floersheim, G.L.; Eberhard, M.; Tschumi, P.; Duckert, F. Effects of penicillin and silymarin on liver enzymes and blood clotting factors in dogs given a boiled preparation of *Amanita phalloides*. *Toxicol. Appl. Pharmacol.*, **1978**, *46*, 455–462.

Fraschini, F.; Demartini, G.; Esposti, D. Pharmacology of Silymarin. *Clin. Drug Inv.*, **2002**, *22*, 51–65.

Freitag, A.F.; Cardia, G.F.E.; Da Rocha, B.A.; et al. Hepatoprotective effect of silymarin (*Silybum marianum*) on hepatotoxicity induced by acetaminophen in spontaneously hypertensive rats. *Evid. Based Compl. Altern. Med.*, **2015**, *2015*, 1–8.

Gadgoli, C.; Mishra, S.H. Antihepatotoxic activity of p-methoxy benzoic acid from *Capparis spinosa*. *J. Ethnopharmacol.*, **1999**, *66*, 187–192.

Gao, Q.; Zhao, X.; Yin, L.; Zhang, Y.; Wang, B.; Wu, X.; Zhang, X.; Fu, X.; Sun, W. The essential oil of *Artemisia capillaris* protects against CCl_4-induced liver injury in vivo. *Rev. Bras. Farmacog.*, **2016**, *26*, 369–374.

Garg, H.S.; Bhandari, S.P.S.; Tripathi, S.C.; Patnaik, G.K.; Puri, A.; Saxena, R.; et al. Antihepatotoxic and immunostimulant properties of iridoid glycosides of *Scrophularia koelzii*. *Phytother. Res.*, **1994**, *8*(4), 224–228.

Ghosh, N.; Ghosh, R.; Mandal, V.; Mandal, S.C. Recent advances in herbal medicine for treatment of liver diseases. *Pharm. Biol.*, **2011**, *49*(9), 970–988.

Ghosh, A.; Sil, P.C. Anti-oxidative effect of a protein from *Cajanus indicus* L against acetaminophen-induced hepato-nephrotoxicity. *J. Biochem. Mol. Biol.*, **2007**, *40*(6), 1039–1049.

Grace-Lynn, C.; Chen, Y.; Latha, L.Y.; Kanwar, J.R.; Jothy, S.L.; Vijayarathna, S.; Sasidharan, S. Evaluation of the hepatoprotective effects of lantadene A, a pentacyclic triterpenoid of Lantana plants against acetaminophen-induced liver damage. *Molecules*, **2012**, *17*(12), 13937–13947.

Grespan, R.; Aguiar, R.P.; Giubilei, F.N.; Fuso, R.R.; Damiao, M.J.; Silva, E.L.; et al. Hepatoprotective effect of pretreatment with *Thymus vulgaris* essential oil in experimental model of acetaminophen-induced injury. *Evid. Based Complement. Altern. Med.*, **2014**, *2014*, Art. no. 954136, doi: 10.1155/2014/954136

Gupta, S.C.; Kismali, G.; Aggarwal, B.B. Curcumin, a component of turmeric: From farm to pharmacy. *Biofactors*, **2013**, *39*, 2–13.

Gutiérrez-Ruiz, M.C.; Bucio, L.; Correa, A.; Souza, V.; Hernández, E.; Gómez-Quiroz, L.E.; et al. Metadoxine prevents damage produced by ethanol and acetaldehyde in hepatocyte and hepatic stellate cells in culture. *Pharmacol. Res.*, **2001**, *44*, 431–436.

Gutierrez, R.M.P.; Anaya Sosa, I.; Hoyo Vadillo, C.; Victoria, T.C. Effect of flavonoids from *Prosthechea michuacana* on carbon tetrachloride induced acute hepatotoxicity in mice. *Pharm. Biol.*, **2011**, *49*(11), 1121–1127.

Hafez, M.M.; Al-Harbi, N.O.; Al-Hoshani, A.R.; Al-Hosaini, K.A.; Shrari, S.D.; Al-Rejaie, S.S.; et al. Hepatoprotective effect of rutin via IL-6/STAT3 pathway in CCl_4-induced hepatotoxicity in rats. *Biol. Res.*, **2015**, *48*, Art. no. 30, doi: 10.1186/s40659-015-0022-y.

Hahn, G.; Lehmann, H.D.; Kurten, M. Zur pharmakologie und toxikologie von silymarin, des antihepatotoxischen wirkprinzipes aus *Silybum marianum* (L.) Gaertn. *Arzneimittelforschung,* **1968**, *18*, 698–704.

Han, S.; Wang, C.; Cui, B.; Sun, H.; Zhang, J.; Li, S. Hepatoprotective activity of glucosyloxybenzyl succinate derivatives from the pseudobulbs of *Pleione bulbocodioides*. *Phytochemistry*, **2019,** *157*, 71–81.

Hase, K.; Kasimu, R.; Basnet, P.; Kadota, S.; Namba, T. Preventive effect of lithospermate B from *Salvia miltorhiza* on experimental hepatitis induced by carbon tetrachloride or Dgalactosamine/lipopolysaccharide. *Planta Med.*, **1997a**, *63*(1), 22–26.

Hase, K.; Li, J.; Basnet, P.; Xiong, Q.; Takamura, S.; Namba, T.; Kadota, S. Hepatoprotective principles of *Swertia japonica* on galactosamine/lipopolysaccharide-induced liver injury in mice. *Chem. Pharm. Bull.*, **1997b**, *45*(11), 1823–1827.

Hawas, U.W.; Soliman, G.M.; El-Kassem, L.T.A.; Farrag, A.R.H.; Mahmoud, K.; León, F. A new flavonoid C-glycoside from *Solanum elaeagnifolium* with hepatoprotective and curative activities against paracetamol-induced liver injury in mice. *Z Naturforsch C. Biosci.*, **2013**, *68*(1–2), 19–28.

Hikino, H.; Kiso, Y.; Kinouchi, J.; Sanada, S. Antihepatotoxic actions of ginsenosides from *Panax ginseng* roots. *Planta Med.*, **1985**, *51*, 62–64.

Hoskeri, H.J.; Krishna, V.; Vinay Kumar, B.; Shridar, A.H.; Ramesh Babu, K.; Sudarshana, M.S. *In vivo* prophylactic effects of oleanolic acid isolated from chloroform extract of *Flaveria trinervia* against ethanol induced liver toxicity in rats. *Arch. Pharm. Res.*, **2012**, *35*(10), 1803–1810.

Hsouna, A.B.; Dhibi, S.; Dhifi, W.; Mnif, W.; Nasr, A.B.; Hfaiedh, N. Chemical composition and hepatoprotective effect of essential oil from *Myrtus communis* L. flowers against CCl_4-induced acute hepatotoxicity in rats. *RSC Adv.*, **2019**, *9*(7), 3777–3787.

Huang, Q.; Li, Y.; Zhang, S.; Huang, R.; Zheng, L.; Wei, L.; He, M.; Liao, M.; Li, L.; Zhuo, L.; Lin, X. Effect and mechanism of methyl helicterate isolated from *Helicteres angustifolia* (Sterculiaceae) on hepatic fibrosis induced by carbon tetrachloride in rats. *J. Ethnopharmacol.*, **2012**, *143*(3), 889–895.

Huang, Q.F.; Zhang, S.J.; Zheng, L.; He, M.; Huang, R.B.; Lin, X. Hepatoprotective effects of total saponins isolated from *Taraphochlamys affinis* against carbon tetrachloride induced liver injury in rats. *Food Chem. Toxicol.*, **2012**, *50*, 713–718.

Hung, W.L.; Yang, G.; Wang, Y.C.; Chiou, Y.S.; Tung, Y.C.; Yang, M.J.; Wang, B.N.; Ho, C.T.; Wang, Y.; Pan, M.H. Protective effects of theasinensin A against carbon tetrachloride-induced liver injury in mice. *Food Funct.*, **2017**, *8*(9), 3276–3287.

Hwang, Y.P.; Choi, C.Y.; Chung, Y.C.; Jeon, S.S.; Jeong, H.G. Protective effects of puerarin on carbon tetrachloride-induced hepatotoxicity. *Arch. Pharm. Res.*, **2007**, *30*, 1309–1317.

Hyun, C.; Jun, J.Y.; Song, E.K.; Kang, K.H.; Baek, H.Y.; Ko, Y.S.; Kim, Y.C. Bakuchiol: A hepatoprotective compound of *Psoralea corylifolia* on tacrine-induced cytotoxicity in Hep G2 cells. *Planta Med.*, **2001**, *67*(8), 750–751.

Igarashi, K.; Mikami, T.; Takahashi, Y.; Sato, H. Comparison of the preventive activity of isorhamnetin glycosides from atsumi-kabu (red turnip, *Brassica campestris* L.) leaves on carbon tetrachloride-induced liver injury in mice. *Biosci. Biotechnol. Biochem.*, **2008**, *72*(3), 856–860.

Jadon, A.; Bhadauria, M.; Shukla, S. Protective effect of *Terminalia belerica* Roxb. and gallic acid against carbon tetrachloride-induced damage in albino rats. *J. Ethnopharmacol.*, **2007**, *109*(2), 214–218.

Jain, M.; Kapadia, R.; Jadeja, R.N.; Thounaojam, M.C.; Devkar, R.V.; Mishra. S.H. Hepatoprotective potential of *Tecomella undulata* stem bark is partially due to the presence of betulinic acid. *J. Ethnopharmacol.*, **2012**, *143*, 194–200.

Janbaz, K.H.; Gilani, A.H. Evaluation of the protective potential of *Artemisia maritima* extract on acetaminophen- and CCl_4-induced liver damage. *J. Ethnopharmacol.*, **1995**, *47*(1), 43–47.

Janbaz, K.H.; Saeed, S.A.; Gilani, A.H. Protective effect of rutin on paracetamol- and CCl_4-induced hepatotoxicity in rodents. *Fitotrapia*, **2002**, *73*(7–8), 557–563.

Janbaz, K.; Saeed, S.; Gilani, A. Studies on the protective effects of caffeic acid and quercetin on chemical-induced hepatotoxicity in rodents. *Phytomedicine,* **2004**, *11*, 424–430.

Jeong, H.G.; You, H.J.; Park, S.J.; Moon, A.R.; Chung, Y.C.; Kang, S.K. Hepatoprotective effects of 18-beta-glycyrrhetinic acid on carbon tetrachloride-induced liver injury: inhibition of cytochrome P450 2E1 expression. *Pharmacol. Res.*, **2002,** *46*, 221–227.

Ji, L.; Jiang, P.; Lu, B.; Sheng, Y.; Wang, S.; Wang, Z. Chlorogenic acid, a dietary polyphenol, protects acetaminophen-induced liver injury and its mechanism. *J. Nutr. Biochem.*, **2013**, *24*, 1911–1919.

Jia, X.-Y.; Zhang, Q.-A.; Zhang, Z.-Q. et al. Hepatoprotective effects of almond oil against carbon tetrachloride induced liver injury in rats. *Food Chem.*, **2011**, *125*, 673–678.

Jiang, W.; Guo M.-H.; Hai X. Hepatoprotective and antioxidant effects of lycopene on non-alcoholic fatty liver disease in rat. *World J. Gastroenterol.*, **2016**, 22(46), 10180–10188.

Jiang, W.-P.; Huang, S.-S.; Matsuda, Y.; et al. Protective effects of tormentic acid, a major component of suspension cultures of *Eriobotrya japonica* cells, on acetaminophen-induced hepatotoxicity in mice. *Molecules*, **2017**, *22*(5), Art. no. 830.

Joshi, U.P.; Mishra, S.H. *In vitro* antioxidant and hepatoprotective activity of isolated compounds from *Pistacia integerrima. Aus. J. Med. Herbalism,* **2010**, *22*, 22–34.

Junnila, M.; Rahko, T.; Sukura, A.; Lindberg, L.A. Reduction of carbon tetrachloride-induced hepatotoxic effect by oral administration of betaine in male Wistar rats: A morphometric histologic study. *Vet. Pathol.*, **2000**, *37*(3), 231–238.

Kapil, A.; Koul, I.B.; Banerjee, S.K.; Gupta, B.D. Anti-hepatotoxic effects of major diterpenoid constituents of *Andrographis paniculata. Biochem. Pharmacol.*, **1993**, *46*(1), 182–185.

Kapil, A.; Koul, I.B.; Suri, O.P. Antihepatotoxic effects of chlorogenic acid from *Anthocephalus cadamba. Phytother. Res.*, **1995**, 9(3), 189–193.

Khan, R.A.; Khan, M.R.; Sahreen, S. Carbon tetrachloride-induced hepatotoxicity: protective effect of rutin on p53, CYP2E1 and the antioxidative status in rat. *BMC Complement. Altern. Med.*, **2012**, *12*, Art. no. 178.

Kim, H.Y.; Kim, J.K.; Choi, J.H.; Jung, J.Y.; Oh, W.Y.; Kim, D.C.; Lee, H.S.; Kim, Y.S.; Kang, S.S.; Lee, S.H.; Lee, S.M. Hepatoprotective effect of pinoresinol on carbon tetrachloride-induced hepatic damage in mice. *J. Pharmacol. Sci.*, **2010**, *112*(1), 105–112.

Kim, K.H.; Kim, Y.H.; Lee, K.R. Isolation of quinic acid derivatives and flavonoids from the aerial parts of *Lactuca indica* L. and their hepatoprotective activity *in vitro*. *Bioorg. Med. Chem. Lett.*, **2007**, *17*(24), 6739–6743.

Kinoshita, S.; Inoue, Y.; Nakama, S.; Ichiba, T.; Aniya, Y. Antioxidant and hepatoprotective actions of medicinal herb, *Terminalia catappa* L. from Okinawa Island and its tannin corilagin. *Phytomedicine*, **2007**, *14*(11), 755–762.

Kiso, Y.; Tohkin, M.; Hikino, H. Assay method for antihepatotoxic activity using galactosamine-induced cytotoxicity in primary-cultured hepatocytes. *J. Nat. Prod.*, **1983**, *46*, 841–847.

Kiso, Y.; Ogasawara, S.; Hirota, K.; Watanabe, N.; Oshima, Y.; Konno, C.; Hikino, H. Antihepatotoxic principles of *Artemisia capillaris* buds. *Planta Med.*, **1984**, *50*, 81–85.

Kumar, K.J.S.; Liao, J.W.; Xiao, J.H.; Vani, M.G.; Wang, S.Y. Hepatoprotective effect of lucidone against alcohol-induced oxidative stress in human hepatic HepG2 cells through the up-regulation of HO-1/Nrf-2 antioxidant genes. *Toxicol. In Vitro,* **2012,** 26, 700–708.

Kumaravelu, P.; Dakshinamoorthy, D.P.; Subramaniam, S.; Devaraj, H.; Devaraj, N.S. Effect of eugenol on drug-metabolizing enzymes of carbon tetrachloride intoxicated rat liver. *Biochem. Pharmacol.*, **1995**, *49*(11), 1703–1707.

Lattanzio, V.; van Sumere, C.F. Changes in phenolic compounds during the development and cold storage of artichoke (*Cynara scolymus* L.) heads. *Food Chem.*, **1987**, *24*, 37–50.

Lee, H.S.; Jung, K.H.; Hong, S.W.; Park, I.S.; Lee, C.; Han, H.K.; Lee, D.H.; Hong, S.S. Morin protects acute liver damage by carbon tetrachloride (CCl_4) in rat. *Arch. Pharm. Res.*, **2008**, *31*, 1160–1165.

Lee, I.C.; Kim, S.H.; Baek, H.S.; Moon, S.H.; Kim, Y.B.; Yun, W.K.; et al. Protective effects of diallyl disulfide on carbon tetrachloride-induced hepatotoxicity through activation of Nrf2. *J. Environ. Toxicol.,* **2015**, *30*(5), 538–540.

Lee, K.J.; Choi, J.H.; Jeong, H.G. Hepatoprotective and antioxidant effects of the coffee diterpenes kahweol and cafestol on carbon tetrachloride-induced liver damage in mice. *Food Chem. Toxicol.,* **2007**, 45, 2118–2125.

Lettiron, P.; Labbe, G.; Degott, C.; Berson, A.; Fromenty, B.; Delaforge, M.; et al. Mechanism for the protective effects of silymarin against carbon tetrachloride-induced lipid peroxidation and hepatotoxicity in mice. *Biochem. Pharmacol.*, **1990**, *39*, 2027–2034.

Leung, J.; Bonis, P.A.; Kaplan, M.M. Colchicine or methotrexate, with ursodiol, are effective after 20 years in a subset of patients with primary biliary cirrhosis. *Clin. Gastroenterol. Hepatol.*, **2011**, *9*, 776–780.

Lijie, Z.; Ranran, F.; Xiuying, L.; Yutang, H.; Bo, W.; Tao, M. Soyasaponin Bb protects rat hepatocytes from alcohol-induced oxidative stress by inducing heme oxygenase-1. *Pharmacogn. Mag.,* **2016,** *12,* 302–306.

Lim, H.K.; Kim, H.S.; Choi, H.S.; Oh, S.; Choi, J. Hepatoprotective effects of bergenin, major constituent of *Mallotus japonicus* on CCl_4 intoxicated rats. *J. Ethnopharmacol.*, **2000**, *72*, 469–474.

Lin, C.C.; Lee, H.Y.; Chang, C.H.; Namba, T.; Hattori, M. Evaluation of the liver protective principles from the root of *Cudrania cochinchinensis* var. *gerontogea*. *Phytother. Res.*, **1996**, *10*(1), 13–17.

Lin, C-C.; Shieh, D-E. *In vivo* hepatoprotective effect of baicalein, baicalin and wogonin from *Scutellaria rivularis*. *Phytother. Res.*, **1996**, *10*, 651–654.

Lin, C.C.; Shieh, D.E.; Yen, M.N. Hepatoprotective effect of fractions Ban-zhi-lian of experimental liver injuries in rats. *J. Ethnopharmacol.*, **1997**, *56*, 193–200.

Lin, C.-N.; Tome, W.-T. Antihepatotoxic Principles of *Sambucus formosana. Planta Med.*, **1988**, *54*(3), 223–224.

Lin, G.S.; Luo, D.D.; Liu, J.J.; et al.; Hepatoprotective effect of polysaccharides isolated from *Dendrobium officinale* against acetaminophen-induced liver injury in mice via regulation of the Nrf2-Keap1 signaling pathway. *Oxid. Med. Cell. Longev.*, **2018**, *2018*, Art. no. 6962439.

Lin, J.; Zhao, J.; Li, T.; Zhou, J.; Hu, J.; Hong, Z. Hepatoprotection in a rat model of acute liver damage through inhibition of CY2E1 activity by total alkaloids extracted from *Rubus alceifolius* Poir. *Int. J. Toxicol.*, **2011**, *30*(2), 237–243.

Liu, G.; Liu, X.; Zhang, Y.; Zhang, F.; Wei, T.; Yang, M.; Wang, K.; Wang, Y.; Liu, N.; Cheng, H.; Zhao, Z. Hepatoprotective effects of polysaccharides extracted from *Zizyphus jujube* cv. Huanghetanzao. *Int. J. Biol. Macromol.*, **2015**, *76*, 169–175.

Loguercio, C.; Andreone, P.; Brisc, C.; et al. Silybin combined with phosphatidylcholine and vitamin E in patients with nonalcoholic fatty liver disease: A randomized controlled trial. *Free Radical Bio. Med.*, **2012**, *52*, 1658–1665.

Mahmoud, M.F.; Hamdan, D.I.; Wink, M.; El-Shazly, A.M. Hepato-protective effect of limonin, a natural limonoid from the seed of *Citrus aurantium* var. *bigaradia*, on *D*-galactosamine-induced liver injury in rats. *Naunyn Schmiedebergs Arch. Pharmacol.*, **2014**, *387*, 251–261.

Mahmoud, A.M.; Hozayen, W.G.; Ramadan, S.M. Berberine ameliorates methotrexate-induced liver injury by activating Nrf2/HO-1 pathway and PPARγ, and suppressing oxidative stress and apoptosis in rats. *Biomed. Pharmacother.*, **2017**, *94*, 280–291.

Maiti, K.; Mukherjee, K.; Murugan, V.; Saha, B.P.; Mukherjee, P.K. Enhancing bioavailability and hepatoprotective activity of andrographolide from *Andrographis paniculata*, a well, known medicinal food, through its herbosome. *J. Sci. Food Agric.*, **2010**, *90*, 43–51.

Makni, M.; Chtourou, Y.; Fetoui, H.; Garoui, E.M.; Boudawara, T.; Zeghal, N. Evaluation of the antioxidant, anti-inflammatory and hepatoprotective properties of vanillin in carbon tetrachloride treated rats. *Eur. J. Phar.*, **2011**. *668*(1–2), 133–139.

Manna, P.; Sinha, M.; Sil, P.C. Protection of arsenic-induced hepatic disorder by arjunolic acid. *Basic Clin. Pharmacol. Toxicol.*, **2007**, *101*, 333–338.

Mansour, M.A. Protective effects of thymoquinone and desferrioxamine against hepatotoxicity of carbon tetrachloride in mice. *Life Sci.*, **2000**, *66*, 2583–2591.

Matsuda, H.; Morikawa, T.; Fengming, X.; Ninomiya, K.; Yoshikawa, M. New isoflavones and pterocarpane with hepatoprotective activity from the stems of *Erycibe expansa*. *Planta Med.*, **2004**, *70*(12), 1201–1209.

Mastuda, H.; Samukawa, K.; Kubo, M. Antihepatic activity of Ginsenoside Ro. *Planta Med.*, **1991**, *57*, 523–526.

Mehrzadi, S.; Fatemi, I.; Esmaeilizadeh, M.; Ghaznavi, H.; Kalantar, H.; Goudarzi, M. Hepatoprotective effect of berberine against methotrexate induced liver toxicity in rats. *Biomed. Pharmacother.*, **2018**, *97*, 233–239.

Mereish, K.A.; Bunner, D.L.; Ragland, D.R.; Creasia, D.A. Protection against microcystin-LR-induced hepatotoxicity by silymarin: Biochemistry, histopathology and lethality. *Pharm. Res.*, **1991**, *8*, 273–277.

Moghaddamm, E.Z.; Azami, K.; Minaei-Zangi, B.; Mousavi, S.Z.; Sabzevari, O. Protective Activity of *Fumaria vaillantii* extract and monomethyl fumarate on acetaminophen induced hepatotoxicity in mice. *Int. J. Pharmacol.*, **2012**, *8*(3), 177–184.

Mohamed, M.A.; Marzouk, S.A.; Moharram, F.A.; El-Sayed, M.M.; Baiuomy, A.R.

Phytochemical constituents and hepatoprotective activity of *Viburnum tinus*. *Phytochem.*, **2005**, *66*(23), 2780–2786.

Morales-López, J.; Centeno-Álvarez, M.; Nieto-Camacho, A.; López, M.G.; Pérez-Hernández, E.; Pérez-Hernández, N.; Fernández-Martínez, E. Evaluation of anti-oxidant and hepatoprotective effects of white cabbage essential oil. *Pharm. Biol*, **2017**, *55*, 233–241.

Moreira, P.R.; Maioli, M.A.; Medeiros, H.C.D.; Guelfi, M.; Pereira, F.T.V.; Mingatto, F.E. Protective effect of bixin on carbon tetrachloride-induced hepatotoxicity in rats. *Biol. Res.*, **2014**, *47*(1), Art. no. 49. doi: 10.1186/0717-6287-47-49

Morikawa, T.; Pan, Y.; Ninomiya, K.; Imura, K.; Matsuda, H.; Yoshikawa, M.; Yuan, D.; Muraoka, O. Acylated phenylethanoid oligoglycosides with hepatoprotective activity from the desert plant *Cistanche tubulosa*. *Bioorg. Med. Chem*, .**2010**, *18*(5), 1882–1890.

Morita, T.; Jinno, K.; Kawagishi, H.; Arimoto, Y.; Suganuma, H.; Inakuma, T.; Sugiyama, K. Hepatoprotective effect of myristin from nutmeg (*Myristica fragrans*) on lipopolysaccharide/d-galactosamine-induced liver injury. *J. Agric. Food Chem.*, **2003**, *51*, 1560–1565.

Mourelle, M.; Franco, M.T. Erythrocyte defects precede the onset of carbon tetrachloride-induced liver cirrhosis: Protection by silymarin. *Life Sci.*, **1991**, *48*, 1083–1090.

Mourelle, M.; Muriel, P.; Favari, L.; Franco, T. Prevention of carbon tetrachloride-induced liver cirrhosis by silymarin. *Fundam. Clin. Pharmacol.*, **1989**, *3*, 183–191.

Muntoni, S.; Rojkind, M.; Muntoni, S. Colchicine reduces procollagen III and increases pseudocholinesterase in chronic liver disease. *World J. Gastroenterol.*, **2010**, *16*, 2889–2894.

Muriel, P.; Mourelle, M. Prevention by silymarin of membrane alterations in acute carbon tetrachloride liver damage. *J. Appl. Toxicol.*, **1990**, *10*, 275–279.

Muriel, P.; Garciapiña, T.; Perez-Alvarez V.; Mourelle, M. Silymarin protects against paracetamol-induced lipid peroxidation and liver damage. *J. Appl. Toxicol.*, **1992**, *12*, 439–442.

Nabavi S.F.; Nabavi S.M.; Habtemariam S.; et al. Hepatoprotective effect of gallic acid isolated from *Peltiphyllum peltatum* against sodium fluoride-induced oxidative stress. *Ind. Crops Products*, **2013**, 44, 50–55.

Nagaraja, Y.; Krishna, V. Hepatoprotective effect of the aqueous extract and 5-hydroxy, 7,8,2¢ trimethoxy flavone of *Andrographis alata* Nees. in carbon tetrachloride treated rats. *Achievements Life Sci.*, **2016**, *10*, 5–10.

Natanzi, A.R.E.; Mahmoudian, S.; Minaeie, B.; Sabzevari, O. Hepatoprotective activity of phloretin and hydroxychalcones against acetaminophen induced hepatotoxicity in mice. *Iran. J. Pharm. Res.*, **2011**, *7*, 89–97.

Nguyen, N.T.; Banskota, A.; Tezuka, Y.; Quan, L.T.; Nobukawa, T.; Kurashige, Y.; Sasahara, M.; Kadota, S. Hepatoprotective effect of taxiresinol and (7′R)-7′-hydroxylariciresinol on d-galactosamine and lipopolysaccharide-induced liver injury in mice. *Planta Med.*, **2004**, *70*(1), 29–33.

Nili-Ahmadabadi, A.; Alibolandi, P.; Ranjbar, A.; Mousavi, L.; Nili-Ahmadabadi, H.; Larki-Harchegani, A.; Ahmadimoghaddam, D.; Omidifar, N. Thymoquinone attenuates hepatotoxicity and oxidative damage caused by diazinon: An *in vivo* study. *Res. Pharm. Sci.*, **2018**, *13*(6), 500–508.

Obidah, W.; Garba, G.K.; Fate, J.Z. Protective effects of *Bixa orellana* seed oil on carbon tetrachloride induced liver damage in rats. *Rep. Opin.*, **2011**, *3*(1), 92–95.

Oh, H.; Kim, J.S.; Song, E.K.; Cho, H.; Kim, D.H.; Park, S.E.; Lee, H.S.; Kim, Y.C.., Sesquiterpenes with hepatoprotective activity from *Cnidium monnieri* on tacrine-induced cytotoxicity in Hep G2 cells. *Planta Med.*, **2002**, *68*(8), 748–749.

Oh, H.; Ko, E.K.; Jun, J.Y.; Oh, M.H.; Park, S.U.; Kang, K.H.; Lee, H.S.; Kim,

Y.S. Hepatoprotective and free-radical scavenging activities of prenylflavonoids, coumarins and stilbene from *Morus alba*. *Planta Med.*, **2002**, *68*, 932–934.

Oh, H.; Kim, D.H.; Cho, J.H.; Kim, Y.C. Hepatoprotective and free-radical scavenging activities of phenolic petrosins and flavonoids isolated from *Equisetum arvense*. *J. Ethnopharmacol.*, **2004**, *95*, 421–424.

Ohta, Y.; Ohashi, K.; Matsura, T.; Tokunaga, K.; Kitagawa, A.; Yamada, K. Octacosanol attenuates disrupted hepatic reactive oxygen species metabolism associated with acute liver injury progression in rats intoxicated with carbon tetrachloride *J. Clin. Biochem. Nutr.*, **2008**, *42*(2), 118–125.

Oliveira, F.A.; Chaves, M.H.; Almeida, F.R.C.; Lima, R.C.P.; Silva, R.M.; Maia, J.L.; Brito, G.A.C.; Santos, F.A.; Rao, V.S. Protective effect of α- and β-amyrin, a triterpene mixture from *Protium heptaphyllum* (Aubl.) March. trunk wood resin, against acetaminophen-induced liver injury in mice. *J. Ethnopharmacol.*, **2005**, *98*, 103–108.

Osman, S.M.; El-Haddad, A.E.; El-Raey, M.A.; El-Khalik, S.M.A.; Koheil, M.A.; Wink, M. A new octadecenoic acid derivative from *Caesalpinia gilliesii* flowers with potent hepatoprotective activity. *Pharmacog. Mag.*, **2016**, *12*, Art. no. S332.

Özbek, H.; Ugras, S.; Bayram, I.; Uygan, I.; Erdogan, E.; Ozturk, A.; Huyut, Z. Hepatoprotective effect of *F. vulgare* essential oil: A carbon tetrachloride induced liver fibrosis model in rats. *Scand. J. Lab. Anim. Sci.*, **2004**, *31*, 9–17.

Özbek, H.; Ugras, S.; Dülger, H.; Bayram, I.; Tuncer, I.; Öztürk, G.; Öztürk, A. Hepatoprotective effect of *Foeniculum vulgare* essential oil. *Fitoterapia,* **2003,** *74*(3), 317–319.

Pardede A.; Adfa M.; Juliari Kusnanda A.; Ninomiya M.; Koketsu M. Flavonoid rutinosides from *Cinnamomum parthenoxylon* leaves and their hepatoprotective and antioxidant activity. *Med. Chem. Res.*, **2017**, 26(9), 2074–2079.

Park, E.J.; Jeon, C.H.; Ko, G.; Kim, J.; Sohn, D.H. Protective effect of curcumin in rat liver injury induced by carbon tetrachloride. *J. Pharm. Pharmacol.*, **2000**, *52*, 437–440.

Park, E.J.; Zhao, Y.Z.; Young, H.K.; Jung, J.L.; Dong, H.S. Acanthoic acid from *Acanthopanax koreanum* protects against liver injury induced by tert-butyl hydroperoxide or carbon tetrachloride *in vitro* and *in vivo*. *Planta Med.*, **2004**, *70*(4), 321–327.

Parveen, R.; Baboota, S.; Ali, J.; Ahuja, A.; Vasudev, S.S.; Ahmad, S. Effects of silymarin nanoemulsion against carbon tetrachloride-induced hepatic damage. *Arch. Pharm. Res.*, **2011**, *34*(5), 767–774.

Pattanayak, S.; Nayak, S.S.; Panda, D.P.; Dinda, S.C.; Shende, V.; Jadav, A. Hepatoprotective activity of crude flavonoids extract of *Cajanus scarabaeoides* (L.) in paracetamol intoxicated albino rats. *Asian J. Pharm. Biol. Res.*, **2011**, *1*(1), 22–27.

Pradhan, S.C.; Girish, C. Hepatoprotective herbal drug, silymarin from experimental pharmacology to clinical medicine. *Indian J. Med. Res.*, **2006**, *124*, 491–504.

Prasad, S.; Kalra, N.; Shukla, Y. Hepatoprotective effects of lupeol and mango pulp extract of carcinogen-induced alteration in Swiss albino mice. *Mol. Nutr. Food Res.*, **2007**, *51*, 352–359.

Pavanato, A.; Tuñón, M.J.; Sánchez-Campos, S.; Marroni, C.A.; Llesuy, S.; González-Gallego, J.; Marroni, N. Effects of quercetin on liver damage in rats with carbon tetrachloride-induced cirrhosis. *Dig. Dis. Sci.*, **2003**, 48, 824–829.

Qiusheng, Z.; Xiling, S.; Xubo; Meng, S.; Changhai, W. Protective effects of luteolin-7-glucoside against liver injury caused by carbon tetrachloride in rats. *Pharmazie,* **2004**, *59*, 286–289.

Quan, J.; Piao, L.; Wang, X.; Li, T.; Yin, X. Rossicaside B protects against carbon tetrachloride-induced hepatotoxicity in mice. *Basic Clin. Pharmacol. Toxicol.*, **2009**, 105, 380–386.

Rajalakshmi, G.; Jothi, K.A.; Venkatesan, R.; Jegatheesan, K. Hepatoprotective activity of *Andrographis paniculata* on paracetamol induced liver damage in rats. *J. Pharm. Res.*, **2012**, *5*, 2983–2986.

Ramakrishnan, G.; Raghavendran, H.R.B.; Vinodhkumar, R.; Devaki, T. Supression of N-nitrosodiethylamine induced hepatocarcinogenesis by silymarin in rats. *Chem. Biol. Interact.*, **2006**, *161*, 104–114.

Ramellini, G.; Meldolesi, J. Liver protection by silymarin: *In vitro* effect on dissociated rat hepatocytes. *Arzneimittelforschung*, **1976**, *26*, 69–73.

Rani, M.; Emmanuel, S.; Sreekanth, M.R.; Ignacimuthu, S. Evaluation of in vivo antioxidant and hepatoprotective activity of *Cassia occidentalis* Linn. against paracetamol-induced liver toxicity in rats. *Int. J. Pharm. Pharm. Sci.*, **2010**, *2*, 67–70.

Rao, G.M.M.; Rao, C.V.; Pushpangadan, P.; Shirwaikar, A. Hepatoprotective effects of Rubiadin, a major constituent of *Rubia cordifolia. J. Ethnopharmacol.*, **2006**, *103*, 484–490.

Rao, M.S.; Asad, B.S.; Fazil, M.; Sudharshan, R.; Rasheed, S.; Pradeep, H.; Aboobacker, S.; Thayyil, A.; Riyaz, A.; Mansoor, M.; Aleem, M.; Zeeyauddin, K.; Narasu, M.L.; Anjum, A.; Ibrahim, M. Evaluation of protective effect of *Sapindus mukorossi* saponin fraction on CCl_4-induced acute hepatotoxicity in rats. *Clin. Exp. Gastroenterol.*, **2012**, *5*, 129–137.

Rašković, A.; Milanović, I.; Pavlović, N.; Ćebović, T.; Vukmirović, S.; Mikov, M. Antioxidant activity of rosemary (*Rosmarinus officinalis* L.) essential oil and its hepatoprotective potential. *BMC Complement. Altern. Med.*, **2014,** *14*(1), 225–234.

Rasool, M.K.; Sabina, E.P.; Ramya, S.R.; Preety, P.; Patel, S.; Mandal, N.; Mishra, P.P.; Samuel, J. Hepatoprotective and antioxidant effects of gallic acid in paracetamol-induced liver damage in mice. *J. Pharm. Pharmacol.*, **2010**, *62*, 638–643.

Renganathan, A. *Pharmaco dynamic properties of andrographolide in experimental animals.* M.D. thesis. Pharmacology. Pondicherry: Jawaharlal Institute of Postgraduate Medical Education and Research, Pondicherry, **1999**.

Renugadevi, J.; Prabu, S.M. Cadmium-induced hepatotoxicity in rats and the protective effect of naringenin. *Exp. Toxicol. Pathol.*, **2010**, *62*, 171–181.

Salamone, F.; Galvano, F.; Marino, A.; et al. Silibinin improves hepatic and myocardial injury in mice with non-alcoholic steatohepatitis. *Digest. Liver Dis.*, **2012a**, *44*, 334–342.

Salamone, F.; Galvano, F.; Cappello, F.; et al. Silibinin modulates lipid homeostasis and inhibits nuclear factor kappa B activation in experimental non-alcoholic steatohepatitis. *Transl. Res.*, **2012b**, *159*, 477–486.

Saleem, S.; Shaharyar, M.A.; Khusroo, M.J.; Ahmad, P.; Rahman, R.U.; Ahmad, K.; Alam, M.J.; Al-Harbi, N.O.; Iqbal, M,; Imam, F. Anticancer potential of rhamnocitrin 4'-β-D-galactopyranoside against N-diethylnitrosamine-induced hepatocellular carcinoma in rats. *Mol. Cellular Biochem.*, **2013**, *384*(1–2), 147–153.

Saleh, D.O.; Abdel Jaleel, G.A.; El-Awdan, S.A.; Oraby, F.; Badawi, M. Thioacetamide-induced liver injury: Protective role of genistein. *Can. J. Physiol. Pharmacol.*, **2014**, *92*(11), 965–973.

Salmi, H.A.; Sarna, S. Effect of silymarin on chemical, functional, and morphological alterations of the liver. A double-blind controlled study. *Scand. J. Gastroenterol.*, **1982**, *17*(4), 517–521.

Samojlik, I.; Lakić, N.; Mimica-Dukić, N.; Đaković-Švajcer, K.; Božin, B. Antioxidant and hepatoprotective potential of essential oils of coriander (*Coriandrum sativum* L.) and caraway (*Carum carvi* L.) (Apiaceae). *J. Agric. Food Chem.*, **2010**, *58*(15), 8848–8853.

Sangale, P.; Patil, R. Hepatoprotective activity of alkaloid fractions from ethanol extract of *Murraya koenigii* leaves in experimental animals. *J. Pharm. Sci. Pharmacol.*, **2017**, *3*, 28–33.

Sangi, S.; El-feky, S.A.; Ali, S.S.; Ahmedani, E.I.; Tashtoush, M.H. Hepatoprotective effects of oleuropein, thymoquinone and fruit of *Phoenix dactylifera* on CCl_4 induced hepatotoxicity in rats. *World J. Pharm. Pharm. Sci.*, **2014**, *3*, 3475–3486.

Saraswat, B.; Visen, P.K.; Agarwal, D.P. Ursolic acid isolated from *Eucalyptus tereticornis* protects against ethanol toxicity in isolated rat hepatocytes. *Phytother. Res.*, **2000**, *14*(3), 163–166.

Saraswat, B.; Visen, P.K.; Dayal, R.; Agarwal, D.P.; Patnaik G.K. Protective action of ursolic acid against chemical induced hepatotoxicity in rats. *Indian J. Pharmacol.*, **1996**, *28*, 232–237.

Saraswat, B.; Visen, P.K.S.; Patnaik, G.K.; Dhawan, B.N. Effect of andrographolide against galactosamine-induced heaptotoxicity. *Fitoterpia*, **1995**, *66*, 415–420.

Saraswat, B.; Visen, P.K.; Patnaik, G.K.; Dhawan, B.N. Protective effect of picroliv active constituent of *Picrorhiza kurroa*, against oxytetracycline-induced hepatic damage. *Indian J. Exp. Biol.*, **1997**, *35*, 1302–1305.

Saravanan, R.; Viswanathan, P.; Pugalendi, K.V. Protective effect of ursolic acid on ethanol-mediated experimental liver damage in rats. *Life Sci.*, **2006**, *78*(7), 713–718.

Sarkar, C.; Bose, S.; Banerjee, S. Evaluation of hepatoprotective activity of vasicinone in mice. *Indian J. Exp. Biol.*, **2014**, *54*, 705–711.

Sarkar, K.; Ghosh, A.; Kinter, M.; Mazumder, B.; Parames, C. Purification and characterization of a 43 kD hepatoprotective protein from the herb *Cajanus indicus* L. *Protein J.*, **2006**, *25*(6), 41–421.

Schriewer, H.; Weinhold, F. The influence of silybin from *Silybum marianum* (L.) Gaertn. on *in vitro* phosphatidylcholine biosynthesis in rat livers. *Arzneimittelforschung*, **1973**, *29*, 791–792.

Shakya, G.; Manjini, S.; Hoda, M.; Rajagopalan, R. Hepatoprotective role of kaempferol during alcohol- and ΔPUFA-induced oxidative stress. *J. Basic Clin. Physiol. Pharmacol.*, **2014**, *25*, 73–79.

Sharma, A.; Chakraborti, K.K.; Handa, S.S. Antihepatotoxic activity of some herbal formulations as compared to silymarin. *Fitoterapia*, **1991**, *62*, 229–235

Shi, Y.; Zhang, L.; Jiang, R.; Chen, W.; Zheng, W.; Chen, L.; Tang, L.; Li, L.; Li, L.; Tang, W.; Wang, Y.; Yu, Y. Protective effects of nicotinamide against acetaminophen-induced acute liver injury. *Int. Immunopharmacol.*, **2012**, *14*, 530–537.

Shim, S.B.; Kim, N.J.; Kim, D.H. β-Glucuronidase inhibitory activity and hepatoprotective effect of 18β-glycyrrhetinic acid from the rhizomes of *Glycyrrhiza uralensis*. *Planta Med.*, **2000**, *66*(1), 40–43.

Shimoda, H.; Tanaka, J.; Kikuchi, M.; Fukuda, T.; Ito, H.; Hatano, T.; et al. Walnut polyphenols prevent liver damage induced by carbontetrachloride and D-galactosamine: hepatoprotective hydrolysable tannins in the kernel pellicles of walnut. *J. Agric. Food Chem.*, **2008**, *56*, 4444–4449.

Shukla, B.; Visen, P.K.S.; Patnaik, G.K.; Tripathi, S.C.; Srimal, R.C.; Dayal, R.; et al. Hepatoprotective activity in the rat of ursolic acid isolated from Eucalyptus hybrid. *Phytother. Res.*, **1992**, *6*(2), 74–79. doi.org/10.1002/ptr.2650060205

Siegers, C.P.; Fruhling, A.; Younes, M. Influence of dithiocarb, (+)-catechin and silybine on halothane hepatotoxicity in the hypoxic rat model. *Acta Pharmacol. Toxicol.* (Copenh), **1983**, *53*, 125–129.

Simandi, B.; Deak, A.; Ronyani, E.; Yanxiang, G.; Veress, T.; Lemberkovics, E. et al. Supercritical carbon dioxide extraction and fractionation of Fennel oil. *J. Agric. Food Chem.*, **1999**, *47*, 1635–1640.

Singh, B.; Saxena, A.K.; Chandan, B.K.; Agarwal, S.G.; Anand, K.K. In vivo hepatoprotective activity of active fraction from ethanolic extract of *Eclipta alba* leaves. *Indian J. Physiol. Pharmacol.*, **2001**, *45*(4), 435–441.

Singh, B.; Saxena, A.K.; Chandan, B.K., Bhardwaj, V.; Gupta, V.N.; Suri, O.P.; Handa, S.S. Hepatoprotective activity of indigtone: A bioactive fraction from *Indigofera tinctoria* Linn. *Phytother. Res.*, **2001**, *15*(4), 294–297.

Singh, M.; Sasi, P.; Gupta, V.H. et al. Protective effect of curcumin, silymarin and N-acetylcysteine on antitubercular drug-induced hepatotoxicity assessed in an in vitro model. *Hum. Exp. Toxicol.*, **2012**, *31*, 788–797.

Song, E.K.; Kim, J.H.; Kim, J.S.; Cho, H.; Nan, J.X.; Soku, D.H.; Ko, G.J.; Oh, H.; Ki, Y.C. Hepatoprotective phenolic constituents of *Rhodiola sachalinensis* on tacrine-induced cytotoxicity in Hep G2 cells. *Phytother. Res.*, **2003**, *17*(5), 563–565.

Sotelo-F′elix, J.I.; Martínez-Fong, D.; Muriel De la Torre, P. Protective effect of carnosol on CCl_4-induced acute liver damage in rats. *Eur. J. Gastroenterol. Hepatol.*, **2002**, *14*(9), 1001–1006.

Sumathi, T.; Nongbri, A. Hepatoprotective effect of Bacoside-A, a major constituent of *Bacopa monniera* Linn. *Phytomedicine,* **2008**, *15*, 901–905.

Sung, S.H.; Won, S.Y.; Cho, N.J.; Gkim, C.Y. Hepatoprotective flavonol glycosides of *Saururus chinensis* herbs. *Phytother. Res.*, **1997**, *11*(7), 500–503.

Sunitha, S.; Nagaraj, M.; Varalakshmi, P. Hepatoprotective effect of lupeol and lupeol linoleate on tissue antioxidant defence system in cadmium-induced hepatotoxicity in rats. *Fitoterapia*, **2001**, *72*, 516–523.

Syamasundar, K.V.; Singh, B.; Thakur, R.S.; Husain, A.; Yoshinobu, K.; Hiroshi, H. Anti-hepatotoxic principles of *Phyllanthus niruri* herbs. *J. Ethnopharmacol.*, **1985**, *14*(1), 41–44.

Tan, J.; Shenghua, L.; Zeng, J.; Wu, X. Antioxidant and hepatoprotective activities of total flavonoids of *Indocalamus latifolius. Bangladesh J. Pharmacol.*, **2015**, *10*, 779–789.

Tan, S.; Lu, Q.; Shu, Y.; Sun, Y.; Chen, F.; Tang, L. Iridoid glycosides fraction isolated from *Veronica ciliata* Fisch. protects against acetaminophen-induced liver injury in mice. *Evid. Based Complement. Altern. Med.*, **2017**, *2017*, 1–11, doi.org/10.1155/2017/6106572

Thakur, S.K. Silymarin-A hepatoprotective agent. *Gastroenterol. Today,* **2002**, *6*, 78–82.

Tirkey, N.; Pilkhwal, S.; Kuhad. A.; Chopra, K. Hesperidin, a citrus bioflavonoid, decreases the oxidative stress produced by carbon tetrachloride in rat liver and kidney. *BMC Pharmacol.*, **2005**, *5*, 2–12,

Tran, Q.L.; Adnyana, I.K.; Tezuka, Y.; Harimaya, Y.; Saiki, I.; Kurashige, Y.; Tran, Q.K.; Kadota, S. Hepatoprotective effect of majonoside R2, the major saponin from Vietnamese ginseng (*Panax vietnamensis*). *Planta Med.*, **2002**, *68*, 402–406.

Tran, Q.I.; Adnyana, I.K.; Tezuka, Y.; Nagaoka, T.; Tran, Q.K.; Kadota, S. Triterpene saponins from Vietnamese ginseng (*Panax vietnamensis*) and their hepatoprotective activity. *J. Nat. Prod.,* **2001**, *64*, 456–461.

Tripathi, S.C.; Patnaik, G.K.; Dhawan, B.N. Hepatoprotective activity of picroliv against alcohol-carbon tetachloride induced damage in rat. *Indian J. Pharmacol.,* **1991**, *23*, 143–148.

Tyutyulkova, N.; Gorantcheva, U.; Tuneva, S.; Chelibonova-Lorer, H.; Gavasova, E.; Zhivkov, V. Effect of silymarin (carsil) on the microsomal glycoprotein and protein biosynthesis in liver of rats with experimental galactosamine hepatitis. *Meth. Find. Exp. Clin. Pharmacol.,* **1983**, *5*, 181–184.

Uchida N. S.; Silva-Filho S. E.; Aguiar R. P.; et al. Protective effect of *Cymbopogon citratus* essential oil in experimental model of acetaminophen-induced liver injury. *Am. J. Chin. Med.,* **2017**, 45(3), 515–532.

Uchida N.S.; Silva-Filho S.E.; Cardia G.F.E.; et al. Hepatoprotective effect of citral on acetaminophen-induced liver toxicity in mice. *Evid.-Based Complement. Altern. Med.* **2017**, 2017, doi: 10.1155/2017/1796209.

Upadhyay, K.; Gupta, N.K.; Dixit, V.K. Development and characterization of phyto-vesicles of wedelolactone for hepatoprotective activity. *Drug Dev. Ind. Pharm.,* **2012**, *38*(9), 1152–1158.

Vaidya, A.B.; Antarkar, D.S.; Doshi, J.C.; Bhatt, A.D.; Ramesh, V.; Vora, P.V.; Perissond, D.; Baxi, A.J.; Kale, P.M. *Picrorhiza kurroa* (Kutaki) Royle ex Benth as a hepatoprotective agent—experimental & clinical studies. *J. Postgrad. Med.,* **1996**, *42*(4), 105–108.

Valan, M.F.; Britto, A.J.; Venkataraman, R. A brief review: Phytoconstituents with hepatoprotective activity. *Int. J. Chem. Sci.*, **2010**, *8*, 1421–1432.

Valezuela, A.; Barria, T.; Guerra, R.; Garrido, A. Inhibitory effect of the flavonoid silymarin on the erythrocyte hemolysis induced by phenylhydrazine. *Biochem. Biophys. Res. Commun.*, **1985**, *126*, 712–718.

Valenzuela, A.; Garrido, A. Biochemical bases of the pharmacological action of flavanoid silymarin and its structural isomer silibinin. *Biol. Res.,* **1994**, *27*, 105–112.

Valenzuela, A.; Guerra, R. Differential effect of silybin on the Fe^{2+} -ADP and t-butyl hydroperoxide-induced microsomal lipid peroxidation. *Experientia,* **1986**, *42*, 139–141.

Valenzuela, A.; Guerra, R.; Garrido, A. Silybin dihemisuccinate protects rat erythrocytes against phenylhydrazine-induced lipid peroxidation and hemolysis. *Planta Med.*, **1987**, *53*, 402–405.

Valenzuela, A.; Lagos, C.; Schmidt, K.; Videla, L.A. Silymarin protection against hepatic lipid peroxidation induced by acute ethanol intoxication in the rat. *Biochem. Pharmacol.,* **1985**, *34*, 2209–2212.

Van Puyvelde, L.; Kayonga, A.; Brioen, P.; Costa, J.; Ndimubakunzi, A.; De Kimpe, N.; Schamp, N. The hepatoprotective principle of *Hypoestes triflora* leaves. *J. Ethnopharmacol.,* **1989**, *26*(2), 121–127.

Vimal, V.; Devaki, T. Hepatoprotective effect of allicin on tissue defense system in galactosamine/endotoxin challenged rats. *J. Ethnopharmacol.,* **2004**, *90*(1), 151–154.

Visen, P.K.; Saraswat, B.; Dhawan, B.N. Curative effect of picroliv on primary cultured rat hepatocytes against different hepatoxins: An in vitro study. *J. Pharmacol. Toxicol. Meth.,* **1998**, *40*, 173–179.

Visen, P.K.S.; Shukla, B.; Patnaik, G.K.; Kaul, S.; Kapoor, N.K.; Dhawan, B.N. Hepatoprotective activity of picroliv, the active principle of *Picrorhiza kurroa*, on rat hepatocytes against paracetamol toxicity. *Drug Dev. Res.*, **1991**, *22*, 209–219.

Visen, P.; Shukla, B.; Patnaik, G.K.; Tripathi, S.C.; Kulshreshtha, D.K.; Srimal, R.C.; Dhawan, B.N. Hepatoprotective activity of *Ricinus communis* leaves. *Pharm. Biol.,* **2008**, *30*(4), 241–250.

Visen, P.K.S.; Shukla, B.; Patnaik, G.K.; Chanda, R.; Singh, V.; Kapoor, N.K.; Dhawan, B.N. Hepatoprotective activity of picroliv isolated from *Picrorrhiza kurroa* against thioacetamide toxicity on rat hepatocytes. *Phytother. Res.,* **1991**, *5*, 224–227.

Visen, P.K.S.; Shukia, B.; Patnaik, G.K.; Dhawan, B.N. Andrographolide protects rat hepatocytes against paracetamol-induced damage. *J. Ethnopharmacol.,* **1993**, *40*(2), 131–136.

Wafay, H.; El-Saeed, G.; El-Toukhy, S.; Youness, E.; Ellaithy, N.; Agaibi, M.; Eldaly, S. Potential effect of garlic oil and silymarin on carbon tetrachloride-induced liver injury. *Aust. J. Basic Appl. Sci.,* **2012**, 6, 409–414.

Wang, C.-J.; Wang, J.-M.; Lin, W.-L.; Chu, C.-Y.; Chou, F.-P.; Tseng, T.-H. Protective effect of *Hibiscus anthocyanins* against tert-butyl hydroperoxide-induced hepatic toxicity in rats. *Food Chem. Toxicol.,* **2000**, 38(5), 411–416.

Wang, M.; Grange, L.L.; Tao, J. Hepatoprotective properties of *Silybum marianum* herbal formulation on ethanol induced liver damage. *Fitoterpia,* **1996**, *67*, 167–171.

Wang, X.; Liu J.R.; Zhang X.H.; et al. Seabuckthorn berry polysaccharide extracts protect against acetaminophen induced hepatotoxicity in mice via activating the Nrf-2/HO-1-SOD-2 signaling pathway. *Phytomedicine,* **2018**, 38, 90–97.

Wang, Y.; Han, T.; Xue, L.M.; Han, P.; Zhang, Q.Y.; Huang, B.K.; et al. Hepatotoxicity of kaurene glycosides from *Xanthium strumarium* L. fruits in mice. *Pharmazie*, **2011**, *66*, 445–449.

Wu, J.-B.; Lin, W.-L.; Hsieh C.-C.; Ho, H.-Y.; Tsay, H.-S.; Lin, W.-C. The hepatoprotective activity of kinsenoside from *Anoectochilus formosanus. Phytother. Res.*, **2007**, *21*, 58–61.

Wu, Y.; Li, L.; Wen, T.; Li, Y.Q. Protective effects of echinacoside on carbon tetrachloride-induced hepatotoxicity in rats. *Toxicology,* **2007**, 232, 50–56.

Wu, Y.; Yang, L.; Wang, F.; Wu, X.; Zhou, C.; et al. Hepatoprotective and antioxidative effects of total phenolics from *Laggera pterodonta* on chemical-induced injury in primary cultured neonatal rat hepatocytes. *Food Chem. Toxicol.,* **2007**, *45*(8), 1349–1355.

Xin, Y.; Wei, J.; Chunhua, M.; Danhong, Y.; Jianguo, Z.; Zonqqi, C.; et al. Protective effects of Ginsenoside Rg1 against carbon tetrachloride-induced liver injury in mice through suppression of inflammation. *Phytomedicine,* **2016**, *23*, 583–588.

Xiong, Q.; Hase, K.; Tezuka, Y.; Tani, T.; Namba, T.; Kadota, S.K. Hepatoprotective activity of phenylethanoids from *Cistanche deserticola. Planta Med.*, **1998**, *64*, 120–125.

Yadav, N.P.; Chanda, D.; Chattopadhyay, S.K.; Gupta, A.K.; Pal, A. Hepatoprotective effects and safety evaluation of coumarinolignoids isolated from *Cleome viscosa* seeds. *Indian J. Pharm. Sci.,* **2010**, *72*(6), 759–765.

Yadav, N.P.; Pal, A.; Shanker, K.; Bawankule, D.U.; Gupta, A.K.; et al. Synergistic effect of silymarin and standardized extract of *Phyllanthus amarus* against CCl_4-induced hepatotoxicity in *Rattus norvegicus. Phytomedicine,* **2008**, *15*(12), 1053–1061.

Yan, H.M.; Xia, M.F.; Wang, Y.; Chang, X.X.; Yao, X.Z.; Rao, S.X.; Zeng, M.S.; Tu, Y.F.; Feng, R.; Jia, W.P.; Liu, J.; Deng, W.; Jiang, J.D.; Gao, X. Efficacy of berberine in patients with non-alcoholic fatty liver disease. *PLoS One,* **2015**, 10, e0134172.

Yang, H.; Sung, S.H.; Kim, Y.C. Two new hepatoprotective stilbene glycosides from *Acer mono* leaves. *J. Nat. Prod.,* **2005**, *68*(1), 101–103.

Yang, J.; Li, Y.; Wang, F.; Wu, C. Hepatoprotective effects of apple polyphenols on CCl_4-induced acute liver damage in mice. *J. Agric. Food Chem.*, **2010**, *58*(10), 6525–6531.

Yang, L.; Wang, C.Z.; Ye, J.Z.; Li, H.T. Hepatoprotective effects of polyprenols from *Ginkgo biloba* L. leaves on CCl_4-induced hepatotoxicity in rats. *Fitoterapia,* **2011**, *82*(6), 834–840.

Yang, S.Y.; Hong, C.O.; Lee, G.P.; Kim, C.T.; Lee, K.W. The hepatoprotection of caffeic acid and rosmarinic acid, major compounds of *Perilla frutescens*, against *t*-BHP-induced oxidative liver damage. *Food Chem. Toxicol.*, **2013**, *55*, 92–99.

Yoshikawa, M.; Morikawa, T.; Kashima, Y.; Ninomiya, K.; Matsuda, H. Structure of new dammarane-type triterpene saponins from the flower buds of *Panax notoginseng* and hepatoprotective effects of principal ginseng sap saponin. *J. Nat. Prod.,* **2003**, *66*, 922–924.

Yoshikawa, M.; Ninomiya, K.; Shimoda, H.; Nishida, N.; Matsuda, H. Hepatoprotective and antioxidative properties of *Salacia reticulata*: Preventive effects of phenolic constituents on carbontetrachloride-induced liver injury in mice. *Biol. Pharm. Bull.*, **2002**, *25*(1), 72–76.

Yoshikawa, M.; Xu, F.; Morikawa, T.; Ninomiya, K.; Matsuda, H. Anastatins A and B, new skeletal flavonoids with hepatoprotective activities from the desert plant *Anastatica hierochuntica*. *Bioorg. Med. Chem. Lett.*, **2003**, *13*(6), 1045–1049.

Yu, H.; Zheng, L.; Yin, L.; Xu, L.; Qi, Y.; Han, X.; Xu, Y.; Liu, K.; Peng, J. Protective effects of the total saponins from *Dioscorea nipponica* Makino against carbon tetrachloride-induced liver injury in mice through suppression of apoptosis and inflammation. *Int. Immunopharmacol.*, **2014**, 19, 233–244.

Yu, J.; Wang, Y.; Qian, H.; Zhao, Y.; Liu, B.; Fu, C. Polyprenols from *Taxus chinensis* var. *mairei* prevent the development of CCl_4-induced liver fibrosis in rats. *J. Ethnopharmacol.*, **2012**, *142*, 151–160.

Yue, S.M.; Kang, W.Y. Lowering blood lipid and hepatoprotective activity of amentoflavone from *Selaginella tamariscina* in vivo. *J. Med. Plant. Res.*, **2011**, *5*, 3007–3014.

Zakaria, Z.A.; Rofiee, M.S.; Somchit, M.N.; Zuraini, A.; Sulaiman, M.R.; Teh, L.K.; et al. Hepatoprotective activity of dried- and fermented-processed virgin coconut oil. *Evid. Compliment. Alt. Med.*, **2011**, *2011*, 1–8.

Zhang, M.; Pan, L.; Jiang, S.; Mo, Y. Protective effects of anthocyanin from purple sweet potato on acute carbon tetrachloride-induced oxidative hepatotoxicity fibrosis in mice. *Food Agric. Immunol.*, **2015**, *27*, 157–170.

Zhang, Y.X.; Hai, J.; Cao, M.; et al. Silibinin ameliorates steatosis and insulin resistance during non-alcoholic fatty liver disease development partly through targeting IRS-1/PI3K/Akt pathway. *Int. Immunopharmacol.*, **2013**, *17*, 714–720.

Zhang, Z.F.; Fan, S.H.; Zheng, Y.L.; Lu, J.; Wu, D.M.; Shan, Q.; Hu, B. Troxerutin protects the mouse liver against oxidative stress-mediated injury induced by D-galactose. *J. Agric. Food Chem.*, **2009**, *57*(17), 7731–7736.

Zheng, Q. S.; Sun, X. L.; Xu, B.; Li, G.; Song, M. Mechanisms of apigenin-7-glucoside as a hepatoprotective agent. *Biomed. Environ. Sci.*, **2005**, *18*, 65–70.

Zheng, Y-F.; Wei, J-H.; Fang, S-Q.; Tang, Y-P.; Cheng, H-B.; Wang, T-L.; et al. Hepatoprotective triterpene saponins from the roots of *Glycyrrhiza inflata*. *Molecules*, **2015**, *20*, 6273–6283.

Zhou, C.M.; Yin, S.J.; Yu Z.F.; et al. Preliminary characterization, antioxidant and hepatoprotective activities of polysaccharides from Taishan *Pinus massoniana* pollen. *Molecules*, 2018, 23(2), Art. no. 281.

Zhu, X.; Guo, X.; Mao, G.; Gao, Z.; Wang, H.; He, Q.; Li, D. Hepatoprotection of berberine against hydrogen peroxide-induced apoptosis by upregulation of Sirtuin 1. *Phytother. Res.*, **2013**, 27, 417–421.

CHAPTER 5

Ethnomedicinal Plants for Hepatoprotection in India

ABSTARCT

In recent times, medicinal plants have received much needed attention as sources of bioactive substances used to treat wide variety of diseases and disorders of major body organs including liver as a hepatoprotective and antioxidants. Ethnobotany is a study of how people of particular cultures and regions make use of the plants in their local environments. Liver is the heaviest gland of the body and plays the major role in metabolic activities and bio-chemical conversions. Hepatic disease is a basic collective term of conditions, diseases, and infections that affect the cells, tissues structures, or functions of the liver. The present review is an attempt to review the usage of ethnomedicinal plants for hepatoprotection in India, utilizing among the different tribal culture in India by using scientific studies. This may be useful to researchers who are working in the area hepatopharmacology and therapeutics.

INTRODUCTION

The simplest definition of ethnobotany is provided by the word itself: *ethno* (people) and *botany* (science of plants). In essence, it is a study of how people of particular cultures and regions make use of the plants in their local environments. These uses can include as food, medicine, fuel, shelter, and in many cultures, in religious ceremonies. In 1895, during a lecture in Philadelphia, a botanist named John Harshberger, provided the first definition of *ethnobotany* as the study of *how* native tribes used plants for food, shelter, or clothing (Harshberger, 1896). One of the best-known modern ethnobotanists was Richard Evans Schultes identified the field of ethnobotany as an interdisciplinary field, combining botany, anthropology, economics, ethics, history, chemistry, and many other areas of study.

Ethnobotanists need to be prepared to ask the following questions:

1. What are the fundamental ideas and conceptions of people living in a particular region about the plant life surrounding them?
2. What effect does a given environment have on the lives, customs, religion, thoughts, and everyday practical affairs of the people studied?
3. In what ways do the people make use of the local plants for food, medicine, material culture, and ceremonial purposes?
4. How much knowledge do the people have of the parts, functions, and activities of plants?
5. How are plant names categorized in the language of the people studied, and what can the study of these names reveal about the culture of the people?

The modern system of medicine still lack in providing suitable medicament for a large number of disease conditions, in spite of tremendous advances made in the discovery of new compounds. A few of these diseases are hepatic disorders, viral infections, AIDS, rheumatic diseases (Mohammad, 1994), etc. The available therapeutic agents only bring about symptomatic relief without any influence on the curative process, thus causing the risk of relapse and the danger of untoward effects. A large number of populations suffer due to various reasons from hepatic diseases of unknown origin. The development of antihepatotoxic drugs being a major thrust area has drawn the attention of workers in the field of natural product research.

India is a vast country with greatest emporia of plant wealth and represents a colorful mosaic of about 563 tribal communities which possess considerable knowledge regarding use of plants for livelihood, healthcare and other proposes through their long association with forestry inheritance and experiences (Bala et al., 2011).

ETHNOBOTANY OF HEPATOPROTECTIVE PLANTS

In this chapter, medicinal plants being used for curing and prevention of liver diseases by different ethnic tribes is reviewed. Information is drawn from the published papers in different journals, books, and theses. We do not claim to have included all the existing tribal communities of India information about ethnobotanical use of medicinal plants but we rather focused on information easily accessible. A list of plants used by the tribal people prepared which gives the scientific name, part of the plant used, name of the tribal community, local name of the plant, uses and/or mode of use. The precision of botanical identification in this review depends on that from original sources.

TABLE 5.1 Ethnobotanical Literature Review of Indian Medicinal Plants Used for Treatment of Various Liver Disease/Disorders

S. No	Plant Name	Part Used	Tribe/Local Community	Local Names (State)	Uses/Mode of Use (Reference)
1	*Abrus precatorius* L.	Roots	Mahadevkoli, Thakur, and Katkari	(Konkan Region, Maharashtra)	The decoction of roots has been used for jaundice (Naikade and Meshram, 2014)
2	*Abrus precatorius* L.	Seeds	Local healers	Gunja/Ratti (Arunachal Pradesh)	Seeds are used for jaundice and hemoglobin uric bile (Rama et al., 2012)
3	*Acacia catechu* (L.f.) Willd.	Leaf, bark	Local people and traditional healers	Karintaali (Wayanad, Palakkad, Idukki districts, Kerala)	Hepatoprotective (Girish and Pradhan, 2012)
4	*Acacia nilotica* (L.) Willd.	Whole plant/ flower	Local people and traditional healers	Babool (Chhattisgarh)	Used for the treatment of jaundice (Janghel et al., 2019)
5	*Acacia nilotica* (L.) Willd.	Bark, gum	Local people and traditional healers	Karuvelam (Coimbatore, Tamil Nadu)	Hepatoprotective (Janghel et al., 2019)
6	*Acanthus ilicifolius* L.	Leaves	Local people and traditional healers	Arrumulli (Thrissur)	Hepatoprotective (Babu et al., 2001)
7	*Achillea millefolium* L.	Flower, leaf	Local people and traditional healers	Achchilliya (Wayanad, Palakkad, Idukki districts, Kerala)	Hepatoprotective (Girish and Pradhan, 2012)
8	*Achras sapota* L.	Leaf, flower	Local people and traditional healers	Pilaghanti (Chhattisgarh)	Used for the treatment of jaundice and liver tumor (Janghel et al., 2019)
9	*Achyranthes aspera* L.	Leaf, root, whole plant	Local medical practitioners/healers	Puthkanda (Shivalik Hills of Northwest Himalaya)	Decoction of rootis used for jaundice (Devi et al., 2014)
10	*Achyranthus aspera* L.	Leaves	Traditional healers	Uttareni (Parnasala Sacred Grove Area Eastern Ghats of Khammam District,Telangana)	Jaundice: Tender leaves along with the tender leaves of *Careya arborea, Mimosa pudica,* and *Ziziphus mauritiana* are crushed to paste and the paste along with cow milk is administered for 7 days (Srinivasa et al., 2015)
11	*Aconitum palmatum* D. Don	Root	Traditional healers	Prativishaa (Sikkim, Nepal)	Hepatoprotective (Valvi et al., 2016)

TABLE 5.1 *(Continued)*

S. No	Plant Name	Part Used	Tribe/Local Community	Local Names (State)	Uses/Mode of Use (Reference)
12	*Aconitum rotundifolium* Kar. & Kir.	Whole plant	Larje, Amchis	Bonkar, Pongtha, Vashi (Lahaul–Spiti region Western Himalaya)	Whole plant juice extract is taken orally with equal volume of water to cure jaundice (Koushalya, 2012)
13	*Adenanthera pavonina* L.	Leaves	Local people	Rakta Kanchana (Himalayan Tract)	Hepatoprotective (Mujahid et al., 2013)
14	*Adhatoda vasica* Nees	Flower, leaves	Local people of Taindol	Adusa (Taindol, Jhansi, Uttar Pradesh)	Leaves extract used for jaundice (Jitin, 2013).
15	*Adhatoda zeylanica* Nees	Leaf	Local healers	Vasaka (Arunachal Pradesh)	Fresh leaf juice is administered for one month for jaundice (Rama et al., 2012)
16	*Adiantum lunulatum* Burm.f.	Leaves	Yenadi	Hamsa padi (Andhra Pradesh)	Leaves are grinded to get juice and used in jaundice (Bala et al., 2011)
17	*Adiantum venustum* D.Don	Whole plant	Gujjars, Bhoris Bakerwals	Kakbai (Bandipora district, Jammu and Kashmir)	Dried fronds are crushed to obtain powder. Powder is added to a glass of water and kept as such overnight. The extract is given next day early in the morning for the treatment of cough, jaundice and stomach ailments (Parvaiz et al., 2013)
18	*Adina cordifolia* Hook. f. ex Brandis	Root	Local people and Ayurvedic practioners	Haridru	Hepatoprotective (Sharma et al., 2012)
19	*Aegiceras corniculatum* (L.) Blanco	Stem	Local people and traditional healers	Halasi (Maharastra, Andhra Prasesh)	Hepatoprotective (Kshirsagar et al., 2011)
20	*Aegle marmelos* (L.) Corr.	Leaf	Malayali tribes	Vilvam (Shevaroy Hills, Tamil Nadu)	The leaf juice 50 mL mixed with cow's milk used to cure jaundice (Alagesaboopathi, 20015)
21	*Aegle marmelos* (L.) Corr.	Leaf	Mahadevkoli, Thakur, and Katkari	Bel (Konkan Region, Maharashtra)	Leaf powder is given along with goat milk for jaundice (Naikade and Meshram, 2014)

TABLE 5.1 *(Continued)*

S. No	Plant Name	Part Used	Tribe/Local Community	Local Names (State)	Uses/Mode of Use (Reference)
22	*Aegle marmelos* (L.)Correa	Fruit	Local people of Taindol	Bel (Taindol, Jhansi, Uttar Pradesh)	Used for jaundice (Jitin, 2013)
23	*Aerva lanata* (L.) Juss.	Roots	Local people	Kali-Bui (Shekhawari region, Rajasthan)	Root extract is given to patients of liver congestion and jaundiece (Rishikesh and Ashwani, 2012)
24	*Agave americana* L.	Leaves	Local people of Taindol	Kantala (Taindol, Jhansi, Uttar Pradesh)	Used for jaundice (Jitin 2013)
25	*Agave americana* L.	Pulp	Local medical practitioners/healers	Ramban (Shivalik Hills of Northwest Himalaya)	The pulp of plant is taken orally as blood purifier, constipation, acidity, as liver tonic. (Devi et al., 2014)
26	*Alangium salvifolium* (L.f.) Wang.	Root, twig, bark	Local traditional healers	Ankula (Nawarangpur District, Odisha)	Root bark paste (5 g) mixed with 7 blackpeppers (*Piper nigrum)* is administered twice a day for 10 days against hepatitis (Dhal et al., 2014)
27	*Allium cepa* L.	Bulb	Local people	Erra ulligadda (Andhra Pradesh)	Hepatoprotective (Kumar et al., 2013)
28	*Aloe barbadensis* L.	Leaf	Local healers	Ghritkumari (Arunachal Pradesh)	Fresh leaf juice is administered for one month (Rama et al., 2012)
29	*Aloe vera* L.	Leaves	Local people and traditional healers	Catevala (Wayanad, Palakkad, Idukki districts, Kerala)	Hepatoprotective (Girish and Pradhan, 2012)
30	*Alstonia scholaris* R. Br.	Bark	Local healers	Saptaparna/Sitan Gachha (Arunachal Pradesh)	Pieces of bark are worn in a garland for curing jaundice (Rama et al., 2012)
31	*Alternanthera pungens* L.	Whole plant	Local people	Katua Shak (Amarkantak region, Madhya Pradesh)	Used for liver and spleen complaints (Achuta et al., 2010)
32	*Alternanthera sessilis* (L.) R.Br.	Leaves	Irula tribals	Ponaganikerai (Kalavai village, Vellore district, Tamil Nadu)	Used in jaundice (Natarajan et al., 2013)

TABLE 5.1 *(Continued)*

S. No	Plant Name	Part Used	Tribe/Local Community	Local Names (State)	Uses/Mode of Use (Reference)
33	*Alternanthera sessilis* (L.) R.Br.	Whole plant	Fisherman community	Ponnanganni (Pudhukkottai District, Tamil Nadu)	The whole plant is used to treat diarrhea, skin disease, dyspepsia, hemorrhoids, liver problems (Rameshkumr and Ramakritinan, 2013)
34	*Amaranthus spinosus* L.	Roots	Kondh, Sabra, Naik tribes	Kanta Marish (Khordha forest division of Khordha District,Odisha)	A handful of dried roots made into fine powder. 3–5 g powder is given twice a day with sufficient water to treat jaundice (Mukesh et al., 2014)
35	*Amaranthus tricolor* L.	Roots	Local people and Ayurvedic practioners	Maarisha-rakta (Throughout India)	Hepatoprotective (Aneja et al., 2013)
36	*Amomum subulatum* Roxb.	Seed	Ethnic and rural people in Eastern Sikkim Himalayan Region	Elaichi (Sikkim)	Liver tonic (Trishna et al., 2012)
37	*Amomum subulatum* Roxb.	Seeds	Hakims, priests, tribal people	Elaichi (Eastern Sikkim Himalaya region)	Stomachic, heart, and liver tonic (Trishna et al., 2012)
38	*Amorphophallus paeoniifolius* L.	Tubers	Ayurvedic practioners	Elephant-foot yam (Throughout India)	Liver tonic (Hurkadale et al., 2012)
39	*Anacardium occidentale* L.	Fruit, leaf, bark	Local people and traditional healers	Kapamava (Wayanad, Palakkad, Idukki districts, Kerala)	Hepatoprotective (Girish and Pradhan, 2012)
40	*Andrographis affinis* Nees	Leaf	Malayali tribes	Kodikkurundhu, Keeripparandai (Shevaroy Hills, Tamil Nadu)	Leaf paste mixed with cow's milk used in liver ailments. Leaf extract 50 mL with mixed with Buffalo curd given internally a day for one week to cure jaundice (Alagesaboopathi, 20015)
41	*Andrographis alata* (Vahl) Nees	Leaf	Malayali tribes	Periyanangai (Shevaroy Hills, Tamil Nadu)	Leaf extract 25 mL mixed with hot water is given orally twice a day for seven to ten days in jaundice (Alagesaboopathi, 2015)
42	*Andrographis echioides* (L.) Nees	Whole plant	Chenchu	Dontarala aku (Nallamalais, Andhra Pradesh)	Whole plant water extract used to cure liver disease (Sabjan et al., 2014)

TABLE 5.1 *(Continued)*

S. No	Plant Name	Part Used	Tribe/Local Community	Local Names (State)	Uses/Mode of Use (Reference)
43	*Andrographis lineata* Nees	Root	Malayali tribes	Periyanangai (Shevaroy Hills, Tamil Nadu)	Root paste (25 g) given along with cow's milk for four to six days to treat enlargement of liver. Leaf juice 50 mL mixed with cow's milk and taken orally twice a day for 5 days in liver diseases (Alagesaboopathi, 2015)
44	*Andrographis macrobotrys* Nees	Leaf, root	Malayali tribes	–	Fresh leaf juice is given orally thrice a day for one week to treat liver disorders. The root powder mixed with goat's milk and taken orally to treat jaundice (Alagesaboopathi, 2015)
45	*Andrographis neesiana* Wight	Leaf	Malayali tribes	–	Fresh leaf ground with water and the paste is given orally to cure jaundice (Alagesaboopathi, 2015)
46	*Andrographis ovata* C.B. Clarke	Leaf	Malayali tribes	–	Leaf extract (25 mL) mixed with common salt and *Piper nigrum* (black pepper) three times a day for 7 days to treat liver ailments (Alagesaboopathi, 2015)
47	*Andrographis paniculata* (Brum.f.) Nees	Whole plant	Local tribe	Kalmegh; Chirata (Oraon) (Jalpaiguri district, West Bengal)	Whole plant or leaf extract used as liver tonic (Bose et al., 2015)
48	*Andrographis paniculata* (Brum.f.) Nees	Leaf	Koch, Meich, Rava, Munda, Santhal, Garo, Oraon	(Coochbehar district, West Bengal)	Leaf extract to treat jaundice (Tanmay et al., 2014)
49	*Andrographis paniculata* (Brum.f.) Nees	Leaf, root	Chenchu	Nelavemu (Nallamalais, Andhra Pradesh)	Leaf or root extract filtered and administered for liver diseases. This plant leaf extract especially used by chenchu tribals in Bairluty tribal hamlet to get rid of hangover symptoms (Sabjan et al., 2014)

TABLE 5.1 *(Continued)*

S. No	Plant Name	Part Used	Tribe/Local Community	Local Names (State)	Uses/Mode of Use (Reference)
50	*Andrographis paniculata* (Burm. f.) Nees	Leaves	Local healers	Kalmegha/Chirata	Leaves and young twigs are smashed and made paste; 20–30 g paste taken three times daily after meal for 2–3 weeks to cure symptoms (Rama et al., 2012)
51	*Andrographis paniculata* (Burm.f.) Nees	Leaves	Yerukala,Yanadi,Sugali	Nelavemu (Sheshachala hill range of Kadapa District, Andhra Pradesh)	Decoction of leaves cures jaundice (Rajagopal et al., 2011)
52	*Andrographis paniculata* (Burm.f.) Nees	Whole plant	Malayali tribes	Siriyanangai, Periyanangai, Nilavembu (Shevaroy Hills, Tamil Nadu)	The decoction of the whole plant mixed with goat's milk is given two times a day for 7–10 days for jaundice and live complaints (Alagesaboopathi, 2015)
53	*Andrographis serpyllifolia* Wight	Leaf	Malayali tribes	Kaatuppooraankodi, Siyankodi, Thutuppoondu. (Shevaroy Hills, Tamil Nadu)	Decoction of the leaf juice 50 mL mixed with cow's milk and drink to treat liver related stomach pain (Alagesaboopathi, 2015)
54	*Andropogon muricatus* Retz. (*Vetiveria zizanioides*)	Whole plant	Local people and traditional healers	Ramaccam (Tamil Nadu, Kerala)	Hepatoprotective (Girish, 2012)
55	*Anethum sowa* Roxb. ex Flem.	Seeds	Local people and traditional healers	Shataahvaa (Throughout India)	Hepatoprotective (Kshirsagar et al., 2011)
56	*Anogeissus latifolia* Wall. ex Bedd.	Flower, gum	Local people	Dhava	Hepatoprotective (Pradeep et al., 2009)
57	*Aphanamixis polystachya* (Wall.) Parker	Leaf	Local people and traditional healers	Harinhara (*Calcutta)*	Hepatoprotective (Gole and Dasgupta, 2002)

TABLE 5.1 *(Continued)*

S. No	Plant Name	Part Used	Tribe/Local Community	Local Names (State)	Uses/Mode of Use (Reference)
58	*Apium graveolens* L.	Fruit	Local pecle	Shalari, callari (Throughout India)	Hepatoprotective (Girish and Pradhan, 2012)
59	*Aquilaria agallocha* Roxb.	Bark, root, and leaves	Traditional healers	Agar (Assam)	Hepatoprotective (Alam et al., 2017)
60	*Aralia elata* (Miq.) Seem.	Leaves	–	–	Hepatoprotective (Girish, 2012)
61	*Ardisia paniculata* Roxb.	Roots	Local healers	–	Root in combination with those of *Smilax ovalifolia* and *Bridelia tomentosa* are crushed and boiled in water and drunk @ 1 cup (100 mL) twice daily for jaundice (Rama et al., 2012)
62	*Areca catechu* L.	Young inflorescence	Kani, Kurumar, Kurumbar, Paniyan (Adakkamaram).	Adakkamaram (Kerala)	The pounded mass of young inflorescence about 5–10 g mixed in goat's milk is given orally twice daily for 7 days to treat jaundice (Asha and Pushpangadan, 2002)
63	*Arenga wightii* Griff.	Young inflorescence, fruit husk	Paniyan (Ayasthingu), Kani (Kudappana)	Ayasthingu, Kudappana (Kerala)	Fresh toddy obtained from the young inflorescence is given internally for jaundice; expressed juice of the fruit husk is also given to treat jaundice (Asha and Pushpangadan, 2002)
64	*Argemone maxicana* L.	Latex, oil, root, bark, leaf seed	Local people and traditional healers	Pili Kateri (Chhattisgarh)	Used for the treatment of jaundice (Janghel et al., 2019)
65	*Argemone mexicana* L.	Latex, root	Gond, Kol, Baiga, Panica, Khairwar, Manjhi, Mawasi, Agaria	Bharbhanda, Kateli (Rewa district, Madhya Pradesh)	Latex used in jaundice (Achuta et al., 2010)
66	*Argemone mexicana* L.	Latex, stems, shoots	Local medical practitioners/healers	Barbhand (Shivalik Hills of Northwest Himalaya)	Stems extraction is used in diabetes and that of leaves is used for jaundice, malaria and fever (Devi et al., 2014)

TABLE 5.1 *(Continued)*

S. No	Plant Name	Part Used	Tribe/Local Community	Local Names (State)	Uses/Mode of Use (Reference)
67	*Argemone mexicana* L.	Whole plant	Local people and traditional healers	Pillibhutti, Satyanasi (Ambala district, Haryana)	Plant juice is used orally; 2–3 spoons daily for one week to cure jaundice (Vashistha and Kaur, 2013).
68	*Argemone mexicana* L.	Leaf	Local healers	Bhant/Satyanasi (Arunachal Pradesh)	Decoction of leaf is used in jaundice (Rama et al., 2012)
69	*Argemone mexicana* L.	Seeds	Malayali tribes	Perammathandu (Shevaroy Hills, Tamil Nadu)	Seed powder is taken with hot water internally twice a day for one week to treat jaundice (Alagesaboopathi, 2015)
70	*Aristolochia indica* L.	Roots	Kani (Kattukka ooli),Kurumbar (Garudakodi), Muthuvan (Cheriya arayan)	Kattukkamooli,Garudakodi, Cheriya arayen (Kerala)	Paste of tender roots along with cow's milk administered internally for 5 days to treat jaundice (Asha and Pushpangadan, 2002)
71	*Arnica montana* L.	Flower, leaf	Local people	Leopards-bane (Puducherry)	Hepatoprotective (Girish and Pradhan, 2012)
72	*Artemisia scoparia* Waldst. & Kitam.	Whole plant	Local people and traditional healers	Danti (Maharastra, Punjab)	Hepatoprotective (Kshirsagar et al., 2011)
73	*Artocarpus chaplasha* Roxb.	Ripe fruits	Herbal practitioners and elders	Unem (Chungtia village, Nagaland)	Eaten either raw or as an infusion taken orally twice a day for liver problems. (Kichu et al., 2015)
74	*Asparagus racemosus* Willd	Roots	Kani (Shathavai), Kurumbar, Mannan (Kilavari)	Sathavari, Kilavari (Kerala)	Expressed juice of freshroot (20–30 mL) given internally twice daily for 7 days to treat jaundice. Roots roasted and taken on an empty stomach in the morning to treat liver disorders (Asha and Pushpangadan, 2002)
75	*Asparagus racemosus* Willd.	Root	Mahadevkoli, Thakur, and Katkari	(Konkan Region, Maharashtra)	Decoction obtained from the root has been used to cure jaundice (Naikade and Meshram, 2014)

TABLE 5.1 *(Continued)*

S. No	Plant Name	Part Used	Tribe/Local Community	Local Names (State)	Uses/Mode of Use (Reference)
76	*Asparagus racemosus* Willd.	Tuberous root	Ethnic and Rural people, Hakims, priests	Kurilo (Sikkim)	Decoction used as for diabetes, jaundice, urinary disorder (Trishna et al., 2012)
77	*Asparagus recemosus* Willd.	Roots	Local tribe	Sharanoi (Kumaun region, Uttarakhand)	Dried root powder is used to cure liver disorders (Gangwar et al., 2010)
78	*Asplenium adiantoides* C. Chr.	Whole plant	Local healers	–	Plant decoction is used in jaundice (Rama et al., 2012)
79	*Asteracantha longifolia* Nees	Whole plant	Local healers	Talmakhana (Arunachal Pradesh)	Plant extract and decoction of leaves is used in jaundice (Rama et al., 2012)
80	*Asteracantha longifolia* Nees	Leaves, seeds	Local people	Talimkhana (Ajara Tahsil, Kolhapur District, Maharashtra)	Used in jaundice (Sadale and Karadge, 2013)
81	*Avena sativa* L.	Whole plant	Local people	Vilaaythi jaw (Maharastra)	Hepatoprotective (Girish and Pradhan, 2012)
82	*Averrhoa carambola* L.	Fruit	Local healers	Kamrakh (Arunachal Pradesh)	3–4 slices of the fruit are taken for jaundice or juice of crushed fruit is taken orally for jaundice @ 1/2–1 cup (50–100 mL) three times daily (Rama et al., 2012)
83	*Azadirachta indica* A. Juss.	Bark	Malayali tribes	Vembu (Shevaroy Hills, Tamil Nadu)	Decoction of the bark mixed with sugar is given internally in jaundice. (Alagesaboopathi, 2015)
84	*Azadirachta indica* A. Juss.	Leaves	Mahadevkoli, Thakur, and Katkari	(Konkan Region, Maharashtra)	Young leaves are fried with salt and powder given with milk (Naikade and Meshram, 2014)
85	*Azima tetracantha* Lam.	Leaves, root	Local people	Kundali (Orissa, West Bengal)	Hepatoprotective (Ekbote et al., 2010)
86	*Balanites aegyptiaca* (L.) Delile,	Bark	Ayurvedic practioners	Ingudi (Rajasthan, Gujarat, Madhya Pradesh, and Deccan)	Hepatoprotective (Jaiprakash et al., 2003)

TABLE 5.1 *(Continued)*

S. No	Plant Name	Part Used	Tribe/Local Community	Local Names (State)	Uses/Mode of Use (Reference)
87	*Baliospermum montanum* (Willd.) Muell	Root	Kondareddis of Khammam dist., Telangana	Konda Amudam (Andhra Pradesh, Telangana)	Root decoction (three teaspoons) administered daily once a week (Reddy et al., 2008)
88	*Barringtonia acutangula* (L.) Gaertn.	Leaf, fruits	Local people and Ayurvedic practioners	Nichula (Assam and Madhya Pradesh)	Hepatoprotective (Mishra et al., 2011)
89	*Bauhinia acuminata* L.	Bark	Gond, Kol, Baiga, Panica, Khairwar, Manjhi, Mawasi, Agaria	Sivamalli (Rewa district, Madhya Pradesh)	Root bark decoction is used for treating inflammation of liver (Achuta et al., 2010)
90	*Bauhinia variegata* L.	Bark	Local people	Kachnar (Rajkot, Gujarat)	Hepatoprotective (Gupta et al., 2015)
91	*Begonia laciniata* Roxb.	Roots	Yenadi	Hooirjo (West Bengal), Teisu (Nagaland)	A decoction of the root is given for liver diseases and fever (Ganapathy et al., 2013)
92	*Benincasa hispida* (Thunb.) Cogn.	Fruit	Local healers	Petha/Chaulkumhra (Arunachal Pradesh)	Boiled extract of fruit is given in stomach ulcer and jaundice (Rama et al., 2012)
93	*Benincasa hispida* (Thunb.) Cogn.	Fruit	Local communities	Torbot (Manipur)	About a glass of extracted and filter juice is mixed with 2–3 teaspoon of sugar candy and a spoon of honey for the treatment of jaundice patient, thrice daily (Rajesh et al., 2012)
94	*Berberis aristata* DC.	Root, stem	Local healers	Daru Haridra (Arunachal Pradesh)	Root and stem decoction is taken orally for jaundice (Rama et al., 2012)
95	*Berberis aristata* DC.	Root, bark	Ethnic and rural people in Eastern Sikkim Himalayan Region	Chutro (Sikkim)	Used in jaundice, malaria, fever, and diarrhea. It is also used externally to cure eye disease (Trishna et al., 2012)

TABLE 5.1 *(Continued)*

S. No	Plant Name	Part Used	Tribe/Local Community	Local Names (State)	Uses/Mode of Use (Reference)
96	*Berberis asiatica* DC.	Bark, stem, wood, and fruits	Local tribe	Rasanjana, Daruhaldi, Kilmora (Kumaun region, Uttarakhand)	The roots are used for curing diabetes and jaundice (Gangwar et al., 2010)
97	*Bergenia stracheyi* (Hook. f. & Thoms.) Engl.	Rhizome and leaves	Local elderly people, hermits, shepherds, Vaids, Gujjars and Gaddies	Shamlot (Tribal Pangi Valley, Chamba, Himachal Pradesh)	Rhizome paste is given internally to cure jaundice (Dutt et al., 2014)
98	*Bixa orellana* L.	Leaves	Local healers	Sinduriya (Arunachal Pradesh)	Leaves are useful in jaundice (Rama et al., 2012)
99	*Boerhavia diffusa* L.	Root	Local healers	Punarnava (Arunachal Pradesh)	Root is used in various ways for jaundice (Rama et al., 2012)
100	*Boerhavia diffusa* L.	Leaf, root	Gond, Kol, Baiga, Panica, Khairwar, Manjhi, Mawasi, Agaria	Santan, Santh, Gadapurena, Punarnava (Rewa district, Madhya Pradesh)	Root decoction used in treatment of jaundice (Achuta et al., 2010)
101	*Boerhavia diffusa* L.	Roots	Chenchu	Atuka Mamidaku (Nallamalais, Andhra Pradesh)	Root paste mixed with water used in liver diseases (Sabjan et al., 2014)
102	*Boerhavia diffusa* L.	Root, leaf	Local medical practitioners/healers	(Shivalik Hills of Northwest Himalaya)	The decoction of leaves is used for jaundice (Devi et al., 2014)
103	*Boerhavia diffusa* L.	Root	Malayali tribes	Mukkurattai (Shevaroy Hills, Tamil Nadu)	The root powder mixed with cow's milk is used in jaundice (Alagesaboopathi, 2015)
104	*Boerhavia diffusa* L.	Whole plant	Mahadevkoli, Thakur, and Katkari	(Konkan Region, Maharashtra)	Fresh whole plant material is boiled in water along with sugar, half cup of the decoction is given to the patient thrice a day for 2–3 weeks (Naikade and Meshram, 2014)
105	*Boerhavia erecta* L.	Root	Fisherman community	Padarmookirattai (Pudhukkottai District, Tamil Nadu)	Used to treat jaundice, enlarged spleen, and gonorrhea. (Rameshkumar and Ramakritinan, 2013)

TABLE 5.1 *(Continued)*

S. No	Plant Name	Part Used	Tribe/Local Community	Local Names (State)	Uses/Mode of Use (Reference)
106	*Brassica juncea* L.	Seeds	Mahadevkoli, Thakur, and Katkari	(Konkan Region, Maharashtra)	Alum 40 g (White mineral salt) + Brassica seed 3 g made in to the paste and eaten along with fruit of banana twice a day for jaundice (Naikade and Meshram, 2014)
107	*Brassica oleracea* L.	Foliage	Herbal practitioners and elders	Pandacobi (Chungtia village, Nagaland)	The fresh juice of the foliage is consumed to treat jaundice (Kichu et al., 2015)
108	*Bridelia monoica* (Lour.) Mec.	Root	Local healers	Karagnalia (Arunachal Pradesh)	The root in combination with the root of *Smilax ovalifolia* and *Ardisia paniculata* are rubbed on grindstone and the paste is collected in a cup of water. The mixture is boiled and taken orally for jaundice (Rama et al., 2012)
109	*Bridelia montana* (Roxb.) Willd.	Bark	Sugali, Yerukala, Yanadi	Sankumanu (Salugu Panchayati of Paderu Mandalam, Visakhapatnam, District, Andhra Pradesh)	Used in jaundice (Padal et al., 2013b)
110	*Bridelia stipularis* (L.) Bl.	Leaves	Local healers	–	Leaves are used for jaundice (Rama et al., 2012)
111	*Bryophyllum pinnatum* (Lam.) Oken	Leaf	Local people and traditional healers	Patharchata (Chhattisgarh)	Used for the treatment of jaundice (Janghel et al., 2019)
112	*Bupleurum falcatum* L.	Fruit	–	Jangli jeera	Hepatoprotective (Girish and Pradhan, 2012)
113	*Butea monosperma* (Lam.) Taub.	Gum	Chenchus, Lambadi and Yerukulas	Moduga (Prakasam District, Andhra Pradesh)	A red astringent gum or resin from stem is administered in the treatment of Jaundice and diarrhea (Mohan and Murthy, 1992)
114	*Caesalpinia bonduc* L.	Root bark	Local tribe	Kalaachikottai (Athinadu Pachamalai Hills of Eastern Ghats in Tamil Nadu)	Root bark, with garlic, pepper, kasakasa cure jaundice (Anandkumar et al., 2014)

TABLE 5.1 *(Continued)*

S. No	Plant Name	Part Used	Tribe/Local Community	Local Names (State)	Uses/Mode of Use (Reference)
115	*Cajanus cajan* (L.) Millsp.	Leaf	Koch, Meich, Rava, Munda, Santhal, Garo, Oraon	Coochbehar district, West Bengal	Leaf decoction for jaundice (Tanmay et al., 2014)
116	*Cajanus cajan* (L.) Millsp.	Leaf	Local tribe	Arhar; Tauri kalai (Koch): Jehu (Garo); Kokhleng (Mech) (Jalpaiguri district, West Bengal)	Leaf decoction beneficial for jaundice (Bose et al., 2015)
117	*Cajanus scarabae-oides* (L.) Thours.	Whole plant	Local people and Ayurvedic practioners	Rantur	Hepatoprotective (Pattanayak et al., 2011)
118	*Caladium hortulanum* Bridsey	Corm	Local people	Elephant ear (Kanyakumari district, Tamil Nadu)	Hepatoprotective (Shazhni et al., 2018)
119	*Calotropis gigantea* (L.) Dryand	Leaves	Yanadi	Tellajilledu (Kavali, Nellore district, Andhra Pradesh)	Leaves tied with black thread and tied around neck for jaundice patients (Swapna, 2015)
120	*Camellia sinensis* (L.) O.Kuntze	Leaves, flower	Local people	Nallateyaku (Throughout India)	Hepatoprotective (Girish and Pradhan, 2012)
121	*Canavalia gladiata* (Jacq.) DC.	Fruits	Local villagers	Abai, Ghevada (Kolhapur district, Maharashtra)	The root is ground in cow urine and adminis-tered internally for consecutive days is said to cure enlargement of liver (Jadhav et al., 2011)
122	*Canavalia gladiata* (Jacq.) DC.	Roots	Chenchu	Adavi thamba (Nallamalais, Andhra Pradesh)	Root paste (20 grams) given along with rice gravel for 2–3 days to cure enlargement of liver (Sabjan et al., 2014)
123	*Capparis spinosa* L.	Shoots	Larje, Amchis	Martokpa, Rutoka (Lahaul–Spiti region, Western Himalaya)	Green shoots are cut and dried in shade and powdered. The powder is taken orally twice a day to cure liver pain in jaundice (Koushalya, 2012)

TABLE 5.1 *(Continued)*

S. No	Plant Name	Part Used	Tribe/Local Community	Local Names (State)	Uses/Mode of Use (Reference)
124	*Capsicum annuum* L.	Fruit	Local healers	Mircha (Arunachal Pradesh)	10–15 g splitted fruit without seed are kept in 100–150 mL water for 3–4 hours and after removing the fruit, water is taken orally for jaundice (Rama et al., 2012)
125	*Carissa opaca* Stapf ex Haines	Leaf, root	Local medical practitioners/healers	Garnu (Shivalik Hills of Northwest Himalaya)	Leaf and root decoction is used against asthma, jaundice (Devi et al., 2014)
126	*Carica papaya* L.	Fruit, root, bark	Local people and traditional healers	Papita (Chhattisgarh)	Used for the treatment of jaundice (Janghel et al., 2019)
127	*Carthamus tinctorius* L.	Fruit	Local healers	–	Fruit juice is used for cure of jaundice (Rama et al., 2012)
128	*Casearia graveolens* Dalz.	Roots, leaves	Gond, Halba and Kawar	Arni (Darekasa hill range, Gondia district, Maharashtra)	Root and leaves extract mixed with warm water and given bath to child in jaundice (Chandrakumar et al., 2015)
129	*Cassia angustifolia* Vahl	Whole plant	Local people and traditional healers	Charota bhaji, Markandik (Chhattisgarh)	Used for the treatment of jaundice (Janghel et al., 2019)
130	*Cassia fistula* L.	Bark, leaves	Yenadi,	Rela (Andhra Pradesh)	Used in the treatment of jaundice in the form of powder (Bala et al., 2011)
131	*Cassia fistula* L.	Flowers	Malayali tribes	Konnai	Powdered flower is used to cure liver ailments (Alagesaboopathi, 2015)
132	*Cassia fistula* L.	Fruit	Herbal practitioners	Rela-kayalu, Relarala, Reylu, Suvarnam (Vizianagaram District, Andhra Pradesh)	One spoon of fruit pulp is administered with sugarcane juice to cure jaundice (Padal et al., 2013c)
133	*Cassia fistula* L.	Fruit	Sugali, Yerukala, Yanadi	Rela (Salugu Panchayati Of Paderu Mandalam, Visakhapatnam, District, Andhra Pradesh)	Used in jaundice (Padal et al., 2013b)

TABLE 5.1 *(Continued)*

S. No	Plant Name	Part Used	Tribe/Local Community	Local Names (State)	Uses/Mode of Use (Reference)
134	*Cassia italica* Lam.	Leaves, fruits	Tribal people Kailasagirikona Forest, Chittor district, Andhra Pradesh	–	About 5–10 g paste is prepared from leaves and fruits are administered orally for a month to recover from jaundice (Pratap et al., 2009)
135	*Cassia occidentalis* L.	Leaf	Traditional healers	Kasintha (Parnasala Sacred Grove area Eastern Ghats of Khammam District, Telangana)	Jaundice: Ten spoonfuls of leaf juice mixed with butter milk is given thrice a day for 7 days (Srinivasa et al., 2015)
136	*Cassia tora* L.	Leaves, seeds	Fisherman community	Oosithagarai (Pudhukkottai District, Tamil Nadu)	Skin diseases, dandruff, constipation, cough, hepatitis, fever, and hemorrhoids (Rameshkumar and Ramakritinan, 2013)
137	*Cassytha filiformis* L.	Whole plant	Local people and Ayurvedic practioners	Amarvalli (Throughout India)	Hepatoprotective (Raj et al., 2013)
138	*Centella asiatica* (L.) Urb.	Laeves	Koragar (Kudagon), Kurumbar (Kumara), Kani (Varambil), Hill Pulayar Kodakan)	Kudagan, Kumara, Varambil, Kodakan (Kerala)	Expressed juice of fresh leaves (20–30 mL) given internally twice daily for seven days to treat jaundice (Asha and Pushpangadan, 2002)
139	*Ceratopteris siliquosa* (L.) Copel.	Whole plant	Muthuvan and Kani (Shirunagal)	Shirunagal (Kerala)	Decoction of the whole plant taken internally (25–30 mL) twice daily to treat jaundice and other ailments (Asha and Pushpangadan, 2002)
140	*Ceriops decandra* (Griff.) Ding Hou	Bark, leaf	Local people and traditional healers	Goran (Tamil Nadu)	Used as cure for ulcer and hepatitis (Gnanadesigan et al., 2017)
141	*Chenopodium album* L.	Whole plant, root, fruit	Local medical practitioners/healers	Bhadhu (Shivalik Hills of Northwest Himalaya)	Decoction of roots is effective against jaundice (Devi et al., 2014)
142	*Chenopodium album* L.	Whole plant	local people and traditional healers	Bathu sag (Ambala district, Haryana)	Root is used in jaundice and urinary problems (Vashistha and Kaur, 2013)

TABLE 5.1 *(Continued)*

S. No	Plant Name	Part Used	Tribe/Local Community	Local Names (State)	Uses/Mode of Use (Reference)
143	*Chloroxylon sweitenia* DC.	Root bark	Local tribe	Vambarai (Athinadu Pachamalai Hills of Eastern Ghats in Tamil Nadu)	Root bark is used for jaundice cure (Anandkumar et al., 2014)
144	*Chrozophora plicata* (Vahl) A. Juss. ex Spreng.	Aerial part	Local people and traditional healers	Sahadevi (Chhattisgarh)	Used for the treatment of jaundice (Janghel et al., 2019)
145	*Chrozophora tinctoria* L.	Aerial part	Local people and traditional healers	Kukronda (Chhattisgarh)	Used for the treatment of jaundice (Janghel et al., 2019)
146	*Cicer arietinum* L.	Seeds	Local people and traditional healers	Chana, Channa, Channo (Barmer district, Rajasthan)	The tribals of Barmer district eat roasted seeds to cure jaundice. (Singh and Pandey,1998)
147	*Cichorium intybus* L.	Leaves	Gujjars, Bhoris Bakerwals	Kasni/Wari Hundh (Bandipora district, Jammu and Kashmir)	Leaves are cooked and given to fresh mothers to cure body weakness, loosening of joints, body muscular pains, frequent bleeding, as appetizer and liver tonic (Jadhav et al., 2011)
148	*Cichorium intybus* L.	Seeds, root, and leaves	Local tribe	Kasni (Kumaun region, Uttarakhand)	Herb is taken internally to cure liver disorders, spleen problems (Gangwar et al., 2010)
149	*Cissampelos pareira* L.	Root	Local healers	Ambashtha/Patha (Arunachal Pradesh)	Root is placed in water for overnight and the extract is taken orally for jaundice. (Rama et al., 2012)
150	*Cissmpelos pareira* L.	Leaves	Chenchus, Lambadi, and Yerukulas	(Prakasam district, Andhra Pradesh)	The juice obtained from crushing the leaves is mixed with black pepper and is given to cure jaundice (Mohan and Murthy, 1992)

TABLE 5.1 *(Continued)*

S. No	Plant Name	Part Used	Tribe/Local Community	Local Names (State)	Uses/Mode of Use (Reference)
151	*Citrullus colocynthis* (L.) Schrad.	Roots	Chenchus, Lambadi, and Yerukulas	Chedupuccha (Prakasam District, Andhra Pradesh)	Crushed roots and fruits are boiled with water and the decoction is given for drinking to cure jaundice and urinary diseases (Mohan and Murthy, 1992)
152	*Citrus aurantifolia* (Christm.) Wingle, *C. medica* L., *C. reticulata*, *C. sinensis* (L.) Osbeck	Fruit	Local healers	–	Fruit administered during jaundice (Rama et al., 2012)
153	*Citrus hystrix* DC	Leaves	Local people and traditional healers	Kaf-fir lime (Mayiladuthurai, Nagai district, Tamil Nadu)	Hepatoprotective (Abirami et al., 2015)
154	*Citrus maxima* L.	Leaves	Local people and traditional healers	Aruttiramacu (Mayiladuthurai, Nagai district, Tamil Nadu)	Hepatoprotective (Abirami et al., 2015)
155	*Citrus medica* Salisb.	Fruit, leaves	Local people of Taindol	Bara Nimbu (Taindol, Jhansi, Uttar Pradesh)	Used for jaundice (Jitin, 2013)
156	*Cleome viscosa.* L.	Leaf	Malayali tribes	Naaivelai (Shevaroy Hills, Tamil Nadu)	Fresh leaf juice mixed with hot water is used in jaundice (Alagesaboopathi, 2015)
157	*Clerodendrum indicum* (L.) O.Kuntze	Root	Local healers	Vanabhenda (Arunachal Pradesh)	Root is soaked in water for overnight and fresh extract is taken orally for 7 - 15 days for jaundice (Rama et al., 2012)
158	*Clitorea ternatea* L.	Seed	Local people and traditional healers	Aparajita (Chhattisgarh)	Used for the treatment of liver swelling and jaundice (Janghel et al., 2019)
159	*Coccinia grandis* (L.) Voigt	Leaves	Irula tribals	Kovai (Kalavai village, Vellore district, Tamil Nadu)	Used in jaundice (Natarajan et al., 2013)

TABLE 5.1 *(Continued)*

S. No	Plant Name	Part Used	Tribe/Local Community	Local Names (State)	Uses/Mode of Use (Reference)
160	*Coccinia grandis* (L.) Voigt	Leaf	Koal, Panika, Bhuriya, Kharvar, Gaund	Kularu (Renukot forest division, Sonbhadra, Uttar Pradesh)	Fresh leaf juice is used in the treatment of jaundice (Anurag et al., 2012)
161	*Coccinia grandis* L.	Leaves	Local people and traditional healers	Aracanviroti (Kadayanallur, Tirunelveli District of Tamil Nadu)	Shows hepatoprotective activity (Janghel et al., 2019)
162	*Cocculus hirsutus* (L.) Diels	Root, whole plant	Local traditional healers	Dahdahiya (Nawarangpur District, Odisha)	Root paste is taken thrice a day for liver dysfunction (Dhal et al., 2014)
163	*Coleus aromaticus* Benth.	Root	Local people and traditional healers	Pashan bhed, patar Bhed (Chhattisgarh)	Used for the treatment of jaundice (Janghel et al., 2019)
164	*Convolvulus pluricaulis* Choisy	Whole plant	Local people and Ayurvedic practioners	Shankhapushpi (Throughout India)	Hepatoprotective (Ravichandra et al., 2013)
165	*Coptis teeta* Wall.		Local healers	Mishmi teeta (Arunachal Pradesh)	Root is soaked in water for overnight and fresh extract is taken orally for 7–15 days for jaundice (Rama et al., 2012)
166	*Corchorus depressus* (L.) Stocks	Whole plant	Local people and traditional healers	Chamkas, Baphuli (Rajasthan)	The decoction of fresh plant is given orally to cure liver disorders (Singh and Pandey,1998)
167	*Cordia dichotoma* Forst. f.	Leaves	Sugali, Yerukala, Yanadi	Banka nakkeri (Salugu Panchayati of Paderu Mandalam, Visakhapatnam, District, Andhra Pradesh)	Used in jaundice (Padal et al., 2013b)
168	*Cordyceps sinensis*	Whole plant	Ethnic and Rural People	Yarcha gombuk (Eastern Sikkim Himalaya region)	Rejuvenates liver and heart (Trishna et al., 2012)
169	*Coriandrum sativum* L	Seeds., leaf	Local healers	Dhanyaka/Dhaniya (Arunachal Pradesh)	Leaf and fruits are taken orally during jaundice (Rama et al., 2012)

TABLE 5.1 *(Continued)*

S. No	Plant Name	Part Used	Tribe/Local Community	Local Names (State)	Uses/Mode of Use (Reference)
170	*Coriandrum sativum* L.	Seeds	Gujjars, Bhoris Bakerwals	Daniwaal (Bandipora district, Jammu and Kashmir)	Seed decoction is given to cure jaundice, drying of mouth and headache (Parvaiz et al., 2013)
171	*Costus speciosus* (Koenig ex Retz.) Sm.	Rhizome	Local healers	Kebuk/Keun (Arunachal Pradesh)	Fresh juice of rhizome is taken orally (Rama et al., 2012)
172	*Crataeva nurvala* Buch-Ham.	Leaves	Herbal practitioners and elders	Kongkawa (Chungtia village, Nagaland)	Boiled leaves are consumed as liver tonic (Kichu et al., 2015)
173	*Crepis flexuosa* (Ledeb.) Benth. ex C.B.Clarke	Whole plant	Larje, Amchis	Homa sili (Lahaul–Spiti region Western Himalaya)	Fresh juice of the plant is mixed with equal amount of water is taken regularly once a day to cure jaundice until cured (Koushalya, 2012)
174	*Crinum amoenum* Roxb. ex Ker.-Gawler	Root	Local tribe	Astachatur (Rava) (Jalpaiguri district, West Bengal)	Root used to treat jaundice (Bose et al., 2015)
175	*Cryptolepis buchanani* Roem. & Schult.	Latex	Chenchus, Lambadi, and Yerukulas	Adavipalathiga (Prakasam District, Andhra Pradesh)	The milky latex mixed with water is given to cure hepatic troubles (Mohan and Murthy, 1992)
176	*Cucumis sativus* L.	Fruit	Local healers	Kheera (Arunachal Pradesh)	Fresh fruit is administered during jaundice (Rama et al., 2012)
177	*Cucumis trigonus* Roxb.	Fruit	Local people and traditional healers	Bhakura (Beed, Maharashtra)	Hepatoprotective (Patil et al., 2011)
178	*Cuminum cyminum* L.	Fruit	Kani and Kurumbar (Jera) Kani, aniyan, Irular, Kurumar, Kurichar (Jeeragam)	Jera Jeerakom, Cheerakam (Kerala)	Decoction of the fruit is given in jaundice to purify blood and as diuretic (Asha and Pushpangadan, 2002)
179	*Curculigo orchioides* Gaertn.	Roots	Gond and Madiya	Kali-musli (Markanda Forest Range of Gadchiroli District, Maharashtra)	Used for Jaundice (Chavhan and Aparna, 2015)

TABLE 5.1 *(Continued)*

S. No	Plant Name	Part Used	Tribe/Local Community	Local Names (State)	Uses/Mode of Use (Reference)
180	*Curculigo orchioides* Gaertn.	Tubers	Local villagers	Kali-musali (Kolhapur district, Maharashtra)	Tubers cut and shade dried, in that add equal amount of sugar and one glass milk mix well to make thick mucilage, this mixture is used in asthma, jaundice and diarrhea (Jadhav et al., 2011)
181	*Curculigo orchioides* Gaertn.	Rhizome	Local healers	Kali Musali (Arunachal Pradesh)	Rhizome is prescribed in piles, jaundice, asthma, diarhoea, and gonorrhoea, considered demulcent, tonic (Rama et al., 2012)
182	*Curculigo orchioides* Gaertn.	Root	Kondareddis of Khammam dist.	Adavi taadi, Naela tadi (Andhra Pradesh, Telangana)	Tuberous root paste (12 g) administered daily twice for three days (Reddy et al., 2008)
183	*Curcuma angustifolia* L.	Rhizome	Local people and traditional healers	Tikhur (Chhattisgarh)	Used for the treatment of jaundice (Janghel et al., 2019)
184	*Curcuma domestica* Valeton	Rhizome	Kani, Kurichar, Paniyan Kattunaikan, Koragar, Hill Pulayar (Manjal)	Manjal (Kerala)	Pounded mass of fresh rhizome, equal to the size of the fruit of *Emblica officinalis* is mixed with calcium hydroxide (equal to the size of the seed of *Adenanthera pavonia* L.), diluted with water and kept overnight in an airtight bottle. Three ounces of the liquid portion from this preparation is given on an empty stomach in the morning for 7 days to treat jaundice and other chronic liver disorders. Daily bath is recommended when this drug is administrered (Asha and Pushpangadan, 2002)
185	*Curcuma longa* L.	Whole plant	Local people of Taindol	Haldi (Taindol, Jhansi, Uttar Pradesh)	Used for jaundice (Jitin, 2013)

TABLE 5.1 *(Continued)*

S. No	Plant Name	Part Used	Tribe/Local Community	Local Names (State)	Uses/Mode of Use (Reference)
186	*Curcuma longa* L.	Rhizome	Local healers	Haridra/Haldi (Arunachal Pradesh)	40–50 g rhizome pounded and made extract; extract is mixed with fruits of *Piper longum* L. and taken daily for 20–25 days during jaundice (Rama et al., 2012)
187	*Curcuma longa* L.	Rhizome	Mahadevkoli, Thakur, and Katkari	Konkan Region, Maharashtra	Paste of rhizome is mixed with cow milk and taken once day for 12–13 days (Naikade and Meshram, 2014)
188	*Cuscuta epithymum* (L.) L.	Whole plant	Yenadi	Sitammapogunalu (Andhra Pradesh,Telangana)	Extract is used in urinary, spleen and liver disorders (Ganapathy et al., 2013)
189	*Cuscuta reflexa* Roxb.	Whole plant	Local healers	Amarvela (Arunachal Pradesh)	Plant juice is taken for 6–7 days for jaundice (Rama et al., 2012)
190	*Cuscuta reflexa* Roxb.	Whole plant	Local tribe	Swarnalata; Alokzori (Oraon) (Jalpaiguri district, West Bengal)	Whole plant juice used to treat jaundice (Bose et al., 2015)
191	*Cydonia oblonga* Mill.	Seeds, fruits, and flowers	Gujjars, Bhoris Bakerwals	Bumchuont (Bandipora district, Jammu and Kashmir)	The seeds also form an important ingredient of a combination of different herbs such as seeds of *Cucumis sativa, Malva neglecta, Foeniculum vulgare*, fruits of *Ziziphus jujuba*, leaves and flowers of *Arnebia benthamii* and fronds of *Adiantum capillus-veneris*. This combination is locally called as "Sharbeth." The composite decoction of "Sherbeth" is given to cure jaundice, cough, cold, chronic constipation, fever and as a good blood purifier (Parvaiz et al., 2013)
192	*Cynara scolymus* L.	Leaves, Flower	Local people and traditional healers	–	Hepatoprotective (Kshirsagar et al., 2011)

TABLE 5.1 *(Continued)*

S. No	Plant Name	Part Used	Tribe/Local Community	Local Names (State)	Uses/Mode of Use (Reference)
193	*Daucus carota* L.	Seeds	Local people and Ayurvedic practioners	Gaajara (Throughout India)	Hepatoprotective (Singh et al., 2012)
194	*Decalepis hamiltonii* Wt. & Arn.	Tuberous roots	Nakkala, Suygali, Yenadi, Yerukala.	Maredu kommulu (Chittoor, Andhra Pradesh)	The root bark is powdered and used in a dose of 3–5 g. 2 times a day to stop jaundice (Jyothi et al., 2011)
195	*Deeringia amaranthoides* (Lamarck) Merrill	Root	Local tribe	Chhorachurisag (Oraon) (Jalpaiguridistrict, West Bengal)	Root is used to treat jaundice (Bose et al., 2015)
196	*Delonix regia* Rafin.	Aerial parts	Ayurvedic practioners	Vadanarayana	Hepatoprotective (Ahmed et al., 2011)
197	*Dendrobium ovatum* (L.) Krenzl.	Whole plant	Yenadi	Nagli (Maharashtra)	Juice of fresh plant used for stomachic, carminative, antispasmodic, laxative, liver tonic (Ganapathy et al., 2013)
198	*Dendrophthoe falcata* (L. f.) Etting.	Leaf	Local people	Bandaaka (Throughout India)	Hepatoprotective (Pattanayak and Priyashree, 2008)
199	*Desmodium biflorum* L.		Kurichar and Kurumbar, Hill Pularyar (Cherupullari), Kani (Nilampullari)	Cheupulladri, Nilampullari (Kerala)	The whole plant is pounded well and 10 g of this is mixed with goat's milk, given on an empty stomach in the morning and evening for five days to treat jaundice and other liver-complaints (Asha and Pushpangadan, 2002)
200	*Desmodium laxiflorum* DC.		Local healers	–	50–100 gs roots are crushed and boiled in water used for jaundice (Rama et al., 2012)
201	*Desmodium triflorum* (L.) DC.	Leaves	Traditional healers	Tripaadi (Throughout India)	Hepatoprotective (Valvi et al., 2016)

TABLE 5.1 *(Continued)*

S. No	Plant Name	Part Used	Tribe/Local Community	Local Names (State)	Uses/Mode of Use (Reference)
202	*Desmostachya bipinnata* (L.) Stapf		Local healers	Kusha (Arunachal Pradesh)	50 mL extract mixing with powder of three fruits of *Piper longum* is taken daily for 10–15 days to cure jaundice (Rama et al., 2012)
203	*Dioscorea bulbifera* L.	Tubers, bulbs	Local people	Kadu karanda (Ajara Tahsil, Kolhapur District, Maharashtra)	Used in jaundice (Sadale and Karadge, 2013)
204	*Diplazium esculentum* (Retz.) Sw.	Leaf	Local people	Dhekia Saag (Dibrugarh, Assam)	Hepatoprotective (Junejo et al., 2018)
205	*Drynaria quercifolia* (L.) J. Smith	Rhizome	Local tribe	Marappannakizhangu, Attukalkizhangu (Western Ghats, Kerala)	As a vegetable as well as hepatoprotective drug especially in chronic jaundice (Anuja et al., 2018)
206	*Ecbolium viride* (Forssk.)	Roots	Local people and traditional healers	Nakka toka (Tirupathy, Andhra Pradesh)	Hepatoprotective (Cheedella et al., 2013)
207	*Ecipta prostrata* (L.) L.	Leaves	Malayali tribes	Karisalankanni (Shevaroy Hills, Tamil Nadu)	Leaf juice mixed with cow's milk is given internally twice a day for one week to cure jaundice (Alagesaboopathi, 2015)
208	*Eclipta alba* (L.) Hassk.	Leaves	Malayali tribes	Majalkarisalankanni (Shevaroy Hills, Tamil Nadu)	Decoction of leaves mixed with hot water used in liver disorders (Alagesaboopathi, 2015)
209	*Eclipta prostrata* (L.) L.	Leaves	Chenchu	Gunta-galijaraku (Rudrakod Sacred Grove, Nallamalais, Andhra Pradesh)	Leaves are squeezed to get leaf sap and a teaspoonful of sap is given in orally, thrice in a day for 15 days to cure jaundice (Rao and Sunitha, 2011)
210	*Eclipta prostrata* (L.) L.		Local healers	Bhringaraja (Arunachal Pradesh)	20–30 g paste of the whole plant mixed with salt is taken once daily for 15–20 days to cure jaundice (Rama et al., 2012)

TABLE 5.1 *(Continued)*

S. No	Plant Name	Part Used	Tribe/Local Community	Local Names (State)	Uses/Mode of Use (Reference)
211	*Eclipta prostrata* (L.) L.	Whole plant	Local people	Karisalanganni (Ariyalur District, Tamil Nadu)	The powder of *Eclipta prostrata, Leucas aspera* and *Phyllanthus niruri* are mixed with butter milk and taken orally to cure jaundice (Satishpandiyan et al., 2014)
212	*Eclipta prostrata* (L.) L.	Whole plant	Irula tribals	Manjal (Kalavai village,Vellore district, Tamil Nadu)	Used in jaundice (Natarajan et al., 2013)
213	*Elaeagnus caudata* Schlecht	Stem bark	Local healers	–	200–250 g stem bark and fruit of the species are pounded and boiled in water; 100 mL extract mixed with *Piper longum* is taken daily for 2–3 weeks to cure jaundice and other liver troubles (Rama et al., 2012)
214	*Elettaria cardamomum* Maton	Seed	Many Kerala tribes	Kerala	Seed extract is used for jaundice (Asha and Pushpangadan, 2002)
215	*Embelia ribes* Burm. f.	Fruits	Local people	Vaividungalu	Hepatoprotective (Girish and Pradhan, 2012)
216	*Emblica officinalis* Gaertn.	Bark, fruits	Local tribe	Aonla, Aonwala (Kumaun region, Uttarakhand)	Bark decoction is used for treating diarrhea, dysentery, Cholera and jaundice (Gangwar et al., 2010).
217	*Emblica officinalis* Gaertn.	Stems	Local healers	Amlaki (Arunachal Pradesh)	100–50 g pith of young branches are boiled in cow milk and made extract. 100 g of extract is taken one time daily in the early morning for 20 days to get relief from jaundice (Rama et al., 2012)
218	*Emblica officinalis* Geartn.	Fruit	Malayali tribes	Nelli (Shevaroy Hills, Tamil Nadu)	Fruit is consumed orally to control jaundice (Alagesaboopathi, 2015)
219	*Erythrina variegata* L.	Stem bark	Local healers	Mura (Arunachal Pradesh)	Stem bark is used for jaundice (Rama et al., 2012)

TABLE 5.1 *(Continued)*

S. No	Plant Name	Part Used	Tribe/Local Community	Local Names (State)	Uses/Mode of Use (Reference)
220	*Erythroxylum monogynum* Roxb.	Leaf	Chenchu	Devadari, Dadiri (Rudrakod Sacred Grove, Nallamalais, Andhra Pradesh)	Leaf juice is administered for jaundice (Rao and Sunitha, 2011)
221	*Euphorbia fusiformis* Buch.-Ham. ex D.Don	Root, rhizome	Malayali tribes	Ilai kalli (Chitteri hills, Tamil Nadu)	Traditionally used treat liver disorders (Anusuya et al., 2010)
222	*Euphorbia geniculata* Ortega	Aerial parts	Gondand Madiya	Bada dudhi (Markanda Forest Range of Gadchiroli District, Maharashtra)	Used for curing jaundice (Chavhan and Aparna, 2015)
223	*Euphorbia hirta* L.	Whole plant	Local people of Taindol	Dudhi (Taindol, Jhansi, Uttar Pradesh)	Used for curing jaundice (Jitin, 2013)
224	*Euphorbia hirta* L.	Whole plant	Local medical practitioners/healers	Doddak (Shivalik Hills of Northwest Himalaya)	Decoction of the whole plant is used in cough, asthma, bronchitis, jaundice, and digestive problems (Devi et al., 2014)
225	*Euphorbia hirta* L.	Leaf, latex	Herbalists, farmers, spiritualist	Dudhi (Yamuna Nagar district of Haryana)	Leaf juice is used to treat jaundice, fever, fungal infection, and syphilis (Parul and Vashistha, 2015)
226	*Exacum tetragonum* Roxb.	Whole plant	Local healers	Chireta (Arunachal Pradesh)	Plant juice or decoction is used thrice a day for jaundice (Rama et al., 2012)
227	*Feronia limonia* (L.) Sw.	Stem	Traditional healers	Elaga (Throughout India)	Hepatoprotective (Jain et al., 2012)
228	*Ficus glomerata* Roxb.	Fruit	Many Kerala tribes	Kerala	Used in jaundice (Asha and Pushpangadan, 2002)
229	*Ficus hispida* L.	Leaves, fruits, bark	Local people and traditional healers	Gobla (Chhattisgarh)	Used for the treatment of jaundice (Janghel et al., 2019)

TABLE 5.1 *(Continued)*

S. No	Plant Name	Part Used	Tribe/Local Community	Local Names (State)	Uses/Mode of Use (Reference)
230	*Ficus microcarpa* L. f.	Bark, leaf	Local people and Ayurvedic practioners	Plaksha (West Bengal, Bihar, New Delhi)	Hepatoprotective (Kalaskar and Surana, 2011)
231	*Ficus racemosa* L.	Roots, leaves	Kani, Kurichar, Mannan (Athi)	Kerala	Tender roots and leaves are powdered to make a paste and 5–6 g taken daily 5 days to treat jaundice.Tender roots powdered well and boiled in Goat's milk and given for liver complaints (Asha and Pushpangadan, 2002)
232	*Ficus religiosa* L.	Seed, latex, bark	Local people of Taindol	Pipal (Taindol, Jhansi, Uttar Pradesh)	Used for curing jaundice (Jitin, 2013)
233	*Ficus semicordata* Buch.-Ham.	Leaf	Local healers	Padhotado (Arunachal Pradesh)	Leaf decoction in combination with that of *Byttneria pilosa* Roxb. and *Phyllanthus fraternus* and the bark of *Callicarpa arborea* is taken orally for get relief from jaundice (Rama et al., 2012)
234	*Flacourtia indica* Burm.f	Leaf, fruit	Gond, Kol, Baiga, Panica, Khairwar, Manjhi, Mawasi, Agaria	Rakatsonk, kateyya (Rewa district, Madhya Pradesh)	Fruit juice is given in the morning in liver problems (Achuta et al., 2010)
235	*Flacourtia montana* J. Grah	Leaf	Traditional healers	Malayalam, Chalirpazham, Kattuloika, Painellicka, Vayyamkaitha (Alappuzha district, Kerala)	Hepatoprotective (Joshy et al., 2016)
236	*Foeniculum vulgare* Mill.	Whole plant	Gujjars, Bhoris Bakerwals	Bodiyaan (Bandipora district, Jammu and Kashmir)	Used in jaundice (Parvaiz et al., 2013)
237	*Fumaria indica* (Haussk.) Pugsely	Whole plant	Local medical practitioners/healers	Pitpapar (Shivalik Hills of Northwest Himalaya)	Powder of whole plant is used for jaundice (Devi et al., 2014)

TABLE 5.1 *(Continued)*

S. No	Plant Name	Part Used	Tribe/Local Community	Local Names (State)	Uses/Mode of Use (Reference)
238	*Fumaria officinalis* L.	Seeds, flowers	Local people	(Nilgiris and Salem, Tamil Nadu)	Hepatoprotective (Girish and Pradhan, 2012)
239	*Garcinia pedunculata* Roxb.	Fruits	Local healers	Thekera (Arunachal Pradesh)	Young fruits are prescribed in jaundice besides use as stimulant, emetic, diuretic, pulmonary and renal troubles (Rama et al., 2012)
240	*Gardenia jasminoides* Ellis	Whole plant	Local healers	Gandharaja (Arunachal Pradesh)	Plant is considered an indigenous medicine for cough, rheumatism, anemia and jaundice (Rama et al., 2012)
241	*Garuga pinnata* Colebr.	Bark	Local tribe	Jum, Tinn, Kharpat, Nil bhadi; Rosuni (Rava) (Jalpaiguri district, West Bengal)	Bark is used in jaundice (Bose et al., 2015)
242	*Gentiana moor-croftiana* G. Don	Whole plant	Larje, Amchis	Santik (Lahaul–Spiti regionn Western Himalaya)	Fresh plant parts are crushed and the juice obtained is taken in empty stomach to cure jaundice (Koushalya, 2012)
243	*Gentiana tubiflora* (Wall. ex G.Don) Griseb.	Whole plant	Larje, Amchis	Chatik (Lahaul–Spiti regionn Western Himalaya)	Juice of the whole plant is mixed with equal quantity of water and about half glass is taken orally during morning hours with empty stomach to cure jaundice (Koushalya, 2012)
244	*Geranium pratense* L.	Whole plant	Larje, Amchis	Podh-Lo, Tapan (Lahaul–Spiti region, Western Himalaya)	About one spoon of plant powder is taken orally with water to cure jaundice (Koushalya, 2012)
245	*Glycosmis arborea* (Roxb.)DC.	Root	Local tribe	Ashshewra (Jalpaiguri district, West Bengal)	Root powder used to treat fever, hepatopathy, eczema, skin diseases, wounds, liver disorder (Bose et al., 2015)
246	*Glycosmis pentaphylla* (Retz.) Correa	Whole plant	Local healers	–	Boiled extract of plant is administered in jaundice (Rama et al., 2012)

TABLE 5.1 *(Continued)*

S. No	Plant Name	Part Used	Tribe/Local Community	Local Names (State)	Uses/Mode of Use (Reference)
247	*Glycyrrhiza glabra* L.	Root	Local people	Athimathuram (Punjab, Jammu and Kashmir, and South India)	Hepatoprotective (Girish and Pradhan, 2012)
248	*Gyrocarpus asiaticus* Willd.	Aerial parts	Traditional healers	Tanikior Nalla Poliki (Thoothukudi district, Tamil Nadu)	Hepatoprotective (Kanthal et al., 2017)
249	*Hedychium spicatum* Buch. Ham. ex Smith	Rhizome	Local tribe	Kapoor kachri (Kumaun region, Uttarakhand)	The powder of root is useful in the treatment of liver complaints (Gangwar et al., 2010)
250	*Hedyotis auricularia* L.	Whole plant	Local healers	–	Decoction of plant is used for 20–25 days or till complete relief (Rama et al., 2012)
251	*Hedyotis corymbosa* L.	Whole plant	Local people and traditional healers	Davana Patta (Chhattisgarh)	Used for the treatment of jaundice (Janghel et al., 2019)
252	*Helicteres isora* L.	Whole plant	Local tribe	–	Whole plant used to treat jaundice (Bose et al., 2015)
253	*Helminthostachya zeylanica* Hook.	Rhizome	Local tribe	Dinshabalindo (Meich); Nagdhup (Rava) (Jalpaiguri district, West Bengal)	Rhizome used to treat jaundice (Bose et al., 2015)
254	*Hemidesmus indicus* (L.) R.Br.	Roots	Mahadevkoli, Thakur, and Katkari	Konkan Region, Maharashtra	Root powder given along with honey once a day for jaundice (Naikade and Meshram, 2014)
255	Herbal formula-1	–	Mahadevkoli, Thakur, and Katkari	Konkan Region, Maharashtra	*Cynodon dactylon*, *Phyllanthus amarus* and *Piper nigram*. Mixture of three plants leaves, fruits made into a juice and given to patient to cure the jaundice (Naikade and Meshram, 2014)

TABLE 5.1 *(Continued)*

S. No	Plant Name	Part Used	Tribe/Local Community	Local Names (State)	Uses/Mode of Use (Reference)
256	Herabal formula-2	Leaves	Mahadevkoli, Thakur, and Katkari	Konkan Region, Maharashtra	*Eclipta alba* L., *Phyllanthus amarus* Schum & Thonn., *Leucas aspera* (Willd.) Link. Leaves of above three plants are ground and extract is given for jaundice (Naikade and Meshram, 2014).
257	Herabal formula-3	Stems, fruits	Mahadevkoli, Thakur, and Katkari	Konkan Region, Maharashtra	*Musa paradisiaca* L., *Lablab purpureus* L. Interior stem portion and fruits legume plants are prepared as a vegetable.curry and given along with diet for jaundice (Naikade and Meshram, 2014)
258	Herabal formula-4	Leaf	Mahadevkoli, Thakur, and Katkari	Konkan Region, Maharashtra	*Cynodon dactylon* (L.) Pers., *Phyllanthus amarus* Schum. & Thonn. Leaf extracts of above two plants are mixed and given with water (Naikade n grum and Meshram, 2014)
259	*Herniaria glabra* L.	Whole plant	Local people and traditional healers	Kashmir to Kumaon	Hepatoprotective. (Kshirsagar et al., 2011)
260	*Hibiscus vitifolius* L.	Root	Traditional healers	Adavi patthi (Throughout India)	Hepatoprotective (Samuel et al., 2012)
261	*Hibiscus lampas* Cav.	Roots	Kani, Kurumbar, Kuruchar (kolukatta)	Kolukatta (Kerala)	Expressed juice of the fresh roots (10–15 mL) is administered internally for 7 days for treating jaundice (Asha and Pushpangadan, 2002)
262	*Hippophae rhamnoides* Linn.	Fruit	Local people and Ayurvedic practioners	Dhurchuk (Himalayas)	Hepatoprotective (Geetha et al., 2008)
263	*Hippophae tibetana* Schltdl.	Berries	Larje, Amchis	Chha-Tuan (Lahaul–Spiti regionn Western Himalaya)	Dried berries are crushed and boiled in water and decoction obtained is taken as a tea to cure jaundice (Koushalya, 2012)

TABLE 5.1 *(Continued)*

S. No	Plant Name	Part Used	Tribe/Local Community	Local Names (State)	Uses/Mode of Use (Reference)
264	*Holarrhena pubescens* (Buch.-Ham.) Wallich ex Don	Root, bark	Local people and traditional healers	Kutaja (Travancore, Assam, and Uttar Pradesh)	Hepatoprotective (Girish and Pradhan, 2012)
265	*Homalomena aromatica* Schott	Rhizome	Local healers	Sugandhamantri (Arunachal Pradesh)	Rhizome juice is taken orally in jaundice and other liver complaints. (Rama et al., 2012)
266	*Hordeum vulgare* Linn.	Fruit	Local people and traditional healers	Barley (Uttar Pradesh, West Bengal, Bihar, Madhya Pradesh, Rajasthan, Haryana, Punjab, Himachal Pradesh, and Jammu and Kashmir)	Hepatoprotective (Girish and Pradhan, 2012)
267	*Houttuynia cordata* Thunb.	Leaves	Local healers	Masundhuri (Arunachal Pradesh)	3–4 fresh leaves are eaten twice daily in case of jaundice. It is also used as condiment.Plant extract is also used for stomach complaints and jaundice (Rama et al., 2012)
268	*Hoya pendula* R. Br.	Root	Sugali, Yerukala, Yanadi	Pala thiga (Salugu Panchayati of Paderu Mandalam, Visakhapatnam, district, Andhra Pradesh)	Used in jaundice (Padal et al., 2013b)
269	*Hygrophila salicifolia* Nees	Whole plant	Local healers	Talmakhana, Minjonjo (Arunachal Pradesh)	Pounded whole plant is taken orally for stomach complaints and jaundice (Rama et al., 2012)
270	*Hygrophila auriculata* (K. Schum.) Heine	Roots, leaves, seeds	Local people and traditional healers	Talmakhana Kamtakalya (Chhattisgarh)	Used for the treatment of jaundice and other liver related disorders (Janghel et al., 2019)
271	*Impatiens henslowiana* Arn.	Flowers	Kani, Mannan (Manja thetchi), Kurichar, Paniyan, Muthuvan (Nedimoorkhan)	Perumthumba (Kerala)	Expressed juice of flowers and tender leaves is used as nasal drops in severe conditions of jaundice (Asha and Pushpangadan, 2002)

TABLE 5.1 *(Continued)*

S. No	Plant Name	Part Used	Tribe/Local Community	Local Names (State)	Uses/Mode of Use (Reference)
272	*Indigofera tinctoria* L.	Aerial parts	Local people and traditional healers	Gouli (Jammu Tawi)	Hepatoprotective (Singh et al., 2001)
273	*Inula cappa* (Buch- Ham.ex D. Don) DC.	Leaves	Local healers	–	The leaves are crushed with those of *Plantago asiatica* and *Lobelia angulata* and the juice is taken orally 10 mL twice a day for jaundice (Rama et al., 2012)
274	*Ionidium suffruticosum* Ging.	Root	Local people	Ratna-purush (Delhi to Bengal)	Hepatoprotective (Vuda et al. 2012)
275	*Ipomoea aquatica* Forssk.	Aerial parts	Local healers	Kalmi (Arunachal Pradesh)	Aerial part is taken orally for jaundice cure (Rama et al., 2012)
276	*Ixora coccinea* L.	Roots	Many Kerala tribes (perumthumba)	Manjathetchi, Nedimoorkhan (Kerala)	Powdered mass of fresh roots (10–15 g) in cold water are given for 7 days (thrice daily) to treat jaundice (Asha and Pushpangadan, 2002)
277	*Jatropha curcas* L.	Fruits, seeds, leaves	Local peoples of taindol	Taindol, Jhansi, Uttar Pradesh	Used for jaundice (Jitin, 2013)
278	*Jatropha gossypifolia* L.	Leaf, seed	Local people and traditional healers	Rakta-Vyaaghrairanda (Tamil Nadu)	Hepatoprotective (Kshirsagar et al., 2011)
279	*Juniperus communis* L.	Berries	Local people	Hapushaa (Kumaon Westwards)	Hepatoprotective (Singh et al., 2016)
280	*Lactuca runcinata* DC.	Aerial parts	Traditional healers	Undirachakam (Thoothukudi district, Tamil Nadu)	Hepatoprotective (Kanthal et al., 2017)
281	*Lagenaria siceraria* (Mol.) Standl.	Whole plant	Local healers	Lao, Lauki (Arunachal Pradesh)	Plant decoction mixed with sugar is taken as protective agent during jaundice. (Rama et al., 2012)

TABLE 5.1 *(Continued)*

S. No	Plant Name	Part Used	Tribe/Local Community	Local Names (State)	Uses/Mode of Use (Reference)
282	*Lagerstroemia speciosa* (L.) Pers.	Root bark	Local healers	Ajar (Arunachal Pradesh)	Decoction of root bark is given in jaundice and enlargement of spleen (Rama et al., 2012)
283	*Lantana camara* L.	Whole plant	Herbal practitioners and elders	Aiangketba naro (Chungtia village, Nagaland)	Plant decoction is taken orally to treat jaundice (Kichu et al., 2015)
284	*Lawsonia inermis*	Roots	Godaba	Manjuati (Semiliguda block, Koraput district, Odisha)	The root is crushed and taken with water of raw rice to cure jaundice (Raut et al., 2013)
285	*Lawsonia inermis* L.	Stem bark	Local healers	Menhadi/Jetuka (Arunachal Pradesh)	Stem bark is pounded in water and taken in jaundice and spleen enlargement (Rama et al., 2012)
286	*Leucas aspera* (Willd.) Link	Leaf	Malayali tribes	Thumbai (Shevaroy Hills, Tamil Nadu)	Fresh leaf juice is taken with water orally thrice a day for five days to liver diseases (Alagesaboopathi, 2015)
287	*Leucas aspera* (Willd.) Link	Leaf	Mahadevkoli, Thakur, and Katkari	(Konkan Region, Maharashtra)	Leaf paste applied on head to cure Jaundice (Naikade and Meshram, 2014)
288	*Leucas aspera [Syn.: L. plukenetii* (Roth) Spreng.]	Whole plant	Local healers	Tumakusir (Arunachal Pradesh)	30–40 g paste of the whole plant is taken three times in meal for 2–3 weeks to cure jaundice. (Rama et al., 2012)
289	*Leucas aspera [Syn.: L. plukenetii* (Roth) Spreng.]	Leaf	Koch, Meich, Rava, Munda, Santhal, Garo, Oraon	(Coochbehar district, WestBengal)	Leaf juice used in jaundice (Tanmay et al., 2014)
290	*Leucas cephalotus* L.	Juice	Local people and traditional healers	Gubha (Chhattisgarh)	Used for the treatment of jaundice (Janghel et al., 2019)
291	*Limonia acidissima* L.	Leaves, fruits	Local communities	Vilam pazham (Jawadhu hills, Thiruvannamalai district of Tamil Nadu)	The fruit is much used in India as a liver and cardiac tonic (Ranganathan et al., 2012)

TABLE 5.1 *(Continued)*

S. No	Plant Name	Part Used	Tribe/Local Community	Local Names (State)	Uses/Mode of Use (Reference)
292	*Litsea monopetala* (Roxb.) Pers.	Stem bark	Local healers	Rogu (Arunachal Pradesh)	The stem bark is taken 2 times a day for one week orally to cure jaundice associated with hepatitis as follows: The bark is ground with the bark of *Vitex peduncularis*, 3 leaves of *Piper betel*, 4 clones of *Allium sativum*, 2–3 grains of *Piper nigrum* (Golmarich) and added 2 spoonfuls (10 g) of sugar. The paste is made into pills and taken orally (Rama et al., 2012)
293	*Lobelia alsinoides* Lam.	Whole plant	Ayurvedic practitioners	Cheriya manganari (Travancore part of Kerala)	Used for the treatment of jaundice (Binitha et al., 2019)
294	*Ludwigia adscendens* (L.) Hara	Young twigs	Local healers	Talijuri (Arunachal Pradesh)	200–300 g of young twigs are smashed and boiled in water; 100 mL extract is taken orally for 15–20 days to cure jaundice (Rama et al., 2012)
295	*Luffa acutangula* (L.) Roxb.	Fruit, tendrils, seed	Local people and traditional healers	Torae (Chhattisgarh)	Used for the treatment of jaundice (Janghel et al., 2019)
296	*Luffa echinata* Roxb.	Fruit	Local people and traditional healers	(Uttar Pradesh, Bihar, Bengal, and Gujarat)	Hepatoprotective (Girish and Pradhan, 2012)
297	*Lumnitzera racemosa* Willd.	Bark	Local people and traditional healers	Cinakkantal, katarkantal, thipparathai, tipparathai, tipparutti, tippiruttai (Tamil Nadu)	Hepatoprotective (Gnanadesigan et al., 2011)
298	*Macrothelypteris torresiana* (Gaudich.) Ching	Aerial parts	Traditional healers	Sword fern (Godavari district, Andhra Pradesh)	Hepatoprotective (Mondal et al., 2017)
299	*Madhuca indica* Gmel.	Leaves	Local people and traditional healers	Mahua (Bhopal, Madya Pradesh)	Hepatoprotective (Janghel et al., 2019)

TABLE 5.1 *(Continued)*

S. No	Plant Name	Part Used	Tribe/Local Community	Local Names (State)	Uses/Mode of Use (Reference)
300	*Madhuca longifolia* Mach	Bark, flowers	Local people and traditional healers	Mahua (Chhattisgarh)	Used for the treatment of jaundice (Janghel et al., 2019)
301	*Mahonia nepaulensis* DC.	Root	Local healers	–	Root juice is pounded and two tea spoon juice is given in jaundice and other liver disorders (Rama et al., 2012)
302	*Mallotus philippensis* Muell.-Arg.	Fruit	Local people and traditional healers	Rohini (Chhattisgarh)	Used for the treatment of jaundice (Janghel et al., 2019)
303	*Malus domestica* Borkh.	Fruits	Gujjars, Bhoris Bakerwals	Maharaji Treil (Bandipora district, Jammu and Kashmir)	Fruits are harvested and stored at some warm place for 15–20 days so as to ripe completely. Ripe fruits are eaten to cure dyspepsia, diabetes, jaundice, urinary problems, loss of appetite, and to remove phlegm from the chest, quench the thirst and dissolve the body fats (Jadhav et al., 2011)
304	*Mangifera indica* L.	Fruit	Local people and traditional healers	Aam (Uttar Pradesh., Punjab, Maharashtra, Andhra Pradesh, West Bengal and Tamil Nadu)	Hepatoprotective (Girish and Pradhan, 2012)
305	*Manilkara hexandra* (Roxb.) Dubard.	Fruits	Gond and Madiya	Khirani (Markanda Forest Range of Gadchiroli district, Maharashtra, India)	Used for curing jaundice (Chavhanand Aparna, 2015)
306	*Marrubium vulgare* L.	Whole plant	Traditional healers	Paharigandana (Srinagar, Jammu, and Kashmir)	Hepatoprotective (Akther et al., 2013)
307	*Maytenus emarginata* (Willd.) Ding Hou	Root bark, leaves	Herbal practitioners	Danti, Pedda chintu (Vizianagaram district, Andhra Pradesh)	10 to 15 leaves with sugar cube taken orally two times for 7days to cure jaundice (Padal et al., 2013c)

TABLE 5.1 *(Continued)*

S. No	Plant Name	Part Used	Tribe/Local Community	Local Names (State)	Uses/Mode of Use (Reference)
308	*Melochia corchorifolia* L.	Leaves	Local people	Chunch (Kumaon to Sikkim, Gujarat)	Hepatoprotective (Rao et al., 2013)
309	*Melothria perpusilla* (Blume) Cogn.	Whole plant	Local communities	Lamthabi (Manipur)	Vegetative parts of this plant is boiled with sugar candy in water and given in patient of Jaundice, Kidney infection (Rajesh et al., 2012)
310	*Mentha arvensis* L.	Whole plant	Local people and traditional healers	Pudina (Chhattisgarh)	Used for the treatment of jaundice (Janghel et al., 2019)
311	*Mentha spicata* L. emend. Nath	Leaf	Local people and traditional healers	Pudinaa (Punjab, Uttar Pradesh and Maharashtra)	Hepatoprotective (Girish and Pradhan, 2012)
312	*Mimosa pudica* L.	Leaves, root	Local people and traditional healers	Lajwanti (Chhattisgarh)	Used for the treatment of jaundice (Janghel et al., 2019)
313	*Momordica charantia* L.	Fruit, leaf	Local medical practitioners/healers	Karela (Shivalik Hills of Northwest Himalaya)	Fruits extract is used for diabetes, jaundice and cholera (Devi et al., 201[illegible])
314	*Momordica charantia* L.	Leaves	Local healers	Karela (Arunachal Pradesh)	The leaves are boiled with that of *Benincasa hispida* (Mainpawl) in the proportion of 5:100 gs, and the extract is taken orally against jaundice (Rama et al., 2012)
315	*Momordica charantia* L.	Fruits	Mahadevkoli, Thakur, and Katkari	Konkan Region, Maharashtra	Fruits are dried and fried given with normal diet for jaundice (Naikade and Meshram, 2014)
316	*Momordica charantia* L.	Leaves	Yenadi	Kakara (Andhra Pradesh)	Leaves juice used in curing jaundice (Bala et al., 2011)
317	*Momordica subangulata* Bl.	Fruits	Muthuvan Kurichar, Kani and Paniyan (kattupaval)	Kattupaval (Kerala)	Fresh juice of tender fruits used for curing jaundice (Asha and Pushpangadan, 2002)

TABLE 5.1 *(Continued)*

S. No	Plant Name	Part Used	Tribe/Local Community	Local Names (State)	Uses/Mode of Use (Reference)
318	*Morinda pubescens* Smith	Stem bark	Chenchu	Maddi (Rudrakod Sacred Grove, Nallamalais, Andhra Pradesh)	Fresh stem bark is crushed and stained in a glass of water throughout the night. The infusion is given in orally in the morning for 7 days for curing jaundice (Rao and Sunitha, 2011)
319	*Morinda tinctoria* Roxb.	Leaves	Local people and traditional healers	Mannanunai (Pagadappadi village, Tamil Nadu)	Hepatoprotective (Subramanian et al., 2013)
320	*Moringa concanensis* Nimmo ex Dalz. & Gibson	Leaves	Gond tribe	Munaga (Adilabad district, Andhra Pradesh)	Leaves are boiled along with pulses and taken as food in anemia and jaundice (Murthy, 2012)
321	*Moringa concanensis* Nimmo ex Dalz. & Gibson	Leavess	Local tribes	Jangli saragavo (R.D.F. Poshina forest range of Sabarkantha district, North Gujarat)	Half tea spoonful paste is prepared from the leaves and applied over the surface of body for a week to relief from jaundice (Patel and Patel, 2013)
322	*Moringa oleifera* Lam.	Stem bark	Many Kerala tribes	Kerala	Uesd in jaundice (Asha and Pushpangadan, 2002)
323	*Morus alba* L.		Local healers	Tuda (Arunachal Pradesh)	Root decoction is taken orally for 10–12 days for cure of jaundice (Rama et al., 2012)
324	*Murdannia japonica* (Thunb.) Faden	Root	Local tribe	–	Root used to treat jaundice (Bose et al., 2015)
325	*Murraya koenigii* L.	Fruits, leaves, seeds	Irula tribals	Karuveppilai Kalavai village,Vellore district, Tamil Nadu)	Used in jaundice (Natarajan et al., 2013)
326	*Musa paradisiaca* L.	Stem	Mahadevkoli, Thakur, and Katkari	Konkan Region, Maharashtra	Interior stem portion is dried and powder is given with honey for jaundice (Naikade and Meshram, 2014)

TABLE 5.1 *(Continued)*

S. No	Plant Name	Part Used	Tribe/Local Community	Local Names (State)	Uses/Mode of Use (Reference)
327	*Mussaenda frondosa* L.	Leaves	Local healers	–	Juice of fresh leaves is good for jaundice. Root decoction is taken orally for 10–12 days. 100 gs. pounded roots are boiled in water and made extract, 100 mL extract is orally taken daily for 15 days to reduce jaundice (Rana et al., 2012)
328	*Myristica fragrans* Houtt.	Friut	Many Kerala tribes	Kerala	Used to cure janundice (Asha and Pushpangadan, 2002)
329	*Nardostachys jatamansi* DC.	Rhizome	Traditional healers	Balchhar (New Delhi)	Hepatoprotective (Ali et al., 2000)
330	*Naregamia alata* Wight & Arn.	Whole plant	Kurumbar, Kani, paniyan, (Nilachar) Mannan, Kurumar, Muthuvan (Nilavepu)	Nilachara, Nilavepu (Kerala)	Whole plant extract is used for curing jaundice (Asha and Pushpangadan, 2002)
331	*Nerium indicum* Mill.	Flower	Loal people	Chandni (North Bengal)	Hepatoprotective (Dey et al., 2015)
332	*Nerium oleander* L.	Flowers	Traditional healers	Kaner (Kota, Rajasthan)	Hepatoprotective (Singhal and Gupta, 2012)
333	*Nymphaea stellata* Willd.	Flowers	Local people and traditional healers	Nilotpala (Throughout India)	Hepatoprotective (Girish and Pradhan, 2012)
334	*Ocimum gratissimum* L.	Leaves	Local people	Banjara (Rajkot, Gujarat)	Hepatoprotective (Gupta et al., 2015)
335	*Ocimum sanctum* L.	Whole plant	Local people and traditional healers	Tulasi (Throughout India)	Hepatoprotective (Girish and Pradhan, 2012)
336	*Oldenlandia corymbosa* L.	Whole plant	Traditional healers	Vermela–vemu (Parnasala sacred grove area, Eastern Ghats of Khammam District, Telangana)	The fresh plant extract is given in jaundice and other liver complaints. The decoction is given in low fever with gastric problems (Srinivasa et al., 2015)

TABLE 5.1 *(Continued)*

S. No	Plant Name	Part Used	Tribe/Local Community	Local Names (State)	Uses/Mode of Use (Reference)
337	*Operculina turpethum* (L.) Silva Manso	Root	Local people and traditional healers	Trivrta (Throughout India)	Hepatoprotective (Girish and Pradhan, 2012)
338	*Oroxylum indicum* (L.) Kurz	Bark	Local healers	Shyonaka/Bhatgila (Arunachal Pradesh)	Half kg. crushed bark is boiled in water and 100 mL extract is taken thrice daily for 2–3 weeks to cure jaundice (Rama et al., 2012)
339	*Oroxylum indicum* (L.) Kurz	Fruit, seed, bark	Local tribe	Sona, Kanaidingi; Hatipanjara, Totala (Oraon); Dagduya (Munda); Jamblaophang (Rava); Kharukhandai (Meich) (Jalpaiguri district, West Bengal)	Paste of hydrated fruit or seed or bark applied in stomach pain, chest pain, used as appetizer, and for jaundice. (Bose et al., 2015)
340	*Orthosiphon grandiflorus* Boldingh.	Leaves	Local people and traditional healers	Mutri-Tulasi (Manipur, Naga and Lushai hills, Chota Nagpur, Western Ghats)	Hepatoprotective (Kumar et al., 2011)
341	*Oxalis corniculata* L.	Leaf	Gaddi tribe	Amblu (Bharmour, Himachal Pradesh)	Leaf juice is useful in liver problems (Dutt et al., 2011)
342	*Parkia roxburghii* G. Don	Fruits	Local people	Yongchak (Bishnupur, Manipur)	Hepatoprotective (Sheikh et al., 2016)
343	*Parmelia perlata* (Huds.) Ach.	Whole thallus	Local people and Ayurvedic practioners	Shaileya (Kashmir hills and the Himalayas)	Hepatoprotective (Shailajan et al., 2014)
344	*Peltigera canina* Willd.	Whole plant	Local healers	–	Plant juice is recommended for cure of jaundice and other liver disorders (Rama et al., 2012)
345	*Pergularia daemia* (Forssk.) Chiov.	Leaf	Malayali tribes	Vealiparuthi (Shevaroy Hills, Tamil Nadu)	Leaf juice is mixed with cow's milk and drink to treat jaundice and liver problems (Alagesaboopathi, 2015)

TABLE 5.1 *(Continued)*

S. No	Plant Name	Part Used	Tribe/Local Community	Local Names (State)	Uses/Mode of Use (Reference)
346	*Peristrophe bicalyculata* (Retz) Nees	Whole plant	Local people and traditional healers	Kak jangha (Chhattisgarh)	Used for the treatment of jaundice (Janghel et al., 2019)
347	*Phaseolus mungo* L.	Seeds	Local people	Maasha (Karnataka)	Hepatoprotective (Nitin et al., 2012)
348	*Phyllanthus amaras* Schum.& Thonn.	Whole plant	Local healers	Bhumyamlaki (Arunachal Pradesh)	Plant juice as well as powder of dried plant is taken orally with water for 10–12 days to treat liver problems (Rama et al., 2012)
349	*Phyllanthus amarus* Schum. & Thonn.	Fruit	Traditional healers	Nelausiri (Parnasala sacred grove area, Eastern ghats of Khammam district,Telangana)	Jaundice: Plant paste mixed with curd 3 spoonfuls is given orally twice a day for 7 days (Srinivasa et al., 2015)
350	*Phyllanthus amarus* Schum. & Thonn.	Entire plant	Gond tribe	Nela usiri (Adilabad district, Andhra Pradesh)	Entire plant powder along with pepper powder is administered for jaundice (Murthy, 2012)
351	*Phyllanthus amarus* Schum. & Thonn.	Whole plant	Malayali tribes	Kellanelli (Shevaroy Hills, Tamil Nadu)	Decoction of the whole plant mixed with cow's milk and taken two times a day for one week to manage jaundice and liver complaints (Alagesaboopathi, 2015)
352	*Phyllanthus amarus* Schum. & Thonn.	Whole plant	Yandi	Nela vusiri (Chittoor district, Andhra Pradesh)	Oral administration of whole plant paste, about 10 g once daily for one week, cures jaundice (Ganesh and Sundarsanam, 2013)
353	*Phyllanthus amarus* Schum. & Thonn.	Leaves	Chenchu	Nela usirika (Nallamalais, Andhra Pradesh)	Leaves mixed with curd given orally for jaundice—3 spoonfuls twice a day for 7 days (Sabjan et al., 2014)

TABLE 5.1 *(Continued)*

S. No	Plant Name	Part Used	Tribe/Local Community	Local Names (State)	Uses/Mode of Use (Reference)
354	*Phyllanthus amarus* Schum. & Thonn.	Root, fruit	Local people	Keela nelli (Ariyalur District, Tamil Nadu)	Roots and fruits are crushed and mixed with goat's milk. The mixture is taken orally to cure jaundice and liver problems (Satishpandiyan et al., 2014)
355	*Phyllanthus amarus* Schum. & Thonn.	Leaf	Mahadevkoli, Thakur, and Katkari	Konkan Region, Maharashtra	Leaf juice 2 spoonful + cow milk is given early in the morning for jaundice (Naikade and Meshram, 2014)
356	*Phyllanthus emblica* L.	Fruits, roots	Local tribe	Amlaki, Amlab (Jalpaiguri district, West Bengal)	Fresh fruit and root paste used to treat jaundice (Bose et al., 2015)
357	*Phyllanthus emblica* L.	Fruits	Mahadevkoli, Thakur, and Katkari	Konkan Region, Maharashtra	Dried fruit and seeds of *Punica granatum* L. are grounded together along with sugar and made into powder, two-three teaspoons of the powder are dissolved in one cup of water and taken orally twice a day for three weeks (Naikade and Meshram, 2014)
358	*Phyllanthus fraternus* Webst.	Whole plant	Many Kerala tribes	Kerala	Used to cure liver problems (Asha and Pushpangadan, 2002)
359	*Phyllanthus niruri* L.	Whole plant	Yanadi	Nela usiri (Andhra Pradesh)	Juice used in curing jaundice (Bala et al., 2011)
360	*Phyllanthus urinaria* L.	Fresh leaves	Local people and traditional healers	Bhuiamla (Chhattisgarh)	Used for the treatment of jaundice and other liver related disorders (Janghel et al., 2019)
361	*Physalis minima* L.	Leaf	Chenchu	Buda busara (Nallamalais, Andhra Pradesh)	Leaf extract (15–20 mL) mixed with buffalo curd or sheep milk (100 mL) taken daily two times for 21days to cure jaundice (Sabjan et al., 2014)

TABLE 5.1 *(Continued)*

S. No	Plant Name	Part Used	Tribe/Local Community	Local Names (State)	Uses/Mode of Use (Reference)
362	*Physalis minima* L.	Leaves	Traditional healers	Budima (Parnasala sacred grove area, Eastern Ghats of Khammam district,Telangana)	Leaves used to cure jaundice (Srinivasa et al., 2015)
363	*Picrorhiza kurroa* Benth.	Root	Local healers	Katuki (Arunachal Pradesh)	Root extract/decoction is used in jaundice for 12 days (Rama et al., 2012)
364	*Pimpinella anisum* L.	Whole plant	Local people and traditional healers	Anisoon (Uttar Pradesh., Punjab, Assam, Orissa)	Hepatoprotective (Kshirsagar et al., 2011)
365	*Piper longum* L.	Fruit	Local people and traditional healers	Pippali (Throughout India)	Hepatoprotective (Girish and Pradhan, 2012)
366	*Piper nigrum* L.	Fruits	Rural people	Jalook (Ahoms, Chutiyas and Deuris villages of Chenijan, Jorhat district, Assam)	Fruits are crushed and mixed with two teaspoonful leaf extracts of Bel, *Aegle marmelos* Corr. (Rutaceae) is given to take orally thrice daily (Sonia et al., 2011)
367	*Plumbago zeylanica* L.	Root	Local people and traditional healers	Chitraka (Throughout India)	Hepatoprotective. (Girish and Pradhan, 2012)
368	*Podophyllum hexandrum* Royle	Roots, rhizomes	Local people and traditional healers	Giriparpata (from Kashmir to Sikkim)	Hepatoprotective (Girish and Pradhan, 2012)
369	*Polygala arvensis* Willd.	Whole plant	Malayali tribes	Milakunanka (Shevaroy Hills, Tamil Nadu)	Decoction of the whole plant is given internally twice a day for one week to cure liver disorders (Alagesaboopathi, 2015)
370	*Polygala glabrum* Willd.	Stem	Local healers	Maradu (Arunachal Pradesh)	Stem decoction is used in curing jaundice and related disorders (Rama et al., 2012)
371	*Polygonum glabrum* Willd	Root	Local healers	Bihagni, Bhilamgori (Arunachal Pradesh)	Root juice is given in jaundice and other disorders (Rama et al., 2012)

TABLE 5.1 *(Continued)*

S. No	Plant Name	Part Used	Tribe/Local Community	Local Names (State)	Uses/Mode of Use (Reference)
372	*Polygonum tortuosum* D. Don	Aerial parts	Larje, Amchis	Nayalo (Lahaul–Spiti regionn Western Himalaya)	Powder obtained from aerial parts is consumed orally with water to cure jaundice. (Koushalya, 2012)
373	*Portulaca oleracea* L.	Whole plant	Local people and traditional healers	Badinoni (Chhattisgarh)	Used for the treatment of jaundice (Janghel et al., 2019)
374	*Portulaca oleracea* L.	Leaves, seeds	Local people and traditional healers	Badi noni (Throughout India)	Hepatoprotective (Kshirsagar et al., 2011)
375	*Prunus armeniaca* L.	Kernels	Local people and traditional healers	Chihri (Northwestern Himalayas)	Hepatoprotective (Kshirsagar et al., 2011)
376	*Psidium guajava* L.	Fruit	Local healers	Madhurika, Amrud (Arunachal Pradesh)	Juice from one fruit, 1/4 liter goat milk and root of *Sida cordifolia* are mixed together thoroughly. The preparation is administered orally. Three doses is sufficient and will result in disappearance of symptoms like clear urine and removal of yellowness from the eyes of the patients (Rama et al., 2012)
377	*Pterocarpus marsupium* Roxb.	Bark	Malayali tribes	Vengai (Shevaroy Hills, Tamil Nadu)	Decoction of the bark is taken as liver tonic (Alagesaboopathi, 2015)
378	*Pterocarpus santalinus* L.	Bark	Malayali tribes	Semmaram, Chandana vengai (Shevaroy Hills, Tamil Nadu)	Decoction of the bark is given orally twice a day for 5 days in jaundice (Alagesaboopathi, 2015)
379	*Pueraria lobata* (Willd.) Ohwi	Root	Local people and traditional healers	Vidaari (Eastern Himalayas, Assam and Khasi Hills)	Hepatoprotective (Girish and Pradhan, 2012)
380	*Punica gratum* L.	Fruit, leaf, whole plant	Local people and traditional healers	Anar (Chhattisgarh)	Used for the treatment of jaundice (Janghel et al., 2019)

TABLE 5.1 *(Continued)*

S. No	Plant Name	Part Used	Tribe/Local Community	Local Names (State)	Uses/Mode of Use (Reference)
381	*Raphanus sativus* L.	Leaf, root	Gond, Kol, Baiga, Panica, Khairwar, Manjhi, Mawasi, Agaria	Mooli (Rewa district, Madhya Pradesh)	Juice of roots and leaves is given for curing jaundice (Achuta et al , 2010)
382	*Raphanus sativus* L.	Leaf, root	Local medical practitioners/healers	Muli (Shivalik Hills of Northwest Himalaya)	Extract of leaves and root is used for jaundice (Devi et al., 2014)
383	*Raphanus sativus* L.	Roots, leaves	Local people	Muli (Solan, Himachal Pradesh)	Young leaves taken for curing jaundice (Mamata and Sood, 2013)
384	*Rheum emodi* Wall. ex Meissn.	Roots	Local people and traditional healers	Amlaparni (Subalpine Himalayas)	Hepatoprotective (Kshirsagar et al., 2011)
385	*Rhizophora mucronata* Lam.	Bark	Local people	Kamo (Bengal)	Hepatoprotective (Ravikumar and Gnanadesigan, 2012)
386	*Rhodiola imbricata* Edgew.	Rhizome	Local people	Arctic root (Western Himalayas, Leh–Ladakh)	Hepatoprotective (Senthilkumar et al., 2014)
387	*Ricinus communis* L.	Leaves	Yanadi	Amudam (Andhra Pradesh)	Leaves paste used in jaundice (Bala et al., 2011)
388	*Ricinus communis* L.	Leaf	Gond, Halba and Kawar	Erandi (Darekasa hill range, Gondia district, Maharashtra)	Juice is prepared and given orally to cure jaundice (Chandrakumar et al., 2015)
389	*Ricinus cummunis* L.	Leaves	Yerukala, Yanadi, Sugali	Patcha amudam (Sheshachalam hill range of Kadapa District, Andhra Pradesh)	Decoction of leaves cures jaundice (Rajagopal et al., 2011)
390	*Rosa chinensis* Jacq.	Flowers	Local people and traditional healers	Taruni-Kantaka (Kannauj, Kanpur and Hathras)	Hepatoprotective (Girish and Pradhan, 2012)
391	*Rosa webbiana* Wall. ex Royle	Fruit	Larje, Amchis	T-siya, Seva, Shanab, susli (Lahaul–Spiti region, Western Himalaya)	Ripened fruit juice is taken with water to cure jaundice and impotency (Koushalya, 2012)

TABLE 5.1 *(Continued)*

S. No	Plant Name	Part Used	Tribe/Local Community	Local Names (State)	Uses/Mode of Use (Reference)
392	*Rubia cordifolia* L.	Tubers	Konda dora tribe	Mangala katthi (Visakhapatnam district, Andhra Pradesh)	Tuber ground into paste with that of *Mirabilis jalapa* and made into pills. One pill each is administered daily thrice with water on empty stomach for jaundice (Padal et al., 2013a)
393	*Rumex crispus* L.	Roots	Local people and traditional healers	Chukka (Mt. Abu)	Hepatoprotective (Kshirsagar et al., 2011)
394	*Rumex nepalensis* Spreng.	Roots	Hakims, priests, tribal people	Halhaley (Eastern Sikkim Himalaya region)	Dried or fresh extract used orally in hepatitis, loss of hair (Trishna et al., 2012)
395	*Rumex vesicarius* Linn.	Whole plant	Local people and Ayurvedic practioners	Chukra (Tripura, West Bengal and Bihar)	Hepatoprotective (Tukappa et al., 2015)
396	*Saccharaum officinarum* L.	Whole plant	Koal, Panika, Bhuriya, Kharvar, Gaund	Ganna (Renukot forest division, Sonbhadra, Uttar Pradesh)	Used in curing jaundice (Anurag et al., 2012)
397	*Saccharum officinarum* L.	Stem	Local medical practitioners/healers	Ganna (Shivalik Hills of Northwest Himalaya)	Extract of stem is used for constipation and jaundice (Devi et al., 2014)
398	*Sapindus emarginatus* Vahl	Leaves	Yerukala,Yanadi,Sugali	Kunkudu (Sheshachalam hill range of Kadapa District, Andhra Pradesh)	Extraction of leaves cures jaundice (Rajagopal et al., 2011)
399	*Sapindus mukorossi* Gaertn.	Fruits	Tradional healers	Ritha (Throughout India)	Hepatoprotective (Ibrahim et al., 2008)
400	*Scorzonera divaricata* Turcz.	Leaves, shoots	Larje, Amchis	Thunpu (Lahaul–Spiti regionn Western Himalaya)	Decoction of leaves and shoots prepared at low temperature is prescribed orally to cure jaundice (Koushalya, 2012)
401	*Scutia myrtina* Kurz.	Whole plant	Local people and Ayurvedic practioners	Cheemaat, Gariki (Mahabaleshwar southwards and Orissa)	Hepatoprotective (R.S.Kumar et al., 2011)

TABLE 5.1 *(Continued)*

S. No	Plant Name	Part Used	Tribe/Local Community	Local Names (State)	Uses/Mode of Use (Reference)
402	*Sesbania grandiflora* (L.) Pers.	Leaf	Koch, Meich, Rava, Munda, Santhal, Garo, Oraon	Coochbehar district, West Bengal	Extract of leaves used to cure jaundice (Tanmay et al., 2014)
403	*Sida cordifolia* L.	Leaves, root	Local people and traditional healers	Balaa (Throughout India)	Hepatoprotective (Girish and Pradhan, 2012)
404	*Silybum marianum* (L.) Gaertn.	Seeds	Local people and traditional healers	Western Himalayas	Hepatoprotective (Girish and Pradhan, 2012)
405	*Solanum dulcamara* L.	Berries	Local people and traditional healers	Agnidamani	Hepatoprotective (Kshirsagar et al., 2011)
406	*Solanum nigrum* L.	Fruits	Bhil, Meena, Sahariya	Makoe (Baran district from Rajasthan)	Fruits are eaten thrice in a day in jaundice (Rishikesh and Ashwani, 2012)
407	*Solanum nigrum* L.		Local healers	Kakamachi (Arunachal Pradesh)	Plant juice is administered with juice of equal quantity of *Phyllanthus amarus* and *Aloe barbadensis* for curing jaundice (Rama et al., 2012)
408	*Solanum nigrum* L.	Leaf	Malayali tribes	Manathaka (Shevaroy Hills, Tamil Nadu)li	The leaf juice mixed with *Piper nigrum* (black pepper) and drink to treat liver ailments (Alagesaboopathi, 2015)
409	*Solanum nigrum* L.	Whole herb	Local tribe	Makoi (Kumaun region, Uttarakhand)	Decoction of leaves is used for liver problem (Gangwar et al., 2010)
410	*Solanum surattense* Burm.f.	Root bark	Traditional healers	Errivanga (Parnasala sacredgrove area, Eastern ghats of Khammam district,Telangana)	Jaundice: Root bark pound with stem bark of *Moringa oleifera*. 3 g of the paste had given orally once a day for 6 days (Srinivasa et al., 2015)

TABLE 5.1 *(Continued)*

S. No	Plant Name	Part Used	Tribe/Local Community	Local Names (State)	Uses/Mode of Use (Reference)
411	*Solanum xanthocarpum* Schrad. & Wendl.	Whole plant	Local people	Kantakari (Taindol, Jhansi, Uttar Pradesh)	Used for curing jaundice (Jitin, 2013)
412	*Solena heterophylla* Lour.	Roots	Local healers	Amtamoola (Arunachal Pradesh)	Fresh roots inhaled reduce jaundice; also the fresh roots are cut into pieces and tied with root of *Plumbago zeylanica* and rhizome of *Curcuma longa* then worn around neck for 15 to 20 days to reduce jaundice (Rama et al., 2012)
413	*Sonchus oleraceus* L.	Leaves	Local people and traditional healers	Bakri booti (Ambala district, Haryana)	Plant extract mixed with clove is taken orally to cure liver diseases, particularly enlarged liver (Vashistha and Kaur, 2013)
414	*Sphaeranthus hirtus* Willd.	Herb	Local people and traditional healers	Gorakmundi	Hepatoprotective (Kshirsagar et al., 2011)
415	*Sphaeranthus indicus* L.	Fruit, whole plant	Local people and traditional healers	Molal Phaji, Mundi (Chhattisgarh)	Used for the treatment of jaundice (Janghel et al., 2019)
416	*Spinacia oleracea* L.	Seeds	Local healers	Palangshak (Arunachal Pradesh)	Seeds boiled in water and extract is taken for 7–10 days. (Rama et al., 2012)
417	*Spondias pinnata* (L.f.) Kurz.	Drupes and leaves	Herbal practitioners and elders	Pakho (Chungtia village, Nagaland)	Ripe drupes are eaten raw or the juice is taken orally as a liver tonic and appetizer. (Kichu et al., 2015)
418	*Suaeda maritima* (L.) Dumort	Whole plant	Traditional healers	Nila Vumarai (Tamil Nadu)	Hepatoprotective (Ravikumar et al., 2011)
419	*Swertia chirata* Buch.-Ham. ex Wall.	Whole plant	Local tribe	Bhucharitta, Kariyata, Chirata (Kumaun region, Uttarakhand)	Flowers, stem and roots are used in asthma, jaundice and anemia (Gangwar et al., 2010)

TABLE 5.1 *(Continued)*

S. No	Plant Name	Part Used	Tribe/Local Community	Local Names (State)	Uses/Mode of Use (Reference)
420	*Swertia chirata* Buch.-Ham. ex Wall.		Local healers	Chireta (Arunachal Pradesh)	Plant juice or decoction after boiling in water for 3–4 hours is taken orally one teaspoon full, thrice a day for 7–8 days for the treatment of jaundice (Rama et al., 2012)
421	*Swertia densifolia* (Griseb.) Kashyapa	Leaves, flowers	Local people and traditional healers	Shailaja (From Konkan to Kerala)	Hepatoprotective (Girish and Pradhan, 2012)
422	*Symplocos racemosa* Roxb	Stem, bark	Local traditional healers	Ludho (Nawarangpur District, Odisha)	Stembark decoction with honey (3:2) is given to childern below 10 years against liver complaints (Dhal et al., 2014)
423	*Tabernaemontana divaricata* (L.) R. Br.	Roots	Local healers	–	100- 200 gs roots are crushed and boiled in water; extract is taken three times to cure jaundice (Rama et al., 2012)
424	*Tabernaemontana divaricata* (L.) R.Br.	Root	Mahadevkoli, Thakur, and Katkari	Konkan Region, Maharashtra	Root powder is boiled in water and the extract is given thrice a day for 2–3 weeks (Naikade and Meshram, 2014)
425	*Tacca aspera* Roxb.	Embarked stem	Local people and traditional healers	Barahi kand (Chhattisgarh)	Used for the treatment of jaundice (Janghel et al., 2019)
426	*Tagets erecta* L.	Flowers	Local people and traditional healers	Genda Phul (Cuttack, Orissa)	Hepatoprotective (Giri et al., 2011)
427	*Tamarindus indica* L.	Fruit, bark	Local people	Imli (Taindol, Jhansi, Uttar Pradesh)	Used for the treatment of jaundice (Jitin, 2013)
428	*Tamarix indica* Roxb.	Whole plant	Local people and traditional healers	Jhaavuka (West Bengal, Bihar, Orissa and South India)	Hepatoprotective (Girish and Pradhan, 2012)
429	*Taraxacum officinale* Wigg.	Root, rhizome	Local medical practitioners/healers	Dulal, Dudhi (Shivalik Hills of Northwest Himalaya)	Decoction of rhizome is used for treatment of jaundice (Devi et al., 2014)

TABLE 5.1 *(Continued)*

S. No	Plant Name	Part Used	Tribe/Local Community	Local Names (State)	Uses/Mode of Use (Reference)
430	*Tecomella undulata* (G. Don) Seem.	Bark	Local people and traditional healers	Rohitaka (Himalayas)	Hepatoprotective (Girish and Pradhan, 2012)
431	*Tephrosia purpurea* L.	Roots	Chenchu	Vempalaku (Nallamalais, Andhra Pradesh)	Root extract (10 mL) mixed with a pinch of salt for Liver related stomach pain (Sabjan et al.,2014)
432	*Tephrosia purpurea* L.	Whole plant, roots	Gond, Kol, Baiga, Panica, Khairwar, Manjhi, Mawasi, Agaria	Sarpokha (Rewa district, Madhya Pradesh)	Decoction is given in liver disorders (Achuta et al., 2010)
433	*Terminalia arjuna* (Roxb.) Wight & Arn.	Bark	Local people and traditional healers	Dhananjaya (Throughout India)	Hepatoprotective (Girish and Pradhan, 2012)
434	*Terminalia bellirica* (Gaertn.) Roxb.	Fruit	Local people and traditional healers	Adamarutha, tani (Kerala)	Hepatoprotective (Kuriakose et al., 2017)
435	*Terminalia catappa* L.	Bark	Fisherman community	Nattu Vathamaram (Pudhukkottai District, Tamil Nadu)	It is used against liver diseases (Rameshkumar and Ramakritinan, 2013)
436	*Terminalia chebula* Retz.	Fruit	Many Kerala tribes	Kerala	Used to cure jaundice (Asha and Pushpangadan, 2002)
437	*Terminalia chebula* Retz.	Fruit, bark	Yerukala,Yanadi, Sugali	Karakkaya (Sheshachalam hill range of Kadapa district, Andhra Pradesh)	The decoction of bark cures fractures, ulcers, asthma, cough and jaundice (Rajagopal et al., 2011)
438	*Thespesia lampas* Lav.	Leaf, fruit juice, root, bark, seed	Local people and traditional healers	Amagong (Chhattisgarh)	Used for the treatment of jaundice (Janghel et al., 2019)

TABLE 5.1 *(Continued)*

S. No	Plant Name	Part Used	Tribe/Local Community	Local Names (State)	Uses/Mode of Use (Reference)
439	*Thespesia populnea* Corr.	Leaves	Herbal practitioners	Gangaravi, Gangaraya, Gangarenu, Gangirana, Muniganga-ravi (Vizianagaram district, Andhra Pradesh)	Extract of 3–4 fleshy leaves ground with an equal quantity of cow milk. This mixture is taken on empty stomach early in the morning for seven days, it is effective remedy for jaundice (Padal et al., 2013c)
440	*Thymus serpyllum* L.	Leaves	Local tribe	Ajwain (Kumaun region, Uttarakhand)	The herb is given in weak vision, complaints of liver and stomach, suppression of urine and menstruation (Gangwar et al., 2010)
441	*Tinospora cordifolia* (Willd.) Miers	Whole plant	Mahadevkoli, Thakur, and Katkari	Konkan Region, Maharashtra	Infusions of whole plant along with sugar juice are given to patient (Naikade and Meshram, 2014)
442	*Tinospora cordifolia* (Willd.) Miers	Stems	Local healers	Amrita, Guduchi (Arunachal Pradesh)	Fresh stem juice 10 mL or plant soaked in water for overnight taken twice a day for 7 days. (Rama et al., 2012)
443	*Tinospora cordifolia* (Willd.) Miers	Fruit	Yanadi	–	Dry fruits powder, about 5–10 g once daily for one month, with honey is administered orally to relieve from jaundice and burning micturition (Ganesh and Sudarsanam, 2013)
444	*Tinospora cordifolia* (Willd.) Miers	Leaf	Fisherman community	Seendhil (Pudhukkottai District, Tamil Nadu)	Leaf extract is taken orally with equal quantity of honey daily in the morning for jaundice until cure (Rameshkumar and Ramakritinan, 2013)
445	*Trachyspermum ammi* (L.) Sprague	Fruit	Local people	Vamu (Throughout India)	Hepatoprotective (Girish and Pradhan, 2012)
446	*Trianthema portulacastrum* L.	Leaf	Gond, Kol, Baiga, Panica, Khairwar, Manjhi, Mawasi, Agaria	Patharchatta (Rewa dist., Madhya Pradesh)	Leaf juice is given in jaundice (Achuta et al., 2010)

TABLE 5.1 *(Continued)*

S. No	Plant Name	Part Used	Tribe/Local Community	Local Names (State)	Uses/Mode of Use (Reference)
447	*Tribulus terrestris* L.	Leaf	Yanadi	–	Leaf paste, about 10–15 g once daily for a week, is administered orally to relieve from jaundice (Ganesh and Sudarsanam, 2013)
448	*Trichosanthes cordata* Roxb.	Roots	Local people and traditional healers	Ilaru (Himalayas from Garhwal to Sikkim)	Hepatoprotective (Kshirsagar et al., 2011)
449	*Tridax procumbens* L.	Whole plant	Traditional healers	Gaddichamanthi (Parnasala Sacred Grove Area, Khammam District, Telangana)	Jaundice: Plant paste with jaggery is administered in doses of two spoonfuls per day for 7 days (Srinivasa et al., 2015).
450	*Trigonella emodi* Benth.	Flowers, leaves	Larje, Amchis	Rebuksu (Lahaul–Spiti region Western Himalaya)	Flowers and leaves are dried and ground to prepare powder. One spoon of powder is taken twice a day for one week to cure jaundice (Koushalya, 2012)
451	*Trigonella foenum-graecum* L.	Seeds	Local people and traditional healers	Methikaa (Throughout India)	Hepatoprotective (Girish and Pradhan, 2012)
452	*Triticum aestivum* L.	Leaves, seeds	Local people and traditional healers	Ganha (Chhattisgarh)	Used for the treatment of jaundice (Janghel et al., 2019)
453	*Triticum aestivum* L.	Roots	Local people and traditional healers	Gaihoon (Punjab, Haryana, Uttar Pradesh., Madhya Pradesh, Maharashtra, Bihar, and Rajasthan)	Hepatoprotective (Kshirsagar et al., 2011)
454	*Uncaria gambier* Roxb.	Leaves, shoots	Local people and traditional healers	Khadira	Hepatoprotective (Kshirsagar et al., 2011)
455	*Urtica dioica* L.	Whole plant	Traditional healers	Bichu Butti (Ranikhet, Uttarakhand)	Hepatoprotective (Joshi et al., 2015)
456	*Valeriana wallichii* DC.	Whole herb	Local tribe	Samoy (Kumaun region, Uttarakhand)	Hepatoprotective (Gangwar et al., 2010)

TABLE 5.1 *(Continued)*

S. No	Plant Name	Part Used	Tribe/Local Community	Local Names (State)	Uses/Mode of Use (Reference)
457	*Vanda tessellata* (Roxb.) Hook.	Root	Local traditional healers	Malang (Nawarangpur District, Odisha)	Root decoction with common salt (2:1) is administered twice aday for 7 days against hepatitis (Dhal et al., 2014)
458	*Vitex negundo* L.		Malayali tribes	Notchi (Shevaroy Hills, Tamil Nadu)	Decoction of the flowers is given internally twice a day for five days in liver complaints (Alagesaboopathi, 2015)
459	*Withania somnifera* (L.) Dunal (cultivated var.)	Root, leaf	Local people and traditional healers	Ashwagandhaa (Throughout India)	Hepatoprotective (Girish and Pradhan, 2012)
460	*Woodfordia fruticosa* (L.) Kurz	Bark, flowers	Hakims, priests, tribal people	Dhayeroo (Eastern Sikkim Himalaya region)	Dried flower for piles, liver complaints bark for gastric trouble (Trishna et al., 2012)
461	*Woodfordia fruticosa* (L.) Kurz	Dried flower	Ethnic and Rural People	Dhayeroo (Sikkim)	For piles, liver complaints (Trishna et al., 2012)
462	*Xeromphis spinosa* (Thunb.) Keay	Stems, roots	Local healers	–	Decoction of 200–300 g of stem and root bark of the species is mixed with *Piper longum* powder and taken thrice daily for 2–3 weeks to relieve jaundice (Rama et al., 2012)
463	*Zanonia indica* L.	Fruits	Traditional healers	Chirpoti (Khasi hills of Meghalaya, Andamans.)	Hepatoprotective (Valvi et al., 2016)
464	*Zingiber officinale* Rosc.	Rhizome	Local people and traditional healers	Aardraka (Kerala, Andhra Pradesh, Uttar Pradesh, West Bengal, Maharashtra)	Hepatoprotective (Girish and Pradhan, 2012)
465	*Ziziphus oenoplia* Mill.	Root, fruits, stem batk	Traditional healers	Laghu-badara (North India)	Hepatoprotective (Valvi et al., 2016)
466	*Zizphus jujuba* Mill.	Fruits	Local people	Ber (Taindol, Jhansi, Uttar Pradesh)	Used for the treatment of liver disorders (Jitin, 2013)

NOTE:

Readers, please note that all material provided herein is for information only and may not be construed as medical advice. Readers should consult appropriate health professionals on any matter relating to their health and well-being. The effectiveness of the flowers and the leaves are at their best, when they are requested and plucked from the plants after they are fully bloomed on the plant. It is preferred not to offer the flowers plucked well in advance, in a bud condition to avoid the ill effects.

CONCLUSION

Many Indian plants described above have been traditionally used individually or in combination for the treatment of variety of liver diseases and disorders. This review hopefully will help to find out the effective and safe plant for the treatment of liver disorders. However, the active ingredients in the herbal formulations are not well defined. It is therefore important to know the active component and their molecular interactions, which will help to analyse the therapeutic efficacy of the product so that it can be standardized and commercialized by pharma companies. Preclinical and clinical investigation of traditional medicinal plants for hepatoprotective activity may provide valuable leads for the development of safe and effective drugs.

KEYWORDS

- **ethnobotany**
- **hepatoprotective plants**
- **local community**
- **tribal community**
- **mode of use**

REFERENCES

Abirami, A.; Nagarani, G.; Siddhuraju, P. Hepatoprotective effect of leaf extracts from *Citrus hystrix* and *C. maxima* against paracetamol induced liver injury in rats. *Food Sci. Human Wellness* **2015**, *4*(1), 35–41.

Achuta, N.S.; Sharad, S.; Rawat, A.K.S. An ethnobotanical study of medicinal plants of Rewa district of Madhya Pradesh. *Indian J. Trad. Knowl.* **2010**, *9*(1), 191–202.

Ahmed, J.; Nirmal, S.; Dhasade, V.; Patil, A.; Kadam, S.; Pal, S.; ... & Pattan, S. Hepatoprotective activity of methanol extract of aerial parts of *Delonix regia*. *Phytopharmacology* **2011**, *1*(5), 118–22.

Akther, N.; Shawl, A. S.; Sultana, S.; Chandan, B. K.; Akhter, M. Hepatoprotective activity of *Marrubium vulgare* against paracetamol induced toxicity. *J. Pharm. Res.* **2013**, *7*(7), 565–570.

Alagesaboopathi, C. Medicinal plants used for the treatment of liver diseases by Malayali tribes in Shevaroy hills, Salem district, Tamil Nadu, India. *World J. Pharm. Res.* **2015**, *4*(4), 816–828.

Alam, J.; Mujahid, M.; Jahan, Y.; Bagga, P.; Rahman, M. A. Hepatoprotective potential of ethanolic extract of *Aquilaria agallocha* leaves against paracetamol induced hepatotoxicity in SD rats. *J. Tradit. Complement. Med.* 2017, *7*(1), 9–13.

Ali, S.; Ansari, K. A.; Jafry, M. A.; Kabeer, H.; Diwakar, G. *Nardostachys jatamansi* protects against liver damage induced by thioacetamide in rats. *J Ethnopharmacol.* **2000**, *71*(3), 359–363.

Anandakumar, D.; Rathinakumar, S.S.; Prabakaran, G. Ethnobotanical survey of Athinadu Pachamalai hills of Eastern Ghats in Tamil Nadu, South India. *Int. J. Adv. Interdis. Res.* **2014**, *1*(4), 7–11.

Aneja, S.; Vats, M.; Aggarwal, S.; Sardana, S. Phytochemistry and hepatoprotective activity of aqueous extract of *Amaranthus tricolor* Linn. roots. *J. Ayurveda Integr. Med.* **2013**. *4*(4), 211.

Anuja, G. I.; Shine, V. J.; Latha, P. G.; Suja, S. R. Protective effect of ethyl acetate fraction of *Drynaria quercifolia* against CCl_4 induced rat liver fibrosis via Nrf2/ ARE and NFκB signalling pathway. *J. Ethnopharmacol.* **2018**, *216*, 79–88.

Anurag, S.; Singh, G.S.; Singh, P.K. Medico-ethnobotanical inventory of Renukot forest division, Sonbhadra, Uttar Pradesh, India. *Indian J. Nat. Prod. Resour.* **2012**, *3*(3), 448–457.

Anusuya, N.; Raju, K.; Manian, S. Hepatoprotective and toxicological assessment of an ethnomedicinal plant *Euphorbia fusiformis* Buch.-Ham. ex D. Don. *J Ethnopharmacol.* **2010**, *127*(2), 463–467.

Asha, V.V.; Pushpangadan, P. Hepatoprotective plants used by the tribals of Wynadu, Malappuram and Palghat districts of Kerala, India. *Anc. Sci. Life* **2002**, *22*(1), 1–8.

Babu, B. H.; Shylesh, B. S.; Padikkala, J. Antioxidant and hepatoprotective effect *of Acanthus ilicifolius. Fitoterapia* **2001**, *72*(3), 272–277.

Bala, K. M.; Mythili, S.; Sravan, K. K.; Ravinder, B.; Murali, T.; Mahender, T. Ethnobotanical survey of medicinal plants in Khammam district, Andhra Pradesh, India. *Int. J. Appl. Boil. Pharm. Tech.* **2011**, *2*(4), 366–370.

Binitha, R. R.; Shajahan, M. A.; Muhamed, J.; Anilkumar, T. V.; Premlal, S.; Indulekha, V.C. Hepatoprotective effect of *Lobelia alsinoides* Lam. in Wistar rats. *J Ayurveda Integr Med.* **2019.** https://doi.org/10.1016/j.jaim.2019.04.004

Bose, D.; Roy, J.G.; Mahapatra, S.D.; Datta, T.; Mahapatra, S.D.; Biswas, H. Medicinal plants used by tribals in Jalpaiguri district, West Bengal, India. *J. Med. Plants Stud.* **2015**, *3*(3), 15–21.

Chandrakumar, P.; Praveenkumar, N.; Sushama, N. Ethnobotanical studies on the medicinal plants of Darekasa hill range of Gondia district, Maharashtra, India. *Int. J. Res. Plant Sci.* **2015**, *5*(1), 10–16.

Chavhan, P.R.; Aparna, S. M. Ethnobotanical survey of Markanda forest range of Gadchiroli district, Maharashtra, India. *Brit. J. Res.* **2015**, *2*(1), 055–062.

Cheedella, H. K.; Alluri, R.; Ghanta, K. M. Hepatoprotective and antioxidant effect *of Ecbolium viride* (Forssk.) Alston roots against paracetamol-induced hepatotoxicity in albino wistar rats. *J. Pharm. Res.* **2013**, *7*(6), 496–501.

Devi, U.; Sharma, P.; Rana, J.C. Assessment of ethnomedicinal plants in Shivalik hills of Northwest Himalaya, India. *Am. J. Ethnomed.* **2014**, *1*(4), 186–205.

Dey, P.; Dutta, S.; Sarkar, M. P.; Chaudhuri, T. K. Assessment of hepatoprotective potential of *N. indicum* leaf on haloalkane xenobiotic induced hepatic injury in Swiss albino mice. *Chem. Biol. Interact.* **2015**, *235*, 37–46.

Dhal, N.K.; Panda, S.S.; Muduli, S.D. Ethnobotanical studies in Nawarangpur district, Odisha, India. *Am. J. Phytomed. Clin. Therapeut.* **2014**, *2*(2), 257–276.

Dutt, B.; Nath, D.; Chauhan, N. S.; Sharma, K. R.; & Sharma, S. S. Ethno-medicinal Plant Resources of Tribal Pangi Valley in

District Chamba, Himachal Pradesh, India. *Int. J. Bio-Resour. Stress Manag.*, **2014**, 5(3), 416–421.

Dutt, B.; Sharma, S.S.; Sharma, K.R.; Gupta, A.; Singh, H. Ethnobotanical survey of plants used by gaddi tribe of Bharmour area in Himachal Pradesh. *ENVIS Bulletin Himalayan Ecol.* **2011**, *19*, 22–27.

Ekbote, M. T.; Ramesh, C. K.; Mahmood, R.; Thippeswamy, B. S.; Verapur, V. Hepatoprotective and antioxidant effect of *Azima tetracantha* Lam. leaves extracts against CCl_4-induced liver injury in rats. *Indian J. Nat. Prod. Resour.* **2010**, *1*(4), 493–499.

Ganapathy, S.; Ramaiah, M.; Sarala, S.; Babu, P.M. Ethnobotanical literature survey of three Indian medicinal plants for hepatoprotective activity. *Int. J. Res. Ayur. Pharm.* **2013**, *4*(3), 378–381.

Ganesh, P.; Sudarsanam, G. Ethnomedicinal plants used by Yanadi tribes in Seshachalam biosphere reserve forest of Chittoor district, Andhra Pradesh India. *Int. J. Pharm. Life Sci.* **2013**, *4*(11), 3073–3079.

Gangwar, K.K.; Deepali; Gangwar, R.S. Ethnomedicinal plant diversity in Kumaun Himalaya of Uttarakhand, India. *Nat. Sci.* **2010**, *8*(5), 66–78.

Geetha, S.; Jayamurthy, P.; Pal, K.; Pandey, S.; Kumar, R.; Sawhney, R. C. Hepatoprotective effects of sea buckthorn (*Hippophae rhamnoides* L.) against carbon tetrachloride induced liver injury in rats. *J. Sci. Food Agric.* **2008**, *88*(9), 1592–1597.

Giri, R. K.; Bose, A.; Mishra, S. K. Hepatoprotective activity of *Tagetes erecta* against carbon tetrachloride-induced hepatic damage in rats. *Acta Pol. Pharm.* **2011**, *68*(6), 999–1003.

Girish, C.; Pradhan, S. C. Indian herbal medicines in the treatment of liver diseases: problems and promises. *Fundam. Clin. Pharmacol.* **2012**, *26*(2), 180–189.

Gnanadesigan, M.; Ravikumar, S.; Anand, M. Hepatoprotective activity of *Ceriops decandra* (Griff.) Ding Hou mangrove plant against CCl_4 induced liver damage. *J. Taibah Univ. Sci.* **2017**, *11*(3), 450–457.

Gnanadesigan, M.; Ravikumar, S.; Inbaneson, S. J. Hepatoprotective and antioxidant properties of marine halophyte *Luminetzera racemosa* bark extract in CCl_4 induced hepatotoxicity. *Asian Pac. J. Trop. Med.* **2011**, *4*(6), 462–465.

Gole, M. K.; Dasgupta, S. Role of plant metabolites in toxic liver injury. *Asia Pac. J. Clin. Nutr.* **2002**, *11*(1), 48–50.

Gupta, A.; Sheth, N. R.; Pandey, S.; Yadav, J. S.; Joshi, S. V. Screening of flavonoids rich fractions of three Indian medicinal plants used for the management of liver diseases. *Rev. Brasil. Farmacogn.*, **2015**, *25*(5), 485–490.

Harshberger, J. The purpose of Ethnobotany. *Bot. Gaz.* **1896**, *21*, 146–154.

Hurkadale, P. J.; Shelar, P. A.; Palled, S. G.; Mandavkar, Y. D.; Khedkar, A. S. Hepatoprotective activity of *Amorphophallus paeoniifolius* tubers against paracetamol-induced liver damage in rats. *Asian Pac. J. Trop. Biomed.* **2012**, *2*(1), S238-S242.

Ibrahim, M.; Khaja, M. N.; Aara, A.; Khan, A. A.; Habeeb, M. A.; Devi, Y. P.; ... Habibullah, C. M. Hepatoprotective activity of *Sapindus mukorossi* and *Rheum emodi* extracts: in vitro and in vivo studies. *World J. Gastroenterol.* **2008**, *14*(16), 2566.

Jadhav, V.D.; Mahadkar, S. D.; Valvi, S. R. Documentation and ethnobotanical survey of wild edible plants from Kolhapur district. *Recent Res. Sci. Tech.* **2011**, *3*(12), 58–63.

Jain, M.; Kapadia, R.; Jadeja, R.N.; Thounaojam, M. C.; Devkar, R. V.; Mishra, S. H. Protective role of standardized *Feronia limonia* stem bark methanolic extract against carbon tetrachloride induced hepatotoxicity. *Ann. Hepatol.* **2012**, *11*(6), 935–943.

Jaiprakash, B.; Karadi, R. V.; Savadi, R. V.; Hukkeri, V. L. Hepatoprotective activity of bark of *Balanites aegyptiaca* Linn. *J. Nat. Remed.* **2003**, *3*(2), 205–207.

Janghel, V.; Patel, P.; Chandel, S.S. Plants used for the treatment of icterus (jaundice) in Central India: A review. *Ann. Hepatol.* **2019**, *18*(5), 658–672.

Jitin, R. An ethnobotanical study of medicinal plants in Taindol Village, district Jhansi, region of Bundelkhand, Uttar Pradesh, India. *J. Med. Plants Stud.* **2013**, *1*(5), 59–71.

Joshi, B. C.; Prakash, A.; Kalia, A. N. Hepatoprotective potential of antioxidant potent fraction from *Urtica dioica* Linn. (whole plant) in CCl_4 challenged rats. *Toxicol. Rep.* **2015**, *2*, 1101–1110.

Joshy, C.; Thahimon, P. A.; Kumar, R. A.; Carla, B.; Sunil, C. Hepatoprotective, anti-inflammatory and antioxidant activities of *Flacourtia montana* J. Grah. leaf extract in male Wistar rats. *Bull. Faculty Pharm. Cairo Univ.* **2016**, *54*(2), 209–217.

Junejo, J. A.; Gogoi, G.; Islam, J.; Rudrapal, M.; Mondal, P.; Hazarika, H.; Zaman, K. Exploration of antioxidant, antidiabetic and hepatoprotective activity of *Diplazium esculentum*-A wild edible plant from North Eastern India. *Future J. Pharm. Sci.* **2018**, *4*(1), 93–101.

Jyothi, B.; Prasad, G.P.; Sudarsanam, G.; Sitaram, B.; Vasudha, K. Ethnobotanical investigation of underground plant parts form Chittoor district, Andhra Pradesh, India. *Life Sci. Leafl.* **2011**, *18*, 695–699.

Kalaskar, M. G.; Surana, S. J. Free radical scavenging and hepatoprotective potential of *Ficus microcarpa* L. fil. bark extracts. *J. Nat. Med.* **2011**, *65*(3–4), 633.

Kanthal, L. K.; Satyavathi, K.; Bhojaraju, P.; Kumar, M. P. Hepatoprotective activity of methanolic extracts of *Lactuca runcinata* DC. and *Gyrocarpus asiaticus* Willd. *Beni-Suef Univ. J. Basic Appl. Sci.* **2017**, *6*(4), 321–325.

Kichu, M.; Malewska, T.; Akter, K.; Imchen, I.; Harrington, D.; Kohen, J.; ... Jamie, J. F. An ethnobotanical study of medicinal plants of Chungtia village, Nagaland, India. *J. Ethnopharmacol.* **2015**, *166*, 5–17.

Koushalya, N.S. Traditional knowledge on ethnobotanical uses of plant biodiversity: A detailed study from the Indian Western Himalaya. *Biodiv. Res. Conserv.* **2012**, 28, 63–77.

Kshirsagar, A. D.; Mohite, R.; Aggrawal, A. S.; Suralkar, U. R. Hepatoprotective medicinal plants of Ayurveda-A review. *Asian J. Pharm. Clin. Res.* **2011**, *4*(3), 1–8.

Kumar, R. S.; Asokkumar, K.; Murthy, N. V. Hepatoprotective effects and antioxidant role of *Scutia myrtina* on paracetamol induced hepatotoxicity in rats. *J. Compl. Int. Med.* **2011**, *8*(1). doi: 10.2202/1553–3840.1461.

Kumar, C. H.; Ramesh, A.; Kumar, J. S.; Ishaq, B. M. A review on hepatoprotective activity of medicinal plants. *Int. J. Pharm. Sci. Res.* **2011**, *2*(3), 501.

Kumar, K. E.; Harsha, K. N.; Sudheer, V. In vitro antioxidant activity and in vivo hepatoprotective activity of aqueous extract of *Allium cepa* bulb in ethanol induced liver damage in Wistar rats. *Food Sci. Hum. Wellness*, **2013**, *2*(3–4), 132–138.

Kuriakose, J.; Raisa, H. L.; Vysakh, A.; Eldhose, B.; Latha, M. S. *Terminalia bellirica* (Gaertn.) Roxb. fruit mitigates CCl_4 induced oxidative stress and hepatotoxicity in rats. *Biomed. Pharmacother.* **2017**, *93*, 327–333.

Mamta, S.; Sood, S.K. Ethnobotanical Survey for wild plants of district Solan, Himachal Pradesh, India. *Int. J. Environ. Biol.* **2013**, 3(3), 87–95.

Mishra, S.; Sahoo, S.; Rout, K. K.; Nayak, S. K.; Mishra, S. K.; Panda, P. K. Hepatoprotective effect of *Barringtonia acutangula* Linn. leaves on carbon tetrachloride-induced acute liver damage in rats. *Indian J. Nat. Prod. Resour.*, **2011**, *2*(4), 515–519.

Mohammad, A. 1994. *Text book of Pharmacognosy.* SBS publishers, New Delhi. 8–14.

Mohan, R.K.; Murthy, P.V.B. Plants used in traditional medicine by tribals of Prakasam district, Andhra Pradesh. *Anc. Sci. Life*, **1992**, *11(*3), 176–181.

Mondal, S.; Ghosh, D.; Ganapaty, S.; Chekuboyina, S.V.G., Samal, M. Hepatoprotective activity of *Macrothelypteris torresiana* (Gaudich.) aerial parts against CCl_4-induced hepatotoxicity in rodents and analysis of polyphenolic compounds by HPTLC. *J. Pharm. Anal.*, **2017**, *7*(3), 181–189.

Mujahid, M.; Siddiqui, H. H.; Hussain, A.; Hussain, M. S. Hepatoprotective effects of *Adenanthera pavonina* (Linn.) against anti-tubercular drugs-induced hepatotoxicity in rats. *Pharmacogn. J.* **2013**, *5*(6), 286–290

Mukesh, K.; Tariq, A.B.; Hussaini1, S.A.; Kishore, K.; Hakimuddin, K.; Aminuddin; Samiulla, L. Ethnomedicines in the Khordha forest division of Khordha district, Odisha, India. *Int. J. Curr. Microbiol. App. Sci.* **2014**, *3*(1), 274–280.

Murthy, E.N. Ethnomedicinal plants used by Gonds of Adilabad district, Andhra Pradesh, India. *Int. J. Pharm. Life Sci.* **2012**, *3*(10), 2034–2043.

Naikade, S.M.; Meshram, M.R. Ethno-medicinal plants used for jaundice from Konkan region, Maharashtra, India. *Int. J. Pharm. Sci. Invent.* **2014**, *3*(12), 39–41.

Natarajan, A.; Leelavinodh, K.S.; Jayavelu, A.; Devi, K.; Senthil, K.B. A study on ethnomedicinal plants of Kalavai, Vellore district, Tamil Nadu, India. *J. Appl. Pharm. Sci.* **2013**, *3*(1), 99–102.

Nitin, M.; Ifthekar, S. Q.; Mumtaz, M. Evaluation of hepatoprotective and nephroprotective activity of aqueous extract of *Vigna mungo* (Linn.) Hepper on rifampicin-induced toxicity in albino rats. *Int. J. Health Allied Sci.* **2012**, *1*(2), 85.

Padal, S.B.; Butchi, R.J.; Chandrasekhar, P. Traditional knowledge of Konda Dora tribes, Visakhapatnam district, Andhra Pradesh, India. *IOSR J. Pharm.* **2013a**, *3*(1), 22–28.

Padal, S.B.; Chandrasekhar, P.; Vijakumar, Y. Traditional uses of plants by the tribal communities of Salugu Panchayati of Paderu Mandalam, Visakhapatnam district, Andhra Pradesh, India. *Int. J. Comp. Eng. Res.*, **2013b**, *3*(5), 98–103.

Padal, S.B.; Venkaiah, M.; Chandrasekhar, P.; Vijayakumar, Y. Traditional phytotherapy of Vizianagaram district, Andhra Pradesh, India. *IOSR J. Pharm.* **2013c**, *3*(6), 41–50.

Parul, B.; Vashistha, D. An Ethnobotanical study of plains of Yamuna Nagar district, Haryana, India. *Int. J. Inn. Res. Sci. Eng. Tech.* **2015**, *4*(1), 18600–18607.

Parvaiz, A.L.; Ajay, K.B.; Fayaz, A.B. A study of some locally available herbal medicines for the treatment of various ailmentts in Bandipora district of J & K, India. *Int. J. Pharm. Biol. Sci.* **2013**, *4*(2), 440–453.

Patel, H.R.; Patel, R.S. Ethnobotanical plants used by the tribes of R.D.F. Poshina forest range of Sabarkantha district, North Gujarat, India. *Int. J. Sci. Res. Pub.*, **2013**, *3*(2), 1–8.

Patil, K.; Mohammed Imtiaz, S.; Singh, A.; Bagewadi, V.; Gazi, S. Hepatoprotective activity of *Cucumis trigonus* Roxb. Fruit against CCl_4 induced hepatic damage in rats. *Iran. J. Pharm. Res.* **2011**, *10*(2), 295.

Pattanayak, S.; Nayak, S.S.; Panda, D.P.; Dinda, S.C.; Shende, V.; Jadav, A. Hepatoprotective activity of crude flavonoids extract of *Cajanus scarabaeoides* (L.) in paracetamol intoxicated albino rats. *Asian J. Pharm. Biol. Res.* **2011**, *1*(1), 22–27.

Pattanayak, S.; Priyashree, S. Hepatoprotective activity of the leaf extracts from *Dendrophthoe falcata* (L.f.) Ettingsh against carbon tetrachloride-induced toxicity in wistar albino rats. *Pharmacogn. Mag.* **2008**, *4*(15), 218.

Pradeep, H.A.; Khan, S.; Ravikumar, K.; Ahmed, M. F.; Rao, M. S.; Kiranmai, M.,... Ibrahim, M. Hepatoprotective evaluation of *Anogeissus latifolia*: In vitro and in vivo studies. *World J. Gastroenterol.* **2009**, *15* (38), 4816.

Pratap, G.P.; Prasad, G.P.; Sudarsanam, G. Ethnomedical studies in Kilasagirikona Forest Range of Chittoor district, Andhra Pradesh. *Anc. Sci. Life* **2009**, *29*(2), 40–45.

Raj, B.; Singh, S.J.; Samual, V.J.; John, S.; Siddiqua, A. Hepatoprotective and antioxidant activity of *Cassytha filiformis* against CCl_4 induced hepatic damage in rats. *J. Pharm. Res.* **2013**, *7*(1), 15–19.

Rajagopal, R.S.; Madhusudhana, R.A.; Philomina, N.S.; Yasodamma, N. Ethnobotanical survey of Sheshachalam hill range of Kadapa district, Andhra Pradesh, India.

Indian J. Fund. Appl. Life Sci. **2011**, *1*(4), 324–329.

Rajesh, S.Y.; Onita, D.C.H.; Santosh, K.S.; Abujamand, D.C. Study on the ethnomedicinal system of Manipur. *Int. J. Pharm. Biol. Arch.* **2012**, *3*(3), 587–591.

Rama, S.; Rawat, M.S.; Deb, S.; Sharma, D.K. Jaundice and its traditional cure in Arunachal Pradesh. *J. Pharm. Sci. Inn.* **2012**, *1*(3), 93–97.

Rameshkumar, S.; Ramakritinan, C.M. Floristic survey of traditional herbal medicinal plants for treatments of various diseases from coastal diversity in Pudhukkottai district, Tamil Nadu, India. *J. Coastal Life Med.* **2013**, *1*(3), 225–232.

Ranganathan, R.; Vijayalakshmi, R.; Parameswari, P. Ethnomedicinal survey of Jawadhu hills in Tamil Nadu. *Asian J. Pharm. Clin. Res.* **2012**, *5*(2), 45–49.

Rao, B.G.; Rao, Y.V.; Rao, T. M. Hepatoprotective and antioxidant capacity of *Melochia corchorifolia* extracts. *Asian Pac. J. Trop. Med.* **2013**, *6*(7), 537–543.

Rao, B.R.P.; Sunitha, S. Medicinal plant resources of Rudrakod Sacred Grove in Nallamalais, Andhra Pradesh, India. *J. Biodiv.*, **2011**, *2*(2), 75–89.

Raut, S.; Raut, S.; Sen, S.K.; Satpathy, S.; Pattnaik, D. An ethnobotanical survey of medicinal plants in Semiliguda of Koraput district, Odisha, India. *Res. J. Recent Sci.* **2013**, *2*(8), 20–30.

Ravichandra, V. D.; Ramesh, C.; Sridhar, K. A. Hepatoprotective potentials of aqueous extract of *Convolvulus pluricaulis* against thioacetamide induced liver damage in rats. *Biomed. Aging Pathol.* **2013**, *3*(3), 131–135.

Ravikumar, S.; Gnanadesigan, M. Hepatoprotective and antioxidant properties of *Rhizophora mucronata* mangrove plant in CCl_4 intoxicated rats. *J. Exp. Clin. Med.* **2012**, *4*(1), 66–72.

Ravikumar, S.; Gnanadesigan, M.; Inbaneson, S. J.; Kalaiarasi, A. Hepatoprotective and antioxidant properties of *Suaeda maritima* (L.) Dumort ethanolic extract on concanavalin-A induced hepatotoxicity in rats. *Indian J. Exp. Biol.* **2011**, 49(6), 455–460.

Reddy, K.N.; Reddy, C.S.; Raju, V.S. Ethnomedicinal Observations among the Kondareddis of Khammam District, Andhra Pradesh, India. *Ethnobot. Leaflets* **2008**, *12*, 916–926.

Rishikesh, M.; Ashwani, K. Ethnobotanical survey of medicinal plants from Baran district of Rajasthan, India. *J. Ethnobiol. Trad. Med.* 2012, 117, 199–203.

Sabjan, G.; Sudarsanam, G.; Dharaneeshwara R.D.; Muralidhara Rao, D. Ethnobotanical crude drugs used in treatment of liver diseases by Chenchu tribes in Nallamalais, Andhra Pradesh, India. *Am. J. Ethnomed.* **2014**, *1*(3), 115–121.

Sadale, A.N.; Karadge, B.A. Survey on ethnomedicinal plants of Ajara Tahsil, district Kolhapur, Maharashtra, India. *Trend. Life Sci.* **2013**, *2*(1), 21–25.

Samuel, A.J.S.J.; Mohan, S.; Chellappan, D.K.; Kalusalingam, A.; Ariamuthu, S. *Hibiscus vitifolius* (Linn.) root extracts shows potent protective action against anti-tubercular drug induced hepatotoxicity. *J. Ethnopharmacol.* **2012**, *141*(1), 396–402.

Sathishpandiyan, S.; Prathap, S.; Vivek, P.; Chandran, M.; Bharathiraja, B.; Yuvaraj, D.; Smila, K. H. Ethnobotanical study of medicinal plants used by local people in Ariyalur district, Tamil Nadu, India. *Int. J. Chem. Tech. Res.* **2014**, *6*(9), 4276–4284.

Senthilkumar, R.; Chandran, R.; Parimelazhagan, T. Hepatoprotective effect of *Rhodiola imbricata* rhizome against paracetamol-induced liver toxicity in rats. *Saudi J. Biol. Sci.* **2014**, *21*(5), 409–416.

Shailajan, S.; Joshi, M.; Tiwari, B. Hepatoprotective activity of *Parmelia perlata* (Huds.) Ach. against CCl_4 induced liver toxicity in Albino Wistar rats. *J. Appl. Pharm. Sci.* **2014**, *4*(2), 70.

Sharma, A.; Sangameswaran, B.; Jain, V.; Saluja, M. S. Hepatoprotective activity of *Adina cordifolia* against ethanol induce hepatotoxicity in rats. *Int. Curr. Pharm. J.* **2012**, *1*(9), 279–284.

Shazhni, J.A.; Renu, A.; Vijayaraghavan, P. Insights of antidiabetic, anti-inflammatory and hepatoprotective properties of antimicrobial secondary metabolites of corm extract from *Caladium x hortulanum*. *Saudi J. Biol. Sci.* **2018**, *25*(8), 1755–1761.

Sheikh, Y.; Maibam, B.C.; Talukdar, N.C.; Deka, D.C.; Borah, J.C. In vitro and in vivo anti-diabetic and hepatoprotective effects of edible pods of *Parkia roxburghii* and quantification of the active constituent by HPLC-PDA. *J. Ethnopharmacol.* **2016**, *191*, 21–28.

Singh, V.; Pandey, R. P. *Ethnobotany of Rajasthan, India*. Scientific Publishers. **1998**.

Singh, H.; Prakash, A.; Kalia, A.N.; Majeed, A.B.A. Synergistic hepatoprotective potential of ethanolic extract of *Solanum xanthocarpum* and *Juniperus communis* against paracetamol and azithromycin induced liver injury in rats. *J. Trad. Complement. Med. 2*, **2016**, *6*(4), 370–376.

Singh, B.; Saxena, A.K.; Chandan, B.K.; Bhardwaj, V.; Gupta, V.N.; Suri, O.P.; Handa, S.S. Hepatoprotective activity of indigtone—a bioactive fraction from *Indigofera tinctoria* Linn. *Phytother. Res.* **2001**, *15*(4), 294–297.

Singh, K.; Singh, N.; Chandy, A.; Manigauha, A. In vivo antioxidant and hepatoprotective activity of methanolic extracts of *Daucus carota* seeds in experimental animals. *Asian Pac. J. Trop. Biomed.* **2012**, *2*(5), 385–388.

Singhal, K.G.; Gupta, G.D. Hepatoprotective and antioxidant activity of methanolic extract of flowers of *Nerium oleander* against CCl_4-induced liver injury in rats. *Asian Pac. J. Trop. Med.* **2012**, *5*(9), 677–685.

Sonia, S.G.S.; Neeliand, S.S.; Jalapure. Ethnobotanical survey of some of the herbs used in Jorhat district, Assam. *Int. J. Curr. Pharm. Res.* **2011**, *3*(4), 53–54.

Srinivasa, R.D.; Shanmukha, R.V.; Prayaga, M.P.; Narasimha, R.G.M.; Venkateswara, R.Y. Some ethnomedicinal plants of Parnasala Sacred Grove area, Eastern Ghats of Khammam district, Telangana, India. *J. Pharm. Sci. Res.* **2015**, *7*(4), 210–218.

Subramanian, M.; Balakrishnan, S.; Chinnaiyan, S.K.; Sekar, V.K.; Chandu, A.N. Hepatoprotective effect of leaves of *Morinda tinctoria* Roxb. against paracetamol induced liver damage in rats. *Drug Inv. Today* **2013**, *5*(3), 223–228.

Swapna, B. An ethnobotanical survey of plants used by Yanadi tribe of Kavali, Nellore district, Andhra Pradesh, India. *J. Sci. Innov. Res.* **2015**, *4*(1), 22–26.

Tanmay, D.; Amal, K.P.; Santanu, G.D. Medicinal plants used by tribal population of Coochbehar district, West Bengal, India—an ethnobotanical survey. *Asian Pac. J. Trop. Biomed.* **2014**, *4* (1), S478-S482.

Trishna, D.; Shanti, B.M.; Dipankar, S.; Shivani, A. Ethnobotanical survey of medicinal plants used by ethnic and rural people in Eastern Sikkim Himalayan Region. *African J. Basic Appl. Sci.* **2012**, *4*(1), 16–20.

Tukappa, N.K.A.; Londonkar, R.L.; Nayaka, H.B.; Kumar, C.B.S. Cytotoxicity and hepatoprotective attributes of methanolic extract of *Rumex vesicarius* L. *Biol. Res.* **2015**, *48*(1), 19.

Valvi, A. R.; Mouriya, N.; Athawale, R.B., & Bhatt, N.S. Hepatoprotective Ayurvedic plants–a review. *J. Complement. Integr. Med.* **2016**, *13*(3), 207–215.

Vashistha, B.D.; Kaur, M. Floristic and ethnobotanical survey of Ambala district, Haryana. *Int. J. Pharm. Biol. Sci.* **2013**, *4*(2), 353–360.

Vuda, M.; D'Souza, R.; Upadhya, S.; Kumar, V.; Rao, N.; Kumar, V.; ... & Mungli, P. Hepatoprotective and antioxidant activity of aqueous extract of *Hybanthus enneaspermus* against CCl_4-induced liver injury in rats. *Exp. Toxicol. Pathology.* **2012**, *64*(7–8), 855–859.

CHAPTER 6

Polyherbal Formulations

Herbal formulation having one or more herbs in specified quantity prescribed to achieve specific therapeutic activity. The active phytoconstituents present in individual plants are sometime ineffective to achieve the desirable therapeutic scale. When combining the multiple herbs in a particular ratio, the mixture may give a better therapeutic effect. This chapter mainly focused on various polyherbal formulations used as a hepatoprotective.

INTRODUCTION

Plant-based medication for liver disorder has been used in India, China, and other Asian and some African countries for a long time. Now, it has also popularized in the whole world. In recent years, main attention has been focused on the development of hepatoprotective natural occurring components and their mechanism of action. More than 600 commercial herbal products and their active components have been claimed for their hepatoprotective activity (Girish and Pradhan, 2012). Almost 160 phytochemical compounds obtained from more than 100 plants with liver protection properties have already been investigated (Saleem et al., 2008). In latest research, it has been investigated that drug–herbal combination treatment is an alternative medication in the market. Modulation in metabolic enzyme and active transporters by herbal photochemical fractions may affect therapeutic outcomes of coadministered allopathic medicines due to changes in their pharmacokinetic profiles.

A combination of different plant extracts as polyherbal medication is a new approach in the treatment of liver disease; for example, Liv. 52 (Himalaya Drug Co. product), Livergen (Standard Pharmaceuticals), Tefroliv (TTK Pharma Pvt. Ltd), etc., are generally used in India for hepatotoxicity treatment. Multiherbal formulation is widely used by different pharmaceutical companies for the treatment of liver toxicity. Some polyherbal drugs are given as follows.

BIHERBAL COMBINATION

Medicinal plants are used either as mono-, bi-, or multiherbal. Ezeonwu and Dahiru (2013) evaluated the protective effect of a biherbal formulation of *Ocimum gratissimum* and *Gongronema latifolium* aqueous leaf extracts in acetaminophen-induced liver and kidney toxicity in rats. Oral administrations of 200, 300, and 500 mg/kg body weight (BW) of the biherbal extract, 500 mg/kg BW each of mono extract, and 100 mg/kg BW silymarin (reference drug) were administered to laboratory animals for 10 days. A suspension of acetaminophen at 750 mg/kg BW was administered once every 72 h to induce toxicity. Biochemical analysis showed a significant increase in the levels of alanine transaminase (ALT), aspartate aminotranferase (AST), alkaline phosphatase (ALP), total bilirubin, creatinine (CRT), urea, cholesterol (CHOL), and triglyceride (TG) with a reduction in total protein (TP) levels of rats administered acetaminophen only. There was a dose-dependent reversal of these changes in rats pretreated with 200, 300, and 500 mg/kg BW of the biherbal extract. The biherbal extract showed higher protective activity than the mono extracts of *Ocimum gratissimum* and *Gongronema latifolium*. The biherbal extract's activity at 500 mg/kg BW was comparable with 100-mg/kg BW silymarin. The study demonstrated that the aqueous biherbal extract of *Ocimum gratissium* and *Gonglonema latifolium* possessed better protective activity against the mono extracts in acetaminophen-induced toxicity. This possibly suggested the combined use of the plants in traditional medicine.

BIHERBAL FORMULATION (*Aerva lanata AND Achyranthes aspera*)

The combined hepatoprotective effect of biherbal ethanolic extract (BHEE) was evaluated by Chaitanya et al. (2012) against paracetamol-induced hepatic damage in albino rats. Ethanolic extract from the leaves of *Aerva lanata* and leaves of *Achyranthes aspera* at a dose level of 200 and 400 mg/kg BW was administered orally once for three days. Substantially elevated serum marker enzymes such as serum glutamic oxaloacetic transaminase (SGOT), serum glutamic-pyruvic transaminase (SGPT), and ALP due to paracetamol treatment were restored toward normal. Biochemical parameters such as TP, tuberculosis (TB), total CHOL, TGs, and urea were also restored toward normal levels. In addition, BHEE significantly decreased the liver weight of paracetamol-intoxicated rats. Silymarin at a dose level of 25 mg/kg used as a standard reference also exhibited significant hepatoprotective activity against paracetamol-induced hepatotoxicity.

POLYHERBAL MIXTURE 1

Rachmawati et al. (2014) investigated the effect of polyherbal combination on the hepatoprotective effect against liver dysfunction due to antituberculosis drugs. A clinical study was conducted on patients diagnosed with TB, who were coadministered for four weeks with a polyherbal mixture (*Phyllanthus niruri, Curcuma xanthorrhiza*, and *Curcuma longa*) together with TB multitherapy. The authors found that the polyherbal mixture prevents the rise in ALT with respect to those treated with anti-TB drugs alone; thus, the authors concluded that this polyherbal mixture possesses a hepatoprotective effect (Rachmawati et al. 2014).

POLYHERBAL FORMULATION-2

A polyherbal formulation (PHF) comprising of *Gongronema latifolia, Ocimum gratissimum*, and *Vernonia amygdalina* demonstrated significant hepatoprotective activities by attenuating the increase in serum hepatic enzyme levels after carbon tetrachloride (CCl_4) treatment compared to the toxin control group and increasing the levels of serum catalase (CAT), glutathione peroxidase (GPx), glutathione (GSH), glutathione-*S*-transferase (GST), superoxide dismutase (SOD), and TP and significantly decreasing lipid peroxidation compared to the toxin control group (Iroanya et al., 2012).

POLYHERBAL FORMULATION-3

Dandagi et al. (2008) explored the hepatoprotective activity of various extracts of *Ferula asafoetida, Momordica charantia*, and *Nardostachys jatamansi* against experimental hepatotoxicity. Polyherbal suspensions were formulated using extracts showing significant activity and evaluated for both physicochemical and hepatoprotective activity in comparison with Liv. 52 as standard. Petroleum ether (60–80°), chloroform, benzene, ethanol, and aqueous extracts of *Ferula asafetida, Momordica charantia*, and *Nardostachys jatamansi* were evaluated for hepatoprotective activity against CCl_4-induced liver toxicity in Wistar rats. Polyherbal suspensions were prepared by the trituration method using a suspending agent and other excipients. Formulation F3 has shown significant hepatoprotective effect by reducing the elevated serum enzyme levels such as SGOT, SGPT, and ALP. These biochemical observations were supplemented by histopathological examination of liver sections. Experimental data suggested that treatment with formulation F3 enhances the recovery from CCl_4-induced hepatotoxicity. From these results, it may be concluded that the F3 formulation (containing chloroform, petroleum ether, and aqueous extracts of *Ferula asafoetida*, petroleum ether and ethanol extracts of *Momordica charantia*, and petroleum ether and ethanol extracts of *Nardostachys jatamansi*) demonstrated significant hepatoprotective activity, which might be due to combined effect of all these extracts.

POLYHERBAL FORMULATION-4

Dinesh et al. (2014) evaluated the hepatoprotective effect of PHFs (containing *Boerhavia diffusa, Phyllanthus amarus, Tinospora cordifolia, Euphorbia hirta*, and *Wedelia chinensis*) on paracetamol-induced hepatic-damaged experimental animals. PHFs pretreatment showed normal morphological parameters signs and significant effect of serum enzymes, TP, and bilirubin levels.

POLYHERBAL FORMULATION-5

Khan et al. (2008) evaluated the hepatoprotective effects of mixture of *Berberis lycium, Galium aparine,* and *Pistacia integerrima* in CCl_4-treated rats. ALT, AST, and ALP activities were significantly increased in hepatocurative groups as compared to the normal control and decreased as compared to the CCl_4-treated rated and the hepatopreventive group. The results of this study indicate that a mixture of *Berberis lycium, Galium aparine,* and *Pistacia integerrima* has hepatoprotective effects. These medicinal plants have more effect as curative agents rather than preventive agents.

POLYHERBAL FORMULATION-6

Mistry et al. (2012) explored the hepatoprotective effects of PHF against CCl_4-induced hepatotoxicity in rats. The liver damage was confirmed by estimation of elevated levels of SGOT, SGPT, ALP, and serum bilirubin. PHF pretreatment (300 mg/kg/bw) significantly reduces the CCl_4-induced elevated serum levels of SGOT, SGPT, ALP, and serum bilirubin.

POLYHERBAL FORMULATION-7

Gumaa et al. (2017) evaluated the hepatoprotective activity of water extract mixtures of rhizomes of *Zingiber officinale*, *Curcuma longa, Glycyrrhiza glabra*, the bark of *Cinnamomum zeylanicum,* and the calyces of *Hibiscus sabdariffa* using CCl_4-induced liver injury in rats. The water extract mixtures were administered for 10 days; on the 10th day, all rats were challenged with CCl_4 except control group animals. AST, ALT, and albumin levels were determined to prove the hepatoprotective effect. The enzyme activities were significantly increased in CCl_4-treated rats. The four water extract mixtures exhibited significant protective effect against CCl_4-induced hepatoxocity and nephrotoxicity by decreasing the levels of serum markers, especially AST and CRT, respectively. In contrast, the serum lipid profiles were slightly improved; HDL-CHOL significantly increased in all the water extract mixtures used.

AQUEOUS EXTRACT FORMULA (AEF) DERIVED FROM *Artemisia capillaris, Lonicera japonica,* AND *Silybum marianum*

Yang et al. (2013) evaluated the hepatoprotective effects of an AEF derived from *Artemisia capillaris, Lonicera japonica,* and *Silybum marianum* (ratio 1:1:1) by its antioxidant properties and its attenuation of CCl_4-induced liver damage in rats. The antioxidant analyses revealed that the AEF showed higher 2,2-diphenyl-1-picrylhydrazyl (DPPH) radical and superoxide anion radical scavenging activities as well as ferric reducing antioxidant power (FRAP) and Trolox equivalent Antioxidant capacity compared with the individual herbs, suggesting a synergism in antioxidation between the three herbs. The animal experiments showed that the CCl_4 treatment increased serum ALT and AST activities, but decreased TG and GSH levels as well as GPx, SOD, and CAT activities. However, AEF administration can successfully lower serum ALT and AST activities, restore the GSH level, ameliorate or restore GPx and CAT activities, and improve SOD action depending on AEF dosage. Histological examination of liver showed that CCl_4 increased the extent of bile duct proliferation, necrosis, fibrosis, and fatty vacuolation throughout the liver, but AEF can improve bile duct proliferation, vacuolation, and fibrosis and restore necrosis.

AMALKADI GHRITA (AG)

AG, a PHF, was evaluated by Achliya et al. (2004) for its hepatoprotective activity against CCl_4-induced hepatic damage in rats. The hepatoprotective activity of AG was evaluated by measuring levels of serum marker enzymes such as SGOT, SGPT, ALP, and acid phosphatase (ACP). The serum levels of TPs and bilirubin were also estimated. The histological studies were also carried out to support the above parameters. Silymarin was used as standard drug. Administration of AG (100 and 300 mg/kg, p.o.) markedly prevented CCl_4-induced elevation of levels of serum SGPT, SGOT, ACP, ALP, and bilirubin. The decreased level of TPs due to hepatic damage induced by CCl_4 was found to be increased in the AG-treated group.

The results were comparable to that of silymarin. A comparative histopathological study of liver exhibited almost normal architecture, as compared to the CCl_4-treated group. Hepatoprotective effect of AG is probably due to combined action of all ingredients (Achliya et al., 2004).

ATA-OFA

"Ata-Ofa," a PHF consisting of 21 plant products, including *Zingiber officinale, Tamarindus indica, Khaya senegalensis, Moringa oleifera, Nauclea latifolia, Camellia sinensis, Anacardium occidentale, Aframomum melegueta, Phyllantus amarus, Morinda lucida,* and *Mangifera indica,* was reported (at 5 mg/kg) for in vivo antioxidant, hepatoprotective, and curative effects by its ability to ameliorate CCl_4-induced alterations in biochemical parameters and antioxidants enzymes in intoxicated rat (Atawodi, 2011).

CHUNGGAN EXTRACT (CGX)

CGX is a commercially available hepatotherapeutics consisting of 13 herbal plants, which developed from a traditional Chinese formula, which means "liver cleaning." Samik Pharmaceutical (Seoul, South Korea) manufactured the CGX according to over-the-counter Korean monographs. CGX comprises 5 g each of Artemisiae Capillaris Herba (*Artemisia capillaris* Thunb.), Trionycis Carapax (*Amyda maackii* Payloff), and Raphani Semen (*Raphanus sativus* L.); 3 g each of Atractylodis Rhizoma (*Atractylodes macrocephala* Koidzumi), Poria (*Poria cocos* Wolf), Alismatis Rhizoma (*Alisma orientale* Juzepczuk), Atractylodis Rhizoma (*Atractylodes lancea* DC.), and Salvia Miltiorrhizae Radix (*Salvia miltiorrhiza* Bunge); 2 g each of Polyporus (*Polyporus umbellatus* Fries), Amomi Fructus (*Amomum villosum* Loureiro), and Ponciri Fructus (*Poncirus trifoliata* Rafinesqui); and 1 g of Glycyrrhizae Radix (*Glycyrrhiza uralensis* Fischer) and Aucklandiae Radix (*Aucklandia lappa* Decne.).

Shin et al. (2006) evaluated the therapeutic effect of CGX, a modified traditional Chinese hepatotherapeutic herbal, on the dimethylnitrosamine (DMN)-induced chronic liver injury model in rats. CGX administration restored the spleen weight to normal after having been increased by DMN treatment. Biochemical analysis of the serum demonstrated that CGX significantly decreased the serum level of ALP, ALT, and AST that had been elevated by DMN treatment. CGX administration moderately lowered lipid peroxide production and markedly lowered hydroxyproline generation caused by DMN treatment in accordance with histopathological examination. DMN treatment induced a highly upregulated expression of TNF-α, TGF-β, TIMP-1, TIMP-2, PDGF-β, and MMP-2. Of these, the gene expression encoding PDGF-β and MMP-2 was still further enhanced two weeks after secession of the four-week DMN treatment and was remarkably ameliorated by CGX administration.

HD-03

HD-03 is a PHF containing plant drugs, which are known for their hepatoprotective properties in the Ayurvedic system of medicine. Mitra et al. (1998) evaluated the herbal formulation HD-03 for its protective effect against diverse hepatotoxic agents viz., paracetamol, thioacetamide, and isoniazid. Treatment with HD-03 led to significant amelioration of toxin-induced changes in the biochemical parameters. Since the protective effect of HD-03 was observed in all three types of intoxication, which are different in their primary mechanism of inducing hepatotoxicity, a protective mode of action of HD-03, not specific to the hepatotoxin, was suggested.

HEPATOPLUS™

Another polyherbal Indian formula known as Hepatoplus™ [the capsule containing *Phyllanthus amarus* (whole plant) 100 mg, *Eclipta alba* (leaves) 50 mg, *Tephrosia purpurea* (leaves) 30 mg, *Curcuma longa* (rhizome) 30 mg, *Picrorhiza kurroa* (root) 20 mg, *Withania somnifera* (root) 100 mg, *Pinus succinifera* (amber) 37.50 mg, *Pistacia lentiscus* (resinous exudates) 25 mg, *Orchis mascula* (seed) 25 mg, and *Cycas circinalis* (flower) 62.50 mg] was evaluated against anti-TB-induced liver toxicity (isoniazid (INH)/rifampicin (RIF) 50 mg/kg each) in Sprague Dawley rats at a dose of 50 and 100 mg/kg for 30 days. Liv. 52 (each tablet contains *Capparis spinosa* 32 mg, *Cichorium intybus* 32 mg, *Mandur bhasma* 32 mg, *Solanum nigrum* 32 mg, *Terminalia arjuna* 32 mg, *Cassia occidentalis* 16 mg, *Achillea millefolium* 16 mg, and *Tamarix gallica* 16 mg) was used as positive control at 100 mg/kg. Hepatoplus™ exhibited a good protection against oxidative damage by increasing antioxidant defenses (GPx, CAT, SOD, and GSH) and also restored serum levels of liver enzymes; although the hepatoprotective effect was dose dependent, the effect observed was similar to the control Liv. 52 (Sankar et al., 2015a).

This PHF (100 mg/kg) administered during 30 days in Sprague Dawley male was effective against oxidative damage generated by anti-TB-induced hepatotoxicity, decreasing lipid peroxidation (LPO) levels and DNA damage by one half in the liver cells. Additionally, the gene expression of caspases and oxidases such as CYP2E1 was regulated by the coadministration of Hepatoplus; the activity and levels of antioxidant defenses such as GPx, CAT, SOD, and GSH were increased in the hepatocytes of rats treated with the polyherbal remedy. In this case, the authors used Liv. 52 (100 mg/kg) as positive control (Sankar et al., 2015b). In another study, rats treated with Hepatoplus™ revealed normal architecture of liver cells and demonstrated the in vitro hepatoprotective effect of the same PHF on liver cell lines against oxidative-induced damage through a mixture of INH/RIF (30 ng/mL). The cell line was treated with three concentrations of Hepatoplus™ (50, 100, and 200 ng/mL). Liver cell lines treated with Hepatoplus™ were protected from oxidative stress and maintained a normal antioxidant profile and liver marker enzymes in a dose-dependent manner. The authors concluded that Hepatoplus™ protects hepatocytes from OS and apoptosis (Sankar et al., 2015c).

HEPAX

Hepax is a PHF that consists of *Plumbago zeylanica (Chitraka), Picrorrhiza kurroa* (Katuka), *Piper nigrum* (Maricha), *Zingiber officinale* (Ardraka), *Sodii carbonas impura* (Sajikakshara), *Phyllanthus emblica* (Amalaki), *Terminalia chebula* (Haritaki), *Calcii oxidum* (Chuna), and *Potassii carbonas impura* (Yavakshara). Devaraj et al. (2011) evaluated the hepatoprotective potential of Hepax, a PHF, against three experimentally induced hepatotoxicity models in rats. Administration of hepatotoxins (CCl_4, paracetamol, and thiocetamide) showed significant morphological, biochemical, and histological deteriorations in the liver of experimental animals. Pretreatment with Hepax had significant protection against hepatic damage by maintaining the morphological parameters (liver weight and liver weight to organ weight ratio) within normal range and normalizing the elevated levels of biochemical parameters (SGPT, SGOT, ALP, and TB), which were evidently shown in histopathological study. The Hepax has highly significant hepatoprotective effect at 100 and 200 mg/kg, *p.o.*, on the liver of all the three experimental animal models.

HERBAL EXTRACTS FROM MEXICO

Various hepatoprotective herbal products from plants are available in Mexico, where up to 85% of patients with liver disease use some form of complementary and alternative medicine. Using a model of CCl_4-induced hepatotoxicity in rats, Cordero-Pérez et al. (2013) evaluated the effects of commercial herbal extracts most commonly used in the metropolitan area of Monterrey, Mexico. The most commonly used herbal products were Hepatisan® capsules, Boldo capsules, Hepavida® capsules, Boldo infusion, and milk thistle herbal supplement (80% silymarin). None of the products tested was hepatotoxic according to transaminase and histological analyses. AST and ALT activities were significantly lower in the Hepavida+CCl_4-treated group as compared with the CCl_4-only group. AST and ALT activities in the silymarin, Hepatisan, and Boldo tea groups were similar to those in the CCl_4 group. The CCl_4 group displayed submassive confluent necrosis and mixed inflammatory infiltration. Both the Hepatisan+CCl_4 and Boldo tea+CCl_4 groups exhibited ballooning degeneration, inflammatory infiltration, and lytic necrosis. The silymarin+CCl_4 group exhibited microvesicular steatosis. The Hepavida+CCl_4- and Legalon+CCl_4-treated groups had lower percentages of necrotic cells as compared with the CCl_4-treated group; this treatment was hepatoprotective against necrosis.

HIMOLIV (HV)

HV is PHF for different liver diseases. Effects of this multiherbal medicine were evaluated on CCl_4- and paracetamol-induced liver toxicity (Bhattacharyya et al., 2003). Aqueous extract of almost 25 plants, mainly *Picorrhiza kurroa, Boerhavia diffusa, Tinospora cordifolia, Andrographis paniculata,* and *Phyllanthus emblica,* was investigated for its hepatoprotective properties in different experimental models. This herbal treatment is useful for biliary disorder, liver diseases, and jaundice via reduction of serum marker enzymes level. All these research studies support that HV can modulate antioxidant pathway by which it can regulate hepatoprotection from different drugs.

JIGRINE

Jigrine, a polypharmaceutical herbal formulation containing aqueous extracts of 14 medicinal plants developed on the principles of Unani system of medicine, is used for liver ailments. The effects of oral pretreatment with Jigrine (0.5 mL and 1.0 mL/kg for seven days) were studied by Vivek et al. (1994) on hepatic damage induced by alcohol and CCl_4 and also with paracetamol in rats. Biochemical parameters such as SGOT, SGPT, serum bilirubin, plasma prothrombin time, and tissue lipid peroxides were estimated to assess the liver function. Alcohol-CCl_4 and paracetamol treatment produced an increase in serum transaminases, bilirubin, plasma prothrombin time, and lipid peroxides in liver. These effects were progressively reduced by pretreatment doses of "Jigrine."

These biochemical observations were supplemented by histopathological examination of liver sections. The activity of Jigrine was also comparable to Liv. 52, a known Ayurvedic hepatoprotective formulation.

Karunakar et al. (1997) evaluated the antihepatotoxic activity of Jigrine against alcohol and CCl_4-induced hepatotoxicity in rats. The study confirms the antihepatotoxic activity of Jigrine, which may be attributed to the membrane stabilizing and antioxidant property of this medicine.

Aftab et al. (1999) evaluated the hepatoprotective potential of jigrine pretreatment on thioacetamide-induced liver damage in rats. Thioacetamide administration increased the AST, ALT, Na+, and K+ levels in serum and TBARS in liver. Pretreatment of the rats with jigrine or silymarin reduced the thioacetamide-induced elevated levels of the above indices. Thioacetamide administration decreased the concentration of GSH in rat liver. Jigrine pretreatment at both the doses and silymarin restored the GSH levels to near normal values in the liver of rats treated with thioacetamide. Similarly, the elevation in liver TBARS levels due to thiocetamide was prevented by silymarin and jigrine pretreaments. Thioacetamide administration induced centrilobular necrosis. Pretreatment of the rats with jigrine at both the doses showed mild protection, while silymarin showed marked protection on pretreatment.

The hepatoprotective potential of jigrine post-treatment at the dose of 0.5 mL/kg per day p.o. for 21 days was evaluated against thiocetamide-induced liver damage in rats (Aftab et al., 2002). Biochemical parameters such as AST and ALT in serum and TBARS and GSH in tissues were estimated to assess liver function. Data on the biochemical parameters revealed hepatoprotective potential of jigrine post-treatment against thioacetamide-induced hepatotoxicity in rats. Silymarin used as reference standard also exhibited significant hepatoprotective activity on post-treatment against thioacetamide-induced hepatotoxity in rats. The biochemical observations were supplemented with histopathological examination of rat liver sections.

Jigrine was evaluated by Najmi et al. (2005, 2010) for its hepatoprotective activity against galactosamine-induced hepatopathy in rats. Biochemical parameters such as AST, ALT, and urea in serum, TBARS and GSH in liver, and whole blood GSH were estimated to assess liver function. DPPH-free radical scavenging activity of jigrine was also evaluated. Biochemical data exhibited significant hepatoprotective activity of jigrine against galactosamine. Silymarin used as reference standard also exhibited significant hepatoprotective activity against galactosamine. The biochemical observations were supplemented with histopathological examination of rat liver sections. Histopathological evaluation showed marked improvement in the livers of jigrine- and silymarin-treated animals.

KABIDEEN

Kabideen is a polyherbal pharmacopoeial formulation mainly prescribed for the management of liver disorders. It comprises 21 ingredients: *Achillea millefolium, Fumaria officinalis, Cassia occidentalis, Cichorium intybus, Chenopodium album, Cuscuta reflexa, Cucumis sativus, Cucumis melo, Solanum nigrum, Rheum emodi, Nardostachys jatamansi, Aquillaria agallocha, Ochrocorpus longifolius, Zataria multiflora, Smilax regelli, Alpinia galanga, Swertia chirata, Rosa damascena, Nymphaea alba, Butea frondosa, Agrimonia eupatoria,* and *Cichorium intybus.*

Zameer et al. (2015) evaluated the hepatoprotective effect of Kabideen Syrup in rats treated with CCl_4. Hepatic damage as evidenced by a rise in the levels of AST, ALT, ALP, bilirubin CHOL, and MDA decreased level of TP in serum. Liver showed a tendency to attain near normalcy in animals coadministered with Kabideen (50 mL/kg) significantly reduced the serum ALT, AST, ALP, and CHOL levels and increased the TP levels when compared to the CCl_4 group. The histopathological findings showed a significant difference between the Kabideen (50 mL/kg) and CCl_4-treated groups.

KAMILARI

Rajesh and Latha (2004) studied the hepatoprotective efficacy of Kamilari, a polyherbal preparation in CCl_4-induced liver dysfunction in albino rats by determining different biochemical parameters in serum and tissues. In serum, the activities of enzymes such as AST, ALT, and ALP and the concentrations of protein and bilirubin were evaluated. The concentrations of total lipids, CHOL, TGs, and phospholipids were studied in serum and different tissues. Here, a dose-dependent study was conducted, and oral administration of Kamilari at a dose of 750 mg/kg BW significantly reduced the toxic effects of CCl_4.

KHAMIRA GAOZABAN AMBRI JADWAR OOD SALEEB WALA (KGA)

Akhtar et al. (2013) evaluated hepatoprotective formulation KGA ex vivo antioxidant and in vivo hepatoprotective properties against CCl_4 toxicity in albino rats. First, phytochemical analysis of test preparation was conducted to estimate its total phenolic and flavonoid contents.

Then, their antioxidant activity was determined by various tests and compared it with standard ascorbic acid and rutin. Afterward, hepatoprotective activity was studied against CCl_4-induced liver damage by determining SGOT, SGPT, ALP, total CHOL, bilirubin, and TP contents in the serum of rats before and after treatment. This suggests that the hepatoprotective activity of formulation is possibly attributed to its free radical scavenging properties.

LIV. 52

In the absence of reliable liver protective drugs in the allopathic system of medicine, many herbs have been claimed to play an important role in the management of various liver disorders. Numerous medicinal plants and their formulations are used for liver disorders in ethnomedical practices and in the traditional system of medicine in India. Liv. 52 is such a herbal formulation of The Himalaya Drug Company, India. Plants that are used in formulation of Liv. 52 medicine are *Achillea millefolium, Capparis spinosa, Cassia occidentalis Cichorium intybus, Solanum nigrum, Tamarix gallica, Terminalia arjuna, Eclipta alba, Phyllanthus niruri, Boerhavia diffusa, Phyllanthus embilica, Fumaria officinalis, Terminalia chebula, Tinospora cordifolia,* and *Andrographis paniculata.* A number of research articles have been published in its favor including experimental as well as clinical studies.

A study was conducted by Chauhan et al. (1994) on Wistar male rats to assess the effect of chronic alcohol administration on blood ethanol and acetaldehyde levels and their modification by Liv. 52, an Ayurvedic formulation. Following chronic ingestion of 6% alcohol (v/v) through a water feeding bottle for 42 days, the administration of an acute oral dose of 5 mL of 5% alcohol (v/v) resulted in a significant decrease in blood ethanol and significant increase in acetaldehyde levels, respectively, at 30 min. Larger doses (i.e., 5 mL of 10% and 15%, respectively) failed to demonstrate an effect of accelerated ethanol metabolism because of saturation kinetics. In the second experiment, temporal observations on blood ethanol and acetaldehyde levels were done at 1, 3, and 4 h. Following chronic ingestion of 6% alcohol for 42 days, stimulation of Phase I and suppression of Phase II metabolism was reproduced at all the time points. As compared with placebo, treatment with Liv. 52 in the last two weeks reversed the blood ethanol and acetaldehyde levels at all the time points, and they were comparable to the levels observed on day 1 (basal readings). In the third experiment, chronic exposure to 6% alcohol for 180 days showed further deterioration in ethanol metabolism as compared to 42-day exposure. In a chronic model of 180 days, Liv. 52 normalized the blood ethanol and acetaldehyde levels in a dose-related manner.

The mechanism of protective effects of Liv. 52 on ethanol-induced hepatic damage has been investigated by Sandhir and Gill (1999). The results indicate that Liv. 52 treatment prevents ethanol-induced increase in the activity of the enzyme gamma-glutamyl transpeptidase. Concomitantly, there was also a decrease in ethanol-accentuated lipid peroxidation in liver following Liv. 52 treatment. The activity of antioxidant enzymes, SOD, GPx, and the levels of GSH were decreased following ethanol ingestion. Liv. 52 treatment was found to have protective effects on the activity of SOD and the levels of GSH.

The efficacy of herbal medicine Liv. 52 (consisting of Mandur basma, *Tamarix gallica* and herbal extracts of *Capparis spinosa, Cichorium intybus, Solanum nigrum, Terminalia arjuna,* and *Achillea millefolium*) on liver cirrhosis outcomes was compared with the placebo for six months in 36 cirrhotic patients referred to Tehran Hepatic Center (Huseini et al., 2005). Patients treated with Liv. 52 for six months had significantly better Child-Pugh score, decreased ascites, and decreased serum ALT and AST. In placebo-administered patients, all the clinical parameters recorded at the beginning of the study were not significantly different than after six months. This protective effect of Liv. 52 can be attributed to the diuretic, anti-inflammatory, antioxidative, and immunomodulating properties of the component herbs.

Mitra et al. (2008) investigated whether ethanol and Liv. 52 could modulate PPARgamma and TNFalpha induction in human hepatoma cells, HepG2. The study with RT-PCR and confocal microscopy experiments showed that ethanol (100 mM) induced suppression of PPARgamma expression in HepG2 cells. The ethanol-induced PPAR-gamma suppression was abrogated by Liv. 52. Moreover, Liv. 52 also induced upregulation of PPAR-gamma mRNA in liver cells as compared to the untreated cells. Furthermore, 100-mM ethanol has also induced TNFalpha gene expression in HepG2 cells, and interestingly, Liv. 52 abolished ethanol-induced TNFalpha. The study also showed that Liv. 52 alone down-regulated TNFalpha expression in HepG2 cells. Taken together, these findings suggest that Liv. 52 is capable of attenuating ethanol-induced expression of TNFalpha and abrogating ethanol-induced suppression of PPARgamma in liver cells. These results indicate that Liv. 52-induced PPARgamma expression and concomitant suppression of ethanol-induced elevation of TNFalpha in HepG2 cells suggest the immunomodulatory and hepatoprotective nature of Liv. 52.

Girish et al. (2009) evaluated the hepatoprotective activity of Liv. 52 in CCl_4-induced liver toxicity in mice. When pretreated with Liv. 52 at a dose of 5.2 mL/kg BW, the CCl_4-induced changes were significantly reversed. The pretreatment with Liv. 52 can prevent acute liver damage induced by CCl_4 only at a higher dose. Therefore, it was suggested that the dose adjustment of Liv. 52 may be necessary for its optimal effect in human liver disease.

Dange (2010) conducted a meta-analysis on the efficacy and short- and long-term safety of Liv. 52 as a hepatoprotective in TB patients receiving anti-TB drugs, as published in eight controlled clinical trials. A meta-analysis of eight clinical studies conducted between 1970 and 1992 in 689 tubercular patients receiving antitubercular treatment (ATT) along with Liv. 52 or placebo was taken up for this study. Duration of the treatment varied from four weeks to one year. Children below the age of two years received 10–20 drops of Liv. 52 three to four times daily. Children in the age group of 2–5 years received 20 drops of Liv. 52 three times daily. Children aged >5 years and adults received one to two teaspoonsful of Liv. 52 syrup three times daily or one to two tablets of Liv. 52 two to three times daily. Improvement in the various parameters of hepatotoxicity, such as hepatomegaly, anorexia, weight gain, general well-being, and liver function test (ALT), and improvement in the ultrasonographic findings of the hepatobiliary system were taken into consideration.

Results of the meta-analysis showed a significant improvement in hepatotoxicity in patients receiving anti-TB drugs and Liv. 52. Significant improvements were observed in associated symptoms such as anorexia, weight gain, hepatomegaly, and general well-being. The protective effect of Liv. 52 against hepatotoxic reaction caused by ATT was further substantiated by a significant reduction in ALT values and alleviation of gastrointestinal symptoms due to hepatitis. No adverse effects were reported or observed due to Liv. 52 during the study period, and the compliance to the drug therapy was good. Therefore, from the above findings, it can be concluded that Liv. 52 acts as a hepatoprotective in the hepatotoxicity of TB patients receiving ATT.

Sapakal et al. (2008) compared the hepatoprotective activity of two marketed formulations: Liv. 52 and Livomyn. The formulations showed significant hepatoprotective effect by reducing elevated serum enzyme levels such as SGPT, SGOT, bilirubin content (direct and total), and

TP. These biochemical observations were supplemented by weight and histopathological examination of liver sections. Various pathological changes such as steatosis, centrilobular necrosis, and vacuolization observed in rats treated only with CCl_4, but the groups treated with the CCl_4, and hepatoprotective formulations were producted to a moderate extent from such pathological changes. Silymarin was used for positive control. It was concluded from study that Liv. 52 has shown more significant hepatoprotective activity against CCl_4-induced hepatotoxicity in rats in comparison to Livomyn.

The hepatoprotective effect of Liv. 52 against oxidative damage induced by *tert*-butyl hydroperoxide (t-BHP) in HepG2 cells was evaluated by Vidyashankar et al. (2010) in order to relate in vitro antioxidant activity with cytoprotective effects. Cytotoxicity was measured by MTT assay. Antioxidant effect of Liv. 52 was determined by DPPH assay, FRAP assay, and lipid peroxidation and measurement of nonenzymic and antioxidant enzymes in HepG2 cells exposed to t-BHP over a period of 24 h. The results obtained indicate that t-BHP induced cell damage in HepG2 cells as shown by significant increase in lipid peroxidation as well as decreased levels of reduced GSH. Liv. 52 significantly decreased toxicity induced by t-BHP in HepG2 cells. Liv. 52 was also significantly decreased lipid peroxidation and prevented GSH depletion in HepG2 cells induced by t-BHP. Therefore, Liv. 52 appeared to be important for cell survival when exposed to t-BHP. The protective effect of Liv. 52 against cell death evoked by t-BHP was probably achieved by preventing intracellular GSH depletion and lipid peroxidation.

LIVACTINE

Livactine is a herbal formulation, which contains extracts of nine medicinal plants of which some plants are *Boerhavia diffusa, Tinospora cordifolia, Andrographis paniculata,* and *Phyllanthus emblica* (Syn.: *Emblica officinalis*). Mayuran et al. (2010) evaluated the antihepatotoxic activity of Livactine against CCl_4- and paracetamol-induced toxicity in rats. Livactine showed significant dose-dependent hepatoprotective effect by reducing elevated serum enzyme levels when compared to that of Liv. 52.

LIVERGEN (GREEN WORLD GROUP, CHINA)

Livergen, product of Multiherbal Green world Group, China, contains extract of different plant such as *Eclipta alba, Picrorhiza kurroa, Carum copticum, Tamarix gallica, Ipomoea turpethum, Plumbago zeylanica, Solanum nigrum, Cichorium intybus, Solanum nigrum, Achillea millefolium. Tinospora cordifolia, Fumaria officinalis, Terminalia chebula*, etc. (Arsul et al., 2007). This herbal drug can significantly reduce enzyme level of SGOT, SGPT, ALP, CHOL, and bilirubin and significantly increase TP level. The report suggested that this hepatoprotection activity is due to phenolic and flavonoid content of plant extracts (Arsul et al., 2007). This drug may protect hepatic cells via resorting the normal architecture of liver parenchyma, and pretreatment of this drug may control rise of all liver marker enzymes when 5.2-mL/kg CCl_4 has been given to mice for induction of hepatic damage (Girish et al., 2009).

LIV-FIRST

Liv-First is a polyherbal Ayurvedic proprietary medicine, which is used as a hepatoprotective agent. Each capsule contains *Eclipta alba* (100 g), *Andrographis paniculata* (50 mg), *Tinospora cordifolia* (100 mg), *Picrorhiza kurroa* (75 mg), *Boerhavia diffusa* (50 mg), and *Berberis aristata* (50 mg). Lima et al. (2010) investigated the hepatoprotective activity of Liv-First against CCl_4-induced hepatotoxicity in rats. Liv-First (16.3 mg/kg, p.o.) was used to screen the hepatoprotective activity in experimental animals by administration of CCl_4 (1 mL/kg, i.p). Silymarin (25 mg/kg, p.o.) was used as the standard. CCl_4 administration in rats elevated the levels of SGPT, SGOT, ALP, and bilirubin. Administration of Liv-First significantly prevented this increase. The activity of antioxidant enzymes in the CCl_4-treated group was decreased, and these enzyme levels were significantly increased in Liv-First-treated groups. Histopathological studies revealed that the concurrent administration of CCl_4 with the extract exhibited protection of the liver tissue, which further evidenced the above results.

LIVINA

Livina comprises of *Picrorhizha kurroa, Phyllanthus niruri, Andrographis paniculata, Cichorium intybus, Tephrosia purpurea, Solanum dulcamara, Crinum asiaticum, Alstonia scholaris* (50 mg each), *Holarrhena antidysenterica, Tinospora cordifolia, Terminala chebula,* and *Hygrophila schullii* (Syn.: *Asteracantha longifolia*) (25 mg each). The capsule (500 mg) with a polyherbal preparation known as Livina was tested in a clinical trial in TB patients. Two capsules of Livina were administered twice daily after meals for six months. The levels of SGOT, SGPT, and ALP were lower in the Livina-treated group with respect to the placebo group at weeks 4 and 8 after treatment. The authors suggest that Livina was efficacious against hepatic dysfunction caused by anti-TB in patients with TB (Gulati et al., 2010).

Darbar et al. (2009) showed that Livina, a polyherbal liquid formulation, was effective in blunting ethanol induced enhanced activities of SGOT, SGPT, ALP, level of serum bilirubin (both total and direct), blood urea nitrogen (BUN), serum total CHOL, and liver weight loss and in reducing ethanol-induced lipid peroxidation. Results of hepatocellular damage caused by ethanol and its recovery by Livina suggest that it might be considered as a potential source of natural hepatoprotective agent. Darbar et al. (2011) found that the supplementation of Livina significantly reduced the damaging effects on liver by paracetamol. Histopathological changes (congestion of central vein, centrilobular necrosis, and sinusoidal congestion) induced by paracetamol were also reduced to a moderate extent in Livina-treated mice.

LIVOBOND

Satyapal et al. (2008) evaluated the hepatoprotective activity of Livobond, a polyherbaal formulation, against CCl_4-induced hepatotoxicity in male Sprague–Dawley rats. The toxic effects of CCl_4 in the Livobond-treated group were controlled significantly by restoration of the levels of serum bilirubin, proteins, and enzymes compared to the CCl_4-treated and silymarin-treated groups Histopathological studies further confirmed the hepatoprotective acivity of Livobod.

LIVOKIN (HERBO-MED, KOLKATA, INDIA)

Livokin (Herbo-Med, Kolkata, India) contains *Andrographis paniculata, Apium graveolens, Berberis lyceum, Carum copticum, Cichorium intybus, Cyperus rotundus, Eclipta alba, Ipomoea turpethum, Oldenlandia corymbosa, Picrorhiza kurroa, Hygrophila spinosa, Plumbago zeylanica, Solanum nigrum, Tephrosia purpurea, Terminalia arjuna, Terminalia chebula,* and *Trigoenella foenum-graecum.*

Girish et al. (2009) evaluated the hepatoprotective activity of Livokin in CCl_4-induced liver toxicity in mice. When pretreated with Livokin at a dose of 5.2 mL/kg BW, the CCl_4-induced changes were significantly reversed. The pretreatment with Livokin can prevent acute liver damage induced by CCl_4 only at a higher dose. Therefore, it was suggested that the dose adjustment of Livokin may be necessary for its optimal effect in human liver disease.

LIVPLUS

Livplus is a PHF containing *Eclipta alba, Phyllanthus niruri, Cichorium intybus, Picrorhiza kurroa, Boerhavia diffusa, Berberis aristata,* and *Andrographis paniculata.* Maheshwari et al. (2015) investigated the hepatoprotective effect of Livplus against CCl_4-induced hepatotoxicity in rats. Treatment with Livplus significantly reduced the elevated levels of ALT, AST, ALP, bilirubin (direct and total), gamma-glutamyl transferase, total CHOL, and TGs and increased levels of TP compared to CCl_4 control rats. The treatment with Livplus also showed a significant increase in GSH contents, SOD, and CAT activity and a decrease in MDA levels compared to CCl_4 control rats.

LIVSHIS

Livshis is product of Southern Health Improvement Samity, West Bengal, India. Main composition and ingredients of this medication are *Aloe barbadensis* Mill., *Andrographis paniculata* Wall. ex Nees., *Asteracantha longifolia* Nees., *Berberis chitria* Lindl., *Fumaria parviflora* Lam., *Phyllanthus fraternus* Webs., and *Picrorhiza kurroa* Royle ex Benth. Hepatoprotective effect of Livshis has been investigated on the CCl_4-induced male albino rat model and shows a significant decrease in liver damage marker enzymes and TB, urea, uric acid, CRT, and blood urea nitrogen, but toxicity profiling and mechanism of action of this polyherbal medicine has not been investigated yet. Ethanol, water, and methanol extract of these plants has shown hepatoprotection activity in different other experiments.

Bera et al. (2011, 2012) investigated the hepatoprotective activity of the PHF Livshis in CCl_4-induced hepatotoxic male albino rats and included an assessment of the toxicity of the formulation. Hepatic antioxidant enzyme activities diminished significantly, and hepatic lipid peroxidation rates were elevated in CCl_4-treated animals that were pretreated with distilled water. The activities of serum toxicity marker enzymes and serum liver function test biosensors increased significantly in animals pretreated with distilled water, whereas these biosensors were significantly protected in Livshis pretreated, CCl_4-treated animals. Animals, treated with Livshis alone, did not exhibit any significant variation in levels of hematological and renotoxicity markers compared to controls. In an acute toxicity study, there were no toxic symptoms up to the dose level of 3200 mg/kg BW. They concluded that Livshis is safe for long-term treatment for hepatic protection at doses of 50 mg/kg BW.

LONGYIN DECOCTION

The Longyin decoction was modified and prepared according to the classic prescription of traditional Chinese medicine "Oriental wormwood Decoction" in Waitai miyao written by Wang Tao of Tang Dynasty. It includes *Capillaries, Gentian, Gardenia, Bupleurum root,* and *Licorice*. *Gentian* refers to the root and rhizome of *Gentiana manshurica* Kitag., *Gentiana scabra* Bge., *Gentiana triflora* Pall., or *Gentiana rigescens* Franch. ex Hemsl. *Capillaries* was whole herb or aboveground parts of *Artemisia scoparia* Waldst. et Kit. or *Artemisia capillaris* Thunb., *Gardenia* is the dried ripe fruit of *Gardenia jasminoides* Elli. *Bupleurum root* is the dried root of *Bupleurum chinense* DC. or *Bupleurum scorzonerifolium* Willd. *Licorice* refers to the dried root and rhizome of *Glycyrrhiza uralensis* Fisch., *Glycyrrhiza inflata* Bat. or *Glycyrrhiza glabra* L. *Long dan* (*Gentian), and Yin chen (Capillaries)* functions to treat jaundice according to the Chinese herbal book, while *Zhi zi* (*Gardenia), chai hu (Bupleurum root)*, and *gancao* (*Licorice*) work together to assist *Long dan* and *Yin chen*. Take the first words of the two main herbs; hence, the formula name is Longyin decoction. Wang et al. (2013) evaluated the hepatoprotective activity in poultry liver injury model induced by CCl_4. An herbal formula, Longyin decoction, was prepared for hepatoprotection test on chicken acute liver injury models. The pathologic changes of the liver were observed, and the activities of ALT and AST were, respectively, detected to evaluate the hepatopro tective effects of Longyin decoction on chickens. The chicken acute liver injury model was successfully established by injecting CCl_4 via pectoral muscle. The best dose of CCl_4 inducing chicken liver injury was 4.0 mL/kg BW. The results of qualitative determination by HPTLC showed that the components of Longyin decoction contained *Gentian, Capillaries, Gardenia,* and *Bupleurum root.* In the high-dose Longyin group and the middle-dose Longyin group, the pathological changes of the damaged liver were mitigated, and the activities of ALT and AST in serum were reduced significantly. Longyin decoction has obvious hepatoprotective effect on acute liver injury induced by $CCl_{4.}$

MAJOON-E-DABEED-UL-WARD (MD)

MD is a Unani herbal formulation containing *Nardostachys, Pistacia lentiscus, Crocus sativus, Bambusa bambos, Cinnamomum zeylanicum, Cymbopogon jwarancusa, Asarum europaeum, Saussurea hypoleuca, Gentiana olivieri, Cuscuta reflexa, Rubia cordifolia, Coccus lacca* (insect), *Cichorium intybus,*

Apium graveolens, Aristolochia donga, Commiphora opobalsamum, Aquilaria agallocha, Syzygium aromaticum, Elettaria cardamomum, and *Rosa damascena*. Shakya and Shukla (2011) evaluated the hepatoprotective effect of MD against acetaminophen (APAP; 2 g/kg, p.o.)-induced liver damage. The latter was evidenced by elevated levels of AST, ALT, serum ALP, lactate dehydrogenase (LDH), bilirubin, albumin, urea, and CRT in experimental animals. Increased levels of lipid peroxidation were associated with a concomitant decline in reduced GSH levels, adenosine triphosphatase (ATPase), and glucose-6-phosphatase (G-6-Pase) caused by APAP treatment. Treatment with MD (250, 500, and 1000 mg/kg, p.o.) reversed the altered levels of AST, ALT, SALP, LDH, bilirubin, albumin, urea, and CRT in a dose-dependent manner. Significant restoration was found in LPO and GSH content, and metabolic enzymes (ATPase and G-6-pase) were seen after therapy.

Myristica fragrans, Astragalus membranaceus, AND *Poria cocos* (MAP)

Yimam et al. (2016) described the potential use of MAP, a standardized blend comprising three extracts from MAP, in ameliorating acetaminophen (APAP) and CCl_4-induced acute liver toxicity models in mice. MAP administered at doses of 150–400 mg/kg showed statistically significant and dose-correlated inhibitions of serum ALT in the APAP and CCl_4 models. Moreover, MAP resulted in up to 75.7%, 60.9%, and 33.3% reductions in serum AST, bile acid, and TB, respectively. Mice treated with oral doses of composition of MAP at 300 mg/kg showed significant reduction in hepatocyte necrosis when compared with vehicle control. Unexpected synergistic protection of liver damage was also observed.

OCTOGEN (PLETHICO PHARMACEUTICALS LTD., INDORE, INDIA)

Octogen contains Arogyavardini rasa and *Phyllanthus niruri*. Girish et al. (2009) evaluated the hepatoprotective activity of Octogen in CCl_4-induced liver toxicity in mice. When pretreated with Octogen at a dose of 5.2 mL/kg BW, the CCl_4-induced changes were significantly reversed. The pretreatment with Octogen can prevent acute liver damage induced by CCl_4 only at a higher dose. Therefore, it was suggested that the dose adjustment of Octogen may be necessary for its optimal effect in human liver disease.

PANCHAGAVYA GHRITA (PG)

Achliya et al. (2003) investigated the hepatoprotective activity of PG against CCl_4-induced hepatotoxicity. Administration of PG (150–300 mg/kg, p.o.) markedly prevented CCl_4-induced elevation of levels of serum SGPT, SGOT, ACP, and ALP. The results are comparable to that of silymarin. A comparative histopathological study of liver exhibited almost normal architecture, as compared to control group.

SAIKOKEISHITO EXTRACT

Saikokeishito extract is a traditional Japanese herbal medicine, i.e., Kampo medicine, which is composed of nine herbs such as *Bupleuri radix*, *Pinelliae tuber*, *Scutellariae radix*, *Glycyrrhizae radix*, *Cinnamomi cortex*, *Paeonia radix*, *Zizyphi fructus*, *Ginseng radix*, and *Zingibers rhizome*. Ohta et al. (2007) examined whether Saikokeishito extract (TJ-10) exerts a therapeutic effect on alpha-naphthylisothiocyanate (ANIT)-induced liver injury in rats through attenuation of enhanced neutrophil infiltration and oxidative stress in the liver tissue. TJ-10 is a spray dried material of Saikokeishito extract. In rats treated once with ANIT (75 mg/kg, i.p.), liver injury with cholestasis occurred 24 h after treatment and progressed at 48 h. When ANIT-treated rats orally received TJ-10 (0.26, 1.3, or 2.6 g/kg) at 24 h after the treatment, progressive liver injury with cholestasis was significantly attenuated at 48 h after the treatment at the dose of 1.3 or 2.6 g/kg. At 24 h after ANIT treatment, increases in hepatic lipid peroxide and reduced GSH contents and myeloperoxidase activity occurred with decreases in hepatic SOD and GSH reductase activities. At 48 h after ANIT treatment, these changes except for reduced GSH were enhanced with decreases in CAT, Se-GPx, and glucose-6-phosphate dehydrogenase activities. TJ-10 (1.3 or 2.6 g/kg) postadministered to ANIT-treated rats attenuated these changes found at 48 h after the treatment significantly. These results indicate that TJ-10 exerts a therapeutic effect on ANIT-induced liver injury in rats possibly through attenuation of enhanced neutrophil infiltration and oxidative stress in the liver tissue.

STIMULIV (FRANCO-INDIAN PHARMACEUTICALS)

Stimuliv syrup and tablets are polyherbal production of Franco-Indian Pharmaceuticals Mumbai, a combination of herbal plants *Fumaria parviflora, Eclipta prostrata, Andrographis paniculata, Phyllanthus niruri, Picrorhiza kurroa, Tephrosia purpurea, Tinospora cordifolia, Aloe vera, Azadirachta indica*, and *Plumbago indica* (Girish et al., 2009). Hepatoprotective properties of these plants have been shown by different researchers on different models such as CCl_4, galactosamine, paracetamol, etc. Components of these plants such as phorbol ester, phenolic, flavonoids, and alkaloid have hepatoptoective activity via mechanism of mitochondrial dysfunction and hepatocytes protection against toxins.

TENG-KHIA-U

Teng-Khia-U is a folk medicine of Taiwan, derived from the entire plants of *Elephantopus scaber*, *E. mollis,* and *Pseudoelephantopus spicatus*. The hepatoprotective effects of water extracts of these three plants against beta-D-galactosamine and acetaminophen-induced acute hepatic damage were determined by Lin et al. (1995) in rats. The results indicated that the SGOT and the SGPT levels caused by D-GalN and APAP decreased after treatment with crude extracts of Teng-Khia-U. The pathological changes of hepatic lesions, caused by D-GalN and APAP, improved following treatment with the drug extracts mentioned above.

TEPHROLIV (TTK PHARMA PVT. LTD., CHENNAI, INDIA)

Tephroliv contains *Andrographis paniculata, Eclipta alba, Ocimum sactum, Phyllanthus niruri, Picrorhiza kurroa, Piper longum, Solanum nigrum, Tephrosia purpurea,* and *Terminalia chebula*. Girish et al. (2009) evaluated the hepatoprotective activity of Tephroliv in CCl_4-induced liver toxicity in mice. When pretreated with Tephroliv at a dose of 5.2 mL/kg BW, the CCl_4-induced changes were significantly reversed. The pretreatment with Tephroliv can prevent acute liver damage induced by CCl_4 only at a higher dose. Therefore, it was suggested that the dose adjustment of Tephroliv may be necessary for its optimal effect in human liver disease.

VASAGUDUCHYADI KWATHA

Vasaguduchyadi Kwatha is a compound Ayurvedic formulation having a mixture of eight medicinal plants (*Adhatoda vasica* Nees, *Tinospora cordifolia* (Willd.) Miers., *Emblica officinalis* Gaertn., *Terminalia chebula* Retz., *Terminalia bellerica* Roxb., *Swertia chirayita* (Roxb.) Karsten., *Picrochiza kurroa* Royle., and *Azadirachta indica* A.Juss.) explained in Astangahridaya for the treatment of liver diseases, especially for *Kamala* (jaundice) and *Panduroga* (anemia). This formulation is extensively used by physicians to treat various liver diseases, especially in alcoholic liver disease.

Kotecha et al. (2015) evaluated, in the same hepatotoxicity model, the protective activity of aqueous extract of Vasaguduchyadi Kwatha on liver injury induced by the administration of anti-TB drugs (INH 27 mg/kg, RIF 54 mg/kg, and PZD 135 mg/kg) in Wistar rats during 60 days. The results showed that at a 5.04-mL/kg dose, the polyherbal remedy decreased liver enzyme (ALT and AST) and bilirubin levels in serum, as well as attenuated hepatocellular necrosis and led to a reduction of inflammatory cell infiltration (Kalpu et al. 2015).

ZHI-ZI-DA-HUANG

Wang et al. (2009) investigated the hepatoprotective effect of Zhi-Zi-Da-Huang decoction (ZZDHD) and its two fractions (one is extracted with diethyl ether as a solvent from the water extract and is called ZD-DE for short; the other is the remained aqueous fraction after extracted with diethyl ether and is abbreviated as ZD-AQ) against acute alcohol-induced liver injury in rats. The high dose of ZZDHD exhibited a significant protective effect by reversing the biochemical parameters and histopathological changes. ZD-DE and ZD-AQ demonstrated different protective actions in biochemical examination. Partly assayed indexes were ameliorated after administrated the media dose of ZZDHD. High-performance liquid chromatography analysis indicated that ZZDHD contained flavonoids, anthraquinones, and iridoids, which might be the active chemicals.

In the market, several polyherbal hepatoprotective tablets are available in India, which are given in Table 6.1.

TABLE 6.1 List of Some Commercially Available Polyherbal Hepatoprotective Tablets in India

S. No	Brand name	Composition	Manufacturer
1	Arosi	Silymarin, L-ornithine-L-aspartate	Avincare
2	Culiv Plus Tablets	*Kutki (Picrorhiza kurroa), Amla (Phyllanthus emblica), Makoy (Solanum nigrum), Kalmegh (Andrographsi paniculata), Sharpankha (Tephrosia purpurea), Daru Haldi (Berberis aristata), Bhringraj (Eclipta alba), Bhui Amla (Phyllanthus niruri), Harad (Terminalia chebula), Kumari (Aloe barbadensis), Punernava (Boerhavia diffusa), Shigru (Moringa pterygosperma), Kasni (Cichorium intybus), Guduchi (Tinospora cordifolia) Viavidang (Embelia ribes), Mullethi (Glycyrrhiza glabra), Mandur bhasma Chiraita (Swertia chirata)*	Ayusearch Drugs & Laboratories

TABLE 6.1 *(Continued)*

S. No	Brand name	Composition	Manufacturer
3	Hepa	Silymarin, Calcium Pantothenate, Choline Bitartrate, Coenzyme Q 10, Ferrous Fumarate, Folic Acid, Inositol, L-Carnitine, L-Glutalthione, L-Ornithine, Pyridoxine, Riboflavin, Sodium Selenate, Vit B1, Vit B12,Vit C, Vit D3,Vit E, Zinc, N-Acetyl Cysteine	Venus Remedies
4	Hepabert	Silymarin, L-Ornithine, L-Aspartate, Pyridoxine, Folic acid	Hilbert Healthcare
5	Hepanit tablet	Silymarin, L-ornithine-L-Aspartate	Mac Organics
6	Hepa-Preserve	*Picrorhiza kurroa* (Katuka), *Phyllanthus niruri* (Bhumyamalaki), *Boerhavia diffusa*, *Andrographis paniculata* (Kalmegh)	Medlife Essentials
7	Herbal Hills Liv-First	*Tinospora cordifolia, Andrographis paniculata, Eclipta alba, Phyllanthus niruri, Boerhavia diffusa, Tephrosia purpurea, Picrorhiza kurroa, Cichorium intybus, Terminalia arjuna, Achillea millefolium, Tamarix gallica*	Isha Agro Developers Private Limited
8	Herbal Liv 7	*Eclipta alba, Picrorhiza kurroa, Andrographis paniculata, Boerhavia diffusa, Terminalia arjuna, Solanum nigrum,*Revand Chini, *Fumaria officinalis, Phyllanthus niruri*	Juventus Spanish Remedies
9	Herbitars	Ptychotis ajowan, *Emblica officinalis, Terminalia chebula, Eclipta alba, Swertia chirata*	J & J Dechane Laboratories Private Limited
10	L38 tablets	Bhumi Amla (*Phylanthus niruri*); Bhringraj (*Eclipta alba*); Kutki (*Picrorhiza kurroa*); Giloy (*Tinospora cordifolia*); Kalmegh (*Andrographis paniculata*); Makoy (*S. nigrum*); Punarnava (*Boerhavia diffusa*); Arjuna (*Terminalia arjuna*); and Daruhaldi (*Berberis aristata*).	Himalaya Herbal Healthcare Products, Indi
11	L52 tablets	Caper brush (Himsra—*Capparis spinosa*); Wild chicory (Kasani—*Cichorium intybus*); Mandur Bhasma (Calx—Ferric oxide); Black Night Shade (Kakimanchi—*Solanum nigrum*); Arjuna (*Terminalia arjuna*); Negro Coffee (Kasamarda—*Cassia occidentalis*); Yarrow (Biranjasipha—*Achillea millefolium*) and Tamarisk (Jhavuka—*Tamarix gallica*)	Himalaya Herbal Healthcare Products, India
12	Liv. 52	Himsra (*Capparis spinosa*), Kasani (*Cichorium intybus*), Mandura bhasma, Kakamachi (*Solanum nigrum*), Arjuna (*Terminalia arjuna*), Kasamarda (*Cassia occidentalis*), Biranjasipha (*Achillea millefolium*), Jhavuka (*Tamarix gallica*)	The Himalaya Drug Company

TABLE 6.1 (Continued)

S. No	Brand name	Composition	Manufacturer
13	Liv. 52 DS	Himsra (*Capparis spinosa*), Kasani (*Cichorium intybus*), Mandura bhasma, Arjuna (*Terminalia arjuna*), Kasamarda (*Cassia occidentalis*), Biranjasipha (*Achillea millefolium*), Jhavuka (*Tamarix gallica*), Kakamachi (*Solanum nigrum*)	The Himalaya Drug Company
14	Liv-Ayu	Arjun, Bhirngraj, Bhumi Amla, Galo, Kadu, Mandoor Bhasma, Sapankho, Satodi, Shingu, Sudh Kuchla, Tulsi, kalmegh, kasni	Ayushastra Wellness Private Limited
15	Liver Care	Milk Thistle and multi vitamin	Vasuki Fitness
16	Liver Care Tab	Milk thistle, Dandelion, Turmeric, Artichok	Biotrex Nutraceutical
17	Liveril	Silymarin, Calcium Pantothenate, Choline Bitartrate, Coenzyme Q 10, Folic Acid, Inositol, Iron, L-Carnitine, L-Glutalthione, L-Ornithine, Selenium, N-Acetyl Cysteine, Vit A, Vit B1,Vit B12,Vit B2,Vit B6,Vit C, Vit D3, Zinc	Meyer Organics Pvt. Ltd.
18	Liv-First	Kalmegh, Bhuiamla, Bhringraj, Tulsi, Sharpunkha, Punarnava, Guduchi	Isha Agro Developers Private Limited
19	Livo Care Tab	Mandur bhasma, *Boerhavia diffusa,* Nishoth, Sonth, *Piper longum,* Kali Mirch, *Embelia ribes, Cedrus deodara, Plumbago zeylanica,* Pushkarmool, *Curcuma longa,* Daruhaldi, *Baliospermum montanum,* Harad, Baheda, *Emblica officinalis, Piper retrofractum,* Indrajav, *Picrorhiza kurroa* rhizome, Nagarmotha	Immunorich Marketing Private Limited
20	Livomap tablets	Punarnava (*Boerhaavia diffusa*), Nimb (*Melia azedarach*),Tikta patola (*Trichosanthes cucumerina*), Shunthi (*Zingiber officinale*), Katuki (*Picrorhiza kurroa*), Guduchi (*Tinospora cordifolia*), Devdaru (*Cedrus deodara*), Haritaki (*Terminalia chebula*), Varuna (*Crataeva religiosa*), Shigru (*Moringa oleifera*), Daruharidra (*Berberis aristata*), Afsantin (*Artemisia absinthium*), Sharapunkha (*Tephrosia purpurea*), Bhumiamalaki (*Phyllanthus niruri*)	Maharishi Ayurveda.
21	Livomyn tablet	*Andrographis paniculata, Phyllanthus niruri,* Triphala, *Cichorium intybus, Boerhavia diffusa, Amoora rohituka, Eclipta alba,Adhatoda vasica, Zingiber officinale, Berberis aristata, Tinospora cordifolia, Tephrosia pupurea, Fumaria officinalis, Embelia ribes, Coriandrum sativum, Aloe barbadensis, Picrorrhiza kurroa*	Charak Pharmaceuticals Pvt.Ltd

TABLE 6.1 *(Continued)*

S. No	Brand name	Composition	Manufacturer
22	Livonex	*Picrorrhiza kurroa, Tephrosia purpurea, Phyllanthus emblica*	AyurLeaf Herbals
23	Livo-Pure	Curcumin, Kalmegh, Silymarine & *Phyllanthus niruri*	RV Herbals (Maulik Remedies)
24	Livotrit Tablets	Arogyvardhini Rasa, Mandur Bhasma, *Boerhavia diffusa, Eclipta alba, Andrographis paniculata*	Zandu Ayuveda
25	Medilib Liver Tab	Rohatak chall, Punarnava, Makoi, Kasani, Triphala, Trikuta, Mandoor bhasma	Medito Pharmaceutical Company
26	Milk Thistle Complex	Beta Carotene, vitamin B1, B2, B3, B6, B12, B5, Dandelion Root, Artichoke Leaf Powder, Milk Thistle extract, Curcumin extract	Aarkios Health Private Limited
27	Natural Orange Livobeet Ayurvedic/ Herbal Tablet	*Phyllanthus niruri, Taraxacum officinale, Trigonella foenum-graecum, Glycyrrhiza glabra, Boerhavia diffusa, Capparis spinosa, Cichorium intybus, Terminalia arjuna, Andrographis paniculata*	Orange Organic Pharma
28	Nitlor Plus	Silymarin, L-ornithine-L-aspartate	BMW Pharmaco India Pvt Ltd
29	Nitoliv	Silymarin, L-ornithine-L-aspartate	Biomax Biotechnics
30	Setu Liver Lift	Silymarin, Glutathione, Alpha Lipoic Acid, and Nac (*N*-Acetyl-Cysteine), Beetroot, Turmeric, Selenium and Biotin	Prakruti Products Pvt. Ltd.
31	Shamliv	Silymarin, Lecithin	Shamsri Pharma Pvt. Ltd
32	SILYBON-70	Silymarin	Micro labs
33	Silymarin Plus	Vitamin C, Vitamin E, Silymarin, Inositol, Choline	Source naturals
34	Stivil	*Arogyavardhini Ras, Punarnava mandur, Phyllanthus niruri, Andrographis paniculata, Melia azedarach, Solanium nigrum, Aloe vera, Eclipta alba, Tephrosia purpurea, Tinospora cordifoia, Glycyrrhiza glabra*	Vital Care Private Limited
35	ZycoLiv	Bhumyamalki (*Phyllanthus niruri*), Yestimadhu (*Glycyrrhiza glabra*), Rohitka (*Tecomella undulata*), Punernava (*Boerhavia diffusa*), Pittpapra (*Fumaria indica*), Daruhaldi (*Berberis aristata*), Pipplimool (*Piper longum*), Bhrungraj (*Eclipta alba*), Arjuna (*Terminalia arjuna*), Kadu (*Piccorhiza kurroa*), Makoy (*Solanum nigrum*), Amla (*Emblica officinalis*)	Zoic Pharmaceuticals

KEYWORDS

- biherbal formulation
- polyherbal formulation
- polyherbal mixture
- hepatoprotective

REFERENCES

Achilya, G.S.; Kotagale, N.R.; Wadodkar, S.G.; Dorle, A.K. Hepatoprotective of panchegava ghrita against carbon tetrachloride in rats. *Indian J. Pharmacol.* **2003**, *35*, 308–311.

Achliya, G.S.; Wadodkar, S.G.; Dorle, A.K. Evaluation of hepatoprotective effect of Amalkadi Ghrita against carbon tetrachloride-induced hepatic damage in rats. *J. Ethnopharmacol.* **2004**, *90*, 229–232.

Aftab, A.; Pillai, K.K.; Najmi, A.K.; Ahmad, S.J.; Pal, S.N.; Balani, D.K. Evaluation of hepatoprotective potential of jigrine post-treatment against thioacetamide induced hepatic damage. J. Ethnopharmacol. **2002**, *79*(1), 35–41.

Aftab, A.; Pillai, K.K.; Shibli, J.A.; Balani, D.K.; Najmi, A.K.; Renuka, M.; Abdul, H. Evaluation of the hepatoprotective potential of jigrine pretreatment on thioacetamide induced liver damage in rats. *Indian J. Pharmacol.* **1999**, *31*, 416–421.

Akhtar, M.S.; Asjad, H.M.M.; Bashir, S.; Malik, A.; Khalid. R.; Gulzar, F.; Irshad, N. Evaluation of antioxidant and hepatoprotective effects of Khamira Gaozaban Ambri Jadwar Ood Saleeb Wala (KGA). *Bangladesh J. Pharmacol.* **2013**, *8*, 44–48.

Arsul, V.A.; Wagh, S.R.; Mayee, R.V. Hepatoprotective activity of livergen, a polyherbal formulation against carbon tetrachloride induced hepatotoxicity in rats. *Int. J. Pharm. Pharm. Sci.* **2011**, *3*, 228–231.

Atawodi, S.E. In vivo antioxidant, organ protective, ameliorative and cholesterol lowering potential of ethanolic and methanolic extracts of "Ata-Ofa" polyherbal tea (A-Polyherbal). *Int. J. Res. Pharm. Sci.* **2011**, *2*, 473–482.

Bera, T.K.; Chatterjee, K.; De, D.; Ali, K.M.; Jana,K.; Maiti, S.; Ghosh, D. Hepatoprotective activity of Livshis, a polyherbal formulation in CCl_4-induced hepatotoxic male Wistar rats: A toxicity screening approach. *Genomic Med. Biomarkers Health Sci.* **2011**, *3*(3–4), 103–110.

Bera, T.K.; Chatterjee, K.; Jana, K.; Ali, K.M.; De, D.; Maiti, S.; Ghosh D. Antihepatotoxic effect of "Livshis," a polyherbal formulation against carbon tetrachloride-induced hepatotoxicity in male albino rat. *J. Nat. Pharm.*, **2012**, *3*, 17–24.

Bhattacharyya, D.; Mukherjee, R.; Pandit, S.; Das, N.; Sur, T.K. Prevention of carbon tetrachloride induced hepatotoxicity in rats by Himoliv, a polyherbal formulation. *Indian J. Pharmacol.* **2003**, *35*, 183–185.

Chaitanya, D.A.K.; Siva Reddy, C.; Manohar Reddy, A. Hepatoprotective effect of biherbal ethanolic extract against paracetamol-induced hepatic damage in albino rats. *J. Ayurveda Integr. Med.* **2012**, *3*(4), 198–203.

Chauhan, B.L.; Mohan, A.R.; Kulkarni, R.D.; Mitra, S.K. Bioassay for evaluation of the hepatoprotective effect of Liv-52, a polyherbal formulation, on ethanol metabolism in chronic alcohol exposed rats. *Indian J. Pharmacol.* **1994**, 26, 117–120.

Cordero-Pérez, P.; Torres-González, L.; Aguirre-Garza, M.; Camara-Lemarroy, C.; Guzmán-De La Garza, F.; Alarcón-Galván, G.; Cantú-Sepúlveda, D. Hepatoprotective effect of commercial herbal extracts on carbon tetrachloride-induced liver damage in Wistar rats. *Pharmacogn. Res.*, **2013**, *5*(3), 150–156.

Dandagi, P.M.; Patil, M.B.; Mastiholimath, V.S.; Gadad, A.P.; Dhumansure, R.H. Development and evaluation of hepatoprotective polyherbal formulation containing some indigenous medicinal plants. *Indian J. Pharm. Sci.* **2008**, *70*(2), 265–268.

Dange, S.V. Liv. 52 in the prevention of hepatotoxicity in patients receiving antitubercular drugs: A metaanalysis. *Indian J. Clin. Practice.* **2010**, *21*, 81–86.

Darbar, S.; Bhattacharya, A.; Chattopadhyay, S. Antihepatoprotective potential of livina, a polyherbal preparation on paracetamol induced hepatotoxicity: A comparison with silymarin. *Asian J. Pharm. Clin. Res.* **2011**, *4*(1), 72–77.

Darbar, S.; Chakraborty, M.R.; Chattarjee, S.; Ghosh, B. Protective effect of *Livina*, a polyherbal liquid formulation against ethanol induced liver damage in rats. Anc. Sci. Life. **2009**, *28*(3), 14–17.

Devaraj, V.C.; Krishna, B.G.; Viswanatha, G.L.; Kamath, J.V.; Kumar, S. Hepatoprotective activity of Hepax—A polyherbal formulation. Asian Pac. J. Trop. Biomed. **2011**, *1*(2), 142–146.

Dinesh, K.; Sivakumar, V.; Selvapriya, B.; Deepika, E.; Sadiq, A.M. Evaluation of hepatoprotective polyherbal formulations contains some Indian medicinal plants. *J. Pharmacogn. Phytochemistry* **2014**, *3*, 1–5.

Ezeonwu, V.U.; Dahiru, D. Protective effect of bi-herbal formulation of *Ocimum gratissimum* and *Gongronema latifolium* aqueous leaf extracts on acetaminophen-induced Hepato-nephrotoxicity in rats. *Am. J. Biochem.* **2013**, *3*, 18–23.

Girish, C.; Koner, B.C.; Jayanthi, S.; Rao, K.R.; Rajesh, B.; Pradhan, S.C. Hepatoprotective activity of six polyherbal formulations in paracetamol induced liver toxicity in mice. *Indian J. Med. Res.* **2009**, *129*, 569–578.

Girish C.; Pradhan S.C. Hepatoprotective activities of picroliv, curcumin, and ellagic acid compared to silymarin on carbontetrachloride-induced liver toxicity in mice. *J. Pharmacol. Pharmacother.* **2012**, 3(2), 149–155.

Gulati, K.; Ray, A.; Vijayan, V.K. Assessment of protective role of polyherbal preparation Livina, against anti-tubercular drug induced liver dysfuntion. *Indian J. Exp. Biol.* **2010**, *48*(1), 318–322.

Gumaa, S.A.; Hassan, E.M.; Khalifa, D.M. Hepatoprotective effect of aqueous extracts of some medicinal plant mixtures on CCl_4-induced liver toxicity. *IOSR J. Pharm. Biol. Sci.* **2017**, *12*(1), 43–52.

Huseini, H.F.; Alavian, S.M.; Heshmat, R.; Heydari, M.R.; Abolmaali, K. The efficacy of Liv-52 on liver cirrhosis patients: A randomized, double-blind, placebo-controlled first approach. *Phytomedicine* **2005**, *12*, 619–624.

Iroanya, O.; Okpuzor, J.; Adebesin, O. Hepatoprotective and antioxidant properties of a triherbal formulation against carbon tetrachloride induced hepatotoxicity. *IOSR J. Pharm.* **2012**, *2*, 130–136.

Karunakar, N.; Pillai, K.K.; Hussain, S.Z.; Rao, M.; Balani, D.K.; Imran, M. Further studies on the antihepatotoxic activity of jigrine. *Ind. J. Pharmacol.* **1997**, *29*, 222–227.

Khan, M.A.; Khan, J.; Ullah, S.; Malik, S.A.; Shafi, M. Hepatoprotective effects of *Berberis lycium*, *Galium aparine* and *Pistacia integerrima* in carbon tetrachloride (CCl_4)-treated rats. *J. Postgrad. Med. Inst.* **2008**, *22*, 19–25.

Kotecha, K.N.; Ashok, B.K.; Shukla, V.J.; Prajapati, P.; Ravishankar, B. Hepatoprotective activity of Vasaguduchyadi Kwatha—A compound herbal formulation against antitubercular drugs (isoniazid + rifampicin + pyrazinamide) induced hepatotoxicity in albino rats. *Pharma Sci. Monit.* **2015**, *6*(3), 73–84.

Lima, T.B.; Suja, A.; Jisa, O.S.; Sathyanarayanan, S.; Remya, K. S. Hepatoprotective activity of LIV-first against carbon tetra chloride-induced hepatotoxicity albino rats. *Int. J. Green Pharm.* **2010**, *4*, 71–74.

Lin, C.-C.; Tsai, C.-C.; Yen M.-H. The evaluation of hepatoprotective effects of Taiwan folk medicine 'Teng-Khia-U', *J. Ethnopharmacol.* **1995**, *45*, 113–123.

Maheshwari, R.; Pandya, B.; Balaraman, R.; Seth, A.K.; Yadav, Y.C.; Sankar, V.S. Hepatoprotective effect of Livplus-A polyherbal formulation. *Pharmacogn. J.* **2015**, 7(5), 311–316.

Mayuren, C.; Reddy, V.V.; Priya S.V.; Devi, V.A. Protective effect of Livactine against CCl_4 and paracetamol induced hepatotoxicity in adult Wistar rats *North Am. J. Med. Sci.*, **2010**, *2*, 491–495.

Mistry, S.; Dutt, K.R.; Jena, J. Formulation and evaluation of hepatoprotective polyherbal formulation containing indigenous medicinal plants. *Int. J.Appl. PharmaTech Res.* **2012**, *3*, 5–11.

Mitra, S.K.; Varma, S.R.; Godavarthi, A.; Nandakumar, K.S. Liv-52 regulates ethanol induced PPARc and TNF an expression in HepG2 cells. *Mol. Cell Biochem.* **2008**, *315*, 9–15.

Mitra, S.K.; Venkataranganna, M.V.; Sundaram, R.; Gopumadhavan, S. Protective effect of HD-03, a herbal formulation, against various hepatotoxic agents in rats. *J. Ethnopharmacol.* **1998**, *63*, 181–186.

Najmi, A.K.; Pillai, K.K.; Pal, S.N.; Aqil, M. Free radical scavenging and hepatoprotective activity of jigrine against galactosamine induced hepatopathy in rats. J. Ethnopharmacol. **2005**. *97*(3), 521–525.

Najmi, A.K.; Pillai, K.K.; Pal, S.N.; Aqil, M.; Sayeed, A. Hepatoprotective and behavioral effects of jigrine in galactosamine-induced hepatopathy in rats. Pharm. Biol. **2010**, *48*(7), 764–769.

Ohta, Y.; Kongo-Nishimura, M.; Hayashi, T.; Kitagawa, A.; Matsura, T.; Yamada, K. Saikokeishito extract exerts a therapeutic effect on alpha-naphthylisothiocyanate-induced liver injury in rats through attenuation of enhanced neutrophil infiltration and oxidative stress in the liver tissue. *J. Clin. Biochem. Nutr.* **2007**, *40*, 31–41.

Rachmawati, E.; Nurrochmad, A.; Puspita Sari, I. Assessment of hepatoprotective effect of polyherbal combination of *Phyllanthus niruri* (meniran), *Curcuma xanthorrhiza* (wild ginger), and *Curcuma longa* (turmeric) against liver dysfunction due to anti-tuberculosis drugs. In *Proceeding of the International Conference Pharmaceutical Care New Development of Pharmaceutical Care in a Pharmacogenetic and Pharmacogenomic Approach.* University of Muhammadiyah Malang, Malang, Indonesia; **2014**, p. 4.

Rajesh, M.G.; Latha, M.S. Preliminary evaluation of the antihepatotoxic activity of Kamilari, a polyherbal formulation. *J. Ethnopharmacol.* **2004**, *91*, 99–104.

Sandhir, R.; Gill, K. Hepatoprotective effects of Liv. 52 on ethanol-induced liver damage in rats. *Indian J. Exp. Biol,* **1999**, *37*, 762–766.

Sankar, M.; Rajkumar, J.; Devi J. Hepatoprotective activity of hepatoplus on isonaizid and rifampicin induced hepatotoxicity in rats. *Pak. J. Pharm. Sci.* **2015b,** *28*(3), 983–990.

Sankar, M.; Rajkumar, J.; Sridhar, D. Hepatoprotective activity of hepatoplus on isoniazid and rifampicin induced liver damage in rats. *Indian J. Pharm. Sci.* **2015a**, *77*(1), 556–562.

Sankar, M.; Rajkumar, J.; Sridhar, D. Effect of hepatoplus on isoniazid and rifampicin induced hepatotoxicity in liver cell lines. *Int. J. Pharm. Pharm. Sci.* **2015c**, *7*(5), 215–219.

Sapakal, V. D.; Chadge, R. V.; Adnaik, R. S.; Naikwade, N. S. and Magdum, C. S. Comparative hepatoprotective activity of Liv-52 and Livomyn against carbon tetrachloride-induced hepatic injury in rats. *Int. J. Green Pharm.* **2008**, *43*, 79–82.

Satyapal, U.S.; Kadam, V.L.; Ghosh, R. Hepatoprotective activity of Livobond a polyherbal formulation against CCl_4-induced hepatotoxicity in rats. *Int. J. Pharm.* **2008**, *4*, 472–476.

Shakya, A.K.; Shukla, S. Evaluation of hepatoprotective efficacy of Majoon-e-Dabeed-ul-ward against acetaminophen-induced liver damage: A Unani herbal formulation. *Drug Dev. Res.* **2011**, *72*, 1–7.

Shin, J.-W.; Son, J.-Y.; Oh, S.M.; Han, S.H.; et al. An herbal formula, CGX, exerts hepatotherapeutic effects on dimethylnitrosamine-induced chronic liver injury model in rats. *World J. Gastroenterol.* **2006**, 12(38), 6142–6146.

Vidyashankar, S.; Mitra, S.K.; Nandakumar, K.S. Liv-52 protects HepG2 cells from oxidative damage induced by tert-butyl hydroperoxide. *Mol. Cell Biochem.* **2010**, *333*, 41–48.

Vivek, K.; Pillai, K.K.; Hussain, S.Z.; Balani, D.K. Hepatoprotective activity of Jigrine on liver damage caused by alcohol, carbon tetrachloride and paracetamol in rats. *Indian J. Pharmacol.* **1994**, *26*, 35–40.

Wang, H.; Feng, F.; Zhuang, B.-Y.; Sun, Y. Evaluation of hepatoprotective effect of Zhi-Zi-Da-Huang decoction and its two fractions against acute alcohol-induced liver injury in rats. *J. Ethnopharmacol.* **2009**, *126*, 273–279.

Wang, C.; Zhang, T.; Cui, X.; Li, S.; Zhao, X.; Zhong, Z. Hepatoprotective effects of a Chinese Herbal Formula, Longyin Decoction, on Carbon-Tetrachloride-induced liver injury in chickens. *Evid. Based Complement. Alternat. Med.* **2013**, *2013*, 392743.

Yang, C.C.; Fang, J.Y.; Hong, T.L.; Wang, T.C.; Zhou, Y.E.; Lin, T.C. Potential antioxidant properties and hepatoprotective effects of an aqueous extract formula derived from three Chinese medicinal herbs against CCl_4-induced liver injury in rats. *Int. Immunopharmacology* **2013**, *15*(1), 106–113.

Yimam, M.; Jiao, P.; Hong, M.; Jia, Q. Hepatoprotective activity of an herbal composition, MAP, a standardized blend comprising *Myristica fragrans*, *Astragalus membranaceus,* and *Poria cocos*. *J. Med. Food.* **2016**, *19*(10), 952–960.

Zameer, M.; Rauf, A.; Qasmi, I.A. Hepatoprotective activity of a unani polyherbal formulation "kabideen" in CCl_4 induced liver toxicity in rats. *Int. J. Appl. Biol. Pharm. Technol.* **2015**, *6*(1), 8–16.

Index

C

D

E

F

G

H

I

J

K

L

N

O

P

Q

R

S

T

U

V

W

X

Z